cology and Field

SECOND EDITION

Robert Leo Smith

VIRGINIA UNIVERSITY

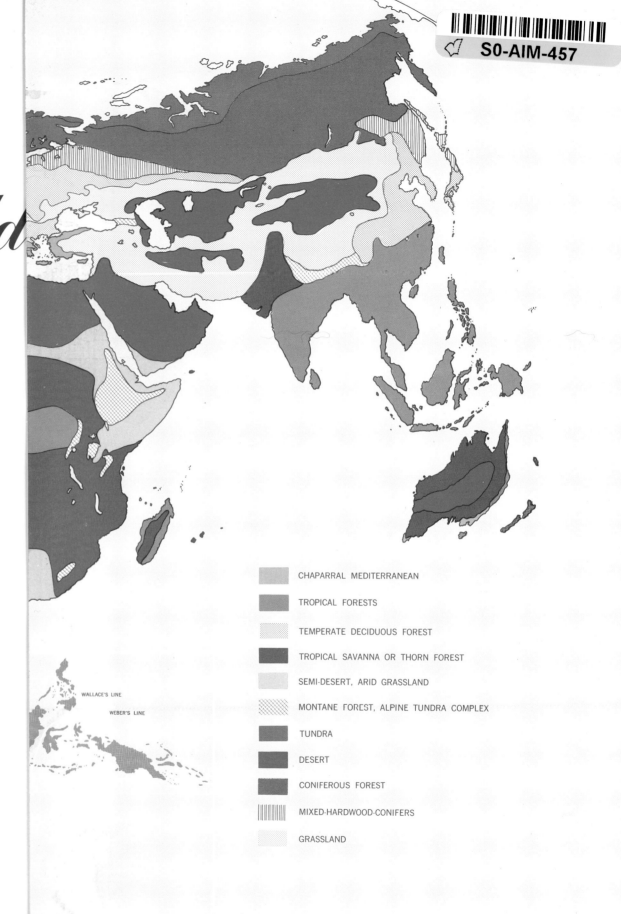

CHAPARRAL MEDITERRANEAN

TROPICAL FORESTS

TEMPERATE DECIDUOUS FOREST

TROPICAL SAVANNA OR THORN FOREST

SEMI-DESERT, ARID GRASSLAND

MONTANE FOREST, ALPINE TUNDRA COMPLEX

TUNDRA

DESERT

CONIFEROUS FOREST

MIXED-HARDWOOD-CONIFERS

GRASSLAND

WALLACE'S LINE

WEBER'S LINE

Biology

HARPER & ROW, PUBLISHERS
New York, Evanston,
San Francisco, London

Sponsoring editor: Joseph Ingram
Special projects editor: Carol J. Dempster
Project editor: Holly Detgen
Designer: Gayle Jaeger
Production supervisor: Robert A. Pirrung

Ecology and Field Biology, Second Edition

Library of Congress Cataloging in Publication Data
Smith, Robert Leo.
    Ecology and field biology.
    Bibliography: p.
    1. Ecology. I. Title.
QH541.S6 1974   574.5   73–13305
ISBN 0–06–046334–1

to Alice

Contents

In the first edition of *Ecology and Field Biology* I wrote: "Prefaces are written to be read, although I suspect that more readers flip past them than have read them." In the eight years this text has been on the market I have received a number of letters from readers telling me that they read the preface. I still believe that in many books the preface is often the most absorbing part. In technical books and textbooks, the preface is about the only place where the author can give any real glimpse of himself and tell his reasons for writing the book, how he thinks it should be used, what problems he had, and who assisted him. So, as in the first edition, this is being written to be read before the reader ventures too far into the book.

I have retained the title *Ecology and Field Biology* for the second edition. I was tempted to drop the words *field biology* but didn't. Field biology emphasizes the field approach to ecology that is an important characteristic of this text.

A lot has happened in ecology since the appearance of the first edition. In 1966 ecology was a poor second cousin to molecular biology in academic programs. But after 1970, when the environmental problems reached home and demanded attention, ecology became almost a household word, although misused and poorly understood. Colleges that had never offered ecology courses before added them to their curricula. New ecology books began to flood the market. Many new ecology texts have appeared since 1970, as well as scores of shorter paperback texts. Lacking a choice in 1966, instructors now face a diversity of texts.

Among the current ecology texts, several approach the subject from an evolutionary viewpoint. In fact, some fall somewhat short of being ecology texts but are quite strong on population biology and population genetics. Others take an historical approach to ecology, still others a population approach. Some of the texts are strictly advanced, some elementary. Except for one text, applied ecology is largely ignored; most are devoted to theory. At least two attempts have been made to provide an adequate text for both graduate and undergraduate levels by suggesting that some chapters be utilized and others skipped, depending upon the level desired.

In *Ecology and Field Biology*, I have not attempted to write for an audience of two levels. This book is written as an introductory text to ecology at the sophomore and junior level, although graduate students and others may find it a useful reference. It is designed as a first course for majors and a terminal course for nonmajors. I believe that both should be exposed to the same material. An examination of current texts leads me to conclude that some authors believe that majors should be introduced directly into pure theory, a debatable point. Majors, I believe, should be exposed to theory early, but at the same time they should develop an appreciation of how theory can be applied to the numerous ecological problems facing us today politically, socially, and ecologically. Nonmajors should not concentrate only on the applied ecology without having some understanding of underlying ecological theory.

In this revision I have attempted to balance theoretical ecology with the applied, quantitative with descriptive and qualitative. I have tried not to be dogmatic but to point out areas where considerable controversy exists. Readers will discover that this edition contains considerably more theory and some mathematics. The math is relatively simple, and the examples can be followed easily. I have brought all sections up to date and rewritten most of the book. I have added new examples, expanded the material on populations, broken the chapter on energy and biogeochemical cycles into four new chapters. I am surprised that the subject of biogeochemical cycles so relevant to air and water pollution receives such little attention in most current ecology texts. Yet my first edition was criticized for its inadequate treatment of that subject. I have added a new chapter on ecosystem approach to resource management and one on organisms and the environment—really a chapter of physiological ecology, a topic most requested by readers.

Readers will also note a major change in organization of the text. A criticism of the first edition was its failure to integrate sufficiently the various areas of ecology discussed, ecosystem, community, population, and behavior. Some instructors failed to use the behavioral ecology material at all. To tighten the organization, I developed the revision around a central theme, the ecosystem, which is introduced in Chapter 2 and is followed throughout the text. A secondary theme is evolution and natural selection. This theme might not appear as obvious as it does in some evolution-oriented texts, but the theme is there; and a chapter is devoted to the subject of natural selection. The first edition of *Ecology and Field Biology* was the first general ecology text to include evolution and natural selection as a part of ecology.

One may criticize the book for being too long for a one-semester course in ecology. That was a persistent objection to the first edition. (I am happy to see that all the major ecology texts that have appeared are approximately the same number of pages.) I gave this criticism serious consideration in the revision and ended up with an even larger book in spite of deletion of considerable material. In revising the text I was faced with two alternatives. One was to reduce the book considerably to a traditional one-semester size of 450 pages. If I did that, the result would have been a book with a minimum amount of material, a skeleton of ecological principles with many areas overlooked. I would have to ignore most of the applied aspects of ecology. Then the book could be criticized as being too thin, a point made about many of the short paperback texts used in some ecology courses. The second alternative was to increase the size of the book, a necessary step if I were to maintain a balanced picture of ecology. I chose the latter alternative, although it meant some sacrifice, such as reduction in the size of some of the illustrations. The book is as large as it can grow. What will be cut in future editions is up to the readers. I would appreciate your suggestions.

In spite of it all, the book basically is not too large for a one-semester course. A text should be an aid and not a crutch for a course. It should not only provide the basic material around which lectures are built but also provide supplementary reading for the student (after all, humanities courses demand that students not only digest a text but a number of additional readings as well). *Ecology and Field Biology* is large enough to allow the instructor a great deal of flexibility. The instructor can explain theory in the lecture; the student will find sufficient material in the text to amplify the lecture material.

In some schools ecology is a lecture and discussion course. This text should fit such courses admirably. It provides the instructor and the student with necessary additional material to develop topics for discussion, as well as a key to finding new sources of informa-

Preface

tion. In other schools ecology is a lecture and laboratory course. In such a course the descriptions of the various ecosystems provide the students with an insight into the communities and ecosystems that might be involved in the laboratory. The appendixes (a popular section of the book I had considered dropping to save space) provide a guide to necessary techniques to use in the lab. (In my own course the laboratory is devoted to a study of the forest ecosystem. The various vegetational layers are sampled by different techniques, the composition and structural layers analyzed, species diversity of layer determined, populations of small mammals and forest soil invertebrates sampled, distribution of organisms analyzed, soil nutrients determined, etc., and the whole tied together in a comprehensive report.) If the instructor requires individual projects, then *Ecology and Field Biology* provides the student with the basic information and techniques needed to undertake independent field studies. This versatility would not be possible in a shorter text, nor is this versatility available in other ecology texts now on the market. In addition, the lists of journals reasonably complete at the time of compilation, general references, suggested readings, and general bibliographies provide a guide to available literature that should be invaluable to instructor and student alike.

This edition is different from other ecology texts in another important way. It recognizes that man is the dominant ecological force on earth. With one exception: Other ecology texts mention man but consider ecological theory, nature, and natural processes apart from man; yet virginal nature no longer exists. Even the most remote wilderness areas are affected by air pollution, by protection from fire, by pesticides, and by other intrusions of man. Several texts do devote the terminal chapter or so to man and ecology. In this text man is integrated into all of the discussions.

The manner in which the text is used is basically the instructor's own decision. The text approaches ecology from the ecosystem to the organisms. Some instructors may want to reverse that approach and start with the organism and work toward the ecosystem. In that event, the instructor might want to begin with Chapters 1 and 2, move to Chapter 7, then 15 and 16, and work backward through Chapters 10 through 14, and 3 through 9, and end with Chapter 20.

The book is divided into six parts. Part I is an introduction. Chapter 1 is a brief look at the history and nature of ecology, while Chapter 2 sets the stage for the material to come. Part II considers the ecosystem and man's place in it, ecosystem function, environmental influences, the nature of the community and ecosystem development. Part III considers the populations that make up the ecosystem. Chapter 10 deals with basic demographic aspects of population; Chapter 11, the interaction among members of the same population; Chapter 12 with interactions among populations of different species with special emphasis on competition, predation, and a special type of predation—exploitation by man. These chapters are followed by one on social behavior in populations and one on natural selection and evolution. Part IV looks at the individual in the ecosystem, particularly its response to its environment. Part V deals with a diversity of ecosystems, terrestrial, fresh water, and marine. Part VI and Chapter 20 look at the application of ecological principles or the lack of their application to resource management. Considered are agriculture, range management, forestry, and wildlife management. Two controversial areas, clear-cutting and hunting, are examined in some detail. The chapter ends with a short introduction to systems analysis and its application in ecology.

The Appendixes, too, have been reorganized. Appendix A provides an annotated bibliography on the use of statistics in ecology. Appendix B is devoted to the sampling of animal and plant populations. Appendix C is concerned with sampling community attributes, including primary and secondary productivity, species diversity, association between species, and population dispersion. Appendix D includes techniques for measuring a number of environmental variables.

I have written this book to be read and used. I hope that I have been able to infuse into *Ecology and Field Biology* some enthusiasm for the subject and some feeling for the natural world and man's place in it. By necessity the reader will find in these pages some of the dull textbook stuff; but in it, too, I hope the user will find a feeling for the world outdoors. If that too can become as much a part of ecology as theory, mathematics and computers, then perhaps posterity also will be able to study ecology.

Robert Leo Smith

The author of a textbook depends upon a number of people, mostly those researchers whose long hours in the laboratory and field have provided the raw material out of which textbooks are fashioned. Aside from these, there are a number of people who must be singled out individually. The idea and the encouragement for writing the book in the first place must go to Dr. F. Reese Nevin, State University of New York at Plattsburgh. Among those whose comments and criticisms were important in the development of the first edition were Professor Arnold Benson, Dr. Willem van Eck, Dr. Warren Chase, Dr. David E. Davis, Drs. Robert and Millicent Ficken, Dr. Henry Tompson, Dr. Harold A. Mooney, Dr. J. T. Enright, and Dr. Monte Lloyd.

For the second edition, Dr. Willem van Eck of West Virginia University reviewed the material on soils again. Dr. David E. Reichle, Oak Ridge National Laboratory, reviewed the material on biogeochemical cycling. Drs. William Kodrich, Robert Moore, and John Williams of Clarion State College, Clarion, Pa., and Dr. Dale E. Birkenholz of Illinois State College, Normal, Ill., reviewed the entire manuscript for readability, classroom usefulness, and accuracy. For their pointed comments and helpful suggestions I am more than grateful. Dr. Norman Kowal of West Virginia University not only reviewed the material on systems ecology, but he also provided me with many insights into systems analysis. His recent decision to study medicine is ecology's loss and medicine's gain. Dr. James Kroll of Stephen F. Austin College provided me with suggestions and reviewed the material on ecological physiology. Thanks also go to a former graduate student of mine, Jerry Moore of the pesticides division of the Environmental Protection Agency. He not only provided many sound suggestions for the revision, but also obtained a number of photos and supplied me with material and references on pesticides.

Between the publication of the first edition and the writing of the second I received many helpful comments from readers, most of whom will have to go unnamed. I particularly appreciate the comments of those who responded to a questionnaire. Suggestions were passed along by Dr. Thomas Pauley, Salem College; Dr. Paul Hafer, State University of New York

Acknowledgments

at Potsdam; Dr. Richard Hartman, University of Pittsburgh; and Dr. Richard W. Coleman, Upper Iowa College. Dr. James Monro, New South Wales Institute, Australia, clarified the *Cactoblastis* story and provided additional information.

For photographs used in the first edition and reused here I am indebted to Dr. George Ammann, George Harrison, Dr. L. W. Gysel, Dr. Glenn Sanderson, and Dr. Wendell Swank. For new photos I thank George Trimble of the U.S. Forest Service; Dr. Harry Wilkes, University of Massachusetts, Boston; Dr. Norman Kowal, West Virginia University; Dr. Richard Johnston, University of Kansas; Dr. S. Oden, University of Helsinki; Pioneer Corn Company, Tipton, Ind.; the U.S. Fish and Wildlife Service; The Soil Conservation Service; the U.S. Forest Service; and the Rockefeller Foundation, New York City. Thanks also go to the Wildlife Society, The Ecological Society of America, and the various publishers and authors who permitted me to redraw or adapt illustrations from their publications. They are credited in the captions and bibliography.

The majority of drawings (graphics excluded) were done by Ned Smith, a nationally known wildlife artist who did a superb job for the first edition. New drawings were done by my son, Robert Leo, Jr., a student of graphic design and a developing outdoors and wildlife illustrator. He did the task well under heavy pressure of deadlines and classes.

The book would not have arrived at the present stage without the constant encouragement, advice, and prodding from Al Abbott, Carol Dempster, and Holly Detgen of Harper & Row. Dan Cooper, field representative of Harper & Row, did a great job of passing along suggestions from users of the text and of acting as a liaison between the Clarion State College reviewers and me. Drs. Dave Samuel and Ed Michael of the wildlife biology section of the Division of Forestry, West Virginia University, took over some of my work with students when time was closing in. Finally, I wish to acknowledge the assistance and encouragement of my wife, Alice, and children who had to endure my many hours of preoccupation with the revision and all that it involved.

Ecology and Field Biology

Introduction

This is an age of ferment and change. A little more than two-score years ago Charles Lindberg bravely crossed the Atlantic alone in a single-engine plane. Today we have landed upon and explored the surface of the moon and have sent photographic probes to Mars. In the early 1920s radio was just beginning to get its feet on the ground; today pictures along with voices are being transmitted around the world by way of communication satellites. We have probed the secrets of the atom and unleashed its awesome power. In so doing we have changed the direction of world history and the destiny of mankind. A ferment also has developed in biology. The probing of the secrets of the cell, the discovery of DNA, and advances in biochemistry and genetics have revolutionized biology and have produced profound implications for the future of man.

While all these advances were taking place, other changes were also at work. Man's environment was deteriorating, and although warnings had been sounded, few paid any attention. Population was increasing explosively. Technological advances were destroying the environment at an accelerating rate. Nitrogenous wastes and excessive phosphorus were draining from farmlands and urban areas, causing eutrophication of natural waters and lowering water quality. DDT and other hydrocarbons, PCB, and mercury and other heavy metals were accumulating in some species of animals, impairing their reproduction or making them unfit as human food. The nonreturnable bottle, nonbiodegradable plastic, and forced obsolescence of appliances and automobiles were creating massive solid-waste problems and littering the countryside. Demand for increased electricity brought about by industry and the public acceptance of air conditioning and all-electric homes increased sulphur dioxide and particulate pollution of the air. It upturned midwest farmland and Appalachian Mountains for cheap strip-mined coal. A rapid multiplication of automobiles poured increasing amounts of nitrous oxides and lead into the atmosphere, which under proper conditions produce choking photochemical smog. Roads slashed through open country, and urban and suburban expansion

ate into the hinterlands and farmlands. Wilderness areas and wild places were disappearing at an accelerating pace, and the increased interest in outdoor recreation placed intolerable pressures on state and national parks. Even the oceans were not spared, as man's debris and chemicals were deadening the seas. Suddenly the public began to awaken to the fact that planet earth was in trouble, and suddenly they became aware of ecology.

In the late 1960s the general public was hardly aware of the term *ecology*. As a topic of interest ecology stirred little public discussion, and as a science it had none of the glamor of molecular biology. By 1970 ecology had become a household word, but it was misunderstood, misused, and equated with environmental science. Too many failed to understand that ecology refers to the interrelations of an organism with its environment and that this includes man. They only vaguely realized that the relationship is two-way, that just as the environment has an impact on an organism, so an organism has an impact on its environment. But at least a great majority became aware of the environment. And the shattering view of earth from outer space forced on us the realization that the earth is finite and that what it is and what it contains are all we have.

Because it deals with life, ecology has been considered a part of biology. A quarter of a century ago and earlier the major introductory path to biology was through natural history, or as it was more popularly known, nature study. This was a time when people were just awakening to the world about them. Nature had ceased to be an enemy. The fields were cleared, the forest subdued, and there was even becoming a danger that many common animals—gray squirrels, beaver, deer, wild turkeys, and ducks—were on the border of extinction. The conservation movement was building up full steam in the 1930s, and nature study was a part of nearly every school curriculum, even though more often than not it was poorly taught. Too often it consisted only of coloring bird pictures and writing paragraphs about them. But at least youngsters became aware that birds existed, that they were colorful and interesting, and that they were something more than living targets for BB guns. It was a time when John Burroughs was popular, the Reed *Bird Guides*

Ecology: its meaning and scope

were the last word in field guides, and the Comstock *Handbook of Nature Study* was the bible of natural history.

Out of this background of close contact with nature and an interest in life, the biologists developed. But as the country became more urbanized and less rural, people lost this contact with nature. Interest in biology from a field approach declined, and research biologists became more concerned about the functioning of an organism than about its relationship to its environment. Modern biologists appeared at the doorways of chemistry, physics, and mathematics—disciplines not immediately related to the living environment. They looked upon biology as beginning and ending with a group of chemical compounds, and they thought that the answer to life lay within the realm of the physical sciences.

Part of the reason for the swing away from natural history lies in biology itself. For a long time traditional biology started and ended with the naming of organisms. Biology as taught in schools and colleges was an endless repetition of the study of types of organisms. It was largely descriptive, weak in quantitative data, and it lacked the strong conceptual foundation that so marked physics, chemistry, and mathematics. Even at the popular level, the mass of amateur naturalists who started out watching birds or collecting insects rarely got beyond the identification stage. They made little or no attempt to understand the organism, to find how it really lived or what its function was in nature. Even professionals fell into this trap, or at least they confined their work to descriptive biology. As a result natural history, once a rigorous subject, lost its position among the sciences and became equated with emotionalism and superficiality. But the ecological revolution of 1970 ended all that.

With the environmental awareness of the 1970s, interest in natural science began to revive. Suburban man has become acutely aware of his environment, and there is a new impetus to study the natural world. Books on natural history and ecology have become popular sellers; even the Reed *Bird Guides* and Comstock and Burroughs are back in vogue. Environmental study has returned to some classrooms; interest in wildlife and forestry has increased. Public outcries, wise or unwise, have been voiced against hunting and against environmental destruction by timber-cutting

5

practices, highways, dams, power plants, and strip mining. Many people are seeking a closer contact with the natural world. Some, especially the young, seek to return to the earth by establishing rural communes and attempting a subsistence agricultural way of life. Industry for the first time is finding itself on the defensive. Its uncontested right to pollute the air and water and to destroy the landscape for profit is being challenged.

Thus natural history evolved into ecology and ecology into a science that has entered the public consciousness. Where the old focal point was kinds of organisms, the new focal point is the nature of living systems. Just as molecular biology attempted to probe the secrets of living systems at the cellular level, so ecology probes the secrets of living systems at the levels of the organism, the populations, and the ecosystem.

The term *ecology* was coined by the German zoologist Ernst Haeckel, who called the "relation of the animal to its organic as well as its inorganic environment" *ökologie*. The origin of the word is the Greek *oikos*, meaning "household" or "home" or "place to live." Thus ecology deals with the organism and its place to live. Basically this is the organism's environment; so ecology might well be called environmental biology. That word *environment*, like sin, covers a multitude of things. For one thing the environment includes the organism's surroundings. It also includes for the individual organism those of its own kind, as well as organisms of other kinds. There are relationships between individuals within a population and with individuals of different populations. Animals react in a social sort of way, in various behavior patterns. Because all organisms have become adapted to the environment and are always adjusting to a changing environment, natural selection and evolution become a part of ecology.

Because of its far-flung involvements with so many fields, ecology, call it what you will, is often regarded as a generality rather than a speciality. Indeed one ecologist, A. MacFadyen, in his book *Animal Ecology: Aims and Methods* (1963)* wrote:

The ecologist is something of a chartered libertine. He roams at will over the legitimate preserves of the plant and animal biologist,

* Full information for sources can be found in the Bibliography.

the taxonomist, the physiologist, the behaviourist, the meteorologist, the geologist, the physicist, the chemist, and even the sociologist; he poaches from all these and from other established and respected disciplines. It is indeed a major problem for the ecologist, in his own interest, to set bounds to his divagations.

This statement nicely emphasizes that ecology is a multidisciplinary science. It has to be to reach the heart of the problems of environmental biology.

It is difficult to trace ecology back to any clear beginnings. The Greek scholar Theophrastus, a friend and associate of Aristotle, wrote of the interrelation between organisms and their environment. But modern impetus to the subject probably came from the plant geographers Humboldt, De Candolle, Engler, Gray, and Kerner. They described the distribution of plants, and in so doing raised some questions that have not been answered yet.

Out of the roots of plant geography developed another subject of study, the plant community, which became *community ecology*. The study of the plant community developed in two regions, western Europe and the United States. In Europe Braun-Blanquet (1932) and others concerned themselves with the composition, structure, and distribution of plant communities. In America, such plant ecologists as Cowles (1899), Clements (1916, 1939), and Gleason (1926) studied the development and dynamics of plant communities. While these investigators were studying plants, Shelford (1913, 1937), Adams (1909), and Dice (1943) in America and Elton (1927) in England were investigating the interrelations of plants and animals.

At the same time an interest in dynamics of populations was developing. The theoretical approaches of Lotka (1925) and Volterra (1926) stimulated the experimental approaches by biologists. In 1935 Gause investigated the interactions of predators and prey and the competitive relationships between species. At the same time Nicholson studied intraspecific competition. Later the work of Andrewartha and Birch (1954) and the field studies of Lack (1954) provided a broader foundation for the study of the regulation of populations. The discovery of the role of territory in bird life by H. E. Howard in 1920 led to further studies by Nice in the 1930s and 1940s. Out of such studies came

the field of *behavioral ecology*. In the 1940s
and 1950s Lorenz and Tinbergen developed
concepts of instinctive and aggressive behavior.
The role of social behavior in the regulation
of populations was explored in some depth by
Wynne-Edwards (1962) in England.

The writings of Malthus (1798), who
called attention to expanding populations and
limited food supply, led indirectly to *evolutionary studies* and the *new systematics*, for
it was from Malthus' essay on populations
that Darwin received the first inspiration for
his theory of evolution. Out of the work of
Darwin (1859) and the genetic theories of
Mendel (1866), S. Wright (1931), and others
grew the field of *population genetics*, the study
of evolution and adaptation.

For a number of years the dynamics of
population and community ecology occupied
the full attention of most ecologists. But early
investigations of the physical environment of
organisms, stemming largely from the work
of Leibig (1840), led to *ecoclimatology* and
physiological ecology. Out of such studies in
the aquatic environment developed the field
of *ecological energetics*. In 1920 the German
limnologist Thienemann introduced the concept of trophic levels in terms of producers
and consumers. Two American limnologists,
Birge and Juday in the 1940s, through their
measurements of the energy budgets of lakes,
developed the idea of primary production. Out
of their studies came the trophic-dynamic
concept of ecology. Introduced by Lindeman
in 1942, this concept marked the beginning
of modern ecology. Out of Lindeman's study
came further pioneering work on energy flow
and energy budgets by Hutchinson and H. T.
and E. P. Odum in the 1950s in America.
Early work on the cycling of nutrients was
done by Ovington (1957) in England and
Australia and by Rodin and Bazilevic (1967)
in Russia.

Modern ecology is focused upon the concept of the *ecosystem*, the major functional
unit consisting of interacting organisms and
all aspects of the environment. About this
concept revolve a number of basic principles.
Even in the 1970s they have not become
standardized into the kinds of basic laws that
one finds in genetics or mathematics, but in
time they will be. As an introduction to the
material to come, some principles might be
stated as follows.

1. The ecosystem is a major ecological unit. It contains both abiotic and biotic components through which nutrients are cycled and energy flows.

2. To accomplish those cycles and flows ecosystems must possess a number of structured interrelationships between soil, water, nutrients, producers, consumers, and decomposers.

3. The function of ecosystems is related to the flow of energy and the cycling of materials through the structural components of the ecosystem.

4. The total amount of energy that flows through a natural system depends upon the amount fixed by plants or producers. As energy is transferred from one feeding level to another, a considerable portion is lost for further transfer. This limits the number and mass of organisms that can be maintained at each feeding level.

5. Ecosystems tend toward maturity; in so doing they pass from a less complex to a more complex state. This directional change is called succession. Early stages are characterized by an excess of potential energy and a relatively high energy flow per unit of biomass. In mature ecosystems there is less waste and less accumulation of energy because the energy flows through more diverse channels.

6. When an ecosystem is exploited and that exploitation is maintained, the maturity of the ecosystem declines.

7. The major functional unit of the ecosystem is the population. It occupies a certain functional niche that is related to the population's role in energy flow and cycling of nutrients.

8. Relationships among populations create new functional niches, so that the accumulation of species in an ecosystem, and the increase in maturity, are to some extent self-reinforcing processes.

9. A functional niche within a given ecosystem cannot be simultaneously and indefinitely occupied by a self-maintaining population of more than one species.

10. Both the environment and the amount of energy fixation in any given ecosystem are limited. When a population reaches the limits imposed by the ecosystem, its numbers must stabilize or, failing this, decline (often sharply) from disease, strife, starvation, low reproduction, and so on.

11. Changes and fluctuations in the environment (exploitation and competition, among others) represent selective pressures upon the population to which it must adjust. Organisms that cannot adjust disappear, perhaps decreasing for a time the maturity of the ecosystem.

12. The ecosystem has historical aspects: the present is related to the past, and the future is related to the present.

Ecology developed along two routes—one plant, the other animal. Plant ecologists focused their attention on the relationships of plants to other plants, and largely ignored the influence of animals on the plant community. Animal ecologists studied population dynamics and behavior. However, because animals depend upon plants for both food and shelter, animal ecologists also considered the relationships between animals and plants. This is especially true in the applied areas of ecology such as range management, wildlife management, and to some extent forestry.

Both plant and animal ecology are artificially divided into two parts, *synecology* and *autecology*. (For a good example of this division in plant ecology see R. F. Daubenmire's *Plants and Environment: A Textbook of Plant Autecology*, 1959, and *Plant Communities: A Textbook of Plant Synecology*, 1968). Autecology is concerned with the study of the interrelations of individual organisms with the environment. Synecology is concerned with the study of groups of organisms—the community. Both have developed independently, although a knowledge of both is necessary for the understanding of the individual, population, or ecosystem. Autecology is experimental and inductive. Synecology is philosophical and deductive. Autecology, because it is concerned with the relationship of an organism to one or more variables such as light or salinity, is easily quantified and subject to experimental design both in the laboratory and in the field. Synecology is largely descriptive and not easily subject to experimental design. Autecology has borrowed techniques from physics and chemistry. With the development of such tools as computers and radioactive tracers synecology has entered a strong experimental phase.

Synecology is often subdivided into aquatic,

terrestrial, and marine ecology. Terrestrial ecology, subdivided further into such areas as forest ecology, grassland ecology, and desert ecology, is concerned with terrestrial ecosystems—their microclimate, soil chemistry, nutrient and hydrological cycles, and productivity. Because they are biologically controlled and subject to much wider environmental fluctuation, terrestrial ecosystems are more difficult to study than aquatic ecosystems. Because aquatic ecosystems are affected more by the physical environment and have more stable environmental conditions (such as temperature), considerable attention is paid to chemical and physical characteristics. But as will be seen in the pages to come it is very difficult to separate aquatic from terrestrial ecology. Because one relates to the other in the form of inputs and outputs of nutrients and water, one cannot be studied without some references to the other.

Autecology and synecology aside, the most recent approach is that of systems ecology. That part of ecosystem ecology is concerned with the analysis and understanding of the function and structure of ecosystems by the use of applied mathematics such as advanced statistical techniques, mathematical models, and computer science. Involved in systems ecology is the construction of models that represent the real system for the purpose of experimentation. To be valid the model has to mimic the real system at least over some restricted range, include the important variables, and be subject to mathematical expression. Models can be constructed to provide a simplified description of a system or to predict changes over time. One of the most commonly used is the compartment model, which essentially describes the input, flow, and output of energy and material through an ecosystem (see Figs. 5–17 and 6–9). The compartment is broken down into subcompartments representing subprocesses. These subcompartments become experimental components subject to experimental analysis.

Because ecologists are working with living systems, the variables are numerous and often highly complex. The tools and techniques used by the physical scientists cannot be easily applied; nor are the results as precise as those obtained in physics and chemistry. But in spite of these problems the ecologist does use the tools of chemistry, physics, and mathematics. Physical and chemical measurements are taken to measure the various parameters of the environment. These may range from simple chemical determinations of the various elements to the use of such sophisticated apparatus as infrared gas analyzers, recording spectrophotometers, and microbomb calorimeters. The use of statistical procedures such as correlation, multiple regression, and matrix algebra, and the application of modern algebra, calculus, and computer science to mathematical models simulating field conditions are providing new insights into population interactions and ecosystem functioning.

Research has been stimulated by the development of new tools that enable ecologists to explore new areas. With the use of electronic equipment and biotelemetric techniques, ecologists can sample and measure plant and animal populations without destroying them. Radioisotopes enable investigators to follow the pathways of nutrients through ecosystems and to determine the time and extent of transfer. Laboratory microcosms—samples of both aquatic and soil microecosystems taken from natural ecosystems—are useful in determining rates of nutrient cycling and other parameters of ecosystem functioning. Such modern tools are encouraging new advances in ecology that are necessary for the continuance of life on earth. Thus ecology, so long criticized as purely descriptive, is gaining quantitative data and conceptual strength and is rapidly becoming one of the most important and critical areas of modern science.

During the 1960s biologists were divided into camps. In one were the molecular biologists, who were concerned with the basic biochemical structure of life. To them much of biology centered on chemistry and physics. These areas of study—cellular physiology, biochemistry, and biophysics—were regarded as somewhat glamorous; they had a sort of bandwagon attraction. In the other camp were the environmental biologists—the ecologist, ethologists, taxonomists, evolutionists, and pedologists. They were rather low in status, in spite of the mounting environmental crisis. Because of the complexity of the problems, often beyond experimental control, ecology was viewed by many as almost a nonscience. Now in the 1970s ecology has emerged as the "science of survival" and as such is beginning to suffer from the bandwagon effect so characteristic of the molecular sciences in the 1960s.

The division between the two areas of biology has been unfortunate, for each can offer much to the other. Many of the discoveries of molecular biology are relatively meaningless unless they can be viewed from the vantage point of the population and the ecosystem. And some of the problems at the population and ecosystem levels can be answered only with the help of molecular biology. In reality biology is a gradient from the molecular level to the cell to the organism, to the ecosystem, and each segment blends into the others.

Now in the 1970s ecology is being described as the new science, and is involved in social, political, and economic issues. This involvement came about primarily because the public was aware of environmental problems and was seeking answers to the problems of pollution, overpopulation, and degraded environments. Many treat the issues as if they are new, as if ecology and what the science and its spokesmen have to say are very contemporary. Actually ecologists have been saying the same things for many years, but few paid any attention. In 1864 George Perkins Marsh's dramatic book *Man and Nature* called attention to the effects of poor land use on man's environment, yet it had little impact. Today it is a popular paperback. Aldo Leopold wrote his famous essay "The Conservation Ethic" back in 1933. It was reprinted under the title "The Land Ethic" in his posthumous book *A Sand County Almanac*, which was read largely by people interested in wildlife conservation. It was not until 1970 that the book was widely distributed as a paperback and became a bible of the ecology movement.

Rachel Carson probably did more than anyone to call attention to environmental problems. Since the publication of her book *Silent Spring*, we have become increasingly aware of the chemical poisons and other pollutants that are being cycled through the ecosystem. Once castigated as more fiction than fact, Carson's dire predictions became only too true. As urban developments eat away at the countryside, more and more people are becoming concerned about open spaces and are beginning to accept the work of such ecological land planners as Ian McHarg. The problem of population growth, once whispered about, is now widely debated in public. *The Population Bomb* by Paul Erlich and *Too Many* by George Bergstrom are widely read paperbacks; zero population growth is a goal toward which many in the United States have become committed. Population control, a subject once taboo, is now the topic of numerous books and newspaper and magazine articles. The public has become aware that man can exceed the limits of his environment, that he can deplete and contaminate it.

In spite of the widespread acceptance of the facts, we have had little real success in improving the environment or preventing its continued contamination. Population growth has been slowed only in some parts of the world. Industry has reluctantly accepted the fact that it must reduce pollution, but efforts to do so have been impeded by such unpopular yet inevitable results as higher costs, employment layoffs, and decreased production. Efforts to reduce pollution by pesticides and heavy metals are resisted by both industry and agriculture. The increasing concentration by agriculture upon single-crop farming has developed simplified ecosystems that create many difficulties and dangers. Pressures still mount against parklands and other semiwild places. Wildlife species decline in the face of "progress." Roads still cut across the landscape to make room for more cars to move faster from one place to another. Subdivisions still close in on prime agricultural lands while in more undeveloped regions forest and grasslands are being destroyed to make room for more agricultural lands.

This discourse could be continued, but enough has been written to press home the fact that the future of human life on earth demands more knowledge about ecosystems than we now possess. For the first time in the history of the earth man has become the completely dominant organism, changing the earth and its vegetation almost at will, with little regard for the consequences. It is little wonder then that some of today's most intellectually challenging problems are found in the area of ecology.

magine yourself in a spacecraft far from earth. Through the window the planet appears as a bluish ball suspended in the black void of space. Its surface is kaleidoscopic, patterns of white, blues, reds, and greens constantly changing across its surface as it turns and as swirls of clouds move across it. It hangs alone, self-contained, dependent on the outer reaches of space only for the energy of sunlight. It is close enough to the sun to be warmed by its radiant energy, yet it does not become overheated or overcooled; it is protected from damaging radiation by an atmosphere unlike that possessed by any of its sister planets.

As the spacecraft approaches, it enters the outer fringes of the earth's atmosphere, the *exosphere* (Fig. 2-1), which extends some 2,200 mi from the planet. The dominant element is hydrogen, and the temperature ranges from 200° to 10,000° F. In this region the atoms and particles are so far apart that they rarely collide, and friction is negligible. High temperatures are determined by the radiation of solar energy from an object, not by the temperature of the surrounding gases.

Further into the exosphere the spacecraft enters a subregion characterized by helium. Passing out of the helium zone, it moves into the *ionosphere*, the outer reaches of which are some 250 mi from the earth. The temperatures are still high, reaching 200° F, and the upper limits are characterized by a layer of oxygen largely ionized into ozone by ultraviolet radiation. As the spacecraft travels closer to earth, it passes through the so-called D region (Fig. 2-2) in which the nitrogen increases. Within the lower subregion, which is some 125 mi out from the earth, the ratio of oxygen to nitrogen, largely in the form of nitric oxide, is nearly 1 : 1. The closer the spacecraft comes to earth, the greater the ratio of nitrogen to oxygen. Eventually the craft moves through a nitrogen layer, mostly nitric oxide ionized by sunlight. This region is known as the *thermosphere*. There the temperature decreases as the distance from the sun increases.

At this point our spacecraft enters the first of the inner envelopes of the earth's atmosphere, the *mesopause*, which is within 55 mi of the earth's surface. This is a region of cold temperatures: −120° F. As the spacecraft descends through the mesopause, the temperature of the atmosphere increases. About 30 mi. out, the craft enters the *stratosphere*, a layer where the temperature is fairly constant at −75° to −45° F in the outer reaches, increasing to 30° F at its lower boundary. Roughly coinciding with the lower stratosphere is a layer often called the *ozonesphere*. There the sunlight ionizes oxygen to ozone by photochemical dissociation. The ozone layer is most heavily concentrated at 60,000 to 80,000 ft. The density of the layer is so low that if it were brought to the pressure and temperature of sea level, it could be compressed into a layer 0.1 in. thick. Yet this layer is sufficient to absorb solar radiation completely, so that it becomes the primary source of heat for the stratosphere. If not for this layer of ozone, the temperature of the lower boundary of the stratosphere would be nearer −70° than 30° F. This would have a pronounced effect on life on earth.

Once through the ozone layer, the spacecraft passes through a layer of sulfates into the *troposphere*, the transition point where the temperature drops rapidly from 30° to nearly −70° F. The troposphere hangs at 5 mi over the earth at the poles, nearly 10 mi over the equator. As the features of the earth come rapidly into view, the temperature gradually increases, and the craft is immersed in a sunlit ocean of air.

Solar radiation

The amount of solar energy per unit time incident on a surface at the mean distance from the earth to the sun is estimated at 2.0 cal cm^{-2} min^{-1}. This value, called the *solar constant*, varies only slightly during the year. Of this quantity about 1.0 cal cm^{-2} min^{-1} reaches the earth's surface. The amount of solar radiation that reaches a place depends upon the atmosphere, moisture, concentration of ozone and dust particles, altitude, and location on the planet. As the altitude above sea level increases, the amounts of water vapor, dust, air molecules, and carbon dioxide in the path of the sunlight diminish. Thus at higher altitudes solar radiation is more intense.

In passing through the atmosphere, the energy of certain wavelengths is absorbed, so the spectrum of energy that reaches the earth's surface is different from that of the incident solar radiation (Fig. 2-3). Ultraviolet wavelengths are nearly all removed by the atmosphere. Molecules of atmospheric gas scatter the shorter wavelengths, giving a bluish color to the sky and causing the earth to shine out in space. Water vapor scatters radiation of all wavelengths, so that an atmosphere with much water vapor is whitish; thus the grayish appearance of a cloudy day. Dust scatters long wavelengths to produce the reds and yellows in the atmosphere. Because of the scattering of solar radiation by dust and water vapor, part of the solar radiation reaches the earth as diffuse light from the sky, called *skylight*. Infrared radiation that reaches the earth and is felt as heat (sensible heat) is absorbed, and a portion is reradiated back as far infrared (4 to 100 μ). What we see as light is visible radiation, which can be broken down into the spectrum that ranges from violet to red (Fig. 2-3).

Of the solar radiation that penetrates the earth's atmosphere, 27 percent is reflected back by clouds, dust, and aerosols, and 2 percent by the earth. Thus 29 percent of the solar radiation is returned as short wave radiation, and is lost as far as energy exchange is concerned. Clouds absorb another 14 percent. Therefore, of the solar radiation that reaches the atmosphere, only 57 percent reaches the earth's surface. Of this roughly 38 percent is direct sunlight and 19 percent is scattered light. The amount of each at any one place varies daily and seasonally, depending upon cloud cover. On a cloudy day most of the sunlight is skylight, diffused and evenly distributed throughout the visible wavelength to give a gray color to the clouds. The solar radiation absorbed as short wave energy by the earth and reradiated as long-wave energy is responsible for heating the earth's atmosphere. Long wave radiation, prevented from escaping into space by the ozone layer of the stratosphere, warms the upper atmosphere. Thus the atmosphere receives most of its heat directly from the earth and only indirectly from the sun.

Sunlight does not strike the earth uniformly. Because of the earth's shape and its tilt on the axis, the sun's rays strike it more directly on the equatorial than on the polar regions (Fig. 2-4a). The lower latitudes are heated far more than the polar ones. Air heated at the equatorial regions rises until it eventually

Space to organism: an overview

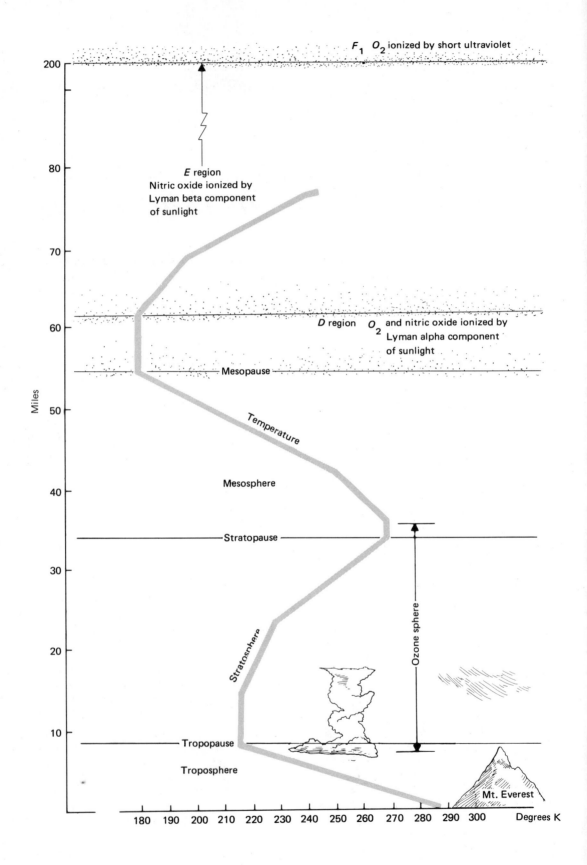

reaches the stratosphere, where the temperature no longer decreases with altitude. There the air whose temperature is the same or lower than that of the stratosphere is blocked from any further upward movement. With more air rising the air mass is forced to spread out north and south toward the poles. As the air masses approach the poles, they cool, become heavier, and sink over the arctic regions. This heavier cold air then flows toward the equator, displacing the warm air rising over the tropics.

Winds

If the earth were stationary and without irregular land masses and oceans, the foregoing would be the unmodified circulatory pattern of the atmosphere. The earth, however, spins on its axis from west to east. This spinning produces a Coriolis force that deflects the winds and prevents a simple direct return flow of air from the poles to the equator. In the Northern Hemisphere, the winds are deflected to the right; in the Southern Hemisphere to the left. Friction produced by the earth's surface modifies the simple flow pattern. The result is a series of belts of prevailing east winds in the polar regions, the polar easterlies, and near the equator, the easterly trade winds. In the middle latitudes is a region of west winds known as the westerlies. These belts break the simple flow of air toward the equator and the flow aloft toward the poles into a series of cells (Fig. 2-4a).

The flow is divided into three cells in each hemisphere. The air that flows up from the equator forms an equatorial zone of low pressure. The equatorial air cools, loses its moisture, and descends. By the time the air has reached 30° latitude it has lost enough heat to sink, forming a cell of semipermanent high pressure encircling the earth and a region of light winds known as the horse latitudes. The air, warmed again, picks up moisture and rises once more. Some of it flows toward the poles, some toward the equator. Meanwhile, at each pole another cell builds up and flows outward in a southerly and northerly direction to meet the rising warm air at approximately 60° latitude, flowing toward the poles. This convergence produces a semipermanent low-pressure area at about 60° N and S latitude.

Land heats and cools more rapidly than the

FIGURE 2-2 [OPPOSITE]
The atmosphere surrounding the earth in turn consists of a series of layers. As the spaceship returns to earth it passes through the heterosphere, in which oxygen and nitric oxides are ionized by short ultraviolet radiation. The heterosphere can be subdivided into the ionosphere (E and F regions) and the thermosphere. In the ionosphere the value of the temperature and even the definition of temperature are uncertain. The temperatures of material objects are determined by radiative equilibrium conditions. In the thermosphere, molecules of air are so widely spaced that high-frequency audible sounds are not carried by the atmosphere, and the concept of the speed of sound loses meaning. Below the thermosphere the spaceship passes through the mesosphere, the stratosphere, and the troposphere.

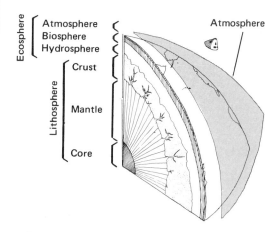

FIGURE 2-1
The planet earth viewed in cross section is layered. Enveloping the earth is a gaseous layer, the atmosphere. The outer layer of the earth itself is the biosphere, the relatively thin zone where the land, sea, and sky meet and where life exists. Below the crust and extending to the surface of the earth is the hydrosphere, the water system. The atmosphere, the hydrosphere, the biosphere, and the crust make up the ecosphere. The bulk of the earth, the lithosphere, consists of the crust, the mantle, and the outer and inner cores.

15

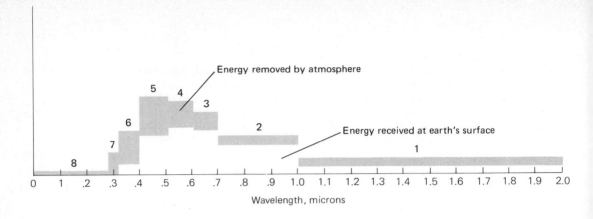

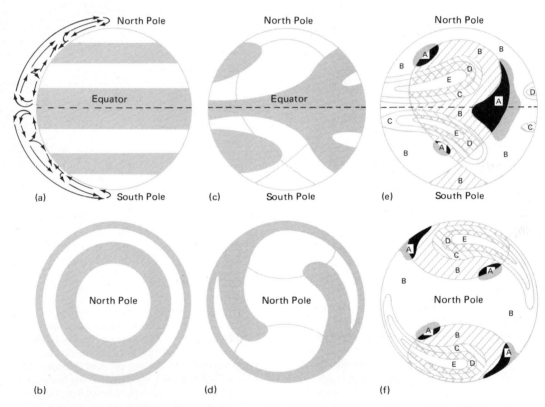

The letters and shadings in parts (e) and (f) represent the climatic conditions of the earth according to Thornthwaite's classification:

Climatic region	Vegetation type	Code
Superhumid	Rainforest	A
Humid	Forest	B
Subhumid	Grassland	C
Steppe	Semiarid	D
Desert	Arid	E

oceans. Because the masses of land and water on the earth are not uniform, the surface of the earth experiences uneven heating and cooling. At any given time the temperature changes are much greater over continental areas than over oceans. Oceans act as heat reservoirs; the continents affect the circulation. In winter the west coasts of the continents are warmer than the east coasts because the air reaching the west coasts has traveled over warmer ocean areas. The interaction of winds and heating produces more or less permanent high-pressure cells known as the Subtropical Highs in the Atlantic and Pacific Oceans; winds and cooling produce low-pressure cells such as the Aleutian and Icelandic Lows. The highs are more pronounced during the summer months, the lows in the winter months.

Also produced are the monsoon winds, dry winds that blow from continental interiors to the oceans in summer and winds heavy with moisture that blow from the oceans to the interiors in winter. Last, there are moving air masses with their cyclonic and anticyclonic frontal systems. These major air circulations are responsible for the changing swirls or cloud patterns seen over the earth from space.

These circulatory patterns and the moisture regimes they influence are responsible for the forests, grasslands, and deserts, which in turn reflect the climatic pattern of the planet (Fig. 2-4).

Ocean currents

The turning of the earth, solar energy, and the winds produce the currents that tie all the seas together. Just as the rotation of the earth deflects the movement of the air, so it deflects the currents of the ocean to the right in the Northern Hemisphere and to the left in the Southern Hemisphere. These deflections are further accented by the continental land masses.

In the equatorial regions there are two broad streams running east to west, the North and South Equatorial Currents (Fig. 2-5). Flowing east between them is the narrow Equatorial Counter Current. The currents in the Northern Hemisphere flow first in a northerly direction, and turn directly east and then south to complete the circle. In the Southern Hemisphere the currents are reversed. The water flows south, then east and north. The

17

FIGURE 2-3 [OPPOSITE TOP]
Energy in the solar spectrum before and after depletion by the atmosphere. Note the variations in that depletion in the different wavelengths. (From Reifsnyder and Lull, 1965.)

FIGURE 2-4 [OPPOSITE BOTTOM]
If the earth were uniform, without irregular masses of lands and oceans, the general circulation of the atmosphere and the rainfall belts would appear as in (a), looking down on the equator, and (b), looking down on the pole. Because of the tilting of the earth on its axis, the rays of the sun fall more directly on the equatorial than on the polar regions, and the lower latitudes are heated more than the polar ones. Air heated at the equator expands, rises, and flows poleward. The total weight of the air at the poles increases and creates high pressure on the ground. At the same time the outflow of air around the equator causes low pressure near the ground. The difference in pressure causes the flow of air near the ground southward from the poles to the equator. The earth, however, is spinning on its axis from west to east. This causes a deflection of the winds, which prevents a simple direct return flow of air from the poles to the equator. This deflection results in a series of belts of winds that break the flow of air on the ground toward the equator, and the flow aloft toward the poles into a series of three cells in each hemisphere, as indicated by the arrows. This influences rainfall. Shaded portions are areas of maximum rainfall; unshaded portions are dry areas.

The earth is not uniform. Masses of water and land are unequal, which prevents the development of rain belts that correspond to the belts of ascending air in the low and high middle latitudes. Land heats and cools more rapidly than water; the oceans act as heat reservoirs, and the continents affect the pattern of circulation of air. Parts (c) and (d) show a generalized rainfall pattern modified by oceans and continents. Again, the shaded portions are areas of maximum rainfall; unshaded areas are dry. The loop formed by the black line in (c), looking down on the equator, is a generalized continental area. It roughly represents North and South America in the Western Hemisphere and Europe, Asia, and Africa in the Eastern Hemisphere. In (d), looking down on the North Pole, the continents are represented by an egg-shaped loop in each hemisphere. The dry areas form an S in the Northern Hemisphere, the center being at the pole. Parts (e) and (f) carry the circulation pattern one step further. The rainfall varies greatly over the earth from less than 1 in. to over 900 in. This variation is reflected in the vegetation pattern, which in turn reflects climate. (From U.S.D.A. 1941 Yearbook, Climate and Man.)

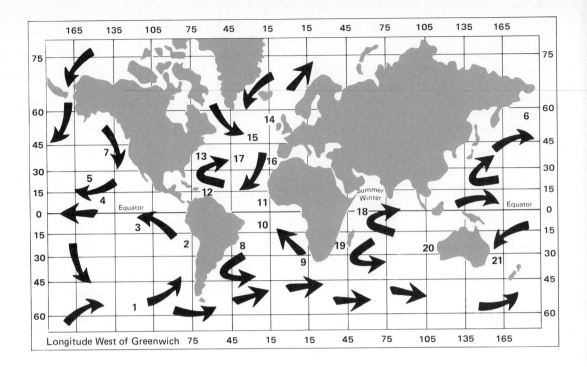

Longitude West of Greenwich 75 45 15 15 45 75 105 135 165

Southern Hemisphere also possesses another great current, the Antarctic Drift, which moves from west to east around the Antarctic continent (Fig. 2-5).

Feeding out of these are other currents. In the North Atlantic, the westerly flowing North Equatorial Current is joined by a part of the South Equatorial Current. The two waters move north through the Straits of Florida. Past the Bahamas this stream picks up waters from another branch of the North Equatorial Current. Together these two become the Gulf Stream, which flows northeast to meet the cold south-flowing Labrador Current in the vicinity of the Grand Banks of Newfoundland. The waters in several branches flow south along the European coast to create the great central eddy, the Sargasso Sea, an area of still and slowly turning water.

In the South Atlantic, part of the South Equatorial Current that does not join the Gulf Stream flows south along the South American coast as the warm Brazil Current. It joins the Benguela Current, which brings cold water out of the Antarctic Drift along the west coast of Africa. In the South Pacific, the Humboldt or Peru Current carries subantarctic water from the Antarctic Drift north along the South American coast. Winds drive the surface waters away from the coast; this water is replaced by cold water upwelling from the deeper layers.

The North Equatorial Current in the Pacific flows west to the Philippines, where most of it moves northeast to Japan. The Japan Current splits into two branches, one of which flows south to Hawaii and merges into the easterly North Pacific Current. The northern branch mixes with the cold Oyashio Current, which forms a mass of subarctic water below the Aleutians and Alaska. This water flows east as the Aleutian Current, one part of which branches west to the Gulf of Alaska; the other part flows south along the United States to join the California Current. It in turn merges with the North Equatorial Current, completing the great circle.

The biosphere

As the spacecraft more closely approaches the earth, the blues and greens and reds sharpen into broad outlines of mountains, deserts, plains, and seas. The colors reflect the blanket of life that wraps the earth. This thin covering occupies the narrow interface of land, air, and water known as the *biosphere* (Fig. 2-1). It does not extend far into the atmosphere or very deeply into land or sea. To this interface life on the planet is largely confined.

Supporting the biosphere is the *lithosphere* (Fig. 2-1) which consists of the crust, the mantle, and the outer and inner cores. For life the most important layer is the crust, and only a very thin portion of that. Beyond serving as a substrate, the crust is a primary source of nutrients. Of the elements and minerals it contains, the ones that are important to the biosphere are not necessarily the ones found in greatest abundance. For example, silica, the most abundant mineral in igneous rocks, comprises only 0.15 percent of living plants. Calcium and potassium, which make up a little more than 3 percent each of igneous rocks, make up 0.5 and 0.3 percent of terrestrial plants and 1.38 and 2.2 percent, respectively, of man.

The ecosystem

As the spaceship comes even closer to the earth, the deserts, mountains, flatlands, and seas focus into expanses of grasslands, forests, croplands, rivers, lakes, estuaries, and oceans (Fig. 2-6). Each is physically and biologically different, and each is inhabited by different organisms that are well adapted to the environment in which they are found. Yet in spite of their differences, each functions in the same fashion. Energy is fixed by plants and transferred to animal components. Nutrients are withdrawn from the substrate, deposited in the tissues of plants and animals, cycled from one feeding group to another, released by decomposition to the soil, water, and air, and then recycled. The deserts, forests, grasslands, and seas are not independent of each other. Energy and nutrients in one find their way to another, so that ultimately all parts of the earth are interrelated, each comprising a part of the total system that keeps the biosphere functioning. The forests, grasslands, desert and tundras, lakes, rivers, and seas—the total planet—all are *ecosystems* (Fig. 2-7a, b).

The word *ecosystem* was coined by A. G. Tansley (1935) in an article that appeared in *Ecology:*

The more fundamental conception is . . . the whole *system* [in the sense of physics] including not only the organism-complex, but also the whole complex of physical factors forming what we call the environment. . . . We cannot separate them [the organisms] from their special environment with which they form one physical system. . . . It is the system so formed which . . . [provides]

FIGURE 2-6
The landscape is a mosaic of topographical patterns, forests, grasslands, old fields, flowing waters. Much of this mosaic is the result of man's attempts to adapt the natural environment to his own social and economic life.

the basic units of nature on the face of the earth. . . . These *ecosystems*, as we may call them, are of the most various kinds and sizes.

The prefix *eco* means environment. A system, according to Webster, is "an aggregation or assemblage of objects joined in a regular interaction or interdependence." A system is more simply defined by Churchman (1968) as "a set of parts coordinated to accomplish a set of goals." The goal of a living system such as the tissues and organs that make up the animal body is a coordinated effort to cause a flow of energy through the system and thus maintain life.

A system implies homeostasis and feedback. Every system consists of sets of different elements. Each set can exist in many different states. The state of one set influences or determines the state of another. As the condition of one set is affected, it in turn affects other sets. In the end the state of the last influences the state of the first. Thus the sets form a complex of coordinated units that are linked together by reciprocal influences that make up a feedback loop. The loop may be positive or negative, stabilizing or disruptive, depending upon the condition of each set. A stable system is one that responds to stimuli by developing forces to restore it to its original condition. In natural ecosystems this stability may mean that the system can survive many changes but still preserve a similar structure. Or stability may imply persistence: the system remains much the way it was.

Consider two ecosystems: the forest and the pond. The sun shining on the open pond warms the shallow water and supplies energy for photosynthetic activity of microscopic plants. These tiny plants in turn support a variety and abundance of minute animal life. Both provide food for young sunfish, tadpoles, and aquatic insects. The insects are eaten by adult sunfish, frogs, and birds. The sunfish and frogs become food for bass and heron. Cattails, reeds, and waterlilies growing along the pond shore furnish food and shelter for muskrats, nesting sites for ducks and red-winged blackbirds, and support for aquatic insects, snails, and flatworms. If the water of the pond is drained, all pond life is destroyed. If the cattails and reeds are covered by fill, the blackbirds, muskrats, and many aquatic insects disappear. If insect life is destroyed, the food supply of frogs and sunfish is eliminated, and this in turn affects the bass and heron. Remove the bass, and the sunfish population may become so large that the fishes' growth will be stunted. Thus all the organisms of the pond depend not only upon clean water but also, directly or indirectly, upon one another for their well-being and existence.

The forest on the slope is quite different from the pond, yet there are many similarities. Trees and other plants capture and channel energy from the sun to other members of the forest. Deer browse on leaves and twigs; earthworms and other soil organisms consume fallen leaves. Insects feed on leaves and plant juices. Woodland mice eat seeds and insects, and they in turn become food for weasels and hawks. The forest provides shelter for many forms of life and modifies the wind and temperature. The forest vegetation depends upon those organisms that break down organic matter and return the minerals back to the soil. When trees are cut or burned, the forest inhabitants disappear and are replaced by other organisms. If deer become too numerous, they overbrowse the forest and destroy young trees and food and shelter for other animals. Just as in the pond, all forest organisms depend, directly or indirectly, on each other for their existence.

All ecosystems, terrestrial or aquatic, have four basic components—the abiotic environment, the producers, the consumers, and the decomposers. The producers, largely green plants, make up the *autotrophic* element, which fixes the energy of the sun and manufactures food from simple inorganic substances. The consumers and decomposers make up the *heterotrophic* element, which utilizes the food stored by the autotrophs, rearranges it, and finally decomposes the complex materials into simple, inorganic compounds again. These two functional elements are arranged into strata, or layers, in the ecosystem. Autotrophic metabolism is greatest in the upper stratum, where light is most available. In the forest this is in the canopy; in the pond it is in the sunlit surface waters where small microscopic plants are concentrated. Heterotrophic activity is most intense where organic matter accumulates. In terrestrial ecosystems this is in the upper layer of the soil; in aquatic ecosystems it is in the sediment.

The producers and consumers in the ecosystem can be arranged into several feeding groups, each known as a *trophic* (feeding) *level* (see Chapter 3). Each trophic level contains at any one time a certain amount of living material composed of a number of kinds of organisms. This is the *standing crop,* most often expressed as the number per unit area or *biomass* (living weight) per unit area. The standing crop consists of a great diversity and number of plants and animals, which collectively make up the species structure of the ecosystem.

The community

The spacecraft is on earth. As one emerges from it, he is aware of the most conspicuous part of the ecosystem, the *biotic community* (Fig. 2-8). It is a naturally occurring assemblage of plants and animals that live in the same environment, are mutually sustaining and interdependent, and are constantly fixing, utilizing, and dispensing energy.

Among the plants and animals of the community only a comparatively few species are found in abundance, either in numbers or in biomass (Preston, 1948; MacArthur, 1960). The common species often are considered the *dominants,* and in the community they may exert some influence over other organisms. In the forest, for example, certain trees of the upper canopy influence the amount of light and moisture reaching the ground and the type of shelter offered other plants and animals. Dominants also affect the soil structure and its chemical composition and in turn the organisms that live in the forest soil.

Although the dominants do influence the physical conditions for many organisms, their influence on the actual distribution of associated species is open to question (Whittaker, 1962). For the most part the plant dominants are widely distributed and thus contain across their range a number of different *ecotypes*—ecological variants adapted to local conditions. Although ecotypes are very similar in appearance and although they have the same modifying influences over the internal environment of the community, their associated species may be quite different. For example, in the southern Appalachians beech forest and oak forest grow at several elevations. The dominants are the same at each elevation but the associated species in the understory are different. The American beech dominates certain communities at low elevations. A second

21

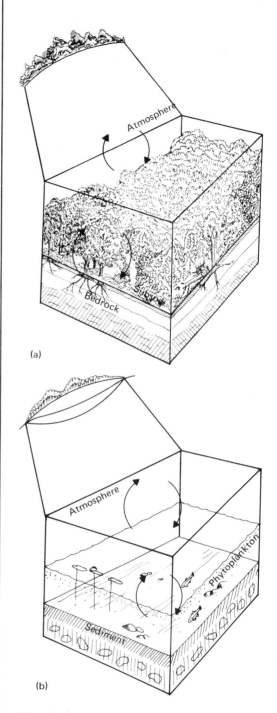

(a)

(b)

FIGURE 2-7
As the spaceship approaches earth the biosphere becomes more distinct. Two types of ecosystems make up the biosphere, the terrestrial (a) and the aquatic (b). Viewed as three-dimensional, both are characterized by a photosynthetic layer, in which energy is fixed, and a layer of decomposition (the forest floor and the aquatic mud), where the materials are recycled back through the plants. Both ecosystems also possess interfaces between the atmosphere and the productive surfaces, and between the decomposition layer and the productive layer, as indicated by the arrows.

beech community growing above 3,000 ft and separated by about 2,000 ft from the lowland population is floristically very different (Whittaker, 1956). On the other hand, some dominant species have a very narrow ecological tolerance, and in this case the associated organisms are the same from one place to another.

Occasionally an animal comprises the dominant mass of the community; this is the case for the coral reef, found in clear, warm waters of tropical and subtropical regions. Coral reefs are built up by anthozoan organisms related to jellyfish and sea anemones. These animals secrete hard, compact, stony skeletons, or cups, in which they live, one generation secreting new cups upon the old. In the chasms and crevices of coral ledges live protozoans, sponges, clams, starfish, octopuses, brilliantly colored fish, and myriads of other forms of life.

Dominance in a community may be the result of the coaction between two or more species, particularly between plants and animals. Prairie dogs can produce and maintain a short-grass stage in the mixed prairie, especially if they have some supplemental help in the form of grazing by cattle and buffalo (Koford, 1958). In earlier days the buffalo on the western plains preferred the taller grasses over the short-grass species. Consequently, over an extensive area the short grass replaced the tall almost entirely (Larson, 1940). The activity of beavers, damming streams and cutting woody vegetation, will eliminate one community and develop and maintain another (Fig. 2-9).

Physical features also may directly control the community. Examples of this are mud flats along a river, rock outcrops on a mountainside, or a body of water or marine bottom where wind and water are controlling elements.

The diversity of species within a community reflects in part the diversity in the physical environment. The greater the variation in the environment, the more numerous are the species because there are more microhabitats available and more niches to fill. Communities possessing a high number of different species usually have complex trophic structures.

There is least diversity in the simpler, less stable ecosystems (A. G. Fischer, 1960; Elton, 1958) such as the tundra and agricultural ecosystems. In tropical regions where the climate is stable, many ecological communities have evolved to maturity. Within them has developed a large number of different species, none of them really abundant. Because each species occupies a rather specialized niche (think, for example, of the wide spectrum of grazing animals on the African plains), the food demands of the population are more widely and evenly distributed. At higher latitudes, in areas of less stable and more severe climates, and in youthful landscapes, the trophic structure is simpler and the species are fewer, but they reach much higher population densities, as a rule (see Chapter 8).

Within any community the diversity is greatest in small organisms. There are many more kinds of insects than there are birds, more birds than mammals, and more small mammals than large ones, irrespective of their position in the food chain. Because they are small, these organisms have become adapted to the conditions afforded by the small, diversified microenvironments found within the community. Each species uses the environment in such a way that it is unavailable to the others (see Chapter 12—the competitive exclusion principle). The total number of individuals of all species together, however, is essentially constant.

The relative abundance of various species within a community depends in part on the species themselves. Some are strictly opportunistic (MacArthur, 1960), common when conditions are highly favorable and scarce when this is not so. No equilibrium is established within the population. Examples of this include diatom blooms in polluted water, many invertebrate populations, and plant populations in the very early stages of succession. Community structure is reflected best by those species that develop some sort of population equilibrium. Among these, relative abundance is important. Some are very common, some are very rare, and between the two is a whole range of intermediates. These latter species, appearing in moderate abundance, are more numerous than either extreme (Preston, 1948).

The population

The many groups of different organisms that make up the biotic community consist of populations of plants and animals (Fig. 2-8). The term *population*, like so many others, has come to mean many things. Considered ecologically, a population is a group of interbreeding organisms of the same kind occupying a particular space. Each population is a structural component of an ecosystem through which energy and nutrients flow. The population is a self-regulating system. It exists because new matter continuously replaces what it inevitably loses. Within an ecosystem stability is related in part to the regulation of populations.

A population can be considered in several different ways. For instance, it is a demographic unit. It is characterized by *density*, the number of organisms occupying a definite unit of space. It possesses a certain *age structure*, the ratio of one age class to another. It has a *birth rate* and a *death rate*. It experiences the movement of new individuals into the population—*immigration*—and loses others through *emigration*. The additions to the population by births and immigrations in relation to losses from death and emigration determine the rate of population growth. A population may fluctuate rather widely at times, but if averaged over a long period its numbers rise and fall about some point or long-term mean. Thus populations are characterized by some regulatory mechanism that tends to keep their size somewhere within the limits of the environment. The regulatory mechanism may be contained within the population and be related to density (density dependent); fluctuations may be environmentally induced (density independent); or fluctuation might result from the interaction of the two.

Because a population is composed of interbreeding organisms, it can also be considered a genetic unit, a collection or pool of genes rather than a group of individuals. Each individual carries a certain combination of genes, or genetic information. The sum of all genetic information carried by all individuals of an interbreeding population is the *gene pool*. In sexual reproduction the genes possessed by individuals are shuffled, sorted, and recombined during meiosis and gamete formation. When members of a population interbreed, gametes combine to form a new set of genotypes. At each generation the gene pool is randomly reconstituted. Local or semi-isolated genetic populations are known as *demes*. Gene flow

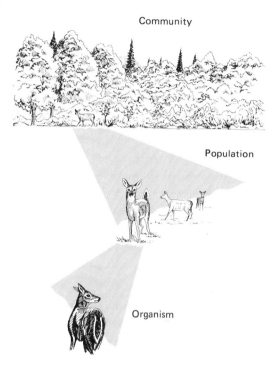

FIGURE 2-8
The living component of the ecosystem is the community that consists of interacting plants and animals occupying the same site. The community in turn consists of populations of plants and animals, and the population of each species consists of individual organisms. In addition to investigating the structure and functions of ecosystems a study of ecology also involves the study of the interaction of organisms with their environment (physiological ecology), of the intraspecific and interspecific relationships of populations (population ecology), and of interaction of plants and animals and their environment (community ecology).

between the demes comes about by immigration and emigration.

The population is also an evolutionary unit. Evolution involves changes in the gene pool and physical expressions of genetic constitution. These changes are caused by selective pressures brought to bear by the environment upon individuals of the population. If the population contains enough variability to allow the gene pool to be altered when necessary, then the population responds by becoming more adapted to the environment. Selection (anything that produces a systematic heritable change in a population) acts as a cybernetic system. Information from the environment is transmitted back to the gene pool, which responds with changes in its variability.

Evolutionary changes involve adaptiveness, or the ability of individuals to live and reproduce in a given environment. Within a population certain individuals are more able to live within a certain range of environmental conditions than others. In other words they are more adapted. Thus better-adapted individuals on the average have more offspring than those not so well adapted, who may not survive long enough to reproduce. As a result less-adapted types decrease, and more-adapted types increase. Subsequently certain variations within the gene pool increase at the expense of other variations and bring about adaptations within the population. Because the environment is always changing, the population, through the survival of certain individuals, is constantly changing adaptive characteristics. The rate at which a population responds to environmental changes determines the ability of that population to survive.

These various features separate one kind of population from another into groups or species recognizable as relatively distinct morphological and physiological types. The *species* is a group of interbreeding individuals living together in a similar environment, reproductively isolated from other groups. The individuals recognize each other as potential mates. They interact with other species in the same environment but do not mate with them. They are a genetic unit in which each individual holds for a short period of time a portion of the contents of an intercommunicating gene pool.

One population becomes genetically isolated or independent of other populations by the process of *speciation*. This is accomplished in most organisms by an interaction of heritable variation, natural selection, and spatial isolation. Once a species becomes established, it remains distinct from other species by *isolating mechanisms*, which inhibit different species from interbreeding successfully. These include morphological characteristics, behavioral traits, habitat selection, and genetic incompatibility.

The population is not simply an assemblage of independent individuals. Individuals interact with each other, communicate in some fashion, mate, and have disputes over mates, territory, and food. This relationship between individuals is expressed as *social behavior*. It involves expressions of aggressiveness and submission, of dominance and subdominance among individual members of a population. It results in the formation of a social hierarchy that influences the ability of an individual to acquire a mate or obtain food. It results in the parceling out of resources through territoriality. Because social behavior involves competitive relationships between members of a population, it has important consequences for the survival and reproductive success of the individual. In this manner social behavior influences and is influenced by natural selection. Social behavior also facilitates the maintenance of an independent existence of one species from another because certain behavior patterns are characteristic of one species only. At the same time it reduces competition for resources among species.

Populations occupy specific places within the community. The place where a population lives and its surroundings, both living and nonliving, is its *habitat*. The woodthrush inhabits a deciduous forest; the wood sorrel grows in acid, humus-rich soil of the cool, deep woods. Even within a given community the distribution of certain organisms may be quite localized because of microdifferences in moisture, light, and other conditions. These localized areas are *microhabitats*.

More than just occupying space, the population of each species in the community performs some function. What the organism does—or, to say it somewhat anthropomorphically, its occupation—in the community is called its *niche*. Some species occupy a very broad ecological niche. They may feed on many kinds of food, plant and animal; or if strictly herbivorous, they may feed on a wide variety of plants. Other organisms occupy highly specialized niches. The woodcock, for example, possesses a sensitive, flexible bill adapted for probing in the soft earth for earthworms, which make up the bulk of the bird's diet. Organisms have arrived at their respective niches through long periods of evolution. Because no two species in the community occupy the same niche (this is discussed in Chapter 8), each more or less complements the other. This reaches a high refinement among the grazing animals of the African plains. Because each kind feeds on preferred food plants, between them they utilize all forms of vegetation from ephemeral annuals and herbs to acacia trees (Darling, 1960; Talbot and Talbot, 1963b). Giraffes feed chiefly on trees, rhinos on brush, wildebeest on grass. Even among animals living on the same class of food, the diets are complementary. Red oats grass is the major forage species for the wildebeests, the topis, and the zebras. The wildebeests feed chiefly on the short, fresh leaves of this grass. Zebras feed on it when it is more mature, when the leaves are over 4 inches long, but they avoid this grass when it is dry. The topis, however, prefer the grass dry.

Just as ecosystems have some basic components, so are the niches they contain basic and repetitive. Through natural selection, different animals (although some may belong to the same genera) in widely separated ecosystems occupy similar niches and perform similar tasks. The mountain lion of North America, for example, feeds on the deer; the African lion feeds on plains antelope and the wildebeest. Animals that have the same occupation in different ecosystems are termed *ecological equivalents*.

The organism

Although populations collectively make up the community, one is most aware of the individual organism (Fig. 2-8), which reflects the morphological and physiological characteristics of the species. One identifies the population or the species and thus the composition of the community by the individuals representing it. The individual organism expresses the genetic variations found in the population. It is the unit upon which natural selection acts. It responds to environmental stimuli, and the collective responses

FIGURE 2-9
A beaver dam across a small stream flooded out acres of forest and created a pond habitat. The presence of certain animals such as the beaver can have a pronounced effect on the nature of local environments. (Photo by author.)

of all individuals become the responses of the population.

The individual reflects the adaptation of the population or the species to the environment in which it is found. The ability of plants and animals to live in the desert results from adaptations to heat and a limited water supply. The ability to exist in the Arctic results from adaptations of plants to make maximum use of a short growing season and of vertebrate animals to conserve metabolic heat. The timing of daily and seasonal activities of populations of plants and animals is mediated through the photoperiodic responses of individuals.

Responses of animals to the environment are more observable than those of plants because animal reactions to sudden changes involve behavior. Behavior is intrinsic to the ecology of an animal. It is involved in the success of the animal in caring for itself, in seeking appropriate shelter, in obtaining food, in escaping enemies, in courtship and mating, and in caring for the young. It is a mechanism that, in part, reduces competition between animals and controls population densities.

The behavior of an animal, like its structure, is the result of natural selection. Perhaps more often than not, structure and behavior evolved together, structure influencing behavior, and behavior in turn influencing the development of structure. To respond to environmental changes, physical and social, the animal must first receive a stimulus from the environment through its sensory system—sight, taste, smell, touch, hearing. The stimulus is transmitted to the motor organs, the muscular system, and finally the motor organs respond. The kind of sensory apparatus the animal possesses, the complexity and organization of its central nervous system, and the type and development of motor organs determine the manner in which the animal is able to respond to its environment. Thus how an animal perceives and reacts to the world about it is limited by what its eyes can see, its ears can hear, and its taste, hearing, and smell can respond to. Because of this, the world appears different to other animals than it appears to man. What is visible and important to them may be unperceivable to humans. Each animal lives, as Jacob von Uexkull put it, in its own phenomenal or self-world of perception and response, its *Umwelt*.

But the responses are not always completely controlled by structure. Species, and even individuals within a species, that appear to be structurally similar and live in the same physical environment may react differently to similar stimuli. Aggressiveness or tameness, vigor in courtship or defense, and the ability to learn are not wholly associated with structure. Often behavioral patterns in the individual are genetically determined, and their appearance, retention, and further development are influenced by the forces of natural selection. Behavior becomes a mechanism for survival, and through a period of time natural selection favors the best-adapted behavior pattern for the species.

Man in the biosphere

As they approach the earth in the spacecraft, the occupants must be impressed with the imprint of man on the planet. No matter what ecosystem one observes, he cannot escape the presence of that one dominant organism. For miles his domination is revealed by the layers of dust and smog that blanket much of the earth. As one gets closer one can see the results of the ameboid growth of human populations stretching across the land. Large areas of natural vegetation have been transformed into patterns of croplands; free-flowing rivers have been dammed to create enormous lakes. Mountains and hills have been pared away for the exploitation of minerals; refuse and debris of modern civilization litter the planet. In places the landscape still remains wild; in places it has been humanized; and in other places it has been ravished.

This is the visible impact of man. Not so noticeable are the more subtle changes he has brought about. He has introduced excessive quantities of certain elements such as mercury, lead, and cadmium into the biogeochemical cycles. He has added synthetic materials such as chloronated hydrocarbons to which organisms had never been exposed.

In his nearly 2 million years of occupancy of the earth, man has been molding the planet to his own designs. For much of his existence he was part of the natural scheme of things. He was a part of the nutrient cycle, and like the other consumer organisms with which he shared the earth, he received his portion of natural energy flow. As his technology developed and his mastery over the environment increased, man destroyed natural ecosystems of which he was part and replaced them with simplified ecosystems. Their components became domesticated plants and animals shaped to man's needs and without whose protection and care they could not exist. Man directed the flow of energy toward his own ends by reducing or eliminating plant and animal competitors. His interference interrupted nutrient cycles, so he supplemented or replaced natural cycling with fertilizers. To increase yields even more he added energy input from draft animals, and later replaced them with power supplied by fossil fuels. When the bulk of human population became concentrated away from the source of food supply, man had to develop mechanisms and institutions for the distribution of foods. This process moves the nutrients in food far from their source and introduces excessive wastes into aquatic ecosystems. This results in the eutrophication of natural waters, which degrades aquatic ecosystems and pollutes water for man's own uses. He preyed upon fish and grazing animals to the point of extinction and he reduced or exterminated other species by destroying their habitats.

Now man in his domination of the earth finds himself on the verge of an ecological catastrophe. At the same time that he is causing his environment to deteriorate and is exhausting exploitable resources, he is allowing his population to push upward to the ultimate carrying capacity of the earth, and there is no place into which it can expand. The result is that man the organism may find it difficult to adapt to an environment he is rapidly changing. However, there are indications that he is slowly beginning to realize that he is not separate from nature but a part of it. He is discovering that he is a functional (even if malfunctioning) unit of ecosystems and that his activities must be in harmony with their functioning. His artificial and natural ecosystems, modified under management, must benefit to the fullest from functional processes common to natural ecosystems. To accomplish this man must develop an understanding of and respect for natural ecosystems. If man is to exist he has to work within the framework of ecosystems of which he is a part. In other words, he needs to develop an ecosystem approach to the management of the earth.

The ecosystem

The sunlight that floods the earth is a source of two forms of energy that keep the planet functioning: *heat energy*, which warms the earth, heats the atmosphere, drives the water cycle, and provides currents of air and water; and *photochemical energy*, which is utilized by plants in photosynthesis, fixed into carbohydrates and other compounds, and becomes fuel for the cool-burning cellular furnaces of living organisms.

The nature of energy

The solar energy incident upon the outer atmosphere of the earth that eventually penetrates the atmosphere and reaches the earth's surface has wavelengths of 0.1 to 10.0 μ (Fig. 3-1). Of this approximately 10 percent is ultraviolet (wavelengths of 0.1 to 0.4 μ), about 45 percent is visible light (wavelengths of 0.4 to 0.7 μ), and the remaining 45 percent is infrared or long wave radiation (wavelengths of 0.7 to 10.0+ μ). What we see as light is visible radiation, broken down into a spectrum that ranges from violet to red.

The transfer of radiational energy involves particles called *quanta*. The quantum units of radiant energy are called *photons*. Radiant energy is also expressed in terms of the *einstein*, the energy of 1 mole (6×10^{23}) of photons. The energy of a photon varies with the frequency of the radiation. A low frequency (long wavelength radiation) possesses less energy than a high frequency. Thus within the range of visible light, 1 einstein of red light with a wavelength of 0.67 μ possesses 42,000 cal, whereas 1 einstein of blue light with a wavelength of 0.47 μ has an energy value of 60,000 cal.

Heat transfer of radiation is measured in several units. For ecologists the *gram calorie* is the most convenient unit of energy. Because ecologists are concerned with energy flow for a region, they measure energy per square centimeter or square meter. For example, 1 g cal/cm² is known as a *langley*. It is a unit of irradiance. Ecologists measure illumination in terms of *lux* (1 lumen/m²). In English units, 1 lumen/ft² is known as a *footcandle*.

Radiation, whether visible light or other wavelengths, is either absorbed, reflected, or transmitted. Natural opaque bodies absorb only a portion of the radiant energy incident upon them. The fraction that is absorbed is termed the *absorptivity*, and it can vary from 0 for the perfect reflector to 1 for the perfect absorber. The fraction that is reflected is termed *reflectivity*. The sum of the two has to equal 1. Part of the energy that is absorbed is reradiated back to the environment. This reradiation, or loss of heat, is termed the object's *emissivity*. It is, of course, directly related to absorptivity. If the object is not opaque, then a fraction of light incident upon it will be transmitted through the object, and the sum of the absorptivity and reflectivity will not equal 1.

Because the transfer of radiation involves the movements of photons of energy from one point to another, energy can be said to flow. This energy flow is known as a *flux*. Energy flows only if there exists an energy source and an energy sink. Without a sink for the flow of thermal energy, the sun could not be an energy source. The earth receives energy from the sun, absorbs part of it, and gives up the energy as heat to a sink, outer space. Without the dispersal of heat energy into a sink, energy cannot flow.

The flow of energy is mediated at the molecular level. Characteristically thermal energy is distributed rapidly among all molecules in the system without necessarily causing any chemical reaction. The effect of thermal energy is to set the molecules into a state of random motion and vibration. The hotter the object, the more its molecules are moving, vibrating, and rotating. These motions tend to spread from a hot body to a cooler one, transferring energy from one to the other. The energy of light waves, on the other hand, especially the red and blue wavelengths, causes electronic transitions within atoms and molecules, called excitations. These excitations can lead to photochemical reactions. The energy of light, or photons, sends one electron of a pair of bonding electrons to a higher state, or orbit. Uncoupled from its partner, it is free to be involved in photochemical reactions.

Energy flow in ecosystems

Energy and the laws of thermodynamics

There are two kinds of energy, *potential* and *kinetic*. Potential energy is energy at rest. It is capable of and available for work. Kinetic energy is due to motion, and results in work. Work that results from the expenditure of energy can both store energy (as potential energy) and arrange or order matter without storing energy.

The expenditure and storage of energy is described by two laws of thermodynamics. The first law states that energy is neither created nor destroyed. It may change forms, pass from one place to another, or act upon matter in various ways, but regardless of what transfers and transformations take place, no gain or loss in total energy occurs. Energy is simply transferred from one form or place to another. When wood is burned the potential present in the molecules of wood equals the kinetic energy released, and heat is evolved to the surroundings. This is an *exothermic* reaction. On the other hand, energy from the surroundings may be paid into a reaction. Here, too, the first law holds true. In photosynthesis, for example, the molecules of the products store more energy than the reactants. The extra energy is acquired from the sunlight, but again, there is no gain or loss in total energy. When energy from outside surroundings is put into a system to raise it to a higher energy state, the reaction is *endothermic*.

Although the total amount of energy involved in any chemical reaction, such as burning wood, does not increase or decrease, much of the potential energy stored in the substance undergoing reaction is degraded during the reaction into a form incapable of doing any further work. This energy ends up as heat, serving to disorganize or randomly disperse the molecules involved, thus making them useless for further work. The measure of this relative disorder is named *entropy*.

The second law of thermodynamics makes an important generalization about energy transfer. It states that when energy is transferred or transformed, part of the energy assumes a form that cannot be passed on any further. When coal is burned in a boiler to produce steam, some of the energy creates steam that performs work, but part of the energy is dispersed as heat to the surrounding air. The same thing happens to energy in the

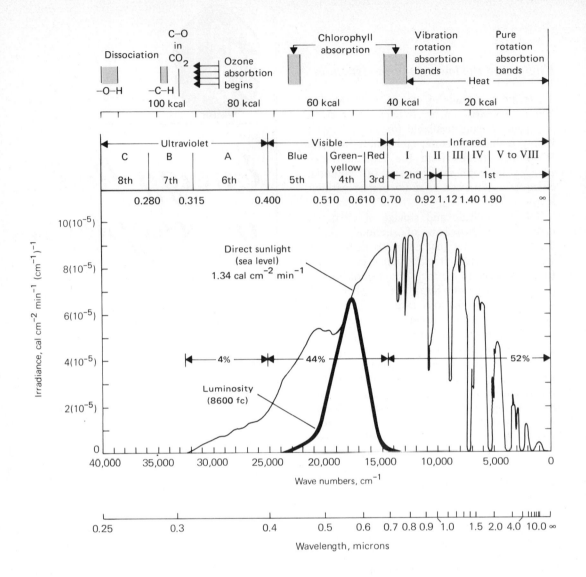

ecosystem. As energy is transferred from one organism to another in the form of food, a large part of that energy is degraded as heat and as a net increase in the disorder of energy. The remainder is stored as living tissue. But biological systems seemingly do not conform to the second law, for the tendency of life is to produce order out of disorder, to decrease rather than increase entropy.

The second law applies to isolated and closed systems approaching thermodynamic equilibrium. However, because the isolated system is an ideal state that can never be achieved, in reality equilibrium thermodynamics applies to *closed systems*, those in which energy, but not matter, can be exchanged between the system and its surroundings. There is a third kind of system, an *open system*, in which both energy and matter can

be exchanged between system and surroundings. A closed system tends toward a state of minimum free energy (energy available to do work) and maximum entropy, whereas an open system maintains a state of higher free energy and lower entropy. In other words, the closed system tends to run down; the open one does not. As long as there is a constant input of matter and free energy to the system and a constant drainoff of entropy (in the form of heat and waste), the system maintains a steady state. Thus life is an open system maintained in a steady state.

Energy enters the biosphere as visible light that is stored in energetic covalent bonds during photosynthesis. From that point biochemical changes involve a series of rearrangements of matter into compounds of less chemical potential energy. These chemical rearrangements are accompanied by the production of heat that eventually goes into the energy sink. This loss of heat is accompanied by a loss of carbon dioxide, water, and nitrogenous compounds, which are recycled through the biosphere. Although some energy is irrevocably lost from the biosphere, some of it is stored in the system. The more organized the system, the longer the energy is stored.

Photosynthesis

The familiar formulation of photosynthesis,

$$6CO_2 + 6H_2O + energy \rightarrow C_6H_{12}O_6 + 6H_2O + 6O_2$$

is incomplete. Whereas the intermediate compounds of photosynthesis do include glucose and the by-products of water and oxygen, the products of photosynthesis also include free amino acids, proteins, fatty acids and fats, vitamins, pigments, and coenzymes. All of these substances probably are synthesized in the chloroplasts by reactions involving photoelectron transport and photophosphorylation. The synthesis of various products may take place in different parts of the plant or under different environmental conditions. Mature leaves of certain species of plants may produce only simple sugars, whereas young shoots and rapidly developing leaves may produce fats, proteins, and other constituents.

Photosynthesis is carried out primarily by cholorophyll-bearing vascular plants, both aquatic and terrestrial. Other minor contributions are made by photosynthetic bacteria that use, instead of water, hydrogen, hydrogen sulfide, and various organic compounds as electron donors. Although photosynthesis is a complicated process, the details of which are still largely conjectured, the essential features can be outlined briefly.

The light energy that strikes a green leaf is absorbed by the pigments in the chloroplasts. The most important of these pigments is chlorophyll a, which has a red light absorption of less than 680 nm. When a molecule of chlorophyll a absorbs a photon of light energy, a low-energy electron is lifted to a higher energy state. This excitation energy forms a strong oxidant and a weak reductant. The strong oxidant splits 2 molecules of water, which donate electrons, e^-. The excitation

FIGURE 3-1 [OPPOSITE]
Spectral distribution of solar radiation showing the segmentation of the spectrum involved in plant processes. Shown at the top are basic photochemical and radiation processes that may occur with the absorption of radiation by plants and other objects. The quantum content, expressed on the scale as kcal, is indicated for each frequency of incident radiation. (From Gates, 1965.)

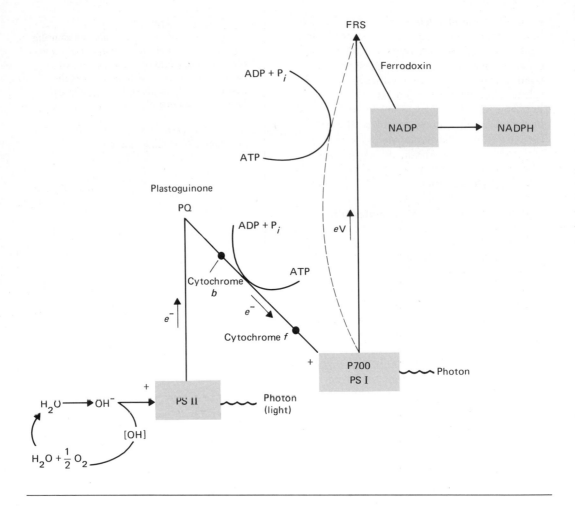

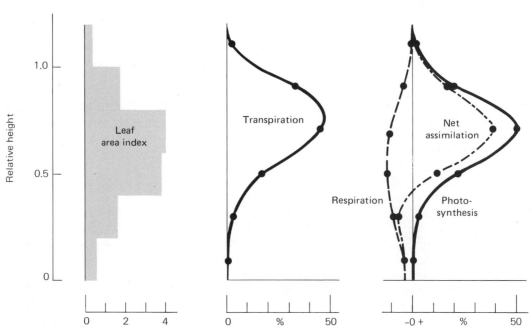

energy also initiates the reduction of an electron carrier, plastoquinone (PQ) (see Fig. 3-2). This reduction triggers a series of energy-releasing reactions, involving the two other electron carriers, cytochrome *b* and cytochrome *c*. The energy from these reactions causes the synthesis of ATP. From the cytochrome carriers the excited electron gets passed on to a specially bound chlorophyll *a*, P 700, which has a red light absorption of about 700 nm. This is a chlorophyll at a lower energy state, which acts as an energy sink. This series of reactions belongs to photosystem II.

When the reduced P 700 chlorophyll, a part of photosystem I, is excited by a photon, it forms a strong reductant and a weak oxidant. The excited electron possesses enough energy to pass to another electron carrier, FRS. In another series of energy-releasing reactions, FRS passes the electron to ferredoxin and then to NADP (nicotinamide adenine dinucleotide diphosphate), reducing it to NADPH. The energy released is used to form more ATP. The weak reductant from photosystem II is oxidized by the weak oxidant from photosystem I. Supplying the electrons to replace those lost from photosystem II are $OH^{e}-$ radicals of the water. Both systems are now ready for further energy-fixing processes. The OH radicals that remain combine to form water and O_2.

The rate of photosynthesis in an ecosystem is limited by both the environmental conditions and the nature of community structure. These limits include such environmental variables as light intensity, temperature, moisture, atmospheric gases, soil nutrients, and competition, and such plant variables as leaf area and the geometry of the vegetational canopy. To further complicate the picture, all of these variables interact with each other.

The intensity of light that reaches the plant is influenced by the local light climate, the modification of that light by the structure of the vegetation, and the position of the leaf in the vegetational profile (Fig. 3-3). The relationship of the leaf to light interception can be described as the leaf area index (*LAI*). This is the ratio of leaf area per unit of ground area. For some plants there is an optimum *LAI*, the point at which there is minimal shading of one leaf by another (Pearce et al., 1965). As the LAI increases beyond this value, there is a decrease in photosynthesis. Other plants do not show this re-

FIGURE 3-2 [OPPOSITE TOP]
A model of the photochemical reactions of photosynthesis. This model emphasizes the fact that energy flow in the ecosystem begins with the photochemical reactions in plants, a fact recognized but not often fully appreciated. (Adapted from Levine, 1969, and Fogg, 1970.)

FIGURE 3-3 [OPPOSITE BOTTOM]
Vertical distribution of respiration, net assimilation, and transpiration as a function of the distribution of plant matter and energy. (From Baumgartner, 1968. Reprinted by permission of Unesco. © Unesco 1968.)

TABLE 3-1
Comparative yields and photosynthetic efficiencies of some herbaceous communities

Community	Location	Growing season (days)	Dry matter (g/m²/day)	Mean daily insolation (langleys)	Efficiency (%)	Reference
Corn	Minnesota	92	11	500	2.1	Ovington and Lawrence, 1967
Tall-grass prairie	Colorado	140	8	590	1.2	Moir, 1969
Meadow-steppe	West Siberia	100	15	400	3.4	Rodin and Bazilevic, 1965
1-year weed field	New Jersey	120	19.5	540	3.8	Botkin and Malone, 1968
Short-grass steppe	Colorado	115	22	590	3.4	Moir, 1969
Theoretical maximum yield	—	100	77	500	12.0	Loomis and Williams, 1963

sponse (Loomis et al., 1967). The angle of the leaf influences light interception. Leaves that are perpendicular to incoming light (i.e., horizontal leaves) intercept the most light, but a number of layers of horizontal leaves reduce the amount of light reaching lower leaves (Loomis et al., 1967). Leaves growing at an angle require a much higher LAI to intercept the same quantity of light as horizontal leaves. As their LAI increases above a certain value, the angled leaves carry on photosynthesis at a faster rate. Such an adaptation in leaf position to obtain maximum photosynthetic rate and still possess a high LAI is characteristic of corn, grasses, beets, turnips, and other row crops (Loomis et al., 1967).

For an individual leaf, at low light intensity, photosynthesis increases linearly as the light intensity increases. At higher intensities this rate increases more slowly until a maximum rate is achieved (Kok, 1965). For most plants full sunlight is far beyond the maximum light needed to achieve the highest photosynthetic rate. In fact, damaging photoinhibition may take place at this intensity, so that the light is used inefficiently. Individual leaves perpendicular to full sunlight are light-saturated at one-tenth the intensity of full sunlight. Leaves angled 30° to full sunlight, however, receive only half the intensity of full sunlight. Plants, however, appear to adjust their total leaf surface to the light intensity available. Leaves in most plants are arranged for greatest efficiency of light utilization (Went and Sheps, 1969).

The limitations of temperature and moisture on photosynthesis become most pro-nounced during dry periods. Plants have an optimum temperature for photosynthesis, which may vary with the developmental stages. Early growth stages usually have a higher optimum temperature than later stages. Lack of moisture, called *water stress*, can reduce photosynthesis markedly (Moir, 1969). The decline is attributable to stomatal closing and the curling of leaves, responses designed to reduce transpirational loss of water. At the same time, this response reduces the amount of carbon dioxide diffusion into the leaf, depleting the supply of carbon dioxide at the site of the chloroplast fixation. It also reduces the transport of latent heat away from the leaf, increasing the leaf temperature, which in turn intensifies the lack of moisture.

EFFICIENCY OF PHOTOSYNTHESIS

The photosynthetic efficiency of converting the energy of the sun to organic matter can be assessed from two viewpoints: (1) the number of quanta of energy required for the evolution of a molecule of oxygen and (2) the ratio of calories per unit area of harvested vegetation to the total solar radiation intercepted. From the quanta viewpoint, photosynthesis is a rather efficient process. To release 1 mole of oxygen and fix 1 mole of carbon dioxide, a green plant needs an estimated 8 einsteins of energy. The photons absorbed have wavelengths in the red region, 1 einstein of which amounts to 40 kcal. Since 8 einsteins of energy are used per mole of oxygen, 320 kcal of light energy are needed to release 1 mole of oxygen. For each mole of oxygen evolved, approximately 120 kcal of energy are fixed. This corresponds to an efficiency of approximately 38 percent. Efficiencies calculated for isolated chloroplasts and for some algae amount to 21 to 33 percent (Kok, 1965; Wassink, 1968; Bassam, 1965).

When considered from the standpoint of calories stored in relation to light energy available, the efficiency is considerably less. The usable spectrum, 0.4 to 0.7 μ wavelengths, is only about half the total energy incident upon vegetation. Highest short-term efficiency measured over a period of weeks of active growth may amount to 12 to 19 percent (Wassink, 1968). In most instances, however, photosynthetic efficiency is computed either on an annual or a growing-season basis (Table 3-1). Mean photosynthetic efficiency of a Puerto Rican tropical rain forest has been estimated at 7 percent (H. T. Odum, 1970). Sugarcane fields in Java have an efficiency of around 1.9 percent (Hellmers and Bonner, 1959). The photosynthetic efficiency of land areas is about 0.3 percent, of the ocean about 0.13 percent. Total yields of solar energy on earth amount to about 0.15 to 0.18 percent (Wassink, 1968).

A sampling of ecological efficiencies is given in Table 3-1. These figures cannot be accepted with a great deal of confidence, nor are they really comparable. All of them involve a number of assumptions in their determination. There is no standard method for determining photosynthetic efficiency. In many cases the radiant energy incident upon the vegetation studied was estimated from tables or from measurements taken in the region but not in the study area. Temperatures, moisture, nutrient status of the soil, and other variables are not held constant. Estimates are often based on the total energy available for the year rather than for the growing season alone. This lowers the estimates considerably. If based on the growing season, estimates might be as high as 10 percent instead of 1 to 3 percent. Error is also introduced if energy fixation is based on estimates of community peak standing crops rather than on the determination of peak standing crop of individual species within the community. Because different species reach peak standing crop at different periods of the growing season, estimates made only at one period would seriously underestimate photosynthetic efficiency.

37

TABLE 3-2
Net primary production and plant biomass for major ecosystems and for the earth's surface

| | Area ($10^2 km^6$) | Net primary productivity, per unit area (dry g/m^2/year) | | World net primary production (10^9 dry ton) | Biomass per unit area (dry kg/m^2) | | World biomass (10^9 dry ton) |
		Normal range	Mean		Normal range	Mean	
Lake and stream	2	100–1,500	500	1.0	0–0.1	0.02	0.04
Swamp and marsh	2	800–4,000	2,000	4.0	3–50	12	24
Tropical forest	20	1,000–5,000	2,000	40.0	6–80	45	900
Temperate forest	18	600–3,000	1,300	23.4	6–200	30	540
Boreal forest	12	400–2,000	800	9.6	6–40	20	240
Woodland and shrubland	7	200–1,200	600	4.2	2–20	6	42
Tropical savanna	15	200–2,000	700	10.5	0.2–15	4	60
Temperate grassland	9	150–1,500	500	4.5	0.2–5	1.5	14
Tundra and alpine	8	10–400	140	1.1	0.1–3	0.6	5
Desert scrub	18	10–250	70	1.3	0.1–4	0.7	13
Extreme desert, rock, and ice	24	0–10	3	0.07	0–0.2	0.02	0.5
Agricultural land	14	100–4,000	650	9.1	0.4–12	1	14
Total land	149		730	109.0		12.5	1,852.0
Open ocean	332	2–400	125	41.5	0–0.005	0.003	1
Continental shelf	27	200–600	350	9.5	0.001–0.04	0.01	0.3
Attached algae and estuaries	2	500–4,000	2,000	4.0	0.04–4	1	2
Total ocean	361		155	55.0		0.009	3.3
Total for earth	510		320	164.0		3.6	1,855

Source: R. H. Whittaker, 1970, *Communities and Ecosystems.* Reprinted with permission of Macmillan Publishing Co., Inc. © Copyright Robert H. Whittaker, 1970.

Primary production

The flow of energy through the community starts with the fixation of sunlight by plants, a process that in itself demands the expenditure of energy. Initially plants rely on the food stored in the seed for energy until their own production machinery is working. Once mobilized, the green plant begins to accumulate energy. Energy accumulated by plants is called *production*, or more specifically, *primary production*, because it is the first and basic form of energy storage in an ecosystem. The rate at which energy is stored by photosynthetic activity is known as *primary productivity*. All of the sun's energy that is assimilated—i.e., total photosynthesis—is *gross primary production*. Like other organisms, plants require energy for reproduction and maintenance. The energy required for these needs is provided by a reverse of the photosynthetic process, *respiration*. The energy remaining after respiration and stored as organic matter is *net primary production*, or plant growth. Thus net production for a single plant or a plant community can be described by the formula

net primary production=
    gross production — autotrophic respiration

The basic unit of energy used in ecological work is the calorie; production is usually expressed in kilocalories per square meter per year (kcal/m^2/year). However, production may also be expressed as dry organic matter in grams per square meter per year (g/m^2/year). If either of these two measures is employed to estimate efficiencies and other ratios, the same unit must be used for both the numerator and denominator of the ratio. Only calories can be compared with calories, dry weight with dry weight.

Net primary production accumulates over a period of time as plant biomass. Part of this accumulation is turned over seasonally through decomposition. Part is retained over a longer period as living material. The amount of this accumulated living organic matter found on a given area at a given time is the *standing crop biomass*. Biomass is usually expressed as grams dry weight of organic matter per square meter (g/m^2), calories per square meter (kcal/m^2), or some other appropriate measure per unit area. Thus biomass differs from productivity, which is the rate at which organic matter is created by photosynthesis. Biomass present at any given time is not the same as production.

Biomass varies seasonally and even daily. In grasslands and old-field ecosystems much of the net production is turned over every year. The standing crop of living material in an old field in Michigan amounted to about 4×10^3 kg/ha (kilograms per hectare) in late summer, compared to 80 kg/ha in late spring. But at this time the standing crop in dead matter was nearly 3×10^3 kg/ha. (Golley, 1960). The above-ground biomass of a tall grass prairie that included both living and dead material was approximately twice that of the standing crop, the living material added during the growing season (Kucera et al., 1967). The above-ground biomass has a turnover rate of approximately 2 years, the below-ground biomass of roots a turnover rate of 4 years. In a salt marsh the standing crop in autumn was 9×10^3 kg/ha; in winter it was just one-third of this. In a forest ecosystem a considerably greater proportion of the net production is tied up as wood. In an oak–pine forest, leaves, fruits, flowers, dead wood and bark contributed 342 g/m^2/year to the organic horizon, and the roots 311 g/m^2/year, for a total of 653 g/m^2 or about 58 percent of the net primary production (Woodwell and Marpels, 1968).

Annual net production in woodlands increases with age up to a certain point and then declines. Mean annual net primary production of a Scots pine plantation was a little over 12×10^3 kg/ha at 30 years of age; it then declined slightly (Ovington, 1961). The plantation, however, achieved maximum production at the age of 20, when it amounted to 22×10^3 kg/ha. Woodlands apparently achieve their maximum annual production in the pole stage, when the dominance of the trees is the greatest and the understory is at a minimum. The understory makes its greatest contribution during the juvenile and mature stages of the forest. As age increases, more and more of the production is needed for maintenance, and very little gross production is left for growth.

Just as net production of a given plant community declines with age, so net production relative to gross production declines from a young ecosystem such as a weedy field or an agricultural crop to a mature plant community such as a forest. As plant communities ap-

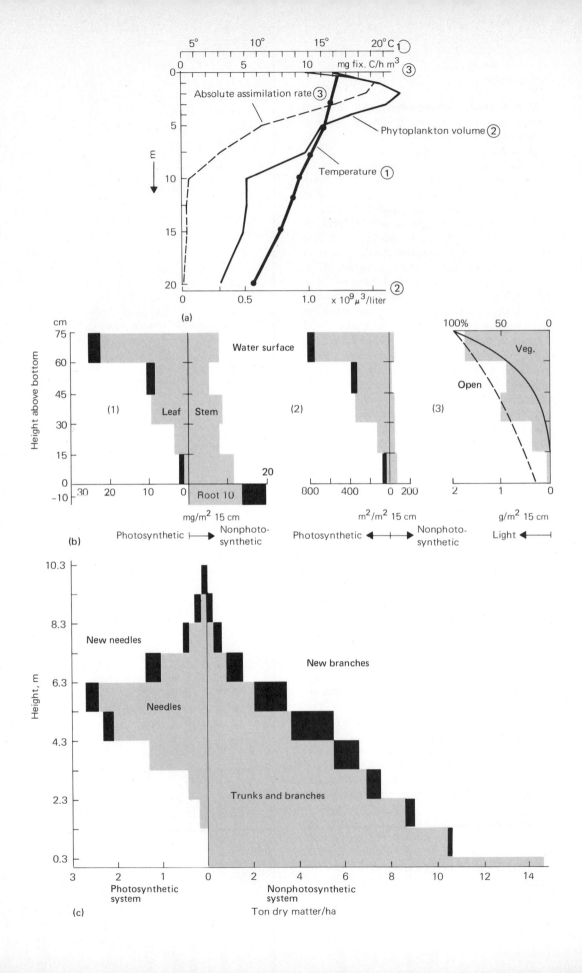

(a)

Absolute assimilation rate ③

Phytoplankton volume ②

Temperature ①

(b)

Water surface

(1) Leaf Stem

Root 10

mg/m² 15 cm

Photosynthetic → Nonphoto-synthetic

(2)

m²/m² 15 cm

Photosynthetic ←→ Nonphoto-synthetic

(3) Veg.

Open

g/m² 15 cm

Light ←

Height above bottom

(c)

New needles

New branches

Needles

Trunks and branches

Photosynthetic system

Nonphotosynthetic system

Ton dry matter/ha

Height, m

proach a stable or steady-state condition (see Chapter 9), more of the gross production is used for the maintenance of biomass and less goes into newly added organic matter. Thus the ratio of gross production to biomass declines through time (Fig. 3-4).

Productivity of ecosystems may vary considerably, not only between the same types of ecosystems, but also within the same ecosystem from year to year. Productivity is influenced by such factors as nutrient availability, moisture (especially precipitation), temperature, length of the growing season, animal utilization, and fire. For example, the herbage yields of a grassland may vary by a factor of eight between wet and dry years (Weaver and Albertson, 1956). The effects of drought may be further intensified by grazing. Overgrazing of grasslands by cattle and sheep, or the defoliation of forests by such insects as the saddled prominent and gypsy moth can seriously reduce net production. Fire in grasslands may result in increased productivity if moisture is normal, but if precipitation is low, fire can reduce productivity (Kucera et al., 1967). An insufficient supply of nutrients, especially nitrogen and phosphorus, can limit net productivity, as can the mechanical injury of plants, atmospheric pollution, and the like.

Gross and net productivities vary widely among different plant communities. Variations of net production for a variety of plant communities are summarized in Table 3-2. For a mature Temperate Zone hardwood forest, net production ranges somewhere between 1,200 and 1,500 g/m²/year (Whittaker, 1965), whereas other woodlands may have a net production of 400 to 1,000 g/m²/year. Shrublands such as heath balds and tall grass prairie have net productions of 700 to 1,500 g/m²/year (Whittaker, 1963; Kucera et al., 1967). Desert grasslands produce about 200 to 300 g/m²/year, whereas deserts and tundra range between 100 and 250 g/m²/year (Rodin and Bazilevic, 1967). Net production of the open sea is generally quite low. The productivity of the North Sea is about 170 g/m²/year; for the Sargasso Sea, 180 g/m²/year. However, in some areas of upwelling, such as the Peru Currents, net production can reach 1,000 g/m²/year.

Some communities have a consistently higher production. A *Spartina* salt marsh, for example, may have a net productivity of 3,300 g/m²/year, sugarcane 1,725 g/m²/year, hy-

FIGURE 3-5 [OPPOSITE]
Vertical distribution of production and biomass in aquatic and terrestrial communities. (a) Volume and vertical density of phytoplankton and the rate of energy assimilation in Lake Constance, Germany. (After Elster, 1966. Originally published by the University of California Press; reprinted by permission of The Regents of the University of California.) (b) A Potamogeton crispus (pondweed) community. (1) Biomass is divided into leaf, stem, and root; solid area represents winter buds. (2) Concentration of chlorophyll in the plant community. (3) Leaf area and light profile in the community. The solid lime is light in the community; the broken line represents light in the open water. (After Ikusima, 1965.) (c) Structure and productive systems of a pine–spruce–fir forest in Japan. (After Monsi, 1968. Reprinted by permission of Unesco. © Unesco 1968.)

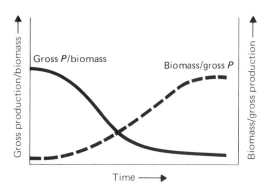

FIGURE 3-4
A model showing the change in ratios between gross community photosynthesis and biomass, or production efficiency, and between biomass and gross community photosynthesis, or maintenance efficiency, in time. Note the high production efficiency early in time, and the accumulation of biomass later. (After G. D. Cooke, 1967.)

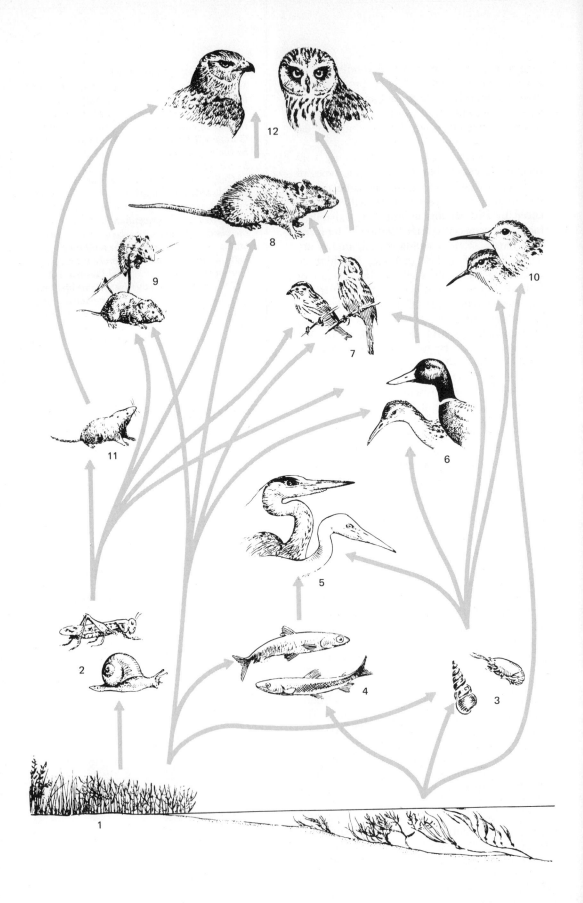

brid corn 1,000 g/m²/year, and some tropical crops 3,000 g/m²/year. Such high productivity usually results from an additional energy subsidy to the system. This subsidy may be a warmer temperature, greater rainfall, circulating or moving water that carries food or additional nutrients into the community, or in the case of agricultural crops the use of fossil fuel for cultivation and irrigation, the application of fertilizer, and the control of pests.

Within the structural profile of the ecosystem there is a vertical distribution of biomass. The vertical distribution of leaf biomass or the concentration of plankton influences the penetration of light, which in turn influences the distribution of production in the ecosystem. The region of maximum productivity in the aquatic ecosystem is not the upper sunlit surface (strong sunlight inhibits photosynthesis) but some depth below, depending upon the clarity of the water and the density of the plankton growth. As depth increases, the light intensity decreases until it reaches a point at which the light received by the plankton is just sufficient to meet respiratory needs, and production equals respiration (Fig. 3-5a). This is known as the compensation level. In the forest ecosystem a similar situation exists. The greatest amount of photosynthetic biomass as well as the highest net photosynthesis is not at the very top but at some point below maximum light intensity (Fig. 3–5b, c). In spite of wide differences in plant species and in the types of ecosystems, the vertical profiles of biomass of the various ecosystems appear to be quite similar.

Food chains

Net production theoretically represents the energy available either directly or indirectly to the consumer organisms and is the base upon which the rest of life on earth depends. This energy stored by plants is passed along through the ecosystem in a series of steps of eating and being eaten known as a *food chain*.

Food chains are descriptive. When worked out diagrammatically, they consist of a series of arrows, each pointing from one species to another, for which it is a source of food. In Fig. 3-6, for example, the marsh vegetation is eaten by the grasshopper, the grasshopper is consumed by the shrew, the shrew by the

FIGURE 3-6 [OPPOSITE]

A midwinter food web in a Salicornia *salt marsh (San Francisco Bay area). Producer organisms, (1) terrestrial and salt marsh plants, are consumed by herbivorous terrestrial invertebrates, represented by the grasshopper and the snail (2). The marine plants are consumed by herbivorous marine and intertidal invertebrates (3). Fish, represented by smelt and anchovy (4), feed on vegetative matter from both ecosystems. The fish in turn are eaten by first-level carnivores, represented by the great blue heron and the common egret (5). Continuing through the web we have the following omnivores: clapper rail and mallard duck (6); savanna and song sparrows (7); Norway rats (8); California vole and salt-marsh harvest mouse (9); the least and western sandpipers (10). The vagrant shrew (11) is a first-level carnivore; the top carnivores (second level) are the marsh hawk and the short-eared owl (12). (Food web adapted from R. F. Johnston, 1956a.)*

marsh hawk or the owl. Thus we have a relationship that can be written:

marsh grass → grasshopper →
shrew → marsh hawk

But as Fig. 3-6 indicates, no one organism lives wholly on another; the resources are shared, especially at the beginning of the chain. The marsh plants are eaten by a variety of birds, mammals, and fish; and some of the animals are consumed by several predators. Thus food chains become interlinked to form a *food web*, the complexity of which varies within and between ecosystems.

At each step in the food chain a considerable portion of the potential energy is used for maintenance and lost as heat. As a result, organisms in each trophic level pass on less energy than they receive. This limits the number of steps in any food chain to four or five. The longer the food chain, the less energy there is available for the final members.

HERBIVORES

Feeding on plant tissues is a whole host of plant consumers, the *herbivores*, which are capable of converting energy stored in plant tissue into animal tissue. Their role in the community is essential, for without them the higher trophic levels could not exist. The English ecologist Charles Elton, in his classic little book *Animal Ecology*, suggested that the term *key industry* be used to denote animals that feed on plants and are so abundant that many other animals depend upon them for food.

Only the herbivores are adapted to live on a diet high in cellulose. Modification in the structure of the teeth, complicated stomachs, long intestines, a well-developed cecum, and symbiotic flora and fauna enable these animals to use plant tissues. For example, ruminants, such as deer, have a four-compartment stomach. As they graze, these animals chew their food hurriedly. The material consumed descends to the first and second stomachs (the rumen and reticulum), where it is softened to a pulp by the addition of water, kneaded by muscular action, and fermented by bacteria. At leisure the animals regurgitate the food, chew it more thoroughly, and swallow it again. This time the mass again enters the rumen, where the coarse particles remain behind for

further bacterial digestion. The finer well-chewed material is pulled into the reticulum and from there forced by contraction into the third compartment, or omasum. There the material is further digested and finally forced into the abomasum, or true glandular stomach.

The digestive process in ruminants relies heavily on bacterial fermentation in the rumen, reticulum, and omasum. Millions of microorganisms attack various digestive materials such as cellulose, starch, pectin, and hemicellulose sugars and convert part of them to short-chain volatile fatty acids. These are rapidly absorbed through the wall of the rumen into the bloodstream and are oxidized to form the animal's chief source of energy. Part of the material is converted to methane and lost to the animal and part remains as fermentation products. Many of the microbial cells involved are digested in the abomasum to recapture still more of the energy and nutrients. In addition to fermentation, the bacteria also synthesize B-complex vitamins and essential amino acids.

Another outstanding group of herbivores are the lagomorphs—the rabbits, hares, and pikas. In contrast to ruminants, these herbivores have a simple stomach and a large cecum. During digestion, part of the ingested plant material enters the cecum, and part enters the intestine to form dry pellets. In the cecum the ingested material is attacked by microorganisms and is expelled into the large intestine as moist soft pellets surrounded by a proteinaceous membrane. The soft pellets, much higher in protein and lower in crude fiber than the hard fecal pellets, are reingested (coprophagy) by the lagomorphs. The amount of feces recycled by coprophagy ranges from 50 to 80 percent. The reingestion is important because it provides bacterially synthesized B vitamins and ensures a more complete digestion of dry material and a better utilization of protein (see McBee, 1971).

CARNIVORES

Herbivores are the energy source for *carnivores*, the flesh-eaters. Those organisms that feed directly upon the herbivores are termed first-level carnivores or second-level consumers. Usually they are larger and stronger than their prey and more or less solitary in habit.

First-level carnivores represent an energy source for the second-level carnivores. Still higher categories of carnivorous animals feeding on secondary carnivores may exist in some communities. As the trophic level of carnivores increases, their numbers decrease and their fierceness, agility, and size increase. Finally, the energy stored at the top carnivore level is utilized by the decomposers.

As a group the carnivores are well adapted to a diet of flesh. Hawks and owls have sharp talons for holding prey and have hooked beaks for tearing flesh. Mammalian carnivores have canine teeth for biting and piercing. Cheek teeth are reduced, but many forms have sharp-crested shearing or carnasial teeth.

OMNIVORES

Not all consumers can be fitted neatly into each trophic level, for many consumers do not confine their feeding to one level. The red fox feeds on berries, small rodents, and even dead animals. Thus it occupies herbivorous and carnivorous levels, as well as acting as a scavenger. Some fish feed on both plant and animal matter. The basically herbivorous white-footed mouse also feeds on insects, small birds, and bird eggs. Many species, including the white-footed mouse, are cannibalistic. They eat the flesh of their own kind not only in an extreme food shortage, as one might like to believe, but frequently as a way of supplementing their diet. The food habits of many animals vary with the seasons, with stages in the life cycle, and with the size and growth of the organism (Fig. 3-7). Consumers that utilize both plant and animal matter are called *omnivores*.

DECOMPOSERS

In traditional texts the *decomposers* make up the so-called final feeding group. Actually the decomposers are much more than that. Their basic function is to release the nutrients contained in the plant and animal biomass back into the mineral cycles (see Chapter 5). The work of the decomposers then is the opposite of that of the producers, who fix nutrients and energy into plant biomass. The process of decomposition involves a complex series of food chains in which the organisms of decay utilize the energy and materials of dead plant and animal matter.

There are two groups of decomposer organisms, macroorganisms and microorganisms.

Micropogon undulatus, Atlantic Croaker

Food categories

Young	Juvenile	Adult
Fishes		
Macrobottom animals		
Microbottom animals		
Zooplankton		
Phytoplankton		
Vascular plant material		
Organic detritus and undetermined organic material		

FIGURE 3-7
Trophic spectra for young, juvenile, and adult stages of the Atlantic croaker from Lake Pontchartrain. The young fish subsist chiefly on zooplankton, the juveniles on organic debris, the adults on bottom animals and fish. (Graph after Darnell, 1961. © The Ecological Society of America.)

45

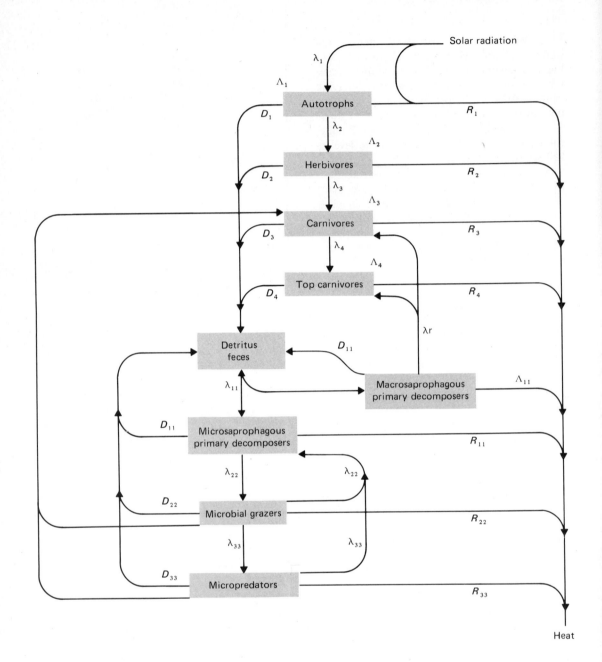

The macroorganisms include small detritus-feeding animals such as mites, earthworms, millipedes, and slugs in the terrestrial ecosystems, and crabs, mollusks, and worms in the aquatic ecosystems. The microorganisms are the bacteria and fungi. The detritus-feeders act as reducer–decomposers. They ingest organic matter, break it down into smaller pieces, mix it with soil, excrete it, spread spores, break down microbial antagonisms, and even add substances to the material that stimulate microbial growth. The same reducer–decomposers also consume bacteria and fungi associated with the detritus, as well as protozoans and small invertebrates that cling to the material. Some of these decomposers are microbial grazers, feeding on bacteria and fungi. These grazers may reduce bacterial and fungal populations, inhibiting the effects of increased population density and accelerating the division of soil microbes, thus speeding up microbial activity. The reducer–decomposers then reduce detritus to smaller pieces, condition the material for microbial action, and at the same time utilize the energy and nutrients stored in the biomass of bacteria and fungi.

The final breakdown of detritus is accomplished by bacteria and fungi. These bacteria may inhabit the digestive tracts of the reducer–decomposers, producing the enzymes necessary for the digestion of the material. Bacteria and hyphae of fungi secrete enzymes necessary to carry out the specific chemical action on the detritus. These organisms then absorb a portion of the product as food, and the unconsumed material is available as food for other organisms. In addition the bacteria and fungi become food for the microbial grazers. In this manner food chains involving decomposers reach up into the traditional herbivore–carnivore food chains (Fig. 3-8). In fact, the decomposers, so frequently considered as some distant feeding group unclassifiable in the general scheme of food chains, actually function as herbivores or carnivores, depending upon whether their food is plant or animal material. At any rate, the end result of decomposition is the transformation of mineral matter of low solubility into nutrients available for plants.

OTHER FEEDING GROUPS

Several other feeding groups are also involved in energy transfer. *Parasites* are one; most parasites spend a considerable part of their life cycle living on, or in, and drawing their nourishment from, their hosts without actually killing them. Other parasites, the *parasitoides*, draw nourishment from the host until the host dies. At this point the parasitoid transforms into another stage of its life cycle, becoming independent of the host. Functionally, parasites are specialized carnivores, or in the case of plant parasites, herbivores. *Scavengers* are animals that eat dead plant and animal material. Among these are termites and various beetles that feed in dead and decaying wood, and crabs and other marine invertebrates that feed on plant particles in water. Botflies, dermestid beetles, vultures, and gulls are only several of many animals that feed on animal remains. Depending on what they eat, the scavengers are either herbivores or carnivores. *Saprophytes* are plant counterparts of scavengers. They draw their nourishment from dead plant and animal material, chiefly the former. Because they do not require sunlight as an energy source, they can live in deep shade or dark caves. Examples of saprophytes are fungi, Indian pipe, and beech drops. The majority of these are herbivorous; others, such as the entomophagous fungi, feed on animal matter.

MAJOR FOOD CHAINS

Within any ecosystem there are two major food chains, the grazing food chain and the detritus food chain (Fig. 3-8). Because of the high standing crop and the relatively low harvest of primary production, the most important food chain in most terrestrial and shallow-water ecosystems is the detritus chain. In deep-water aquatic ecosystems, with their low biomass, rapid turnover of organisms, and high rate of harvest, the grazing food chain is the dominant one.

The amount of energy shunted down the two routes varies among communities. In an intertidal salt marsh, less than 10 percent of living plant material is consumed by herbivores and 90 percent goes the way of the detritus-feeders and decomposers (Teal, 1962). In fact most of the organisms of the intertidal salt marshes obtain the bulk of their energy from dead plant material. Fifty percent of the energy fixed annually in a Scots pine plantation is utilized by decomposers (Fig. 3-9). The remainder is removed as yield or is stored in tree trunks (Ovington, 1961). In some communities, particularly undergrazed grasslands, unconsumed organic matter may accumulate, and the materials might remain out of circulation for some time, especially when conditions are not favorable for microbial action. The decomposer or detritus food chain receives additional materials from the waste products and dead bodies of both the herbivores and carnivores.

There are a number of supplementary food chains in the community, including the parasitic and saprophagic. Parasitic food chains are highly complicated because of the life cycle of the parasites. Some parasites are passed from one host to another by predators in the food chain. External parasites (ectoparasites) may transfer from one host to another. Other parasites are transmitted by insects from one host to another through the blood stream or plant fluids. Food chains also exist among parasites themselves. Fleas that parasitize mammals and birds are in turn parasitized by a protozoan, *Leptomonas*. Chalcid wasps lay eggs in the ichneumon or tachinid fly grub, which in turn is parasitic on other insect larvae. In these parasitic food chains, the members, starting with the host, become progressively smaller and more numerous with each level in the chain.

Food chains involving saprophages may take two directions (Fig. 3-8), toward the carnivores or toward microorganisms. The role of these feeding groups in the final dissipation of energy has already been mentioned. But they also are food to numerous other animals as well. Slugs eat the larvae of certain Diptera and Coleptera, which live in the heads of fungi and feed on the soft material. Mammals, particularly the red squirrel and chipmunks, eat woodland fungi. Dead plant remains are food sources for springtails and mites, which in turn are eaten by carnivorous insects and spiders. These in turn are energy sources for insectivorous birds and small mammals. Blowflies lay their eggs in dead animals and within 24 hours the maggot larvae hatch. Unable to eat solid tissue, they reduce the flesh to a fetid mass by enzymatic action in which they feed on the proteinaceous material. These insects are food for other organisms.

FLOW OF ENERGY THROUGH
A NATURAL FOOD CHAIN

Food chains in nature are rather difficult to study, but one that has been rather carefully worked out (Golley, 1960) involves old field vegetation, meadow mice, and least weasels (Fig. 3-10). The mice were almost exclusively herbivorous, and the weasels lived mainly on mice. The vegetation converted about 1 percent of the solar energy into net production, or plant tissue. The mice consumed about 2 percent of the plant food available to them, and the weasels about 31 percent of the mice. Of the energy assimilated, the plants lost about 15 percent through respiration, the mice 68 percent, and the weasels 93 percent. The weasels used so much of their assimilated energy in maintenance that a carnivore preying on weasels could not exist.

In a more general sense, in the transformation of energy through the ecosystem, the energy is reduced in magnitude by 10 from one level to another. Thus if an average of 1,000 kcal of plant energy were consumed by herbivores, about 100 kcal would be converted to herbivore tissue, 10 kcal to first-level carnivore production, and 1 kcal to second-level carnivores. The amount of energy available to second- and third-level carnivores is so small that few organisms could be supported if they depended on that source alone. For all practical purposes, each food chain has from three to four links, rarely five. The fifth link is distinctly a luxury item in the ecosystem.

From the example of the plant–meadow mouse–weasel food chain it is rather obvious that all of the production of green plants is not consumed by the herbivores, nor are all of the herbivores utilized by the carnivores. In fact two-thirds to three-fourths of the energy stored by photosynthesis in a grassland ecosystem, ungrazed by domestic animals, is returned to the soil as dead plant material, and less than one-fourth is consumed by herbivores (Hyder, 1969). Of the quantity consumed by herbivores, about half is returned to the soil as feces. In the forest ecosystem only about 2.5 percent of the annual leaf production is consumed by insects, exclusive of energy consumed by sap-feeders (Van Hooke, 1971). Grasshoppers consume just 2 percent of the net production available to them (Smalley,

49

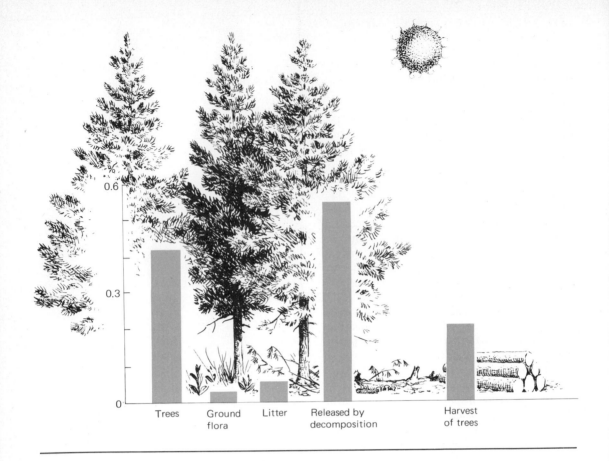

Trees	Ground flora	Litter	Released by decomposition		Harvest of trees

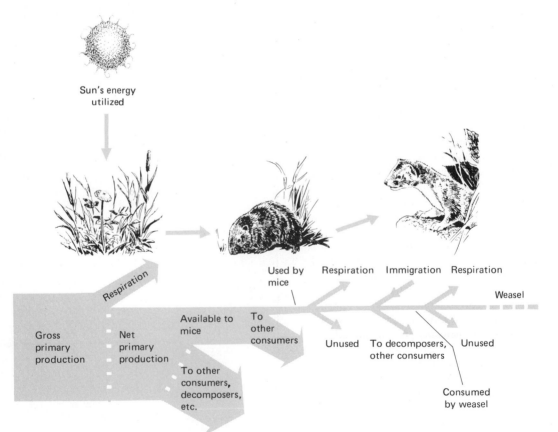

Sun's energy
utilized

Respiration

Used by mice   Respiration   Immigration   Respiration

Weasel

Gross primary production

Net primary production

Available to mice

To other consumers

Unused

To decomposers, other consumers

Unused

To other consumers, decomposers, etc.

Consumed by weasel

1960), and mice about 7 percent (Golley, 1960). No predator, however skillful, completely exterminates its prey. The weasel in the old-field ecosystem consumed 30 percent of the energy available to it (Golley, 1960). A mountain lion population in Idaho utilized less than 4 percent of the deer and elk population available to it (Hornocker, 1970). All of this unutilized production, primary and secondary, as well as the inedible and undigested portions, go to the decomposers.

Secondary or consumer production

Net production is the energy available to the heterotrophic components of the ecosystem. Theoretically at least, all of it is available to the grazers or even to the decomposers; but rarely is it all utilized in this manner. The net production of any given ecosystem may be dispersed to another food chain outside of the ecosystem by man, wind, or water. For example, about 45 percent of the net production of a salt marsh is lost to estuarine water (Teal, 1962). Much of the living material is physically unavailable to the grazers—they cannot reach many plants. The organic matter of live organisms is unavailable to decomposers and detritus-feeders, and that of dead plants may not be relished by grazers. The amount of net production available to herbivores may vary from year to year and from place to place. The quantity consumed varies with the type of herbivore and the density of the population. Once consumed, a considerable portion of the plant material, again depending upon the kind of plant involved and the digestive efficiency of the herbivore, may pass through the animal's body undigested. A grasshopper assimilates only about 30 percent of the grass it consumes, leaving 70 percent available to the detritus–decomposer food chain (Smalley, 1960). Mice, on the other hand, assimilate about 85 to 90 percent of what they consume (Golley, 1960; R. L. Smith, 1962).

Energy, once consumed, either is diverted to maintenance, growth, and reproduction, or is passed from the body as feces and urine (Fig. 3-11). The energy content of the feces will be transferred to the detritus food chain. Part of the energy will be lost through urine, and depending upon the nature of the organism, the loss of energy can be variable

FIGURE 3-9 [OPPOSITE TOP]
The fate of the 1.3 percent of solar energy assimilated as net production by a Scots pine plantation 23 years old. (After J. D. Ovington, 1961.)

FIGURE 3-10 [OPPOSITE BOTTOM]
Energy flow through a food chain in an old-field community in southern Michigan. The relative sizes of the blocks suggest the quantity of energy flowing through each channel. Values are in calories per hectare per year. (Based on data from Golley, 1960.)

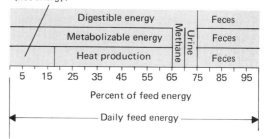

FIGURE 3-11
Relative values of the end products of energy metabolism in the white-tailed deer. Note the small amount of net energy gained (body weight) in relation to that lost as heat, gas, urine, and feces. The deer is a first-level consumer or herbivore. (After Cowan, 1962.)

51

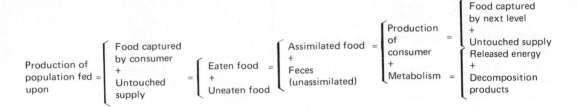

and often quite high (Coo et al., 1952). Another portion is lost as fermentation gases. Of the energy left after losses through feces, urine, and gases, part is utilized as *heat increment*, which is heat required for metabolism above that required for basal or resting metabolism. The remainder of the energy is "net" energy, available for maintenance and production. This includes energy involved in capturing or harvesting food, muscular work expended in the animal's daily routine, and energy needed to keep up with the wear and tear on the animal's body. The energy used for maintenance is lost as heat. Maintenance costs, highest in small, active, warm-blooded animals, are fixed or irreducible. In small invertebrates energy costs vary with temperature; and a positive energy balance exists only within a fairly narrow range of temperatures. Below 5° C spiders become sluggish and cease feeding, and have to utilize stored energy to meet metabolic needs. At approximately 5° C assimilated energy approaches energy lost through respiration. From 5° to 20.5° C spiders assimilate more energy than they respire. Above 25° C, their ability to maintain a positive energy balance declines rapidly (Van Hooke, 1971). Energy left over from maintenance and respiration goes into the production of new tissue, fat tissue, growth, and new individuals. This net energy of production is *secondary production*. Within secondary production there is no portion known as gross production. What is analogous to gross production is actually assimilation. Secondary production is greatest when the birth rate of the population and the growth rates of the individuals are the highest. This usually coincides, for obvious reasons, with the time when net primary production is also the highest.

This scheme is summarized in Fig. 3-12. It is applicable to any consumer organism, herbivore or carnivore. The herbivore represents the energy source of the carnivore; and like the plant food of the herbivore, not all of the energy contained in the body of the herbivore is utilized by the carnivore. Part

of it, such as hide, bones, and internal organs, is unconsumed, and the same metabolic losses can be accounted for. At each transfer considerably less energy is available for the next consumer level.

The energy budget of a consumer population can be summarized by the following formula:

$$C = A + FU$$

where C is the energy ingested, A is the energy assimilated, and FU is the energy lost through feces, urine, gas, and other rejecta.

The term A can be further refined as

$$A = P + R$$

where P is secondary production and R is energy lost through respiration. Thus

$$C = P + R + FU$$

or secondary production is

$$P = C - FU - R$$

Just as net primary production is limited by a number of variables, so is secondary production. The quantity, quality (including nutrient status and digestibility), and availability of net production are three limitations. So is the degree to which primary and available secondary production are utilized.

The latter can be examined from the viewpoint of two different ratios. One is the ratio of assimilation to ingestion, A/C. This is the measure of the efficiency of the consumer population in extracting energy from the food it consumes. The other is the ratio of productivity to assimilation, P/A. This is a measure of the efficiency with which the consumer population incorporates assimilated energy into new tissue, or secondary production.

The ability of the consumer population to convert the energy it consumes varies with the species and the type of consumer. Vertebrates utilize about 98 percent of their assimilated energy in metabolism and only about 2 percent in net production; invertebrates utilize 79 percent in metabolism. Thus invertebrates convert a greater proportion of their assimilated energy into biomass than vertebrates. A clearer distinction can be made if one considers, as Engelmann (1968) does, the animal world divided into two broad energetic groups, the thermoregulators, or homiotherms, and the nonregulators, or poikilotherms. In this context the nonregulators are more efficient producers than the thermoregulators. However, a major difference exists in their assimilation efficiency. Poikilotherms have an efficiency of about 30 percent in digesting food, whereas homiotherms have an efficiency of around 70 percent. Thus the poikilotherm has to consume more calories than a homiotherm to obtain sufficient energy to meet the needs of maintenance, growth, and reproduction.

Trophic levels

An ecosystem consists of numerous food chains. Because no organism lives wholly on another, the resources are shared, especially at the beginning of the chain. Thus food chains become interwoven into food webs, whose complexity varies within and between ecosystems. If all organisms that obtain their food in the same number of steps (that is, all those that feed wholly or in part on plants, wholly or in part on herbivores, and so on) are superimposed, the structure can be collapsed into a series of single points representing the trophic or feeding levels of the ecosystem. Thus each step in the food chain represents a trophic level. Animals at the lower level, such as the grasshopper and the snail, may occupy a single trophic level. But most of the animals at the higher levels participate simultaneously in several trophic levels, as do the vole and the song sparrow (Fig. 3-6). Thus the first trophic level belongs to the producers, the second to the herbivores, the third to the first-order carnivores, and so on.

Although trophic levels as considered by most ecological references do not include the decomposer and side chains, they logically should. As already mentioned, decomposers and parasites should be considered as herbivores or carnivores, depending upon the nature of their food source. Decomposers feeding on dead plant material, as well as bacteria

FIGURE 3-12 [OPPOSITE]
A model of components of energy flow in secondary production. (After Rafes, 1967.)

occupying the rumen of ungulate animals, should be considered functional herbivores. Decomposers feeding on the bodies of animals should be considered carnivores, and so on. In this manner, all the various steps in energy transfer in an ecosystem can be placed in some trophic level.

Ecological pyramids

By adding all of the biomass or living tissue contained in each trophic level and all of the energy transferred between levels, one can construct pyramids of biomass and energy for the ecosystem (Fig. 3-13b, c).

The pyramid of biomass indicates by weight or other measurement of living material the total bulk of organisms or fixed energy present at any one time—the standing crop. Because some energy or material is lost in each successive link, the total mass supported at each level is limited by the rate at which energy is being stored below. In general the biomass of the producers must be greater than the biomass of the herbivores they support, and the biomass of the herbivores must be greater than that of the carnivores. This usually results in a gradually sloping pyramid for most communities, particularly the terrestrial and shallow-water ones, where the producers are large and characterized by an accumulation of organic matter, where the life cycles are long, and where there is a low rate of harvesting.

This trend, however, does not hold for all ecosystems. In such aquatic ecosystems as lakes and open seas, primary production is concentrated in the microscopic algae. These algae have a short life cycle, multiply rapidly, accumulate little organic matter, and are heavily exploited by herbivorous zooplankton. At any one point in time the standing crop is low. As a result, the pyramid of biomass for these aquatic ecosystems is inverted: the base is much smaller than the structure it supports.

When production is considered in terms of energy, the pyramid indicates not only the amount of energy flow at each level, but more important, the actual role the various organisms play in the transfer of energy. The base upon which the pyramid of energy is constructed is the quantity of organisms produced per unit time, or, stated differently, the rate at which food material passes through the food chain. Some organisms may have a small biomass, but the total energy they assimilate and pass on may be considerably greater than that of organisms with a much larger biomass. On a pyramid of biomass these organisms would appear much less important in the community than they really are.

Energy pyramids are always sloping because less energy is transferred from each level than was paid into it. This is in accord with the second law of thermodynamics. In instances where the producers have less bulk than the consumers, particularly in open-water communities, the energy they store and pass on must be greater than that of the next level. Otherwise the biomass that producers support could not be greater than that of the producers themselves. This high energy flow is maintained by a rapid turnover of individual plankton, rather than an increase of total mass.

Another pyramid commonly found in ecological literature is the pyramid of numbers (Fig. 3-13a). This pyramid was advanced by Charles Elton (1927), who pointed out the great difference in the numbers of organisms involved in each step of the food chain. The animals at the lower end of the chain are the most abundant. Successive links of carnivores decrease rapidly in number until there are very few carnivores at the top. The pyramid of numbers often is confused with a similar one in which organisms are grouped into size categories and then arranged in order of abundance. Here the smaller organisms again are the most abundant; but such a pyramid does not indicate the relationship of one group to another.

The pyramid of numbers ignores the biomass of organisms. Although the numbers of a certain organism may be greater, their total weight, or biomass, may not be equal to that of the larger organisms. Neither does the pyramid of numbers indicate the energy transferred or the use of energy by the groups involved. And because the abundance of members varies so widely, it is difficult to show the whole community on the same numerical scale.

Model of energy flow

The concept of energy flow in ecological systems is one of the cornerstones of ecology.

The model was first developed by Raymond Lindeman in 1942. It is based on the assumptions that the laws of thermodynamics hold for plants and animals, that plants and animals can be arranged into feeding groups or trophic levels, that at least three trophic levels—plants, herbivore, and carnivore—exist, and that the system is in equilibrium.

In the Lindeman trophic–dynamic model, the energy content or standing crop of any trophic level is designated by the Greek capital lambda, Λ. This letter is followed by a numerical subscript to denote trophic level. Thus Λ_1 represents the energy content of the producers, Λ_2 that of the herbivores, and so on. And Λ_n indicates any designated trophic level. Because energy is continuously entering and leaving (the dynamic aspect), the contribution of energy through time from one trophic level to another can be indicated by a lowercase lambda, λ. λ_1 represents the proportion of energy the organisms on any one trophic level, Λ_n, receive from the trophic level below. The loss of heat or respiration, R, plus the energy passed on to the next trophic level, λ_{n+1}, is symbolized as $\lambda_{n'}$. Thus for any trophic level designated as Λ_n, $\lambda_{n'}$ represents $\lambda_{n+1} + R$, and λ_{n+1} is the amount of energy passed on to the next trophic level, Λ_{n+1}.

From this model one can write a generalized formula for energy flow from one trophic level to another:

$$\frac{\Delta \Lambda_n}{\Delta t} = \lambda_n - \lambda_{n'}$$

What this equation states is that the rate of change of the energy content of a trophic level is equal to the rate at which energy is assimilated minus the rate at which energy is lost from it. In the formula, λ_n is positive and represents the contribution of energy from the previous trophic level, Λ_{n-1}, and $\lambda_{n'}$ represents the sum of energy lost from Λ_n.

In the Lindeman model, the amount of energy λ_n transferred to the next trophic level Λ_{n+1} is considered true productivity. The assumption, of course, is that all of this production must represent the gross primary production at the first trophic level and assimilation at the several consumer levels. As used in many studies, this has not been the case for the model because assimilation by decomposers at each level has been ignored, and obviously all of the production has not been assimilated

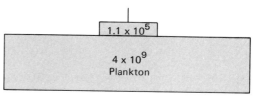

(a) Pyramid of numbers

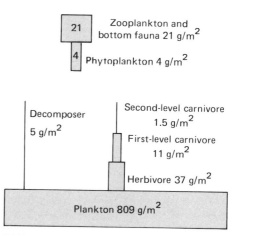

(b) Pyramids of biomass

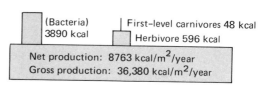

(c) Pyramid of energy

FIGURE 3-13
Ecological pyramids. (Data from W. E. Pequegnat, 1961; H. T. Odum, 1957; Harvey, 1950; Teal, 1962.)

55

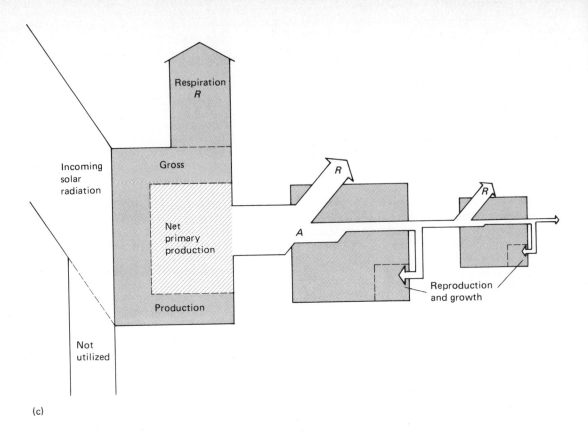

(c)

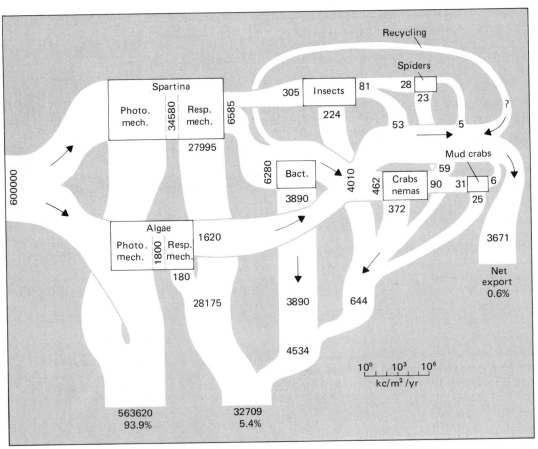

(d)

by those organisms usually considered to be involved in the various trophic levels.

Whether at the producer or consumer level, energy flow through the ecosystem is mediated at the level of the individual (Fig. 3-14a). A quantity of energy of food is consumed. Part of it is assimilated, and part is lost as feces, urine, and gas. Part of the assimilated energy is used for respiration, and part is stored as new tissue which can be utilized to some extent as an energy reserve. Part is used for growth or the production of new individuals.

The model of energy flow through a population, Figure 3-14b, obviously exhibits many of the characteristics of the model for the individual, but with some additions. Growth in the individual becomes changed in standing crop. Part of the biomass goes to predators and parasites, and there are gains and losses of energy and biomass from and to other ecosystems. The population boxes can be fitted into several trophic levels and linked to form a model of energy flow through the ecosystem. In the model, illustrated in Fig. 3-14c, there is no attempt to separate out the detritus-feeders and the decomposers (see Figs. 3-8 and 3-14d). Contrary to what occurs in the Lindeman model, they must fall into one of several trophic levels when the food web is collapsed.

To provide for the grazing and detritus paths of energy transfer, Wiegert and Owens (1971) propose a somewhat different model of energy flow in which transfers through the two channels beyond the autotrophs are considered (Fig. 3-15). They define all organisms utilizing living material as biophages, and all organisms utilizing nonliving matter as saprophages. First-order biophages utilize living plants and are the traditional herbivores, whereas first-order saprophages feed on dead plant material as well as organic material egested by first-order biophages. First-order biophages in turn are utilized at death by second-order saprophages, which are really functional carnivores. In turn, first-order saprophages may be utilized by second-order biophages. Although this model, like others, cannot separate out organisms that occupy several trophic levels, it does modify the Lindemann model so that the decomposers are broken down into functional components.

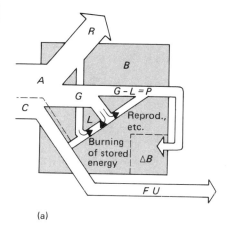

(a)

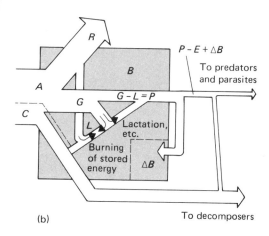

(b)

FIGURE 3-14 [ABOVE AND OPPOSITE]
Models of energy flow. (a) *Energy flow through the individual organism. Note the losses and portion of energy accumulated as growth.*
(b) *Energy flow through the population.*
(c) *Energy flow through the ecosystem. This model considers the decomposers as occupying one of the several trophic levels and does not regard them as a separate trophic pathway.*
(d) *Energy flow through a Georgia salt marsh. (From Teal, 1962.)*

Ecological efficiencies

The Lindeman model stimulated the study of energy flow through populations and communities and has given rise to various estimations of ecological efficiencies to the point that a great deal of confusion exists over the real meaning of the concept. Ecological efficiency is the ratio of any of the various parameters of energy flow in or between trophic levels, populations, or individual organisms (Kozlovsky, 1968). (Efficiencies of individual organisms are more physiological than ecological.)

Although the value of many ecological efficiencies might be dubious, a few are of real interest. Before they can be compared, the parameters involved must be defined. One parameter, ingestion, I, is the quantity of food or energy taken in by an organism, be it a consumer, producer, or saprophage. In the case of producers I is the amount of light available to or absorbed by the photosynthetic pigments. Assimilation, A, is the amount of food absorbed in the alimentary canal by consumers, the absorption of extracellular products by decomposers, and the energy fixed by plants in photosynthesis. Respiration, R, is all of the energy lost in metabolism and in activity, including loss in urine. Net productivity, NP, is the energy accumulated, and represents that left over after respiration: $NP = A - R$. Production, P, is that portion of productivity a trophic level passes on to the next trophic level. It is that portion actually available and does not include losses to decomposers (if decomposers are not considered a part of the trophic level), losses to other systems, or increases and decreases to the standing crop. In this context production must be considered yield.

One useful ecological efficiency is *assimilation efficiency*. The general formula is

$$\frac{A_n}{I_n}$$

Among producers this would be

$$\frac{\text{energy fixed by plants}}{\text{light absorbed}}$$

For consumer levels it would be

$$\frac{\text{food absorbed (assimilation)}}{\text{food ingested}}$$

A second is *ecological efficiency*. This is the ratio:

$$\frac{P_n}{I_n}$$

It reads

$$\frac{\text{energy passed to } n + 1}{\text{energy ingestion at } n}$$

Other efficiencies include *ecological growth efficiency*:

$$\frac{NP_n}{I_n} \quad \frac{\text{net production at } n}{\text{ingestion at } n}$$

tissue growth efficiency:

$$\frac{NP_n}{A_n} \quad \frac{\text{net production at } n}{\text{assimilation at } n}$$

trophic-level production efficiency:

$$\frac{A_n}{NP_{n-1}} \quad \frac{\text{assimilation at } n}{\text{net production at } n - 1}$$

utilization efficiency:

$$\frac{I_n}{NP_{n-1}} \quad \frac{\text{ingestion at } n}{\text{net production at } n - 1}$$

Another useful ratio is that of respiration to assimilation:

$$\frac{R_n}{A_n}$$

These efficiencies vary among species, populations, and trophic levels. Growth efficiencies among larger animals appear to be less than among small animals, and greater among young animals than older ones. Assimilation efficiencies appear to be higher among carnivores than among herbivores; but at the same time respiration in proportion to both ingestion and assimilation also increases at higher trophic levels. As a result, net productivity and production decrease in proportion to ingestion at carnivore levels. Although it may vary widely, the ratio of assimilation between one trophic level and another, A/A_{n-1}, is about 10 percent, and the net productivity of one level compared to the net productivity of the previous level is virtually constant.

BODY SIZE

Body size imposes a limit on energy flow through a given population. There is no linear correlation between energy flow and body weight or surface area per se, but there is a correlation with the metabolically effective body weight. Thus an increase of 100 percent in body weight means a 70 percent increase in metabolic weight. As the body size increases, the neuroendocrine system, which controls metabolism, increases proportionally with body surface rather than with weight. Thus metabolic rate per gram of weight rises exponentially as the weight of the individual declines.

Size acts in still another way. It has considerable influence on the direction a food chain takes because there are upper and lower limits to the size of food an animal can capture. Some animals are too fleet to be caught; others are large enough to defend themselves successfully. Some foods are too small to be collected economically because it takes too long to secure enough to meet the animal's metabolic needs. Thus the size of a carnivore's prey is determined by the upper limit by the predator's strength and on the lower limit by the animal's opportunity to secure enough prey to meet its needs.

There are exceptions, of course. By injecting poisons, spiders and snakes kill prey much larger than themselves; wolves hunting in packs can kill an elk or a caribou. The idea that food chains involve animals of progressively larger sizes is true only in a very general way. In the parasitic chain the opposite situation exists. The larger animals are at the base and, as the number of links increases, the size of the parasites becomes smaller. The only animal that can deal with food of any size is man.

SIZE OF THE STANDING CROP

The size of the standing crop influences the capacity to produce. A pond with too few fish, or a forest with too few trees, does not have the capacity to utilize the energy available. On the other hand, too many fish or too many trees mean less energy available to each individual. This lowers the efficiency of use. The size and composition of the standing crop vary with the region. It may consist of many species, as in a coral reef, or few, as in the tundra. This has a considerable influence on the storage and transfer of energy through the community.

A large standing crop, however, is not syn-

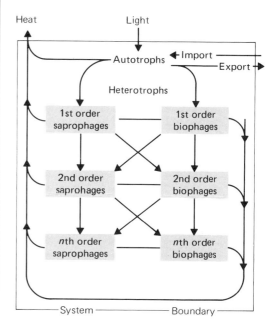

FIGURE 3-15
A model of two-channel energy flow, which separates energy flow into two pathways, that utilized by organisms that feed on living material and that utilized by organisms that feed on nonliving material. (After Wiegert and Owen, 1971.)

TABLE 3-3
Community energy balance sheets for an autotrophic and heterotrophic community

AUTOTROPHIC COMMUNITY: THE SALT MARSH[a]

Input as light	600,000 kcal/m²/year
Loss in photosynthesis	563,620, or 93.9%
Gross production	36,380, or 6.1% of light
Producer respiration	28,175, or 77% of gross production
Net production	8,205 kcal/m²/year
Bacterial respiration	3,890, or 47% of net production
First-level consumer respiration	596, or 7% of net production
Second-level consumer respiration	48, or 0.6% of net production
Total energy dissipation by consumers	4,534, or 55% of net production
Export	3,671, or 45% of net production

HETEROTROPHIC COMMUNITY: TEMPERATE COLD SPRING[b]

Organic debris	2,350 kcal/m²/year, or 76.1% of available energy
Gross photosynthetic production	710 kcal/m²/year, or 23.0% of available energy
Immigration of caddis larvae	18 kcal/m²/year, or 0.6% of available energy
Decrease in standing crop	8 kcal/m²/year, or 0.3% of available energy
Total energy dissipation to heat	2,185 kcal/m²/year, or 71% of available energy
Deposition	868 kcal/m²/year, or 28% of available energy
Emigration of adult insects	33 kcal/m²/year, or 1% of available energy

Source: [a] Teal, 1962; [b] Teal, 1957.

onymous with productivity. Instead, the reverse may be true. Production of bottom fauna in a pond amounted to nearly 17 times the standing crop when fish were present. In the absence of fish, the production rate of fish food decreased and finally stopped with a larger standing crop. Thus the standing crop of bottom fauna in the presence of fish was depressed, but production increased. In ponds with fish, the annual production amounted to 811 lb of fish food and 181 lb of fish per acre. This represents an efficiency of 18 percent in energy conversion from bottom fauna to fish (Hayne and Ball, 1956).

Community energy budgets

Because energy flow involves both inputs and outputs, the efficiency of an ecosystem can be estimated by measuring the quantity of energy entering the community through various trophic levels and the amount leaving. Thus a balance sheet for energy and production with debit and credit sides (Table 3-3) can be drawn up for a community. Few communities have been studied intensively enough to present such a broad picture, but there are some available. One is a salt marsh, an autotrophic community (Teal, 1962), and another a cold spring, a heterotrophic community (Teal, 1957), whose major energy source was plant material fallen into the water.

Of the energy transformed by organisms in the spring, 76 percent entered as leaves, fruit, and branches of terrestrial vegetation. Photosynthesis accounted for 23 percent, and 1 percent came from immigrating caddisfly larvae. Of this total input, 71 percent was dissipated as heat, 1 percent was lost through the emigration of adult insects, and 28 percent was deposited in the community. In the salt marsh the producers themselves were the most important consumers, for plant respiration accounts for 70 percent of gross production, an unusually high figure. (Scots pine, for example, utilizes only about 10 percent in respiration.) Plants are followed by the bacteria, which utilize only one-seventh as much as the producers. Primary and secondary consumers come in a poor third, using only one-seventh as much energy as the bacteria.

In spite of energy budget sheets, these examples point out how incomplete and fragmentary is our knowledge of energy flow through ecosystems. To understand something of energy flow through ecosystems, one has to know something of energy flow through populations within the ecosystem and then relate this information to the flow of energy from one trophic level to another. Herein lies the weakness of the model. To date, energy budgets of several ecosystems are based in part on assumptions rather than on known values of energy flow. Too little is known of energy flow through any population to draw a clear picture of energy flow through an ecosystem. And what knowledge we do possess often is unreliable. If this knowledge is to be used, one has to assume that energy flow through a population is constant, or if fluctuating, at least predictable. But probably it is not because energy flow through populations is variable, depending upon ecological conditions. For example, growth, and thus energy storage by salmonid fish, is related to the size of food (Paloheimo and Dickie, 1966). When particle size is small, fish expend more energy obtaining food, and a smaller proportion is used in growth. Variations in temperature affect the rate of food assimilation, and variations in salinity and nitrogenous wastes in the water affect both energy turnover and efficiency of utilization.

Within any ecosystem, animal populations may have a pronounced influence on the rate of energy fixation of plants. Overgrazing and overbrowsing reduce the amount of primary production in grassland and forest ecosystems. Overexploitation of a prey species either by man or by predators can affect the amount of secondary production. The nutrient composition of solids or plants can limit energy fixation and storage in both primary and secondary production. Lack of boron in the soil, for example, can severely depress the growth of alfalfa; and the low nutrient status of a soil, and thus of the plants, can affect the production of animals. We lack enough detailed studies of trophic efficiency to give us any clear picture of the structure of ecosystems or to strongly support the Lindemann model. Because they involve populations rather than trophic levels, studies to date neither support nor refute the model of energy flow through the ecosystem. But the concept of energy flow is valuable as a guideline for future studies and as a basis for understanding some of the relationships and interactions of modern man and his environment.

Summary

A basic functional characteristic of the ecosystem is the flow of energy. The energy of sunlight is fixed by the autotrophic com-

ponent of the ecosystem, the green plants, as primary production. This energy is then available to the heterotrophic component of the ecosystem of which the herbivores (plant eaters) are the primary consumers. The herbivores in turn are a source of food for the carnivores. At each step or transfer of energy in the food chain, a considerable amount of potential energy is lost as heat, until ultimately the amount of available energy is so small that few organisms can be supported at that source alone. The animals high on the chain often utilize several sources of energy, including plants, and thus become omnivores. All food chains eventually end with the decomposers, chiefly organisms that reduce the remains of plants and animals into simple substances. Energy flow in the ecosystem may take two routes: one goes through the grazing food chain, the other through the detritus food chain, in which the bulk of production is utilized as dead organic matter by the decomposers.

The loss of energy at each transfer limits the number of trophic levels, or steps, in the food chain to four or five. At each level the biomass usually declines; if the total weight of individuals at each successive trophic level is plotted, a sloping pyramid is formed. In certain aquatic situations, however, where there is a rapid turnover of small aquatic consumers, the pyramid of biomass may be inverted. Energy, however, always decreases from one trophic level to another and is pyramidal.

The ratios of energy flow in or between trophic levels of natural communities, in or between populations of organisms, and in or between individual organisms, are ecological efficiencies. Because efficiencies are dimensionless, any number of ratios can be determined. Among some of the most useful are assimilation efficiencies, growth efficiencies, and utilization efficiencies.

Although knowledge of energy flow in ecosystems is fragmentary and difficult to come by, the concept of energy flow is a valuable guide for the study and understanding of both ecosystem functioning and the relationship of man to his environment.

 s Aldo Leopold (1949) wrote,

There are two spiritual dangers of not owning a farm. One is the danger of supposing that breakfast comes from the grocery, and the other that heat comes from the furnace. To avoid the first danger one should plant a garden, preferably where there is no grocer to confuse the issue. . . . To avoid the second, he should lay a good split of oak on the andirons, preferably where there is no furnace, and let it warm his shins while a February blizzard tosses the trees outside.

Relatively few people in the United States plant gardens any more, and most have succumbed to the first spiritual danger. To the mass of our people breakfast does come neatly boxed from the grocery and nowhere else. Modern man has mentally divorced himself from the food chain. Nevertheless he is a part of the food chain, a link in the flow of energy through the biosphere. He has moved out of a natural food chain into an artificial one that is manipulated toward his own ends and dependent upon an input of energy in addition to that currently supplied by the sun.

The hunter food chain

Man has lived as a hunter–gatherer for over 99 percent of the 2 million years he has lived on earth. He began the domestication of plants and animals only 10,000 years ago. And it was only 2,000 years ago that agriculturalists and pastoralists replaced the hunter over half the earth. Today only scattered pockets of hunting man survive, yet the hunting way of life has been the most successful and persistent adaptation of man. It has dominated human evolution, and it enabled man to colonize the earth. Agriculture, on the other hand, has influenced less than 1 percent of human history.

Throughout his long history as a hunter and gatherer, man was a part of the natural ecosystem in which he existed, competing with other animals for the same resources. He was wholly dependent upon the productivity of his habitat, and he acquired his food by muscular energy. Although he had no con-

Man, energy, and the food chain

trol over scarcity and abundance, he was able to exploit a wide range of plants and animals, which tended to stabilize his food supply. To obtain his food he had to move about the countryside from areas of scarcity to areas of abundance. Like the animals with which he lived, he expanded his hunting territory when prey was scarce, and was able to contract the size of the home range when game was abundant.

Hunting man, like most terrestrial warm-blooded vertebrates, evolved behavioral patterns that resulted in efficient spatial distribution. If the food resources were concentrated seasonally and regionally and the diet was specialized, then the group the diet supported was large, perhaps a hundred or more. If the food resources consisted of a wide variety of plants and animals, the size of the groups changed seasonally with the abundance of food and the region they occupied. Gregarious tendencies were limited in part by the food resources. In the Pleistocene epoch mean densities rarely exceeded one man per square mile.

Of the hunting societies, the purest is that of the Eskimo. He is a specialized hunter existing in a specialized natural ecosystem, the Arctic. Eskimos (at least before their extensive contact with western technology) have been characterized by a high degree of mobility, a diet almost exclusively of meat, and a social organization adapted to the exploitation of an animal source of food.

Eskimos live in a severe environment. Half the year, summer, is daylight; the other half, winter, is darkness. Temperatures range from $-50°$ F in winter to $50°$ F in summer, with an occasional maximum of $80°$. For part of the year the land is exposed and there is a brief period of greening. Most of the year the land is snow-covered, and the open water is frozen. To meet the energy demands needed to survive, the Eskimo requires somewhere around 3,000 cal/day, as well as an additional input of energy to warm his shelter and to feed his dogs. The source of this energy is the wild game that shares the arctic ecosystem with him.

To exist the Eskimo exploits the full range of the environment: from sea ice to open water, from tidal flats and sea cliffs to open tundra. To secure needed energy he hunts a wide range of arctic game. Seals, both common and bearded, are the mainstay of his

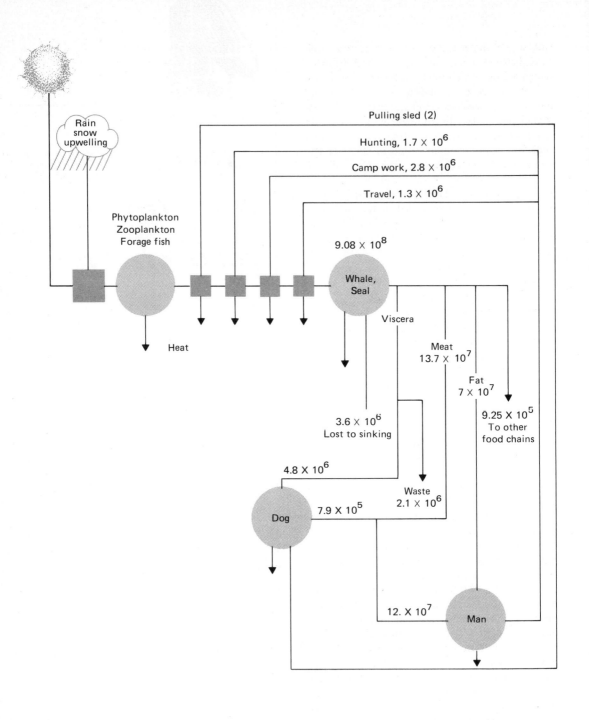

Rain
snow
upwelling

Pulling sled (2)

Hunting, 1.7×10^6

Camp work, 2.8×10^6

Travel, 1.3×10^6

Phytoplankton
Zooplankton
Forage fish

9.08×10^8

Whale,
Seal

Viscera

Heat

Meat
13.7×10^7

Fat
7×10^7

9.25×10^5
To other
food chains

3.6×10^6
Lost to sinking

4.8×10^6

Dog

7.9×10^5

Waste
2.1×10^6

$12. \times 10^7$

Man

existence, providing food, oil for heating, and skins for clothing, shelter, and boats. The beluga whale is also an important item, along with the caribou, the arctic char, eider ducks, and other nesting sea birds, which are sources of eggs, meat, and down.

The Eskimo expends most of his energy hunting. In pretechnological times, he moved about seasonally, reducing that expenditure. During the fall and winter, when the arctic world is locked in ice, the Eskimo lives a sedentary life, his energy needs met by meat obtained and cached in the fall, when animal life is abundant. In the spring, when thaw sets in and animal life becomes active, the Eskimo leaves his permanent settlement, and follows the food supply, which includes fish and birds and their eggs. When fall comes he returns to his winter settlement. Thus the greatest output of energy on the part of the Eskimo takes place during the spring, summer, and early fall, when energy sources are most abundant. Because hunting activity is reduced in winter, energy expenditure is minimal, most of it directed toward maintenance and social activities.

The Eskimo portrays the close relationship of man, the hunter, to natural ecosystems. His energy needs are met by the natural food chains of which he is a part (Fig. 4-1). His input of energy into the system is muscular, and like a predatory animal, he expends the greatest amount of his energy on hunting. The Eskimo, like members of most hunting cultures, recognizes his close dependence on natural food chains. All living beings, the Eskimo believes, have a soul, and he seeks a friendly relation with the animals he hunts, if not physically then spiritually. Eskimos take care never to offend the soul of the animals they kill.

With the introduction of technology into Eskimo society, the hunting culture and energy relationships are changing. Kerosene replaces seal oil for heat; the snowmobile and gasoline replace dog teams and wild meat. The rifle and ammunition replace the spear and harpoon; the motor boat replaces the kayak. Wooden winter houses replace the igloo and the tent. Canned goods and carbohydrates now supply part of the caloric intake; money becomes a medium of energy exchange; and crafts and periodic employment are a new expenditure of energy. Technology becomes the source of well-being,

FIGURE 4-1 [OPPOSITE]
Energy flows through an Eskimo hunting society. Sunlight, precipitation, ocean currents, and upwellings furnish energy and nutrients for food chains extending to seals and whales, staple foods of the Eskimo hunting society. Other energy inputs are those supplied by Eskimos and their dogs involved in hunting, as well as societal activities. Note the dependence on animal foods. Technology, canned foods, the use of gasoline as a fuel, and snowmobiles rather than dog sleds as a means of transportation are changing Eskimo culture and energy relationships (Quantities in kcal.) (Based on data from W. B. Kemp, 1971.)

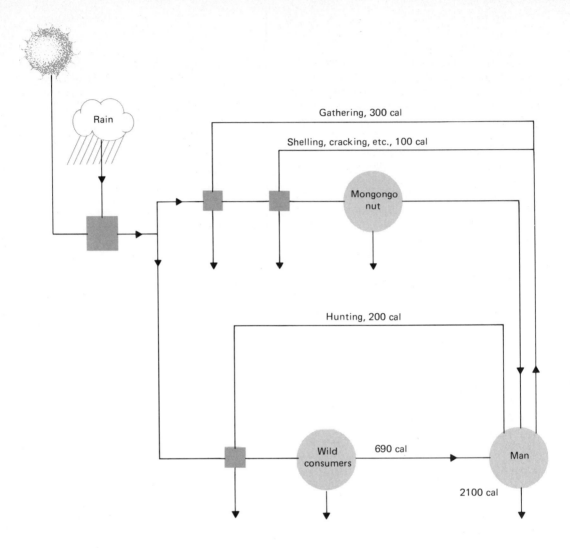

Gathering, 300 cal

Shelling, cracking, etc., 100 cal

Rain

Mongongo
nut

Hunting, 200 cal

Wild
consumers

690 cal

Man

2100 cal

and the ritual control of the forces of nature and the food supply are disappearing. The Eskimo as hunter is being partially dislodged from the natural food web and is taking more direct control of his energy system.

At the other end of the hunter spectrum are the true hunter–gatherers, generalists rather than specialists, living in generalized ecosystems. Such people usually form small localized groups. They know the country intimately, they are less nomadic than the specialized hunters, and they subsist on a wide range of wild plants and animals.

One example of the generalized hunter–gatherers is the !Kung Bushmen of the Kalahari (Lee, 1966) (Fig. 4-2). Unlike hunter–gatherers who lived in more moderate environments such as the eastern woodlands of North America or on the plains, the Bushman lives in a harsh desert environment. The springs are hot and dry, the summers hot and rainy,

and the winters cool and dry. Summer daytime temperatures range from 60° to 100° F, winter temperatures from 30° to 78° F. Rainfall averages 6 to 10 in. The loose soil, in spite of its porosity, supports an abundance of vegetation.

The !Kung Bushmen live in small settlements within a mile of a permanent well, although they move during the summer rainy season when seasonal pools of water appear. The men spend the days hunting game, and the women forage for nuts, berries, roots, and melon. (The basic item in the diet, however, is the mongongo nut.) Although group members hunt and gather independently, they share the food at the end of the day. There is little if any exchange of food between groups, and little food is stored. Most of it is consumed within 48 hours.

The basic foraging strategy is to collect and eat the most desirable foods that are the least distance from standing water. As the distance of the food from standing water increases, energy costs increase. The maximum cost rises sharply beyond a 12-mi round trip. If the cost is too great the alternative is to stay nearer camp and exploit foods less desirable in terms of taste, ease of collecting, and abundance. During the dry season, the younger, more mobile members make the long trips to the nut forests. The older members stay nearer home and collect the less desirable foods. During the rainy season, when seasonal pools of water are abundant, the entire camp moves when the nearby supply of nuts is exhausted. The cost of nuts during the rainy season never exceeds a 6-mi round trip.

From the point of view of energy, the per capita caloric requirement of the Bushman is 2,140/cal/day. Of this, meat furnishes 37 percent of the diet by weight and 690 cal, mongongo nuts 33 percent and 1,260 cal, and other foods 30 percent and 190 cal. Based on activities of hiking for food and nut cracking, the calorie requirement for males is estimated by Lee (1966) to be 2,250/day. The per capita food requirement of the group is about 1,975 cal. Thus the societal energy needs per person exceeds each person's food demands by 165 cal. A portion of these extra calories is utilized by the hunting dogs. This food consists largely of leftovers, the quantities of which vary with the abundance of food brought into camp. The food is so distributed that each age and sex group fulfills its respective caloric needs.

Contrary to popular opinion, man the hunter–gatherer did not live on the edge of catastrophe, nor did he spend the greater part of his time seeking food. Studies of the few remaining hunting cultures indicate that he worked less than we do, that this quest for food, except under dire circumstances, took up only a small portion of his time, and that leisure was abundant. His wants were restricted, and because of his need to move, wealth was a burden. The nature and distribution of food, along with the mobility required, precluded the establishment of any large permanent settlement.

Early agriculture

Of the preagricultural peoples, the hunter–gatherers occupying the forest and grassland

FIGURE 4-2 [OPPOSITE]
Energy flow through a hunting–gathering culture, the !Kung Bushman of the Kalahari. In contrast to the Eskimo culture, only a portion of the energy base comes from meat. Most of the energy input from man involves the gathering and preparation of manango nut. (Based on data from Lee, 1966.)

margins, where plants and animals were most diverse, established small and relatively permanent settlements. For thousands of years such hunter–gatherers of western Asia lived by intensively collecting the seeds of wild wheat, barley, wild rye grass, wild flax, and large-seeded legumes such as vetch and chickpea. They also had access to dates in the lowlands, acorns and almonds in the foothills, and grapes, apples, and pears in the mountains.

Hunting man undoubtedly was aware that seeds thrown away subsequently sprouted. Open disturbed areas about his settlements and piles of debris provided the bare soil and freedom from competition needed by the weedy plants that furnished food. Requiring soils rich in nitrogen and growing only in disturbed areas, such plants quickly colonized an expanding number of such sites. They became in effect weeds about the habitations of man.

Wild wheats and barley were plant opportunists and as such possessed certain adaptations characteristic of agricultural weeds today. The seeds were able to survive long dry spells in well-baked soil. When rains came the large seeds with abundant food reserves responded quickly, sprouted, and grew. As the seeds ripened, the hunter–gatherers harvested them, probably selecting the larger-seeded forms. Undoubtedly the next step was to plant the seeds deliberately as a source of food. Planting such seeds in these man-created habitats exposed the plants to new selective pressures as well as removing certain others. Protection allowed new deviants and mutations to survive. When they benefited man, he kept them even though they no longer were suited to grow as wild plants (see Helbaek, 1959; Ucko and Dimbleby, 1969).

Two types of agriculture emerged: *seed culture* (the planting of seeds) and *vegeculture* (the planting of root crops). Seed culture developed in subtropical regions of cool winters, dry summers, and wet autumns and springs. It arose independently in Mexico and in the Fertile Crescent of Iran. The former was characterized by corn, the latter by wheat and barley, all developed from wild types (Fig. 4-3) (see J. R. Harris, 1967; Helbaek, 1959; P. C. Mangelsdorf, 1958). Vegeculture arose in the tropical lowlands of South America and in southeastern Asia and Africa, regions also characterized by wet and dry seasons. Vegeculture included such starch-rich plants as yucca, sweet potato, yams, taro, and cassava. In the Andean highlands, where there is a cool temperate climate, another root crop, the potato, flourished in the wild state. Early development of vegeculture probably began by colonization and harvesting of root crops without replanting.

Early agriculture was polycultural rather than monocultural. Involved were an assortment of crops that were in many ways functionally interdependent. To achieve this interdependence, man in his agricultural efforts manipulated rather than transformed natural ecosystems. He altered selected components without modifying overall structure, even when his own domesticated livestock were involved.

Early agriculture was swidden, or slash-and-burn, agriculture. Swidden agriculture, still widely practiced in tropical regions, involves the clearing of land by cutting and burning the natural vegetation to open the soil for planting (Fig. 4-4). It varies widely from region to region, country to country, and culture to culture. It may involve year-long or seasonal cropping of the land, fertilization or no fertilization. No attempt is made to plow: the farmers use a digging stick or hoe to plant the seeds or cuttings directly into the soil. Within 2 to 3 years the original plot is abandoned and the land is allowed to revert to and remain in wild vegetation for some time before it is cultivated again. For every acre under cultivation some 15 to 20 other acres must lie fallow, especially if a plot is reopened only once every 20 to 25 years.

Ecologically such agriculture is closely tied to the functioning of natural ecosystems. Although the native vegetation on the plot is destroyed and the material burned, the mineral matter is returned to the soil. The vegetation, even though domesticated and cultivated, is diversified and grows in layers, or strata. A number of different crops are grown together, from roots to grains and fruits. The plot is in cultivation for a short time only. It is not open enough to be exposed to heavy erosion, nor is it so depleted of nutrients that the growth of invading natural vegetation is inhibited. By allowing the natural vegetation to reclaim the area, the swidden agriculturalists restore fertility to the plot.

Essentially there are two types of swidden agriculture, *milpa* and *conuco*. Milpa is based on seed crops, especially the combination of corn, beans, and squash, the three forming a sort of symbiotic relationship. Corn is dominant, grows tall, and claims most of the sunlight. It also supports the bean stalks in their climb to the sunlight. The bean, a legume, returns nitrogen to the soil. The lower stratum of the field is claimed by squash and pumpkin, whose vines cover the ground. The result is a protective cover of vegetation that makes full use of sunlight and moisture and is an adequate substitute for the natural cover that existed before. However, the milpa system, because of its nutrient-demanding crops, simple stratification, and open canopy, increases the opportunity for invasion by weeds and also wastes soil resources (D. R. Harris, 1969). Thus there is a need for more shifts to temporary clearings, which in itself encouraged expansion of that type of agriculture.

Conuco, a system of root-crop cultivation, is ecologically more stable. It often involves the preparation of earthen mounds in which stem cuttings and other vegetative parts are planted. The diversity of plants tends to be greater. In fact the Hanunoo of the Philippines plant as many as 50 different kinds of plants. The fields are weeded by hand, by hoe, or by machete. As the various crops ripen, they are harvested and often replanted, resulting in continuous year-long production. Because it provides an even greater ground cover and is not so demanding of soil nutrients as the protein-rich crops of milpa, conuco cultivation can be maintained on fixed plots for many years. In river-bank, seashore, and savanna-edge ecotones, which have an assured supply of animal protein, conuco agricultural systems provided a sort of easy life that did not demand the foresight or labor of grain or milpa agriculture. It was grain farming that created the agricultural societies of Europe, Egypt, India, and China.

Accompanying the development of agriculture was another major intervention into the natural functions of the ecosystem: irrigation. Even before he learned how to use draft animals, man discovered that he could supplement natural rainfall with water held back in impoundments. He held back the water that naturally flowed through rivers and streams to the sea and diverted it as needed through canals and ditches to his croplands. Irrigation as an agricultural technique is nearly as old as seed agriculture. Its early develop-

69

(a)

(b)

FIGURE 4-3

These photos contrast three stages in the development of corn. (a) Teosinte, progenitor of modern corn, grows at the edge of a cultivated plot in Mexico. (Photo by H. Wilkes.) (b) A field of two kinds of modern corn. On the right are rows of open pollinated corn, replaced as a commercial crop by hybrid corn, growing on the left. (Photo courtesy Pioneer Corn Co., Tipton, Indiana.)

ment was in the Fertile Crescent and in the Nile Valley, where grain agriculture evolved. It enabled those early agriculturists to grow food successfully in a relatively dry climate. It permitted the development of a complex agricultural system capable of supporting a growing population and a rising civilization. Later the Incas, the Mayas, the Indians of southwestern North America, and the peoples living along the rivers of Asia also developed irrigation techniques. Today irrigation, with water provided both by impoundments and wells, is used not only to expand agriculture to land that would otherwise not be tillable but also to intensify agriculture on lands already cultivated.

As man was domesticating certain wild plants, he was also domesticating certain animals, perhaps more by trial and error than by purposeful forethought. In the same region east of the Mediterranean and in southwestern Iran where wild emmer and barley grew, wolves, sheep, goats, pigs, and cattle also lived. What prompted man to domesticate these animals is a question that probably never will be answered. Whatever the reason, man did assume leadership and dominance over certain animals, particularly sheep and cattle. He thus changed from a hunter of animals to a domesticator, a major behavioral shift in his cultural and ecological adaptation. This practice eventually freed him from dependence on wild game as a source of animal protein. To maintain his animals he displaced wild grazing animals with his domesticates and eliminated their predatory enemies. In time he harnessed the animals as additional input of energy in the production of crops.

From the domestication of herding animals came a new way of life for some men, that of pastoralism, a form of agriculture in which a livelihood is gained from the care of large herds of domestic animals. They are kept for food products, such as milk, meat, and blood; for salable products such as wool; or as instruments of production, as in the case of the horse, which was used by the North American Plains Indians to hunt buffalo. Basically, pastoralism represents a cultural adjustment to a semiarid grassland ecosystem that can support grazing animals but is poorly suited for cultivated crops.

Pastoralism involves the mutual dependence of man and animal. The livestock provide food, fuel in the form of dried dung, clothing, housing, a means of transport, and items for trade. In return the herdsman provides the animals protection, sometimes shelter from climate extremes, postnatal care of young, and in some instances winter food. Both man and animal extract more than they return to grassland ecosystems. In too many instances livestock overgraze grassland (inducing water and wind erosion) and destroy vegetation around waterholes and sometimes even the waterhole itself. Dung, which might return some fertility to the grassland, is collected and used as fuel.

When man achieved the domestication of animals, he more completely severed his dependence upon natural ecosystems as an energy base. As early as 9000 B.C. the Neolithic people of Mesopotamia had shifted their dependence from wild to domestic animals. This shift altered man's attitude and relations toward his natural environment. No longer requiring wild animals for food, and secure with an energy base supported by domestic plants and animals, yet retaining his biological inheritance as a hunter, man began to kill animals indiscriminately and wastefully, something the cultural hunter rarely did. Many of the larger mammals were killed to protect both crops and livestock, but massive slaughters were the result of purposeful overkill in the name of sport and entertainment. Egyptian pharoahs and early rulers in southwestern Asia had animals driven into compounds for easy kill. The coliseum game of the decadent Roman Empire consumed thousands of elephants, cattle, bears, lions, and other animals and were a major cause of their eventual extinction in North Africa.

Accompanying this change in attitude toward wild animals was the continued encroachment of agriculture on the habitats of wild animals and the constant competition between domestic livestock and wild ungulates for forage. This intense competition still continues, especially on the African and Australian savannas and on the rangelands of the western United States.

In the course of his domestication of animals, man has adapted and changed wild animals to suit his religious, economic, and recreational needs. He has recognized the utilitarian value of mutants and deviants that could never have survived under natural conditions. By exposing these animals to new and often artificially applied selective pressure and by protecting them, man has been able to increase the usefulness of animals. Some produce specialized products such as milk and wool, and others are drafted for work and war. Dogs are used for protection, hunting, food, or household pets.

An example of a primitive type of general agriculture is the native cattle-keeping system in Uganda. Cattle harvest the primary productivity of the range, and man utilizes secondary production in the form of blood, milk, and meat. Various crops, however, supply basic parts of his energy supply. Man's energy input involves both tillage and caring for the cattle (Fig. 4-5).

The domestication of two animals, cattle and horses, provided man with a new source of power. At first horses and oxen were used primarily to pull chariots and carts. In time oxen, or in southeast Asia water buffalo, were hitched to wooden plows that often were little more than heavy sharpened sticks. The harnesses were inefficient and did not take full advantage of the animals' muscle power. Inadequate foot protection for the animals further limited their efficiency. However, draft animals still could perform ten times as much tillage in a day as a man, and their use significantly increased crop yields (Fig. 4-6). Energy flow in such a primitive system is exemplified by Indian rice culture on small plots of ground. The cattle supply power and milk. In turn they consume roughage unusable by man and recycle nutrients in the form of manure (Fig. 4-7).

The efficiency of harnesses was improved, iron horseshoes were developed, and the iron moldboard plow was invented, thus enabling agricultural man to break wet, heavy soils. This opened up new areas for food production (e.g., all of northern Europe and eventually the western grasslands of the United States), and it reduced the need for keeping large acreages of land lying fallow. The increasing use of draft animals and subsequent invention of such horse-pulled equipment as reapers, cultivators, mowing machines, and multiple gangplows permitted the rapid expansion of agriculture with limited amounts of manpower.

A growing agricultural technology required fewer people on the land and in fact produced a surplus of rural people. This agricultural revolution, with its reduced need for manpower, fostered the Industrial Revolution,

FIGURE 4-4
Cleared garden plots in the Philippines surrounded by montane forest. This form of swidden agriculture is called "kaingin." The usual pattern has been for the "kaingineros" (gardeners who practice shifting agriculture) to clear patches of forest, raise kamote for 2 or 3 years, and then abandon the garden because of soil depletion and the growth of low, dense vegetation (see Fig. 9-11). (Photo by N. E. Kowal.)

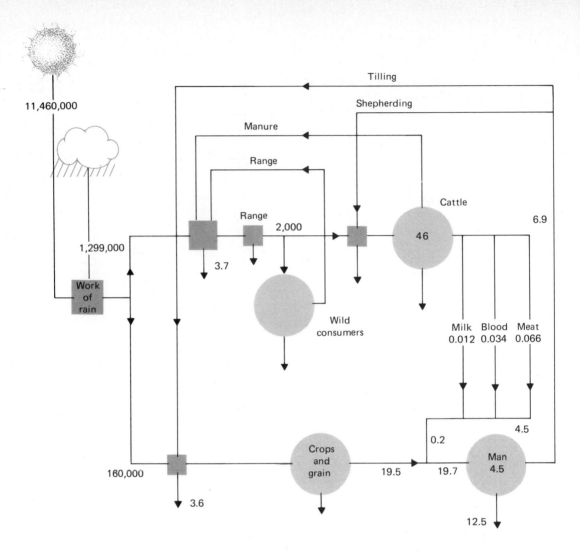

which utilized surplus rural labor. The Industrial Revolution also harnessed a new source of energy, fossil fuels, and produced the steam and internal combustion engines.

Breakthroughs in science added to agricultural productivity. The discovery by von Leibig that the addition of nutrients could renew the fertility of the soil led to the use of fertilizers to increase crop production without increasing the acreage under cultivation. The work of Mendel, de Vries, and other geneticists enabled plant breeders to tailor plants and animals to regional needs. Plants were bred to be more tolerant of cold, more resistant to drought and disease, more responsive to fertilizers, and higher-yielding. Cows were bred for high milk production, chickens for maximum egg production. Meat animals were bred for high gains, rapid maturity, and a minimum of fat (vegetable oils replaced animal fats as human food). The broiler industry, for example, has so advanced that 2.5 kg of

feed produces 1 kg of living weight of chicken. Advances in nutrition, the use of artificial protein sources such as urea, controlled feeding, the restriction of chickens to the laying battery and broiler house, feedlot feeding of cattle, and barn feeding of dairy cattle all hold to a minimum the amount of energy required for activities other than the production of eggs, milk, and meat. Modern agricultural plants and animals have no protective mechanisms left; they are unable to fend for themselves and depend wholly upon an energy subsidy supplied by industrialized agriculture.

Modern agriculture

All these developments brought an end to traditional agriculture, particularly after World War II. Before that time farms were relatively small, fields were planted with a wide diversity of crops, pastures were usually separated by hedgerows and interspersed with woodlands. The land supported a diversity of plant and animal life even though species composition was heavily influenced by agricultural practices. After World War II, the explosive increase in the use of tractors, self-propelled combines, and other large machines made small farms inefficient and uneconomical (Fig. 4-8). Farms were combined, and hedgerows were removed to make larger fields, a process just now taking place in England. The variety of crops characteristic of semi-subsistence and livestock farming were replaced by large acreages of single crops, for example, grass, corn, wheat, soybeans. Monoculture, characterized by low diversity of life, replaced polyculture. Tractors replaced horses, releasing vast acreages (in the United States alone, 70 million acres) once needed to support draft animals for food production and other uses. Food production became a large business, and the grains, meats, and wastes that once were recycled locally are shipped to other parts of the country and the world. Waste products once returned to land now create solid waste problems.

In many ways modern agricultural production has become divorced from natural ecological systems. The chemical and mechanical orientation of modern agriculture depends upon a heavy input of energy and a large quantity of material inputs such as chemical fertilizer and pesticides. In fact far more energy is put into the production of crops than is harvested. Food production greatly depends on an energy subsidy from fossil fuels. Without it agriculture would have to revert to semisubsistence levels. As the input of fossil fuels into agricultural systems increased, yields also increased (Fig. 4-9). In 1870 1 American farmer could produce food enough to feed 5 people. Today, in combination with other inputs, 1 farmer can produce enough food to feed 43 people.

Among natural ecosystems, organisms don't worry about distribution systems. The animal goes to the source of food (many aquatic organisms allow the currents to carry the food to them). Primitive man sought out sources of food and even followed his food supply. In subsistence agriculture as practiced in our pioneer days and as still practiced in most underdeveloped nations, food is consumed at the point where it is raised.

FIGURE 4-5 [OPPOSITE]
Energy flow in a primitive herding and cropping system in Uganda. Energy inputs from sun and rain flow through native rangeland and through cultivated crops. Range forage supports cattle and wild herbivores which in turn recycle some nutrients and return some energy through body wastes. Inputs of energy into the system by man involves cultivation of crops and tending of cattle. Energy flow to man includes meat, milk, and blood, but the bulk of energy comes from cultivated crops. (Quantities in kcal.) (Adapted from H. T. Odum, 1971, Environment, Power, and Society, Wiley, New York.)

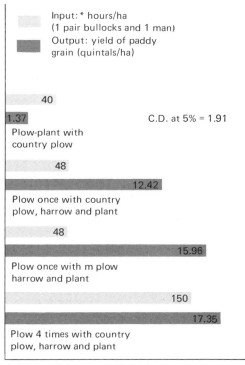

FIGURE 4-6
Increased yields that can be gained from additional energy inputs including increased intensity of soil management are demonstrated by increased output of grain from additional input of human labor and technology. Maximum output relative to input resulted from a simple procedure of replacing a primitive plow with a moldboard plow. (From U.S. Panel of World Food Problems, 1967.)

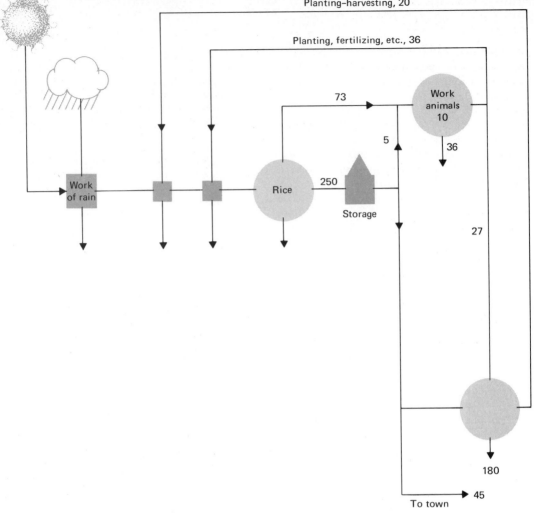

Planting–harvesting, 20

Planting, fertilizing, etc., 36

Work animals 10

Work of rain

Rice

Storage

To town

But a large urban society requires that food be transported to it. Unless a distribution system is available, high food production is of no consequence.

Nowhere in the world is there a more elaborate food distribution system than in the United States. The marketing system must assemble products from farms across the country, sort them for uniformity, and combine them into units suitable for handling and shipping. Food travels through a host of middlemen, each step involving further input of energy from fossil fuel (Fig. 4-10). To preserve wholesomeness, quality, and freshness the process must be continuous. Transportation must be rapid and storage space must be available. Only in this way can regular flow of food be ensured throughout the year.

The expansion of high-yield agriculture is primarily an accomplishment of the western world. It was made possible by a series of

revolutionary developments. The first was the domestication of plants; the second was the introduction of irrigation; another was the harnessing of draft animals. Still others were the exchange of crops from one country to another, the application of the principles of genetics to the improvement of plants and animals, the invention of such labor-saving machinery as the reaper, the development of refrigeration and methods of preserving food, the use of tractors and input of fossil fuel into food production, and the development of food distribution pathways and applied research and education in food production.

In recent years the technology of high-yield agriculture has been introduced into Asia, Africa, and Latin America. This technological change, known as the green revolution, involves among other things the development of crops adapted to climate and heavy fertilization (Fig. 4-11). But the green revolution is not without its ecological problems. Because of the lack of transportation facilities, roads, and the needed political and social structures, food stuffs cannot be distributed to the whole population. Unknown are the effects of short-term high-yield cropping obtained by heavy chemical fertilization on soils. Only recently have agricultural fields been subject to such intensive farming. We cannot compare the effects of this type of farming with the less intensive type that has kept soils productive for thousands of years. Intensified farming on limited acres requires strong measures to control both pests and diseases. The use of pesticides has already created worldwide ecological problems. There is a question whether chlorinated hydrocarbons can continue to be used without disastrous results to both animals and man. The effectiveness of the green revolution may rest with the development of ecologically safe pesticides. There is a problem of water scarcity. To provide needed water storage, rivers have been dammed, creating more ecological problems. The fertility of the lands along the Nile depended on deposition of fertile silt. Now the silt is deposited in the dam. Replacement of simple age-old irrigation practices with large-scale projects and slow, warm water in miles of irrigation ditches has created ideal conditions for certain species of snails that serve as secondary hosts for schistosomiasis or bilharziasis, a growing health problem. No one as yet can predict or foresee the ecological consequences of cultivating more, often marginal, land.

But ecological problems are not the only ones being created. There are also a number of social, economic, and political problems. As the yields of grain increase, there is a corresponding need to expand the capacity to harvest, thresh, dry, and store the grain. There is a need to produce fertilizers that require industrial expansion and capital outlay. Marketing facilities need to be improved and in some places, developed. The food must be distributed from grain-producing areas to urban concentrations. Poor distribution and marketing systems produce local gluts of grain that depress prices. Because of poverty there is often no demand for the food produced because demand is based on purchasing power. Involved technology decreases the number of rural people needed to raise crops, and these landless groups are being squeezed in large cities. Because of the economic demands of

FIGURE 4-7 [OPPOSITE]
Energy flow in a primitive plow agriculture as practiced in the rice culture in India. Energy flow is channeled through rice. Inputs of energy into the system come from man's hand labor and the work of draft animals. A portion of the grain is fed to the work animals, but the bulk is used as human food. It differs from the cattle economy in the lack of any utilization of meat of herbivores, the independence from a natural ecosystem, and the input of animal work into the system. (Quantities in kcal.) (Adapted from H. T. Odum, 1971, Environment, Power, and Society, Wiley, New York.)

FIGURE 4-8
Modern agriculture is subsidized by a heavy input of fossil fuels, not only to power the machinery, but also to manufacture machinery and to process and distribute foods to consumers. (Photo courtesy U.S. Department of Agriculture.)

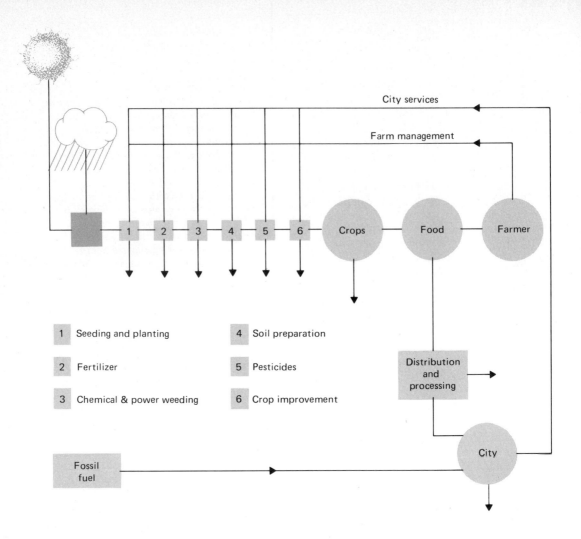

1	Seeding and planting	
2	Fertilizer	
3	Chemical & power weeding	

4	Soil preparation	
5	Pesticides	
6	Crop improvement	

large-scale grain production and the need to buy fertilizer and pesticides, the small poor farmer with a lack of credit is being forced out of production.

Man's place in the food chain

Although man has been growing food for thousands of years, and although he has applied the knowledge of plant breeding and genetics to the improvement of food sources, he still relies on the basic crops that have been with him since the early days of agriculture.

The most significant change in man's place in the food chain came when he switched from animals to vegetables as a source of food. Today, on the average, 88 percent of man's global food energy supply comes from plants. It is significant that in only four countries, Canada, the United States, Australia, and New Zealand does man obtain more calories from meat, milk, and eggs than from starchy

foods (Brown and Finsterbusch, 1972). The remainder of the world to varying degrees lives on the first consumer level.

The dominant staple foods of mankind are wheat and rice. Wheat is the principal staple in almost all high-income countries. It is the world's most widely cultivated and traded cereal, supplies one-fifth of man's calorie intake, and provides a third of mankind with food. Rice is the principal food of the world's poor. Although a staple food in only 16 countries, those lands contain half the world's population. Providing one-fifth of man's calorie intake, rice yields twice as much per acre as wheat. This higher yield supports a much higher population.

The third-ranking staple food is corn. The leading grain of 14 countries, including most of Latin America and many African countries, corn supplies about 5 percent of man's energy needs.

An assortment of plants provide the rest of the energy. Millet, sorgum, and rye provide about 7 percent of man's energy intake. Starchy roots, particularly potatoes, sweet potatoes, and yams provide another 5 percent; cassava (popularly known in developed nations as tapioca) supplies about 2 percent, mainly in several South American and African countries. Fruits, sugar, and vegetables supply another 9 percent; fats and oils provide a little less than 9 percent; and livestock products and fish supply the remaining 12 percent (Brown and Finsterbusch, 1972).

The poorer nations of the world, particularly those in southeastern Asia, live close to primary production. Cereal crops are consumed directly; primary production is channeled directly to humans. In richer countries more and more of the grains are fed to livestock, and man moves up one step in the trophic level. Corn that provides the staple diet in Latin America is converted to pork and beef in the United States and Canada. Thus the direction of the food chain differs in various parts of the world. In monsoon Asia the major link is rice to man; in Latin America it is corn to man, and in Europe and North America it is wheat to man. Peoples of poorer regions subsist on monotonous diets of cereal grains. In Africa the poorest groups live on even cheaper starchy food, such as potatoes, sweet potatoes, and yams. In the richer nations the diet becomes more diversified and there is greater dependence on animal protein. In North America, 31 percent of man's daily food energy is in the form of meat, milk, eggs, and fish. In western Europe the proportion is about 22 percent, in Oceania it reaches 36 percent. This is in sharp contrast to eastern Europe and Russia, where animal products contribute only 15 percent, and Asia, where the proportion drops to 5 percent.

In spite of the seeming abundance of food in some parts of the world, increased food production, and technological developments in agriculture, much of the world goes hungry. Although there has not been a major famine (except for the man-caused one in Biafra) since World War II, the bulk of the world's population, including a surprising number in affluent nations, are inadequately nourished, suffering from the ills caused by malnutrition.

Man has certain nutritive requirements. He needs carbohydrates—sugars and starches —as an energy source for daily activity, for

FIGURE 4-10 [OPPOSITE]
Energy flow through a modern agricultural production system. Note the heavy input of fossil fuels into the system to do work once done by man. A considerable input of energy is used not only for management but also for services to provide materials necessary for management such as (1) seeding and planting, (2) fertilization, (3) herbicides and cultivation, (4) soil preparation, (5) insecticides, and (6) development of new varieties of crops. (Adapted from H. T. Odum, 1971, Environment, Power, and Society, Wiley, New York.)

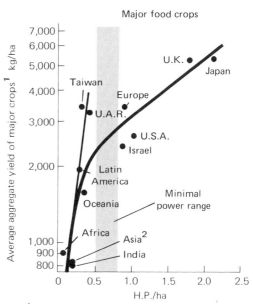

[1]Cereals, pulses, oilseeds, sugar crops (raw sugar), potatoes, cassava, onions, and tomatoes
[2]Excluding mainland China

FIGURE 4-9
The input of fossil fuels to increase production of agricultural crops is emphasized in this graph. Note the sharp rise in yield with the input of minimum horsepower and the steady rise as increasing amounts of fossil fuels are paid into production. (From U.S. Panel of World Food Problems, 1967.)

TABLE 4-1
The potential productivity of earth and the population it could support

North latitude (degrees)	Land surface in ha ($\times 10^8$)	Number months above 10°C	Carbohydrate/ha yr in kg ($\times 10^3$)	No allowance for urban and recreational needs		750 m^2/man for urban and recreational needs		Agricultural land %
				m^2/man	No. men ($\times 10^9$)	m^2/man	No. men ($\times 10^9$)	
70	8	1	12	806	10	1556	5	52
60	14	2	21	469	30	1219	11	38
50	16	6	59	169	95	919	17	18
40	15	9	91	110	136	860	18	13
30	17	11	113	89	151	839	20	11
20	13	12	124	81	105	831	16	10
10	10	12	124	81	77	831	11	10
0	14	12	116	86	121	836	17	10
−10	7	12	117	85	87	835	9	10
−20	9	12	123	81	112	831	11	10
−30	7	12	121	83	88	833	9	10
−40	1	8	89	113	9	863	1	14
−50	1	1	12	833	1	1583	1	53
Total	131				1022		146	

Source: From deWit, 1968, in A. San Puetro et al. (eds.), *Harvesting the Sun,* Academic, New York.

metabolism and body maintenance and growth, and for storage for future energy demands. Carbohydrate intake greater than immediately needed is stored as liver glycogen and fat. Fats, needed in lesser quantity, serve as a condensed reserve of energy. They are an important structural component of cell membranes and are essential for various reactions in intermediary metabolic reactions. Proteins, the structural elements of life and the principal constituents of organs and soft tissues, are needed in continuous and liberal supply for body growth and maintenance throughout life. They are especially important to the growing young. The thousands of different proteins all consist of different arrangements of amino acids, of which eight (nine in the young) are essential. The quality of proteins varies considerably, determined by the ratio of the 8 essential amino acids in the food. In addition there are vitamins, essential nutrients needed in small amounts. Without them growth, metabolism, general health and vigor, and reproduction may be impaired. Humans and other vertebrates also require some 17 essential minerals.

Man's metabolic energy demands can be met by many different combinations of proteins, carbohydrates, and fats (proteins and fats are also sources of carbohydrates). Of all the nutrients required by man, proteins are the scarcest and the most expensive. They come from both plants and animals, but the most useful are of animal origin. The conversion cost is high. Four pounds of plant protein must be fed into poultry for every pound of egg protein, the most beneficial of all for man. One pound of beef protein requires the consumption of 14 to 16 lb of protein by the beef animal. Protein in plants, which is the source of most animal protein, varies widely. To obtain the daily requirements of protein from rice, which provides 60 percent of the energy to half the world's population, an adult would need to eat 2 lb daily. To obtain sufficient protein from corn one would have to consume 1.75 lb daily; for wheat the figure is 1.25 lb. And even then the protein would not be of the right quality because corn and wheat do not contain enough of the essential amino acids.

Other commonly consumed plants provide even less protein. Two that form an important part of the diet in much of the world, potatoes and cassava, are extremely low in protein. The potato contains only 2 percent, and to obtain the minimum daily requirement from that source alone one would have to eat 8 lb daily. Cassava made into tapioca contains even less. Twenty-five pounds would have to be eaten daily to meet protein requirements.

Some plants are rich in high-quality protein, especially the pulses—beans, peas, and related plants. One cup of soybean concentrate, 170 g, supplies the daily requirements of protein, vitamins, and minerals. Dried beans and peas contain 22 to 26 percent protein, yet in underdeveloped nations the full possibilities of these protein-rich plants are not realized. Production is concentrated in high-yield crops such as rice, potatoes, cassava, bananas, and other carbohydrate-rich plants. The accepted standard for protein intake is 60 g/person/day. Many people in Latin America, Africa, and East Asia receive less than 7 g/day.

Man, food, and the future

The number of organisms supported at any position in a food chain theoretically depends upon the limits of the energy supply available. For man the supportable levels of populations depend upon the limits of photosynthesis.

deWit (1967) has estimated this maximum sustainable population (Table 4-1). The calculations are based on a number of selected variables: (1) a leaf scattering coefficient of .3; (2) a photosynthetic rate of 20 kg c/ha/hour; (3) a leaf area index of 5; (4) a leaf display similar to that of young grass or grain; (5) photosynthesis taking place on a clear day; (6) latitudinal variations. He further assumes that each human requires 10^6 kcal/year. Thus the number of people who can exist on a hectare equals the total photosynthesis of the hectare expressed in tons. Based on these assumptions, the productivity of the earth is calculated for each 10° of latitude.

But man also needs space. To meet this need, deWit adds 750 m² per person to support urban needs, recreation, and the like. The calculation cannot take into account the areas of land not suitable for agriculture, urban use, and so on. Of the earth's total land area only about 15 percent is available for agricultural purposes.

The calculations are based on the idea that man will occupy the first consumer level in the food chain. If he desires animal protein at a rate of 200 g/day, then about 5,000

FIGURE 4-11
The "green revolution" centers about the development of new grains that are more productive, yet adapted to local ecological conditions. New wheats that will not "lodge" or fall over from weak stalks must be developed to take increased nitrogen fertilization. Punjab Agricultural University scientists compare their new three-gene wheat with a stand of lesser straw strength that was lost 2 weeks before harvest. (Photo courtesy Rockefeller Foundation, New York.)

kcal of forage will be needed per 500 kcal of meat, requiring twice the amount of land for agricultural purposes. If the need for space is doubled then the population would have to be halved from 146 billion to 73 billion. Of course all of this is highly theoretical, pointing out how many people under ideal photosynthetic conditions the earth might possibly support.

On a more realistic level, the supportable population, in terms of energy, depends upon how well man takes care of the earth. To date his care of the planet has not aimed toward increasing productivity. The basic support system of agriculture is the soil. Ever since he began to cultivate the earth, man has caused a deterioration of the soils in most places, even destroying such ancient and once prosperous civilizations as the Sumerian and the Mayan. In spite of technological ability, erosion continues. In the United States thousands of acres of farmland are lost each year to highways and urban expansion. Every mile of freeway demands 24 acres of ground and every interchange 80 acres. Three-fourths of the land invaded by urban expansion has been high-quality cropland (Fig. 4-12). In fact in the last decade almost 15 million acres in the United States have been shifted from farming to nonfarming uses. Agricultural activities are increasing the amount of deserts and wasteland. The Sahara Desert of Africa and the Thar Desert

of western India are growing, and new wastelands have been created in the dry lands of the USSR. As land is withdrawn from agriculture, few new lands are available for cultivation. Today virtually all of the land that can be cultivated economically is under cultivation. The remainder is marginal, requiring expensive and superior technology.

In the tropics the soil itself imposes a severe handicap to the development of agriculture. Tropic soils are lateritic (see Chapter 7), that is, they are acidic, high in iron, nickel, and aluminum, and low in silica and organic matter. Although the soil supports dense forests, the high temperature and humidity cause a rapid breakdown and recycling of organic matter (see Chapter 5). The forest protects the soil from erosion and slows soil degeneration. As farms for conventional agriculture are cleared, the soil is exposed to air, and in a few years it bakes to a pavement-hard laterite, which is difficult to break. In fact the lateritic soils are used to make bricks in tropical countries. To develop agriculture in tropical lands, much more sophisticated methods than those of temperate climates are needed.

The vast desert areas of the world are viewed as potential cropland. The success in Israel is an example of how deserts can be made to produce. But the great limitation is water. Desert cropland depends upon irrigation, much of the water for which is literally mined from exhaustible underground storage. To circumvent this there has been growing interest in desalting seawater for irrigation purposes. But three overwhelming problems limit desalination. One is the need for a source of energy. Involved are a large capital investment and high operating costs that to be met must be shared by concurrent urban or industrial development. The second problem involves the cost and techniques of desalination itself. The plants in operation today, as in southern California, are uneconomical. Along with poor economy goes the problem of disposal of hot, bitter brine, which could cause adverse ecological consequences. A third problem is the transport of water from the desalting plants along the sea to croplands inland and at higher elevations. Once pumped to the general area the water has to be stored and then distributed to farms and fields. Currently the cost to produce irrigation water from seawater is greater than the value of water to agriculture. In fact irrigation water in general costs more than its value to agriculture. It becomes economical only because the cost of the water is subsidized by urban development that the water development projects stimulate.

Nations conceive grandiose schemes to bring new lands under cultivation by irrigation. Often these create more problems than they solve. The high Aswan Dam in Egypt holds back the floodwaters of the Nile for irrigation use during the dry seasons, but it also holds back the silt that refertilized the river bottom lands that supported a rich and viable agriculture. The loss of nutrients that flowed into the estuary of the Nile has virtually eliminated sardine fishing in the eastern end of the Mediterranean. Irrigation ditches carrying water to desert soils support increasing populations of the snail that is host to the parasitic blood fluke that causes schistosomiasis when it burrows into the flesh of people working in water-soaked soil. There are plans to turn the Mekong River into a series of pools like the Tennessee River. The fertility of the Mekong River Valley comes from the annual flooding and deposition of silt, which would be trapped behind the dam. Large-scale forest clearings that would have to accompany such river development would increase the laterization process in that region.

Irrigation causes other problems. Of the water that is diverted from rivers to crop fields, part is used by the plant, part is lost through evaporation from the plant and soil, and part percolates downward to form groundwater. Unless there is some sort of drainage scheme that includes the cycling of groundwater for irrigation, the groundwater accumulates. In time it raises the water table to the root zone, causing waterlogged soil, which crop plants cannot tolerate. In addition the rising water also brings salts from the deeper layers of the soil and deposits them on the surface. Through time the surface soil becomes so salty that it cannot grow plants. Waterlogged and salty soils are then abandoned. This problem, recognized by ancient agriculturists in the Near East and South who were forced to abandon vast areas of land once devoted to agriculture, still remains a critical one in such countries as Egypt, India, and Pakistan.

Much of the recent increase in world food production has come about not by increasing land under cultivation but by increasing yield from land already cultivated. The yield increases are the result of input of fossil fuel, labor, and fertilizer into croplands and the use of improved crop varieties. The replacement of open pollinated varieties of corn with hybrids and the use of herbicides have increased the yield of corn per acre in the United States nearly threefold since 1940, and the yield of rice per acre in Japan has more than doubled in that time (Fig. 4-13). In these instances the maximum production per acre has been achieved. As nonrecurring sources of productivity are exhausted and as inputs of fertilizer and cultural techniques no longer yield significant returns, increases in the rate of yield per acre begin to slow. This results in a typical S-shaped growth curve (Fig. 4-14). Although yield increases are continuing at a rapid rate in some countries, they are slowing for some crops, such as wheat in the United States and rice in Japan. Eventually all crop production will reach this upper-yield plateau.

The problem then becomes one of attempting to achieve new plateaus. This might come about through the development of new food plants. Of the 700,000 plants in the world, nearly 80,000 are edible, but only 3,000 are considered crop plants, and of these only 300 are in abundant use. Only 12 species or genera supply man with 90 percent of his food. New sources of food may be found among unexploited plants.

FEED CONVERSION EFFICIENCY

Obtaining adequate protein is one of man's most serious food problems. Man has two sources, animals and plants; the highest-quality protein comes from animal sources. The consumption of animal protein is directly related to affluence (Table 4-2). As more nations and people become more affluent, the demand for animal protein increases. But the production of animal protein is expensive in terms of energy and protein input. For example, 7.3 kg of digestible feed protein and 7 kcal of energy are needed to produce 1 kg of protein for man. Among animals there are various conversion efficiencies (the amount of feed needed to produce a unit of live weight). Beef and milk cows, lambs, and pigs are about 35 percent efficient, which has changed little during the past 25 years (Byerly, 1967). The greatest improvement in efficiency has come about in poultry, which under ideal conditions will convert 1 lb of feed to 1 lb live weight of broiler.

The efficient production of animal protein involves the input of high-quality protein from fish meal and oilseeds such as soybeans and cottonseed, which are also adequate sources of protein for man. To increase the availability of both animal and plant proteins for human consumption, nonprotein nitrogen sources such as urea may be fed to ruminant animals without decreasing the production of milk and meat. The plant and animal proteins saved are then available for human consumption. As yields of cereal grains increase, the protein content declines. Efforts are needed to improve the protein yield of cereal grains. The proteins of cereals can be fortified by adding amino acids and by the development of new protein foods from oil seeds (Altschul, 1967). If populations continue to expand at present rates, the world population may reach a point where most of mankind cannot afford the luxury of animal protein. Plant proteins will have to be channeled directly to man.

FOOD FROM THE SEA

As the world's population increases, man tends to look to the sea for food. But the sea has only a limited potential. This may seem rather surprising when one considers that the sea contains 90 percent of the world's vegetation and fixes from 18 to 20 billion tons of carbon a year. The productivity of the sea varies greatly. The open sea fixes approximately 50 g C/m², shallow waters and estuaries twice as much, and areas of upwelling perhaps six times as much. Occupying only 1 percent of the oceans, the regions of upwelling supply half the world's ocean fish production; the coastal waters and estuaries furnish the other half. In terms of fishing, the open sea is a watery desert.

Fish production has increased steadily over the years, reflecting intensified fishing effort and improvement in fishing gear and methods. It is the one resource in which the increase in production is higher than the rate of population growth. Today ocean fisheries supply 18 percent of the world's protein. Although fish protein is the major supplement to plant protein in much of the world, little of the increased protein production is used to fight protein deficiency. The poorer countries lack the purchasing power and the distribution facilities to utilize it. As a result most fish protein is used to supplement livestock and poultry feeds in the richer countries of the world.

Based on the photosynthetic capacity of the sea, 240 million tons of fish is the maximum production, of which 100 million tons could be harvested on a sustained-yield basis (Ryther, 1969). To make the most of fish production man will have to move to a lower trophic level. As it is now, the harvest comes from the upper trophic levels. Flounder and herring are on level three; tuna and halibut on the fourth. Every 1.5 oz of tuna fish requires 500 lb of phytoplankton. To obtain a larger harvest, man will have to utilize the less palatable, less attractive species that now are thrown away. Yet a product of those unpalatable fish, fish flour or fish protein concentrate, has not been widely accepted, for no way has been found to incorporate it successfully into other foods (Holden, 1971).

The productivity of the sea is further limited by man himself. A major problem is overexploitation. The rapid increase in fish yields has been at the expense of the stock of the fish. Of the 30 major fish stocks, 14 have been overfished, and some have been eliminated as commercial species. Their populations are too low to be harvested economically. Among these are the Pacific sardine, the Atlanto-Scandinavian herring, the Berents Sea cod, the menhaden, and the haddock. Because the ocean is extraterritorial, no enforceable in-

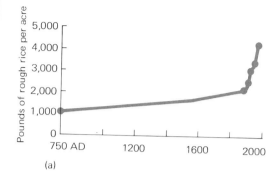

(a)

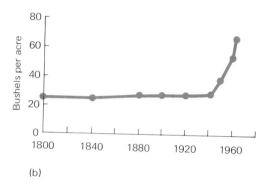

(b)

FIGURE 4-13
Increases in yield of rice in Japan (a) *and corn in the United States* (b) *show a sharp increase in recent times. Expansion in rice production resulted largely from the use of fertilizers and the hand planting of rice in rows. Expansion of corn production resulted from the replacement of open-pollinated varieties by hybrid corn and the use of herbicides, reducing competition from weeds.* (*From U.S. Department of Agriculture.*)

TABLE 4-2
Affluence and world consumption and distribution of protein

World area	Availability of animal protein (g/person/day)	Distribution of world supply (%) per capita	
		Animal	Plant
North America	76	27	47
Europe	33	13	11
South America	26	12	11
Africa	9	4	7
Asia	2	3	5
Oceania	130	41	19

Source: G. C. Anderson, 1973.

ternational regulations exist to control the take. Each nation takes as much as possible, with no regard for the future. The estimated 100-million-ton production was based on a sustained-yield management.

Pollution endangers much of the productive coastal waters. Although the region of upwelling may escape for the time any serious pollution, the coastal waters are becoming heavily polluted from sewage and industrial wastes, and are being destroyed by dredging and filling. These coastal waters are the nursery ground for many of our most valuable species—the flounder, the striped bass, and the bluefish, as well as the oyster, clam, shrimp, and other shellfish. Highly vulnerable to pollution, the coastal waters are rapidly declining as the nursery of commercial species. Unless abated, pollution can eventually destroy half the world's ocean fisheries.

The sea provides approximately 80 percent of the world's fish catch. Providing another 5 percent is fish farming, or mariculture (Fig. 4-15). Considered by some to be at least a partial solution to the protein scarcity of many countries (Webber, 1968), it holds a promise for the future (Bardach, 1968). Currently it is practiced largely in the Asian countries, where the Indian carp, gray mullet, milkfish, tilapia, shrimp, and prawns are reared, and in the United States, where catfish and oyster farming are increasing significantly. Through the efforts of the Food and Agricultural Organization, interest in fish farming is spreading to many countries. Out of the total production of over 3 million tons of food fish, 2.6 million come from the nine countries of Asia and the Far East. Among the Western nations the United States is the leading producer. In the Philippines, Indonesia, and India 1 million ha are devoted to fish production; another 8.2 million ha can be reclaimed for fish culture. The most extensive developments involve the capture of estuarine species close inshore. The fry are transferred to ponds prepared by diking lowland coastal flats. A major problem is close control of the salinity of the pond. Production runs from 500 to 2,000 lb/acre. In the United States some 30,000 acres of land are devoted to catfish farming, producing over 40 million tons of channel and blue catfish in 1968. Trout culture adds another 212 million lb. In experimental stages are methods of farming salmon and off-bottom methods of oyster culture. If pollution of estuarine waters can be controlled, fish farming should expand rapidly. The Food and Agricultural Organization estimates that by 1985 aquaculture will produce around 20 million tons of high-quality fish and shellfish for human consumption, a third of the current total world fishery production.

Suggested is the scheme of increasing food production from the sea by recycling human wastes (Ryther, et al., 1972). The algae would be utilized by oyster cultured on rafts. The feces and pseudofeces of oysters would support an invertebrate bottom fauna, which in turn could feed a fish population. Seaweeds would utilize the nitrogenous compounds excreted by animals. Algae are also considered a potential source of food. But algae cultured on a large scale requires too high an input of fossil fuel, and is too artificial a method for economical mass production. Such systems are in an experimental stage.

Summary

Through his history man has changed his relationships in the food chain. For over 99 percent of the 2 million years he has lived on earth, he existed as a hunter–gatherer. As such he was a part of the natural ecosystem in which he existed. His energy input consisted largely of the muscular energy needed to hunt game and to gather seeds and fruits. About 10,000 years ago man began to adapt the food chain and energy sources to his own needs. In certain regions of the world, notably Mexico and western Asia, hunter–gatherers discovered ways of raising plants whose seeds and roots they once collected wild. From this developed the growing of such grain crops as wheat and corn and such root crops as yams and cassava.

Early agriculture was polycultural, and man manipulated rather than transformed natural ecosystems. To achieve greater crop production man had to increase his input of energy. In time he domesticated some animals, used oxen and horses as sources of power, and developed irrigation techniques to supplement natural rainfall.

Following the Industrial Revolution, man employed new techniques to increase food production. Polyculture was replaced by monoculture. Such equipment as plows, reapers, cultivators, and mowing machines

decreased human labor on farms, but demanded more input of energy, first from
draft animals and later from fossil fuels.
Techniques of fertilization, development of
hybrid varieties, and advances in animal nutrition made agriculture highly artificial and
placed it largely outside of natural food
chains and biogeochemical cycles. As a result,
nutrients are transported to distant points,
and most are lost to ecosystems through waste
disposal and sewage effluents. Maintenance
of the agricultural ecosystems depends upon
heavy input of fossil fuels for production and
distribution of food stuffs and for protection
from pests and fertilization.

Expansion of high-yield agriculture is
largely an accomplishment of the Western
world. Technological agriculture—including
mechanization, fertilization, pest control,
and the development of high-yielding hybrid
varieties of wheat, corn, and rice—is being
introduced into underdeveloped countries to
increase their food energy base. Although
these developments are improving the food
supply, they are creating a number of social,
economic, and political problems.

The development of agriculture moved
man away from an animal diet to a plant diet.
Today 80 percent of his global food energy
supply comes from plants, largely rice, wheat,
and corn. Man lives at the second trophic level.
Needed, however, is a source of animal protein, globally in short supply. Many look to
the sea for sources of protein, but the sea
will never become a major source of food, for
it has only a limited potential. In the future
man will have to depend upon terrestrial
agriculture for his food. To do so, he will
have to develop new food plants from unexploited wild plants.

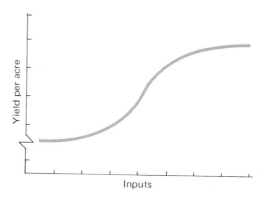

FIGURE 4-14
S-shaped yield curves result when nonrecurring
or "one-shot" sources of increased production
(such as the replacement of open-pollinated by
hybrid varieties) are reduced until eventually the
rate of increased productivity slows and
levels off. Further inputs of resources do not result
in increased yields. (From U.S. Department of
Agriculture.)

FIGURE 4-15
A growing agricultural industry throughout the
world, but particularly in Asia and the United
States, is aquaculture, or fish farming. In Asian
countries such species as milkfish are commonly
reared. In the United States catfish farming is
expanding rapidly. Aquaculture requires intensive
management. Here workers remove a female
catfish from a spawning pen prior to the removal
of catfish fry to rearing ponds. (Photo courtesy
U.S. Fish and Wildlife Service.)

he existence of the living world depends upon the flow of energy and the circulation of materials through the ecosystem. Both influence the abundance of organisms, the metabolic rate at which they live, and the complexity of the community. Energy and materials flow through the community together as organic matter; one cannot be separated from the other. But the flow of energy is one way; once used by the ecosystem, it is lost for any further transfer. Nutrients, on the other hand, recirculate. An atom of carbon or calcium may pass between the living and the nonliving many times, or it may even be exchanged between ecosystems. The continuous round trip of materials, paid for by the one-way trip of energy, keeps ecosystems functioning.

Important nutrients in the ecosystem

Living organisms require at least 30 to 40 elements for their growth and development. Most important of these are carbon, hydrogen, oxygen, phosphorus, potassium, nitrogen, sulfur, calcium, iron, magnesium, boron, zinc, chlorine, molybdenum, cobalt, iodine, and flourine.

Molecular *oxygen*, O_2, which makes up 21 percent of the earth's atmosphere, is a by-product of photosynthesis. Three major pools of oxygen are carbon dioxide, water, and molecular oxygen, all of which interchange atoms with one another. Other sources include nitrate and sulfate ions, which release oxygen upon their decomposition and supply oxygen for a number of living organisms. Molecular oxygen is a building block of protoplasm and is necessary for biological oxidation, in which it serves as a hydrogen acceptor, producing water. Plants and animals use large amounts of oxygen in respiration. In terrestrial habitats the supply of oxygen is rarely inadequate for life, except at high altitudes (Chapter 17), deep in the soil, and in soils saturated with water. In aquatic communities, however, oxygen may be limited. Here the supply comes from photosynthesis and from the diffusion of atmospheric oxygen into the water, the rate of which is proportional to the surface area of water exposed to the air. The total quantity of oxygen that water can hold at saturation varies with temperature, salinity, and depth. It is depleted by the respiration of aquatic plants and animals. Oxygen content in aquatic environments fluctuates daily from a low at night, when respiration is greatest, to a high at midday, when photosynthesis is the highest.

Carbon is the basic constituent of all organic compounds. In the ecosystem it occurs as carbon dioxide, carbonates, fossil fuel, and as part of living tissue. Unlike oxygen, carbon dioxide, which makes up only 0.03 percent of the atmosphere, seems to be limited in terrestrial environments. Plant physiologists have demonstrated that a 10 percent increase in the atmosphere results in a 5 to 8 percent increase in photosynthesis (see Bonner, 1962). The amount of carbon dioxide in natural waters is highly variable, for it occurs in free and combined states. The pH of aquatic media and of the soil has a pronounced influence on the proportions of the two types. Carbon dioxide combines with water to form a weak carbonic acid, H_2CO_3, which dissociates:

$$CO_2 + H_2O \rightleftharpoons H_2CO_3 \rightleftharpoons H^+ + \\ HCO_3^- \rightleftharpoons H^+ + CO_3^{2+}$$

Carbon dioxide in solution and carbonic acid make up free carbon dioxide; the bicarbonate (HCO_3^-) and the carbonate (CO_3^{2+}) ions are the combined forms. The presence of bicarbonate and carbonate ions in soil or water helps to buffer or maintain a certain pH of that medium. This is significant in ecological systems because an increase in soil pH lowers the availability of most nutrients to plants. The amount of carbon dioxide fluctuates during the day but in a manner opposite to daily oxygen fluctuations. The carbon dioxide concentration is lowest around midday and highest at night, when only respiration is taking place.

Nitrogen makes up 78 percent of the atmosphere as molecular nitrogen, N_2, but most plants can utilize it only in a fixed form, such as in nitrites and nitrates (the exceptions are nitrogen-fixing bacteria and blue-green algae. Most of the nitrogen in the soil is found in organic matter. Nitrates leached from the soil and transported by drainage water are an important source of nitrogen for aquatic communities. During the summer the nitrogen supply in these may be utilized completely by phytoplankton, and nitrates may disappear from the surface water. As a result phytoplankton growth, or "bloom," is reduced greatly in late summer. Nitrates build up again in winter. On the other hand, nitrogenous wastes draining from agricultural fields, sewage disposal plants, and other sources often overload aquatic ecosystems with nitrogen. This results in massive plankton growth and other undesirable changes in community structure (see Chapter 6).

All life processes depend upon nitrogen. Even chlorophyll is a nitrogenous compound. Nitrogen is a building block of protein and a part of enzymes. It is needed in an abundant supply for reproduction, growth, and respiration.

Oxygen, carbon, and nitrogen are considered the energy elements and are needed in relatively large quantities. Other elements and compounds, called macro- and micronutrients, are needed in smaller amounts. The macronutrients include calcium, magnesium, phosphorus, potassium, sulfur, sodium, and chlorine. The micronutrients are the trace elements. These include copper, zinc, boron, manganese, molybdenum, cobalt, vanadium, and iron. Some are essential to all organisms; others appear to be essential only to animals. If micronutrients are lacking, plants and animals fail as completely as if they lacked nitrogen, calcium, or any other major element.

Two elements needed in appreciable quantities are *calcium* and *phosphorus*, which are closely associated in the metabolism of animals. Together the two make up 70 percent of the total ash present in the animal body and nearly 1 percent of the wet body mass. In animals calcium is necessary for proper acid–base relationships, for the clotting of blood, for contraction and relaxation of the heart muscles, and for the control of fluid passage through the cells. It gives rigidity to the skeleton of vertebrates and is the principal component in the exoskeleton of insects and the shells of mollusks and arthropods. A number of mollusks and bivalves are restricted to hard water because there is insufficient calcium in soft water to harden the shells. In plants calcium is especially important in combining with pectin to form calcium pectate, a cementing material between cells. Plant roots need a supply of this element at the growing tips to develop normally.

Phosphorus not only is involved in photosynthesis, but also plays a major role in energy

Biogeochemical cycles in the ecosystem

transfer within the plant and animal. It is a major component of the nuclear material of the cell, where it is involved in cellular organization (DNA) and in the transfer of genetic material. Animals require an adequate supply of calcium and phosphorus in the proper ratio, preferably 2 : 1 in the presence of vitamin D. An inadequate supply of either may limit the nutritive value of other elements. The lack of either in animals is associated with rickets, a condition involving the improper calcification of bones. When the supply of phosphorus in plants is low, growth is arrested, maturity delayed, and roots stunted. An excessive intake of calcium can inhibit the assimilation of other mineral elements.

Accompanying calcium and phosphorus is *magnesium*. In animals most of it is in the bones. It is an integral part of chlorophyll, without which no ecosystem could function; and the element is active in the enzyme systems of plants and animals, especially those enzymes transferring phosphate from ATP to ADP. Low intake of magnesium by grazing ruminants causes a serious disease, grass tetany, which may result in death.

Potassium is utilized in large quantities by plants, and if it is readily available and growing conditions are favorable, the uptake (in crop plants at least) may be above that of their average total requirements (Reitemeier, 1957). The formation of sugar and starches in plants, the synthesis of proteins, normal cell division and growth, and carbohydrate metabolism in animals all depend upon an adequate supply of potassium. Unlike calcium, phosphorus, and magnesium, potassium occurs in body fluids and soft tissues. In animals it is readily absorbed metabolically, and excess over needs is immediately excreted. Because plants usually contain an adequate supply, deficiencies rarely occur among grazing animals.

Iron and *manganese* are involved in the production of chlorophyll. Iron is part of complex protein compounds that serve as activators and carriers of oxygen and as transporters of electrons. Over half the iron present in the animal body is in the hemoglobin of the blood. Lack of iron results in anemia. A low level of manganese in animals can result in malformation of bones, in delayed sexual maturity, and in impared reproduction.

Boron and *cobalt* are two micronutrients whose deficiency effects, the one in plants and

the other in animals, are notable. Some 15 functions have been ascribed to boron, including pollen germination, cell division, carbohydrate metabolism, water metabolism, maintenance of conductive tissue, and translocation of sugar in plants. Plants with boron deficiency are stunted both in leaves and roots. This condition is most common in the croplands of eastern and central North America, where vegetation is continuously being removed. Without cobalt, an element not required by plants, animals become anemic and waste away. Cobalt deficiency is most pronounced in ruminants such as deer, cattle, and sheep, which require the element for the synthesis of vitamin B_{12} by bacteria in the rumen. The quantity of cobalt required is very small. An acre of grassland supporting seven sheep needs to supply only 0.01 oz/year.

In addition to producing deficiency symptoms when undersupplied, some micronutrients can be toxic when they are in excess. Among these are *copper* and *molybdenum*. Molybdenum acts as a catalyst in the conversion of gaseous nitrogen into a usable form by free-living, nitrogen-fixing bacteria and blue-green algae. But high concentrations of molybdenum cause "teart disease" in ruminants such as cattle and deer. This disease is characterized by diarrhea, debilitation, and permanent fading of the hair color. In plants copper is concentrated in the chloroplasts, where it affects the photosynthetic rate, is involved in oxidation–reduction reactions, and acts as an enzyme activator. When present in excess, copper interferes with phosphorus uptake in plants, depresses iron concentration in the leaves, and reduces growth. Copper can interact with molybdenum effectively to "tie-up" the copper and produce a copper deficiency. In animals this may cause poor utilization of iron. As a result, iron concentrates in the liver, anemia develops, and calcification of bones decreases.

Sulfur, like nitrogen, is a basic constituent of protein, and many plants may utilize as much of this element as they do phosphorus. The sulfur supplied by rainwater and by organic matter in soil is sufficient to meet plant needs. Considerable quantities are released to the atmosphere in industrial areas and carried to the soil by rainwater. Excessive sulfur can be toxic to plants. Exposure for only an hour to air containing 1 ppm (part per million) of sulfur dioxide is sufficient to kill vegetation. For this reason countrysides near smelters are often denuded of vegetation. Plants, especially in arid and semiarid country, also are affected by high concentrations of soluble sulfates in soils, which limit the uptake of calcium.

Sodium and *chlorine* are on the borderline between micro- and macronutrients. They are required in minute quantities by plants but in much greater quantities by animals. (Sodium can substitute for potassium to satisfy the plant's need for that element.) Both sodium and chlorine are indispensable to vertebrate animals. Obtained from common salt, these elements are important for the maintenance of acid–base balance of the body, the total osmotic pressure of extracellular fluids, and for the formation and flow of gastric and intestinal secretions.

Zinc usually is abundant in the soil, but it may be unavailable to plants. Zinc may exist as insoluble compounds in the soil when the pH is around 7. Zinc is needed in the formation of auxins in plant growth substances, is a component of several plant enzyme systems, and is associated with water relations in plants. In animals, zinc functions in several enzyme systems, especially the respiratory enzyme carbonic anhydrase in red blood cells, where it plays an essential role in the elimination of carbon dioxide. Insufficient zinc in the diet of mammals can cause a dermatitis known as parakeratosis.

Iodine and *selenium* are two other micronutrients of note. Iodine, the natural deficiency of which occurs in the soils of the northeast region of North America, the Andes Mountains of South America, and the mountainous parts of Europe, Asia, and Africa, is involved in thyroid metabolism. In animals lack of iodine results in goiter, hairlessness, and poor reproduction. Selenium, closely related to vitamin E in its function, is required to prevent white muscle disease in the newborn ruminant. The amount required is on the order of 0.1 mg/kg of ration. The border line between requirement level and toxicity level is very narrow. Too much selenium results in the loss of hair, sloughing of hooves, liver injury, and death by starvation.

All of the essential nutrients and many others besides, including a number of manmade materials such as chlorinated hydrocarbons, flow from the nonliving to the living and back to the nonliving parts of the ecosystem in a more or less circular path known as the *biogeochemical cycle* (*bio* for living, *geo* for water, rocks, and soil, and *chemical* for the processes involved). Some of the material is returned to the immediate environment almost as rapidly as it is removed; some is stored in short-term nutrient pools such as the bodies of plants and animals or the soil and sediment of lakes and ponds; and some may be tied up chemically or buried deep in the earth in long-term nutrient storage pools before being released and made available to the biota. Between the easily accessible and the relatively unavailable there exists a slow but steady interchange.

The important roles in all nutrient cycles are played by green plants that organize the nutrients into biologically useful compounds, by the organisms of decay that return them to their simple elemental state, and by air and water that transport the nutrients between the abiotic and the living components of the ecosystem. Without these there would be no cyclic flow of nutrients.

The role of decomposers

In the November woods the leaves of summer lie brown and withered (Fig. 5-1). They had formed the green canopy of the forest, and in the course of a summer they pumped water and minerals from the soil, traded oxygen for carbon dioxide in the atmosphere, and trapped the energy of the sun by photosynthesis converting it into carbohydrates and other energy-storing compounds. What energy the trees did not need for their own maintenance and production of fruit and seed, the leaves sent back to the trunk, branches, and roots, to be stored as woody growth.

As the days shortened, the process slowed and finally stopped. A delicate layer of cells formed between the leaf and the twig, causing the stem of the leaf to fall off. As many as 10 million leaves/acre may flutter to the ground in fall in our hardwood forests, an equivalent of 2,000 to 3,000 lb.

It is the activity of the organisms of the forest floor that ensures that next spring will be green. Unless the tons of leafy material and other debris are disposed of, the forest would smother in its own litter. And unless the mineral matter it contains is released to the soil to be recycled to the trees again,

forest plants would face a shortage of nutrients necessary to life. Thus in the leaves underfoot is an almost unbelievable activity as billions of organisms reduce the debris to chemical elements and mix the organic matter with the soil.

In effect what the organisms of decay accomplish is the reversal of the process of photosynthesis—the reduction of organic matter to the inorganic compounds—carbon dioxide, water, and oxygen—from which it was synthesized and the utilization of the energy this matter contains. Although the process of decomposition sounds deceptively simple, it is a complex of many processes, including fragmentation, mixing, change in physical structure, ingestion, egestion, concentration, and action of enzymes, all accomplished by a wide diversity of organisms.

The organisms involved in this process are the decomposers. Too often the term is misleading, for it implies that all the organisms involved function in the same manner, all somehow mysteriously converting organic matter back to inorganic compounds. Actually the world of the decomposers is one of complex food webs, involving herbivory, carnivory, parasitism, and to a very limited extent, producers. In one way or another all contribute to the breakdown of organic matter.

There are both micro- and macroscopic organisms involved in decomposition. There may be over 100 billion of them in a square foot of the first 3 in. of a hardwood forest soil. Approximately 40 percent of these organisms are bacteria, about 50 percent are microscopic fungi; 5 to 9 percent may be protozoans; and 0.05 may be true fungi (mushrooms). Animals that can be seen by the naked eye, the macrofauna, account for only 0.000004 percent of the total population. Most are *saprophages*, organisms that exist on dead matter. The macrofauna can be further subdivided into such groups as mycetophages (those that feed on fungi), saproxylophagi (the wood-feeders), scatophagi (those that feed on animal excreta), and so on.

Bacteria, which range in size from forms invisible to the naked eye to relatively large, clubbed, stalked, and branching cells, are the organisms most commonly associated with decomposition. They can be divided into two broad groups, autotrophic and heterotrophic. The autotrophic forms are commonly considered producers, but they differ considerably from true autotrophs, the green plants. These chemosynthetic bacteria are something of heterotrophs, also. They can use the oxidation of simple inorganic compounds such as ammonia, nitrates, and sulfides as the source of energy for conversion of carbon dioxide into living material. In utilizing this source of energy they are aiding in decomposition.

The heterotrophic bacteria are true decay organisms, and they can be classified in any number of ways. They may be *aerobic*, requiring oxygen as the electron acceptor; or they may be *anaerobic*, in which case they can carry on their metabolic functions without oxygen by using some inorganic compound as the oxidant. Anaerobic bacteria are commonly found in aquatic muds and sediments and in the rumen of ungulate herbivores. Many are facultative anaerobes: when oxygen is present they use it, but in its absence they can utilize inorganic compounds as their energy source.

Bacteria can also be grouped according to their nutritional requirements, which reflect their specialization. Some can grow on very simple media such as glucose and nitrate salts; some require a number of amino acids; others need vitamin B_{12}; some need both amino acids and vitamins; and some require soil and root exudates that as yet are unidentified.

Considered as being somewhere between bacteria and fungi are the filamentous Actinomycetes, or ray fungi, very closely associated with the soil. They are slower decomposers and attack material not readily broken down by true bacteria.

Major decomposers of plant materials are the *fungi*, whose hyphae penetrate plant and animal matter. Like bacteria, microscopic fungi can be grouped according to their role in decomposition: saprophytic sugar fungi, lignin decomposers, coprophilious fungi, and so on.

The bacteria, actinomycetes, and fungi produce enzymes necessary to carry out specific chemical reactions. Enzymes are secreted into the plant and animal material, some of the products are absorbed as food, and the remainder is left for other organisms to utilize. Once one group has exploited the material to its capabilities, another group of bacteria and fungi move in to continue the processes. Thus a succession of microorganisms occurs in the detritus until the material is finally reduced to elemental nutrients.

FIGURE 5-1

In autumn in the deciduous forest, the leaves of summer fall to the ground, where they are subject to decomposition by fungi and soil fauna. The winter woods are far from lifeless.

Feeding on the bacteria and fungi is another group, the *microbial consumers*. These consist of such diverse groups as the nematodes, collembolans or springtails, larval forms of beetles (Coleoptera), flies (Diptera), and mites (Acarina). Nematodes feed heavily on bacteria, fungi, and algae in the soil. Their impact on the microbiotic world is not known. They may so reduce the microbial population that they delay ordinary decomposition. Or perhaps they promote microbial activity by preventing the microbial population from overproducing and by maintaining them at their level of maximum productivity or rate of division. Thus nematodes and other grazers may prevent aging and senescence of the bacterial populations. Collembolans and mites may have the same effect on fungi. If they do reduce the fungal population, they also stimulate its growth by dispersing fungal spores. By consuming fungi these microbial grazers may hasten recirculation of nutrients by releasing the nutrients locked up in the biomass of microbial populations.

The macrofauna make up the final group of decomposer organisms, the *reducer–decomposers*. They feed on plant and animal remains and on fecal material. This group includes in terrestrial ecosystems such organisms as earthworms, millipedes, isopods, and the larvae of beetles and flies. In aquatic and semiaquatic situations, mollusks, crabs, and scuds are prominent among the detritus-feeders.

Prominent in the terrestrial environment is the earthworm, one of the few reducers that can feed directly on fallen leaves before they are softened by initial microbial action. Prior to feeding on a leaf the earthworm moistens a part of the leaf's surface with enzymes so that it is partially digested before the leaf enters the worm's digestive tract. The earthworm functions not only as a consumer of detritus, but also mixes the material with mineral matter and buries it deeper in the soil.

Other reducers also contribute directly to decomposition by digesting some of the material. Isopods and saprophagous insects feed on plant remains, mollusks, and millipedes; other Dilopoda digest significant amounts of structural polysaccharides. Certain beetle larvae are prominent agents in the breakdown of dung, and dipterous insects play an important role in the breakdown of carrion. The amount of vegetable detritus introduced by excretions of herbivores is considerable, and dung beetles further reduce this material. Some beetles are specialists, feeding only on the dung of certain animals. The amount of organic material processed by dung beetles may amount to a fourth of the vegetable matter consumed by livestock (Ghilarov, 1970).

The major contribution of the reducers to decomposition is largely indirect. They break up detritus into small particles that are vulnerable to microbial attack; and material passing through the reducer's intestines is more easily decayed. In this way they accelerate decomposition.

The importance of the reducer's role has been demonstrated in several experiments. These employed nylon litter bags to exclude reducers from the litter sample or used naphthalene, which drives away arthropods and macrofauna but does not inhibit the activity of bacteria and fungi. Witkamp and Olsen (1963) placed white oak leaves, some confined in litter bags, some unconfined, in pine, oak, and maple stands in November. By the following June both the unconfined and confined leaves showed a similar loss in weight. But after June the unconfined leaves showed a sudden increase in breakdown. Before June both types of leaves lost weight by the breakdown of easily decomposable substances through the action of microflora and -fauna. Because the unconfined leaves were broken into fragments by the reducer–decomposers as well as birds, mice, wind, and rain, they were more available to microorganisms for further decay. Experiments by Kurcheva (1964), Edwards and Heath (1963), and Witkamp and Crossley (1966) show that suppression of activities of the reducer-decomposers results in a marked slowdown in the rate of microbial decomposition. In the absence of saproxylophagous invertebrates, the decomposition of wood is slowed by half. Not only do these organisms physically fragment the substrate but they also inoculate it with fungi and bacteria (Ghilarov, 1970). Without the activity of the reducers, nutrient elements could stagnate in the litter, bound energy in the ecosystem would increase, and both primary and secondary productivity would decrease.

Thus the decomposition of plant and animal material involves a complicated succession and interaction of different organisms, each of which contributes a part.

THE PROCESS OF DECOMPOSITION

In a popular sense decomposition is associated with death—the withered vegetation lying on the ground or an animal carcass lying in a field. With vegetation, the decomposition begins long before the potential detritus falls to the ground or to the bottom of a pond. In many ways decomposition begins when any herbivore consumes vegetation. Not only does the animal extract minerals and nutrients from the plant for its own nutrition, but it also deposits a substantial portion as partially decomposed fecal material. A portion of what is not consumed is left as fragments exposed to microbial action. In this manner, all animals are both consumers and decomposers.

Microbial decomposition of plant leaves begins while the leaves are still on the trees. Living plants produce varying quantities of exudates that support an abundance of surface microflora. These organisms feed on the exudates and on any cellular material that sloughs off. The same exudates account for the source of nutrients leached from the leaves during a rain. In tropical rain forests leaves are heavily colonized by bacteria, actinomycetes, and fungi (Ruinen, 1961).

While microbes are utilizing exudates of the leaves, other organisms are utilizing organic material from the roots. The soil region immediately surrounding the roots, known as the rhizosphere, and the root surface itself, known as the rhizoplane, support a whole host of microbial feeders on root litter and root exudates. The latter may consist of simple sugars, fatty acids, and amino acids. In fact some 10 sugars, 21 amino acids, 10 vitamins, 11 organic acids, 4 nucleotides, and 11 miscellaneous compounds have been identified in the rhizosphere. Obviously not all such exudates occur in the rhizosphere of all plants. The absence of certain exudates in the rhizosphere can influence the quantitative and qualitative differences in the microflora of rhizospheres of different plant species (F. E. Clark, 1969a, b).

When the plant body becomes senescent, it is invaded by microbial decomposers. If moisture is sufficient, the fungi colonize decaying culms of grass plants. A favorite point of invasion is the internode where the culm is attached to the stem. The species of fungal flora involved is influenced by the distance of the internodes and culms from the ground (the closer to the ground the more humid the habitat) and by the species of grass (Hudson and Webster, 1958; J. Webster, 1956–1957). Fungi infect pine needles 5 to 6 months before the needles fall (Burges, 1963). Destruction of the palisade layers of deciduous leaves by leaf miners opens up the affected leaves to microbial attack while they are still hanging on the tree. But the bulk of decomposition does not take place until the dead vegetation comes in contact with the soil.

Once on the ground, plant debris is subject to attack by microbes, bacteria, yeasts, actinomycetes, and fungi. The rate at which these organisms feed on the debris depends upon moisture and temperature. Higher temperature favors more rapid decomposition, and continuous moisture is more favorable than alternate wetting and drying. The microbial mass is largely fungal mycelium. Among the first to invade the material are the sugar fungi and nonspore-forming bacteria that rapidly utilize the readily decomposable organic compounds in this material such as sugars. The rapidity of this utilization has been demonstrated by Stewart (1966). He found that 50 percent of the glucose carbon added to the soil was mineralized in 3 days; 70 percent in 10 days; and 80 percent in 24 days. Once the glucose is utilized the debris is invaded by spore-forming bacteria and myxobacteria that feed on the cellulose. Although less easily digested, cellulose carbon nevertheless disappears rather rapidly, 65 percent being utilized within a 24-day period (Stewart et al., 1966). Lignin is the least decomposable—only 30 percent of it is utilized in a 60-day period (Mayandon and Simonart 1959a).

As the bacteria and fungi work on the plant debris, they assimilate the nutrients and incorporate them into living matter. At this point these nutrients are still unavailable for recycling. This is known as nutrient *immobilization*. The amount of mineral matter that can be tied up by microbes varies greatly, but many exhibit luxury consumption (use in excess of need) of such nutrients as potassium, calcium, and nitrogen. This in itself can affect primary production.

Both bacteria and fungi are short-lived. They die or are consumed by microbial grazers. This death and consumption, as well as the leaching of soluble nutrients from the decomposing substrate, releases minerals contained in the microbial and detritus biomass. This process, known as *mineralization*, makes nutrients available for use by primary producers and microbes. Thus a cycle of immobilization and mineralization takes place within the soil. Nutrients are temporarily immobilized in microbial tissue. As microbes die, the nutrients are released or mineralized, and become available for uptake again. Microbial uptake occurs simultaneously with mineralization. The amount of nutrients available for primary producers depends in part on the magnitude of uptake by the microbial decomposers.

The process of decomposition is aided materially by the fragmentation of detritus and by litter-feeding invertebrates. They consume the cuticle, mine the parenchyma, eat holes in leaves, all serving to open up the material to microbial invasion. The action of such litter-feeders as millipedes, earthworms, and isopods may increase exposed leaf area to 15 times its original size (Ghilarov, 1970). Because the net assimilation of plant debris by litter-feeders is on the average less than 10 percent, a great deal of the material passes through the gut of these saprophages. They utilize only the easily digested proteins and carbohydrates. Mineral matter in the material is often concentrated. This fecal matter is readily attacked by microbes. Some litter-feeders such as earthworms enrich the soil with vitamin B_{12}. In addition they mix organic matter with soil, thus bringing the material in contact with other microbes. Although the mineral pool in the contained biomass and the contribution to energy flow (van der Drift, 1971) by the reducers is relatively small (about 4 to 8 percent), they still make a major but indirect contribution to decomposition.

As the amount of energy decreases with time, the least decomposable material, largely derived from lignin, is left behind as humus. Humus is a structureless, dark-colored, chemically complex material whose characteristic constituents are humin, a complex of unchanged plant chemicals, and other organic compounds such as fulvic acid and humic acid. The latter is derived from lignin and plant flavonoids, which undergo degradation and conjugation with amino compounds, carbohydrates, and silicate materials. Because of the chemical complexity of the material, this process is accomplished in part by such fungi as *Penicillium* and *Aspergillus* and such bacteria as *Streptomyces* and *Pseudomonas*. Decomposition proceeds so slowly that the amount of organic matter in soil changes little each year. Annual loss by decomposition is balanced by the formation of new humic material. Carbon-dating indicates that humus in podzol soils has a mean residence time of 250 ±60 years, that in chernozems 870 ±50 years (Campbell et al., 1967).

In terrestrial ecosystems bacteria and fungi take the major role in decomposition. The same cannot be said for aquatic ecosystems. Although the role of bacteria in aquatic environments is poorly understood, evidence suggests that bacteria, and to a limited extent fungi, act more as converters than as regenerators of nutrients, whereas phytoplankton and zooplankton take a major part in the cycling of nutrients.

Phytoplankton, macroalgae, and zooplankton furnish dissolved organic matter, with algae being main contributors. Phytoplankton and other algae excrete quantities of organic matter at certain stages of their life cycles, particularly during rapid growth and reproduction. During photosynthesis the marine algae *Fucus vesticulosus* produces as exudate on the average of 42 mg C/100 g dry weight of algae/hour. Total exudate accounts for nearly 40 percent of the net carbon fixed (Sieburth and Jensen, 1970). Johannes (1968) points out that 25 to 75 percent of the regeneration of nitrogen and phosphorus takes place in the presence of microorganisms by autolysis and solution rather than by bacterial decomposition. In fact 30 percent of the nitrogen contained in the bodies of zooplankton is lost by autolysis within 15 to 30 minutes after death, too rapidly for any significant bacterial action to occur.

Bacteria, phytoplankton, and zooplankton utilize inorganic nutrients as well as such organic nutrients as vitamin B_1 (necessary for the growth of both phytoplankton and zooplankton) and organic sources of nitrogen and phosphorus. In effect they tend to concentrate these nutrients by incorporating them into their own biomass. Important in this concentration of nutrients are the bacteria. Dissolved organic matter is a substrate for the

95

growth of bacteria. Both dissolved and colloidal matter condense on the surface of air bubbles in the water, forming organic particles on which bacteria flourish (R. T. Wright, 1967; Riley, 1963). Fragments of cellulose supply another substrate for bacteria. Bits of plant detritus, bacteria, and phytoplankton are consumed by both bacteria and planktonic animals (Strickland, 1965). As in terrestrial ecosystems, the utilization of these organic nutrients by bacteria results both in the increase and the immobilization of nutrients. Bacteria can use nutrients such as phosphorus in excess of their need. Such luxury consumption can reduce the supply of available nutrients to phytoplankton, thus reducing algal blooms.

Bacteria are consumed by ciliates and zooplankton that in turn excrete nutrients in the form of exudates and fecal pellets in the water. Zooplankton, too, in the presence of an abundance of food, consume more than they need, and can reduce microbial population. In the presence of abundance, zooplankton will excrete half or more of the ingested material as fecal pellets, which make up a significant fraction of suspended material. These pellets are attacked by bacteria that utilize the nutrients and growth substances they contain. Thus the cycle starts over again.

Aquatic muds are largely anaerobic habitats. Fungi are absent, and the decomposer bacteria are largely facultative anaerobes. Incomplete decomposition often results in the accumulation of peat and organic muck. But the particulate matter from dead vegetation and animals nevertheless supports a rich bacterial flora. Plant fragments are colonized by bacteria, and both become food for snails and mollusks. Newell (1965) points out that the snail *Hydrobia* feeds on the detritus found in the mud flats. Its fecal pellets are devoid of nitrogen but rich in carbohydrates, suggesting that the snail cannot digest cellulose and other complex carbohydrates. If the pellets are held in filtered seawater, the nitrogen content quickly rises. The rise in nitrogen is accomplished by the growth of marine bacteria that colonize on the fecal material and utilize nitrogen dissolved in seawater. The fecal pellets, enriched with nitrogen in the form of bacterial protein, are reingested by the snail. The snail then digests the bacterial bodies, and the resultant fecal pellets again

are devoid of nitrogen. Recolonization of fecal pellets is repeated.

Thus bacteria function primarily to concentrate nutrients rather than to release them to the environment by decomposition. That task is accomplished largely by the algae, zooplankton, and detritus-feeding animals that release certain nutrients and metabolites into the water by physical and chemical breakdown of plant and animal tissues, releasing cellular contents and excreting organic matter into the water.

Influences on decomposition. The rate of decomposition is influenced by a number of environmental and biotic variables. Among these are moisture, temperature, exposure, altitude, type of microbial substrate, and vegetation. Both temperature and moisture greatly influence microbial activity by affecting metabolic rates. Alternate wetting and drying and continuous dry spells tend to reduce both activity and populations of microbes. Slope exposure, especially as it relates to temperature and moisture, and type of vegetation can increase or decrease decomposition. Witkamp (1963) found that bacterial counts from north-facing slopes were nine times higher in hardwoods than in coniferous stands; but counts from hardwood and coniferous stands on south slopes did not differ. This was undoubtedly due to drier conditions on the south slopes. The species composition of leaves in the litter had the greatest influence. Easily decomposable and highly palatable leaves from such species as redbud, mulberry, and aspen support initially higher populations of decomposers than litter from oak and pine, for example, which is high in lignin. Earthworms have a pronounced preference for such species as aspen, white ash, and basswood, take with less relish and do not entirely consume sugar maple and red maple, and do not eat red oak at all (Johnston, 1936). In a European study (Lindquist, 1942) earthworms preferred the dead leaves of elm, ash, and birch, ate sparingly of oak and beech, and did not touch pine or spruce needles. Millipedes likewise show a species preference (van der Drift, 1951). Thus decomposition of litter from certain species proceeds more slowly than litter from others. On easily decomposable material initially high populations of microbes decline with time as energy is depleted. But on more resistant oak and pine litter, initially low population densities increase as decomposition proceeds (Witkamp, 1963).

The water cycle

Leonardo de Vinci wrote, "Water is the driver of nature." Perceptive as he was, even he could not have appreciated the full meaning of his statement based on the scientific knowledge of his time. Without the cycling of water, biogeochemical cycles could not exist, ecosystems could not function, and life could not be maintained. Water is the medium by which materials make their never-ending odyssey through the ecosystem.

THE STRUCTURE OF WATER

Because of the physical arrangement of its hydrogen atoms and hydrogen bonds, liquid water consists of branching chains of oxygen tetrahedra. The physical state of water, whether liquid, gas, or solid, is determined by the speed at which hydrogen bonds are being formed and broken. Heat increases that speed; hence weak hydrogen bonds cannot hold molecules together as they move faster. The thermal status of water in a liquid state is such that hydrogen bonds are being broken as fast as they form. At low temperatures the tetrahedral arrangement is almost perfect; when water freezes, the arrangement is a perfect lattice with considerable open space between ice crystals, and thus a decrease in density. For this reason ice floats. As the temperature of the frozen water is increased, this molecular arrangement becomes looser and more diffuse, resulting in random packing (because of the continuous breaking and reforming of hydrogen bonds) and contraction of molecules. The higher the temperature, the more diffuse the pattern becomes, until the whole structure (and the hydrogen bonds) breaks down and the water melts. Upon melting, water contracts, and its density increases up to a temperature of $3.98°$ C. Beyond this point, the loose arrangement of molecules means a reduction in density again. The existence of this point of maximum density at approximately $4°$ C is of fundamental importance to aquatic life.

Seawater behaves somewhat differently. The density of seawater (salinity of 24.7 o/oo and higher), or rather its specific gravity relative to that of an equal volume of pure water (sp. gra. = 1) at atmospheric pressure, is correlated with salinity. At 0° C the density of seawater with a salinity of 35 0/oo is 1.028. The lower its temperature, the greater becomes the density of seawater; the higher its temperature, the lower the density. No definite freezing point exists for seawater. Ice crystals begin to form at a temperature that varies with salinity. As pure water freezes out, the remaining unfrozen water is even saltier and has an even lower freezing point. Ultimately a solid block of ice crystals and salt is formed.

PHYSICAL PROPERTIES

Specific heat. Water is capable of storing tremendous quantities of heat with a relatively small rise in temperature. It is exceeded in this only by liquid ammonia, liquid hydrogen, and liquid lithium. Thus water is described as having a high *specific heat*, the number of calories necessary to raise 1 g of water 1° C. The specific heat of water is given the value of 1.

Latent heat. Not only does water have a high specific heat, but it also possesses the highest heat of fusion and heat of evaporation—collectively called *latent heat*—of all known substances that are liquid at ordinary temperatures. Large quantities of heat must be removed before water can change from a liquid to a solid, and conversely it must absorb considerable heat before ice can be converted to a liquid. It takes approximately 80 cal of heat to convert 1 g of ice to a liquid state when both are at 0° C. This is equivalent to the amount of heat needed to raise the same quantity of water from 0° to 80° C.

Evaporation occurs at the interface between air and water at all ranges of temperature. Here again considerable amounts of heat are involved; 536 cal are needed to overcome the attraction between molecules and convert 1 g of water at 100° C into vapor. This is as much heat as is needed to raise 536 g of water 1° C. When evaporation occurs, the source of thermal energy may come from the sun, from the water itself, or from objects in or around it. Rendered latent at the place of evaporation, the heat involved is returned to actual heat at the point of condensation (see

Chapter 7). Such phenomena play a major role in worldwide meteorological cycles.

Viscosity. The viscosity of water also is high because water molecules interact with neighboring molecules by forming hydrogen bonds. Viscosity can be visualized best if one imagines or observes liquid flowing through a glass tube or clear plastic hose. The liquid moving through the tube behaves as if it consisted of a series of parallel concentric layers flowing over one another. The rate of flow is greatest at the center; but because of the amount of internal friction between layers, the flow decreases toward the sides of the tube. This same phenomenon can be observed along the side of any stream or river with uniform banks. The water along the streamside is nearly still, whereas the current in the center may be swift. This resistance between the layers is called *viscosity*.

This lateral or laminar viscosity is complicated by another type, eddy viscosity, in which water masses pass from one layer to another. This creates a turbulence both horizontally and vertically. Biologically important (see Chapter 18), eddy viscosity is many times greater than laminar viscosity.

Viscosity is the source of frictional resistance to objects moving through the water. Since this resistance is 100 times that of air, animals must expend considerable muscular energy to move through the water.

Surface tension. Within all substances, particles of the same matter are attracted to one another. Water is no exception. Molecules of water below the surface are symmetrically surrounded by other molecules. The forces of attraction are the same on one side of the molecule as on the other. But at the water's surface, the molecules exist under a different set of conditions. Below is a hemisphere of strongly attractive similar water molecules; above is the much smaller attractive force of the air. Since the molecules on the surface, then, are drawn into the liquid, the liquid surface tends to be as small as possible, taut like the rubber of an inflated balloon. This is *surface tension*. In the aquatic ecosystem it is a barrier to some organisms, and a support for others. It is the force that draws liquids through the pores of the soil and the conducting networks of plants. Aquatic insects and plants have evolved structural adaptations that prevent the penetration of water into the

tracheal systems of the former and the stomata and internal air spaces of the latter.

DISTRIBUTION OF WATER

Although one views water as something of a
local phenomenon, such as a stream or autumn rains, it forms a single worldwide resource distributed in land, sea, and atmosphere, and unified by the hydrological cycle.
It is influenced by solar energy, by the currents of the air and oceans, by heat budgets,
and by water balances of land and sea.
Through historical time the balance of free
water has remained relatively stable although
the balances between land and sea have fluctuated. According to the Russian hydrologist
Shinitnikov (Kalinin and Bykov, 1969), at
present we are passing from a humid period in
earth's history to a dry one, in which the land
areas are losing water at the rate of 105 mi³/
year, and the oceans are gaining that amount,
rising on the average of 0.05 in./year.

Oceans cover 71 percent of the earth's surface (Table 5-1). With a mean depth of 3.8
km (2.36 mi), they hold 93 to 97 percent of
all the earth's waters (depending on the estimate used). Thus fresh water usable by man
represents only 3 percent of the planet's water
supply. Of the total fresh water on earth, 75
percent is locked up in glaciers and ice sheets,
enough to maintain all the rivers of the world
at their present rate of flow for the next 900
years. If oceans contain 97 percent of the
world's water, then nearly 2 percent of the
remainder is tied up in ice. This leaves less
than 1 percent of the world's water available
fresh. Fresh-water lakes contain 0.3 percent
of the fresh-water supply, and at any one time,
rivers and streams contain only 0.005 percent
of it. Soil moisture accounts for approximately
0.3 percent. Another very small portion of
the earth's water is tied up in living material.

More stable is the groundwater supply,
which accounts for 25 percent of our fresh
water. Groundwater fills the pores and hollows within the earth just as water fills pockets and depressions on the surface. Estimates,
necessarily rough and inaccurate, place renew-
able and cyclic groundwater at 7×10^6 km³
(Nace, 1969), or approximately 11 percent
of the fresh-water supply. Some of the groundwater is "inherited," as in aquifers in desert
regions, where the water is thousands of years

TABLE 5-1
World's water resources

Resource	Volume (W) (in thousands of km³)	Annual rate of removal (Q) (in thousands of km³) and process		Renewal period (T = W/Q)
Total water on earth	1,460,000	520,	evaporation	2,800 years
Total water in the oceans	1,370,000	449,	evaporation	3,100 years
		37,	difference between precipitation and evaporation	37,000 years
Free gravitational waters in the earth's crust (to a depth of 5 km)	60,000	13,	underground run-off	4,600 years
Of which, in the zone of active water exchange	4,000	13,	underground run-off	300 years
Lakes	750		—	
Glaciers and permanent snow	29,000	1.8,	run off	16,000 years
Soil and subsoil moisture	65	85,	evaporation and underground run-off	280 days
Atmospheric moisture	14	520,	precipitation	9 days
River waters	1.2	36.3,[a]	run-off	12(20) days

Source: G. P. Kalinin and V. D. Bykov, *Impact of Science on Society,* 19:2. © Unesco 1969.
Note: Average error is probably 10–15 percent.
[a] Not counting the melting of Antarctic and Arctic glaciers.

old. Because inherited water is not rechargeable, heavy use of these aquifers for irrigation and other purposes is mining the supply. In the foreseeable future the supply could be exhausted. A portion of the groundwater, approximately 14 percent, lies below 1000 m. Known as fossil water, it is often saline, and does not participate in the hydrological cycle.

The atmosphere, for all its clouds and obvious close association with the water cycle, contains only 0.035 percent fresh water. Yet it is the atmosphere and its relation to land and oceans that keeps the water circulating over the earth.

THE CYCLING OF WATER

Outside it is raining. The rain strikes against the window, and runs down the panes and walls and into the ground. It disappears into the grass, drips from the leaves of trees and shrubs, and trickles down the trunks. In a while the rain will stop, and the windows and walls will dry. The entire episode—the spring shower, the infiltration into the ground, the throughfall in the trees and bushes, the runoff from the walks, the evaporation—epitomizes the water cycle on a local scale (Fig. 5-2).

Precipitation is the driving force of the water cycle. Whatever its form, precipitation begins as water vapor in the atmosphere. When air rises it is cooled adiabatically (see Chapter 2), and when it rises beyond the temperature level at which condensation takes place, clouds form. The condensing moisture coalesces into droplets 1 to 100 μ in diameter and then into rain droplets with a diameter of approximately 1,000 μ (1 mm). Where temperatures are cold enough, ice crystals may form instead. Particulate matter smaller than 10 mμ in the atmosphere may act as nuclei on which water vapor condenses. At some point the droplets or ice crystals fall as some form of precipitation. As the precipitation reaches the earth, some of the water reaches the ground directly, and some falls on vegetation, on litter on the ground, and on urban structures and streets. It may be stored, hurried off, or in time infiltrate the soil.

Because of interception, which can be considerable, various amounts of water never reach the ground and evaporate into the atmosphere. Grass in the Great Plains may intercept 5 to 13 percent of the annual precipitation (average loss, 7.9 percent) (Corbett and Crouse, 1968), and litter beneath a stand of grass may intercept 2.8 to 8.0 percent of annual precipitation (average loss, 4.3 percent) (Corbett and Crouse, 1968). Precipitation striking forest trees or forest canopy must penetrate the crowns before it reaches the ground. A forest in full leaf in the summer can intercept a significant portion of a light summer rain. The amount of rainfall intercepted depends upon the type and the age of the forest (Fig. 5-3). In general conifers intercept more rainfall on an annual basis than do hardwoods. Mature pine stands, for example, will intercept 20 percent of a summer rainfall and 14 to 18 percent of the annual precipitation, whereas an oak forest will intercept 24 percent of a summer rain and 11 percent of annual precipitation (Lull, 1967).

In summer a relatively greater proportion of rainfall is intercepted during a light shower than during a heavy rain. During a light shower the rain does not exceed the storage capacity of the canopy. The water held by the leaves subsequently evaporates or runs down the leaves, stems, twigs, and trunk as stemflow. The water then enters the soil in a relatively narrow band around the base of a tree. In winter hardwood or deciduous trees intercept very little precipitation.

In urban areas a great portion of the rain falls on roofs and sidewalks, which are impervious to water. The water runs down gutters and drains to be hurried off to rivers. The percentage of rainfall striking impervious surfaces varies with the nature of the urban area. In downtown sections, areas impervious to rain may be 100 percent. In residential areas imperviousness varies with the size of the lot. For lots of about 6,000 ft^2 about 80 percent of the area is impervious; for lots 6,000 to 10,000 ft^2, 40 percent; and for those 15,000 ft^2 and over, 25 percent (Antoine, 1964). In suburbia, with its expanse of lawns, impervious areas are smaller. On a 1.8-acre lot about 92 percent is lawn, 8 percent pavement and roof. A more typical 0.4-acre lot has 24 percent of its area in roof and pavement (Felton and Lull, 1963). The largest paved areas in the suburbs are shopping centers, which require three to four times as much area for parking as for stores.

City streets may intercept, store, and eventually lose by evapotranspiration 0.04 to 0.10 in. of water, and buildings 0.04 in. (Viessman, 1966). Estimates place interception by urban areas at about 16 percent, approximately the same amount intercepted by forest trees in summer leaf. Residential areas, surprisingly,

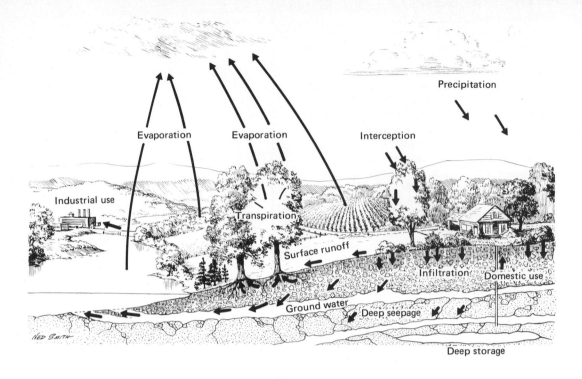

intercept less; lawn grass 2 in. high intercepts 0.01 in. of water, and storage in leaf depression is about 0.04 in. (Felton and Lull, 1963). Total annual interception by residential areas comes to about 13 percent of a 45-in. rainfall (Lull and Sopper, 1969).

The precipitation that reaches the soil moves into the ground by infiltration. The rate of infiltration is governed by soil, slope, type of vegetation, and the characteristics of the precipitation itself. In general the more intense the rains, the greater the rate of infiltration, until the infiltration capacity of the soil, determined by soil porosity, is reached. Because water moves through the soil by the action of two forces, capillary attraction and gravity, soils with considerable capillary pore space have initial rapid infiltration rates, but the pores rapidly fill with water. Water will infiltrate longer into soils with a high proportion of noncapillary pore space. Vegetation that tends to roughen the surface retards water flow, and holds surface flow and allows it to move into the soil. Slope, impeding layers of stone or frozen soil, and other conditions influence the rate at which water moves into the soil.

During long wet spells and heavy storms, the soil may become saturated; or intense rainfall or rapid melting of snow can exceed the infiltration capacity of the soil. At this point water becomes overland flow. In places it becomes concentrated into depressions and rills, and the flow changes from sheet flow to channelized flow, a process that can be observed even on city streets as water moves in sheets over the pavement and becomes concentrated into streetside gutters. Again the amount of runoff that affects the erosion of soil depends upon slope, texture of the soil, soil moisture conditions, and the type and condition of vegetation.

In the undisturbed forest, infiltration rates usually are greater than intensity of rainfall, and surface runoff does not occur. In urban areas infiltration rates may range from 0 to a value exceeding the intensity of rainfall on certain areas where soil is open and uncompacted. Because of low infiltration, runoff from urban areas might be as much as 85 percent of the precipitation (Lull and Sopper, 1969). Because they are so compacted by frequent tramping and mowing, lawns have a low infiltration rate. Felton and Lull (1963) have demonstrated that water infiltrates in lawns at an average rate of 0.01 in./minute, compared to 0.58 in./minute in forest soil.

Water entering the soil will percolate or seep down to an impervious layer of clay or rock to collect as groundwater. From here the water finds its way into springs, streams, and eventually to rivers and seas. A great portion of this water is utilized by man for domestic and industrial purposes, after which it reenters the water cycle by discharge into streams or into the atmosphere.

A part of the water is retained in the soil. The portion held by capillary forces between the ground-soil particles is capillary water. Another portion adheres as a thin film to soil particles. This is hygroscopic water and is unavailable to plants. The maximum amount of water that a soil can hold after gravitational water is drained away is called field capacity.

The capacity of soils to retain and store water varies considerably. Because sandy soils have few fine clay or silt particles filling the pore space, they have less surface area to which the water can cling; and the pores are so large that the weight of the water forces it to run down and out of the soil. Fine-textured soils retain water for a longer time in rather large quantities. Because the surface area on which the water can cling is greater and the size of the pores is smaller, water is slowly discharged into the subsoil. Sandy soils may store 10 to 15 percent of their weight in water, clay 50 to 70 percent. Water retention is increased by the humus, or organic-matter, content of the soil. Humus may retain 100 to 200 percent of its own weight in water. For each inch of humus, about 0.8 in. of water is stored for as long as 2 days after a rain, and is slowly discharged into streams (Lull, 1967). Storage in urban areas obviously is considerably less, amounting to nearly 0 (on paved areas) to 10 percent. Much of this storage goes into sewer manholes and basements.

The water remaining on the surface of the ground, lying in soil depressions in the surface layers and collected in vegetation, as well as water of the surface layers of streams, lakes, and oceans evaporates, a process by which more water molecules leave a surface than enter it. The rate at which water moves back into the atmosphere is governed by the vapor-pressure deficit of the atmosphere.

As surface soils dry out, evaporation from them ceases because there exists a dry barrier through which little soil water moves. At this point major water losses from the soil take place through the leaves of plants. Plants take in water through the roots and they lose it through the leaves. As long as sufficient water is available the leaves remain turgid and the openings of the stomata are maximal, which permits an easy inflow of carbon dioxide into the leaf, but at the same time permits a large leakage of water. This leakage will continue as long as moisture is available for roots in

103

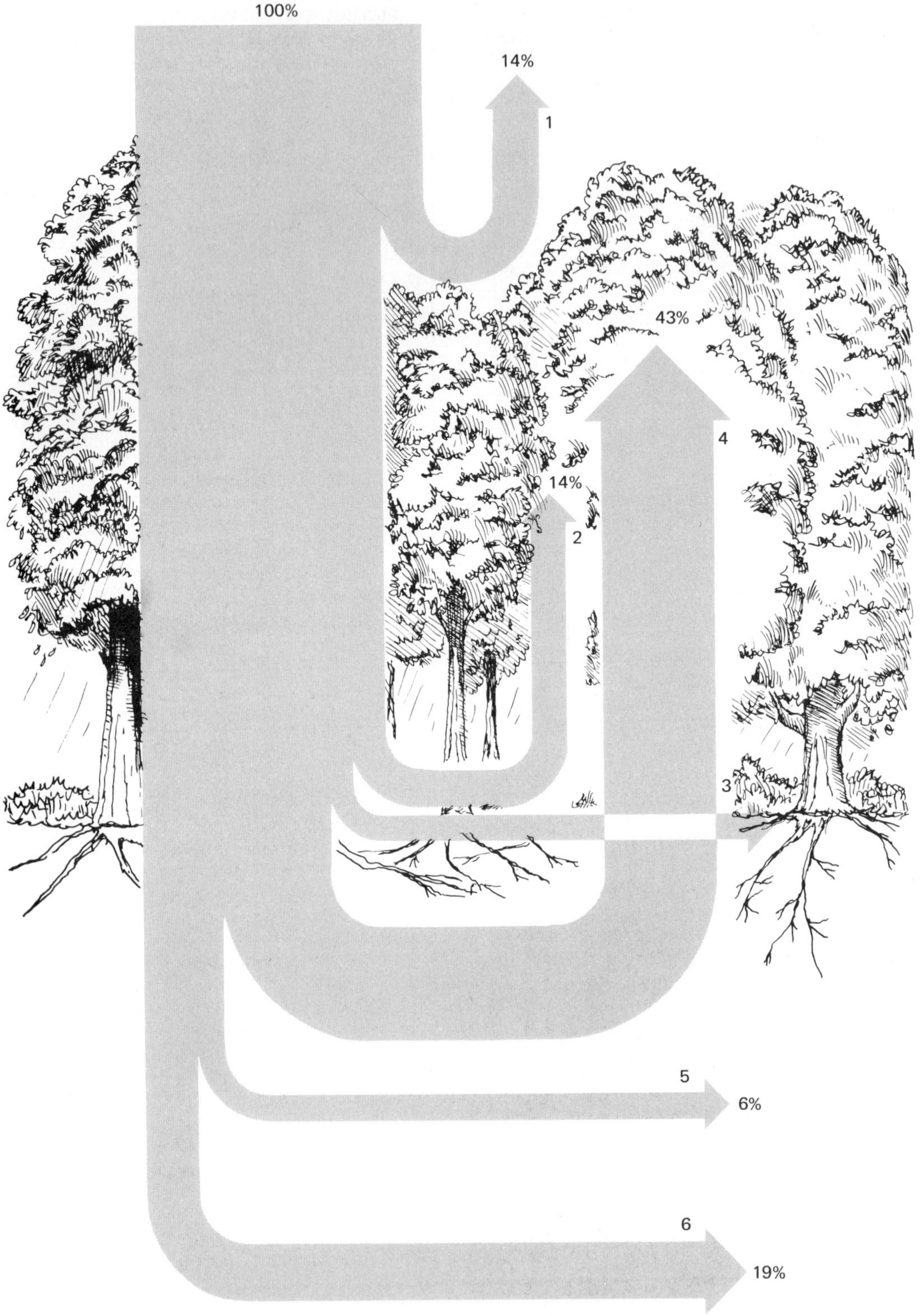
100%
14%
1
43%
4
14%
2
3
5
6%
6
19%

the soil, as long as the roots are capable of removing water from the soil, and as long as the amount of energy striking the leaf is enough to supply the necessary latent heat of evaporation. Thus plants can continue to remove water from the soil until the capillary water is exhausted. Some plants, known as phanerophytes, have roots that can reach and tap groundwater. Annual evapotranspiration in northeastern forests may range from 23 to 29 in. of water. In urban areas evapotranspiration may range from 4 to 5 in.

The temperate deciduous forest and northeastern urbanized areas represent two environmental extremes involved in water cycling. In comparison to the forest, urbanized areas are characterized by reduced interception, less infiltration, much smaller soil moisture storage, less evapotranspiration, and reduced water quality. Such areas also exhibit increased overland or surface flow, increased runoff, and increased peak flows of streams and rivers. Once evaporated into the atmosphere or carried away by surface runoff, the water involved in the local hydrological cycle enters the global water cycle.

THE GLOBAL WATER CYCLE

The molecules of water that fell in the spring shower might well have been a part of the Gulf Stream a few weeks before and perhaps spent some time in the Amazon tropical rain forest some time before that. The local storm is simply a part of the mass movement and circulation of water about the earth, a movement suggested by the changing cloud patterns over the face of the earth. The atmosphere, oceans, and land masses form a single gigantic water system that is driven by solar energy. The presence and movement of water in any one part of the system affects the presence and movement in all other parts.

The atmosphere is one key element in the world's water system. At any one time the atmosphere holds no more than a 10- to 11-day supply of rainfall in the form of vapor, clouds, and ice crystals. Thus the turnover rate of water molecules is rapid. Because the source of water in the atmosphere is evaporation from land and sea, there are global differences in the amount of evaporation and the amount of moisture in the atmosphere at any given point around the globe. Evaporation at lower latitudes is considerably greater than

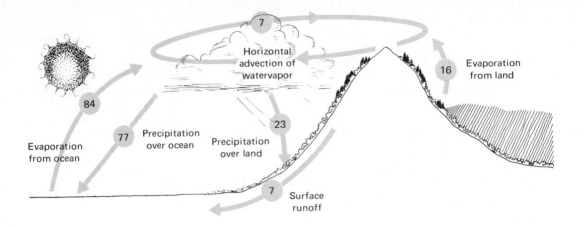

Horizontal advection of watervapor 7

Evaporation from land 16

Evaporation from ocean 84

Precipitation over ocean 77

Precipitation over land 23

Surface runoff 7

evaporation at higher latitudes, reflecting the greater heat budgets produced by the direct rays of the sun. Evaporation is greater over oceans than over land, for not only is there more free water to evaporate, but oceans also contain well over 90 percent of the world's water. Oceans provide 85 percent of evaporation—considerably more than they receive in return from precipitation (Fig. 5-4). Land areas contribute 16 percent of the annual evaporation, yet they intercept more water than evaporates from them.

Moisture in the atmosphere moves with the general circulation of the air. Air currents, hundreds of kilometers wide, are in fact giant unseen rivers moving above the earth. Only a part of this moisture falls as precipitation in any one place. For example, in a year's time the United States receives an unequally distributed 6,000 km³ of precipitation, but the liquid equivalent of water vapor passing over the country is 10 times that much. Atmospheric precipitation for the earth as a whole is approximately 100 cm, and the average resident time for a water molecule in the atmosphere is approximately 10 days.

Variations in evaporation are also reflected in precipitation variations. As described in Chapter 2, air moves about the earth in giant swirls. In the equatorial areas the trade winds move moisture-laden air toward the equator, where it is warmed. The hot air over the equator rises, cools, and drops its moisture as rain. Thus the equatorial regions are areas of maximum precipitation. The air that rises over the equator descends earthward in two subtropical zones around 30° N and 30° S latitude. As the air descends, it warms, and picks up moisture from land and sea. The highest annual losses to evaporation occur in the subtropics of the western North Atlantic and Northern Pacific, or the Gulf Stream, and

the Kinshio Currents. North of this are two more zones of ascending air and low pressure that produce the west-coast areas of maximum rainfall. In high latitudes the air descends again in the polar regions, where it remains dry.

Global detention of precipitation varies with region and season. In the Northern Hemisphere the maximum detention occurs in March and April, when snow-cover is still on the ground and ice still covers ponds and lakes. In the tropics maximum detention time is in October, the time of the summer rains and monsoons. But whatever the location or season, the residence time of water on land is 10 to 120 days.

Precipitation on land that is excess over evaporation is eventually carried to the sea by rivers. Rivers are the prime movers of water over the globe and carry many more times the amount of water their channels hold. By returning water to the sea they tend to balance the evaporation deficit of the oceans. Sixteen major rivers discharge 13,600 cm annually, 45 percent of all water carried by rivers. Adding the next 50 largest rivers brings the total to 17,600 cm, 60 percent of all water discharged to the sea.

Evaporation, precipitation, detention, and transportation maintain a stable water balance on the earth. Of the estimated 100 cm of water that falls annually on earth, 100 cm is returned by evaporation. The oceans receive 112 cm of precipitation a year but lose 135 cm in evaporation, for a net loss of 23 cm annually. Continental areas pick up 73 cm in precipitation, lose 31 cm annually, and return 31 cm as runoff. There is a wide variation among continents (Budyko, 1963). South America, for example, receives 135 cm/year of precipitation, loses 86 cm in evaporation, and contributes 49 cm/year in runoff. The

Australian continent receives 47 cm and evaporates 41 cm, leaving 6 cm as runoff. North America picks up 67 cm in precipitation, 36 cm evaporates, and 31 cm runs off. These values are close to those for Europe.

In its global circulation the water also influences the heat budgets of the earth. As already suggested in the distribution of precipitation, the highest heat budgets are in the low latitudes, the lowest in the polar regions, and a balance between incoming and outgoing cold and hot is achieved at 38° to 39° latitude. Excessive cooling of higher latitudes is prevented by the north and south transfer of heat by the atmosphere in the form of sensible and latent heat in water vapor and by warm ocean currents.

Examined from a global point of view, the water cycle emphasizes the close interaction between the physical and geographical environments of the earth. Thus the water problem often considered in local terms is actually a global problem, and local water management schemes can affect the planet as a whole. Problems result not because an inadequate amount of water reaches the earth, but because it is unevenly distributed, especially relative to human population centers. Because man has strongly interjected himself into the water cycle, the natural usable water resources have decreased, and water quality has declined. The natural water cycle has not been able to compensate for man's deteriorating effects on water resources.

TYPES OF BIOGEOCHEMICAL CYCLES

There are two types of biogeochemical cycles, the *gaseous* and the *sedimentary*. In gaseous cycles the main reservoir of nutrients is the atmosphere and the ocean. In sedimentary cycles the main reservoir is the soil and the sedimentary and other rocks of the earth's crust. Both involve biological and nonbiological agents, both are driven by the flow of energy, and both are tied to the water cycle.

Gaseous cycles

Because gaseous cycles are closely linked to the atmosphere and the ocean, they are pronouncedly global and involve those compounds (or substances) of which we are most consciously aware: oxygen, carbon dioxide, and nitrogen.

THE OXYGEN CYCLE

Oxygen, the by-product of photosynthesis, is involved in the oxidation of carbohydrates with the release of energy, carbon dioxide, and water. Its primary role in biological oxidation is that of a hydrogen acceptor. The breakdown and decomposition of organic molecules proceeds primarily by dehydrogenation. Hydrogen is removed by enzymatic action from organic molecules in a series of reactions and is finally accepted by the oxygen, forming water.

Oxygen is very active chemically. It can combine with a wide range of chemicals in the earth's crust and is able to react spontaneously with organic compounds and reduced substances. Thus oxygen, necessary for life, can also be toxic, as it is to anaerobic bacteria. Higher organisms evolved a system to protect themselves from the toxic effects of molecular oxygen. This system involves organelles called preoxisomes, which are contained within the cells and which produce peroxide. The oxidative reactions they mediate result in the production of hydrogen peroxide, which in turn is used through the mediation of other enzymes as an acceptor in oxidizing other compounds. The energy evolved is not utilized by the cells.

At some time higher organisms evolved elaborate mechanisms to ensure a supply of oxygen. Some animals obtain oxygen by diffusion through the skin from air or water. Other animals evolved lungs and gills; plants evolved stomata. Organisms evolved an array of catalysts such as iron-containing molecules, the cytochromes and copper-containing enzymes, and cytochrome oxidases to mediate the transfer of hydrogen to oxygen molecules. In higher organisms this oxidative system is contained in the mitochondria of the cell, which acts as a low-temperature furnace where organic molecules are slowly burned with oxygen and the energy evolved is used to form the high-energy bonds of ATP.

The major supply of free oxygen that supports life is in the atmosphere. There are two significant sources of atmospheric oxygen. One is the photodisassociation of water vapor in which most of the hydrogen released escapes into outer space. If the hydrogen did not escape, it would oxidize and recombine with the oxygen. The other source is photosynthesis, active only since life began on earth. Because photosynthesis and respiration are cyclic, involving both the release and utilization of oxygen, one would seem to balance the other, and no significant quantity of oxygen would accumulate in the atmosphere. At some time in the earth's history the amount of oxygen introduced into the atmosphere had to exceed the amount used in the decay of organic matter and that tied up in the oxidation of sedimentary rocks. Part of the atmospheric oxygen represents that portion remaining from the unoxidized reserves of photosynthesis—coal, oil, gas, and organic carbon in sedimentary rocks. The amount of stored carbon in the earth suggests that 150×10^{20} g of oxygen has been available to the atmosphere, over 10 times as much as now present, 10×10^{20} g (F. S. Johnson, 1970). The main nonliving oxygen pool consists of molecular oxygen, water, and carbon dioxide, all intimately linked to each other in photosynthesis and other oxidation–reduction reactions, and all exchanging oxygen with each other. Oxygen is also biologically exchangeable in such compounds as nitrates and sulfates, utilized by organisms that reduce them to ammonia and hydrogen sulfide.

On the surface the oxygen cycle might appear to be quite simple. But because oxygen is so reactive, its cycling is quite complex. As a constituent of carbon dioxide, it circulates freely throughout the biosphere. Some carbon dioxide combines with calcium to form carbonates. Oxygen combines with nitrogen compounds to form nitrates, with iron to form ferric oxides, and with many other minerals to form various other oxides. In these states oxygen is temporarily withdrawn from circulation. In photosynthesis the oxygen freed is split from the water molecule. This oxygen is then reconstituted into water during plant and animal respiration. Part of the atmospheric oxygen that reaches the higher levels of the troposphere is reduced to ozone (O_3) by high-energy ultraviolet radiation.

There is some concern that world oxygen will be depleted by the increased burning of fossil fuels and by a decreased biomass of photosynthetic plants caused by deliberate destruction and pollution. The earth is enveloped in a cloak of oxygen approximately

60,000 moles/m². To this, photosynthesis adds 8 moles/m² annually. Most of the oxygen released to the atmosphere by photosynthesis is consumed by animals and bacteria. The bulk of the remainder is utilized by the oxidation of geologic materials. Only a very small fraction, about 1 part per 10,000, escapes oxidation and remains in the atmosphere. Thus the amount of oxygen in the atmosphere remains rather stable and resistant to short-time changes of 100 to 1,000 years. If we were to completely burn our fossil fuel reserves, we would still use up less than 3 percent of our oxygen reserve. And if photosynthesis should cease and all organic matter were decomposed, we would still have a larger reserve of molecular oxygen on which to draw (Broecker, 1970). However, before oxygen is depleted, some other catastrophe might wipe out man. The major problem is not oxygen depletion, but the accumulation of harmful gases and dusts in the world's atmosphere.

THE CARBON CYCLE

Because it is a basic constituent of all organic compounds and a major element involved in the fixation of energy by photosynthesis, carbon is so closely tied to energy flow that the two are inseparable. In fact the measurement of productivity (see Appendix C) is commonly expressed in terms of grams of carbon fixed per square meter per year. The source of all the fixed carbon both in living organisms and fossil deposits is carbon dioxide, CO_2, found in the atmosphere and dissolved in the waters of the earth. To trace its cycling to the ecosystem is to redescribe photosynthesis and energy flow (see Chapter 3).

Carbon, together with oxygen and hydrogen in the presence of sunlight, is converted to a simple carbohydrate, glucose, $C_6H_{12}O_6$. Involved in the production of 1 kg of glucose is 3.7×10^3 cal of energy, equivalent to 4.35 kwh of energy, 1.47 kg of carbon dioxide, and 0.6 kg of water. Because plants continually carry on respiration during both day and night, a considerable portion of glucose is oxidized to yield carbon dioxide, water, and energy:

$$\tfrac{1}{6}C_6H_{12}O_6 + O_2 \rightarrow CO_2 + H_2O + \text{energy}$$
$$30\,g + 32\,g \rightarrow 44\,g + 18\,g$$

Plants utilize approximately 40 percent of the gross primary production for maintenance,

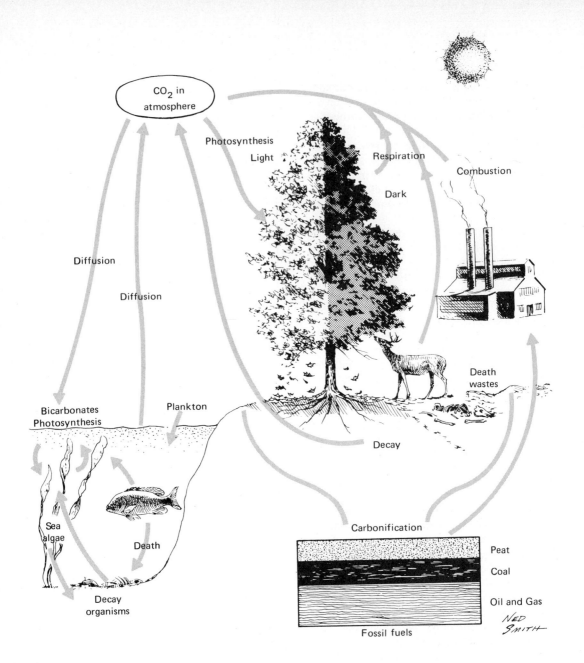

CO$_2$ in atmosphere

Photosynthesis

Light

Respiration

Dark

Combustion

Diffusion

Diffusion

Death wastes

Bicarbonates
Photosynthesis

Plankton

Decay

Sea algae

Death

Carbonification

Peat

Coal

Oil and Gas

Decay organisms

Fossil fuels

NED SMITH

to transpire water, and to convert first-level glucose to higher carbohydrates. To produce 1 kg of carbohydrate material the plant requires 6.2×10^3 cal of energy, equivalent to 7.2 kwh. From 180 g of glucose the plant produces 162 g of cellulose and 18 g of water. Each kilogram of cellulose produced then removes 1.6 kg of carbon dioxide from the atmosphere.

Once produced by the plant, the polysaccharides and fats synthesized from glucose and stored as tissue are utilized by plant-feeding animals that digest and synthesize the carbon compounds into others (Fig. 5-5). Meat-eating animals feed on the herbivores, and the car-

bon compounds again are redigested and resynthesized into other forms. Some of the carbon is returned by these organisms directly because carbon dioxide is a by-product of respiration of both plants and animals. Some becomes incorporated into the bones of land animals and the exoskeletons of invertebrates, especially such marine forms as Foraminifera.

Carbon contained in animal wastes and in the protoplasm of plants and animals is released eventually by assorted decomposer organisms. Rate of release depends upon environmental conditions such as soil moisture, temperature, and precipitation. In tropical forests most of the carbon in plant remains is quickly recycled, for there is little accumulation in the soil. The turnover rate of atmospheric carbon over a tropical forest is about 0.8 year (Leith, 1963). In drier regions such as grasslands, considerable quantities of carbon are stored as humus. In swamps and marshes, where dead material falls into the water, organic carbon is not completely mineralized, and is stored as raw humus or peat, and is circulated only slowly. The turnover rate of atmospheric carbon over peat bogs is somewhere on the order of 3 to 5 years (Leith, 1963).

Similar cycling takes place in the freshwater and marine environments. Phytoplankton utilizes the carbon dioxide that has been diffused into the upper layers of water or is present as carbonates, and converts it into carbohydrates. The carbohydrates so produced pass through the aquatic food chains. The carbon dioxide produced by respiration is re-utilized by the phytoplankton in the production of more carbohydrates. Under proper conditions a portion is reintroduced into the atmosphere. Significant portions of carbon bound as carbonates in the bodies of shells, snails, and foraminifers become buried in the bottom mud at varying depths when the organisms die. Isolated from biotic activity, that carbon is removed from cycling and becomes incorporated into bottom sediments, which through geological time may appear on the surface as limestone rocks or as coral reefs. Other organic carbon is slowly deposited as gas, petroleum, and coal at an estimated global rate of 10 to 13 $g/m^2/$year.

The cycling of carbon as carbon dioxide involves its assimilation and respiration by plants, consumption in the form of plant and animal tissue by animals, its release through their respiration, the mineralization of litter and wood, soil respiration, accumulation of carbon in a standing crop, and withdrawal into longer-term reserves such as humus and peat fossil deposits (see Fig. 5-5).

Diurnal and seasonal patterns. If you were to measure the concentration of carbon dioxide in the atmosphere above and within a forest on a summer day as Woodwell and Dykeman (1966) did, you would discover that it fluctuates throughout the day. At daylight, when photosynthesis begins, plants start to withdraw carbon dioxide from the air, and the concentration declines sharply (Fig. 5-7). By afternoon, when the temperature is increas-

FIGURE 5-5 [OPPOSITE]
The carbon cycle in the ecosystem.

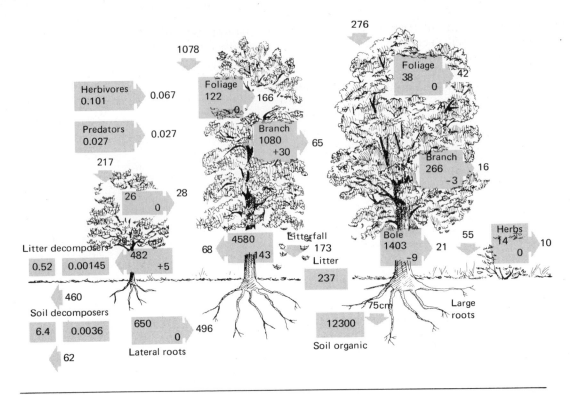

1078

Herbivores
0.101 → 0.067

Predators
0.027 → 0.027

Foliage
122   166
0

Branch
1080
+30   → 65

276

Foliage
38   42
0

Branch
266   16
-3

217

26   28
0

Litter decomposers
0.52   0.00145   482   28
+5

68   4580   143

Litterfall
173
Litter
237

Bole
1403   21
-9

55
Herbs
14   10
0

Soil decomposers
6.4   0.0036

460

62

Lateral roots
650   496
0

75cm

Large
roots

Soil organic
12300

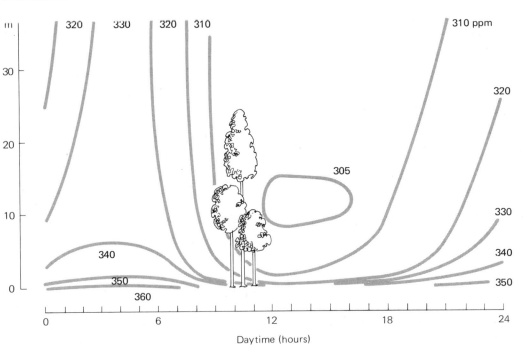

320   330   320   310   310 ppm

320

305

330

340

340

350

360   350

30

20

10

0

0   6   12   18   24

Daytime (hours)

ing and the humidity is decreasing, the respiration rate of plants is increased, the assimilation rate of carbon dioxide declines, and the concentration of carbon dioxide in the atmosphere increases. By sunset the light phase of photosynthesis ceases (see Chapter 3), carbon dioxide is no longer being withdrawn from the atmosphere, and its concentration in the atmosphere increases sharply. A similar diurnal fluctuation takes place in aquatic ecosystems.

The diurnal pattern of carbon dioxide production varies through the vertical profile of the forest (Fig. 5-7). During the day the amounts of carbon dioxide in the atmosphere may vary little from one level to another, but at night the accumulation is greatest nearest the ground. Because the soil and litter layers are the sites of decomposition, microbial respiration and thus the rate of carbon dioxide evolution would be high. Added to microbial respiration is the transfer of carbon assimilated by the plants from the leaves to the roots and into the soil, where it disappears as carbon dioxide. Similar profiles occur in other types of vegetation, including grasslands and agricultural crops (see Huber, 1960).

Not only is there a diurnal pattern of carbon dioxide concentration in the atmosphere, but there is also an annual course in the production and utilization of carbon dioxide. This seasonal change relates both to temperature and to the dormant and growing seasons. In the spring, when land is greening and phytoplankton is actively growing, the daily production of carbon dioxide is high. As measured by nocturnal accumulation in spring and summer, the rate of carbon dioxide production may be two to three times higher than winter rates at the same temperature (Fig. 5-8). The transition from lower to higher rates increases dramatically about the time of the opening of buds and falls off just as rapidly about the time the leaves of deciduous trees start dropping in the fall.

The global carbon dioxide cycle. Like water, the carbon budget of the earth is closely linked to the atmosphere, the land, the oceans, and the mass movements of air around the earth. Just as most of the world's water is tied up in the oceans, most of the world's carbon, some 99.9 percent of it, is tied up in the land mass. Deposits of carbon in the earth's crust amount to around 2.7×10^{16} metric tons largely in the form of carbonates;

FIGURE 5-6 [OPPOSITE TOP]
Carbon cycle in a mesic hardwood forest ecosystem at Oak Ridge, Tennessee. From left to right: the trees represent the understory, dominant tulip poplar and other overstory trees. Structural components of the ecosystem have been abstracted as compartments with major fluxes. Vertical arrows represent photosynthetic fixation. Lateral arrows indicate respiratory losses. Units of measure are grams of carbon per square meter and grams of carbon per square meter per year for compartment increments and fluxes. (From Reichle et al., 1972; courtesy Oak Ridge National Laboratory.)

FIGURE 5-7 [OPPOSITE BOTTOM]
Carbon dioxide distribution in a mixed forest. (From Baumgartner, 1968, after Miller and Rusch, 1960. Reprinted by permission of Unesco. © Unesco 1968.)

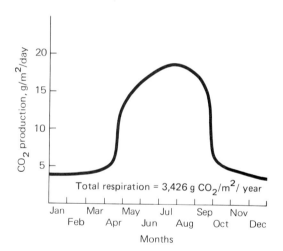

FIGURE 5-8
Production of carbon dioxide fluctuates throughout the year, as data from the Brookhaven pine forest show. The respiration rates expressed in grams of carbon dioxide per square meter per day and plotted by the month show a high increase in production during the summer, when photosynthesis and decomposition are the highest. Based on respiration rates during inversions. Rates corrected to mean monthly temperatures. (From Woodwell and Dykeman, 1966. © 1966 American Association for the Advancement of Science.)

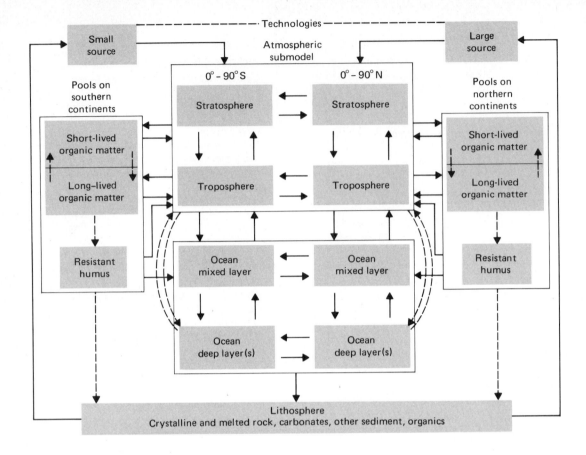

a much smaller fraction is in the form of hydrocarbons and carbohydrates. The oceans contain 0.1 percent and the atmosphere 0.0026 percent in the form of carbon dioxide. As with the water cycle, the atmosphere is the major coupling mechanism in the cycling of carbon between land and sea.

The atmosphere on an annual average contains approximately 320 ppm of carbon dioxide. This is equivalent to 2.6×10^{12} tons of carbon dioxide or 7×10^{11} metric tons of carbon. Since 1850 the quantity of carbon dioxide in the atmosphere has been increasing from the burning of fossil fuels, carbon fixed by photosynthesis millions of years ago. In 1970 man injected into the atmosphere around 2×10^{10} tons of carbon dioxide, or 5.4×10^7 tons of carbon. Over the past 100 years the carbon dioxide content of the atmosphere has increased from 290 to 320 ppm, most of the rise occurring in the past decade (F. S. Johnson, 1970).

Although the average figures suggest it, the atmosphere is not a homogenous pool of carbon dioxide. Like the sea and the land, the atmosphere has compartments or reservoirs between which there is circulation. Recognized atmospheric compartments are the stratosphere and troposphere, separated by the tropopause, and the Northern and Southern Hemispheres (Fig. 5-9) (Nydal, 1968). This pool of carbon dioxide is involved in exchanges with the land mass and the sea.

The carbon cycle in the sea is nearly a closed system. Phytoplankton assimilates carbon dioxide from the surface layers of the water and releases both oxygen and carbon dioxide back to the system. Phytoplankton is consumed by invertebrates and fish, which release carbon dioxide into the water. Eventual decay of organic matter in the lower water and shallow bottoms replaces the carbon dioxide utilized by the phytoplankton. A small fraction of the organic matter reaches the deep layers, which subsequently become enriched with carbon dioxide, especially in the form of carbonates and bicarbonates. A portion becomes incorporated in bottom deposits and is removed from circulation. Another portion enters into a long-term circulation and through upwellings eventually returns to surface layers (Fig. 5-9). A transfer of carbon dioxide across the interface of air and water couples the atmosphere and the ocean.

The exchanges between land masses and the atmosphere are nearly in equilibrium. Photosynthesis by terrestrial vegetation, including that of the tundra, removes about 1.5×10^{10} tons of carbon each year; plant decay returns 1.7×10^{10} tons (Table 5-2). Forests are the main consumers, fixing about 36×10^9 tons of carbon per year, twice as much as all other types of terrestrial vegetation combined (Olson, 1970). Forests also are the major reservoirs of the living carbon pool, containing approximately 482×10^9 tons. This is two-thirds of the amount present in the earth's atmosphere, which is estimated as 640×10^9 billion tons.

The equilibrium of carbon dioxide exchange between land, sea, and atmosphere has been disturbed by the rapid injection of carbon dioxide into the atmosphere by man. This input, already mentioned, which amounts to approximately 2×10^{10} tons/year, has resulted in an increase of carbon dioxide concentration in the atmosphere by 7.5×10^9 tons/ year. Because this amount is only one-third of the annual input, nearly two-thirds must go into the ocean or into increased growth of terrestrial vegetation. Plant growth increases directly with an increased carbon dioxide concentration.

If nearly two-thirds of the carbon dioxide injected into the atmosphere disappears from the atmosphere annually, then the exchange rate between the atmosphere, sea, and land must be fairly rapid. The measurement of the rate of disappearance of the radioactive isotope ^{14}C indirectly produced by nuclear testing in 1963 indicates the rate of turnover. The ^{14}C became well mixed in the atmosphere within several years after the test. The decline thereafter was rapid. According to the estimates of Nydal (1968) the redistribution of excess ^{14}C over natural levels suggests an exchange time from the troposphere into the ocean and biosphere of 4.0 ± 1.0 years. A similar amount of carbon dioxide would have been injected into the atmosphere from land masses and the sea.

The global cycling of carbon dioxide shows the same seasonal variations as the cycle exhibited in the forest (Fig. 5-8). In January the concentration of carbon dioxide at the

FIGURE 5-9 [OPPOSITE]
This flow diagram of carbon cycling in the biosphere shows the interconnected subsystems of the atmosphere, oceans, and main continental organic matter pools. The last two have contributed to fossil fuels over geologic time, while man is releasing carbon from this sink at an accelerating rate. Each subsystem has a complex substructure (for example, see Fig. 20-8). (Courtesy Oak Ridge National Laboratory.)

TABLE 5-2
Estimated atmospheric carbon dioxide balance (1970)

	Exchange in units of 10^{10} tons/year	
	Carbon	Carbon Dioxide
INPUTS TO ATMOSPHERE FROM		
Tropical ocean	4.0	14.8
Plant decay	1.7	6.3
Man (fuel burning)	0.54	2.0
Volcanic action	0.01 (variable)	0.05
Total	6.25	23.15
OUTPUTS FROM ATMOSPHERE TO		
Northern oceans	1.6	6.0
Southern oceans	2.6	9.6
Arctic tundra	0.5	1.8
Other plants	1.35	5.0
Total	6.05	22.4
DISTRIBUTION OF MAN INPUTS TO ATMOSPHERE		
Atmosphere gain	0.2	0.75
Biosphere (land) gain		
Lumber	0.02	0.07
Forests, soil and peat deposits	0.1	0.
Ocean gain	0.22	0.8
Total	0.54	2.00

Source: Prepared by A. Watt.

North Pole is near 313 ppm, at the South Pole 318 ppm. As spring in the north progresses, the concentration of carbon dioxide from 20° N latitude to the North Pole declines sharply in July and reaches its lowest concentration in August. By October the level increases again. The decline during the arctic summer suggests a heavy and rapid removal of carbon dioxide from the atmosphere by photosynthetic activity of arctic plant life. The scrubbing effect is at a maximum in August, when arctic ecosystems are most active and polar ice is at its minimum. A similar but much less pronounced seasonal change takes place at the South Pole.

THE NITROGEN CYCLE

Nitrogen is an essential constituent of protein, which is a building block of all living material. It is also the major constituent of the atmosphere, comprising about 79 percent of it. The paradox is that in its gaseous state, N_2, abundant though it is, nitrogen is unavailable to most life. Before it can be utilized it must be converted to some chemically usable form. Getting it into that form comprises a major part of the nitrogen cycle.

Nitrogen is chemically versatile, capable of entering into a number of interesting reactions. This versatility is due to its variety of valence states, or bonding character. In its most highly oxidized form (most electron-donating toward other atoms) nitrogen shares the five electrons in its outer shell with other atoms, giving it a valence of $+5$. In its most reduced state (most electron-accepting from other atoms) it has a valence of -3. In the nitrate ion, an example of nitrogen in its most highly oxidized form, N shares five electrons with oxygen. In the ammonium ion, representative of its most highly reduced form, N borrows three electrons from hydrogen. Between these two extremes of nitrogen compounds there is a total valence change of eight electrons.

To be used, the free molecular nitrogen has to be fixed, and fixation requires an input of energy. In the first step molecular nitrogen, N_2, has to be split into two atoms:

$$N_2 \rightarrow 2N$$

This step requires an input of 160 kcal for each mole (28 g) of nitrogen. The free N atoms then must be combined with hydrogen to form ammonia, with the release of about 13 kcal of energy:

$$2N + 3H_2 \rightarrow 2NH_3$$

This fixation comes about in two ways. One is by high-energy fixation such as cosmic radiation, meteorite trails, and lightning that provide the high energy needed to combine nitrogen with the oxygen and hydrogen of water. The resulting ammonia and nitrates are carried to the earth in rainwater. Estimates (Eriksson, 1952) suggest that less than 8.9 kg N/ha is brought to the earth annually in this manner. About two-thirds of this comes as ammonia and one-third as nitric acid, H_2NO_3.

The second method of fixation is biological. This amounts to 100 to 200 kg N/ha, roughly 90 percent of the fixed nitrogen contributed to the earth each year. This fixation is accomplished by symbiotic bacteria living in association with leguminous and root-noduled nonleguminous plants, by free-living aerobic bacteria, and by blue-green algae. In agricultural ecosystems, the nodulated legumes of approximately 200 species are the preeminent nitrogen fixers. In nonagricultural systems some 12,000 species of plants, from free-living bacteria and blue-green algae to nodule-bearing plants, are responsible for nitrogen fixation.

Legumes, the most conspicuous of the nitrogen-fixing plants, have a symbiotic relationship with members of the bacterial genus *Rhizobium*. Rhizomia are aerobic, gram-negative, non-spore-forming rod-shaped bacteria that exist in the immediate surroundings of the plant roots, called the rhizosphere. Here the bacteria, stimulated by secretions from the legumes, multiply. The secretions, together with enzymes secreted by the legumes in response to the exudates of the bacteria, loosen the fibrils of the root hair walls. Swarming rhizobia enter the root hair tips and penetrate the inner cortical cells. This is usually accomplished with the aid of an infection thread, which consists of rhizobia imbedded in mucilage and surrounded by the cell wall of the host. The bacteria, singly or in groups, are released by a pinching off of the thread into the host cell. Here they multiply and increase in size, resulting in swollen infected cells in which hemoglobin develops. These cells make up the central tissue of the nodules. Inside the nodule the bacteria change from rod-shaped bacteria of infection threads to a nonmotile bacteroid condition associated with nitrogen fixation.

In addition to the legumes there are a large number of nonleguminous nodule-bearing plants, most of them associated with the early or pioneering stages of succession when the soil is usually low in nitrogen. Among the angiosperms are such plants as *Alnus, Ceanothus, Shepherdia, Elaeagnus,* and *Myrica.* Although their nodules differ structurally from those of legumes, they function in much the same manner. The plants may contribute as much nitrogen to wildlands as leguminous crops contribute to agricultural ecosystems. Nodules are not necessarily confined to the roots. Some plants, such as the African genera of *Rubiaceae* and *Pavetta,* bear nodules on the leaves.

Also contributing to the fixation of nitrogen are free-living soil bacteria. The most prominent of the 15 known genera are the aerobic *Azotobacter* and the anaerobic *Clostridium* (see Mishustin and Shilnikova, 1969). *Azotobacter,* although distributed worldwide, is not found in all soils. It prefers soils with a pH of 6 to 7, that are rich in mineral salts and low in nitrogen. Somewhat characteristic of cultivated soils, *Azotobacter* occurs in relatively low numbers from a few to 20,000/g of soil (Jenny, 1958). One of its outstanding physiological characteristics is the possession of the highest respiratory rate of all kinds of living matter (Jenny, 1958). This rate is understandable when one considers that nitrogen fixation requires a heavy input of energy. To fix 12 to 20 mg of nitrogen, *Azotobacter* utilizes 3.7 kcal of energy.

The anaerobic *Clostridium* is ubiquitous, found in nearly all soils. Although it prefers a neutral pH, it is more tolerant of acidic conditions than *Azotobacter.* In its fixing of nitrogen *Clostridium* leaves a considerable fraction of energy behind as fermentation products such as butyric and acetic acid. For 3.7 kcal of energy utilized, *Clostridium* produces 2 to 3 mg of nitrogen.

Both genera produce ammonia as the first stable product and, like the symbiotic bacteria, they require molybdenum as an activator and are inhibited by an accumulation of nitrates and ammonia in the soil. However, recent experimental work (Mishustin and Shilnikova, 1969) suggests that in agricultural ecosystems *Azotobacter* is an effective fixer of nitrogen only in the presence of added fertilizer. Its role seems to be primarily as a growth stimulator.

Blue-green algae are another important group of largely nonsymbiotic nitrogen fixers. Of the some 40 known species the most common are in the genera *Nostoc* and *Calothrix*, found both in soil and aquatic habitats. Blue-green algae are well adapted to exist on the barest requirements for living. They are often pioneers on bare mineral soil. Especially successful in waterlogged soil, they appear to be nitrogen fixers in rice paddies of Asia (Singh, 1961), where studies indicate that they annually fix 30 to 50 kg N/ha. Blue-greens are perhaps the only fixers of nitrogen over a wide range of temperatures in aquatic habitats from arctic and antarctic seas to fresh-water ponds and hot springs. In the hot springs of Yellowstone, the blue-greens, which are responsible for the bluish-green color, fix nitrogen at a temperature of 55° C (W. D. P. Stewart, 1967). As with *Azotobacter*, blue-greens also require molybdenum for nitrogen fixation.

Recently certain lichens (*Collema tunaeforme* and *Peltigera rufescens*) were also implicated in nitrogen fixation (Henriksson, 1971). Lichens with nitrogen-fixing ability possess nitrogen-fixing blue-green species as their algal component.

Nitrogen fixed by symbiotic and nonsymbiotic microorganisms in soil and water is one source of nitrogen. Another source is organic matter. The wastes of animals broken down by decomposition release nitrates and ammonia into the ecosystem. All of these nitrogenous products are involved in another phase of the nitrogen cycle: the processes of nitrification, denitrification, and ammonification.

In ammonification the amino acids are broken down by decomposer organisms to release energy. It is a one-way reaction. Ammonium, or the ammonia ion, is directly absorbed by plant roots and incorporated into amino acids, which are subsequently passed along through the food chain. Wastes and dead animal and plant tissues are broken down to amino acids by heterotrophic bacteria and fungi in soil and water. Amino acids are oxidized to carbon dioxide, water, and ammonia, with a yield of energy:

$$CH_2NH_2COOH + 1\tfrac{1}{2}O_2 \rightarrow$$
$$2CO_2 + H_2O + NH_3 + 176 \text{ kcal}$$

Part of the ammonia is dissolved in water, part is trapped between soil sols, and some is trapped and fixed in both acid clay and certain base-saturated clay minerals near the point where first broken down or introduced into the soil (Nommik, 1965).

Nitrification is a biological process in which ammonia is oxidized to nitrate and nitrite, yielding energy. Two groups of microorganisms are involved. *Nitrosomonas* utilize the ammonia in the soil as their sole source of energy because they can promote its oxidation to nitrite ions and water:

$$NH_3 + 1\tfrac{1}{2}O_2 \rightarrow HNO_2 + H_2O + 165 \text{ kcal}$$
$$HNO_2 \rightarrow H^+ + NO_2^-$$

Nitrite ions can be oxidized further to nitrate ions in an energy-releasing reaction. This energy left in the nitrite ion is exploited by another group of bacteria, the *Nitrobacter*, which oxidize the nitrite ion to nitrate with a release of 18 kcal of energy (M. Alexander, 1965):

$$NO_2^- + \tfrac{1}{2}O_2 \rightarrow NO_3^-$$

Thus nitrification is a process in which the oxidation state (or valence) of nitrogen is increased. *Nitrosomonas* oxidizes 35 mols of nitrogen for each mol of CO_2 carbon assimilated; *Nitrobacter* oxidizes 100 mols.

The activity of nitrifying bacteria is influenced by environment and the chemical nature of the substrate. Both types function best at a neutral or alkaline pH and are inhibited in acid conditions. They require an adequate supply of magnesium, phosphorus, and iron. The optimum temperature for the process lies somewhere between 35° and 30° C. Oxygen is needed, of course, and its deficiency in wet and waterlogged soil can inhibit the process. In general the greatest nitrifying activity takes place when soils are at half to two-thirds of their water-holding capacity.

Although nitrification is generally considered a beneficial process, it may not always be so. Nitrification involves the conversion of a slowly leached cationic form of nitrogen into a readily leached anionic form, NO_2^-. If quantities of NO_2 are large enough, and sufficient water percolates through the soil, the nitrate can be removed faster than it can be taken up by the roots, a situation that results in the eutrophication of water. An abundance of nitrate also leads to increased losses of gaseous nitrogen. Nitrates are a necessary substrate for denitrification, in which nitrogen in the nitrate form is reduced to the gaseous form.

The denitrifiers, represented by fungi and the bacteria *Pseudomonas*, are facultative anaerobes—they prefer an oxygenated environment. But if oxygen is limited they can use nitrate instead of O_2 as the hydrogen acceptor and can release N_2 in a gaseous state as a by-product:

$$C_6H_{12}O_6 + 4NO_3^- \rightarrow 6CO_2 + H_2O + 2N_2$$

Like nitrification, denitrification takes place under certain conditions: a sufficient supply of organic matter, a limited supply of molecular oxygen, a pH range of 6 to 7, and an optimum temperature of 60° C.

CYCLING OF NITROGEN

With the basic and necessary processes described, the nitrogen cycle (Fig. 5-10, Table 5-3) can be followed briefly. The sources of inputs of nitrogen under natural conditions are the fixation of atmospheric nitrogen, additions of inorganic nitrogen in rain from such sources as lightning fixation and fixed "juvenile" nitrogen from volcanic activity, ammonia absorption from the atmosphere by plants and soil, and nitrogen accretion from windblown aerosols, which contain both organic and inorganic forms of nitrogen. In terrestrial ecosystems, nitrogen, largely in the form of ammonia or nitrates, depending upon a number of variable conditions, is taken up by plants, which convert it into amino acids. The amino acids are transferred to consumers, which convert them to different types of amino acids. Eventually their wastes (urea and excreta) and the decay of dead plant and animal tissue are broken down by bacteria and fungi into ammonia. Ammonia may be lost as gas to the atmosphere, may be acted upon by nitrifying bacteria, or may be taken up directly by plants. Nitrates may be utilized by plants, immobilized by microbes, stored in decomposing humus, or leached away. This material is carried to streams, lakes, and eventually the sea, where it is available for use in aquatic ecosystems. There nitrogen is cycled in a similar manner, except that the large reserves contained in the soil humus are largely lacking. Life in the water contributes organic matter and dead organisms that undergo decomposition and subsequent release of am-

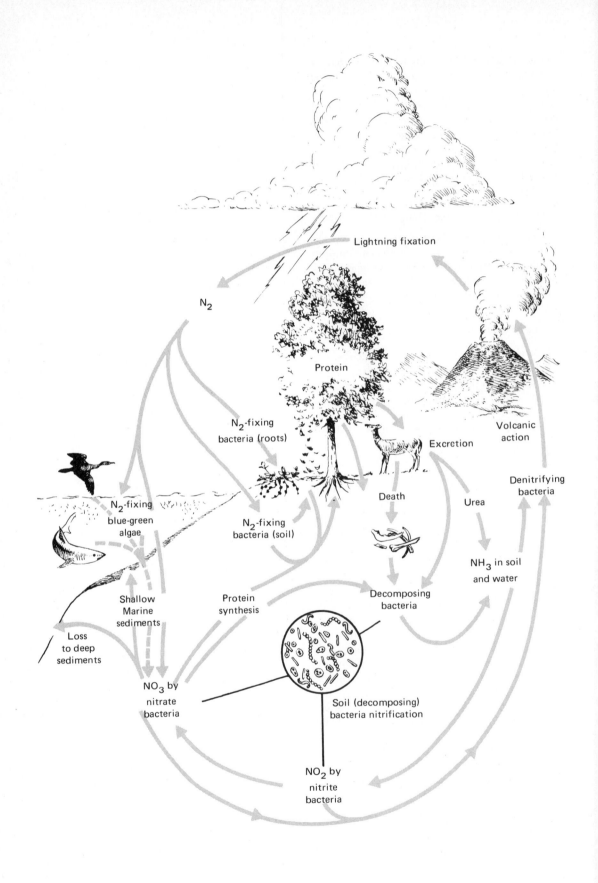

Lightning fixation

N₂

Protein

N₂-fixing
bacteria (roots)

Excretion

Volcanic
action

N₂-fixing
blue-green
algae

N₂-fixing
bacteria (soil)

Death

Urea

Denitrifying
bacteria

Shallow
Marine
sediments

Protein
synthesis

Decomposing
bacteria

NH₃ in soil
and water

Loss
to deep
sediments

NO₃ by
nitrate
bacteria

Soil (decomposing)
bacteria nitrification

NO₂ by
nitrite
bacteria

monia and ultimately nitrates. Atmospheric nitrogen is fixed by a number of blue-green algae. Tracer studies with ^{15}N, a short-lived nonradioactive isotope, show that in marine ecosystems, ammonia is recycled rapidly and preferentially by phytoplankton (Dugdale and Goering, 1967). As a result there is little ammonia in most natural waters, and nitrate is utilized only in the virtual absence of ammonia. In addition to biological cycling there are small but steady losses from the biosphere to the deep sediments of the ocean and to sedimentary rocks. In return there is a small addition of "new" nitrogen from the weathering of igneous rocks and juvenile nitrogen from volcanic activity.

Under natural conditions nitrogen lost from ecosystems by denitrification, volatilization, leaching, erosion, windblown aerosols, and transportation out of the system is balanced by biological fixation and other sources. Both chemically and biologically, terrestrial and aquatic ecosystems constitute a dynamic equilibrium system in which a change in one phase affects the other.

Man's intrusion into the system has either upset the steady state or has shifted the system into a new steady state. Cultivation of grasslands, for example, has resulted in a steady decline in the nitrogen content of the soil (Fig. 5-11) (Jenny, 1933). The mixing and breaking up of the soil exposed more organic matter to decomposition and decreased the amount of root material, thus decreasing new additions of organic matter. And the removal of nitrogen through harvested crops or grazing caused additional losses. Harvest of timber results in a heavy outflow of nitrogen from the forest ecosystem not only in timber removed but also in short-term nitrate losses from the soil. Normal loss from a hardwood-forest ecosystem in New Hampshire was less than 21 kg/ha/year. After cutting, outflow was more than 60 kg/ha/year. Nitrate concentrations exceeded pollution standards for over a year. This outflow equals the normal annual turnover of nitrogen in the ecosystem before the forest was cut (Likens et al., 1969) and emphasizes the tight cycling and steady-state conditions in the temperate forest ecosystem.

Heavy addition of commercial fertilizer, especially in the form of anhydrous ammonia, increases the amount of nitrogen in cropland ecosystems. Unless properly applied, a considerable portion of this fertilizer may be lost

FIGURE 5-10 [OPPOSITE]
Nitrogen cycle in the ecosystem.

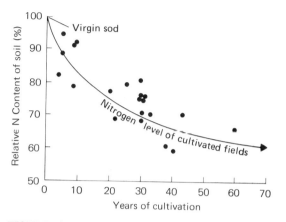

FIGURE 5-11
When the prairie soil was broken and cultivated, the nitrogen level in the soil declined sharply, and eventually stabilized at a much lower level of equilibrium. In effect, agriculture mined the nitrogen reserves of the soil. (After Jenny, 1933.)

TABLE 5-3
Budget for the nitrogen cycle

	Land		Sea		Atmosphere	
	rate/year	% error	rate/year	% error	rate/year	% error
INPUT						
Biological nitrogen fixation	—	—	10	50	—	—
Symbiotic (31)	14	25	—	—	—	—
Nonsymbiotic (31)	30	50	—	—	—	—
Atmospheric nitrogen fixation (31)	4	100	4	100	—	—
Industrially fixed nitrogen fertilizer (31)	30	5	—	—	—	—
N-oxides from combustion	14	25	6	25	20	25
Return of volatile nitrogen compounds in rain	?	—	?	—	—	—
River influx (31)	—	—	30	50	—	—
N_2 from biological denitrification (31)	—	—	—	—	83	100
Natural NO_2	—	—	—	—	?	—
Volatilization (HN_3)	—	—	—	—	?	—
Total input	92+		50		103+	
STORAGE						
Plants (31)	12,000	30	800	50	—	—
Animals (31)	200	30	170	50	—	—
Dead organic matter (31)	760,000	50	900,000	100	—	—
Inorganic nitrogen (31)	140,000	50	100,000	50	—	—
Dissolved nitrogen (31)	—	—	20,000,000	10	—	—
Nitrogen gas (31)	—	—	—	—	3,800,000,000	3
$NO + NH_4$ (25)	—	—	—	—	Less than 1	50
$NH_3 + NH_4$ (17)	—	—	—	—	12	50
N_2O (33)	—	—	—	—	1,000	50
Total storage	912,200		21,000,970		3,800,001,013	
LOSS						
Denitrification (31)	43	—	40	100	—	—
Volatilization	?	—	?	—	—	—
River runoff (31, 32) (includes enrichment from fertilizers)	30	50	—	—	—	—
Sedimentation (31)	—	—	0.2	50	—	—
N_2 in all fixation processes	—	—	—	—	92	50
NH_3 in rain (17)	—	—	—	—	Less than 40	50
NO_2 in rain	—	—	—	—	?	—
N_2O in rain	—	—	—	—	?	—
Total loss	73		40.2		132+	

Source: Robert F. Inger et al. (eds.), 1972, *Man in the Living Environment*, University of Wisconsin Press, p. 76. © 1972 by the Board of Regents of the University of Wisconsin System.
Note: All numbers are in millions of metric tons. The error columns list plus-or-minus probable errors as a percentage of the estimate.

as nitrate nitrogen to the groundwater. Cultivation itself is an added source of nitrate nitrogen (Stewart et al., 1968). Nitrate nitrogen appears to accumulate in tilled soil at a rate three times that of accumulation in grassland soils. It decreases the amount of nitrogen being immobilized by microorganisms, thus making organic matter once inaccessible to decomposition accessible. Grass may adversely affect nitrification by inhibiting the growth of nitrifying bacteria. This can be caused by competition from grass roots and other heterotrophs for ammonium ions (Robinson, 1963),

and by secretion of toxic substances (Rice, 1964, 1965). Automobile exhausts and industrial combustions add nitrous oxides to the atmosphere which are carried to the soil and water in precipitation. Additional inputs of nitrogen into the natural cycle from industrial nitrification and from large-scale cultivation of nitrogen-fixing legumes may be at a higher rate than the total amount being denitrified. The difference represents the nitrate buildup in soils and groundwater and the eutrophication of rivers, lakes, and estuaries.

Sedimentary cycles

Mineral elements required by living organisms are obtained initially from inorganic sources. Available forms occur as salts dissolved in soil water or in lakes, streams, and seas. The mineral cycle varies from one element to another, but essentially it consists of two phases: the salt-solution phase and the rock phase. Mineral salts come directly from the earth's crust by weathering. The soluble salts then enter the water cycle. With water they move through the soil to streams and lakes and eventually reach the seas, where they remain indefinitely. Other salts are returned to the earth's crust through sedimentation. They become incorporated into salt beds, silts, and limestones; after weathering they again enter the cycle.

Plants and many animals fulfill their mineral requirements from mineral solutions in their environments. Other animals acquire the bulk of their minerals from plants and animals they consume. After the death of living organisms the minerals are returned to the soil and water through the action of the organisms and processes of decay.

There are as many different kinds of sedimentary cycles as there are elements. Two may serve as examples. One is sulfur, the cycling of which is something of a hybrid between the gaseous and the sedimentary because it has reservoirs not only in the earth's crust but also in the atmosphere. Phosphorus, on the other hand, is wholly sedimentary—it is released from rock and deposited in both the shallow and deep sediments of the sea.

THE SULFUR CYCLE

Sulfur, like nitrogen, is an essential part of protein and amino acids and is characteristic of organic compounds. It exists in a number of states: elemental sulfur, S; sulfides, with a valence of -2; sulfur monoxide, with a valence of $+2$; sulfites, with a valence of $+4$; and sulfates, with a valence of $+6$. Of these the three that are important in nature are elemental sulfur, sulfides, and sulfates.

The sulfur cycle is both sedimentary and gaseous. It involves a long-term sedimentary phase in which sulfur is tied up in organic and inorganic deposits, is released by weathering and decomposition, and is carried to terrestrial and aquatic ecosystems in a salt

123

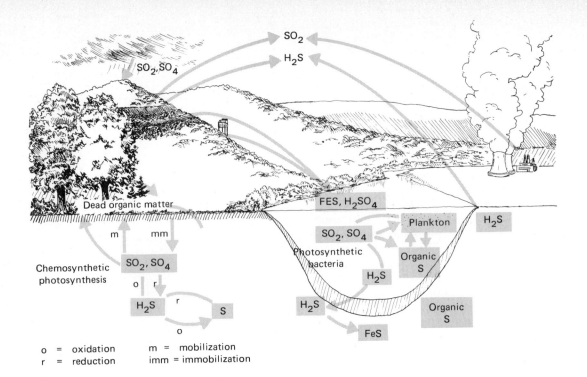

o = oxidation   m = mobilization
r = reduction   imm = immobilization

solution. On the other hand, a considerable portion of sulfur is cycled in a gaseous state, which permits its circulation on a global scale.

Sulfur enters the atmosphere from several sources: the combustion of fossil fuels, volcanic eruption, the surface of oceans, and gases released by decomposition. Initially sulfur enters the atmosphere as hydrogen sulfide, H_2S, which quickly oxidizes into another volatile form, sulfur dioxide, SO_2. Atmospheric sulfur dioxide, soluble in water, is carried back to earth in rainwater as weak sulfuric acid, H_2SO_4. Whatever its source, sulfur in a soluble form is taken up by plants and is incorporated through a series of metabolic processes, starting with photosynthesis, into such sulfur-bearing amino acids as cystine. From the producers the sulfur in amino acids is transferred to the consumer groups (Fig. 5-12).

Excretions and death carry sulfur in living material back to the soil and to the bottoms of ponds, lakes, and seas where the organic material is acted upon by bacteria, releasing the sulfur as hydrogen sulfide or sulfate. One group, the colorless sulfur bacteria, both reduces hydrogen sulfide to elementary sulfur and oxides it to sulfuric acid. The green and purple bacteria, in the presence of light, utilize hydrogen sulfide as an oxygen acceptor in the photosynthetic reduction of carbon dioxide. Best known are the purple bacteria found in salt marshes and in the mud flats of estuaries. These organisms are able to carry the oxida-

tion of hydrogen sulfide as far as sulfate, which may be recirculated and taken up by the producers or may be used by sulfate-reducing bacteria. The green sulfur bacteria can carry the reduction of hydrogen sulfide only to elementary sulfur.

Sulfur, S^{2-}, in the presence of iron and under anaerobic conditions, will precipitate as ferrous sulfide, FeS_2. This compound is highly insoluble under neutral and alkaline conditions and is firmly held in mud and wet soil. Some ferrous sulfide is contained in sedimentary rocks overlying coal deposits. Exposed to the air in deep and surface mining, the ferrous sulfide oxidizes and in the presence of water produces ferrous sulfate and sulfuric acid:

$$2FeS_2 + 7O_2 + 2H_2O \rightarrow 2FeSO_4 + 2H_2SO_4$$

Other reactions in nature may be more nearly the following:

$$12FeSO_4 + 3O_2 + 6H_2O \rightarrow$$
$$4Fe_2(SO_4)_3 + 4Fe(OH_3) \downarrow$$

In this manner sulfur in pyritic rocks, suddenly exposed to weathering by man, discharges heavy slugs of sulfur, sulfuric acid, ferric sulfate, and ferrous hydroxide into aquatic ecosystems. These compounds destroy aquatic life and have converted hundreds of miles of streams and rivers in eastern United States to highly acidic water.

Global cycling. Sulfur circulates globally through the world's hydrological cycle and through the atmosphere as a gas. Eriksson (1963) has calculated that in the past approximately 76 g S/cm² of the earth's surface has been released from igneous rocks. The sea alone contains 246 g S/cm² of the earth's surface; and sedimentary rocks contain another 261 g/cm². Thus of these 500 g S/cm², 76 g/cm² came from igneous rocks. The remainder either was in the atmosphere or was released by volcanic eruptions.

The global circulation of sulfur is complex (Fig. 5-13). As with nitrogen, oxygen, and other gaseous cycles, the biosphere plays an important role. At the same time the sulfur cycle operates within the sedimentary cycle. Sources of sulfur include the weathering of rocks, especially pyrites, erosional runoff, industrial production, and decomposition of organic matter. The bulk of sulfur appears first as a volatile gas, hydrogen sulfide. In the hydrosphere, the soil, and the atmosphere hydrogen sulfide is oxidized to sufides and sulfates, the forms in which sulfur is most readily circulated. The atmosphere contains sulfate particles, sulfur dioxide, and hydrogen sulfide. The latter is most abundant over continents. The concentration of sulfur as hydrogen sulfide in the unpolluted atmosphere is estimated at 6 g/m³; as sulfur dioxide at 1 g/m³. Part of the sulfur in the atmosphere is recirculated to land and sea by rainwater. The concentration of sulfur dioxide in rain falling over land has been estimated as 0.6 mg/liter, and over sea as 0.2 mg/liter, excluding sea spray.

It is almost impossible to estimate the biological turnover of sulfur dioxide because of the complicated cycling within the biosphere. Ericksson estimates that net annual assimilation of sulfur by marine plants is on the order of 130 million tons. Added to the anaerobic oxidation of organic matter, this brings the total to an estimated 200 million tons. Both industrially emitted sulfur and fertilizer sulfur are eventually carried to the sea; therefore these two sources probably account for the 50-million-ton annual increase of sulfur in the ocean. The balance sheet of sulfur in the global cycle is summarized in Table 5-4.

THE PHOSPHORUS CYCLE

The phosphorus cycle differs from the sulfur cycle in that the element is unknown in the atmosphere, and none of its known compounds

FIGURE 5-12 [OPPOSITE]
The sulfur cycle. Major sources are burning of fossil fuels and acid mine water from coal mines.

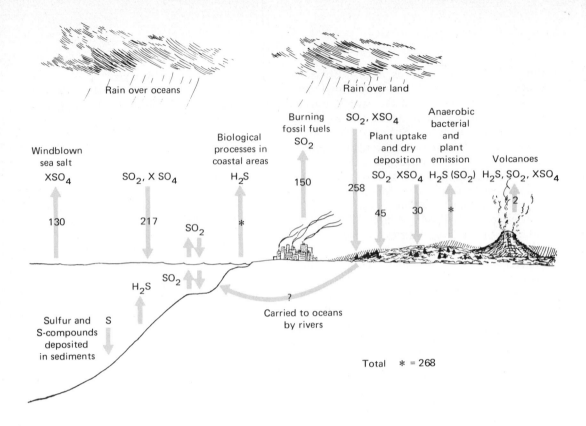

Rain over oceans

Rain over land

Windblown
sea salt

XSO_4

130

SO_2, X SO_4

217

SO_2

H_2S

Biological
processes in
coastal areas

H_2S

*

Burning
fossil fuels

SO_2

150

SO_2, XSO_4

258

SO_2

Plant uptake
and dry
deposition

SO_2 XSO_4

45

Anaerobic
bacterial
and
plant
emission

H_2S (SO_2)

30

*

Volcanoes

H_2S, SO_2, XSO_4

2

SO_2

Sulfur and
S-compounds
deposited
in sediments

S

Carried to oceans
by rivers

?

Total   * = 268

have any appreciable vapor pressure. Thus phosphorus can follow the hydrological cycle only part-way from land to sea (Fig. 5-14). Under undisturbed natural conditions, phosphorus is in short supply. It is freely soluble only in acid solutions and under reducing conditions. In the soil it becomes immobilized as phosphates of either calcium or iron. Even superphosphate applied to croplands may be rapidly converted to unavailable inorganic compounds. Its natural limitation in aquatic ecosystems is emphasized by the almost explosive growth of algae in water receiving heavy discharges of phosphorus-rich wastes.

The main reservoirs of phosphorus are rock and natural phosphate deposits, from which the elements are released by weathering, by leaching, by erosion, and by mining for agricultural use. Some of it passes through terrestrial and aquatic ecosystems by way of plants, grazers, predators, and parasites; and it is returned to the ecosystem by excretion, death, and decay. In terrestrial ecosystems organic phosphates are reduced by bacteria to inorganic phosphates. Some are recycled to plants, some become immobilized as unavailable chemical compounds, and some are immobilized by incorporation into bodies of microorganisms. Some of the phosphorus of terrestrial ecosystems escapes to lakes and

seas, both as organic phosphates and as particulate organic matter.

In marine and fresh-water ecosystems, the phosphorus cycle moves through three compartments: particulate organic phosphorus, dissolved organic phosphates, and inorganic phosphates. Inorganic phosphates in the water are taken up rather rapidly, often in as short a time as 5 minutes, by all forms of phytoplankton. The phosphorus in the phytoplankton may be ingested by zooplankton or detritus-feeding organisms. Zooplankton in turn may excrete as much phosphorus daily as is stored in their biomass (Pomeroy et al., 1963). By excreting phosphorus it is instrumental in keeping the cycle going. More than half of the phosphorus zooplankton excrete is inorganic phosphate, which is taken up by phytoplankton. In some instances 80 percent of this excreted phosphorus is sufficient to meet the needs of the phytoplankton population. The remainder of the phosphorus in aquatic ecosystems is in organic compounds that may be utilized by bacteria, which fail to regenerate much dissolved inorganic phosphate. Bacteria are consumed by the microbial grazers, which then excrete the phosphate they ingest (Johannes, 1968). Part of the phosphorus is deposited in shallow sediments and part in the deep. In the ocean some of the latter may be recirculated by upwelling, which brings the phosphates from the unlighted depths to the photosynthetic zones, where it is taken up by phytoplankton. Part of the phosphorus contained in the bodies of plants and animals is deposited in shallow sediments, and part in the deeper ones. As a result the surface waters may become depleted of phosphorus, and the deep waters become saturated. Because phosphorus is precipitated largely as calcium compounds, much of it becomes immobilized for long periods in the bottom sediments. Upwelling returns some of it to the photosynthetic zones, where it is available to phytoplankton. The amount available is limited by the insolubility of calcium phosphate.

In the intertidal salt marshes of the southeastern United States, marsh grass, *Spartina alterniflora*, withdraws phosphorus from subsurface sediments. When *Spartina* dies the abundant animal population in marsh and adjacent waters and tidal creeks use the detritus as food. The excretions of these animals add phosphorus to the water, which is taken up by the phytoplankton and eventually by the zooplankton (Pomeroy et al., 1969).

One of the detritus-feeders is the ribbed mussel, *Modiolus dimissus*. It plays a major role in the turnover of phosphorus through the ecosystem (Kuenzler, 1961). To obtain its food, which consists of small organisms as well as particles rich in phosphorus suspended in the tidal waters, the mussels must filter great quantities of seawater. Some of the particles they ingest, but most are rejected and deposited as sediment on the intertidal mud. These particles, rich in phosphorus, are then retained in the salt marsh instead of being carried out to sea. Each day the mussels remove from the water one-third of the phosphorus found in suspended matter. Or to state it a little more precisely, every 2.6 days the mussels remove as much phosphorus as is found on the average in the particles in the water. The particulate matter deposited on the mud is utilized by the deposit-feeders, which release the phosphate back to the ecosystem. Thus the ribbed mussel, although of little economic importance to man and relatively unimportant as an energy consumer in the salt marsh, plays a major role in the cycling and retention of phosphates in the salt marsh. The worth of an animal cannot always be measured in terms of economic values or ignored because it contributes little to energy

FIGURE 5-13 [OPPOSITE]
The sources and sinks of atmospheric sulfur compounds. Units are 10^6 tons calculated as sulfate per year. (After Kellogg et al., 1972. © 1972 American Association for the Advancement of Science.)

TABLE 5-4
The yearly budget of sulfur in nature

Item	Atmosphere To	Atmosphere From	Lithosphere (Sediment. rocks) To	Lithosphere (Sediment. rocks) From	Pedosphere To	Pedosphere From	Hydrosphere (oceans) To	Hydrosphere (oceans) From
River discharge						80	80	
Weathering				15	15			
Fertilizers				10	10			
Precipitation		165			65		100	
Sea spray	45							45
Dry deposition		200			100		100	
Sedimentation			15					15
Industrial	40			40				
Increase in sea			50[a]					50
Balance								
Soils—atmosphere	110					110		
Oceans—atmosphere	170							170
Total	365	365	65	65	190	190	280	280
Specification								
As SO$_4$ sulfur	45	165			90	80	180	95
As SO$_2$ sulfur	40	200			100		100	
As H$_2$S sulfur	280					110	170	
As other forms of sulfur			65	65				15

Source: E. Erikisson, 1963. Copyright by American Geophysical Union.
Note: Values in million tons.
[a] For the balance this has to be treated as an item borrowed by the ocean from the lithosphere.

flow. It may serve some other important ecological function in the community that remains to be discovered.

Phosphorus, like other elements, may take diverse pathways through an ecosystem. Its movements through the several levels can be followed by the use of [32]P (phosphorus 32) as a tracer. By adding carefully regulated amounts of ^{32}P, Ball and Hooper (1963) followed the movements of this element through a trout-stream ecosystem. The uptake was rather rapid, the material traveling from 453 to 11,263 yd downstream, depending upon conditions, before being finally removed. Microscopic plants (periphyton) growing on the rocks and other substrate, and three species of larger plants, *Potamogeton*, the alga *Chara*, and water moss *Fontanalis*, were responsible for most of the uptake. Maximum amounts of radioactive phosphorus appeared in plant tissues shortly after the material passed through the area; and the rate of loss was the greatest shortly thereafter. Losses decreased with time for almost 15 to 20 days, when equilibrium was achieved. This suggested that the plants were recycling phosphorus. Concentration of phosphorus in consumer organisms reflected

differences in both metabolic turnover rates and in food relationships. Small filter feeders, especially the black fly larvae, reached the highest level of concentration in the shortest time, followed by the animals that scraped periphyton from the rocks. At the same time these organisms lost ^{32}P quite rapidly. The material persisted longer in such omnivorous feeders as the scud and the caddisfly. Phosphorus-32 appeared the latest and was retained the longest in the invertebrate and vertebrate predators. The investigators found considerable variation in the uptake and retention of ^{32}P by different plants and animals from year to year. This suggested that the major differences in the cycling of radioactivity was related to the way ^{32}P distributed itself between soluble and particulate phases in the streamwater.

Seasonal changes in phosphorus sources have been described for a marine ecosystem (Ketchum and Corwin, 1965; Ketchum, 1967). In the prebloom period 28 percent of the phosphorus supply necessary for phytoplankton production was supplied by the inorganic fraction in solution in the euphotic zone or upper water layer and 72 percent was supplied by vertical mixing and transport of nutrients from deeper water. During the period of bloom 86 percent of the phosphorus came from the inorganic phosphorus dissolved in solution, 12 percent from vertical mixing, and 2 percent from regeneration by biological cycling. When the system was nearly in equilibrium, 43 percent of the phosphorus was supplied by regeneration and 57 percent by vertical mixing. The distribution of the phosphorus (Table 5-5) under the three environmental situations tells something about the cycling of phophorus. During the prebloom period 13 percent of the phosphorus assimilated was located on the particulate matter about equally divided between the upper and lower layers; during the bloom, 92 percent was located in particulate matter, of which about 47 percent remained in the upper water layers; and during the steady-state condition only 3 percent was in particulate form. Thus during the bloom stage much of the phosphorus used is tied up in organic matter and only by a rapid turnover of phytoplankton populations could the phosphorus requirements be met. The importance of turnover is further emphasized by the fact that under steady-state conditions only a small part of the phosphorus assimilated in production is in living matter—most of it is supplied by regeneration and vertical mixing from the deep.

The phosphorus requirements of marine phytoplankton can be met only by biological regeneration of nutrient and vertical mixing from the deep. Regeneration can come about only by a rapid recycling in the aquatic ecosystem, which in coastal waters may occur six to ten times a year.

Turnover period, the time in which a given quantity of an element is utilized, also shows a seasonal pattern. During the summer the turnover time for phosphorus among the several lakes studied by F. H. Rigler (1956) ranged on the average from 0.9 to 7.5 minutes. As fall and winter approached, turnover time slowed down. Average times ranged from 7 minutes to 7 days. In marine ecosystems these turnover rates are slower. The average turnover time in summer seas ranges from 1 to 56 hours and in winter, 30 to 169 hours (Pomeroy, 1959).

As with other biogeochemical cycles, man's activities have altered the phosphorus cycle. Because the cropping of vegetation depletes the natural supply of phosphorus in the soil,

TABLE 5-5
Sources of supply of phosphorus under three environmental conditions

	Prebloom	Bloom	Steady state
Total phosphorus cycle, in mg at/m^2/day	0.89	1.65	0.54
Source			
Removal, %	28	86	0[a]
Vertical mixing, %	72	12	57
Regeneration, %	0	2	43

Source: B. H. Ketchum, 1967.
[a] Inorganic phosphate increased in the euphotic zone by 0.22 mg at m^2/day. This amount was also supplied by vertical mixing.

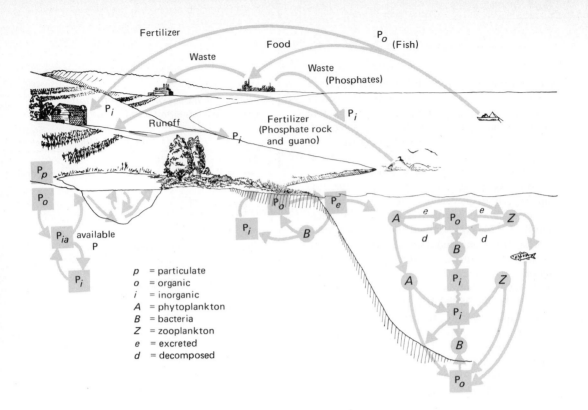

Fertilizer

Food

P_o (Fish)

Waste

Waste
(Phosphates)

P_i

Runoff

Fertilizer
(Phosphate rock
and guano)

P_i

P_i

P_p

P_o

P_{ia} available
P

P_i

P_o

P_e

P_i

B

A P_o Z

B

A P_i Z

P_i

B

P_o

e e

d d

p = particulate
o = organic
i = inorganic
A = phytoplankton
B = bacteria
Z = zooplankton
e = excreted
d = decomposed

phosphate fertilizers must be added. The source of phosphate fertilizer is phosphate rock. Because of the abundance of calcium, iron, and ammonium in the soil, most of the phosphate applied as fertilizer becomes immobilized as insoluble salts. In 1968, for example, 50 percent more fertilizer was added to cropland than was lost to the oceans from global runoff from all sources.

Part of the phosphorus used as fertilizer is removed in crops when harvested. Transported far from the point of fixation, this phosphorus in vegetables and grain eventually is released as waste when foodstuffs are processed or consumed. Concentration of phosphorus in wastes of food processing plants and of feedlots adds a quantity of phosphates to natural waters. Greater quantities are supplied by urban areas, where phosphates are concentrated in sewage systems. Sewage treatment is only 30 percent effective in removing phosphorus, so 70 percent remains in the effluent and is added to the waterways. In the equatic ecosystems the phosphorus is taken up rapidly by the aquatic vegetation, resulting in a great increase in biomass. Unless new input of phosphorus continues, as it would through sewage effluents, the phosphorus is lost to the sediments. Eventually all phosphorus mobilized by man becomes immobilized in the soil or in the bottom sediments of ponds, lakes, and seas.

Phosphorus requirements for agricultural production depend largely on the utilization of natural deposits of phosphate rock and to a lesser extent upon the harvest of fish and guano deposits. Because so much of the phosphate applied to soil is eventually immobilized in deep sediments and because the activity of phytoplankton seems inadequate to keep phosphorus in circulation, more of the element is being lost to the depths of the sea than is being added to terrestrial and fresh-water aquatic ecosystems (Hutchinson, 1957).

Nutrient budgets

Constantly, nutrients are being removed or added by natural and artificial processes (Fig. 5-15). In woodland, shrub, and grassland ecosystems, nutrients are returned annually to the soil by leaves, litter, roots, animal excreta, and bodies of the dead. Released to the soil by decomposition, these nutrients again are taken up first by plants and then by animals. In fresh water and the sea, the remains of plants and animals drift to the bottom, where decomposition takes place. The nutrients again are recirculated to the upper layers by the annual overturns and by upwellings from the deep.

THE TEMPERATE FOREST

The cycle, however, is not a closed circuit within an ecosystem. Nutrients are continuously being imported, as well as carried out of ecosystems. Appreciable quantities of plant nutrients are carried in by rain and snow (Emanuelsson, Eriksson, and Egner, 1954) and by aerosols (White and Turner, 1970). In western Europe, at least, the weight of nutrients supplied by these sources is roughly equivalent to the quantity removed in timber harvest (Neuwirth, 1957). A small quantity of the nutrients carried to the forest by rain and snow is absorbed directly through the leaves, but this hardly offsets the quantity leached out. Rainwater dripping down from the canopy is richer in calcium, sodium, potassium, phosphorus, iron, manganese, and silica than rainwater collected in the open at the same time, although less rainwater reaches the forest floor (Tamm, 1951; Madgwick and Ovington, 1959). Throughfall of rain in an oak woodland accounted for 17 percent of the nitrogen, 37 percent of the phosphorus, 72 percent of the potassium, and 97 percent of the sodium added by the canopy to the soil. The remainder was supplied by fallen leaves (Carlisle et al., 1966). These nutrients leached from the foliage are taken up in time by the surface roots and translocated to the canopy. Such localized nutrient cycles may require only a few days.

For some elements, the amount carried in by aerosols may exceed that carried in by the rain. Estimates of the annual income of nutrients to an English mixed deciduous woodland by airborne particles were 125.2 kg Na/ha, 6.3 kg K/ha, 4.2 kg Ca/ha, 16.2 kg Mg/ha, and 0.34 kg P/ha. The income from aerosols was greater for elements known to occur in droplets, such as sodium, potassium, and magnesium. The income of calcium, associated with terrestrial sources, and phosphorus, associated with biological activity, was greater in rainfall (White and Turner, 1970).

These little nutrient cycles can be followed by means of radioactive tracers. By inoculating white oak trees with 20 microcuries of ^{134}Cs (cesium-134), Witherspoon and others (1962) were able to follow the gains, losses, and transfers of this radioisotope. About 40 percent of the ^{134}Cs inoculated into the oaks in April moved into the leaves by early June

FIGURE 5-14 [OPPOSITE]
The phosphorus cycle in aquatic and terrestrial ecosystems.

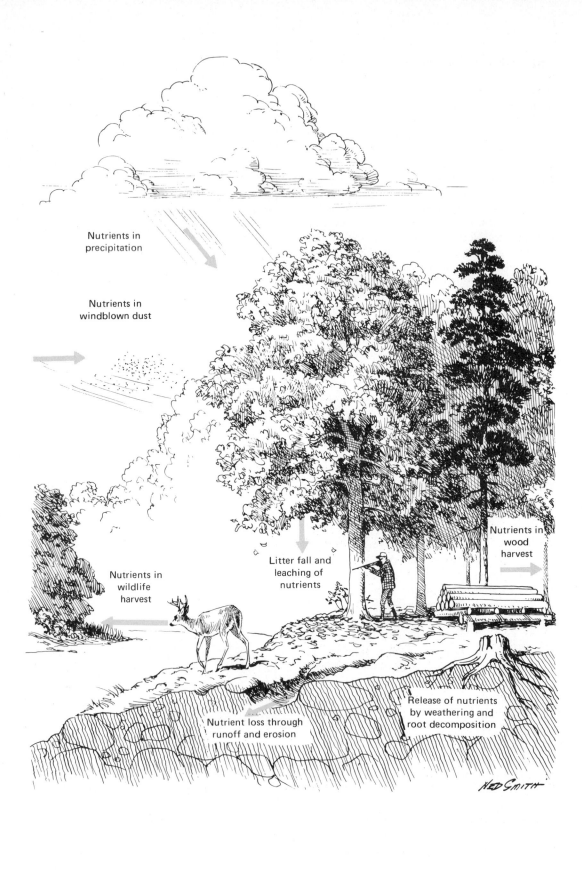

Nutrients in
precipitation

Nutrients in
windblown dust

Litter fall and
leaching of
nutrients

Nutrients in
wood
harvest

Nutrients in
wildlife
harvest

Nutrient loss through
runoff and erosion

Release of nutrients
by weathering and
root decomposition

NED SMITH

(Fig. 5-16). When the first rains fell after inoculation, leaching of radiocesium from the leaves began. By September this loss amounted to 15 percent of the maximum concentration in the leaves. Seventy percent of this rainwater loss reached mineral soil; the remaining 30 percent found its way into the litter and understory. When the leaves fell in autumn, they carried with them twice as much radiocesium as was leached from the crown by rain. Over the winter, half of this was leached out to mineral soil. Of the radiocesium in the soil, 92 percent still remained in the upper 4 in. nearly 2 years after the inoculation. Eighty percent of the cesium was confined to an area within the crown perimeter and 19 percent was located in a small area around the trunk. This suggests that cesium distribution in the soil was greatly influenced by leaching from rainfall and stem flow.

The use of ^{137}Cs to trace the pathway of mineral cycling in the forest was carried one step further in a study of cesium cycling in a tulip poplar (*Liriodendron tulipifera*) stand (Olsen et al., 1970). Dominant trees were tagged with ^{137}Cs. The fate of the tracer was continuously monitored for 5 years, and its pathway through the forest ecosystem was traced as illustrated in Fig. 5-17. The leaves were richest in cesium in the spring and poorest in the fall. The reason for the decline in the leaves was the movement of ^{137}Cs from the foliage to the woody tissue when the leaves were mature and senescent. There was a cumulative loss of the tracer from the foliage both from rainfall and litter fall, but the greatest transfer to the soil was through the roots. The ^{137}Cs accumulated in the litter and organic layers of the forest floor in the upper 10 cm of the mineral soil. From here the cesium was picked up by decomposers, soil organisms, and litter-feeding arthropods and their predators.

To trace the pathway of a single radioactive tracer element through an individual or through some part of the ecosystem is one thing. To attempt to measure the input and outflow of the mineral elements through the various biotic components of an ecosystem, and come up with a budget for the ecosystem, is quite another question. Such a study is loaded with difficulties, but some estimates for the primary producer component can be made by sampling a number of trees of the various species, by determining the nutrient distribu-

FIGURE 5-15 [OPPOSITE]
A generalized nutrient cycle in a forest ecosystem. Input includes nutrients carried in by precipitation and windblown dust, through litter fall, and through the release of nutrients by weathering and root decomposition. Outgo is through wood harvest, wildlife harvest, runoff, erosion, and leaching.

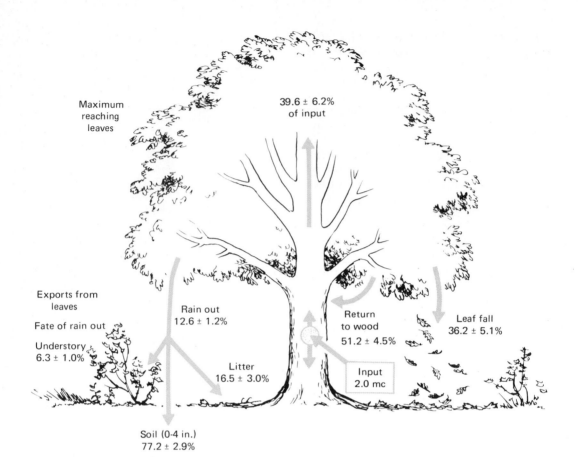

Maximum
reaching
leaves

39.6 ± 6.2%
of input

Exports from
leaves

Fate of rain out

Understory
6.3 ± 1.0%

Rain out
12.6 ± 1.2%

Return
to wood
51.2 ± 4.5%

Leaf fall
36.2 ± 5.1%

Litter
16.5 ± 3.0%

Input
2.0 mc

Soil (0-4 in.)
77.2 ± 2.9%

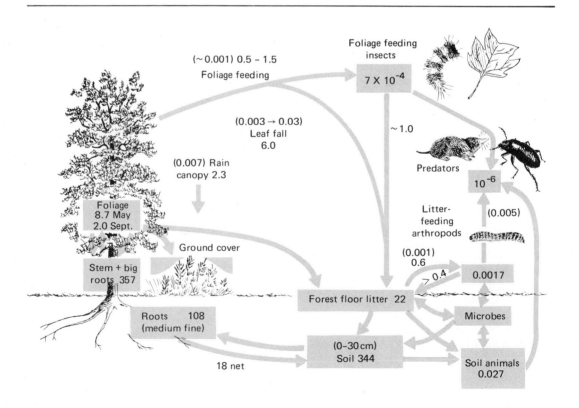

(~0.001) 0.5 – 1.5
Foliage feeding

Foliage feeding
insects

7×10^{-4}

(0.003 → 0.03)
Leaf fall
6.0

~1.0

(0.007) Rain
canopy 2.3

Predators

10^{-6}

Foliage
8.7 May
2.0 Sept.

Litter-
feeding
arthropods

(0.005)

Ground cover

(0.001)
0.6

Stem + big
roots 357

> 0.4

0.0017

Forest floor litter 22

Microbes

Roots   108
(medium fine)

(0–30 cm)
Soil 344

Soil animals
0.027

18 net

tion within the biomass, and by measuring mineral content of leaf wash and stemflow. Transfers from the forest soil to the roots can be measured by collecting samples of soil solution with tension lysimeters. The uptake by forest vegetation can be estimated by sampling the mineral content of current year's increment of growth, including foliage, branches, and bole. A nutrient budget, summarized in Table 5-6, shows that 10 to 15 percent of the nutrients held by the Douglas fir is located in the roots and stump below ground level. Mineral accumulation in the bark is relatively high. The understory plants contain a minimal proportion of the nutrients.

The budget also points out that the internal cycle is more important to nutrient budget than additions to or losses from precipitation. A portion of the nutrients is stored in the tree, then returned to the forest floor by way of litter fall, stemflow, and leaf wash. Ten percent of nitrogen, 50 percent of phosphorus, 71 percent of potassium, and 22 percent of calcium are returned by this route. The annual accumulation of calcium, phosphorus, potassium, and nitrogen is considerable both in trees and the forest floor. The nutrient cycle can continue only by withdrawing elements from the soil at the annual rate of 34.6 kg N/ha, 6.3 kg P/ha, 19.9 kg K/ha, and 11.5 kg Ca/ha. The time required for total soil depletion at this rate is summarized in Table 5-7. This depletion is speculative because it does not take into account change through time in the transfer rate between components of the ecosystem or the addition of elements by the way of nitrogen fixation, mineral solubility, and accumulation of organic matter (Cole et al., 1969). This budget suggests that Douglas fir is an accumulator plant (Fortescue and Martin, 1970). Accumulator plants are important in the ecosystem because they can remove elements from the soil in quantities large enough to upset the nutritional balance in the ecosystem.

A more extensive nutrient budget is one determined for an oak wood in Belgium by Duvigneaud and Denaeyer-DeSmet (1970), in which more than one species of tree is involved. The budget is presented pictorially in Fig. 5-18, and a summary of the annual budget appears in Table 5-8. Again there is considerable retention relative to total uptake.

These and other budgets, such as those de-

FIGURE 5-16 [OPPOSITE TOP]
The cycle of ^{134}Cs in white oak as an example of nutrient cycling through plants. The figures are the average of 12 trees at the end of the 1960 growing season. (After Witherspoon et al., 1962; courtesy Oak Ridge National Laboratory.)

FIGURE 5-17 [OPPOSITE BOTTOM]
The biogeochemical cycle of ^{137}Cs in a tulip poplar (Liriodendron tulipifera) forest ecosystem. The trees of this forest were originally tagged with a total of 467 microcuries of ^{137}Cs in May 1962. Continuous inventory followed the seasonal and annual distributions and fluxes of ^{137}Cs among the biotic and abiotic components of the ecosystem. Data in this figure are for the 1965 season. Numbers in the boxes are microcuries of ^{137}Cs per square meter of ground surface. Arrows indicate the pathways of ^{137}Cs transfer between compartments. Numbers by the arrows are estimates of annual fluxes; numbers in parentheses are the averaged transfer coefficients ($days^{-1}$) of ^{137}Cs for the seasons during which the process occurs. (Courtesy Oak Ridge National Laboratory.)

TABLE 5-6
Budget of selected nutrients for second-growth Douglas fir forest

Erosystem component	N (kg/ha)	% of total	P (kg/ha)	% of total	K (kg/ha)	%of total	Ca[a] (kg/ha)	%of total
TREES								
Total	320	9.7	66	1.7	220	44.6	333	27.3
Distribution								
Foliage	102	31.9	29	43.9	62	28.2	73	21.9
Branches	61	19.1	12	18.2	38	17.3	106	31.8
Wood	77	24.0	9	13.6	52	23.6	47	14.1
Bark	48	15.0	10	15.2	44	20.0	70	21.0
Roots	32	10.0	6	9.1	24	10.9	37	11.2
SUBVEGETATION								
Total	6	0.2	1	0.1	7	1.4	9	0.7
FOREST FLOOR								
Total	175	5.3	26	0.6	32	6.5	137	11.2
SOIL								
Total	2,809	84.8	3,878	97.6	234	47.5	741	60.8
INPUT BY PRECIPITATION	1.1		trace		0.8		2.8	—
UPTAKE BY FOREST FROM SOIL								
Total	38.8		7.2		29.4		24.4	—
Distribution								
Foliage	24.3		4.7		16.2		17.8	—
Branches	4.2		0.8		2.7		2.6	—
Trunk	10.3		1.7		10.5		4.0	—
RETURN TO FOREST FLOOR								
Total	15.3		0.6		15.0		15.7	—
Distribution								
Litter fall	13.6		0.2		2.7		11.1	—
Stem flow	0.2		0.1		1.6		1.1	—
Leaf wash	1.5		0.3		10.7		3.5	—
LEACHED FROM FOREST FLOOR	4.8		0.95		10.5		17.4	—
LEACHED BEYOND ROOTING ZONE	0.6		0.02		1.0		4.5	—
ANNUAL ACCUMULATION IN ECOSYSTEM								
Forest	23.5		6.6		14.4		8.7	—
Forest floor	11.6		−0.4		5.3		1.1	—
Soil	−34.6		−6.3		−19.9		−11.5	—

Source: D. W. Cole et al., 1967.
[a] Exchangeable components only.

veloped for an oak forest at Brookhaven (Whittaker and Woodwell, 1969), of a Scots pine forest (Ovington, 1959), and of a northern hardwoods forest (Likens et al., 1969) suggest several characteristics of mineral cycling in forest ecosystems. The accumulation of nutrients in the rooting zone of the soil and in the tree biomass making them unavailable slows down the biological cycle of minerals. Thus the rate of biological cycling can be maintained only by the pumping of nutrients from deep soil reserves, and the weathering of parent rock. Nutrient cycling is influenced by the climate, the nature of soil, and the nature of the plants occupying the area, including their ability to pump, transport, store, and return certain mineral elements Seasonal variation in utilization, storage, and retention complicate the cycling. Except for a sharp

peak in mineral uptake, which occurs when the productivity of the forest seems to be at its peak, age seems to have little influence on mineral cycling.

The gain to the ecosystem from precipitation, extraneous material, and mineral weathering is offset by losses. Water draining from forests carries away more mineral matter than is supplied by precipitation (Viro, 1953). In the Brookhaven oak–pine forest, for example, total input from precipitation and dust of the four *cations*, K^+, Ca^{2+}, Mg^{2+}, and Na^+ amounted to 2.48 g/m^2, whereas losses to the water table ranged from 3.68 to 5.16 g/m^2. Losses from the Hubbard Brook Forest almost equaled the input (Table 5-9). As already indicated, considerable quantities of nutrients are tied up in forest trees. Assuming that all twigs, litter, and other parts are left behind to decay in the forest, the harvest of oak timber from the 89-year-old European stand would remove 51 percent of accumulated nutrients potassium, calcium, magnesium, nitrogen, sulfur, and phosphorus; and if the 36-year-old Douglas fir stand were cut, 20 percent of nitrogen, calcium, phosphorus, and potassium would be taken from the system.

In addition to losses brought about by the actual removal of timber from the ecosystem, there are additional losses from erosion and leaching. Instability resulting from the harvesting of a forest shows up in the water losses. When an experimental watershed on the Hubbard Brook Experimental Forest in New Hampshire was clear-cut, evapotranspiration decreased 68 percent and runoff increased 40 percent. As one might predict, most of the increase in flow occurred during the spring and summer, when loss of water through the leaves would have been the greatest.

An increased amount of water passing through and out of the ecosystem also means an increased amount of nutrients lost from the clear-cut forest ecosystem. Because the water is no longer being withdrawn from the soil by the trees during the growing season, more water is able to percolate through the soil, dissolving the nutrients and carrying them in solution. With the death of the trees, the amount of roots and root surface has been reduced. This in turn means that smaller portions of the nutrient in the soil water will be removed by the roots from the leaching water. At the same time there is relatively little vegetation left to take up the nutrient.

Because a change in the physical environment alters the organismal balance, it can have a pronounced effect on nutrient depletion. Loss of vegetation and the subsequent increased surface temperatures may speed up action of decomposers and organisms. This enables them to convert the organic matter rapidly into mineral form, matter that is then dissolved in the soil water. The amount of calcium ions, magnesium ions, sodium ions, and potassium ions lost by subsurface runoff in the Hubbard Brook Forest were 9, 8, 3, and 20 times greater respectively than the amount lost from undisturbed forests (Likens et al., 1969). There was an even greater loss of nitrogen. Once the vegetation was removed, ammonia was converted rapidly to nitrate. This process caused the formation of both ni-

TABLE 5-7

Utilization of N, P, K, Ca from soil by second-growth Douglas fir

Element	% yearly uptake[a]	Static supply (year)[b]	Cyclic supply (year)
N	1.4	73	125
P	0.2	537	582
K	12.5	8	12
Ca	3.3	30	64

Source: D. W. Cole et al., 1967.
[a] Yearly uptake divided by total in ecosystem; includes input by precipitation and return by litter stem flow, and wash.
[b] Years: Number of years supply will last.

TABLE 5-8

Annual element balance (kg/ha) of the Belgium mixed-oak forest

	K	Ca	Mg	N	P	S
Retained	16	74	5.6	30	2.2	4.4
Returned	53	127	13	62	4.7	8.6
Uptake	69	201	18.6	92	6.9	13

Source: Data from P. Duvigneaud and S. Denaeyer-DeSmet, 1970.

TABLE 5-9

Input and losses of nutrients of the Hubbard Brook Forest expressed in kilograms per hectare

	Ca	Mg	Na	K
Input	3.0 ± 0	0.7 ± 0	1.0 ± 0	2.5 ± 0
Output	8.0 ± 0.5	2.6 ± 0.06	5.9 ± 0.3	1.8 ± 0.1
Loss	5.0 ± 0.5	1.9 ± 0.06	4.9 ± 0.3	0.7 ± 0.1

Source: Data from G. E. Likens et al., 1967.

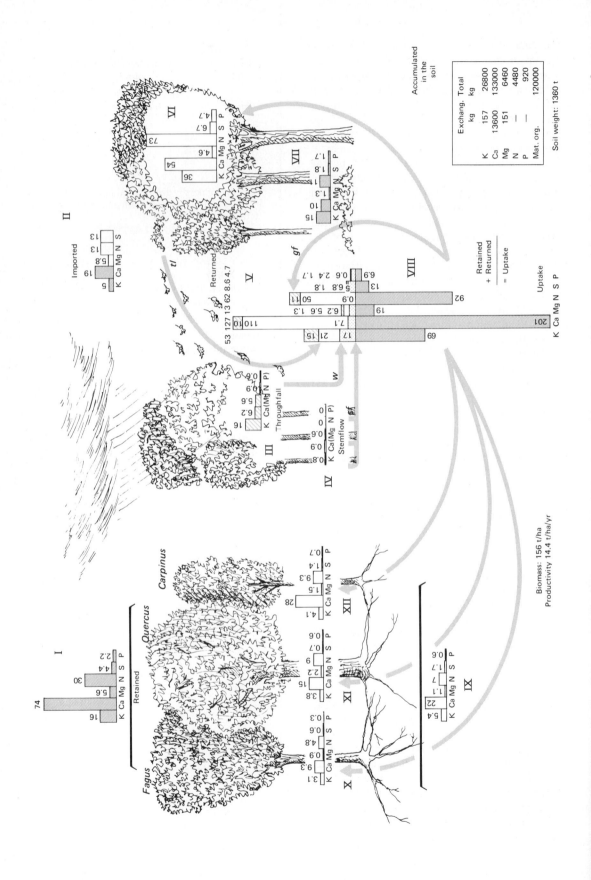

trate and hydrogen ions. The greater concentration of hydrogen ions increased the acidity (lowered the pH) of the soil, making the calcium, magnesium, sodium, and potassium more soluble. Once dissolved, they were carried away from the ecosystem. The amount of nitrogen carried from the clear-cut forest in the form of nitrates was 100 times greater than the nitrogen loss of the undisturbed forest.

Once there was some regrowth of vegetation, these nutrient losses were halted; minerals once again accumulated in vegetation, and were held in the system by tight internal cycling. The losses on the Hubbard Brook Forest clear-cutting were intensified and prolonged by the use of the herbicide Bromacil, which inhibited the immediate regrowth of vegetation. Under normal conditions the herbaceous plants, saplings, and sprout growth that come in after cutting restore the nitrogen cycle (Marks and Bormann, 1972).

Budgets of nutrients vary widely among ecosystems. A number of budgets are described by Rodin and Bazilevic (1967). An average uptake of nitrogen from the soil into organic matter ranges on the average from 1.8 g/m² for the cold desert to 42.7 g/m² for a wet tropical forest. Both cold and aridity limit the uptake of nitrogen and other nutrients. Annual loss of nitrogen as litter is complete for nonwoody vegetation to 26.1 g/m² in wet tropical forests. Retention in biomass may amount to 4 to 6 g/m²/year in forest ecosystems. An interesting contrast in budgets of nitrogen exists between two tropical ecosystems, which points out the wide variation in nutrient budgets that exist under somewhat the same light and moisture conditions. One ecosystem is the Middle Asian tuagi (a Russian term), an irrigated, fertilized vegetational complex of river bottomland that consists of grass, shrubs, and lianas. The other ecosystem is wet tropical forest. Both ecosystems have the advantage of high radiation and plenty of moisture, the former by both irrigation and flooding, the latter by rainfall. The differences in the budgets of nitrogen reflect the difference in vegetation (Table 5-10). The tuagi takes up four times as much nitrogen as the wet tropical forest and returns over six times as much through litter fall.

Nutrient budgets naturally reflect the type of ecosystem, the climate, the vegetation, and the differences in the accumulation of organic matter on the top of mineral soil, all of which

FIGURE 5-18 [OPPOSITE]

Annual mineral cycling (in kilograms per hectare) of macronutrients in a mixed oak wood ecosystem at Virelles, Belgium. Retained: in annual wood and bark increment of roots and aerial parts of each species (total is hatched). Returned: by tree litter (tl), ground flora (gf), washing and leaching of the canopy (w), and stem flow (sf). Imported: by incident rainfall (not included). Macronutrients contained in crown leaves when fully grown (July) are shown on the right side of the figure in italics; these amounts are higher (except for calcium) than those returned by leaf litter. Values for magnesium, nitrogen, and phosphorus in throughfall and stemflow. Exchangeable and total element content in the soil are expressed on air-dry soil weights of particles < 2 mm. (From Duvigneaud and Denaeyer-DeSmet, 1970.)

TABLE 5-10

Nitrogen budget tropical rain forest vs. tuagi (g/m²)

	Tropical rain forest	Tuagi
Biomass	50,000	11,100
Litterfall	2,500	7,650
N stored in biomass	294	188
N Uptake	42.7	161.0
N Returned in litterfall	26.1	161.0

Source: Data from Rodin and Bazilevic, 1967.

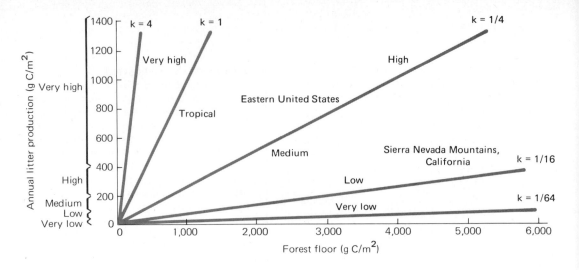

affect the rates of decomposition. Differences in decomposition in relation to the amount of forest litter in different forest ecosystems are summarized in Fig. 5-19. The rate parameter k evaluates the effectiveness of microbial activity on the breakdown of litter. The graphs illustrate that the most rapid cycling takes place in the warm, moist tropical forest, the lowest in pine forest, where the litter is characterized by high content of lignin and resins.

NUTRIENT CYCLING IN THE TROPICAL FOREST

In the temperate forest, bacteria and fungi, the main agents of decay, release the nutrients directly into the mineral soil, where they are subject to leaching. Feeder roots of the trees are woven into the matrix of mineral soil, and large volumes of litter are returned to the forest floor. In the tropical forest the volume of litter added is less. A thin layer of about 1 to 2 cm is about all that accumulates. Feeder roots of tropical trees are concentrated in the well-aerated upper 2 to 15 cm of humus, and only a few penetrate the upper layer of mineral soil (Cornforth, 1970). Associated with the roots are mycorrhizal fungi, which are symbiotically associated with the roots (hence the name mycorrhize, or fungus root). Some mycorrhizal fungi are ectotrophic: they live around the roots. Others are endotrophic, and live partly within the cells of the roots. Although their role as symbionts is not clearly defined, it appears that the mycorrhiza are capable of digesting organic matter and passing phosphates and other minerals from the soil to the roots. During the process the mycorrhiza also extract sugar and growth substances from the roots.

The roots of tropical forest trees support an abundance of endotrophic mycorrhizal fungi that attach the feeder roots to dead organic matter by hyphae and rhizomorph tissue. As a result nutrient cycling in the tropical forest appears to be through the mycorrhiza, where, according to the theory proposed by Went and Stark (1968),

. . . fungi cycle nutrients directly from dead organic matter to the living roots with only a minimum leakage into the mineral soil. The minerals lost from the closed organic system are partly reclaimed by roots and partly leached. There is probably a very slow replenishing of minerals from the sand or clay soil. . . . In bacterial decay, minerals are made soluble, released into the soil, and then taken up by the roots. In direct mineral cycling, the minerals remain tied up in living or dead organic matter, are transferred through hyphae from dead branches or leaves to living roots and very little mineral is made soluble and moved into the soil. This would explain why the feeder roots are mainly restricted to the humus layer and why there are more mycorrhizal roots in poorer soils.

If indeed the mineral cycling in tropical forests is that direct, then one can explain why tropical soils lose their fertility so rapidly under cultivation. When the forest is cleared the mycorrhiza are destroyed, mineral cycling ceases, the organic matter is quickly depleted, and the mineral-poor soil quickly loses through leaching what nutrients it did contain. The lush tropical vegetation is supported not by a rich mineral soil but by an efficient direct recycling of nutrients by mycorrhizal fungi from dead organic matter back to living plants again.

This theory is given some support by the observations of Cornforth (1970) in Trinidad. There, large areas of natural tropical evergreen seasonal forests were cleared and burned to plant Caribbean Pine. Not only were the nutritional reserves lost through burning and erosion, but the pines introduced were much less efficient in the cycling of nutrients, even though litter accumulated beneath the trees. As a result, the reserves of nitrogen, phosphorus, and potassium decreased sharply, and only calcium and potassium showed some increase. This decline in nutrient cycling can reflect the loss of the mycorrhizal fungi associated with the natural tropical forest. (Also, see Jordan and Kline, 1972).

MODELS OF NUTRIENT CYCLES

Biogeochemical cycling is inseparable from energy flow and is in fact powered by it, as illustrated in the model, Fig. 5-20. Nutrients are pumped through the system and sped on their circular path by the action of photosynthesis. Then they are made available for recycling by the action of decomposers, largely microscopic, that carry on complex chemical reactions that set the nutrients free. Nutrients follow the same pathways as energy through the food chain, passing from one trophic level to another. The difference is that a given unit of energy is ultimately dissipated, whereas nutrients to varying degrees are recycled. That is, nutrients can be returned to their original chemical form. Some of the nutrients are involved in short-term cycles, some are tied up in organic storage temporarily, and some are locked up in deep sediments and rocks.

Because the medium in which all nutrients are eventually carried through the ecosystem is water, nutrient cycles are also inseparably tied to the water cycle. Foliar leaching, rainfall, percolation through the soil, decomposition, uptake transport of nutrients in and out of an ecosystem—in fact the whole nutrient cycle—could not function without water. It is the movement of water, too, that ties the terrestrial ecosystems to the aquatic and in so doing relates the local ecosystems to the global ecosystem.

Summary

Materials flow from the living to the non-living and back to the living parts of the eco-

FIGURE 5-19 [OPPOSITE]
Estimated decomposition rates (k) of litter in evergreen forests based on the ratio of annual litter production to steady-state accumulation on the forest floor. Note that litter production in tropical forests is high but with relatively little accumulation. This contrasts with low decomposition and high accumulation in such evergreen forests as ponderosa pine that occupy colder regions. (From Olson, 1963.)

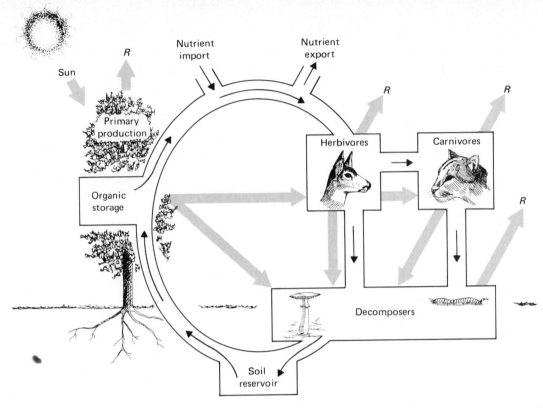

Sun

R

Nutrient import

Nutrient export

R

R

Primary production

Herbivores

Carnivores

R

Organic storage

Decomposers

Soil reservoir

system in a perpetual cycle. By means of these cycles plants and animals obtain nutrients necessary for their well-being. Plants and animals require some of these elements, the macronutrients, in relatively large quantities. Others, the trace elements, or micronutrients, are needed in lesser and often only minute quantities. Yet without them plants and animals will fail, as if they lacked one of the major nutrients. On the other hand, micronutrients in too great a quantity can be toxic.

Involved in nutrient cycling are the decomposer organisms. Most are saprophages, which exist on dead matter. True decay organisms are the heterotrophic bacteria and the fungi; but making organic matter more available to them are the reducer-decomposers. They feed on plant and animal remains and on fecal material. Feeding on the bacteria and fungi are the microbial grazers, which reduce microbial populations and thus influence microbial activity. In terrestrial ecosystems bacteria and fungi take the major role in decomposition; in aquatic ecosystems bacteria and fungi act more as converters, whereas phytoplankton and zooplankton take a major part in the cycling of nutrients.

There are two kinds of cycles, the gaseous represented by the carbon, oxygen, and

nitrogen cycles, and the sedimentary, represented by the phosphorus cycle. The sulfur cycle is a combination of the two. The carbon cycle is so closely tied to energy flow that the two are inseparable. It involves the assimilation and respiration of carbon dioxide by plants, its consumption in the form of plant and animal tissue, its release through respiration, the mineralization of litter and wood, soil respiration, and accumulation of carbon in standing crop and withdrawal into long-term reserves. The carbon dioxide cycle exhibits both diurnal and annual curves. The equilibrium of carbon dioxide exchange between land, sea, and atmosphere has been disturbed by its rapid injection into the atmosphere by man from burning of fossil fuels, but nearly two-thirds is removed from the atmosphere by land vegetation and the sea.

The nitrogen cycle is characterized by fixation of atmospheric nitrogen by nitrogen-fixing plants, largely legumes and blue-green algae. Involved in the nitrogen cycle are the processes of ammonification, nitrification, and denitrification.

The sedimentary cycle involves two phases, salt solution and rock. Minerals become available through the weathering of the earth's crust, enter the water cycle as salt solutions, take diverse pathways through the ecosystem, and ultimately return to the sea or back to the earth's crust through sedimentation. The phosphorus cycle is wholly sedimentary, with reserves coming largely from phosphate rock. Much of the phosphate used as fertilizer becomes immobilized in the soil, but great quantities are lost in detergents and other wastes carried by sewage effluents.

The sulfur cycle is a combination of the gaseous and sedimentary cycles because it has reservoirs in both the earth's crust and the atmosphere. It involves a long-term sedimentary phase in which sulfur is tied up in organic and inorganic deposits, is released by weathering and decomposition, and is carried to terrestrial and aquatic ecosystems in a salt solution. A considerable portion of sulfur is cycled in gaseous state, which permits its circulation on a global scale. Sulfur enters the atmosphere from the combustion of fossil fuel, volcanic eruption, the surface of the ocean, and gases released by decomposition. Entering the gaseous cycle initially as hydrogen sulfide, sulfur in this form quickly oxidizes to sulfur dioxide. Soluble in water, it is carried back to earth as weak sulfuric acid. Whatever the source, sulfur is taken up by plants and corporated into sulfur-bearing amino acids, later to be released by decomposition.

Thus important roles in mineral cycles are played at one end by green plants, which take up the materials, and at the other end by decomposers, which release the materials for reuse, and by air and water, in which the return trips are made.

FIGURE 5-20 [OPPOSITE]
Nutrient cycles and energy are closely interrelated, as this model indicates. It also stresses the fact that energy flow is unidirectional, whereas nutrient flow is cyclic. (After R. L. Smith, 1972, Ecology of Man, Harper & Row, New York.)

143

Ever since man discovered fire, adapted to a village way of life, and developed a technology, he has to varying degrees had an impact on biogeochemical cycles. As populations increased and as technology became more sophisticated and demanding of fossil fuels, his intrusions into the cycles became more intense (Fig. 6-1). He injected into the biosphere natural substances such as nitrogen, sulfur, mercury, and lead compounds in quantities greater than the system is able to handle. He further upset biogeochemical systems by throwing in such unnatural substances as chlorinated hydrocarbons that circulate through local and global ecosystems. These materials, injected into the biosphere in quantities so great that they affect the functioning of ecosystems and have an adverse effect on plants, animals, and man, are collectively considered pollutants.

Some pollutants enter the biogeochemical cycle by way of the water cycle; others enter a gaseous cycle and circulate for a time in the atmosphere; still others enter the soil, where they may become incorporated into plants and animals, become chemically bound in the soil, or enter into a water or gaseous cycle. Whatever their mode of entry into biogeochemical cycles, they circulate among the air, water, and sediments and move through the food chains. What is an air pollutant one day may become a water pollutant the next.

A major source of pollution is the particulate and gaseous matter released to the air by the burning of fossil fuels. Out of this comes a variety of emissions: (1) fine particles (less than 100 μ in diameter), which include carbon particles, metallic dusts, tars, resins, aerosols, solid oxides, nitrates, and sulfates; (2) coarser particles (over 100 μ), largely carbon particles and heavy dust that is quickly removed by gravity from the air; (3) sulfur compounds; (4) nitrogen compounds; (5) oxygen compounds; (6) halogens; and (7) radioactive substances.

At least five major fuel-burning sources pour these pollutants into the air. Automobiles are the greatest source, producing nearly two-thirds of the carbon monoxide and one-half of the hydrocarbons and nitrous oxides. These latter two pollutants are responsible for photochemical smog. Electrical power plants burning fossil fuels, particularly coal, produce two-thirds of the sulfur oxides. Industrial processors such as metallurgical plants and smelters, chemical plants, petroleum refineries, pulp and paper mills, and synthetic rubber manufacturing plants are responsible for about one-fifth of the air pollution. Heating plants of homes, apartments, schools, and industrial buildings are the fourth largest source of air pollution. The transportation industry, exclusive of private cars, including railroads, ships, and aircraft all contribute the same type of pollutants as cars.

Other sources of pollution, relatively minor in quantity but often important because of the substances they release, are the construction industry and agriculture, which is responsible for pesticides, dust from agricultural practices, and burning of fields. Nature, too, adds a fraction: pollen, hydrocarbons released by vegetation, and dust from deserts and volcanic activity. From an ecological viewpoint these are natural occurrences that the normal biogeochemical cycles are able to handle if not overburdened by man. Many of these natural substances become pollutants only when they interfere with man's well-being, as do pollen-caused allergies.

Once injected into the atmosphere, pollutants enter biogeochemical cycles by various routes. Fine particulate matter and gaseous substances may be carried by atmospheric currents to points far removed from the source. A portion reaches land as dry fallout, which may enter various nutrient cycles and food chains through water and soil. Other contaminants react chemically and photochemically with each other and produce such secondary pollutants as sulfuric acid and ozone. Aerosols and other forms of fine particulate matter act as condensation nuclei and return to earth as rainfall.

Although atmospheric pollutants, particularly DDT, sulfur, and heavy metals, enter easily into the water cycle and become a source of water pollution, the major source of water pollution is waste disposal in its broadest sense. For centuries rivers and lakes have been used as dumping grounds for human sewage and industrial wastes of every conceivable kind, many of them highly toxic. Added to this are materials leached and transported from land by water percolating through the soil and running off its surface to aquatic ecosystems.

Man and biogeochemical cycles

One outcome of this pollution has been the destruction of aquatic life by direct poisoning from toxic chemicals, pesticides, acid mine drainage, and oil spills. Another outcome has been the excessive enrichment of aquatic ecosystems. Because of this a term once in the vocabulary of ecologists only, *eutrophication*, is becoming commonplace (see Chapter 9). *Eutrophic* means well nourished; eutrophication refers to the natural and artificial addition of nutrients to lakes, streams, and estuaries, and to the effects of this addition. Like the material injected into the atmosphere, these materials too enter the biogeochemical cycles.

Carbon dioxide and carbon monoxide

Carbon dioxide, a natural constituent of the atmosphere, is increasing at a rate of 0.02 percent annually, 0.7 ppm in 320 ppm. A by-product of the burning of fossil fuel, it is not necessarily a pollutant. It produces adverse physiological effects only at very high levels. Approximately half of the input remains in the atmosphere; the other half is removed by the oceans and by plants and animals. The main concern about increased carbon dioxide is its possible effects on the earth's temperature.

If one dismisses carbon dioxide as a pollutant, then the next most abundant and widely distributed air pollutant is carbon monoxide (Jaffe, 1971). The major atmospheric source is the gasoline engine. Lesser amounts come from the burning of coal. Worldwide, the annual input amounts to 27.17×10^6 ton, 95 percent of which is produced in the Northern Hemisphere. The concentration of carbon monoxide over the Northern Hemisphere is 0.1 to 0.2 ppm, but in urban areas, local concentrations build up to much higher levels—50 to 100 ppm. In heavy automobile traffic, drivers may be exposed to 650 ppm. The concentration in slow-moving traffic on an expressway may reach 140 ppm. At the level of 30 ppm, normal concentration for some urban atmospheres, carbon monoxide binds up about 5 percent of the hemoglobin of the blood.

Unlike carbon dioxide, carbon monoxide is not easily removed from the atmosphere. A very stable gas, it remains as a monoxide for a long time. Atmospheric carbon monoxide may be scrubbed by atmospheric circulation, by photochemical oxidation to carbon dioxide, by plants and soil bacteria, and by oceanic absorption.

Nitrogen

Man's intrusion into the nitrogen cycle takes two pathways. One is the release of nitrogen in the form of various oxides into the atmosphere and into the gaseous cycle. The other is the input of nitrates into the salt-solution phase of the cycle. One contaminates the air; the other pollutes rivers, lakes, and estuaries.

NITROGEN OXIDES

The major sources of nitrogen oxides are the automobile and power plants that burn fossil fuels. The major type of nitrogenous air pollutants is nitrogen dioxide, NO_2. In the atmosphere nitrogen dioxide is reduced by ultraviolet light to nitrogen monoxide and atomic oxygen:

$$NO_2 \rightarrow NO + O$$

Atomic oxygen reacts with oxygen to form ozone:

$$O_2 + O \rightarrow O_3$$

Ozone, in a never-ending cycle, then reacts with nitrogen monoxide to form nitrogen dioxide and oxygen, thus closing the cycle:

$$NO + O_3 \rightarrow NO_2 + O_2$$

This cycle illustrates only a few of the reactions that nitrogen oxides undergo or trigger. In the presence of sunlight, atomic oxygen from nitrogen dioxide also reacts with a number of reactive hydrocarbons to form radicals. These radicals then take part in a series of reactions to form still more radicals that combine with oxygen, hydrocarbons, and nitrogen dioxide. As a result, nitrogen dioxide is regenerated, nitrogen monoxide disappears, ozone accumulates, and there form a number of secondary pollutants, such as formaldehyde, aldehydes, and peroxyacytnitrates, known as PAN (see Am. Chem. Soc., 1969). All of these collectively form photochemical smog.

Nitrogen dioxide and the secondary pollutants are harmful both to man and plants. Nitrogen dioxide, a pungent gas that produces

a brownish haze, causes nose and eye irritations at 13 ppm and pulmonary discomfort at 25 ppm. Ozone irritates the nose and throat at 0.05 ppm, and at 0.1 ppm it causes dryness of the throat, headaches, and difficulty in breathing. All of these, along with other contaminants such as sulfur dioxide and carbon monoxide, usually act synergistically. The total effect is much greater than individual effects alone. Thus air pollution has been positively correlated with increase in asthma, bronchitis, emphysema, and lung cancer (Lave and Seskin, 1970).

Ozone, PAN, and nitrogen dioxide severely injure many forms of plant life. PAN is extremely toxic. By destroying some of the lower epidermal cells of the leaves and by damaging the chloroplasts, it interferes with the plant's metabolic processes. Although the mechanism has not been defined, PAN also interferes with enzymes important in providing the energy necessary to split the water molecules in photosynthesis (see Treshow, 1970).

A number of important leafy vegetable plants are extremely sensitive to PAN. Among them are spinach, endive, and tobacco. Other sensitive crops are oats, alfalfa, beets, beans, corn, celery, and peppers. Affected plants usually show some form of glazing, silvering, or bronzing in irregular patches on the undersides of the leaves (see Hill and Heggestad, 1970).

Ozone is especially prevalent in the air near populated regions. Reactions to it vary among plants. Highly sensitive tobacco is flecked with white lesions, bean leaves show stippling and bleached areas, and the leaves of woody plants may have reddish-brown lesions. Plants especially sensitive to ozone include white and ponderosa pines, alfalfa, oats, spinach, tobacco, and tomato. Nitrogen dioxide too causes direct injury to plants. Symptoms of damage from nitrogen dioxide are white or brown collapsed lesions or tissues between the veins and near the margins of the leaves. These symptoms closely resemble reactions to both ozone and sulfur dioxide.

NITRATES

Natural sources of nitrates are the fixation of nitrogen in water and soil, microbial activity in the soil, and the subsequent leaching of nitrates to aquatic ecosystems. Added to these natural sources are the often excessive inputs by man from domestic sewage and from agricultural croplands and feedlots. Addition of excessive quantities of nitrogen and the associated element phosphorus has caused an unnatural enrichment, or cultural eutrophication, of lakes, rivers, and estuaries (see Chapter 9).

In recent years the distortion of the nitrogen cycle by agriculture has been an area of controversy (Commoner, 1970). The intensified use of nitrogenous fertilizers has been blamed for stress to the natural nitrogen cycle, causing increased eutrophication and hazards to health from nitrite poisoning. As a result efforts are made to restrict the use of nitrogen fertilizers on agricultural crops. Much of the controversy results from a lack of understanding of the nitrogen cycle and the role of nitrates in the ecosystem.

In a stable undisturbed ecosystem the nitrogen inputs and outputs are somewhat balanced, with varying amounts of nitrogen tied up in organic matter. Nitrogen often is a limiting nutrient, and because only minimal amounts are leached (see Table 6-1), aquatic ecosystems are nutrient-poor, or oligotrophic. When man cut the forests and broke the virgin prairie, he stimulated the breakdown of organic matter and loss of nitrogen through leaching (see Chapter 5). For example, after 100 years of cultivation, some 1.75 billion tons of nitrogen have been lost from the upper 40 in. of soil in the United States. The potential nitrate concentration in waters leaching through a prairie soil due to 100 years of cultivation is enormous. Stout and Buran (1967) estimate that if the prairie soil initially contained 0.1 percent organic nitrogen in the top foot, one-fourth of this nitrogen would be lost in 100 years of cultivation, and if only one hundredth of it leached as nitrate in 12.6 in. of water annually, the percolate would contain 28.6 ppm of nitrate nitrogen. Water leached from agricultural lands today may run no higher (Viets, 1971b).

One cannot generalize about the role of chemical fertilizers in nitrate losses through leaching. It is difficult to distinguish among the nitrate sources—organic matter, chemical fertilizers, and degradation products. The amount of nitrates lost from the soil by way of percolating water varies with the type of crop, the rate and time of fertilizer application, the soil permeability, the ratio of precipitation or irrigation to evaporation, hydrol-

FIGURE 6-1
One aspect of modern man's relationship to global ecosystems is the excessive amount of sulfur, nitric oxide, carbon dioxide, and particulate matter injected into the global biogeochemical cycle. Present in excessive quantities, these materials become pollutants. (Photo by John R. Shrader; courtesy Environmental Protection Agency.)

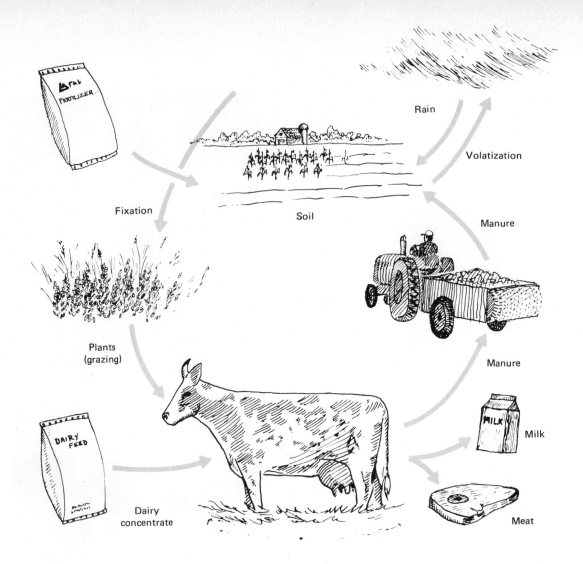

Rain

Volatization

Fixation

Soil

Manure

Plants
(grazing)

Manure

Dairy
concentrate

Milk

Meat

ogy of the area, the portion of the watershed in crops, and the general climate. Greater leaching of nitrates takes place in the humid eastern United States than in the drier Midwest. In dryland agriculture, excessive nitrates are carried to the depth of free water movement (Power, 1970). In times of excessive precipitation the nitrates may be deposited below the root zone and subsequently be available for leaching when moisture conditions permit it.

The amount of nitrates available for leaching depends upon the portion of nitrates available in excess of crop usage. In rare instances crops may utilize 95 percent of applied nitrogen (Allison, 1966). Utilization of 70 to 90 percent is fairly common, but a more usual nitrogen recovery is 50 to 60 percent. Thus in many agricultural areas losses to aquatic ecosystems can be considerable (Table 6-2). But losses of nitrogen from agricultural fields show no consistency as losses

from forest ecosystems do. Losses are as varied as agriculture itself. They may run as low as 3.8 lb/acre (Jaworski and Helling, 1970) to as high as 50 lb/acre (Bigger and Corey, 1969).

Loss of nitrates from agricultural soils is a natural process, and because nitrate is a form in which nitrogen is most effectively used by plants, the source of nitrogen, organic or inorganic, is of little consequence. There is little reason to believe that the amount of nitrate leached per unit of crop production would be less if nitrogenous sources other than commercial fertilizers were used. Failure to apply nitrogenous fertilizers on agricultural lands would result in a decline in productivity (see Viets, 1971b), in a deterioration of soil structure, and a decline in the water-holding capacity of the soil, its organic-matter content, and its productive capacity (Aldrich, 1972).

The often heavy loss of nitrates from agricultural ecosystems can be reduced by restricting the application of nitrates to the quantity usable by the crop, by applying the fertilizer as a side dressing throughout the growing season close to the time of maximum demand by the crop, and by planting a cover crop of small grains or grass to utilize the nitrates that form in the fall after the row crops are harvested.

Another source of agricultural nitrate pollution is animal waste. Waste produced by domestic animals in the United States is equivalent to that of a human population of 1.9 billion. Fifty percent of this animal waste is produced in concentrated livestock operations (Wadleigh, 1968). This does not mean, however, that the impact of livestock waste on water quality is equal to four to eight times that of the human population. A considerable portion is dispersed over the land, unlike human waste which is poured directly into natural waters.

Only recently has animal waste become an environmental problem. Prior to changes in the management of livestock, animal waste was recycled to croplands and the nutrients eventually returned to the animals as feed (Fig. 6-2). Inputs of feed and fertilizer, including manure, were converted into outputs of milk and meat. The system was ecologically sound, and losses decreased in proportion to the size of the farm because there was a greater area over which the wastes could be spread (Fig. 6-3).

But today manure piles up. In the dairy region of the northeast and northcentral United States, the cost of spreading manure on the fields exceeds its value as a fertilizer. In the Midwest beef cattle once fattened in pastures are now confined to feedlots handling 16,000 head and involving huge capital investments (see Viets, 1971a). In 1940 40 percent of the beef came from feedlots, in 1968, 67 percent. Areas of feedlot concentrations are the panhandles of Texas and Oklahoma, the Corn Belt, eastern Oklahoma to North Dakota, and southern California and Arizona. The feces in feedlots contain thousands of chemical compounds that come from feeds, animal metabolism, the symbiotic microbial metabolism, and the putrefaction upon exposure on the feedlot surface. The environmental significance of these compounds is unknown. About 25 tons/acre of nitrogen is excreted in a moderate-density feedlot. Of this about half is urine, which is quickly

FIGURE 6-2 [OPPOSITE]
Nitrogen cycle in a well-managed agricultural system involving livestock is complete with nitrogenous wastes returned to the fields and the nitrogen recycled through crops and cattle. Inputs into the cycle are nitrogenous concentrates of feed and a limited amount of fertilizer. Outputs include milk and meat in the form of veal and older unproductive cows. This contrasts with a feedlot system of handling beef cattle, in which nitrogenous wastes are stored in huge refuse piles. (After C. R. Frink, 1970.)

TABLE 6-1
Estimates of sources of plant nutrients entering Wisconsin surface waters

Source	N (%)	P (%)
Municipal treatment facilities	24.5	55.7
Private sewage systems	5.9	2.2
Industrial wastes	1.8	0.8
Rural sources		
Manured lands	9.9	21.5
Other cropland	0.7	3.1
Forest land	0.5	0.3
Pasture, woodlot, and other lands	0.7	2.9
Groundwater	42.0	2.3
Urban runoff	5.5	10.0
Rainfall on water areas	8.5	1.2

Source: C. R. Frink, 1971.

TABLE 6-2
Estimates of sources of plant nutrients entering the Potomac River estuary

Source	N (%)	P (%)
Agricultural runoff	31	8
Forest runoff	16	4
Urban runoff	2	1
Wastewater discharge	51	87

Source: C. R. Frink, 1971.

hydrolyzed to ammonia. Nitrates are produced on or near the feedlot surface by microbial action, although nitrification can be inhibited by high pH. In a study encompassing a feeding period of 203 days, the manure contained 70 percent of the nitrogen excreted, some nitrogen entered the soil (see Table 6-3), and the rest was either denitrified or escaped as ammonia. The waste of feedlots is either left to decay anaerobically beneath a caked and almost impervious outer surface or is stored in large piles covering acres of ground. Because of low rainfall in the feedlot areas, feedlot wastes contribute an insignificant amount of nitrates to aquatic ecosystems, although local groundwater may be contaminated. In more humid areas such wastes would create a much greater environmental problem. The major problem of feedlots is air pollution from dust, odors, and ammonia.

The solution to the problem of feedlot and dairy waste is to return the solid waste to the land in appropriate amounts for the maximum crop production. Estimates suggest that 1 acre of cropland could safely handle 11.7 tons of dry manure per year for the production of feed grains. Still, manure would only be sufficient to fertilize one-tenth of the croplands in the United States. Such spreading has to be done when the ground is not frozen; otherwise runoff would carry nitrogen and phosphorus to surface waters.

A third source of nitrate pollution is human waste, particularly in the form of sewage. In spite of the magnitude of water pollution from agricultural sources, human effluents contribute even heavier loads. Municipal sewage treatment facilities contribute one-fourth of the nitrogen and one-half of the phosphorus found in the surface waters of Wisconsin (Corey et al., 1967, cited by Frink, 1971). Groundwater from rural areas contributes one-half of the nitrogen and only 2 percent of the phosphorus. In Connecticut domestic wastes contribute as much nitrogen as animal waste and twice as much phosphorus. Because 60 percent of domestic waters are treated by sewage plants, the soil filter is bypassed and the nitrogen is discharged directly into waterways. In the Potomac River, estuary waste water accounts for 51 percent of the nitrogen, compared to 31 percent from agricultural runoff (Table 6-2) (Jaworski and Helling, 1970).

As with agricultural wastes, the potential solution is to return human waste to the land, where the nitrates can be utilized by the vegetation and the phosphorus can be tied up in the soil. Sewage sludge can be used as a fertilizer, and sewage effluents can be sprayed on crops and forest lands. Groundwater can then be purified by cycling it through a crop (Parizek et al., 1967). This is an ecologically sounder solution than diluting sewage effluents in streams.

Phosphorus

Phosphorus is perhaps more intimately involved in the eutrophication of water than nitrogen. Of the three nutrients required for aquatic plant growth, potassium is usually present in excess, nitrogen is supplemented by fixation, and phosphorus tends to be precipitated in the sediments and cannot be supplemented naturally (Am. Chem. Soc., 1969). Thus phosphorus is usually limiting, and in the presence of a luxury supply, algae respond with luxurious growth.

Most of the phosphorus enrichment of aquatic ecosystems comes from sewage disposal plants (Table 6-1). Primary sewage treatment removes only 10 percent of the total phosphorus. Secondary treatment removes only 30 percent at best, and even some of this ultimately finds its way into streams. Feedlots may contribute some from runoff. However, phosphorus has such a strong affinity to the soil particles that the amount coming from agricultural lands is insignificant. Thus sewage, with its heavy load of phosphate detergents, contributes nearly all of the phosphorus reaching lakes and rivers (see Chapter 9).

The impact of phosphorus entering estuarine and coastal ecosystems may differ from that in fresh-water ecosystems. Ryther and Dunstan (1971) point out that nitrogen rather than phosphorus limits algal growth and eutrophication in coastal marine waters. Coastal algae have available twice as much phosphorus as they need. The reason can be found in the ratio of nitrogen to phosphorus in the contribution from land. Incoming pollution has a ratio of nitrogen to phosphorus of just over 5 : 1. Phytoplankton in coastal waters assimilates nitrogen and phosphorus in a ratio of 10 : 1. By the time the nitrogen is depleted, half of the phosphorus still remains.

If half of the phosphorus in sewage comes from detergents and if this source were eliminated entirely, the ratio of nitrogen to phosphorus in the coastal waters would be 10 : 1, and the amount of phosphorus would still be excessive. If phosphate detergents were replaced by nitrogenous ones, as has been suggested, the problem of eutrophication in estuarine waters would be further compounded.

Sulfur

To produce energy to keep our technological civilization running, we have to burn fossil fuels, and in doing so we inject into the environment a heavy dose of sulfur. Globally each year we pour into the atmosphere 147 million tons of sulfur dioxide. Seventy percent of it comes from the combustion of coal. Once in the atmosphere, sulfur dioxide does not remain in the gaseous state, but reacts with the moisture to form sulfuric acid (see Chapter 5).

Sulfuric acid in the atmosphere has a number of effects. It is irritating to respiratory tracts in concentrations of a few parts per million. In a fine mist or absorbed in small particles, it can be carried deeply into the lungs to attack sensitive tissue. High concentrations of sulfur dioxide (over 1,000 m²/m³) have implicated this pollutant as a prime cause of many air pollution disasters characterized by higher than expected death rates and increased incidence of bronchial asthma. Among such disasters are the Meuse Valley in Belgium in 1930, Donora, Pennsylvania, in 1938, London in 1952, and New York and Tokyo in the 1960s.

Although sulfur compounds in the atmosphere are only one of a number of pollutants, sulfur concentrations are a measure of serious air pollution. There is a strong correlation between sulfur content of the air and human health. Studies have shown a correlation between atmospheric concentrations of sulfur trioxide and absences from work due to respiratory disease in five different cities (Lave and Seskin, 1970). During Asian flu epidemics there was a 200 percent increase in illness in cities with polluted air and only 20 percent increase in those with a relatively unpolluted air. In Nashville, Tennessee, the death rates of infants under 1 year of age could be correlated with atmospheric concentra-

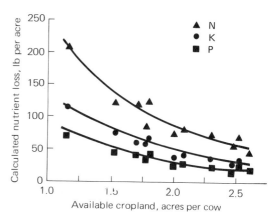

FIGURE 6-3

Estimated losses of nutrients from dairy farms in the northeastern United States. Phosphorus and potassium usually accumulate in the soil in unavailable forms, but nitrogen may be lost to waterways. Note that as the intensity of management increases (that is, the number of cows per acre), the loss of nutrients increases. (From C. R. Frink, 1971.)

TABLE 6-3
Fate of nitrogen removal from a Midwest feedlot

Source of loss	Percentage
Cleaning and hauling	59.1
Winter (snow) runoff	10.1
Spring and summer (rain) runoff	1.3
Denitrification and ammonia	29.5(?)

Source: Data from Gilbertson et al., 1970.

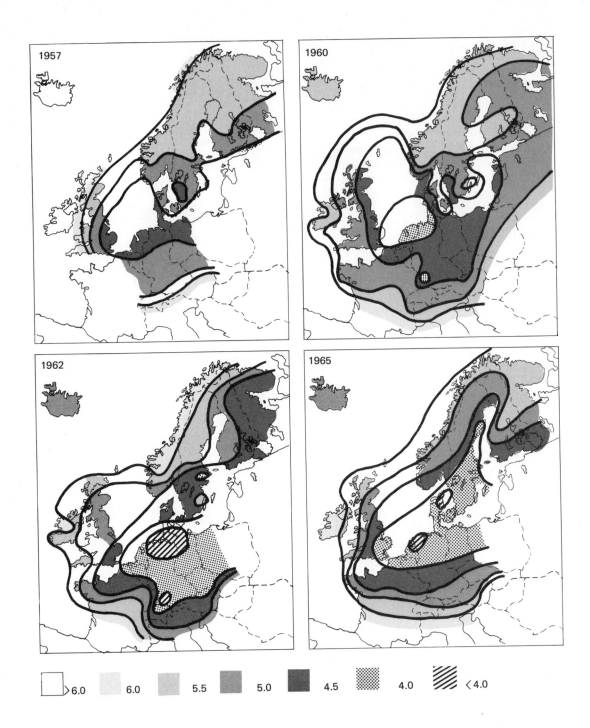

tions of sulfur trioxide. In another study involving 114 Standard Metropolitan Statistical Areas in the United States, the total death rate was significantly related to the minimum level of sulfate pollution, to population density, and to the percentage of people over 65, the ones most likely to feel the effects of long-term sulfur pollution. The same minimum atmospheric concentration of sulfates is also a significant explanatory variable in infant death rates.

Plants exposed to atmospheric sulfur are injured or killed outright. Exposure of plants to sulfur dioxide with as low a concentration as 0.3 ppm for 8 hours can produce both acute and chronic injury. Acute injury is characterized by dead tissue between the veins and along the margins of plant leaves; chronic injury is marked brownish-red or blackish areas in the blade of the leaf.

Injury to plants is caused largely by acidic aerosols during periods of foggy weather, during light rains, or during periods of high relative humidity and moderate temperatures. Pines are more susceptible than broadleaf trees and react by partial defoliation and reduced growth.

Not only does the emission of sulfur into the atmosphere cause injury to plants and problems in public health, but it also produces acid rainfall over parts of the earth. The problem was first noticed in Scandinavia, downwind from the industrial centers of Britain and the Ruhr Valley (Fig. 6-4). Acidity of the rainfall has increased 200-fold since 1966; pH values as low as 2.8 have been recorded (Oden and Ahl, 1970). This acid rainwater is increasing the acidity of Scandinavian streams, interfering with salmon reproduction and destroying salmon runs. It reduces forest growth and increases the amount of calcium and other nutrients leached from agricultural soils.

Current studies of chemistry of surface water and rainfall in areas in northeastern United States that are also downwind from major industrial centers show that the acidity of rainfall has increased 10 to 100 times and that the acidity of streams has increased in recent years (N. H. Johnson et al., 1972). Rain and snowfall at the Hubbard Brook Experimental Forest in New Hampshire have had a pH of 4.1, with weekly samples as low as 3.0. Contemporary rainfall over much of New

FIGURE 6-4 [OPPOSITE]
Trends in the pH of rainfall over northern Europe from 1957 to 1965 are shown on these maps. Note the increasing acidity over northern Europe, the source of which is the industrial heart of Europe. (From S. Oden, "Nederborden forsurning-ett generellt hot mot ekosystemem." I Mysterud [red.] Forunensning og biologish miljovern. Universit.)

England has a pH of 4.1 to 4.4, compared to a more natural pH of 5.7.

As a result, anions in the streamwater of the northeastern United States are changing from bicarbonates to sulfates. Such waters are detrimental to fish and other aquatic life. Thus the burning of high-sulfur fuels can easily convert nonacid streams into acid waters without the input of mine acid drainage and the like. Because of the global air currents, even a stream relatively remote from industrial centers is not protected from an input of sulfuric acid.

Heavy metals

Bringing biogeochemical cycles forcibly to the attention of the public has been the accumulation of heavy metals in the biosphere. Among these are lead, cadmium, and mercury. Although always present naturally at low levels in the environment, concentrations have been increased markedly by the activities of man. Like other elements, heavy metals enter into natural cycles of the ecosystem. As a result their concentrations have been measurably increased in the upper trophic levels.

LEAD

The major input of lead into the biosphere comes from the burning of lead alkyl additives in automobile fuels. In 1968 more than 2.3×10^8 kg of lead was emitted to the atmosphere by automobiles alone (Chow and Earl, 1970). Minor sources are metal-smelting plants and agricultural areas where lead arsenate is used as an orchard spray. Introduced as a fine aerosol, lead eventually falls out either in precipitation or in dust to contaminate the soil. Once in or on plants, lead enters the food chain (Chow, 1970). Plant roots take up lead from the soil and leaves take it up from contaminated air or from particulate matter on the leaf surface. There is evidence that microbial systems are also capable of taking up and immobilizing substantial quantities of lead (Tornabene and Edwards, 1972). Because long-term increases in concentrations of atmospheric lead result in appreciably higher concentrations of lead in the blood of people who breath this lead-contaminated air, this pollutant is a potential

health problem (Goldsmith and Hexter, 1967).

MERCURY

The danger of mercury-contaminated ecosystems has been tragically emphasized by the deaths and impaired lives of the Japanese who live in the villages around Minamata Bay and along the Agano River. The fish in these waters, a major part of these people's diet, became contaminated with methylmercury when neighboring industrial plants discharged mercurial wastes into these waters. When the people of Minamata Bay ate contaminated fish, the methylmercury was assimilated and transported by the blood to the brain. Toxic accumulations resulted in severe neurological symptoms and death. Called Minamata Disease, mercury poisoning is characterized by blindness, deafness, incoordination, and intellectual deterioration. In some instances mercury can be transferred from a pregnant woman through the placenta to the fetus, even though the mother herself exhibits no symptoms of mercury poisoning.

The toxic properties of mercury have been recognized for centuries, but only in recent times has the element accumulated in the ecosystem. The discovery of mercury concentrations in wild birds and fish signaled its widespread distribution throughout the global ecosystem. High mercury contents in the feathers and beaks of grain-eating birds and their predators in Sweden gave the first scientific evidence that the terrestrial environment was contaminated (Berg et al., 1966; Borg, 1969). The source of contamination was the alkyl mercury used as a fungicide on seeds. It was consumed directly by seed-eating birds and passed on to their predators. In Sweden 41 percent of seed-eating birds and 67 percent of predatory birds contained 2 mg/kg and over of mercury. However, since mercurial fungicides on seeds has been banned, the levels of mercury in seed-eating birds have declined.

More highly contaminated than terrestrial environments are aquatic ecosystems, which receive heavy local influxes of mercurial wastes from the manufacturers of chlorine, caustic soda, and electrical equipment, from pulp and paper mills, and from the burning of fossil fuels (Joensuu, 1972). The problem in aquatic environments is compounded by the chemical nature of mercury. It is unique in its ability to form stable compounds with organic radicals. These compounds can be classified into three groups: (1) alkyl, as represented by methylmercuric hydroxide; (2) alkoxyalkyl, such as methoxyethylmercuric hydroxide; and (3) aryl, such as phenylmercuric acetate. Mercury is usually discharged into the aquatic environment as inorganic mercury compounds or as phenylmercury.

In the aquatic environment mercury may precipitate out as a highly insoluble sulfide. Under anaerobic conditions this compound may remain in the sediments for an indefinite period. Under aerobic or partially anaerobic conditions it can oxidize to the sulfate form and be subject to methylation (Fig. 6-5). In the bottom sediments microbes capable of synthesizing vitamin B_{12} can transform inorganic, aryl, and alkoxyalkyl compounds into both monomethyl and dimethyl forms. The less volatile monomethyl is usually formed under acidic conditions; the highly volatile dimethyl form is favored under neutral and alkaline pH. Methylmercury dissolves in the water. The more volatile dimethylmercury evaporates into the atmosphere and enters the global cycle (Fig. 6-5). Fish in the methyl mercury-contaminated water pick up this pollutant either by the food chain or by diffusion across the gills.

The concentration of methylmercury in fish inhabiting locally contaminated water may be much too high. In Sweden average concentrations range from 0.2 to 5.0 mg/kg; a level of 0.05 mg is considered acceptable. Not only do these levels have a direct and often lethal effect on fish, but mercury is passed along to consumer organisms, and when that consumer is man, the results, as already discussed, can be tragic.

Because mercury is a natural substance, it has probably always been present at some level in oceanic and fresh-water fish. In spite of bans on tuna and swordfish because of higher than permissible levels in their flesh, levels have changed little in time. Mercury levels of museum specimens of tuna and swordfish caught 64, 93, and 25 years ago are in the same range as specimens taken in recent years, all above permissible levels. This suggests that mercury levels now found in wide-ranging ocean fish come from natural sources rather than from man-made pollutants (Weiss et al., 1971). In fact one authority

155

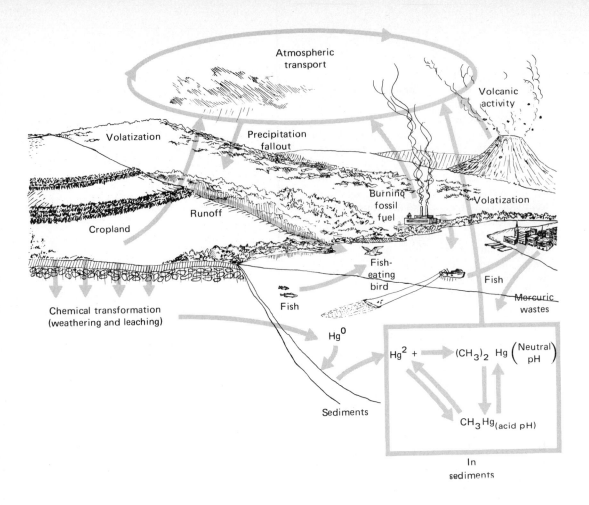

Atmospheric
transport

Volcanic
activity

Volatization

Precipitation
fallout

Volatization

Runoff

Cropland

Burning
fossil
fuel

Fish-
eating
bird

Fish

Fish

Mercuric
wastes

Chemical transformation
(weathering and leaching)

Hg^0

$Hg^{2+} \longrightarrow (CH_3)_2 \; Hg \left(\begin{matrix} \text{Neutral} \\ \text{pH} \end{matrix} \right)$

Sediments

$CH_3 Hg_{\text{(acid pH)}}$

In
sediments

(A. L. Hammond, 1972) estimates that if the total amount of mercury processed by man since 1900 were introduced and well mixed in the world's oceans, the concentration would be increased only by 1 percent. In fresh-water ecosystems the story is different. D'Itri (1971) analyzed museum specimens of fish caught in the St. Clair River System before it received mercurial pollution and compared concentrations of mercury in those fish with ones recently caught. Concentrations of mercury in muskellunge and sea lamprey increased threefold from 1965 to 1970; concentrations in sturgeon, pike, sauger, small-mouthed bass, walleye, and white crappie increased five times during this period.

On a global basis the cycling of mercury appears to present no particular hazard, although as yet relatively little is known about the mercury cycle. Inorganic mercury and methylmercury are highly volatile at ordinary temperatures, and are interjected into the atmosphere as an aerosol or vapor or both. Estimates place the atmospheric burden of mercury as 80,000 metric tons, but little is

known of actual concentrations. The few measurements taken in the United States are well below the threshold limit of 0.1 mg/m³ for metallic vapor and 0.01 mg/m³ for organic mercury.

The major dangers of mercury are concentrations in local aquatic environments, where it accumulates in fish, and in terrestrial ecosystems, where it is introduced as a fungicide on seeds. Its greatest hazard to health is limited to those people for whom fish is the major part of the diet. For wildlife, mercury is a hazard to seed-eating birds feeding directly on treated seeds and to predatory birds at the top of the aquatic food chain. For both man and animals, intake of mercury can induce chromosome breakage and other genetic risks not fully understood.

Hydrocarbons

Associated with oxides of nitrogen in the photochemical production of smog are the hydrocarbons. Like sulfur dioxide and the oxides of nitrogen, hydrocarbons originate from the combustion of coal, oil, natural gas, gasoline, and wood, as well as from the natural decay of vegetation. Man-made emissions of hydrocarbons amount to about 90 million tons/year; natural emissions amount to 1.6 billion tons/year. In addition, man-made emissions are in areas occupied by man and are in contact with other pollutants. Of greatest concern are the unsaturated hydrocarbons—olefins and the benzene group. Olefins react easily with other chemicals, so they enhance smog formation. The aromatic hydrocarbons appear to be cancer-producing.

A different type of contamination from hydrocarbons is oceanic pollution by oil. Although accidental oil spills cause the most localized damage and receive the publicity, they are really a minor source of this type of pollution. Ninety percent of oil pollution comes from the normal operation of ocean shipping, refineries, petrochemical plants, submarine oil wells, and the fallout of airborne hydrocarbons from automobiles and industry. In 1967 normal tanker operations contributed an estimated 530,000 tons of oil to the sea; non-tankers, another 500,000 tons. Offshore oil production contributed 100,000 tons; refineries and petrochemical plants another 300,000 tons. Accidental spills added only 200,000 tons. Industrial and automotive waste oils and grease add some 500,000 tons, even though most of it is dumped on land. The emission of petroleum hydrocarbons is about 90 million tons/year, 40 times the amount of substances entering the ocean from other sources. If only 10 percent of these hydrocarbons reached the oceans, the total contamination from this source would be five times the direct influx from ship and land sources.

Hydrocarbons, largely emitted by the industrialized Northern Hemisphere, are dispersed slowly by the wind. Once in the ocean, they are diluted and dispersed by natural mixing, and they eventually disappear. A portion is lost through microbial action; another part by oxidation, evaporation, and deposition in bottom sediments. But such action is slow, and depending upon location, character, and concentration, the hydrocarbons can be quite destructive to marine ecosystems. They can poison or kill marine life, disrupt ecosystems by destroying juvenile forms of life and other links in the food chain, and interfere with the

FIGURE 6-5 [OPPOSITE]
Mercury cycle in nature. The portion of the cycle involving methylation is shown in the aquatic insert.

communication and sensory systems of animals.

Airborne halogenated hydrocarbons enter the oceans through an air–water interface. This interface, only a fraction of an inch thick (Duce et al., 1972), forms a slick-like layer on the ocean's surface. Incorporated in the surface film are such natural substances as fatty acids and such man-introduced materials as oils from leakages and spills, lead, nickel, copper, and chlorinated hydrocarbons. Concentrations of these materials in the top 10 to 15 cm are 1.5 to 50 times greater than concentrations in water 20 cm below the surface (Duce et al., 1972). The concentration of a chlorinated hydrocarbon, dieldrin, in the top millimeter of water off the Florida coast was 0.1 ppm compared to a concentration of 1.0 ppt in the underlying water (Seba and Cochrane, 1969). If this concentration were uniform over the entire ocean area between 15° and 50° N latitude, the total quantity of hydrocarbons and associated compounds in the upper meter would be equivalent to 70,000 tons. These same materials often cling to small particles, droplets, and tarry lumps that settle to the bottom and eventually return to the top through the motions of water (Horn, Teal and Backus, 1970). These slicks, often detectable only by optical means, concentrate toxic hydrocarbons and heavy metals on the surface where small larval stages of fish and plant and animal plankton may well extract and further magnify the materials through the food chain.

CHLORINATED HYDROCARBONS

Of all of man's intrusions into biogeochemical cycles, none has done more to call attention to nutrient cycling than the widespread application of DDT. Since World War II this pesticide has been used in huge quantities to control disease-carrying and crop-destroying insects. As early as 1946 Clarence Cottam called attention to its damaging effects on ecosystems and nontarget species. But the impact of pesticides on ecosystems remained obscured until Rachel Carson's book *Silent Spring* exposed the dangers of hydrocarbons. The detection of DDT in the tissues of animals in the Antarctic, far removed from any applied source of the insecticide, emphasized the fact that DDT does indeed enter the global biogeochemical cycle and that it becomes dispersed around the earth.

DDT. DDT (along with other chlorinated hydrocarbons) has certain characteristics that enable it to enter global circulation. It is highly soluble in lipids or fats, and not very soluble in water. Therefore it tends to accumulate in the lipids of plants and animals. It is persistent and stable under environmental conditions. It undergoes little degradation (largely from DDT to DDE) and has a half-life of approximately 20 years. It has a vapor pressure high enough to ensure direct losses from plants. It can become adsorbed by particles or remain as a vapor; in either state be transported by atmospheric circulation. It can return to land and sea with rainwater.

The major input of DDT to the biosphere is its manufacture and subsequent application to croplands, forest, and marshes for insect control. In 1963 the maximum production of DDT was 8.13×10^{10} g. Production declined sharply in 1970. The major areas of application have been and continue to be humid temperate and tropical regions.

Insecticides are applied on a large scale by aerial spraying. Half or more of a toxicant applied in this manner is dispersed to the atmosphere and never reaches the ground. If the vegetation cover is dense, only about 20 percent reaches the ground. In a massive spraying of DDT over 66,000 acres of forest in eastern Oregon, only about 26 percent reached the forest floor (Tarrant, 1971).

On the ground or on the water's surface, the pesticide is subject to further dispersion. There is apparently little movement of DDT from the surface soil to the subsoil. In the Oregon study the concentration of DDT in the prespray samples was 0.006 ppm at a depth of 0 to 3 in. Twelve months after spraying, the concentration was 0.029 ppm; 36 months later it was back to 0.006 ppm again. There was no significant gain in the subsoil levels. Input from litterfall likewise declined from 11.32 ppm at 0 to 6 months after spraying to 3.08 ppm 3 years later. Throughfall precipitation contained insignificant amounts. Woodwell et al. (1971) estimate that agricultural soils of the United States contain 0.168 g/m² (1.50 lb/acre) of DDT, and nonagricultural soils 0.0045 g/m². Mean lifetime of DDT in the soil is about 4.5 years. Pesticides reaching the soil are lost through volatilization, chemical degradation, bacterial decomposition, run-off, and the harvest of organic matter, which can amount to about 1 percent of the total DDT used on the crop.

In flowing water, pesticides are subject to further distribution and dilution as they move downstream. Insecticides released in oil solutions penetrate to the bottom and cause mortality of fish and aquatic invertebrates (see reviews in Pimentel, 1971*a*; and cope, 1971). Trapped in bottom rubble and mud, the insecticide may continue to circulate locally and kill for some days.

In lakes and ponds, emulsifiable forms of DDT tend to disperse through the water, but not necessarily in a uniform way. DDT in oil solutions tends to float on the surface and move about in response to the wind. Eventually the pesticides reach the ocean, where they may concentrate on the surface slicks in which the pesticide is concentrated 10,000 times greater than in lower waters. These slicks, which attract plankton, are carried across seas by ocean currents. In the oceans part of the DDT residues may circulate in the mixed layer. Some may be transferred to below the thermocline to the abyss. More may be lost through the sedimentation of organic matter.

Although considerable amounts of DDT and other pesticides are transported by water, they are relatively insignificant from a viewpoint of global circulation (Fig. 6-6). Estimates based on concentration of pesticides in river water and annual runoff amount to about 0.1 percent of the amount of DDT produced per year in the United States.

The major transport of pesticide residue takes place in the atmosphere. Not only does the atmosphere receive the bulk of pesticidal sprays (well over 50 percent of that applied), but it also picks up that fraction volatilized from soils, vegetation, and water. The vapor pressure of DDT is such (1.5×10^{-7} mm Hg at $20°$ C) that the equilibrium concentration is about 2 ppb by weight. If DDT remained as a vapor alone, the saturation capacity of the atmosphere to the troposphere would hold as much DDT as produced to date. But the capacity of the atmosphere to hold DDT is increased greatly by the adsorption of residues to particulate matter. Thus the atmosphere becomes a large circulating reservoir of DDT and other chlorinated hydrocarbons.

159

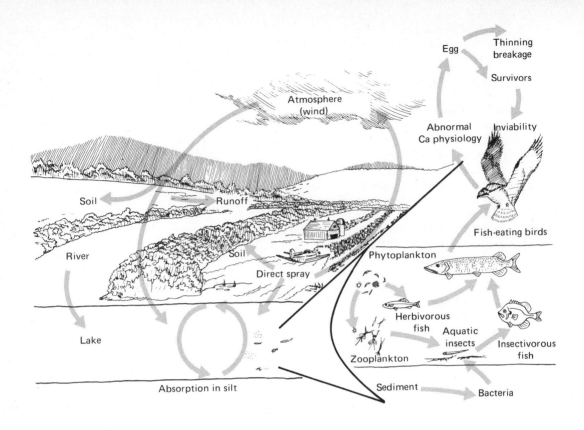

Residues are removed from the atmosphere by chemical degradation, diffusion across the air–sea interface, and mostly by rainfall and dry fallout (SCEP, 1970). The mean concentration of DDT in rainfall in England in 1966–1967 was 80 ppt. Rainfall in southern Florida between June 1968 and May 1969 contained an average of 1,000 ppt of DDT residues. If the total annual precipitation over the world's oceans contained an average of 80 ppt, a total of 2.4×10^4 metric tons of DDT would be carried to the oceans annually, about one-fourth of the world's annual production. Pesticides carried to land and sea are subject to volatilization and subsequent return to the atmosphere.

Although the quantity of residues of DDT and other chlorinated hydrocarbons may be relatively small, amounting to 1/30 or less of the amount produced each year, the concentrations still are sufficient to have a deleterious impact on marine, terrestrial, and fresh-water ecosystems. DDT and related compounds tend to concentrate in the lipids of living organisms where they undergo little degradation (see Menzie, 1969; Bitman, 1970; and Peakall, 1970).

The high solubility of DDT in lipids allows the magnification of its concentration through the food chain. Although only a portion of the food is ingested by consumer organisms, most of the DDT contained in the food is retained in the fatty tissues of the consumer. Because it breaks down slowly, DDT accumulates to high and even toxic levels. DDT so concentrated is passed on to the next trophic level. The carnivores on the top level of the food chain receive massive amounts of pesticides.

There are a number of examples of species concentrating this pollutant. Eastern oysters held in flowing seawater containing only 0.1 ppb of DDT for 40 days concentrated the pesticide some 70,000 times that contained in the water (Butler, 1964). Four species of algae in water containing 1 ppm for 7 days concentrated the pesticide 227-fold (Vance and Drummond, 1969). *Daphnia*, a genus of zooplankton, concentrated DDT 100,000-fold during a 14-day exposure in water containing 0.5 ppb of DDT (Preuster, 1965). Slugs and worms in a cotton field concentrated DDT 18 and 11 times that of the level in the soil. Pesticides concentrated by first-level consumers are then passed on to second-level consumers, carnivores, who in turn magnify the pesticide even more. The concentration of DDT in a Long Island salt marsh sprayed for mosquito control over a period of years was 13 lb/acre. Total residue ranged from

0.04 ppm in plankton, 0.16 ppm in shrimp; 0.28 in eel; 2.07 ppm in predacious fish; and 75 ppm in ringbilled gulls. This contrasts with the actual concentration of DDT in the water —0.00005 ppm.

Thus the ring-billed gull, a predacious bird near the top of the food chain, contained a level of DDT a million times greater than in the water (Woodwell et al., 1967). In Clear Lake, California, the concentration of DDT in the tissue of organisms over that of water was 265-fold in plankton, 500-fold in small fishes, 85,000-fold in predacious fishes and 80,000 fold in grebes, fish-eating birds (Rudd and Genelly, 1956).

High concentrations of DDT in the tissues often results in death or impaired reproduction and genetic constitution of organisms. Laboratory populations of zooplankton, shrimps, and crabs are killed outright by exposure to DDT in parts per billion. The continuous exposure of shrimp to DDT in 0.2 ppb, a concentration that has been detected in waters flowing into shrimp nursery areas, killed the entire population in less than 20 days (SCEP, 1972). A residue level of 5 ppm in the lipid tissues of the ovaries of fresh-water trout causes 100 percent die-off of the fry, which pick up lethal doses as they utilize the yolk sac. High levels of DDT are correlated with the decline of such fish as sea trout and California mackerel. DDT and its degradation product DDE interferes with calcium metabolism in birds. Chlorinated hydrocarbons block ion transport by inhibiting the enzyme ATPase, which makes available the needed energy. This reduces transport of ionic calcium across membranes and can cause death of organisms. DDE also inhibits the enzyme carbonic anhydrase (Bitman, 1970; Peakall, 1970), essential for the deposition of calcium carbonate in the eggshell and for the maintenance of a pH gradient across the membrane of the shell gland.

There is evidence that DDT can depress the growth of certain commercial vegetables, can cause significant changes in macro- and microelements in the above-ground parts of plant tissues (see Pimentel, 1971a), and can reduce the productivity of phytoplankton (Wurster, 1968). In most instances, however, the concentrations necessary to affect these changes were much higher than normally applied or higher than might occur in natural ecosystems.

FIGURE **6-6** [OPPOSITE]
Cycle of DDT and other chlorinated hydrocarbon pesticides. The initial input comes from spraying. A relatively large portion fails to reach the ground and is carried on water droplets and particulate matter through the atmosphere.

POLYCHLORINATED BIPHENYLS

Another recently discovered widespread contaminant of the environment is polychlorinated biphenyl (PCB), a widely used industrial chemical. Never intended for release into the environment, PCBs have been found in rainwater, in human tissues, and in many species of birds and fish. Like DDT, PCBs have an affinity for fatty tissue, degrade slowly, and accumulate in the food chain. Used as dielectric fluids in capacitors and transformers, in plastics, solvents, and printing inks, the production of PCBs has grown steadily since its introduction in 1930 to a high of 34,000 tons in 1970. Production and use are now declining since the discovery that it is an environmental contaminant.

How PCB gets into the environment is largely unknown, but some sources may be sewage outfalls and industrial disposal. Rivers are a means of transport, but the atmosphere appears to be the major mode (see A. L. Hammond, 1972). The pathway of PCBs is similar to that of DDT. In many areas residues of PCBs in fish are higher than that of DDT; but typical residues range from 0.01 to 1.0 ppm both in ocean fish and in fresh-water fish away from heavily polluted industrial areas. The osprey and cormorant, predatory fish-eating birds, have concentrations that range from 300 to 1,000 ppm. Traces have been found in human tissue. Like DDT, PCBs appear to cause thinning of egg shells, deformities in newly hatched birds (Hays and Risebrough, 1972), and reduction in growth rates of certain marine diatoms (Mosser et al., 1972).

Radionuclides

Ever since the atomic bomb violently ushered in the atomic age, the impact of nuclear radiation on life on earth has been of major concern. Although the threat of atomic weapons is always present, of more immediate concern is the effect of much lower levels of irradiation on life. One source of radionuclides is the nuclear power plant (Fig. 6-7).

Pressurized water reactors, commonly used in nuclear power plants, release low levels of radioactivity to the air and to the condenser water. Radioactive contamination of water from nuclear power plants comes from several sources. When water passes through the intense neutron flux of the reactor, trace elements in the water are activated, producing radioisotopes. Added to this are radioactive corrosive products from the surface of metal cooling tubes. Except for tritium, most of the radioisotopes are removed in a radioactive waste removal process. Those left have a short half-life and rapidly decay below detection levels.

Radioactive materials that enter the water become incorporated in bottom sediments and circulate between mud and water. Some become absorbed by organisms and concentrated at various trophic levels. In fact a sort of equilibrium is established between retention in the biomass and in the bottom sediment, the water, daily input, and decay (Jennings and Osterberg, 1969). For example, when a strike shut down the Hanford water reactor on the Columbia River for 40 days in 1966, a rapid and extensive decline in the concentration of radionuclides, particularly ^{51}Cr, ^{52}P, and ^{75}Zn, took place in both fish and phytoplankton (Renfro and Osterberg, 1969; Watson et al., 1969). When the plant resumed operations, the old equilibrium was reestablished within 2 to 3 weeks. Although the levels of radioactivity in the biota were too low to cause any short-range biological damage, the biological effects of long-term exposure to low levels of radiation are unknown.

A second source of radiation pollution is gaseous effluents. The most important source of radioactive pollutants, however, is not the nuclear power station itself, but fuel reprocessing plants, and the combustion of fossil fuels in conventional power plants. Although gaseous emissions are contained 45 days in a holding system, two radioactive elements, the so-called noble gases, krypton and xenon, are still present and are released to the atmosphere. Of the two, ^{85}Kr is more important, and the total global content comes from nuclear reactors of all types. At present the level of exposure of man to krypton is less than 0.1 percent of the combination from background cosmic radiation and natural radioactivity. But if nuclear power production increases as predicted, then the exposure due to ^{85}Kr could increase to 2 milliroentgens or 1000 times that of tritium and about 1 percent of the radiation protection guides recommended by national and international standards groups (J. H. Wright, 1970).

Some radionuclides introduced into biogeochemical cycles in recent years are ^{60}Co,

^{65}Zn, ^{90}Sr, ^{137}Cs, ^{131}I, ^{32}P, and ^{85}Kr. These radioactive materials are by-products of both weapons testing and nuclear reactors. When the uranium atom is split or fissioned into smaller parts, it produces, in addition to tremendous quantities of energy, a number of new elements, or fission products, including strontium, cesium, barium, and iodine. Some of these fission products last only a few seconds; others can remain active for several thousand years. Although they are not essential for life, these radioactive elements enter the food chain and become incorporated in living organisms. In the same atomic reaction some particles with no electrical charges, called neutrons, get in the way of high-energy particles. Nonfission products are the result, and they include the radioisotopes of such biologically important elements as carbon, zinc, iron, and phosphorus, which are useful in tracer studies. Both fission and nonfission products are released to the atmosphere by nuclear testing and by wastes from nuclear reactors unless carefully handled. Later these return to the earth along with rain, dust, and other material as atomic fallout. In the case of large-weapons testing in the atmosphere, this fallout can be worldwide. Once they reach the earth the isotopes enter the food chain and become concentrated in organisms in amounts that exceed by many times the quantities in the surrounding environment. This, in effect, produces local radiation fields in the tissues of plants and animals.

One of the most important radioactive materials released into the biogeochemical cycle is ^{90}Sr. A common component of atomic fallout and atomic wastes, it decays slowly (half-life 28 years). Ecologically it behaves like calcium and follows it in the cycling of materials. Strontium-90 is rapidly absorbed by plants through the foliage and the roots; when taken in with food by animals and man, it becomes concentrated in the bone and other areas of calcium deposition.

Strontium-90 enters animal tissue most easily through the grazing food chain, especially in regions with high rainfall and with low levels of calcium and other mineral nutrients in the soil. Sheep grazing on the English moors, where soils are acid and the rainfall high, have a higher concentration of ^{90}Sr than sheep grazing in drier, more fertile areas. A more striking example involves the lichen–

FIGURE 6-7
Nuclear power may be the source of energy for the future if safe breeder reactors can be developed. A conventional nuclear power plant, Connecticut Yankee, began generating electricity in 1968 and in 1972 became the first single nuclear unit to have generated more than 20 billion kilowatt hours. (Photo courtesy Westinghouse Electric.)

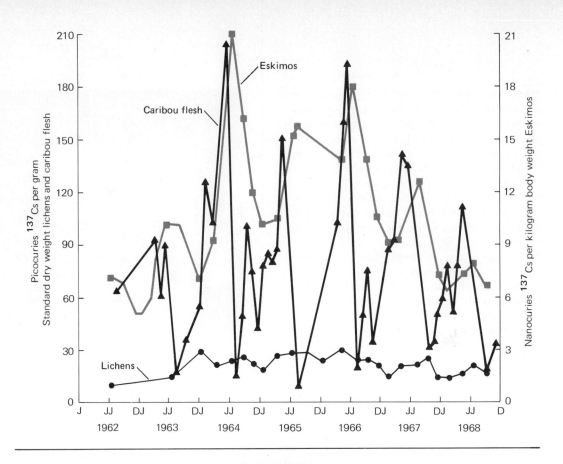

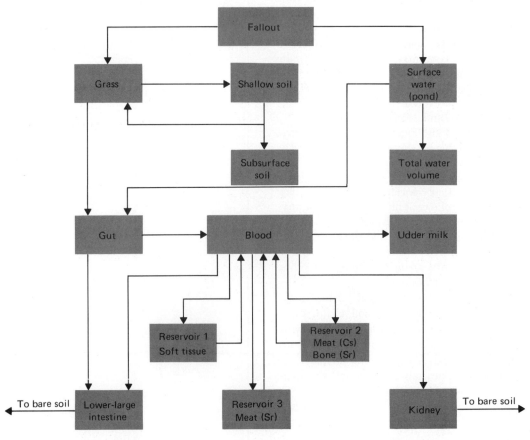

caribou–man food chain in the arctic regions of the world. The Far North has been subject to rather heavy atomic fallout and a dominant plant, the lichen, absorbs virtually 100 percent of the radioactive particles falling upon it. From lichens, the main contaminants, ^{90}Sr and ^{137}Cs, travel up the food chain from caribou and reindeer to carnivores and man. Humans who depend upon caribou and reindeer for their protein already have accumulated from one-third to one-half of the permissible amounts (Palmer et al., 1963; W. C. Hanson, 1971). Caribou are a major source of other radionuclides for the northern Alaska natives, who kill them during the northward migration in spring and stockpile the meat for late spring and early summer food. Because caribou feed all winter on lichens, their flesh in spring contains three to six times as much ^{137}Cs as it does in fall. Thus there is a corresponding rise in the ^{137}Cs level in the Eskimo, which often increases by 50 percent (Fig. 6-8). This level decreases when the people change to a diet of fish in summer and fall.

With the decline in nuclear weapons testing, the amounts of ^{90}Sr and ^{137}Cs are also declining. But the knowledge that radionuclides can be taken up and channeled through a food chain has led to further intensive study of the behavior of radionuclides in ecosystems.

Although radionuclides became more concentrated at higher trophic levels in the simple food chains of the arctic ecosystems, such behavior is not universal. Retention time and concentration of radionuclides vary with the isotope, the species of organism and its physiology, the food chain, the ecosystem, and the mechanisms governing the rates of transfer involved.

Dispersion of radionuclides in terrestrial ecosystems comes from gaseous, particulate, or aerosol deposition and from liquid and solid wastes. As illustrated in the arctic ecosystem, the most important mechanism for radionuclide contamination of the terrestrial environment is interception of particulate deposition and foliar absorption. Contaminants then enter a short-term food chain. Subsequent retention, decay, and weathering may result in some reduction in intercepted radioactivity. Additional input into vegetation may come by way of uptake of radionuclides from soil and litter. Both pathways can result in significant accumulations of such fission products as ^{90}Sr,

FIGURE 6-8 [OPPOSITE TOP]
Cesium–137 concentrations in lichens, caribou flesh, and Eskimos at Anaktuvuk Pass, Alaska, during the period 1962–1968. Note the relationship between the concentration of ^{137}Cs in the caribou flesh and the amount in Eskimos. As the concentration in caribou declined seasonally so did the concentration in humans. (From W. C. Hanson, 1971.)

FIGURE 6-9 [OPPOSITE BOTTOM]
Models of transfer of strontium and cesium through fallout to man. The radionuclides pass from fallout to cows to man. (Courtesy Oak Ridge National Laboratory.)

TABLE 6-4
Aquatic and terrestrial food-chain concentration of elements

Trophic level[a]	Element concentration factors[b]												
	Ca	Sr	K	Cs	Na	Co	Zn	Mn	Ru	Fe	P	Ra	I
AQUATIC													
Water	1.0	1.0	1.0	1.0	1.0	1.0	1.0	1.0	1.0	1.0	1.0	1.0	1.0
Algae and higher plants	1–400	10–3,000		50–25,000		2,500–6,200	140–33,500	700–35,000	80–2,000	2,400–200,000	36,000–50,000	0.5	60–200
Invertebrates — S	16	10–4,000		60–11,000		325		6,000–140,000	130		2–100,000	0.5	20–1,000
— H		1		600			150			125	2,000		
— C				800									
Fish — O	1	1	300–2,500	125–6,000						10,000	3,000–100,000	0.5	
— C	0.5–300	1–150	400–2,700	640–9,500			4–40					1.5	25–50
TERRESTRIAL													
Plants	1.0	1.0	1.0	1.0	1.0	1.0	1.0	1.0	1.0	1.0	1.0	1.0	
Invertebrates — S	0.1–18	0.1	3.5	0.2	17	0.4			0.4		11		
— H	0.1	0.1	3.0	0.3–0.5	21	0.5			1.2		17		
— C	0.1		2.0	0.1–0.5	27	0.3					18		
Mammals — H		0.5–4.5		0.3–2.0					0.4	0.8		0.01	0.5
— O				1.2–2.0						0.2		0.2	
— C				3.8–7.0								0.1	

Source: D. E. Reichle et al., 1970.

[a] S = saprovore (detritus-feeder); H = herbivore; C = carnivore; O = omnivore.
[b] Ratio of element level in consumer to element level in food-chain base, with base value normalized at 1.0.

^{137}Cs, and ^{131}I. From the plants the radionuclides move through the ecosystem by way of the food chain, but there can be no generalization about that movement. Strontium and cesium usually accumulate relative to whole-body levels at higher trophic levels, in vertebrate food chains (Fig. 6-9), whereas cobalt, ruthenium, and iodine do not. Some, such as iodine, may concentrate in certain tissues. In arthropod food chains, potassium, sodium, and phosphorus accumulate, whereas strontium and cobalt do not (Reichle et al., 1970) (see Table 6-4).

Radionuclides contaminate aquatic ecosystems largely through waste from nuclear power plants and from the nuclear processing industry. Bottom-dwelling insects and fish downstream from the source may be exposed to chronic low-level radiation. Under such conditions organisms approach equilibrium with environmental levels. As in terrestrial ecosystems, concentrations of radionuclides in food chains vary with the system and the species involved. In aquatic environments ^{32}P concentrations increase at higher trophic levels; ^{90}Sr becomes concentrated in the bones of fish and in invertebrates possessing calcareous exoskeletons. Yet the concentration in fish flesh is less than that in plants and in invertebrates. For example, cobalt-60 concentrations in algae (plants in which physical absorption is greater than biological concentration) are 2,500 to 6,200 times that of the surrounding water; concentrations in fish are only 25 to 50 times that of the water. Thus the concentration of radionuclides does not necessarily increase consistently through the food chain as does the concentration of chlorinated hydrocarbons. In many situations the concentration of radionuclides decreases at higher trophic levels. Aquatic organisms tend to concentrate radionuclides the same as they do the stable element.

In general ^{137}Cs, related biogeochemically to potassium, does not appear to increase at higher trophic levels in aquatic ecosystems. However, the fact that fish do accumulate some ^{137}Cs in their bones, and clams ^{90}Sr in their shells, serves as a monitor of low-level radiocontamination of the environment. Because they continuously feed on particulate matter in the aquatic environment and accumulate certain radionuclides, clams are excellent indicator species. Since there is no turnover of radionuclides deposited in the growth rings

167

of the shell, their shells are a record of the radionuclide contamination in their environment (D. J. Nelson, 1962).

In spite of considerable study, we still know little about uptake, assimilation, distribution in tissues, turnover rates, and equilibrium levels of radionuclides for many taxonomic groups. We know even less about the role of the environment in the cycling of radionuclides through various ecosystems. As the number of nuclear power plants and the necessity of waste disposal increase, our knowledge of the behavior of radionuclides needs to be more sophisticated. The knowledge of hazards must extend not only to man but also to the biota upon which he depends.

Summary

Man has overburdened biogeochemical cycles by adding more natural substances than the ecosystem is able to handle. He has further upset the cycles by introducing into the biosphere synthetic materials with which the system is unable to cope. Such substances, known as pollutants, may enter biogeochemical cycles by way of the water cycle. Some enter the gaseous cycle and circulate among air, water, and sediments. A certain number, including lead, mercury, and chlorinated hydrocarbons, move through the food chain. Among sources of these pollutants are automobiles, power plants, industry, sewage disposal, and agriculture.

Carbon dioxide, although hardly a pollutant, is increasing in the atmosphere at the rate of 0.2 percent annually. One of the most abundant pollutants is carbon monoxide, the source of which is the automobile. Heaviest concentrations are in urban areas. A very stable gas, it remains as a monoxide in the atmosphere for a considerable time.

Man's major intrusion into the nitrogen cycle involves inputs of nitrogen dioxide into the atmosphere and nitrates into the aquatic ecosystems. The major sources of nitrogen dioxide are automobiles and burning of fossil fuels. Nitrogen dioxide is reduced by ultraviolet light to nitrogen monoxide and atomic oxygen. These react with hydrocarbons in the atmosphere to produce a number of pollutants, including ozone and PAN. These make up photochemical smog, a pollutant harmful to plants and animals. Excessive quantities of nitrates are added to aquatic ecosystems by improper use of nitrogen fertilizer on agricultural crops, by animal wastes, and by sewage effluents. The latter accounts for the largest source. More closely involved in the eutrophication of aquatic ecosystems is phosphorus, most of which comes from sewage effluents.

A by-product of our technological civilization is sulfur dioxide. Injected into the atmosphere, it becomes a major pollutant, affecting and even killing plant growth, causing respiratory afflictions in man and animals, and producing acid rainfalls over parts of the world.

Industrial use of such heavy metals as lead and mercury, always present at low levels in the biosphere, has significantly increased their occurrence. Both pose potential and actual health problems, especially because they enter the food chain. Currently the major dangers of mercury are local rather than global.

Of more serious consequence globally are the chlorinated hydrocarbons. Used in insect control, these pesticides have contaminated the global ecosystem and entered food chains. Because they become more concentrated at higher trophic levels, the chlorinated hydrocarbons affect the predacious animals most adversely. Fish-eating birds are endangered because chlorinated hydrocarbons interfere with reproductive capability.

Also introduced are radioactive materials from testing nuclear weapons and from nuclear power plants. Some radioisotopes can enter and become concentrated in food chains. In many situations concentrations of radionuclides decrease at higher trophic levels. As man pours more pollutants into the biogeochemical cycles, we need to know more about the hazards they present.

Environmental influences

I n 1840 a German organic chemist, the foremost of his day, Justus von Liebig, published a book, *Organic Chemistry and Its Application to Agriculture and Physiology*. He described in it his analyses of surface soil and plants; and he set forth this simple statement, revolutionary for his day: "The crops on a field diminish or increase in exact proportion to the diminution or increase of the mineral substances conveyed to it in manure." Essentially what he said was that each plant requires certain kinds and quantities of nutrients or food materials. If one of these food substances is absent, the plant dies. And if it is present in minimal quantities only, the growth of the plant will be minimal. This became known as the *law of the minimum*.

Continued investigation through the years disclosed that not only nutrients, but other environmental conditions, such as moisture and temperature, also affected the growth of plants. Later, animals were found to be limited by food, water, temperature, and humidity. Eventually the law of the minimum was extended to cover all environmental requirements of both plants and animals.

Later studies showed that too much, as well as too little, of a substance or condition limits the presence or success of an organism. Organisms, then, live within a range between too much and too little, the limits of tolerance. This concept of maximum substances or conditions limiting the presence or success of organisms was incorporated by Shelford in 1913 into the *law of tolerance* (Fig. 7-1).

Modern ecologists, however, recognize that organisms actually are limited by a number of conditions, and often by an interaction between them. An organism, for example, may have a wide range of tolerance for one substance or condition and a narrow range for another, and thus be limited by the item for which it possesses a narrow range of tolerance. It follows then that an organism that exhibits a wide range of tolerance for all environmental influences would be widely distributed. In some cases when one condition is not optimum for the species, the limits of tolerance are reduced in others. On the other hand some organisms may utilize an item in surplus as a substitute for another that is deficient. Some

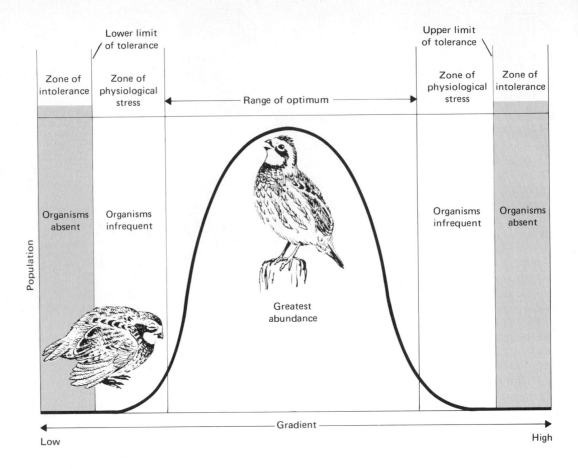

Lower limit of tolerance

Upper limit of tolerance

| Zone of intolerance | Zone of physiological stress | Range of optimum | Zone of physiological stress | Zone of intolerance |

Organisms absent

Organisms infrequent

Organisms infrequent

Organisms absent

Greatest abundance

Population

Gradient

Low

High

plants respond to sodium when the potassium supply is inadequate (Reitemeier, 1957). The maximum and minimum levels of tolerance for all environmental conditions of a species vary seasonally and geographically, Examples of the extensions of this law will be found in the following discussion of some of the items important to organisms.

Nutrients and plant life

Plants require some 16 essential elements. Not all plants require these elements in the same quantities or in the same ratios, but all plants do require a certain minimal amount of essential ions for growth, and the requirements are specific. Each plant species has a specific ability to exploit the nutrient supply in some manner that may not be duplicated by other species. This enables different plants growing in the same environment to exploit slightly different nutrient sources. Shallow-rooted plants, for example, may utilize nutrient supply on the upper surface soil, whereas those with deep tap roots may draw on deeper supply of nutrients.

A wide variation exists among and within species in their ability to take up nutrients. Some species growing on soils poor in nutrients have become adapted to low nutrient levels, whereas species growing on more fertile soil have become adapted to higher levels of nutrition. In one experiment two strains of the clover *Trifolium repems*, one growing on soils rich in phosphorus and the other growing on soils low in phosphorus, were grown in cultures of varying levels of that element. Plants from the soils rich in phophorus increased their growth as the level of phosphorus was increased. Plants from low-phosphorus soils exhibited a slight decline in growth. However, the concentration of phosphorus in the tissues of the plants from the low-phosphorus soils increased with the increasing level of phosphorus. Plants from the rich soil did not show this increase.

The nutrient levels of the soil have a pronounced influence on the distribution of plants. Nowhere has this been better demonstrated than on the long-term grassland fertilizer trial plots at the Rothamsted Experimental Farm in England. Among the plots, established in 1856, are some containing natural vegetation that probably resembles the original. The unfertilized, unmanaged plots contain some 60 species of higher plants (Thurston, 1970) representing not only every plant found in all other plots, but some restricted to the natural plots as well. Species diversity on the plots was high, and no one species was clearly dominant. The vegetation was short and the yield of hay low. On plots that received applications of phosphorus, potassium, sodium, and magnesium, but no nitrogen, legumes became dominant at the expense of other species. The addition of nitrogen discouraged the legumes, reduced their growth, and encouraged the grasses.

A demonstration of the effectiveness of the limitations of nutrient supply on vegetation is provided by Willis (1963) in his study of the addition of mineral nutrients to sand dune soils. Dry dune pastures, characterized by a closed mossy turf or relics of marram grass, and lichen pasture, characterized by the reindeer lichens *Cladonia* and the moss *Hypnum*, were treated with additions of nitrogen and phosphorus. The result was a pronounced change in vegetational composition, a reduction of the number of species, and an increase in the height and yield of vegetation. On the dry dune pasture and lichen pastures the two grasses red fesque, *Festuca rubra*, and bluegrass, *Poa pratensis* ssp *subcaerulea*, became strong dominants. A few of the large dicotyledonous plants such as hawksbeard, *Crepis capillaris*, and groundsel, *Senecio jacobaea*, were eventually reduced to minor importance, and rosette-type plants and the byrophytes were completely eliminated. On wetter areas bentgrass, *Agrostis astolonifera*, became dominant when complete nutrients were added. When nutrients except nitrogen were added, the vegetation changed little. The addition of nutrients except phosphorus resulted in a change in the floristic composition of the vegetation. Certain species of *Carex* and *Juncus* became dominants, suggesting that such plants have low phosphorus requirements and are favored by the addition of nitrogen and potassium. Interestingly, the addition of minor elements resulted in no significant changes in growth, beyond the addition of major elements. These experiments demonstrated that the open character of old pasture fields and other similar areas is due largely to severe deficiencies of nitrogen and phosphorus.

Shifts and changes in species composition

FIGURE 7-1 [OPPOSITE]
The law of tolerance, illustrated graphically.

reflect the competitive abilities of plants under different nutrient regimes, as well as their ability to exploit a rich source of nutrient or to get along on a poor source. Legumes, which are nitrogen-fixers, surpass grasses only when the nitrogen content of the soil is low (Wolf and Smith, 1964). When nitrogen fertilizer is added to the system, grasses dominate.

Soil acidity can affect nutrient uptake of plants, so it also has a strong influence on plant distribution. Many plants fall into two broad categories, those which require a relatively high pH of the soil (usually over 6.5), called calcicoles, and those that can tolerate a low pH, the calcifuges. More specifically, the characteristics of a calcifuge, the opposites of which would describe a calcicole, are (1) the ability to grow where available phosphorus is low and the bulk of the nutrient is immobilized by aluminum ions at the root surface; (2) the ability to grow in a low concentration of calcium or in the face of impaired calcium uptake and translocation; (3) the possession of specific sites within the cytoplasm where aluminum may accumulate harmlessly; and (4) the ability to chelate aluminum or to precipitate it at the cell surface. It is this ability to grow in the presence of toxic ions, especially aluminum, that sets the calcifuge plants apart from the calcicoles. Calcifuge plants are especially sensitive to calcium. On calcareous soils they suffer from lime chlorosis, a disease in which the roots and leaves become stunted and the leaves bleached. This sensitivity may be due to the plant's mechanism for chelating aluminum, which also has an affinity for iron at higher pH. On the other hand, calcicole plants are restricted to soils of a high pH because of their susceptibility to aluminum toxicity, which begins to show up at a pH of 4.5. Free aluminum accumulates on the surface of the root and in the root cortex, and it interacts with phosphorus to form highly insoluble compounds. Thus between the two types there exists an inverse relationship between aluminum toxicity and lime chlorosis.

Other ions such as sodium, manganese, chlorine and sulfur also cause specific toxic effects. Toxicity of sodium and chlorine ions are noticeable in the alkaline soils of western United States. Just as plants exhibit adaptive tolerances to aluminum, so they show adaptive tolerances to salt conditions. As salt stress increases, plant communities tend to become restricted to those species able to exist in a highly alkaline or salty environment.

The limiting influence of nutrients on production is emphasized when fertilizer is applied to soils and ponds. Responses of trees to an application of nitrogen or a complete fertilizer (containing nitrogen, phosphorus, and potassium) often are surprising. The application of 60 to 100 lb of nitrogen and 100 lb of calcium/acre increased the stand volume increment of white and scarlet oak growing on sandy loam soil in central Pennsylvania (Ward and Bowersox, 1970) more than 40 percent. This increase in growth was associated with increased nitrogen content of foliage and litter after leaf fall. Fertilization of ponds and small lakes is followed by an increase in the photosynthetic rate. Fish in such waters grow faster because of an improved food supply. When fertilizer was applied to a 120-acre unstratified lake on Kodiak Island, Alaska, the rate of photosynthesis in a 10-day period was increased by a factor of 2.5 to 7 (Nelson and Edmondson, 1955).

Nutrients and animal life

Because all animals depend directly or indirectly on plants for food, plant quantity and quality affect the well-being of animals. In the face of a limited quantity or shortage of foods, animals either suffer from acute malnutrition, are forced to leave the area, or starve. In other situations the quantity of food available may be sufficient to allay hunger, but the low quality affects reproduction, health, and longevity.

A certain correlation appears to exist between the presence and absence of certain nutrients and the abundance of certain animals. Throughout most of its range in North America, the introduced ring-necked pheasant is most closely associated with high-calcium soils. Outside of such areas its abundance declines, and the range is considered marginal. This apparent correlation has stimulated a number of investigative studies (Anderson and Stewart, 1969; Chambers et al., 1966; J. A. Harper, 1964), but the results are inconclusive. A similar correlation may exist between the population density of microtine rodents and the relative abundance of sodium in the soil (Amman and Emlem, 1968), but this has been challenged by Krebs et al. (1971).

On the other hand, the spatial distribution of elephants in the Wankie National Park in central Africa is closely correlated with the sodium content of the drinking water (Weir, 1972). The numbers of elephants were highest at waterholes with a high sodium content. Fewer appeared where the water had a lower sodium content if the soil contained water-soluble sodium. Few elephants were found where little sodium was available in soil or water, in spite of the abundance of browse. Reasons for such a relationship are unclear, but when elephants ingest sodium-rich soil and water they also ingest quantities of calcium, magnesium, potassium, and other minerals.

In addition to nutrients in the soil, the presence or absence of certain vitamins and hormone-like substances in plants can influence the well-being of animals. The presence of estrogens in such plants as alfalfa, birdfoot trefoil, and ladino clover has been related to reproductive difficulties of herbivores feeding on such plants (see Samuel, 1967).

Other studies show a more general relationship between the level of soil fertility and the health of animals. Studies of the relationship between soil and the rabbit in Missouri showed no significant difference in body weight of rabbits collected from soils of contrasting fertility (C. E. Williams, 1965), but there was a positive relationship between soil fertility and fecundity (Williams and Caskey, 1965). Rabbits from areas of highly fertile soils had significantly larger litters. Analyses indicated that this difference in fecundity resulted from contrasts in soil fertility, not differences in latitude, age, or stress.

The size of deer, their antler development, and their reproductive success all relate to nutrition (Fig. 7-2). Only deer obtaining high-quality food grow large antlers; deer on diets low in calcium, phosphorus, and protein are stunted in growth, and the bucks develop only thin spike antlers (French et al., 1955). Reproductive success of does is highest where food is abundant and nutritious. On the best range in New York State, 1.71 fawns on the average were born for each doe. On a poor range, however, the average fawn production was only 1.06 for each doe (Cheatum and

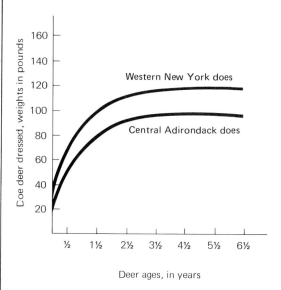

Deer ages, in years

FIGURE 7-2
Differences in the dressed weights of doe deer from a good range in western New York to a poor range in the central Adirondacks. (After Severinghaus and Gottlieb, 1959.)

Severinghaus, 1950). A higher proportion of well-fed young does, including fawns, conceive than those occupying range where food is low in quality and quantity (Taber, 1953).

Some animals appear to be able to select the most nutritious forage available. Free-ranging deer feeding on farm crops selected forages that gave them 12 percent more lipids (ether extract), 38 percent more calcium, and 34 percent more phosphorus (Swift, 1948). In California, mule deer concentrate on vegetation growing in pockets of deep soil scattered throughout regions of poor soil (Taber and Dasmann, 1958). Among the black-tailed deer of Washington, trailing blackberry (*Rubus nacropetalus*) makes up to 30 percent of the diet. When nutritionists determined the nutrient content of the plants they found that it had the highest nitrogen-free extract (which comprises sugars, starches, and hemicellulose) and the lowest crude fiber of all the forage species.

If deer show some preference for the highest-quality forage, this preference is due to improved palatability rather than any conscious recognition of those plants that have the greatest nutritive value. The sense of smell probably plays an important role in selection (Longhurst et al., 1968). Certain volatile substances in different plants not only give each plant a certain aroma, but also inhibit the growth of rumen bacteria. The least palatable plants happen to be those with the volatile substances most inhibiting to bacterial growth in the rumen.

Red grouse, hares, and rabbits inhabiting the Scottish moors also show a preference for the most nutritious foods (Miller, 1968). Some plots of heather were fertilized; others were not. In winter, but not summer, grouse preferred heather with a high level of nitrogen. In winter hares and rabbits selected heather with a high content of nitrogen and phosphorus; in summer they selected for high nitrogen content only.

The need for quality foods differs among herbivores. Ruminant animals, for example, can subsist on rougher or lower-quality forage because the rumen can synthesize such requirements as vitamin B_1 and certain amino acids from simple nitrogen compounds. Thus both the caloric content and the nutrient status of a certain food item might not reflect its real nutritive value for the ruminant. Non-ruminant herbivores require more complex proteins and may carry bacterial symbionts in the cecum. Seed-eating herbivores exploit the concentration of nutrients in the seed. Such herbivores are not likely to destroy their habitat, but when seeds are scarce or absent these animals must leave the area or starve. Such animals rarely have a dietary problem.

Among the carnivores, quantity is more important than quality. Carnivores rarely have a dietary problem because they consume other animals that have stored up and resynthesized proteins and other nutrients from plants. What they fail to obtain from an animal source, carnivores can obtain by consuming some plant concentrate such as berries and fruits.

A necessary nutrient for all organisms is protein. Essential for growth, proper development, weight gain, milk secretion, and other functions, a liberal continuous supply is needed throughout life, particularly during the period of rapid growth. Among the grazing herbivores protein requirements are essentially the same for all, both domestic and wild. The white-tailed deer requires 13 to 16 percent protein for optimum growth and a minimum of 6 to 7 percent for maintenance. If crude protein levels in deer forage fall below this, rumen functions are seriously impaired (Dietz, 1965). Deer fed on diets containing only 7 percent protein developed more slowly and had lower survival of fawns than deer fed more protein (Murphy and Coates, 1966). Omnivorous species such as wild pigs and most rodents require at least 15 percent protein in their diet, and lagomorphs require protein levels of 15 to 20 percent. Seed- and insect-eating bobwhite quail need up to 28 percent protein (Nestler, 1949).

Among the herbivores, the nutritive values of the food and the nutritive requirements of the animals follow a seasonal cycle. Because it has been rather widely studied, the diet and nutrition of the deer serves as an excellent example. Throughout early and mid-winter deer have to rely on dormant plants, dried vegetation, and woody twigs, all characterized by minimum protein. In late winter and early spring, when the protein content increases, the protein requirements of the deer for lactation and growth of fawns also reaches its maximum. After flowering and setting seed, the protein contents of plants decline, but by having available a variety of various plants, deer on good range can meet their protein needs.

As long as the deer population is not so great as to tax the available food supply, deer are able to meet minimum protein requirements throughout the winter. But if numbers are high, deer are forced to consume low-quality forage that cannot supply the protein needed, even though the animal may be getting sufficient food to fill the rumen. The deer has a rumen with a small capacity and a high metabolic rate (26 kcal/kg/day). To meet its energy demands, the deer requires highly nutritious, easily digested food. When the diet changes from one high in carbohydrates and crude protein to one high in crude fiber, the food has to remain longer in the rumen to be digested. A slower rate of passage of food through a digestive tract together with low protein can produce malnutrition.

Malnutrition then can affect deer herds in a number of ways. Some effects are death, increased sterility, reduced ovulation, weak fawns, and failure of does to permit the fawns to suckle. Thus the quality of food available to the herbivore can change its population and influence energy flow through the ecosystem.

Temperature

Environmental temperatures fluctuate both daily and seasonally. Temperatures of any one area will vary from sunlight to shade and from daylight to dark. Surface temperatures of soil may be 30° C higher in the sunlight than in the shade and up to 17° C higher during the day than during the night. On the desert this spread may be as high as 40° C. Temperatures on tidal flats exposed to the sun may rise to 38° C; in a few hours these same flats are covered by water at 10° C. Seasonal fluctuations may be just as extreme. In North Dakota, where the annual mean temperature is between 5° and 9° C (36° to 44° F), yearly temperatures fluctuate from a low of −43° C (−56° F) in winter to 49° C (120° F) in summer. In the mountainous eastern state of West Virginia, where the mean annual temperature is 12° C (54° F), temperatures range from −37° to 44° C (−34 to 108° F).

The ability to withstand extremes in temperature varies widely among plants and animals, but there are temperatures above and below which no life can exist. A temperature of 52° C is about as high as any animal can withstand and still grow and multiply. Among plants, however, some hot-spring algae can live in water as warm as 73° C under favorable conditions (Brock, 1966), and some arctic algae can complete their life cycles in places where the temperatures barely rise above 0° C. Nonphotosynthetic bacteria inhabiting hot springs can actively grow at temperatures greater than 90° C (Bott and Brock, 1969).

An animal's response to wide ranges in temperature is influenced by its physiology. Invertebrates, fish, amphibians, and reptiles have no internal mechanism for temperature regulation, and their body temperatures vary with external conditions. Such animals are called *poikilothermal*. These animals are active during the warm seasons; during winter, most of them are dormant, except for fish and social insects such as the honeybee. By constant vibration of heavy wing muscles, the bees are able to maintain temperatures within the hive a few degrees above the air outside. Most poikilothermal animals become inactive when the temperature of their surroundings goes below 8° C (43° F) or rises to 42° C (108° F).

Birds and mammals can, within limits, maintain constant body temperatures, regardless of temperature variations of air and water. Such animals are termed *homeothermal*. Their life processes are adjusted to function at the animal's normal temperature, averaging a little less than 38° C (100° F) in mammals and 3 to 4° higher in birds. If its temperature control fails, the animal dies. Although these animals cannot withstand serious temperature changes within the body, they are active during all seasons of the year and are independent of temperature change in the external environment.

All living organisms apparently have a temperature range outside of which they fail to grow or reproduce. Within the favorable range, organisms have an optimum or preferred temperature, at which they best maintain themselves. The optimum temperature may vary within the life cycle or with the process involved. Optimum temperature for photosynthesis is lower than that for respiration. If the temperature goes much above the optimum for photosynthesis the plant may not be able to balance energy fixation with respiration. The seeds of many plants will not germinate, and the eggs and pupae of some insects will not hatch or develop normally until chilled.

Brook trout grow best at 13° to 16° C, but the eggs develop best at 8° C.

Temperature influences the speed and success of development of poikilothermic animals. In general complete development of eggs and larvae is more rapid in warm temperatures. Trout eggs, for example, develop four times faster at 15° C than at 5° C. Similar examples can be found among insects. The chironomid fly, *Metriocnemus hirticollis*, requires 26 days at 20° C for the development of a full generation, 94 days at 10° C, 153 days at 6.5° C, and 243 days at 2° C (Andrewartha and Birch, 1954).

TEMPERATURE AND DISTRIBUTION

Because the optimum temperature for the completion of the several stages of the life cycle of many organisms varies, temperature imposes a restriction on the distribution of species. Optimal temperatures for some species are so different from those of others that the animals cannot inhabit the same area. Some organisms, particularly plants that are growing under suboptimal temperatures, cannot compete with the surrounding growth, a situation that would not exist under optimal conditions.

Generally the range of many species is limited by the lowest critical temperature in the most vulnerable stage of its life cycle, usually the reproductive stages. Although the Atlantic lobster will live in water with a temperature range of 0° to 17° C, it will breed only in water warmer than 11° C. The lobster may live and grow in colder water, but a breeding population never becomes established there.

A classic example of temperature limitation on animal distribution is found among four species of ranid frogs (J. A. Moore, 1949*a*). The wood frog breeds in late March, when water temperature is about 10° C. Its eggs can develop at temperatures as low as 2.5° C. Larval stages transform in about 60 days. This frog ranges into Alaska and Labrador, further north than any other North American amphibian or reptile. The meadow frog breeds in late April, when water temperatures are about 15° C, and the larvae require around 90 days to develop. As a result the northern limit of its range is southern Canada. The southernmost species of the three, the green frog, does not breed until the water is about

25° C, and the eggs will develop at 33° C, a lethal temperature for the others. Its eggs, however, will not develop at all until the temperature exceeds 11° C. The range of the green frog extends only slightly above the northern boundary of the United States.

A number of examples can be found among plants. The northern limit of the sugar maple closely parallels the 35° F mean annual isotherm. The paper birch, a cold-climate species, is found as far north as the 53° F July isotherm and seldom grows naturally where the average July temperature exceeds 70° F. The distribution of the black spruce follows a similar pattern. Its southern distribution is approximately the same as that of the paper birch, and its northernmost outliers are seldom far from the mean July isotherm of 51° F. In southeastern North America the northern limit of the loblolly pine is set by winter temperature and rainfall.

Some plants, such as blueberries, will not flower or fruit successfully unless chilled. Other plants may grow in a region during the summer, but are unable to reproduce or grow to normal size because the twigs freeze back in winter or are killed by late spring frosts. These plants are restricted in their natural range to areas where temperatures are favorable for growth and reproduction.

ATMOSPHERIC TEMPERATURES

Another important aspect of environmental temperature is the daily heating and cooling of the air masses.

The earth's atmosphere is a gas. When the pressure of a gas changes, the volume and temperature also change. Any column of air above the earth has a measurable mass and exerts a pressure proportional to its height. The atmospheric pressure at high altitudes is therefore lower than that at low altitudes. If a mass of air rises it moves into a region of lower pressure, and its temperature decreases at the rate of 5.5° F/1,000 ft. This rate of change in the temperature with the rate of change in height is called the *environmental lapse rate*. If the air mass moves downward to lower elevations, the air is compressed and warmed at this same rate. Any process in which heat is neither gained from nor lost to the surroundings is called an *adiabatic process*. The temperature change that accompanies a pressure change is called a *dry adiabatic change*.

177

If a mass of air is cooled at the dry adiabatic rate, it will rise; if it becomes immersed in warmer air, it will fall to a level where the surrounding air has the same temperature. If an air mass descends but the region it encounters is cooler and denser, it will rise to its original level. When this condition prevails the atmosphere is said to be stable. If a parcel of air moves up and tends to remain at the new level, the air mass is termed neutral. A layer of such air in which the temperature increases with height is extremely stable. Such a layer is called an *inversion*. But if the air mass moves up or down on its own accord and continues to rise or fall, the atmosphere is unstable.

If a parcel of air moving up and down condenses its moisture as it cools, it gains some heat of condensation. Thus its rate of temperature change is less than a dry adiabatic rate. The rate varies with the amount of water vapor in the atmosphere. Usually it is somewhere between 2° and 5° F/1,000 ft. This is known as the *moist adiabatic rate*. The surrounding air is judged to be stable, neutral, or unstable by comparing its lapse rate with the moist adiabatic rate. In mountain country it is not uncommon for the air to change its lapse rate from dry to moist as the air cools sufficiently to condense its moisture.

Instability of the atmosphere is created by differential heating of the earth and lower atmosphere. The earth by day is heated by short-wave solar radiation (Fig. 7-3a), which is absorbed in different amounts over the land depending on vegetation, slope, soil, season and so on. Lower layers of the atmosphere, heated by the earth's surface by radiation, conduction, and convection, rise in small volumes; colder air falls. This turnover of the atmosphere produces a turbulence that is increased by winds.

After the sun goes down, the earth begins to lose heat (Fig. 7-3b). This nighttime radiational cooling is most pronounced on calm clear nights. The layer of surface air is cooled while the air aloft remains near daytime temperatures. This can form a surface inversion in which temperature increases rather than decreases with altitude. In mountainous or hilly country, cold dense air flows down slopes and gathers in the valleys. The cold air then is trapped beneath a layer of warm air (Fig. 7-4). Such inversions trap impurities and other air pollutants. Smoke from industry and other

heated pollutants rise until their temperature matches the surrounding air. Then they flatten out and spread horizontally. As pollutants continue to accumulate, they may fill the entire area with smog. Such inversions are most intense if the atmosphere is stable. Inversions break up when surface air is heated during the day to create vertical convections up through the inversion layer, or when a new air mass moves in.

Similar but more widespread inversions occur when a high-pressure area stagnates over a region. In a high-pressure area the air flow is clockwise and spreads outward. The air flowing away from the high must be replaced, and the only source for replacement air is from above. Thus surface high-pressure areas are regions of sinking air movement from aloft, called *subsidence*. When high-level winds slow down, cold air at high levels in the atmosphere tends to sink. As this air moves from the high altitude and low atmospheric pressure to a lower altitude and higher atmospheric pressure, it heats up at the dry adiabatic rate. The sinking air becomes compressed as it moves downward, and as it warms it becomes drier. As a result, a layer of warm air develops at a higher level in the atmosphere (Fig. 7-5). Rarely reaching the ground, it hangs several hundred to several thousand feet above the earth, forming a subsidence inversion. Such inversions tend to prolong the period of stagnation and increase the intensity of pollution. The subsidence inversion that brings about our highest concentrations of pollution is often accompanied by lower-level radiation inversions.

Along the west coast of the United States, and occasionally along the east coast, the warm seasons often produce a coastal or marine inversion. In this case cool, moist air from the ocean spreads over low land. This layer of cool air, which may vary in depth from a few hundred to several hundred thousand feet, is topped by warmer, drier air, which also traps pollutants in the lower layers.

THERMAL POLLUTION

A new dimension is being added to aquatic ecosystems. It goes by a variety of names: thermal pollution, thermal loading, thermal enrichment. It has been with us in some form for years. Most dams, even beaver dams, raise the temperature of the streams and rivers they retain. Irrigation practices and industrial and steam electrical plants have been returning heated waters back to rivers and lakes. In recent years, however, with the tremendous increase in electrical power plants and their need for cooling water, thermal pollution has become a problem of great concern. The term *thermal pollution* must be used with discrimination. Heat becomes a pollutant only when it is inimical to the interests and needs of man, including adverse effects on the aquatic ecosystem.

Creating the greatest amount of thermal pollution are steam electric stations. In producing electricity by steam generation, whether by coal or nuclear power, about 60 to 70 percent of the energy is lost as heat. The ideal way to get rid of the heat is to release it to the atmosphere. Because water can store large quantities of heat with a small temperature increase, it is most economical to use water as the coolant and to allow it to discharge the heat to the air. In this process fresh water from a river or lake generally is circulated through the condensers and back to the river or lake again, without precooling. More expensive cooling towers, of either the wet or evaporative type or the nonevaporating or dry type (see R. L. Smith, 1972), reduce water temperature, permit the recirculation of the same water through the plant, and reduce thermal pollution.

The amount of water used for industrial cooling has increased tremendously in the past decade, and most of the increase is directly attributable to greater demand for and production of electricity. In 1958 steam-generating utilities required 90 billion gal of water daily for cooling purposes. In 1965 the electric power industry used 128 billion gal of water/day. By the year 2000, if no other methods of cooling are introduced and if nuclear power becomes the common source of power, the cooling water requirements will be 1,250 billion gal/day. If 30 percent of the power-generated waste heat is discharged to seawater cooling systems, 875 billion gal of cooling water/day will have to be supplied by inland fresh-water sources. The average daily runoff in the United States is about 1,200 billion gal/day, so the power industry would need nearly three-quarters of the total runoff for cooling purposes (J. H. Wright, 1970). This is a nationwide average. Locally, the problems could be much greater, especially during the

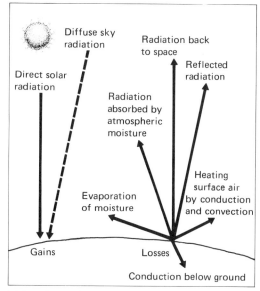

(a) Daytime surface heat exchange

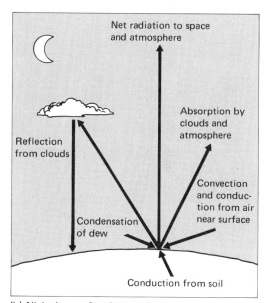

(b) Nighttime surface heat exchange

FIGURE 7-3
(a) *Solar radiation that reaches the earth's surface in the daytime is dissipated in several ways, but heat gains exceed heat losses.* (b) *At night there is a net cooling of the earth's surface, although some heat is returned by various processes.* (*Adapted from* Fire Weather, *USDA Agricultural Handbook 360.*)

45°

50°

45°

40°

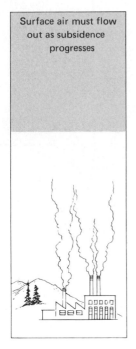

Surface air must flow out as subsidence progresses

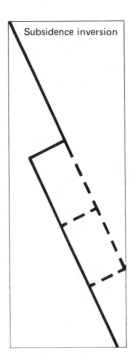

Subsidence inversion

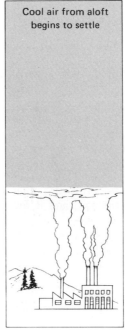

Cool air from aloft begins to settle

Warm, very dry air approaches the surface

summer, when the flow is naturally low. In some Delaware River watersheds, power plants utilize 150 percent of the fresh-water flow by reusing it. Power plants can increase water temperatures 10° to 30° F above ambient temperatures. In summer the discharges may reach 100° to 115° F. Except for local situations, thermal pollution is not yet a major problem, but with the increased demands for power and the extensive construction of a number of steam electric plants on the same waterways, it is a threat to the aquatic ecosystems.

Heating in natural waters places a stress on aquatic life. Through long periods of evolution, poikilothermal aquatic organisms have become adapted to a particular temperature range. Each species has an optimum for each stage in its life cycle. Within limits these organisms are able to adjust to higher or lower temperatures (see Chapter 15). Under natural conditions with a slow rise and fall in temperature, fish and other organisms gradually adjust to changing environmental temperatures. But if a new temperature regime is imposed quickly, it can produce thermal shock and death.

Thermal shock is an extreme result of thermal pollution. There are other effects, often more insidious, that result from introducing heated effluents into the waterway. The metabolic processes of organisms are affected (see Chapter 15). Because the body temperatures of poikilothermic animals are influenced by environmental temperatures, a rise in environmental temperature increases the metabolic rates of fish and aquatic invertebrates. This in turn increases their oxygen demands. At the same time high temperatures decrease the oxygen content of the water when the organisms need it the most. For fish the situation is further aggravated by the reduced affinity of hemoglobin for oxygen. Increased oxygen demand, its decreased availability, and the reduced efficiency for obtaining it cause severe physiological stress, and even death. In summer a temperature rise of only a few degrees may cause 100 percent mortality among fish and invertebrates, especially those living near the southern extremities of their ranges.

Temperature tolerances vary with the species. Cold-water fish, represented by such salmonids as the brook trout (*Salvelinus fontanilis*) and such invertebrates as the stonefly (*Plecoptera*), are very sensitive to warm water.

FIGURE 7-4 [OPPOSITE TOP]
Topography plays an important role in the formation and intensity of nighttime inversions. At night air cools next to the ground, forming a weak surface inversion in which the temperature increases with height. As the cooling continues during the night, the layer of cool air gradually deepens. At the same time cool air descends downslope. Both cause the inversion to become deeper and stronger. In mountain areas the top of the night inversion is usually below the main ridge. If air is sufficiently cool and moist, fog may form in the valley. Smoke released in such inversions will rise only until its temperature equals that of the surrounding air. Then smoke flattens out and spreads horizontally just below the thermal belt. (Adapted from Fire Weather, *USDA Agricultural Handbook 360.)*

FIGURE 7-5 [OPPOSITE BOTTOM]
The descent of a subsidence inversion may be followed by successive temperature measurements as shown by the dashed lines. As the more humid air flows outward, the drier air aloft is allowed to sink and warm adiabatically. (Adapted from Fire Weather, USDA Agricultural Handbook *360.)*

The brook trout rarely occurs in water whose temperatures exceed 20° C. The optimum temperature range for its food source, the stonefly, is 10° to 15° C (Gaufin, 1965). Warm-water species are much more tolerant of high temperatures. The preferred temperature of large-mouth bass (*Micropterus dolomieui*) is 27° C; for bass acclimatized to 30° C the lethal temperature is 35.4° C. *Trichoptera* or caddisfly larvae occur in water up to 35° C (Robach, 1965), whereas the majority of the Tendipedidae or midge larvae live in water 30° to 32.6° C. (Curry, 1965).

The temperature of the water can be well within the range of tolerance for one species and still have an adverse effect on its population if it limits or eliminates the food supply. In the Patuxent River estuary of Chesapeake Bay the juvenile striped bass, *Roccus saxatilis*, feeds heavily on the opossum shrimp, *Neomysis americana*. This shrimp, a northern latitude species, is at the southern extremity of its range in the Chesapeake Bay. It is highly intolerant of high temperatures, and cannot survive above 31° C (Mihursky and Kennedy, 1967). This temperature occasionally is reached and even exceeded in the waters of Chesapeake Bay.

The behavior of fish can also be changed or modified by heated waters. Warmer water in winter may cause spawning out of season. Most aquatic organisms reproduce when a given temperature level is reached after a period of temperature change. If the temperature change, normally a seasonally dependent environmental phenomenon, is altered by introducing heated water, then the fish may spawn out of season or may experience a protracted spawning season (Naylor, 1965). Mayflies and other aquatic insects may hatch early and emerge into an environment too cold for them to mate or lay eggs.

Altered temperatures interfere with the migratory behavior of fish. Anadromous fish, highly sensitive to temperature, have rigid time schedules for upstream and downstream migrations. High temperatures act as a barrier both to the spawning of adults and to the downstream migration of young. Heated water also interferes with the timing of migration and may actually inhibit movements altogether. Some fish, such as the bluefish, may remain in the plume of heated water instead of leaving the area for naturally warmer water (A. C. Jensen, 1970).

Artificially warming natural waters can change the structure and functioning of aquatic ecosystems. Results of a number of studies suggest the impact heavy thermal pollution can have on aquatic environments. Most species of fish are eliminated from the zone of maximum heat in summer (Trembley, 1965), and the diversity, numbers, and biomass of invertebrate fauna of the riffles are reduced substantially (Coutant, 1970; Warringer and Brehmer, 1966). As temperatures rise, diatoms, which thrive in cool water, are replaced by green algae, and at high temperatures by blue-green algae (see Coutant, 1970). In fresh-water streams such cold-water species as trout, dace, and stoneflies are replaced by large-mouth bass and carp, and even these fish disappear if water temperatures exceed 30° C.

Disease, parasitism, and predation may be influenced by changing temperature. Heated water may bring new predators into the area, and it may influence the ability of the predators to capture prey, or of the prey to escape their predators. Experimental brook trout living in 63° F water were comparatively slow in capturing minnows, and at 70° F they were incapable of capturing them. Thus, increased temperatures favored the prey species and were detrimental to the trout, which literally starved to death in the midst of plenty. Warming of the Columbia River by nuclear power plants encouraged a once rare but deadly bacterial disease of fish, columnaris, to flourish (Stroud and Douglas, 1968). It has caused extensive mortality among salmon ascending the river.

Extension or contractions of natural range and the introduction and spread of exotic species are other possible effects of thermal pollution (Naylor, 1965). In marine environments where certain cold-water species such as the American lobster and Ipswich clam reach their southernmost distribution and certain warm-water species are at the northernmost part of their range, heated effluents can tip the scales in favor of the warm-water species, enabling them to breed and expand their range, something they were unable to do at former temperatures. Northern species would be forced to retreat into cooler waters. On the other hand, in areas where the ranges of such species do not overlap there may be a scarcity of replacement species. In such situations the niche remains vacant unless some species

adapted to warmer water are deliberately introduced or accidentally carried there by currents or on the hulls of boats (Naylor, 1965).
Often such colonizers are species that are able
to live on pilings and other structures located
near discharges of heated effluents.

A rise in water temperature speeds up the
rate of decomposition of organic matter. This
increase in bacterial action lowers oxygen
supplies, especially in summer months. Great
loads of nitrogen and phosphorus poured into
cool water may produce a minimal immediate
biological effect, but a rise in temperature in
the presence of these nutrients can trigger
massive blooms of algae.

Moisture

Moisture relationships within an ecosystem
closely relate to the distribution of rainfall.
Mean annual precipitation is influenced by
topography and mass air movements (Fig.
7-6). Mountains along the east and west
coasts of North America interfere with the
even sweep of the winds off the oceans across
the continent and intercept the moisture they
contain. This causes excessive moisture on the
windward side of the mountains and local and
regional rainshadows on the leeward side.

Seasonal distribution of rainfall is more important than average annual precipitation. A
world of difference exists between a region
receiving 50 in. of rain rather evenly distributed throughout the year and a region in
which nearly all of the 50 in. falls during a
several-month period. In this latter situation,
typical of tropical and subtropical climates,
organisms must face a long period of drought.
An alternation of wet and dry seasons influences the reproductive and activity cycles of
organisms as much as light and temperature
in the temperate regions. For example, the
coming of the dry season to the African plains
initiates the migrational movements of the
large herbivores. In southwestern United States
the lack of winter rains fails to stimulate the
growth of green vegetation, needed by Gambel's quail as a source of vitamin A. This in
turn inhibits the reproductive activity of the
bird, which is reflected in reduced numbers
of offspring.

The moisture content of the air is usually
expressed as *relative humidity*. This is the
percent of moisture relative to the amount of

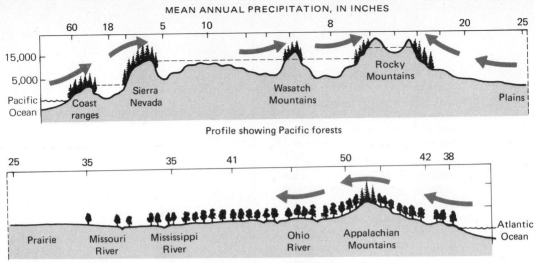

MEAN ANNUAL PRECIPITATION, IN INCHES

Profile showing Pacific forests

Profile showing Atlantic forests

water that the air holds at saturation at the existing temperature. As the air warms up, the relative humidity drops (as long as the moisture content of the air is constant), because warm air can hold more moisture than cool air. As the relative humidity drops, the vapor pressure deficit—the difference between the partial pressure of water at saturation and the prevailing vapor pressure of the air—increases, and increased evaporation takes place.

The relative humidity varies during a 24-hour period. Generally it is lower by day and higher by night. During the day relative humidity under a closed canopy is higher than it is on the outside, and it is lower than the outside during the night. Over normal surfaces relative humidity during the day usually increases with height because of the decrease in temperature with height. This contrasts with absolute humidity, which decreases with height. But if a subsidence inversion occurs, especially at night, the relative humidity decreases upward to the top of the inversion, then changes little or increases only slightly.

In any one area relative humidity varies widely from one spot to another, depending upon the terrain. Variations in humidity are most pronounced in mountain country. Low elevations warm up and dry out earlier in the spring than high elevations, and soil moisture becomes more depleted later in summer. The daily range of humidity is greater in the valley and the least at high elevations. Because daytime temperatures decrease with altitude, as does the dew point, relative humidities are greater on the tops of mountains than in the valleys. As nighttime cooling begins, the temperature change with altitude is reversed. Cold air rushes downslope and accumulates at the bottom. Through the night, if additional cooling occurs, the air becomes saturated with moisture and fog or dew forms by morning. These differences in humidity can produce vegetative differences on mountain slopes. They are most pronounced on the slopes of the mountains along the Pacific coast.

Temperature and wind both exert a considerable influence on evaporation and relative humidity. An increase in air temperature causes convection currents. This sets up an air turbulence, which mixes surface layers with drier air above. Wind movements associated with cyclonic disturbances also mix moisture-laden air with drier air above. As a result, the vapor pressure of the air is lowered and evaporation from the surface increases.

MEETING MOISTURE PROBLEMS

Both excessive and deficient moisture can be detrimental to organisms, but many are adapted to meet moisture extremes in one way or another. Because many invertebrates and amphibians must avoid dry air to prevent desiccation, they are found in moist or aquatic situations and are active chiefly at night, when humidity is highest. Many invertebrates are quite sensitive to moisture variations and will leave an area of low humidity to go to one where the moisture is more favorable. A reverse movement to a drier situation occurs among animals that are sensitive to high humidity. The reaction, however, of animals to moisture is influenced by the condition of the animal, the temperature, the light intensity, and the like.

Survival during long periods of dry weather or existence in arid regions is achieved by organisms in various ways. Some plants of the dry country are small annuals that complete their life cycles in a very short time during and immediately after the rainy season, before the soil dries out. They survive the dry period as seeds. Other plants of the arid regions may be succulents, possessing reduced intercellular space and enlarged vacuoles in which water accumulates during the rainy season. They may also have shallow root systems, which enable them to absorb the maximum amount of water during the rainy season. Other plants either have deep root systems that can tap deep moisture supplies or they go dormant during the dry season. Mesquite, for example, has a root system that penetrates to a depth of 175 ft (W. S. Phillips, 1963). Or the plants may have evergreen leaves that are heavily waxed and resistant to dessication.

Plants in regions with greater moisture reduce transpiration and survive the physiologically dry period of winter by shedding leaves. Waxy secretions on buds and twigs also help reduce water loss. Plants react to dry conditions by rolling, curling, or folding their leaves to reduce the unit area exposed to drying air and sun. Long periods of dry weather that result in soil drought reduce plant growth, cause die-back of plants or outright death. Drought-injured plants are susceptible to attack by insects and are highly susceptible to fire. Drought can also influence the composition of plant communities. Weaver and Albertson (1956) found that during the 1933–1939 drought on the plains, buffalo grass either disappeared entirely from some ranges or was reduced to small scattered patches. Its more resistant associate, blue grama, was never killed uniformly and persisted. When the drought ended, buffalo grass responded rapidly to moisture and became the dominant plant.

Animals of the desert confine their activities to the period between sunset and sunrise, when the humidity is the highest and the temperature the lowest. Desert amphibians burrow into beds of temporary streams where the moisture is high and remain there during the day or during extensive dry spells. Desert mammals such as the kangaroo rat and many insects that live in dry situations possess physiological mechanisms that enable them to conserve metabolic water (see Chapter 15).

FIGURE 7-6 [OPPOSITE]
The influence of moist air currents on forest distribution in the East and the West along the thirty-ninth parallel of north latitude. (Adapted from Climate and Man, *1941 USDA Yearbook of Agriculture.)*

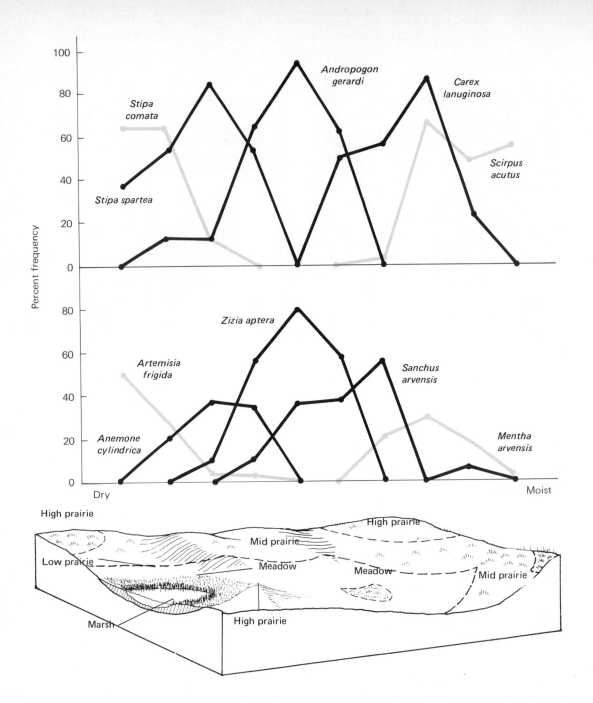

Percent frequency

100

80 — *Stipa comata*

60 — *Stipa spartea*

Andropogon gerardi

Carex lanuginosa

Scirpus acutus

Zizia aptera

Artemisia frigida

Sanchus arvensis

Anemone cylindrica

Mentha arvensis

Dry — Moist

High prairie

Mid prairie

Low prairie

Meadow

Meadow

High prairie

Mid prairie

Marsh

High prairie

High prairie

Many animals go into a dormant state during dry periods. The flatworm (*Phagocytes vernalis*), which occupies ponds that dry up during the summer, encysts. Other aquatic or semiaquatic organisms retreat deep into the soil until they reach groundwater level. Still others become dormant during the dry season. Many insects undergo diapause, just as they do when confronted with unfavorable temperatures.

Moisture also influences the speed of development and even the fecundity of some insects. If the air is too dry, the eggs of some

locusts and other insects may become quiescent. There is an optimum humidity at which nymphs develop fastest. Some insects lay more eggs at certain relative humidities than above or below that point.

Too much water may be as detrimental as too little. High water tables result in shallow-rooted trees that are easily toppled by the wind and are sensitive to drought and frost. Terrestrial plants subject to prolonged flooding, particularly during the growing season, will die from the lack of oxygen about the roots. This happens frequently to trees in areas where floods are caused by beavers. Heavy rain and prolonged wet spells cause widespread death among mammals and birds from drowning, exposure, and chilling. Excessive moisture and cloudy weather kill insect nymphs, inhibit insect pollination, and spread parasitic fungi, bacteria, and viruses among both plants and animals.

MOISTURE AND PLANT DISTRIBUTION

Moisture or the lack of it can have a major influence on the distribution of plants on both a geographic and local basis. For example the western red cedar and the western hemlock grow where the average annual rainfall in western North America is around 32 in. (Little, 1971). The influence of moisture on the local distribution of plant communities is well illustrated by the grassland vegetation of the central plains. In a study of the prairie, meadow, and marsh vegetation of Nelson County, North Dakota, Dix and Smeins (1967) divided the soils into ten drainage classes from excessively drained to permanent standing water. They determined the indicator species for each drainage class and then divided the vegetational display into six units corresponding to the drainage pattern (Fig. 7-7). The uplands fell into high prairie, mid-prairie, and low prairie, and the lowlands into meadow, marsh, and cultivated depressions. High prairies dominated the excessively drained areas, and they were characterized by stands of needle-and-thread grass, western wheatgrass, and prairie sandweed. The mid-prairie, considered to be the climax or true prairie, was dominated by big and little bluestem porcupine grass and prairie dropseed. Low prairie on soils of moderate moisture was characterized by big bluestem, little bluestem, yellow Indian grass, and muhly. Lowlands that occupied soils in which the drainage was sluggish and the water table within the rooting depth of most plants were characterized by canary grass, sedge, and *Scolochloa festucacea*. Meadows on even wetter soils were dominated by northern reed grass, wooly sedge, and spikerush. Marshes that contained permanently standing water contained stands of reed, cattails, and tule bulrush. Cultivated depressions were usually colonized by spikerush and water plantain.

Although each drainage class supported a characteristic stand of vegetation, no one spe-

FIGURE 7-7 [OPPOSITE]
Prairie vegetation forms a mosaic that is influenced by topography and drainage regimes. Below is a hypothetical block diagram of North Dakota landscape, showing the relative positions of vegetational units. Uplands: high, mid, and low prairie. Lowlands: meadow and marsh. Above are distributional curves of selected species along a drainage gradient regime. Excessive drainage = high prairie. (After Dix and Smeins, 1967. Reproduced by permission of the National Research Council of Canada from the Canadian Journal of Botany, 45:21–58.)

cies was associated solely with another species. Each species had its own set of optimal drainage requirements and behaved independently of other species. At any particular site the drainage conditions influenced the combination of dominant plants (see Chapter 8). Because of this each community blended into the other. The only sharp breaks came where the drainage differences were sharp and severe.

In a similar manner soil moisture influences the distribution of woody plants in the northern Rocky Mountains. In northern Idaho, warm, dry lowland soils, droughty in summer, support Idaho fesque and snowberry. As soil moisture improves upslope, ponderosa pine and Douglas fir replace the former plants; on midslope grand fir appears, and still higher western red cedar, western hemlock, and subalpine fir. On the upper parts of south-facing slopes where the soil may become as droughty as that of the lowlands, wheat grass and Idaho fesque replace woody plants (Daubenmire, 1968a).

INTERACTION OF TEMPERATURE AND MOISTURE

A close interaction exists between temperature and moisture in terrestrial environments; and the two determine in large measure the climate of a region and the distribution of vegetation. Low-moisture conditions are more extreme when temperatures are high or low. Moisture in turn influences the effects of temperature, a fact observable to everyone. Cold is more penetrating when the air is moist, and high temperatures are more noticeable when relative humidity is high.

A climograph is a plot of the mean monthly temperatures against the mean monthly relative humidities or precipitation, and gives a composite picture of the climate of an area (Fig. 7-8). Twelve dots for the year connected together form an irregular polygon, which can be compared with that for another area for similarity or difference in fit. By use of this, climates can be compared much more easily than they can by tables. Such climatographs are useful to contrast or compare one region or one year with another. Often this is done to determine the suitability of an area for the introduction of exotic animals, particularly game birds.

East-west zonation of vegetation follows a pattern of moisture distribution more than that of temperature. If temperature alone controlled plant distribution in North America, the vegetation zones would be in broad belts running east and west. Only in the far north do the vegetation zones (tundra and the coniferous forest) stretch in these directions. Below this, vegetation is controlled by precipitation and evaporation, the latter influenced considerably by temperature. Because available moisture becomes less from east to west, vegetation follows a similar pattern, with belts running north and south. Humid regions along both coasts support natural forest vegetation. This zone is broadest in the east. West of this eastern forest region is a subhumid zone, where precipitation is less and evaporation is higher. Here the ratio of precipitation to evaporation is about 60 : 80 percent, and the land supports a tall-grass prairie. Beyond this is semiarid country, where the precipitation–evaporation ratio is 20 : 40 percent; it supports a short-grass prairie. To the west of this and on the lee of the mountains is the desert.

In mountainous country, both east and west, vegetation zones reflect climatic changes on an altitudinal gradient (Fig. 7-9). These belts often duplicate the pattern of latitudinal vegetation distribution. In general the belts include the land about the mountain base that has a climate characteristic of the region. Next is a higher montane level, which has greater humidity and temperatures that decrease as altitude increases. Here the forest vegetation changes from deciduous to coniferous. Beyond this is a subalpine zone that includes coniferous trees adapted to a more rigorous climate than the montane species. Above this is the alpine or tundra zone, where the climate is cold and cloudy. Here trees are replaced by grasses, sedges, and small tufted plants. Between the alpine and the subalpine lies the krummholz, a land of stunted trees. On the very top of the highest mountains is a land of perpetual ice and snow.

Wind

Because the atmosphere is constantly in motion, winds are a continual and highly variable influence on the environment. Winds, especially those near the earth's surface, are strongly affected by the topography and by local heating and cooling.

Winds that move with the leading edge of the air mass or frontal system or are carried by the general circulation winds aloft are *general winds*. In the mountains of western North America the circulation patterns of winds are interrupted by *foehns*, dry downslope winds. Their development depends upon a strong high-pressure system on the windward side of the mountains and a correspondingly well-situated low-pressure system on the leeward side.

In addition to broad general winds, there are local convective winds caused by temperature differences within a locality. The most familiar of these are the land and sea breezes experienced along ocean shores and large inland lakes and bays. Sea breezes flow inland from the water and bring in cool, moist marine air, often accompanied by morning fog. Land breezes at night are the reverse of daytime sea breezes. Air in contact with land becomes cooler at night, gains in density, and flows from land to water.

In mountainous topography, local winds can be exceedingly complex. Differences in heating of air over mountain slopes and canyon and valley bottoms result in several wind systems. Because of the larger heating surface, air in mountain valleys and canyons becomes warmer during the day. Similarly, a larger cooling area of the valleys causes a reversal of this situation at night. The resulting pressures causes a flow of air upslope by day and downslope by night. Combined valley and upslope winds exit at the ridgetops by day. As the slopes go into the shadows of late afternoon, the slope and valley winds shift direction and become downslope winds.

Local winds affect soil moisture and humidity and have an important bearing on forest fire conditions and local drought situations. Drying action of high, warm winds in winter when soil moisture is low or unavailable causes drought. The wind removes humid air about the leaves and increases transpiration. Losing more water than they are able to absorb, evergreens in particular dry out and their foliage turns brown.

Plants that normally grow tall become low and spreading when high winds are frequent and regular, a situation characteristic of the timberline. On the high wind-swept ridges, cushion plants with small, uniform, crowded branches are most common. Because of constant desiccation, cells of plants growing in

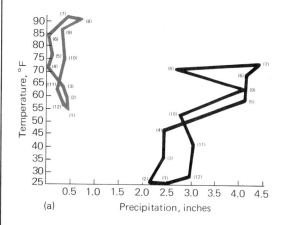

(a)

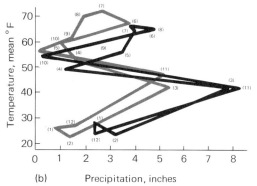

(b)

FIGURE 7-8

Temperature-moisture climographs. (a) A climograph comparing two very different regions, the desert and the eastern deciduous forest. Note how the hot dry desert climate differs graphically from the cool, temperate, moist climate of the east. These 12-sided polygons give a picture of temperature and moisture conditions and permit the comparison of one set of conditions with another. (Data: mean temperature and precipitation, 1941–1950, for Yuma, Arizona, and Albany, New York.) (b) A climograph comparing conditions on the rain-shadow side and the high rainfall side in the Appalachian Mountains of West Virginia.

these places never expand to normal size, and all organs are dwarfed. Terminal branch shoots are killed by desiccation, by blasting of ice particles, and, along the ocean, by the effects of salt spray. As a result the terminals are replaced by strong laterals, which form a mat close to the ground.

Strong and persistent winds blowing from a constant direction bend the branches of trees around to the windward side until, like a weathervane, they point out the direction of the prevailing wind. Often the wind may kill all the twig-forming buds, so that no limbs develop on the windward surface.

Shallow-rooted trees and trees with brittle woods, such as the willows, cottonwoods, and maples, are thrown or broken by strong winds. Windthrow is most prevalent among trees growing in dense stands that, through logging or natural damage, are suddenly exposed to the full force of the wind. Hurricanes and other violent windstorms sweeping over forested areas may uproot and break trees across a considerable expanse of land.

Hurricanes and strong storms cause deaths among animals also, and often carry individuals far from their normal environment and set them down elsewhere. Winds are an important means of dispersal for seeds and small animals such as spiders, mites, and even snails. (Andrewartha and Birch, 1954; Darlington, 1957).

Wind may also play a secondary role in the distribution of small mammals. The deeper accumulation of litter and snow in areas sheltered from the wind support more small mammals than exposed areas (Vose and Dunlap, 1968).

Light

Light influences ecosystems in two ways: it affects photosynthetic activity, and it influences the daily and seasonal activity patterns of plants and animals. The influence of light depends upon its intensity, quality, and duration. The photosynthetic rate of a plant leaf increases linearly with increasing light intensity up to the point of light saturation, usually one-tenth to two-tenths of full sunlight; after this point the rate remains the same. The efficiency of the photosynthetic process, however, declines steadily with increasing light intensity. Thus, a leaf exposed to full sunlight is not very efficient at utilizing light energy; at the

best it utilizes about 5 percent. At low light intensity the photosynthetic rate is lower, but efficiency increases and may approach that of 20 percent (Bonner, 1962). This fact, however, must not be interpreted as meaning that expected plant yields decrease under high light intensity. Actually the reverse is true; the greater the light intensity, the higher the yield, because more light reaches the lower leaves and even the lower layers of chlorophyll within the leaf.

The light intensity at which plants can no longer carry on sufficient photosynthesis to maintain themselves is called the *compensation intensity*. This is the point at which photosynthesis balances respiration, where plants are just able to replace material loss in respiration night and day. Few green plants can live where light intensity is less than 1 percent full sunlight, or about 100 footcandles.

Some plants are more shade-tolerant than others. Sugar maple, white cedar, and hemlock successfully exist under a dense forest canopy at low light intensities, but they do not attain normal growth there. In aquatic situations, light penetration is influenced by turbidity either from debris or from phytoplankton growth. Light stimulates the development of a dense phytoplankton growth at the surface, which prevents the light from reaching deeper water. This limits the growth of rooted aquatics.

Visible light that penetrates water becomes limited more and more to a narrow band of blue light at wavelengths of about 0.5 μ as the water depth increases from 0.1 to 100 m. This is in part the reason why water of deep, clear lakes looks blue. Eventually blue light is filtered out and the remaining green light is poorly absorbed by chlorophyll. Depths at which green light occurs are occupied by red algae, which possess supplementary pigments that enable them to utilize the energy of green light.

Periodicity and biological clocks

One aspect of communities with which everyone is familiar is rhythmicity, the recurrence of daily and seasonal changes. Bird song signals the arrival of dawn. Butterflies, dragonflies, and bees become conspicuous; hawks seek out prey, and chipmunks and tree squirrels become active. At dusk, light fades and daytime animals retire, the blooms of waterlilies and other flowers fold, and animals of the night appear. The fox, the raccoon, the flying squirrel, the owls, and the moths take over niches occupied by others during the day. As the seasons progress, day length changes and with it other conspicuous activities. Spring brings the migrant birds and initiates the reproductive cycles of many animals and plants. In fall the trees of temperate regions become dormant, insects and herbaceous plants disappear, the summer-resident birds return south, and winter visitors arrive. On the ocean shore the tide rises and falls about 50 minutes later each day and affects all life on the edge of the sea. Underlying these rhythmicities are the movements of the earth relative to the sun and the moon. The earth's rotation on its axis results in the alternation of night and day. The tilt of the earth's axis, along with its annual revolution around the sun, produces the seasons.

DAILY PERIODICITY:
THE CIRCADIAN RHYTHMS

Life evolved under the influences of daily and seasonal environmental changes, so it is natural that plants and animals would have some rhythm or pattern to their lives that would synchronize them with fluctuations in the environment. For years biologists have been intrigued over the means by which organisms kept their activities in rhythm with the 24-hour day, including such phenomena as the daily pattern of leaf and petal movements in plants, the emergence of insects from pupal cases, the sleep and wakefulness of animals (Fig. 7-10). At one time biologists thought that these rhythmicities were entirely exogenous, that is, that the organisms responded only to external stimuli such as light intensity, humidity, temperature, and tides. Laboratory investigations, however, indicate that this is not the complete answer.

At dusk in the forests of North America a small squirrel with silky fur and large, black eyes emerges from a tree hole. With a leap the squirrel sails downward in a long sloping glide, maintaining itself in flight with broad membranes stretched between its outspread legs. Using its tail as a rudder and brake, it makes a short, graceful, upward swoop that lands it on the trunk of another tree. This is *Glaucomys volans* the flying squirrel, perhaps

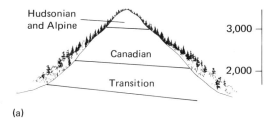

(a)

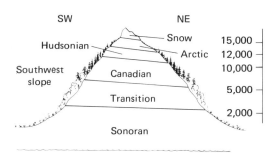

(b)

FIGURE 7-9
*Altitudinal zonation in mountains. (a) Mount Marcy in the Adirondacks of New York. The forest on the lower slope is Transition, consisting chiefly of northern hardwoods. The forest on the middle slope is Canadian, with paper birch, red spruce, and balsam. The upper slope is Hudsonian and Alpine, characterized by dwarf spruce and willows and heaths. In the southern Appalachians, the northern hardwoods are replaced by oaks and hickory; the northern hardwoods replace the spruce, and the spruce replaces the willows and stunted spruce.
(b) A generalized Western mountain. The Sonoran zone is characterized by grassland and shrubby vegetation—chaparral, juniper, etc.; the Transition by oaks and, higher up, lodgepole pine. The Canadian zone contains lodgepole pine, Englemann spruce, red fir, silver fir; the Hudsonian, mountain hemlock, western white pine; the Arctic-Alpine, willows, etc. Note that the life zones extend higher on the southwest slope than on the northeast slope. Elevations are approximate only.*

191

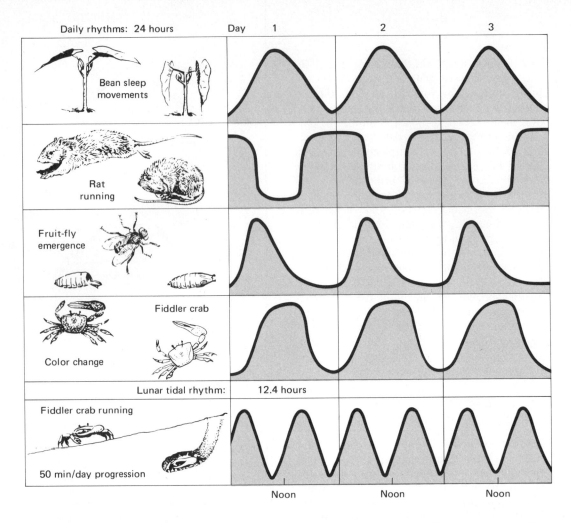

Daily rhythms: 24 hours   Day   1   2   3

Bean sleep movements

Rat running

Fruit-fly emergence

Fiddler crab

Color change

Lunar tidal rhythm: 12.4 hours

Fiddler crab running

50 min/day progression

Noon   Noon   Noon

the commonest of all our tree squirrels. But because of its nocturnal habits, this mammal is seldom seen by most people. Unless it is disturbed, the flying squirrel does not come out by day. It emerges into the forest world with the coming of darkness; it returns to its nest with the first light of dawn.

If the flying squirrel is brought indoors and kept under artificial conditions of night and day, the animal will confine its periods of activity to darkness, its periods of inactivity to light. Whether the conditions under which the animal lives are 12 hours of darkness and 12 hours of light, or 8 hours of darkness and 16 hours of light, the onset of activity always begins shortly after dark. The squirrel's day-to-day activity forms a 24-hour period. This correlation of the onset of activity with the time of sunset suggests that light has a regulatory effect on the activity of the squirrel.

But the photoperiodism (response to changing light and darkness) exhibited by the squirrel is not quite so simple. There is more

to it than the animal becoming active because darkness has come. If the squirrel is kept in constant darkness, it still maintains a relatively constant rhythm of activity from day to day (DeCoursey, 1961). But in the absence of any external time cues, the squirrel's activity rhythm deviates from the 24-hour periodicity exhibited under light and dark conditions. The daily cycle under constant darkness varies from 22 hours, 58 minutes, to 24 hours, 21 minutes, the average being less than 24 hours (most frequent 23:50 and 23:59) (DeCoursey, 1961). The length of the period maintained under a given set of conditions is an individual characteristic. Because of the deviation of the average cycle length from 24 hours, each individual squirrel gradually drifts out of phase with the day–night changes of the external world (Fig. 7-11). If the same animals are held under continuous light, a very abnormal condition for a nocturnal animal, the activity cycle is lengthened, probably because the animal, attempting to avoid run-

192

ning in the light, delays the beginning of its activity as much as it can.

Nocturnal animals possess a rhythm of activity that under field conditions exhibits a periodicity of 24 hours. Moreover, when these organisms are brought into the laboratory and held under constant conditions of light, darkness, and temperature, away from any external time cues, they still exhibit a rhythm of activity of approximately 24 hours. Because these rhythms approximate but seldom match the periods of the earth's rotation, they are called *circadian* (from the Latin *circa*, about, and *dies*, day). The period of the circadian rhythm, the number of hours from the beginning of activity one day to the beginning of activity on the next, is referred to as free-running. In other words the activity period exhibits a self-sustained oscillation under constant conditions. The length of this period is usually a function of the intensity of light provided under constant conditions. Increasing the light intensity causes a lengthening of the free-running period in organisms active by night and a shortening of the period in organisms active by day (Hoffman, 1965). Circadian rhythms apparently are internally driven or endogenous, are affected very little by temperature changes, are insensitive to a great variety of chemical inhibitors, and are innate, not learned from or imprinted upon the organism by the environment.

The innate character of the circadian rhythm is demonstrated by observations of several animals. When fruit flies, *Drosophila*, are kept under constant conditions from the larval stage on, they will still emerge from the pupae with a regular circadian rhythm (Bünning, 1935a). In fact *Drosophila* reared for 15 generations under continuous dim light still retain their capacity to emerge from the pupae according to a circadian rhythm (Bünning, 1935a). Eggs of chickens and lizards kept under constant conditions produce animals that later show regular circadian cycles (Aschoff and Meyer-Lohmann, 1954).

Thus many plants and animals are influenced by two periodicities, the internal circadian rhythm of approximately 24 hours and the external environmental rhythm, usually a precise 24-hour rhythm. If the activity rhythm of the organism is to be brought into phase, or synchrony, with the external one, then some environmental "time-setter" must adjust the endogenous rhythm with that of the outside

FIGURE 7-10 [OPPOSITE]
Examples of rhythmic phenomena experimentally demonstrated to persist under constant conditions in the laboratory, illustrating diagrammatically the natural phase relationships to external physical cycles. (Redrawn by permission from F. A. Brown, 1959, "Living Clocks," Science, *130:1537.)*

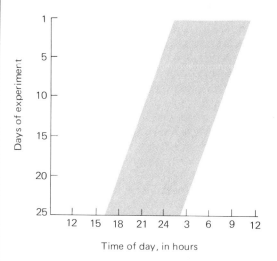

Time of day, in hours

FIGURE 7-11
Drift in the phase of the activity rhythm of a flying squirrel held in continuous darkness at 20°C for 25 days. (Adapted by permission from P. J. DeCoursey, 1960, Cold Spring Harbor Symposia on Quantitative Biology, *25:51.)*

world. The most obvious time keepers, cues, synchronizers, or *Zeitgebers* (Aschoff, 1958) are temperature and light. Of the two, light is the master *Zeitgeber*. It brings the circadian rhythm of many organisms into phase with the 24-hour photoperiod of the external environment.

Although one might have difficulty proving in the field the role of light in synchronizing the circadian rhythm with the environment, it can be demonstrated in the laboratory (see Bruce, 1960). If an animal or plant is kept under constant conditions, such as continuous darkness or continuous light, the circadian rhythm drifts out of phase with natural light and dark and eventually may fade away. The length of time required for this to happen depends upon the organism and the conditions of light and darkness. The activity rhythm of rodents may continue for several months in constant darkness. Other rhythms, such as the leaf movements of plants, may fade much more quickly. Once a rhythm has faded, a new one can be started by some exposure to light or dark. This may be the interruption of continuous darkness by a short flash of light, the interruption of continuous light by darkness, the change from continuous darkness to continuous light, or vice versa. With some organisms, a change in temperature may start a new rhythm.

Pittendrigh (1954) in an early experiment demonstrated that light was a synchronizer in *Drosophila*. In their natural habitat, fruit flies emerge from their pupal cases at about dawn. When they are allowed to lay their eggs in the laboratory and the eggs are left to develop, all under continuing darkness, the flies emerge from the pupal cases at random through the day. But if larvae hatched from such eggs are subjected to only a single flash of light during the period of controlled darkness, the height of emergence of adult flies from pupal cases will take place at approximately the same hour of day that they were previously exposed as larvae to the flash of light (Fig. 7-12). A single flash of light during the larval period was sufficient to establish a rhythm of emergence of adult flies.

The activity rhythm of some vertebrates shows a similar entrainment to light–dark cycles. The flying squirrel, both in its natural environment and in artificial day–night schedules, synchronizes its daily cycle of activity to a specific phase of the light–dark cycle.

This was demonstrated in a series of experiments by DeCoursey (1960a, 1961). Flying squirrels were held in constant darkness until their circadian rhythms of activity were no longer in phase with the natural environment. Then they were subjected to a light–dark cycle that was out of phase with their free-running period. If the light period fell in the animals' subjective night, it caused a delay in the subsequent onset of activity. Synchronization took place in a series of stepwise delays, until the animals' rhythms were stabilized with the light–dark change (Fig. 7-13). If the light period fell at the subjective dawn or at the end of the dark period (when the animals' activity period was about to end), it caused an advance of activity toward the dusk period. And if the light fell in the animals' inactive day phase, it had no effect. The flying squirrels do not need to be exposed to a whole light–dark cycle to bring about a shift in the phase of the activity rhythm. A single 10-minute light period is sufficient to cause a phase shift in the locomotory activity, provided it is given during the squirrel's light-sensitive period (DeCoursey, 1960b).

The chaffinch, a bird active during the day, responds more rapidly to changes in the light–dark cycle. Birds held in 12 hours of light and 12 hours of dark were active during the light period (Fig. 7-14). After they were entrained to this light–dark cycle, the birds were placed in a constant dim light of 0.4 lux. The birds were active for a shorter length of time than they were before, their circadian activity rhythm (measured from the beginning of one period of activity to the beginning of another) was slightly more than 24 hours and drifted out of phase with the original 12 : 12-hour cycle (Aschoff and Wever, 1962; Aschoff, 1965). On the thirty-third day the birds were exposed to a new cycle of 12 hours of light, 12 of darkness. To this the birds became entrained immediately. Fifteen days later the same birds were placed under continuous light of 120 lux. The birds became active earlier each "day" and were active for a longer length of time; but their free-running period was somewhat shorter than 24 hours (Fig. 7-14a). This latter phenomenon has resulted in a circadian rule that with increasing intensities of illumination the circadian rhythm is shortened in diurnal animals, lengthened in nocturnal ones (see Aschoff, 1962).

For the circadian rhythm to function as a time-measuring device, it must be and is temperature-independent; that is, it is unaffected by a normal range of temperatures. However, a 24-hour temperature cycle can entrain the circadian activity rhythm of some organisms. One is the lizard *Lacerta sicula*. When this animal was kept in a constant light and a sinusoidal 24-hour temperature cycle, its activity cycle became entrained to the one of temperature (Fig. 7-15). The phase of the entrained rhythm depended upon the free-running period that the individual animal exhibited under the constant conditions in which it was held immediately before or after exposure to the temperature cycle (Hoffman, 1963).

There is some experimental evidence that temperature can influence the morning awakening time of birds (Enright, 1966). Seasonal changes in average temperature and day length may have opposite influences on awakening time. House finches held in the laboratory under constant day length and subjected to a cold environmental temperature (2° C) awakened later relative to dawn than house finches kept at 20° C. House finches subjected to constant temperature but changing day length also awakened later relative to dawn. Thus as the days of spring begin to lengthen, the warming temperatures would counteract the minimal anticipation of dawn induced by longer day length. Environmental temperatures and the accompanying reproductive maturation in the normal breeding cycle can counteract the seasonal influence of day length and cause a wide variation of awakening time among many species of birds.

Light and dark may be the *Zeitgebers* that control the phase of an organism's circadian activity rhythm, but the rhythms may relate more directly to other aspects of the environment, which, ecologically, are more significant to the organism than light and dark per se. The transition from day to night, for example, is accompanied by such environmental changes as a rise in humidity and a drop in temperature. Woodlice, centipedes, and millipedes, which lose water rapidly when exposed to dry air, spend the day in a fairly constant environment of darkness and dampness under stones, logs, and leaves. At dusk they emerge, when the humidity of the air is more favorable. In general, these animals show an increased tendency to escape from light as the

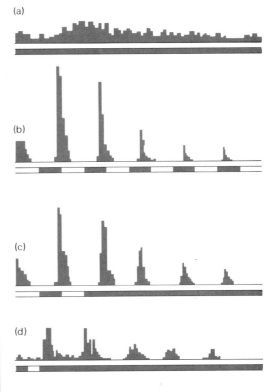

FIGURE 7-12
Daily rhythm of emergence from pupae by
Drosophila. *(a) From cultures held under*
continuous darkness emergence is random.
(b) From cultures held under light-dark
conditions at 21°C. (c) From cultures held
under light-dark conditions and then transferred to
constant darkness at 21°C (note that the rhythm
tied to the original dawn is maintained in darkness).
(d) From cultures held in continuous darkness
and then exposed to a single light stimulus of
4 hours. (From C. S. Pittendrigh, 1954, Proc.
Nat. Acad. Sci., 40:1020, 1023.)

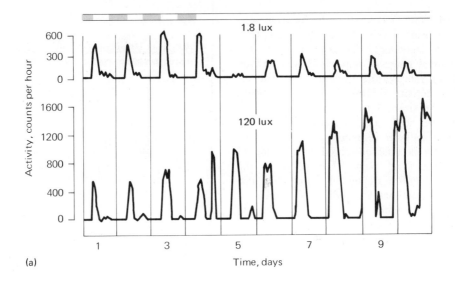

(a)

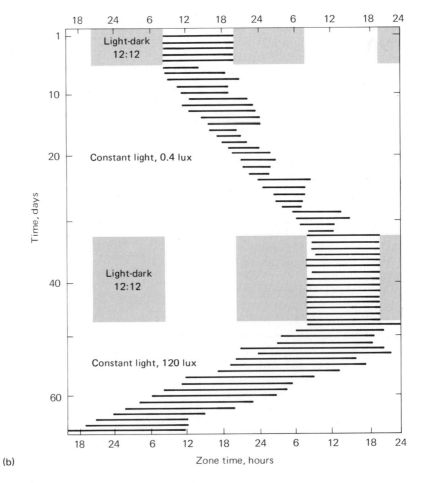

(b)

length of time they spend in darkness increases. On the other hand, their intensity of response to low humidity decreases with darkness. Thus they come out at night into places too dry for them during the day; and as light comes they quickly retreat to their dark hiding places (Cloudsley-Thompson, 1956, 1960). Among some animals the biotic rather than the physical aspects of the environment may relate to the activity rhythm. Deer undisturbed by man may be active by day, but when they are hunted and disturbed they become strongly nocturnal. Predators must relate their feeding activity to the activity rhythm of the prey. Moths and bees must visit flowers when they are open and provide a source of food. And the flowers must have a rhythm of opening and closing that coincides with the time when the insects that pollinate them are flying. The entrainment of the phase of its activity rhythm to a natural light–dark cycle means more to an organism than simply an adjustment to a precise 24-hour period. More important, the entrainment serves to time the activities of plants and animals to a day–night cycle in a manner that is appropriate to the ecology of the species.

The possession of a circadian rhythm that can be entrained to environmental rhythms provides plants and animals with a biological clock, which probably is an integral part of cellular structure. With this, organisms can not only determine the time of day, they can also use the clock for time measurement. The clock, as already suggested, is not simply an hourglass or stopwatch. It does not start on some given signal, such as dawn, run until stopped by another signal, such as darkness, and then start up again on another. The clock runs, or oscillates, continuously, but it must be regulated or reset by environmental signals. It is this latter characteristic of the clock that makes the 24-hour photoperiod possible. The environmental rhythm of daylight and dark is the signal by which the biological clock is set to the correct local time each day.

CELESTIAL ORIENTATION

Some organisms, particularly arthropods, birds, and fish, utilize their time sense as an aid to find their way from one area to another. To orient themselves, the animals use the sun (or in some cases the moon and the stars) as a compass. To do this, they utilize

FIGURE 7-14 [OPPOSITE]
(a) *Activity rhythm of two chaffinches* (Fringilla coelebs). *The birds were kept three days in an artificial light-dark cycle and thereafter in continuous illumination, one in light with an intensity of 1.8 lux and the other in light of 120 lux. Ordinate, perch-hopping activity recorded by means of print-out counters.* (b) *Activity rhythm of a chaffinch in a light-dark cycle of 12 hours of light and 12 hours of darkness and in continuous illumination with an intensity of 0.4 lux and of 120 lux. Note how rapidly the birds became entrained to the light-dark cycle, and the direction of phase drift in the two illuminations. (Redrawn by permission from J. Aschoff, 1965, "Circadian rhythms in man,"* Science, 148:1428. © 1965 American Association for the Advancement of Science.)

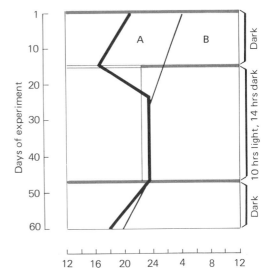

FIGURE 7-13
Diagrammatic representation of the synchronization for flying squirrels with circadian rhythm in constant darkness of less than 24 hours to a cycle of 10 hours of light, 14 hours of darkness. In squirrel A, the rephasing light fell during its subjective night, synchronization was accomplished by a stepwise delay, and the onset of activity was stabilized shortly after light-dark change. In squirrel B, the light fell in the subjective day, and the free-running period continued unchanged until the onset drifted up against the "dusk" light change. This prevented it from drifting forward by a delaying action of light. When returned to constant darkness, the onset of activity continued a forward drift. (Redrawn from P. J. DeCoursey, 1960, Cold Spring Harbor Symposia on Experimental Biology, 25:52.)

both their biological clock and observations on the azimuthal position of the sun in relation to an established direction. The azimuth is the angle between a fixed line on the earth's surface and a projection of the sun's direction on the surface. Using the sun as a reference point involves some problems for the animals because the sun moves. The target angle changes throughout the day. If the animal is to use the sun as a reference, it must be able to correct for the movement. This the animal seems to accomplish.

Hoffman (1959) trained starlings during natural daylight to a certain direction of the compass by feeding the birds at that position. Then he exposed the starlings to an artificial day–night cycle that began and ended 6 hours behind the natural one. The birds then searched for food 90° to the right of the training direction (Fig. 7-16). Thus the birds were able to adjust for a change in time and sought the food by allowing for the change in the position of the sun. The birds continued to choose the direction they maintained in the artificial day, even when held under constant light and nearly constant temperature. This same ability also exists in fish (Braemer, 1959; Hasler, 1960), turtles (Carr, 1962), and lizards (K. Fischer, 1960), as well as in such invertebrates as bees (von Frisch, 1954), wolf spiders (Papi, 1955), and sand hoppers (Papi, Serretti, and Parrini, 1957).

For animals that migrate over great distances, the use of the sun compass must be further refined to compensate for the changes in azimuthal speed of the sun, which varies during the day according to the season and latitude. As the migrating animals pass rapidly from one region to another, they encounter changing day lengths with an increasing azimuthal angle speed around noon; and some eventually reach regions where the sun culminates in the north instead of the south. Exactly how the animals correct for these changes is unknown, but they may take into account the height of the sun and the rate of azimuthal change, together with a sense of time to estimate longitude (Matthews, 1959, 1961). Whether the sun compass is a sufficient or even tenable explanation is open to question.

Directional finding by birds may be influenced by magnetic fields of the earth. This theory was advanced by Yeagley in 1947 and was thoroughly rejected. But lately the theory

has been revived. Recent experiments suggest that European robins will change their directional preferences when presented with a change in the direction of the magnetic field (Wiltschko and Hock, 1972).

ANNUAL RHYTHMS

As the seasons turn, the daily periods of daylight and darkness change. The activities of plants and animals are geared to this seasonal rhythm of night and day. The flying squirrel starts its daily activity with nightfall and maintains this relation through the year. As the short days of winter turn to the longer days of spring, the squirrel begins its activity a little later each day (Fig. 7-17). The commencement of bird song follows the dawn, but not until the light reaches a certain intensity, which varies with the species (Fig. 7-18).

The song of birds is associated with the reproductive cycle, which also follows a seasonal pattern. For most birds the height of the breeding season is spring; for deer the mating season is the fall. Brook trout spawn in the fall; bass and bluegills in late spring and summer. The trilliums and violets bloom in the short days of early spring before the forest leaves are out and while an abundance of sunlight reaches the forest floor. Asters and goldenrods flower in the shortening days of fall. Most animals and plants of temperate regions have reproductive periods that closely follow the changing day lengths of the seasons, or the photoperiod.

PHOTOPERIODISM IN PLANTS AND ANIMALS

Based on photoperiodic responses, plants can be classed as short-day, long-day, and day-neutral. Day-neutral plants are those whose flowering is not affected by day length but rather is controlled by age, number of nodes, previous cold treatment, and the like. Short-day and long-day plants both are influenced by the length of day. When the period of light reaches a certain portion of the 24-hour day, it inhibits or promotes a photoperiodic response. The length of this period so decisive to the response is called the *critical day length*. It varies among organisms but usually falls somewhere between 10 and 14 hours. Throughout the year plants (and animals) "compare" this time scale with the actual length of day or night. As soon as the actual length of day or night is greater or smaller than the critical day length, the plant may flower or cease to flower, expand its leaves, and lengthen its shoots. Short-day plants are those whose flowering is stimulated by day lengths shorter than the critical day length. Long-day plants are those whose flowering is stimulated by day lengths longer than a particular value. These latter usually bloom in late spring and summer.

In the cotton fields of the southern United States lives the pink cotton bollworm, the larva of a tiny moth. Except for a few hours directly after hatching, the larva spends its life in the flower buds or bolls of cotton. At the fourth larval instar stage, the insect goes into *diapause*, a stage of arrested growth over winter. The onset of diapause comes in late August; but not until near the autumnal equinox, September 21, when the night becomes equal to or longer than the day, does the number of diapausing larvae sharply increase. By the end of October virtually all the larvae are in diapause. Through the winter, the larva remains in this state of arrested development; then in late winter, as the days begin to lengthen, the insect comes out of diapause and continues its growth. The emergence from diapause reaches its maximum just after the spring equinox, when the days are just slightly longer than those that induced diapause.

When the larvae of the pink bollworm were exposed to regimes of light and dark in the laboratory, the insect would go into diapause only when the light phase of the 24-hour day was 13 hours or less (Adkisson, 1966). If the larvae were exposed to a light period of 13.25 hours, the insect was prevented from going into diapause. So precise is the time measurement in the insect that a quarter-hour difference in the light period determines whether the insect goes into diapause or not. Once growth is arrested, the exogenous rhythms do not cease but continue through the diapause until the day length becomes longer than 13 hours. Diapause terminates most rapidly under photoperiods of 14 hours, less rapidly at 16 and 12. Thus to the pink bollworm, the shortening days of late summer and fall forecast the coming of winter and call for diapause; and the lengthening days of late winter and early spring are the signals for the insect to resume development, pupate, emerge as an adult, and reproduce.

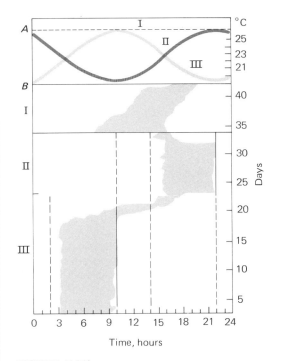

FIGURE 7-15
Activity rhythm of a lizard (Lacerta sicula) *held in constant conditions and in a 24-hour temperature cycle.* (A) *The temperature regimes under which the lizard was held: I, constant temperature; II, III, 24-hour temperature cycles.* (B) *Activity cycles of the lizard held under the three temperature regimes. The Roman numerals I, II, and III correspond to the temperature cycles in A. The broken vertical lines indicate the time of minimum temperatures; the solid line the maximum temperature. I, activity in constant temperature. Note the drift to the left. This drift was halted when the lizard was placed in temperature regime II. After a short stepwise rephasing to the right, the onset of activity became synchronized with the temperature minimum and was prevented from drifting forward by the delaying action of the temperature maximum. When the temperature cycle was reversed, in III, the lizard again became synchronized by a longer series of stepwise advances until the forward advance was stopped by the minimum temperatures.* (Adapted from Hoffman, 1963.)

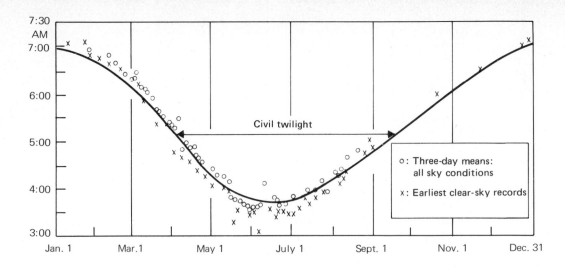

That increasing day length increases gonadal development and spring migratory behavior in birds was experimentally demonstrated some 40 years ago when Rowan (1925, 1929) forced juncos into the reproductive stage out of season by artificial increases in day length. Results of subsequent experimental work with a number of species have shown that the reproductive cycle is under the control of an exogenous seasonal rhythm of changing day lengths and an endogenous physiological response timed by a circadian rhythm.

After the breeding season, the gonads of birds studied to date have been found to regress spontaneously. This is the *refractory* period, a time when light cannot induce gonadal activity, the duration of which is regulated by day length (see Farner, 1959, 1964*a*, *b*; Wolfson, 1959, 1960). Short days hasten the termination of the refractory period; long days prolong it. However, Hamner (1968) makes the suggestion, based on experimental work with house finches, that the refractory period consists of two distinct periods of physiological states. One is an absolute refractory period, the length of which is independent of photoperiod. This is followed by a relative refractory period when photosensitive birds will not respond to day lengths equal to or shorter than the one to which they were previously exposed. As winter approaches, and the natural day shortens, there is a continual temporal readjustment of the birds' timing mechanism so that the progressively shorter days become photoperiodically stimulatory. After the refractory phase is completed, the *progressive* phase begins in late fall and winter. During this period the birds fatten, they migrate, and their reproductive organs increase in size. This process can be speeded up by exposing the

bird to a long-day photoperiod. Completion of the progressive period brings the bird into the *reproductive* stage.

A similar photoperiodic response exists in the cyprinid fish, the minnows. The annual sexual cycle among the minnows consists of a sexually inactive period, followed by the reproductive period (Harrington, 1959). Underlying this is an intrinsic sexual rhythm that consists of a long responsive period, alternating with a shorter refractory period. Long days occurring within the responsive period start the prespawning period, characterized by mating and territorial behavior. The prespawning period ends with the laying of the first eggs. The subsequent spawning period is consummatory and ends sometime before the days shorten to the critical length that initiated the prespawning activity. Following this is the refractory, or postspawning period, in which light fails to stimulate gonadal development. The prespawning period, from about mid-November to mid-July, is the phase in which the annual sexual period is timed.

Seasonal cycles of photoperiodism influence the breeding cycles of many mammals (Fig. 7-19). In northeastern United States the flying squirrel has two peaks of litter production, the first in early spring, usually April, and the second in late summer, usually August. To produce litters in April, the flying squirrel must be in breeding condition in January and February. Muul (1969) investigated the responses of gonadal development to changing photoperiod under laboratory conditions. He found that in the flying squirrel the descent of the testes into the scrotum (in nonbreeding condition, the testes are held in the body cavity) occurred in January under short-day and long-night conditions. An accelerated decrease in day length hastened the descent. The experimental animals held under natural photoperiod came into reproductive condition and produced litters at the same time as squirrels in the wild in absence of temperature cues. After the maximum day length in summer, the testes regressed and remained in that condition. If the photoperiod was altered so that the animal's minimal day length came 2 months early and then increased, the testes descended 2 months early.

Muul subjected one group of squirrels to a photoperiod 6 months out of phase with the natural world. Squirrels were exposed in July to a photoperiod characteristic of January. Their testes descended in July and regressed in January, a time when the testes of males exposed to a natural photoperiod were descending. Likewise females produced litters out of phase in January, 6 months later than births observed in nature. Thus in the flying squirrel the testes of the male descend and ovulation in females takes place when day length increases from 11 to 15 hours, and ovulation ceases and testes regress when the photoperiod decreases.

The food-storing behavior of the flying squirrel in fall also appears to be photoperiodically controlled (Muul, 1965). Squirrels held in the laboratory under seasonal photoperiods and controlled constant temperature exhibited an intensity of food storing similar to that of animals held under natural conditions. Squirrels exposed to seasonal temperatures and a controlled photoperiod of 15 hours of daylight showed no intense food-storage activity through the winter. When the light was reduced to 12 hours in March, there was a marked rise in food storage. Another group was held under constant temperature and a controlled photoperiod of 15 hours of daylight, which was reduced to 13 hours in mid-December. Within a week this group increased food-storage activity sharply and continued it from January through March. In still another experiment, the squirrels were subjected in mid-October to a photoperiod typical of mid-November. The intensity of food storage increased more rapidly than normal and reached an equivalent of that of mid-November under natural conditions. Further decreases in the length of day increased the performance of squirrels. By the beginning of November, when the squirrels were subject to a photoperiod equivalent to that of

FIGURE 7-18 [OPPOSITE]
Daybreak song of the cardinal from February 26, 1944, to April 20, 1948, in relation to the beginning of civil twilight. (From **Leopold** *and Eynon, 1961.)*

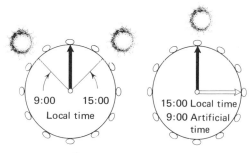

FIGURE 7-16
Change in direction by a starling trained to feed from a dish oriented to the south. After the bird was held in an artificial day 6 hours behind local time for a period of 12 to 18 days, it was tested under a natural sun at a time when its height was the same as expected under artificial conditions. The orientation of the bird changed 90° clockwise from the training direction. (Adapted from Hoffman, 1965.)

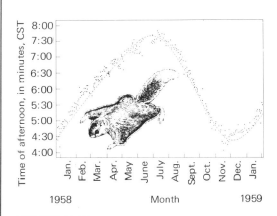

FIGURE 7-17
Onset of running wheel activity for one flying squirrel in natural daylight conditions throughout the year. The graph is the time of local sunset through the year. (With permission from P. J. Decoursey, 1960, Cold Spring Harbor Symposia on Quantitative Biology, 25:50.)

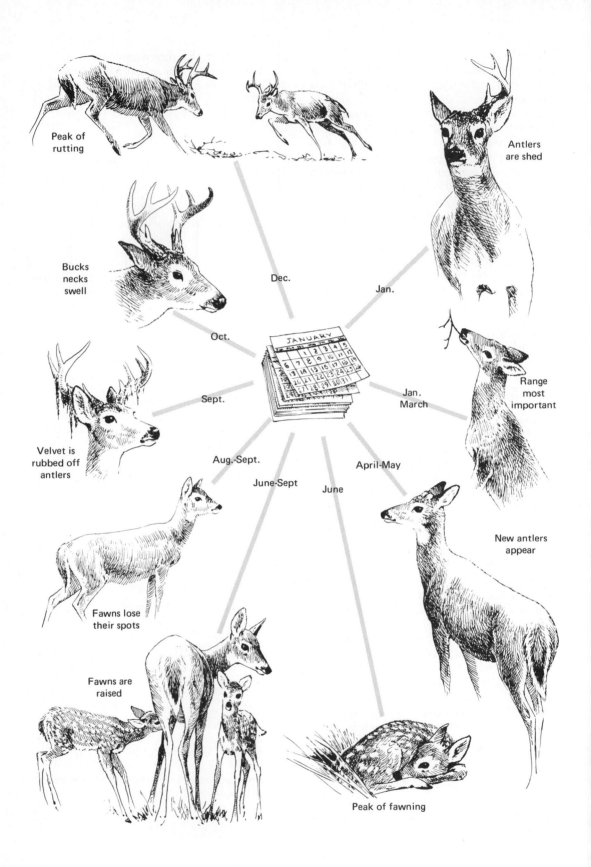

Peak of rutting

Antlers are shed

Bucks necks swell

Dec.

Jan.

Oct.

Sept.

Jan. March

Range most important

Velvet is rubbed off antlers

Aug.-Sept.

April-May

June-Sept

June

New antlers appear

Fawns lose their spots

Fawns are raised

Peak of fawning

late December, the storage peak was maximum. Immediately a long day of 15 hours of light and 9 hours of dark, equivalent to midsummer conditions, was imposed on the squirrels. Some squirrels showed a sudden decrease in storage intensity; others showed a gradual decrease. But among the squirrels held under natural conditions, storage of food was still increasing (Fig. 7-20). These experiments demonstrate that the food-storage activity of the flying squirrel is photoperiodically controlled. Such a control synchronizes exploratory and storing behavior with a ripening of the mast crop—nuts and acorns—and prevents a premature harvest.

PHYSIOLOGICAL MECHANISMS: THE BÜNNING MODEL

The photoperiodic responses of plants and animals are not dependent upon the length of day as such, nor even on the length of night. Instead, what seems to be involved is a circadian rhythm of sensitivity to light as the inducing or inhibiting agent. Current experimental evidence suggests that a time-measuring process starts at the beginning of the light period or the beginning of the dark. This induces, after a certain length of time from the beginning of light or dark, a sensitive stage that responds specifically to light.

When plants are held under short-day and long-night conditions, the short-day plants are stimulated to flower, and the long-day plants are inhibited. When day length is increased, flowering is inhibited in the short-day and stimulated in the long-day plants (Fig. 7-21). If the dark period of the short-day and long-day plant is interrupted, each reacts as if it had been exposed to a long day. The long-day plant flowers; the short-day plant does not.

A similar response occurs in animals. The diapause of insects, a short-day phenomenon, is inhibited by light breaks in the dark period (Bünning and Jaerrens, 1960). The breeding period of the ferret can be initiated by exposing the animal to 12 hours of light each day for a month if 1 hour of light is given from midnight to one o'clock (Hart, 1951). This contrasts with 18 hours of light required daily if the animal is exposed to light in a continuous period. In fact, 6 hours of light are sufficient to bring the ferret into breeding condition if the cycle includes 4 hours of continuous light and 20 hours of darkness

FIGURE 7-19 [OPPOSITE]
The seasonal cycle of the white-tailed deer, which begins with the breeding season in the fall. The annual breeding cycle of the white-tailed deer is attuned to the decreasing day length of fall.

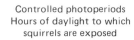

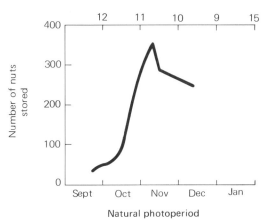

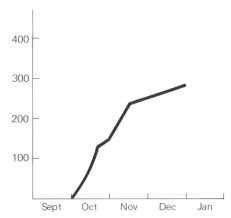

FIGURE 7-20
Food-storage activity of flying squirrels held under controlled photoperiods compared to natural photoperiod. (I. Muul, 1965. Reprinted by permission from Natural History Magazine, *March, 1965. Copyright © The American Museum of Natural History, 1965.)*

interrupted between the seventeenth and nineteenth hours (J. Hammond, 1953).

That such a circadian rhythm of light sensitivity exists in birds has been demonstrated in a series of experiments by Hamner (1963). House finches were held under six experimental light regimes, all involving 6 hours of daylight: short-day cycles of 24 hours (6L/ 18D), 48 hours (6L/42D), and 72 hours (6L/66D); and long-day cycles of 12 hours (6L/6D), 36 hours (6L/30D), and 60 hours (6L/54D). The finches held under the first three responded as though they had received a short-day treatment and exhibited no enlargement and maturation of the testes and no production of sperm. The birds held on the other three treatments responded as if under long-day conditions. The testes enlarged and spermatogenesis began. Since all the birds were subject to the same length of light but to varying lengths of darkness, the experiments showed that neither the length of the light period nor the length of the dark was critical. The results indicate that the house finch has an endogenous rhythm with a periodicity of about 24 hours. During the long dark period the birds endogenously reached light-sensitive states about 24 hours apart. When light is given at the proper phase of the rhythm, gonadal enlargement and maturation take place, but when light is given at the light-insensitive phase of the rhythm, no response occurred. Thus photoperiod-controlled reproductive cycles of birds are timed by a circadian rhythm of light sensitivity.

Further experimentation (Hamner, 1968) suggests that this response has a rhythm of two approximately 12-hour phases of different sensitivity to light. The basic rhythm can be phase-shifted by and entrained to artificial lighting cycles. Thus as days shorten, the timing mechanism readjusts to increase the absolute duration of the second sensitive phase of the rhythm. As the days of spring begin to lengthen, additional light interacts with the readjusted sensitive phase of the photoperiodic clock and stimulates the growth of the testes.

How this rhythm of light sensitivity in plants and animals might function is suggested in a hypothesis advanced by Dr. Edwin Bünning of Germany to explain the short-day and long-day reactions of plants. According to the Bünning model (Fig. 7-22), the circadian rhythm of light sensitivity goes through at least two half-cycles of more or less opposite sensitivity to light: a tension or photophil phase, in which development is favored by light, and a relaxation or scotophil phase, in which development is inhibited. Each phase of the rhythm is about 12 hours long, but to function, the light phase must precede the dark. Short-day plants are considered scotophil in the second half of the cycle; if exposed to light at that time, their flowering is inhibited (Fig. 7-21). In the same half-cycle, long-day plants are photophil; their flowering is stimulated. In addition, light sensitivity itself is rhythmic and seems to fluctuate in half-cycles within the light and dark periods. There is a time of maximum sensitivity during the dark (Fig. 7-23), which in some organisms is about 14 to 16 hours from the beginning of the main light period. In the flying squirrel the sensitive periods are at the beginning of the subjective night and at subjective dawn (DeCoursey, 1960a, b). Light exerts its maximum effect in the pink bollworm either 10 hours after dark or 10 hours before dawn (Adkisson, 1964). Additional light offered a few hours after the beginning of the light period has a stimulatory effect on the flowering of short-day plants (Bünning, 1964). Among many organisms the beginning of the previous light period may have a stronger influence on the timing of the point of maximum sensitivity than the beginning of the dark. At any rate both light and dark periods of a certain duration are required for a proper response. These rhythms in light sensitivity are endogenous. Light controls or sets the biological clock, and the clock in turn controls the light sensitivity of the organism.

Although the circadian rhythm is often considered as a single oscillation of approximately 24 hours, the fact is that the activity periods of most animals exhibit two peaks, occasionally more. The first, representing major activity, is followed by a second peak, which is much smaller, more variable in its position in time, and limited to about a half hour. The two peaks are separated by a trough of relative inactivity. In many animals this period is usually associated with some environmental condition adverse to the organism, such as high temperature or low humidity. Because the period of inactivity in nocturnal animals comes after midnight and in diurnal animals after noon, and the first peak of activity is associated with the dim light of dawn and dusk, some consider these peaks to be directly caused by environmental stimulus.

If this were so, the peaks should disappear under constant conditions of temperature and humidity and under artificial cycles of light and dark, with a sharp change between light and dark. Aschoff (1966) subjected three different species of finch to light cycles interposed with artificial twilight and then light cycles with a sharp change from light to dark. Under both regimes the activity pattern remained the same (Fig. 7-24). The birds were further subjected to light and dark cycles without twilight and then to constant illumination. Even under constant light the bimodal pattern of activity remained, indicating that the rhythm has two peaks that are endogenous, that are a persistent property of the circadian systems, and that do not depend on any concurrent change in environmental stimulus. Regardless of the length of the activity time, the period between the two peaks remains proportionately the same.

In all cases the responses of the organism are mediated by hormones and enzymes. Exactly how the hormonal and enzymatic systems function in periodism is the subject of much current research. A discussion of this is beyond the scope of the book, but for an elementary and enlightening discussion see Hastings, 1970.

SOME OTHER ASPECTS OF PERIODICITY

Seasonal periodicities in addition to activity also exist in the lives of plants and animals. Annual coloration of leaves in fall is a seasonal response brought on within the leaf by chemical changes. Animal coloration is influenced by photoperiod. The summer coat of the snowshoe hare is brown; in winter it is white. The white hair makes the animal inconspicuous against the snow and may aid it to escape detection by predators (Fig. 7-25). Seasonal changes in light indirectly affect the food habits of animals because the availability is directly related to the activity and reproductive cycle of prey species, and the flowering and fruiting of plants (Fig. 7-26).

The endogenous rhythms considered so far are ones that are correlated with the environment; there are also endogenous rhythms that are not so synchronized. Often these are temperature-dependent and speed up their frequencies with increasing temperatures. Among

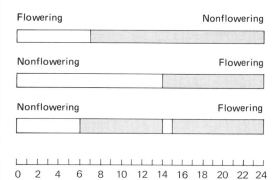

FIGURE 7-21
The time of flowering in long-day and short-day plants, as influenced by photoperiod.

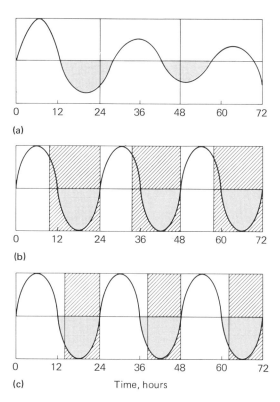

FIGURE 7-22
The Bünning model. Oscillations of the clock cause an alternation of half cycles with quantitatively different sensitivities to light (white versus black). (a) The free-running clock in continuous light or continuous darkness tends to drift out of phase with the 24-hour photoperiod. (b) The short day. (c) The long day. Short-day conditions allow the dark to fall into the white half-cycle; in the long day, the light falls into the black half-cycle. (Redrawn by permission from Bünning, 1960, Cold Spring Harbor Symposia in Quantitative Biology, 25:253.)

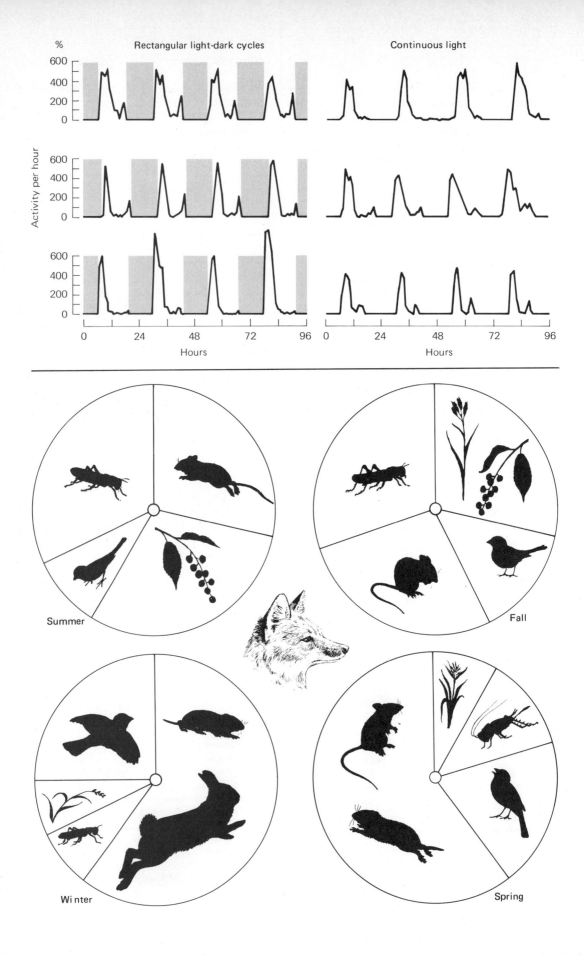

Summary

Fall

Winter

Spring

FIGURE 7-24 [OPPOSITE TOP]
Activity pattern of three greenfinches (Chloris
chloris L.) *kept first in artificial light-dark cycles,
thereafter in continuous illumination. Ordinate:
Deviation of hourly activity from mean activity
per 24 hours, the average being expressed as
100 percent. (From J. Aschoff, 1966. © The
Ecological Society of America.)*

FIGURE 7-26 [OPPOSITE BOTTOM]
*The food habits of many animals, such as the
red fox, are influenced by seasonal periodicities.
The timing of flowering and the onset of breeding
activities of animals influence the availability of
foods through the year. Note the prominence of
fruits and insects in summer and fall, and rodents
in spring and winter. (Based on data from
Scott, 1955.)*

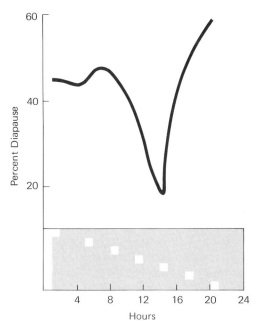

FIGURE 7-23
*Effects of light offered in the first half and
second half cycle of about 12 hours each in
Pieris brassicae. Additional light of 1 hour
promotes diapause in this cabbage butterfly if
offered during the first half of the cycle, but
inhibits diapause during the second half of the
cycle. Diapause is a short-day response. (Adapted
from Bünning and Joerrens, 1960.)*

FIGURE 7-25
*The white color of the snowshoe hare in winter
and the brown color in summer are regulated by
changing day length. As the nights lengthen
and the days shorten, the pelage changes from
brown to white, regardless of snow conditions
and temperature. (Photo courtesy Pennsylvania
Game Commission.)*

such rhythms are the beating of the heart, the respiratory rate, and the transmission of nervous impulses. The timing of these rhythms involves internal feedback. The respiratory rate, for example, increases as the carbon dioxide content of the blood increases, the information being fed to the lungs by way of the central nervous system.

CIRCADIAN RHYTHMS IN MAN

Man, in common with other animals, possesses circadian rhythms (see Aschoff, 1965; Sollberger, 1965). Some of these endogenous rhythms are influenced by man's activities and the imposition of his will. Others show up as truly circadian, the peak coming about once every 24 hours. Volunteer subjects kept without timepieces, in complete isolation from the outside world but allowed to set their own rhythms of eating and sleeping, of lights on and lights off, showed circadian rhythms in the calcium and potassium contents and volume of urine excretion, in body temperature, and in sleep and wakefulness. These rhythms ranged from 23.6 hours to 25.8 hours, with an average period of about 25 hours (Fig. 7-27). In the absence of any time cues, these rhythms showed a steady drift against clock time. Because man is a diurnal animal, the peaks of many of his endogenous rhythms come during the day (Fig. 7-28). If his activity cycle is shifted to night, in general he is less efficient and may suffer psychologically and physiologically. He will live in a conflict situation between two tendencies in the circadian rhythm: to remain in phase with the normal environment as entrained by social *Zeitgebers* or to shift to a new work–rest cycle.

The jet and space age has complicated the system. Space travel can require days and weeks of short work–rest schedules under abnormal day–night situations within one 24-hour period. These abnormal cycles occur when astronauts pass from dark to light on every orbit of the earth or when they travel to the moon. Even if one flies from Paris to New York, he will be out of phase with local time because his circadian clock will still be running on Paris time. Entrainment to new local time requires 2 or 3 days during which the traveler may experience periods of sleeplessness and hunger out of phase with those around him. A traveler headed east experi-

ences a shorter day, one going west a longer day. It takes longer to adapt from an easterly flight than from a westerly one. As man shortens travel time between distant points in the world and as he invades outer space, he will have to take his endogenous circadian rhythms into account.

Ionizing radiation

A new dimension has been added to ecological influences in the nuclear age: high-energy radiation. The source is fallout from atomic blasts, radioactive wastes, and nuclear reactors. Involved are high-energy short-wavelength radiations, known as ionizing radiations. They are so called because they are able to remove electrons from some atoms and attract them to other atoms, producing positive and negative ion pairs. Two ecologically important ionizing radiations, alpha and beta, are corpuscular. They are streams of atomic or subatomic particles that have a definite mass, and travel only a short distance through the air and water. Once stopped, they transfer their energy to the object they strike, producing locally a large amount of ionization. They have their greatest biological effect when ingested, absorbed, or internally deposited. The other form is ionizing electromagnetic or gamma radiation, which has a short wavelength, travels a great distance, and penetrates matter easily. The biological effects of electromagnetic radiation depend upon the number and energy of the rays and the distance of the organism from the source. Because this radiation is penetrating, it can produce its effects without being taken inside.

Increasing attention is being given to the effects of gamma radiation on ecosystems. When an unshielded nuclear reactor was built in a forested valley near Atlanta, Georgia, a group of biologists under Robert B. Platt took advantage of the situation to observe the effects of radiation in the forest ecosystem. After measuring the amounts of radiation at increasing distances from the site, they revisited the area to follow the changes taking place. Gamma radiation and neutrons (large, unchanged particles that, although they themselves do not emit radiation, cause ionization by bumping atoms out of normal position) devastated the vegetation, killing pines several hundred feet away, reducing the reproductive capacity of more distant plants, and altering the normal sequence of succession.

At the Brookhaven National Laboratory on Long Island, additional studies were made on the effects of ionizing radiation on two ecosystems, an oak–pine forest and an old field (Woodwell, 1962; Sparrow and Woodwell, 1963). A source of gamma radiation was suspended in a tower so that it could be raised or lowered into a lead container by a winch, from a safe distance. The container shielded the source and allowed movement into the area. The radioactive source for the forest was ^{137}Cs and for the old field, ^{60}Co. Both produced daily exposure rates that ranged from several thousand roentgen/day (Table 7-1) at a distance of a few meters to 2 roentgen at 130 m. The highest exposures available, continued daily through the winter and into the spring, devastated both communities, but there was a wide difference in the resistance of the vegetation (Figs. 7-29, 7-30).

None of the higher plants of the forest community survived 360 roentgen/day. Most trees of all species were killed by 60 roentgen/day over a 6-month period; and major damage from chronic exposure at 20 to 30 roentgen/day appeared in a couple of weeks. Pitch pine was the most sensitive to radiation. Growth was inhibited by 1 or 2 roentgen/day, and the trees were killed by 20 to 30 roentgen, about half the exposure that caused comparable damage to oaks. In Georgia, Platt (1963) found that loblolly pine receiving doses of 2,000 or more rad turned brown and in a few weeks were dead. As pines at a greater distance from the reactor accumulated 7,000 rad, they too died, within 90 to 120 days. At the end of 2 years all pines within 2,000 ft of the reactor were dead, yet the hardwoods were not noticeably affected.

On the other hand, an old-field community, dominated by pig weed and crab grass, exposed to the same amount of radiation, was five times as hardy as pine. The basic plant population was essentially unaltered at exposures of less than 300 roentgen/day, although growth, form, and reproductive capacity were affected by daily exposures of 100 to 300 roentgen. Of the old-field plants crabgrass was the most resistant, surviving exposures of more than 2,000 roentgen. At Brookhaven, mosses and lichens survived total

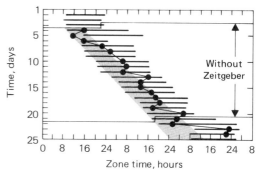

FIGURE 7-27
Circadian rhythm of activity and urine excretion in a human subect held for 3 days under normal conditions, then for 18 days in isolation, and finally again under normal conditions. Black bars indicate times of being awake; circles, maximum of urine excretion; mean values of period for onset and end of activity and for urine maxima. (From J. Aschoff, 1966. © 1966 American Association for the Advancement of Science.)

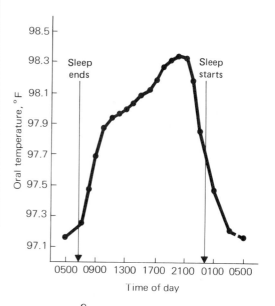

FIGURE 7-28
Circadian rhythm of temperature in man, based on the measurements of 70 young men with normal work activity. Note the decline in body temperature at night after onset of sleep. (From W. P. Colquhoun, 1971, Biological Rhythms and Human Performance, Academic, New York.)

TABLE 7-1
Ionizing radiation terminology used in text

ROENTGEN (R)	The original unit of radiation. It is the amount of X- or gamma radiation that is produced in 1 cm³ of standard dry air ionization equal to 1 electrostatic unit of charge. It is used to describe the radiation field to which one may be exposed.
RAD (RADIATION ABSORBED DOSE)	The quantity of radiation that delivers 100 ergs of energy to 1 g of substance (almost equivalent to a roentgen when referred to body tissue).
REM (ROENTGEN EQUIVALENT, MAN)	A biological (rather than physical) unit of radiation damage, the quantity of radiation that is equivalent in biological damage to 1 rad of 250-kilovolt peak X-rays.
CURIE (C)	A measure of radioactivity. One curie is the amount of any radioactive nuclide that undergoes 37 billion transformations a second.
HALF-LIFE	The average time required for half the atoms of an unstable nuclide to transform.
DOSE	A measure of energy actually absorbed in tissue by interactions with ionizing radiation.
EXPOSURE	A measure of X- or gamma-radiation at any point, used to describe the energy of the radiation field outside the body.

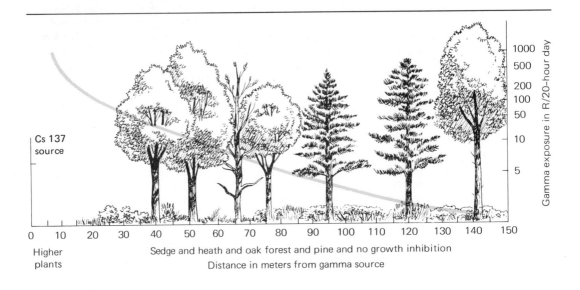

Higher plants

Sedge and heath and oak forest and pine and no growth inhibition

Distance in meters from gamma source

exposures of more than 200,000 roentgen (Woodwell, 1962, 1967).

The forest is more sensitive to radiation than any other natural ecosystem and tends to revert to an earlier stage of succession. In recent studies (Woodwell, 1962, 1967; Woodwell and Whittaker, 1968; F. McCormick, 1969) disturbance fell into five well-defined zones of modification of vegetation: (1) a devastated zone receiving more than 200 roentgen/day, where all woody and most herbaceous plants were killed; (2) a sedge zone receiving 200 to 150 roentgen/day, where the sedge *Carex pensylvanica* became the dominant plant; (3) a shrub zone receiving 40 to 150 roentgen/day, where the tree canopy had been reduced by 50 percent and the dominant plants were huckleberry and blueberry; (4) oak forest receiving 16 to 40 roentgen/day in which the pitch pine was killed but the oaks and understory were undisturbed; (5) oak–pine forest receiving less than 12 roentgen/day in which no mortality occurred. Thus the strata are removed layer by layer, with all but the herbaceous layer gone (Fig. 7-30). Upright forms are at a particular disadvantage, whereas plants with underground stems and buds survive. In effect radiation changes the structure of the ecosystem and alters species diversity. The changes are somewhat suggestive of the effects of fire or other severe environmental damage (Brayton and Woodwell, 1966). The gradient of vegetational life that forms shows a reduction in the stature of dominants, and in total coverage and stratal complexity parallels the gradient one finds in vegetation from favorable to severe environments. Such gradients are not found in irradiated forests only. A similar one is produced in the vegetation of old fields

(Woodwell, 1967). The zones include in order: total kill, groundsel, crab grass, panic grass, and pigweed.

Also affected by irradiation are diversity of species and productivity. Diversity and abundance of species in irradiated communities decrease as exposures to irradiation increase, the extent of the decline depending upon the nature of the community. Decline in the diversity of species in an oak–pine forest sets in at an exposure to 50 roentgen/day, in an old field at 100 roentgen/day, and in a lichen community at 200 to 300 roentgen/day.

The effects of irradiation on net productivity are not so clear. In general there is a decline in productivity beyond the area of total kill. In the old-field community, net productivity increased through exposures up to 1,000 roentgen/day, then declined to 0 (Woodwell, 1967). This was probably brought about by the replacement of an open herb community by a mat of crabgrass. Exposure of a forest community over 6 months to 40 roentgen/day destroyed the dominant species and reduced net production by half.

Radiation damage is closely correlated with the amount of energy absorbed by the chromosomes (Sparrow and Woodwell, 1963). Large chromosomes are more susceptible to damage than small ones, and organisms with few choromosomes are more subject to damage than those with many. Thus polyploidy —the occurrence of extra sets of chromosomes in the nucleus—adds resistance to radiation damage. The amount of energy absorbed is also influenced by the period of exposure. Slowly dividing cells, irradiated over an extended period of time, are subject to greater total exposure before the next division than those rapidly dividing. Rapidly dividing tissues can replace damaged cells with undamaged ones. But the reproductive stages are highly susceptible to radiation damage. Dose rates far below those required for severe inhibition of growth or for lethal effects will cause the complete failure of sexual reproduction. On the other hand, dormant seeds are highly resistant. The seeds of pitch pine, one of the most susceptible of plants, will withstand a radiation dosage of 12,000 roentgen, whereas the tree itself will succumb to 27 roentgen/day.

The life form of the plant, independent of the chromosome size and number, influ-

FIGURE 7-30 [OPPOSITE]
Pattern of radiation damage to an oak–pine forest in 1962 after about 6 months exposure to radiation. (From Woodwell, 1967. © 1967 American Association for the Advancement of Science.)

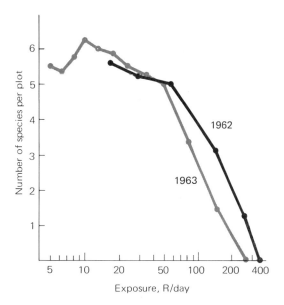

FIGURE 7-29
Species vary in their ability to withstand radiation. As this graph indicates, as the intensity or dosage of radiation is increased, the number of species that survive decrease. (From Woodwell, 1967. © 1967 American Association for the Advancement of Science.)

ences the radiosensitivity of the organisms. Because perennial grasses and ferns have some shoot meristems at and below ground level, they are partially or completely shielded from radiation (see Chappel, 1963). Hemicryptophytes and cryptophytes can survive intensities of radiation that are lethal for plants with a similar nuclear volume and chromosome number but whose shoots are more fully exposed.

Of all the animal groups, insects are the most radioresistant, tolerating exposures from 1,000 to 10,000 roentgen, depending upon the species. When plants are injured by radiation herbivorous insects can intensify the damage. At Brookhaven, leaftiers, leaf rollers, leaf beetles, and loopers were more resistant than their white oak host. Without any apparent increase in absolute numbers they ate most of the remaining foliage on the trees (Woodwell, 1962, 1967). Among soil invertebrates the oribatid mites and deep-soil collembolans were more resistant to irradiation than centipedes, surface collembolans, and parasitic mites (Edwards, 1969). Thus changes in host–parasite and predator–prey relationships can influence the population dynamics of insects and soil invertebrates.

Mammals are the most sensitive to radiation, succumbing to exposures of 200 to 1,000 roentgen. The effects of irradiation, however, are variable and difficult to generalize. No apparent relationship exists between radiation resistance and taxonomic groups or adaptations to stress of rigorous environments. Dunaway and his associates (1969) exposed six species of mice, two species of shrews, and the laboratory mouse to ^{60}Co radiation. Estimates of LD_{50} (lethal dose—the amount of radiation needed to kill 50 percent of the population) ranged from 525 to 1,069 rad. Two species of *Peromyscus*, the golden mouse and the white-footed mouse, were the most resistant. The most sensitive was the marsh rice rat. The cotton rat, a resistant species, had a 91 percent survival when exposed to 500 roentgen and released to a field enclosure, and a 25 percent survival when exposed to 1,200 rad (Pelton and Provost, 1969). An environmental dose of 5,369 rad delivered over a 15-day period resulted in the disappearance of 38 percent of the resident individuals of two species of *Peromyscus*. The white-footed mice living in an open field all disappeared, but some of the beach mice, a

deep-burrowing species, and some of the woodland white-footed mice survived. Their habitat shielded the latter two species considerably, reducing their radiation 88 percent. Mammals receiving sublethal doses show typical effects. One is a shortened life span (French et al., 1969). A very characteristic effect is the graying of the pelage (Schnell, 1963; Dunaway et al., 1969). Other symptoms include conjunctivitis, ataxia, a general reduction of activity, loss of appetite, diarrhea, lack of aggressiveness and alertness, and impairment of motor reflexes and equilibrium.

The effects of radiation on fish are difficult to evaluate because they vary with different stages of the life cycle. Gametes and eggs at the one-cell stage are very sensitive (Rice and Angelovic, 1969). Irradiation results in a slowing of the rate of increase in length and weight, in retarded development of young, and in increased abnormalities.

Nestling bluebirds, more resistant to gamma radiation than chickens, die when exposed to 3,000 roentgen (Willard, 1963). The estimated LD_{50} for a 6-day-old nestling was about 2,500 roentgen; however, birds receiving a sublethal dose were so severely stunted that their chances of survival were poor. All nestlings that survived irradiation developed normal fledgling behavior and attempted to leave the nest, even though growth and feather development were stunted. Nestlings that received 800 to 900 roentgen when 2 days old were able to leave the nest with normal nestmates; but because they were weakened and possessed subnormal flying ability, their survival was doubtful. Nestlings receiving 1,500 roentgen were able to crawl out of the nest box but were unable to fly even when they attempted to do so. Thus even though the nestling birds received sublethal doses that affected growing tissues, the radiation did not impair the brain centers responsible for inherent behavior.

Soil

Soil is the site where nutrient elements are brought into biological circulation by mineral weathering. It also harbors the bacteria that incorporate atmospheric nitrogen into the soil. Roots occupy a considerable portion of the soil. They serve to tie the vegetation to the soil and to pump water and its dissolved minerals to other parts of the plant for photosynthesis and other biochemical processes. Vegetation in turn influences soil development, its chemical and physical properties and organic-matter content. Thus the soil acts as a sort of pathway between the organic and mineral worlds.

SOIL FORMATION

Soil is the collection of natural bodies at the earth's surface that is composed of mineral and organic matter and is capable of supporting plant growth. Soil formation begins with the weathering of rocks and their minerals. Exposed to the combined action of water, wind, and temperature, rock surfaces peel and flake away. Water seeps into crevices, freezes, expands, and cracks the rock into smaller pieces. Accompanying and continuing long after this disintegration is the decomposition of the minerals themselves. Water and carbon dioxide combine to form carbonic acid, which reacts with calcium and magnesium in the rock to form carbonates. These slightly soluble carbonates either accumulate deeper in the rock material or are carried away, depending upon the amount of water passing through. Primary minerals that contain aluminum and silicon, such as feldspar, are converted to secondary minerals, such as clay. Because iron is especially reactive with water and oxygen, iron-bearing minerals are prone to rapid decomposition. Iron remains oxidized in the red ferric state or may be reduced to the gray ferrous state. Fine particles, especially clays, are shifted or rearranged within the mass by percolating water and on the surface by run-off, wind, or ice. Eventually the rock is broken down into loose material. This may remain in place, but is more often than not lifted, sorted, and carried away. Material transported from one area to another by wind is known as loess; that transported by water as alluvial, lacustrine, (lake), and marine deposits; and that by glacial ice as till. In a few places soil materials come from accumulated organic matter such as peat. Materials remaining in place are called residual.

This mantle of unconsolidated material is called the regolith. It may consist of slightly weathered material with fresh primary minerals or it may be intensely weathered and consist of highly resistant minerals such as quartz. Because of variations in slope, climate, and native vegetation, many different soils can develop in the same regolith. The thickness of the regolith, the kind of rock from which it was formed, and the degree of weathering affect the fertility and water relations of the soil.

Eventually plants root in this weathered material. Frequently intense weathering goes on under some plant cover, particularly in glacial till and water-deposited materials, which already are favorable sites for some plant growth. Thus soil development often begins under some influence of plants. They root, draw nutrients from mineral matter, reproduce, and die. Their roots penetrate and further break down the regolith. The plants pump up nutrients from its depths and add them to the surface, and in doing so recapture minerals carried deep into the material by weathering processes. By photosynthesis, plants capture the sun's energy and add a portion of it, as organic carbon—approximately 18 billion metric tons, 1.7×10^{17} kcal—to the soil each year. This energy source, the plant debris, enables bacteria, fungi, earthworms, and other soil organisms to colonize the area.

The breakdown of organic debris into humus is accomplished by decomposition and finally mineralization. Higher organisms in the soil —millipedes, centipedes, earthworms, mites, springtails, grasshoppers, and others—consume fresh material and leave partially decomposed products in their excreta. This is further decomposed by microorganisms, the bacteria and fungi, into various compounds of carbohydrates, proteins, lignins, fats, waxes, resins, and ash. These compounds then are broken down into simpler products such as carbon dioxide, water, minerals, and salts. This latter process is called *mineralization*.

The fraction of organic matter that remains is called humus. It is a stage in the decomposition of soil organic matter and therefore it is not stable. New humus is being formed as old humus is being destroyed by mineralization. The equilibrium set up between the formation of new humus and the destruction of the old determines the amount of humus in the soil.

Activities of soil organisms, the acids produced by them, and the continual addition of organic matter to mineral matter produce profound changes in the weathered material. Rain falling upon and filtering through the accumulating organic matter picks up acids and minerals in solution, reaches mineral soil, and

sets up a chain of complex chemical reactions. This continues further in the regolith. Calcium, potassium, sodium, and other mineral elements, soluble salts, and carbonates, are carried in solution by percolating water deeper into the soil or are washed away into streams, rivers, and eventually the sea. The greater the rainfall the more water moves down through the soil and the less moves upward. Thus, high precipitation results in heavy leaching and chemical weathering, particularly in regions of high temperatures. These chemical reactions tend to be localized within the regolith. Organic carbon, for instance, is oxidized near the surface, whereas free carbonates precipitate deeper in the rock material. Fine particles, especially clays, also move downward. These localized chemical and physical processes in the parent material result in the development of layers in the soil, called *horizons*, which impart to the soil a distinctive *profile*. Within a horizon, a particular property of the soil reaches its maximum intensity and away from this level decreases gradually in both directions. Thus each horizon varies in thickness, color, texture, structure, consistency, porosity, acidity, and composition.

In general soils have four major horizons: an organic, or O, horizon and three mineral horizons, the A, characterized by major organic-matter accumulation, by the loss of clay, iron, and aluminum, and by the development of granular, crumb, or platy structure; the B, characterized by an alluvial concentration of all or any of the silicates, clay, iron, aluminum, and humus, alone or in combination, and by the development of blocky, prismatic, or columnar structure; and the C, material underlying the two horizons, either like or unlike the material from which the soil is presumed to have developed. Below all this may lie the R horizon, the consolidated bedrock. Because the soil profile is essentially a continuum, often there is no clear-cut distinction between one horizon and another. Horizon subdivisions (Fig. 7-31) are indicated by arabic numbers, e.g., O_1, O_2, A_1, A_2, etc.; lowercase letters are used to indicate significant qualitative departures from the central concept of each horizon; e.g., A_{2g} or B_t.

The O horizon, once designated as L, F, H, or A_o and A_{oo}, is the surface layer, formed or forming above the mineral layer and composed of fresh or partially decomposed organic material, as found in temperate forest soils. It is usually absent in cultivated soils. This and the upper part of the A horizon is the region where life is most abundant. It is subject to the greatest changes in soil temperatures and moisture conditions and contains the most organic carbon. And it is the site where most or all decomposition by organisms takes place (see Chapter 5).

THE ORGANIC HORIZON

Of all the horizons of the soil, none is more important or ecologically more interesting than the forest floor, or organic horizon. A close relationship exists between litter and humus, and the environmental conditions in the forest community—the internal microclimate of the soil, the moisture regime, its chemical composition, and its biological activity. The forest floor plays a dominant role in the life and distribution of many forest plants and animals, in maintenance of soil fertility, and in many of the soil-forming processes. The nature and quality of the forest organic layer depends in part on the kind and quality of forest litter. And the fate of that litter and the development of the A horizon are conditioned by the activity of microflora and soil animals. In fact many active humus forms undergo initial breakdown in the bodies of animal organisms. To complete the circle the composition and density of the soil fauna are influenced by the litter. Thus as Bernier (1961) put it, the forest humus is both "a consequence and a cause" of local ecological conditions.

The importance of the organic layer was stressed early in the history of ecology. Charles Darwin in his famous work of 1881, *The Formation of Vegetable Mould Through the Action of Worms, with Observations on Their Habits,* pointed out the influence of these animals on the soil. About the same time, in 1879 and 1884, the Danish forester P. E. Müller described the existence of two types of humus formation in the temperate forest soil, which he called mull and mor. Not only did he observe differences in vegetation, soil structure, and chemical composition, but he discovered differences in their fauna also. Müller considered mull and mor as biological, rather than purely physiochemical, systems and regarded the fauna present as aiding their formation. Others have regarded mull and mor from the physical and chemical point of view, with little regard for the biological

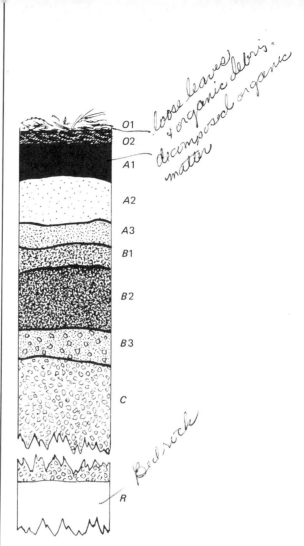

FIGURE 7-31
A generalized profile of the soil. Rarely does any one soil possess all of the horizons shown. O_1: Loose leaves and organic debris. O_2: Organic debris partly decomposed or matted. A_1: A dark-colored horizon with a high content of organic matter mixed with mineral matter. The A horizon is the zone of maximum biological activity. A_2: A light-colored horizon of maximum leaching. Prominent in podzolic soils; faintly developed or absent in chernozemic soils. A_3: Transitional to B, but more like A than B. Sometimes absent. B_1: Transitional to B but more like B than A. Sometimes absent. B_2: A deeper-colored horizon of maximum accumulation of clay minerals or of iron and organic matter; maximum development of blocky or prismatic structure or both. B_3: Transitional to C. C: The weathered material, either like or unlike the material from which the soil presumably formed. A glei layer may occur, as well as layers of calcium carbonate, especially in grasslands. R: Consolidated bedrock.

mechanisms involved. Actually both are the result of interaction of all three.

Mor, characteristic of dry or moist acid habitats, especially heathland and coniferous forest, has a well-defined unincorporated and matted or compacted organic deposit resting on mineral soil. It results from an accumulation of litter that is slowly mineralized and remains unmixed with mineral soil. Thus there is a sharp distinction or break between the O and the A horizons. Slow though mineralization may be, it is the manner in which this process proceeds that distinguishes mor from other humus types. The main decomposing agents are fungi, both free-living and mycorrhizal, which tend to depress soil animal activity and to produce acids; nitrifying bacteria may be absent. Vascular cells of leaves disappear first, leaving behind a residue of mesophyll tissue. Proteins within the leaf litter are stabilized by protein-precipitating material, making them, in some cases, resistant to decomposition. Because of limited volume, poor space, acidity, type of litter involved, and the nature of its breakdown, mor is inhabited by a small biomass of the smaller soil animals. These organisms have little mechanical influence on the soil but live instead in an environment of organic material cut off from the mineral soil beneath.

Mull, on the other hand, results from a different process. Characteristic of mixed and deciduous woods, on fresh or moist soils with a reasonable supply of calcium, mull possesses only a thin scattering of litter on the surface, and the mineral soil is high in organic content. All organic materials are converted into true humic substances, and because of animal activity, these are inseparably bound to the mineral fraction, which absorbs them like a dye. There is no sharp break between the O and the A horizons. Because of less acidity and a more equitable base status, bacteria tend to replace fungi as the chief decomposers, and nitrification is rapid. Soil animals are more diverse and possess a greater biomass, a reflection of more equitable distribution of living space, of oxygen, of food and moisture, and of a smaller fungal component. This faunal diversity is one of mull's greatest assets, because the humification process flows through a wide variety of organisms with differing metabolisms. Not only do these soil animals fragment vegetable debris and associate it with mineral particles, thus enhancing micro-

bial and fungal activity, but they also incorporate the humified material with mineral soil. This constant interchange of material takes place from the surface to the soil and back again. Plants extract nutrients from the soil and deposit them on the surface. Then the soil plants and animals reverse the process.

Between the two extremes, the mull and mor, is moder, the insect mull of Müller. In this humus type, plant residues are transformed into the droppings of small arthropods, particularly collembolans and mites. Residues not consumed by the fauna are reduced to small fragments, little humified and still showing cell structure. The droppings, plant fragments, and mineral particles all form a loose, netlike structure held together by chains of small droppings. In acid moder, the shape of the droppings is destroyed by the washing action of rainwater; and under more extreme conditions, humus leached from the droppings acts as a binding substance to form a dense, matted litter approaching a mor. At the other end of the spectrum, the border line between moder and mull, the droppings of large arthropods, which are capable of taking in considerable quantities of mineral matter with food, are common. However, moder differs from the mull in its higher organic-matter content, restricted nitrification, and a more or less mechanical mixture of organic components with the mineral, the two being held together by humic substances, but yet separable. In other words, the organic crumbs are deficient in mineral matter in contrast to mull, in which the mineral and organic parts are inseparably bound together.

Litter source plays a decisive role in its decomposition and in the direction that this takes, whether to mull, mor, or moder. Litter from plants growing on nutrient-rich soil under favorable temperature and moisture conditions encourages the development of an active microfauna and bacteria able to mineralize the material rapidly. Beech and sugar maple, which grow on soils of a wide range of fertility, will produce mull on rich soils and moder on poor soils (Bernier, 1961). Some plant litters, such as pine and spruce needles, are high in lignins, which resist decomposition and inhibit the decomposition of cellulose (Lutz and Chandler, 1946). Tannin, common in oak litter, may prevent the rapid breakdown of protein and often gives rise to raw humus. Litter types influence the abundance and composition of soil fauna, which play an active role in humus formation, and the lack of which results in reduced humification. Different plants, different parts of plants, and even particular plant tissues are attacked selectively by various soil animals. Soil animals mix and bind organic matter with mineral and influence aggregate formation in the soil. The interplay of these two, vegetation and fauna, may result in the reversibility of mull and mor. As vegetation changes either by succession or by the establishment of other plant species, the soil fauna changes and with it the type of litter breakdown. Chemical treatments with fertilizers or pesticides, which may increase nutrient status or perhaps kill off certain soil organisms, likewise may result in a change in the type of forest litter decomposition.

PROFILE DIFFERENTIATION

The differentiation of the soil profile into horizons, and the nature of the soil material, its content and distribution of organic matter, its color, and its chemical and physical characteristics are influenced over large areas by the combined action of vegetation and its prime determinant, climate (Fig. 7-32). Thus the soil beneath native grassland differs from that beneath native forest.

Grassland vegetation developed in the subhumid-to-arid and temperate-to-tropical climates of the world—the plains and prairies of North America, the steppes of Russia, and the veldts and savannas of Africa. Dense root systems may extend many feet below the surface. Each year nearly all of the vegetative material aboveground and a part of the root system are turned back to soil as organic residue. Although this material decomposes rapidly the following spring, it is not completely gone before the next cycle of death and decay begins. The humus then becomes mixed with mineral soil by the action of the soil inhabitants, developing a soil high in organic matter. The humus content is greatest at the surface and declines gradually with depth. Because the amount of rainfall in grassland regions generally is insufficient to remove calcium and magnesium carbonates, these are removed only down to the average depth that the percolating water reach. The high calcium content of the surface soil is maintained by grass, which absorbs large quantities from lower

Chernozem

Tundra

Prairie (Brunizem)

Podzol

Mountain
soils

Chestnut
and Brown

Gray-brown
Podzolic

Sierozem
and
Desert

Red-yellow
Podzolic

Laterite

horizons and redeposits them on the surface. Likewise there is little loss of clay from the surface layer. This process of soil development is called *calcification*; the soil itself is called *pedocal*.

Soils developed by calcification have a distinct A horizon of great thickness and an indistinct B horizon, characterized by an accumulation of calcium carbonate. The A horizon is high in organic matter and nitrogen even in tropical and subtropical regions.

Forests are the dominant vegetation in the humid regions. Here the cycle of organic matter accumulation differs from that of the grassland. Only part of the organic matter, leaves, twigs, and some trunks, is turned over annually. Leaves, which are the largest source of organic matter, and vegetation of the ground layer remain on the surface. Dead roots add little to soil organic matter because they die over an irregular period of time and are not concentrated near the surface. Because only the leaves are returned regularly to the soil and much of the mineral matter and energy is tied up in trunk and branches, most of the currently available nutrients turned back to the soil come from annual leaf fall. The amount of nutrient return varies with the species composition of the forest because trees differ in the nutrient content of their leaves. For example basswood, quaking aspen, hickories, American elm, and flowering dogwood contain more calcium in their leaves and return more calcium to the soil than sugar maple, red maple, yellow birch, and red oak (Lutz and Chandler, 1946). And the latter return more than beech, red pine, white pine, and hemlock.

Rainfall in forested regions is sufficient to leach away many elements, especially calcium, magnesium, potassium, iron, and aluminum. Because trees generally return an insufficient amount of bases back to the surface soil, it becomes acid, although the degree of acidity will vary, depending upon the forest composition and its site. Some forests in the southern Appalachians, particularly those containing yellow poplar and basswood and growing on north and northeast slopes, have rather high, often neutral pH in surface horizons, even though they grow on soils weathered from acid sandstone (unpublished data from van Eck and Smith). Increased acidity may cause the dispersion and downward movement of organic and clay colloids. A soil developed by

FIGURE 7-32 [OPPOSITE]

The great soil groups. This map of North America shows the general distribution of the important zonal, or great soil, groups of the continent and points out the general relation of soils to vegetation and climate. The majority of the soils illustrated here (the exception is the tundra) are those that develop on well-drained sites. In the humid regions bases do not accumulate in the soils because of the leaching processes associated with high rainfall. Podzol soils, or spodosols, characterized by a very thin organic layer on top of a gray, leached soil lying over a dark-brown horizon, generally develop in a cool, moist, climate under coniferous forests. Under deciduous forest in a cool, temperate, moist climate develop the gray-brown podzolic soils or alfisols. These differ from podzols in that the leaching is not so excessive and beneath the organic layer is a horizon of grayish-brown leached soil. The red and yellow soils, or ultisols, occur in a warm-temperate moist climate of southeastern North America. Developed through podzolization with some laterization, yellow soils are characterized by a grayish-yellow leached horizon over a yellow one; the red by a yellowish-brown leached soil over a deep-red horizon. In the tall-grass country with a temperate, moist climate is prairie or brunizem soils or mollisols, the result of calcification. It is very dark brown in color, grading through lighter brown with depth. West of this lies the chernozem, another mollisol, black soil high in organic matter, some 3 to 4 ft deep, which grade into lime accumulations. They developed under tall- and mixed-grass prairie. Closely related are the chestnut and brown soils, also mollisols, dark brown and grading into lime accumulations at 1 to 4 ft. These soils developed under mixed- and short-grass prairie. In desert regions are sierozem and desert soils or aridisols. They are pale grayish in color, low in organic matter, and closely underlain with calcareous material. Lateritic soils, or oxisols, typical of tropical rain forest where decomposition is rapid, have a thin organic layer over a reddish leached soil. In high mountains are a variety of soils, here vaguely classified as mountain soils, or entisols. Many of them are stony and lack any well-developed horizons. Tundra soils are variable, but the common one is a glei, subject to considerable disturbance from frost action and underlain with a permanently frozen substrate.

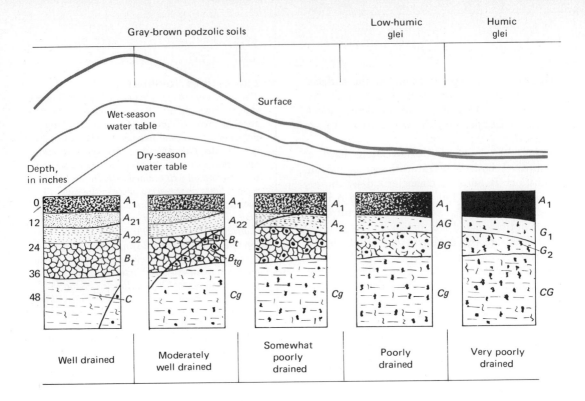

this process is called podzolic and the process *podzolization*. The name comes from the Russian, meaning "ash beneath," and refers to the leached horizon of strongly podzolized soils. The latter are characterized by a white-colored A2 horizon. A brilliant, yellow-brown B horizon results from accumulations of iron and aluminum compounds and humus. Iron accumulations in some podzol soils may act as a cement, creating a hardpan layer in the B horizon. This layer, called ortstein, impedes the free circulation of air and water. Soils developed by this process are referred to as *pedalfer*.

In the humid subtropical and tropical forested regions of the world, where rainfall is heavy and temperatures high, the soil-forming process is much more intense. Because temperatures are uniformly high, the weathering process in these regions is almost entirely chemical, brought about by water and its dissolved substances. The residues from this weathering—bases, silica, alumina, hydrated aluminosilicates, and iron oxides—are freed. Because precipitation usually exceeds evaporation, the water movement is almost continuously downward. With only a small quantity of electrolytes present in the soil water because of continual leaching, silica and aluminum silicates are carried downward, while sesquioxides of aluminum and iron remain behind. The reason for this is that these sesquioxides are relatively insoluble in pure rainwater, but the silicates tend to be precipitated as a gel in solutions containing humic substances and electrolytes. If humic substances are present they act as protective colloids about iron and aluminum oxides and prevent their precipitation by electrolytes. The end-product of such a process is a soil composed of silicate and hydrous oxides, clays, and residual quartz, deficient in bases, low in plant nutrients, and intensely weathered to great depths. Because of the large amount of iron oxides left, these soils possess a variety of reddish colors and they generally lack distinct horizons. Below, the profile is unchanged for many feet. The clay has a stable structure and unless precipitated, iron is hardened into a cemented laterite. It is very pervious to water and is easily penetrated by plant roots. This soil-forming process is termed *laterization* or *latosolization*.

Arid and semiarid regions have relatively sparse vegetation. Because plant growth is limited by low rainfall, there is very little organic matter and nitrogen in the soil. Light precipitation results in slightly weathered and slightly leached soils high in plant nutrients. Their horizons usually are faint and thin. In these regions occur areas where soils contain excessive amounts of soluble salts, either from parent material or from the evaporation of water draining in from adjoining land. The infrequent rain penetrates the soil, but soon afterward evaporation at the surface draws the salt-laden water upward. The water evaporates, leaving saline and alkaline salts at or near the surface, to form a crust, or *caliche*.

Calcification, podzolization, and laterization are processes that take place on well-drained soil. Under poorer drainage conditions a different soil-development process is at work. The slope of the land determines to a considerable extent the amount of rainfall that will enter and pass through the soil, the concentration of erosion materials, the amount of soil moisture, and the height at which the water will stand in the soil (Fig. 7-33). The amount of water that passes through or remains in the soil determines the degree of oxidation and breakdown of soil minerals. The iron in soils where water stays near or at the surface most of the time is reduced to ferrous compounds. These give a dull gray or bluish color to the horizons. This process, called *gleization*, may result in compact structureless horizons. Gley soils are high in organic matter because more organic matter is produced by vegetation than can be broken down by humification, which is greatly reduced because of the absence of soil microorganisms. On gentle to moderate slopes, where drainage conditions are improved, gleization is reduced and occurs deeper in the profile. As a result the subsoil will show varying degrees of mottling of grays and browns. On hilltops, ridges, and steep slopes, where the water table is deep and the soil well drained, the subsoil is reddish to yellowish-brown from the oxidized iron compounds.

In all, five drainage classes are recognized (Fig. 7-33). (1) well-drained soils are those in which plant roots can grow to a depth of 36 in. without restriction due to excess water; (2) on moderately well-drained soils, plant roots can grow to a depth of 18 to 20 in. without restriction; (3) in somewhat poorly drained soils, plant roots cannot grow beyond a depth of 12 or 14 in.; (4) poorly drained soils are wet most of the time, and usually are characterized by the growth of alders, willows, and sedges; (5) on very poorly drained soils, water stands on or near the surface most of the year.

Soils and time. The weathering of rock material; the accumulation, decomposition, and mineralization of organic material; the loss of

FIGURE 7-33 [OPPOSITE]
Effect of drainage on the development of gray-brown podzolic soils. Wetness increases from left to right. This diagram represents the topographic position the profiles might occupy. Note that the strongest soil development takes place on well-drained sites where weathering is maximum. The least amount of weathering takes place on very poorly drained soils where the wet-season water table lies above the surface of the soil. G or g indicates mottling; t indicates translocated silicate clays. (In part after Knox, 1952.)

minerals from the upper surface, and gains in minerals and clay in lower horizons; and horizon differentiation all require considerable time. Well-developed soils in equilibrium with weathering and erosion may require 2,000 to 20,000 years for their formation. But soil differentiation from parent material may take place in as short a time as 30 years. Certain acid soils in humid regions develop in 100 years because the leaching process is speeded by acidic materials. Parent materials heavy in texture require much longer to develop into "climax" soils, because of impeded downward flow of water. Soils develop more slowly in dry regions than in humid ones. Soils on steep slopes often remain young regardless of geological age because rapid erosion removes the soil nearly as fast as it is formed. Floodplain soils age little through time because of the continuous accumulation of new materials. Young soils are not as deeply weathered as and are more fertile than old soils because they have not been exposed to the leaching process as long. The latter tend to be infertile because of long-time leaching of nutrients without replacement from fresh material.

SOIL MORPHOLOGY

In the field, differences among soils and among horizons within a soil are primarily reflected by variations in texture, arrangement, structure, and color. The *texture* of a soil is determined by the proportion of different size soil particles. It is partly inherited from parent material and partly a result of soil-forming processes. The particles are classified on the basis of size into gravel, sand, silt, and clay. Gravel consists of particles larger than 2.0 mm. Sand ranges from 0.05 to 2.0 mm, is easily seen, and feels gritty. Silt consists of particles from 0.002 to 0.05 mm in diameter, which scarcely can be seen by the naked eye, feels and looks like flour. Clay particles, too fine to be seen even under the ordinary microscope, are colloidal in nature. Clay controls the most important properties of soils, including plasticity and exchange of ions between soil particles and soil solution.

Most soils are a mixture of these various particles. Based on the proportions of the various particles contained in them, soils can be grouped into 12 textural classes that fall into 5 textural groups: (1) coarse-textured soils are loose, consist mainly of sand and gravel, and

when moist will form a cast if squeezed that will fall apart when touched. They retain very little moisture and supply some plant nutrients. (2) Moderately coarse soils from sandy loam to very fine sandy loam, if squeezed when moist will form a cast that requires careful handling to avoid breaking. (3) Medium-textured soils are mixtures of sand, silt, and clay high enough to hold water and plant nutrients. (4) Moderately fine-textured soils are high in clay. Moderately sticky and plastic when wet, they may form a crust on the surface if organic matter is low. They have a high moisture-holding capacity. (5) Fine-textured soils contain more than 40 percent clay, may be sticky and plastic when wet and hold considerable water and plant nutrients, but may have restricted internal drainage. Soil texture may vary through the profile. The *B* horizon is usually finer-textured than those above and below because the clay particles may have been lost from the surface and deposited deeper in the soil.

Soil particles are held together in clusters or shapes of various sizes, called *aggregates* or *peds*. The arrangement of these aggregates is called *soil structure*. Like texture, there are many types of soil structure. Soil aggregates may be classified as granular, crumblike, platelike, blocky, subangular blocky, prismatic, and columnar (Fig. 7-34). Structureless soil can be either single-grained or massive. Soil aggregates tend to become larger with increasing depth. Structure is influenced by texture, the plants growing on the soil, other soil organisms, and the soil's chemical status.

Color has little direct influence on the function of a soil, but considered with other properties, it can tell a good deal about the soil. In fact it is one of the most useful and important characters for the identification of soil. In temperate regions dark-colored soils generally are higher in organic matter than light-colored ones. Well-drained soil may range anywhere from very pale brown to dark brown and black, depending upon the organic-matter content. But it does not always follow that dark-colored soils are high in organic matter. In warm temperate and tropical regions, the dark clays may have less than 3 percent organic matter. Red and yellow soils are the result of iron oxides, the bright colors indicating good drainage and good aeration. Other red soils obtain their color from parent material and not from soil-forming processes. Well-drained yellowish sands are white sands containing a small amount of organic matter and such coloring material as iron oxide. Red and yellow colors increase from the cool regions to the equator. Quartz, kaolin, carbonates of lime and magnesium, gypsum, and various compounds of ferrous iron give whitish and grayish colors to the soil. The grayest are permanently saturated soils in which the iron is in the ferrous form. Imperfectly and poorly drained soils are mottled with various shades of yellow-brown and gray. The colors of soils are determined with the use of standardized color charts.

SOIL CLASSIFICATION

Each combination of climate, vegetation, soil material, slope, and time results in a unique soil, the smallest repetitive unit of which is called a *pedon*. Soils even within a small area may vary considerably. Changes in slope, drainage, and soil material account for local differences among soil individuals. These soil individuals are roughly equivalent to the lowest category in the soil classification system—the *soil series*. The present soil taxonomic system consists of orders (Table 7-2), suborders, great groups (Fig. 7-32), families, and series. Soil series are named after the locality in which they were first described. For example, the Ovid series in New York was named after the town of Ovid; the Miami after the Miami River in western Ohio. Like the species among plants and animals, soil series are defined in terms of the largest number of differentiating characteristics and they occur in fairly limited areas. Higher categories of taxonomy combine series into larger groupings distinguished by even fewer differentiating properties. At the highest level of classification, one differentiates only among the classes that roughly correspond with broad climatic zones (Fig. 7-32).

Every soil series has neighboring soil series with different properties, into which it grades abruptly or gradually. If these several soils found side by side have developed from the same soil material but differ mainly in natural drainage and slope, they are said to form a *catena* (see Fig. 7-33). When soils are mapped, unlike soils for reasons of scale or practical use may be grouped together into *associations*. When occurring in inseparable patterns, soils are mapped as *complexes* (Fig. 7-35). In de-

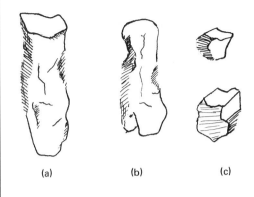

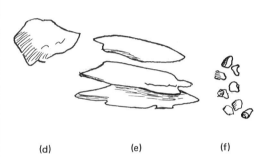

FIGURE 7-34
Some types of soil structure: (a) *prismatic;* (b) *columnar;* (c) *angular blocky;* (d) *subangular blocky;* (e) *platelike;* (f) *granular.* (*Adapted from USDA Handbook No. 18.*)

TABLE 7-2

New soil orders[a] and approximate equivalents in old classification[b]

Order	Formative syllable	Derivation and meaning	Approximate equivalents
1. Entisol	ent	coined from recent	Azonol soils, and some low humic gley soils
2. Vertisol	ert	L. *verto*, for inverted	Grumusols
3. Inceptisol	ept	L. *inceptum*, for young	Ando, sol brun acids, some brown forest, low humic gley, and humic gley soils
4. Aridisol	id	L. *aridus*, for arid	Desert, reddish desert, sierozem, solonchak; some brown and reddish-brown soils, and associated solonetz
5. Mollisol	oll	L. *mollis*, for soft	Chestnut, chernozem, brunizem (prairie) rendzinas, some brown, brown forest, and associated solonetz and humic gley soils
6. Spodosol	od	Gk. *spodos*, for ashy	Podzols, brown podzolic soils, and groundwater podzols
7. Alfisol	alf	Coined from Al-Fe	Gray-brown podzolic, gray wooded soils, noncalcic brown soils, degraded Chernozem, and associated planosols and half-bog soils
8. Ultisol	ult	L. *ultimus*, for last	Red-yellow podzolic soils, reddish-brown lateritic soils of the U.S., and associated planosols and half-bog soils
9. Oxisol	ox	Fr. *oxide*, for oxidized	Lateritic soils, latosols
10. Histosol	ist	Gk. *histos*, for organic	Bog soils

Source: [a] Soil Survey Staff, Soil Classification, A Comprehensive System, Soil Conservation Service, Washington, D.C., 1960. [b] M. Baldwin, C. E. Kellogg, and J. Thorp, Soil Classification, in Soils and Men, 1938 U.S.D.A. Yearbook of Agriculture, pp. 979–1001, Washington, D.C., 1938.

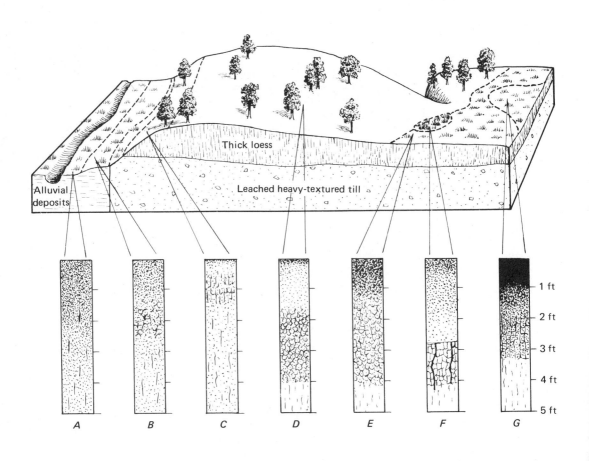

tailed mapping, a soil series can be subdivided into *types*, *phases*, and *variants*.

Fire

For many years ecologists ignored the importance of fire as an environmental influence. After regarding fire as destructive and as largely an act of man, ecologists now recognize fire as an important ecological force that is a part of the natural environment along with moisture, temperature, wind, and soil.

Three conditions are necessary for fire to assume ecological importance: (1) an accumulation of organic matter sufficient to burn; (2) dry weather conditions to render the material combustible; and (3) a source of ignition. Two sources of ignition are lightning and man.

Globally certain regions possess conditions conducive for burning and the spread of fires set off by lightning. One condition is a fire climate, a dry period in which fuel built up during wetter times can burn. The wetter the growing season and the longer and hotter the dry season, the more chances there are for lightning-set fires. Africa is ideally located for the development and occurrence of thunderstorms and lightning (J. Phillips, 1965; Batchelder, 1967). North America is literally swept with waves of electricity (Komarek, 1964; 1966). Southern and western Australia experience hot summers with low humidity and drying westerly winds (Cochrane, 1968). Long before the evolution of man, fires periodically swept these regions (T. M. Harris, 1958). Thus each is characterized by vegetation that evolved under the influence of fire: the grasslands of North America, the chaparral of southwestern United States, the maquis of the Mediterranean, the South African fynbos (J. Phillips, 1965), the African grasslands and savannas (Batchelder, 1967), the southern pinelands of the United States, and the even-aged stands of coniferous forests of western North America.

Since the Cenozoic era began, lightning has been a primary source of ignition. Annually 16 million thunderstorms occur on the earth (McCann, 1942), causing an average of 100 lightning strokes to the ground per second, 24 hours a day for 365 days a year (Komarek, 1968).

FIGURE 7-35 [OPPOSITE]
Effect of topography and native vegetation on the soil. This diagram shows the normal sequence of eight representative soil types from the Mississippi to the uplands in Illinois. The drawing also illustrates how bodies of soil types fit together in the landscape. Boundaries between adjacent bodies are gradations or continuums, rather than sharp lines.

The lower part of the diagram pictures the profiles of seven of the soils, showing the color and thickness of the surface horizon and the structure of the subsoil. Note how the natural vegetation that once covered the land (trees for forest, grass clumps for grass) influenced surface color. The diagram also shows how topographic position and distance from the bluff influence subsoil development.

Profile A is a bottomland soil (Sawmill) formed from recent sediments and has not been subjected to much weathering. Profile B (Worthen) on the foot slope also developed from recent alluvial material and shows little structure. Profile C (Hooper) on the slope break developed from a thick loess on top of leached till, while the soil on the bottom of the slope developed directly from the till. Profile D is an upland soil (Seaton) formerly covered with timber. It possesses a light surface color and lacks structure, the result of a rapid deposition of loess during early soil formation, holding soil weathering to a minimum. Profile E represents an upland soil (Joy) developed under grass. Note the dark surface and lack of structure, again the result of rapid deposition of loess. Profile F (Edgington) is a depressional wet spot. Extra water flowing from adjacent fields increased the rate of weathering, resulting in a light-colored grayish surface and subsurface and a blocky structure to the subsoil. This indicates a strongly developed soil. The depth of subsoil suggests that considerable sediment has been washed in from the surrounding area. Profile G (Sable) represents a depressional upland prairie soil. Note deep, dark surface and coarse, blocky structure. Abundant grass growth produced the dark color. (After Veale and Wascher, 1956.)

Lightning storms are not universally ac-companied by precipitation. In the temperate regions weather patterns at the end of a drought are often characterized by cloud thunderstorms with no rain. The same is true at the beginning and the end of dry periods in the tropical regions. When lightning strokes hit the ground, the dry material is kindled, and fires are set off. In the western United States 70 percent of the forest fires are caused by dry lightning during the summer. Because of the seasonal nature of lightning, fires so caused are most numerous during the summer in the midst of the growing season, when they would have the greatest impact as a selection force.

As a selective pressure, fire has a pronounced impact on the ecosystem. It reduces dead and dry organic matter to soluble ash and releases phosphorus, calcium, potassium, and other elements for rapid recycling, stimulating new growth. Although the flush of new vigorous growth is attributed to an increased availabil-ity of nutrients, this may not necessarily be true for all ecosystems. Daubenmire (1968b) suggests that the response of grasslands to fire may reflect the release of new shoots from the competition of old tillers and increased root activity. The response of animals to vegeta-tion on burns may be due to the increased availability of both tender shoots and biomass previously unavailable because litter and old stems were a hindrance to grazing. Consider-able nitrogen may be lost to the atmosphere, but unless the litter is converted to white ash, some nitrogen will remain in the charred lit-ter, increasing the total nitrogen content of the surface soil. Further increases in nitrogen come about by a marked increase in nitrogen-fixing legumes following a fire.

Fire exposes the mineral soil, stimulates the germination of certain seeds, and may en-courage erosion, changing the character of the site. In the western United States erosive forces favor such species as the knobcone pine (Vogl, 1967). Hot fires may heat the soil sufficiently to kill the roots of some plants and change the nature of the soil faunal community. The dark color of burned-over lands absorbs more heat from the sun, warming up the soil in temperate regions and reducing the depth of permafrost in arctic regions.

Fire modifies the vegetational community (Fig. 7-36). Crown and severe surface fires in a forest can destroy all existing vegetation. Light and moderate surface fires may destroy only the undergrowth, kill some thin-barked fire-sensitive trees by heating the cambium, and damage others. Heat-damaged trees are susceptible to attack from insects and disease (Fig. 7-37). Plants resistant to fire are char-acterized by underground stems and buds, aerial parts that die off annually, dormant underground buds at the soil surface, or thick fire-resistant bark. Fire-tolerant species are also characterized by an intolerance of shade and a requirement of mineral soil and full sunlight for the germination of seeds and growth of seedlings. Other plant species, though not fire-resistant as individual plants, require fire in their life cycles to release seeds from cones or provide a seedbed of mineral soil. Among such species are jack pine, knob-cone pine, white pine, red pine, paper birch, aspen, and eucalyptus.

Fire can free grass from woody competi-tion and maintain grass as dominant vegeta-tion. However, in some communities fire may have no effect on the composition of vege-tation. Through long evolution in the pres-ence of fire, many species have become fire-adapted. Communities containing such spe-cies may experience no loss of species, but with the stand opened up, opportunistic spe-cies may invade the area, increasing the rich-ness of the community. In such a manner fire can diversify the mosaic of vegetation over the landscape.

When man appeared on the scene, fire be-came an even more powerful influence on vegetation, for man added a new dimension to it. Whereas lightning fires are random and often periodic, man-made fires are often de-liberately set in an attempt to modify or change the environment. Fires became more numerous through the years, and their pat-tern was and is adjusted to the season, to agricultural calendars, and to religious be-liefs. Fires were set to clear ground for agri-cultural use, to improve conditions for hunt-ing, to develop grass and shrubby vegetation attractive to game, to improve forage for graz-ing, to open up the countryside, to reduce enemy cover, to develop areas for wild fruits and berries and other desirable plants, and to make travel easier. Other fires were set simply for excitement or revenge or to burn trash.

Whatever the reason, most fires set by man burned in the nongrowing seasons of fall and spring, when damage can be much more severe.

As man spread from the fire-evolved grasslands and savannas to the more humid forested areas, he introduced fire into less fire-resistant vegetational areas, such as hardwood forests. Further, he caused fires when the forests were the most inflammable—fall and early spring. Indian fires undoubtedly produced the open heaths of the northeast United States and the glades of Kentucky, developed oak stands in the central hardwoods (R. L. Smith, in press), and maintained large areas of blueberries (Thompson and Smith, 1970). Lumbering operations left massive piles of debris that fed extremely hot fires that swept across much of the logged-over country. In many places fire burned into the deep layers of organic matter and peat to rock and mineral soil, destroying any opportunity for former forest types to return (Fig. 7-36).

Because of the destructiveness of man-made fires, man has moved to the other extreme, the exclusion of all fire, which can have as deleterious effect on the ecosystem as too frequent or too hot fires. The lack of periodic fires allows the accumulation of trash and litter. Then if fires do start, they are much hotter and more destructive than more frequent cool fires. Exclusion of fire alters the composition of forest stands by eliminating the fire-resistant species, by permitting the dominance of fire-sensitive species, and by encouraging the spread of woody vegetation into grasslands. It inhibits the regeneration of fire-resistant species and causes deterioration of forage and range, stagnation of forest stands, and decline in the habitats of certain wildlife.

Carefully used, fire can be an important tool in the regulation and manipulation of vegetation. It can be used to improve forage stands and increase net productivity of grasslands. It can reduce the hazard of destructive forest fires by removing litter before it accumulates to a great degree, and can improve seed bed for regeneration of certain forest types and tree species. It can be used to improve wildlife habitats, to maintain certain fire-controlled ecosystems, to maintain the naturalness of wilderness areas, and even to improve the esthetics of the natural landscape.

FIGURE 7-36
Forest fires, by destroying existing vegetation, modify the nutrient and hydrologic cycles, set back succession to an earlier stage, and eliminate some species and encourage others. Fire is a powerful ecological force in molding local ecosystems. (Photo courtesy U.S. Forest Service.)

FIGURE 7-37
Fire can damage some fire-resistant trees, but the scars remain. In the Appalachian hardwood forest, fire often results in "cat-face" logs and butt rot, which causes an economic loss to the timber industry. (Photo by author.)

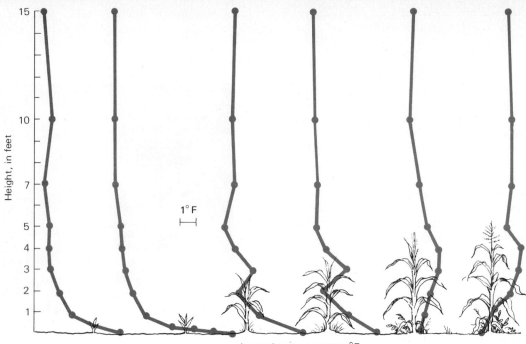

Increasing temperature, °F

Microclimates

When the weather report states that the temperature is 75° F and the sky is clear, the information may reflect the general weather conditions for the day. But on the surface of the ground in and beneath the vegetation, on slopes and cliff tops, in crannies and pockets, the climate is quite different. Heat, moisture, air movement, and light all vary radically from one part of the community to another to create a whole range of "little" or "micro" climates.

On a summer afternoon the temperature under calm, clear skies may be 82° at 6 ft, the standard level of temperature recording. But on or near the ground—at the 2-in. level—the temperature may be 10° higher; and at sunrise, when the temperature for the 24-hour period is the lowest, the temperature may be 5° lower than the standard level (Biel, 1961). Thus in the middle eastern part of the United States, the temperature near the ground may correspond to the temperature at the 6-ft level in Florida, 700 mi to the south; and at sunrise the temperature may correspond to the 6-ft-level temperature in southern Canada. Even greater extremes occur above and below the ground surface. In New Jersey, March temperatures about the stolons of clover plants ½ in. above the surface of the ground may be 72° F (21° C), while 3 in.

below the surface the temperature about the roots is 30° F (−1° C) (Biel, 1961). The temperature range for a vertical distance of 3.5 in. is 40° F (20° C). Under such climatic extremes most organisms exist.

The chief reason for the great differences between the ground level and 6 ft high is solar radiation. During the day the soil, the active surface, absorbs solar radiation, which comes in short waves as light, and radiates it back as long waves to heat a thin layer of air above (Fig. 7-38). Because air flow at ground level is almost nonexistent, the heat radiated from the surface remains close to the ground. Temperatures decrease sharply in the air above this layer and in the soil below. The heat absorbed by the ground during the day is reradiated by the ground at night. This heat is partly absorbed by the water vapor in the air above. The drier the air, the greater is the outgoing heat and the stronger is the cooling of the surface of the ground and the vegetation. Eventually the ground and the vegetation are cooled to the dewpoint, and water vapor in the air may condense as dew. After a heavy dew a thin layer of chilled air lies over the surface, the result of rapid absorption of heat in the evaporation of dew.

By altering wind movement, evaporation, moisture, and soil temperatures, vegetation influences or moderates the microclimate of an area, especially near the ground. Temperatures at the ground level under the shade are lower than those in places exposed to the sun and wind. Average maximum soil temperatures at the 1-in. level below the surface in a northern hardwoods forest and an aspen–birch forest in New York State were 60° F (15.5°C) from mid-May to late June and 68° F (20° C) (absolute maximum 78° F) from early July to mid-August (Spaeth and Diebold, 1938). Mean differences between maximum temperatures for the forest and adjacent fields ranged from 9° to 24° F. Maximum soil temperatures at 1 in. in open chestnut–oak forests were 66° F (19° C) from mid-May to late June and 75° F (24° C)(absolute maximum, 93° F) from July to mid-August. On fair summer days a dense forest cover reduces the daily range of temperatures at 1 in. by 20° to 30° F, compared with the temperature in the soils of bare fields.

Vegetation also reduces the steepness of the temperature gradient and influences the height of the active surface, the area that intercepts the maximum quantity of solar insolation. In the total absence of or in the presence of very thin vegetation, temperature increases sharply near the soil; but as the plant cover increases in height and density, the leaves of the plants intercept more solar radiation (Fig. 7-39). The plant crowns then become the active surface and raise it above the ground. As a result temperatures are highest just above the dense crown surface and lowest at the surface of the ground. Maximum absorption of solar radiation in tall grass occurs just below the upper surface of the vegetation, whereas in the short grass maximum temperatures are at the ground level (Waterhouse, 1955). As the grasses grow taller the level of maximum temperature falls into and rises with the upper level of the grass stalks until the temperature eventually reaches an approximate equilibrium with the air above. (Among broad-leafed plants daily maximums occur on the upper leaf surfaces.) At night minimum temperatures are some distance above the ground because the air is cooled above the tops of plants and the dense stalks prevent the chilled air from settling to the ground.

Within dense vegetation air movements are reduced to convection and diffusion (Fig. 7-40). In dense grass and low plant cover, complete calm exists at ground level. This calm is an outstanding feature of the micro-

FIGURE 7-39 [OPPOSITE]
Vertical temperature gradients at midday in a cornfield from the time of seeding stage to the time of harvest. Note the increasing height of the active surface. (Adapted from Wolfe et al., 1949.)

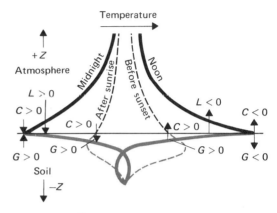

FIGURE 7-38
Idealized temperature profiles in the ground and air for various times of day, and the transport of heat by convection, C; by conduction, G; and by latent heat of evaporation or condensation, L. (From D. M. Gates, 1962, Energy Exchange in the Biosphere, Harper & Row, New York.)

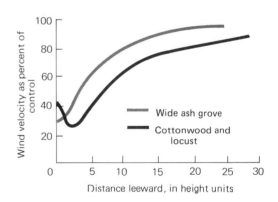

FIGURE 7-40
Comparative wind velocity at 16 in. above the ground to the leeward of an open ash grove and a cottonwood and locust shelter-belt. The curve for the ash grove illustrates the effectiveness when a stand of trees is open, allowing the wind to sweep under the crowns. Zero is the lee face of the barrier. (After Stoeckeler, 1962.)

climate near the ground because it influences both temperature and humidity and creates a favorable environment for insects and other animals.

Vegetation also deflects wind flow up and over its top (Fig. 7-41). If the vegetation is narrow, such as is a windbreak or a hedgerow, the microclimate on the leeward side may be greatly affected. Deflection of wind produces an area of eddies immediately behind the vegetation, in which the wind speed is low and small particles such as seeds are deposited. Beyond this is an area of turbulence, in which the climate tends to be colder and drier than normal. If some wind passes through the barrier and some goes over it, no turbulence develops, but the mean temperature behind the barrier is high in the morning and lower in the afternoon.

Humidity differs greatly from the ground up. Because evaporation takes place at the surface of the soil or at the active surface of plant cover, the vapor content (absolute humidity) decreases rapidly from a maximum at the bottom to atmospheric equilibrium above. Relative humidity increases above the surface, since actual vapor content increases only slowly during the day whereas the capacity of the heated air over the surface to hold moisture increases rather rapidly. During the night little difference exists above and on the ground. Within the growing vegetation, however, relative humidity is much higher than above the plant cover. In fact near-saturation conditions may exist.

Soil properties, too, enter the microclimatic picture. In a soil that conducts heat well, considerable heat energy will be transferred to the substratum, from which it radiates to the surface at night. On such soils, surface temperatures are lower by day and higher by night than the surface temperatures of poorly conducting soils. This influences the occurrence of frost. Moist soils are better conductors of heat than dry soils. Light-colored sandy soils increase reflection and reduce the rate at which heat energy is absorbed. Dark soils, on the other hand, absorb more heat.

NORTH AND SOUTH SLOPES

The microclimatic variations throughout a given area, then, result from differences in slope, soil, and vegetation. The greatest microclimatic differences exist between north and south slopes. South-facing slopes receive the most solar energy, which is maximal when the slope grade equals the sun's angle from the zenith point. North-facing slopes receive the least energy, especially when the slope grade equals or exceeds the angle of sun-ray inflection. At latitude 41° N (about central New Jersey and southern Pennsylvania) midday insolation on a 20° slope is, on the average, 40 percent greater on south slopes than on north during all seasons. This has a marked effect on the moisture and heat budget of the two sites. High temperatures and associated low vapor pressures induce evapotranspiration of moisture from the soil and plants. The evaporation rate often is 50 percent higher, the average temperature higher, the soil moisture lower, and the extremes more variable on south slopes. Thus the microclimate ranges from warm, xeric conditions with wide extremes on the south slope to cooler, less variable, more mesic conditions on the north slope. Xeric conditions are most highly developed on the top of south slopes where air movement is the greatest, while the most mesic conditions are at the bottom of the north slopes (Fig. 7-42). In the central and southern Appalachians, north slopes are steeper and include many minor microreliefs—small depressions and benches created largely by the upheaved roots of thrown trees. South slopes are longer and less steep, because of long-term downward movement of the soil. The whole north–south slope complex is the result of a long chain of interactions: the solar radiation influences moisture regimes, the moisture regime influences the species of trees and other plants occupying the slopes. The species of trees in turn influence mineral recycling, which is reflected in the nature and chemistry of the surface soil and the nature of the herbaceous ground cover.

Tree cover on the north-facing and south-facing slopes of the New Jersey hills does not reflect the microclimates of the two slopes as well as the herbaceous layer, according to Cantlon (1953), but differences do show up. In New Jersey yellow poplar does not occupy the south slope nor black oak the north. Flowering dogwood grows on both slopes, but is most abundant on the north. Rhododendron, spicebush, and maple-leaf viburnum occupy the north slopes. Ferns and mosses characterize the ground layer of the north slopes; grasses and sedges, the south.

The differences apparent between the north and south slopes in the New Jersey hills are much more pronounced in the mountains of the central Appalachians, where the land is strongly dissected by a dendritic drainage pattern, the slopes are extreme—ranging up to 85 and 90 percent—and the valleys are very narrow. In places, particularly in the Cumberland Plateau, the hills arise abruptly from narrow stream bottoms and the ridges drop off sharply. In these situations the plant communities of the north and south slopes often are sharply defined at the ridge top, the xeric south-slope vegetation meeting the more mesic vegetation of the upper north slope (Fig. 7-43).

The lower north slopes of the Cumberland Plateau support a forest dominated by beech, sugar maple, and yellow poplar, with black birch, basswood, black walnut, and butternut as associates. This gradually gives way to a red oak–white oak–basswood forest on the upper slopes. The umbrella tree and the redbud are characteristic of the understory.

The lower south slope that lies within the shelter and shadow of the north slope is very similar to the lower and middle north slope and often contains the most abundant hemlock stands. This bottom slope forest soon changes to oak forest on the middle south slope, where black oak and scarlet oak are dominant. The upper south slope supports a forest of chestnut oak, pitch pine, and shortleaf pine. The characteristic understory tree is sourwood. Flowering dogwood and sassafras are common to both slopes, but dogwood is the most abundant on the north and sassafras on the south. Red maple, an irregular understory tree on the north slope, is abundant on the south. Red oak, occurring in all strata, is most abundant on the north slope; black locust and shadbush are generally distributed on the south slope.

Even more pronounced differences exist underfoot in the herbaceous layer. Forty-eight species of herbs and ferns are exclusive to the north slopes in the Cumberland Plateau; 61 are exclusive to the south slope (Smith and van Eck, unpublished data).

Similar vegetational changes are characteristic of north-facing and south-facing slopes in the chaparral of the North Coast Range of California. South slopes—warm, dry, and thin-soiled—are covered with chamise, buckbush, and yerba santa. The cool, moist, north slopes with deep fertile soil support deerbush,

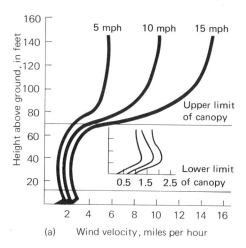

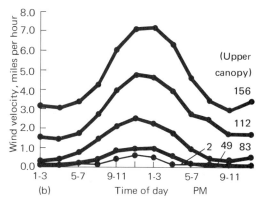

FIGURE 7-41
Comparative wind velocities. (a) *Distribution of wind velocities with height as affected by the timber canopy of coniferous forests for wind velocities of 5, 10, and 15 mi/hour measured 142 ft above the ground. (After Fons, 1940.) (b) The average wind velocity during a June day (based on 1938, 1939, 1940 dates) inside a coniferous forest with a cedar understory in northern Idaho. Note decrease in velocity near the ground. (After Gisborne, 1941.)*

eastwood manzanita, interior live oak, and California laurel (Taber and Dasmann, 1958).

Being mobile, few if any animals are typical of north or south slopes, as far as we now know. Deer tend to use south slopes more heavily in winter and early spring and north slopes more heavily in summer (Taber and Dasmann, 1958). In the central Appalachians the red-backed vole, throughout its range normally an inhabitant of the cool, fir, spruce, aspen, and northern hardwoods forests, is restricted in its local distribution to forested, mesic, north-facing slopes (Smith and Violet, in press). Studies of the north- and south-facing slopes of ant mounds in Denmark (Haarløv, 1960) suggest that some soil invertebrates may be confined to either of the two slopes. The steep north sides of the mounds were completely covered with *Festuca ovina*; the gentle south slopes had an open cover of *Achillea* but were dominated by moss. Similar to the large slopes, the south-facing sides of the mounds had the highest maximum temperature and the greatest fluctuations. The soil fauna common to the north slopes were *Pyemotidae* sp. and *Tectocepheus alatus*; common to the south slopes were *Speleorchestes termitophilus* and *Passalogetes perforatus*. Of these only the mite *Passalogetes* was confined to one slope, the south, alone. This species, intolerant of humidity, can exist only in a dry habitat.

VALLEYS AND FROST POCKETS

The widest climatic extremes occur in valleys and pockets, areas of convex slopes, and low concave surfaces. These places have much lower temperatures at night, especially in winter, and much higher temperatures during the day, especially in summer, and a higher relative humidity. Protected from the circulating influences of the wind, the air becomes stagnant. It is heated by insolation and cooled by terrestrial radiation, in sharp contrast to the wind-exposed, well-mixed air layers of the upper slopes. In the evening cool air from the uplands flows down the slope into the pockets and valleys to form a lake of cool air. Often when the warm air in the valley comes into contact with the inflowing cold air, the moisture in the warm air may condense as valley fog.

A similar phenomenon takes place on small concave surfaces. Like the larger valley, these concave surfaces radiate heat rapidly on still, cold nights and cold air flows in from surrounding higher levels. On such sites the air temperature near the ground may be 15° lower than the surrounding terrain. This results in a temperature inversion (see Atmospheric Temperatures). Because low ground temperatures in these areas tend to result in late spring frosts, early fall frosts, and a subsequent short growing season, these depressions are called frost pockets. The pockets need not be deep. Minimum temperatures in small depressions only 3 to 4 ft deep were equivalent to those of a nearby valley 200 ft below the general level of the land (Spurr, 1957). Such variations in temperature due to local microrelief can strongly influence the distribution and growth of plants. Tree growth is inhibited; and because the low surfaces more often than not accumulate water as well as cold air, such sites may contain plants of a more northern distribution. Frost pockets may also develop in small forest clearings. The surface of the tree crowns channels cold air into the clearings as terrestrial radiation cools the layer of air just above.

MICROCLIMATE OF THE CITY

Ever since man became an urban dweller, he has not only altered the natural environment, but he has also altered the atmosphere, creating a distinctive urban microclimate.

The urban microclimate is a product of the morphology of the city and the density and activity of its occupants. In the urban complex, stone, asphalt, and concrete pavement and buildings with a high capacity for absorbing and reradiating heat replace natural vegetation with low conductivity of heat. Rainfall on impervious surfaces is drained away as fast as possible, reducing evaporation. Metabolic heat from masses of people, and waste heat from buildings, industrial combustion, and vehicles raise the temperature of the surrounding air. Industrial activities, power production, and vehicles pour water vapor, gases, and particulate matter into the atmosphere in great quantities. The effect of this storage and reradiation of heat is the formation of a heat island about cities (Fig. 7-44) in which

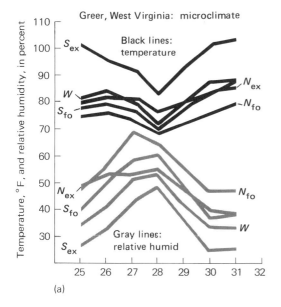

Greer, West Virginia: microclimate

(a)

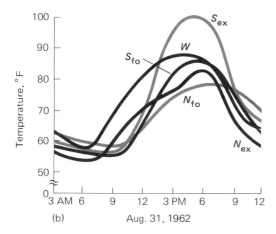

(b) Aug. 31, 1962

W : Standard weather station on ridge
N_{fo} : Microclimate station on forested north-facing slope
N_{ex} : Microclimate station on exposed north-facing slope
S_{fo} : Microclimate station on forested south-facing slope
S_{ex} : Microclimate station on exposed south-facing slope

FIGURE 7-42
(a) *Daily maximum temperature, minimum relative humidity, as recorded by four weather stations measuring microclimate and a single standard weather station during a week in August, 1962. Note extremes recorded at exposed sites* (S_{ex}), *differences in readings of microclimate stations, standard station.* (b) *Temperatures recorded at the five stations in August on a sunny day. Temperatures recorded at the exposed site showed the greatest variation. Contrast this with the forested north-facing slope.* (*Data courtesy Dr. W. A. van Eck.*)

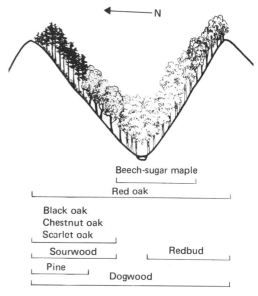

TREE DISTRIBUTION, CUMBERLAND MTS.

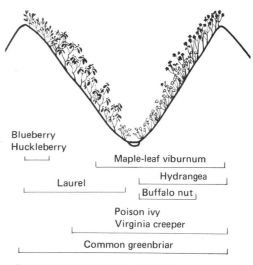

SHRUB DISTRIBUTION, CUMBERLAND MTS.

FIGURE 7-43
The influence of microclimate on the distribution of trees and shrubs on a north-facing and south-facing slopes in the Cumberland Plateau of southwestern West Virginia.

233

the temperature may be 6° to 8° C higher than the surrounding countryside (see Landsberg, 1970; SMIC, 1971).

Heat islands are characterized by high temperature gradients about the city. Highest temperatures are associated with areas of highest density and activity; temperatures decline markedly toward the periphery of the city (Fig. 7-45). Although detectable throughout the year, heat islands are most pronounced during summer and early winter and are more noticeable at night than during the day when heat stored by pavements and buildings is reradiated to the air. The magnitude of the heat island is influenced strongly by local climatic conditions such as wind and cloud cover. If the wind speed, for example, is above some varying critical value, a heat island cannot be detected.

During the summer the buildings and pavement of the inner city absorb and store considerably more heat than does the vegetation of the countryside. In cities with narrow streets and tall buildings the walls radiate heat toward each other instead of toward the sky. At night these structures slowly give off heat stored during the day. Although daytime differences in temperature between the city and the country may not differ noticeably, nighttime differences become pronounced shortly after sunset and persist through the night. The nighttime heating of the air from below counteracts radiative cooling and produces a positive temperature lapse rate while an inversion is forming over the countryside. This, along with the surface temperature gradient, sets the air in motion, producing "country breezes" to flow into the city.

In winter solar radiation is considerably less because of the low angle of the sun, but heat accumulates from human and animal metabolism, from home heating, power generation, industry, and transportation. In fact heat contributed from these sources is 2.5 times that contributed by solar radiation. This energy reaches and warms the atmosphere directly or indirectly, producing more moderate winters in the city than in the country.

Urban centers influence the flow of wind. Buildings act as obstacles, reducing the velocity of the wind up to 20 percent of that of the surrounding countryside, increasing its turbulence, robbing the urban area of the ventilation it needs, and inhibiting the movement of

cool air in from the outside. Strong regional winds, however, can produce thermal and pollution plumes, transporting both heat and particulate matter out of the city and modifying the rural radiation balance a few miles downwind (Clarke, 1969; Oke and East, 1971).

Throughout the year urban areas are blanketed with particulate matter, carbon dioxide, and water vapor. The haze reduces solar radiation reaching the city by as much as 10 to 20 percent. At the same time, the blanket of haze absorbs part of the heat radiating upward and reflects it back, warming both the air and the ground. The higher the concentration of pollutants, the more intense is the heat island.

The particulate matter has other microclimatic effects. Because of the low evaporation rate and the lack of vegetation, relative humidity is lower in the city than in surrounding rural areas. But the particulate matter acts as condensation nuclei for water vapor in the air, producing fog and haze. Fogs are much more frequent in urban areas than in the country, especially in winter (Table 7-3).

Another consequence of the heat island is increased convection over the city. Updrafts, together with particulate matter and large amounts of water vapor from combustion processes and steam power, lead to increased cloudiness over cities and increased local rainfall both over cities and over regions downwind. An evidence of weather modification by pollution is the increase in precipitation and stormy weather about La Porte, Indiana, downwind from the heavily polluted areas of

TABLE 7-3
Climate of the city compared to the country

Elements	Comparison with rural environment
Condensation nuclei and particles	10 times more
Gaseous admixtures	5–25 times more
Cloud cover	5–10 percent more
Winter fog	100 percent more
Summer fog	30 percent more
Total precipitation	5–10 percent more
Relative humidity, winter	2 percent less
Relative humidity, summer	8 percent less
Radiation, global	15–20 percent less
Duration of sunshine	5–15 percent less
Annual mean temperature	0.5°–1.0° C more
Annual mean wind speed	20–30 percent less
Calms	5–20 percent more

Source: Adapted from H. E. Landsberg, 1970.

Chicago, llinois, and Gary, Indiana, and close to moisture-laden air over Lake Michigan. Since 1925 there has been a 31-percent increase in precipitation, a 34-percent increase in thunderstorms, and a 240-percent increase in the occurrence of hail (Changnon, 1968).

Summary

The physical and chemical conditions of the environment influence the well-being and distribution of plants and animals and the functions of the ecosystem. The kind and quantity of elements and nutrients available for circulation in the biogeochemical cycle affect the growth and reproduction of plants and animals that vary in their requirements and tolerances for different elements. Some, the macronutrients, are required in relatively large quantities by all living organisms. Others, the trace elements, or micronutrients, are needed in lesser and often only minute quantities; yet without them plants and animals will fail, as if they lacked one of the major nutrients. Organisms live within a rather limited range of environmental temperatures and moisture conditions, outside of which they fail to reproduce or grow. Temperature and moisture, acting together, determine in large measure the climate of a region and the distribution of plant and animal life. Wind increases evaporation and is a major agent in the dispersal and distribution of plant and animal life.

Light is essential to plants; without it ecosystems could not function. Light also influences the reproductive and activity cycles of plants and animals. An almost universal feature of life is an internal physiological biological clock, whose basic structure is probably chemical and is involved in the makeup of the cell. It is free-running under constant conditions, with an oscillation or fluctuation that has its own inherent frequency. For most organisms, the inherited clock deviates more or less from 24 hours. Under natural conditions this clock is set or entrained to 24 hours by external time cues, or Zeitgebers, which synchronize the activity of plants and animals with the environment. Because the most dependable external time setter is light and dark—day and night—most of the selected species studied so far are en-

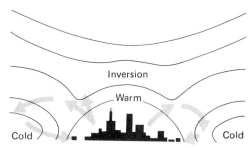

FIGURE 7-44
An idealized scheme of nighttime circulation above a city in clear, calm weather. A heat island develops over the city. At the same time a surface inversion develops in the country. This results in a flow of cool air toward the city, producing a country breeze in the city at night. Dashed lines are temperature isotherms; the arrows represent wind. (From H. Landsberg, 1970. © 1970 American Association for the Advancement of Science.)

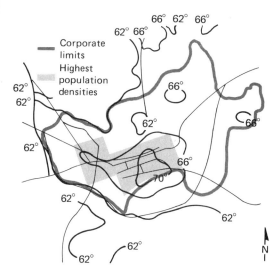

FIGURE 7-45
Thermal pattern of night air in a small city, Chapel Hill, N.C. Note that the highest temperatures are inside the corporate limits, where the population and the activity are the greatest. (From Kopec, 1970.)

trained to a 24-hour photoperiod. The onset
and cessation of activity are usually synchronized with dark and dawn, the response depending upon whether the organisms are
diurnal (light-active) or nocturnal (dark-
active). The biological clock is useful not
only as a means to synchronize the daily
activities of plants and animals with night
and day, consistent with the ecology of the
species, but also to time the activities with the
seasons of the year. The possession of a self-
sustained rhythm with approximately the same
frequency as that of environmental rhythms
enables organisms to "predict" such advance
situations as the coming of spring. It brings
plants and animals into a reproductive state
at a time of year when the probability for the
survival of offspring is the highest; it synchronizes within a population such activities as
mating and migration, dormancy and flowering. The acquisition and refinement of a
physiological timekeeper, geared to Zeitgebers
that provide organisms with distinct and
species-specific or population-specific synchronization with the environment, is a result of
natural selection. The secrets of the clock, an
understanding of how it works and where it
resides in the organism, have yet to be
discovered.

Gamma radiation is a new ecological influence of the nuclear age. Not only can it
kill animal life; it can destroy complex ecosystems as well and convert the areas into
simple ecosystems, dominated by radioresistant plants and animals, the latter chiefly
insects.

Whatever their classification, whatever the
major or minor differences between them,
all soils are the base for terrestrial ecosystems.
Soil is the site of decomposition of organic
matter and of the return of mineral elements
to the nutrient cycle. It is the home of animal
life, the anchoring medium for plants, and
their source of water and nutrients. Soil begins with the weathering of rocks and minerals,
which involves the leaching out and the carrying away of mineral matter. Its development
is guided by slope, climate, original material,
and native vegetation. Plants rooted in the
weathering material further break down the
substrata, pump up nutrients from their
depths, and add all-important organic material. This material, through decomposition
and mineralization, is converted into humus,
an unstable product that is continuously be-

ing formed and destroyed by mineralization.
As a result of the weathering process, accumulation and breakdown of organic matter, and
the leaching of mineral matter, horizons or
layers are formed in the soil. Of these there
are four: the O, or organic, layer; the A
horizon, characterized by an accumulation of
organic matter and a loss of clay and mineral
matter; the B horizon, in which mineral
matter accumulates; and C, the underlying
material. These horizons may be further divided into subhorizons. Of all the horizons,
none is more important than the humus layer,
which plays a dominant role in the life and
distribution of plants and animals, in the
maintenance of soil fertility, and in much of
the soil-forming process. Humus usually is
grouped into three types: mor, characteristic
of acid habitats, whose chief decomposing
agents are fungi; mull, characteristic of deciduous and mixed woodlands, whose chief
decomposing agents are bacteria; and moder,
which is highly modified by the action of
soil animals.

Soil profile development is influenced over
large areas by the vegetation and climate. In
grassland regions, calcification is the chief
soil-forming process, in which calcium accumulates at the average depth reached by
percolating water. In forest regions, podzolization, involving the leaching of calcium,
magnesium, iron, and aluminum from the
upper horizon and the retention of silica, takes
place. In tropical regions, laterization, in
which silica is leached and iron and aluminum
oxides are retained in the upper horizon,
is the major soil-forming process. Gleization
takes place in poorly drained soils. Organic
matter decomposes slowly and iron is reduced
to the ferrous state.

Differences between soils and between
horizons within soils are reflected by variations in texture, structure, and color. Each
combination of climate, vegetation, soil material, slope, and time results in a unique
soil, of which the smallest repetitive unit is
the pedon. Soil individuals equivalent to the
lowest category in the soil classification system are the soil series. These may be further
categorized into families, great soil groups,
suborders, and orders.

Another important environmental influence
is fire, only recently given due recognition.
Lightning-set fires have been a part of the
natural environment since the emergence of

terrestrial plant life. Many plant communities, such as grasslands and certain forest types, evolved under a regime of fire. With the coming of man fire became an even more important influence because it became more frequent, occurred during the fall and spring rather than the summer, and was often deliberately set to modify or change the environment. Because of the destructiveness of man-caused fires, man moved to another extreme, the exclusion of fires, which also has adverse ecological effects. Properly handled, fire is an important tool in the regulation and manipulation of vegetation.

Not only do all of the foregoing environmental conditions influence life over a region, they also influence the distribution of life on a much smaller scale. Organisms occupying the same general habitat may be living under completely different conditions because of variations in the microclimate or microenvironment. Most pronounced are the environmental differences between ground level and the upper strata and between north-facing and south-facing slopes. Other microclimates exist over urban areas. A city is characterized by the presence of a heat island. Compared to surrounding rural areas, a city has a higher average temperature, particularly at night, more cloudy days, more fog, more precipitation, a lower rate of evaporation, and lower humidity.

he most visible part of the eco-system is the biotic, the living portion of vegetation and animal life. The assemblage of plants and animals in any given ecosystem or, to be more restrictive, in any given physical environment is a *community*. Thus a community can be considered not only as the combination of plant and animal populations that comprise the biotic portion of a given ecosystem such as a forest, but also as the assemblage of organisms that inhabit a fallen log, an acorn, or even a tiny pool of water held in the hollow of a tree. The former is a major or *autotrophic* community, independent of others and requiring from the outside only the energy of the sun. The latter is a minor or *heterotrophic* community dependent upon the major community for its energy source. Thus the community is something more than a loose assemblage of independent organisms. The plants and animals that make up the community are interdependent, living together in some orderly fashion, forming a functional unit of the ecosystem.

Although the community usually consists of a certain combination of species, similar communities (an oak–hickory forest, for example) need not consist of exactly the same assemblage of species, which does create some problems in defining a community. Botanists use the term *association* for a plant community possessing a definite floristic composition, as a unit of vegetation that is comparable to a species. They use the word *community* only in a very general sense. Zoologists and many ecologists apply the term equally to a specific assemblage or some general groupings. Because of the problems, some ecologists question whether definitive communities actually exist or whether or not they are simply abstractions resulting from human attempts to arrange communities, like species, into neat categories.

But the general community does exist, and regardless of its species composition it does possess certain general attributes such as dominance, niche, species diversity, structure, stability, development, and a metabolic role in the functioning of the ecosystem. Both major and minor communities can be looked at in terms of these attributes.

Dominance

In a general sort of way the nature of a community is controlled either by physical or abiotic conditions such as the substrate, the lack of moisture, and wave action, or by some biological mechanism. Biologically controlled communities are often influenced by a single species or by a group of species that modify the environment. These organisms are called *dominants*.

It is not easy to delineate what constitutes a dominant species or, in fact, how to determine the dominant species. The dominants in a community may be the most numerous, possess the highest biomass, preempt the most space, make the largest contribution to energy flow or mineral cycling, or by some other means control or influence the rest of the community.

In a practical sense, some ecologists have given the role of the dominant organisms to those that are numerically superior. But numerical abundance alone is not sufficient. A species of plant, for example, can be widely represented and yet exert little influence on the community as a whole. In a forest the small or understory trees can be numerically superior, yet the nature of the community is controlled by a few large trees that overshadow the smaller ones. In such a situation the dominant organisms are not the ones with the greatest number, but the ones with the greatest biomass or those that preempt most of the canopy space. Other ecologists use biomass or basal area as a measure of dominance: the dominant organism may be relatively scarce yet by its activity control the nature of the community. The predatory starfish *Piaster*, for example, preys on a number of similar species and thereby reduces competitive interactions between them so that these different prey species can coexist (Paine, 1966). If the predator is removed, a number of prey species disappear and one of them becomes a dominant. In effect the predator controls the nature of the community, and so must be regarded as the dominant.

The concept of dominance involves certain implications. In the first place the dominant species may not be the most essential species in the community from the standpoint of energy flow and nutrient cycling, although they often are. Dominant species achieve their status by occupying niche space that might

The ecosystem and the community

potentially be occupied by other species in the community. For example, when the American chestnut was eliminated by blight from the oak-chestnut forests, the chestnut's position was taken over by other oaks and hickory. Although dominants frequently shape populations of other trophic levels, dominance necessarily relates to species occupying the same trophic level. If a species or a small group of species is to achieve dominance, it must relate to a total population of species, all of which possess similar ecological requirements. One or several become dominants because they are able to exploit the range of environmental requirements more efficiently than other species in the same trophic level. The subordinate species exist because they are able to occupy the niche or portions of it that the dominants cannot effectively occupy. Dominant organisms then are generalists capable of utilizing a wide range of physiological tolerances. The subdominants tend to be more specialized in their environmental requirements and more narrow in their physiological tolerances.

The degree of dominance expressed by any one species appears to depend in part on the position the community occupies on a physical or chemical gradient. At one particular point on a moisture gradient, species A and species B may be the dominants. As the gradient becomes drier, species B may assume a subdominant position in the community, and its place might be taken by a third species, C. Nutrient enrichment can change the structure of the community. Lakes receiving excessive sewage discharges shift from a diverse assemblage of nutrient-thrifty diatoms to a few blue-green algae that are able to exploit a nutrient-rich system (see Edmundson, 1970).

To determine dominance ecologists have used several approaches. One can measure relative abundance of the species involved, comparing the numerical abundance of one species to the total abundance of all species (see Appendix B). Or one can measure relative dominance, which is a ratio of the basal area occupied by one species to total basal area; or one can use relative frequency as a measure. Often all three of these measurements are combined to arrive at an Importance Value.

Niche

The consideration of dominance as the ability of certain species under some sets of environmental conditions to more successfully exploit a site than other associated species suggests that the community can also be regarded as an aggregation of niches.

The English ecologist Charles Elton (1927) considered the niche as the fundamental role of the organism in the community—what it does, its relation to its food and to its enemies. On the other hand the American ecologist Joseph Grinnell considered the niche to be a subdivision of the environment occupied by a species. The modern concept of the niche involves the combination of both, the functional role of an organism in the ecosystem as well as its position in time and space (Hutchison, 1959).

In a way the community can be considered as an aggregation of many environmental and functional features, each of which can be represented as a coordinate in an infinite-dimensional space (Hutchison, 1959), the *hypervolume*. For any one species the hypervolume would be the upper and lower limits of all environmental variables in which the species can persist. For the community the hypervolume would consist of all the hypervolumes of its constituent species (Fig. 8-1). In any one community a given species free from any sort of interference from another species will occupy some fraction of the community hypervolume. This is the *fundamental niche* of the species. But rarely does one species of a given trophic level alone make up a community. Other species are also there. Competitive relationships between them force the original species to constrict the portion of the hypervolume it occupies. The amount of the hypervolume actually occupied by a species in the face of competition from others is its *realized niche*.

The concept of the niche is closely associated with the concept of competitive exclusion and competitive relationships among species. In fact the two in many ways are synonymous (see Chapter 12). Basically what the competitive exclusion theory says is that no two species can occupy the same niche in a given environment. Either one must become extinct or one or both must, through natural selection, diverge into different niches.

In a theoretical sense each species occupies its own distinctive niche within the community hypervolume measured by such parameters as climatic tolerances, chemical environment, food size, required living space, and position in vertical structure of the community. The centers of each species' niches are distributed through the community niche hypervolume. The niche can be considered as the fraction of the community's resource that the species utilizes, and the fraction of community productivity contributed by that species. Considering the whole niche hypervolume of each species, overlap may occur. The niche of each species is such that areas of competition are not critical and the area of differences between niches is sufficient to allow coexistence.

Niche segregation, which for simplicity is usually measured in terms of one to several variables, is often based on the utilization of space. Territorial animals, for example, parcel out bits of a given area of suitable habitat among relatively few individuals (see Chapter 11). Space demands become most critical among those species exhibiting interspecific territoriality. Space can be divided into feeding niches. Organisms have restricted areas of the vertical profile in which they forage. Among a group of North American warblers (see Chapter 12) some restrict their feeding areas to the canopy, some to the outer branches, A similar division exists among the titmice of Europe (Lack, 1971). The division of feeding space may exist even between the sexes of the same species. The male red-eyed vireo, for example, gleans its insectivorous food in the upper canopy, the female in the lower canopy and nearer the ground, with only about a 35 percent overlap (Fig. 8-2) in feeding areas between the two (Williamson, 1971). Although similar foods may be utilized, each secures the insects from different levels. Similar sexual differences in foraging areas exist between males and females of several of the woodpeckers (Ligon, 1968). For birds, niches may be separated by food size dictated, perhaps, by the size of the bill (see Lack, 1971). Morphological differences influencing food procurement may even exist between sexes of a given species. A pronounced sexual difference in bill size exists between some male and female woodpeckers such as the Arizona woodpecker, *Dendrocopos arizonae*. Differences in length of bill is correlated with differences in foraging behavior. The male seeks food on the trunk, the female on the branches (Ligon, 1968). Thus definition of niche to date is restricted to a very few environmental and functional measurements.

Some clue to the nature of niche relationships may be obtained from patterns of species abundance. Theoretically the way species divide up the hypervolume should be reflected in abundance of species. There are several hypotheses relating to species abundance, the random-niche-boundary hypothesis, the niche-preemption hypothesis, and the log-normal distribution of species.

The random-niche hypothesis (MacArthur, 1960) suggests that boundaries of the niche hypervolumes of several species can be treated as points cast at random on a line. The points then represent the niche boundaries. The length of each contiguous, nonoverlapping segment represents the niche size. If the segments, representing the importance value of the species (the percentage of total density, biomass, etc. of all species in a community that a single species represents), are plotted in sequence from the longest to the shortest, then a curve like that of Fig. 8-3 A will result. This theory is the discredited and obsolete broken-stick model (see Hairston, 1969), which except for a very few situations, is not realistic. The curve is approached only by some small samples of taxonomically related animals with stable populations and relatively long life cycles occupying a small homogenous community such as nesting birds in a forest.

The niche-preemption hypothesis supposes that the most successful or dominant species preempts the most space, the next most successful claims the next largest share of the space, and the least successful occupies what little space is left. If the relative importance of the species is plotted in species sequences on a log scale, the result is a straight line (Fig. 8-3), and the distribution of the species forms a geometric series. Such a distribution is achieved only by a few plant communities containing relatively few species and occupying severe environments such as a desert. In most plant and animal communities, species overlap in the use of space and resources.

The log-normal hypothesis (Preston, 1962) supposes that the niche space occupied by a species is determined by a number of conditions such as resources, space, microclimate, and other variables that affect the success of one species in competition with another. The relative importance of each species is determined by the way the various variables affect each species. This results in a bell-shaped or normal distribution of importance values (Fig.

241

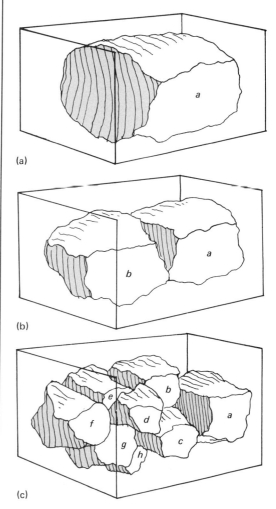

(a)

(b)

(c)

FIGURE 8-1
An abstraction of the hypervolume of the niche. The open box represents the hypervolume. In (a) the hypervolume is occupied by a highly adaptive generalized species. That portion of the hypervolume occupied by the species represents its fundamental niche. In the presence of a competitor the species is forced to retreat to that portion of the hypervolume in which it is most highly adapted, leaving another portion of the niche to its competitor (b). The two species now occupy realized niches. With more species present (c) capable of highly exploiting narrow ranges of the fundamental and environmental aspects of the niche, the realized niches become even more narrow. Eventually the hypervolume is occupied by a diversity of species.

8-4). Again, as in the random-niche theory, a line can be divided into segments representing the ranges in importance values, and the ranges are placed in frequency distribution classes. The central range of importance values will have the most segments or species in it, whereas fewer segments will fall into the ranges on either side. Again the important species will be few in number, and many will be of intermediate importance. If this distribution is plotted logarithmetically, the curve produced will fall somewhere between the random-niche-boundary and the niche-preemption curves (Fig. 8-3). The log-normal distribution most closely approximates the distribution of importance values obtained from communities rich in species. All three approaches, however, are inadequate, and a way to define the boundaries of the niche still remains elusive.

Species diversity

Whatever the ecosystem may be—a grassy plain, a rain forest, or an ocean—one universal characteristic is that the tropics support many more kinds of plants and animals than do the regions to the north and south. If one were to leave from the tropics and travel north through the Temperate Zone to the Arctic, one would find the numbers of plants and animals decreasing on a latitudinal gradient. Species of nesting birds might approach 1,395 in Colombia, drop to 1,100 in Panama, to 143 in Florida, to 118 in Newfoundland, and to 56 in Greenland (A. G. Fischer, 1960). One can follow the same pattern in mammals (Simpson, 1964), fish (Lowe-McConnell, 1969), lizards (Pianka, 1967), and trees (Monk, 1967). But the gradient in diversity is not restricted to a latitudinal one from the tropics to the Arctic. Small or remote islands have fewer species than large islands and those nearer continents (MacArthur and Wilson, 1967). From east to west in North America, the number of species of breeding land birds increases (MacArthur and Wilson, 1967) as does that of mammals (Simpson, 1964) (Fig. 8-5) amphibians and reptiles (Keister, 1971). Mountain areas in general support more species than flatlands, and peninsulas fewer species than adjoining continental areas. Fields of annual weeds support fewer

species than grasslands, and grasslands are not as rich in species as the forest.

Because this pattern of gradient in abundance of species has stirred considerable interest among ecologists over the years, many ideas have been generated to describe and explain it. Such terms as *species abundance*, *species dominance*, *species equibility*, *species importance*, *species diversity*, and *species richness* have arisen. Although these terms are all related, they are not equivalent. Diversity has become equated with richness or the abundance of different species. Species importance implies dominance yet it can relate to a species' role in the productivity of an ecosystem, which may or may not relate to abundance. To all of these, different ecologists have applied the term *species diversity*, which as a result has become almost a meaningless concept.

It is generally accepted that species diversity is a function of the number of species present in a given area (species richness) and of the evenness with which the individuals are distributed among the species; this is expressed as the ratio of observed diversity to maximum diversity (see Lloyd and Ghelardi, 1964; Pielou, 1966, 1971). Thus a community that contains a few individuals of many species would have a higher diversity of species than a community containing the same number of individuals and even the same number of species but with most of the individuals confined to a few species. In the temperate and arctic regions, a few species are common, the rest are rare, and species diversity is low. In the tropics individuals are more evenly divided among all species and the species diversity is relatively high.

In order to quantify species diversity for purposes of comparison, a number of indices have been proposed (see Appendix C). One of the most widely used is the Shannon-Weiner Index which is based on information theory.

The Shannon-Wiener Index measures diversity by

$$H = -\sum_{i=1}^{s} (p_i)(\log_2 p_i)$$

where

H = the diversity index
s = the number of species

i = the species number

p_i = proportion of individuals of the total sample belonging to the ith species

This formula takes into consideration the relative abundance of species. The biological significance of this index and others (see Appendix C) has been questioned (Hurlbert, 1971).

All the indices assume that the more abundant a species is, the more important it is to the community. But the more abundant species are not necessarily the most important or the most influential. In communities embracing organisms possessing a wide range of sizes, the importance of fewer but larger individuals may be underestimated and the more common species are weighted more heavily than the many rare species. Thus one of the distinctive failures of the indexes is the inability to distinguish between the abundant and important species. Often there is little agreement among the several indexes that are supposed to measure both richness and evenness. Some ecologists claim that the index is misused, because communities with different species composition are not intrinsically arrangeable in linear order on a diversity scale. Thus the diversity index may be of little ecological importance unless one is interested

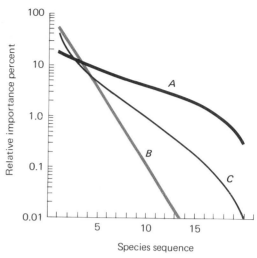

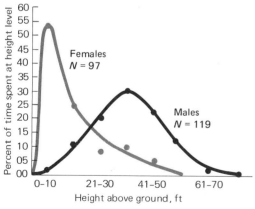

FIGURE 8-2

Separation by height of foraging male and female red-eyed vireos (Vireo olivaceus). Mean height for males, 37.1 ft; standard deviation; 12; S.E. 1.0; range 9–75. Mean height for females, 14.2 ft; standard deviation; 10.8, S.E. 1.1; range 2–50. (From Williamson, 1971. © The Ecological Society of America.)

FIGURE 8-3

A number of hypotheses have been suggested to delineate the niche. These hypotheses may be illustrated by means of graphs in which the importance values of the species (percentage of all species that that particular species represents) expressed as total density, total net productivity, or some other measurement is plotted against the sequence of species. In the random-niche-boundary hypothesis, represented in (A), boundaries are located at random positions in the hyperspace. In a geometric series, represented in (B), the size of the niche hypervolume is determined primarily by the ability of dominant species to preempt part of the niche space, leaving the less successful species to occupy what is left. In a log-normal distribution (C), the niche space occupied by a species is determined by a large number of variables that affect the competitive abilities of the several species. (After Whittaker, 1965. © 1965 American Association for the Advancement of Science.)

only in probability of a certain interspecific encounter.

Despite the disagreement as to how to best measure the diversity gradient, it is generally agreed that it exists. Many hypotheses have been proposed to explain why the tropics should hold a greater abundance of species than the temperate region or why one island should hold more than another. Many of these hypotheses are similar, but not identical.

The *time theory* (A. G. Fischer, 1960; Simpson, 1964) proposes that diversity relates to evolutionary time. Thus older communities hold a greater diversity than young communities. Communities in the tropics evolve and diversify faster than temperate or arctic communities, in part because the environment is more constant and climatic catastrophes less likely. There is some evidence to support this. For example, fossil planktonic Foraminifera from the Cretaceous period in the Northern Hemisphere show a gradient in species abundance from the tropics to the Arctic similar to the found in living Foraminifera of today (Stehli et al., 1969).

Considering a shorter time scale, a species requires a certain amount of time to disperse into unoccupied areas of suitable habitat. Because there has not been enough time for many species to move into temperate zones, these areas are unsaturated as to the total number of species they now support. Many cannot move until barriers to dispersal are broken; others are already moving out of the tropics into temperate zones, as both the natural spread of the cattle egret in North America from Africa by way of South America and the northward spread of the armadillo indicate.

The theory of *spatial heterogenity* (Simpson, 1964) holds that the more complex and heterogenous the physical environment, the more complex and diverse will its flora and fauna be. The greater variation in topographic relief, the more complex the vertical structure of the vegetation, and the more types of microhabitats the community contains, the more kinds of species it will hold. This theory is supported by the fact that the more complex the vertical stratification of a community, the more species of birds it holds (MacArthur, 1972; Pearson, 1971).

The *climatic stability theory* (A. G. Fischer, 1960; Connell and Orias, 1964) holds that the more stable the environment, the more species will be present. Through evolutionary time, the tropics, of all regions of the earth, has probably remained the most constant and has been relatively free from severe environmental conditions that could have catastrophic effects on a population. Under tropical conditions, selection is strongly influenced by the competition of individuals against members of other species. At higher latitudes selection is influenced more by severe environment and by competitive relations of individuals with others of the same species.

Another interesting hypothesis closely related to climatic stability is the *productivity theory* advanced by Connell and Orias (1964). In brief, this hypothesis proposes that the level of diversity of a community is determined by the amount of energy flowing through the food web. The rate of energy flow is influenced by the limitation of the ecosystem and by the degree of stability of the environment.

If one assumes a hypothetical increase in the stability of the physical environment, then with increasing environmental stability, less energy is required for regulatory activities and more energy enters into net production. Increased net productivity can support larger populations. Larger populations maintain greater genetic variety and increase the opportunity for interspecific association. Greater productivity per unit area permits the less mobile animals to become even more sedentary, and the species tend to be broken into may semi-isolated populations, which may bring about greater intraspecific genetic variety. As a result speciation is favored, especially if semi-isolated segments are exposed to new environments. Any new species that arise would tend to be more specialized and initially to have smaller populations.

In the early stages of the evolution of a community, positive feedback would increase the rate of speciation, resulting in a faster cycling of nutrients and an increase in net productivity. As the number of species increases, the food webs become more complex and the community becomes more stabilized (see Chapter 3). But the tendency toward overspecialization and the smaller population per species would tend to decrease community stability and act as a negative feedback on the whole system.

The productivity theory in effect says that the more food produced, the greater the diversity. While perhaps true in a general sense, there are too many exceptions. In some aquatic ecosystems, increased productivity results in a decrease in diversity, a situation that results from enrichments of the system with sewage and other nutrients. Marine benthic regions of low productivity have a higher abundance of species than areas of high productivity (Sanders, 1968).

The *competition theory* (Dobzhansky, 1951; C. B. Williams, 1964) states that in environments of high physical stress, such as the Arctic, with its frigid cold, and the temperate regions with their wide fluctuations in annual temperatures, selection is controlled largely by the physical variables. In more benign climatic regions biological competition becomes more important in the evolution of species and in the specialization of niches.

Out of his studies of animals of the rocky intertidal zones, Paine (1966) has proposed the *predation theory*. Because more predators and prey exist in the tropics than elsewhere, the predators tend to hold down the prey species to such a low level that competition among prey species is reduced. This theory supposes that more diverse communities support an increased proportion of predators, that predators are very efficient at regulating the abundance of prey (see Chapter 12), and that these assumptions are true at all trophic levels including the primary-producer level. Janzen (1971) suggests that this indeed does happen in the tropics. Because seed-eating animals tend to cluster about the seed-producing tree, seed mortality is heaviest about the seed source, whereas the probability that seeds will be overlooked increases with the distance from the tree.

Sanders (1968) has combined the environmental stability hypothesis and the time hypothesis into still another one, the *stability–time hypothesis*. This assumes that two contrasting types of communities exist, the physically controlled and the biologically controlled.

In the physically controlled communities organisms are subjected to physiological stress due to fluctuating physical conditions. The organisms in time evolve adaptive mechanisms to meet these conditions. But at least some of the time the organisms are subject to severe physiological stress, and the probabilities of reproductive success and survival are low. As a result diversity is low.

Low-diversity environments fall into three categories: (1) new environments in which

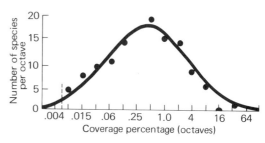

FIGURE 8-4

A bell-shaped curve of importance values resulting from a log-normal distribution of plant species. The importance value in this example is determined by the percent of ground surface covered by the species. It is represented on the horizontal scale (logarithmic) by octaves in which the species are grouped by doubling the units of percentage of cover. The largest number of species occurs in the middle octaves. (After Whittaker, 1965. © 1965 American Association for the Advancement of Science.)

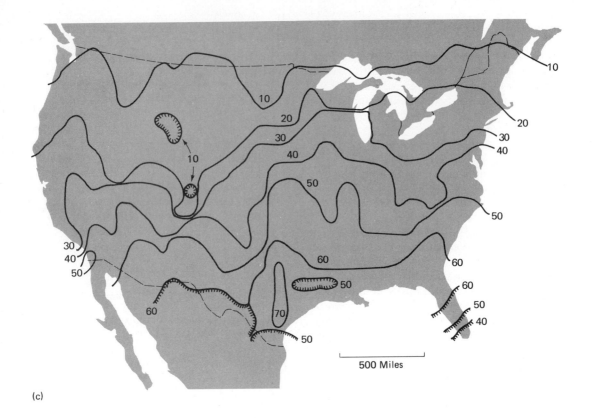

(c)

500 Miles

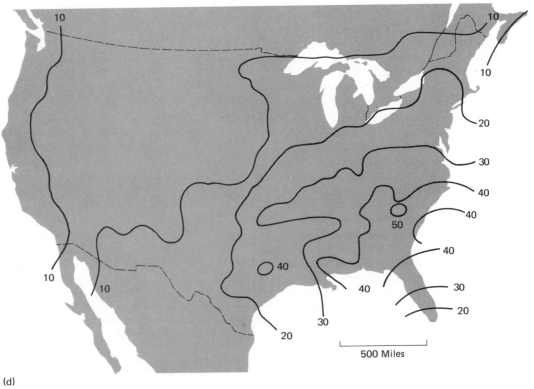

(d)

500 Miles

the number of organisms colonizing the area are increasing but are subject to environmental stress; (2) severe environments in which a slight environmental change such as an increase in temperature or salinity can eliminate life altogether; and (3) unpredictable environments in which the environmental properties vary widely and unpredictably about some mean value. A wide fluctuation from the mean can severely stress the population.

In the biologically controlled community the physical conditions are relatively uniform over long periods of time and are not critical in controlling the species. Evolution proceeds along the lines of interspecific competition, one species adapting to the presence of the other and sharing the resources with it. The environment is more predictable, the physiological tolerances of the organisms are low, and diversity is high.

But there is no such entity as a wholly physically controlled or biologically controlled community. Rather the community is influenced by the interaction of the two. In situations where physiological stress has been low, biologically accommodated communities evolved. As physiological stress increases due to increasing fluctuations in the physical environment the community changes from a biologically controlled to a physically controlled one. The number of species diminishes gradually along the gradient of stress. When the point is reached where stress is too severe, no species exist (Fig. 8-6).

However intriguing these hypotheses may appear, it is difficult to test any of them in the field, or even to put them into mathematical models. But in spite of this it is apparent that the diversity of species can be related to a number of such variables as the structure of the habitat, the diversity of microhabitats, the nature of the physical environment, climate and protection from its adverse effects, the availability of food, nutrient supply, and time.

Community structure

All communities have a physical structure and biological pattern. The structure can be subdivided into more or less distinct layers both on a vertical and horizontal plane. The stratification of terrestrial communities largely reflects the life forms of plants—the character-

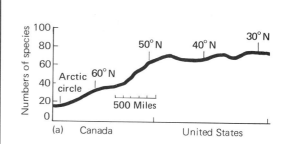

(a)   Canada   United States

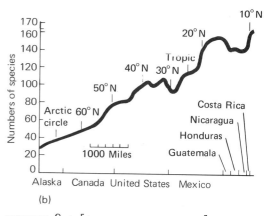

(b)

FIGURE 8-5 [ABOVE AND OPPOSITE]
The diversity of animals across a continent is influenced by temperature and moisture reflected in latitudinal and altitudinal variations. (a) Species densities of North American mammals from the Arctic to the Mexican border and (b) from the Pacific to the Atlantic across the middle of the United States. (After Simpson, 1964.) (c) The most pronounced latitudinal variations occur among the reptiles and amphibians. Being poikilothermic and ectothermic, reptiles have their greatest density in the hot desert regions and lower latitudes of North America. (From Kiester, 1971.) (d) Being not only poikilothermic but also highly sensitive to moisture conditions, the amphibians reach their richest diversity in the central Appalachians, and decrease northward, southward, and westward. They are the lowest in species in the dry and the cold regions of the continents. (From Kiester, 1971.)

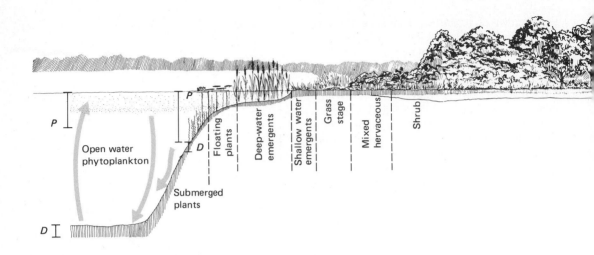

Open water
phytoplankton

P

D

Floating
plants

Submerged
plants

Deep-water
emergents

Shallow water
emergents

Grass
stage

Mixed
hervaceous

Shrub

istic appearance of vegetation, such as size, branching, and leaves. This is an expression of genetic characteristics, often modified by the environment. The general appearance of a community and the stratification of animal life as well are determined largely by the life form of the dominant plants. A well-developed forest ecosystem, for example, has several layers of vegetation (Fig. 8-7), each of which provides a habitat for animal life in the forest. From top to bottom there is the canopy, the understory, the shrub layer, the herb or ground layer, and the forest floor.

The canopy, which is the major site of food production, has a major influence on the rest of the forest. If it is fairly open, considerable sunlight will reach the lower layers, and the shrub and the understory tree strata will be well developed. If the canopy is closed, the shrub and the understory trees and even the herbaceous layers will be poorly developed.

The understory consists of tall shrubs such as the witch hazel, understory trees such as the dogwood and hornbeam, and younger trees. Some are the same as those in the crown, others are of different species. Species that are unable to tolerate the shade and competition will die; others will eventually reach the canopy after some of the older trees die or are harvested.

The shrub layer differs with various types of forest. In oak forest on south-facing slopes, blueberries are most characteristic; in the cove forests grow buffalo nut, hydrangea, and rhododendron. In the northern hardwood forests, witch hobble, maple-leaf viburnum,

and striped maple are common. The nature of the herb layer depends upon the soil moisture conditions, slope position, density of the overstory, and aspect of the slope, all of which vary from place to place through the forest. The final layer, the forest floor, has already been discussed (in Chapter 7) as the site where the important process of decomposition of the forest litter takes place and where the nutrients are released to the nutrient cycle again.

The variety of life in the forest is directly related to the number and development of its layers. If certain layers are absent, then the animals they normally shelter and support also are missing. Thus, a well-developed forest supports a rich diversity of life that often goes unappreciated.

Other ecosystems have similar if not as highly stratified structures. Grasslands have the herbaceous layer, the ground or mulch layer, and the root layer. The latter is more pronounced in grasslands than in any other ecosystem.

Aquatic ecosystems such as the lakes and oceans have strata determined by light penetration, temperature profile, and oxygen profiles (Fig. 8-7). Well-stratified lakes have a layer in summer of freely circulating surface water, the epilimnion; a second layer, the metalimnion, characterized by a thermocline (a very steep and rapid decline in temperature); the hypolimnion, a deep cold layer of dense water about 4° C, often low in oxygen; and a layer of bottom mud (see Chapter 18; Fig. 18-1). In addition, two other structural layers are recognized based on light penetra-

tion—an upper zone, roughly corresponding to the epilimnion, dominated by plant plankton and the site of photosynthesis; and a lower layer in which decomposition is most active. The lower layer roughly corresponds to the hypolimnion and the bottom mud.

Ecosystems, whether terrestrial or aquatic, have similar biological structures. They possess an autotrophic layer concentrated where the light is most available, which fixes the energy of the sun and manufactures food from inorganic substances. In the forest this layer is concentrated in the canopy, in the grassland the herbaceous layer, and in the lake and sea the upper layer of water. Ecosystems also possess a heterotrophic layer, which utilizes the food stored by autotrophs, transfers energy, and circulates matter by means of herbivory, predation in the broadest sense, and decomposition.

Stratification of terrestrial communities reflects largely the life form of plants—the characteristic appearance of vegetation such as size, branching, and leaves. This is an expression of genetic characteristics, often modified by environment. The general appearance of a community and the stratification of animal life are determined largely by the life form of the dominant plants. Because of this, the life form has been widely used to describe the plant structure of a community. Among several classification systems proposed, the most widely used is that of Raunkaier, which is based largely on overwintering parts (Table 8-1).

A life-form classification is very useful for comparing communities. All species in a re-

FIGURE 8-7 [OPPOSITE]
A vertical section view of communities from aquatic to terrestrial illustrates the general features of each. Both are structurally similar in that the zone of decomposition and regeneration is in the bottom stratum and the zone of energy fixation is in the upper stratum. In succession from aquatic through the terrestrial stages, stratification and the complexity of the community become greater. In the aquatic community there is little storage of materials in biomass. In the terrestrial communities biomass storage increases as ecosystems become more mature. (P, production; D, decomposition.)

Gradient of physiological stress
→

Predominantly biologically accommodated   Predominantly physically controlled   Abiotic

Stress conditions beyond adaptive means of animals

FIGURE 8-6
A bar-graph representation of the stability-time hypothesis in which species numbers diminish continuously along a stress gradient. The greatest diversity occurs among the predominately biologically accommodated communities.
(From Sanders, 1968. © 1968 by The University of Chicago Press.)

TABLE 8-1
Raunkaier's life forms

Name	Description
Therophytes	Annuals survive unfavorable periods as seeds. Complete life cycle from seed to seed in one season.
Geophytes (Cryptophytes)	Buds buried in the ground on a bulb or rhizome.
Hemicryptophytes	Perennial shoots or buds close to the surface of the ground; often they are covered with litter.
Chamaephytes	Perennial shoots or buds on the surface of the ground to about 25 cm above the surface.
Phanerophytes	Perennial buds carried well up in the air, over 25 cm. Trees, shrubs, and vines.
Epiphytes	Plants growing on other plants; roots up in the air.

TABLE 8-2
An example of an analysis of life-forms spectra of two plant communities:
a New Jersey pine barren and a Minnesota jack pine forest

Basis of spectrum	Community	Number of species	Percentage				
			Ph	Ch	He	G	Th
Species list	New Jersey	19	84.2	0	10.5	5.2	0
	Minnesota	63	23.8	4.7	60.3	7.9	3.1
Cover	New Jersey	19	98.1	0	1.9	0	0
	Minnesota	63	11.8	2.5	55.6	28.7	1.4

Source: Stern and Buell, 1951.

TABLE 8-3
Life forms of aquatic plants

Relation to substratum			Type	Example
Free (not rooted or anchored)			Natantia	*Lemna minor, Ceratophyllum demersum*
Rooted in soil	Emersed, at least in part	Broad leaved	Foliacea	*Sagittaria latifolia, Pontederia cordata*
		Narrow, tubular, or linear leaved	Junciformia	*Scirpus validus, Juncus nodosus*
		Floating leaved	Nymphoidea	*Nymphaea odorata, Nymphoides lacunosum*
	Submersed (at most a few floating leaves)	Long, leafy stems and/or leaves ribbonlike	Vittata	*Potamogeton richardsonii, Vallisneria americana*
		Leaves much reduced, crowded at base	Rosulata	*Isoetes braunii, Lobelia dortmanna*
		Annuals	Annua	*Najas flexilis, Potamogeton pusillus*
Adnate or epipyhtic			Adnata	*Podostemon cerato-phyllum, Fontinalis* spp.

Source: Dansereau, 1945.

gion or community can be grouped into several classes, and the ratio between them can be expressed as a percentage (Table 8-2), giving a biological spectrum of the area (Fig. 8-8). Hemicrytophytes, for instance, would be most abundant in grasslands and old fields, threophytes in the desert and weed communities. In woodlands of eastern North America geophytes, such as the dogtooth violet, are most characteristic of the forest in spring, whereas hemicryptophytes, such as asters, are common in autumn.

A useful classification system for aquatic communities, developed by Dansereau (1945, 1959), is well correlated with the horizontal zonation of communities along streams and

the shores of lakes and ponds (Table 8-3). Vertical gradients in aquatic environments are aptly expressed as changes in the physical environment, involving light, temperature, and oxygen. These three govern to a large extent the distribution of life and biological activity in deep water.

Each vertical layer in the community is inhabited by its own more or less characteristic organisms. Although considerable interchange takes place between several strata, many highly mobile animals confine themselves to only a few layers, particularly during the breeding season. Occupants of the vertical strata may change during the day or the season. Such changes reflect daily and seasonal variations in humidity, temperature, light, acidity, oxygen content of water, and other conditions, or the different requirements of the organisms for the completion of their life cycles. For example, D. L. Pearson (1971) found that birds occupying the upper strata of a tropical dry forest in Peru moved to lower strata during the middle of the day for several reasons: to secure food (insects moved to lower levels), to escape heat stress, to escape a high degree of solar radiation, and to conserve moisture.

Horizontal stratification or zonation (Fig. 9-5) is caused primarily by differences in climatic or edaphic conditions, which retard or inhibit rooted vegetation. This type of stratification is most conspicuous around ponds and bogs (see Chapter 18, Fig. 18-8). Stratification increases the number and variety of places in which the organisms can live in a given area. Highly stratified communities generally contain a wider variety of animal life than those with only a few layers.

Ecotone

The zone where two or more different communities meet and integrate is an *ecotone* (Fig. 8-9). This zone of intergradation may be narrow or wide, local (e.g., a zone between a field and a forest) or regional (e.g., the transition between forest and grassland).

Three types of ecotones are recognizable (Daubenmire, 1968a). One is an abrupt transition, the result of an abrupt change in such environmental conditions as soil type or soil drainage. The second is a sharp transition brought about by a plant interaction, particularly competition. The third type is a blending of two or more adjacent vegetational types. In the second and third types the superior plant competitors on one side of the ecotone extend as far out as their ability to maintain themselves allows. Beyond this competitors of the adjacent community take over. As a result the ecotone exhibits a shift in dominance of the conspicuous species of both communities. The ecotone contains not only species common to the communities on both sides; it also may include a number of highly adaptable species that tend to colonize such transitional areas. Because of this the variety and density of life is often the greatest in such areas (Leopold, 1933). This phenomenon has been called the *edge effect*.

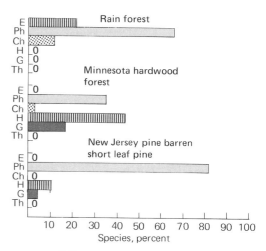

FIGURE 8-8

Life form spectra of a tropical rain forest (adapted from Richards, 1952), a Minnesota hardwood forest (data from Buell and Wilbur, 1948), and a New Jersey pine barren (data from Stern and Buell, 1951).

FIGURE 8-9

A broad geographical ecotone where the western coniferous forest in the Black Hills of South Dakota meets the short grass plains. (Photo by author.)

A number of different studies point this out. A breeding-bird census in Illinois showed 14 species in the forest interior and 22 on the forest edge (V. R. Johnston, 1947). A census on a 130-acre area in Pennsylvania showed that the greatest density of birds was in orchards, temporary thickets, slashings, and mixed conifers and hardwoods. Areas of uniform vegetation, such as large meadows or pure conifer stands, had low populations (Edeburn, 1947). If meadows were strip-cropped, the bird populations increased from 48 pairs/100 acre in large meadows to 93 pairs in meadows broken by strips of grain, and from 10 pairs on large grain fields to 27 pairs/100 acre on strip-cropped grain fields (Good and Dambach, 1943).

A number of different game species are considered occupants of the edge, and in fact require a diversity of community and ecotone during the year. For example, the ruffed grouse requires new forest openings with an abundance of herbaceous or low shrubs, dense sapling stands, pole stage stands for nesting cover, and mature forest for winter food. Because the ruffed grouse spends its life in an area of 10 to 20 acres, this amount of land must provide all of its seasonal requirements (Gullion, 1970).

Nature of the community

Although the definition of a community is rather straightforward—an assemblage of species occupying a given area—the nature of the community has been the object of study and dispute for years. Is a community such as an oak–hickory forest a real entity that is definable, describable, and constant from one stand of oak–hickory to another? Or is it something of an abstraction, a collection of different populations that exist together because they have similar environmental requirements? To the last question one group of ecologists says yes, and another group says no.

The composition of any one community is determined in part by the species that happen to be distributed on the area and can grow and survive under prevailing conditions. Seeds of many plants may be carried by the wind and animals, but only those adapted to grow in the habitat where they are deposited will take root and thrive. The element of chance also is involved. One adapted species may colonize an area and prevent others equally as well adapted from entering. Wind direction and velocity, size of the seed crop, disease, insect and rodent damage all influence the establishment of vegetation. Thus the exact species that settle an area and the number of individual species that succeed are situations that seldom if ever are repeated in any two places at any two times. Nevertheless there is a certain pattern, with more or less similar groups recurring from place to place. Only a relatively small group of species are potential dominants because a limited number are well adapted to the overall climate and soils of the region they occupy.

Communities are often regarded as distinct natural units or associations, especially for practical reasons of description and study, but more often than not community boundaries are hard to define. Some, such as ponds, tidal beaches, grassy balds, islands of spruce and fir within a hardwood forest, old fields and burns, have sharply defined boundaries. Here the vegetational pattern is *discontinuous*. Most often, however, one community type blends into another. The species comprising the community do not necessarily associate only with one another but are found with other species where their distribution overlaps (Figs. 8-10 and 8-11). Some organisms will succeed only under certain environmental conditions and tend to be confined to certain habitats. Others tolerate a wider range of environmental conditions and are found over a wider area. Species shift in abundance and dominance, because of change in altitude, moisture, temperature, and other physical conditions. One species may be dominant in one group, an associated species in another. This sequence of communities showing a gradual change in composition is called a *continuum* (Curtis, 1959). Each community is similar to but slightly different from its neighbor, the difference increasing roughly as the distance between them. Even when the dominant plants change completely, the community may integrate in the understory vegetation. The continuum is much like a light spectrum. The end colors, red and blue, and other primary colors in the middle are distinguishable, but the boundaries grade continuously in either direction. Eventually the continuum must end, when environmental conditions favor a completely different group of organisms. A community in such a gradient can be described as a discrete area or

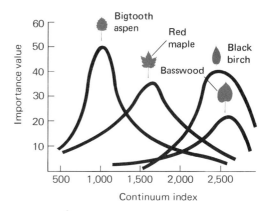

point in the continuum, the point being defined by some given criteria.

The distribution of species along an environmental gradient is not confined to plants alone. The same phenomena also have been found in the case of insects (Whittaker, 1952) and birds (Bond, 1957) (Fig. 8-11).

Involved in the concept of the community are two opposing philosophies. One is the organismic approach developed by Clements (1916) and supported by such proponents as Daubenmire (1966, 1968a) and Langford and Buell (1969). The organismic viewpoint regards the community as a sort of superior organism, the highest stage in the organization of the living world—rising from cell to tissue, organs, organ systems, organism, population, community. The whole is more than the sum of its parts. Just as tissues have certain characteristics and functions above and beyond those of the cells that comprise them, so the community has characteristics and functions above and beyond the various populations it embodies. The distribution and abundance of one species in the community are determined by the species' interaction with others in the same community. Species making up the floristic community are organized into discrete groups. Groups of stands similar to one another form associations. Stands of one association are clearly distinct from stands of other associations. The community acts as a unit in seasonal activity, in competition with other communities, in trophic functions, and in succession.

The individualistic approach was advanced by Gleason (1926) and further developed by Curtis (1959), McIntosh (1958, 1967), and Whittaker (1962, 1965, 1967, 1970a). It emphasizes the species rather than the community as the essential unit in the analysis of interrelationships and distribution. Species respond independently to the physiological and biotic environment according to their own genetic characteristics. They are not bound together into groups of associates that must appear together. Instead, when species populations are plotted along an environmental gradient, long or short, the resulting graph suggests a normal or bell-shaped curve (Fig. 8-12). The curves of many species overlap in a heterogenous fashion. Thus the vegetation and its associated animal life exhibit a gradient or continuum from one extreme (e.g., dry conditions) to another (wet conditions). In

FIGURE 8-10
Distribution of some forest tree species on a continuum index. (Adapted from Curtis and McIntosh, 1951. © The Ecological Society of America.)

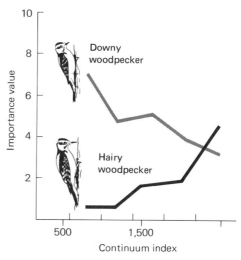

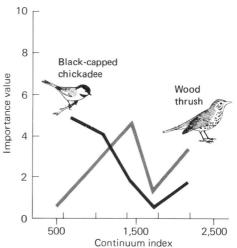

FIGURE 8-11
Distribution of several bird species along a vegetational gradient. (Adapted from Bond, 1957. © The Ecological Society of America.)

this view, the community is regarded as a collection of populations of species requiring the same environmental conditions. It is a continuous variable, not an integrated unit.

The individualist proponents grant that community discontinuities do exist, but that these reflect environmental discontinuities, such as breaks in soil type or sharp changes in moisture and salinity. If stands are as distinct as the organismic group considers them to be, then the boundaries between associations should be distinct. But they rarely are. Except where man or some environmental catastrophe such as fire has interfered, it is rare for the vegetation to be a mosaic of discontinuous units. Instead, boundaries between units are more or less diffuse. If separate stands of an association are similar, all such associations should have similar distributions, and plants that comprise such associations should have distributions that coincide locally and over continental limits. If the association is a natural unit, then the species that comprise it should be bound together by obligate interrelationships. But most species are not obligatory.

The organismic school argues that stands studied by gradient analysis are disturbed stands or ones that are not in equilibrium. If undisturbed, all stands develop to an endpoint in a few hundred years; and stands heading for an endpoint will naturally show a continuum of species. Besides, the technique of gradient analysis forces data into a continuum. These gradients are usually based on one variable such as moisture; but it is impossible to restrict a continuum to one variable alone because many interacting variables influencing plant distribution are involved. The continuum approach assumes that all species are equal when in fact some species are dominant.

The future may reveal that each is partly right and partly wrong. The species that collectively make up the biotic community may respond individually to the environment; yet each community, in ways perhaps not clearly definable by human criteria, still operates as a functional unit, especially in regard to the flow of energy and the cycling of materials.

Naming the community

Although the major current concept of the community is that of a continuum, the plant and animal life of any large area are so com-

plex that they must be separated into subdivisions. Thus the aggregation of organisms in any given locality or habitat must be regarded as a unit if the community is to be studied, described, or compared with similar community stands in other habitats. To give order to the study of communities, some system of classification is needed, even though the communities of a region often cannot be placed in discrete units.

There are a number of approaches to community classification, each arbitrary and each suited to a particular need or viewpoint (H. C. Hanson, 1958). The most widely used classification systems are based on physiognomy, species composition, dominance, and habitat.

Physiognomy, or general appearance, is a highly useful method of naming and delineating communities, particularly in surveying large areas, and as a basis for further subdivision of major types into their component communities. Because animal distribution is most closely correlated to the structure of vegetation and not the species composition (cf., for birds, MacArthur and MacArthur, 1961), classification by physiognomy will relate both the animals and the vegetations of an area. Communities so classified are usually named after the dominant form of life, usually plant, such as the coniferous or deciduous forest, sagebrush, short-grass prairie, and tundra. A few are named after animals, such as the barnacle-blue mussel (*Balanus–Mytilus*) community of the tidal zone. One, of course, may grade into the other; so even here the classification may be based on arbitrary, although specific, criteria.

In areas where the habitat is well defined, physiography is used to classify and name communities. Examples of such are sand dunes, cliffs, tidal mud flats, lakes, ponds, and streams.

Finer subdivisions are often based on species composition, a system which works much better with plants alone than with animals or with both. Such a classification requires at first a detailed study of the individual community. Such a system also involves a number of concepts: frequency (the regularity with which a species is distributed throughout a community), dominance, constancy, presence, and fidelity (see Appendix B).

A group of stands in which more or less the same combination of species occurs can be classified as the same community type, named after the dominant organisms or the ones with the highest frequency. Examples of such are the *Quercus–Carya* association, or oak–hickory forest, the *Stipa–Bouteloua* association, or mixed prairie, and the animal-dominated *Balanus–Mytilus*, or barnacle-blue mussel, community of the tidal zone. European ecologists have developed the floristic classification with emphasis on dominance, constancy, and diagnostic species. They group communities into classes, orders, alliances, and associations (for a complete discussion on this, see Poore, 1962; Whittaker, 1962).

The floristic system is modified when the stands are treated as a continuum. The community complex of a major physiognomy is subdivided by species composition and correlated with an environmental gradient arbitrarily divided into five segments: wet, wet mesic, mesic, dry mesic, and dry (Curtis, 1959). Thus the deciduous forest in Wisconsin has been divided into southern and northern hardwoods and northern forest. These are further divided in a moisture gradient. The southern hardwoods, for example, include the dry southern hardwoods with bur, black or white oak as the dominants; the dry mesic with red oak or basswood; the mesic with sugar maple and beech; the wet mesic with silver maple, elm, and ash; and the wet mesic with willow or cottonwood. Such a system recognizes the influences of habitat on community composition. Detailed studies on animal distribution may reveal similar influences of animal composition. A shift of species composition of birds occurs in the southern forests of Wisconsin (Bond, 1957). In the dry stands the most important species are the scarlet tanager, rose-breasted grosbeak, cardinal, blue jay, black-capped chickadee, downy woodpecker, and the red-headed woodpecker. In the dry-mesic segment, however, the wood thrush, least flycatcher, blue-gray gnatcatcher, redstart, yellow-throated vireo, veery, and ruby-throated hummingbird are the dominant species.

A large problem in community classification is to arrive at a system that will embrace animals as well as plants. Communities distinguished by plant composition indicate little about the animals of the community because animal distribution cannot be correlated with plant species distribution. As a result animal and plant communities usually are studied separately, which unfortunately obscures the

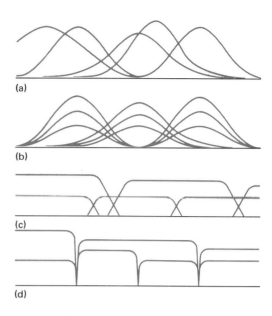

(a)

(b)

(c)

(d)

FIGURE 8-12
Four models of species distribution along environmental gradients. (a) The abundance of one species on an environmental gradient is independent of the others. Thus the association of several species along the gradient changes with the response of the individual species to that gradient. (b) The abundance of one species is associated with the abundance of another. The two or more species are always found in association with each other. (c) The distribution of one species is independent of another in an environmental gradient, but the abundance and distribution of each species is sharply restricted at some point on the gradient by interspecific competition. (d) The distribution of a species is sharply restricted by a change in some environmental variable. This is characteristic of an ecotone.

wholeness of the community and limits our understanding of its functions.

To escape this dilemma in part, the distribution of animals can be related to the life form of plants and types of vegetation. This results in a more inclusive classification, which embraces several plant communities but includes all animal life associated with them; this classification is called the *biome*. The biome is a broad ecological unit characterized by the distinctive life forms of the climax species, plant or animal. The biome is further subdivided into smaller units, distinguished by uniformity and distinctness in species composition of the climax and its successional stages. Thus the life form of plants is given greater emphasis, rather than the taxonomic composition, which in the final analysis plays the most important role in dominance.

This brings up another concept: fidelity, or the "faithfulness" of a species to a community type. Species with low fidelity occur in a number of different communities and those with high fidelity in only a few. Seldom if ever are the latter found away from certain other plant and animal associates. The greater the ratio of the constant species to the total number of species, the more homogeneous is the community and the more sharply can it be delineated. Often, however, this simply reflects a group of species unable to grow successfully under a wide range of ecological conditions or with other species. Species, in general, can be grouped as *exclusive*, those completely or almost completely confined to one kind of community; *characteristic* (including the selective and preferential species of plant ecologists), those that are most closely identified with a certain community; and *ubiquitous* (or indifferent), those which have no particular affinity to any community. The characteristic species high in constancy and dominance are the ones that really characterize the community type.

Recognition of characteristic species often offers problems, and some decision is necessary as to how abundant or constant a species must be in order to be characteristic. This is a greater problem among animals than among plants because dominance and comparative abundance of the latter offer a firmer base for a decision. To be characteristic, an animal species should be conspicuous and occur in at least 50 percent of all samples taken in the community (Thorson, in Hedgpeth, 1957). In a study of breeding birds in Ontario (Martin, 1960), a species was considered characteristic of a community type if at a population level of 1 to 9 pairs/100 acre (40 ha), it was three times more abundant there than in any other community. Species with a population density of 10 to 100 pairs/100 acre were considered characteristic if twice as abundant in one type of vegetation over another. If a species had a density of more than 100 pairs/100 acre, a 50 percent difference was considered adequate.

Summary

However the community may be classified, or whatever methods may be employed to distinguish one community from another, the basic concept remains unchanged. A biotic community is a naturally occurring assemblage of plants and animals living in the same environment, mutually sustaining and interdependent, constantly fixing, utilizing, and dissipating energy.

Communities are organized about dominant species, especially in the Temperate Zone. The dominants may be the most numerous, possess the highest biomass, preempt most of the space, or make the largest contribution to energy flow. But the dominant species may not necessarily be the most important species in the community.

Each organism in the community occupies a particular functional niche, at which it arrived by a long process of natural selection and evolution. The more niches there are to occupy, the more specialized the occupants, the more complex the community, the greater the diversity of species, and the more stable the ecosystem.

Species diversity implies both a richness in the number of species and equability in the distribution of individual among the species. A number of hypotheses have been proposed to explain species diversity. The time theory proposes that diversity relates to the time available for speciation and dispersal. The spatial heterogenity theory relates diversity to the physical complexity of the environment. The climatic stability theory holds that the more stable the environment the more species will be present. The productivity theory suggests that the level of diversity is determined by the amount of energy flowing through the food chain. The competition theory says that competition is keener in the tropics—the organisms are specialized and thus more can coexist. The predation theory states that predators reduce competition and permit more species to coexist. And the stability-time hypothesis suggests that diversity is related to stress gradient, with diversity lowest in environments subject to environmental stress, and highest in those where the environment is benign.

All communities exhibit some form of layering or stratification, which largely reflects the life form of the plants and which influences the nature and distribution of animal life in the community. Communities most highly stratified offer the richest variety of animal life, for they contain a greater assortment of microhabitats and available niches.

There are two opposing views concerning the nature of the community. According to the organismic school, communities are integrated units that have definable boundaries. The individualistic school argues that the community is a collection of populations that require the same environmental conditions. The makeup of any one community is determined in part by the species that happen to be distributed on the area and can grow and survive under prevailing conditions. The exact species that settle on an area and the number that survive are rarely repeated in any two places at the same time, but there is a certain recurring pattern of more or less similar groups. Rarely can different groups of communities be sharply delimited, for they blend together to form a continuum. The area where two major communities meet and blend together is an ecotone.

bandoned cropland is a common sight in agricultural regions, particularly in areas once covered with forests. No longer tended, the lands grow up in grasses, goldenrod, and other herbaceous plants. Only the most unobservant would fail to notice that in a few years these same weedy fields are invaded by "brush"—blackberries, sumac, and hawthorns, followed by fire cherry, pine, and aspen. Many years later this abandoned cropland supports a forest of maple, hickory, oaks, or pines. Thus over a period of years, one community replaces another until a relatively stable forest or climax finally occupies the area (Fig. 9-1) and ends successional progress.

The changes involved in the return of the forest were not haphazard, but orderly, and barring disturbance by man or natural events, the reappearance of the forest was predictable. This orderly and progressive replacement of one community by another until a relatively stable community occupies the area is called *ecological succession*. The whole series of communities, from grass to shrub to forest, that terminate in a final stable community is called a *sere*, and each of the changes that take place is a *seral stage*. Each seral stage is a community, although temporary, with its own characteristics, and it may remain for a very short time or for many years. Some stages may be completely missed or bypassed. This happens when old fields grow up immediately into forest trees, but even in these situations the young trees form a sort of shrub community.

The succession that took place on the old field is called *secondary* because it proceeded from a state in which other organisms already were present. Secondary succession arises on areas disturbed by man, animals, or natural forces such as fires, wind storms, and floods. Its development may be controlled or influenced by the activities of man or animals, domestic or wild. Succession that takes place on areas devoid or unchanged by organisms is called *primary* (Fig. 9-2).

The idea of vegetational change is not recent. The fact that one aggregation of plants eventually is replaced by another has been noted for many years (see Spurr, 1952). Back in 1863 Anton Kerner described in a fascinating book, *Plant Life of the Danube Basin*, the formation of meadow from swamp, forest regeneration, and the forest edge, and he explained the "genetical relationship of plant formations," as he called succession. Within the pages of this old book lay the field of plant sociology in an embryonic state. In America, Henry David Thoreau knew about succession and wrote about it. Then in 1899 Henry Cowles published his classic description of plant succession on the sand dunes of Lake Michigan. Sixteen years later, the pioneer plant ecologist, Frederick Clements, published a book, *Plant Succession*, which became the foundation of a system of studying and describing plant communities that colors ecological thinking today.

The nature of succession

Succession is characterized by progressive changes in species structure, organic structure, and energy flow. It involves a gradual and continuous replacement of one kind of plant and animal by another until the community itself is replaced by another that is more complex. It is brought about by a modification of the physical environment by the organisms themselves. As the environment is modified by their own life activities, the organisms make the habitat unfavorable for themselves. They are gradually replaced by a different group of organisms more able to exploit the new environment.

In the early stages of succession, dominants may replace one another from year to year and the number of species, and often individuals, increases. If the density of species is plotted against time, a sort of step diagram results. Later, changes in species structure take place more gradually. Because some organisms have wider tolerances or occupy more generalized niches, they may persist over a longer period of time (Fig. 9-3), forming a continuum from one stage to another.

Studies of successional communities, especially in microcosms or laboratory cultures of algae and natural aquatic ecosystems, have led H. T. Odum and Pinkerton (1955), Cooke (1967), Margalef (1968), E. P. Odum (1969, 1971), and others to develop a model of succession (Fig. 9-4). According to the model, early stages of succession are characterized by relatively few species, low biomass, and a largely extrabiotic source of nutrients. The ratio between gross production and biomass is high and the production is greater than respiration $(P > R)$ (Table 9-1). Energy is channeled through relatively few pathways to many individuals of a few species and production per unit is high. The food chains are short, linear, and largely grazing. The mature stages in succession are characterized by many species, high biomass, and a nutrient source largely organic in nature. Although production may be high, the ratio between gross production and biomass is low and production equals respiration $(P = R)$. Energy is channeled down many diverse pathways and shared by many units. Food chains are complex and largely detrital.

As the developmental stages pass from immature to mature, both stratification and diversity increase, and niches change from broad, general ones to narrow, specialized ones. Both the increased diversity of species and the increased number of niches are brought about by increased stratification. As the strata become more developed, diversity changes within them. In early successional forest communities, such as longleaf and slash pine, the diversity of species is higher in the seedling and sapling classes than in the canopy class. In the mature stages such as mixed southern hardwood, diversity is more uniformly distributed among the three layers (Monk, 1967).

Diversity can become a measure of ecosystem maturity and stability. According to the model, diversity increases with maturity (Fig. 9-4). By this measure, tundra and croplands would be young, immature, or unstable ecosystems, and the tropical forest would be mature and stable. Another point is that as diversity increases, dominance decreases. As more and more species make up an ecosystem and the energy is more finely shared, there are fewer true dominant species.

Finally, diversity is related to stability. The more diverse the species composition, the more stable the ecosystem. Thus the most efficient ecosystems are ones in which biomass accumulates in a large number of individuals of very few species. Because of this, the ecosystem is very sensitive to disturbance, and a heavy interference with a major pathway of energy flow can disrupt the ecosystem. Plants are more susceptible to exploitation by heterotrophs, and much the same holds true for dependent consumers. Disturbance at one trophic level is quickly felt at other levels. In

Ecosystem development

(a)

(b)

(c)

(d)

Thallus vegetation: pioneer stage

Spruce-fir community

Tall-shrub community

Mixed upland herb community

species-rich ecosystems, biomass accumulates in a few individuals of many species. Disturbance to one component of the food chain would not have a seriously disturbing effect on the total ecosystem. The mature system, in effect, has a high degree of stability. Thus the price paid for stability is decreased production, whereas the price for increased production is instability.

There are other changes involved in succession. Organisms that comprise the younger stages of succession usually have shorter, simpler life cycles (for example, annual plants) and usually are smaller in size than those characteristic of later stages. This generalization may hold true for plants, but the application to animal life is open to some question. Because of the severe and more fluctuating environment, species of the early developmental stages possess a high rate of increase (r) (see

Chapter 10). Many young are produced, but mortality is high. Selection is for a high birth rate (r selection) because replacements are needed, and the organisms possess means for long-range dispersal. In the more mature stages both mortality and the production of young are lower, and selection is for the maintenance of the population at carrying capacity (K selection) (see Chapter 10). In this instance selection favors individuals that can most effectively utilize the resources. The life span of the organisms is long, and usually the organisms possess poor means of long-range dispersal. Because the life span of the organisms are relatively long, few replacements are needed to maintain the population at carrying capacity.

Although this concept of succession has been widely accepted by ecologists, it has some shortcomings. Largely developed from experimental and field work in aquatic systems, it has not been tested adequately in terrestrial systems, and may not be wholly applicable. In the annual grasslands of California, for example, dominance declines as the diversity of grassland species increases, as predicted by the model, and biomass is no longer concentrated in the top species (McNaughton, 1968). Likewise, as diversity increases, production declines. But the stability of grasslands is unrelated to production, diversity, or dominance. The diversity of species, too, may not necessarily approach a maximum in the mature stable stages. In some instances diversity may actually decline as the successional stages reach maturity. For example, in the upland forests of Wisconsin, species diversity declines after 100 years of forest succession (Loucks, 1970).

Another problem in terrestrial ecosystems is the question of stability itself. A mature oak and hickory or Appalachian cove hardwoods forest may approach stability, but there is no evidence that production equals respiration. Even in so-called old-climax forests, new wood is still being added annually so that P still is actually greater than R. Some data on mineral cycling (Rodin and Bazilevic, 1967) indicate that in some mature ecosystems nutrients retained by plant biomass do not equal the difference between plant uptake and plant return, as the model predicts.

The idea of succession is most applicable in biologically controlled ecosystems. Many major ecosystems, however, are controlled by

FIGURE 9-1 [OPPOSITE TOP]
Successional changes in an old field over 20 years. (a) The field as it appeared in 1942, when it was moderately grazed. (b) The same area as it appeared in 1963. (c) A close view of the rail fence in the left background of (a). (d) The same area 20 years later. The rail fence has rotted away and white pine and aspen grow in the area. (Photos by author.)

FIGURE 9-2 [OPPOSITE BOTTOM]
Primary succession in the subalpine zone in the Wasatch Mountains, Utah. Here the early stages include the trees; the climax is a mixed herb community. Note the changes in soil depth from a rocky surface with fine soil only in crevices to a well-defined solum essentially free from rocks. (Based on data from Ellison, 1954.)

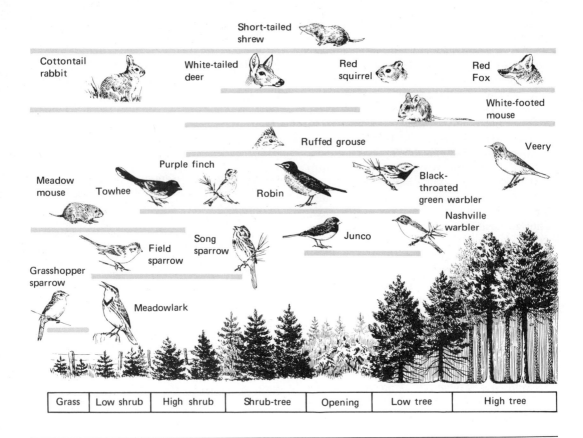

| Grass | Low shrub | High shrub | Shrub-tree | Opening | Low tree | High tree |

TABLE 9-1
Trends in ecological succession

Attribute	*Stage in ecosystem development*	
	Young	*Mature*
Biomass	Small	Large
Gross production/community respiration	Greater or less than 1	Approaches 1
Gross production/biomass	High	Low
Biomass supported per unit of energy flow	Low	High
Food chains	Short, grazing	Long, complex; detritus
Stratification	Less	More
Species diversity	Low	High (?)
Niche specialization	Broad	Narrow
Feeding relations	General	Specialized
Size of individuals	Smaller	Larger (?)
Life cycles	Short, simple	Long, complex
Population control mechanisms	Physical	Biological
Fluctuations	More pronounced	Less pronounced
Mineral cycles	Open	More or less closed
Role of detritus	Unimportant	Important
Stability	Low	High
Potential yield to man	High	Low

Source: Adapted from E. P. Odum, 1969, and R. Margalef, 1968.

physical conditions such as high temperature and low rainfall in the desert, or strong currents in streams and estuaries, which restrict the kinds and numbers of organisms living there. For example, Tilly (1968) found that the community of a cold spring supported largely by detritus was immature, yet was a stable ecosystem. It was maintained in that condition by restricted space, a heavy flow of water, and a low but constant temperature. In such physically controlled environments, succession may never occur.

Aquatic succession

The development of a pond into a mesic forest can be observed in a limited area, often about one pond alone. The first step in succession is the *pioneer stage*, characterized by a bottom barren of plant life. Such a stage can be found in newly formed, man-made ponds and lakes. The earliest forms of life to colonize the area are plankton, which may become so dense as to cloud the water. This plankton consists of microscopic algae and animal life, which upon death settle to the bottom to form a layer of muck.

If the plankton growth becomes rich enough, the pond may support other forms of life—bluegills, green sunfish, and largemouth bass, and small caddis flies that build cases of sand and feed on microorganisms living on the bottom.

The developing layer of loose, oozy material on the pond bottom creates a substrate for rooted aquatics, such as the branching green alga, *Chara*, the pondweeds and waterweeds (Fig. 9-5). These plants bind the loose bottom sediment into a firmer matrix and add materially to the deposition of bottom organic matter. Organisms common to the barren pond bottom cannot exist in the changed conditions of the submerged vegetation stage. The caddis flies of the pioneering stage are replaced by other species able to creep over submerged vegetation and build cases from plant material. Dragonflies, mayflies, and small crustaceans appear.

Rapid addition of organic matter on the bottom reduces water depth and provides nutrients for more demanding plants. Floating aquatics, roots embedded in the bottom muck and leaves floating on the water's surface, invade the pond. Because these plants shut out

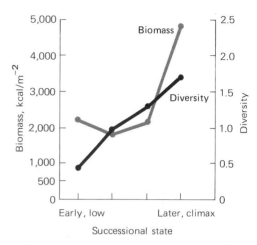

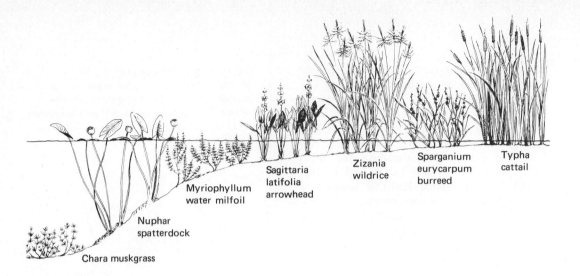

Chara muskgrass

Nuphar spatterdock

Myriophyllum water milfoil

Sagittaria latifolia arrowhead

Zizania wildrice

Sparganium eurycarpum burreed

Typha cattail

the light from the pond depths, they tend to eliminate the submerged aquatic growth. This is the *floating-aquatics stage*, one in which faunal living space is increased and diversified. Hydras, frogs, diving beetles, and a host of new insects capable of utilizing the undersurfaces of floating leaves appear.

Seasonal fluctuations in water levels alternately expose and cover the increasingly shallow bottom about the edge of the pond. Lacking the bouyancy and protection of water, the weak and soft-tissued floating plants cannot exist in the changing environment. Emergent plants—the cattails, sedges, bulrushes, and arrowheads, firmly anchored in the bottom muck by spreading fibrous roots and rhizomes—occupy the area. Because the leaves rise above and lack the protection of water, they possess flexible leaves and wandlike stems that bend easily before the wind and water. Animals of the floating stage are replaced by those that inhabit the jungle of emergent plant stems. Gill-breathing snails give way to the lung breathers. Different species of mayflies and dragonflies spend their nymphal stages on submerged stems and climb to the surface when they are ready to emerge as adults. Redwinged blackbirds, ducks, and muskrats become common to the area. As the oxygen supply of the water decreases, because of the increasing quantities removed through respiration by organisms of decay breaking down the accumulating organic matter, only animals with low oxygen requirements can exist. Bullheads replace sunfish and annelid worms colonize the bottom muck.

Since the dense root system and the annual deposition of leaf growth add great quantities of organic matter to the bottom, the substrate builds up rapidly after the emergents have appeared. Much of the old open-water area now is covered by sedges, cattails, and associated plants to form a *marsh*. As the bottom rises above the groundwater level, the remnant of the open pond dries up in summer. It has now become a *temporary pond*, which contains only those organisms that can withstand drying in summer and freezing in winter.

Drainage improves as the land builds higher. The emergents disappear, the soil lies above the water table, and organic matter, exposed to the air, decomposes more rapidly. Meadow grasses, accompanied by land animals, invade to form a marsh meadow in forested regions and a prairie in the grass country. In forested areas and in certain topographical situations between the prairie and the forest, alders, willows, and buttonbushes colonize the site. If the marsh is invaded directly by woody plants, the marsh-meadow stage never develops. Shrubs give way to trees—aspen, elm, red and silver maples, and white pine. Root systems, limited by high water tables, spread horizontally instead of vertically in the soil. The substrate is rapidly but unevenly built up by the accumulation of fallen trunks and by upturned roots and soil. As the forest floor becomes drier and the crown closes, seedlings of intolerant forest trees are unable to develop; but seedlings of sugar maple, beech, hemlock, spruce, and cedar, able to grow in low light intensities, dominate the understory and subsequently replace the intolerant trees. These trees tolerate the environmental conditions they create, so the forest cover becomes stabilized.

SUCCESSION AND EUTROPHICATION

Aquatic succession is considered natural eutrophication, and in classical limnology natural eutrophication is considered an aging process. Inflowing waters carry silt, which builds up bottom sediments and fills in the basins of lakes, ponds, and estuaries. Nutrients carried in from surrounding watersheds stimulate the growth of phytoplankton. The increased production of phytoplankton increases total biological productivity and gradually causes major changes in lakes and ponds. Phytoplankton becomes concentrated in the upper layer of water, giving it a murky green cast. This turbidity reduces light penetration and restricts biological productivity to the surface waters. Zooplankton feeds on phytoplankton and bits of organic matter and it in turn become food for fish. Algae unconsumed by phytoplankton, as well as inflowing organic debris and remains of rooted plants, drift to the bottom, where bacteria convert the dead matter to inorganic substances. The activity of these decomposers depletes the oxygen supply of the bottom sediments and the benthic waters to the point that the region is unable to support aerobic forms of life. The numbers and biomass of organisms are high although the diversity of species is often low.

This contrasts with oligotrophy, or poor nourishment. Oligotrophic lakes are relatively impoverished in nutrients, especially nitrogen and phosphorus, and nutrients normally added by inflow are quickly taken up by the phytoplankton. The density of the algal growth is low, and light easily penetrates to considerable depth. The water is clear and appears blue to blue-green in the sunlight. The oxygen profile (see Chapter 18) is nearly the same at all depths and the bottom fauna is well developed. Although the numbers of organisms may be low, the diversity of species is high. Fish life is dominated by members of the salmon family. When nutrients in moderate amounts are added to oligotrophic lakes they are rapidly taken up and circulated. As increasing quantities are added, the lakes and ponds begin to change from oligotrophic to mildly eutrophic to eutrophic conditions.

From the developmental point of view, limnologists consider eutrophic bodies of water as mature or aged and oligotrophic lakes as young ecosystems. Natural succession is assumed to proceed from an oligotrophic to

FIGURE 9-5 [OPPOSITE]
Zonation of vegetation about ponds and along river banks. Note the changes in vegetation with water depth. (Drawings based on Dansereau, 1959.)

a eutrophic state. This is accompanied by a filling in of the basin and subsequent conversion to land. Yet the characteristics of oligotrophic and eutrophic bodies of water do not conform to those characteristic of young and mature ecosystems (see Table 9-1). Eutrophic lakes possess a relatively low diversity of species, a high ratio of primary production to biomass, a circulation of nutrients, and a high rate of exploitation, all characteristics of immature ecosystems. Oligotrophic bodies of water have a high diversity of species, low ratio of production to biomass, and the nutrients are tightly circulated and quantities are tied up in biomass, all characteristics of mature ecosystems.

The terms *eutrophic* and *oligotrophic* were introduced by the German biologist C. A. Weber in 1907, when he applied them to the evolution of peat bogs. In his studies he found that the upper layers of the peat bogs were nutrient-poor and the deeper layers reflected a much richer supply. Thus, the development of a bog from a lake proceeded from a eutrophic to an oligotrophic state (Rodhe, 1969; Hutchinson, 1969). Somehow and somewhere the terms used in the context of lake succession became reversed. Perhaps it might be advisable to separate the terms from ecosystem development.

Lakes and other bodies of water in many ways passively exploit terrestrial ecosystems. The characteristics of streams, lakes, and estuaries are heavily influenced by their drainage basins and the sediments they receive. Hutchinson (1969) suggests that it might be more useful to stop thinking in terms of eutrophic and oligotrophic water types and instead to consider rivers, lakes, and estuaries and their watersheds as eutrophic or oligotrophic systems. A northern bog lake is oligotrophic because its watersheds, the terrestrial ecosystems, are nutrient poor. A lake fed by drainage from urban and agricultural watersheds is eutrophic because the watersheds are eutrophic.

Terrestrial succession

A similar sequence takes place on dry areas. Barren areas, whether they are natural primary sites, such as rock and sand dunes, or disturbed areas, such as abandoned cultivated fields or roadbanks, are a sort of natural biological vacuum eventually to be filled by living organisms. Plants and animals that colonize such sites comprise the pioneer communities. On primary sites no soil exists initially and successive communities can become more complex only as soils develop. Bare, disturbed areas, the secondary sites, have some sort of soil present already. Both, however, are characterized by full exposure to the sun, violent fluctuations in temperature, and rapid changes in moisture conditions.

PRIMARY SUCCESSION

Rocks and cliffs are common terrestrial primary sites. Bare rocks, on fully exposed cliffs and slopes, are colonized by crustose lichens, such as the widespread gray-green *Parmelia conspersa* and the black moss, *Grimmia*. *Cladonia* lichens may invade mats of black moss, followed by the hair-cap moss *Polytrichium*, or even vascular plants and grass (Oosting and Anderson, 1937, 1939). As yet there is no evidence that lichens either modify the rock or aid in the accumulation of fine soil or organic matter that can be colonized by other plants. As long as conditions are such that soil or organic debris cannot accumulate from some external source, crustose lichens and a few mosses persist indefinitely.

Wherever some soil has accumulated in crevices, recesses, and depressions, plants can take root. Usually these are a chance assortment of adapted species from the surrounding countryside plus a few distinctive rock species. From such areas a thin soil mat may spread outward and eventually cover the bare rock surface, but on more exposed sites the destructive forces of wind and water tear away the soil and plant cover and expose bare rock again. In soil pockets and crevices where the soil is deeper, woody plants become established.

The sand dune is another severe primary site. A product of pulverized rock, sand is deposited by wind and water. Where deposits are extensive, as along the shores of lakes and oceans and on inland sand barrens, sand particles may be piled up in long windward slopes to form dunes (Fig. 9-6) that move before the wind and often cover forests and buildings in their path, until stabilized by plants. With high surface temperatures by day and cold temperatures at night, dunes are rigorous environments for life to colonize. Grasses are the most successful pioneer and binding plants. When these, and such associated plants as beach pea, have stabilized the dunes, at least partly, mat-forming shrubs invade the area.

From this point the vegetation may pass from pine to oak, or to oak directly. The low fertility of the dunes favors plants with low nutrient requirements. Because these plants are inefficient in cycling nutrients, especially calcium, soil fertility remains low. Because of this infertility and the low moisture reserves in the sand, oak is rarely replaced by more moisture-demanding or more nutrient-demanding trees (Olson, 1958). Only on the more favorable leeward slopes and in depressions where the microclimate is more moderate and where moisture can accumulate does succession lead to a mesophytic growth of such trees as sugar maple, basswood, and red oak. Because these trees recycle nutrients more efficiently, effectively shade the soil, and add to litter accumulation, they aid in the rapid improvement of nutrient and moisture conditions. On such sites a mesophytic forest may become established without going through the pine and oak stages (Olson, 1958).

Deposits left by receding glaciers also are often nutrient-poor sites. In Alaska, newly exposed raw glacial till is invaded first by mountain avens (*Dryas*), whose feathery seeds are carried to the site by the wind (Lawrence, 1958). Tolerant of adverse environmental conditions present there, the avens seeds germinate quickly to form a horizontal evergreen mat, green in spring and red bronze in fall. This mat reduces soil erosion and begins to build up organic matter. Mountain avens is followed closely by Sitka alder, a rapidly growing, upright shrub, capable of reproducing abundantly by 7 years of age. Willow and cottonwood seedlings also appear, but on the nitrogen-impoverished soil they grow only as prostrate plants with yellowish leaves. But the alders, and to some extent the avens, possess in their roots large nodules of nitrogen-fixing bacteria, and nitrogen compounds fixed in alder roots leak out into the soil. Alder leaves, in which nitrogen constitutes up to 3 percent of their dry weight, add considerably more of this nutrient to the soil. In fact alder thickets 5 years old and 5 ft tall add to the soil each autumn 140 lb N/acre, an amount that increases as the thicket grows older. Because the rate of nitrogen accumulation is high and the cool, moist climate prevents the utiliza-

(a)

(b)

FIGURE 9-6
Sand dunes along the northeastern Atlantic coast. (a) *Grass on a dune along the Virginia coast.* (b) *The fore dune in the Cape Breton coast is claimed by marra grass while the older dunes support a growth of white spruce.* (*Photos by author.*)

tion of nitrogen faster than it is fixed, nitrogen is stored in the young soil, more than a ton an acre in the upper 18 in. of leaf mold and soil in 70 years following the ice recession. As a result cottonwoods and alders are stimulated to vigorous, upright growth.

As the alder thicket matures and can grow no taller, cottonwoods and hemlocks rapidly emerge and eventually shade out the alder. One hundred and seventy years, more or less, after the ice has melted, the alder is gone, incorporated into the forest floor. The nitrogen reserves are depleted and a carpet of moss and deep litter covers the forest floor.

SECONDARY SUCCESSION

Secondary succession is most commonly encountered on abandoned farmland and noncultivated ruderal sites such as fills, spoil banks, railroad grades, and roadsides, all artificially disturbed and frequently subject to erosion and settling movements.

The species most likely to colonize such places are the so-called weeds, a catch-all name that means something a little different to nearly everyone, depending upon his personal interests. Although virtually undefinable, weeds usually have two characteristics in common. They invade areas modified by human action; in fact a few are confined to such artificially modified habitats. And they are exotics, not natives, to the region. Once native species invade the area, these "weed" plants eventually disappear.

Annual, biennial, or perennial, all plants that settle initially on disturbed areas possess a great tolerance for soil disturbance and partial defoliation. Their seeds remain viable for a long time and may remain in the soil for a number of years until the conditions are right for germination. Some weeds require an open seedbed and exposed mineral soil for germination. Their rapid and successful colonization is aided by an efficient means of dispersal. Some have seeds that are light and carried by the wind; others spread by underground rhizomes. Vigorous, these pioneer plants grow rapidly under favorable conditions; in less favorable habitats they will set seed even when small (Sorensen, 1954).

In spite of this, these plants cannot maintain their dominance very long—3 to 4 years at the most—if all disturbance ceases. Short life cycles, advantageous at first, are not adaptable to conditions imposed by incoming plants with long life cycles, ones that begin growth earlier in the spring and persist throughout the summer.

The species composition of pioneering communities is highly variable. Infinite combinations exist, depending upon the seed source, the cultural practices prior to abandonment of the land, and moisture and soil conditions. Broad differences exist between the first-year communities that spring up in small grain fields and those that appear in cultivated row-crop fields. Because land in row crops has large areas of exposed soil, the number of annuals and biennials, such as ragweed, all of which do well in the hot weather of midsummer, is high. In the absence of cultivation, herbs are able to establish themselves in small grain fields unseeded to grass. In these the number of annuals and biennials is greatest the last year of use and decreases rapidly thereafter (Fig. 9-7) (Beckwith, 1954). Moist soils rich in plant nutrients support such plants as burdock, catnip, and nettle; poorer, drier soils grow shepherd's-purse, chicory, and ragweed. On railroad grades and yards, fresh cinders, high in sulfur compounds and other toxic materials, compose a substrate suggestive of the saline soils of the western plains (Curtis, 1959). Here western plants carried eastward on rail cars may become established, among them western wheatgrass and sunflowers. In addition there is a wide assortment of exotic weeds.

Thus succession, whether primary or secondary, starts with the colonization of the area by pioneer species—plants able to grow on a substrate low in nutrients and organic matter in an excessively wet or excessively dry environment, as well as able to withstand very bright sunlight and, on land, wide variations of surface temperature. The number of plants colonizing and surviving on the site are few at first, but as conditions improve, more of them occupy the area. Through the deposition of organic matter and shading of the surface, they reduce the surface evaporation and modify the environment enough to permit more demanding plants to invade the area. Better adapted to utilize the nutrients available, the new arrivals eventually take over and crowd out the pioneers by shading and by vigorous growth. Some new arrivals may take hold not because of changed environmental conditions but in spite of them. Many plants of an advanced seral stage could well be the pioneers had they chanced to colonize the area earlier. Some pioneer species such as crabgrass and sunflower may produce chemicals that inhibit their own growth (Rice, 1965) paving the way for invasion by grasses that are not affected by toxins of weeds (Parenti and Rice, 1969). Grasses in turn may inhibit nitrogen-fixing bacteria, slowing succession to the next stage (Rice, 1965). But eventually the stage is set for other plants and animals to invade and colonize the area and for a different community to develop. Thus one community replaces another until a stable community occupies the site.

Although succession has been described as a series of discrete steps leading to a final stage, the change is gradual, continuous, often variable, and in part controlled by the community itself. Stages may be skipped, telescoped, or extended. Often a seral stage may be prolonged by soil conditions or by a temporary climatic condition, such as drought; or succession may be accelerated by the omission of a stage, or by the combination of two or more stages into one. Later stages may be independent of earlier ones. Virginia pine and yellow poplar, for example, can and often do invade open sites normally colonized by pioneering herbs and grasses if a seed source is available. Succession may not necessarily proceed to a theoretical end point such as a forest. Shrub communities, such as alder, may be stabilized for long periods of time without proceeding toward the final stage.

Under certain conditions succession may regress. Man, fire, floods and the deposition of soil may alter conditions and set succession back to an earlier stage. In the prairie grasslands the constant burrowing of rodents and ants creates new bare areas, which are invaded by annuals. As the mounds and disturbed areas "mature," annuals are replaced first by plants with more exacting needs, then finally by climax species. This microsuccession is an excellent example of "gap-phase" replacement. Although the grassland community as a whole is relatively stable, it may be undergoing violent fluctuations from disturbance on very small areas within.

In certain topographical situations, a seral stage may be destroyed by the vegetation itself and succession turned at an angle from its original direction. Sphagnum moss growing on poorly drained sites in Alaskan spruce forests

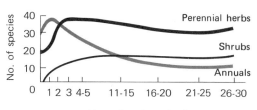

FIGURE 9-7
The succession of vegetation on cultivated fields. Note the early rise and rapid decline of annuals. (After Beckwith, 1954. © The Ecological Society of America.)

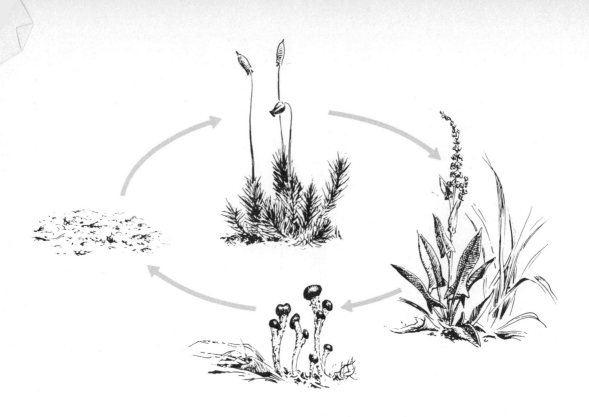

becomes so soggy that soil aeration is impaired, trees die, and logs and stumps become buried under a blanket of peat. As the forest dies, the sphagnum moss of the forest is replaced by sphagnum of the open. Seeds of hemlock and spruce may sprout in the mat, but they soon die for lack of nutrients and root aeration (Heilman, 1966). The result is the famous northern muskeg (Heinselman, 1963). In spite of the fact that the muskeg appears to be a regression, species diversity does not decline (Fig. 9-8) with the replacement of the forest community (Reiners et al., 1970). The loss in diversity contributed by trees and tall shrubs is compensated by an increased diversity of low shrubs and herbs of the muskeg.

CYCLIC REPLACEMENT

To determine whether a particular stage of vegetation is successional may be difficult. Succession may be confused with another type of vegetational and habitat change that occurs in cycles. Each successive community or phase is related to the other by orderly changes in an upgrade and downgrade series. These phasic cycles were first recognized and described in the Scottish heaths (S. A. Watt, 1947). There Scotch heather represents the peak of the upgrade series. After its death, a lichen, *Cladonia silvatica*, becomes dominant

and covers the dead heather stems. Eventually the lichen disintegrates to expose bare soil, the last of the downgrade series. The bare soil is colonized by bearberry, to initiate the upgrade series. Then heather reclaims the area and dominates again. There are other shorter phasic cycles, one involving heather, lichen, and bearberry and another involving heather, lichen, bare soil, and back to heather.

A similar cycle, frequently initiated by ants or ground squirrels, occurs in old-field communities in Michigan (Evans and Cain, 1952). Here bare areas at the bottom of the downgrade series are invaded by mosses, to start the upgrade series (Fig. 9-9). The mosses are invaded by Canada bluegrass and dock. The accumulation of dead culms and stems of these plants are covered by lichens of several species. Together the organic debris and the lichens crowd out the grass. Rain, frost, and wind destroy the lichens, and bare soil is exposed and left open to invasion by mosses again.

THE CLIMAX

Eventually succession slows down and an equilibrium or steady state with environment is more or less achieved. Theoretically, at least, this last sere is mature, self-maintaining, self-reproducing through developmental stages, and permanent; the vegetation is tolerant of the environmental conditions imposed by itself. This terminal community is characterized by an equilibrium between production and respiration, a diversity of species, a well-developed spatial structure, and complex food chains. This final stable community of the sere is the climax community, and the vegetation supporting it is the climax vegetation.

There are three theoretical approaches to the climax. One is the monoclimax theory developed largely by Clements. This theory recognizes only one climax, determined solely by climate, no matter how great the environmental conditions are at the start. All seral communities in a given region, if allowed sufficient time, will ultimately converge to a single climax and the whole landscape will be clothed with uniform plant and animal community. All communities other than the climax are related to the climax by successional development and are recognized as subclimax, postclimax, disclimax, and so on.

Another approach is the polyclimax theory (Tansley, 1935). The climax vegetation of a region consists not of just one type, but a mosaic of vegetational climaxes controlled by soil moisture, soil nutrients, topography, slope exposure, fire, and animal activity (see Daubenmire, 1968a, b).

The third and perhaps most acceptable approach is the climax-pattern hypothesis (Whittaker, 1963; McIntosh, 1958; Selleck, 1960). The composition, species structure, and balance of a climax community are determined by the total environment of the ecosystem and not by one aspect. Involved are the characteristics of each species population, their biotic interrelationships, availability of flora and fauna to colonize the area, the chance dispersion of seeds and animals, and the soils and climate. The mosaic of climax vegetation will change as the environment changes; and the climax community represents a pattern of populations that corresponds with and changes with the pattern of environmental gradients, intergrading to form ecoclines (see Chapter 17). The central and most widespread community in the pattern is the prevailing or climatic climax. It is the community that most clearly expresses the climate of the area. This mean of the prevailing climaxes relates the community to the climate in major ecoclines and provides a regional geographic pattern to vegetation (Fig. 17-1).

For all practical purposes the climax is

FIGURE 9-9 [OPPOSITE]
Cyclic replacement in an old field community. Moss—dock—lichen—bare ground.

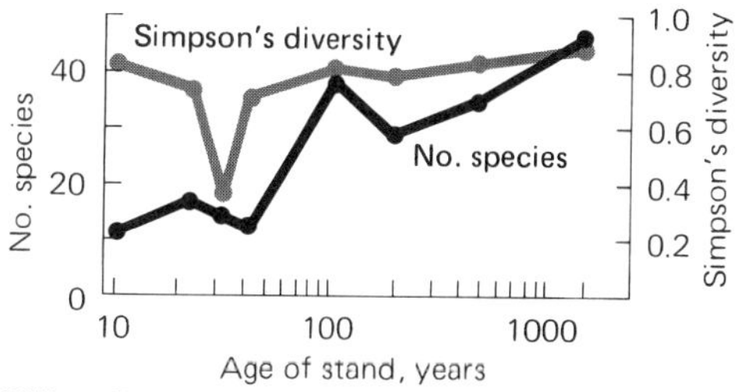

Willow-*Dryas*-Alder-Spruce——Hemlock——Muskeg

FIGURE 9-8
Changes in the number of species and species diversity in a successional sequence of vegetation at Glacier Bay, Alaska. (After Reiners, Worley, and Lawrence, 1971. © The Ecological Society of America.)

rarely if ever attained. In fact there is considerable question if any such thing really exists. True, the late stages in succession are relatively stable; they do exist for some time, and the trend, regardless of whether it started out from xeric or wet sites, is toward a mesophytic condition. But, that succession ends here is not a proved fact. Even in the so-called climax communities, stability is never really achieved. Self-destructive biological changes are continually taking place, even though slowly. Trees grow old and die and may be replaced by new trees of a different species. Cyclic replacement of microcommunities within the overall community may be taking place on a large or small scale. Replacement or recycling of nutrients may be lagging; too much may be tied up in woody vegetation, and the whole metabolism of the community may be slowing down. Thus the idea of the climax appears to be changing; the word takes on a different meaning. It is rapidly coming to mean those more or less stable and long-lived communities that develop late in succession in absence of disturbance. What comes after, how slowly or in what direction they will change, only time can tell.

Heterotrophic succession

Within each major community and dependent upon it for an energy source are a number of microcommunities. Dead trees, animal carcasses and droppings, plant galls, tree holes, all furnish a substrate on which groups of plants and animals live, succeed each other, and eventually disappear, becoming in the final stages a part of the nutrient base of the major community itself. In these instances, succession is characterized by early dominance of heterotrophic organisms, maximum energy available at the start, and a steady decline in energy as the succession progresses.

An acorn supports a tiny parade of life from the time it drops from the tree until it becomes a part of the humus (Winston, 1956). Succession often begins while the acorn still hangs on the tree. The acorn may be invaded by insects, which carry to the interior pathogenic fungi fatal to the embryo. Most often the insect that invades the acorn is the acorn weevil, *Curculio rectus*. The adult female burrows through the pericarp into the

embryo and deposits its eggs. Upon hatching, the larvae tunnel through to the embryo and consume about half of it. If fungi—*Penicillium* and *Fusarium*—invade the acorn simultaneously with the weevil, or alone, they utilize the material. The embryo then turns brown and leathery and the weevil larvae become stunted and fail to develop. This represents the pioneer stage.

When the embryo is destroyed, partially or completely, by the pioneering organisms, other animals and fungi enter the acorn. Weevil larvae leave the acorn through an exit hole, which they cut through the outer shell. Through this hole, fungi feeders and scavengers enter. Most important is the moth *Valentinia glandenella*, which lays its eggs on or in the exit hole, mostly during the fall. Upon hatching, the larvae enter the acorn, spin a tough web over the opening, and proceed to feed on the remainder of the embryo and the feces of the previous occupant. At the same time several species of fungi enter and grow inside the acorn, only to be utilized by another occupant, the cheese mites, *Tryophagus* and *Rhyzozhyphus*. By the time the remaining embryo tissues are reduced to feces, the acorn is invaded by cellulose-consuming fungi. The fruiting bodies of these fungi, as well as the surface of the acorn, are eaten by mites and collembolans and, if moist, by cheese mites too. At this time predaceous mites enter the acorn, particularly *Gamasellus*, which is extremely flattened and capable of following smaller mites and collembola into crevices within the acorn. Outside on the acorn, cellulose and lignin-consuming fungi soften the outer shell and bind the acorn to twigs and leaves on the forest floor.

As the acorn shell becomes more fragile, holes other than the weevils' exits appear. One of the earliest appears at the base of the acorn where the hilum falls out. Through this, larger animals such as centipedes, millipedes, ants, and collembolans enter, although they contribute nothing to the decay of the acorn. The amount of soil in the cavity increases and the greatly softened shell eventually collapses into a mound and gradually becomes incorporated into the humus of the soil.

Thus microcommunities illustrate the major concept of succession: that the change in the substrate is brought about by the organisms themselves. When organisms exploit an environment, their own life activities make the habitat unfavorable for their own survival, and instead create a favorable environment for different groups of organisms. Those responsible for the beginning of succession are all quite specialized for feeding in acorns, the later forms are less so, and the final group are generalized soil animals, such as earthworms and millipedes.

Influences on succession

Over most of the settled parts of North America and other continents, extensive areas of original climax vegetation no longer exist. Succession either has been set back, modified, or arrested by man. Even new "climax" communities have been modified by the elimination of some plants through lumbering and overgrazing.

FIRE

Fire, whether in climax vegetation or in a seral stage, sets succession back, influences the species composition, and shapes the character of a community. It long has played an important role in vegetational development. Eighty to 90 percent of the virgin northern coniferous forests can be traced to a post-fire origin. Charcoal is universal in the organic layer of the prairie (Heinselman, 1971). More than 95 percent of the virgin forests of Wisconsin were burned during the five centuries before the land was settled (Curtis, 1959). These fires not only enabled such species as yellow birch, hemlock, pines, and oaks to persist, but also, from an ecological viewpoint, were normal and necessary to perpetuate these forests (Maisurow, 1941; Curtis, 1959). The open ponderosa pine forests of southwestern United States evolved under the influence of natural fires. Forty years of fire exclusion from these forests have resulted in a thick growth of young pines, stagnation of stands, elimination of grass, and detrimental changes in the forest community (C. F. Cooper, 1960). Even-aged stands of Douglas fir, western white pine, and longleaf pine, red pine, and jack pine usually result from a fire-prepared seed bed. In Alaska, on the other hand, fires have converted white spruce stands into treeless herbaceous or shrub communities of fireweed and grass or dwarf birch and willow (Lutz, 1956), whose growth is so thick that forest trees cannot become established.

The effects of fire depend upon its intensity, the age of the plants affected, soil moisture at the time of burn, season of burn, health of the herbaceous plants, and the frequency of drought. Crown and severe surface fires can reestablish pioneer conditions and initiate secondary succession. But deep peat burns can convert forests to stagnant openings. Fire can inhibit the growth of shrubs in grasslands and stimulate the growth of grass; it can promote the growth of shrubs in forests and stimulate shrubby growth in chaparral. Fire can reduce the numbers of certain species, increase the numbers of others, and eliminate some altogether. It produces the vegetational pattern on which the animal component of ecosystems depends. Fire can produce successional stability by destroying the mature stages and initiating their redevelopment on the same site. In this manner fire recycles certain vegetational types such as chaparral and jack pine. Exclusion of fire from fire-dependent ecosystems will alter the type of vegetation that will ultimately claim the site. What type of vegetation might develop is largely unknown to science and will be in a sense unnatural. (Heinselman, 1971).

The plants that grow after a fire originate from several sources. Some are carried in by the wind or by animals. The former are well represented by aspens, paper birch, pine, and some herbs; the latter by fleshy-fruited plants. Others, such as oaks, bracken fern, and some perennial grasses sprout from fire-resistant roots. Because the seeds of such trees as beech, birch, and hemlock are destroyed by heat, sprout trees—red oak, black oak, white oak, and scarlet oak—dominate the area, and other species are reduced to a minimum. In this way fire can change the future composition of a forest (for a detailed discussion, see Ahlgren and Ahlgren, 1960).

Fire is a useful tool to control succession and to maintain economically more valuable seral stages. Prescribed burning—fire under control—is used as a management tool to eliminate hardwood understory beneath southern pines and thus perpetuate these intolerant trees, to prevent the encroachment of woody growth into grassland, to develop openings and browse for wildlife, and to maintain certain shrub communities, such as blueberries.

LUMBERING

Removal of a forest, especially by clear-cutting, turns the land back to an earlier stage of succession (Fig. 9-10). Unless followed by fire or badly disturbed by erosion and logging activities, the cutover area rapidly fills in with herbs and shrubs—blackberries, sumac, and dense thickets of sprout growth and seedlings of trees. The area passes quickly through the shrub stage to an even-aged pole forest (trees 4 to 8 in. diameter breast height). As the pole forest matures, the crown closes and shuts out the sunlight from the forest floor. This eliminates all but the most tolerant herbaceous and woody plants and reduces the stratification. Through time, trees in a better competitive position, those with larger crowns and more vigorous root systems, crowd out the subdominant trees and the area gradually passes into a mature and more highly stratified forest.

Man can modify the forest in many ways to meet his requirements. Early in the life of a new forest, trees of undesirable species and poor form, which in time would retard the growth of more valuable ones, can be removed. This improves (economically, at least) the composition of the stand and the quality of the trees. Later, the maximum growth of crop trees can be encouraged by thinning. The increased space between the trees stimulates crown expansion and increases growth.

Many of the most valuable and desirable timber trees exist in the lower seral stages instead of the climax. To maintain and reproduce this seral stage often is a problem. Stands of pine, balsam fir, and some spruces, aspen, and yellow poplar are maintained by clear-cutting the mature trees to expose the ground to sunlight. Only under this condition will intolerant seedlings survive.

The other extreme in management is selection cutting. In this process single trees or groups of trees are removed based on their position in the stand and their possibilities of future growth. Because successional changes do not follow, the forest remains structurally the same. Intolerant trees are excluded as well as the wildlife species associated with them; and the proportion of such tolerant trees as sugar maple is increased.

Mismanagement of the forest, such as clear-cutting excessively large areas, erosion initiated by poor layout of logging roads, and poor slash disposal can limit the rate of succession and delay the return of the original vegetation. Highgrading, taking the best and leaving the poorest, not only eliminates certain tree species from the future forest, but also tends to leave genetically inferior trees to supply seed for the future.

GRAZING AND BROWSING

Grazing by domestic and wild animals may arrest succession or even reverse it. Buffalo herds controlled the height of grass by grazing on the taller-grass species. This permitted the short grasses to assume dominance. Plant clipping by prairie dogs tends to decrease the proportion of annual grasses and forbs, to increase perennial forbs, and to influence the relative area of ground cover by each plant species (Koford, 1958). In prairies with both tall- and short-grass species, prairie dogs can develop and maintain a short-grass prairie. Overgrazing of grasslands by domestic stock results in denudation and erosion of the land. In the rangelands of the southwestern United States, overgrazing reduces the organic mat and thus the incidence of fire. Because of this, the reduced competition from grass, and dispersal of seeds through cattle droppings, mesquite and other unwanted shrubs rapidly invade the area (W. S. Phillips, 1963; Humphrey, 1958; Box et al., 1967) (Fig. 9-11).

The effects of overgrazing are well illustrated on the Wasatch Plateau in Utah. There the virgin subalpine meadows were so overgrazed in the 1880s and 1890s by cattle and sheep that the area was changed to a virtual desert. In spite of management efforts, accelerated soil erosion still continues (Ellison, 1954). The overgrazed meadows were taken over by annuals and early-withering perennials. In moderately grazed meadows, however, vegetation reacted differently to sheep and cattle grazing. Grazing by sheep tended to favor development of grasses and low shrubs, whereas grazing by cattle resulted in dominance by forbs, because grasses were suppressed.

If a relatively large number of cattle are allowed to graze in eastern woodlands year after year, they will in time destroy the forest. First, young trees and shrubs and lower limbs of taller trees are browsed, opening up the forest. The forest floor, unprotected by vegetation, is exposed to the wind; the soil dries out and weeds invade the woodland. In time

275

FIGURE 9-10
Lumbering, like fire, has an impact on forest ecosystems. Nutrients stored in the trunks of trees are removed, light and water regimes are altered, cover is removed, and stratification is destroyed. The cycle of successional development begins over again. The altered habitat attracts new species of animal life and eliminates for the time species dependent upon structural forest vegetation. (Photo courtesy U.S. Forest Service.)

FIGURE 9-11
The spread of a species and successional development is often the result of an interaction between plants and animals. Here a severely overgrazed range is invaded by mesquite, a deep-rooted shrub of semiarid regions. The removal of competing grasses by cattle provided the site. The seeds of mesquite dropped by cattle hastened the plant's spread. (Photo courtesy U.S. Forest Service.)

the understory completely disappears; some of the dominant trees die, both from natural causes and from the effects of reduced soil moisture and excessive exposure to sun and wind; and sufficient light reaches the forest floor to encourage grass. As the canopy is reduced to 50 percent or less, opened in part by the dying tops in old trees, the sod cover becomes complete. Eventually most of the trees die, and the once-forested land is converted to an open field with a tight sod cover (Day and DenUyl, 1932).

Overpopulations of wild grazing and browsing animals also influence community succession. In many parts of eastern North America, white-tailed deer, overstocked because of the failure to harvest does as well as bucks, and because of the lack of natural predators, have destroyed forest reproduction and developed a browse line, the upper limits on the trees at which a deer can reach food. The effects of browsing vary widely between forests, depending upon the type and the region. In the Ottawa National Forest, for example, white spruce was little browsed by deer, whereas yellow birch, basswood, hemlock, white cedar, and aspen were eagerly sought and have been greatly reduced or eliminated (Graham, 1958). In mixed northern hardwood–hemlock stands, only sugar maple and red maple have been able to survive the effects of browsing and probably will dominate the future forest. Basswoods and hemlock were completely eliminated. Mule deer and elk in the Rocky Mountain National Park prevent the invasion of forest openings by trees and shrubs. Extensive studies indicate that only in areas protected from deer by fencing did aspen, willow, and other woody invaders grow (Fig. 9-12).

In Africa the elephant has a pronounced effect on the nature of the ecosystem. When elephant populations are in balance with forage supplies and their movements are not restricted, the elephant has an important role in creating and maintaining the forest. When elephants are overpopulated, the feeding habits of the animal combined with destructive fires can devastate the flora, fauna, and soils. Elephant depredations on trees act as a catalyst to fires, which are the primary cause of converting forests to grasslands (Wing and Buss, 1970).

CULTIVATION

Cultivated plant communities are simple, highly artificial, and consist mainly of introduced species well adapted to grow on disturbed sites. Such vegetation cannot survive or perpetuate itself without constant interference and assistance from man. Because of the very simple and homogeneous ecosystem involved, tillage brings with it new pests, such as the Japanese beetle, which spreads rapidly and is destructive to both cultivated and natural vegetation. Tillage also disturbs the structure of the soil by mixing the upper strata and by exposing it to erosion.

Accelerated erosion (in contrast to geological or natural erosion) is a result of cultivation and also poor management. Stripped of its protective vegetation and litter by plow, axe, and grazing, soil is removed by wind and water faster than it can be formed. As the upper layers of humus-charged, granular, high-absorptive topsoil are removed, humus-deficient, less stable, less absorptive, and highly erodable layers beneath are exposed. If the subsoil is clay, it absorbs water so slowly that heavy rains produce a highly abrasive and rapid runoff.

The intensity of water erosion is influenced by slope, the kind and condition of soil, land use, and rainfall. The least conspicuous type of water erosion is *sheet erosion,* a more or less even removal of soil over the field. Because it takes place slowly, the change in the color of soil from dark to light and the appearance of subsoil and bedrock often goes unnoticed for a long time. When runoff tends to concentrate in streamlets instead of moving evenly over a sloping field, the cutting force is increased and *rill erosion,* which produces small channels down slope, results. On land areas where concentrated water cuts the same rill long enough or where runoff is concentrated in sufficient volume to cut deeply into the soil, highly destructive gulleys are formed (Fig. 9-13). *Gully erosion* often begins in furrows running up and down hill, in wheel ruts, livestock trails, and logging skid trails in the woods.

Bare soil, finely divided, loose and dry as it often is after tillage, is ripe for wind erosion. The forward velocity of the wind well above the soil surface is much higher than that near the surface, where it approaches zero. Just above the surface, wind movement is influenced by surface irregularities. These conditions produce eddies with an upward velocity two to three times the forward velocity near the surface. The eddies dislodge the most erodable grains of soil occupying the most exposed positions on the surface and move them along a short distance on the surface. Suddenly the grains shoot upward in a jumping movement, the height of which is influenced by size, density of the particle, and the surface. When they strike the ground again, they rebound and dislodge still other particles, forcing them to jump into the air. The impact of the particles on the surface initiates still another movement, surface creep. And very fine particles of dust, bounced by larger particles, rise high enough to be picked up by the wind and carried on as dust clouds. Often dust particles may be lifted high in the atmosphere and carried for hundreds of miles.

Erosion by wind or water impoverishes the land. It carries away organic layers, exposes the subsoil, depletes nutrients, changes soil structure, deposits soil elsewhere, increases runoff, and causes land ruin and abandonment. Land abandoned because of mismanagement is usually so degraded that natural vegetation has difficulty colonizing the area. Erosion worsens, gullies deepen, and conditions become progressively worse, unless drastic steps are taken to stop erosion and restore vegetation.

In tropical regions cultivation associated with fire can permanently change vegetation. Kowal (1966) studied in detail the effects of fire and shifting cultivation in the Philippines. Two types of cultivation are practiced: one is kaingin, or slash-and-burn (see Chapter 4). Practiced in the past on gentle slopes, kaingin caused few long-time environmental effects. Because of the topography, absence of hoeing, absorbent soil organic matter, and absence of fires, erosion was minimal. Once abandoned, the plots reverted to montane forest. But on steep slopes soil erosion and slides so deteriorated the site that xerophytic pines and grasses seeded in. Frequent fires, often only 1 to 5 years apart, encouraged continued erosion of the soil. Pines and grasses were never replaced by broad-leaved evergreen forest (Fig. 9-14). Recurring fires in the pine forests slowly ate into the montane forest, which eventually became restricted largely to stream depressions and areas remote from cultivation. Today kaingin is replaced by per-

(a)

(b)

FIGURE 9-12
Influence of elk and deer on vegetation. (a) *When this exclosure was built, aspen saplings were numerous on the hillside. Elk, deer, and beaver have eliminated all those that are not within the exclosure.* (b) *Old scars on dying aspen resulted from elk feeding on the bark. (Photos courtesy L. W. Gysel and* Journal of Forestry.)

manent gardens that involve working the soil with hoe, intensive fertilization, and permanency, a result of increasing demands for truck crops by an expanding urban population. With permanent gardens the equilibrium between agriculture and the soil is broken. Severe erosion occurs even on gentle slopes. The destruction of the montane forest is even more extensive, and if and when abandoned, the land rarely returns to montane forest because of the deterioration of the soil and frequency of fires on the mountain slopes. Abandoned land is claimed by pines and grasses, which in turn may be cleared again for permanent gardens.

IONIZING RADIATION

Intense radiation such as that produced by unshielded nuclear reactors, by atomic and hydrogen bomb blasts, and by areas of fallout—up to 100 mi from the blast site—can have a pronounced effect on the plant community (see Chapter 7). Early successional stages, simple in structure, are more resistant to high levels of chronic gamma radiation than more complex systems, such as the forest. Early pioneers, the mosses and lichens, can withstand exposures of over 200,000 roentgens. Grasses, hemicryptophytes, and cryptophytes that have shoots and buds protected by the soil, and annuals, whose dormant seeds are radioresistant, can withstand months of exposures of radiation between 100 and 300 roentgens. Radiosensitive dominant species would be replaced by normally infrequent but radioresistant species. As seral communities become more complex, their radiosensitivity increases, and radioactivity that would hardly alter simpler ecosystems would cause major damage. Pine, the most sensitive of plants because of the low number of chromosomes (40), would be destroyed by chronic irradiations of 20 to 60 roentgens, and the hardwoods would undergo transformation of leaves that reduces the thickness of the canopy. If radiation rises above 360 roentgens, the hardwood forest would be destroyed and the area would revert to a simple community of radioresistant grasses, annuals and perennials, and low shrubs (Woodwell, 1963).

If a complex ecosystem is subject to low exposure, the growth of sensitive species is temporarily inhibited. The effect would be comparable to that of wind or frost damage.

Recovery would be rapid, and the direction of succession would not be changed. At higher levels of dosage, radiation would cause selective inhibition or mortality of sensitive populations. Its impact on succession would be similar to that following logging, grazing, or a catastrophic storm. Under intense radiation the basic structure of the ecosystem and the capacity for the site to support future life might be reduced. The effect would be similar to that of destroying a tropical rain forest by burning. The loss of nutrients can limit succession and delay indefinitely the reestablishment of vegetation (Woodwell and Sparrow, 1965). Long-continued disturbance favors vigorous organisms that reproduce rapidly and asexually.

POLLUTION

The settlement of a region by man subjects the aquatic ecosystems to increased inputs of natural and unnatural substances. Agriculture, road construction, and building have clogged rivers and streams with incalculable tons of silt and filled lakes, dams and estuaries. Man has added a heavy load of nutrients, especially nitrogen, phosphorus, and organic matter from sewage and industrial effluents. He has poured in thousands of different chemicals and wastes, including pesticides, which natural ecosystems are ill-adapted to handle.

One of the outcomes of this has been excessive nutrient enrichment of aquatic ecosystems. This enrichment has been changing oligotrophic lakes to eutrophic, and eutrophic lakes to hypereutrophic conditions. This accelerated enrichment has been called *cultural eutrophication* (Hasler, 1969).

Cultural eutrophication has produced significant biological changes in many lakes and estuaries. The tremendous increase in nutrients stimulates a dense growth of planktonic algae, dominated by blue-green forms, and rooted aquatics in shallow water. This upsets normal food chains. The herbivores, principally grazing zooplankton, are unable to consume the bulk of the algae as they normally would do. Abnormal quantities of unconsumed algae as well as rooted aquatics die and sink to the bottom. On the bottom the aerobic decomposers are unable to reduce organic matter to inorganic matter, and they perish from the depletion of oxygen. They are replaced by anaerobic organisms that only incompletely decompose the organic matter. Partially decomposed sediments build up the bottom, and sulfate-reducing bacteria release hydrogen sulfide that can poison benthic waters. These chemical and environmental changes cause major shifts in the plant and animal life of the affected aquatic ecosystem.

An example is a major shift that resulted in part from cultural eutrophication in Lake Erie (Beeton, 1965, 1969). In 1930, before it was subject to enormous quantities of pollutants, the dominant bottom organism was the burrowing mayfly *Hexagenia*. In 1961 the mayfly in the western part of the lake was virtually extinct, replaced by tubificid worms. Between 1911 and 1963, phytoplankton production increased significantly. Spring and fall blooms increased in intensity and duration. Dominant phytoplankton in the spring bloom in western Lake Erie shifted from *Asterionella* to an exotic species *Melospira binderana*, not reported in the United States prior to 1961. Now the dominant species, it comprises in certain areas of the lake as much as 99 percent of the total phytoplankton.

Changes in fish populations were as pronounced as those in the invertebrate fauna and the phytoplankton. Oligotrophic species were replaced by mesotrophic, and eventually the mesotrophic species by eutrophic. Although the fish populations were reduced by overexploitation and the invasion of the lake by two marine species, the sea lamprey and the alewife, the decline and extinction of certain species throughout the Great Lakes occurred both in the absence and the presence of the marine fish and commercial fisheries.

The first to go were the salmonids, the Atlantic salmon and the lake trout (S. H. Smith, 1972), followed by the coregonines— whitefish, lake herring and deep-water cisco, whose populations declined in the 1940s. As these species declined, the mesotrophic percid fish—walleyes, sauger, and yellow perch—became the dominants. They too are at the end of succession as the walleyes and saugers have declined, leaving the yellow perch. These mesotrophic fish are being replaced by carp, white bass and freshwater drum (Fig. 9-15). Lake Ontario has a similar history. The three other Great Lakes, are in various stages of decline from the greatest to the least, Huron, Michigan, and Superior.

Similar and even more disastrous examples of biological change can be found in many

FIGURE 9-13
Gully erosion. This gully began as a path used by cows. Another cow path cutting across the photograph looks like an open furrow. (Photo courtesy Soil Conservation Service.)

FIGURE 9-14
An abandoned swidden agricultural plot in the Philippines reclaimed by sprout growth and pine. Swidden agriculture has been responsible for the spread of pine forests in the Philippines. (Photo by N. Kowal.)

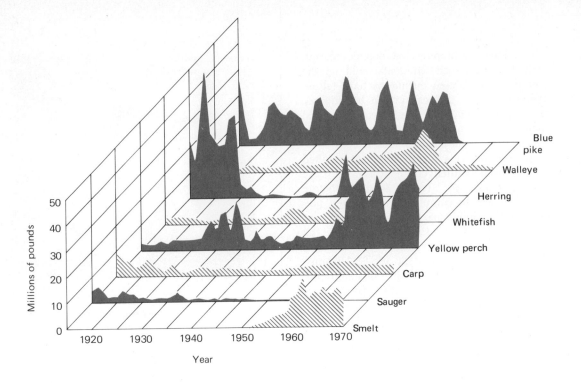

lakes (Edmondson, 1969). These changes affect the value of water for man. Murky green water and algal scums make lakes and ponds undesirable for recreational use. The phytoplankton and products of decomposition impart an unsatisfactory flavor to the water and interfere with its proper treatment. The rotting mass of vegetation and the die-off of fish create objectional odors.

Cultural eutrophication has been attributed largely to the increase of nitrates and phosphorus added to the water from many sources. Both are important but of the two, phosphorus is considered more important because it is an element in short supply naturally. Because the principal sources of phosphorus are human sewage (each human produces about 1.5 to 4 lb/year), agricultural drainage (each cow produces about 1.5 lb/year), and detergents, some consideration must be given to its removal from sewage effluents.

Its removal can have dramatic effects. From 1941 to 1968 Lake Washington received the effluent from sewage plants servicing Seattle. In 1932 the lake was oligotrophic. After effluents were poured into it, sediments and their organic content, phosphorus, and nitrogen all increased, and the waters supporting a heavy phytoplankton population were turbid. From 1963 to 1969 the sewage effluent was diverted from the lake. After the first diversion, con-

ditions in the lake began to improve steadily. The phosphorus concentration has declined, the abundance of phytoplankton has decreased, and the summer transparency of the water has increased (Edmundson, 1970). Although the nitrogen level is still high, the reduction of phytoplankton suggests that phosphorus is the limiting nutrient.

SURFACE MINING

During the past decade surface mining for coal has increased to the point that it supplies 40 percent of the nation's coal. Once confined to the mountainous areas of Appalacia and the anthracite region of eastern Pennsylvania, surface mining has spread to the corn fields of Ohio and Illinois, the desert lands of Arizona, and the mountains of Wyoming, Montana, and Alberta. Except for those who live in or have visited coal-producing regions, few appreciate the environmental destruction caused by strip-mining. The impact and the magnitude of damage vary with the region.

Basically there are two types of strip-mining, area and contour. Area surface mining is done on relatively level land and on mountain tops (Fig. 9-16). A trench or box cut is made through the overlying rock (overburden) to expose a portion of the coal seam. When that coal is removed, a parallel cut is made, and the material (spoil) is deposited in the cut previously excavated. The final cut leaves an open trench equal to the overburden, and the coal bed is recovered. Reclamation laws now require that this final cut also be filled. Contour strip-mining is done on mountain slopes. The operator bulldozes a haul road to the seam and proceeds to remove the overburden from the bed, starting at the outcrop and working around the hillside, carving out a long broad trench. Left on the inside is a high wall often as high as 100 ft. On the outside is the overburden, shoved over the side of the hill to create a precipitous downslope. Some mountains contain one seam of coal. Others may contain up to four seams. The strip-mine operator starts on the lowest seam first, and when that is mined, he moves on to the next seam until he reaches the topmost. When the coal has been removed, the mountainside has been converted into a series of level terraces joined by rocky outslopes, like giant steps (Fig. 9-17).

Whatever the method, strip-mining does violence to the land. Undisturbed, unweathered rock strata are broken and brought to the surface, where the material is subject to rapid weathering. Nutrients that were once slowly leached and recycled are now subject to rapid weathering and chemical action. Elements that are necessary in small amounts become toxic to plants when released in great amounts (see Chapter 7). Carried away in high concentration by water coming off the stripped slopes, these elements reduce water quality downstream. Dissolved solids coming from stripped sites can be 12 times greater than from unmined sites, and sulfites can be 50 times greater (Collier, 1962). The duration of major effects of these materials can be expected to last at least 10 years. In fact, although the amount of leaching of toxic materials decreases rapidly after mining operations, the half-life of toxic materials in the mountains of West Virginia and Kentucky is estimated to be 125 years for sulfur, 100+ years for magnesium, and 1,000+ years for aluminum (Vimmerstedt and Struthers, 1968).

The spoils are highly unstable. The mixture of soil and rock that makes up the overburden is often so deposited that the outer slope of the spoils is steeper than the original slope. Water seeping into and percolating

FIGURE 9-15 [OPPOSITE]
The commercial production of Lake Erie fishes, 1915 to 1970, Canadian and United States landings combined (1 lb = 0.434 kg), shows the effects of accelerated cultural eutrophication, increased water temperatures (of 1.1°C.), severe oxygen depletion, as well as exploitation and influx of new species. This is an excellent example of the impact of man on the succession of fish life from oligotrophic to eutrophic species. (From W. L. Hartman, 1972; by permission of the Journal of the Fisheries Research Board of Canada.)

FIGURE 9-16
Huge shovels capable of digging 400,000 lb of earth in a single bite swing 180° and deposit the load 325 ft away from the digging point at heights up to 120 ft. Used in area mining, these machines remove one cut of coal and pile the overburden on the surface. On the next cut the overburden is deposited in the trench of the first cut. The ecological impacts are enormous. (Photo by author.)

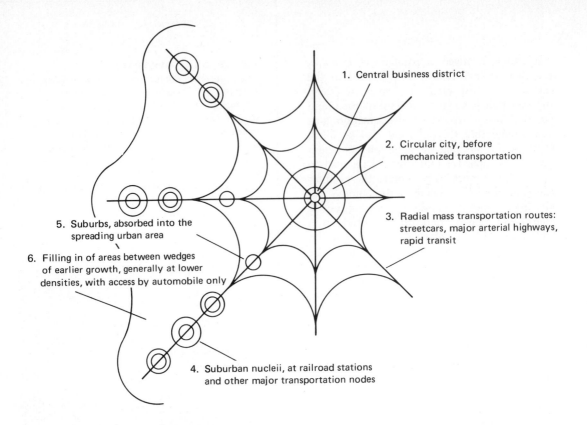

1. Central business district

2. Circular city, before mechanized transportation

3. Radial mass transportation routes: streetcars, major arterial highways, rapid transit

5. Suburbs, absorbed into the spreading urban area

6. Filling in of areas between wedges of earlier growth, generally at lower densities, with access by automobile only

4. Suburban nucleii, at railroad stations and other major transportation nodes

through the spoils wets the clays and shales. The slipperiness of the clays, the weight of the spoils, and gravity combine to cause massive landslides that block mountain roads and dam mountain streams. Water rushing down steep slopes carries with it tons of sediment that are washed far downstream. Sediment yields from spoil banks may go as high as 1,000 times the amount that comes from undisturbed forested slopes, and the erosion rate is five to ten times as much (Collier, 1962).

Strip-mining alters the ground–water regime. Water tables once deep in the underlying rock strata are exposed and flow freely to the newly created surface. Large quantities of water that would have been taken up by the trees and lost to the atmosphere as evapotranspiration are added to the amount of runoff. During periods of heavy storms in spring and summer, water flows from strip-mine sites with great force, intensifying the height and damage from flash floods and washing out stream channels and narrow flood plains below.

Because of public pressure, eastern coal states now require extensive reclamation ef-

forts that involve slope reduction, backfilling, leveling, and burying of toxic materials, and revegetation. Revegetation on many spoils is nearly impossible because of the acid and toxic material of the overburden, the extremely unstable slope, and severe environmental conditions. Even though toxic materials are covered, water moving through the material still carries a burden of toxic elements. Revegetation in mountainous areas in particular requires heavy fertilization and liming before vegetation can be established. The plant cover established is largely short-lived grasses such as weeping love grass and Kentucky fesque. Unless this plant cover receives annual fertilization and some management, it will deteriorate, exposing the materials to erosion again. Further grass cover is unable to stabilize the outslopes of the spoils, and the area is still subject to slides (Fig. 9-18). On more level terrain revegetation is more successful. Unless managed as grazing lands, some care is needed to insure the maintenance of a grassy cover until natural succession can take over. In the desert lands of the west, there is no known method of reclamation to date, for the aridity makes revegetation extremely difficult.

FIGURE 9-17
In the mountainous regions of Appalachia, coal is removed by contour strip mining which results in the destruction of the mountain environment. (Photo by author.)

FIGURE 9-20 [OPPOSITE]
The expansion of urban areas is similar to succession in old fields. As the forest moves out into old fields, so do urban and suburban areas move into the countryside, their spread following avenues of transportation, as suggested by this model. (From Mayer, 1969. Reproduced by permission from the Association of American Geographers Commission on College Geography, Resource Paper #7, "Spatial Expression of Urban Growth.")

FIGURE 9-19
Many areas once settled by man have been abandoned and the forest returned. Here an old graveyard has been reclaimed by white pine. (Photo by author.)

FIGURE 9-18
In contour strip mining the overburden is pushed over the mountainside, burying the vegetation and creating steep unstable slopes difficult if not impossible to revegetate and subject to slides and erosion. (Photo by author.)

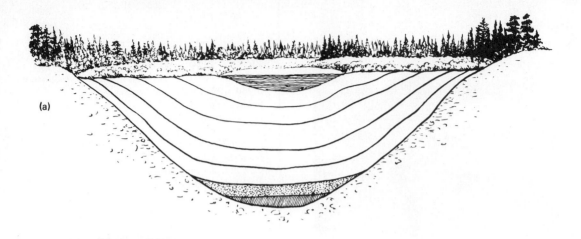

(a)

INDUSTRIALIZATION AND URBANIZATION

No land-use change is more complete and final than industrialization and urbanization, a climax type in human succession. Natural vegetation is destroyed by man and replaced by ecologically permanent bare areas of concrete, asphalt, and steel. But even here a diversified group of animals and some plants are able to exist. Norway rats, common rock pigeons, starlings, English sparrows, cockroaches, and flies are common to this environment, as well as some grasses, algae, and other plants able to gain a foothold in cracks in concrete and in vacant lots. Nighthawks have substituted the artificial canyons created by tall buildings for natural cliffs. Fumes from factories, coke ovens, and smelters destroy the vegetation of surrounding areas. Even after the cause has been eliminated, many years are required before the vegetation begins to return. Pollution of streams by sewage, industrial wastes, and siltation eliminates oxygen-demanding fish like trout; these are replaced by carp and bullheads, able to adapt to polluted conditions. Dam construction for power drowns terrestrial communities and converts part of the river community to a deep lake. Migrant fish, particularly the salmon, may be blocked from reaching their spawning grounds in the headwater streams.

Human settlement of an area from past to present has undergone a sort of succession. The first to live in or penetrate a region, the pioneers, are hunters and trappers, who, aside from harvesting animals, leave little mark on the land. They are followed by a subsistence farming or grazing culture, which can completely change a natural community. Some

plants and animals may be destroyed, succession set back to an earlier and more economically productive stage, and new animals introduced. If the land is too poor or too abused to support human society economically, the land at this stage may revert back to natural vegetation. Traces of old settlements and abandoned land can be found throughout the country (Fig. 9-19). Ghost towns, old stone and rail fences, house and barn foundations, lilac bushes, wells and springs hidden back in the woods, all attest to former human occupancy of the land.

Industrial and urban settlements are the climax stages of human succession. The tremendous growth of suburban settlements onto fertile farmland marks well this type of succession. Just as there are successional trends in the various later stages of forest development, so there are successional trends in the stages of urban development. The urban community begins as a small central core and grows outward into the surrounding countryside. Just as the forest invades an old field, so this outward expansion or invasion takes over the surrounding country.

Initially the center of the city is the most desirable place to locate. But because all resident and associated establishments cannot locate there, a sorting process takes place, resulting in the segregation of both functional units and social units based on socioeconomic status, culture, and race (Mayer, 1969). As the city grows, the pressure for other central locations forces outward expansion. This expansion or invasion does not take place at equal rates because of resistance (competition) from land-use functions, people, and transportation facilities. As one zone exerts

(b)

pressure on the adjacent outer zone, it eventually replaces the outer zone. At the same time, the outer zone tends to invade the next or colonize a new area. Thus one successional stage replaces another (Fig. 9-20).

With the passage of time, the mature community deteriorates, and the zones about the central core become less desirable. Buildings deteriorate, industrial and commercial complexes become less efficient and far removed from the source of productivity—the young vigorous communities on the outer zones or the earlier stages of succession. There new central core areas develop, determined by transporation links, shopping centers, and industrial parks. As these new core areas become firmly established they in turn initiate their own successional forces and patterns (Fig. 9-20). In time they will be absorbed by the expanding urban area.

Meanwhile, in the original central core, homes once occupied by higher socioeconomic classes have been abandoned to new immigrants, the original resident having moved to outer zones. As the new inhabitants of the core better their socioeconomic status, they too move out to the adjacent zones while their occupants move further out from the original city. Eventually the structure of the inner city become unattractive, the density of the population declines, and in time the core becomes a blighted area unable to pay any sort of economic rent. Subsidized public intervention becomes necessary. Just as fire or some other natural event tends to renew the aged, declining natural community, so urban renewal represents an attempt to put the core city back to an earlier, more productive stage of succession.

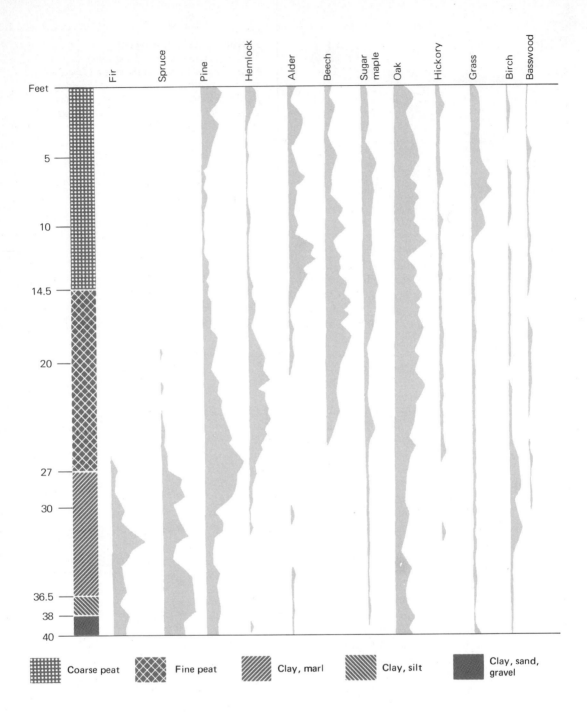

Past communities

Community changes and succession take place over a relatively short period of time and under similar climatic conditions. Communities also changed throughout geological periods of time, changes usually associated with shifts in climate. The emergence of mountains, sinking and rising of seas, and glaciation all influenced the climate and other environmental conditions of a given area. Many plant and animal species became extinct; one type of climax community was replaced by another.

The records of these past communities, their animals and plants, lie buried as fossils. From these, plant and animal associations of the past can be determined and in a broad way the climatic changes that brought about the gradual destruction of one stable community and the emergence of another. Such interpretation is based on the assumption that organisms of the past that had a structure similar to those living today also possessed similar ecological requirements. Thus one assumption is that if modern palms and broad-leaf evergreens are tropical plants, then their ancient prototypes also lived in a tropical climate. The study of this past relationship of ancient flora and fauna to their environment is called *paleoecology*.

Of particular interest to the paleoecologist are the climatic and vegetational changes that followed the advance and retreat of glaciers. As the glaciers moved south in several advances, vegetation was destroyed and the relief or physiognomy changed radically. The climate about the edges of the glacier supposedly was rather cold and the growing conditions optimum only for tundralike vegetation. Recent evidence, however, suggests that the climate, particularly along the southern edges of the glacial advance, may not have been as cold as once thought and that some forest-tree species quickly invaded the glacial-drift country following deglaciation.

Changes in postglacial vegetation and climate are recorded in the bottoms of lakes and bogs. As the glaciers retreated, they left scooped-out holes and dammed-up rivers and streams, which filled with water to form lakes. Organic debris accumulated on the bottoms to form peat (Fig. 9-21). In the peat were embedded fossil pollen and spores and small invertebrates that blew in from adjacent vegetation, settled on the water, and sank. Microscopic examination of samples of organic bottom deposits obtained at regular intervals reveals fossil pollen. Identification of various genera is based on comparisons of fossil pollen to that of the same genera growing today. The relative abundance of pollen of several genera indicates the predominate vegetation of the specified depth of deposition (Fig. 9-22).

Generally five vegetational or climatic periods following glaciation are recognized. First spruce and fir were dominant, suggesting a cool, moist period. Following this was a period when pine was dominant, with an intermix of oak, suggesting a drier, warm climate. Then an invasion of hardwoods took place. In the first stage of this invasion, the climate apparently was humid and warm, for hemlocks, oaks, and beech were the dominant trees, but a later warm, dry period favored the oaks. This was followed by a return to pine and a reappearance of hemlock and spruce, which indicated a climatic shift to a cool, moist period. A sequence of postglacial forest vegetation from an area near the edge of the glaciers' furthest advance south is shown in Fig. 9-22.

Results of pollen investigation can only indicate trends in vegetation and climate through the past. At present it is impossible to determine the exact structure and composition of prevailing vegetation at any one time period. Inaccuracies are caused by variations in pollen production. Because some tree species produce more pollen than others, these would appear more abundant than they really were. Some pollen might have been carried a considerable distance by the wind and perhaps even mixed deeper by earthworms. Many pollen grains can be identified only to genera and not species. Thus the oaks cannot be separated out to give a clue to the particular type of oak forest. And finally the possibility exists that some of the species had environmental requirements different from their modern-day counterparts.

Summary

With the passing of time, natural communities change. Old fields of today return to forests tomorrow; weedy fields in the prairie country revert to stable grassland. This gradual change from one community to another is called succession. It is characterized by a progressive change in species structure, an increase in biomass and organic-matter accumulation, and a gradual balance between community production and community respiration. This change is brought about by the organisms themselves. As they exploit the environment, their own life activities make the habitat unfavorable for their own survival. But in doing so, they create an environment for a different group of organisms. Eventually, however, an equilibrium, or steady state with the environment, is more or less achieved. This stage, usually called the climax, is self-maintaining and usually long-lived, as long as it is free from disturbance.

But relatively few communities are free from disturbance, and the greatest cause of disturbance is man himself. He has greatly modified natural communities the world over. Fire, more often man-caused than not, sets back succession to an earlier stage and changes the composition and even the type of community. In cutting the forest for timber, man most often has tended either to remove the trees completely, or to take the best and leave the poorest in quality and the genetically inferior to reproduce the future forest. Overgrazing of forests and grassland by both domestic stock and mismanaged wild grazing animals has resulted in the denudation of grasslands and serious disturbance and even destruction of forests. To provide food for himself, man has cleared away natural vegetation and replaced it with simple, highly artificial communities of cultivated species, adapted to grow on disturbed sites. This has brought about an explosion of insect pests, accelerated erosion of unprotected soil, and changed the nature of successional communities that follow abandonment.

Nowhere is land change more complete than in industrial and urban areas, a climax-type of human succession that has its own developmental stages. This succession is accompanied by air and water pollution from industrial and human wastes. The enrichment of aquatic ecosystems causes cultural eutrophication, characterized by a decline in species diversity and the replacement of oligotrophic species by eutrophic species. Demands for fossil fuels encourage the growth of surface mining that completely destroys natural communities. Attempts to restore the areas are not wholly successful.

In recent years man has added a new dimension to successional change: nuclear radiation

*from atomic testing and nuclear reactors.
Such radiation can have a powerful effect both
on community composition and directional
change. Some plants and most animals are
killed by radiation; a few plants are radioresistant. Under intense radiation the basic
community structure is destroyed and the
ecosystem is simplified and unstable.*

*Perhaps the most outstanding character-
istic of natural communities is their dynamic
nature. They are constantly changing through
time, rapidly in early stages of development,
more slowly in the later stages. Even the
seemingly most stable natural communities
may change through time. That they have
in the geological past is shown by fossil
records in the earth. What changes will take
place in the future, and how rapidly or how
slowly they will come, only time can answer.*

Populations
and ecosystems

P. ILLNER. SC.

The ecosystem, whatever its type, consists in part of populations of living organisms. Considered ecologically, a population is a group of organisms of the same species occupying a particular space. But more than this, a population is one unit through which energy flows and nutrients are cycled. It is also a self-regulating system that helps maintain stability within the ecosystem. A population has a birth rate and a death rate, a growth form, density, age structure, and a numerical dispersion in time and space. The study of the numbers of organisms and what determines their abundance and distribution is called *population ecology*. As one might imagine, even in its simplest form, the laboratory population, this field is extremely complex.

Density

The size of a population in relation to a definite unit of space is its *density*. Every 10 years the Census Bureau counts the number of people living in the United States. Wildlife and fishery biologists determine the number of fish and game in areas with which they are concerned. A forester cruises a timber stand to determine the volume and number of trees. These people may express their results as so many thousand per square mile, per acre, or per square meter. This is *crude density*. But populations do not occupy all the space within a unit because all is not suitable habitat. A biologist might estimate that 500,000 deer live in a 40,000 mi² area. The deer, however, will not utilize all the land within this area because of human habitation and land-use practices. A sample of soil may contain 2 million arthropods/m². These arthropods do not inhabit the entire substrate; they live only in the pore spaces in the soil. No matter how uniform a habitat may appear, it is in fact heterogeneous. There are microdifferences in moisture, light, temperature, exposure, to mention a few conditions. Each organism occupies only those areas that can adequately meet its requirements, resulting in a patchy distribution. To be accurate, the density of organisms should refer to this amount of area available as living space. This would be *ecological density*.

Ecological densities are rarely estimated be-cause it is difficult to determine what portion of a habitat represents living space. In Wisconsin, the densities of bobwhite quail were expressed as the number of birds per mile of hedgerow rather than as birds per acre (Kabat and Thompson, 1963). This is an ecologically more realistic estimate. Determining just what area is available to one particular kind of organism and just how suitable it is represents one of the most important problems in modern ecology.

The density of organisms on any area varies. It may change with the seasons, with weather conditions, with food supply, and with many other influences. There is, however, an upper limit to the density of a population within a unit area, imposed by size and trophic level. Generally the smaller the organism, the greater its abundance per unit area. A 100-acre forest will support a greater number of woodland mice than deer. A forest stand will contain many more trees 2 to 3 in. dbh (diameter breast height) than trees 12 to 14 in. dbh. The larger its size or the higher its position on the trophic levels, the less is the numerical density of an organism. For biologists, determining the density of organisms is a major problem in field studies of animal and plant populations (see Appendix B).

Patterns of population distribution

How a population is distributed within a given unit of space has considerable influence on ecological density as well as the estimation of density. The distribution of populations may be random, uniform, or clumped; the clumps in turn may be distributed uniformly or randomly (Fig. 10-1).

Individuals of a population are distributed at random if the position of each individual is independent of the others. Very rarely are members of a population distributed at random. Some invertebrates of the forest floor, particularly spiders, appear to be randomly distributed (Cole, 1946; Kuenzler, 1958) as do the clam *Mulinia lateralis* of the intertidal mud flats of the northeast coast of North America. Bergerud and Manuel (1969) found that winter aggregations of moose in central Newfoundland were randomly distributed, whereas the moose within the groups were not (Fig. 10-2).

When organisms are more evenly spaced

Populations: demographic units

than they would be if they were distributed by chance, they are uniformly distributed. Regular distribution results from intraspecific competition. Among animals territoriality might produce uniform distribution if environmental conditions are relatively homogenous. Uniform distribution happens among plants when competition is severe for crown and root space, as among forest trees, and for moisture, as among desert plants (Beals, 1968). Most common is clumped (sometimes called contagious) distribution. Clumping produces aggregations, the result of response by plants and animals to habitat differences, daily and seasonal weather and environmental changes (Fig. 10-3), reproductive patterns, and social behavior.

Among plants, aggregations often are influenced by the nature of the seed or other propagative means. Nonmobile seeds, such as those of oak and cedar, are clumped near the parent plant or where seeds are placed by animals. Mobile seeds are more widely distributed, but even here the number of individuals is greater near the parent plant (Fig. 10-4) or along natural barriers such as fencerows, where seeds are dropped by the deflection of the wind. Runners or vegetative propagation also produce clumping of plants as do differences in the germination of seeds, survival of seedlings, and competitive relationships.

Some animal aggregations are purely accidental, brought about by environmental conditions, as aquatic organisms washed ashore in concentrations along the beachline. Individuals may be drawn together by a common source of food, water, or shelter, or by favorable moisture conditions. Other aggregations reflect a varying degree of social tolerance. Moths attracted to light, earthworms congregated in a moist pasture field, barnacles clustered on a rock, all reflect a low level of social interaction. The individuals do not aid one another, yet they do not prevent other members of the same species from sharing the conditions that brought them together.

Aggregations on a higher social level reflect some degree of response by population members. Prairie chickens congregate on "leks" for communal courtship. Birds flock together for migration flights; with geese these flocks are organized with a leader. Elk band together in herds with some social organization, usually with a cow as the head (Altmann, 1952). Birds may congregate on feeding grounds away from territorial sites, yet show intolerance for

293

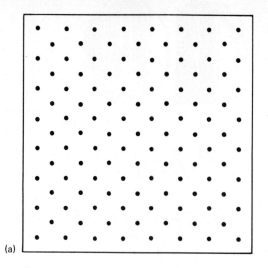

(a)

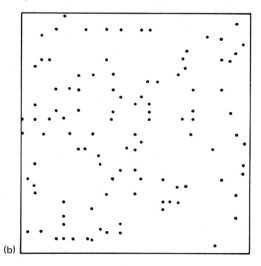

(b)

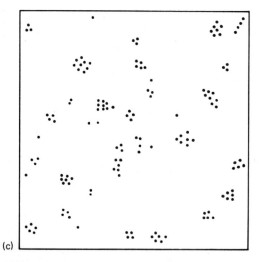

(c)

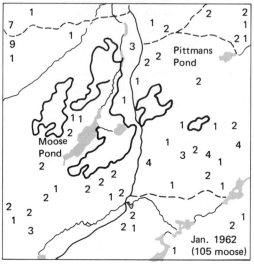

Jan. 1962
(105 moose)

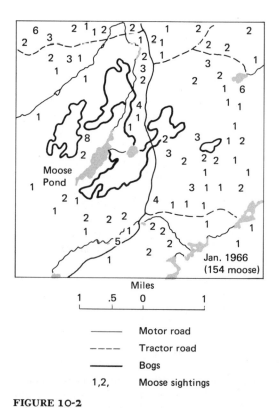

Jan. 1966
(154 moose)

Miles

1   .5   0   1

——————— Motor road

--------- Tractor road

━━━━━━━ Bogs

1,2,   Moose sightings

FIGURE 10-1
*Patterns of distribution. Note the difference
between* (a) *uniform,* (b) *random, and*
(c) *clumped.*

FIGURE 10-2
*Random aggregate distribution of moose in
the Lower Noel Paul River of Newfoundland.
(From Bergerud and Manuel, 1969.)*

FIGURE 10-3

Aggregate distribution of elephants in the Kibale Forest. (a) Distribution of elephants during the December–February dry season showing definite central orientation in swampy basins. The south block is generally deserted. (b) Distribution during the March–May rainy season showing a general north and east movement in the Central Block. Elephants moved into and were widely dispersed throughout the South Block. (c) Distribution in June–August dry season showing north movement out of South Block. Moderate concentrations in swampy areas. (d) Distribution during the September–November rainy season showing western orientation and wide dispersion from north to south. (From Wing and Buss, 1970.)

Number of droppings per chain (66') of transect:

$0 = \ .01 - \ .05$
$1 = \ .06 - 1.4$
$2 = 1.5 - \ 2.4$
$3 = 2.5 - \ 3.4$
$4 = 3.5 - \ 4.4$
$5 = 4.5 - \ 5.4$
$6 = 5.5 - \ 6.4$
$7 = 6.5 - \ 7.4$
$8 = 7.5 - \ 8.4$
$9 = 8.5 - \ 9.4$
$10 = 9.5 - 10.4$
$N = 0$

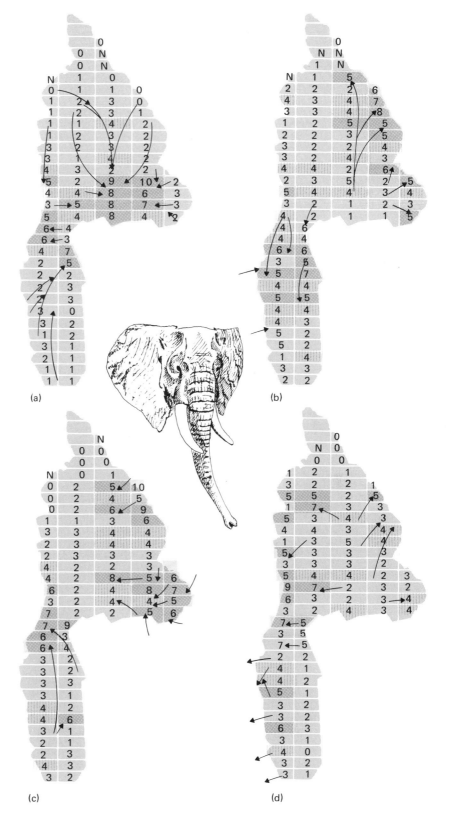

(a)

(b)

(c)

(d)

each other about the nest. Aggregations of the highest social structure are found among insect societies, such as ants and termites, where individual members are organized into social castes according to the work performed.

Aggregations result in competitive or cooperative situations. White-tailed deer gather in severe winter weather. Often too many are gathered in an area of limited food supply. After low browse is eliminated, only the larger, stronger deer are able to browse the higher limbs; the smaller, weaker deer die of starvation (Fig. 20-6). Fewer deer in the yard would have extended the food supply and permitted more to survive the winter. On the other hand, aggregation may improve group survival. An animal may have a greater chance of survival if it is one of a large organized band or flock rather than solitary or in a small group. Pronghorned antelope herds of more than 15 will stand ground when attacked by wolves and bunch into a defense unit, whereas smaller herds will stampede and scatter (Leopold, 1933). Schooling of fish apparently serves as a protection against predation (Brock and Riffenburgh, 1960). Any increase in the size of the school over and above the quantity a predator can consume on a single encounter reduces the frequency of predator–prey encounters. Fish appear to condition water with a growth-promoting substance secreted in the slime of skin glands. Fish raised in such water grow much better than those reared in water frequently changed. Green sunfish in groups of four grew faster than single individuals in one-fourth the volume of water (Allee et al., 1948). Covies of bobwhite quail mass in a tight circle at night during winter, some even perching on top of another. This maintains a high temperature on cold nights. Small covies and lone individuals succumb to low temperatures.

Among most aggregating organisms there is an optimum density, illustrated graphically in Fig. 10-5. Among some species the optimum population is the smallest possible; among others, it lies between too many and too few.

Clumping is also characteristic of human populations. Man is nonrandomly distributed over the earth, with the greatest aggregations in eastern Asia, the Indian subcontinent, and along the coastal regions of the other continents. Within a given region populations are clustered in urban areas. Within these urbanized areas, the distribution of populations follows several patterns. In one type there is a

sharp gradient in density from the core toward the periphery (Berry et al., 1963) (Fig. 10-6). or thinly settled margin. In another type the population is more generally dispersed, and a sharp gradient in density is lacking (Fig. 10-7). Many urban aggregations are characterized by a differential change in population distribution through time. There is an inverse relationship between population density and growth rates (Newling, 1966). The population tends to decline in the central urban area and to increase in the outer units. This decline in density in the central urban area and expansion on the periphery results in a series of waves of urban expansion (Fig. 10-8).

DETERMINING TYPES OF DISTRIBUTION

Accurately distinguishing between random, uniform, and clumped distribution is difficult. To detect randomness versus nonrandomness requires the judicious use of certain statistical tools such as the negative binomial, Poisson distribution, and the chi-square goodness-of-fit. It also demands careful sampling because the size, shape, and number of the sample quadrats can influence the determination. For example, if a nonrandomly distributed population is sampled by random quadrats, the distribution will appear random if very small or large quadrats are employed, but nonrandom if intermediate-size quadrats are used (Grieg-Smith, 1964). Populations characterized by a low unit density but with high density patches require a relative large number of quadrats of some intermediate size in order to detect departure from randomness. It is not a simple matter to detect the type of distribution characteristic of a given population under given environmental conditions.

Age ratio

Unless a population consists of seasonal breeders with nonoverlapping generations, such as annual plants and animals, it will be characterized by a certain age structure. The birth rates and death rates of a population are influenced considerably by the age distribution within it (see sections following on natality and mortality). Even the density of a population is without much meaning until age structure is considered.

Populations can be divided into three eco-

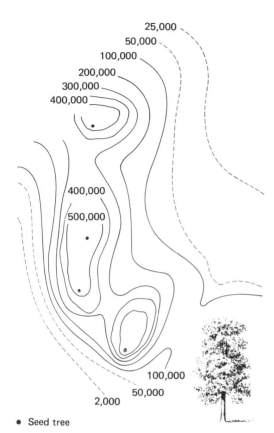

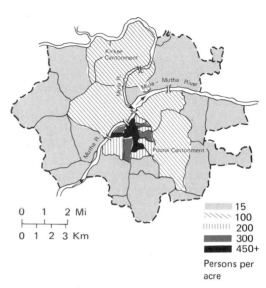

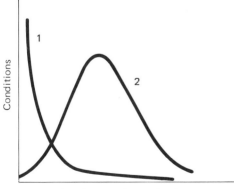

• Seed tree

FIGURE 10-4
Pattern of annual yellow poplar seedfall. Lines show equal seeding density. The seed is wind carried. (After L. G. Engle, 1960, Yellow Poplar Seedfall Pattern, Central States For. Expt. Stat. Note 143.)

FIGURE 10-6
The population density of Greater Poona, India, in 1961 showing a sharp density gradient on distribution of human populations. (From Brush, 1968. Reprinted from the Geographical Review, Vol. 58, 1968, copyrighted by the *American Geographical Society of New York.)*

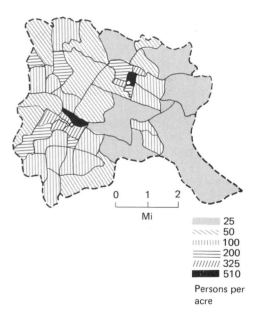

FIGURE 10-5
Optimum density for some organisms is the smallest possible; for others, somewhere between the two extremes.

FIGURE 10-7
The population density of Bangalore, India, in 1961. There is no sharp gradient in the distribution among individuals of the population. (From Brush, 1968. Reprinted from the Geographical Review, Vol. 58, 1968, copy-*righted by the American Geographical Society of New York.)*

logical periods—prereproductive, reproductive, and postreproductive. The ratio of these three ages may vary, although the density may remain the same. Theoretically at least, populations in a constant environment where emigration and immigration are minimal or nonexistent tend toward a stable age distribution. The proportion of individuals in different age classes becomes constant when mortality equals natality (see Table 10-13). Stable age distributions are approximated by any population in a steady environment regardless of whether the population is increasing, decreasing, or holding steady in size. For any given environmental condition, each population has its own stable age distribution. If the stable situation is disrupted by any cause, such as natural catastrophe, exploitation, or emigration, the age composition will tend to restore itself upon the return of normal conditions.

Because reproduction is limited to certain ages and mortality is most prominent in others, the ratio of age groups is important to the future of the population. These ratios are best analyzed when the data on age structure are presented in the form of an age pyramid (Fig. 10-9). The pyramid is a bar graph of the ratio of one age group by sex to another in the population (expressed in percentages). The shape of the pyramid tells much about the present and the future of the population (Fig. 10-10). A pyramid with a broad base and a narrow, pinched top indicates an inflated young-age group and a very small old-age group. Such a pyramid indicates a youthful population in which increasing numbers will enter the reproductive period. It foretells a rapid population growth. In a narrow pyramid the ratios of one age class to another are about the same. However the dependent age groups are rather large in relation to the reproductive (and working) population. Such a pyramid suggests an aging population with no significant population growth. A narrow-based pyramid describes a population with a large reproductive age group, but declining young. This reflects a decreased birth rate and predicts a small reproductive age class in the future. A pyramid with a marked indentation in the reproductive age period has a relatively large part of its population in the pre- and postreproductive age periods, which in human populations implies an added economic burden on the working population. Thus age distributions of human populations have important implications for the economic, so-

cial, and political life of the country or region in which the population resides. If the number of young are disproportionately larger than the number of reproductive adults, a greater input of money and energy is needed to support and educate the young. If too few young enter the population, that population ages with a growing entrenchment of conservative philosophies and declining community vigor.

The aging of a regional population is illustrated by a series of age pyramids for a five-county section of Appalachia. At one time isolated, the region was subject to a rapid population expansion through immigration while the coal industry was expanding. This industry collapsed economically and the laborers were replaced by machines to counteract the economic decline. The changing environment is reflected in the age pyramids (Fig. 10-11) describing the locale over a 40-year period. In 1930 the age pyramid was distorted by a large cohort in the 34-to-54 age group, the adult immigrant group. The dependency ratio, the number of those under 15 and over 65 in relation to the working group, was 43.2 percent. In 1940 the pyramid assumed the shape of a growing but not rapidly expanding population, and the dependency increased to 65 percent. By 1960 the age pyramid changed radically. The dependency ratio jumped to 85 percent, and there was a distinct pinch in the 15-to-34 year class, which reflects the emigration of the younger working-age groups to secure better opportunities. In 1970 the age pyramid, with its narrow base of young-age classes and nearly vertical sides to the older working-age classes, indicates an aging, declining population. The pyramid of 1970 reflects the heavy emigration of the reproductive age groups, and a loss of young indicated by the 1960 age pyramid. From 1960 to 1970 the dependency ratio dropped 17 percentage points to 67 percent.

Changes in age-class distribution reflect changes in production of young, their survival to maturity, and the influence of the ratios of the three classes to the population. Hunting and fishing an unexploited population immediately alter the life expectancy of members of the population. In general the young and the old age groups are more vulnerable than others. Fish are selected by the size of the gill net. Such exploitation causes age ratios to shift, because of changes in age-specific death rates (Fig. 10-12). This in turn affects the population's birth rate. Age ratios are influenced continually by management subsequently applied to the exploited population. In populations in which life expectancy for the oldest ages is reduced, a higher proportion falls into the reproductive class, increasing the birth rate. Conversely, if life expectancy is extended, a greater proportion of the population falls into the postreproductive age, reducing the birth rate. This is true both of natural and human populations.

Examples in human populations have been worked out by Keyfitz and Flieger (1971). They compare the percentage of people under 15 in a population having the same death rate as the United States had in 1967 and the same birth rate as Venezuela had in 1965 with the actual percentage of the population under 15 in the United States in 1967. The percentage under 15 years old in the United States in 1967 was 24.5 and in the hypothetical population 48.5. As in Venezuela, the effect of the high birth rate is to raise the proportion of those under 15 by 24 percentage points. Conversely, if one compared the percentage of a population under 15 in the United States with a hypothetical population having U.S. birth rates and Venezuela's death rate, the percentages are 24.5 for the United States and 23.9 for the hypothetical population, a 0.6 percentage-point difference. The difference between Venezuela's 47.7 percent under 15 to the United States' 24.5 percent is due to the 24-percentage-point difference in fertility and only 0.6 percent to mortality.

Basically a population can do three things: increase, decrease, or remain stable. The ratio of young to adults in a relatively stable population is approximately 2 : 1 in most populations of game animals. A normally increasing population should have an increasing number of young; a decreasing population a decreasing number of young. Within this framework there are a number of variations. A population may be decreasing, yet show an increasing percentage of young. Or a population might be stable with a decreasing percentage of young. By combining information on population density, age ratios, and reproduction, a biologist can correlate changes in population structure with habitat changes and ecological and human influences.

Age structure is rarely employed in the study of plants because in many instances plants do not lend themselves to that type of analysis. Information on age structure is most widely

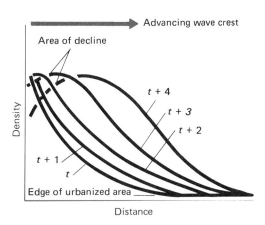

FIGURE 10-8
Population density as a function of population expansion. This diagram of waves of human population expansion in urbanized areas is typical of the expansion of any animal population with its advancing wave-line crest and its area of decline. (From Mayer, 1969. Reproduced by permission from the Association of American Geographers Commission on College Geography, Resource Paper #7, "Spatial Expression of Urban Growth.")

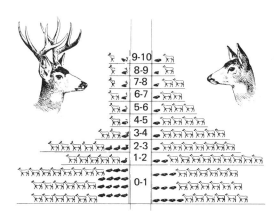

FIGURE 10-9
Age pyramid of a black-tailed deer population in California chaparral. This pyramid is broken down by sex, showing sex ratio by age classes. The black skulls indicate mortality in each group. (Redrawn from Taber and Dasmann, 1958, California Fish and Game Bulletin No. 8.)

299

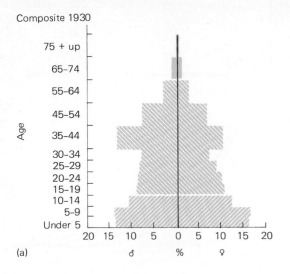

Composite 1930

(a)

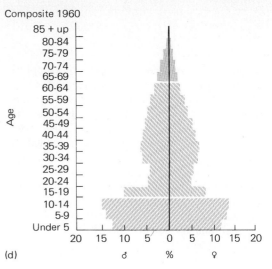

Composite 1960

(d)

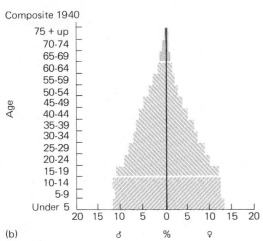

Composite 1940

(b)

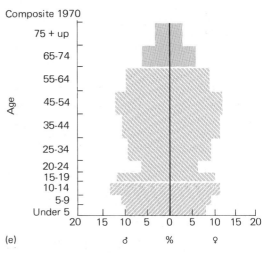

Composite 1970

(e)

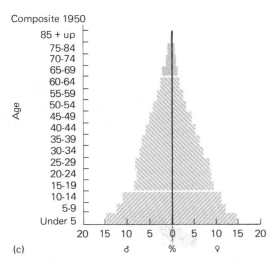

Composite 1950

(c)

FIGURE 10-11

A series of population pyramids for a coal-mining region in Appalachia from 1930 to 1970. The pyramid for 1930, (a), reflects the immigration of people into the mining region. A heavy base of young suggests a future expansion of the population. The pyramid for 1940, (b), is a "normal" one but suggests that a population increase may come about in the future. The population expansion as predicted has been partly achieved as the profile of the pyramid for 1950, (c), becomes slightly pinched at the top. The 1960 pyramid, (d), shows a major change. The middle has become pinched. The pyramid reflects the heavy emigration of the younger elements of the reproductive age group. This emigration results in a high ratio of dependent young and old to the economically active. It also reflects a declining population. The extent of that decline is indicated in the age pyramid for 1970, (e). The population is nearly equally distributed over all age groups. The nearly vertical pyramid is typical of an aging and declining population. (Pyramids prepared by the author from U.S. Bureau of Census data for selected counties in southern West Virginia.)

employed in forestry. Forest stands are divided into two general classes, even-aged stands and uneven-aged stands. Even-aged forests consist of trees all of the same age class and come about by clear-cutting, fire, or disease and insect attacks. Uneven-aged stands consist of trees of a variety of ages. Analysis of most stands, however, is based on diameter classes rather than age because basal area and volume, the parameters of greatest interest to foresters, are functions of diameter, not age. There is little correlation between age and diameter, and the largest trees in the stand are not necessarily the oldest.

Kerster (1968) attempted to determine the age structure of a plant population. He worked up the age structure of the prairie forb, blazing star, which can be aged by collecting (destructively) all the plants from sample quadrats and counting the annual rings on the freshly cut surfaces of the corms. Some of these plants had rosettes only; others had rosettes bearing one or more flowering spikes. Thus the plants could be sorted into age categories and the reproductive members could be separated from the nonreproductive individuals.

The "age pyramids" developed from the data are informative and contrast two populations (Fig. 10-13). The maximum age attained was 35 years. The scarcity of plants over 20 years was mainly due to a hot prairie fire in the past and secondarily to dessication and insect attacks. There is a greater variation in size of adjacent-year classes than one finds in animal populations, due to variations in the germination of seeds and the survival of seedlings. Reproduction or flowering starts at about 9 years of age, and each "adult" reproduces rather regularly every other year. Average generation span is equal to the average age of the flowering plant. One population appeared to be senescent. The other showed recruitment of young, or resurgence.

Sex ratio

Among mammals the sex ratio at birth is often slightly in favor of males. By adulthood it may favor either male or female, depending upon species, mortality, and habitat. A study of muskrat populations in Wisconsin showed that the sex ratio is about equal at birth, but after only 3 weeks the ratio shifts to about 140 males to 100 females (Beer and Traux, 1950). There

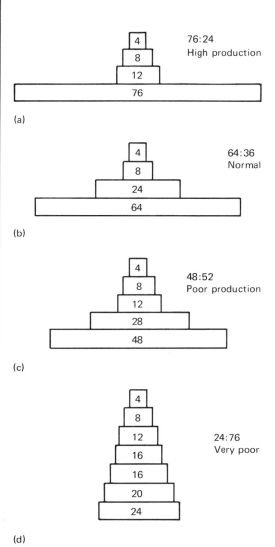

FIGURE 10-10
Theoretical age pyramids: (a) *a population with high production of young;* (b) *normal production;* (c) *low production;* (d) *very poor production. Note how the pyramids flatten out with high production of young. Pyramid* (d) *is typical of an aging population. These figures are based on the assumption that the life table is the same from* (a) *through* (d). (*After M. Alexander, 1958.*)

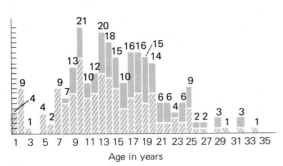

Age in years	Age in years
'65 '63 '61'59'57 '55 '53'51'49'47'45'43'41'39'37'35'33'31	'65 '63 '61 '59 '57 '55 '53 '51 '49 '47 '45 '43
(a) Year	(b) Year

is little or no change after this until the following breeding season. Then mortality from intraspecific strife brings the sex ratio back to equality again. Sex ratios of gray squirrels in West Virginia (Uhlig, 1955) and fox squirrels in Michigan (D. L. Allen, 1943) show a predominance of females, ranging from 51 to 60 percent of the population. Among white-tailed deer, the males outnumber the females at birth and maintain their predominance the first year. By adulthood the ratio shifts strongly in favor of females, as much as 42 males per 100 females (Dahlberg and Guettinger, 1956). This apparently reflects the selective hunting of bucks (see Fig. 10-9).

A similar pattern has been observed in elk (Flook, 1970). The sex ratio of fetuses from elk in the western Canadian national parks was 113 males to 100 females. The losses of males did not exceed those of females before 1.5 years of age. Between 1.5 and 2.5 years, the ratio of males to females dropped abruptly and continued to do so until the ratio remained at about 85 : 100, although in places the ratio dropped as low as 37 : 100. The greatest decline in the number of males occurred between the ages of 7 and 14. In contrast, the decline of the females was less rapid, the oldest being 21 years. The loss of males was attributed to the forced dispersal of subdominant males into unfavorable habitat, and to the loss of energy reserves due to reproductive activities (rut) prior to the winter, making them vulnerable to severe malnutrition in late winter and spring. The effect of this loss was more space and food

for females and young, sustaining a high rate of increase. Among humans, the males, too, exceed the females at birth, but as age increases the ratio swings in favor of the females. In 1965 the ratio of males to females based on a stable age distribution in the United States was age 0 to 4 years, 104 : 100; age 40 to 44, 100 : 100; age 60 to 64, 88 : 100; age 80 to 84, 54 : 100. The lower mortality of females was characteristic of both advanced and underdeveloped countries.

Males tend to be excessive among birds, particularly adult waterfowl (Bellrose et al., 1961). Fall and winter sex ratios of prairie chickens show a preponderance of males in adult groups (Ammann, 1957). Apparently there is some selective mortality of females of this species at an early age, which may be counterbalanced by a selective mortality of males during and prior to the breeding season.

The sex ratios of a population of some species may change with a change in population density. Snyder (1962) removed females from a population of woodchucks and then compared the sex ratio of the progeny to that of a control group in which the sex ratio was not disturbed. In the experimental area the sex ratio of captured young males to females was 100 to 222; the ratio on the control areas was 100 young males to 97 females. The sex ratio of the fetuses in the experimental area was also heavily in favor of the females, in contrast to the control, where male ratio was higher. Increasing populations of gray squirrels and fox squirrels are associated with a predominance of females (D. L. Allen, 1943; Uhlig, 1955). Low populations of cottontail rabbits are associated with large populations of males (Linduska, 1947). Among polygamous species, such as the ring-necked pheasant and white-tailed deer, a large proportion of males can be killed off without affecting the population. This is considered in harvesting deer and pheasants, where the males can be distinguished from the females. But a continual harvest of males from such populations may permit females to increase to the point that the numbers grow beyond the carrying capacity of the range. In such cases it is necessary to harvest both sexes until a favorable sex ratio is established.

In any population of sexually reproducing animals in which a group of all males contribute as much to the ancestry of future generations as a group of all females, the sex ratio is the result of natural selection (R. A. Fisher, 1929; MacArthur, 1961; Kolman, 1960). To rear young to independence, the parents must expend as much energy, time, and nutrients on the male fraction as the female fraction of the population. Other things being equal, the sex that is in short supply has the greater value per individual and will tend to be favored by selection pressures. For example, if males in a population suffer heavy mortality during the period of parental care, they are more expensive per individual to produce because an expenditure is involved not only for those that survive but also for those that die prematurely. Thus there is a greater expenditure for each male reared, but less for each male born than for females at corresponding stages. Natural selection would tend to favor more males at birth. Because of a higher death rate, they would become less numerous, and by the end of the period of parental care, the sex ratio would be the result of differential mortality. In the end approximately the same expenditures would be involved for both the male and the female groups, but this equal expenditure does not imply equal sex ratios.

Mortality

One great influence on population size is mortality. Like natality, mortality varies with age groups. In fact mortality is distributed according to age even more dramatically than births.

Mortality, which begins even in the uterus and the egg, can be expressed either as the

FIGURE 10-13 [OPPOSITE]
An age pyramid for the plant Liatris aspera. *Numbers atop bars indicate numbers in year classes. Shaded portions of bars indicate plants not flowering in 1965, the year of collection. Unshaded portions of bars indicate flowering specimens. The relatively low numbers of young plants suggest recruitment is not adequate to maintain the population.* (b) *An age pyramid for another population of* Liatris aspera. *The low number of plants 7 to 11 years old suggests population senescence, followed recently by a resurgence inferred from large numbers in the seedling through 6-year-old classes.* (From Kerster, 1968.)

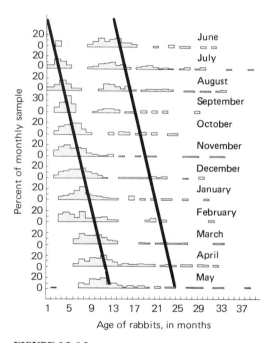

FIGURE 10-12
Monthly changes in the age composition of a cottontail rabbit population in east central Illinois. The black lines follow one age group. Note how rapidly the proportion of an age class declines as that age group grows older. (From Lord, 1961c; courtesy Illinois Natural History Survey.)

TABLE 10-1
Human life table, United States, 1966

		Male					Female		
Age	l_x	d_x	q_x	e_x		l_x	d_x	q_x	e_x
0	1000	26	.02576	66.75		1000	20	.01997	73.86
1	974	4	.00405	67.51		980	3	.00338	74.36
5	970	2	.00253	63.78		977	2	.00183	70.61
10	968	3	.00260	58.93		975	1	.00153	65.74
15	965	7	.00730	54.08		973	3	.00295	60.83
20	958	10	.00992	49.46		971	3	.00357	56.01
25	949	9	.00938	44.93		967	4	.00440	51.20
30	940	10	.01088	40.33		963	6	.00632	46.41
35	930	14	.01520	35.74		957	9	.00911	41.69
40	916	21	.02345	31.26		948	13	.01391	37.05
45	894	33	.03716	26.94		935	20	.02104	32.53
50	861	51	.05956	22.88		915	28	.03082	28.18
55	810	75	.09216	19.16		887	40	.04501	23.99
60	735	97	.13260	15.84		847	56	.06601	19.99
65	637	124	.19505	12.86		791	84	.10673	16.22
70	513	137	.26772	10.35		707	114	.16147	12.84
75	376	132	.35064	8.22		593	146	.24644	9.81
80	244	115	.47188	6.33		447	170	.38102	7.17
85+	129	129	1.00000	4.75		276	276	1.00000	5.05

Source: Data from Keyfitz and Flieger, 1971.

TABLE 10-2
Human life table, Mauritius, 1966

		Male					Female		
Age	l_x	d_x	q_x	e_x		l_x	d_x	q_x	e_x
0	1000	69	.06942	59.48		1000	58	.05848	63.71
1	931	26	.02824	62.89		942	27	.02917	66.64
5	904	8	.00929	60.69		914	7	.00787	64.61
10	896	5	.00564	56.23		907	3	.00361	60.10
15	891	6	.00682	51.54		904	7	.00734	55.31
20	885	6	.00646	46.87		897	13	.01416	50.70
25	879	7	.00815	42.16		884	13	.01512	46.39
30	872	9	.01067	37.49		871	15	.01724	42.06
35	863	19	.02151	32.86		856	17	.02026	37.75
40	844	22	.02646	28.53		839	16	.01898	33.48
45	822	33	.04051	24.23		823	24	.02952	29.08
50	788	59	.07459	20.14		798	30	.03792	24.89
55	730	84	.11548	16.54		768	50	.06513	20.76
60	645	114	.17648	13.36		718	68	.09432	17.02
65	531	129	.24349	10.67		650	97	.14928	13.52
70	402	138	.34330	8.29		553	115	.20811	10.44
75	264	121	.45652	6.32		438	141	.32172	7.50
80	143	72	.50084	4.62		297	143	.48035	4.85
85+	72	72	1.00000	1.89		154	154	1.00000	2.01

Source: Data from Keyfitz and Flieger, 1971.

probability of dying or as a *death rate* (D. E. Davis, 1960). The death rate is the number of deaths during a given time interval divided by the average population, and is an instantaneous rate. The probability of dying is the number that died during a given time interval divided by the initial population, the number alive at the beginning of the period. The complement of the latter is the probability of living, the number of survivors divided by the initial number in the group. The number of survivors is more important to the population than the number dying, so mortality is better expressed as survival, or as *life expectancy*, the average number of years that members of a population have left to live.

To obtain a clear and systematic picture of mortality and survival, a life table can be constructed. The life table, first developed by students of human populations and widely used by actuaries of life insurance companies, is simply an account book of deaths (Tables 10-1 and 10-2). It consists of a series of columns, headed by standard notations, each of which describes mortality relations within a population when age is considered. It always begins with a certain size group, usually 1,000, at birth or hatching. The columns include x, the units of age; l_x, the number of animals in a cohort that survive to age x, or the probability at birth of an individual surviving to age x; d_x, the fraction of a cohort that dies during the age interval $x, x + 1$, or if expressed as a decimal of 1, the probability at birth of dying during the interval $x, x + 1$. If l_x and d_x are converted to proportions, that is, if the number of animals that died during the interval $x, x + 1$, is divided by the number of animals alive at the beginning of age x, the result is an age-specific death rate, q_x. Two additional columns are L_x, the average years lived, and T_x, total years lived. These two columns are used to calculate e_x, the life expectancy at the end of each interval.

At one time data for life tables could be obtained only for laboratory animals and humans. As census methods and age-determination techniques became more and more refined, sufficient data necessary for at least an approximate life table could be acquired for other species (Deevey, 1947; Hickey, 1952; Caughley, 1966). It is difficult to obtain information on mortality and survival of wild animals. Mortality can be determined by finding the ages at death of a large number of animals born at the same time (Table 10-3). This procedure could involve marking or banding a considerable number of animals. Such a method provides information for the d_x column. One can record the ages at death of animals marked at birth, but not necessarily born during the same season or year. Data from several years and several cohorts are pooled to provide the information for the d_x column. Another approach is to determine the age at death of a representative sample of carcasses of the species concerned. Age is obtained by examining wear and replacement of teeth in deer, from growth rings in the cementum of teeth of ungulates and carnivores, from annual rings in the horns of

TABLE 10-3
Dynamic-composite life table for squirrels marked as nestlings, 1956–1964 (sexes combined)

Year marked	Nestlings marked	Year of return							
		1957	1958	1959	1960	1961	1962	1963	1964
1956	40	8	4	3	2	0	0	0	0
1957	138		60	30	28	13	9	4	3
1958	229			61	26	12	10	7	3
1959	193				58	26	19	12	9
1960	162					19	13	8	6
1961	99						4	1	1
1962	82							18	6
1963	80								25

Age (x)	Total known alive	Maximum available for recapture	Known live per 1,000 available
0–1	1,023	1,023	1,000.0
1–2	253	1,023	247.3
2–3	106	943	112.4
3–4	71	861	82.5
4–5	43	762	56.4
5–6	25	600	41.7
6–7	7	407	17.2
7–8	3	178	16.9

Age (x)	Known survival (l_x)	Apparent mortality (d_x)	Proportional mortality (q_x)	Average years lived (L_x)	Total years lived (T_x)	Life expectancy (e_x)
0–1	1,000.0	752.7	0.753	538.9[a]	989.6	0.99
1–2	247.3	134.9	0.545	179.9[b]	450.7	1.82
2–3	112.4	29.9	0.266	97.4	270.8	2.41
3–4	82.5	26.1	0.316	69.5	173.4	2.10
4–5	56.4	14.7	0.261	49.0	103.9	1.84
5–6	41.7	24.5	0.588	29.4	54.9	1.32
6–7	17.2	0.3	0.017	17.1	25.5	1.48
7–8	16.9	16.9	1.000	8.4	8.4	0.50

[a] $n = \dfrac{1,000\,(l - e^i)}{i}$; $\overline{q_x} = 0.635$.

[b] $n = \dfrac{l_x + (l_{x+1})}{2}$.

Source: Barkalow et al., 1970.

mountain sheep, and from the lens of the eye in rabbits. Such information also goes into the d_x column. Recording the ages at death of a sample of a population wiped out by some catastrophe could provide data for an l_x series. Life tables derived from the aging of animals taken during a hunting season provides information for the l_x column because the sample came from a living population. But the data often are biased in favor of older age classes, especially if the data are collected between breeding seasons.

There are two kinds of life tables. One is the *cohort* or *dynamic* kind, which records the fate of a group of animals all born at the same time. Because biologists often find it difficult to obtain a sufficiently large cohort to follow through a period of time, some use a more practical (but questionable) approach. They mark a number of new-born animals over several years, and after following the fate of each year class, treat all of the marked animals as a single cohort. This was the approach taken in Table 10-3 in the construction of the life table for gray squirrel.

Another example of a dynamic life table is the one Lowe (1969) constructed for a cohort of red deer 1 year of age in 1957. He censused the population of red deer on the Isle of Rhum, Scotland, in 1957. Between 1957 and 1966 he recorded the age, sex, and number of deer that had been shot or died. He found that these deer accounted for nearly 92 percent of the animals alive in 1957, and he knew the number of deer on Rhum in 1957, so he was able to determine the age structure of the population at that time. In addition, because he knew the number of red deer 1 year of age in the population during 1957, he had a single cohort that he could follow through the 9 years of his study. The result was the dynamic life table presented in Table 10-4.

The other type is the *time-specific* life table, in which the mortality of each age class is recorded over a given period of time, usually a year. This table is constructed from a sample of animals of each age class (taken in proportion to their numbers) in a population and the age at death. This type of life table assumes that the birth and death rates are constant and that the population is stationary. An example of a time-specific life table is another one for the red deer. In this instance the table (Table 10-5) was constructed from data on the age structure of the population in the year 1957. The l_x schedule shows the probability of surviving from year 1 rather than from birth. Each age class represents a composite of many cohorts. The differences in the number of animals in adjacent age groups is due to differences in cumulative mortality. The time-specific life table can represent no more than a crude generalization of the probable age structure of a population at the time the data were collected.

Life tables as constructed portray a *stationary age distribution*. This distribution results when a population does not change in size and its age structure is constant with time. This concept, which was developed in human demography, is most applicable to species that have no seasonally restricted period of birth. But it can be applied to a population of species of seasonal breeders if one defines stationary age distribution as that which does not vary either in numbers or age structure at a particular time of year. A stationary age distribution for this kind of population is the distribution of ages at a given time of year. For restricted seasonal breeders, age structure can be sampled over that period and the number of live births determined by observing the number of females pregnant and caring for young. If the age distribution is sampled half-way between breeding seasons, as is often done with deer and other game animals, neither the l_x or the d_x series can be accurately established, but the

TABLE 10-4

Dynamic life table for red deer on Isle of Rhum, 1957

x (age, years)	l_x (survivors at beginning of age class x)	d_x (deaths)	e_x (further expectation of life, years)	1000 q_x (mortality rate/1000)
		STAGS		
1	1000	84	4.76	84.0
2	916	19	4.15	20.7
3	897	0	3.25	0
4	897	150	2.23	167.2
5	747	321	1.58	430.0
6	426	218	1.39	512.0
7	208	58	1.31	278.8
8	150	130	0.63	866.5
9	20	20	0.5	1000
		HINDS		
1	1000	0	4.35	0
2	1000	61	3.35	61.0
3	939	185	2.53	197.0
4	754	249	2.03	330.2
5	505	200	1.79	396.0
6	305	119	1.63	390.1
7	186	54	1.35	290.3
8	132	107	0.70	810.5
9	25	25	0.5	1000

Source: V. P. W. Lowe, 1969.

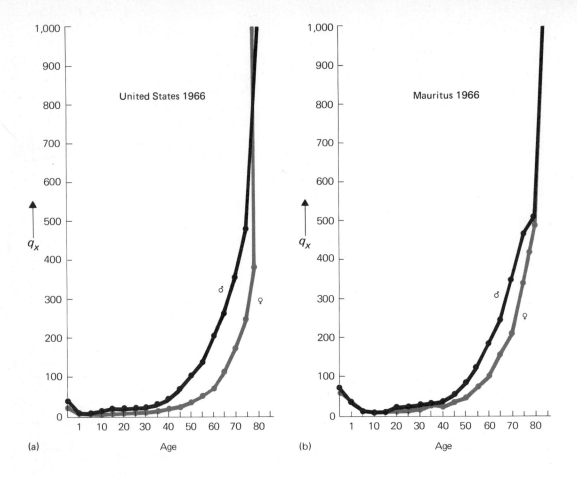

(a)   Age   (b)   Age

q_x values can be determined for the age intervals $x + 0.5$, $x + 1.5$ (Caughley, 1966). The stationary age distribution should not be confused with stable age distribution (see next section). Stable age distribution results when a population increases at a constant rate while its survivorship and fecundity rates are constant.

From the life tables data curves and survivorship curves can be plotted. Mortality curves plot mortality (q_x) against age. In general these exhibit a juvenile phase, in which the rate of mortality is high, and a postjuvenile phase, during which the rate decreases to a minimum and then rises with increasing age (Figs. 10-14, 10-15). For populations of mammals a roughly J-shaped curve results (Fig. 10-16).

Because q_x is a ratio of the number dying during an age interval to the number alive at the beginning of the period (or surviving the previous age period), it is independent of the size of the younger age classes. Thus it is freer of the biases inherent in the l_x column and survivorship curves, which plot l_x against age. Most life tables of wild populations are subject to bias because the 1-year age class is not ade-

quately represented. This error in the frequency of the 1-year age class distorts each succeeding l_x and d_x value. If the first values are inaccurate, all succeeding ones are inaccurate. But if the first values of q_x are wrong, the error does not affect the other values. For this reason mortality curves are more informative than survivorship curves, which indicate the rate of mortality indirectly by the slope of the line.

Survivorship curves can be plotted two ways. One is to plot the data from the survivorship or l_x column against time with the time interval on the horizontal coordinate and survivorship on the vertical coordinate (Fig. 10-17). The other method is to plot the data with the time interval scaled as percent deviation from the mean length of life (Fig. 10-18). The validity of the survivorship curves depends upon the validity of the life table and the l_x column. Because life tables and thus survivorship curves depict the nature of a population at a specific place during a specific time rather than a standard population, survivorship curves are useful to compare the population of one area, time, or sex with the population of another (Fig. 10-19 and Tables 10-6 and 10-7).

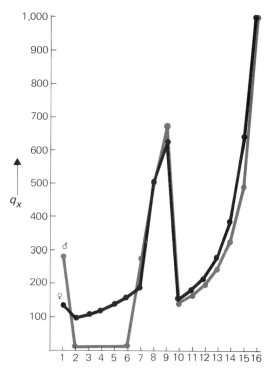

FIGURE 10-14 [OPPOSITE]
*Mortality curves for human populations in the
United States* (a) *and Mauritius* (b) *in 1966.
Although the two countries are different
demographically with Mauritius having a young,
rapidly growing population, the mortality curves
follow a typical J-shape characteristic of mammals.*

TABLE 10-5
*Time-specific life tables for red deer on
Isle of Rhum, 1957*

x (age, years)	l_x (survivors at beginning of age class x)	d_x (deaths)	e_x (further expectation of life, years)	100 q_x (mortality rate/1000)
		STAGS		
1	1000	282	5.81	282.0
2	718	7	6.89	9.8
3	711	7	5.95	9.8
4	704	7	5.01	9.9
5	697	7	4.05	10.0
6	690	7	3.09	10.1
7	684	182	2.11	266.0
8	502	253	1.70	504.0
9	249	157	1.91	630.6
10	92	14	3.31	152.1
11	78	14	2.81	179.4
12	64	14	2.31	218.7
13	50	14	1.82	279.9
14	36	14	1.33	388.9
15	22	14	0.86	636.3
16	8	8	0.5	1000
		HINDS		
1	1000	137	5.19	137.0
2	863	85	4.94	97.3
3	778	84	4.42	107.8
4	694	84	3.89	120.8
5	610	84	3.36	137.4
6	526	84	2.82	159.3
7	442	85	2.26	189.5
8	357	176	1.67	501.6
9	181	122	1.82	672.7
10	59	8	3.54	141.2
11	51	9	3.0	164.6
12	42	8	2.55	197.5
13	34	9	2.03	246.8
14	25	8	1.56	328.8
15	17	8	1.06	492.4
16	9	9	0.5	1000

Source: V. P. W. Lowe, 1969.

FIGURE 10-15
*Mortality curves for red deer, male and female,
on the Isle of Rhum. Note the sharp rise in
mortality between the sixth and tenth years
that breaks the J-shaped curve. Compare this
with the survivorship curve for the same
population, Fig. 10-19. (Data from Lowe, 1969.)*

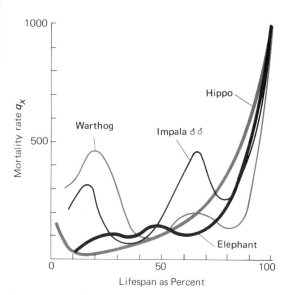

FIGURE 10-16
*Mortality curves for some selected African
herbivores. Only the hippo and the elephant
follow a good J-shaped mortality curve. The
impala and the wart hog have two periods of
high mortality. (From Spinage, 1972. © The
Ecological Society of America.)*

TABLE 10-6
Life table typical of dense gypsy moth populations in Glenville, N.Y.

x	l_x	$d_x f$	d_x	$100q_x$
Age interval	Number alive at beginning of x	Factor responsible for d_x	Number dying during x	d_x as percent of l_x
Eggs	250	Parasites	50.0	20
		Other	37.5	15
		Total	87.5	35
Instars I–III	162.5	Dispersion, etc.	113.8	70
Instars IV–VI	48.7	Parasites	2.4	5
		Disease	29.2	60
		Other	12.2	25
		Total	43.8	90
Prepupae	4.9	Desiccation, etc.	0.5	10
Pupae	4.4	Parasites	1.1	25
		Disease	0.7	15
		Calosoma larvae	0.9	20
		Other	0.4	10
		Total	3.1	70
Adults	1.3	Sex (S : R = 30 : 70)	0.9	70
Adult ♀♀	0.4	—	—	—
Generation	—	—	249.6	99.84

Source: R. W. Campbell, 1969.

TABLE 10-7
Life table typical of sparse gypsy moth populations in Glenville, N.Y.

x	l_x	$d_x f$	d_x	$100q_x$
Age Interval	Number alive at beginning of x	Factor responsible for d_x	Number dying during x	d_x as percent of l_x
Eggs	450	Parasites	67.5	15
		Other	67.5	15
		Total	135.0	30
Instars I–III	315	Dispersion, etc.	157.5	50
Instars IV–VI	157.5	Parasites	7.9	5
		Disease	7.9	5
		Other	118.1	75
		Total	133.9	85
Pre-pupae	23.6	Desiccation, etc.	0.7	3
Pupae	22.9	Vertebrate predators	4.6	20
		Other	2.3	10
		Total	6.9	30
Adults	16.0	Sex (S : R = 65 : 35)	5.6	35
Adult ♀♀	10.4	—	—	—
Generation	—	—	439.6	97.69

Source: R. W. Campbell, 1969.

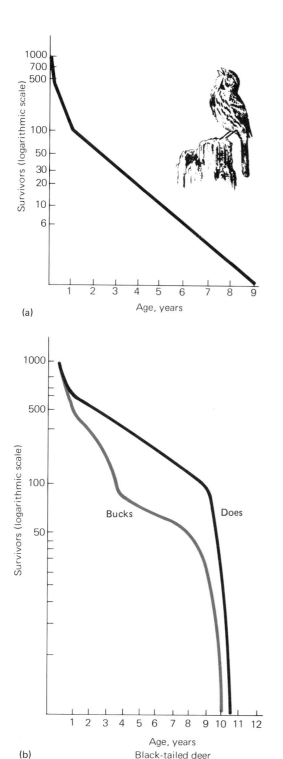

(a)

(b)

Age, years
Black-tailed deer

FIGURE 10-17
(a) *Survivorship curves for the song sparrow
(from Johnston, 1956) and* (b) *the black-tailed
deer* (*Taber and Dasmann, 1957*). *The difference
between the survivorship curves of bucks and
does reflects the hunting seasons for bucks only.*

311

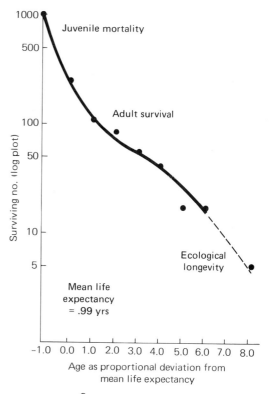

FIGURE 10-18
*The survivorship curve for 1,023 known-age
gray squirrels.* (*From Barkalow, 1970.*)

Survivorship curves are often classified into three hypothetical types (Fig. 10-20). If mortality rates are constant at all ages, the survivorship curve will be a diagonal line (if plotted on semilog paper). Such a curve is characteristic of hydra and adult stages of many birds and rodents. If mortality rates are extremely high in early life, as in oysters, fish, and many invertebrates, the survivorship curve is concave. When individuals tend to live out their physiological life span and when there is a high degree of survival throughout life and a heavy mortality at the end of a species' life span, the curve is strongly convex. Such a curve is typical of rotifers and of man, mountain sheep, and other mammals.

These are conceptual models; actual survivorship curves do not necessarily resemble them. But they do serve as a model for comparison with actual survivorship curves. Most survivorship curves are intermediate between two of the models. Consider, for example, the series of survivorship curves of the population of Sweden from the middle of the eighteenth century to the present time (Fig. 10-21). During the early eighteenth century the mortality of young was high and life expectancy relatively low. The survivorship curve fell somewhere between A and B (Fig. 10-20). With advances in medicine and modifications of the environment, the survivorship curve reflected a shift in mean life expectancy toward maximum and more nearly resembled A.

Mortality of plants has not received the same kind of conceptual treatment as mortality of animals. Plants do not lend themselves easily to a life table and survivorship approach to mortality. Survivorship in plants can be expressed as percentage germination of seeds, and mortality and survival of seedlings. Some studies consider the percent mortality of trees and shrubs caused by drought, disease, and outbreaks of insects.

The closest approach to life tables for plants is the yield table (Tables 10-8 and 10-9) developed in commercial forestry. Like the life table, the yield table considers age classes and number of trees in each class, and has additional columns to give diameter, basal area, and volume. Yield tables indicate mortality of trees by the shrinkage of numbers. But as numbers decline through competition or removal, basal area and biomass increase. Thus mortality does not necessarily reflect a declining population but rather a maturing one. Like life tables,

yield tables are not constant for a species but are constructed for different site classes. That takes into consideration the different environmental conditions under which the species grows.

Natality

The greatest influence on population increase is usually natality, the production of new individuals in the population. Natality often is described in two ways, as *maximum* or *physiological natality* and *realized natality*. Physiological natality represents the maximum possible births under ideal environmental conditions, the biological limit. This is rarely achieved in wild populations, so it is of little value to the field biologist. It is somewhat theoretical but nevertheless is a useful yardstick against which to compare realized natality. Realized natality is the amount of successful reproduction that actually occurs over a period of time. It reflects the type of breeding season —continuous, discontinuous, or strongly seasonal; the number of litters or broods a year; the length of gestation or incubation; and so on. It is influenced by environmental conditions, by nutrition, and the density of the population.

Maximum natality varies with the species and even within the species. The bluegill, a fish common in lakes and ponds, produces 4,000 to 61,000 fry from a single nest. The pelagic codfish produce up to 61 million eggs/female. The ground-nesting bobwhite quail lays about 15 eggs/clutch; the tree-nesting mourning dove lays only 2 eggs/clutch. The white-tailed deer has 1 or at the most twin fawns once a year; the white-footed mouse has a litter of 4 to 6 young 4 times a year. Lack (1947) suggests that the birth rates of animals are adjusted through natural selection (see Chapter 14) to the maximum rates at which the offspring are destined to reach adulthood. For some species this is the largest number of young the parents can rear successfully. For others this may be the maximum number the parents can feed, or brood and protect. Some animals, such as many fish and amphibians, simply lay their eggs in relatively unprotected masses. With these, the number of eggs laid probably represents that which the animal is physiologically able to lay. Here, too, selection would favor those that laid sufficient eggs to assure some survivors.

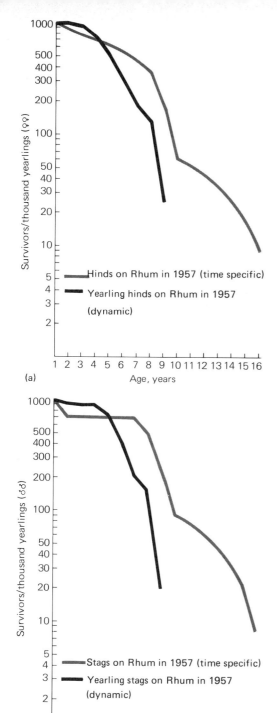

(a)

(b)

FIGURE 10-19
Survivorship curves for the red deer on the Isle of Rhum, as constructed from both time-specific and dynamic life tables. See text for discussion. (a) Survivorship curves for hinds; (b) survivorship curves for stags. Contrast the survivorship curves with those of the black-tailed deer. Both hinds and stags are harvested, which holds the survivorship curves of the two sexes at about the same level. (From Lowe, 1969.)

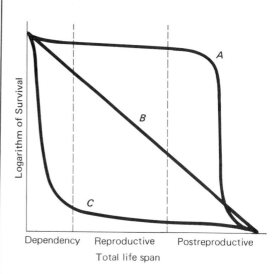

FIGURE 10-20
The three basic types of survivorship curves. The vertical scale may be graduated arithmetically or logarithmically. If graduated logarithmically, the slope of the lines will show the following rates of change: (A) curve for animals living out the full physiological life span of the species; (B) curve for animals in which the rate of mortality is fairly constant at all age levels— a more or less uniform percentage decrease in the number that survive; (C) curve for animals with high mortality early in life. (After M. Alexander, 1958.)

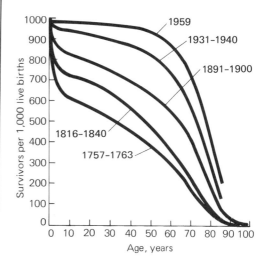

FIGURE 10-21
Survivorship curves for the population of Sweden over several centuries. Note that as health conditions improved and the standard of living increased over the years, the survivorship curves began to approach the physiological life span, changing from concave to convex. (From Clark, 1967, Population Growth and Land Use. By permission of Macmillan, London and Blasingstoke, and St. Martin's Press, Inc.)

TABLE 10-8
Yield tables for Douglas fir on fully stocked acre, total stand

Age (years)	Site index 100			Site index 200		
	Trees (per acre number)	Av. DBH (in.)	Basal area (ft²)	Trees (per acre number)	Av. DBH (in.)	Basal area (ft²)
20	4,150	1.8	76	571	5.7	101
30	1,800	3.4	114	350	9.0	154
40	1,090	4.9	143	240	12.2	195
50	764	6.3	165	176	15.3	224
60	580	7.6	182	138	18.2	248
70	468	8.8	197	113	20.9	268
80	394	9.9	210	97	23.3	285
90	347	10.8	220	84	25.6	299
100	311	11.6	229	75	27.6	312
110	281	12.4	238	69	29.4	323
120	259	13.2	245	63	31.1	332
130	240	13.9	251	59	32.7	341
140	224	14.5	257	55	34.3	350
150	211	15.1	263	51	35.8	357
160	200	15.7	268	48	37.2	364

Source: McArdle et al., 1949.

TABLE 10-9
Life table of an eastern white pine plantation, simulated value for a 1-acre plot

x	l_x	d_xF	d_x	$100q_x$	$100r_x$
Age interval (years)	No. of trees alive at start of x	"Mortality" factor	No. of trees "dying" during x	d_x as percent of l_x	d_x as percent of initial number
0–1	1200	Improper planting	80	6.7	6.7
		Mice	30	2.5	2.5
		Subtotal	110	9.2	9.2
1–3	1090	Pales weevil	40	3.7	3.3
		White-pine weevil	20	1.8	1.7
		Unknown	20	1.8	1.7
		Subtotal	80	7.3	6.7
3–20	1010	White-pine weevil	400	39.6	33.3
		Pine leaf aphid	50	5.0	4.2
		White-pine blister rust	50	5.0	4.2
		Unknown	75	7.4	6.2
		Subtotal	575	57.0	47.9
20–40	435	White-pine weevil	40	9.2	3.3
		Ips beetles	35	8.1	2.9
		White-pine blister rust	30	6.9	2.5
		Root rot	20	4.6	1.7
		Suppression	25	5.7	2.1
		Unknown	25	5.7	2.1
		Partial cutting(s)	60[a]	13.8	5.0
		Subtotal	235	54.0	19.6
40–80	200	Gypsy moth	15	7.5	1.2
		Ips beetles	10	5.0	0.8
		White-pine blister rust	10	5.0	0.8
		Chlorotic dwarf	8	4.0	0.7
		Lightning	2	1.0	0.2
		Suppression	8	4.0	0.7
		Unknown	4	2.0	0.3
		Partial cutting(s)	18[a]	9.0	1.5
		Subtotal	75	37.5	6.2
80	125	Total	1075	—	89.6

Source: W. E. Waters, 1969. [a] Number of trees cut in addition to those killed or deformed by other agents.

A similar situation is found in plants. Seed production in annual plants such as dandelions, and perennial plants, such as cattail, is enormous. Because these seeds are dispersed by the wind, large numbers are needed to ensure some survival. In such forest trees as oak, seed production is relatively low and irregular, with seed crops being produced only once every several years. The seeds are heavy, protected, and are usually dispersed by animals to favorable sites.

Species such as the codfish and the dandelion, which produce large numbers of seeds, eggs, or offspring that experience low survival are known as r strategists. They exhibit high reproduction over longevity and high survival, and possess the ability to make use of temporary habitats and exploit relatively uncompetitive situations. Species that have fewer offspring develop into larger individuals, are more competitive, and take good care of their young are known as K strategists. They are characterized by stable populations of relatively long-lived individuals. The implications of this will become apparent later.

Natality, measured as a rate, may be expressed either as crude birth or specific birth rate. *Crude birth rate* is expressed in terms of population size, as for example 50 births/1,000 population/year. *Specific birth rate*, more accurate in its interpretation, is expressed relative to a specific criterion, such as age or condition. The usual form is the number of offspring produced per unit time by females of age x, since reproductive success varies with age. Females of reproductive age are divided arbitrarily into age classes, and the number of births for each age class is tabulated. From this, an age-specific schedule of births can be constructed (Table 10-10). Because population increases are a function of the female, the age-specific birth schedule can be modified further by determining only the mean number of females born in each female age group. This information is known as the *gross reproductive rate*, in contrast with the *net reproductive rate*, the number of females left during a lifetime by a newborn female or the mean number of females born in each female age group. Because it is calculated by multiplying the gross reproductive rate by survival of each age class, it includes adjustments for mortality of females in each age group. In human demography, the net reproductive rate is usually modified into a *fertility rate*, the number of births/1,000

TABLE 10-10

Gross reproductive rates (GRR) and net reproductive or replacement rates, R_0, of an unexploited gray squirrel population

Age (x)	Age structure	Survival series l_x[a]	Productivity m_x[b]	$l_x m_x$[c]
0–1	530	1.000	0.05	0.050
1–2	134	0.253	1.28	0.324
2–3	56	0.116	2.28	0.264
3–4	39	0.089	2.28	0.203
4–5	23	0.058	2.28	0.132
5–6	12	0.039	2.28	0.089
6–7	5	0.025	2.28	0.057
7–8	2	0.022	2.28	0.050
			GRR = 15.01	R_0 = 1.169

Source: Barkalow et al., 1970.
[a] l_x = survival series.
[b] m_x = number of female progeny produced per female.
[c] $l_x m_x$ = number of female offspring born to surviving individuals at time t_0.

TABLE 10-11
Life history table, hypothetical squirrel population

Age	*Year*										
	0	1	2	3	4	5	6	7	8	9	10
0	20	27	37.2	48.87	60.1	77.25	98.99	126.28	161.02	205.7	262.71
1	10	6	8.1	11.16	14.66	18.03	23.17	29.69	37.88	48.31	61.71
2	0	5	3	4.05	5.58	7.33	9.02	11.58	14.84	18.94	24.16
3	0		4	2.4	3.24	4.46	5.86	7.21	9.26	11.87	15.15
4	0			2.4	1.44	1.54	2.67	3.51	4.32	5.55	7.12
5	0				1.92	1.15	1.55	2.13	2.81	3.45	4.44
Total	30	38	52.3	68.88	86.94	110.16	141.26	180.4	230.13	293.19	375.29
λ	0	1.27	1.38	1.32	1.27	1.26	1.28	1.28	1.28	1.27	1.28

women, 15 to 40 years of age. That compensates for differences in sex ratios and age structures.

The term *reproductive rate* applies to animal populations. In plants reproduction may be quantified as pounds of seeds per acre and number of seeds per pound, percentage germination under different environmental conditions, and seedling survival. Vegetative reproduction is difficult to handle because one may not be able to distinguish individual plants.

The reproductive rate is measured on the basis of mature females and omits the males and immature females. Useful for comparative purposes, the reproductive rate summarizes information on the frequency of pregnancy, the number of females born, and the length of the breeding season. Because the simplest methods of obtaining reproductive rates involve the counting of embryos, placental scars, the number of eggs and unfledged young in birds, the reproductive rate incorporates a measure of mortality of the original group of ova. In field work with wild populations, the reproductive rate may have little more than academic interest because the important point is the number of offspring per unit female that survive to fledgling stage, or independence.

Population growth

Possessing sufficient data to construct a life table and knowing the age-specific fecundity, one can determine some characteristics of population growth. By summing the m_x values of all ages one obtains the gross productive rate, the number of female young a female who lived through all age groups would be expected

to produce. By multiplying the age-specific birth rate, m_x, by survivorship, l_x, and summing the products, one obtains the net reproductive rate, R_0 (see Table 10-10). If $R_0 = 1$, then the birth rate equals the death rate and the population is replacing itself or remaining stable. If the value is greater than 1, the population is increasing; and if it is less than 1, it is decreasing. From the same information one can also chart the growth of the population. This is illustrated in the life history table (Table 10-11). To construct this table one needs to obtain one more parameter, p_x, the proportion of animals surviving each age class. This is expressed as $1 - q_x$. The value p_x, like q_x, is independent of the survivorship of earlier age classes, is dependent only on the value of the previous age class, and expresses survivorship as a rate. Survivorship, l_x, is an absolute value, not a ratio.

The life table for a hypothetical population is presented in Table 10-12. For brevity, only the major columns are presented: q_x, p_x, l_x, m_x, and $l_x m_x$, and for ease in calculations the l_x column is expressed as a decimal fraction of 1 rather than in terms of 1,000. The population is started with 10 females aged 1 year that will produce a litter the year of introduction, year 0. With that information the life history table can be easily constructed.

During year 0, the initial population of 10 1-year-old females gives birth to 20 females, which are added to age class 0 of that year. The survival of these two age groups is obtained by multiplying the number of each of the approximate p_x value. Since the p_x of the females in age 1 is 0.5, 5 individuals ($10 \times 0.5 = 5$) survive to year 1 (age 2). The p_x value of age 0 is 0.3, so only 6 of the 20 in this age group in year o survive $(20 \times 0.3 = 6)$ to year 1 (age 1). Survivorship is tabulated year by year diagonally down the table to the right through the various age groups. In year 1 the 6 1-year-olds and the 5 2-year-olds together contribute 27 young to age class o. The m_x value of the 6 1-year-olds is 2, so they produce $6 \times 2 = 12$ offspring. The 5 2-year-olds (m_x value of 3.0) produce 15 offspring. The steps for determining the number of offspring in year t, N_{to}, is given by the equation

$$N_{to} = \sum_{x=1} N_{tx} m_x$$

where N_{to} is the number in the population at the given year t, and N_{tx} is the number in any year x prior to year t_0.

Thus for year o the calculation is

$$N_0 = (10)(2) = 20$$

and for succeeding years

$$N_1 = (6)(2) + (5)(3) = 27$$
$$N_2 = (8.1)(2) + (3)(3) + (4)(3) = 37.2$$

and so on. Thus the offspring are obtained by multiplying the number in each age group by the m_x value for that age and summing these values for all ages.

From such a life history table one can also calculate age distribution (Table 10-13). The age distribution for any one year can be obtained by dividing the number in each age group by the total population size for that year. The general equation is

$$c_{tx} = \frac{N_{tx}}{\sum\limits_{y=0}^{\infty} \int N_{ty}}$$

where

c_{tx} is the proportion of age group x in year t
N_{tx} is the number in age group x in year t
N_{ty} is the sum of all age groups, y, in year t

By comparing the age distribution of the hypothetical population in year 3 with the population in year 6, one observes that the population has attained a stable age distribution by the year 6. From that year on, the proportions of each age group in the population remain the same year after year.

The rate at which the population grows can be expressed as a graph of the numbers in the population against time. The graph of the

TABLE 10-12
Life table, hypothetical squirrel population

Age	q_x	p_x	l_x	m_x	$l_x m_x$
o	.7	.3	1.0	o	o
1	.5	.5	0.3	2.0	.60
2	.2	.8	0.1	3.0	.30
3	.4	.6	0.08	3.0	.24
4	.2	.8	0.05	3.0	.15
5	o	o	0.03	o	o

$$m_x = GRR = 11.0$$
$$l_x m_x = R_0 = 1.29$$

TABLE 10-13
Approximation of stable age distribution

Age	Year										
	0	1	2	3	4	5	6	7	8	9	10
0	.67	.71	.71	.71	.69	.70	.70	.70	.70	.70	.70
1	.33	.16	.15	.16	.17	.16	.16	.16	.16	.16	.16
2		.13	.06	.06	.06	.07	.06	.06	.06	.06	.06
3			.08	.03	.04	.04	.04	.04	.04	.04	.04
4				.03	.02	.02	.02	.02	.02	.02	.02
5					.02	.01	.01	.01	.01	.01	.01

growth of the hypothetical population indicates that it is growing at an exponential rate (Fig. 10-22).

The rate of population growth also can be determined by dividing the size of the population in year t by the total of the previous year, $t - 1$. For year 1 this is $38/30 = 1.27$. The rate of increase for each of the years is given in Table 10-11. Once the population has reached a stable age distribution, the rate of growth becomes constant, a law of population growth proven by Sharp and Lotka in 1911.

This eventual rate of increase attained with a stable age distribution is the finite rate of increase λ. Because λ was obtained from N_t/N_{t-1} where N_t is the population size in year t and N_{t-1} the size of the population of the previous year, it follows that the population size for any one year is $N_t = \lambda N_{t-1}$. Thus for k years of growth of a population with a stable age distribution, $N_t = \lambda^k N_{t-k}$, which describes the population growth as exponential. The finite rate of increase is rarely expressed as λ, but rather as the natural logarithm, $\log_e \lambda = r$. Thus $\lambda = e^r$ where e is the base of natural logarithms. The value r is known as the *intrinsic rate of increase*, the *maximum rate of increase*, and the *Malthusian parameter*. For the finite rate of increase of the hypothetical population, 1.28, the natural log of λ is 0.253.

The example of population growth derived from the construction of a life history table based on age-specific fecundity and age-specific mortality points out a rather obvious fact: births represent additions to a population and death subtractions from it, and the rate of increase, r, of a population equals births minus deaths. When births exceed deaths, in the hypothetical population, the population increases; when deaths exceed births, the population decreases. Thus r can also be expressed by the simple formula $r = b - d$.

RATE OF INCREASE

The rate of increase, r, when based on life-table functions and stable age distribution, is the maximum rate at which a population can increase in an unlimited environment and is expressed as r_m or r_i (Caughley and Birch, 1971). It is not, as one is often led to believe, a constant for the species; rather it varies with the environmental conditions in which a population exists. The environment in which a population of a species has its highest r_m is considered the optimal environment for the species. The index r_m can be determined only when a population is newly introduced into an environment or by fitting a curve to the growth of the population after a portion of that population has been artificially reduced. The index r_m can be useful in the study of wild population while the population is in the lower portion of the growth curve. It can be used to compare the differences in growth rates of two populations of the same species under different environmental conditions, and can serve as an indicator of the suitability of the environment and of the relationship between populations and their environments.

Another assessment of the rate of increase is r_s. This is the measurement of r based on life tables and fecundity tables at any given density with a stable age distribution. It must be calculated by the formula derived by the eighteenth-century French mathematician Leonhard Euler:

$$\Sigma\, l_x m_x e^{-rx} = 1$$

This formula states that the growth of a population is the sum of survivorship times female productivity (the $l_x m_x$ column of the life table) times the exponential, e^{-x} of r multiplied by age. Thus $l_x m_x e^{-rx}$ for the year 2 would be $l_x m_x e^{-r2}$. The number of offspring born at any time t

time *t* equals the contribution made by individuals 2 years ago, and so on. Although there is no real way to solve for *r*, the simplest, yet still laborious way, is to substitute values by trial and error into the equation until the right side balances with the left. This becomes rather involved, and the statistic has little meaning in the field (Caughley and Birch, 1971).

As the hypothetical life history table suggests, life history events, especially the physiological aspects, influence *r* (Cole, 1954*a*). In general, *r* is increased in a continuously reproducing population by a reduction of the age at which a female first produces young, the average size of the litter, and longevity. Of these the age when a female produces her first young probably has the greatest influence on *r*. It takes two young to replace the parents, but the consequences of producing more than two young per lifetime varies greatly with the age at which reproduction takes place. Offspring from young females enter the reproduction period while the mother may still be producing young herself. In contrast, the offspring of those whose first reproduction came much later in life would not become a part of the reproductive population until the mother was in late reproductive or postreproductive stage. Thus the female who produces early begins to compound interest in the population much sooner. In man, for example, if a first birth occurred when the mother was 13, an average of 3.5 children per female would contribute as much to the future growth of a population as an average of six children would if the first birth came when the mother was 25 years old (Cole, 1954*a*). The effect of increased litter size per female lifetime is very small if the initial reproductive age is very young. Again using man as an example, if human females on the average produced their first offspring at the age of 20 and had a total of five children spaced at 1-year intervals, they would be contributing almost half as much to the population growth as would hypothetical females capable of living forever producing a child every year. On the other hand, if the reproductive life of the female is short, or if initial reproduction does not occur until a relatively late age, an increase in longevity would significantly increase *r*.

Implied in the example is the idea of reproductive value, the reproductive contribution of a female aged *x* compared with a female aged *x* + *t*, if both begin their initial production of

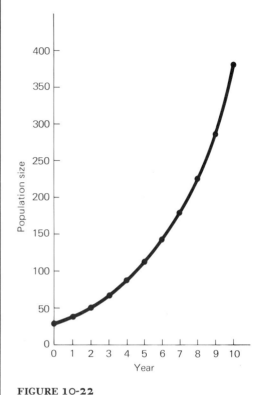

FIGURE 10-22
The exponential growth curve of the hypothetical squirrel population discussed in the text.

319

young at two different age levels. The statistician R. A. Fischer (1929) defined the reproductive value as

$$\frac{v_x}{v_0} = \frac{e^{rx}}{l_x} \Sigma e^{-ry} l_y m_y .$$

where

y = all ages the female passes through
v_x = reproductive value at age x
v_0 = reproductive value at birth

Because v_0 is usually made to equal 1, the term v_0 can be dropped in the working equation.

The equation means that the reproductive value of a female aged x is the ratio of number of female offspring produced at a given moment by a female aged x to the number of females of age x at the given moment.

Solving for the reproductive value again requires the determination of r by trial and error. Although somewhat laborious to determine, the statistic has implications in the management of exploitable populations. The maximum yield (see Chapter 13) can be obtained by removing females of lowest reproductive value and allowing them to be replaced by females with a higher reproductive value. Reintroduction of a species into a depopulated area or the colonization by a species of a new area can succeed only if the females involved possess a high reproductive value.

In general the reproductive value for all females just born is 1. The value gradually rises to a peak—for humans the peak age is fifteen—and then declines. The value can also vary among populations of females of the same species. For example Keyfitz and Flieger (1971) have calculated the reproductive values of females from various countries of the world. Reproductive values for four countries for the ages 15 and 30 are compared in Table 10-14. In the United States the reproductive values of a female aged 15 is nearly five times that of a woman aged 30. In Mexico a 15-year-old is only two times more productive than a 30-year-old. This suggests that it is the younger females that contribute most heavily to the population growth of the United States (and other developed countries) whereas in Mexico (and other underdeveloped countries) reproduction continues in the older age classes. The difference in the reproductive values of females aged 15 is not that great between developed and under-developed countries. But the reproductive value of a woman 30 years old in Mexico is four times as great as that of a 30-year-old woman in the United States.

THE GROWTH CURVE

If a population were suddenly presented with an unlimited environment, as can happen when an animal is introduced into a suitable but unoccupied habitat, it would tend to expand geometrically. Assuming there were no movements into or out of a population and no mortality, then birth rate alone would account for population changes. Under this condition population growth would simulate compound interest. If you refer to a mathematics handbook, you will learn that if interest is compounded annually then

$$A = P(1 + r)^n$$

where

A is the new amount at some given time
P is the original amount or principal
r is the rate of interest expressed as a decimal
n is the number of years

If the interest is compounded several times a year, then

$$A = P\left(1 + \frac{r}{q}\right)^{nq}$$

where q is the number of times interest is compounded during the year.

If interest is compounded continuously, then q approaches infinity. Letting r/q equal x, the expression can be written

$$A = P(1 + x)^{rn/x}$$

When q approaches infinity, x approaches 0, and from calculus it can be shown that

$$\lim_{x \to 0} (1 + x)^{1/x} = e$$

where e is the base of natural logarithms, which is approximately 2.7183. Thus if interest is compounded continuously, the expression reads

$$A = Pe^{rn}$$

In the symbols used in population ecology and already introduced, A, the amount, at some time, t, becomes N_t; P, the principal becomes N_o, the initial population; r becomes the rate of increase; and t is the units of time. Then our compound formula will read

$$N_t = N_0 e^{rt}$$

the expression for logarithmic population growth, the accumulation of compound interest on the population. Changed to a logarithmic form,

$$\log_e N_t = \log_e N_0 + rt$$

If r is positive and the conditions remain the same, the exponential growth will increase faster and faster, like interest rapidly accumulating.

But movement in and out of the population does occur and death is a reality, so the birth rate alone does not account for population changes. However, by subtracting deaths from births, one can use the adjusted or age-specific birth rate as a measure of population increase. The same formula then can be used to describe the growth of a population in which deaths are accounted for and for which the environmental conditions are for all practical purposes constant and the resources excessive.

The rate of growth at first is influenced by heredity or life history features, such as age at the beginning of reproduction, the number of young produced, survival of young, and length of reproductive period. Regardless of the initial age of the colonizers, the number of animals in the prereproductive category would increase because of births, whereas those in the older categories for a time would be stationary. As the young mature, more would enter into the reproductive stage and more young would be produced, as has already been demonstrated in the hypothetical population example. If the number of animals are plotted against time, the points will fall into an exponential growth curve defined by the foregoing formula (see Fig. 10-22). If the logs of the numbers of animals are plotted against time, the points fall into a straight line.

But the environment is not unlimited, nor does age distribution remain stable; and rarely do populations long maintain a constant rate of increase. As a population increases, detrimental effects of increased density begin to inhibit the growth until it reaches some asymptotic level or the carrying capacity, the maximum number that can be supported in a given habitat.

The inhibition and slowing down of the

TABLE 10-14
Reproductive values, selected countries

	Reproductive value			
Age	United States	Finland	Mexico	Mauritius
15	1.129	1.034	1.945	1.783
30	0.227	0.212	0.927	0.690

Source: Data from Keyfitz and Flieger, 1971.

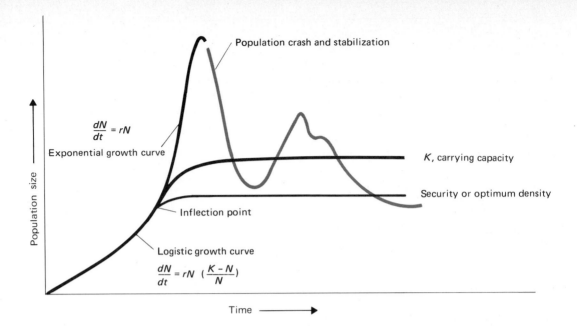

Population crash and stabilization

$\frac{dN}{dt} = rN$

Exponential growth curve

K, carrying capacity

Security or optimum density

Inflection point

Logistic growth curve

$\frac{dN}{dt} = rN \left(\frac{K - N}{N} \right)$

Population size

Time

growth rate can be described mathematically by taking the exponential equation, $N_t = N_o e^{rt}$ and inserting into it some variables to describe the effects of density:

$$N = \frac{K}{1 + e^{a-rt}} \quad \text{where} \quad a = \frac{r}{K}$$

and where K represents the upper limit or carrying capacity. This same equation is often written

$$\frac{dN}{dt} = rN \frac{K - N}{K}$$

In words the equation says that the rate of increase of a population is equal to the potential increase of the population times the proportion of the carrying capacity of the habitat that is still unexploited. This formula was developed in 1838 by the French mathematician Verhulst as a model of population growth in a limited environment. In 1920 Pearl and Reed, in a classic paper, plotted the growth of the population of the United States by years and fitted it to the curve described by the Verhulst equation. Since then the model has been known as the Verhulst–Pearl equation.

The Verhulst–Pearl equation describes a logistic or sigmoid growth curve. The rate of increase is slow at first, then accelerates until it reaches a maximum (Fig. 10-23). As density increases, the rate slows, marked by the inflection point in the curve. As the population reaches the carrying capacity, the curve flattens out. Or to state it more precisely, as $K - N$ approaches 0, dN/dt also approaches 0. Accord-

ing to the model, when N is low or near 0, rN approaches the exponential value and N increases, and the value of $(K - N)/K$ declines toward 0. When $N = K = 0$ the population is said to have reached equilibrium.

For example, give K a value of 1,000, allow r_o to equal 2, and the initial population N to stand at 10. Then for population levels of N — 200, 500, 800, 1,000, and 1,200 the following values of dN/dt will result:

$$\frac{dN}{dt} = rN \left(\frac{K - N}{K} \right)$$

$$20 \left(\frac{1000 - 200}{1000} \right) = 20 \,(0.80) = 16$$

$$20 \left(\frac{1000 - 500}{1000} \right) = 20 \,(0.50) = 10$$

$$20 \left(\frac{1000 - 800}{1000} \right) = 20 \,(0.20) = 4$$

$$20 \left(\frac{1000 - 1000}{1000} \right) = 20 \,(0.00) = 0$$

$$20 \left(\frac{1000 - 1200}{1000} \right) = 20 \,(-0.20) = -4$$

As N approachs K, dN/dt decreases; and when $N = K = 0$, dN/dt also equals 0. If N exceeds K then dN/dt is negative and N declines or approaches K from above. If N is less than K, then N climbs up toward K. In effect the movement of N away from K so affects the rate of growth that the population tends to return to an equilibrium level defined by K. The equation indicates that the parameters r and K are independent of each other: that as N increases, births decrease and deaths increase;

that when $N = K$, births = deaths; that changes in population respond to changes in density. In other words population size is density regulated.

The logistic equation has been criticized as unrealistic (F. E. Smith, 1952; Slobodkin, 1962), for it involves a number of oversimplifications. The equation is based on the assumption that all animals in the population have an equal chance to give birth or to die, to eat or to be eaten, which eliminates the concept of age structure. It suggests that all animals respond instantaneously to changes in the environment, when in fact adjustments often lag; that when a population approaches extinction, its rate of increase is the highest; and finally that a constant upper limit to the population exists in any particular situation.

The idea of equilibrium, which involves density dependence, is not universally accepted. Indeed there is a great deal of debate concerning the effects of density-dependent and density-independent mechanisms in population regulation. Part of the argument results from the observations that populations rarely exhibit any real equilibrium. They show constant fluctuation between some upper and lower limits, and where the asymptote falls depends on how one defines the carrying capacity.

Carrying capacity can be interpreted in four different but overlapping ways (Dasmann, 1964). One level is represented by K in the logistic equation, the limit imposed by the ability of the environment to provide food to support the population at bare survival level. There is not enough food for optimum body growth, vigor, and good health. The population flirts with disaster. A small change in the weather or a failure of food supply can be catastrophic. Another level is optimum density, which implies that the population has adequate food, water, and shelter. Individuals approach the maximum in body size and growth, and in health and vigor. Rarely do populations remain at this level. Wild animals controlled by hunting or predation may hold at the optimum level, as may a population controlled by territoriality or some form of birth control. In man such levels may have been maintained by hunting–gathering cultures. The optimum level represents the inflection on the growth curve.

A third way of interpreting carrying capacity is through security density. It is based on Paul Errington's (1946) concept of the threshold of security. If a population density is above a

FIGURE 10-23 [OPPOSITE]
The growth curve, showing variations.

certain level, there is instability in the population. Excess animals may be driven out or they may become very vulnerable to predation. Security density in some instances can correspond either to tolerance density, as in the case of territorial animals, or to optimum density. A fourth interpretation is tolerance density. This is saturation density: the level at which intraspecific tolerance or the lack of it among animals of the same species permits no further increase. The degree of crowding tolerated usually sets the upper limit of population density. It is most marked in territorial species in which units of space are parceled out and defended by members of the population. Among territorial species, tolerance density may be the same level as optimum density. A similar type of density may occur among nonterritorial species that exhibit social hierarchy. The subordinate animals may be driven out or prevented from feeding or reaching shelter. Under such conditions tolerance density might be nearly at the same level as subsistence density.

The growth of a number of laboratory and natural populations is roughly sigmoid, particularly when introduced into unfilled environments. The early phase of population growth follows the exponential or J-shaped curve. Some populations grow exponentially up to a point where some sudden limiting force as weather, lack of space, or depletion of food causes the population to drop sharply. From the low point the population may recover to undergo another phase of exponential growth, it may decline to extinction, or it may recover and fluctuate about some level below the high level attained. The J-shaped growth curve is characteristic of some insects and is often characteristic of game birds and mammals introduced to a new and unfilled environment (Fig. 10-23). The population grows exponentially, overshoots what might be a normal carrying capacity, then suddenly declines and more or less stabilizes at some much lower level. Such has been the history of the ring-necked pheasant populations in the Dakotas and on Point Pelee Island (Stokes, 1955), the eruptive growth and crash of the reindeer population on St. Matthew Island (Klein, 1968), and the irruption of the thar in New Zealand (Caughley, 1970). Riney (1964) has postulated that a herbivore population liberated and left undisturbed in a new environment will exhibit a single eruptive oscillation. The curve consists of four stages (Caughley, 1970): (1) the initial increase stage, the period between the establishment of the population and the attainment of the initial peak; (2) the initial stabilization period at peak population; (3) the decline; and (4) the post-decline stage in which the population adjusts to a lower population level (Fig. 10-24). As the population approaches the asymptote, either downward or upward, the growth curve departs from the logistic and fluctuates between some upper and lower limits.

The growth of the world's human population is exponential (Fig. 10-25). It began with an unusually long period in which the net increase of population was very low. This initial low rate of growth was probably due to a hunting–gathering way of life that characterized a long period of man's history. Dependent upon only what the natural ecosystems could provide, man was limited by the restricted energy source available. Under the conditions in which he lived, man occupied an environment with a relatively low carrying capacity. The eventual development of agriculture enabled man to exploit new resources, to make more efficient use of others, and in effect increase the carrying capacity of his habitat. Population increased steadily, rising from perhaps 16 million 6,000 years ago to 113 million 200 years ago. Man's habit of clustering in urban developments promoted the spread of disease and outbreaks of plagues. Human populations experienced local deteriorations of habitat through water and air pollution and local shortages of food. Combinations of disease, pollution, and starvation kept the world population at a still relatively low level with local increases and declines common around the globe. Then came the Industrial Revolution in Europe about 300 years ago, followed by a medical revolution, advances in agriculture and food preservation, improved transportation, and discovery of new lands available for colonization. The overall result was to increase the carrying capacity of the earth, to raise the value of K, to provide an "unlimited environment" in which the human population could expand. The rate of increase accelerated, and still remains high. The annual rate of increase for the world population is approximately 2 percent/year; some regions are experiencing a 4-percent annual increase.

Today man has about filled the finite earth, yet he is still seeking ways to increase its carrying capacity. But when a carrying capacity is increased, a population has a way of filling it.

Thus as the capacity of the earth has been increased, so has the population. Because the earth, like the local habitats of natural organisms, is finite, there is a limit to the numbers of man it can support. The question is whether man will check his own population growth at a level the earth can support or whether he will experience the same kind of sharp and catastrophic decline experienced by other animals introduced into a new environment.

As indicated by the growth curves, the rate of increase or population growth is at a maximum where there are neither too many nor too few individuals. When the population falls below or exceeds this point, the rate of increase falls off. Increasing sparseness is associated with a reduction in rate of increase; the population may become so low that the rate of increase r becomes negative and the population dwindles to extinction. The heath hen is a classic example. Formerly abundant in New England, this eastern form of the prairie chicken was driven eastward by excessive hunting to Martha's Vineyard, off the Massachusetts coast, and to the pine barrens of New Jersey. By 1880 they were restricted to Martha's Vineyard. Two hundred birds made up the total population in 1890. Conservation measures increased the population to 2,000 by 1917. But that winter, a fire, gales, cold weather, and excessive predation by goshawks reduced the population to 50. Numbers rose slightly by 1920, then declined to extinction in 1925. The last bird died in 1932.

There are several causes for the decline of sparse populations. When only a few animals are present, the females of reproductive age may have a small chance of meeting a male in the same reproductive condition. Many females remain unfertilized, reducing average fecundity. A small population also faces the prospect of increased death rate. The fewer the animals, the greater may be the individual's chance of succumbing to predation.

Overcrowded populations, on the other hand, often have a reduced rate of increase because of strife among individuals. This results in increased death rates, poor reproduction, aberrant maternal behavior increasing the death rate among the young and physiological derangement typified by low level of blood sugar and liver glycogen. Disease and parasites spread rapidly through dense populations, often dramatically reducing numbers.

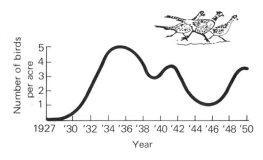

FIGURE 10-24
A growth curve for the ring-necked pheasant population on Point Pelee Island, Ontario. (Redrawn from Stokes, 1955.)

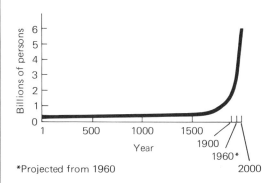

*Projected from 1960

FIGURE 10-25
Growth of the world population. (U.S. Department of Agriculture.)

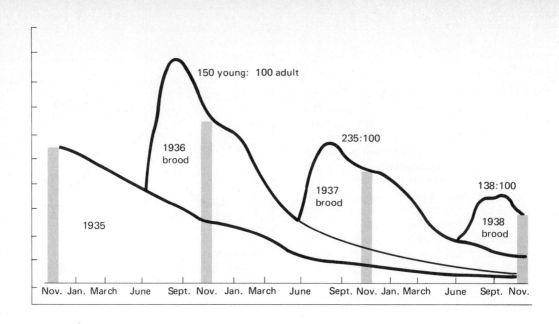

150 young: 100 adult

1936 brood

235:100

1937 brood

138:100

1935

1938 brood

Nov. Jan. March June Sept. Nov. Jan. March June Sept. Nov. Jan. March June Sept. Nov.

TABLE 10-15

Life equation of a black-tailed deer population on a 36,000-acre area from 1949 to 1954

		Males			Females			
	Type of gain or loss	Adults	Yearlings	Fawns	Fawns	Yearlings	Adults	Total
1949	Prehunting population	312	140	456	475	274	1,003	2,690
	Legal hunting kill	204	3	—	—	—	—	− 207
	Crippling loss and							
	illegal kill	10	11	8	8	8	38	− 83
	Winter losses	13	19	304	229	40	135	− 740
1950	Prefawning population	85	107	144	238	226	860	1,660
	Fawning season gain	192	144	+707	+589	238	1,098	2,968
	Summer mortality	2	1	221	83	12	52	− 371
	Prehunting population	190	143	486	506	226	1,046	2,597
	Legal hunting kill	125	25	71	86	43	160	− 510
	Crippling loss	6	12	21	26	13	50	− 128
	Winter losses	2	6	80	61	10	37	− 196
1951	Prefawning population	57	100	314	333	160	799	1,763
	Fawning season gain	157	314	+617	+515	333	959	2,895
	Summer mortality	2	3	311	195	17	48	− 576
	Prehunting population	155	311	306	320	316	911	2,319
	Legal hunting kill	96	9	—	—	—	—	− 105
	Crippling loss	5	10	3	3	6	15	− 42
	Winter loss	2	8	89	67	9	42	− 217
1952	Prefawning population	52	284	214	250	301	854	1,955
	Fawning season gain	336	214	+762	+601	250	1,155	3,318
	Summer mortality	3	2	436	260	12	58	− 771
	Prehunting population	333	212	326	341	238	1,097	2,547
	Legal hunting kill	178	80	62	64	55	205	− 644
	Crippling loss	9	20	19	20	17	70	− 161
	Winter loss	5	6	71	54	8	30	− 174
1953	Prefawning population	141	106	174	203	158	786	1,568
	Fawning season gain	247	174	+608	+506	203	944	2,672
	Summer mortality	2	2	338	225	10	47	− 624
	Prehunting population	245	172	270	281	193	897	2,058
	Legal hunting kill	85	26	38	36	18	70	− 273
	Crippling loss	4	9	8	8	7	27	− 63
	Winter loss	5	7	71	53	7	29	− 172
1954	Prefawning population	151	130	153	184	161	771	1,568

Source: Adapted from E. R. Brown, 1961.

LIFE EQUATIONS

A picture of the limitations of the growth of a population, seasonal gains and losses, and other important events occurring throughout the year can be summarized in a life-equation table (Table 10-15). Since slight changes in reproduction, survival, or sex ratios can influence the rate of increase considerably from year to year, wildlife biologists, especially, find the information summarized in the life equation highly useful.

The life equation is a modification of the life table. The life table is a mathematical expression of the dynamic processes only, the vital statistics of a population. The life equation, on the other hand, illustrates changes within the population. The life equation involves a census or inventory of an identifiable population. Age in the life equation is referred to as stages in the life history of a population within one breeding cycle, instead of within a day, a month, or a year.

Life equations, like life tables, begin with a given population, usually 1,000, broken down into sex and age categories. If a game animal is involved, the table begins with a prehunting population. Hunting losses then are subtracted, according to sex and age, leaving a posthunting population. The number left after winter losses comprise the prebreeding season population. To this is added the breeding-season gains. Finally a new prehunting-season population estimate is obtained 1 year later.

This tabular data of gains and losses can be presented as a sort of survival curve (Fig. 10-26). Curves of several years can be joined to illustrate numerical changes over a period of years. In addition, changes in age structure can be indicated by showing the proportions of several age classes (Fig. 10-9).

Life equations are not very accurate nor are they intended to be. Some categories in the equation cannot be measured accurately; other items are based more on estimates than on facts. At the same time other information might be quite correct. Properly constructed, the life equation shows the magnitude of population losses to several causes and where and when the heaviest losses occur. The life equation also indicates the extent and importance of production of young to the future of the population, gaps in knowledge of population behavior of the species involved, and the most important research problems for future study.

FIGURE 10-26 [OPPOSITE]
Changes in a population of California quail. The bars indicate the ratio of mature to immature birds in November. This is a diagrammatic life equation. (After J. T. Emlen, 1940.)

327

Summary

All living organisms exist in groupings of the same species or populations. These populations occupy a particular space, have a density and age structure, a birth rate and a death rate, a growth form, and numerical dispersion in time and space. The size of a population in relation to a definite unit of space is its density. Population density depends upon the number of individuals added to the group and the number leaving, the difference between the birth rate and the death rate, and the balance between emigration and immigration. Birth rate usually has the greatest influence on the addition of new individuals. Mortality is a reducer and is greatest in the young and old. It is usually stated in terms of survivorship rather than as a rate of mortality. Age and sex ratios influence the natality and mortality rates of populations. Reproduction is limited to certain age classes; mortality is most prominent in others. Changes in age-class distribution bring about changes in the production of young. The sex ratio tends to be balanced between males and females but may change with a change in population density. If mortality is high in one sex before the animals are on their own, the sex in short supply tends to be favored by selective pressures.

When births exceed deaths, the population increases; and the difference between the two represents the rate of increase, r. In an unlimited environment, an animal population expands geometrically, a phenomenon that occurs when a small population is introduced into an unfilled habitat. But because the environment is limited, population growth eventually slows down and arrives at some point of equilibrium with the carrying capacity of the habitat, about which it fluctuates. What the carrying capacity is or what the equilibrium point might be is something of an abstraction. Carrying capacity can be considered in terms of four different but overlapping types: subsistence density or K, optimum density, security density, and tolerance density.

Populations are distributed in some kind of pattern over an area. Territoriality results in uniform distribution. Most organisms exhibit a clumped or contagious distribution, which results in aggregations. Some aggregations reflect a degree of sociality on the part of the population members, which may lead to cooperative or competitive situations.

No population increases indefinitely. Eventually it has to arrive at some kind of equilibrium point about which it fluctuates. The equilibrium point is a long-term constant mean or average density appropriate to the environmental conditions. It is the point to which populations tend to return if disturbed. Because it has to be determined by hindsight, based on census data collected in years or seasons past, the term may or may not have any real meaning. Thus the equilibrium point, defined as density at which $r = 0$, may be more statistical than biological. There is, however, sufficient evidence in the literature (for example, see Tanner, 1966) to support the idea that populations do fluctuate between some upper and lower limits, and that these limits do tend to give some stability to the population. The nature of the processes that regulate the numbers of plants and animals is a major problem of population ecology.

Basically population numbers are influenced by some force outside the population itself or by some force generated within the population. The former, extrinsic in nature, often is termed density-independent because the action or effect is constant, regardless of the size of the population. The latter, intrinsic in nature, is called density-dependent, because the intensity of action varies with the density of the population. In natural situations, however, it is difficult, if not impossible, to separate the effects of each, for the two acting together influence nearly all population fluctuations.

Attempts to answer the question "What determines the number of organisms in a natural population?" have resulted in two widely divergent views. One school, represented by those who have worked mainly with insect populations, argues that the number of animals is determined independent of density by some outside force, particularly weather; that environmental forces act upon the rate of increase during the short reproductive period, causing either a rapid rate of increase or decline; that an equilibrium point, which implies a constant environment, is nonexistent; that as populations decline, the remaining find greater accessibility to favorable habitat and survive (Andrewartha and Birch, 1954; Ehrlich and Birch, 1967).

Intraspecific population regulation

The other school argues that populations are regulated through density-dependent mechanisms. It rejects the idea that density-independent influences do anything more than impose their effects on density-dependent mechanisms (Solomon, 1957; Nicholson, 1958; Lack, 1966). If a population is controlled by some density-independent influence, and environmental conditions change more or less at random, then the population can vary without limit. Eventually by chance it could vary to extinction or to prodigious numbers. This does not happen, so some mechanism must exist to keep a population fluctuating within some upper and lower limits. The point about which this fluctuation takes place is the density at which $r = 0$, and this density ultimately becomes the long-term mean density of the population. Deviations from the mean density are corrected by changes in r that direct the population upward or downward toward the mean. Because regulations are not strict and a time lag exists in the response of the population, populations always fluctuate from year to year or season to season.

This school suggests that the plant or producer populations are limited by available resources; that herbivores do not usually deplete their food resources, therefore are not food-limited and are regulated by predation; that carnivores and decomposers are limited by available resources and intraspecific competition; that although population densities are directly affected by such density-independent influences as weather, the highest densities a population can reach are those set by resources (Hairston et al., 1960; Slobodkin et al., 1967). If a population is not limited by resources, then it must exist at some lower density, controlled by some regulating mechanism or by predation.

From the argument, which colors all papers written on the subject, one conclusion is clear: the limitations on the number of organisms cannot be explained by one simple mechanism; population regulation is complex and the causes unknown. But if the debate has had any results, it has at least stimulated investigations into the nature of population regulation.

Density-independent influences

Density-independent influences on the rate of increase cannot regulate population growth, because regulation implies a homeostatic feedback that functions with density. But density-inde-

329

pendent influences can have a considerable impact on the increase and decrease of a population. They may so reduce a population that density-dependent mechanisms do not come into play, or they may interact with density-related influences to bring about population changes. A cold spring may kill the flowers of the oak, causing the failure of the acorn crop. Because of the failure, squirrels may experience widespread starvation the following winter. Although starvation relates to the density of the squirrels and the available food supply, weather was the major cause of the decline. In general, population fluctuations influenced by annual and seasonal changes in the environment tend to be irregular and correlated with variations in moisture and temperature.

Examples of pronounced density-independent influences on population size can be found among insects. One example is the walnut aphid, an important pest in the walnut orchards of California. The aphid population changes are influenced by (1) the age of the walnut leaflet, which determines the maximum number of aphids that can feed on a leaflet at any one time; (2) the amount of prior aphid feeding, which can damage leaves and reduce their carrying capacity; (3) predation; and (4) temperature (Sluss, 1967).

Sharp declines in the population are associated with high temperatures, especially when the temperature exceeds 100° F over a period of several days. This results in a J-shaped curve, characteristic when a density-independent influence suddenly restricts the rapid growth of a population (Fig. 11-1a). Because of the high density prior to the period of high temperature, the aphids had significantly reduced the food supply and damaged the leaves and thus were unable to resurge after the temperatures became lower. Predation, combined with high temperatures, modified the J-shaped curve (Fig. 11-1b). The population of aphids grew more slowly and peaked at a lower level, but still experienced the sharp decline. Because the population prior to the decline was high enough to reduce food and affect leaf condition, these aphids, too, were unable to resurge. In situations where predation was relatively heavy, aphid populations reached carrying capacity more slowly, did not reach such high densities, and did not experience the same drastic decline at high temperatures. Because the leaves were not unfavorably conditioned by aphid attack,

the insects were able to make some comeback. Because their food supply was not decimated, the predatory beetles remained in the area and were able to respond to any new increase in the aphid population (Fig. 11-1c).

The effect of weather on vertebrate populations is less dramatic but nevertheless important. Data gathered over a 22-year period indicated that winter losses of bobwhite quail fluctuated directly with the number of months the ground remained covered with more than 3 in. of snow (Fig. 11-2). Losses ranged as low as 4 percent during winters having less than 1 month of snow cover to 80 percent during winters having a ground cover of snow lasting 3 months (Kabat and Thompson, 1963). However, only the most severe winter had any apparent direct effect on the population size the following fall. More important in controlling the population was the interrelation between winter losses, spring density, and summer and fall losses. When the spring population was low because of heavy winter mortality, the quail approached their full reproductive potential (intrinsic rate of increase). This resulted in a high summer gain and an increased fall population. Conversely, when the spring population was high, an intolerance developed among birds that significantly affected breeding behavior. Summer gains were suppressed and the fall population leveled off. There exists over the years a fairly stable level of population size about which the population fluctuates (Fig. 11-3). An adverse winter may cause a sharp decline in the population one year, but from this the population quickly recovers.

Populations of white-tailed deer in the northern parts of its range exhibit a somewhat similar response to severe winters (Fig. 11-4). Following severe winters (those that have had 60 days of 15 in. of accumulated snow, or 50 or more days of 24 in. of accumulated snow), populations of deer in the Adirondack Mountains decrease dramatically. The losses are due largely to the inability of deer to obtain sufficient food. During periods of prolonged snow cover, low-growing winter food is buried. The expenditure of energy required by a fawn to move through 15 in. of snow and of an adult to move through 20 in. or more is greater than the energy provided by available food and stored fat reserves of the deer. Often deer are unable to reach food supplies outside of the wintering area. The result is an exceptionally high mortality of fawns.

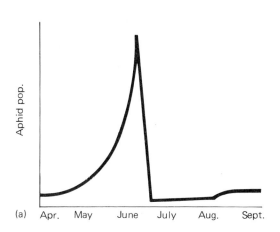

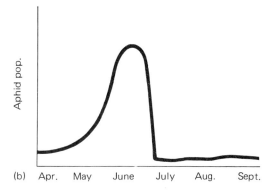

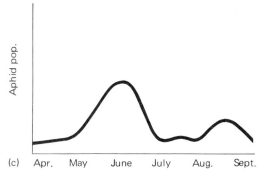

FIGURE 11-1

Three generalized types of aphid population patterns found in northern California walnut orchards indicates: (a) domination by temperature and leaflet condition; (b) similar domination, but the predatory action of coccinellid beetles is superimposed; and (c) domination by the predatory action of coccinellid beetles. (From Sluss, 1967. © The Ecological Society of America.)

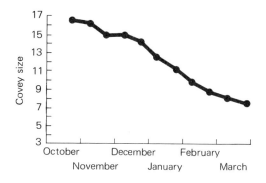

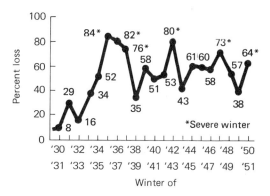

FIGURE 11-2

Annual winter loss, expressed as a percentage of bobwhite quail at Prairie du Sac, Wisconsin, showing the effects of severe winters. The years run from 1929 to 1951, inclusive. The inset shows the shrinkage of quail covies at Prairie du Sac during the winter. The graph is based on an average of 15 years, 1937–1951. (After Kabat and Thompson, 1963.)

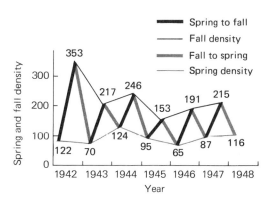

FIGURE 11-3

Spring and fall densities, 1942 through 1947, of bobwhite quail at Prairie du Sac, Wisconsin. Note the seasonal fluctuation of population density from spring to fall and back to spring again. Fall densities are fairly stable in this territorial species. (After Kabat and Thompson, 1963.)

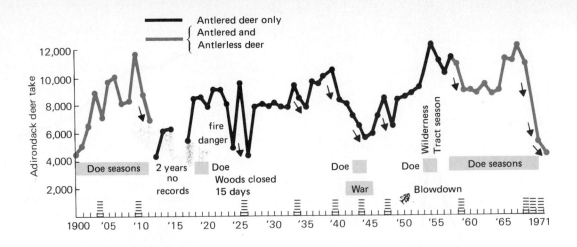

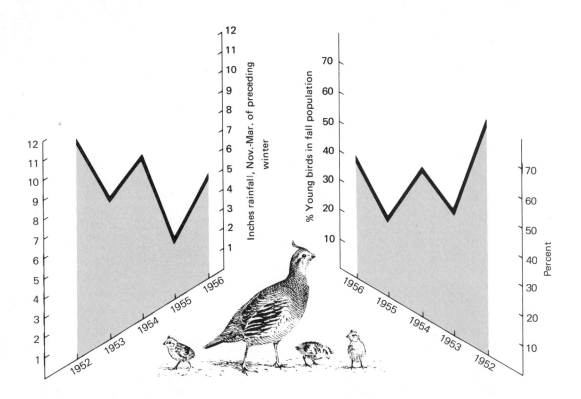

For example, during the severe winters of 1968–1969, 1969–1970, and 1971, characterized by prolonged deep snows and cold temperatures, approximately 5 out of every 6 fawns died (Severinghaus, 1972). These heavy winter kills were reflected in the deer populations the following years.

Approximately 53 percent of the variation in annual productivity of ringnecked pheasants in Wisconsin is associated with variations in spring breeding. Much of the remaining variability is associated with temperature prior to

nesting (Wagner et al., 1965; Wagner and Stokes, 1968). This association is strictly density-independent, influenced by habitat and acting through the physiology of the bird. A correlation exists between r and the temperatures of late April and early May (Fig. 11-5). Mean temperatures in excess of 53° F result in increased productivity; temperature much below 53° F result in decreased productivity. The decrease may be due to stress brought on by the hen's increased need for energy to maintain her body temperature at the expense of reproductive physiology. Decreased production is reflected in egg dropping, nest abandonment, and the laying of eggs in dump nests.

Some of the most definitive examples of density-independent influences on the population growth of vertebrates exists between precipitation and the rate of increase of certain birds and rodents of the desert. The productivity of Gambel's quail is, in order of importance, a function of (1) soil moisture in late April; (2) proportion of females over 1 year old; and (3) seasonal rainfall from September to April (Francis, 1970). The close relationship between soil moisture and population increase results from the dependence of the quail on annual forbs for food in late winter. The amount of green vegetation available depends upon the amount of rainfall in winter. If the critical rains come, green vegetation is abundant, and the quail produce many young. If rainfall is scant, annual forbs fail to develop, and production of quail is low or nonexistent (Fig. 11-6). A similar relationship exists between desert rodents and rainfall. Although water and green vegetation are not necessary for adult survival, females fail to reproduce if they are not available (Beatley, 1969).

Too much or too little moisture can temporarily reduce populations. Drought can be a time of crisis for muskrat populations, as water levels of marshes become low (Errington, 1939, 1943, 1951, 1963). On the other hand, late spring floods drive muskrats out of burrows and lodges and drown the young. When wet years turn marshes into lakes, muskrats are forced to live in danger close to shore and to seek shelter in woodchuck burrows, grain shocks, culverts, and cavities beneath tree roots (Errington, 1937b). After conditions return to normal, the population rises to its previous level, provided the habitat itself was not destroyed.

Heavy rains in May and June appear to be important in determining hatching success and juvenile survival among ringnecked pheasants (Stokes, 1955), as well as rabbits, quail, and other animals. Survival of the spruce budworm, which causes enormous damage in northern forests by consuming the flowers and leaves of spruce, balsam fir, and other conifers, is influenced by an interaction of temperature and moisture (Greenbank, 1956). Overwinter larval survival is high when temperatures fluctuate between freezing and thawing. Wet summers are inimical to active stages of budworms. Out-

FIGURE 11-4 [OPPOSITE TOP]
The fluctuations of the Adirondack deer herd in New York appear to be closely related to the severity of winters. Hard winters result in heavy mortality, decline in reproduction, and a sharp decline in the population. Horizontal hatches refer to severe winters. (After C. S. Servinghaus, The Conservationist 27(1972): 30–31.)

FIGURE 11-6 [OPPOSITE BOTTOM]
The relationship of winter rainfall to age composition of a Gambel's quail population in southern Arizona the following year. (After Sowls, 1960.)

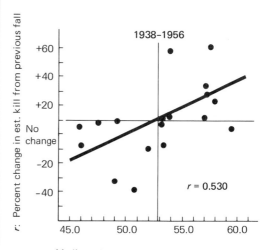

Madison mean temperature, April 21–May 11

FIGURE 11-5
Correlation between Madison mean temperatures for the period April 21–May 11, and the percentage change from one fall to the next in the estimated Wisconsin pheasant kill, 1938–1956. (From Wagner et al., 1965.)

breaks usually start at times when the summers are clear and dry.

Man exerts one of the greatest influences on population rate of increase. Market hunting for hides, meat, and feathers has exterminated or greatly reduced many species over the world. Man's ability to change environments has benefited some species, reduced others. Destruction of virgin forests in the South exterminated the ivory-billed woodpecker; cutting of the northern beech forests combined with market hunting caused the extinction of the passenger pigeon. Drainage of potholes in the prairie regions has seriously reduced waterfowl populations.

Clearing of the forests and subsequent cultivation of land in eastern North America increased the habitat for open field birds and species of the edge. It permitted the eastward spread of such prairie life as the dickcissel and the coyote.

Clear-cutting a forest can produce major biotic changes and have a pronounced impact on *r* among the populations involved. The harvest of old-growth Douglas fir stimulates the population growth of Oregon juncos (Hagan, 1960) and deer mice (Gashwiler, 1970). As the cutover areas pass into weed and brush stages, the populations of these species exhibit a high rate of increase. On the other hand, such animals as red-breasted nuthatches, red-backed voles, Douglas squirrels, and flying squirrels disappear.

Man's widespread use of pesticides affects populations in a density-independent way. Numerous studies indicate that the chlorinated hydrocarbon pesticides cause the death of organisms other than the target species by contact or by ingestion. (Clawson, 1958–1959; G. J. Wallace, 1959; L. B. Hunt, 1960; E. G. Hunt and Bischoff, 1960; Rudd, 1964). They lower fecundity by interfering with the reproductive process or survival of young. Chlorinated hydrocarbons interfere with the calcium metabolism of birds, causing the thinning of eggshell and subsequent loss of embryos (Bitman, 1970; Keith et al., 1970). Chlorinated hydrocarbons concentrate in the eggs of fish to cause heavy mortality of fry (Burdick et al., 1964). Pesticides also interfere with density-regulating processes. Often they eliminate predators, parasites, and competitors of undesirable species so that the pests against which the control was originally directed resurge in greater numbers (Muir, 1965).

Density-dependent influences

Among many organisms the mechanisms for the regulation of populations within the limits imposed by the environment are largely density-dependent. Through such mechanisms organisms avoid the hazard of overexploiting their environment and even bringing about their own extinction. It appears that the regulation of populations is a homeostatic process. As a population approaches a certain density, some density-related effects act to reduce the population. If the population falls below a certain level, density-dependent processes fade, and the population builds up. In other words the rate of increase of a population is inversely related to its density. As density increases, births decline or mortality increases, or both; *r* becomes negative and population declines. As density decreases, births increase, mortality declines and *r* becomes positive. When $r = 0$, the population will not change. Such a feedback system results in oscillations, or regular fluctuations. In general these oscillations will move about an equilibrium density, the level at which the production of offspring precisely compensates for the loss of adults by death. Any departure from this level brings compensating reaction or self-regulation into play, which ceases as soon as equilibrium density is obtained again. A time lag is involved in these oscillations: organisms take a significant time to grow up and join the reproductive segment of the population and an equilibrium density is rarely attained.

In one way or another the limitations on the density of a population ultimately work through the intraspecific competition that results when a number of animals utilize a common resource that is in short supply. Of the various components of the environment, the two that are apt to be in short supply in relation to population density are space and food. Competition for them may result in aggressiveness, stress, starvation, or emigration, but in the last analysis some element of this competition is involved.

Food

The role of food as a limiting influence on population growth and stability has been hotly debated (see Hairston et al., 1960). Because an energy source is essential for the maintenance of a population, the food supply, directly or indirectly, independently or associated with other limited resources, should function in some manner in the regulation of populations.

In a long-term experiment involving the sheep blowfly, *Lucilia cuprina*, Nicholson (1954) demonstrated the influence of intraspecific competition in a population. Although the experimental population lacked all the complex interactions one would expect to find in nature, the work does show what might happen.

In one experiment, Nicholson fed to a culture of blowflies containing both adults and larvae a daily quantity of beef liver, plus an ample supply of dry sugar and water for the adults. The number of adults in the cages varied with violent oscillations (Fig. 11-7). When the population of adults was high, the flies laid such a vast number of eggs that the resulting larvae consumed all the food before they were large enough to pupate. As a result, no adult offspring came from the eggs laid during that period. Through natural mortality, the number of adults progressively declined, and few eggs were laid. Eventually a point was reached where the intensity of larval competition was so reduced that some of the larvae secured sufficient food to grow to a size large enough to pupate. These larvae in turn gave rise to egg-laying adults—after a developmental period of about 2 weeks. Meanwhile the population continued to decline, further reducing the intensity of larval competition and permitting an increasing number of larvae to survive. Eventually the adult population again rose to a very high level and the whole process started over again.

Competition for limited food held this blowfly population in a state of stability and prevented any indefinite increase or decrease. But the time lag involved between the addition of egg-laying adults by the way of larval survival to the declining population resulted in an alternate over- and undershooting of the equilibrium position, causing an oscillating population density.

In a second part of the experiment, Nicholson supplied the adults with a surplus of suitable food, which was unavailable to the larvae. As a result of the enormous quantity of eggs laid by the adults, larval competition intensified and eventually the density of adults decreased in a manner comparable to the other experiment (Fig. 11-8).

In another variation, the larvae were supplied with a surplus of food, and the adults were

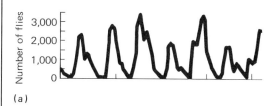

(a)

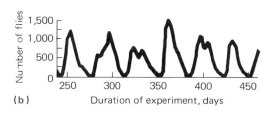

(b)

FIGURE 11-7

Fluctuations in the number of adult blowflies, Lucilia cuprina, *in cultures subjected to constant conditions but restricted to a daily quota of food: 50 g of food in* (a), *25 g in* (b). (*Redrawn with permission from Nicholson, 1957,* Cold Spring Harbor Symposia on Quantitative Biology *22:156.*)

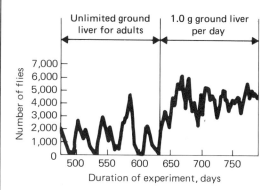

FIGURE 11-8

Fluctuations produced by restricting the daily quota of ground beef for adult sheep blowflies to 1.0 g after a period of ample supply, in a population governed by larval competition for a daily quota of 50 g of meat. (Redrawn by permission from Nicholson, 1957, Cold Spring Harbor Symposia on Quantitative Biology, 22:158.)

given a constant daily quota of protein food. Again the adult population oscillated. The adults produced a high number of eggs that, because of the lack of larval competition, nearly all developed into adults. The adults competed intensively for a limited amount of food. Lacking sufficient protein for the production of eggs, the adults laid fewer eggs; and for the lack of replacements the adult population declined. Competition was gradually relaxed to a point where some of the flies obtained enough protein to produce eggs. After a 2-week lag the adult population began to build up again.

From these results Nicholson felt that the magnitude of the oscillations would be reduced if the larvae and the adults competed for a limited quantity of food not available to the other. This assumption was confirmed experimentally. Under the conditions described not only were the fluctuations slight, but they lost their periodicity, and the mean population level was nearly quadrupled (Fig. 11-8).

In these competitive situations, the larvae and the adults were seeking food, the rate of supply of which was not influenced by the activity of the flies. In effect the resource, or food available, could be subdivided into many small parts to which the competitors, the larvae and the adult flies, had general access. The individuals "scrambled" for their food, which under gross crowding resulted in wastage. Each competitor got such a small fraction of the food that it was unable to survive and take a part in sustaining the population. "Scramble" competition tends to produce violent oscillations in the population not caused by environmental fluctuations. It limits the average density of the population far below that which the food supply could support if there were no wastage.

In contrast to the scramble type of competition is the "contest" type, in which each successful individual claims a supply of requisites sufficient for self-maintenance and reproduction. Unsuccessful individuals are denied access to food or space by the successful competitors, so the deleterious effects of shortage are confined to a fraction of the population. This either eliminates or greatly reduces the wastage of food, permits the maintenance of a relatively high population density, and prevents violent oscillations in numbers.

In the blowfly experiment, food was directly limiting, and its availability had a pronounced effect on the growth and size of the population.

The response of the population was characterized by sharp fluctuations brought on by starvation or the lack of nutrients needed for reproduction.

In a laboratory experiment involving two species of fruit flies, *Drosophila pseudoobscura* and *D. melanogaster*, Ayala (1968a) demonstrated that the growth and size of their populations were controlled by an interaction between limited space and food. The upper limit of population size was controlled by space. As the carrying capacity (which varied with the two temperatures involved, 25° and 19° C) increased, so did interference among flies. This interference decreased longevity. The rate of increase or productivity of the population, however, was influenced by food.

It is not difficult to assess the role of food in limiting population growth and size in a laboratory situation, where variables can be controlled. It is much more difficult to do so in a natural situation, where food may interact with other limiting influences. There are field situations, however, that do illustrate that food, either directly or indirectly, influences both fecundity and mortality.

The sycamore aphid feeds on the phloem sap of the sycamore leaves. Nymphal aphids feeding on young and senescent leaves tap a food rich in amino nitrogen and develop into large, heavy adults with a high reproductive rate (Dixon, 1970). This results in a spring and autumn increase in the population. Because the aphids do not need to move, they do not interfere with one another and can tolerate a high density. In summer, when the mature leaves are low in amino nitrogen and are the only source of food, the populations of the aphid decline, for the limitation of food interacts with population density. Reproduction in summer continues only if the population density is low enough to prevent individuals from disturbing one another or to inhibit movements to a new food source. Individuals in a dense summer population so disturb one another that reproduction is curtailed. The interaction is further complicated by such environmental conditions as temperature, solar radiation, and wind, which can reduce the suitability of many leaves on the sycamore as aphid habitat. These conditions limit the availability of the food resource even though there is an apparent abundance.

Quantity and quality of food have a pronounced effect on the populations of North

American deer. The mule deer and the white-tailed deer, inhabitants of the transitory early successional stages, do not appear to have developed a self-regulating mechanism characteristic of such ungulates as the Uganda kob and the mountain sheep that inhabit a relatively stable vegetation type. As a result, populations can increase dramatically, particularly in the absence of predation, and population decline is brought about by some interaction with the food supply (Klein, 1970a; Caughley, 1970). As the population increases to high levels, both the quantity and quality of available food decline. In periods of stress, such as a hard winter, starvation, especially among fawns, becomes common. But more commonly, too many deer survive and exist on inadequate nutrition. Poor nutrition results in the repression of growth, delayed sexual maturity, lower conception rates, increased intrauterine mortality, increased fawn mortality, and an age structure distorted toward older age classes. But if food quantity and quality are high, fertility is maximum, mortality is low, and the rate of increase is high (Fig. 11-9) unless the population is held down by hunting or predation. Eventually a high population results in the overutilization of the food supply.

Overutilization of food supply is implicated in a rodent-vegetation food supply hypothesis developed first by Lack (1954) and further refined by Pitelka (1957a,b; 1964) and Schultz (1964). The hypothesis relates the 4-year cycle of the Arctic lemming to ecosystem function (Fig. 11-10). According to the hypothesis the forage of the lemming is rich in calcium and phosphorus, but because of high production of forage much of these and other mineral elements are tied up in organic matter. Grazing, burrowing, and tunneling by lemmings reduce the organic insulation of the soil, and the soil thaws deeper. The next year the forage production is low, there is poor cycling of nutrients, and the lemming population is now under stress from a low and nutrient-poor food supply. Nutrient depletion affects reproduction and lactation, and the lack of adequate protective cover exposes the population to excessive predation. A combination of the two causes a rapid decline of the population. The lower population permits the vegetation to recover and organic matter to accumulate, but the forage quality is still low. By the fourth year the plants have fully recovered, protective cover has re-

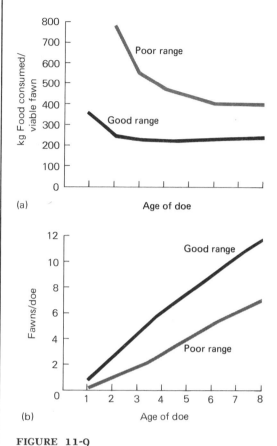

(a)

(b)

FIGURE 11-9

(a) *Production efficiency, and* (b) *the number of fawns produced during an 8-year life span by a typical doe on good and poor ranges.* (*From Short, 1972.*)

turned, nutrient cycling has speeded up, and nutrients accumulate in the plants. The nutrient content of the forage is high enough to stimulate reproduction and support a large population again.

Although the reasoning seems valid, the hypothesis is not fully supportable. Population densities of such rodents as the seed-eating omnivorous deer mice have been maintained by supplemental feeding (Bendell, 1959; M. H. Smith, 1971), as have the populations of pond snails (Eisenberg, 1970). But added food has failed to have any influence on vegetation-consuming microtine rodents such as the meadow mouse. Supplemental feeding of a low density of the California vole failed to produce either a rapidly expanding population or to prevent the decline of the population to low numbers (Krebs and Delong, 1965).

Food in itself may not act as a limitation on human population growth on a worldwide basis, nor have local food shortages reduced the population growth of modern man. But widespread starvation brought about by local food shortages has occurred throughout history. It has had an impact on local population densities and has forced emigrations. Among some of the more famous famines have been the ones in ancient Rome in 436 B.C., in Russia in 1600, in India in 1769–1770 and 1790, in China in 1878, and in Biafra in 1970. The potato famine in Ireland in 1840 killed more than 1 million and forced another million to emigrate.

But food affects human populations in a more insidious way. Although physically man exhibits a considerable metabolic tolerance, he can and does suffer from nutritional stress. Inadequate diet increases the frequency and severity of disease, causes such nutritional diseases as rickets, retards maturation, saps vitality of adults, and causes a rise in infant mortality. Poor nutrition results in multiple vitamin deficiencies. These problems are widespread in underdeveloped countries, where the deficiencies are aggravated by restricted carbohydrate diets of cereals with low fat and protein content. Protein deficiencies in the lower southern latitudes of the earth produce a syndrome known as kwashiorkor. It is characterized by grossly retarded growth, delayed maturity, depigmentation, anemia, and high infant mortality. Deprivation of adequate protein in the young, common among the poor in even the most highly developed countries, impairs the central nervous system during the early years of life and causes mental retardation.

Behavior

Intraspecific competition expresses itself in the social behavior of animals as the degree of tolerance and intolerance between individuals of the same species. Social behavior appears to be a mechanism that regulates populations by limiting the number of animals to a habitat, food supply, and reproductive activity and excluding the surplus (Wynne-Edwards, 1962, 1965). This right to food, the right to space, and the right to reproduce are the greatest competitive situations in which an individual animal can engage. It is a contest-type of competition in which violence is minimized and in which conventionalized display is used to intimidate rivals (Chapter 13).

A tendency exists among many animals to band together in flocks, herds, schools, or loose colonies. Many of these animal groups, especially the invertebrates, have some form of social organization based on intraspecific aggressiveness and intolerance and on the dominance of one individual over another. Two opposing forces are at work. One is the mutual attraction toward one another. At the same time the animals exhibit a degree of social intolerance, a negative reaction against crowding. Each individual occupies a position in the group based on dominance and submissiveness. In its simplest form there is an alpha individual, which is dominant over all others, a beta individual, which is dominant over all but alpha, and finally the totally subdominant individual, the omega. *Social rank* (hierarchy) is determined by fighting, bluffing, and threat at initial encounters between any given pair of individuals or at a series of such encounters. Once social rank is established, it is maintained by habitual subordination of those in lower positions, reinforced by threats and occasional punishment meted out by those of higher rank. Such organization results in social harmony by stabilizing and formalizing intraspecific competitive relationships and by resolving disputes with a minimum of fighting and consequent waste of energy.

Territoriality exists when an individual animal for a period of time defends a given area that is not shared with rivals. Its defense involves definitive behavioral patterns (see Chapter 13) that evoke escape and avoidance in rivals. As a result territorial individuals tend to occur in a more or less regular pattern of distribution.

An animal defends the entire area of his territory. *Home range* is the area in which an animal normally lives and it is not necessarily associated with any particular type of aggressive behavior. Among some species, home range and territory are the same. Among others only a part or none of the range is defended. The two sexes may have the same or different home range; however, these overlap and the home range of the male may embrace those of several females.

The home range is highly variable, even within a species. Seldom is it rigid in its use, its size, its establishment. The home range may be compact, continuous, or broken into two or more disconnected parts reached by trails and runways. Irregularities in the distribution of food and shelter produce corresponding irregularities in the home range and in the frequency of animal visitation. The animal does not necessarily visit every part daily. Its movements may be restricted to runways or its activities may be concentrated in the most attractive parts. These centers of activity vary with the season, the age of the occupant, and the intraspecific and even interspecific competition. Variations in home range are associated with the species, sex, and age of the animal, with the season, and with such ecological conditions as available food, cover, and intraspecific strife.

Territory can be classified according to activities that take place within them (Nice, 1941). Common among songbirds and among some mammals such as the muskrat is the mating, nesting, and feeding territory (Fig. 11-11). The males claim a territory and defend it by fighting or in the case of birds by song and display. The territorial defender usually is invincible on his own area. Late in the breeding season and after it, territorial defense breaks down. A second type is the mating and nesting territory with feeding done elsewhere. This is common among some hawks and black-capped chickadees. Others, such as the woodcock, grouse, and kob, utilize their territory for mating only. The male woodcock maintains an open-country singing and mating territory separate from the feeding area, which is not defended. A few gallinaceous birds, such as

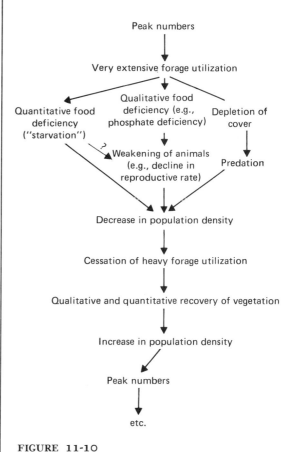

FIGURE 11-10
Pitelka's food supply hypothesis, one of the proposed regulatory mechanisms in an animal population. (From Krebs, 1964.)

the prairie grouse, have elaborate display grounds or leks where they mate promiscuously with females that come to them. Some species defend a nesting site only. Colonial birds such as the puffins and swallows defend territories that are barely large enough to embrace the nest. Their size is determined by the distance the bird can strike with its beak (J. T. Emlen, Jr., 1952). Finally there are feeding and roosting territories, commonly defended in winter. Even within this classification there are many variations. But they all have one feature in common—defense against competitors of the same species.

Territoriality and social hierarchy represent degrees of manifestation of the same basic dominance behavioral pattern. It is difficult to draw a sharp line between the two. A gradient in behavior exists from no social organization at one end to group territoriality at the other. Under one set of environmental conditions a species may be highly territorial; under another, territoriality might disappear into social hierarchy. In any case, social behavior, particularly territoriality, is considered a population-regulating mechanism.

THEORY OF TERRITORIALITY

Because territoriality involves the distribution of animals over a given habitat, a theory of territoriality can be considered from the viewpoint of habitat suitability (Fretwell and Lucas, 1969). The model assumes that each individual will select the optimum habitat, where its chances of survival and success at reproduction will be the highest. If all individuals choose a habitat of highest suitability, then from the point of view of unsettled individuals, all occupied habitats will be equally suitable. The model also assumes that all animals are free to enter any habitat on an equal basis, socially or otherwise. This is of course unrealistic, but is necessary to develop the model. When all of the optimum habitat is filled, then the next most suitable or suboptimal habitat becomes for the remaining individuals the most optimum habitat, and they begin to settle it. Once that is filled, the marginal habitat becomes the most suitable. The model implies that as density increases, the suitability of the habitat decreases. Thus a habitat has its highest suitability when the density is 0 (Fig. 11-12).

Based on this model of habitat suitability some hypotheses of territoriality can be considered. One is that territoriality functions in a density-dependent fashion, forcing some males under conditions of overcrowding into a less suitable habitat, from S_1 to S_2 (Kluijver and Tinbergen, 1953). At low levels of population density only the optimal habitat is occupied; at highest densities, all habitats from optimal to marginal are utilized. The behavior of settled individuals is a cue to unsettled ones that the area is completely populated.

A second hypothesis considers the territory as an elastic disk compressible to a certain size (Huxley, 1945). Territory size decreases as the density increases. As the territory is compressed to a certain limit, the resident resists further compression, and additional settlers are prevented from entering. Because aggressive behavior is not the same among all individuals, the most aggressive have the advantage and the less aggressive individuals are forced to settle somewhere else than in optimum habitat. In birds territoriality operates in a density-dependent fashion only if they are prevented from establishing territory as a result of territorial behavior (J. L. Brown, 1969).

A third hypothesis is that territoriality acts only to space individuals out within a habitat and that density is determined independent of behavior (Lack, 1964). At a certain density individuals separate as much as possible and have nonoverlapping home ranges. The role of territorial behavior is to isolate the male and to strengthen the pair bond.

With these hypotheses in mind, one can now consider territoriality and other social interactions as a regulating mechanism.

SOCIAL BEHAVIOR AND POPULATIONS

If social behavior is to limit populations in a density-dependent fashion, one has to show (Watson and Moss, 1970) that a substantial portion of the population consists of surplus animals that do not breed either because they die, or attempt to breed and fail; that such individuals are prevented from breeding by dominant or territorial individuals; that nonbreeding individuals are capable of breeding if dominant individuals are removed; and that breeding animals are not completely utilizing food and space.

A few field studies have tested this hypothesis. Stewart and Aldrich (1952) counted 148 territorial male birds in a 40-acre spruce forest in Maine from June 6 to June 14, 1947. From

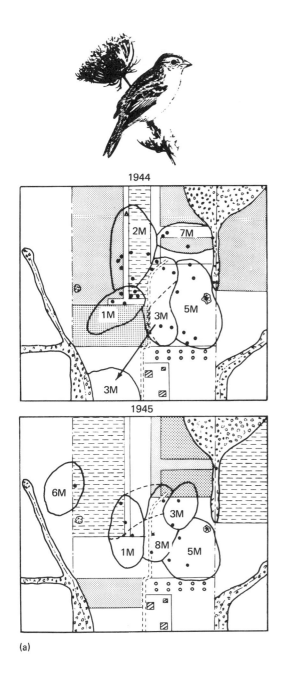

(a)

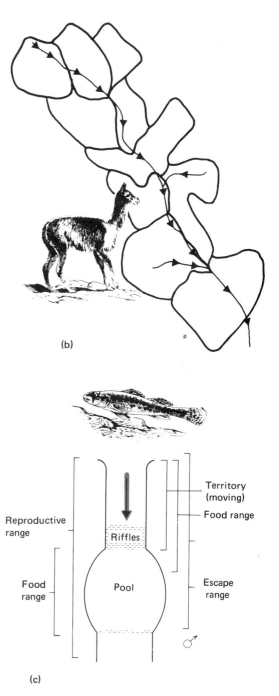

(b)

(c)

FIGURE 11-11

Some examples of territory in vertebrates. (a)
Territory of the grasshopper sparrow. (*From
R. L. Smith,* 1963.) *Note changes between
the two years. Dots indicate singing perches;
open triangles indicate nests.* (b) *Vicuna.*
(*After Koford,* 1957. © *The Ecological Society of
America.*) (c) *Iowa darter.* (*After Winn,* 1958a.
© *The Ecological Society of America.*)

June 15 to July 1, they shot from the area 302 territorial males, 84 adult females, 34 adults for which sex was not determined, and 35 young birds. The large number of males over the census count undoubtedly represented new birds that moved into the area to occupy the vacated territories. Hensley and Cope (1951) returned the following year. The area then was occupied by 154 territorial males. A subsequent removal of all birds they could find this time included 352 territorial males. These studies suggested that the replacements were surplus birds previously prevented from settling in the area. But the results were inconclusive because the investigators had no proof that the birds, largely males, did not or would not have established territories elsewhere. And the females involved in the Hensley and Cope study were not replaced.

A more carefully designed study of 13 experiments running from 1960 to 1966 was done by Watson and Jenkins (1968). It involved the removal of red grouse, which take up territories in autumn and hold them until the following summer. Inhabiting the area are both territorial and nonterritorial birds. When territorial grouse were shot or temporarily removed, territorial cocks that were neighbors of those shot invariably enlarged their territories before they, too, were removed. This suggests that there was previous pressure causing a reduction in the size of the territory. Nonterritorial cocks took over vacated territories and subsequently bred there. This indicates that, if given a chance, nonterritorial birds can hold a territory and breed. The number of territories on the area the following spring rose to the same level as the previous fall before the birds were removed, whether or not all previous birds were shot and regardless of the number of nonterritorial birds available.

In a similar type of experiment, Krebs (1971) removed breeding pairs of great tits from their breeding territory in an English woodland (Fig. 11-13). They were replaced by new birds, largely first-year individuals that moved in from territories in hedgerows, considered suboptimal habitat. The vacated hedgerow territories, however, were not filled, suggesting that a "floating reserve" of nonterritorial birds did not exist. Thus it is territory that limits breeding density, at least in optimal habitat of this bird, rather than food, which Lack (1964) postulated.

One argument against territoriality as a density-dependent regulating mechanism is the year-to-year variation in breeding density. This variation apparently is not related to food supply because abundance of food is not known at time of territorial establishment. But it may be associated with such conditions associated with territorial establishment as pressure from potential settlers, weather conditions, nature of the area, and rate and order of settlement.

The idea is supported by a study of an island population of song sparrows by Tompa (1962, 1964, 1971). Territorial birds claimed the shrubby growth that covered most of the island. The floating population of subdominant birds was restricted to grasslands, the major winter habitat. For 3 years the territorial population held relatively steady at 44 to 47 pairs. In the fourth year the population increased to 69 territorial males. The increase resulted from a spring snowstorm that came at the height of territorial activity. The storm eliminated 70 percent of the territorially active males, and at least 46 percent of the females. But survival was high among the subdominant males still living in winter habitat. These males quickly invaded the optimum nesting habitat. Because the birds established territories almost simultaneously, the area was divided quite uniformly. During the summer 17 males held territories without mates; in the fall migrant females settled in to even up the sex ratio.

The increase, however, was disadvantageous to the population. The high density the following spring resulted in increased social aggressiveness, delay of breeding, hatching failures, and a breaking up of territories. This study, then, indicates that territoriality can act as a limitation on population density as well as provide a degree of stability to the population.

Common to the tall-grass prairie region of midwestern North America is the dickcissel. Zimmerman (1971) has demonstrated that the territorial response of this bird compares to the elastic-disk hypothesis. The optimal habitat is characterized by a tall dense cover of herbaceous vegetation and the presence of song perches. Within this habitat, which is usually fully utilized (Fig. 11-14), territorial size decreases as density increases down to a minimum size of approximately 0.9 acre. This size minimum is reached at a density of 60 to 70/100 acres. At densities greater than this the surplus males are forced to less suitable vegetation. The number of mates attracted by the polygamous males is directly proportional to the volume of

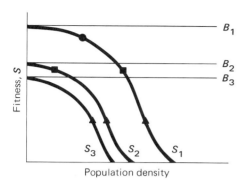

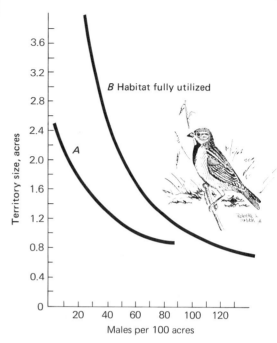

FIGURE 11-12

The relationship between habitat suitability, population density, and expected settling density in three habitats of different quality. B_1, B_2, and B_3 are the suitabilities of all three habitats. The curves S_1, S_2, and S_3 show the *actual fitness of individuals in the habitats at different population densities. At the low population density, indicated by the heavy dot, all populations will settle in habitat 1. At the intermediate density, indicated by the solid square, some individuals will settle in habitat 2. At a high density, indicated by the solid triangle, all three habitats will be settled. (After Fretwell and Lucas, 1969.)*

FIGURE 11-14

Territory as a function of male density in the dickcissel (Spiza americana). If the habitat were fully utilized, the density curve would appear as curve B. (From Zimmerman, 1971.)

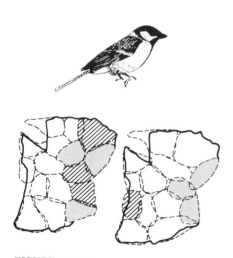

FIGURE 11-13

An example of the replacement of removed birds. Six pairs of great tits (Parus major) were shot between March 19 and 24, 1969 (left, stippled area). Within 3 days, four new pairs had taken up territories in the woods (right, stippled area). There was some expansion of residents' territories during the removal so that after replacement territories again formed a complete mosaic over the wood. (From Krebs, 1971. © The Ecological Society of America.)

vegetation. Males forced into less suitable vegetation attract fewer mates, reflected by a decrease in the sex ratio and a lower density of active nests.

In a similar manner, territoriality appears to stabilize the population of the arctic ground squirrel (Carl, 1971). The breeding population consists of males, restricted by territoriality, and territorial females, held at a relatively constant number by the availability of burrow sites. Mate selection comes about by the overlap of a female's burrow system with a male's territory. In the fall the breeding population holds prehibernation territories. In spring and fall surplus animals from the colony are driven into a refuge as a floating population that cannot breed because of the lack of suitable habitat. The refuge population serves as a reservoir of replacements; but most of the animals are killed off seasonally by environmental changes and by heavy predation by foxes and bears.

How a social hierarchy in the absence of territoriality can reduce the number of breeding pairs is more difficult to answer. Some species that exhibit social hierarchy are polygamous. For example, only a relatively small proportion of male elk and deer are engaged in breeding, but there is no evidence that some female elk and deer exclude other females from the breeding population. But in some species this is not the case.

For example, wolves live in small cooperative hunting groups or packs, which consist of a dominant male, a dominant female, subordinate males and females, peripheral males and females, and juveniles (Woolpy, 1968). The alpha or dominant male is deferred to by all members of the pack; he is the focal point of solicitation and the principal guard of the territory. The alpha female is dominant over all females and most of the males and controls the relationship of the rest of the females to the pack. The alpha male and female and the subordinate males and females form the nucleus of the pack. The peripheral males and females are kept out of it. Social relationships highly restrict fertile mating. The dominant female rather successfully prevents subordinate females from mating with other males. In addition, males exhibit a preference for certain females. The dominant male, preferred by all the females, may himself actually prefer a peripheral female, which the dominant female prevents from mating with him. Fertile mating usually occurs between the alpha male and the alpha female. This behavior places a severe restriction on mating habits and effectively limits the number of young. Perhaps there will be only one litter to a pack, the young being cooperatively raised by a number of adults. Since the optimum density of wolves in relation to an adequate supply of prey is about $1/10$ mi², the restrictive breeding limits population size. In areas where the wolf is heavily persecuted, social restrictions break down. The birth rate is high, with nearly all mature females producing a litter each season (Rausch, 1967).

Social behavior appears to restrict the number of breeding females of the gekkonid lizard *Gehyra variegata* (Bustard, 1969). Behavioral interactions between females result in a definite and rather constant upper limit to the number of female geckos that can inhabit any stump with a single male. The males are territorial and defend the homesite, usually a stump.

Territoriality may regulate populations of some species by restricting the food supply to a certain segment of the population (C. C. Smith, 1968). An example is found among the red squirrels (*Tamiascurius*) that inhabit the coniferous forests of the Pacific Northwest. Individual squirrels, male and female, each defend a territory through the year. Males enter the female's territory during a very short period of 1 to 2 days when she is in heat. At that time the female ceases to defend her territory. The territory centers about the food supply, and the size is adjusted according to its availability, which in general does not exceed three times the territorial owner's yearly requirements. The lower the overall food supply, which consists in part of cones and fungi that can be stored, the larger the territory. Thus territorial behavior allows each individual the optimal conditions for harvesting, defending, and storing a seasonal food supply, so that it will be available throughout the year. Kemp and Keith (1970) observed a similar type of territoriality in the populations of red squirrels in Alberta.

That the size of the territory might be regulated by food is suggested by the correlation between body size and diet, which influences the rate of energy expenditure (McNab, 1963; Schoener, 1968). Large animals tend to have larger home ranges or territories than smaller ones. "Hunters," whether herbivorous or carnivorous, who have to seek their food, have larger ranges relative to their body size than nonhunters. Predators tend to have larger territories than omnivores and herbivores of the same weight. Territory size increases more rapidly with body weight among the predators, which probably reflects a decreasing availability of food for those of increasing weight (Fig. 11-15). The home range of a red fox, for example, is around 1 mi², that of the mountain lion on the average is around 13 mi² (Hornocker, 1970).

TERRITORIALITY IN MAN

In recent years the role of territoriality and aggressive behavior in man has stirred considerable controversy (Ardrey, 1966; Lorenz, 1966; Morris, 1967, 1969; N. Tinbergen, 1968; see R. L. Smith, 1972). Although some aspects of human behavior may suggest territoriality on the part of individuals and nations, and though man may have exhibited group territoriality in the past (Tinbergen, 1968), it is difficult and perhaps dangerous to extrapolate animal territoriality to man without considerable modification.

Modern man doesn't quite measure up to a territorial animal in many respects. As Tinbergen points out, territorial animals, once settled on a territory, attack intruders; and on his territory the owner is virtually invulnerable. Once settled, territorial animals do not seek to expand their territory. In fact they may be forced to compress it to make room for another. But in most cases an animal seeking a territory avoids established owners. Intruders are driven off by hostile behavior that is more ritualistic than combative and rarely leads to the demise of the intruder. Thus a country's defense of its national boundaries is an overt act of territoriality; the invasion of a territory is not.

The premise that man is a territorial animal has never been proven by scientific study, but some insights are gained through the research of sociologists and psychologists. The maintenance of spacing apparently is a feature of human as well as animal behavior, but the nature and degree of spacing differ among different cultures and even different activities. Man may be colonial at work and highly territorial at home from both the respect of property boundaries and even rooms. One's room is usually considered a personal territory. Studies of spacing of students in a library (Sommers, 1971) and patients in old age homes and mental hospitals (Esser, 1965) indicate that each respects a zone of space around a neighbor, and each prefers a certain position at a

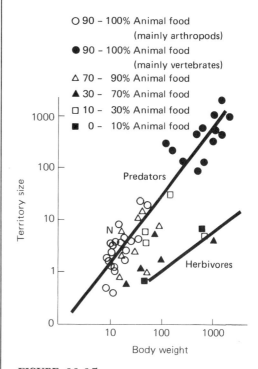

○ 90 – 100% Animal food (mainly arthropods)
● 90 – 100% Animal food (mainly vertebrates)
△ 70 – 90% Animal food
▲ 30 – 70% Animal food
□ 10 – 30% Animal food
■ 0 – 10% Animal food

FIGURE 11-15
The relationship between territory size (acres) and body weight for birds of various feeding categories. N = nuthatch species. (From Schoener, 1968. © The Ecological Society of America.)

table, a certain chair, and so on. Use of territorial mechanisms to ensure man's need for space are being applied to restrict the density in public areas such as campgrounds, which are faced with increasing population pressures. More and more campgrounds are being divided into a finite number of territories to achieve some preconceived idea of optimum density.

Social hierarchy is evident in man at various social and occupational levels. One hierarchy may exist at work, another at home, and another in social relationships. Pecking orders exist not only among inmates of prisons but also among members of industrial, business, and academic institutions. Such a hierarchy, as among animals, reduces the likelihood of conflict (Sommers, 1971) and permits people to coexist in confined areas. Within social groups the suggestion of dominance and subordination exists. The elite are economically dominant, acquire more space, move about more freely, and possess a greater opportunity to escape boredom. Those only partially or marginally integrated into the social structure exhibit various degrees of stress and tension. To further reduce stress from density, man, unlike other species, is able to create a number of occupational and social niches that restrict his interaction to certain individuals.

As with many animals, man appears to need those who lead and those who follow. At one time in history, such a hierarchical arrangement was accepted, particularly in the Middle Ages. But the industrial and scientific revolution caused great upheavals in the social structure, and the Renaissance brought to an end the ideas of tight social hierarchies among men. Man, however, still shows need for some hierarchical arrangement in his society to provide some order to his way of life and to his relationships with others of his own kind, even if hierarchy has lost any meaning for population regulation.

Stress

Associated with behavioral regulation is a physiologic control mechanism. Evidence suggests that social stresses act on the individual through a physiological feedback involving the endocrine system (Fig. 11-16). In vertebrates, this feedback is most closely associated with the pituitary and adrenal glands (Christian,

1963; Christian and Davis, 1964). Increasing populations of mice held in the laboratory resulted in the suppression of somatic growth and curtailment of reproductive functions in both sexes. Sexual maturation was delayed or totally inhibited at high population densities, so that in some populations no females reached normal sexual maturity. Intrauterine mortality of fetuses increased, especially in the fetuses of socially subordinate females. Increased population density resulted in inadequate lactation and subsequent stunting of the nurslings. The same effect appeared again to a lesser degree in the animals of the next generation, even though the parents were not subject to crowding.

Studies of some wild populations under stress seem to support this hypothesis. When the density is high, snowshoe hares suffer from severe physiological stress, typified by a low level of blood sugar and liver glycogen, producing a "shock disease" (MacLulich, 1947). Chitty (1952) found that a decline or crash of a vole population in Wales was characterized by strife during the breeding season, causing the death of young and a psychological derangement among adults. Young that survived from these parents were abnormal from birth. In Germany, F. Frank (1957) found that when the number of voles exceeded supportable density, a population crash occurred. This crash, brought on by psychological stresses induced by crowding and physiological stresses from food shortages, was triggered off by cold, wet weather. The die-off was characterized by the exhaustion of adrenal and pituitary functions, liver degeneration, convulsions, and cannibalism. Before this happened, mass emigration and reduced reproduction took place.

These behavioral and physiological changes take place even in the presence of an abundance of food. Rats living under crowded conditions change their behavior radically (Calhoun, 1952). Individuals form subcolonies and normal hierarchical relationships break down, leading to unstable groups. Females have reduced rates of conception; the suckling young have poor viability and few of them live to have young of their own. Crowding causes a marked increase in aggressive behavior with numerous and often fatal attacks on the young. Subdominant individuals, subjected to excessive punishment, fail to utilize the environment, including food. The growth of the animals is reduced, the adults failing to reach normal weight. Stunted rats characterized by behavioral disturbances result from prenatal impairment.

Confined populations of rabbits were studied by K. Myers and his associates (1971). The rabbits were held at different densities in different living spaces within confined areas of natural habitat. Those living in the smallest space, in spite of decline in numbers, suffered the most debilitating effects. Rates of sexual and aggressive behavior increased, particularly among the females. Reproduction declined, as did the ovulation rate and the number of corpora lutea. Fat about the kidneys decreased, and the kidneys showed inflammation and pitting on the surface. The weight of the liver and spleen decreased, and adrenal size increased. Stressed individuals had adrenals with a high proportion of the gland occupied by the *zona fasciculata-reticularis* and a low proportion by the *zona glomerulosa*. Young rabbits born to stressed mothers showed stunting in all body proportions and in organs. As adults they exhibited such behavioral and physiological aberrations as a high rate of aggressive and sexual activity, and large adrenal glands in relation to body weight. Low body weight, the lack of lipids in the adrenals, a reduced proportion of the *zona glomerulosa*, and poor survival indicated a lack of fitness of such rabbits. On the other hand, rabbits from low-to-medium-density populations showed excellent health and survival.

Among fish there is a marked negative correlation between density and rate of body growth. This relationship, evident to anyone who has observed a pond overstocked with bluegills, is one of the best-established examples of density-dependent relationships in populations (Fig. 11-17). The decline in growth is not attributable only to the lack of food. There are more subtle influences at work, such as the conditioning of the water. The role of conditioners has been studied more intensively in frog populations, among which a high tadpole density apparently inhibits growth. Associated with inhibited growth is an algal cell found in the feces of tadpoles (Richards, 1958). The effects of this inhibitory cell are nonspecific. Water from the environment of crowded tadpoles of one species can inhibit the growth of another (Licht, 1967).

Mechanisms that operate in animal populations do not necessarily also operate in human

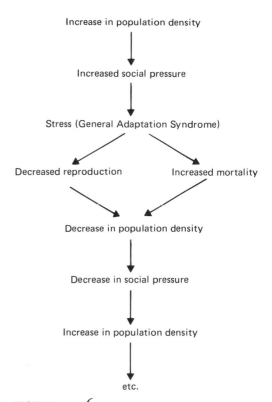

FIGURE 11-16

Christian's stress hypothesis of population regulation, which operates through the General Adaptation Syndrome. (From Krebs, 1964.)

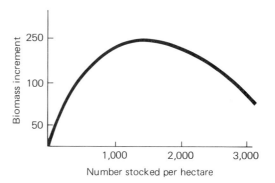

FIGURE 11-17

The growth of fish, as influenced by population density. This graph reflects a scramble type of competition. (From Backiel and LeCren, 1967.)

347

populations. There is evidence, however, that although stress might not regulate human population size, it does have other negative effects on the population. In time of war, when whole populations are living under extreme stress, congenital malformations in the young are more common than in times of peace. Where crowding is the worst, premature births, congenital absence of part of the brain (anencephaly), and infant mortality are the highest (Stott, 1962). There is evidence that emotional disturbances and such stress diseases as ulcers and heart conditions in pregnant women result in children born with gastrointestinal illnesses and other weaknesses that make them especially susceptible to early diseases (Stott, 1962). Children born of stressed mothers exhibit timidity, lack of motivation, excitability, inability to concentrate, and hyperactivity. Under primitive conditions, such children would have little chance of survival.

People living in poor housing are in a situation equivalent to an animal existing in submarginal habitat. This leads, according to Schorr (1963), to pessimism, passivity, poor health, and a high degree of sexual stimulation without a legitimate outlet. Although the evidence is inconclusive, it appears that as crowded conditions increase for an individual, the number of social obligations and the need to inhibit individual desires increase, and privacy becomes more difficult to attain. Individuals react with irritability, weariness, and withdrawal, and they exhibit an inability to plan ahead. In contrast to the situation among animals, increased density in man leads to increased fertility. This reflects hypersexuality and the ability of the human female to conceive over the 12 months of the year. Suboptimal environments also lead to the inability of parents to control their children, to an early abandonment of the home by the young, and to increased delinquency (Galle et al., 1972).

Plant populations also become stressed at high densities. Because plants reproduce both vegetatively and sexually by seed, the response to stress differs between the two modes of reproduction. In many ways the individual plant itself is a population that exhibits an increase in such basic components as leaves, flowers, and runners. As more and more units are added, interference between units increases and so does the demand on resources. As already discussed (see Chapter 10, life tables of plants), productivity of plants per unit area is inde-

pendent of the number of plants over a wide range of densities. The biomass per unit area may be the same whether distributed among many small individuals or concentrated in a few large ones. In this respect plants and fish are similar in their responses of growth to density.

One response, then, of plants to density is the reduction of the growth of individual plants. Botanists call it a plastic response. Individuals adjust their growth form, size, shape, number of leaves, flowers, and production of seeds in a "scramble" fashion to share the limited resources available.

A second response is increased mortality. High mortality is characteristic of the seedling stage, just as high mortality is characteristic of young animals. The size of a population that develops from seed is the function of the availability of seed and sites. If sites are abundant and seeds are plentiful, density-dependent mortality follows. As the number of seedlings increases, the chance that a seed will produce a mature plant decreases. No matter how great the densities involved, all finally converge to a certain number per unit area, a situation most obvious in forest trees (see Table 10-8). As the yield tables suggest, density varies with the size of the plants. A certain size has a similar level of survival at various densities. In some plants, reproduction by seed at high densities is almost nonexistent (Putwain et al., 1968); it is solely by vegetative means. As the density of plants increases, the number of vegetative offspring also declines.

A third response to the stress of density is a hierarchical exploitation of resources and space. A few individual plants become dominant and control the site. The remainder remain stunted and possess poor viability.

Genetic quality

Between 1956 and 1959 the populations of the western tent caterpillar collapsed in southern Manitoba. Wellington (1960), who was studying the population dynamics of this species, noted certain qualitative changes in the behavior of individuals in the caterpillar colonies. Some were active, others sluggish. A colony with active caterpillars built a number of long clavate tents, dispersed them widely, and foraged some distance away from them. A colony with sluggish individuals seldom constructed more than one compact pyramidal tent, and fed nearby. These sluggish colonies were less viable than active colonies in harsh environments and some were too sluggish to survive even in favorable situations.

As the infestation aged, even the active colonies decreased in size and activity. By 1959, fourth-instar colonies along 156 miles of roadway had decreased from 74,000 to 251. In years of minimal density colonies increased in size and vitality, suggesting that the resurgence was due to the elimination of the least viable portion of the population.

The reason behind this, Wellington suggests, is that although the strongest and most active moths may oviposit locally, they are apt to fly some distance away from their birthplace to lay their eggs. On the other hand, less active adults are incapable of sustained flight, and thus all these females able to oviposit must do so near their birthplace. Thus there is a partial emigration of viable stock and a complete retention of the poorer-quality but still viable stock, which hastens the decline of population quality in that locality. As long as the favorable environment permits the survival of poorer and poorer colonies produced by increasingly sluggish resident populations, the latter increase to a point where the sluggish colonies outnumber the active, produced by strong adults that remained or immigrated from the outside. The ultimate result is the presence of numerous colonies too sluggish to reproduce or even to survive under the most favorable circumstances. When they die, the local population decreases abruptly. The decline or "crash" is followed by a sudden recovery, after its least viable portion had been eliminated. In such a manner does colony life lead to the rapid deterioration of local population quality at a time when survival is maximum and the colonies are functioning most effectively.

In his work with voles, Chitty (1960) found that a decline in the number of these rodents can take place even though the environment appears to be favorable. In fact a high density of population is insufficient to start a decline, and a low density is insufficient to stop it. A majority of animals die from unknown causes, males more rapidly than females. However, the adult death rate is not abnormally high during years of maximum abundance. Since Chitty observed that individuals in a declining vole population were intrinsically less viable than their predecessors, any changes in the cause of mortality were insufficient to account for the

349

increased probability of death among the voles.

These studies suggest that as animal populations increase in density, the quality of the population changes. A deterioration in the quality of the population prevents an indefinite increase. The hypothesis suggests that a genetic feedback operates through density pressure, selection pressure, and genetic change within the population. Large increases in the population, brought about by a changing environment, increase the variability in the population, and many inferior genotypes survive. When conditions become more rigorous, these inferior types are eliminated, and the population is reduced, often abruptly (Fig. 11-18).

It has not been proved that the decline in the tent caterpillars or the mice resulted from an increase in inferior genotypes. Although some tests have been tried, the hypothesis has not been adequately tested among natural populations. Tamarin and Krebs (1969), in a study of the genetic quality of a population of the meadow mouse, suggested that aggressive voles have a breeding advantage over others and sire the most progeny. Perhaps certain genotypes are favored over others during different periods of population cycles. Based on studies of the short-tailed vole, the suggestion is made by Chitty and Phipps (1966) that under conditions of mutual interference, some genotypes survive and others do not. But the hypothesis remained untested because of a lack of information on the behavior and genetics of the populations involved. Ayala (1968), who made intensive studies of the population dynamics of fruit flies, suggests that the size of a population living in a certain environment depends upon its genetic constitution. Large populations have greater genetic variability than smaller populations. Such populations could adapt more efficiently to a changing environment. The more fit genotypes would survive, the less fit perish. By the same token, large populations possess the genetic variability that would permit the exploitation of and adaptation to a new environment (see Krebs et al., 1973).

Dispersal

Overcrowding and associated increase in aggressive behavior are major forces in producing dispersal. Social hierarchy and territoriality can force low-ranking individuals to depart and to seek suitable habitat in areas that are unoccupied by higher-ranking individuals or competing species. Individuals forced to leave are largely the maturing young.

Dispersal is a constant phenomenon, regardless of whether a population is dense or sparse. It is most pronounced when densities are high. One-way movement out of a population with no return is *emigration*. It is closely associated with the one-way movement into the population, *immigration*. The emigrants from one region are the immigrants into another.

If populations in an area are fairly stable and are at or near the carrying capacity, dispersal movements have little influence on population density and may only slightly influence age structure. But emigration movements induced by overpopulation and food shortages reduce populations and may greatly influence age structure and the reproductive rate of the remaining population. Immigration into rapidly growing populations can increase the growth rate not only by the addition of new members but also by an increase in the birth rate. Immigration into sparsely populated areas can bolster greatly the reproductive potential of a population and halt the reduction of r, characteristic of very low population densities. This principle is involved in wildlife management, where restocking programs with wild trapped individuals are undertaken to build up low populations; this has been very successful with wild turkey and deer.

The fate of the displaced subordinate individuals is usually death from exposure, from starvation, from competition from others of the same kind or other species already occupying the area, or from predation. If population densities are low, individuals forced from an area have a greater chance to find vacancies in suitable habitats. But if densities are high, emigrants are forced into suboptimal habitats or into traveling relatively great distances to seek new areas (see Errington, 1940, 1951). Once new individuals do become established successfully as breeding units, and occupy an area, the pioneers become the dominant individuals.

In spite of the seeming waste of individuals forced from an area during times when densities are high, the dispersal has important implications for the species. The more animals forced from an area, the more likely will marginal and suboptimal habitats be occupied, the more likely will a species spread into an area previously unoccupied by it. If colonization fails consistently, the chances are that sometime a few dispersing individuals will possess a mutation or genetic constitution that will enable it to adapt successfully to its surroundings.

Once dispersing individuals have established a new population in an unoccupied area, that population serves as a focus for further expansion. As the population grows and achieves a high enough density to encourage significant dispersal, new centers of population become established. This process is characteristic of populations expanding into new habitats. Early pioneers succeed in gaining a foothold; the population then expands and spreads even further. This produces a wavelike spread of the species. The waves of expansion are most pronounced when exotic plants and animals are introduced to a region. Able to exploit new niches, they take hold and spread with extreme rapidity. There are a number of excellent examples, one being the European starling. Introduced into New York City in 1890 and 1891, the bird expanded its range southward and westward until, in 59 years, its range included southern Canada, all of the United States except southern Florida, and northeastern Mexico. Its breeding range extends from British Columbia, Washington, California, Arizona, and Utah to Mississippi and eastward. In addition it has established migratory habits essentially the same as those it had in Europe (Kessel, 1953). More recently, on its own the cattle egret reached the New World from Africa and spread through North and South America (Fig. 11-19). It is now one of the most common egrets in the southeastern United States.

Similar stories could be written for the European pine shoot moth and the gypsy moth, as well as for some North American animals that have migrated into other lands. The gray squirrel and the muskrat have become serious pests abroad. And of course there is the well-known introduction of the European rabbits to Australia. The occupation of a new region by man follows a similar pattern, with early pioneers establishing the outposts, followed by a rapid buildup of the population (Fig. 11-20).

The dispersal of a population, and, if conditions are favorable, its colonization of a new area, involves a number of variables. One is the type of dispersal, active or passive. Most

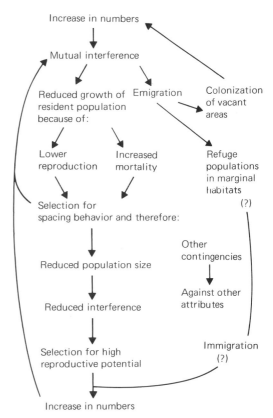

FIGURE 11-18

A modified version of Chitty's hypothesis to explain population fluctuations in small rodents. Density-related changes come about through natural selection. Selection through dispersal is highly important. Animals with highest reproductive potential disperse; those that remain behind are less influenced by population densities. (From Krebs et al., 1973. © 1973 American Association for the Advancement of Science.)

animals, being mobile, are actively dispersed. Parasites may be carried in the guts of animals. Some seeds likewise may be carried in the gut, or in the fur and feathers. But plants, lacking such animal processes as search, migration, and settlement, are for the most part passively dispersed by currents of air and water. The same is true of protozoans, algae, and such invertebrates as spiders.

A second variable is frequency of dispersal. The more frequently a plant has a heavy seed crop or a population reaches a high density, the more often will new propagules be available for colonization, even though losses are enormous. Regardless of the means of dispersal, active or passive, the number of disseminules per unit area decreases with distance, rapidly at first, then more slowly (Fig. 11-21). If one plots the number of seeds against density, the curve approximates exponential decay. The reasons are rather obvious. Surviving exposure, escaping predation, moving through hostile habitat for great distances, deflection of seeds by barriers, loss of a means of transport—all of these and more influence the rate at which emigrants reach a given area.

Just as no species exists alone, so it does not disperse in the absence of other dispersing organisms. Successful colonization requires transport to the area, the arrival of at least one reproductive unit of an unrepresented species, the entry of that species into the community, and its survival for at least several generations. Once the colonists are settled on the area, they face such risks to survival as predation or stiff competition from members of another species that are also colonizing the area. Or they may be unable to survive because of physical or chemical conditions, or be unable to give rise to a population because of biological conditions (see McGuire, 1972).

To successfully survive, the emigrant must find some suitable unoccupied habitat. In many situations these suitable habitats may be considered "islands" surrounded by areas of unsuitable habitat. The island might be an alder thicket surrounded by a broad expanse of drier uplands, a block of pines in an expanse of deciduous forest, a new pond, or even pools of water held in a pocket of leaves, a habitat for algae and protozoans. The colonization of such areas has been theoretically described by MacArthur and Wilson (1963, 1967). As an island or habitat fills up with species, the immigration rate declines, because the number

of species is approaching that found in surrounding source areas. Fewer and fewer immigrants belong to new species. At the same time some species on the island are declining toward local extinction, at a rate that increases as the number of species filling up the island increases. This happens for several reasons. The more species present, the more there are to become extinct. The more populations of species there are on the area, the smaller are the population sizes; and thus the chances of any one becoming extinct through local ecological or genetic changes are increased. The more species present, the greater the chance of increasing interference between them. When the rate of immigration equals the rate of extinction, the number of species in an area comes to an equilibrium (Fig. 11-22). Thus the number of species present in a given area at a given time can influence the chances of successful colonization by a species new to the area.

Migration

Migration is a two-way movement, involving a return to the area originally vacated. Migratory movements occur among highly mobile species, particularly vertebrates and some insects. Migratory movements usually are seasonal or periodic and may or may not follow traditional pathways (Fig. 11-23).

Insect migrations are more widespread than commonly believed, cover tremendous distances, and even involve specific flyways (Johnson, 1969). The most famous insect migrant in North America is the monarch butterfly, which each fall migrates south in huge flocks, roosting at night in selected trees. The return trip the following spring, however, includes overwinter survivors, as well as the progeny of these autumn migrants, which apparently breed on the wintering ground (Beall, 1941*a*, *b*; 1946). In the desert regions of the Middle East and North Africa the migratory movements of locusts have been famous since Biblical times. Locust migrations are governed to a large extent by the wind and temperature (Rainey and Waloff, 1951). Compared to migrations of other insects, locust movements are largely involuntary, but they do show regularity, since they take place in areas where season and trade winds and monsoons occur. These migrations lead the insects only by

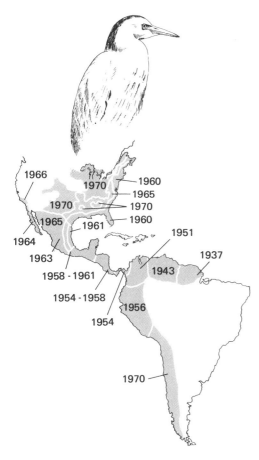

FIGURE 11-19

Unlike the starling and the English sparrow, the cattle egret invaded the New World without the aid of man. It came from Africa and first colonized the northeastern coast of South America. From there it spread northward into North America and southward along the coast of western South America. The cattle egret has been so successful in colonizing the New World because it has been able to occupy a niche unfilled by native herons and egrets. Although it nests in rookeries near other herons, the cattle egret feeds largely on insects in grass, pasture, and cropland. (Map constructed from data from numerous sources.)

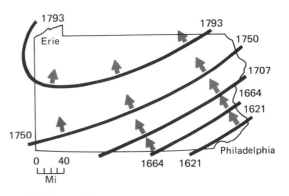

FIGURE 11-20

The spread of human settlements across a region, such as Pennsylvania, follows the same general pattern as the spread of animal populations. (From Florin, 1965.)

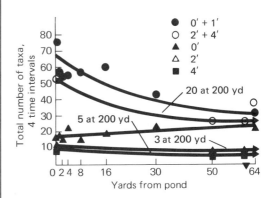

FIGURE 11-21

The number of disseminules of algae, insect larvae, etc., colonizing individuals decreases as the distance from the species pool increases. This graph shows the number of disseminules colonizing small bodies of water held by leaves of plants or in tree holes as a function of distance from a pond. In this case the small bodies of water were small bottles of water set at increasing distances from a pond. (From B. Maguire, Jr., 1971, Ann. Rev. Ecol. and Systematics, 2:441; redrawn by permission.)

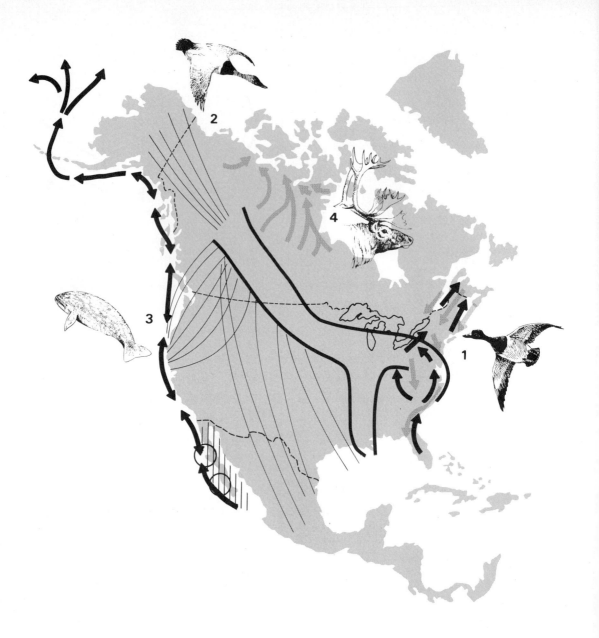

chance to suitable breeding places. When they arrive at a moist area, the locusts develop to sexual maturity and lay eggs (Rainey and Waloff, 1948).

The migratory habit is highly developed among many fish. Some make short journeys; others, such as the salmon and eel, make long, spectacular journeys. Fish that make these long migrations may be *catadromous*, those that migrate from fresh water to the ocean to breed, or *anadromous*, those that return from the ocean to fresh water to breed.

The eel is a catadromous fish that lives in fresh-water streams but migrates to the area of the Sargasso Sea to spawn and then die. The eggs hatch in late winter or early spring

into quarter-inch larvae. The two populations, American and European, both drift with the Gulf Stream to their respective shores. In 1 year's time the American, and in 3 years' time the European eels arrive along the coast, both as 2- to 3-inch larvae. They metamorphose into "glass eels," or elvers, and swarm up streams. In 8 to 10 years the eels mature; then, come autumn, they return downstream to the ocean. The Pacific Coast salmon, anadromous fish, spawn in the cold headwater streams of the northwest river systems. The fry return to the sea, where they mature. They then return up to 1,000 miles inland from the ocean to the home stream where they were spawned. If a population is prevented from reaching its home stream by a dam, the fish will not spawn and the population will perish.

More restricted migrations are made by a number of fish. Lake-dwelling salmonid fish such as the rainbow trout and lake trout may ascend streams to spawn, then return to the lake (Hartman, 1957; Loftus, 1958). On the Pacific Coast, "steelhead" rainbow trout spend their adult life in salt water and migrate upstream to fresh water to spawn. Stream-dwelling trout move upstream to spawn and return downstream to the home area sometime later (Schuck, 1943; Shetter, 1937). Some ocean fish such as tuna and herring also undertake migratory trips.

The first warm spell of spring initiates migration of many amphibians to ponds where they mate and deposit their eggs in the water, and then return to damp retreats beneath logs and stones and in the soil.

The most conspicuous migrants are birds. In autumn, birds flock and move south, to return in the spring. Most of these migrations follow regularly used paths or flyways (Fig. 11-23). The movements and times of departure usually coincide with weather conditions. Southward migrations appear to be timed to start after the passage of a cold front with its flow of continental polar air coming in from a northerly direction. Northward migration in spring usually is timed with the onset of a warm front, with barometric pressure dropping and warm moist air moving up from the Gulf of Mexico and the Caribbean. When a cold front arrives, migratory movements stop, not to resume again until high-pressure areas have passed. In fact radar studies (Drury et al., 1961; Eastwood, 1971) show that a cold front in spring may start immediately a steady reverse southward movement. Radar has revealed a similar reverse migration in autumn, when the birds meet a warm tropical air mass moving north.

Migration among birds does not take place until the urge is released by a stimulus. The proper release of this stimulus requires that the bird be in a proper physiological state. Before migrating in the fall, birds put on a heavy deposit of fat. In spring, the increased photoperiod (see Chapter 7) stimulates the pituitary gland, in turn stimulating gonadal activity. Increasing temperatures reduce energy requirement. This, plus the lengthening day, causes nightly unrest among the birds, which results in increased rate of feeding, be-

FIGURE 11-23 [OPPOSITE]
Migration routes of four vertebrates. (1) Ring-necked duck. Light arrows indicate fall migration; dark arrows, spring migration. (After Mendall, 1958.) (2) Canvasback duck. Fine lines represent minor routes; broad path, the major route. (After R. E. Stewart et al., 1958.) (3) California gray whale. Breeding area is indicated around Baja California. (After Pike, 1962.) (4) Barren-ground caribou. (After Banfield, 1961.)

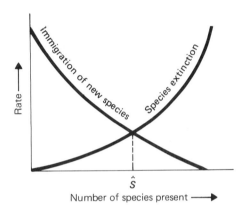

FIGURE 11-22
A model of colonization and extinction, particularly as it applies to island conditions, although the model has validity in other colonization situations. The number of new immigrant species falls and the number of extinctions rises as the number of species on an island increases. The two processes balance, and the island is at equilibrium at the point where the two lines intersect. At this point immigration rate of new species equals extinction rates. When all of the island's habitats are occupied by appropriate species, a sharp rise in the extinction rate occurs. (From R. MacArthur and Wilson, 1963.)

yond the ordinary need for existence. Fat accumulates, providing the reserve needed for northward flight. This nightly unrest, both spring and fall, continues until it reaches a threshold that responds to the environmental change, usually temperature, that triggers migratory movement. Birds not in proper physiological condition fail to migrate (see Farner, 1955; Kendeigh et al., 1960).

The ability of birds and fish to return to the same place they occupied the previous year has been the subject of considerable research, and the final explanation has yet to be learned. Many theories have been advanced to explain direction-finding in certain birds and fish. Those currently supported by research hold that these organisms can determine direction by the sun and compensate for daily change in the sun's position relative to the earth, local time (as indicated by the animal's "internal clock"—see Periodicity), and season (Hochbaum, 1955; Braemer, 1960; Hasler, 1960). Experimental evidence seems to indicate that these animals possess some sort of "computing" mechanism that indicates to them that every 24 hours the sun makes a revolution around the horizon. Even when reared under sunless conditions, fish and starlings still possess the ability to calculate the sun's movement (Kraemer, 1960). When, however, fish trained to compass direction under the sun at Wisconsin were transported to the equator and Southern Hemisphere, they did not adapt to the radically different daily sun movement. Instead they continued to compensate for the azimuth curve of the sun that would have been correct for Wisconsin. (Hasler and Schwassmann, 1960).

Although the sun compass explains direction-finding in the day, how birds navigate at night requires another explanation. Results of experiments carried out by Sauer and Sauer (1955, 1960) suggest that both the azimuth and altitude (bicoordinate celestial navigation) of the starry skies influence the directional choice made by the birds. This presupposes that the birds have the ability to perform actual navigation, an idea that has been severely criticized (Wallraff, 1960).

The findings of the Sauers were partially verified by some careful experimentation by S. T. Emlen (1967, 1972) on the migratory orientation in the indigo bunting. This bird of brushy pastures is a moderately long-distance nocturnal migrant. Emlen used two approaches

to the problem. One was to house the buntings in a planetarium where the seasonal star patterns could be manipulated at will or shut off to simulate a cloudy night. When the buntings in autumnal migratory condition were exposed to a star-filled sky they indeed did orient themselves in a southerly direction as indicated by the star pattern on the panetarium. This happened regardless of whether the planetarium sky was normal or reversed; that is, whether the planetarium skies pointed to true south or not. The birds lost all ability to orient themselves when the stars were not visible. Further experimentation suggested that the birds did not rely on altitude and azimuth for navigation, nor did they employ a means to compensate for time changes as proposed in sun-compass orientation. Birds apparently used additional information provided by numerous other stars and by the patterns of the stars in the sky.

In another set of experiments Emlen manipulated the photoperiod so that one group of buntings was in a physiological state characteristic of the autumnal migratory period and at the same time, another group was in a state characteristic of the spring migratory period. The birds were then exposed simultaneously to the spring planetarium sky. Those in autumn condition oriented southward, those in spring condition, northward. Thus physiological states produce opposite and correct responses to the same stimulus of the stars.

Recently radar observations of migrating birds have provided information that challenges the idea that a star or sun compass is necessary for avian migration. Birds, particularly waterfowl, can continue well-oriented migration under cloudy skies and even within the clouds themselves (Eastwood, 1971). The suggestion has been made (Bellrose, 1967) that the birds use the turbulent structure of the wind as a directional cue. But work with racing pigeons has resurrected a theory (Yeagley, 1947) once discarded that the earth's magnetic field is a source of directional information. When birds carrying small magnets were released under cloudy skies they became disoriented (Wiltschko et al., 1971, 1972). Birds without the magnets were able to orient correctly to home base, however. But as yet no one is able to prove a sensory basis for the detection of magnetic stimuli. Many problems still remain to be solved in the field of animal navigation.

Once the animals arrive in the general home vicinity they find their exact location by random search and probably recognize the area by familiar landmarks, although even this is open to question (Wallraff, 1960). Salmon locate their home stream by a sense of small, the odors serving as guideposts. Young salmon are conditioned to the organic odor of the home stream during the fingerling period. Then when as mature salmon they reach the mouth of the home river, they are stimulated to enter by the characteristic of that water alone. They swim upstream against the current past one tributary, then another, until they detect the strong traces of the home stream and enter it. Occasionally faulty choices are made, but the fish backtrack until the correct spawning or home tributary is located (Hasler, 1960).

Among the reptiles, the journey of young green turtles from their nest far back on the sandy beach to the sea is outstanding (Carr, 1962). Although navigation per se is not involved, the short trip to the oceans seems to be based on a tendency of the turtles to move toward a better-illuminated sky or a horizon free of obstacles.

Migration among North American mammals generally is confined to the more northern forms and to those that spend a part of the year in an area not offering a year-long food supply. The gray whale spends the winters along the bays and lagoons of the lower coast of California, where the young are safe from killer whales, and return, in summer, to the Arctic Ocean (Fig. 11-23). Females, pups, and young males of the fur seal winter as far south as the waters off southern California, while adult males winter south of the Aleutians and in the Gulf of Alaska. As the breeding season approaches, the bulls return to the Pribilof Islands; meanwhile the females migrate 3,000 miles, arriving at the islands a few hours or days ahead of the birth of the young. By November the females with grown pups start the long journey back to southern waters. Caribou migrations are anticipated by the Eskimos and Indians, who depend upon these animals for food. Caribou in the Mount McKinley area of Alaska spend the summer on the grassland of the southern slopes of the Alaskan Range (Murie, 1944). In the winter they shift to the lichen range on the northern side, where the snowfall is lighter. Caribou tend to shift their ranges to new areas after passing over one route a number of years. Nearly as famous are the migrations of elk in the Jackson Hole country, Colorado, and Yellowstone Park. Here elk make regular journeys from the snow-covered highlands in winter to the lowest valleys and sagebrush plains (C. C. Anderson, 1958). They return to the mountains in spring, following the snowline. Mule deer of western United States spend their summers well back in the mountain meadows and along the higher ridges, from 4,000 to 12,000 ft up, and winter in the oak-covered foothills, from 300 to 1,500 feet up. Mule deer of the chaparral country are more restricted but move from south-facing slopes in late spring to cool northern exposures in late summer (Taber and Dasmann, 1958).

This brief discussion of migration as a phase of dispersal has strayed a bit from population regulatory mechanisms, yet migration is very important. The migratory habit enables many species to exploit new food resources at different times of the year or seek more favorable environments. The migratory movements often result in heavy mortality, and survival on the wintering grounds is influenced by a whole new set of conditions. Migratory populations are subject to different sets of environmental stresses; and a limited winter habitat can reduce the number of animals that return to the breeding grounds in spring.

Population fluctuations

Interactions of density-dependent and density-independent influences result in the fluctuations of populations. For the most part, fluctuations are irregular. But the fluctuations of some populations are more regular than one would expect by chance. These are commonly called *cycles*. The two most common intervals between oscillations are 3 to 4 years, typified by lemmings (Fig. 11-24), and 9 to 10 years, typified by the lynx and the snowshoe hare. These cyclic fluctuations are largely confined to the simpler ecosystems such as the northern coniferous forest and the tundra.

Occasionally among the European lemming species, the numbers become so great that mass emigrations take place. These begin as seasonal movements down the mountains from one vegetation type to another, from the pre-

ferred lichens to the willows to the birch forests, and finally to the conifers (Curry-Lindahl, 1962). Breeding as they move slowly down from higher elevations, the lemmings build up to new population highs at lower elevations. If the rodents' slow progress is stopped by a long lake or the meeting of two rivers, their continuing accumulation becomes so great that panic breaks out, resulting in a reckless movement that follows no special direction, that flows north or south, east or west, uphill or down, across rivers and lakes, and sometimes ending disastrously in the sea. The cause of this behavior appears to be caused by endocrine malfunctioning, particularly of the adrenal gland, but this is questioned by Clough (1965a, b).

Further south in the coniferous forest biome the snowshoe hare follows a 9- to 10-year cycle. The hare is the chief food supply of the lynx, so the lynx population is also cyclic, with its peaks and depressions somewhat behind those of the hare, traceable through fur reports back to 1735. The ruffed grouse, too, is supposed to follow a similar cycle; but there is some dispute as to whether any animal south of the tundra is truly cyclic. Some biologists think these are merely statistical pseudocycles.

Cycles do not appear to affect migratory species in the region, only permanent residents; nor do they appear on a continental level. Only local and regional populations are affected, although the 10-year cycle of the snowshoe hare and the lynx are broadly synchronized from Quebec to the Northwest Territory (Keith, 1963).

A number of theories have been advanced concerning cycles. These can be divided into two main schools. One maintains that something in the physical environment, in the ecosystem, or in the population itself is able to cause rhythmic fluctuations. Predation has been singled out as a cause, but predators usually are not abundant enough when rodents are at a peak to bring about a decline. Food shortages, perhaps brought on by overpopulation, have been considered as a cause. Lack (1954) suggests that cycles are caused by a combination of food shortages and predation. Other causes relate to mechanisms already discussed. A vegetation–herbivore hypothesis relates to a food shortage and loss of cover brought about by overpopulation of lemmings, followed by heavy predation and subsequent recovery of the vegetation (Pitelka, 1957*a, b*). Endocrine malfunction has been suggested, as have changes in the quality and behavior of animals. Another school, represented by Cole (1951, 1954*b*) and Palmgren (1949), holds that cycles simply represent random fluctuations of a population. Populations are affected by a variety of environmental forces, and random oscillations are the result. Cycles then simply reflect random oscillations or fluctuations in environmental conditions.

Cyclic or irregular, fluctuations in populations are characterized by certain demographic parameters. Meslow and Keith (1968) followed a snowshoe hare population from its peak in numbers through the decline to a low and its first stages of recovery. The decline, which really began at peak populations, was characterized first by a decline in juvenile survival followed by a decline in adult survival and a decrease in the number of young born per female. The reproductive rate was halved and the age structure became top-heavy with old individuals. In the upswing, reproductive rate and adult survival doubled, and juvenile survival increased. Critical to both the decline and the upswing was juvenile survival.

Decline in a feral house mouse population had similar parameters: a large reduction in weaned young and a cessation of breeding by females (DeLong, 1966). A similar cessation occurred in a declining meadow mouse population, in spite of plenty of food, cover, and absence of predation (Krebs et al., 1971). A declining population can be pushed to extinction by competition and interference (Lidicker, 1966). A feral house mouse population declining from a high density faced competition from a rapidly increasing population of voles. The house mice, experiencing reduced vitality, had a declining reproductive rate that was further reduced by persistent interference from aggressive voles, reduced food supply, and social disintegration caused by increased wandering and annoyance from voles. As a result the local house mouse population became extinct (Fig. 11-25).

GRAPHIC ANALYSIS— REPRODUCTION CURVES

The result of intraspecific competition in all its forms is an increasing rate of mortality or a decrease in the production of young to maturity. Mortality is usually heaviest among the

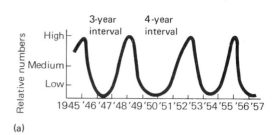

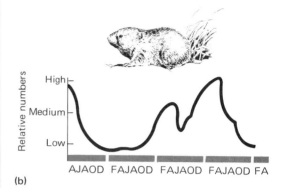

FIGURE 11-24

(a) *A generalized curve of the 3- to 4-year cycle of the brown lemming near Barrow, Alaska.* (b) *A generalized curve of a single oscillation in a short-term cycle showing subordinate fluctuations.* D, December; O, October; J, July; A, April. (*After F. Pitelka, 1957, "Some Characteristics of Microtine Cycles in the Arctic,"* 18th Annual Biology Colloquium, *Oregon State College, Corvallis.*)

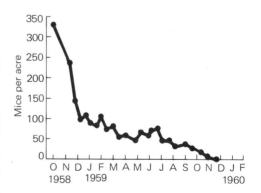

FIGURE 11-25

A graph depicting the decline and extinction of house mouse population. (From Lidicker, 1966. © *The Ecological Society of America.*)

359

young, and it directly influences the recruitment of new reproductive members to the population. The effect of density-dependent mortality upon this recruitment or addition is in part a function of the number of parents that produce the offspring. This relationship can be shown graphically by plotting net reproduction (mature progeny) against the density of the stock that produced them. The resulting graphs are known as *reproduction curves* (Ricker 1954, 1958a, b) (Fig. 11-26).

Reproduction curves consist of two parts: a diagonal 45° line and a domed curve. The 45° line represents the replacement level of the stock, reproduction in which density-dependence is absent. This line is only a guide; the other curve is the actual population. The filial generation tends always to be equal to the parental except for density-independent deflections under such conditions. Such a stock has no mechanism for the regulation of its numbers. Eventually by chance it would decrease to 0. The domed curve plots the actual recruitment in relation to the density of parent stock. As indicated by the sigmoid growth curve, maximum recruitment occurs at some intermediate level of abundance. The ascending limb is steeper than the descending limb. The apex of the domed curve lying above and to the left of the diagonal line represents the maximum replacement reproduction. The curve must cut at least once and usually only once the 45° line and must end below and to the right of it. Where the curve and the diagonal intersect is the point at which the parents are producing just enough progeny to replace current losses from the reproductive units. Basically any curve lying wholly above the 45° line would describe stock that is increasing without limit, something that does not exist in nature. A curve below the 45° line would describe stock headed for extinction.

There are, of course, many kinds of reproduction curves. In the curves in Fig. 11-26 recruitment rises to a maximum as the density of the parental population increases, and then falls off. They differ in the size of maximum recruitment and in the size of parental population necessary to reproduce the maximum. Curve *A* represents a population in which contest competition exists and in which a low density of parental stock is not very productive. The flatter dome and more gently inclined limbs are characteristic of territorial

animals whose populations are divided into partially distinct units, so that the effects of density are not uniformly felt through the stock. Curves *B* and *C*, with a steep right climb, are characteristic of scramble competition. In curves *C* and *D* substantial reproduction takes place only over a narrow range of stock densities considerably below the equilibrium level, the point at which the number of mature progeny equals the number of parents.

An interesting aspect of these reproduction curves is that they generate cycles within a single-species population in a constant environment, just as happened in Nicholson's blowfly experiments. In curves *C* and *D*, for example, the increasing density of stock exceeds and precedes the replacement number. In such a situation a population tends to oscillate in abundance. If the reproduction curve crosses the 45° line with a slope between −1 and −∞, as it does in curve *E*, a population of single-age parental stock will have an irregular but permanent cycle of abundance; if the parental stock is multiple-aged, the cycle is more regular. On the other hand, if the right-hand slope of the curve is between 0 and −1, the cycles are damped and eventually disappear.

SELF-REGULATION IN POPULATIONS

Evidence seems to indicate that some natural populations do possess some form of density-dependent self-regulation that holds the numbers within some upper and lower limits. The direction of change upward and downward is influenced by the size of the population. This regulating mechanism in part involves intraspecific competition over food and space and may be expressed in some form of social behavior such as social hierarchy and territoriality. Reduction in available space or an increase in density relative to available resources may result in undesirable behavioral, physiological, and even genetic changes; in a decline in reproductive success; and in an increase in mortality. These mechanisms often interact and may even be masked by density-independent influences, especially weather. Environmental changes may result in local extinctions (Ehrlich et al., 1972), with the population eventually being reestablished by immigration into the area. Self-regulation also interacts with other pressures, particularly predation, disease, and interspecific competition, all discussed in Chapter 12.

An intriguing question is to what extent self-regulating processes might be involved in human populations, because man does experience stress, behavioral changes, emigration, and immigration, starvation, and disease.

At one time, populations of man as hunter and gatherer were probably subject to regulation similar to that of animal populations. Food shortages, disease, and natural disasters held human populations to some level below the capacity of the environment to support them. Even intraspecific competition over food and space might have been involved, as it is in some aboriginal peoples today (Vayda, 1961; see R. L. Smith, 1972). In the Middle Ages, man, crowded into small towns and cities under unsanitary conditions, suffered heavily from disease and warfare. But after the Industrial Revolution, man escaped many natural adaptations that might have effectively controlled populations, and lost natural mechanisms for population homeostasis. Man lessened the biological aspects of reproduction and imposed upon it cultural and social norms. The regulation of human densities no longer occurs through natural means but rather through cultural processes (Stott, 1962). As a result, the population of man is experiencing a "superexponential" growth rate (Holling, 1966) with all the attributes of a bacterial colony in the first stages of growth when population densities are already high.

At the same time man is experiencing an urban revolution. Never in his history has he existed in such great numbers in an urban environment. In 1900 only one country, Great Britain, was considered urbanized. By 1965 all industrial nations were urbanized. By 1990 half the world's population will live in urban places of all sizes. Man has been undergoing urban industrial evolution for only 200 out of his 10,000 years as an agrarian. It is a new experience, bringing with it new stresses, an intensification of social competition, an emigration from a simple rural life to a complex urban world to which man is not yet adapted. Perhaps responses to these environmental changes and increased densities are beginning to show in the problems of the ghetto, of the poor, of drug addiction and alcoholism, of the violence of the inner city and international crises, of waste disposal and of resource depletion. Sometimes it seems that man must

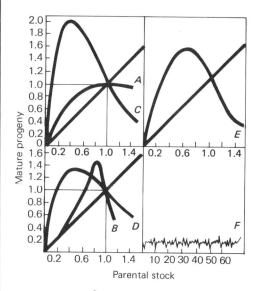

FIGURE 11-26

A set of reproduction curves. In curves A, C, and D reproduction rises to a maximum as the density of parental stock increases, but they differ in the magnitude of maximum reproduction relative to replacement level and in the size of the parental stock that produces the maximum. Curve A is typical of territorial animals, a contest type of competition; curves B and C with a steep right climb are characteristic of scramble competition. B represents a situation in which the lowest stock densities are not very productive. Curve E represents a stock reproduction relationship in which there is an oscillation equilibrium. This curve will generate the cycle shown in the lower righthand square when the population consists of single-age parental stock. The values of the ordinate and the abscissa represent fractions or multiples of replacement level of stock. For the mathematical determination of these values see Ricker, 1958a,b.) (*Based on Ricker, 1958a,b.*)

reach some equilibrium situation character-ized by low mortality, stable age distribution, and a zero growth rate.

The equilibrium can come about by some catastrophic biological process or it can come about by some managed cultural process. Es-sentially cultural regulation involves family planning as a means of population control. It can be accomplished by birth control, by so-cial changes, or by a combination of both. Social changes involve a change in the ideal family size to some low number of children by discouraging early marriages, by encourag-ing life styles that compete against marriages and large families, and by encouraging a nar-rower child-bearing age range.

Because in man, as in animals, population densities are closely related to the availability of resources, another aspect of population stabilization is a move toward a stationary economy. A growing population coupled with a growing economy results in a depletion of natural resources. Because no nation, state, or local community looks with favor on a de-cline in population or a decline in revenues, few view a stationary economy and a stabilized population with enthusiasm.

The stationary state implies a constant stock of physical wealth and a constant stock of people (Daly, 1972). Both involve a low rate of flow through the system. In the econ-omy it means low production, low consump-tion, and high durability. In the population it means a low birth rate, a low death rate, and a high life expectancy. Such a stationary state would require radical changes in the social structure.

Because it is so contrary to economic thought, the stationary state is unacceptable to most economists and politicians. Yet the idea is not new. It was advocated over a cen-tury ago by the economist John Stuart Mill (1857), who wrote:

A population may be too crowded, though all be amply supplied with food and raiment. It is not good for a man to be kept perforce at all times in the presence of his species. . . . Nor is there much satisfaction in con-templating the world with nothing left to the spontaneous activity of nature; with every rood of land brought into cultivation, which is capable of growing food for human beings; every flowery waste or natural pasture plowed up, all quadrupeds or birds which are not

domesticated for man's use exterminated as his rivals for food, every hedgerow or su-perfluous tree rooted out, and scarcely a place left where a wild shrub or flower could grow without being eradicated as a weed in the name of improved agriculture. If the earth must lose that great portion of its pleasantness which it owes to things that the unlimited increase of wealth and population would extirpate from it, for the mere purpose of enabling it to support a larger, but not a happier or a better population, I sincerely hope, for the sake of posterity, that they will be content to be stationary, long before necessity compels them to it.

Perhaps some indication of the regulation of population through cultural processes is the demographic transition exhibited by cer-tain western nations. As European nations became industrialized and urbanized, people saw new economic opportunities and had new aspirations. They found that in cities children were strictly consumers, not producers, ex-pensive to raise and educate. If they and their children were to succeed under urban condi-tions they needed education, special skills, and freedom to move to places of new oppor-tunity. Large families inhibited all of these. Even in rural areas large families were at a disadvantage. Mechanization reduced the number of laborers needed on the farm and required heavier capitalization. Parents lived longer, postponing the time when a son would take over the farm. As farms were subdivided among children, a point came when further subsidivisions were no longer possible economically. Thus urban and rural Europeans began to restrict the size of the family by postponing marriages and by birth control. This was the beginning of the demo-graphic transition from high birth rates and

FIGURE 11-27 [OPPOSITE]
A neo-Malthusian model of failure of economic and demographic transition. A population increase stimulated by improved technology and health services is not accompanied by an increase in productivity in the economy. As population levels increase, the quality of living decreases. As a result the situation in that population never changed. (From H. Frederiksen, 1969. © 1969, American Association for the Advancement of Science.)

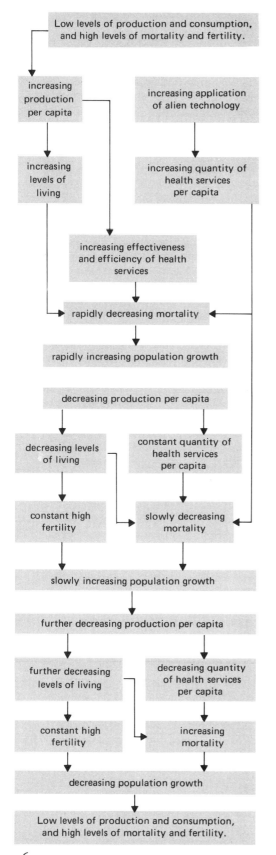

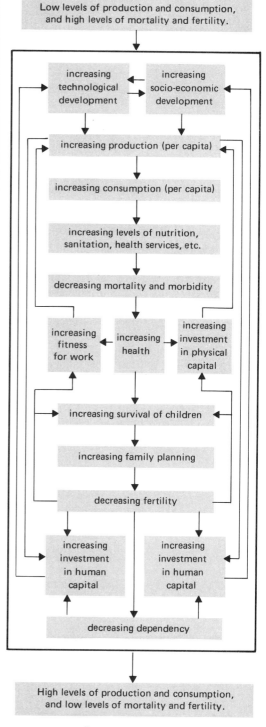

FIGURE 11-28

An alternative model of successful economic and demographic transition from low to high levels of production and consumption, and from high to low levels of mortality and fertility. (From H. Frederiksen, 1969. © 1969, American Association for the Advancement of Science.)

low death rates to low birth rates and low death rates.

As with natural populations, a decline in numbers in one area is often accompanied by an increase elsewhere. But unlike natural populations, an increase in one area can have a pronounced impact on other areas as well. The population increases in underdeveloped nations will have far-reaching consequences on the rest of the world. And it seems unlikely that those nations can pass through the demographic transition before serious political and economic consequences arise.

The interrelation between population and economics, between demographic transition and economic development are illustrated in two feedback models developed by Frederiksen (1969). One, Fig. 11-27 illustrates the failure of economic and demographic transition in which an increasing population faced with high fertility and low mortality experiences decreasing production due to economic stresses. This in turn results in low levels of production and consumption, and in decreasing population growth, characterized by high levels of mortality and fertility. In such a situation high mortality provides no incentive for restraint in family size. The other model, Fig. 11-28, illustrates a transition from low to high levels of consumption and from high to low levels of fertility and mortality. Increased production is accompanied by increased survival of young, an increase in family planning, and a decrease in fertility and dependency.

Summary

The existence of self-regulatory mechanisms in plant and animal populations is a subject of considerable speculation and study. Some biologists maintain that population fluctuations are most affected by some extrinsic or density-independent influences. Others argue that populations fluctuate about some optimum size, and that changes in the density of the population are determined by the numbers of organisms in relation to the optimum. Still others take the position that local populations fluctuate between some upper and lower population levels controlled by the interaction of population density, resource limitations and climate, as well as intraspecific competition and predation.

Density-independent influences such as inclement weather can reduce local populations, even to the point of extinction. But because they are density-independent, they rarely regulate density. Among density-independent influences, man is the most important. By his activities he can increase or exterminate whole populations.

The regulatory process seems to be controlled by some intrinsic, or density-dependent, mechanisms. Among these is intraspecific competition, in which animals compete for some resource often in short supply—food, space, a mate. There are basically two types of competition; the first is scramble, in which the resource is subdivided into many small parts to which all have access. Individuals "scramble" for the resource, which results in wastage. Each individual obtains such a small amount that it is unable to survive. The second is the contest type, in which each successful individual claims a part of the resource and the unsuccessful are denied any access to it. Contest competition shows up in territoriality and social dominance and hierarchy. Such stresses often result in endocrine imbalances, especially in the pituitary–adrenal complex, which results in abnormal behavior and growth and in degeneration and infertility. Also involved in this behaviorally related regulation may be genetic feedback, which results in a deterioration of population quality. These mechanisms produce oscillations in population density, which are damped, or become irregular, through the combined action of density-dependent and density-independent influences.

Social pressure and crowding also result in the movement of members of a population away from their area of birth. This dispersal takes the form of emigration and immigration. These are one-way movements that may result in the establishment of new breeding colonies in unexploited habitats. Migrations are round-trip movements involving a return to the area originally vacated. Often these migrations follow regularly used routes or flyways. Animals find their way along them by some navigational mechanism associated with the biological clock. This clock seemingly enables some animals to orient themselves to the sun, or even the stars.

When oscillations are more regular than one would expect by chance, they are called cycles. The most common intervals between oscillations are 3 to 4 years, typified by the lemmings, and 9 to 10 years, typified by the lynx and the snowshoe hare. Two major groups of theories have been advanced to explain cycles. One is that something in the environment or within the population itself produces the cycles; the other is that they simply reflect random oscillations in environmental conditions.

Human populations respond to population density in ways similar to those of other organisms. A high density can result in emigration, largely for economic reasons; and waves of human settlements are similar to those of other colonizing species. Man is subject to stress, which in its extreme form can result in reproductive impairment and physiological and behavioral changes. He also exhibits a need for some sort of spacing. But he has lost any homeostatic mechanism for population regulation. Control of his numbers now rests with cultural rather than biological processes.

Animal and plant populations of the biotic community exhibit a wide range of relationships to one another. Some populations have little influence on one another, except in the indirect and often distant roles they play in energy exchange. Other populations, such as parasites and their hosts and predators and prey, have a very direct and immediate relationship. From an individual standpoint these relations often are detrimental; from a population standpoint they may act either as a depressant or as a stabilizer of population numbers. Such interactions influence the growth curve of a population.

Mutualism

Situations in which a close and often permanent and obligatory contact exists between two species is called *mutualism*. Such relations are mutually beneficial to both species involved. A classic example is the lichen fungi. The basic fungal structure is the hypha, usually white or colorless. Within the lichen thallus is a thin zone of algae, which usually forms colonies of 2 to 32 cells. The hyphae support the plant while the algae supply the food. Without the algae component the lichen could not survive (see Hale, 1971). Similar relationships exist between algae and lower invertebrates. The hydra has green algae living within its gastrodermal cells (Muscatine and Lenhoff, 1963). The algae photosynthetically incorporates carbon dioxide and about 10 percent of this is released to the hydra, which assimilates it into its cellular components. Evidence suggests that the symbiotic algae are nutritionally significant to the host, but how the material is used is still a subject of study.

Another well-known example of mutualism is the nodule growth of nitrogen-fixing bacteria on the roots of legumes. Less well known are mycorrhizae, the symbiosis of plant roots with the mycelium of fungi. Common to pines, oaks, and beech, mycorrhizae produce shortened and thickened roots that suggest coral. The roots are covered with a dense sheath of fungal hyphae. There are two broad types of mycorrhizae, endotrophic and ectotrophic. Among the endotrophic types, a portion of the hyphae penetrates the host's cells; among the ectotrophic, the hyphae work between the plant cells and extend into a network on the outside of the root. These fungi utilize carbohydrates of the roots and in turn supply the root with inorganic nutrients and auxins.

The ectomycorrhizae are not the most common, but they play a very important role in the temperate and tropical forests. They replace the root hairs lacking in the pines and oaks. The hyphae radiate some distance into the soil and litter surrounding the root, aid in the decomposition of the litter, and absorb and translocate the nutrients, especially nitrogen and phosphorus, from the soil into the root tissue (Zak, 1964; Marx, 1971). They provide the plant with water and produce auxins that retard the elongation of roots and form new ones (Hacskaylo, 1971).

The physiological influence of the fungi on the host and vice versa go beyond supplying nutrients and transferring energy. The mycorrhizae increase the capability of roots to absorb nutrients (Voigt, 1971), provide selective ion absorption and accumulation, mobilize nutrients in infertile soil, render unavailable substances available (particularly those bound up in silicate materials) (Voigt, 1971), and transfer nutrients directly to the conducting systems of higher plants. In addition the mycorrhizae reduce the susceptibility of the host to the invasion of pathogens by utilizing root carbohydrates and other chemicals attractive to pathogens. They provide a physical barrier to pathogens, secrete antibiotics, and stimulate the roots to elaborate chemical inhibitory substances (Marx, 1971). In turn the roots provide a constant supply of simple carbohydrates. A balanced association exists between the two. Any alteration in light or nutrient supply creates a deficiency of carbohydrates and thiamin for the fungi; and interruption of photosynthesis causes a cessation of fruiting of the mycorrhizae. Thus any interruption or imbalance in the continuous supply of metabolites will impair or destroy the association.

One of the outstanding mutualistic relationships between plants and animals is between the fungus-growing attine ants and a slow-growing fungus that cannot survive without them (Martin, 1970). The ants cultivate the fungi, which are fragile, spongelike structures 15 to 30 cm in diameter, consisting of many pieces of leaves held together by a dense growth of mycelium. To cultivate the fungus the ants cut fresh leaves and flowers into small pieces. They scrape, scar, and macerate the pieces that release enzymes and nutrients contained in the leaf cells and provide a large growing surface for the mycelia. On the pieces the ants plant tufts of mycelium, which brings about a rapid colonization of the substrate by the fungus ahead of competitors. But before they plant, the ants drop onto the pieces of leaves liquid droplets of fecal material rich in ammonia, free amino acids, and proteolytic enzymes. That material supplies the fungus with an immediately available source of nitrogen, and also supplies the enzyme that the fungus needs to break down plant protein. The fungus in turn degrades cellulose, making it accessible as a carbohydrate source for the ant. Thus the mutualistic relationship is a biochemical one in which the carbon and nitrogen metabolism of the two organisms are integrated. The ants supply the enzymatic apparatus to degrade the protein so the fungi can utilize it, and the fungi degrade the cellulose to supply food for the ants.

Mutualism is often termed *symbiosis*. Actually, symbiosis means "living together" and includes mutualism, commensalism, and parasitism (see Henry, 1966).

Commensalism

Often a one-sided relationship between two species exists, in which only one benefits and the other is neither benefited nor harmed. Such a relationship is called *commensalism*. Among the commensals are epiphytes, such as orchids. These plants grow in the branches of trees, where they are near light; their roots draw nourishment from the humid air. Other epiphytes are "Spanish moss," which festoons southern live oaks, and "old man's beard" of the northern coniferous forests. All these epiphytes depend upon the trees for support only. They manufacture their own food by photosynthesis. Animal commensals include the barnacles, which attach themselves to the backs of whales and shells of horseshoe crabs. The remoras attach themselves to the bellies of sharks by means of a dorsal fin highly modified into a suction disk on the top of the head. Not only do they obtain a free ride but they

Relationships among populations

also feed on the fragments of the shark's prey. In Arizona the elf owl nests in the hole made by the Gila woodpecker in the stems of Saguaro cactus. The burrows of many animals may be occupied by others. Woodchuck burrows are used by cottontail rabbits; muskrat bank dens by mink.

Amensalism

Contrasting with commensalism is *amensalism,* a situation in which one population definitely inhibits the other while remaining unaffected itself. By so modifying the environment, the organism improves its own chance of survival. Amensalism commonly involves some type of chemical interaction with other organisms. This interaction, by which the organisms of one species affects the well-being and growth of individuals or the population biology of another, is called allelochemic (see Whittaker, 1970; Whittaker and Feeny, 1971). Involved may be inorganic inhibitors such as the production of acids or bases that reduce competition for nutrients, light, and space, especially from pioneering organisms. The production of relatively simple organic toxins is another source of chemical inhibitors. Produced in profusion in natural communities as secondary substances, most remain innocuous; but a few influence community structure. A third type is the antibiotic, a potent antimicrobial agent. It is a substance produced by an organism, which, in low concentration, can inhibit or kill the growth of another organism.

The majority of inhibiting chemicals are produced as secondary substances by plants and released to the soil through the roots or as leaf wash. The suppression of growth through the release of chemicals by a higher plant is known as allelopathy. A commonly observed example is inhibition of growth of certain plants by the walnut tree. This tree produces a nontoxic substance, juglone (Bode, 1958) which is found in its leaves, fruits, and other tissues. The fruit and leaves falling to the ground release this substance to the soil. In the soil, juglone is oxidized to a substance that inhibits the growth of certain understory species and graden plants such as heaths and broad-leafed herbs, and favors others, such as bluegrass and blackberries (M. B. Brooks, 1951).

In the desert shrub community, hard chaparral releases a variety of more or less toxic phenolic compounds to the soil through rainwater. These substances inhibit the germination and growth of the seeds of annual herbs (McPherson and Muller, 1969). Another shrub type, the soft chaparral, which commonly invades grasslands, releases aromatic terpenes such as camphor to the air. These terpenes are adsorbed from the atmosphere onto soil particles. In certain clay soils these terpenes accumulate during the dry season in quantities sufficient to inhibit the germination and growth of herb seedlings (Muller et al., 1968). As a result the invading patches of shrubs are surrounded by belts devoid of herbs, and by a wider belt in which the growth of grassland plants is reduced (W. H. Muller et al., 1968).

Phenolic acids released by exudates and the decay of dead roots have an antibiotic effect on soil microorganisms. For example, the inhibition of nitrogen-fixing bacteria and blue-green algae by the grass *Aristida oligantha* and other old field plants maintains a low concentration of nitrogen in the soil. These plants themselves are tolerant of low nitrogen, and so slow the invasion of the grassland by other species (Wilson and Rice, 1968). Various species of algae produce extracellular byproducts that inhibit the growth of other algal species, usually within the same phylum. Thus a species of algae that begins its development first in a body of water quickly assumes dominance. A sudden crash of this dense algal population is believed to be a result of the inability of algae to tolerate an accumulation of its own extracellular products, a case of autotoxicity.

The inhibitory effect of algae antibiotics may extend beyond algae. A subalpine lake just outside Glacier National Park received drainage from a small farm and barnyard. This raised the nitrogen and phosphorus content enough to support a bloom of *Aphanizomenon flos-aquae.* Each year this bloom develops, mudpuppies die by the scores, while others are sluggish and roll about on the bottom (Prescott, 1960). Another alga, *Anabaena flos-aquae,* has killed Franklin gulls and mallard ducks, which suffered from symptoms of botulism. And the dinoflagellate *Gymnodinium veneficium* has caused the wholesale death of fish by depolarizing nerves and muscle systems (Abbot and Ballentine, 1957).

Molds, actinomycete fungi, and bacteria also produce antibiotics. Soil fungi in particular secrete antibiotics that even in small amounts can inhibit life processes of other susceptible organisms. Such an association between organisms is exploited in medicine, where these chemical substances are used to combat bacterial and virus infections.

Parasitism

It is only a short step from commensalism to *parasitism*, a condition in which two animals live together but one derives its nourishment at the expense of the other. Parasites, strictly speaking, draw nourishment from the tissues of their larger hosts, a case of the weak attacking the strong. Typically, parasites do not kill their hosts as predators do, although the host may die from secondary infection, or suffer from stunted growth, emaciation, or sterility. A strict line cannot be drawn here, since many parasitic larvae of insects draw nourishment from the tissues of their hosts; but by the time of metamorphosis the larvae have completely consumed the soft tissues of the host. These parasites act in much the same way as predators and are discussed later as such; such parasites are used in the biological control of insects.

Parasites exhibit a tremendous diversity in ways and adaptations to exploit their hosts. Parasites may be plants or animals; they may parasitize plants, animals, or both. They may occur on the outside of the host (ectoparasites) or live within the body of the host (endoparasites). Some are full-time parasites; others, only part-time. The latter may be parasitic as adults, free-living as larvae, or the reverse. They have developed numerous ways to gain entrance to their hosts, even to the point of using several hosts to disperse their kind. They have evolved various means of mobility, ranging from free-swimming ciliated forms to dependence upon other animals, and have developed ways of securing themselves to the host to maintain their position. Some, such as the tapeworm, have become so adapted to the host that they no longer require a digestive system. They simply absorb their food directly through their body wall. Parasites that live within the bodies of plants and animals possess cuticles or develop cysts resistant to the digestive enzymatic action of the host.

369

Parasites may be restricted to one host. Some parasites of birds, especially the tapeworm, can live only on one particular order or genera (see Baer, 1951). This fact is utilized by the taxonomist as evidence of relationships between species, genera, or families. Others may be restricted to special habitats within the host (Fig. 12-1). The roundworm lives near the duodenum of the digestive tract, and the soil nematodes live in the rootlets of plants. Some parasites live their entire life cycle on one host; others may require more. An example of this is the brain worm of the white-tailed deer, *Pneumostrongylus tenuis*. Its secondary host is a snail or a slug that lives in the grass. The deer becomes infected by taking up the infected snail or slug while grazing. Inside the stomach of the deer, the larvae leave the snail, puncture the deer's stomach, and enter the abdominal membranes. From here the larvae move to and penetrate the spinal cord, all within approximately 10 days after the initial infection. In the spinal cord the larvae continue their development and within 40 days the worms occupy spaces surrounding the brain. Now adult, the brain worms in these spaces mate and produce eggs which may or may not hatch there. Eggs and larvae pass through the blood stream to the lungs. In the lungs the larvae break into the air sacs, are coughed up, swallowed, and passed out with the droppings. Outside, the larvae must be ingested by a snail or slug in order to continue development to the infective stage, a process that requires about 20 days. As with most parasites and hosts, the deer and the brain worm have achieved a mutual tolerance, and the deer does not suffer greatly from the infection.

Among the plant parasites, rusts require secondary hosts. White pine blister rust, which infects the five-needled pines, enters through and grows in the needles until it reaches the bark, where it forms spindle-shaped cankers. In 2 to 5 years orange-yellow spore-bearing blisters appear on the bark and release millions of spores, which infect only the leaves of gooseberry. Here the fungus forms brownish-yellow spots, in which are produced spores capable of infecting pines only. These spores, unlike those produced on the pine, are delicate, short-lived, and can travel only up to 900 yd. Control of parasites requiring more than one host can be accomplished by eliminating one of the intermediate hosts, such as the gooseberry.

Successful parasitism represents something of a compromise between two living populations. Parasites and hosts that have lived together over a long period of time have developed a sort of mutual toleration with a low-grade, widespread infection, as long as conditions are favorable for both host and parasite. But if conditions become favorable for the parasite or the host, or if a host is exposed to a new parasite against which it has no defense, then conditions worsen for one or the other.

There are two examples of such defenselessness to new parasites among the mammals of eastern North America. At one time the white-tailed deer and the moose occupied nearly exclusive ranges. The deer inhabited the deciduous forest, the moose the northern coniferous forest. When man cut and burned the coniferous forest, aspen and birch replaced the conifers. The deer expanded into the new range, came into closer contact with the moose, and brought to it the brain worm. The moose, experiencing a new parasite, succumbs to the infection.

No longer in demand for fur, lacking predators, and apparently possessing no homeostatic regulatory mechanism, red foxes have greatly increased in many parts of eastern North America. High populations have become decimated by rabies and more recently mange. Mange, caused by a parasitic mite *Scaroptes* and spread by direct contact among foxes, is rapidly becoming a major agent of mortality in fox populations. In some regions it is significantly reducing fox populations. Mange appears to be particularly effective because the spread of the parasite is not density-dependent.

DISEASE

At best, a parasitized animal is not in a state of health, that condition in which all vital processes function harmoniously. Any disturbance to this finely adjusted balance that results in an abnormal function is termed a disease. There are a number of causes of diseases. Parasites are one; physiological stress, nutritional deficiency, and poisoning are others. But most commonly, disease is considered as bacterial, viral, or fungal infection.

Disease organisms either are present in animals at all times, becoming active only when body resistance is lowered; or they are absent and highly virulent only when they enter the body. Single attacks of some diseases confer immunity on the organism, an adaptation that enables the animal to withstand any further attack.

Diseases among plants are caused largely by fungi; and certain diseases, such as chestnut blight, have spread rapidly through a species population, wiped it out, and thus changed community composition. This rapid and fatal spread of a disease is especially characteristic when an exotic infection is introduced into a population that has had no prior adaptation to it.

Several diseases, only partially understood, are prevalent in eastern forests and are potential threats to a number of tree species. Maple, ash, and black cherry are subject to dieback. Beech, particularly in northeastern North America, is subject to a near epidemic of beech bark disease spread by the woolly beech scale. In the oak forests of the upper Mississippi Valley and the southern Appalachians, oak wilt threatens the hardwoods stands. The disease is caused by a fungus, *Ceratocystes fagacearium*, which appears to be spread by root grafts and nitidulid beetles, especially *Glischrochilus* and *Coleopterus* (True et al., 1960). Affected trees have wilted or defoliated tops, a brownish discoloration to the leaves, bark cracks, and pressure cushions caused by fungal mats. Loss of oaks would be more disastrous than the loss of the American chestnut, not only from an economic but also an ecological viewpoint, for a wide spectrum of animal life relies on oaks for food, and the gaps in the plant community would be occupied by lesser species.

Bacteria and viruses are causal agents of important animal diseases. Bacteria may produce localized inflammatory changes in tissue, enter the blood stream, or produce powerful poisons known as toxins. Viruses cause diseases that may be complicated by secondary invaders. Some diseases, both bacterial and viral, may be transmitted from wild to domestic animals, by direct contact or by insect vectors.

One such disease, widespread among rabbits and squirrels, beaver, and muskrats, is tularemia, a bacterial disease transmitted from

FIGURE 12-1
Liver flukes in the white-tailed deer. This is an important parasite of the white-tailed deer, especially in northern herds, where the incidence is very high. (Photo courtesy Michigan Conservation Department.)

animal to animal by the bites of ticks and flies, particularly the rabbit tick, and carried to man by ticks (especially the wood tick, *Dermacentor* spp.) and deerflies, or by direct contact through a break in the skin. Contraction of the disease while cleaning an infected animal is the cause of 90 percent of the cases in man (Yeatter and Thompson, 1952). Infected animals in advanced stages will not move from cover; and if they do, they stagger, stop, and perhaps fall over. Internally, the liver and spleen are swollen and studded with tiny white discolorations.

Possibly the greatest natural drain on waterfowl populations is botulism, or western duck sickness, caused by *Clostridium botulinum*. When conditions are favorable—warm temperature, an organic medium for food, and anaerobic conditions—the spores of *Clostridium* germinate and multiply. In doing so they produce a nerve toxin as a by-product of metabolism. When waterfowl ingest the toxin, found in the decomposing carcasses of aquatic insects and other invertebrates, ducks develop a paralysis of the neck and a looseness of feathers, which is followed by death. The disease is most prevalent during late summer, when water levels are low and exposed plant and animal matter begins to decay.

Viruses are the causal agents of two wildlife diseases important to man, encephalitis, or sleeping sickness, and rabies. Encephalitis is harbored by a host of avian species, from bluebirds, jays, and warblers to herons and ibis and is transmitted by the mosquitoes *Culex* and *Culeseta*. Among birds, only the ringnecked pheasant succumbs to the disease, which is transmissible to both animal and man. Rabies, also highly contagious to man, is a viral disease, which follows the nerves from the infection point to the spinal column and brain before symptoms appear. The symptoms are both behavioral and physical. Infected animals are restless, excitable, exhibit convulsions, wander aimlessly, and show no fear of man. Physically they are emaciated, exhausted, partially paralyzed. Among wild animals, rabies occurs most commonly in coyotes, foxes, raccoons, and skunks, although no species is immune. Foxes, together with domestic dogs, are the primary causes of the spread of the disease. Rabies may reach epidemic proportions, as it did in New York State in 1954. Severe reduction of fox populations

will control the disease, although it can never be entirely eliminated.

Fungal diseases are common among wild animals. Aspergillosis, common in domestic fowl, is caused by a fungus that invades the air sacs of the lungs, developing pneumonia-like symptoms; the infection is found among waterfowl. Fish are susceptible to *Saprolegnia*, or water mold. This whitish growth often begins on lesions caused by some injury, from which it spreads.

Disease, although prevalent, becomes important only when epidemics or epizootics occur. These usually come about when the host populations are high. Disease, when rampant, can reduce populations, but it may not be the primary cause. An increasing population may result in a depression within the animal of antibody formation and other body defenses, an increase in inflammatory processes, and in an increased susceptibility to disease. Thus disease may be the consequence of a high population, rather than a cause of population decline.

Like parasites, diseases, especially introduced ones, can reduce populations, exterminate them locally, or restrict the distribution of the host. Prior to the discovery and settlement of the Hawaiian Islands by Europeans, the avifauna of Hawaii inhabited all parts of the islands from the seashore to the upper limits of vegetation. In 1826 the tropical subspecies of the night mosquito was accidentally introduced. The mosquitos spread rapidly throughout the lowland areas of the high islands and spread bird pox and other unidentified diseases through the lowland bird populations. As a result endemic Drepannid birds disappeared from the lowlands to a 600-m elevation (Warner, 1968). The species of these birds that still exist (a number are extinct) are restricted to regions above 600 m in spite of the availability of habitat at lower elevations. When Drepannid species of birds are carried to a lower elevation, they succumb to malaria or bird pox. Thus the high forests of the Hawaiian Islands are an ecological sanctuary from disease for the birds.

Disease has had its impact on human populations (see Armelagos and Dewey, 1970). Smallpox, intentionally or unintentionally introduced, nearly exterminated some tribes of North American Indians. The Black Death repeatedly cut back the populations of medieval Europe, as much as 25 to 50 percent. Malaria in southern Asia depressed population growth. After DDT was sprayed to control malaria-carrying mosquitos, the human death rate dropped and population growth increased dramatically.

The relationship of diseases and parasites and their hosts is one of evolutionary response. As new defenses are built up by the host, new strains are evolved by the parasite. A dramatic example is the relation of hybrid corn, artificially developed or "evolved" by man, and corn blight. In breeding varieties of hybrid corn, hybridizers succeeded in developing nuclear diversity within a uniform cytoplasm. They did this to utilize factors within the cytoplasm that lead to male sterility, which eliminates the need for detasseling corn in seed fields. One particular corn cytoplasm, Texas male sterile, TMS, has been widely incorporated into many lines. But TMS conferred susceptibility to a virulent race T of southern corn blight (Tatum, 1971). When this blight did appear in the Central Corn Belt in 1969 and in Florida, it became epidemic because 80 percent of field corn hybrids carried the highly susceptible TMS. The remaining hybrids containing normal male-fertile cytoplasm were resistant to race T. Such is the risk man runs with new varieties and races of cereal grains. The problem is to maintain a crop with a high degree of genetic diversity, so that when a disease develops a virulent strain, a genetic bank exists from which resistant forms can be bred.

SOCIAL PARASITISM

Another form of parasitic relationship is social parasitism, in which the parasite foists the rearing of its young onto the host.

Social parasitism in various stages of development is found among some higher vertebrates. Outstanding is egg parasitism among birds, both nonobligatory and obligatory. Nonobligatory egg parasites lay their eggs in nests of other birds, although they have nests of their own. Among these are the pied-billed grebe, black-billed and yellow-billed cuckoos, the bobwhite, and ringnecked pheasant. The ringnecked pheasant is notorious for dumping eggs in the nest of another, and even in dump nests in which the eggs are not incubated. Equally notorious are waterfowl, especially the redhead and ruddy duck. Twenty-one species of ducks are known to lay eggs in nests other than their own (Weller, 1959). An estimated 5 to 10 percent of female redhead ducks are nonparasitic and nest early. All others lay eggs parasitically at one time or another. More than half of these are semiparasites and nest themselves. The remainder are completely parasitic.

Egg parasitism has been carried to the ultimate by the cowbird and European cuckoos, both of which have lost the instinct for nest building, incubating the eggs, and caring for the young. These are obligatory parasites who pass off these duties to the host species by laying eggs in their nests. The cowbird removes one egg from the nest of the intended victim, usually the day before laying, and the next day lays one of her own as a replacement. The young cowbird does not eject its nestmate, a practice carried out by nestling European cuckoos. Another form of social parasitism is robbery or pirating of food from other species. The bald eagle forces the osprey to drop its prey in flight and catches the dropping fish before it hits the water. Jaegars and skuas, dark, hawk-like birds of the sea, force gulls to give up their catch of fish.

Like parasitism and disease, social parasitism can adversely affect a host experiencing its first contact with the social parasite. Such a situation has developed with the Kirtland warbler (Mayfield, 1960), a relict species that inhabits extensive jack pine stands in a compact central homeland of about 100 mi² in northern lower Michigan. Before the white man arrived, Kirtland warbler apparently was isolated by 200 mi of unbroken forest from the parasitic brown-headed cowbird of the central plains, a bird closely associated with grazing animals. When settlers cleared the forest and brought grazing animals with them, the cowbirds spread eastward and northward in the jack pine country. Unlike other birds, Kirtland warblers had no real defense against the cowbird, such as egg ejection, building a new nest over the old, or rearing young successfully along with the cowbird. The warbler has a short nesting season and an incubation period a day or two longer than other songbirds, so that the young cowbird already is out of the egg when the warblers hatch. The result of this parasitism, which involves 55 percent of the nests, is a 36 percent reduction in the number of warbler fledglings produced, below the level attained without cowbird interference. The annual survival rate of adults is about 60 percent from one June to another, so 57 percent of the 1.41 fledglings produced per nest (in contrast with 2.2 under normal conditions) must survive the first year to replace the adult loss, nearly an impossibility. In recent years the situation has not improved, and the Kirtland warbler is faced with extinction.

Predation

No phase of population interaction is more misunderstood (or hotly debated, especially by sportsmen) than predation. Predation in natural communities is a step in the transfer of energy. It is commonly associated with the idea of the strong attacking the weak, the lion pouncing upon the deer, the hawk upon the sparrow. But this idea must be modified, for predation grades into parasitism and vice versa. Between the two exists the broad gray area of the parasitoid and the host, which sometimes is called parasitism and sometimes predation. In this situation, one organism, the parasitoid, attacks the host (the prey) somewhat indirectly by laying its eggs in or on the body of the host. After the eggs hatch, the larvae feed on the host until it dies. Ultimately the effect is the same as that of predation, and the two can be considered the same. The concept of predation has been further extended to include the relationship between plants and herbivores. Grazing herbivores are considered predators upon plants and their impact on plant populations as predation. Thus predation in its broadest sense can be defined as one organism feeding on another living organism.

From the viewpoint of the population ecologist, predation in its actions and reactions is more than just the transfer of energy. It represents a direct and often complex interaction of two or more species, of the eaters and the eaten. The numbers of some predators may depend upon the abundance of their prey, and predation may regulate the population of the prey. The ideas are debatable; certainly the same generalizations cannot apply to all groups of predators and prey.

The influence of predation on population growth of a species received the attention of two mathematicians, Lotka (1925) and Volterra (1926). Separately they proposed formulas to express the relationship between predator and prey populations. They attempted to show that as the predator population increased, the prey decreased to a point where the trend was reversed and oscillations were produced.

Later the biologist G. F. Gause (1934) attempted to prove this experimentally. He reared together under constant environmental conditions a predator population, *Didinium*, a ciliate, and its prey, *Paramecium caudatum*. The predator always exterminated the prey, regardless of the density of the two populations. After the prey was destroyed, the predators died from starvation. Only by periodic introductions of prey to the medium was Gause able to maintain the predator population and prevent it from dying out. In this manner he was able to maintain populations together and produce regular fluctuations in both, as predicted by the Lotka-Volterra models. The predator–prey relations were ones of overexploitation and annihilation, unless there was an immigration from other prey populations.

In another experiment Gause introduced sediment in the floor of the tube. Here the prey could escape from the predator. When the prey was eliminated from the clear medium, the predators died from the lack of food. The paramecia that took refuge in the sediment continued to multiply and eventually took over the medium.

About a decade later, an ecologist, Nicholson, and a mathematician and engineer, Bailey, developed a mathematical model for a host–parasitoid relationship. This model predicted increasingly violent oscillations in single-predator and single-prey populations living together in a limited area with all external conditions constant.

Essentially both the Lotka-Volterra and the Nicholson-Bailey models state that as predator populations increase, they will consume a progressively larger number of prey, until the prey populations begin to decline. As the prey diminishes, the predators are faced with less and less food, and they in turn decline.

375

In time the number of predators will be so reduced by starvation that the reproduction of the prey will more than balance their loss through predation. The prey will then increase, followed shortly by an increase of predators. This cycle, or oscillation, may continue indefinitely. The prey is never quite destroyed by the predator; the predator never completely dies.

The Lotka-Volterra model in many respects is too simple and unrealistic. To improve upon the basic model, Rosenzweig and MacArthur (1963) developed a series of graphic models. These models consider a wider range of outcomes of predator–prey interactions by modifying the zero growth curve of the prey.

The Lotka-Volterra model is graphically depicted in Fig. 12-2. The ordinate H is the number of predators; the abscissa is P, the number of prey. In the area to the right of the vertical line or ordinate, the predators increase; to the left they decrease. In the area below the horizontal line or abscissa the prey increases; above it the prey decreases. The circle of arrows represents the joint population of predators and prey, and the size of the population of each changes with it. If a point or arrow falls in the region to the left of the vertical line, the prey population is not large enough to support the predators, and the predator population declines. If the arrows fall to the left of the vertical and above the horizontal, both populations are declining; in fact the predator population decreases enough to permit the prey population to increase. In this case the arrows would move to the right of the vertical and below the horizontal. The increase in the prey population now permits the predator population to increase, and the arrows now move across the horizontal. As the arrows move across, the predator population increases, depressing the prey population.

The Lotka-Volterra equation overemphasizes the influence of predators on prey populations. Genetic changes, stress, emigration, aggression, and other attributes also influence fluctuations of populations. These influences can be brought into the picture by plotting the growth curve of the prey as convex rather than horizontal (Fig. 12-3). This is biologically realistic because populations at very low densities have a low rate of growth; at very high densities, the growth rate is also depressed. The basic components of increase and decrease remain basically the same.

The predator and prey populations increase if the joint abundance falls inside the convex curve to the right of the vertical. If the joint abundance falls outside of the curve, then the populations of both decline. The prey curve intersects the predator curve at right angles to form four equal quadrants, so the curves produce stable cycles in the populations of predator and prey.

If the predator growth curve is moved to the right (Fig. 12-4), the curves no longer intersect at right angles. The interaction of predators with prey declines. The arrows do not form a closed circle, indicating that population fluctuations become damped, and the system eventually becomes stabilized. If the predator curve is moved to the left (Fig. 12-5), more of the prey growth curve falls within the predator growth curve, and the interaction of predators and prey increases. The arrows now spiral outward, indicating a heavy decline of the prey. This model produces violent oscillations that eventually may result in the extinction of both predator and prey.

All of the models represent situations in which the prey has no refuge from a predator. The growth rates of both the predator and prey populations are a function of the frequency with which the predator comes in contact with its prey. An even more realistic situation removes a portion of the prey population from contact with the predators, and the predator is forced to turn to alternate prey. As the prey population increases in the refuge area, the surplus repopulates the surrounding area (Fig. 12-6). Thus, when the prey is cropped down to a certain level, the predator population declines either from emigration, starvation, or failure to breed. The tawny owl of England (Southern and Lowe, 1968), for example, and the great horned owl of North America (Rusch et al., 1972) fail to breed when the population of their major prey is low. The refuge not only complicates the system, but it also adds stability. This model is more typical of natural situations in which a number of stable predator–prey relationships seemingly exist. Two examples are the muskrat and the mink (Errington, 1957) and the wolf and the moose (Mech, 1969).

The models of the rise and decline of a predator population as its prey population rises and falls indicates that some density-influenced relationship exists between them (Fig. 12-7). The nature of this relationship depends a good deal on whether the prey species is vagile or sedentary, territorial or nonterritorial. Among invertebrates, especially the insects, and nonterritorial animals, intensity of predation is independent of prey density but directly related to predator density. Under such conditions predators may respond to changes in the density of prey in two distinguishable ways. As the density of the prey increases, each predator may take more of the prey or take it sooner. This is a functional response (Holling, 1959, 1961). Or the predators may become more numerous through increased reproduction or immigration, a numerical response.

Functional response. In general the predator or the parasite will take or affect more of the prey, as the density of the prey increases. However, there is also a tendency for the number of prey taken or the number of hosts affected to increase in less than a linear proportion to the total number of host or prey available (Fig. 12-8). The functional response of parasites to hosts is usually curvilinear (Fig. 12-9). The slope decreases as the host density increases, until the curve levels off. When vertebrate predators are involved, the functional response curve is S-shaped. This is caused by the behavior of the predator—its response to hunger, its hunting ability, its frequency of contact with the prey species, and the stimulation it provides.

This latter type of functional response is well illustrated in a study by Tinbergen (1960) of the relation between woodland birds and insect abundance. According to Tinbergen's hypothesis, when a new prey species appears in a given area, its risk of becoming prey is low at first. The birds as yet have not developed a "searching image" for that species. Once the birds have acquired this image for the prey, intensity of predation increases suddenly. However, the relation between the density of the prey species and its percentage in the food of the predator cannot be explained from the probability of encounters alone. When prey density is low, predatory consumption is lower than would be expected on a density-dependent basis. At moderate densities, predation is unexpectedly high. At high densities, predation drops again. Tinbergen explains this on the assumption that when the prey species is low, the birds lack a searching image for it. At high densities, the birds become satiated or stop using that par-

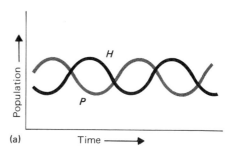

(a)

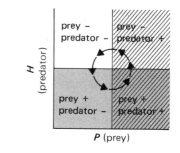

(b)

FIGURE 12-2

The Lotka-Volterra equation produces a predator–prey interaction as shown in (a). The abundance of each is plotted as a function of time. Another way of viewing the Lotka-Volterra equation is shown in (b). This model shows the joint abundances of the species. In this model the zero growth curve for both predator and prey are straight lines that intersect at right angles.

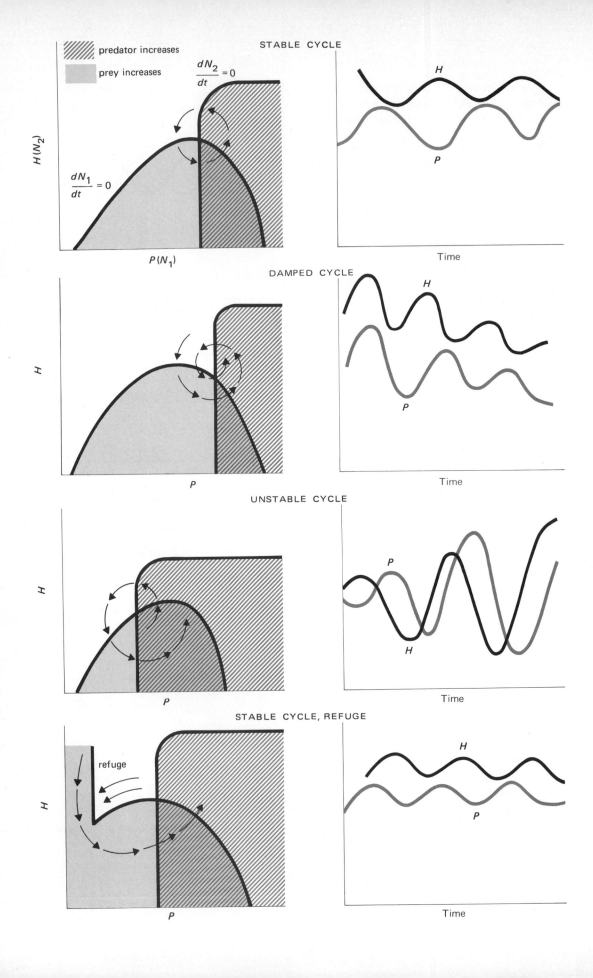

FIGURE 12-3 [OPPOSITE]

In this graph the growth curves of the predator and the prey are drawn more realistically as convex curves rather than straight lines. In this model the prey population can increase if the joint abundances of predator and prey fall inside the bulge. It will decrease if it falls outside the bulge. In this model the position of the growth curves are such that they intersect at right angles, as in Fig. 12-2. This generates a stable cycle. (Redrawn from R. MacArthur and J. Connell, 1966, Biology of Populations, Wiley, New York; after Rosenzweig and MacArthur, 1963.)

FIGURE 12-4 [OPPOSITE]

In this model the predator's zone of increase intersects the descending part of the prey's growth curve, to dampen the oscillations. (Redrawn from R. MacArthur and J. Connell, 1966, Biology of Populations, Wiley, New York; after Rosenzweig and MacArthur, 1963.)

FIGURE 12-5 [OPPOSITE]

In this model the predator's zone of increase intersects the ascending part of the prey's growth curve. This increases the oscillations, producing an unstable or exploding cycle. (Redrawn from R. MacArthur and J. Connell, 1966, Biology of Populations, Wiley, New York; after Rosenzweig and MacArthur, 1963.)

FIGURE 12-6 [OPPOSITE]

In this model the situation is similar to that of Fig. 12-5, except that the prey has a refuge where a portion of the population can escape predation. This limits the oscillations. (Redrawn from R. MacArthur and J. Connell, 1966, Biology of Populations, Wiley, New York; after Rosenzweig and MacArthur, 1963.)

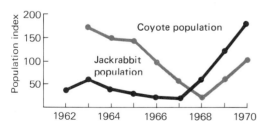

FIGURE 12-7

Coyote and jack rabbit annual population trends for Curlew Valley, Idaho, 1962 to 1970, as shown by population indices. Note that the reproductive rate and changes in coyote population vary with the food base. Since that base depends largely on the jack rabbit population there appears to be a density-dependent relationship between the coyote and jack rabbit populations. (From F. W. Clark, 1972.)

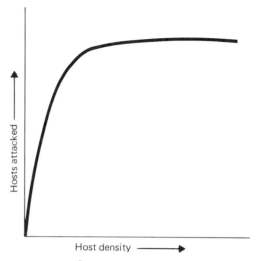

FIGURE 12-8

A functional response curve. The number of prey taken is related to the density of the prey population. This curvilinear relationship is typical of parasite–host relations.

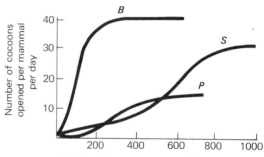

FIGURE 12-9

A functional relationship between small mammal predation and the density of sawfly cocoons. Curve B is the curve for the short-tailed shrew, Blarina; curve S is for the common shrew, Sorex; and P for the white-footed mouse, Peromyscus. (From Holling, 1959.)

ticular searching image when the prey concerned forms more than a critical percentage of the food. This may be the reason for the S-shaped curve of functional response.

Functional responses are expressed graphically as the plot of the number of prey taken per predator per unit time, the dependent variable, against the number of prey. The curves fall into three types (Fig. 12-10): (1) the slope rises at a continuing decreasing rate, indicating a decreasing proportion of the prey being taken; (2) the slope is a straight line, indicating proportional prey mortality throughout; and (3) the slope rises with an increasing gradient, indicating an increasing percentage in prey mortality as prey density increases.

Numerical response. As the density of the prey increases, the numbers of predators may also increase. But the response is not immediate. There is necessarily a time lag between the birth and the appearance of an active predator, and this time lag may prevent the predator or the parasite population from catching up with the prey. As a result an inverse relationship often exists between numerical response and prey density (Figs. 12-11 and 12-12). This is well illustrated in the long-term study of the spruce budworm in Canada (R. F. Morris, 1963). The response of parasites to an increasing density of budworms was sharp at first, then declined rapidly as the pupal density of the budworm increased.

Outstanding as an example of a numerical response to prey density are several warblers, especially the Tennessee, the Cape May, and the bay-breasted, whose abundance is dictated by outbreaks of the spruce budworm. During such periods the population of the bay-breasted warbler has increased from 10 to 120 pair/100 acre (Mook, 1963; Morris et al., 1958); Cape May and bay-breasted warblers have larger clutches during budworm outbreaks than associated warbler species (MacArthur, 1958). In fact the Cape May and possibly bay-breasted apparently depend upon occasional outbreaks of spruce budworms for their continued existence. At these times the two species, because of their extra large clutches, are able to increase more rapidly than other warblers. But during the years between outbreaks, these birds are reduced in numbers and are even extinct locally.

Sparrows prey heavily on the larch sawfly, a serious defoliator of tamarack, in and near the bogs of Manitoba, Canada. These birds were far more important predators of the adult sawfly than the insectivorous warblers and showed a strong numerical response to prey density (Fig. 12-13) (Buckner and Turnock, 1965). The sparrow population increased largely by immigration, which involved family flocks, adults and subadult birds, and premigratory flocks.

Numerical response takes three basic forms (Fig. 12-14): (1) direct response, in which the number of predators per unit area increases as the prey density increases; (2) no response, in which the predator population remains proportionately the same; and (3) inverse response, in which the predator population declines in relation to prey density. To these may be added two other sigmoid curves (Hassell, 1966). One indicates a decrease in the ratio of predators to prey as the density of prey increases; the other indicates that at one point the growth of the predator population is faster than that of the prey (Fig. 12-14).

Functional and numerical responses are defined in terms of the numbers of prey taken per predator, or changes in the number of predators per unit area. Another approach, suggested by Hassell (1966) is to consider the interaction between predator and prey in terms of proportionate changes in predation. This makes it easier to detect the effects of the predator population on the numbers of prey as well as the effects of density of the prey on the behavior of the predators.

The reaction of predators to various prey densities can be considered as a behavioral response to the spatial distribution of prey, measured within a generation of predators. Behavioral response may be individual or aggregative. Individual response, which corresponds to functional response, is the percentage mortality of prey caused by one predator in the region of high density. Aggregative response, which corresponds to numerical response, is one in which the density of the predator is proportionately greater or less in regions of high density.

There may also be an intergenerative relationship between the predator and the prey. This exists when there is a relation between the mean percentage mortality of the prey caused by the predator population and the mean density of the prey over a number of consecutive generations of predators. If the

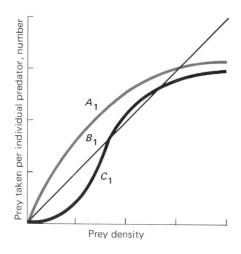

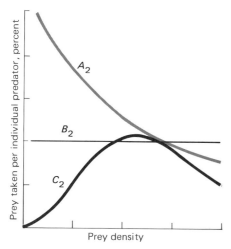

FIGURE 12-10

The three basic types of functional response curves.
(From Hassell, 1966.)

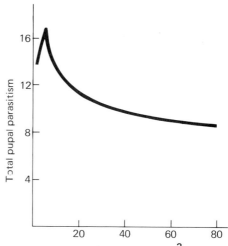

Pupal density (number per 10 ft^2 of foliage)

FIGURE 12-11

Predator and parasitoid populations may show an
inverse density relationship to prey and host
populations. The predators cannot keep up with
the prey or host populations during a time of
outbreak. (After Miller, 1963; in Morris, 1963.)

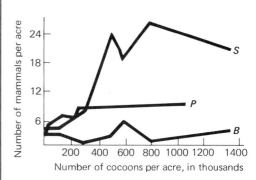

Number of cocoons per acre, in thousands

FIGURE 12-12

Numerical response of small mammalian
predators to a sawfly cocoon population.
The only mammal that shows any strong
response is the shrew, Sorex. (After Holling,
1959.)

points so plotted and joined consecutively form a circle or spiral, it implies a delayed density-dependent response, which results in a predator-induced oscillation of the prey population (Fig. 12-15).

Examples of delayed numerical response of predators to changes in prey population can be found in a number of recent field studies. The population of the great horned owl in a 62-mi² area in Alberta, Canada, increased over a 3-year period, 1966 to 1969, from 10 birds to 18 as the population of its prey, the snowshoe hare, increased sevenfold (Rusch et al., 1972). The proportion of owls nesting increased from 20 percent to 100 percent as the biomass of snowshoe hare in the owls' diet increased from 23 to 50 percent (Fig. 12-16). Coyotes inhabiting a 700-mi² area in Utah increased as the density of black-tailed jackrabbits increased, a species that made up three-fourths of the coyotes' diet (F. W. Clark, 1972). The decline and recovery of a lynx population in central Alberta followed the decline and recovery of the snowshoe hare population (Nellis, 1972). In the great horned owl the numerical response acted through the failure of adults to breed; in the lynx the response acted through mortality of young, not the failure to produce young.

Total response. In analyzing the relationship between predator density and prey density, functional and numerical responses may be combined to give a total response, and predation is plotted as a percentage. If this is done, predation falls into two types. In one the percentage of predation declines continuously as the prey density rises (Fig. 12-17). In the other the percentage of predation rises initially and then declines. The latter results in a dome-shaped curve (Fig. 12-18) produced by the S-shaped functional response to prey density and by direct numerical response.

TYPES OF PREDATION

The predation discussed so far has been recognized by Ricker (1954) as types A and B. In type A, predators of any given abundance take a fixed number of prey species during the time they are in contact, usually enough to satiate themselves. The surplus prey escape. Trout feeding on an evening hatch of mayflies would come under this category. There is a functional response but no numerical response. Type B predation exists if predators of any

given abundance take a fixed fraction of a prey species, as though the prey were captured at random encounters. In other words, the amount of prey eaten is proportional to the abundance of the predator and the abundance of the prey. There is both a functional and a numerical response.

Among many vertebrate populations, the response of predator to prey does not quite follow the functional and numerical response just described. Predators take most or all the individuals of the prey species that are in excess of a certain minimum number, as determined by the carrying capacity of the habitat and social behavior. The prey species compensates for its losses through increased litter and brood size and greater survival of young. For this reason this type of predation, Ricker's type C, is called compensatory. The population level at which predators no longer find it profitable to hunt the prey species has been called the "threshold of security" by Errington (1946). As prey numbers increase above this threshold, the surplus animals are no longer tolerated in the area (see Chapter 11) and become vulnerable to predation. Below the threshold of security, functional response of the predator is very low (Fig. 12-18) and numerical response is nonexistent. Above the threshold, functional response is marked, and numerical response could occur. An outstanding example of compensatory predation is detailed by Errington (1963) in his notable study of the muskrat.

COMPONENTS OF PREDATION

In considering predation, two events come to mind. One concerns a Cooper's hawk that one winter resided in a hemlock grove and so harassed a covey of bobwhite quail that by spring not a single bird remained. Another incident involved two predators working independently side by side, one an alligator and the other a blue heron. The heron, during a relatively short period of observation, was quite successful, securing two fish. The alligator, so situated that he could lunge at fish as they swam by, was unsuccessful. The incidents were so simple that a casual observer might think little of them. Yet each involved a number of components that make up the pattern of predation.

One component is the hunting ability of the predator. There obviously was a wide dif-

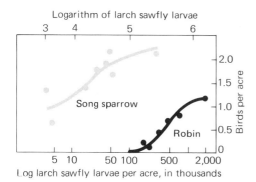

Logarithm of larch sawfly larvae

FIGURE 12-13
Numerical response of song sparrows and robins to larval larch sawfly. (From Buckner and Turnock, 1965. © The Ecological Society of America.)

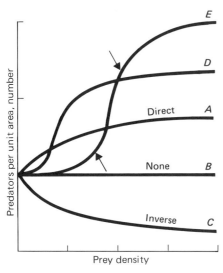

FIGURE 12-14
The basic forms of numerical response. See text for details.

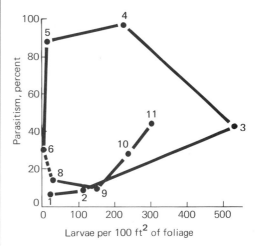

FIGURE 12-15
A model of delayed density-dependent response that results in a predator-induced oscillation of the prey population. (After Solomon, 1964.)

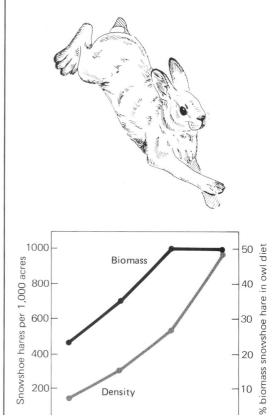

FIGURE 12-16
Spring snowshoe hare densities in hare habitat versus percentage of biomass of snowshoe hare in the diet of great horned owls near Rochester, Alberta. (From Rusch et al., 1972.)

ference in the hunting ability and tactics of the Cooper's hawk—a pursuer, the blue heron —a stalker, and the alligator, which stakes out an ambush. The alligator can afford a low frequency of success because hunting by ambush requires a minimal amount of energy. The heron is a deliberate hunter, seeking suitable pools of water and procuring its prey with a sudden thrust of the beak. The search time may be great but the pursuit time is minimal and the predator can afford to take smaller prey. For the Cooper's hawk the search time is minimal because it usually knows the location of its prey, but its pursuit time is relatively great. In general it must secure relatively large prey.

By their hunting activities predators can be regarded as specialized or generalized. Specialized predators are those adapted to hunt only a few species. They are forced to move when the vulnerability of a staple prey item drops to a point where the predator population cannot support itself. Generalized predators, not so restricted in diet, adjust to other food sources. The horned owl and buteo hawk have a large range of collective prey available. Foxes can shift to a vegetable and carrion diet, should conditions require it. In general, adaptations of predators are such that small rodents are vulnerable to more species of predators than small birds and game.

Hunting ability and success apparently involve the development of a searching image on the part of the predator. Once it has secured a palatable item of prey, the predator finds it progressively easier to find others of the same kind. Thus as the Cooper's hawk and the blue heron remember the appearance and location of their respective prey and hut for them in their own manner, the searching image becomes stronger. The more adept and successful the predator becomes at securing a particular prey, the longer and more intensely it concentrates on the item. In time the numbers of that prey might become so reduced or the population so dispersed over the environment that encounters between the predator and that particular prey are lessened. At some point the searching image for that item will wane and the predator will begin to react to another species.

This fact emphasizes another important aspect of predation, that of the facultative predator and alternate prey. Although the predator may have a strong preference for a particular prey, it can turn in the time of relative scarcity to an alternate, more abundant species that provides more profitable hunting. For example, if rodents are more abundant than rabbits and quail, foxes and hawks will concentrate on them instead of game animals. These alternate prey species then are often called buffer species because they stand between the predator and the game species. If the buffer population is low then the predators will turn to the other source of food, the rabbits and quail, if the populations of these animals are high. This turning by a predator to alternate food has recently been termed switching (Murdoch, 1969). Switching occurs when the more abundant prey species forms a higher proportion of the diet than it does of the food available.

Although the concepts of switching and alternate prey are valid, some predators deliberately seek certain items, no matter how scarce. Peale's falcon, a subspecies of the peregrine falcon found along the northwest Pacific Coast, shows a marked preference for ducks or pheasants and will eat gulls and crows only when necessary (Beebe, 1960). Predators in a California grassland exhibited a distinct preference for *Microtus*, the meadow vole, over *Reithrodontomys*, the harvest mouse, and other rodents even though the alternate prey was abundant. (O. P. Pearson, 1966). In fact the abundance of alternate prey apparently enabled the carnivores to maintain populations high enough to continue predatory pressure on a preferred species (Fig. 12-19). Among the herbivores or plant predators, deer exhibit a pronounced preference for certain species of browse (see Klein, 1970a). Meadow mice often concentrate on the seeds of preferred grasses, even though seeds of other species are abundant and available (Batzli and Pitelka, 1970).

Habitat preferences or overlapping territories can bring predator and prey into close contact, increasing prey risks. Predatory rainbow trout in Paul Lake, British Columbia, move into the shoals when their prey, the redside shiner, are most heavily concentrated there. A complex pattern of seasonal movements of the shiner governed by water temperatures brings prey in contact with the predator in the midsummer. During winter the shiners move into warmer, deep water, where they are isolated from the predatory trout (Crossman, 1959a, b). The herbivorous gastropod *Tegula funebralis* spends the first 5 to 6 years of its life in the high intertidal zone outside of the ecological zone of the predatory starfish, *Pisaster ochraceus*. After 6 years *Tegula* moves into the lower intertidal zone occupied by *Pisaster*. Then vulnerable to predation, *Tegula* makes up 25 to 28 percent of the diet of the starfish (Paine, 1966). Tawney owls whose territories embraced a high percentage of open ground were most successful in securing food when wood mice were abundant. And those whose territories included a light cover of bracken and bramble were most successful when bank voles were abundant (Southern and Lowe, 1968).

Age, size, and strength of the prey influence the direction that predation takes. The various species of predators have certain energy requirements that can be met only by profitable hunting. Predators cannot afford prey that is too small to meet their energy needs unless that prey can be captured quickly and in abundance. Otherwise the energy expended to procure food would exceed the energy obtained. On the other hand, predators have an upper size limitation. The prey may be too big to consume or too difficult or too dangerous to handle. Thus the predator has to select food on the basis of size. Mountain lions, for example, avoid attacking large healthy elk, which they cannot successfully handle, and concentrate instead on deer and young or feeble elk (Hornocker, 1970). The alewife feeds on the larger lake zooplankton and ignores the smaller species, which has an effect on community composition, a point to be discussed later (Brooks and Dodson, 1965). Where a choice in prey is available, the predators will consistently choose the largest food items available consistent with their own size and strength.

It is commonly believed that predatory animals are constantly on the move, seeking to kill. In reality predators, with individual exception, hunt only when it is necessary for them to procure food. Only in that interval of time when the predator is seeking food do predator and prey interact. The frequency of contact between predator and prey is in part determined by the hunger level of the predator. When the predator is satiated, its hunger level is low and there is little reason for pursuit or attack. The duration of the low level of hunger depends upon the maximum food capacity of the gut, the amount of food con-

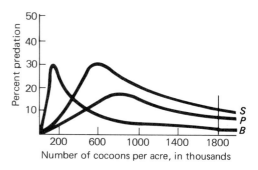

FIGURE 12-17

Total response of predators to prey density expressed as a percentage of predation to prey density. These graphs are based on the graphs in Figs. 12-9 and 12-12. (From Holling, 1959.)

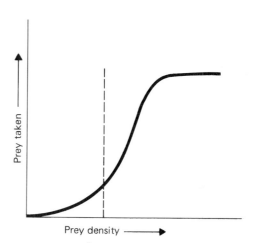

FIGURE 12-18

Compensatory predation as illustrated by a functional response curve. There is no response to the left of the vertical line.

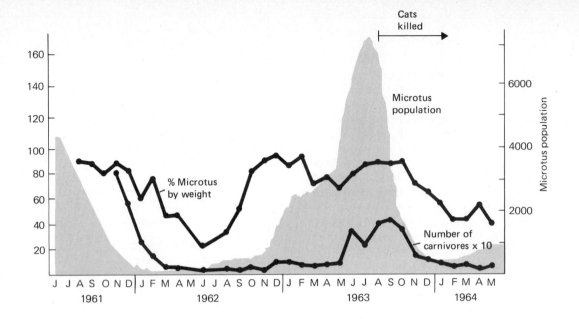

sumed at feeding, the rate of digestion and assimilation, and the need to supply food to offspring. After a certain length of digestive pause the hunger level of the predator rises to a point where it triggers attack (see Holling, 1966; K. E. F. Watt, 1968), and the predator moves out to hunt.

Hunting involves still another variable that is influenced by prey density—the searching rate, the time spent in search and pursuit of food. The searching rate is influenced by the speed of the predator relative to the speed and escape reactions of the prey, to the distance at which predators first notice and attack the prey, and to the proportion of attacks that result in successful capture. The searching rate varies considerably among species and among individuals. Also, the more abundant the prey, the less time is involved in searching. If prey is scarce, the time element may lengthen to a point where the predator cannot support itself.

Once it has captured its prey, the predator must spend additional time in handling and eating it. If the catch does not satisfy the predator, it searches for additional food. When satiated, the predator spends a period of time in digestive pause, during which it is not hungry enough to search and attack.

All of these variables influence the time predators are exposed to prey. This is the amount of time the predators spend in non-feeding activities compared to time spent in feeding activities.

Such variables lend themselves to mathematical modeling of predation. Predation can be broken down to component parts, and the components then can be tested by experimentation. Models are described in detail in Holling, 1966; K. E. F. Watt, 1968; and Morris, 1969.

PREY

The variables discussed so far largely involve the predator. From the viewpoint of the prey there are a number of variables that make up the odds that an animal will be captured by a predator, the prey risk. It is determined in part by the density of the prey population, availability of food and cover, movement, activity, habits, size, age, strength, and escape reactions. In many situations the intensity of predation on a prey species is related to the density of the prey population. Concentrations of prey tend to attract predators (numerical response). The more dense the population of a prey, the more intense is the predatory pressure, provided the prey is an acceptable item. But for an individual in the population, a dense population of prey reduces the probability that a particular individual will fall to a predator.

As discussed already, there is a very strong interaction between predation and cover. Prey that occupy secure habitats with good cover are much less vulnerable to predation. A sunfish hiding in a mass of pondweed is relatively safe, but once the fish moves into clear water, it exposes itself to a number of predacious fish. Young salmon living in clear shallow streams are easy prey for kingfishers and mergansers, whereas those living in deep pools are safe. Meadow mice are secure in heavy cover, but after the spring thaw, when the grass cover has been destroyed by the mice and flattened by winter snows, the mice are exposed to heavy predation (see Pitelka, 1957*b*).

Movement, activity, and habits affect an animal's chances of falling to a predator. During periods of territorial establishment and courtship, passerine birds are more active and less wary, increasing their conspicuousness. Nocturnal animals such as the white-footed mouse are less susceptible to diurnal predators than to owls. During the winter the same mice run along the top of the snow while meadow mice tunnel beneath. The dark pattern against the white makes the white-footed mouse highly vulnerable while the meadow mouse is relatively safe. Concentrations of prey may make them highly vulnerable to predation. Covies of quail are the target of Cooper's hawks in winter; during summer, when the quail are scattered, they are less subject to attack. On the other hand, ringnecked pheasants are relatively secure in winter cover. Spring dispersal takes them into the range of additional predators; and increased daily activity at a time when there is little background cover makes the pheasants more conspicuous (Craighead and Craighead, 1956). Dispersing muskrats are highly vulnerable to predation (Errington, 1957). A similar situation exists with the predation of seeds by granivorous animals. During the predispersal stage, when seeds are concentrated, they are subject to heavy predation. During the postdispersal stage the intensity of predation decreases with an increasing distance from the parent plant as well as with a decreasing density of seeds (Jansen, 1970).

Age, size, and strength vary the prey risk. Although the young, old, and infirm among vertebrate prey appear to possess the greatest prey risk, it is difficult to generalize. Much depends upon the mode of hunting on the part of the predator. Canids that rely on speed to overtake their prey invariably end up taking young and old animals. Hyenas confine their predation of wildebeest to young and old (Schaller, 1972). Cheetas restrict themselves to prey of less than 45 lb, which limits them to gazelles and the young of wildebeest. On the other hand, the big cats, such as the African lion and the mountain lion, hunt by stealth and ambush, and any animal that

FIGURE 12-19 [OPPOSITE]
Relationship between the amount of Microtus, *the meadow vole, by weight in the carnivores' diet compared with the number of carnivores and the number of* Microtus *present. Divide the left scale by 10 to obtain the number of carnivores. (From O. P. Pearson, 1966.)*

places itself in a vulnerable position is liable to attack. Thus the African lion attacks wildebeest in the prime of life (Schaller, 1972), and mountain lions kill mule deer of all ages (Hornocker, 1970).

If predators have evolved means of capturing or securing prey, then the prey, one might expect, must have evolved certain defensive mechanisms against excessive predation. Speed, agility, and escape reactions vary among species and among individuals of the same species. Songbirds and rabbits possess considerable agility and maneuverability. Bobwhite quail, on the contrary, have little speed or agility and must rely on acute awareness of danger, cover, and protective coloration. Some animals react defensively, either collectively or individually, and often succeed in driving off or even injuring or killing the aggressor. Defensive actions are adapted to the predator. A grown muskrat may slash at a mink with the teeth (Errington, 1957). Thompson's gazelle may kill a jackel but flee from a larger predator (Schaller, 1972). Other defensive or escape mechanisms involve structural defenses. Turtles may withdraw within protective shells; porcupines expose sharp quills. Among plants, certain species possess thorns and spines that inhibit browsing by animals. The hard coats of seeds inhibit predation by all but specialists.

Also employed are chemical defenses, and their use is wider than one might suspect. Some plants and animals are toxic to their predators. Toxicity in plants comes from the concentration of such chemicals as selenium, gossypol, cyanogenic glycosides, and saponins. Although these toxins may inhibit predation from a wide range of predators, toxic plants do have their specialized predators. Some are insects that are capable of transferring the toxic substances from the plant to themselves. This in turn protects the plant predator from its enemies. The monarch butterfly, for example, feeds on the milkweed that contains a cardiac glycoside. This substance can cause an illness to birds that eat the monarch. Other chemical defenses include the secretion of offensive substances. Skunks and some arthropods have glandular secretions that they administer to the enemy. These are either burning or ill-smelling fluids sprayed on the enemy or venoms injected into the enemy (T. Eisner, 1970). Other animals, such as the millipede, discharge a secretion that oozes over the surface of the body (T. Eisner, 1970). The effects of these secretions are usually immediate. The ill effects so distract the predator that the prey escapes.

Animals that possess pronounced toxicity and other chemical defenses often possess warning coloration, bold colors with patterns that serve as a warning to would-be predators. The black and white stripes of the skunk, the bright orange of the monarch butterfly, and the yellow and black coloration of many bees and wasps serve notice to their predators, all of whom must have some unpleasant experience with the animal before they learn to associate the color pattern with unpalatability or pain.

Together with these animals, other associated edible species have evolved a similar mimetic or false warning coloration. This phenomenon was described some 100 years ago by the English naturalist H. W. Bates in his observations on tropical American butterflies. This type of mimicry that he described, now called Batesian, is the resemblance of an edible species, the mimic, to an inedible one, the model. Once the predator has learned to avoid the model, it avoids the mimic also. An example among North American butterflies is the mimicry of the palatable viceroy butterfly to the monarch, definitely distasteful to birds (Brower, 1958). Both the model and the mimic are orange in ground color with white and black markings; they are remarkably alike. Yet the viceroy's nonmimetic relatives are largely blue-black in color. Another group of models and mimics involve the inedible pipevine swallowtail and the palatable spicebush and black swallowtails (Fig. 12-20). Tests involving the Florida scrub jay as the predator showed that this bird could not distinguish or confused the color pattern of the mimic with that of the model (Brower, 1958).

Bumblebees and honeybees have their mimics among the flies (Fig. 12-20). Of these perhaps the most interesting is the model bumblebee and the mimic robber fly, *Mallophora bomboides*. Not only does the robber fly benefit from reduced predation, but it also exploits the model as food. This robber fly, which preys on the Hymenoptera by preference, by resembling the bumblebee would not be noticed by the bee until it was too late to defend itself or flee (Brower and Brower, 1962).

A second, less common type of mimicry is

Müllerian, in which both model and mimic are unpalatable. The pooling of numbers between the two species reduces the losses of each, particularly when inexperienced birds are learning to avoid them (Sheppard, 1959).

A more subtle defense is physiological: to so time reproduction that the maximum number of seeds or offspring are produced within one short period of time, thus satiating the predator and allowing a percentage of the reproduction to escape. Species of conifers heavily consumed by squirrels release seeds within one short period of time. To obtain seeds squirrels cut and store green cones. Thus it becomes something of a race between how fast squirrels can harvest cones and how fast the cones can shed the seeds (C. C. Smith, 1970). Schaller (1972) suggests that the peak production of young wildebeest, highly susceptible to predation, is an antipredation mechanism. The young are so abundant at one short period of time that all the predators become satiated. No young are available during the rest of the year.

There are other mechanisms. Plants that escape predation by predator satiation usually produce large amounts of small seeds. All seeds of a seed corp may not be equally available to all its predators. If the predator is something of a specialist it cannot build up its numbers during lean seed years. Thus the population cannot reach a level that might be catastrophic to the seed crop. The seed crop of one species may be timed with an abundant seed crop of another species. Thus predators may be attracted from one seed crop by another. This is a common occurrence among the mast-bearing trees such as the oaks and hickories.

Whatever the mechanism involved, it is apparent that the various predator–prey relations represent a coevolution of predator–prey systems. As the prey evolves an improved defensive mechanism against predation, the predator evolves a new way to meet the change.

An example of an adaptive change in a prey species to meet a new type of predation is the caribou on the island of Newfoundland, effectively isolated from populations on the mainland of Canada. On Newfoundland in the past and on mainland Canada the principal predator of the caribou is the wolf. In response to very efficient predation by the wolf, the caribou evolved ways of keeping ahead of predation. These mechanisms include synchronized calving in which all are

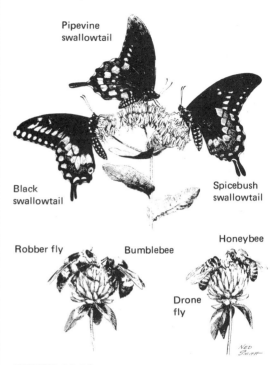

FIGURE 12-20

Mimicry in insects. The model, the distasteful pipevine swallowtail, has as its mimics the black swallowtail and the spicebush swallowtail. The black female tiger swallowtail, not shown, is a third mimic. All these butterflies are found together in the same habitat. The robber fly (left) is a mimic of the bumblebee, on which the robber fly preys. The drone fly, left, is a mimic of the honeybee. Characteristically, the model is much more common than the mimic, which is palatable to insect-feeding toads and birds.

389

born in a couple of weeks, postparturient aggregation of does, and a strong perception of motion as a defense against an open approach by wolves (Bergerud, 1971).

The recent history of the caribou in Newfoundland differs from that of the caribou on the mainland. Predation by wolves, accompanied by heavy hunting pressure from both Indians and white settlers, reduced the caribou in the mid 1800s to 100 to 150 individuals. As the wolves were killed off and the Indians disappeared, the caribou reached peak numbers between 1895 and 1900. Meanwhile a predator, the lynx, was exceedingly rare in the 1800s. After the snowshoe hare was introduced to the island, the populations of both snowshoes and lynx reached a high in 1896. After 1900 the snowshoe hares declined, leaving the abundant lynx without an adequate source of food. The lynx then apparently turned to caribou calves.

The caribou lacked any defense against a predator that travels by night and ambushes its prey. The caribou's habit of congregating during the calving season and keeping its young near cover is a good defense against the wolf but a poor defense against the lynx. In recent years it appears that natural selection in the Newfoundland caribou is in the direction of defensive mechanisms against lynx predation. The caribou, especially on the Avalon Peninsula, exhibit three behavioral traits not common to caribou elsewhere: the caribou have no specific calving grounds, the females do not congregate after calving, and calving is spread over 3 weeks instead of 2. Because the lynx, even though preying on caribou calves, still depends upon the hares, it cannot depart too far from those adaptations effective in catching them. Thus the constraints on the lynx are greater than those on the caribou. This change in behavior is reflected in a much improved rate of increase, 12 percent for the Avalon herd compared to 3 percent elsewhere (Bergerud, 1971).

PREDATOR–PREY SYSTEMS

Evidence in the field suggests that through natural selection, the predator and the prey develop a mutually sustaining system. If no such system evolves or if it is absent at first, prey declines rapidly, as did the caribou in the face of a new predator, the lynx. As long as the females congregated after calving, mortality of young was high. When the females and young dispersed over the range, mortality from lynx predation declined. There have been some attempts to duplicate predator–prey systems in the laboratory.

In an experiment, Huffaker (1958) attempted to learn if an adequately large and complex laboratory environment could be established, in which predator–prey relations would not be self-exterminating. Involved were the six-spotted mite *Eotetranychus sexmaculatus* and a predatory mite, *Typhlodromus occidentalis*. Whole oranges, dispersed on a tray among a number of rubber balls the same size, provided the food (and cover) for the spotted mite. Such an arrangement permitted the experimenter to change both the total food resource available, by covering the oranges with paper and sealing wax to the desired degree, and by the general distribution of oranges among the rubber balls. The experimenter could manipulate conditions to simulate a simple environment where the food of the herbivore was concentrated, or a complex universe where the food was widely dispersed, partially blocked by barriers, and where refuge areas were lacking.

In both situations, the two species found plenty of food available at first for population growth. The density of predators increased as the prey population increased. In the environment where the food was concentrated and the dispersion of the host population was minimal, the predators readily found the prey, quickly responded to changes in prey density, and were able to destroy the prey rapidly. In fact the situation was self-annihilative. On the other hand, in the environment where the primary food supply and thus the host were dispersed, the predator and the prey population went through two oscillations before the predators died out. The prey recovered slowly.

Several important conclusions resulted from this study. For one, predators cannot survive when the prey population is low. Second, a self-sustained predator–prey relationship cannot be maintained without immigration of the prey. Finally, the complexity of prey dispersal and predator-searching relationships, combined with a period of time for the prey population to recover from the effects of predation and to repopulate the areas, had more influence on the period of oscillation than the intensity of predation.

The degree of dispersion and the area employed was too restricted in the experiment to perpetuate the system. But in another experiment, Pimentel, Nagel, and Madden (1963) attempted to provide an environment with a space–time structure that would allow the existence of a parasite–host system. Used in the study was a wasp (*Niasonia vitripennis*) parasite and a fly (*Musca domestica*) host. For the environment the experimenters used a special population cage, which consisted of a group of interconnected cells. A system of 16 cells died out; but a 30-cell system persisted for over a year. Increasing the system from 16 to 30 cells decreased the average density of parasites and hosts per cell and increased the chances for the survival of the system. The lower density was due to the breakup and sparseness of both parasite and host populations. The greater number of individual colonies that remained following a severe decline of the host assured the survival of the system, since these colonies provided a source of immigrants to repopulate the environment. Moreover the amplitude of the fluctuations of the host did not increase with time, as proposed by the model of Nicholson. Apparently the fluctuations were limited by intraspecific competition.

There are in nature a few studied examples that these laboratory experiments support. Perhaps the most famous is the *Cactoblastis* story. Sometime before 1839 the *Opunitia*, or prickly pear, cactus was introduced from America into Australia as an ornamental. As is often the case with introduced plants and animals, the cactus escaped from cultivation and rapidly spread to cover eventually 60 million acres in Queensland and New South Wales. To combat the cacti, a South American cactus-feeding moth, *Cactoblastis cactorum*, was liberated. The moth multiplied, spread, and destroyed the cacti to a point that the plants existed only in small sparsely distributed colonies with wide gaps between them.

But the decline of the prickly pear also meant the decline of the moth. Most of the caterpillars coming from the moths that had bred on the prickly pear the previous generation died of starvation. And because of the low number of moths, not many of the plants were parasitized. In some areas the moth survived; and as the prickly pear increased, so did the moth, until the cactus colony was again destroyed. In other places the moth was absent and the prickly pear spread once more. But sooner or later it was found by the moth and the colony eventually destroyed. Meanwhile seed scattered in new areas established new colonies that maintained the existence of the species. The rate of establishment of prickly pear colonies is determined by the time available for the colonies to grow before they are found by the moth. As a result, an unsteady equilibrium exists between the cactus and the moth. Any increase in the distribution and abundance of the cactus leads to an increase in the number of moths, and a subsequent decline in the cactus. The maintenance of this predator and prey, or more accurately herbivore–plant, system depends upon environmental discontinuity. The relative inaccessibility of the host or prey in time and space limits the number of parasites and predators.

J. Monro (1967), in further investigation of the *Cactoblastis* and *Opunitia* relationship, finds that *Cactoblastis* may conserve food for succeeding generations of moths by limiting its own numbers. The moth accomplishes this by clumping its egg-sticks rather than laying them randomly on prickly pear. At high densities the moth overloads certain plants of prickly pear with eggs, resulting in the destruction of the plant and subsequent starvation of the relatively sedentary larvae. As a result more plants either escape infestation altogether or are subject to a lighter infestation than expected if eggs were laid at random. As the mean density increases, the proportion of eggs wasted by clumping increases. Monro found that this mechanism, which acts most strongly in the center of the range of *Cactoblastis*, conserves food supply for succeeding generations and maintains a constant level of both the food resource and the *Cactoblastis* population.

This in no way alters the events previously described. In dense stands of pear, the clustering of eggs would have little influence on larval survival. As an overloaded plant would collapse, it would fall on its neighbor and the larvae would move to a new source of food. In time, as dense stands were broken up into isolated plants, the starving larvae would be unable to cross the wide gaps. Then the mechanism to regulate the density of the moth would begin to act.

This regulatory mechanism working within a predator–prey relationship differs from an-

other relationship between plants and a herbivore. Among thrips and other insects, whose population growth is limited by some environmental influence, their numbers become large enough to deplete the resource, and there is no advantage to conserve the resource. These predators then exploit the resource to produce the maximum number of offspring. Their numbers in any one generation do not determine the future quantity of food (Monro, 1967).

Genetic feedback and predator–prey systems. Environmental discontinuity, predator density, alternate prey—these and other influences may not be acting alone in the maintenance of an equilibrium, however unstable, in a predator–prey system of nonterritorial animals. In recent years increasing attention has been given to the idea that evolution acting through genetic feedback systems may integrate the herbivore–plant, parasitoid–host, and predator–prey systems in the community.

An example of genetic change functioning to adjust a natural parasite–host system is the myxomatosis outbreak in the European rabbit population in Australia. To control the rabbit, the Australian government introduced myxomatosis in the population. The first epizootic of the disease was fatal to between 97 and 99 percent of the rabbits; the second resulted in a mortality of 85 to 95 percent; the third, 40 to 60 percent mortality (Fenner, 1953). The effect on the rabbit population was less severe with each succeeding epizootic, suggesting that the two populations were becoming integrated and adjusted to one another in the ecosystem. In this adjustment, attenuated genetic strains of virus, evolved by mutation, are tending to replace virulent strains (H. V. Thompson, 1954). Also involved is passive immunity to myxomatosis, conferred to the young born of immune does. Finally a genetic change has occurred in the rabbit population, providing an intrinsic resistance to the disease.

But this is not the whole story. The transmission of the myxomatosis virus is dependent upon *Aedes* and *Anopheles* mosquitos, which feed only on living animals. Rabbits infected with the virulent strain live for a shorter period than those infected with the less virulent strain. Because the latter live for a longer time, the mosquitos have access to that virus for a longer period. This gives the nonvirulent strain a competitive advantage over the virulent. In those regions where the nonvirulent strains have a competitive advantage, the rabbits are more abundant. This means that more total virus is present in those regions than in comparable areas where the virulent strains exist. Thus the virus with the greatest rate of increase and density within the rabbit is not the one selected for. Instead the virus whose demands are balanced against supply has the greatest survival value in the ecosystem.

This example suggests how a genetic feedback mechanism through genetic evolution might integrate the herbivore–plant, the parasite–host, and predator–prey systems in the community. It functions as a feedback system through the dynamics of density pressure, selective pressure, and genetic change in the interacting populations (Pimentel, 1961). In a herbivore–plant system, animal density influences selective pressure on plants. This selection influences the genetic makeup of plants, and the genetic makeup of plants determines the quality of food they provide the herbivores. Because the birth rate and the death rate of herbivores are in part a function of the quality of food they consume, the plants influence animal density. This same action and reaction follow through the food chain. Density influences selection; selection influences genetic makeup, and in turn genetic makeup influences density, all of which result in the evolution and regulation of animal populations.

PREDATION AS A REGULATING MECHANISM

It is almost impossible to generalize about the influence of predation upon a prey population, for as already pointed out, the effect depends in part upon the interaction of many variables. Theoretically at least, a predator can regulate or control a prey population if it can increase its density or effectiveness as the abundance of prey increases, and vice versa. Or to state it somewhat differently, regulation results only if the average risk of each prey individual increases with an increasing density of prey (Nicholson, 1954). As long as the predator or parasitoid feeds only on the interest of the prey population and does not touch the "capital," both predator and prey populations will tend to remain fairly stable. This is typical of compensatory predation. If the predator cuts into the capital stock, then the prey declines. In this situation, in which the predator must have a high rate of increase and an ability to disperse, predation may act to limit or may virtually exterminate local prey populations. The predator population tends to overshoot that of the prey. The decrease in prey results in a sharp decline in the predator population through starvation. If a portion of the host or prey is not available because of environmental discontinuity, then the oscillations will be damped, and under certain conditions the predator–prey system will be self-regulating. On occasions the host or prey do overcome any controlling influence the parasitoids and predators may have upon them and reach outbreak proportions. Under this condition it would be unusual for a parasite or a predator to overtake a prey while it is abundant and cause it to crash.

The impact of predators on prey populations is becoming more and more a subject of investigation. Data are accumulating on the influence of predators on selected prey populations. Waterfowl appear to be particularly vulnerable to predation during the nesting season. The effects of predation have become more pronounced as available breeding habitat shrinks. In the past, large expanses of nesting grounds compensated for predation. In one study, as high as 89 percent of the duck nests were destroyed by predators (Urban, 1970). When predator control was initiated on waterfowl breeding areas, production of young ducklings increased by 60 percent. On the predator control area nesting success was 81 percent; on the untreated area it was 34 percent (Basler et al., 1968).

Cotton rats likewise are vulnerable. Schnell (1968) enclosed 17 different populations of cotton rats in areas of natural habitat. Eight populations were in predator-proof enclosures; nine populations were exposed to predators. A density of about 15 rats/acre was the mean point of inflection on the survival curve and represented the "threshold of security" or predator-limited carrying capacity. Mortality from predators was high above and below this density (Fig. 12-21). Predation had its greatest limiting effect on the population during the nonbreeding period rather than during the breeding season.

Predation may depress populations of ungulates when the predator is one of the highly skilled canids, the wolf or the coyote, or a big cat, such as the mountain lion. In fact predation may be the major regulatory mechan-

ism. Deer, moose, and caribou apparently have not evolved intrinsic population controls. Pimlott (1967) suggests that this may be "because they have had very efficient predators, and the forces of selection have kept them busy evolving ways and means of not limiting their own numbers, but of keeping abreast of mortality factors."

The overworked story of the Kaibab deer herd as an example of the damping effect of predation on ungulate populations is not supportable. Not only are the data unreliable and inconsistent, but another event, the removal of sheep and cattle from the range, may have contributed to the eruption of the deer herd (Caughley, 1970). But other examples exist. Wolves are capable of depressing and holding a moose population below the carrying capacity of the range. Kolenosky (1972) has determined that wolves in east-central Ontario remove 9 to 11 percent of the deer over a 5-month winter period. Predation on the Thompson gazelle by hyenas, wild dogs, and lions is undoubtedly a major regulatory mechanism for that species. On the other hand predation, especially by lions, has little impact on populations of wildebeest, even though the predators kill a large number of these animals. Hornocker (1970) found that predation by mountain lions was inconsequential in determining the numbers of deer and elk. But predation was a powerful force that dampened and protracted severe prey oscillation and that distributed the deer and elk on a restricted and critical range.

Knowledge of the impact of predation on insect pests is essential in the development of biological control. A major predator on the damaging bark beetles is the woodpecker. Koplin (1972), in a study of the impact of three species of woodpecker—northern three-toed, hairy, and downy—on larval spruce beetles, developed a deterministic model of predation. The model incorporates the food requirements of free-living woodpeckers, the average number of prey per woodpecker stomach, population density of woodpeckers and air temperature as inputs, and the number of prey consumed as the output.

The model is

$$PI = \sum_{j=1}^{n} (CR_j \cdot PA_j \cdot d_j)$$

where

PI = predator impact, the predicted number of prey consumed by a population of predators during all time periods of uniform conditions of predator density and average air temperature occurring during a season or year

CR_j = number of prey consumed per predator day

PA_j = number of predators during time periods of uniform population density

d_j = period of time in days when conditions of average predator density and average air temperature are uniform

Each variable is further defined by its own equations, explained in the original paper.

This model was used to predict the impact of each species on endemic, epidemic, and panepidemic populations of larval spruce beetles. Estimates of predatory impact were obtained by measuring the relative survival of larvae inside and outside woodpecker enclosures. The predicted predation compared favorably with the estimates. Both showed the northern three-toed woodpecker to be the most effective predator and the downy the least (Fig. 12-22). The combined impact of all three woodpeckers was least effective in endemic and panepidemic populations and the most effective in epidemic populations. Endemic populations experienced the least predation because of the availability of alternate prey; panepidemic populations because territoriality limited the numerical response of woodpeckers to prey density.

Just as animal predators have an impact on vertebrate and invertebrate prey, so do herbivores or plant predators have an impact on plant populations. Deer can adversely affect such plants as white cedar, birch, and American yew, all quite sensitive to overbrowsing. The American yew is nearly gone in the Adirondacks and central Appalachians. Scattered vigorous stands exist only where populations of deer are low. Grazing of reef fish is so concentrated near the reefs that no sea grasses can sustain viable growth there (Randall, 1965).

In England the rabbit, introduced as a semi-domesticated species in the twelfth century, has been part of the countryside since the mid-1800s. Although some early work by the English ecologist A. Tansley showed that rabbits grazing on the chalk flora produced

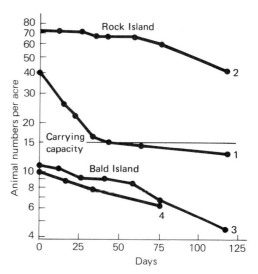

FIGURE 12-21
Comparison of survival curves of cotton rat populations with and without predation. When not subject to predation, populations show low disappearance rates regardless of whether initial densities are high or low. When populations are subject to predation, they show increased mortality rates if the population is high, and reduced rates if they are low. Curves 2 and 3 are populations without predation. (From Schnell, 1969.)

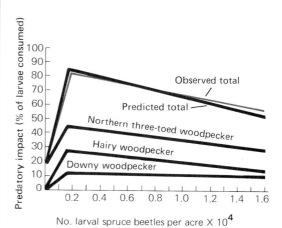

FIGURE 12-22
Relative and total predatory impact of woodpeckers on larval spruce beetles. (From Koplin, 1972.)

a richer diversity of species, the real impact of rabbits was not noticed until after 1954. At that time myxomatosis nearly wiped out the rabbit population, reducing the predatory pressure on the grasslands. With the demise of the rabbits, there was a spectacular increase in the growth of grass and a profusion of flowering by perennial species whose existence had never been recorded. But after the initial flush the number of dicot species declined, the grasslands were dominated by a few species to the exclusion of others, and the number of vegetative shoots of both grasses and dicots were reduced as the plants grew into taller, self-shading vegetation (J. L. Harper, 1969).

CANNIBALISM

A rather special kind of "competitive interference" or predation that exists within a species population or between populations of closely related species is cannibalism. Distasteful as it may seem, cannibalism is a method of population control common to a wide range of animals, including fish, rodents, birds, and primitive human societies, although in the latter (and in today's society) infanticide, abortion, and war are more commonly employed violent methods of intraspecific population control. Of all the methods of population control, cannibalism is the one in which the abundance of the control agent is inseparably linked to that of the population controlled. With cannibalism, an increase in parental stock not only increases the number of eggs laid and young born in a given reproductive season; it is also accompanied by a decreasing survival of young. If the breeding stock is large enough, cannibalism reduces reproduction to practically zero (Fig. 12-23), in spite of the greatly increased production of young. However, cannibalism as a form of population control is probably important only in a minority of populations. Difficult to study in nature, cannibalism is most pronounced as a means of self-regulation in confined populations of flour beetles (Lloyd, 1968; Dawson, 1968). Eggs, pupae, and pre-reproductive adults may experience mortality greater than 90 percent from predation by larvae and mature adults within their own species.

Interspecific competition

When two species in the same community seek the same resources, such as food and space, which are in short supply (in relation to the number seeking it), or interact in a way that affects their growth and survival, they are said to compete with one another. Few concepts have had such an impact on ecological and evolutionary thinking; yet basically the nature of interspecific competition and its effects on the species involved is one of the least known and most controversial fields of ecology.

The idea of competition, or the "struggle between species," was emphasized by Darwin. Later Lotka and Volterra separately developed mathematical formulas describing relationships between two species utilizing the same food source. Essentially these formulas indicate that eventually only one species, the least susceptible to food shortage or most adaptable to changed environmental conditions, will survive.

As with the models of predation, the basic growth curve is involved, but with an added variable to account for the interference of one species on the population growth of another:

$$\frac{dN_1}{dt} = r_1 N_1 \left(\frac{K_1 - N_1 - \alpha N_2}{K_1} \right)$$

and

$$\frac{dN_2}{dt} = r_2 N_2 \left(\frac{K_2 - N_2 - \beta N_1}{K_2} \right)$$

where

$\alpha =$ competitive coefficient for species 2 in relation to species 1
$\beta =$ competitive coefficient for species 1 in relation to species 2

If the system is at equilibrium then

$$\frac{dN_1}{dt} = \frac{dN_2}{dt} = 0$$

In a purely competitive situation there are three possible alternatives. One species can win out over the other, and one species will become locally extinct. Or either of the two species can win, depending upon the ecological variables operative at any one particular time. In such a situation an unstable equilibrium exists between the two species, with one dominant at one particular time, the second species at another. Or it is possible that neither species wins, in which situation they eventually coexist, dividing in some manner the resources between them.

These situations are illustrated graphically in Fig. 12-24. The abcissa represents a set of values of population size for species 1, the ordinate a set of values of population size for species 2. The model proposes that for each species a set of joint values exist along which $dN_1/dt = 0$ and $dN_2/dt = 0$. In other words the population of each species is neither increasing nor decreasing. The values fall in a straight line as illustrated in Fig. 12-24a and b, which happens when $N_1 = K_1 - \alpha N_2$ and $N_2 = K_2 - \beta N_1$. Outside the line the species is decreasing because the population is exceeding the carrying capacity available. Inside the line the population is increasing because the carrying capacity K has not been reached.

Now assume the simple situation in which two species, 1 and 2, occupy the same space as competitors. One species may outcompete the other and eliminate it from the area. This is illustrated in Fig. 12-24c, where species 2 outcompetes species 1. This happens because the area on or above the line K_1, $K_{1/a}$, and below K_2, $K_{2/\beta}$ is below the carrying capacity of species 2 but is above limit of species 1. Under a different set of conditions 1 would win, as illustrated in Fig. 12-24f, in which the area bounded by K_2, $K_{2/\beta}$, and K_1, $K_{1/a}$, is above the carrying capacity of N_2 but below N_1.

In another situation the two species might coexist with their populations in equilibrium. This can happen when the two competing species act as one, possessing such a behavioral trait as interspecific territoriality. Thus as species 1 increases, species 2 decreases and vice versa. The density of each species depends upon some mechanism which slows the growth of species 1 before it becomes so abundant that it can stop or reverse the growth of species 2. The two competing species reach a stable equilibrium point (Fig. 12-24e) when

$$\alpha < \frac{K_1}{K_2} = \beta < \frac{K_2}{K_1}$$

397

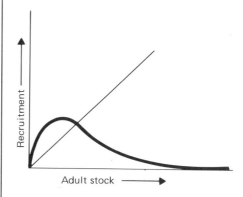

FIGURE 12-23

A reproductive curve illustrating cannibalism. Note how the population dwindles to zero.

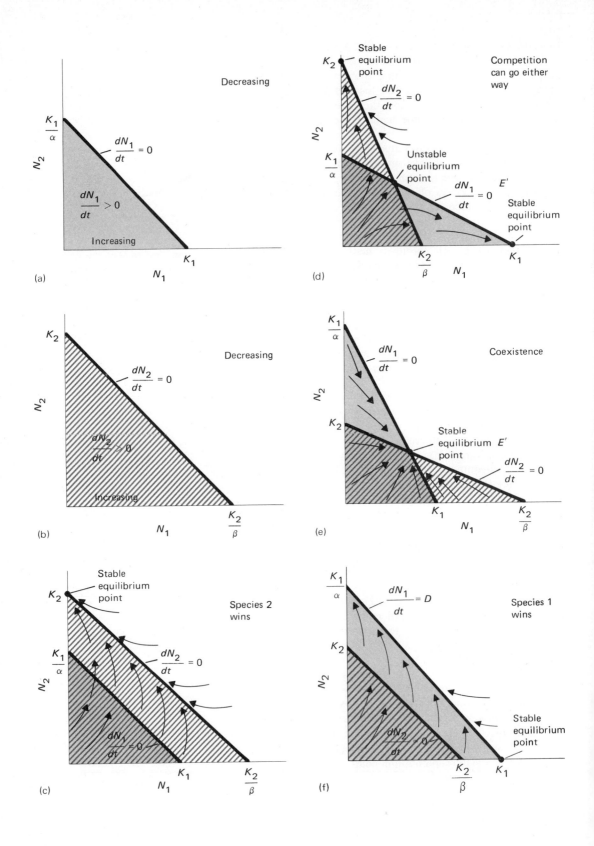

(a)

Decreasing

$\dfrac{K_1}{\alpha}$

$\dfrac{dN_1}{dt} = 0$

$\dfrac{dN_1}{dt} > 0$

Increasing

K_1

N_2

N_1

(b)

K_2

Decreasing

$\dfrac{dN_2}{dt} = 0$

$\dfrac{dN_2}{dt} > 0$

Increasing

$\dfrac{K_2}{\beta}$

N_2

N_1

(c)

Stable equilibrium point

K_2

$\dfrac{K_1}{\alpha}$

$\dfrac{dN_2}{dt} = 0$

$\dfrac{dN_1}{dt} = 0$

K_1

$\dfrac{K_2}{\beta}$

N_2

N_1

(d)

K_2

Stable equilibrium point

$\dfrac{dN_2}{dt} = 0$

$\dfrac{K_1}{\alpha}$

Unstable equilibrium point

$\dfrac{dN_1}{dt} = 0$

E'

Stable equilibrium point

Competition can go either way

$\dfrac{K_2}{\beta}$

K_1

N_2

N_1

(e)

$\dfrac{K_1}{\alpha}$

$\dfrac{dN_1}{dt} = 0$

K_2

Stable equilibrium E' point

$\dfrac{dN_2}{dt} = 0$

Coexistence

K_1

$\dfrac{K_2}{\beta}$

N_2

N_1

(f)

$\dfrac{K_1}{\alpha}$

$\dfrac{dN_1}{dt} = D$

K_2

$\dfrac{dN_2}{dt} = 0$

Species 1 wins

Stable equilibrium point

$\dfrac{K_2}{\beta}$

K_1

N_2

N_1

Species 2 wins

In still another situation, an unstable equilibrium might exist between the two competing species. This is diagrammed in Figure 12-24d, in which either of two species can win. Above the line K_2, $K_{2/\beta}$ species 2 is unable to increase, and above K_1, $K_{1/\alpha}$ species 1 is unable to increase. Where the two lines cross represents an equilibrium point, but unlike the previous case, it is unstable. Instead the equilibrium points are at K_1 and K_2. If the mix of the species is such that the triangle K_2 E' $K_{1/\alpha}$ exists, 1 is above its carrying capacity and 2 is not. 2 will continue to increase and 1 will decrease until it is gone. The reverse situation occurs in triangle K_1, E' $K_{2/\beta}$.

Gause (1934) set out to test these formulas experimentally. He introduced two species of *Paramecium*, *P. aurelia* and *P. caudatum*, in one tube containing a fixed amount of bacterial food. *P. caudatum* died out. The success of *P. aurelia* resulted from its higher rate of increase. But in another experiment *P. caudatum* occupied the same solution with *P. bursaria*. Here both species were able to reach stability because *P. bursaria* confined its feeding to bacteria on the bottom, whereas the other fed on that suspended in the solution. Although the two used the same food supply, they occupied different parts of the culture, thus utilizing food essentially unavailable to the other. Later Crombie (1947) confined two beetle populations, *Trilobium confusum* and *Oryzaephilus surinamensis*, in flour. The *Oryzaephilus* population died out, chiefly because *Trilobium* ate more eggs of *Oryzaephilus* than the latter did of *Trilobium*. In a second experiment he reared another population of two beetles, one *Oryzaephilus* and the other *Rhizopertha dominica*, in a culture of cracked wheat. *Rhizopertha* lived and fed outside the grain and *Oryzaephilus* lived and fed inside the grain. They occupied different niches, so they were able to exist together even though they utilized the same food.

These experiments seemed to bear out in part at least the mathematical models devised by Lotka and Volterra. From this developed the concept that two species with identical ecological niches cannot occupy the same environment. English ecologists called this "Gause's principle," although the idea was far from original with him. More recently the concept has been called the "competitive

FIGURE 12-24 [OPPOSITE]

Models of competition between two species. In (a) and (b), species 1 and 2 will increase in size and come to some equilibrium at some point on the diagonal line. Neither species is competing with the other. In (c), because the zero growth curve for species 2 falls outside the zero growth curve for species 1, species 2 wins, ultimately leading to the extinction of species 1. In (f) species 1 wins. In (d) and (e) the diagonal lines, or growth curves, cross. The equilibrium point represented by their crossing will represent either a stable or an unstable point. The area in which each species is successful is indicated by their respective coded shadings. Note how a slight movement of the diagonal lines upward or downward in (d) and (e) can change the outcome of competition. See the text for details.

exclusion principle" (Hardin, 1960), which can be stated briefly as: "Complete competitors cannot coexist." Essentially this principle means that if two noninterbreeding populations occupy exactly the same ecological niche, if they occupy the same geographic territory, and if population A multiplies even the least bit faster than population B, then A eventually will occupy the area completely and B will become extinct.

Empirical evidence of competition, although often difficult to determine and prove in nature, seems to support competitive exclusion. No two species, as far as is known, possess identical ecological requirements. Each species faces the chance that reproduction and survival will fail and thus permit a competitor to step in. Subtle differences in the ecology of each species, which might escape notice at first, may tip the scales in favor of one species or the other. Park (1948, 1954) studied the relationship of laboratory populations of two flour beetles, *Trilobium castaneum* and *Trilobium confusum*. He found the interaction between the two species to be quite complex. The fate of the two competitors depended considerably upon environmental conditions such as temperature and humidity, upon the presence or absence of parasites, and the fluctuations of the total number of eggs, larvae, pupae, and adults. Often the final outcome of competition was not determined for generations. Frank (1952), working with two water fleas with overlapping niches, *Daphnia pulicaria* and *Simocephalus vetulus*, found that eventually *Daphnia* won out, although *Simocephalus*, better adapted to yeast food, put up a good struggle. In nature, however, under less severe crowding conditions, the two species probably could exist together.

Similar species or members of the same genus are able to exist together in normal numbers in the same community because of ecological adaptations that permit them to occupy different ecological niches, to possess different feeding methods, to utilize different foods, or if the food is the same, to feed at a different time or place. (Fig. 12-25). In other cases competition is avoided because normally there is a superabundance of food. Here competition would be evident only when the shared resource was in short supply.

There are examples. The myrtle warbler, the black-throated green warbler, and the Blackburnian warbler all inhabit the same forest and all have similar feeding habits; but each feeds in a particular part of the forest canopy with a minimum of overlap. Two other species, the bay-breasted warbler and the Cape May warbler, are competitors whose abundance varies with the superabundance of food. The Cape May warbler, in particular, depends upon the spruce budworm. During years between outbreaks of this insect, this warbler is reduced in numbers or eliminated locally (MacArthur, 1958).

Throughout Europe there are nine species of tits, six of which are widespread. Three of these, the blue tit, the great tit, and the marsh tit inhabit the broadleaf woods. The blue tit, the most agile of the three, works high up in the trees gleaning insects, mostly 2 mm in size or smaller, from leaves, buds, and galls. The great tit, which is large and heavy, feeds mostly on the ground and seeks prey in the canopy only when taking caterpillars to feed its young. Its food consists of large insects 6 mm and over, supplemented with seeds and acorns. The marsh tit feeds largely on insects around 3 to 4 mm, which it gleans in the shrub layer and in twigs and limbs below 20 ft above the ground. It, too, feeds extensively on seeds and fruits. In the northern coniferous forests live the coal tit, the crested tit, and the willow tit. The more agile coal tit forages high up in the trees among the needles. There it seeks and stores aphids and spruce seeds. The willow tit consumes a high proportion of vegetable matter and feeds in the few available broadleaf trees. When in the conifers it spends most of its time in the lower parts and on the branches rather than on the twigs. The crested tit is confined mostly in the upper and lower parts of the trees and on the ground, but the bird does not feed in the herb layer. Thus by feeding in different areas and on different size insects, as well as different types of vegetable matter, these species divide the resources among them.

STRATEGY OF COMPETITION

Two strategies are involved in competition, *exploitation* and *interference* (R. S. Miller, 1967).

Exploitation happens when two or more species have access to a limited resource. The outcome is determined by how effectively each of the competitors utilizes the resource. Essentially it is a scramble type of competition between individuals of different species instead of between individuals within a species. For example, in Newfoundland the moose and the snowshoe hare share the same habitat and food (Dodds, 1960). Of the 30 woody plants consumed by the moose, 27 are utilized by the hare. Both browse most heavily on plants under 6 ft tall. When moose feed on balsam fir reproduction on cutover areas, they retard the growth of cover for hares, preventing them from inhabiting the area during the early years after cutting. Birch is the most important hardwood browse for both moose and hares. Competition for this food begins when hares move into the cutover areas. Where forest reproduction is predominantly fir and birch, moose may cause a winter food shortage for hares.

In general when even partial competition exists, the immediate effect is depressed population levels of one or both populations, reduction in rate of increase, or reduction in individual growth rates. Trout, for example, grow more rapidly to a larger size when other trout species and forage species are low or absent (E. L. Cooper, 1959). Shiners in Paul Lake were serious competitors with smaller trout for food, even though they were a food source for larger trout (Crossman, 1959*b*). A rise in the shiner population in this lake caused a drop in production and a decrease in the growth rate of trout. A more extreme response occurs among two species of butterflies, the monarch and the queen monarch in Florida (Brower, 1961). Both are attracted to the flowers of the same milkweed plant, upon which they both feed and lay eggs and on which the larvae feed. The non-migratory queen monarchs, however, avoid some species of milkweed that the migratory monarchs utilize. But after the monarchs have flown northward, the queen monarchs oviposit on this milkweed. Thus the disappearance of the monarchs coincides with a population increase in the queen monarchs.

With the other strategy, interference, one competitor is denied access to the resource, usually by some form of aggressive behavior. For example, the ubiquitous red-winged blackbird and the restricted tricolor blackbird occupy the same range in the lowlands of California and adjacent Oregon and Baja California. Morphologically they are similar; behaviorally they are different. Male red-wings

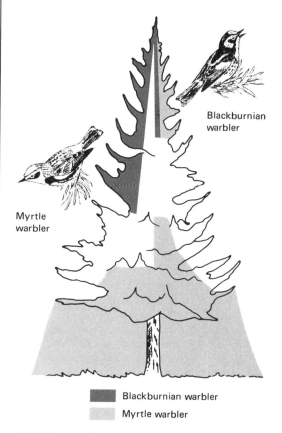

Blackburnian warbler

Myrtle warbler

■ Blackburnian warbler
□ Myrtle warbler

FIGURE 12-25
The feeding niches of two species of warblers, the myrtle and the balckburnian. The shaded areas represent that portion of the trees in which the birds forage over 50 percent of the time. (Adapted from MacArthur, 1958. © The Ecological Society of America.)

establish moderately large territories of 500 to 30,000 ft² early in winter in the California marshes. Within these territories, the females establish subterritories and build their nests. Male red-wings take no part in caring for the nest and young; the females stay partly on the territory and partly on the adjacent dry land. The tricolor is nomadic and is the most colonial of the North American passerines. Colonies of less than 50 nests are rare; and in such optimum habitats as the California rice fields, there may be as many as 200,000 nests. Male territories are about 35 ft², and in them, one to three females construct the nests. The males do not assist in incubation and may even leave the area during the incubation period. But once the eggs have hatched, the males take an active role in feeding the young. Unlike the red-wing, the tricolor feeds off the territory and may range as far as 4 mi from the colony. The colonial system of the tricolor is more demanding of energy than the territorial system of the red-wing. Since the birds have to travel some distance to obtain food for the young, the tricolor requires a rich food supply that can be gathered quickly when the birds arrive. This results in an unpredictable breeding distribution.

Because of early territorial establishment, the red-wings have usually filled the marsh by the time the tricolors start to establish their colonies. When large numbers of the tricolors move into a marsh occupied by red-wings, the male red-wings act aggressively, but the tri-colors simply through superior numbers usually are successful, without offering counteraggression (Orians and Collier, 1963). The red-wings either desert their territories (Fig. 12-26) or if they remain on them, cease to defend them. Because the tricolors do not necessarily breed in the same place year after year, the red-wings do not desert the area but continue to remain aggressive toward the tricolor.

There are other examples that might be noted briefly. When brook trout and rainbow trout occupy the same stream, they treat each other as if they were the same species (M. A. Newman, 1956). The largest fish occupy the deepest and darkest parts of the stream and the smaller ones are forced to live in less favorable parts. Seven species of darters defend territories against other species as well as against individuals of their own species (Winn, 1959a, b).

OUTCOME OF COMPETITION

As predicted by the models and as already illustrated in some of the examples, competition results in the replacement of one species by another, in an unstable equilibrium between species, or in a stable coexistence.

The degree of coexistence depends upon the degree of similarity. If two or more closely related species are quite similar in requirements, and if the population is less than the carrying capacity of the habitat, each has to share a limited environmental resource, and natural selection will favor any new gene in the population that will eliminate or reduce sharing. It will promote the spread of genes that will enable the population to exploit an unfilled ecological niche. In other words, real competitors are dropped along the evolutionary wayside.

Many species tolerate a wide range of habitat conditions, from optimum to marginal. If the population expands, the surplus individuals may settle in marginal habitat. This same habitat also may be marginal for another similar species and be occupied by it. If the new colonizer is a strong competitor, the original occupant is forced to narrow its ecological range by contracting it to the optimum habitat. Here the species remains, until the marginal range again is unoccupied by its competitor, when the species again may expand into the area. Such a situation is called *ecological overlap*.

For example, the fresh-water marshes of western United States are occupied by red-winged and yellow-headed blackbirds. The diets of the two birds apparently are similar enough to preclude the possibility of their exploiting the same simple vegetation in different ways. This limits the population intraspecifically and interspecifically. The total number of birds in an area is limited and territorial conflicts are interspecific. The yellow-headed blackbirds occupy the deeper water areas, whereas the red-wings are restricted to shallow water near shore. The size of the territories of each is influenced by the presence of individuals of the other species. The early-arriving red-wings contract their territories when the yellow-headed blackbirds arrive; and if the red-wings are not there, the yellow-headed blackbirds expand shoreward.

Under optimum conditions, some plants will dominate and suppress their associates.

During moist years the prairie grass, little bluestem, dominates the grassland, but during dry years it almost disappears, while two associates, side-oats grama and blue grama, two grasses of the same genus associated with each other become dominant.

An unstable equilibrium usually results in one species winning out over the other. Which species wins, more often than not, is determined by the environmental conditions under which they are living at any one particular time. In a laboratory experiment Park (1955) observed that the flour beetle *Trilobium castaneum* usually won out over *T. confusum*. But if the two competitors were exposed to the sporozoan parasite *Adelina trilobii*, *T. confusum* usually won.

In some ways unstable equilibrium is more easily observed among plants.

From the time seedlings germinate and develop, the demand for growing space, light, moisture, and nutrients increases. Those plants that utilize resources in short supply most efficiently have the best chance for survival. Plants whose root systems are in the same horizon compete for limited moisture and nutrients. In western North America the shallow-rooted annual grass *Bromus tectorum* grows early in spring and often reduces moisture so much that slow-growing annuals, perenials, and even shrubs are unable to withstand the competition (Holmgren, 1956). In drier regions, plants that develop roots rapidly after germination have the competitive advantage. The weaker plants are overtopped and eventually crowded out by more vigorous and aggressive individuals.

A more usual outcome of competition is the elimination of one species from a given area by another. Each species then is more or less confined to a habitat in which it has the competitive advantage over the other. The marine intertidal zone off the Scottish coast is inhabited by two barnacles, *Chthamalus stellatus* and *Balanus balanoides*. The upper limits of the intertidal zone inhabited by the two species are set by physiological tolerances. Of the two, *Chthamalus* grows higher up on the rocks. When young *Chthamalus* settle in the *Balanus* zone, they seldom survive unless the area is kept free of *Balanus*. The *Balanus* colonize an area in greater numbers and grow so much faster than *Chthamalus* that they undercut, smother, or crush the latter. The *Chthamalus* that do survive are

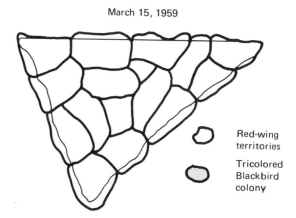

March 15, 1959

Red-wing territories

Tricolored Blackbird colony

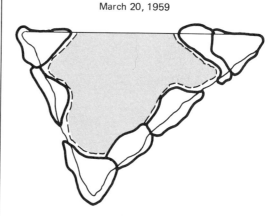

March 20, 1959

FIGURE 12-26
Interaction between redwinged and tricolor blackbirds at Hidden Valley marsh. Ventura County, California, in 1959. The tricolor blackbird is colonial. (After Orians and Collier, 1963.)

small and produce few offspring (Connell, 1961). Thus, the *Chthamalus'* degree of success in colonizing the lower limits is set by competition.

On the eastern slopes of the Sierra Nevada Mountains live four species of chipmunks: the alpine chipmunk, *Eutamias alpinus*, the lodgepole chipmunk *E. speciosus*, the yellow pine chipmunk, *E. amoenus*, and the least chipmunk, *E. minimus*. Each occupies a different altitudinal zone (Fig. 12-27), the line of contact partly determined by interspecific strife. The upper range of *E. minimus* is determined by aggressive interactions with the dominant *E. amoenus* (Heller, 1971). Although *E. minumus* is capable of occupying a full range of habitats from the sagebrush desert to alpine fell fields, in the Sierras it is restricted to the sagebrush habitat. That is somewhat outside of the climatic space (see Chapter 15) of the least chipmunk, but it can exist there by the use of hyperthermia and daytime retreat to burrows (Heller and Gates, 1971). Because the sagebrush habitat is outside of the climatic space of *E. amoenus* it cannot penetrate that part of the range of *E. minimus*. The aggressive behavior of *E. amoenus* also determines the lower limit of *E. speciosus*, restricted to the open coniferous forests. There the abundance of food and cover as well as the secretive habits of the chipmunk do not bring it into aggressive contact with *E. amoenus*. The upper limit of *E. speciosus* is determined by the aggressive behavior of *E. alpinus*, which is limited to the alpine and Hudsonian zones of the Sierras. Thus of the four species the distribution of two are in part influenced by the aggressive behavior of the other two.

In competitive situations aggressive behavior was probably selected for in *E. alpinus* and *E. amoenus* because of a seasonally limited food supply that can be cached and economically defended. Aggressiveness probably was not selected for in *E. minimus* because such activity would not be metabolically feasible in the hot sagebrush desert. Likewise, aggressiveness was not selected for in *E. speciosus* because in the presence of an abundance of food and vegetational diversity there was no adaptative value in doing so (Heller, 1971). Possibly because of pressure from predators, *E. speciosus* remains quiet and rather secretive, in contrast to the noisy activity of the highly aggressive species.

Because of competitive exclusion two species in contact with each other rarely if ever fully occupy their fundamental niche. For at least one and possibly both, the realized niche is less than the fundamental one.

A final type of outcome is a stable equilibrium that is often characterized by interspecific territoriality, in which the area is divided up among all the competing species, each behaving toward the other as the same one. For example Cody (1968) found that three grassland species of birds in Minnesota, Leconte's sparrow, the savannah sparrow, and the bobolink, exhibited interspecific territoriality with no overlap in the territories of any of the birds (Fig. 12-28). In Colorado grasslands, however, four species of birds, the western meadowlark, the horned lark, the lark bunting, and McCowan's longspur, exhibited almost complete overlap in their territories. The degree of overlap in territories of grassland birds was related to the height and vertical density of the vegetation and to the species of birds involved.

Stabilizing situations have been observed in a number of competitive plant situations, largely experimental in nature. Leith (1960) found that mixtures of perennial ryegrass *Lolium perenne* and white clover *Trifolium repens* growing in pastures form moving mosaics. Patches dominated by ryegrass tend to be invaded by clover, and patches of clover by grass. Van den Bergh and De Wit (1960) seeded sweet vernal grass *Anthoxanthum odoratum* and timothy *Phleum pratense* together in field plots in different proportions. They then compared the ratio of tillers of the two species after the first winter with the rations after the second winter (Fig. 12-29). In those plots where sweet vernal grass was in excess, the proportion of timothy increased; and where timothy was in excess, vernal grass increased. This experiment points out that in some mixtures the species in the mix experience more intra- than interspecific interference. In a mix of two species, the existence of a stable equilibrium depends upon frequency-dependent competition.

The competitive exclusion principle is based on the assumption that the competing species themselves and their biotic environment remain genetically constant. Because this is not the case, a situation might arise in which two competitors could coexist. The interspecific competitors may change genetically, so that both could live together and utilize the same food, space, and other necessary resources in the ecosystem (Pimentel et al., 1965). Assume that two species *A* and *B* are fairly evenly balanced, populationwise, but species *A* is slightly superior to species *B*. As the numbers of the stronger species increase, the weaker species declines and becomes sparse. At this point the individual of *A* contends largely with intraspecific competitive selection. Concurrently species *B*, still contending primarily with interspecific competitive pressure, would evolve and improve its ability to compete with the more abundant *A* species. As *B* improves as a competitor, its numbers increase, until finally *B* becomes the abundant species. The original trend is reversed. After many such oscillations, a state of relative stability should result.

Predation, competition, and community structure

The impact on vegetation of grazing by rabbits and the response of the grasses and associated vegetation to a release from grazing suggest that predation coupled with competition may be a major influence on the diversity of species within a community. Further evidence comes from the studies of controlled grazing of artificial and natural stands of grass with sheep the primary grazing animal (Jones, 1933; Milton, 1930). The findings of these studies can be reduced to some basic principles (Harper, 1969). If the predator selects the major dominant, then plant diversity increases. Overgrazing in winter and spring followed by undergrazing in summer produces the maximum floral richness. This is usually the case in natural situations because deer, elk, and others can overexploit their food in winter, but undergraze it in summer. The most species-rich communities are developed by continuous grazing with a

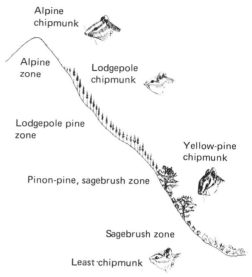

FIGURE 12-27

Transect of the Sierra Nevada, California, latitude 38°N, showing the zonation of dominant vegetation and the altitudinal ranges of the four Eutamias species that inhabit the east slope. (Transect after Heller and Gates, 1971. © The Ecological Society of America.)

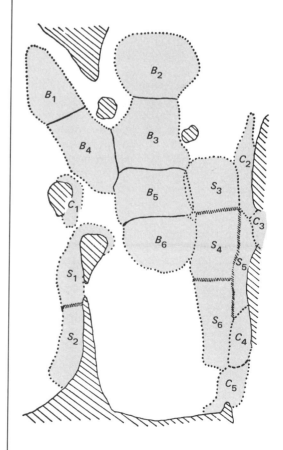

FIGURE 12-28

A map of the territories of several species of birds on a study area in Minnesota. The birds are bobolinks, B; Savannah sparrow, S; LeConte's sparrow, C. Note the lack of overlap, suggesting interspecific territoriality. Undefended territory boundaries are indicated by a dotted line. (From Cody, 1968. © 1968 by The University of Chicago Press.)

405

maintained population of herbivores. If the dominant plant is highly palatable, overgrazing will reduce it and allow other less palatable species to occupy the area. But if the dominant species are unpalatable then grazing only serves to consolidate their dominance. If the herbivore is regulated by its food supply and not by predators, then it can reduce a plant species to a minor component and allow invasion by other species. As the intensity of grazing increases, more and more unpalatable species are grazed, and eventually only wholly unpalatable species remain or move into the area. Thus the complex balance of species in a plant community is sensitive to rather precise control by the feeding activities of the grazing animal.

If plant predators are limited not by food but by predation, then the relationship between predation and community structure becomes more complex, involving not only predation but intra- and interspecific competition as well. Predation may ameliorate the intensity of competition for space and increase species diversity, as demonstrated by the experimental removal of a top carnivore, the starfish *Pisaster* from an intertidal community. Under a normal situation, including predation by *Pisaster*, the mussel *Mytilus californianus*, the barnacle *Balanus cariosus*, and a goosenecked barnacle *Mitella polymerus* form a conspicuous band in the mid-intertidal zone. In areas where the predator *Pisaster* was removed the barnacle *Balanus* occupied 60 to 80 percent of the available space (Paine, 1966). But within a year the *Balanus* themselves were being crowded out by the rapidly growing *Mytilus* and *Mitella*, and the benthic algae disappeared from the lack of space, along with the chitons and larger limpets. Thus the removal of *Pisaster* caused a pronounced decrease in species diversity and an increase in the intensity of interspecific competition. Predation in this case was effective in preventing one or two species from monopolizing a given resource.

The composition of a plant community may be influenced by the competitive advantage of one species over another. Two grasses are common to pastures of Australia, *Bothriochloa ambigua*, of low grazing value, and *Danthonia* spp., a better pasture grass. If the seeds of both are disseminated on a bare area, *Bothriochloa* becomes established as a dominant because of its more vigorous root growth. But

in native pastures already occupied by *Danthonia*, invasion by *Bothriochloa* rarely occurs, unless the former has been weakened by overgrazing (C. W. E. Moore, 1959). In other communities, taller plants may gain the upper hand, receive most of the light, and occupy the greatest root space, and only those plants able to grow and develop in their shade will survive underneath. If the dominant plants are intolerant they will be unable to replace themselves and the more tolerant invaders will assume dominance.

Most plants are not too exacting in soil requirements as long as competition is absent. But this is rarely the case in nature. On deep fertile sites poorer competitors often can coexist only if they can produce more viable seeds and occupy a situation within their range of tolerance. Otherwise the species is swamped. Thus on better sites the inefficient competitor, the intolerant plant, is eliminated. But on poorer sites competitive ability may take second place. The chestnut oak, a prolific seeder, will grow very well on fertile soil; but it is most abundant on poorer sites since it is excluded on the better sites by more vigorous competitors, such as red or white oak.

Population exploitation

A form of highly selective and intensive predation, often not related to the density of either predators or prey, is exploitation by man. Overexploitation of wild populations, especially if coupled with the loss of habitat, has resulted in a serious decline and local if not global extermination. The overharvesting of buffalo, great auk, African ungulates, whales, and many pelagic fish are examples of shortsightedness on the part of man. On the other hand, such valuable wildlife populations as the white-tailed deer are underharvested in many places (see Chapter 20), particularly since their natural predators have been eliminated. In contrast, the objective of the wise exploitation of any natural population is the maintenance of some sort of equilibrium between recruitment and harvest.

BASIC CONCEPTS OF
POPULATION EXPLOITATION

Although some of the terms used in defining exploitation of populations are similar to those used in productivity (see Chapter 3), the meanings are somewhat different. When fishery and wildlife biologists speak of *yield* they refer to the individuals or biomass removed when the population is harvested. *Biomass yield* is the product of the number harvested times the average weight. (Yield may indicate weight without numbers or vice versa.) The *standing crop* is the biomass present in a population at the time it is measured. *Productivity* is the difference between the biomass left in the population after harvesting at time t and the biomass present in the population just before harvesting at some subsequent time, $t + 1$.

The objective of regulated exploitation of a population is *sustained yield*. This means the yield per unit time is equal to productivity per unit time. In its simplest form sustained yield is described by an equation first proposed by E. S. Russell for fishery exploitation. Although the equation was specifically developed for fisheries it is applicable to any exploitable population. With some minor modifications this equation reads:

$$B_{t+1} = B_t + (A_{br} + G_{bi}) - (C_{bf} + M_b)$$

where

$B_{t+1} =$ total biomass of exploitable stock just before harvesting, at time $t + 1$

$B_t =$ total biomass of exploitable stock just after the last harvest, at time t

$A_{br} =$ biomass gained by the younger recruits just grown to exploitable stock

$G_{bi} =$ biomass added by the growth of individuals in both B_t and A_{br}

$C_{bf} =$ any biomass exploitatively removed during the harvest period

$M_b =$ biomass lost from exploitable stock by natural causes during the time t to $t + 1$

Thus the equation stresses the fact that productivity also includes individuals that were born and died and individuals that died during the time interval from one end of one harvest period to the beginning of the next.

The equation is highly simplified. In an unexploited fish population, A_{br}, C_{bf}, and M_b are interdependent. For example, in a stable environment largely undisturbed by man, fish populations appear to be dominated by large species. In turn each species population appears to be dominated by large old fish (Johnson, 1972). When man starts to exploit such a population, significant changes take place.

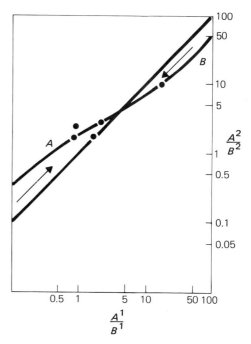

FIGURE 12-29

The graph depicts the competitive relationship between two grasses, timothy (Phleum pratense) and sweet vernal grass (Anthoxanthus). The relationship is expressed as the ratio of the number of tillers of sweet vernal grass (A) and timothy (B) after the first winter, A^1/B^1, and after the second winter, A^2/B^2. In plots in which sweet vernal grass had been in excess, the proportion of timothy increased; where timothy was in excess, sweet vernal grass increased. The point where the line AB intersects the 45-degree slope line represents the equilibrium point under a given set of environmental conditions. (From J. L. Harper, The Journal of Ecology, 55:242–270, 1967, Blackwell Scientific Publication, Oxford; redrawn from Van den Bergh and De Wit, 1960.)

To compensate for the exploitation directed first toward the largest members of the population (organisms that under natural conditions are normally secure from predation), the population exhibits an increased growth rate, a reduced age of sexual maturity, increased number of eggs per unit of body weight, and reduced mortality of small members of the population (Regier and Loftus, 1972). Populations of other vertebrates react in a similar way (see, for example, the white-tailed deer, Chapter 20). Exploitation may also influence behavior of the species. If a particular type of fishing gear or hunting technique that is quite effective during the early stages of exploitation is employed constantly, it gradually becomes less effective with time, possibly because of some conditioning or learning by the members of the population. As the take or catch of the species begins to decline, the exploiters are forced to continually improve their methods of fishing and hunting, a point to be discussed later. Also involved may be an interspecific competition. As both the numbers and the larger members of a population decline, the niche may be occupied by unexploited, highly competitive, and closely related sympatric or introduced species. Thus as the Pacific sardine populations declined, their place was taken by anchovies. If exploitation is carried far enough, then the age classes in the population are too young to carry on reproduction, and the population collapses.

The principle behind sustained yield is to avoid that collapse of the population. The objective is to remove as much of the population each year as possible and still maintain productivity at some desired level year after year. The exploitation and sustained yield of a population is closely dependent upon the rate of increase. The higher the r of a species, the higher will be the rate of exploitation that produces the maximum amount of biomass production. Species characterized by scramble competition (r strategists) have a high wastage of production; to manage a population influenced by density-independent variables such as climate or temperature, the objective should be to reduce wastage by increasing the rate of exploitation. The role of harvesting is to take all individuals that otherwise would be lost to natural mortality. This type of exploitation is described by the expression:

$$\text{maximum yield} = B_t - \min (R_t)$$

where B_t = biomass at time t and $\min (R_t)$ is the minimum number of reproducing individuals left at time t in order to ensure replacement of maximum yield at time $t + 1$ (see K. E. F. Watt, 1968).

Such a population is often difficult to manage because the stock can be severely depleted unless there is repeated reproduction. An example is the Pacific sardine (Murphy, 1966 and 1967), a species in which there is little relationship between breeding stock and the subsequent number of progeny produced. Exploitation of the Pacific sardine population in the 1940s and 1950s shifted the population to young age classes. Prior to exploitation, 77 percent of the reproduction was distributed among the first 5 years of life. In the fished population, 77 percent of the reproduction was associated with first 2 years of life. Thus, the population approached that of a singly reproducing population, subject to pronounced oscillation (Fig. 12-30). Two consecutive years of reproductive failures resulted in a collapse of the population from which it has never recovered.

In populations that are density-dependent regulated (K strategists), productivity can be increased to a certain level by removing sufficient amounts of biomass to reduce intraspecific competition. Once harvest level exceeds that point, the population will decline. The maximum rate at which such a population can be exploited and still sustain maximum production depends upon age structure of the population, the frequency of harvest, and the number left behind after harvest. This type of exploitation is described by

$$\text{maximum } (P_b) = \max [B_{t+1} (X)] - B_t$$

where

P_b = biomass productivity from t to $t + 1$
B_t = biomass at time t
B_{t+1} = biomass at time $t + 1$
X = the various variables that influence biomass production over the time t to $t + 1$

There are a number of examples of such exploitable populations, among them the North Sea plaice, salmon, and many terrestrial vertebrates.

This can be illustrated by means of some reproductive curves, as shown in Fig. 12-31. For any position of the stock to the left of the 45° line there is a rate of exploitation that will

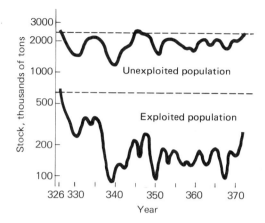

FIGURE 12-30

Simulation of an exploited and an unexploited population of sardines, both subject to the random environmental variation in reproductive success. The dotted line indicates the asymptotic population size. Note how exploitation adds to the instability and how dangerously low the population can get. (From Murphy, 1967. © The Ecological Society of America.)

FIGURE 12-31

Reproductive curves illustrating rates of exploitation. The perpendicular line ac *cuts the 45-degree line at* b *(see text). On curve C, the 82 percent point represents the maximum surplus reproduction and the maximum rate of exploitation possible for this population. Other figures in the population indicate the position on the curve for other rates of exploitation. Segment* ab *(E) is harvest;* bc *is the stock left for recruitment. Note the differences between the two curves. In curve C, $E_2 > E_1$, and $N_2 > N_1$. Under these conditions the greater the standing crop, the greater is the sustained yield. In curve A where $E_2 > E_1$, yet $N_1 > N_2$, a high standing crop does not result in greater sustained yield. A knowledge of parent–progeny relations is essential for the wise exploitation of natural animal populations.*

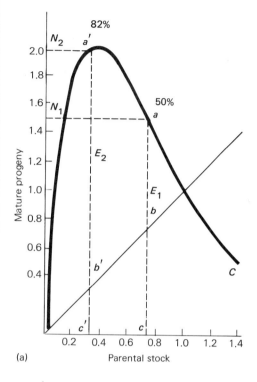

(a)

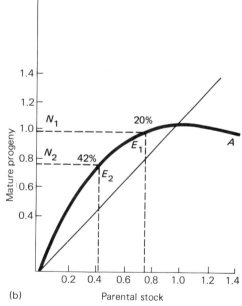

(b)

maintain the stock at that position. Maximum sustained yield does not necessarily require a large standing crop. In the curves in Fig. 12-31 let *a* be any position on the curve and *c* a perpendicular line that cuts the 45° line at *b*. At equilibrium, the portion *bc* of the recruitment must be used for the maintenance of the stock, for *bc = oc*; *ab* can be harvested. There is, however, a limit to exploitation, a limit that is influenced by the inflection point of the curve. For curve *C*, maximum rate of exploitation is about 82 percent; for curve *A*, 42 percent. In these curves the size of the reproductive stock that will give maximum sustained yield will not be greater than half of the replacement of the reproductive population. The greater the area of the reproduction curve above the 45° line, the greater is the optimum rate of reproduction.

So far the discussion has concentrated on commercial fisheries. Another group of exploited animals are those hunted for sport. Most are characterized by contest-type competition and are largely, but not always, density-regulated. The model of exploitation follows that in Fig. 12-31b. Some level of exploitation results in increased productivity because of reduced intraspecific competition (see Chapter 11). Some game animals, such as ruffed grouse and the cottontail rabbit, however, respond to regulated hunting as they do to compensatory predation. Hunting mortality simply replaces natural mortality. If the surplus were not harvested, they would succumb to disease, predation, exposure, and the like. With these species, regulated hunting usually declines sharply as the surplus is removed early in the hunting season. Just as a predator turns to another source of prey when the first source demands too great an expenditure of energy to hunt, so do sport hunters abandon the field when hunting success (animals taken per 100 hours of hunting) drops below a certain level. Sport hunters are unwilling to spend long hours in the field with little chance of success.

Among the ducks a different situation exists. Geis, Smith, and Rodgers (1971) have studied in detail the harvest characteristic and survival of black ducks based on banding returns, harvest characteristics, and other data. The black duck is the major game duck on the east coast of North America. It inhabits coastal marshes, rivers, and ponds, and feeds largely on aquatic invertebrates. By comparing the annual mortality rates with kill rates these investigators discovered that young and female black ducks are more vulnerable to hunting than males. This is apparently due to differences in the speed and timing of migration. The males move out of the production areas earlier, leaving the females and young to absorb the heavy hunting pressure on the breeding grounds. Geis et al. also found that hunting mortality was largely in addition to, rather than in place of, nonhunting mortality (Fig. 12-32). The magnitude of mortality is influenced by hunting regulations. When regulations are liberal, more hunters are in the field, hunting pressure is increased, and the kill is high. Stringent regulations reduce hunting pressure and thus the annual kill.

The continental black duck population has declined considerably. This decline has been attributed to liberal regulations in the mid-1950s. The birds have failed to recover in recent years partly because of an apparent high rate of kill and partly because of poor production. The black duck also is affected by thin eggshells caused by pesticides.

EXPLOITATION AND EXTINCTION

Few natural populations are really managed on a sustained yield basis, largely because enforceable regulatory mechanisms are neither available to control the harvest nor often desired by the exploiters. Exceptions are some game animal populations for which strong state and federal laws regulate the take. Many of these are underharvested rather than overharvested. Where their habitat is maintained, populations of these animals are largely controlled by nature and not man. Because of the lack of international regulations on the take, which could be handled by controlling gear, type of boats, fishing methods employed, and the like, most commercial fish populations are declining.

Regier and Loftus (1972) have reconstructed the probable development and decline of the extensive fresh water fisheries of the Great Lakes. This may serve as an example of the processes involved in overexploitation. Commercial fishing started first in the cove areas and near the mouths of streams. The fishermen utilized seines, hooks, weirs, spears, and the like to take such fish as northern pike, small- and large-mouth bass, and lake whitefish. As the rates of fishing success declined and demand for fish increased, the fishermen, still using the same gear, moved into more exposed areas offshore. Each new site exploited yielded an initial abundance of preferred fish species. As catches of these fish declined, the fishermen turned their attention to less valued species (Fig. 12-33), which in turn provided high catches. As the cove, stream, and shore fisheries became depleted, the fishermen improved their gear and fishing methods. They used poundnets and trot lines in more exposed areas and in deeper water. The fishermen now needed larger boats, more organization, more capital investment. From this they again realized high yields. As these yields subsequently declined, fishermen moved further offshore, where they used gill nets and ultimately trawls and/or purse seines. When they began to understand the migratory movements of different species, the fishermen began to follow the major concentrations everywhere rather than to wait passively in an area until the desired species moved in. Commercial fishermen enhanced their efficiency by periodically improving their gear, vessels, and marketing arrangements. As fish became harder to secure, more capital was invested to harvest the remaining fish. Each improvement was followed by a new burst in yield. As one preferred species declined, fishermen turned to less desirable species, until eventually the populations of one species after another collapsed. The fishing industry faded out of existence.

The whaling industry follows a similar pattern (Fig. 12-34). Early whalers armed with hand lances and harpoons sought only those whales that could be overtaken and killed from small boats. As whales became scarce along the shore, whalers took to the high seas. In the sixteenth century whalers hunted off the Newfoundland coast and Iceland until the stocks failed. The populations of Spitzenbergen and the Davis Straight were next to go. The colorful New England whaling industry, which first exploited the stock of right whales, peaked in the first half of the nineteenth century, but as stocks of slow-moving, easily exploited species were depleted and as petroleum replaced whale oil as a fuel, the New England whaling industry died. Then two developments put international whaling into business. One was the invention of the explosive harpoon in 1865, and the other was

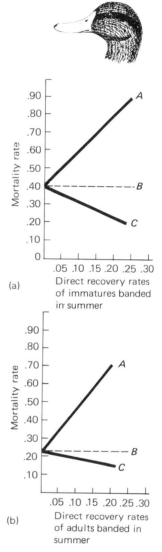

(a) Mortality rate
.90
.80
.70
.60
.50
.40
.30
.20
.10
0

.05 .10 .15 .20 .25 .30
Direct recovery rates
of immatures banded
in summer

(b) Mortality rate
.90
.80
.70
.60
.50
.40
.30
.20
.10

.05 .10 .15 .20 .25 .30
Direct recovery rates
of adults banded in
summer

FIGURE 12-32
Relation between hunting and nonhunting mortality among (a) *immature black ducks and* (b) *adult black ducks* (Anas rubripes) *using a regression of mortality rates on direct recovery rates. A = additional hunting mortality; B = nonhunting mortality replaced by hunting; C = nonhunting mortality. (From Geis et al., 1971.)*

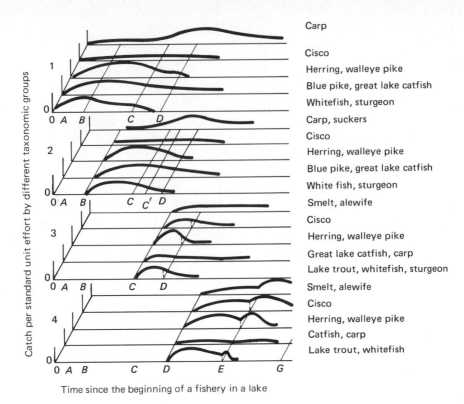

Carp

Cisco

Herring, walleye pike

Blue pike, great lake catfish

Whitefish, sturgeon

Carp, suckers

Cisco

Herring, walleye pike

Blue pike, great lake catfish

White fish, sturgeon

Smelt, alewife

Cisco

Herring, walleye pike

Great lake catfish, carp

Lake trout, whitefish, sturgeon

Smelt, alewife

Cisco

Herring, walleye pike

Catfish, carp

Lake trout, whitefish

Catch per standard unit effort by different taxonomic groups

Time since the beginning of a fishery in a lake

the development of more powerful, faster boats that could overtake the swifter whales. The revitalized industry began to concentrate on the plankton-feeding blue whale and its relatives the fin and the sei whales. Again stocks were overexploited. Blue whale fishery in Norway ended in 1904, followed by a decline of the species off Iceland, the Faeroes, the Shetlands, the Hebrides, and Ireland. When these areas failed, whalers sailing free-ranging factory ships turned to the Antarctic and the Falklands to concentrate on smaller fin whales. The catches rose and the stocks again collapsed. Antarctic whaling was over and most of the whaling nations were out of business. Japan and the USSR developed a whaling industry and hunted the sei, sperm, and other whales wherever they could be found. Investments were made in large factory ships equipped with helicopters and accompanied by catcher boats that captured more whales but with less oil per effort and per ship (Fig. 12-34).

Populations being overexploited exhibit certain easily discernible symptoms (K. E. F. Watt, 1968). Up to a certain rate of exploitation, the stock is able to replace itself. Beyond this critical point certain changes point to impending disaster. Exploiters experience decreased catch per unit effort as well as a decreasing catch of one species relative to the catch of related species. There is a decreasing proportion of females pregnant, due both to sparse populations and to a high proportion of young nonreproducing animals. The species fails to increase its numbers rapidly after harvest. A change in productivity relative to age and age-specific survival shows that the ability of the population to replace harvested individuals has been impaired. An outstanding example is the blue whale (Fig. 12-35). After 1860 the blue whale became the most important commercial species. Catches peaked in 1931 at 150,000 animals and declined to 1,000 to 2,000 in 1963. For the past 40 years the average age of the blue whale caught in the Antarctic has been 6 years, mostly immature females or females carrying their first calf. The species is near extinction.

Summary

The interrelations between populations of different species are as important in population regulation of a species as intragroup relations. In many cases the two relationships are so intertwined that it is difficult to separate them. Disease may decimate a population, yet it usually reaches epidemic pro-

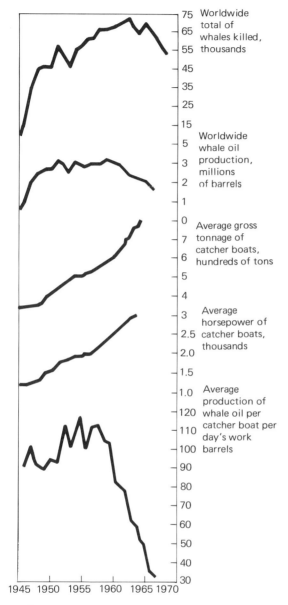

75	Worldwide
65	total of
55	whales killed, thousands
45	
35	
25	
15	
5	Worldwide whale oil
3	production, millions
2	of barrels
1	
0	Average gross
7	tonnage of
6	catcher boats, hundreds of tons
5	
4	
3	Average horsepower of
2.5	catcher boats,
2.0	thousands
1.5	
1.0	Average production of
120	whale oil per
110	catcher boat per day's work
100	barrels
90	
80	
70	
60	
50	
40	
30	

1945 1950 1955 1960 1965 1970

FIGURE 12-34

This graph illustrates the relationship between a declining resource and the intensity of fishing. Since 1945 more and more whales have been killed to produce less oil. At the same time boats have become larger and more powerful, but their efficiency has dropped greatly. (From D. C. Payne, 1968.)

FIGURE 12-33 [OPPOSITE]

This graph depicts the decline of a fishery through overexploitation and eutrophication. Panel 1 shows the history of a stream and cove fishery very near a market center through time for a series of species groups. On the time axis A = 0, the year of first fishing. Panel 2 depicts a cove and stream fishery beginning at year B. This area is some distance removed from both the market center and the fishery begun at time A. Panel 3 illustrates the results when nearshore gear such as poundnets or trot lines are first used at time C near the market center and at time C' some distance removed from that market center. Panel 4 represents the sequence in an offshore gillnet or trawl fishery beginning at time D, with a major technological innovation at time E. This offshore fishery deleteriously affects nearshore and any remaining cove and stream fisheries on species that also occur in offshore waters. (From H. A. Regier and K. Loftus, 1972; by permission of the Journal of the Fisheries Research Board of Canada.)

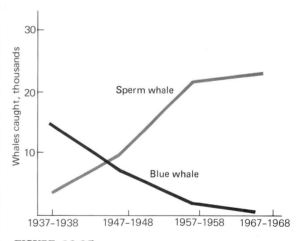

FIGURE 12-35

Since it has been intensively hunted, the great blue whale has slipped toward extinction. (Adapted from J. A. Gulland, ed., 1970, The Fish Resources of the Ocean, FAO Fisheries Tech. Paper no. 97.)

portions only when the population density is high and individuals are under stress. Much the same is true in parasitism, a condition in which two animals live together but one derives its nourishment at the expense of the other. Also rather common is social parasitism, in which one bird, the parasite, lays its eggs to be hatched in the nest of another, the host. Successful parasitism represents something of a compromise between two populations. The hosts and the parasites have developed a sort of mutual toleration with a low-grade, wide infection. Also the products of a long evolutionary process are the predator–prey and parasitoid–host systems. A close relationship exists between both populations of the system. Among many animal populations, especially invertebrates, the intensity of predation is independent of prey density but related to predator density. As the density of the prey increases, the predators may take more of the prey, a functional response; or the predators may become more numerous, a numerical response. These interactions may result in oscillations of both predator and prey populations. If a portion of the prey is not available because of environmental discontinuity, then the oscillations will be damped, and under certain conditions the predator system will be self-regulating. In other predator–prey systems in which the predators take most of the individuals of a prey species that are in excess of a certain minimum number, as determined by the carrying capacity of the habitat and social behavior, the relationship is compensatory. The prey reproductively compensate for their losses; and because only surplus prey is taken, the predators have little regulatory effect on the prey population.

Still another interrelationship exists between populations of different species, interspecific competition. This phenomenon has given rise to the competitive exclusion principle, which states, simply, "complete competitors cannot coexist." If two noninterbreeding populations occupy the same ecological niche, occupy the same geographical area, and if one population multiplies the least bit faster than the other, the former will occupy the area and the latter eventually will disappear. Competition is usually not that complete. Competitive situations between species can result in the replacement of one species by another, in an unstable

equilibrium between species, or in stable coexistence characterized by some division of resources. That is often accomplished by restriction of feeding niches or interspecific territoriality.

Interaction of predation and competition can influence community structure and species diversity. Predation can so reduce populations of interspecific competitors that a number of different competitors can coexist. Interspecific competition unmodified by predation can result in dominance of a few individual species and a reduction in species diversity.

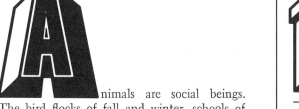

Social behavior in populations

nimals are social beings. The bird flocks of fall and winter, schools of fish, and herds of grazing animals—all are evidence of this. Yet an aggregation is not the only expression of sociality in animals. In fact some groups, such as insects attracted to a light, are not social gatherings at all. For most animals, aggregations at the best are seasonal. In most species, individuals spend much of their lives apart from the rest of their own kind. But at some time or another, especially during the breeding season, even the most solitary animals seek out and, if possible, come in contact with others of the same species. The meeting is formalized; there is some sort of interaction among individuals. It may be the relationship between two rival males at a territorial boundary, between members of a breeding pair, or between parents and offspring. The interactions or joint activities among individuals of a species make up social behavior.

Some basic considerations

Behavior is intrinsic to the ecology of an animal. It is involved in the success of the animal in caring for itself, in seeking appropriate shelter, in obtaining food, in escaping enemies, in courtship and mating, in caring for the young. It is a mechanism that, in part, reduces competition between animals and controls population densities. As Darling (1960) writes concerning the red deer, "social behaviour is an important ecological factor, a part of the environment as large to the deer as society and culture to us. If we interpret one face of ecology as the physiology of community, the inclusion of social behavior in our thinking helps us to understand."

The behavior of an animal, like its structure, is the result of natural selection. Perhaps more often than not, structure and behavior evolved together, structure influencing behavior, and behavior in turn influencing the development of the structure. To respond to environmental changes, physical and social, the animal must first receive a stimulus from the environment through its sensory system —sight, taste, smell, touch, hearing. From here the stimulus is transmitted to the motor organs, the muscular system, and finally the motor organs respond. The kind of sensory apparatus the animal possesses, the complexity and organization of its central nervous system, and the type and development of motor organs determine the manner in which the animal is able to respond to its environment. Thus how an animal perceives and reacts to the world about it is limited by what its eyes are built to see, its ears designed to hear, and its taste, hearing, and smell designed to respond to. Because of this, the world appears different to other animals than it appears to man. What is visible and important to them may be unperceivable to humans. Each animal lives, as Jacob von Uexkull put it, in its own phenomenal or self-world of perception and response, its *Umwelt*.

But the responses are not always completely controlled by structure. Species, and even individuals within a species, that appear to be structurally similar and live in the same physical environment may react differently to similar stimuli. Aggressiveness or tameness, vigor in courtship or defense, the ability to learn, these are not wholly associated with structure. Often behavioral patterns in the individual are genetically determined, and their appearance and retention and further development are influenced by the forces of natural selection. Behavior becomes a mechanism for survival, and through a period of time natural selection favors the best adapted behavior pattern for the species.

The study of behavior of animals in their natural communities grew into the field of *ethology*, which is concerned primarily with determining how the behavioral processes work, how they develop during the life of an animal, and how behavior evolved. Ethology is replete with its own terminology and theory. Much of this is of little direct concern to the ecologist and field biologist. But to understand those aspects of behavior pertinent to the understanding of the ecology of animals, some exposure to ethological concepts is necessary.

Instinctive behavior

In the bright sunlight of an April morning, a male flicker taps out a drumming roll on the bone-gray limb of a dead oak tree. He is answered in the distance by another, and then suddenly he leaves his drumming perch and drops to the ground. Another flicker had flown into the territory. The two males face each other closely on the ground, point their bills into the air, wave them about for four or five seconds, utter a shrill *we-cup*, raise their bright red crests, and display the yellow-colored underwings. After several such dances, the birds fly into different trees. In the quiet of a summer evening a field cricket in a dark crevice in the base of the dead oak tree calls. He occupies the crevice alone and sallies forth for short distances for food and water, to challenge other males, and to court females. He is answered by other males, located in other dark corners, and the evening cricket chorus begins. Nearby female crickets respond to the chirps and travel in the direction of the source.

The response of each to a stimulus was stereotyped and automatic. Each performed as other individuals of their kind would perform under similar circumstances. The response was characteristic of the species and apparently inherited or genetically fixed. It is this internally "programmed" behavior, to use computer-age language, which ethologists refer to as innate or instinctive behavior.

The terms innate and instinct are often dangerous to use, for they are interpreted differently by different people. To some, innate implies that the behavior is "programmed" completely without any environmental influence or control. If an animal held in isolation from birth responds to a stimulus it never experienced in the way that normally reared animals would, then the behavior pattern is considered innate. But it is difficult to prove that no component of the environment influenced the "programming" as the animal developed. To others, the term innate refers to those complete behavioral patterns performed by an animal without prior practice or experience and with all senses fully functional. This latter interpretation is the one implied in this book.

An animal performs a specific innate behavior only when it receives the appropriate external stimulus or cue. Not all of the information gathered by the sense organs of the animal is meaningful. Somewhere between the sense organs and the motor centers that control the behavioral action much of the sensory input is rejected and the remainder is selectively admitted. What stimuli are admitted depends upon the internal condition of the animal. Only when a certain stimulus or stimuli are properly received by the animal is the behavior pattern discharged. This stimulus may be the appearance of a predator or prey, a sex partner, the presence or absence of a species-specific color pattern, morphological structure, or posture; it may be a chemical substance, such as the odor of a stream or a scent emitted by an animal; or it may be a specific call or song. Often the stimulus may be only a small part of the total object perceived, such as the moustache of a flicker (Noble, 1936) (Fig. 13-1), or the red breast feathers of a European robin (Lack, 1953). The essential features of an object or situation that elicit a particular innate behavior pattern are called the *sign stimuli*. And those structures of an animal that serve exclusively to send out sign stimuli and result in a high degree of response are called *releasers*.

Often the stimulus that normally triggers the behavior pattern is not always the one with the greatest triggering function. Supernormal or artificial objects may be chosen over the normal stimulus. The oystercatcher, for example, will choose a giant egg in preference to a normal one and will incubate a supernormal clutch of five eggs over the usual clutch of three (N. Tinbergen, 1952). This leads the way for the evolution of "releasing" structures, such as the conspicuous plumage patterns of birds used in territorial and courtship display.

The first phase of a behavioral pattern usually is spontaneous. Some internal change or stimulus in the animal, such as hunger or hormonal action, causes the animal to seek an external stimulus such as food or a sex partner. When the animal sights food or a potential mate, a series of behavioral acts is set in motion. The preliminary or searching phase is often called *appetitive behavior*. Internally controlled, this behavior in the right situation switches over to external stimuli. If the animal proceeds in a coordinated manner and in the right direction, the activity is self-stimulating and self-reinforcing until the "releaser" for the next stage is found.

An example of such a reaction chain of behavior and its control is the courtship of the queen butterfly (L. P. Brower et al., 1965). The appearance of the female is the

FIGURE 13-1

The moustache of the male flicker (below) is the "releaser" that stimulates territorial defense in the male or courtship behavior in the female.

417

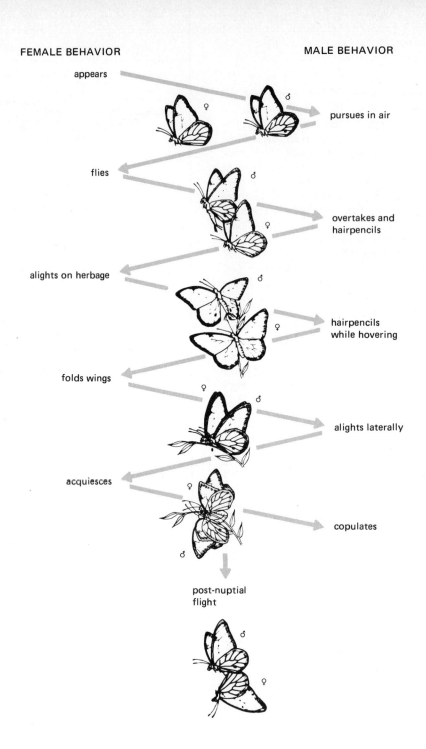

appears

pursues in air

flies

overtakes and
hairpencils

alights on herbage

hairpencils
while hovering

folds wings

alights laterally

acquiesces

copulates

post-nuptial
flight

stimulus that sets off the behavioral pattern or program (Fig. 13-2). When a female queen butterfly comes within the view of the male, he flies after her in aerial pursuit. The female flies off, but eventually the male overtakes her. As he passes a few inches over her back, he extrudes two bundles of hairs on either side of the abdomen. These are the abdominal hair pencils, which unfurled emit a strong, musky perfume. Changing to a bobbing flight, the male rapidly sweeps the hair pencils up and down her head and antennae. She responds to this by alighting in the herbage. The male then continues to "hair pencil" while still maintaining a bobbing flight. In response, the female folds her wings tightly over her back. This stimulates the male to retract the hair pencils and alight alongside of the female. The female acquiesces and the male in turn attempts to copulate. If successful, the male and female engage in a postnuptial flight, in which the male flies off carrying the female suspended at the end of his abdomen. If unsuccessful, the male induces the female to fly up again by hovering over her and striking her on the back. As she flies up, the male pursues her and starts the courtship over again.

Once a behavior is initiated, obviously some internal mechanism must exist to make the animal stop, to prevent the animal from carrying the behavior to an extreme. In the case of the queen butterfly, the behavior program ended with successful mating. Hunting behavior ends with the capture of a prey. And the entire program of feeding ends when the stomach of the animal is full and the animal's desire to eat is reduced. The concluding or *consummatory* act of a behavioral program may act as a sort of a negative feedback mechanism that reduces the effect of a stimulus and brings, if necessary, the behavior to an end.

EXPRESSION OF CONFLICT

The strong stimulation of one behavioral program apparently inhibits all others and allows only one particular behavior pattern to appear in an animal at one time. How this suppression of one pattern by another is accomplished is unknown, but in some way it may be controlled through connections in the central nervous system. On occasions, however, an animal may be strongly stimulated in several ways at once. When neither of the two, or even three, behavior patterns suppresses the other, conflict behavior results. Such conflict behavior often arises when an animal is stimulated both to attack and to escape, and is expressed in the animal as behavior that is not quite appropriate to the situation (Figs. 13-3 and 13-4). For example, the black-headed gull will preen itself or go through the actions of nest building when its brooding drive is thwarted (Moynihan, 1955a). During a fight with another male, the pectoral sandpiper, like other waders, will suddenly turn its head around, put its bill under its scapulars, and act as if it were going to sleep (Hamilton, 1959). Such acts are called *displacement* or *irrelevant* activities. The still controversial theory behind displace-

FIGURE 13-2 [OPPOSITE]
Chain-reaction behavior in the courtship of the queen butterfly. The male behavior is shown on the right and the female behavior on the left. (Drawing courtesy of L. P. Brower and the New York Zoological Society; from Brower et al., 1965.)

(a)

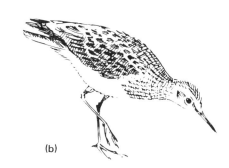

(b)

FIGURE 13-3
Displacement feeding (a) and normal feeding, (b) in the pectoral sandpiper. (After Hamilton, 1959.)

FIGURE 13-4
Grass pulling, a displacement activity in the herring gull. (Based on photographs in N. Tinbergen, 1953.)

ment is that the energy of two opposing tendencies finds its outlet through another behavior pattern. Another theory is that two types of behavior are more or less evenly balanced and cancel each other out so that a third behavioral pattern emerges. A third theory is that if a hierarchy of behavior exists, a subordinate behavioral pattern appears when the dominant ones are absent. Under these circumstances, ethologists argue, such behavior may not be so inappropriate after all (see Iersel and Bol, 1958; Sevenster, 1961; Rowell, 1961; Hinde, 1970).

Another form of conflict behavior is *redirection activity,* a type of behavior elicited when some action directed toward an animal or object is suddenly redirected to another object (N. Tinbergen, 1952). Like displacement, redirection seems to occur when an activity or instinct is thwarted or conflicts with another (see Bastock et al., 1953). When its nest is disturbed by some human intruder, a falcon may direct its attack on some passing bird. Since both attack and escape drives are activated simultaneously, the escape drive may be strong enough to prevent the bird from attacking its adversary. Instead it finds an outlet for the thwarted attack drive through an unprovoked attack on some convenient scape-goat available at the moment.

Some of these displacement and redirection activities have become incorporated in display-behavior patterns, both hostile and courtship; both involve considerable conflict and thwarting.

INSTINCT VERSUS LEARNING

Although much of an animal's behavior may be "internally programmed," it may be "corrected and adjusted" by an external "programmer," experience and learning. Learning in animals enables them to adjust innate behavior patterns to changes in their lives. Individuals modify their behavior patterns as a result of experience and thus adapt themselves to a changing and unstable environment.

Learning is important in perfecting skills, even innate ones. Such behavior as walking, flying, feeding, courtship, nest building, vocal patterns, and others appear in the animal in species-specific form at some stage of maturation; many of them appear suddenly in complete form. But with experience, animals become more adept in performance. For example, nest building in birds is innate, although it improves partly with learning, partly with rising hormone levels. But the bird learns what materials are suitable for nest building, such as the size and type of twigs. Birds do not learn how to fly; they "know" how to fly once they have developed the proper motor coordinations. But what they do learn in flying are such things as a suitable landing place. Animals learn what is and what is not edible, and how to distinguish between social companions and enemies. They also learn not to respond to stimuli that tend to be insignificant in their lives. Invariably the self-protective tendency to flee is triggered by a wide range of generalized stimuli, ranging from moving objects and sudden movements to strange sounds and situations. Unless an animal is able to sort out those situations and movements that are meaningless to its existence, its life would be an impossible mess. Thus animals are able to get used to certain situations and sounds and ignore them.

Learning is a large and complex subject, which for our purposes need not be discussed any further here. For good reviews of the subject see J. P. Scott (1958) and W. H. Thorpe (1963).

Imprinting

The young of many precocial species, such as geese, ducks, and ungulates, possess a tendency during a restricted period after hatching or birth to orient toward or investigate an object, especially a moving one. Ducklings, goslings, and young gallinaceous birds shortly after hatching follow the first moving object to which they are exposed. In nature this obviously is their mother, and after following her for some time, the young will follow no other animal. Incubator-reared ducklings, goslings, and chickens, however, can be induced to follow an artificial or abnormal object, such as a wooden decoy or even a balloon. Once they have followed this for some time, they will continue to do so. This suggests that at the time of birth, the young possess an innate behavioral program that is incomplete and must be supplemented by some influence from the outside world. In other words, the object has to be "imprinted" upon the young before the young will respond to it. This phenomenon has been called *imprinting*.

Imprinting is the rapid establishment of a perceptual preference for an object and seemingly must take place during a limited and definite period of the life cycle (Klopfer and Hailman, 1964; Klopfer, 1964). The term was first suggested by Lorenz (1933) to explain his experiences with incubator-raised graylag geese. These young geese, deprived of a normal mother, attached themselves to Lorenz by following him and finally by accepting him as a substitute parent.

Imprinting presumably is confined to a definite and brief period in the life of an animal and to a particular set of environmental conditions. The "critical period" for imprinting varies among species and is linked to the rate of physical development. Young mallards are susceptible to imprinting somewhere between 13 and 16 hours (Hess, 1959), although this varies among individuals. The moose calf develops a heeling or following response to its dam around 4 days after birth (Altmann, 1960). If the following response is not elicited during this critical period, then the appetitive behavior for following fades. This is perhaps because of an increasing development of a competing fear or fleeing response. As a result, the older the animal becomes, the less likely it will follow strange objects (Hinde, Thorpe, and Vince, 1956). This rapid form of learning is highly adapted to those animals with rapidly developing motor abilities who must establish early and maintain contact with their parents and others of the same species.

Experimental studies have stressed visual imprinting, but there is growing evidence that imprinting also involves other perceptual stimulation. Young wood ducks, reared in a dark tree hole or nest box, are exposed to the call of their mother for a relatively long period before they are exposed to the sight of her. The auditory stimulation apparently plays a major role in the ducklings' recognition of their parent (Gottlieb, 1963). After fluttering to the ground, the young apparently rely on the calls of the mother to lead them to water (Klopfer, 1959). Olfactory imprinting apparently exists among mammals that recognize both species and individuals by body odor. Acceptance of newborn young by fe-

421

male sheep, goats, and cattle seems to depend upon the olfactory and tactile contact with the newborn young for at least a few minutes after birth (Collias, 1956). To separate the young and mother for even as short a time as 1 hour is sufficient to lead to rejection of that young. Immediately after giving birth, a female sheep or goat will accept any young lamb or kid but within a few hours will reject any strange young. This behavioral trait is utilized by some sheep raisers, who attempt to present orphaned young to a foster mother.

For a number of years imprinting has been equated with the following response. But imprinting is much more than a socialization process. It is a genetically programmed learning process that involves not only a parent object but also food (Hess, 1964; Burghardt and Hess, 1966). In imprinting the object to be learned, mother or food, is the reward. Thus the young animal learns the primary object and as a result responds to it. The survival value of imprinting is obvious. Young animals, particularly those that move on their own power shortly after birth, have only a short time in which to learn the primary object. Failure to learn that means the death of the individual.

Social behavior

Socialization, the development of social bonds between and among members of the same species, begins largely in the nest or den, first as a bond between young and parent, then between litter and brood mates. Later the bond between parent and young weakens, and the young show increased ability to recognize and distinguish between individuals of the species. At this stage, the animal begins to exhibit the most commonly observed social behavior patterns—agonistic or aggressive behavior. This is an important form of social behavior, for it regulates population density and is involved in courtship and reproductive behavior.

AGONISTIC BEHAVIOR

Agonistic behavior involves the motivations of attack and escape (Moynihan, 1955*b*). These are usually associated with territoriality and social dominance and frustration (see Marler and Hamilton, 1966; Hinde, 1970).

Under natural conditions hostile displays appear to be instigated by a direct response of one animal to the proximity of another individual. Agonistic behavior involves displays of threat that seem to have evolved because they conferred upon the animal the ability to achieve certain advantages without having to fight for them or risk physical injury. As a result, a wide variety of hostile displays have developed, as well as associated morphological structures. The displays have become ritualized, that is, the movements or components of the display have become exaggerated to increase their efficiency as a signal.

The most common of all ritualized forms of hostility is the intimidation, or threat display, of which most higher animals possess more than one kind. The primary function of the intimidation display is to force the opponent to retreat or flee—in other words, to increase the actual and relative strengths of the opponent's escape drive. Such threat displays are not always successful, especially among territorial species. An outsider may threaten the owner of a territory—an aggressive opponent at any time—but seldom will cause him to flee. The threat display, however, may tend to cause the owner to hesitate before attacking or threatening back.

The reason is that threat display involves two drives activated simultaneously, attack and escape; and the elements of escape are present in many threat displays. The display itself may vary in intensity or exaggeration in the form of movement with the relative strengths of the two drives. The attack definitely is the stronger in most threat displays; in a few the escape drive may be the stronger, and in others the drives may be equal. In some instances, the latter condition may trigger redirection or displacement activities, such as grass pulling by the herring gull.

Threat displays have been derived from a number of sources, both hostile and nonhostile (Moynihan, 1955*c*). Of the hostile sources perhaps the most important are a whole series of unritualized intention movements indicating locomotion, retreat, or avoidance (Fig. 13-5). Others undoubtedly were attack movements, such as pecking at the opponent, flying or charging at another, or actual fighting. Out of this evolved those attack components—the color of bill or head, feather patterns, feather puffing, horns, bristling of body hair, and other morphological structures—that emphasize the visual conspicuousness of threat displays.

It is not uncommon for one threat display to provoke a threat display in return, often the same one (Fig. 13-6). And among birds in particular some threat displays are "designed" to do just that. These are the so-called exemplary displays, as yet little studied. An example is the choking display of gulls, in which the bird or pair of birds lower the breast, bend the legs, point the head downward, and perform rhythmic jerking movements with the head as if they were going to peck at the ground (Moynihan, 1955a). During all this the bird may utter a deep call repeated almost in time with the jerking movement. This threat display usually induces a return choking display. A characteristic of this type of threat display is its infectiousness, for the display invariably causes all birds in the area to undertake identical performances and thus releases a communal suppression of attack of one bird upon another.

Some examples of agonistic display. Agonistic displays vary from species to species, but among families and genera of animals, a number of basic hostile displays appear, interspecific differences being a sort of "variations on a theme." Among passerine birds, in fact among most groups of birds, a common and perhaps universal aggressive posture is the head-forward threat (Fig. 13-7). The body is lowered to the horizontal, head and body are in line, and at the highest intensity, the bill is held open in a gape. Some species, such as the green finch (Andrew, 1961), the grasshopper sparrow (R. L. Smith, 1959), and the redpoll (Dilger, 1960), also vibrate their wings. Another display commonly given is bill raising (Fig. 13-7), in which the bill is raised and held well above the horizontal. It is probably derived from an upward flight intention movement and apparently serves as a distance-increasing display (N. Tinbergen, 1959).

Even mammals possess certain aggressive displays the features of which are possessed in common. The facial expressions of canids —the baring of fangs, the curling of the lips, the erection of the ears, and the movement of the tail—all indicate degrees of aggressiveness in dogs, foxes, and wolves (Fig. 13-8). The aggressive behavior of the black-tailed deer consists of three components (Cowan and Geist, 1961). One is the crouch, which in-

FIGURE 13-5
Unritualized intention movement of flight in the hermit thrush. (After Dilger, 1956.)

FIGURE 13-6
Threat display in underyearling Kamloops trout. (Redrawn from Stringer and Hoar, 1955. © 1955 by The University of Chicago Press.)

423

volves hunching, partial flexing of the hind legs, shoulders, and elbow joints, all tending to lower the animal. The walk is slow, stiff, and stilted, with the head held in line with the body, the ears laid back. Other features are nose licking, circling, and the snort, and finally the rush. The most extreme display consists of lowering the head to bring the antlers into contact with the object of aggression. Similar threat displays are found in the moose (Geist, 1963) and the barren-ground caribou (Pruitt, 1960). In the head-low threat of the moose, the head and neck are in line, the head is lowered toward the ground, the hair on the neck, withers, and rump is raised, the ears are down. From this position, the moose will attack in a short, fast rush. In the agonistic display of the caribou, the muzzle is extended, the nose is curled, the ears are laid back, and the animal advances rapidly toward the antagonist.

APPEASEMENT DISPLAY

Appeasement displays, as one might expect, are almost as common as aggressive displays and are especially characteristic of encounters during the reproductive period. They serve to prevent attack without provoking escape, to reduce the strength of the opponent's attack drive, and to release other specific non-aggressive behavior (Moynihan, 1955*b*). Appeasement results when the escape drive is much stronger than the attack. Like aggressive patterns, appeasement displays have derived from hostile sources, but the patterns involve most strongly the intention movements of escape. Usually appeasement behavior consists of withdrawal or avoidance movements, often specialized to hide offensive weapons, such as beaks, fangs, and any threat releasers. Appeasement displays, unlike intimidation, seldom have colors or structures evolved for this use alone.

A common appeasement display is the submissive pose, varying in appearance but still possessing some basic features in common (Fig. 13-9): among birds, the lowered or indrawn head, the crouched position, the fluffing or ruffling of feathers; among canids, the ears laid back, the eyes looking down, the tail curled between the hind legs (Fig. 13-8); among rabbits, a crouched position with ears laid back, the head pulled in toward the body, and the tail depressed (Marsden and Holler,

1964). Some gulls, such as the black-headed gull, turn their heads away from the opponent (head flagging) to hide the threat stimulus of the red bill and black face (Tinbergen and Moynihan, 1952). Occasionally appeasement displays may be superimposed on threat, and at times attack and escape may integrate, as they do in the "anxiety upright" of the gull (Fig. 13-9). Although basically a threat display, it seldom elicits more than a low escape drive in the opponent.

SOUND AND SCENT

Display also involves sound and scent. Song or other auditory efforts are aggressive among many species of birds. It serves among territorial species as an announcement that a piece of ground has been claimed and warns males of the same species not to trespass. Territorial song is somewhat infectious, for when one male begins to sing, others in the area follow suit. The same song usually serves a dual role of attracting a potential mate and repelling rival males, but this is not always true. The grasshopper sparrow possesses two songs, a territorial, or grasshopper song, the familiar insectlike buzzing song of the species, and a more musical, sustained song, which serves to attract the female. The latter in its entire form consists of two parts, the grasshopper song, followed by a series of sustained trills. The first phase of the song is hostile, necessary in the early period of courtship when the warning function still is of prime importance, but the second phase—often sung alone—is not (R. L. Smith, 1959).

A similar situation possibly exists with the grasshopper sparrow's meadow associate, the meadow grasshopper, who possesses a two-phase song consisting of a series of ticks followed by a buzz. The ticking part of the song seems to be associated with the function of spacing individual males, and the buzzing phase attracts females (R. D. Alexander, 1960).

The chirps and songs of the crickets, cicadas, and grasshoppers, like the songs of birds, are associated either with attraction or repulsion. When sexually responsive male crickets are in close proximity, they frequently spar or fight with each other, using their antennas, forelegs, and mandibles, and kicking with their hind legs (R. D. Alexander, 1961). At the same time distinctive sounds are pro-

AGGRESSION DISPLAYS

Pectoral
sandpiper

Herring
gull

Hermit
thrush

Redpoll

Ring-necked
pheasant

FIGURE 13-7
Agonistic displays among birds. Note the general similarity. Pectoral sandpiper (after Hamilton, 1959); herring gull (based on photos in N. Tinbergen, 1953); hermit thrush (after Dilger, 1956); redpoll (after Dilger, 1960); ring-necked pheasant (after Collias and Taber, 1951).

FIGURE 13-9
Appeasement display in some birds. Herring gull (based on photos and drawings in N. Tinbergen, 1953; Moynihan, 1955b); redpoll (after Dilger, 1956); ring-necked pheasant (after Collias and Taber, 1951).

(a)

(b)

FIGURE 13-8
Aggressive expression (a) *and submissive expression* (b) *in the American wolf, typical of canids. (Suggested by illustration in Schenkel, 1948.)*

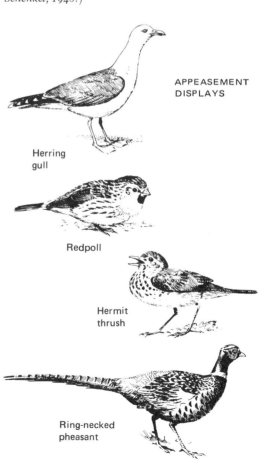

APPEASEMENT
DISPLAYS

Herring
gull

Redpoll

Hermit
thrush

Ring-necked
pheasant

duced by the individuals, which affect the outcome of the encounter. Dominant males usually chirp more frequently after an encounter, while the loser rarely chirps after an encounter. After an encounter the dominant male may continue to chirp, the chirps merging into the calling song.

Mammals, too, have vocalists whose calls are aggressive in function. The choral performances of the howler monkeys of the South American rain forest apparently are important in regulating territorial ranges (Carpenter, 1934). Chimpanzees have a variety of calls that they use in agonistic situations (Lawick-Goodall, 1968). Some calls are strongly aggressive; others signal submissive behavior. The ability to roar and bellow is a common secondary sexual characteristic among ungulates, seals, and walruses. The sounds, usually confined to the rutting season, are aggressive, a challenge to other males. Barking by male sea lions restricts the movements and barking of smaller males (Schusterman and Dawson, 1968). It also serves to identify individuals and to assert dominance.

Scent, too, serves to mark territorial holdings and to warn other males to stay away. In fact communication by scent is widespread throughout the animal kingdom. Like sound, it serves as a means of communication in a variety of ways, from recognition and trail marking to dominance. Known as *pheromones*, these scents are secreted from exocrine glands (Fig. 13-10) as liquids, transmitted as liquids or gases, and smelled or tasted by other animals of the same species. Such scent may release immediate behavioral response or it may alter the physiology of the receiving organisms, usually through the endocrine system, stimulating a new set of behavioral patterns, particularly as related to reproductive behavior (Bronson, 1969).

Pheromones may be simple or complex chemical substances. Among insects, in which their major function appears to be sexual attraction, the pheromones are either a single component or a simple mixture. Among vertebrates pheromones are complex chemicals, which tend to be "personal" odors (Wilson, 1971). Thus different types of pheromones allow vertebrates to recognize individuals, assert dominance, and the like. For example, North American deer are well equipped with a number of suboriferous and sebaceous glands that secrete pheromones (Fig. 13-10). Each

gland secretes its own distinctive scent (Muller-Schwarze, 1971). Scent from the tarsal gland is important in the mutual recognition of sex, age, and individuals. The metatarsal glands, located on the outside of the hind legs, secrete a scent associated with alarm and fear reaction. Scent from the glands in the forehead is used for marking home range; and during aggressive posturing, the black-tailed deer opens fully the orifice of the preorbital gland (Fig. 13-11). Female urine attracts males. And rub-urinating, in which deer of either sex and of all ages occasionally rub their hocks together while urinating on them, serves as a distress signal in fawns and as a threat in adult males and females. Dogs and other canids have various scent posts (the familiar lamp post or fire hydrant) on which they advertise their presence. Shrews may rub abdominal scent glands along the walls of their tunnels to proclaim the burrows occupied (D. A. Pearson, 1946; for some discussion see Ewer, 1968).

Aggressive behavior is most conspicuous among males during the breeding season, although it is not exclusively a male function. The female gray squirrel, for example, becomes highly aggressive during the nesting season and establishes a territory about the den tree, which she defends against all ages and both sexes (Bakken, 1959). The seasonal development of aggressive display is closely associated with hormonal secretions. Secretions of gonadotrophins by the pituitary stimulates the growth of the gonads and the subsequent output of sex hormones. This, plus external stimuli, stimulates aggressive behavior. As the output of sex hormones declines, aggressive behavior too declines, although here also it is difficult to generalize. The decline of song among many birds, for example, is not related to hormones. Song may decline or even cease when a mate has been acquired (Nice, 1943).

EXPRESSION OF AGGRESSIVE BEHAVIOR

Aggressive behavior is expressed most commonly as territoriality and social dominance (see also Chapter 11). Some species are exclusively territorial; some operate within the framework of social hierarchy; others are territorial during the reproductive season only and may exhibit social dominance during the remainder of the year.

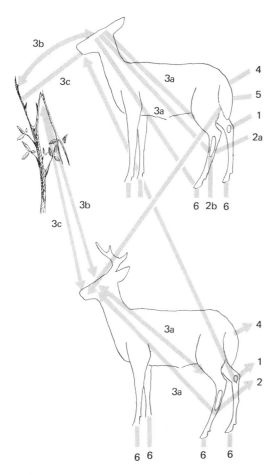

FIGURE 13-11

Details of the head of the black-tailed deer
(Odocoileus hemionus sitkensis) *during the
snort that occurs when the buck is circling
in the aggressive crouch position. Note the
widely opened preorbital gland, curled upper
lip, and bulged neck muscles. The snort is a
sibilant expulsion of air through the closed
nostrils, causing them to vibrate. (After
Cowan and Geist,* 1961.)

FIGURE 13-10

*Odor plays an important role in communication
among animals. Various odors and secretions
from the subcutaneous glands released to the
air and rubbed on ground and twigs have
significant meanings in the world of the deer.
The drawing shows the pathways of social
odors in the deer. Scents of the tarsal organ* (1),
metatarsal gland (2a), *tail* (4), *and urine* (5)
*are transmitted through air. While reclining
the metatarsal gland touches the ground* (2b).
Deer rub their hindleg over forehead (3a),
forehead is rubbed over dry twigs (3b). *Marked
twigs are sniffed and licked* (3c). *Interdigital
glands leave scent on ground* (6). (From
Müller-Schwarze, 1971.)

Territoriality. Territories, discussed in Chapter 11, are defended by the males, occasionally by females, by means of song, aggressive displays, and fighting. Territoriality is most clearly defined in birds; it is next most highly developed among fishes (Greenberg, 1947; Kelleberg, 1958; Winn, 1958*a*, *b*; Gerking, 1959). The males of many species of darters defend a territory that includes reproductive and escape areas, but leave the territory to feed. The territories may be stationary or moving. Salmon maintain in their territories one strongly dominated local station, constant in position, to which the fish returns after a foray. From this station the fish defends an irregular area with a poorly defined boundary.

Territoriality among mamals is not so well defined. An outstanding example is the vicuna, a near relative of the llama, which lives in the high, treeless pastoral zone of the Andes. The male vicuna is the head of a band of females and young. He defends the area they occupy (Fig. 11-11b) against other family males and is intolerant of all other males except the young of his own band (Koford, 1958). Muskrats also appear to defend a territory (Errington, 1937*a*). The Alaskan fur seal establishes territories on rocky shores prior to the arrival of the females. To hold and maintain these territories, males may have to go without food and water for days and live on accumulated body fat. After the bulls acquire harems, territorial defense wanes. The Uganda kob, one of the antelopes of the African savanna, defends a small, fixed territory of 20 to 60 yd in diameter within a central territorial area approximately 200 yd in diameter (Buechner, 1961). Within this central area are 12 to 15 territories; and this area is surrounded by a zone of more widely spaced territories. The territorial ground is situated on a ridge, knoll, or slightly raised area characterized by short grass, good visibility, and proximity to a permanent stream. Females enter the territorial ground through the year for the purpose of breeding. When disturbed by a lion, elephant, or something else, the kob deserts the territorial ground along an established route. Shortly after the disturbance is gone, the kob returns to its own territory.

It is difficult to separate territoriality from social dominance. Depending upon the season and conditions of crowding, territory can grade into social dominance. This behavior is perhaps best expressed in the behavior of feral domestic fowl (McBridge et al., 1969). During the breeding season social behavior may range from extreme territorial to weak social hierarchy, as illustrated by six classes of males. Territorial males with fixed territorial boundaries are dominant to all other males. They restrict their movements to the territory and have females as constant companions. A second group of males possess subordinate territories, small ones defended against even the dominant neighbor. Defense is often unsuccessful, and the dominant neighbor may be able to penetrate the territory but not over the whole extent. These males may have harem flocks. Semiterritorial males roost apart from the dominant males, and make slight defensive reactions against invasion by dominant males. However, the territories of these males are within the territory of a dominant male. They may be attended by females, but not constantly. Subordinate males roost apart from dominant males. The areas they occupy are also usually occupied by females, but the birds do not possess any. Another group of subordinate males roost with a dominant male who does not possess females. Finally there are the runts, which generally leave the territory. Where more than one male occupies an area, territorial succession is a matter of peck-order position. When male domestic fowl are confined to smaller and smaller areas, dominant males adjust from territorial behavior to social dominance. When space is increased, the dominant males become territorial again (M. Schein, pers. comm.).

Territory does not necessarily have to be a fixed area. There is a strong relationship between territoriality and individual distance. Animals commonly keep an area about them free of all other individuals by attacking intruders or moving away from them. This can best be observed, perhaps, by watching a flock of swallows settling on a telephone wire. Cliff swallows, for example, will so disperse themselves that one bird is approximately 4 in. away from its neighbor; any attempt by another bird to reduce this distance is met with a threat display (J. T. Emlen, 1952).

Defended areas and individual distances need not be stationary. Some geese maintain moving territories (Jenkins, 1944). These are definite defended areas that move when family groups move and are maintained in all types of activities throughout the year, except during the reproductive period. A similar type of territory is maintained by the male goose around the female.

Social dominance. Social dominance, which necessarily involves individual recognition of members of the group, usually is firmly fixed in some form of hierarchial arrangement. The simplest of these, first described by Schjelderup-Ebbe (1922) for the domestic chicken, is the straight-line peck order, so named because dominance was indicated by dominant birds pecking the subordinates. There is an alpha bird, which can peck all others, a beta individual, which can peck all but alpha, and finally an omega, which can peck no other (Fig. 13-12). Even within this scheme there are some complexities, common to most flocks, such as triangular hierarchies. These are triplets of individuals whose pair relations are such that the first individual is dominate to the second, the second is dominant over the third, and the third dominates the first. In such a situation an individual of a lower rank can "peck" an individual of higher rank. In some groups peck order is replaced by peck dominance, in which social rank is not absolutely fixed. Threats and pecks are dealt by both members during encounters, and the individual that pecks the most is regarded as the dominant. The position of the individual in the social hierarchy may be influenced by levels of male hormone, strength, size, weight, maturity, previous fighting experience, previous social rank, injury and fatigue, close associates, and environmental conditions.

In flocks made up of both sexes, separate hierarchies may exist—a peck order of males, a peck order of females, and dominance of males over females. Such peck orders are characteristic of flocks of red crossbills (Tordoff, 1954) and ring-necked pheasants (Collias and Taber, 1951). The top-ranking male in a crossbill flock was the most aggressive male (Tables 13-1 and 13-2), yet the low-ranking male was most active in dominating the females, followed closely by the top-ranking male. A similar situation existed in a flock of ring-necked pheasants in which the males dominated the females, who had their own peck order. But at the onset of the breeding season, the dominance of cock over hen declined. Dominance among white-tailed deer

Straight-line peck order

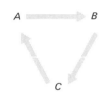
Triangular peck order

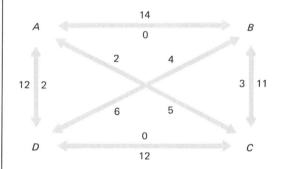
Triangular peck order

FIGURE 13-12
Some examples of peck orders.

in a wintering yard in Wisconsin (Kabat et al., 1953) was one in which the adult bucks dominated other deer, and the does dominated the fawns. Within this scheme, the larger animals, whether buck, doe, or fawn, dominated the smaller animals. Wintering flocks of juncos exhibit no such dominance based on sex (Sabine, 1959). Females dominate the males in some species, such as the redstart (Ficken, 1963), while in other species, such as the snow goose (Jenkins, 1944) and jackdaws (Lorenz, 1931), the females appear to be equal to the male, especially where family ties are strong.

Once social hierarchies are well established within a group, newcomers and subdominant individuals rise in rank with great difficulty. Strangers attempting to join the group either are rejected or, as in the valley quail, are relegated to the bottom of the social order (Guhl and Allee, 1944). New birds in a wild covey remain a few yards behind the main group and do not mingle until after a period of acquaintanceship. Some individuals newly entering the flock then may rise rather rapidly up the hierarchy, whereas others remain unassimilated. Several such individuals then may associate together and by coordinating their behavior maintain a mutual social rank against other members, as do some individuals in howler monkey society (Carpenter, 1934). Removal of the dominant individual or its injury causes a scramble for the alpha position. W. L. Thompson (1960) relates that an injured female house finch lost her alpha position because of her inability to move about and attack as readily as she was attacked, and thus she sank to the bottom of the hierarchy.

Rise in the hierarchy often is related to breeding and sexual activity and hormones. This is particularly true among those species that remain in flocks throughout the year. Breeding condition is important in determining social rank. Male house finches that occupied the bottom rung of the social ladder during the winter but came into breeding condition first rose near to the top of the hierarchy during the reproductive season (Thompson, 1960). The top male, however, was still dominated by his mate. Ring-necked pheasant cocks that occupied the dominant position in winter flocks usually were the first to crow and to establish individual territories in the spring. But subdominant males that came into breeding condition earlier rose higher in the hierarchy (Collias and Taber, 1951). Rise in hierarchy, then, appears to be related to a rise in male hormones.

Mating, too, improves the rank of lower individuals. Both members of a pair of house finches rise in hierarchy, but not always to the same level (Thompson, 1960). In the jackdaw, however, one member of the mated pair rises at the time of pair formation to the same level as the other member (Lorenz, 1931).

A rise in male hormone levels is not the only cause of increased aggressiveness. Shortages of food, space, and mates, among other things, will increase competition and hostility between individuals. The dominant individual has first choice of food, shelter, and space, and subdominant individuals may obtain less than the despots. When shortages are severe, the low-ranking individual may be forced to wait until all others have fed, to take the leavings, if any, to face starvation, or be forced to leave the area.

COOPERATIVE SOCIAL GROUPS

Not all aggregations are based on social dominance. Some groups contrast sharply with those in which considerable intragroup conflict exists, and show no evidence of social rank among individuals. There exists a relatively strong group cooperation in intergroup aggressive situations.

The social structure in two animals, the harvester ant and the prairie dog, are excellent examples. Individual ants, possibly by imprinting, soon learn the odor of their own colony. Foraging ants returning to their own colony may stimulate some aggression on the part of their nest mates, but once the identity is clear, the ants are unmolested. But if an ant enters an alien colony, it is threatened, seized, dragged, and killed, for a strange taste or smell stimulates attack among ants in the colony. The colony acts as a unit; and conflicts between colonies may continue for days, leaving the ground littered with corpses.

Among vertebrates few social organizations are more fascinating than that of the prairie dog, aggregations of which are known as prairie dog towns. Some towns are subdivided by topography or vegetation into wards; further subdivisions reflect prairie dog behavior. Within each ward a group of prairie dogs is united into a cohesive, cooperative unit, known as a coterie, which defends a particular section of the ward against all trespassers (King, 1955). Territories of coteries, which cover less than an acre, may contain from 2 to some 30 members. Breeding coteries usually contain 1 male and 3 or 4 females; nonbreeding coteries may have all males, more males than females, or an equal number of both sexes. No social hierarchy exists, although one male, usually the most aggressive and the strongest defender of the territory, may dominate the rest. All coterie members use the same territory without conflict or the threat of it; and social relations among most members are friendly and intimate, involving grooming, play activities, vocalizations, and the identification kiss—a recognition display in which each individual turns its head and opens its mouth to permit contact with the other. A coterie emphasizes its social unity by the defensive action of its members against invasion, and by the inability of any member of the group to wander beyond the coterie territory without becoming involved in conflict (although members are ready to invade other coteries if the opportunity is present.) The members of the coterie do not drive away the young; instead the young are protected against the antagonism of neighboring coteries. Overpopulation in a coterie may force territorial expansion, in which the defending male of the adjacent coterie is vanquished and driven out; or it may result in social unrest and the eventual emigration of yearlings. Adults may leave perhaps to escape the demands of the pup or to seek more abundant food. Advantages of living in such a group are many, including a limited control over an area of the habitat, increased defense against predators, the prevention of overcrowding, results that can be achieved only by the combined activity of all individuals.

Prairie dogs exhibit little division of labor; but the most highly organized natural societies—those of ants, termites, and some bees and wasps—do. Insect societies can be regarded as families, for they arise from a family unit of parents—the males and the queens—and their offspring. The latter remain with the female and take over the task of nourishing additional young as well as caring for the mother, the queen. In such societies three basic roles exist, the males, the queens, and the workers, each characterized by differences in structure, appearance, and behavior. The

TABLE 13-1
Peck order in male red crossbills, based on 404 encounters

	December 31–January 12 dominates				January 14–March 25 dominates			
	B	A	G	O	B	A	G	O
W	— B	A	G	O	—	B A	G	O
B	—	A	G	O		—	A G	O
A		—	G	O			—	G O
G	W		—	O				— O
O				—				—

Source: Tordoff, 1954.
Note: Each table represents a different method of summarizing dominance in social groups. Table 13-1 is a qualitative summary. Table 13-2 is a quantitative summary showing the number of wins by one individual bird over others in the group.

TABLE 13-2
Social hierarchy of a mixed group of caged male and female house finches in winter

	R	L	GB	WG	GR	WP	Y	O
R		1	2	0	4	2	3	2
L			2	1	2	1	1	1
GB				7	7	4	6	10
WG	2				6	8	8	5
GR			1			3	2	5
WP							8	1
Y								3
O					9			

Source: W. L. Thompson, 1960.
Note: Read from left to right in horizontal rows. The number indicates the number of wins by the bird in question.

431

male's sole role is the fertilization of the reproductive female, or queen. Having accomplished this the male either dies or leaves the group. The queen, usually large in size, lays eggs, from which other members the group develop. The workers among the ants, bees, and wasps are sterile females; among the termites, workers may include male and females, both sterile. The workers, as the name implies, perform the duties of their complicated society. Ants and termites have additional castes, the soldiers, large and equipped with formidable mandibles, who defend the colony. These are males and females in the termites, males in the ants. Defense in bees and wasps, who lack a soldier caste, is a function of the worker. The individuals of these societies have no choice of their roles, and their basic behavior is innate. Individuals have no chance of survival if separated from the social group. Although organically separate, the individuals are inseparably bound to the colony by behavior and physiology.

SOCIAL DOMINANCE AND LEADERSHIP

A wedge of geese flying across the blue sky of spring invariably raises the question of leadership in the mind of the observer. Common belief has it that the wedge is led by an old and experienced bird. This may not be quite true, but evidence seems to indicate that adult geese do lead the migratory flights with others following, although the leadership changes throughout the flight. Where leadership is expressed in groups, this probably is the most common type.

The problem of group leadership in the wild has hardly been investigated. True leadership, as the few studies on sheep, goats, and deer show, implies the ability of an animal to move out ahead of the group, which then follows without the use of force. Dominance is not involved; in most cases only by chance do dominant individuals also become the leaders. Among family groups of geese, the male is the guard and leader of the group, and intragroup despotism is rather weak. Only among certain ungulates, particularly the red deer of Europe (Darling, 1937) and the elk of North America (Altmann, 1956a, b) does true leadership appear to be vested in one individual; and in these species it is an old and experienced female. Each female has two or three followers, usually her own offspring

of past years, their young, and her own young of the year. Her supremacy is never challenged and she is succeeded only after death or after she ceases to drop a fawn each year. Below the leader is a well-defined peck order, in which the next in rank usually brings up and guards the rear of the herd. This is a matriarchal society, extremely cohesive and gregarious. It probably arises from maternal care and the dependence of the young upon the mother for several years. The young run to the mother and are rewarded by being allowed to nurse; they also rush to her side when danger threatens. The young then are habituated to follow the mother; because the older females have the greatest number of young, they are the natural leaders behind which the rest of the herd moves. During the season of rut, the stag of the red deer and the bull of the elk break into the matriarchal herd and establish a harem. The real leadership still rests with the female, especially in the time of danger, when the stag of the red deer may just as often as not retreat until danger is past (Darling, 1937). The male's interest in the herd is strictly egocentric.

Courtship and reproductive behavior

In the bright cold sunlight of a late winter day a flock of mallards is active on the open water of a pond. Suddenly some rear up out of the water, and some appear to be charging at others. Closer observation reveals a pattern to the activity. A number of green-headed males are swimming with their heads drawn in, the feathers ruffled, the body shaking repeatedly (Fig. 13-13). As the tension increases, a few of the drakes rear up out of the water and flick their heads forward. On occasions a drake rears up, arches his head forward, and rakes his bill across the water (Fig. 13-13). Then with his bill pressed to his breast, he slowly sinks back to the water. At times this display may be accompanied by a low courtship call or followed by still another display, in which the male throws his head back in an arched position and jerks it abruptly upward. As the drake turns toward the female, he erects and spreads the tail feathers vertically and lifts the wing coverts to expose the irridescent metallic purple speculum (Fig. 13-13). Then the drake lowers his head, stretches the head and neck forward just above the surface of the water and swims in rapid circles about the female. The female in turn is completely passive. The brown-feathered hens follow a mate or an intended one. With neck arched and head pointed toward the water, the hen moves her head back and forth from front to side away from the drake and toward the females. Often this display is accompanied by short dashes of attack. This courtship display is typical of the mallard. Other surface ducks have similar behavior patterns, but with a number of variations, omissions, or additions to the repertoire.

The ducks represent the courtship behavior of a sexually dimorphic and polygamous group of animals. A somewhat different pattern is common to animals holding territories and to those in which sexual dimorphism is lacking. The female is attracted to the male by song or some other type of sound production, as in birds, frogs, and some insects, or by appearance, as in the case of the three-spined stickleback. The male, in turn, sees an animal of the same species and reacts aggressively. If the intruder happens to be a male or an unreceptive female, the animal flees or threatens back. In the latter situation, a fight, even if a mock one, develops. If the intruder happens to be a receptive female, she remains. She may exhibit some hostility, but eventually she adopts a submissive pose, which tends to reduce or inhibit the male's aggressiveness. When the male's behavior becomes less aggressive, pair formation is accomplished. As times goes on, more generalized sexual elements enter courtship behavior. In contrast, the females of some species become more aggressive after pair formation and may dominate the male. The male's first reaction in courtship behavior then is to attack; the initial reaction of the female is to appease (escape) the male and elicit further courtship. The male in turn must suppress the escape tendencies of the female. Thus all courtship display contains elements of agonistic behavior (attack and escape) and sexual behavior. Progressive changes in the behavior of male and female toward each other depend in part on the stimulus situations presented by the partner: morphological, behavioral, vocal.

The function of courtship display is to attract the female to the male or vice versa. Once oriented to each other, neither may react to the partner's courtship until certain

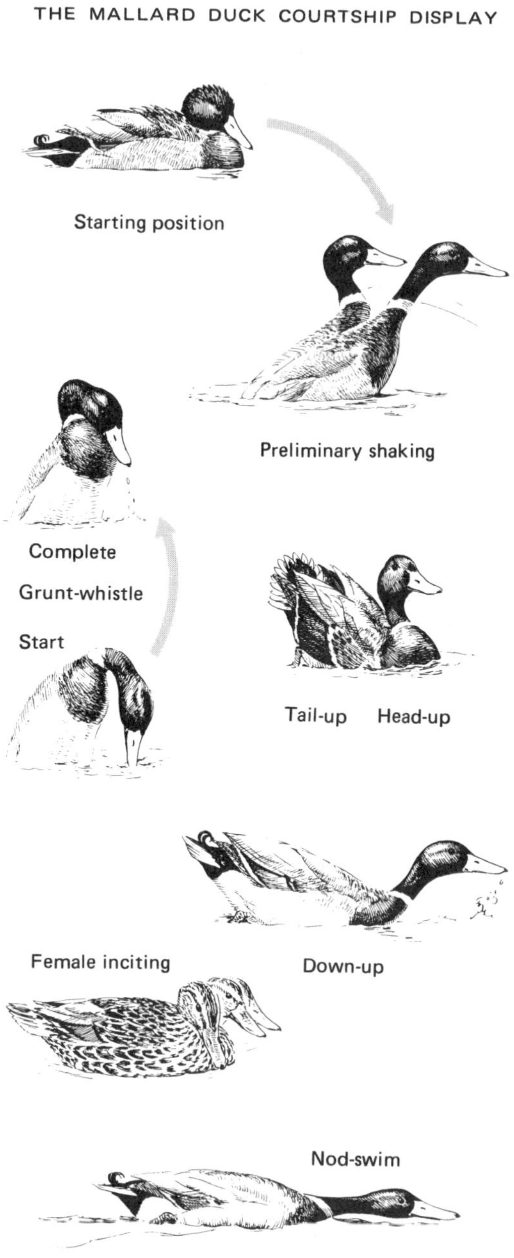

THE MALLARD DUCK COURTSHIP DISPLAY

Starting position

Preliminary shaking

Complete
Grunt-whistle
Start

Tail-up   Head-up

Female inciting   Down-up

Nod-swim

FIGURE 13-13
Courtship display of the mallard. Some of these displays are common to many of the river ducks.

433

TABLE 13-3
Comparison of presumably homologous behavior patterns in the
mergansers (Mergus) *and goldeneyes* (Bucephala)

	Buffle head	Common golden-eye	Barrow's golden-eye	Hooded mer-ganser	Smew	Red-breasted mer-ganser	Common mer-ganser
MALE COURTSHIP							
Upward stretch	×a	×	×	×	×	×	×
Wing flapping	×b	×	×	×	×	×	×
Crest raising	×	×	×	×	×	×	×
Head throw	—	×	×	×	×	—	—
Tail cocking	—	—	—	×	×	×	×
FEMALE COURTSHIP							
Inciting	×	×	×	×	×	×	×
COPULATORY BEHAVIOR							
Drinking by ♂	?	×	×	×	×	×	×
Drinking by ♀	?	×	×	×	×	×	×
Female prone	×	×	×	×	×	×	×
Upward stretch (♂)	—	—	—	×	×	×	×
Preen dorsally (♂)	×	—	—	—	×	×	×
Water twitch (♂)	×	×	×	×	—	—	?
Preen behind wing (♂)	—	×	×	×	?	—	—
Steaming to ♀	—	×	×	×	—	—	—
Flick of wings (♂)	×	×	×	×	×	?	—
Steaming from ♀	—	×	×	×	—	—	—

Source: Johnsgard, 1961.
a × the behavior pattern was observed.
b × the behavior pattern was exceptionally well developed.

aggressive tendencies have been appeased or the reluctance of the male or female has been overcome. Later in courtship the behavior patterns are involved in the proper synchronization of the mating act, as exemplified by the courtship behavior of the queen butterfly (Fig. 13-2). Finally courtship displays serve to assure that males mate with the females of their own species and tend to reduce the error of mating with the wrong species. Thus the displays act as isolating mechanisms.

These functions, orientation, synchronization, persuasion, and appeasement, and reproductive isolation are achieved by species-specific signals or releasers. In fact the most highly developed specific signals and behavior have evolved around courtship, where errors would be most disadvantageous to the species.

Some basic patterns may be common or homologous to a group of animals (see Table 13-3). Among nine species of surface ducks (*Anas*) sympatric to western North America, eight possess many homologous displays (see Fig. 13-13), but there is a striking difference in male plumages. Some displays are almost

universal among passerine birds: bill raising (Fig. 13-14), feather sleeking, feather raising, lowering the bill, vibrating the wing or wings (Andrew, 1961). Interspecific differences, so necessary to reduce the chance of interspecific pairings, set one species apart from the other. These include (1) differences in the color of the plumage, skin, or other body structures, (2) variations in display movements involving differences in relative strengths of the tendencies to attack, to flee from, or to behave sexually toward mates, (3) frequency in occurrence of the several behavior patterns and their intensity, and (4) the production of sound (Hinde, 1959).

In spite of this mistakes do occur and some groups, such as the ducks, are notorious hybridizers (Sibley, 1957; Johnsgard, 1960). This may be the result of two situations. One, common to the periphery of overlapping ranges, is the presence of a female in an area where males of her own species are relatively rare and males of a closely related species similar in appearance or voice are common (see Fig. 14-7). The female will seek a mate;

but the longer she searches in vain, the lower becomes her threshold to respond, until eventually she reacts to suboptimum releasers, usually exhibited by a male of the next most closely related species. In the other situation, a female, because of a mutation or abnormal genetic recombination, reacts unspecifically to male signals.

In either circumstance the resultant hybrids are selected against. In the former situation, either the species is further restricted to ranges where conflicting stimuli do not exist or else male signal characters and the female responses to them are further refined away from the related species. In the latter situation the abnormal birds either would have offspring selected against, or the male would be unable to attract a mate. Hybrids between prairie chickens and sharptailed grouse, for example, carry on mating displays that are a combination or blending of those of both species (Fig. 13-15) with curious results (see Ammann, 1957). The success that these hybrids have in attracting mates is unknown, but it is probably nil. In some areas of Ontario a high percentage, possibly 80 percent, of the prairie chickens show varying degrees of sharptail characteristics. Some biologists suggest that perhaps such hybridization may be the reason for the rapid reduction of prairie chickens in areas where sharptails have become successfully established. If a prairie chicken hen mated with a sharptail cock and produced a brood of hybrids, the reproductive rate of the sharptails would not be reduced but the prairie chicken would suffer the loss of an entire brood.

In closely related sympatric species, strong sexual dimorphism is due in part to the female's choice of her own species, based on innate and learned responses. The males among ducks and prairie grouse, for example, will court the females of any species. Females, on the other hand, unless in some situation as discussed, will mate only with males of their own species, distinguished by color pattern and display movements. Competition for mates also increases the development of signal characters, particularly among polygamous species in which the males with the most pronounced or effective releasers are the most successful in attracting a mate. In such instances the females choose the males for the development of secondary sexual characteristics. The same selective pressures that refined the signal

Flicker

Gray-cheeked thrush

Boat-tailed grackle

FIGURE 13-14
Bill-raising is a courtship display common to many birds.

characters of the male also influence the innate responses of the female. Thus the females may exhibit as much sexual dimorphism as the males, only it is invisible (Dilger and Johnsgard, 1959).

SOURCE OF COURTSHIP DISPLAYS

Because many elements of agonistic behavior enter into courtship and because courtship is so closely related to other aspects of the animal's life, the fact that courtship displays have evolved from other behavior patterns is not surprising. Involved are intention movements, elements of aggressive behavior, ritualized displacement or redirection activities, infantile behavior, and nest building. In fact among some species of birds, such as the wood thrush (Dilger, 1956) and the redstart (Ficken, 1963) there are no special displays associated solely with pair formation. All displays involved are agonistic and appear in other situations. The same is true in many mammals. The courtship display of the male caribou is simply a modified threat pose (Pruitt, 1960). The extended muzzle, the grunts and swift advance express to the does the vigor of the sexual drive in the buck. If the doe is not physiologically receptive, the buck's actions are "interpreted" as antagonistic, and the doe flees; if the doe is sexually receptive, the display fails to release flight. On the other hand, in mountain sheep the ram is much more cautious, and the ewe employs the aggressive behavior (Geist, 1971). Her attacks and butting suggest the behavior of a smaller subordinate ram toward a larger one (Fig. 13-16), except that the female dashes away whereas the young ram does not. Possibly by using aggressive behavior in her courtship, the ewe arouses sexual behavior in the ram by arousing aggressive impulses. This switch from female to male behavior at estrous is also common among the mountain goats and old world deer (Geist, 1971).

Once pair formation—that period from the initial meeting of the male until a bond is formed—has been accomplished, other behavioral elements more sexual in nature appear. Male birds may fluff the feathers of the scapulars, rump, and head, spread the tail, drop and wave the wings, and bow (Figs. 13-17 and 13-18). These latter displays, so common among passerine birds in one form or another, probably are ritualized intention movements of flight, indicating either strong sexual tendencies to fly up and mount the female or tendencies to flee from the female. Common among paired gulls is head bobbing, regarded as solicitation, or precopulatory displays (N. Tinbergen, 1960).

Symbolic nesting and symbolic nest-site selection are common in courtship behavior of birds, from grebes and cormorants to songbirds. Symbolic nesting involves either the manipulation of nesting material by that member of the pair, usually the male, who does not ordinarily help in nest construction, or the unnecessary handling of nesting material by the other member (or both, if male and female together build the nest) prior to actual building.

Among some species, symbolic nest-site selection precedes symbolic building. The male red-winged blackbird crouches near the female, spreads his wings and sings, and then flies to a clump of cattails, to which he clings while holding the wings over his back (Nero, 1956a). If the female follows, he may slowly crawl through the cattails with his wings still partly spread. Then he stops, bows with his beak at his feet, breaks off bits of cattail blades and manipulates them as the female would in building a nest. Symbolic nest building probably represents the vestiges of a true nest-building behavior in a time when the male did assist in nest building. Its possible function is reassurance to the female, for it reappears at times when the female is disturbed during nest building or when the female sits on the nest only irregularly during the egg-laying period.

Begging food by the female and reciprocal feeding of the female by the male are part of the sexual behavior patterns of many species of birds. Begging on the part of the female is regarded as a kind of infantilism, the reappearance of her behavior as a chick. Herring gulls feed the female in response to the head toss, in appearance similar to the food begging of the young (N. Tinbergen, 1960). Terns present a fish to the female, an act that apparently serves in sex recognition. The female may beg several times before the male hands the fish to her, and she may then return it (R. S. Palmer, 1941). Only while incubating does the female accept the fish (N. Tinbergen, 1951). Marsh hawks toss prey to their mates in mid-air prior to incubation. Courtship feeding may serve to reduce or inhibit the aggressive behavior of the male and release sexual behavior, for courtship feeding often precedes or accompanies coition. It also may stimulate and maintain the pair bond. Ethnologists still speculate on the origin of this ritualized feeding, but it appears most likely to have arisen in part from anticipatory feeding (described under Parental Behavior), in which the male prematurely brings to the nest food intended for nestlings not yet hatched. Feeding on the part of the male belongs to the parental behavior, which normally appears long after sexual behavior wanes.

Closely related to courtship feeding is the tidbitting behavior of male gallinaceous birds, among them the domestic chicken (Wood-Gush, 1955; McBride et al., 1969) and the chukar partridge (Stokes, 1961). The male makes incipient pecks toward the ground, especially at conspicuous objects, such as stones, food particles, and feathers, followed by the actual picking up of the objects. At the same time the male gives a tidbitting call to the female, which is similar to the food-finding call of both sexes to the young. Early in the season the female may ignore his calls, but later she breaks off her own activity, runs to the male, and begins to peck at or to pick up the same objects. In a few seconds the male suddenly stands erect, moves off in a stiff-legged step to the rear of the female, and again tidbits. A series of such displays may be followed by coition. This irrelevant activity apparently has an appeasing effect on the hens, who avoid aggressively displaying males. The behavior apparently is ritualized from normal feeding behavior and may be derived from a displacement reaction performed while the cock is undergoing a conflict between sex, attack, or escape or when the sexual drive has been thwarted.

Thwarting of the sex impulse results in irrelevant activities in courtship in animals other than birds. Sexually activated males of the three-spined stickleback and the river bullhead fan their pectoral and caudal fins, movements used in the ventilation of eggs in the nest. These movements appear frequently while the male fish is courting the female that has entered his territory. Over and over the male may swim away from her to the nest he has prepared and do a series of fanning movements even though no eggs are present (van Iersel, 1953; D. Morris, 1954).

One of the functions of courtship is to re-

(a)

(b)

(c)

FIGURE 13-15
(a) *Prairie chicken in display.* (b) *Prairie chicken-sharptailed grouse hybrid displaying (note the blending of the traits of the two parents of the hybrid),* (c) *Sharptail in courtship display. (Photos courtesy Michigan Conservation Department.)*

437

lease the sexual response in the partner through specific signals. The male stimulates the female and in turn the female stimulates the male. The courtship is mutual. When an animal, especially the male, is under strong sexual impulses yet cannot mate because the partner has not given the final signal to release the mating act, a conflict situation arises, which finds its outlet in irrelevant activities.

Some irrelevant activities act as releasers in courtship just as they do in aggressive behavior. The thwarting of the sexual drive in some birds results in preening, which in ducks has become ritualized. The movements or components have become exaggerated to increase their efficiency as a signal. Brightly colored structures, such as the blue wing speculum of the mallard duck, are so located that they are conspicuous during the ritualized displacement activity. The mandarin duck, related to the North American wood duck, strokes a specialized secondary feather, which is extremely broad and bright orange in color, in contrast to the normal, narrow, dark-green secondary feathers. By such ritualized preening the males call attention to the colored feather or feathers, thus making the movement more conspicuous and more species-specific. Ritualization of movements removes them further away from the original source; they become increasingly independent, which obscures the source or instinct from which they were borrowed.

SOLICITATION DISPLAY

Once the pair bond has been established, the need for specific displays diminishes. As a result, the precopulatory or solicitation displays are rather generalized and highly stereotyped, the final signals prior to actual mating. Failure of either male or female to respond to solicitation results in thwarted courtship and often irrelevant activities.

Solicitation displays of female passerine birds (Fig. 13-17) include a horizontally crouched body, wing shivering, and elevation of the tail, often accompanied by a note or series of notes. Somewhat similar displays, particularly the crouched or squatting position, are also characteristic of other birds. These solicitation displays may vary in intensity. In the low-intensity, generalized display of female passerines—but among the blackbirds in

particular—the wings are held out but not quivered and the tail is not elevated. This low-intensity display may be given upon the approach of a soliciting male or be given independently, in absence of the male. Such displays in the redstart often follow periods or bouts of nest building or after the male departs (Ficken, 1963). At high-intensity solicitation display, the tail is cocked at a greater angle, the body is tipped forward, and the display is often accompanied by solicitation notes. Solicitation usually does not appear in the female until the nest-building period, although this varies among species. Most displays prior to this time indicate weak sexual tendencies and are not accompanied by copulation. Actual mating cannot take place unless the drive of both male and female are approximately at the same level, and the responses of both rely closely on the interplay of internal drive and external stimuli.

Recognition of the mate is important to those animals who maintain a long pair bond since it reduces aggression and timidity toward the mate. Some birds, such as the gannet and penguin, the heron and pelican, have rather elaborate nest-relief ceremonies, which are given when one bird returns to relieve the other at the nest. These displays probably have an appeasement value as well as recognition.

Among animals, such as the amphibians, which form no lasting pairs and lack a long-term courtship, the precoital displays are rather intense and serve more to stimulate each member of the pair rather than to attract a mate (Fig. 13-19). Some species of mole salamanders (Ambystomidae) participate in a sort of nuptial dance involving large numbers of individuals, the purpose of which seems to be to stimulate the sexes for the mating activities that follow. The males deposit stump-shaped structures of jelly surmounted by a cap of sperm, the spermatophores, which are picked up by the female with the lips of her cloaca. In some species the male may grasp the female, holding her in firm embrace for a while, after which the two separate, the male to deposit the spermatophore and the female to retrieve it (Fig. 13-19). The behavior of the woodland salamanders (Plethodontidae) includes a series of preliminary movements during which the male rubs his lips, cheeks, mental gland, or side of body on the snout of the female to arouse her (Noble and Bradley, 1933). Later the female in a "tail walk" follows the male, keeping her chin closely pressed against the base of his tail until she picks up the spermatophore.

THE ROLE OF VOICE IN COURTSHIP

Courtship among animals is often characterized by a vocal performance, which may or may not accompany display. Birds and frogs make the most conspicuous use of voice in courtship, although voice is widespread through the animal kingdom. Songs, calls, and other notes may serve to attract females to males or vice versa, to maintain the pair bond, and to stimulate each member of the pair.

Most familiar of all are the songs of birds. These are any vocalizations that when given by one bird repel rivals and attract the opposite sex of the same species, or both (R. L. Smith, 1959; Moynihan, 1962). Song in many species serves a dual function and is essentially hostile in nature. When a female invades the territory of a male, she is greeted in a hostile manner. Some species, such as the chickadee (Odum, 1941) and the grasshopper sparrow (Smith, 1959), have songs that serve primarily to attract the female. Females of some species also possess a song or call that attracts the male. The female grasshopper sparrow possesses a trill, which she sings independently of any vocalization of the male but which is answered by the male either by his own sustained mating song or by flying to her. Recent studies in sound production by fish reveal that even these animals possess some form of vocalization used in courtship. Male satinfin shiners "knock" aggressively when a female enters their territory (Winn and Stout, 1960). If the female remains, the male swims quickly to the potential egg site and vibrates his body. If the female is ready to spawn, she will follow him to the nesting site. Frogs and toads are the most vocal of amphibians, and the pools and roadside ditches of spring resound with their calls. Calling in anurans is restricted largely to males and indicates that they are sexually receptive. The calls play a significant role in the formation of breeding aggregations and in the attraction of receptive females and sexually aroused males to breeding pools. Once on the spawning site, the frogs respond to a complex of auditory, visual, and tactile stimuli, depending upon the species. Calls in

439

FIGURE 13-16
Courtship of the estrous bighorn ewe makes use of contact and aggression as she forcefully butts the ram on the shoulder. Through it all the ram maintains a stiff posture. (After V. Geist, 1971, Mountain Sheep, University of Chicago Press, Chicago.)

Red-wing

FIGURE 13-17
Courtship display of the male redwing blackbird and invitational display of the female. (Drawn from photos in Nero, 1956.)

King rail

FIGURE 13-18
Invitational display of the king rail. This posture is assumed by a mated male upon the close approach of the female. The bird displays the tail and points the bill downward and slowly swings it from side to side. (After Meanley, 1957.)

the singing insects, the grasshoppers, crickets, and cicadas, attract sexually responsive females and in some species stimulate the congregation of males and females. Thus vocalization plays an important role in bringing male and female of the same species together.

Once male and female are near each other, vocalizations serve as sexual recognition. Birds, frogs, and singing insects are able to discriminate between the calls of males of their own species and those of others. That both wide and subtle differences exist between songs is readily apparent from spectrographs of the songs. These differences are important in maintaining reproductive isolation and in enabling close-range orientation of male to female, especially where a number of similar species assemble on a common breeding site, as do the frogs and toads. This discrimination is reinforced by other isolating mechanisms involving visual and tactile cues, as well as the animals' spatial distribution on the breeding area. Depending upon the species, male frogs call from perches at various elevations above ground, on the ground, in exposed or sheltered situations, away from the water, or floating in the water.

Vocalizations play an important role in the stimulation of the sexes during mating activities. Females of many species of birds give a solicitation call or note that releases copulatory behavior in the male. Males of some species, notably the blackbirds, likewise possess solicitation notes that indicate their readiness to mate (Nero, 1956a). These calls are almost always given by grackles during the solicitation display (Selander and Giller, 1961). When sexually responsive male and female crickets and cicadas are at close range, the male produces specialized courtship sounds, which stimulate the female to move forward and walk up the back of the male into position to receive the spermatophore (R. D. Alexander, 1960, 1961). During courtship display the male cod grunts to stimulate the female to display and to swim upward and spawn (Brawn, 1961); and the male satinfin shiner emits a purring sound while he courts the female prior to spawning (Winn and Stout, 1960).

Proper orientation of male to female often is a necessary stimulus to release mating behavior. Under ordinary conditions the stimulus necessary to release copulatory behavior in the male chukar partridge is the sight of

a fully crouched female facing away from him. The female may initiate proper orientation by turning from the male, or the male may move around until he is behind the female. And the female by turning so that the male is no longer behind her can disrupt the male's intentions (Stokes, 1961). A male field cricket, encountering a female face to face, quickly reverses his position and changes his stridulation to courtship sounds, accompanied by a rocking and swaying of his body (Alexander, 1961). This behavior is necessary to induce the female to mount the back of the male, from which position the female picks up the spermatophore released by the male. If the female withdraws upon contact, the male again gives the courtship sounds and initiates distinctive movements with his posterior end directed toward the female's head. If her withdrawal is more pronounced, the male may stop courting, turn around to face the female, and start the preliminaries all over again. If the female leaves, the males of some species will give calls or chirps similar to those produced in aggressive encounters with males.

COMMUNAL COURTSHIP AND HAREMS

Most courtship involves sequences of behavior patterns between single males and single females. But there are species in which courtship is a communal affair. The animals congregate in a special courting area, sexual relations are promiscuous and usually dominated by only a few of the many males present; and the courtship activity tends to have stimulatory effect upon the whole group. The latter effect is not confined to this type of situation alone, since courtship and other social activities of such colonial animals as herring gulls, terns, and herons have a stimulatory effect on the whole colony. Even in the congregations of frogs and insects, the singing of one male stimulates the others to vocalize.

Community display is especially well developed in some species of grouse, notably the black grouse of the Scottish Highlands and the prairie and sage grouse of North America. Early in spring the prairie and sage grouse congregate on their strutting grounds, or leks —called booming grounds for prairie chickens, dancing or parade grounds for the sharptail grouse, and strutting grounds for the sage grouse. The leks are located on areas of open ground, somewhat elevated and so situated that they are visible to the surrounding area. The cocks gather early in the morning and at daybreak the performances begin (for detailed descriptions of the actual performances of each of the three species see Schwartz, 1944; Hamerstrom, 1939; J. W. Scott, 1942, 1950; Hjorth, 1970). Early in the season there is much strutting and challenging with accompanying display, inflation and expellation of air from the air sacs to produce the hooting and booming sound so characteristic of each species, and fighting. Out of this the males acquire locations on the leks and determine dominance. Each dominant bird— the master cock and the most active, aggressive, and vigorous males—has his own accustomed place on the lek, which he must continue to defend from rivals. When the hens arrive, strutting and display by the cocks is most intense, as each endeavors to attract the females to his mating area. Receptive hens wander into the area held by the master cock of their choice.

The master cock of the prairie grouse has two or three subdominant cocks associated with him on the mating area. Except for this master cock, no hierarchy exists and the master cock must ward off his rivals alone. The subcocks and others may gang up on the master cock and steal matings with the hens, especially in the latter part of the mating period when the number of females visiting the leks decreases. The sage grouse master cock, on the other hand, is accompanied by a subcock and several guard cocks who ward off outsiders attempting to break into the mating circle. Nearly 74 percent of all sage grouse matings are accomplished by the master cock, and he may mate with up to 21 birds per morning. When the master cock is satiated, the subcock and later the guard cock may steal some matings. Other males leave the general area to strut and display before the departing females. Among the prairie and sage grouse, the unsuccessful males are for all practical purposes psychological castrates.

A somewhat similar behavior exists among some mammals, in particular the fur, gray, and elephant seals. Dominant males establish territories, which they must constantly defend. Within these territories they rule over harems of females. In the harem of the elephant seal there is a hierarchy of dominance among the females; females of low dominance often are forced to move by females of higher domin-

Spotted newts (courtship)

FIGURE 13-19
Courtship behavior of the common newt.

ance (Bartholomew and Collias, 1962). Elk too have harems, but the bull does not collect a group of females, instead he joins an existing group of cows, a group that also includes calves, yearlings, and extra juvenile males, whom he tolerates to some extent (Altmann, 1956b).

Parental behavior

Courtship and mating in animals inevitably result in the production of young; and in anticipation of the arrival of offspring, behavior gradually becomes oriented toward the care and protection of young. Nest, den, or lodge building among many animals precedes the laying of eggs or the birth of young. The hatching of eggs and birth of young induces behavioral changes appropriate to the care of young, basically those of defense and nourishment.

The degree of parental care given to the young varies widely in the animal kingdom. In a general way the greater the degree of care given to the offspring, the lower is the fecundity of the species. Those invertebrates in which brood protection is highly developed lay relatively few eggs; and those that give the eggs no protection whatsoever produce eggs in the millions (see Thorson, 1950, for example). Parental care is not highly developed among most invertebrates. Some retain eggs within the body until they hatch; others carry the eggs externally. Invertebrate parental care is most highly developed in social ants, bees, and the hunting wasps. The social insects provide all five functions of parental care: defense, food, sanitation, heat, and guidance. Fish and amphibians either lay a few eggs and actively protect both these and the young; or they may lay many eggs and give them no care at all. Parental care is usually poorly developed among the amphibians, although a few salamanders remain with the eggs. Some frogs, notably the male midwife toad, carry the eggs and the subsequent young on their bodies and eventually place them in a suitable environment for further growth. Fish, likewise, may or may not care for the eggs. The cod lays its eggs in the open sea; the trout constructs a gravel nest to ensure proper protection and ventilation of the eggs, but gives no care to the young. Other species, especially those that have highly developed courtship and

mating patterns, build a nest, defend and aerate the eggs, and protect the young. This behavior is typical in the sticklebacks and catfish. Internal fertilization and terrestrial reproduction is fully developed among the reptiles. Relatively few eggs well supplied with yolk may be carried inside the mother's body until they hatch; or they may be placed in nests buried in the ground and given little subsequent care. Crocodiles, however, actively defend the nest and later the young for a considerable period of time.

It is among the homotherms that parental care reaches its highest development and becomes universal. Parental care is most complex among birds, since the young are hatched outside the body. Care must start with the nest, carry through the incubation of eggs and brooding and feeding the young until they become independent. The female plays the major role in the care of eggs and young, but among some animals the males perform this function, and in others both sexes participate. All species defend the eggs, directly or indirectly, and many actively defend the young. Among the mammals, the mother plays the most significant role by carrying the young in the uterus until birth, by nourishing the young with milk, and providing them with heat, sanitation, and guidance after birth. The male may play no part in the care of the young, as in the seals and deer; he may defend them, as does the musk-ox; or he may share in other parental duties by supplying food, for example, as do the wolf and the fox.

INCUBATION BEHAVIOR

Brooding or incubation behavior is that phase of parental activity that provides warmth and shelter for the eggs and young. As one would expect, it is best developed in birds. From the time of the laying of the last eggs, and in some species from the laying of the first, through to the independence of the young, incubation and brooding dominate all other behavior. Sexual behavior is suspended, and even self-feeding, after the young have hatched, is reduced. The duration of the incubation period is influenced by a genetically fixed period of embryonic development and the length of time the young remain in the nest.

The start and continuation of incubation in birds and in those fish that tend the eggs are highly dependent upon nest and eggs, the sign stimuli that induce these animals to settle on the eggs. Effective incubation continues only when there is a proper feedback of the tactile and visual stimuli of the eggs and in some species (perhaps in the females of all) the thermoreceptors of the brood patch (Baerends, 1959). Fanning of eggs by the three-spined stickleback is released by the appearance and fertilization of the eggs. If the eggs are removed, the cycle is broken; at the same time the fanning drive inhibits sexual behavior (van Iersel, 1953).

Interruptions, disturbances, and the lack of proper "releasers" during incubation result in conflict situations and subsequent displacement activities. Building and preening behavior among gulls and terns appear when the number of eggs in the nest is insufficient, when the eggs are abnormal in size and shape, when the temperature is too high or too low, or when the position of the eggs is incompatible with the favorite direction for sitting in the nest (Baerends, 1959). Displacement building and preening are most frequent when resettling or reshifting on the eggs fails to improve the situation, or some disturbance takes place. Building and preening often appear at the same time, but one usually is predominant over the other. Preening is the dominant activity if the bird is reluctant to return to the nest and shows a tendency to escape. If building is dominant, the bird usually returns quickly to the nest. Both patterns appear when the bird, because of the opposing influences of the presence and disturbance of the clutch, is influenced at once to flee and to stay. The ratio of these two drives then determines which of the two, building or preening, appears.

Care and concern for the eggs increase as the incubation period progresses. Birds tend to desert nests less frequently after a disturbance as the time for hatching approaches. Ground-nesting birds, especially geese, gulls, and terns, tend to retrieve any eggs accidently kicked from the nest, although it may take some time for the bird to respond to the egg-out-of-nest stimulus (N. Tinbergen, 1960). Then the bird rolls or attempts to roll the egg with its bill back into the nest. A broken or pecked egg does not release retrieval; rather the bird may eat the egg. No longer a normal shape, the egg has become instead a bit of food. The three-spined stickleback, too, will retrieve eggs that happen to lie outside the

443

nest (van Iersel, 1953). These the male sucks up and inserts back into the nest, but only if the clump of eggs is large enough, at least five or six. Single eggs are eaten. The male also removes or attempts to remove eggs that have become moldy.

The eggs are seldom left unattended among those birds in which both sexes incubate. As one bird arrives to relieve the other at the nest, one of several actions may happen. The signs or signals of broodiness in the mate may stimulate the sitting bird to rise from the nest and allow the other to take over. On the other hand, nest relief may arrive before the sitter is prepared to go. In this case the relief bird, unable to satisfy its brooding urge, may perform some irrelevant activity such as nest building. This activity seems to stimulate the sitting bird to rise. But if all else fails, the relief bird may force the sitter off the nest. Some birds, the herons for example, have a ritualized nest-relief ceremony; others may announce their approach with a call.

CARE OF YOUNG

The hatching or birth of young ushers in another phase of parental behavior, the care of young. The extent and kind of care given to the young is influenced by the maturity of the young at the time of birth. Basically, birds and mammals are either precocial or altricial at birth. Precocial animals are able to move about at or shortly after birth, although some time may elapse before they can fly or move about as adults. Altricial animals are born helpless, naked or nearly so, often blind and deaf. Between the two extremes there is a wide variation in the stage and nature of maturity at birth.

Nice (1962) has classified the maturity at hatching in birds (Table 13-4). Precocial and semiprecocial birds, hatched with eyes open and completely covered with down, are mobile to some degree and leave the nest in a day or so. Most precocial birds are capable of feeding themselves on small invertebrates and seeds; others follow the parents and respond to their food calls; still others are fed by the parents. Semiprecocial birds are able to walk, but because of feeding habits of the parents, they are forced to remain in the nest. Semialtricial birds are hatched with a substantial covering of down and with eyes open or closed, but are unable to leave the nest.

Altricial birds are completely helpless; their eyes are closed and they have little or no down.

A somewhat similar classification could be devised for mammals. Young mice, bats, and rabbits are born blind and naked and thus are altricial. The young of wolves, foxes, dogs, and cats are born with hair and are soon able to crawl about the nest or den, but are blind for several days. These might be called semialtricial. Deer, moose, and other ungulates, as well as horses and pigs, would fall into the semiprecocial category. They are very ungainly on their legs for several days after birth, and during this time they establish a nursing routine. They may be hidden alone by the mother or held in a "pool," characteristic of the elk (Altmann, 1960). The young wait for the dam to return to the hiding place to nurse and to be licked. The most precocial of all mammals are the seals, which might well be a "precocial 4" according to the classification in Table 13-4. Not only is delivery extremely rapid, approximately 45 seconds in the gray seal (Bartholomew, 1959), but movements and vocalization appear very shortly after birth. Newly born fur seals are able to stand up and call from 15 to 45 seconds after birth and are capable of shaky but effective locomotion a few minutes after birth (Bartholomew, 1959). Even while the umbilical cord is still attached, the pups are able to shake off water, bite and nip at each other, and scratch dog-fashion with the hind flippers. They attempt to nurse within five minutes after birth. Although the pups continue to nurse the cows for some time, the cows are protective and attentive toward their young only between parturition and estrus. Thus this behavior is conspicuous only for a few hours to a day after the cow has given birth to the pup.

As a rule parental behavior is not well developed in the poikilotherms, but care for young by the sticklebacks and chiclid fish deserves some comment. At hatching, the young sticklebacks have rather large yolk sacs from which they draw nourishment, and so they tend to remain embedded in or lying on the nest material. After a while the young, moving in a series of jumps, attempt to swim out of the nest. The male tries to catch the jumping and swimming young, chasing them, sucking them into his mouth, and spitting them back into the nest pit. But in most instances the male is unsuccessful (van Iersel, 1953). Young chiclid fish remain near their mother, and for about 6 days return to the mouth of their mother for protection (N. Tinbergen, 1952).

Nourishment of the young is chiefly a function of the female among the mammals, for only she can provide milk. The males defend the young or ignore them. Among the canids, however, the males hunt and bring the prey to the den for the female and later for the female and young when the young have started on solid food. No such behavior exists among the cats.

Among those birds who must feed their young, both male and female participate whether the male assisted in the brooding or not. (There are exceptions, however, such as the hummingbirds.) Exactly how the male knows that the young have hatched when the female alone incubates the eggs is not really known, although changes in the female's behavior probably give the cue. In some species, such as the starling and the prairie warbler, the male may start bringing food to the nest several days to a week prior to the hatching of the young (Nolan, 1958). This is known as anticipatory feeding.

Providing sanitation, guidance, and heat are other parental functions. Brooding provides the necessary warmth for the young while altricial and semiprecocial mammals are sheltered in nests or dens; such precocials as the ungulates have temperature controls of their own from birth. Sanitation is no problem among precocial birds. Altricial birds usually deposit their wastes in fecal sacs, which are carried away by the parents. Mothers of some altricial mammals stimulate excretion in the young by licking their genital and anal regions and then swallowing the excreta.

Guidance by the parents of young who have left the nest and are beginning to acquire some independence occurs among few species of invertebrates other than the social insects, and among only a few species of amphibians, reptiles, or fish. The female caiman of Guiana keeps the young with her until the spring following their birth. Guidance, again, is best developed among birds and mammals. The female ducks, especially the river ducks, stay with the young until they can fly and set the rhythm for such activities as feeding, preening, and resting. The young of diving ducks

TABLE 13-4
Classification of maturity at hatching in birds

FEED SELVES

Precocials
eyes open, down-covered, leave nest first day or two
Precocials 1
independent of parents—e.g., megapodes
Precocials 2
follow parents but find own food—e.g., ducks, shorebirds
Precocials 3
follow parents and are shown food—e.g., quail, chickens

FED BY PARENTS

Precocials 4
follow parents and are fed by them—e.g., grebes, rails

Semiprecocials
eyes open, down-covered, stay at nest though able to walk—e.g., gulls, terns

Semialtricials
down-covered, unable to leave nest
Semialtricials 1
eyes open—e.g., herons, hawks
Semialtricials 2
eyes closed—e.g., owls

Altricials
eyes closed, little or no down, unable to leave nest—e.g., passerines

Source: Nice, 1962.

do not fare so well, since the hen abandons them much earlier. The 2- to 3-week-old ducklings then band together and follow after the 5- to 6-week-old ducklings. Gallinaceous birds are highly dependent on parental guidance for food as well as warmth. Altricial birds also follow the parents for several weeks after they leave the nest (see Nice, 1943). Parental guidance is highly important among the ungulates. The young of moose and elk, for example, may follow the dam until they are yearlings (Altmann, 1960). A close bond exists between dam and calf, and there is considerable vocal communication. The moose cow makes the calf stay within her sight and will retrieve it if it strays. If the dam is killed during the first year, the calf rarely survives the winter. In fact during the rut season a moose cow will leave with her calf at once if the bull intimidates the young animal. Young raccoons and skunks follow the mother on nightly forays, and young wolves join their parents in the hunt. Young gray squirrels follow the female for a while after they have left the nest, respond to her calls, and may even be groomed by her (Bakken, 1959).

A rather strong bond exists between the young of the same brood or litter. Ducklings seemingly need the companionship of their fellows and may do poorly without it. The attachment between members of a brood often outlasts their bond with the parent. A strong bond also exists in broods of gallinaceous birds, but among some shore birds the bond between siblings is not especially strong (Nice, 1962).

As the young mature into adults, the bond between parent and offspring breaks, and agonistic behavior replaces it. This split more often than not is initiated by the parents themselves. The moose cow becomes hostile to the yearling, especially the female offspring, whom the old cow regards as a rival. The yearling bull is tolerated, but if near his mother he becomes a target for the courting bull (Altmann, 1960). Among elk, the yearling cow is tolerated in the herd, but the young bull is chased out of the herd at breeding season by the dominant bull. Not only does he have to face the antagonism of mature cows and the dominant bull; he also has to face other free-roaming, unattached bulls, forcing the animal into a very insecure situation. Once rejected by parents and forced on their own, juvenile ungulates become as

Altmann puts it "a rejected and most erratic non-conforming age group."

Summary

Sociality in animals implies more than the fact that animals stay together. The outstanding characteristic of social behavior is that animals do something together. Their activities are jointly timed and oriented, whether it be fighting between males, relationships between the breeding pair, or flocking. The accomplishment of such joint action requires a means of communication, a "language" between animals. This language may be visual, vocal, or chemical signals; but whatever the displays, the animals understand each other. These behavior patterns are associated with such activities as territoriality, social dominance, pair formation, suppression of agonistic responses between individuals, and the maintenance of reproductive isolation. It is to increase the latter that many of the major differences in display are believed to have evolved. But the distinctive social behavior of a species also may be influenced by the ecology of the animal—its food habits, habitat, avoidance of enemies.

Behavior is essentially a stimulus–response phenomenon. Outside stimuli are selectively admitted, dependent upon the internal state of the animal. The sensory data are processed and integrated within the nervous system and the message sent out to the muscular system for appropriate response.

An animal is born with a number of inherited or innate behavior "programs," which respond to certain external stimuli and which may become modified through learning. Something within the system operates to permit only one behavior program to function at a time. But often two or more behavior programs may be active at once, none of which is dominant. Then conflict behavior, as expressed by irrelevant and redirection activities, results. Animals, however, are not automatic machines, responding only to certain stimuli and nothing else. They are able to modify their behavior through learning, which presumably is most highly developed in mammals. Innate or highly modified through learning, the behavior of an animal enables it to adapt to the changing environment in which it lives.

Population genetics: natural selection and speciation

The community, aquatic or terrestrial, consists of many different locally defined groups of individuals similar in structure and behavior. Individuals within these groups interbreed, oak tree with oak tree, white-footed mouse with white-footed mouse, large-mouth bass with large-mouth bass. Collectively, individuals within each group make up a genetical population, or deme. Beyond one local population may be other similar demes. They may be separated by distance, or they may be only indistinctly separated—the populations may be more or less adjacent and continuous over a wide area. Whatever the situation, hereditary material to a greater or lesser degree passes from one population to the other. Some adjacent demes may interbreed so freely they become essentially one. If a local population of a plant or animal dies out, it will, if conditions are favorable, be replaced by individuals from surrounding populations (see Chapter 11). Individuals that die are replaced by their offspring, so that the population tends to persist down through the years; the inheritable features pass from one generation to the next.

Individuals that make up the deme are not identical. Just as a wide individual variation exists between and among human populations, so the same variation exists among individuals of sexually reproducing plants and animals. This variation is the raw material of natural selection.

The variability evident, and that perhaps not so evident, may be inherited or noninherited. Some of the conspicuous individual variants, such as shortened tails or missing appendages, malformed horns, enlarged muscles, or other features, which are the result of disease, injury, or constant use are not inheritable. These are acquired characteristics, which the early evolutionist Lamarck hypothesized, erroneously, were passed down from one generation to the next. On the other hand some acquired characteristics, environmentally induced, are inherited; or more accurately, the ability of an organism to acquire these characteristics appears to be inherited.

Genetic variation

Of major importance to natural selection and adaptation are genetic or inherited variations within a population, variations that arise from the shuffling of genes and chromosomes, especially in sexual reproduction. That variation exists within a population can easily be demonstrated. All one needs to do is to select some 100 or so specimens from a local population and observe and record the variations of a single character: the tail length of some species of mouse, the number of scales on the belly of a snake, the shapes and sizes of sepals and petals, the rows of kernels on ears of corn. These observations can be tabulated as frequency distributions (Fig. 14-1). Many of the specimens will have the same numerical value; since this is the most common, it is called the *mode*. Others will vary above and below the mode. The frequencies of these values fall off away from the mode—fewer and fewer and fewer individuals are in each class. The frequency distribution of these variable characters tends to follow a bell-shaped curve, the normal curve of probability. In some situations the distribution may deviate from the normal bell-shaped curve. These differences may point out some facts on variation within a population. The variations may be due to heredity, to environment, or, more often than not, to an interaction of both. All tend to produce a bell-shaped distribution of the variations of the character within a population.

Species characteristics and their variants in bisexual organisms are transmitted from the parent to the offspring. Higher plants and animals bridge successive generations by the union of two germ cells, or gametes, the sperm of the male, the egg of the female.

The egg and the sperm carry one of each member of every chromosome pair in a variety of combinations, brought about by random segregation at the time of meiosis. In man, for example, 23 haploid chromosomes (46 diploid, one-half contributed by each parent) may occur in 2^{23} possible combinations, somewhere around 8 million. When egg and sperm unite to form a new individual, the chromosomes recombine in any of a bewildering array of combinations. It is this segregation and recombination of chromosomes and the hereditary information existent for many generations that they carry that is the main immediate source of variation.

Each chromosome carries units of heredity, the genes, the informational units of the DNA molecule. Because chromosomes are paired, genes likewise are paired in body cells. The position a particular gene occupies on a chromosome is known as a locus. Genes occupying the same locus on a pair of chromosomes are alleles, and can be identical or different. During the formation of gametes, the alleles are separated as chromosomes separate; and at the time of fertilization, the alleles, one from the sperm and one from the egg, recombine as the chromosomes recombine. (This is all greatly simplified; the whole story must be sought in books on genetics.)

If genes occur in two forms, A and a, then any individual carrying them can *fall into three possible diploid classes:* AA, aa, and Aa. Individuals in which the alleles are the same, AA or aa, are called homozygous; and those in which the alleles are different, Aa, are heterozygous. The haploid gametes produced by the homozygous individuals are either all A or all a; those by the heterozygous, half A and half a. These can recombine in three possible ways (Table 14-1). Thus the proportion of gametes carrying A and a is determined by the individual genotypes, the genes received from the parents. Eggs and sperm unite at random, enabling the prediction of the proportion of offspring of different genotypes based on parental genotypes.

Assume that a population homozygous for the dominant AA is mixed with an equal number from a population homozygous for the recessive aa. Their offspring, the F_1 generation, then will consist of 0.25 AA, 0.50 Aa, and 0.25 aa (Table 14-1). These proportions are called genotypic frequencies. The gene frequencies, of course, are 0.5 of A and 0.5 of a. This proportion will be maintained through successive generations of a bisexual population (Table 14-2) if at least three conditions exist: (1) reproduction must be random; (2) mutations either must not occur or else they must be in equilibrium, that is, the rate of mutation from A to a is the same as a to A; and (3) the population must be large enough so that changes by chance in the frequency of genes are insignificant.

The equilibrium of these three genotypes can be expressed as a general statistical law, known as the Hardy-Weinberg law. Essentially it is this: If the frequency of allele A is p and the frequency of a is q, where $(p + q) = 1.0$, then $p^2 + 2pq + q^2 = 1.0$, in which p^2 is the genotypic frequency of individuals homozygous for A, q^2 is the frequency of individuals homozygous for a, and $2pq$ is the frequency of the heterozygous, Aa. In the hypothetical population above, the proportion of the genotypes in the F_1 generation will be $(0.5)^2 + 2(0.5 \times 0.5) + (0.5)^2$; and the same tendency can be demonstrated even if the ratio is not the classical Mendelian $1 : 2 : 1$. Imagine a population in which the ratio of A alleles (p) to the a alleles (q) is 0.6 to 0.4. The frequency of the genotypes in the F_1 generation will be 0.36 AA, 0.48 Aa, and 0.16 aa, and the gene frequency will be $(0.6)^2 + 2(0.6 \times 0.4) + (0.4)^2$. From this one can conclude that all succeeding generations will carry the same proportions of the three genotypes (Table 14-3), provided that the assumptions mentioned earlier are fulfilled.

The stated assumptions are never perfectly fulfilled in any real population, so the Hardy-Weinberg law must be considered as wholly theoretical, a distribution against which actual observations can be compared. Nevertheless, the Hardy-Weinberg law is of fundamental importance in theoretical population genetics, since it means that one can state approximately what the genotypic frequencies will be from a knowledge of gene frequencies alone—provided that the population has discrete generations. In a population with overlapping generations, such as the human population, the Hardy-Weinberg law does not hold, even in theory.

Variation in a population seldom is constant from generation to generation. One reason is gene mutation, the ultimate source of genetic variation. Mutation is a change in chromosomes with a genetic effect. The change may involve a multiplication of chromosomes (polyploidy, discussed later), addition to or subtraction from one or more chromosome sets, gross structural changes, translocations, and inversions (all discussed in detail in genetics books). Mutation, which can occur at all gene loci, produces new variation slowly, perhaps one in a million genes. Most common mutations involve the change of one allele into another. Even in a population

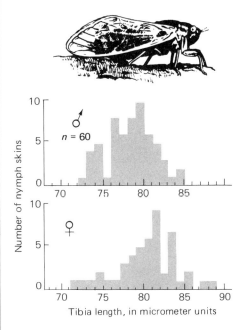

FIGURE 14-1

A histogram showing the frequency distribution of the hind tibia lengths of nymphal exuviae (shed skin) of the periodical cicada, Magicicada septendecim. *(After Dybas and Lloyd, 1962. © The Ecological Society of America.)*

TABLE 14-1
Mixing two homozygous populations

			MALES	
			AA	aa
			.50	.50
			A	a
AA		A	AA	Aa
.50			.25	.25
dd		a	Aa	aa
.50			.25	.25

FEMALES

449

homozygous for A, for example, A eventually will mutate to *a* in some of the gametes; and in a population having both genes, mutations may be forward to *a* or backward to A. If A mutates to *a* faster than *a* to A, then the frequency of allele A increases over the other. This rarely occurs to a point where one of the alleles is lost to the population, for reversibility prevents a long-term or permanent loss. Eventually such mutations arrive at an equilibrium. Even if one allele is lost from the population, it will usually reappear by mutation.

But of more immediate consequence is nonrandomness of reproduction within a population. Not every individual is able to contribute its genetic characteristics to the next generation or to leave surviving offspring. It is this selectivity, this disparity between parents and the rest of the population, that is natural selection.

Before a given individual in a population can contribute to the succeeding generation, it must first survive to reproduce. Survival begins from the time of fertilization through the periods of development, growth, and sexual maturation. Fertilized eggs may fail to develop fully and die, from physiological or environmental causes. Disease, predation, and accidents eliminate those young not quite as swift, as quick, or as strong as their siblings. In such survival, genetic variation plays a key role, for natural selection influences the frequency of alleles in a population. If a mutation arises that places its carrier at a disadvantage, selective pressures eliminate the individual; on the other hand, an advantageous mutation is retained.

An example of such selection can be found among the flies. When DDT was first used as an insecticide against houseflies, the chemical was highly effective and destroyed the bulk of local populations. But among the flies were a few that did not die, that carried a mutation or a certain combination of genes that made them resistant to the spray. Resistance in one strain of flies was due to a recessive gene. Flies homozygous for this gene tolerated a high concentration of DDT, while homozygous dominants and heterozygotes were killed. These flies survived to multiply. Many of their offspring were as resistant to the sprays as the parents; some were even more resistant. The least resistant were selected against; the most highly resistant were retained in the reproductive population. Later applications of DDT continually selected for a combination of genes most resistant to the insecticide. As a result DDT became ineffective in fly control, and newer, stronger sprays were and are required. Eventually these sprays will select resistant strains of flies, which will become adapted to the new environmental conditions. But to acquire this resistance the flies pay a price. In the absence of DDT the flies are inferior competitors to the nonresistant flies, which have a shorter development time (Pimentel et al., 1951). If the spraying is stopped, evolution will be reversed and the resistance will largely disappear from the fly population.

Once they reach reproductive age, more individuals are eliminated from the parental population. The maintenance of genetic equilibrium infers random mating, but mating is not random. Many species of animals, particularly among birds, fish, and some insects (see Chapter 13) have elaborate courtship and mating rituals. Any courtship pattern that deviates from the commonly accepted pattern would be selected against, and the individual and its genes eliminated from the reproductive population. On the other hand, animals possessing a color pattern or movement that accents the typical pattern and increases its signal value or stimulus, especially to the opposite sex, would be selected for. Any new mutations that improved on courtship, mating signals, and ritual would possess a favored position in subsequent generations. Among polygamous species, in particular, the majority of males go mateless, for the females mate with dominant males, that tolerate no interference from younger or less aggressive males (see Chapter 13). States of psychological and physiological readiness also are involved in mate selection. Unless both male and female are at the same state of sexual readiness, mating will not occur.

Neither is fecundity random. Even among human populations it is well known that some parents have many more young than others. The same is true throughout the living world. Some families or lines increase in number through time; others fade away. Obviously those who produce more offspring increase the frequency of their genes in a population and affect natural selection. For example, if individuals with allele A produce 10 offspring to every one produced by those with allele *a*, the proportion of A in the population will increase. There is a limit, however, for natural selection does not always favor fecundity. If an increased number of young per female results in reduced maternal care, survival of offspring may be reduced. This is true particularly among those animals whose chances of individual survival are high. Those organisms whose chances of individual survival are low —for example, ground-nesting game birds and oceanic fish—have become very fecund. A sort of general rule applies to all organisms: high fecundity, low survival; low fecundity, high survival. Natural selection allows a wide range of interplay between these.

GENETIC DRIFT

Sexual reproduction is such that only a few gametes are ever involved in the formation of a new generation. In general, all an individual's genes will be represented somewhere among its gametes, but not in any two of them. Yet on the average, two gametes are all that an individual can leave behind, if the population size is to remain stable. For a heterozygote, *Aa*, there is a 50 : 50 chance that these two gametes will either be **A**, **A** or *a, a*, assuming no natural selection (and with selection the chance is even greater). Thus there is a 50 : 50 chance that a heterozygote will fail to pass on one of its genes. In a whole population, these losses will tend to balance each other, so that the gene frequencies of the filial generation will be a replica, but never an exact one, of the parents' gene frequencies. This is simply the familiar law of averages at work. The larger the population, the more closely will the gene frequencies of each generation resemble those of the previous one; the smaller the population, the greater will be the sampling error, or "genetic drift" (S. Wright, 1931, 1935). If the deme is very small, there is a good chance that the whole population may become homozygous for a particular allele in only a few generations. This is called "genetic fixation." Certain alleles are permanently lost until reintroduced by immigration or mutation. This loss of genetic variability is often maladaptive.

Theoretically, at least, the importance of

TABLE 14-2

Proportions in the F$_2$ generation

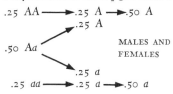

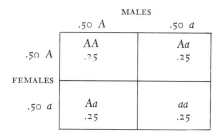

	MALES	
	.50 A	.50 *a*
FEMALES .50 A	AA .25	Aa .25
.50 *a*	Aa .25	*aa* .25

TABLE 14-3

An illustration of the Hardy-Weinberg law

AA	Aa	*aa*
.36	.48	.16

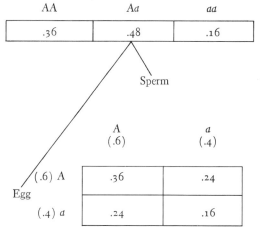

Sperm	A (.6)	*a* (.4)
Egg (.6) A	.36	.24
(.4) *a*	.24	.16

genetic drift in natural selection and evolution may be considerable, especially since most species consist of partially isolated, small populations. Attempts, however, to demonstrate the importance of genetic drift in the field have been inconclusive, owing largely to the difficulties of estimating effective deme sizes in natural populations.

GENETIC ASSIMILATION

From a mathematical point of view, natural selection is regarded as a process that brings about changes in the frequency of genes within a population. But in actual operation, natural selection does not act on genes per se, but on the individual organism, especially as it affects the individual's ability to leave viable offspring. The receiving end of selection, then, is the phenotype, which throughout its development is exposed to the rigors of the environment. Any change in the environment that requires some adaptive change in the species will be lethal unless at least some of the organisms, by some somatic change, are able to weather the period of environmental stress, either until the environment returns to its previous norm or until some appropriate genotypic change occurs. This implies a certain amount of somatic or phenotypic flexibility (Bateson, 1963). This flexibility will involve some of the genotypes in such a way that they will produce a phenotype suitable for the new conditions. If the period is of long enough duration, then the somatic response in the form of acquired characteristics may, under appropriate conditions of selection, be replaced by similar characteristics that are genetically determined. The acquired characters would become "genetically assimilated" (Waddington, 1957). This genetic assimilation, which simulates Lamarckian inheritance, has considerable survival value when organisms must adapt to stress or change that remains constant over a generation. Through genetic assimilation, a species acquires the ability to respond through somatic changes to changes in the environment. Upon the return of the previous environmental norm, the changes in the individual produced in response to specific environmental conditions will follow with a diminution or loss of characteristics (Waddington, 1957). But the ability to respond will be retained genetically.

The species

One has little difficulty distinguishing a robin from a wood thrush or a white oak from a red oak. Each has certain morphological characteristics—ones most useful in field guides—that set them apart from other organisms. Each is an entity, a discrete unit to which has been given a name. This was the way that Carl von Linné, who gave us our system of classification, saw the great number of plants and animals. He, as did others of his day, regarded the many organisms as fixed and unchanging units, the products of special creation. Differences and similarities were based on color pattern, structure, proportion, and other characteristics, and from these criteria the species were described and separated and arranged into groups. Each species was monotypic, that is it contained only those individuals that fairly well approximated the norm or type for the species, the specimens from which the species was described. Some variation was permissible, but these variants were considered accidental, although some slight changes within the species were admitted possible. This is the *morphological species,* a classical concept still alive, useful, and necessary today for classifying the vast number of plants and animals.

Later the studies of Darwin on variation, of Wallace on geographical distribution, and of Mendel on genetics upset the idea of special creation and emphasized that variation was the rule. Naturalists explored new lands, collected new specimens, and observed plants and animals in their natural communities. They found that some species were quite distinct and easily identified by structural or color characteristics; some, such as the song and savanna sparrows, had strong behavioral differences. But among some organisms the distinctions were much hazier. Many animals had only minute but constant morphological differences; some, such as mosquitos, were morphologically indistinguishable as adults and were distinguishable only by the eggs. Others were structurally indistinguishable in all stages of life history but differed in their behavior (see R. D. Alexander, 1962), in ecology (see Dybas and Lloyd, 1962), in biochemistry, as determined by paper chromatography and electrophoresis (Sibley, 1960). These are *sibling species,* defined as "morphologically similar or identical natural populations that are reproductively isolated" (Mayr, 1963).

Examples of sibling species are widespread, especially among the insects. Some of the better-known cases occur in the genus *Drosophila* and in the malaria mosquito complex of Europe. Sibling species can be found in almost every family of beetles and are quite common in the Orthoptera (see Table 14-4). They also occur in snakes, in amphibians, particularly the frogs, in birds to a limited extent, and in fish, especially the whitefish and the salmonids (see Neave, 1944; Ricker, 1940). A characteristic common to all is the lack of a sharp division between ordinary species and the sibling species. The latter are at the far end of a broad spectrum or continuum of increasingly diminishing morphological differences. To the human eye the species may be virtually indistinguishable, but the differences are apparent to the animals themselves.

Biologists also realized, as Darwin observed, that many apparently closely related forms replaced each other geographically, and in doing so intergraded with one another. So smooth and gradual is the transition that it is often difficult to separate precisely one from another. The question "When does a robin cease to be a robin and become something else?" becomes very real and important. The robin, of course, is quite distinct, but the same cannot be said about some other species.

Eventually some biologists realized that the classical morphological or typological species, which emphasizes the individual, was incomplete. It was augmented by the population concept of the species—that species are not fixed, static units but are changing through long periods of time. The anatomical, physiological, and behavioral characteristics that clearly defined a species on a local basis are now regarded as a part of the sum of characteristics found throughout the entire population. The pattern of characteristics can be graphed as a frequency distribution of variants of different characters present in the population at a given time (Fig. 14-2). In other words, the species is multidimensional.

The species has been defined (Mayr, 1942) as "a group of actually or potentially interbreeding populations that are reproductively isolated from other such groups." This concept of the species, then, embodies a group of interbreeding individuals living together in a similar environment in a given region and

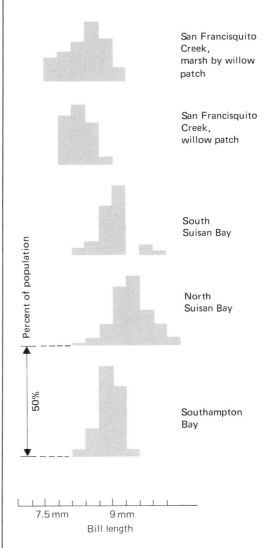

FIGURE 14-2

Geographic variation in the bill length of several populations of song sparrows in the San Francisco Bay area of California. (Adapted from J. T. Marshall, 1948.)

TABLE 14-4
Some differences in the sibling species of the 17-year and 13-year Cicadas (Homoptera, Cicadidae, Magicicada)

	M. septendecim	M. tredecim	M. cassini[a]	M. tredecassini	M. septendecula[a]	M. tredecula
Size	No statistical difference in size between any of these					
Body color	Black above, reddish beneath, appendages reddish; pronotum reddish yellow, prothoracic pleura reddish yellow	Same as M. *septendecim* except radial W in forewing heavily clouded	Black above; almost black below; appendages reddish; pronotum black; prothoracic pleura black		Pronotum black; prothoracic pleura black, reddish or with narrow black apical markings. Tibia reddish all the way to tip. In M. *tredecula* apical tarsal segments reddish all the way to tip. Abdominal sternites with prominent reddish bands.	
Brood	17 years	13 years	17 years	13 years	17 years	13 years
Call	Low-pitched buzzing phrases, fairly even in intensity, ending with a drop in pitch. Phaaaaaaraoh. 1–3 sec.		Rapidly delivered tick series, alternated with high-pitched, sibilant buzzes; noticeable rise and fall in pitch and intensity. Ticks: 2–3 sec; buzz: 1–3 sec.		High-pitched brief phrases in series of 20–40 at rate of 3–5 per sec. Entire call 7–10 sec.	
Chorus	Even, monotonous roaring or buzzing; no regular fluctuations in intensity or pitch. Individual males not synchronized. Chorus most intense in morning.		A shrill, sibilant buzzing, rising and falling in intensity because of synchronization among individual males. Chorus most intense in afternoon.		More or less continuous repeating of short, separated buzzes produced without regular fluctuations in pitch or intensity. Individual males not synchronized. Chorus most intense around midday.	

Source: Data from Alexander and Moore, 1962; M. Lloyd, personal communication.

[a] The only reliable way to distinguish *septendecula* from *cassini*, other than by song, is by morphometric characters. These color characteristics are very inconsistent. Where *cassini* occurs on the edge of its range in the absence of *septendecula*, some character displacement seems to take place. The abdominal sternites of *cassini* have reddish bands as prominent as those of *septendecula*.

under similar ecological relationships. The individuals recognize each other as potential mates. They interact with other species in the same environment. They are a genetic unit in which each individual holds for a short period of time a portion of the contents of an intercommunicating gene pool.

Such a definition of a species is applicable only to bisexual organisms. It is also limited to *sympatric* species, those occupying the same area at the same time. No *allopatric* species, those occupying an area separated by time and space, can be involved because they never have the opportunity to meet other similar species. Thus there is no way of knowing whether they are reproductively isolated or not. The only direct evidence that an individual organism belongs to one species consists of observations in the wild that indicate that the organism is living with a specific population and functioning as a member of it. The individuals tend to possess the same morphological and physiological characteristics because they belong to the same evolutionary population. This, then, is the biological species, in contrast to the static morphological concept—variable, constantly changing, splitting up, reuniting, almost impossible to define precisely.

The concept of the biological species arose mainly in vertebrate systematics, where it has the widest application. The biological species is less generally accepted by botanists, for among plants, as well as among many small invertebrate animals, the biological species, as defined, is inadequate.

The biological species embraces only sexually reproducing organisms. But among plants and some invertebrates, asexual or vegetative reproduction is common. Many of the higher plants, even those in which cross fertilization occurs, reproduce in this fashion and others rely on it, by means of stolons, root runners, bulbs, and corms. The blue flag along the streamside and water lilies in a pond occupy the habitat not primarily through sexually produced seeds but through asexual propagation. In a somewhat similar fashion, the same is true for many annual plants, especially those that occupy extreme environmental situations and pioneering communities. These annuals are self-fertilized and, because they are short-lived, they possess no effective means of vegetative reproduction. Selection has favored self-fertilizing forms most likely to produce a good crop of seeds.

455

Such organisms, which perpetuate their kind outside of sexual reproduction, are called *agamospecies*. They possess, as one may well imagine, very little genetic variability, and they have for the most part lost their capacity to adapt to environmental changes. Variability has been sacrificed for the ability to take the maximum advantage of a given environmental situation. Here they are preeminently successful.

Agamospecies possess for many generations the genetic constitution derived from parent stock. In fact they are the parent stock— genetic individuals on the whole immune to disease, fire, climate, age. These may destroy individual shoots, but separate vegetative parts capable of being carried considerable distances remain. The buffalo grass of the western plains, some botanists believe, probably consists of the same genetic individuals that colonized the plains after the glacial retreat. They possess a sort of immortality.

Any changes in agamospecies are the result either of mutation or occasional reversal to sexual reproduction. Mutations involved are sudden and random changes with seemingly little relationship to the structure and needs of the individual in which they occur. Although small mutations may be tolerated, large ones have very little chance of being beneficial. But if mutations do persist, then they are maintained by the same vegetative reproduction, and a different form of the same species has arisen. Occasionally such plants may revert to sexual reproduction. Pollen from one may be transferred to another and some variability is introduced to the offspring, which propagate themselves either vegetatively or by self-fertilization. Eventually some forms may cross-pollinate and produce hybrids, which in turn are maintained and spread by self-pollination and vegetative reproduction. Hybridization may range from a slight blurring of distinctions between two or more species to the development of a huge complex of forms, in which the original species have become more or less lost. Such natural hybridization is the root of much of the "species problem" in plants and is involved in such "difficult" groups as hawthorns, blackberries, willows, and even some oaks. Certain groups of these are little more than aggregations of hybridizing semispecies (see Benson, 1962).

Agamospecies, however divergent from the biological species, still fit into the morphological species concept. They still can be associated typologically with their sexual relatives, yet they have no certain status.

Geographical variation

Because of the widespread variation of many morphological, physiological, and behavioral characters in a widely distributed species, significant differences often exist among populations of different regions. One local group may differ, more or less, from other local populations; and the greater the distance between populations, the more pronounced the differences become. The geographic variants reflect the environmental selective forces acting on various genotypes, adapting each population to the locality it inhabits.

Geographic variation shows up either as clines, as geographical isolates, or as hybrids. The cline is the result of phenotypic response to environmental selection pressures that vary on a gradient (continuum). The isolate is a population or a group of populations that are prevented by some extrinsic barrier from effecting a free flow of genes with others of the same species (Mayr, 1963). The degree of isolation depends upon the efficiency of the extrinsic barrier, but rarely is the isolation complete. These geographical isolates, or races (or ecological races) (Fig. 14-3), and to some extent the clinal variants, taxonomically make up the *subspecies*, in itself a very thorny problem.

The subspecies is defined by Mayr (1963) as "an aggregate of local populations of a species inhabiting a geographic subdivision of the range of a species, and differing taxonomically from other populations of the species." Here again the population is stressed, for although a whole population can be assigned to a subspecies, it is often very difficult to assign an individual to a subspecies because of the inherent variability within a population and the overlap of the curves of variations of adjacent populations. In reality the subspecies is an artifact of man, set up for practical reasons; it is not a unit of evolution.

The geographical races of a continental species are, more often than not, connected by intermediate forms or intergrades, so that it is virtually impossible to draw a line that will separate them. The differences between the races are greatest at the geographical periphery of the species. Often these differences are so pronouncd that the end races behave as perfectly distinct species. Such is the case in the cline of the meadow frog. The species differs so widely in its physiology on the ends of a north-south gradient that matings of individuals from two geographic extremes, Vermont and Texas and Vermont and Florida, produced young that died before complete development (J. Moore, 1949a, b). Matings from two neighboring locations, however, produced normal offspring.

Sometimes a population of a species has diverged almost too far to be considered a race, yet not far enough to be considered a distinct species. This may result in a so-called ring of races. Sympatric populations that share the same habitat behave like distinct species, yet they are joined by a chain of allopatric races that smoothly intergrade. As a result, it is impossible to draw a line anywhere between the two, although for all practical purposes the end members are different species.

The classical example is the great tit, *Parus major*, of the Eurasian continent. This bird ranges from Ireland to Russia and on through central Siberia to the Pacific without change. The nominate race (the one first named) is *Parus major major*, and it has a dull green

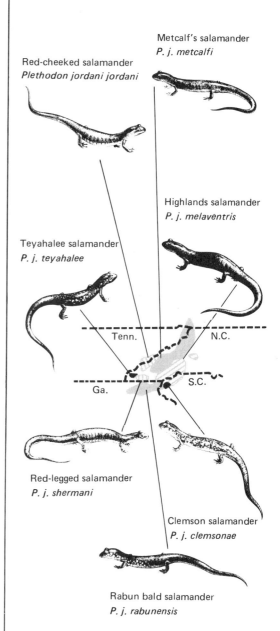

Red-cheeked salamander
Plethodon jordani jordani

Metcalf's salamander
P. j. metcalfi

Highlands salamander
P. j. melaventris

Teyahalee salamander
P. j. teyahalee

Tenn.   N.C.

Ga.   S.C.

Red-legged salamander
P. j. shermani

Clemson salamander
P. j. clemsonae

Rabun bald salamander
P. j. rabunensis

FIGURE 14-3

Geographical races and subspeciation in the Plethodon *salamanders of the Appalachian Highlands. These salamanders of the* jordani *group resulted when the population of the salamander* Plethodon yonahlossee *became separated by the French Broad valley. In the eastern part, the separated population developed into Metcalf's salamander, which spread northeastward, being the only direction in which any group member could find suitable ecological conditions. South, southwest, and northwest the mountains end abruptly, limiting the remaining* jordani. *Metcalf's salamander is the most specialized and ecologically divergent and least competitive. Following separation of Metcalf's salamander, the isolation of Jordan's salamander from* **the** *red-legged and the rest of the group resulted from the deepening of the Little Tennessee River. Remaining members are still somewhat connected, especially around the headwaters of the Little Tennessee. (Information from Hairston and Pope, 1948; Hairston, 1949; distribution map based on data of latter and* Checklist of Amphibians and Reptiles, *1960.)*

back and yellow belly. But in southern Europe some variation becomes evident. The great tit of Spain, Portugal, and North Africa, *P. major excelsus*, has a more vivid yellow on the belly and a reduced amount of white on the outer tail feathers. The variant on the Balearics, Cyprus, Crete, and Greece, *P. m. aphrodite*, has a more grayish back and a paler yellow belly. The birds of Corsica and Sardinia are similar to *aphrodite*, except that they are grayer on the flanks and have been separated as *P. m. corsus*. The Palestinian subspecies, *P. m. terraesanctae*, is paler beneath and more yellowish on the back than any of the others mentioned, a condition even more pronounced in the birds of Persia and North Mesopotamia, which are known as *P. m. blanfordi*. However, all the forms tend to intergrade except where separated by the sea, and all have the same basic coloration of gray back and yellow belly.

In India, Japan, and Manchuria live two other forms of the great tit, whose color variation is so pronounced that by morphological concepts they could easily be called two other species. The Indian bird, *P. m. cinereus*, has a gray back and a white belly; and the Japanese bird, *P. m. minor*, has a green back and a white belly. Both, however, are regarded as subspecies of *Parus major*, for through south-central China and along the coast of Indochina lives another form, *P. m. commixus*, which resembles the gray-backed form except that the tail is gray and the back green tinted; and where this form meets the Indian and Japanese birds, they intergrade. Where they come in contact with the Persian race, the Indian birds intergrade with it and thus connect *P. m. major* of Europe with *P. m. cinereus* and ultimately with *P. m. minor*. The white-bellied, green-backed form of the extreme eastern Asian coast extends northward from Japan to the northern border of Manchuria, the Amur River, where *P. m. minor* meets *P. m. major*. Here they coexist and breed without intermixing as a separate species, although more recent information suggests that hybrid populations are formed (Delacour and Vaurie, 1950). Thus although two races, *Parus major major* and *Parus m. minor*, may behave as separate species in the zone of overlap they are connected with one another by a whole chain of intergrading geographic races. Where does one draw a line between the two? Where would one species, if so classified, begin and

the other end? This is the problem that modern systematists face when they attempt to define a species. This is what is meant by the species problem.

This problem is not so pronounced among insular populations, for here the distribution is discontinuous, and because of natural barriers, zones of intergradation do not exist.

POLYMORPHISM

Geographical variations, especially discontinuous ones, frequently arise within local populations. The occurrence of several distinct forms of a species in the same habitat at the same time is *polymorphism*, which literally means "many forms." This may involve differences in color (such as the gray and black forms of the gray squirrel) and other morphological characters, and in physiology. The important feature about polymorphism is that the forms are distinct and the characteristic involved is discontinuous. There are no intermediates.

Although polymorphism is caused by differences in major genes, some are environmentally induced, the result of the environment modifying the action of genes. This is possible only when two environments, for example, background color, are present at the same time in the same place. Environmentally controlled polymorphism, favoring two or more forms, is the optimal expression of the characters concerned. All intermediates are at a disadvantage and usually are eliminated. The black swallowtail, a common butterfly in eastern North America, and the European swallowtail are good examples. Both swallowtails pupate either on green leaves and stems, or on brown ones. Each possesses two distinct colors to the puparium, one green, the other brown. Through natural selection both have acquired a genetic constitution that produces green pupal color in green environments, brown in brown. Green pupae would be quite conspicuous in winter; butterflies, however, emerge from these in late summer; those in brown pupae do not emerge until the following spring (Sheppard, 1959).

There are times when environmental changes convert a disadvantageous allele or mutant gene to an advantageous one, permitting the latter to spread through the population. During this period, polymorphism will exist, only to disappear when the new advantageous form has completely replaced the original or has so swamped it that the original can be maintained only by recurrent mutation. Such a situation is known as *transient polymorphism*.

An excellent example is industrial melanism. In England, where most observations on transient polymorphism have been made, the physical appearances of more than 70 species of moths are being transformed by environmental changes in the countryside caused by the spread of industrialization. The most famous case is the peppered moth, *Biston betularia*. Before the middle of the nineteenth century the moth, as far as is known, was always white with black speckling in the wings and body (Fig. 14-4). But in 1850, near the manufacturing center of Manchester, a black form of the species was caught for the first time. The black form, *carbonaria* (Fig. 14-4), increased steadily through the years until the black form became extremely common, often reaching a frequency of 95 or more percent in Manchester and other industrial areas. From these places *carbonaria* spread mostly westward into rural areas far from the industrial cities. The black form has come about by the spread of dominant and semidominant mutant genes, none of which are recessive. This increased frequency and spread has been brought about by natural selection. The typical form of the peppered moth has a color pattern that renders it inconspicuous when it rests on lichen-covered tree trunks. But the grime and soot of industrial areas carried great distances over the English countryside by prevailing westerly winds killed or reduced the lichen on trees and turned the bark of the trees a nearly uniform black. The dark form is very conspicuous against the lichen-covered trunk but inconspicuous against the black. A British biologist, H. R. D. Kettlewell (1961; see also Kettlewell, 1965) experimentally demonstrated that not only are the moths eaten by birds but the predation is selective. When melanistic and typical forms were released in woods with lichen-covered trees, the melanistic form was easily seen and subject to heavier predation. In polluted woodlands the typical form bore the brunt of predation. This explained why the typical form has virtually disappeared in polluted country and why it is still common in the unpolluted areas in western and northern Great Britain.

FIGURE 14-4

Normal and melanistic forms of the polymorphic peppered moth, Biston betularia, *at rest on a lichen-covered tree. The spread of the melanic form,* carbonaria, *in industrial areas is associated with improved concealment of black individuals on soot-darkened, lichen-free tree trunks. Away from the industrial areas the normal color is most frequent because black individuals resting on lichen-covered trunks are subject to heavy predation by birds.*

459

TABLE 14-5
Rules correlating variations with environmental gradients (subject to frequent exceptions)

Rule	Statement
Bergmann's rule	Geographic races possessing a smaller body size are found in the warmer parts of the range; races of larger body size are found in the cooler climate.
Allen's rule	The extremities of animals, the ears, tail, bill, etc., are shorter in the cooler part of the range than in the warmer part.
Gloger's rule	Among warm-blooded animals, black pigments are most prevalent in warm and humid areas, reds and yellows in the arid areas, and reduced pigmentation in the cool areas.
Jorden's rule	Fish living in warm waters tend to have fewer vertebrae than those living in cool waters.
—	Races of birds living in the warmer part of the range lay fewer eggs per clutch than those living in the cooler part of the range.

The most common type of polymorphism, however, is stable or balanced polymorphism, a condition in which an apparently optimum proportion of two forms exist in the same habitat and any deviation in one direction or the other is a disadvantage. This can result in several ways. One is when two opposing selective forces so operate that one allele is at a disadvantage when common and at an advantage when rare. The ladybird beetle provides one example of this situation. Individuals of this polymorphic European species possess two color phases, red and black, with a variety of spotting patterns. During the summer, black individuals increase and by fall outnumber the red. During the winter, selective mortality during hibernation is greatest among the black, so that by spring the red form is numerically superior. Thus a balanced selection works in favor of the red phase in winter and the black phase in summer.

Another example is the snow goose *Anser caerulescens*, which is polymorphic and has two clearly defined color phases, a blue one and a white one. The polymorphism appears to be determined by a single pair of alleles. Birds possessing genotype *BB* and *Bb* are blue, whereas the white birds are homozygous, *bb* (Cooke and Cooch, 1968). But there are complications. Dominance appears to be incomplete, and a variable amount of white on the belly of the blue-colored birds indicates a heterozygous individual, whereas a dark-bellied blue-colored bird is homozygous. This polymorphism is further complicated by variations in plumage that are not due to the major alleles for polymorphism but which have both genetic and nongenetic components. In addition there is assortive mating. Males select mates that have a plumage pattern similar to one of their parents, a behavior that could maintain the polymorphism. The goslings apparently become imprinted to the plumage of one of their parents. A similar polymorphism exists among goslings of the related Ross goose, in which the dominant allele confers a gray color to the down and the recessive a yellowish color (Cooke and Ryder, 1971). The gene is not manifested in the adult plumage. Thus mating in terms of the dimorphism appears to be random.

Polymorphism can be extended even further to embrace geographical regions in which each geographical district is characterized by a different proportion of several color phases. This is most pronounced among insular populations, the most widely studied of which are birds. The reef heron of the southwest Pacific, for example, possesses three color phases, gray, white, and mottled. On the Polynesian Tuamotu Islands, white birds make up 50 percent of the population. This color phase comprises only 10 to 30 percent over the rest of its range. White birds are absent or very rare in the Marquesas Islands and in south Australia, southern New Zealand, and New Caledonia. The mottled phase makes up 15 percent of the population from Micronesia to the Solomon Islands and Fiji (Mayr, 1942). Ultimately one form may become dominant or even exclusive over a particular part of the animal's range. And from here it is only a short step to the formation of a well-defined subspecies.

CLINES

In contrast to the discontinuous type of variation exhibited by polymorphic species is the continuous type of variation. Continuous variation is the result of the intergradation of gene pools between local populations. It is most prevalent among organisms with continuous ranges over a continental area.

One of the most familiar of all North American amphibians is the meadow frog, *Rana pipiens*. It has the largest range, occupies the widest array of habitats, and possesses the greatest amount of morphological variability of any North American ranid, and it is the only one successfully established throughout the prairie country. But the variability and adaptability of the meadow frog are orderly, not haphazard. The species embraces a number of temperature-adapted races on a north-south gradient (J. A. Moore, 1949*a*, *b*). When the populations of the north and south extremes are compared, the differences are pronounced, yet between the two extremes, no break in variations occurs. Embryos of southern meadow frogs have an upper-limit temperature tolerance of 4° C above that of the northern embryos, although both survive equally well at low temperatures. Southern forms have smaller eggs and a slower rate of development at low temperatures. In fact so wide are these physiological differences that crossing the northern and southern forms results in defective individuals, although normal hybrids are produced when the meadow frog is crossed with either of the two gopher frogs, *Rana areolata areolata* and *R. a. capito*, and the pickerel frog. This gradual change in a character along a gradient from one area to another is called a *cline*. Clines have distinctive extremes, but because such gradual changes exist from one population to another, it is impossible to group the demes into separate entities. Clines are usually associated with an ecological gradient, such as temperature, moisture, altitude, and light. These changes may take place over a relatively short distance in response to some varying change in ecological conditions (ecocline) or they may take place over a much larger area, as is the case of the meadow frog. This phenomenon has resulted in a number of ecological rules, summarized in Table 14-5.

Some species of plants exhibit clinal gradation, some in size and other structural characteristics, others in time of flowering, in growth, or other physiological responses to the environment. Clinal differences in plants can be demonstrated by transplant studies in which a series of populations from different climates are grown together under one uniform environment in field and greenhouse. Further comparisons of differences can be obtained by growing such a series under several environmental conditions. Such studies have revealed that a number of prairie grasses, among them blue grama, side-oats grama, big bluestem, and switchgrass, flower earlier in northern and western communities and progressively later toward the south and east (McMillan, 1959). The goldenrod, *Solidago sempervirens*, flowers progressively later in the season from north to south along the Atlantic coast. The yarrow *Achillea* blankets the temperate and subarctic Northern Hemisphere with an exceptional number of ecological races. One species, *Achillea lanulosa*, which has been intensively studied (Clausen et al., 1948) occurs at all altitudes in the Sierra Nevada Mountains of California. It exhibits considerable variation, an adaptive response to different climatic environments at various altitudes. Populations at lower altitudes are taller and those at higher altitudes progressively shorter, although considerable variation in height exists within each population.

Until recently clinal variations were measured on such characteristics as wing length and body weight and correlated with environmental gradients such as temperature. Such measurements were often limited to statistical differences among subspecies. Since 1960 the electronic computer has enabled biologists to employ multivariate biometry to measure the relationships of a number of meristic characters simultaneously to several environmental variables. This approach at last permitted biologists to demonstrate the obvious but at one time an unmeasurable fact that geographic variation is not due to the adaptation of a few characters to a single environment. Rather it is a multidimensional process involving the adaptation of many characters to a variety of interdependent environmental variables whose gradients and ranges overlap in a complex way (Sokal and Rinkel, 1963).

One of the best examples is the study of the evolution of the house sparrow in North America by Johnston and Selander (1971).

They studied 16 skeletal characters of 33 samples of North American male and female house sparrows. From correlations of skeletal characters they found that all 16 in both sexes were as geographically variable as such characteristics of skin specimens as wing length. Samples from central and eastern Canada, the Great Plains, and the Rocky Mountains were large; samples from southwestern United States and Texas were intermediates; and samples from the west coast, Gulf Coast, and Mexico had small characteristics. By feeding summaries of gross size variations into the computer, Johnston and Selander obtained a contour diagram of geographical variation in mean values of skeletal sizes (Fig. 14-5). The contoured variations corresponded to the predictions of Bergmann's rule with the larger birds at the higher latitudes. Similarly, geographical variations of such characters as sternum and long bone sizes agreed well with Allen's rule (see Table 14-5).

Not all species exhibit such clinal variations. Some show little or none at all. Ecological rules may be true in a general sort of way, but they are not universal. McNab (1971) argues that for many, if not all, animals the assumption underlying Bergmann's rule is not tenable. This rule assumes that large animals expend less energy to maintain body heat because of a small surface-area-to-volume ratio, and therefore it is more economical for them to live in a colder climate. McNab feels that the north-south gradient in size must be due to something else. He demonstrates that latitudinal variation in body size among carnivores and granivores reflects a change in the size of their prey. A latitudinal change in the size of the available prey is due either to the distribution of the prey or to the distribution of other predators utilizing the same prey species. Allen's rule postulates that animals living in warm climates have small bodies and relatively long appendages for effective thermoregulation at high temperatures. A regression of surface volume ratio of the bill against eight measures of the environment from latitude to four measures of temperature associated thinner bills with higher temperatures, in accordance with Allen's rule (Power, 1970). The regression of bill dimensions against temperature in the red-eyed vireo follows Allen's rule; but in the sympatric Philadelphia vireo, tarsal length increases as temperature decreases, in opposition to Allen's rule (Barlow and Power, 1970). According to Gloger's rule, homeotherm animals are darker in warm and humid regions and lighter in cool areas. Among some insects such as the collembolans, the higher the latitude, the darker the animal (Rapoport, 1969). The nature of the variations in body size, in appendages, and in color may represent an adaptive response to the environment. Species occupying broad ecological niches but possessing high individual tolerance of the environment may show little clinal variation and little tendency to speciate. Those with strong clinal variation and occupying broad ecological niches but showing low individual tolerance of the environment have a strong tendency to speciate (Levins, 1968). The latter is true of the house sparrow, for in a little over 100 years in North America it has evolved into a number of clinal variations comparable to those in its native Europe.

How species arise

The great diversity of living things in the world causes one to wonder how all these species arose. Each is adapted to an ecological niche in the community to which it belongs, and each is genetically independent. The process by which this has come about, by which one form becomes genetically isolated from the other, is *speciation*, the multiplication of species.

Speciation is accomplished in most animals, at least, by an interaction of heritable variation, natural selection, and spatial isolation. Species formation under spatial isolation is described as geographical speciation (see Mayr, 1963).

The first step in geographical speciation is the splitting up of a single interbreeding population into two spatially isolated populations (Fig. 14-3). Imagine for a while a piece of land, warm and dry, occupied by species A. Then at some point in geological time mountains uplift, or land sinks and becomes flooded with water, or some great vegetational catastrophe occurs, which splits and separates by mountain, water, or ecological barrier, a segment of species A from the rest of the population. The newly isolated segment will now become species A', and now occupies an area of cool, moist climate in our imaginary land.

The population of A', because it represents

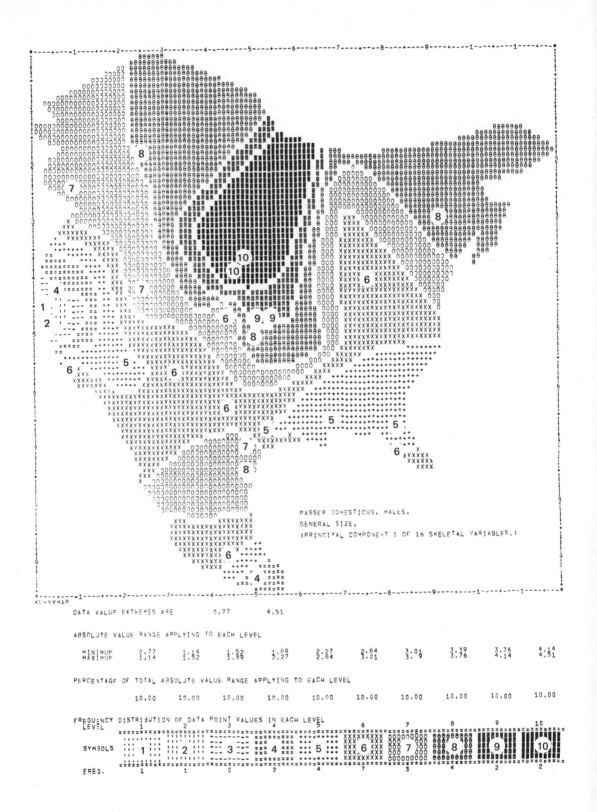

KU-SYMAP

DATA VALUE EXTREMES ARE   0.77   4.51

ABSOLUTE VALUE RANGE APPLYING TO EACH LEVEL

MINIMUM   0.77   1.14   1.52   1.89   2.27   2.64   3.01   3.39   3.76   4.14
MAXIMUM   1.14   1.52   1.89   2.27   2.64   3.01   3.39   3.76   4.14   4.51

PERCENTAGE OF TOTAL ABSOLUTE VALUE RANGE APPLYING TO EACH LEVEL

           10.00   10.00   10.00   10.00   10.00   10.00   10.00   10.00   10.00   10.00

FREQUENCY DISTRIBUTION OF DATA POINT VALUES IN EACH LEVEL

PASSER DOMESTICUS, MALES,
GENERAL SIZE,
(PRINCIPAL COMPONENT 1 OF 16 SKELETAL VARIABLES.)

only a random sample of the population of species A, will possess a slightly different ratio of genetic combinations. The climatic conditions are different; the selective forces are different. Natural selection will favor any mutation or any recombination of already existing genes that will result in a better adaptation to a cool, moist climate. Similar selection for a warm, dry climate will continue in population A on the original land mass. With different selective forces acting upon them, the two populations will tend to diverge. Accompanying this genetic divergence will be changes in physiology, morphology, color, and behavior, resulting in ever-increasing external differences, until A' becomes a geographical race. A', however, is still a part of the species A population, still capable of interbreeding if given the opportunity.

If geographical barriers break down at this point before isolating mechanisms (Dobzhansky, 1947)—those agents that curtail or prevent gene interchange between populations—are fully effective, then interbreeding takes place, and the individuals produced by the cross are fully fertile and viable. If these possess as high a reproductive potential as the parent stock, and if the latter has no other selective advantage, then the two gene pools will merge. The final result will be a population with increased variability over the original population prior to the split.

If the barrier, however, remains, further evolutionary diversification occurs, the two populations become increasingly different, and isolating mechanisms become more fully established. The time eventually is reached when normal interbreeding is no longer possible, even if the two populations do come together. Population A' has now arrived at the species stage.

If the barrier fails at this stage or prior to it, the individuals of the two populations may interbreed and produce hybrid offspring. Such hybrids among animals are less fertile and less viable than the parent stock because they contain discordant gene patterns. Their reproductive potential, if they are fertile at all, is low; they produce fewer offspring. They are, in one, at a selective disadvantage, while any color pattern, voice, behavior, etc., in the parent stock or any mutation or genetic recombination that reinforces reproductive isolation will be selected for. This selection against hybrids continues until gene flow be-

465

tween the two populations has stopped (see Sibley, 1957). Thus species A and new species A′ can invade each other's territory, occupy suitable niches—in our example, a warm, dry environment and a cool, moist environment—and become wholly or partly sympatric (Fig. 14-6). This leads to a diversification of life (see also Ross, 1962).

ISOLATING MECHANISMS

In spring there is a rush of courtship and mating activity in woods and fields, lakes and streams. Fish move into their spawning grounds, amphibians migrate to breeding pools, birds are singing. During this frenzy of activity each species remains distinct. Song sparrows mate with song sparrows, trout with trout, wood frogs with wood frogs, and few mistakes are made even between species similar in appearance. The means through which the many diverse species remain distinct are *isolating mechanisms*. These include any morphological characters, behavioral traits, habitat selection, and genetic incompatibility that enable different species to remain apart.

Isolating mechanisms fall into four broad classes: (1) ecological, which includes habitat and seasonal isolation; (2) ethological, or behavioral; (3) mechanical; and (4) reduction of mating success (Mayr, 1942, 1963).

If two potential mates in breeding condition have little opportunity to meet, they are not likely to interbreed. Habitat selection can effectively reinforce this isolation. From North Africa and the Iberian Peninsula eastward through south-central Europe to Arabia and across central Asia to the Pacific are spread the *Alectoris* partridges, among them the chukar, well known as an exotic game bird in parts of North America (Fig. 14-6). Most of them are allopatric and separated by geographical distance. But in Thrace two very similar forms, the rock partridge, A. *graeca*, and the chukar, A. *chukar*, meet (G. E. Watson, 1962*a, b*). Here they are geographically sympatric. Both species inhabit rock-strewn hillsides and mountainsides with little cover and surface water. But the rock partridge is an alpine form found above 3,000 ft; the chukar lives below 3,000 ft. Otherwise their habitats, habits, and food are the same. In eastern Asia a parallel situation exists. Here the chukar and the great partridge, A. *magna*, both occur but are separated by altitude, the chukar be-

ing found at elevations below 7,000 ft. In central Europe, there is a zone of sympatry between the redleg partridge, A. *rufa*, and the rock partridge. Again altitudinal allopatry separates the two forms. The redleg ranges on the lower, open, cultivated hillsides with their vineyards, cereal fields, and gardens. The rock partridge inhabits mostly the rocky mountain sides between the tree and the snow line. Thus among these partridges geological sympatry exists because of habitat isolation.

Habitat isolation on a very local basis is highly important among frogs and toads (see Bogert, 1960). Different calling and mating sites among concurrently breeding frogs and toads tend to keep the species separated. The barking tree frog typically calls while floating in open water, while sitting under water near shore, or when sitting on woody material projecting from the water. The closely related green tree frog calls from the ground or from low branches of trees or bushes near the breeding pool, but never in the water. Probably this is highly effective in preventing barking tree frog males from clasping green tree frog females. The upland chorus frog and the closely related southern chorus frog breed in the same pools, but ecological preferences tend to separate, partially at least, the calling aggregations of the species. The southern chorus frog calls from concealed positions at the base of grass clumps or among vegetational debris; the upland chorus frog calls from more open situations.

Temporal isolation, differences in the timing of the breeding and flowering seasons, effectively isolates some sympatric species. The American toad, for example, breeds early in the season, whereas the Fowlers toad breeds a few weeks later (W. F. Blair, 1942). Brown trout and rainbow trout may occupy the same streams, but the rainbows spawn in the spring, the brown trout in the fall. Fluctuations in environmental stimuli can time mating seasons. Among the narrow-mouthed toads, *Microhyla olivacea* breeds only after rains, whereas M. *carolinensis* is little influenced by rain (Bragg, 1950). Because temporal isolation is incomplete, call discrimination also is involved (W. F. Blair, 1955), nevertheless some hybridization does occur.

Ethological barriers, differences in courtship and mating behavior, are the most important isolating mechanisms in animals. The males have specific courtship displays to which, in most instances, only females of the same species respond. These displays involve visual, auditory, and chemical stimuli (see Chapter 13). Some insects, such as certain species of butterflies and fruit flies, and mammals possess species-specific scents. Birds, frogs and toads, some fish, and such "singing" insects as the crickets, grasshoppers, and cicadas have specific calls that attract the "correct" mates. Visual signals are highly developed in birds and some fish. Species-specific color patterns, structures, and display, which give rise to a high degree of sexual dimorphism among such bird families as the hummingbirds and ducks, have apparently evolved under sexual selection (Sibley, 1957). Among the insects, the light flashes sent out by fireflies on a summer night are the most unusual visual stimuli. The light signals emitted by various species differ in timing, brightness, and color, which may range from white through blues, greens, yellows, orange, and red (Barber, 1951).

Mechanical isolating mechanisms involve structural differences that make copulation or pollination between closely related species impossible, although evidence for such mechanical isolation among animals is very scarce. Even variations in body size and genitalic differences among insect species do not prevent cross-mating. Differences in floral structures and intricate mechanisms for cross-pollination within the species of many plants present mechanical barriers (see Grant, 1963). If hybrids should occur, especially among the orchids, they would possess such unharmonious combinations of floral structures that these would be unable to function together, either to attract insects to them or to permit the insects to enter the flower.

These three types of isolating mechanisms —ecological, ethological, and mechanical—are significant in that they prevent the wastage of gametes, diminish the appearance of hybrids, and permit populations of incipient species to enter each other's ranges and become partly or wholly sympatric.

The fourth type of isolating mechanism, the reduction of mating success, does not prevent the wastage of gametes, but it is highly effective in preventing crossbreeding. Male gametes of animals that are liberated directly into the water, as is the case of most fish, amphibians, and marine and fresh-water invertebrates, either are unable to fertilize eggs

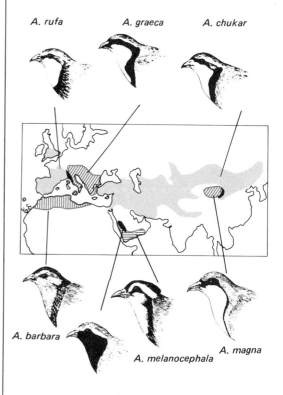

FIGURE 14-6
Sympatry and allopatry in the Alectoris *partridges. The overlap in ranges indicates regions of sympatry and ecological allopatry in the partridges. (Map redrawn from Watson, 1962a.)*

other than of their own kind, produce sterile eggs, or produce juveniles that fail to mature. If hybrids do mature, they may be sterile or at a selective disadvantage and thus eliminated from the breeding population. Among some F_1 *Drosophila* hybrids, the males are sterile and the females fertile only when backcrossed to either parental stock, but the viability of the backcrossed progeny is low (Dobzhansky, 1947). Isolation through sterility is important only in organisms in which ethological isolating mechanisms are poorly developed.

Although the various types of isolating mechanisms have been discussed individually, isolation among species is accomplished by an interaction of several of them. In the coastal sage communities of southern California grow two species of sage, *Salvia apiana* and *S. mellifera*. Although genetically compatible, the two species have established a number of partial barriers to maintain reproductive isolation. *Salvia mellifera* blooms earlier than *S. apiana* (temporal isolation), has smaller flowers (mechanical), and is pollinated by small bees, flies, and butterflies, whereas *S. apiana* is pollinated largely by the larger carpenter bees (ethological) (Epling, 1947). Two mosquitos, *Aedes aegypti* and *A. albopictus*, maintain reproductive isolation by means of at least five barriers: (1) mating behavior involving species differences in flight sounds and differential responses of males; (2) structural incompatibility, the result of differences in male genitalia; (3) sperm inactivation in interspecific crosses; (4) reduced oviposition due to differences in ovulation stimulus provided by substances from the male accessory glands; and (5) genetic incompatibility (Leahy and Craig, 1967).

BREAKDOWN IN ISOLATING MECHANISMS: HYBRIDIZATION

The gaps between sympatric species are absolute. There is no crossing them when the species occupy the same habitat and are adapted to different niches. Between allopatric species, on the other hand, the gaps are relative only. There are no assurances that they are even good separate species; perhaps they are geographical races instead. The final test comes when the geographical barriers are broken and the two or more species in question meet. If the isolating mechanisms are

not highly effective, if the populations are still diverging, then the organisms will hybridize freely. A more or less extensive hybrid zone is formed. Unlike the intergradation of subspecies, which involves a series of intermediate populations no more variable than neighboring populations, hybrid populations in the zone of secondary contact range from character combinations of species A to those of species B. Some hybrids may be indistinguishable from one or the other parent stocks; others may show a high degree of divergence.

If the hybrids are not at a selective disadvantage in competition with the parent populations, the genes of one species will be incorporated with the gene complex of the other. This resulting introgression leads to a swamping of differences between parental forms. There will be a period of increased variability in the rejoined populations, and new adaptive forms will be established. Eventually the variability will decrease to a normal amount, and once again there is a single freely interbreeding population.

Introgressive hybrids are more common in plants than in animals. One of the most conspicuous examples in animals is the golden-wing–blue-wing warbler complex, in which the hybrid forms, the Lawrence and Brewster warblers, are distinctive in color and song. Prairie chicken–sharp-tailed grouse hybrids are rather common, but whether these hybrids are capable of producing offspring among themselves or with either parent species is not known. However, the variety of types (Fig. 14-7) from one extreme to the other suggests that F_1 hybrids crossed with either parent species have reproduced (Ammann, 1957). There are numerous examples among plants. The white oak, for example, hybridizes with seven other members of the white oak group, including the chestnut, post, and overcup oaks. In plants, at least, two species may intergrade not only with each other but with still others to form chains or complex networks of partly segregated and partly interbreeding systems. These hybrids must compete with other plants selected by the environment and probably better adapted than they. Usually these hybrid swarms occupy permanent intermediate habitats somewhat removed from the parents or disturbed habitats where selective advantages to normal inhabitants are removed. If a hybrid population does survive, it may not retain all its members; through selection some members representing various genetic combinations may exploit special niches not suited for either parent.

If selected against, hybridization functions as a source of selection against individuals of both parental species that enter into mixed pairs. Any mechanism reducing the incidence of pairs then is selected for as long as the interaction continues. Any interaction between species that results in deleterious competition or in a wastage of gametes is selected against, reducing diversity in species-specific characters and reinforcing isolating mechanisms (Sibley, 1957).

The North American plains serve or served as an effective barrier to woodland species of animals, a barrier that has existed at least since the Pleistocene, when the ice sheet separated the populations of many animals. Woodlands stretched like fingers along the rivers, but only a few, the Platte in particular, provided a continuous woodland connection across the grasslands. Even these river-bottom forests may have been broken by prairie fires, by floods, and by trampling from buffalo. When the white man settled the plains, he destroyed the woodlands for lumber. But importations of timber from the east, the control of fires, and the control of floods brought riparian woodlands back. Trees and shrubs were planted in shelter belts around towns, farms, fields, and even along the rivers. These changes, which resulted in the development of suitable breeding sites, as well as "islands" of suitable resting places for migrants, permitted the woodland species of birds to move both east and west, to colonize the new habitat, and eventually to come in contact with one another. The red-shafted flicker of the west and the eastern yellow-shafted flicker were two species that spread into an area of contact on the plains, as well as the lazuli bunting and the indigo bunting, the Bullock and the Baltimore orioles, and the black-headed and rose-breasted grosbeaks.

The two grosbeaks illustrate well what happens when two closely related allopatric species that have not yet diverged sufficiently to possess strong isolating mechanisms come into contact (West, 1962). Both the black-headed and rose-breasted grosbeaks have essentially the same habitat preferences, build the same type of frail nest from twigs and rootlets, lay

(a)

(b)

FIGURE 14-7

Sharp-tail grouse, prairie chicken–sharp-tailed hybrids, and prairie chicken specimens. Dorsal and ventral views. (1) Sharptail, adult male; shot Oct. 3, 1941, near Sidnaw, Houghton County, Michigan; (2) prairie chicken–sharptail, adult male; shot Oct. 2, 1946, on Drummond Island, Chippewa County; (3) prairie chicken–sharptail, immature male; shot Oct. 17, 1947, near Ralph, Dickinson County, Michigan; (4) prairie chicken–sharptail, adult female; shot Oct. 28, 1950, near Sharon, Kalkaska County, Michigan; (5) prairie chicken, adult male, from Michigan. No additional data. Specimen no. 2 is in the University of Michigan Museum of Zoology collection. Others are in the Game Division collection, Michigan Department of Conservation. (Photographs from Ammann, 1957; photos courtesy Game Division, Michigan Department of Conservation.)

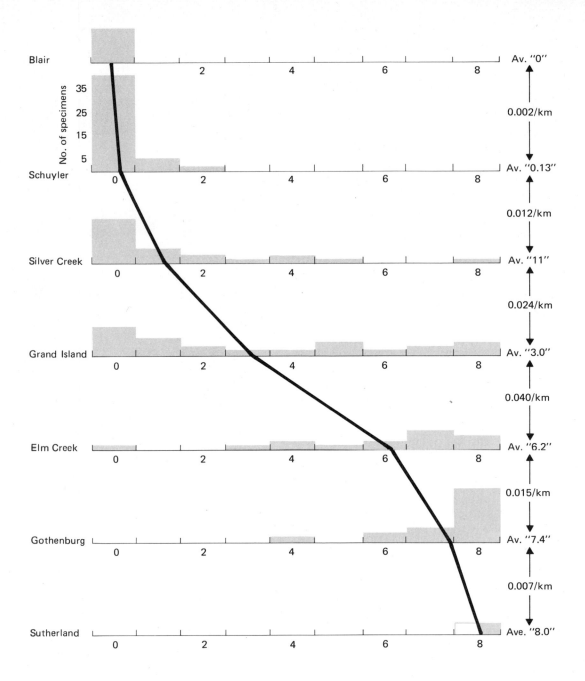

eggs that are almost indistinguishable, possess nearly the same vocalizations, and have nearly identical female plumages. Only in the male plumage do the two species differ widely (Fig. 14-8). On this basis the two have been considered separate species. Where the two come in contact, however, they interbreed, producing a variety of plumage combinations (Fig. 14-9). The variation found in the hybrids suggests that back-crossing, producing second-generation hybrids (F_2), is taking place. Differences in plumage that have evolved in the two under spatial isolation is insufficient to prevent the two from interbreeding. At present there is no indication that any selection against hybrids is operating. Possibly either the differences will be swamped or hybridization will stabilize, with a hybrid zone between the two "pure" populations, since by hypothesis the two species are ecologically too similar to become sympatric.

A similar situation arises in plants when geographic barriers are no longer effective. The eastern red cedar and the Rocky Mountain juniper are related and quite similar in appearance. Where the two come in contact from North Dakota south and east into Kansas and Nebraska, they, like the grosbeaks, hybridize and exhibit to varying degrees the characteristics of both.

NONGEOGRAPHIC SPECIATION: DISRUPTIVE SELECTION

The idea that geographical isolation is necessary for speciation is not universally accepted. There is excellent evidence that racial divergence can come about through disruptive selection, even in the face of a high level of intercrossing (Thoday, 1958; Millicent and Thoday, 1960, 1961; Thoday and Boam, 1959; Thoday and Gibson, 1962).

Within a population, selection may act in three ways. Given an optimum intermediate genotype, it may favor the average expression of the phenotype at the expense of both extremes, in which case selection is *stabilizing.* Or selection may be *directional,* in which one extreme phenotype is favored at the expense of all others. In this case the mean phenotype is shifted toward the extreme, provided that heritable variations of an effective kind are present. The third is *disruptive,* in which both extremes are favored, although not necessarily

FIGURE 14-9 [OPPOSITE]

Histogram of hybrid index scores for samples of grosbeaks from a Platte River transect. The shift per kilometer in average index is given. Localities are about 80 km (50 mi) apart. (From West, 1962.)

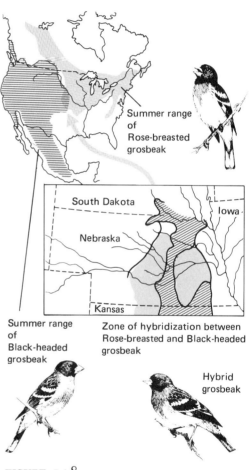

Summer range of Rose-breasted grosbeak

South Dakota
Iowa
Nebraska
Kansas

Summer range of Black-headed grosbeak

Zone of hybridization between Rose-breasted and Black-headed grosbeak

Hybrid grosbeak

FIGURE 14-8

Hybridization in the black-headed and rose-breasted grosbeaks. The map of North America shows the same summer range of both species; inset map shows the zone of hybridization between the two. Figure in the lower right is a hybrid grosbeak showing characters somewhat intermediate between the two parent stocks. (Map based on West, 1962.)

to the same extent, at the expense of the average (Fig. 14-10).

Disruptive selection is most apt to occur in a population living in a heterogeneous environment in which there is a strong selection for adaptability or phenotypic flexibility. Increased competition within the population may select for a closer adaptation to habitat, with the result that the population may subdivide. This would give rise either to a polymorphic situation or to separation into different populations with different characteristics. The latter is most likely to take place in areas where selection is intense and where optimum habitat adjoins or is penetrated by less than optimum habitat. Organisms settling in these habitats will adapt to the local environment. If disruptive selection is strong enough, it will lead to positive assortive mating and eventually genetic divergence of two or more groups.

That local populations can be formed by disruptive selection has been demonstrated experimentally by Thoday and his associates. One experiment involved at the start 80 flies of each sex of a wild strain *Drosophila melanogaster*. From each sex, 8 flies with the highest sternopleural chaeta number and 8 with the lowest were selected. The 32 flies were permitted to mate at random during a 24-hour period, after which the 8 high and 8 low chaeta numbered females were separated. From their progeny, 8 high and 8 low flies of each sex were selected and mated at random. This process was repeated each generation. By the fourth generation of selection, almost all the high flies selected came from progeny of high females; and almost all the low flies from the progeny of low females. By the twelfth generation of selection, the distribution curves did not even overlap. In this experiment, mating was not enforced at any particular pattern and initially probably was entirely random. As the experiment progressed, reproductive isolation developed, limiting hybrids between high and low modes.

This and other experiments demonstrate that it is possible for ecotypes or biological races to diverge under disruptive selection pressures that may be imposed by heterogeneous habitats and that the genetic differences involved are not swamped by random mating. Because other workers have tried similar experiments and failed to get similar results, the idea of disruptive selection is somewhat controversial (see Scharloo, 1971; Thoday and Gibson, 1971).

POLYPLOIDY

Among plants, still another situation arises. New species can arise spontaneously. The most common method by which this takes place is through the alteration of the number of chromosomes. When a diploid cell divides, the chromosomes likewise must divide to produce a full complement in the two daughter cells. In rare instances the rest of the cell does not divide with the nucleus and the chromosomes, and the cell ends up with two times the normal complement of chromosomes. Thus a normal diploid cell becomes a tetraploid ($4n$) cells with more than a normal complement of chromosomes, and the individuals produced by them are called *polyploids*, to distinguish them from the normal diploid state. Polyploids, a type of mutant, usually have fewer but larger cells than the parent stock; often the individuals are larger and tend to be less fertile, because of the physiological upsets and abnormal pairing of chromosomes during the formation of gametes.

There are two types of polyploids, both important from the standpoint of plant speciation—autopolyploids and allopolyploids. The former, autopolyploids, are formed by the doubling of chromosomes in any individual of the species. Thus if the diploid species had AA, BB, CC, and DD chromosomes, its autotetraploid would have AAAA, BBBB, CCCC, and DDDD. When gametes are formed by meiosis, then groups of three and one, four, and two and two may be formed by each type of chromosome. Such irregularity in the chromosomes of the gametes results in few offspring. When a diploid gamete unites with a normal haploid ($1n$) gamete, a hybrid triploid results. Although incapable of sexual reproduction, the triploid can reproduce and spread vegetatively.

The other type is the allopolyploid, which arises something like this. Suppose that species A has four chromosome pairs: AA, BB, CC, DD, and a second species B nearby has chromosome pairs: RR, SS, TT, UU. If the two plants hybridize, the offspring will have a complement of A, B, C, D, R, S, T, U. Because the chromosome pairs are very dissimilar, the hybrid will be infertile. But if during the development of one of the individuals, the chromosome number should be doubled, if it should mutate to a polyploid, then an allopolyploid (allotetraploid) with chromosome pairs *AA, BB, CC, DD, RR, SS, TT, UU* will be formed. Each chromosome now has one definite partner; pairing at meiosis once again is normal, and fertility is improved.

But there are other problems for the allopolyploid. Cross-fertilization with either parental species, both much more abundant, will produce sterile triploid offspring; and because of the small population of allopolyploids, few tetraploids will result, unless the plant is self-fertilized or relies on asexual reproduction. At such a competitive disadvantage, the polyploid is selected against and eventually disappears. If little cross-pollination takes place and asexual reproduction is the normal method of propagation, then the parental stock and the allopolyploid may coexist.

There is another alternative. If the allopolyploid colonizes an area unoccupied by either parent, then the allopolyploid may so completely dominate the area that the parents are excluded. The few that do gain a foothold are at a selective disadvantage, because, being less common, they will receive more pollen from the polyploid than from the diploid stock. Thus the offspring of the parental stock will be sterile triploids, whereas only a few allopolyploid offspring will be in this condition. This is more than a remote possibility because a number of polyploids thrive in areas different from the diploid ancestors. Because plants dispense with sexual reproduction under unfavorable conditions and multiply by rhizomes, bulbs, corms, etc., polyploidy is not at a disadvantage, particularly among perennial herbs. In fact this condition often enables plants to colonize and to tolerate more severe environments. Thus the availability of new ecological niches favors the establishment of polyploidy.

From polyploidy has arisen many of our common cultivated plants—potatoes, wheat, alfalfa, coffee, grasses, to mention a few. It likewise is rather widespread among native plants, in which polyploidy produces a complex of species as in the blackberries. The common blue flag of northern North America is a polyploid and is believed to have originated from two other species, *Iris virginica* and *I. setosa*, when the two, once wide ranging, met during the retreat of the Wisconsin

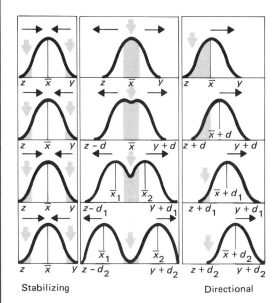

Stabilizing   Directional

FIGURE 14-10

The three main kinds of selection. In stabilizing selection the environment favors the organisms with values close to the populational mean. Consequently little or no change is produced in the population. Disruptive selection favors the extremes and will tend to divide the population in two. Directional selection favors one extreme, and it will tend to move the mean of the population toward that extreme. Directional selection accounts for most of the change observed during evolution. In the figures the curves represent the frequency of organisms with a certain range of values between x and y. The shaded areas are those phenotypes that are being eliminated by selection. The long arrows indicate the direction of evolutionary change and the amount of change. (From O. T. Solbrig, 1970, Principles and Methods of Plant Biosystematics. *Reprinted with permission of Macmillan Publishing Co., Inc. Copyright © 1970 by Otto T. Solbrig.)*

ice sheet. The Sequoia is a relict polyploid, its diploid ancestors having become extinct, as are the willows and the birches. The whole fascinating subject of polyploidy and species formation in plants is a complex one that cannot be pursued any further here. For this one should turn to such sources as Stebbins (1950).

GROUP SELECTION

In general selective pressures are considered as impinging upon individuals of a species. But the idea has been advanced, first by S. Wright (1931, 1935) and more recently by Wynne-Edwards (1963), that a social group or deme is an evolutionary unit. As pointed out previously, most species consist of small populations, typically self-regulating, persistent, strongly localized, and sufficiently (but not completely) isolated to permit some differentiation in sets of gene frequencies. On the other hand, there is sufficient contact between demes to allow a gradual spreading of an advantageous genetic complex to other demes and throughout at least part of the species. But gene flow is not fast enough to prevent local populations from acquiring some characteristics of their own. As a result some demes barely exist or become extinct, and others develop some selective advantage over neighboring demes and produce a greater surplus population. These immigrate into surrounding demes, which in time improve their own selective advantage. This process can spread among inferior demes. In effect the successful take over from the unsuccessful; one group is in sort of passive competition with another. Selection that operates through groups rather than through individuals is known as *group selection* (for discussion see C. B. Williams, 1966).

The resulting higher rate of reproduction through the populations could lead to overpopulation, which would threaten the existence of the group by an exhaustion of resources and poisoning of the environment. But long-time adaptation requires selection at a level that places a priority on a balance between population size and available resources. Involved in the regulation of the population are some kinds of adaptations that belong to and characterize the social group or deme as an entity rather than their members individually. Among many species this regulation is accomplished through territoriality and social dominance.

The idea of group selection is, to say the least, highly controversial, accepted by some geneticists and evolutionary ecologists (see Lewontin, 1965; J. Emlen, 1973) and rejected by others who argue that no characteristic of the group exists that cannot be explained by natural selection at the level of the individual. Proving group selection is difficult. It is most apt to take place among species that are relatively sedentary and show specific breeding site attachments or exist as isolated pockets of populations, and it probably operates largely in relation to dispersal (Van Valen, 1971). It is probably sufficient to say that group selection exists but that its influence is weak, relegated to situations that do not involve biotic adaptations.

ADAPTIVE RADIATION

The direction and degree to which a population diversifies is influenced by the preadaptability of the species population to a new situation, by selective pressures of climate and competition, and by the availability of ecological niches. All species of organisms are adapted to some particular environment, but because the environment is limited, overpopulation can result (see Chapter 10). This in itself is a selective force, for the time finally comes when those individuals that are able to utilize some unexploited environment and resource are at an advantage. Under lessened competition in the environment, these individuals have some opportunity of leaving progeny behind. By eliminating disadvantageous genes, selection will strengthen the ability of the group to utilize the new niche or niches it has occupied.

Not every organism can adapt to a new environment. Before a species can enter a new mode of life, it must first have physical access to the new environment. Unless it is able to reach this new environment, the species cannot exploit it. And having arrived there, the species must be capable of exploiting the niches (preadapted). It must possess a certain level of physical and physiological tolerances to enable it to gain a foothold in the new locality. Here animals, particularly the vertebrates, have an advantage over plants. Most plants, although easily dispersed, are rather exacting in their habitat requirements;

animals, though they find it difficult to cross barriers, are better able to cope with a new environment. Once the organism has established a beachhead, it must possess sufficient genetic variability to establish itself under the selective pressures of climate and competition from other organisms. The adaptations that permit the organism to gain a foothold are only temporary makeshifts, which must be altered, strengthened, and improved by selection before the organism can efficiently utilize the niche. Finally an ecological niche must be available for exploitation. Competition in a new habitat either must be absent or slight enough so that the new invader can survive in its initial colonization. Such niches have been available to colonists of some remote islands, such as the Galapagos, the Hawaiian, and the archipelagos of the South Pacific. The abundant empty niches available in these diversified islands when the first colonists arrived encouraged the rapid evolution of species. Darwin's finches are classical examples of colonization and diversification in an unexploited environment. The original finch population probably consisted of a few chance migrants from South America. Because of the paucity of invaders, on account of the great distance from the mainland, the successful immigrants were able to spread out in many evolutionary directions to exploit the islands' resources (see Suggested Readings at the back of this book).

A similar development took place among the honeycreepers, Drepanididae, in the Hawaiian Islands, which evolved into finchlike, honey eater-like, creeperlike, and woodpeckerlike forms (Fig. 14-11). They so completely occupied the diverse niches on the island that they prevented similar adaptive radiation by later colonists of thrushes, flycatchers, and honey eaters.

The ancestor of the honeycreepers was probably a nectar-feeding coerebidlike bird with an insectivorous diet somewhat similar to *Himatione* (Fig. 14-11h). After colonizing one or two islands, stragglers undoubtedly invaded other surrounding islands. Because each group was under a somewhat different selective pressure, the geographically isolated populations gradually diverged. By colonizing one island after another, and after reaching species level, recolonizing the islands (double invasion) from which they came, the immigrants enriched the avifauna, especially on the larger

475

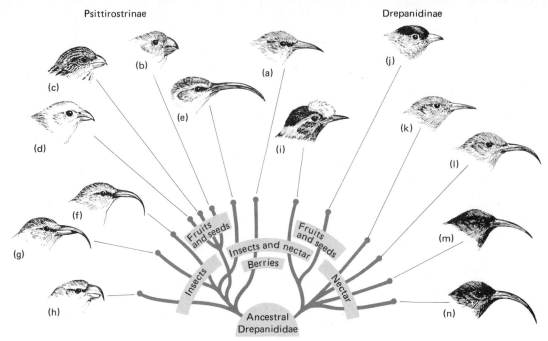

Psittirostrinae

Drepanidinae

(a) (b) (c) (d) (e) (f) (g) (h) (i) (j) (k) (l) (m) (n)

Fruits and seeds

Fruits and seeds

Insects and nectar

Berries

Insects

Nectar

Ancestral
Drepanididae

Adaptive radiation of Hawaiian honey creepers

islands with a more varied ecology. At the same time, competition among sympatric forms placed a selective premium on divergence. The forms then became adapted to the somewhat different ecological niches available, survived by rigid specialization, or one or the other perished. Such divergence of one group into several different forms, each adapted to different ecological niches, able to exploit new environments or to tap a new source of food, is called *adaptive radiation* (Fig. 14-11).

This principle is nicely illustrated by the genus *Hemignathus* (Fig. 14-11e, f, g), all members of which are primarily insectivorous. *Hemignathus obscurus*, whose lower mandible is about the same length as the upper mandible, uses its decurved bill like forceps to pick insects from crevices as it hops along the trunks and limbs of trees. The bill of *H. ludicus* is also decurved, but the lower mandible is much shorter and thickened. The bird uses the lower bill to chip and pry away loose bark as it seeks insects on the trunks of trees. In *H. wilsoni*, the modification is carried even further. The lower mandible is straight and heavy. Holding its bill open to keep the slender upper mandible out of the way, the bird uses the lower mandible to pound, woodpeckerlike, into soft wood to expose insects. The bill of genus *Hemignathus* was specialized at the start to feed on insects and nectar. Although the result of modification through competi-

tion in *H. wilsoni* is grotesque, it is the only species among eight surviving in fair numbers.

The honeycreepers also exhibit parallel evolution, adaptive changes in different organisms with a common evolutionary heritage in response to similar environmental demands. The long, thin, decurved bill of *Hemignathus obscurus* is adapted to a diet of insects and nectar; so too are the bills of several members of the subfamily Drepanidinae (Fig. 14-11m, n).

Similar adaptive radiation took place among the whales. Arising from a carnivorous creodent stock, they eventually radiated into filter feeders, the whalebone whales, whose food is plankton, the great-toothed whales, which feed on deep-sea mollusks, and the fish-eating porpoises and dolphins.

CHARACTER DISPLACEMENT

In areas where two species overlap their ranges, there is a tendency for the differences between them to be accentuated, whereas outside of this area of overlap the differences are weakened or lost entirely (Brown and Wilson, 1956). This divergence, known as *character displacement* may be morphological, ecological, behavioral, or physiological. Differences in feeding habits, in anatomical structures that assist in food gathering in periods of activity, and in nesting sites would reduce competition; differences in reproductive behavior would prevent interspecific hybridization. For example, among the Galapagos finches differences in the depth of bill is exaggerated where the species are sympatric (Lack, 1942). Character displacement, however, is not universal (Mayr, 1970) and perhaps should be considered no more than "a weak rule" (MacArthur and Wilson, 1967). Some (Thielcke, 1966) claim that no known instance of character displacement among birds exists.

CONVERGENCE

Instead of accentuating their differences in zones of overlap, some species reduce their differences, and their characters converge. This apparently happens among species in which reproductive isolation is complete. Selective pressures may tend to favor an increasing resemblance among species as long as reproductive isolation is not upset and the similarities are advantageous to the species (Moynihan, 1968; Cody, 1969). Such convergence can result because sympatric species evolve similar adaptations to the same environment such as background color and because it facilitates social reactions among species.

There are a number of possible situations. Sympatric species may become more or less cryptic against background color. Selection may favor dull or dark-colored forms, which might be a reason for the so-called sibling species. Or the animals may become more conspicuous against background color, a trait that can facilitate flocking among individuals of the same and very similar species such as the herons. Sympatric species may become conspicuous as Batesian or Mullerian mimics. By sharing the same predators they reduce predation on each other. Convergence in color pattern may be selected for if it is advantageous for a species to be a social mimic. Such mimicry to facilitate mixed flocking is common among birds of the mountains of neotropical regions (Moynihan, 1968). For example in parts of the northern Andes most of the more common and conspicuous species in the mixed flocks of the humid Temperate Zone are predominately brilliant blue or blue and yellow. Finally, some species may evolve striking similarities in voice and coloration where they are sympatric and reduced similarity in areas of allopatry. Such convergence

FIGURE 14-11 [OPPOSITE]
Adaptative radiation in the Hawaiian honeycreepers, Drepanididae. Selected representatives of two subfamilies, Drepanidinae and Psittirostrinae, illustrate how the family evolved through adaptative radiation from a common ancestral stock. Both subfamilies show a certain degree of parallel evolution, based on diet. (a) Loxops virens, which probes for insects in the crevices of bark and folds of leaves, as well as feeding on nectar and berries; (b) Psittirostra kona, a seed-eater, extinct; (c) P. cantans, which feeds on a wide diet of seeds, insects, insect larvae and fruit; (d) P. psittacea, a seed-eater, feeding especially on the climbing screw pine, Freycinetia arborea; (e) Hemignathus obscurus, feeding on insects and nectar (see text); (f) H. lucidus, feeding on insects; (g) H. wilsoni, an insect-feeder; (h) Pseudonestor xanthophrys, which feeds on larvae, pupae, and beetles of native Cerambycidae, grips branch by curved upper beak; (i) Palmeria dolei, which feeds on insects and the nectar of Ohio; (j) Ciridops anna, a fruit- and seed-eater, extinct; (k) Himatione sanguinea, which feeds on the nectar of Ohio (Metrosideros) and caterpillars; (l) Vestiaria coccinea, feeding on nectar from a variety of flowers and on loop caterpillars; (m) Drepanis funerea, a nectar-feeder, extinct; (n) Drepanis pacifica, a nectar-feeder, extinct. (Drawings based on Amadon, 1947, 1950, and other sources; evolutionary data from Amadon.)

probably arises in response to interspecific territoriality. The more nearly alike in color and voice, the more they will behave aggressively as one species. Hybridization is prevented by subtle species-specific recognition signals. For example, in southern Mexico live three finches —the collared towhee *Pipilo ocai*, the rufous-sided towhee, *P. erythrophthalmus*, and the chestnut-capped brush finch, *Atlapetes brunneinucha*. The three form two pairs that are interspecifically territorial. The collared towhee and the brush finch colored in green and chestnut, black and white, are so similar in appearance that they can easily be confused in the field. The rufous-sided towhee and the collared towhee have similar songs. Because the brush finch and the rufous-sided towhee differ in both plumage and song, their territories overlap (Cody and Brown, 1970).

Similar convergence exists among plants. Within the chaparral vegetation of California, for example, the dominant plants belong to such diverse families as Ericaceae, Rhamaceae and Rosaceae, yet all are deep-rooted evergreen sclerophyllous shrubs (Mooney and Dunn, 1970). In fact, throughout all areas with a mediterranean-type climate, the vegetation has a similar appearance, dominated by woody evergreen sclerophyllos species. Even though widely separated geographically and possessing different evolutionary histories, the vegetation has converged in both form and function. This has been in response to similar selective forces, including fire, drought, high temperatures, and low rainfall.

COEVOLUTION

In exploitative or feeding situations, interacting species often adjust their relationship to one another. To counter the effects of predation, the prey under the selective pressure of predation evolves some means of defense. Meanwhile, the predators must constantly improve their efficiency in exploiting prey in order to survivie. Each in effect has a reciprocal influence on the evolution of the other. Ultimately both species arrive at some point in which the detrimental aspects of the relationship are balanced with the beneficial. This is *coevolution*. (See Ehrlich and Raven, 1964; Janzen, 1969.)

An example of coevolution is the relationship of the pine squirrels and conifers in the Cascade Mountains of southwestern British Columbia (C. C. Smith, 1970). The squirrels feed principally on the seeds of lodgepole pine and Douglas fir. Because the eastern slopes of the Cascades lie in a rain shadow, the forests there are dry and burn frequently. On these slopes the lodgepole pine is the common species, occurring in even-aged stands. Typical of fire-evolved conifers, this pine has serotinous cones, which remain closed for several years after they are mature. This offers the red squirrel, *Tamiasciurus hudsonicus*, which lives east of the Cascades, a rather constant food supply and stabilizes the squirrel population. The squirrels actually prefer the seeds of the Douglas fir. This tree, however, is not as common, nor does it provide a dependable source of food. In years when the preferred Douglas fir seed crop fails, the squirrels turn to seeds of lodgepole pine. On the western side of the Cascades, moisture is abundant. There the Douglas fir is common and the lodgepole pine is rare. That race of lodgepole does not have serotinous cones, and the resident Douglas squirrel, *T. douglasii* experiences pronounced population fluctuations. Due to the effect of selective pressure from the red squirrel, the lodgepole pine on the eastern slopes has harder texture, and the Douglas fir produces fewer seeds per cone. The harder lodgepole pine in turn exerts a selective pressure for squirrels with stronger jaw musculature. As a result the red squirrels have stronger jaws than the Douglas squirrels on the western slopes. In this situation selective pressures begin with frequent fires that lead to serotinous cones with fewer seeds, which finally selects for squirrels with stronger jaws.

SELECTION AND DOMESTIC PLANTS

By means of constant selective pressures from the environment, coevolution that brings them into genetic equilibrium with their predators and parasites, and relatively unobstructed gene flow, natural populations maintain a degree of genetic diversity. But this is not true with another group of important organisms, man's domestic plants. Because of the outstanding success of modern plant-breeding programs, the genetic resources required for improving our basic food crops are being rapidly destroyed.

When man first domesticated plants he began by harvesting them in the wild. Later he planted what he harvested and eventually ended up with domestic plants. Because he rarely practiced clean cultivation, his fields contained an assortment of desirable and undesirable types from which he selected the more desirable forms. However, the genetic constitution of his crop plants were probably never subjected to stabilization by inbreeding. Because of wild and cultivated species crosses and introgressions, the populations of domestic plants were continually changing in genetic composition.

Wild races, having been around for a long time, reached a genetic equilibrium with their parasites. Genes for virulence in the pathogens were matched by genes for resistance in wild plants so that neither host nor parasite was threatened with extinction. Domestic races that man developed were similar in gene content, possessing a multiple genetic resistance to disease. In the same manner local crops and their wild counterparts were specifically adapted to local environments, further broadening the genetic base. As crops were transported and planted far from centers of origin, new domestic races, adapted to new local environments and genetically balanced with other local races of pathogens, were developed. All of these groups provided a rich genetic store for plant-breeding programs.

In their most useful form, most cultivated plants are polyploids. In the early history of plant domestication their genetic base was never lost by plant breeding. As man learned to exercise higher degrees of selection, planting only those clones or seeds that showed higher yields and favorable qualities, he allowed undesirable types to become extinct or exist as weed species. In modern plant-breeding programs local varieties or landraces were assembled and pure lines selected, usually at the expense of genetic variability. Pure lines were crossed with pure lines to produce hybrids, and inbred lines and controlled crosses were established, especially in corn. Genetic variation was further eroded, and many genes for more general adaptations and resistance to disease were lost. As pure lines were selected for, variety was selected against and with it resistance to disease. As each pure line was attacked by some pathogen, attempts were made to build up resistance in that strain. Then another strain of disease moved in, forcing further changes in gene structure. Entire crop stands, lacking genetic diversity, are open to

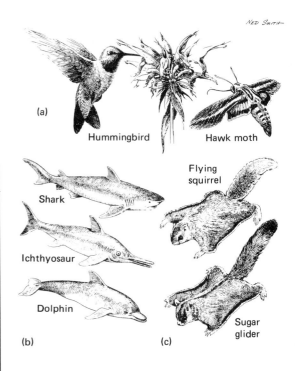

FIGURE 14-12

Convergent evolution. (a) Convergent evolution in two highly dissimilar organisms, the hummingbird and the hawk moth. Both feed on the nectar of flowers; the bird by day, the moth by night. Both have adaptations of the mouth for probing flowers; both have the same rapid, hovering mode of flight. (b) Convergent evolution in three unrelated groups: the prehistoric marine reptile, the ichthyosaur; the modern shark; and the marine mammal, the dolphin. All have the same streamlined shape for fast movement through the water. (c) Convergent evolution in a North American rodent, the flying squirrel, and the Australian marsupial, the sugar glider. Both have a flat, bushy tail and an extension of skin between the foreleg and the hind leg that enables them to glide down from one tree limb to another.

such devastating diseases as the late blight of potato, and southern corn blight.

The demand for tightly controlled genetic strains is confined largely to the United States. Up until World War II at least, such pure strains had little impact on wild, weed, and landrace populations of other countries. But now that new pure lines adapted for more tropical countries have been introduced as part of the "green revolution" into countries that previously harbored the wild types, the last remnant of genetic diversity in food plants is threatened. High-yielding Mexican wheat has been spread to nearly every continent; new wonder rice has been introduced to Asia, hybrid corn to Mexico, modern potatoes to the Andes. With clean cultivation, the expansion of agriculture into uncultivated lands and the overgrazing of wild stocks of corn, potatoes, wheat, and rice are vanishing. With their loss, accelerated by the world's increasing dependence on modern varieties, the genetic diversity of these forms is also lost. The ever-narrowing genetic base of modern food plants, the loss of irreplaceable genetic resources, the evolution of new and virulent plant diseases, as well as pollution, all threaten the future of crop production. For insurance against such a catastrophe, great efforts must be made to maintain centers of variability and to provide complete isolation of modern artificially produced varieties from their wild and landrace progenitors (see Harlan, 1972; Ugent, 1970; Wilkes, 1972; Adams et al., 1971).

Taxonomy and systematics

Taxonomy and systematics are basic to the study of species diversity, structure and function of ecosystems, niches, population genetics, speciation, nature of isolating mechanisms, interactions among populations, comparative morphology and physiology as well as to such practical aspects as biological control of pests, study of domestic plants and animals, resource management, and the construction of field guides. Without the means of attaching some label to organisms, without being able to fit them into a hierarchy, and lacking any knowledge of the relationship of organisms, modern biology would not only be worthless but impossible. This fact is un-appreciated by a great body of modern biologists.

Although taxonomy and systematics are often confused, they are separate yet interdependent. Most taxonomists are also systematists, and all systematists have to be taxonomists. Taxonomy is the theory and practice of classifying organisms, of characterizing species, and arranging them in hierarchies of higher categories. Systematics is the study of the diversity of organisms and any and all relationships among them. It is concerned with organizing the basic knowledge about species and is dedicated to the discovery of differences as well as similarities among species.

Taxonomy had its beginnings with the Greeks, who saw a natural order to the world, which they demonstrated and classified by certain logical procedures. These procedures were defined largely by Aristotle (384–322 B.C.), whose classification scheme was adequate for over 2,000 years. Although he had a number of forerunners, the father of modern taxonomy was Carl von Linnaeus, who in 1753 introduced the system of binomial nomenclature, which gave each plant and animal two names, species and genus. The taxonomy of Linnaeus was concerned with the description of typical individuals and placing it into some category as dictated by its characteristics. Darwin's discoveries during his voyage aboard the *Beagle* presented taxonomy with a number of major problems. To help solve them, taxonomists embraced the theory of evolution and began to ask questions on how different species originated. This approach inevitably led to systematics. The next step was the welding of population genetics and systematics into the "new systematics." Out of new systematics came the concept of the biological species. From then on taxonomy and systematics became highly interrelated. Taxonomists used the concepts of the biological species and evolution upon which to base inferences for classification. The taxonomist usually attempts to place the units and groups in some natural order that will show relationships and phylogeny. This requires considerable study; and for many plant and animal groups a sound phylogenic arrangement is difficult to construct because of a paucity of information, anatomical, genetic, morphological, and behavioral.

In the earlier day of the typological species,

the taxonomist usually based his descriptions on what he considered the typical individual, the type specimen. The modern taxonomist bases his description on measurements and on evaluation of variations within a series of specimens from the population. In fact the typical specimen becomes mean values and ranges of variations. But when he comes to name the organism, the taxonomist still must select one specimen to be the *type* or *voucher specimen*. Because of the concept of the biological species as a complex population that cannot be measured as wholes but only sampled (Darlington, 1970), the samples obviously do not show all of the limits to or ranges of variation. Although statistical analysis is used, the species and its description cannot be handled with mathematical precision. For this reason, among others, interest in taxonomy began to lag behind other areas of biology. But in the mid-1960s new life was infused into taxonomy. Like population biology and molecular biology, taxonomy and systematics have become highly controversial as new concepts challenge the old.

Currently taxonomy and systematics are divided into two camps—the *pheneticists* and the *phyleticists*. The latter are further divided into two subcamps—the *evolutionists* and the *phylogeneticists*. The evolutionists base classification on lines of descent and on the degree of differences. The phylogeneticists (see Henning, 1966) argue that classification should reflect only cladist affinity (the new taxonomy suffers heavily from verbose and obscure terminology designed more to confuse than enlighten); that is, the sequence of branching lines of descent should ignore similarity and difference. In other respects the two camps are similar. The other major group (see Sokal and Sneath, 1963, 1966), the pheneticists or numerical taxonomists, base classification wholly on relative phenetic similarity and ignore descent. By phenetics the numerical taxonomists mean any character that can be measured.

There are basic philosophical differences between the two groups (see Hull, 1970; Michener, 1970). The phyleticists accept, indeed champion, the conventional concept of the biological species. They attempt to obtain the best samples of the population and observe plants and animals in their natural habitat. They sort out and analyze their

481

samples by simple screening and complex statistical analysis, even using techniques employed by numerical taxonomists. They set up a classification system based on man's innate capacity to form group sets and make value judgments. Although the methods may not be mathematically precise, they accept the reality and complexity of the biological species (Darlington, 1970).

Numerical taxonomists do not recognize the biological species, and they reject any application of evolutionary theory to classification. They would replace the biological species with "operational taxonomic units," or OTUs. Taxa are limited or defined so they can be handled numerically and mathematically. The OTUs are various characters drawn from as many parts of the body as possible in an effort to sample total variability in the group. The OTUs are by necessity limited by the current accepted species (which forces the pheneticists to admit that biological species do exist). The species levels of OTUs are described and compared numerically, using component analysis and cluster analysis, and the results are displayed as dendritic phenograms or dendrograms (Fig. 14-13). The weakness is that the dendrogram will vary according to the number of variables used and the coefficients employed. Different workers using the same materials may produce different dendrograms. Thus any formal classification based on dendrograms would be very unstable, if indeed even valid.

The phyleticists accuse the pheneticists, especially the highly theoretical ones, of moving away from reality, of substituting natural situations with purely mathematical and numerical concepts, of replacing the biological species with OTUs without reference to biological organisms or biological significance. On a practical level it would be very difficult for an ecologist to give up the biological species for OTUs and a number in a field guide.

Basically the current conflict between the two approaches to taxonomy is over the question of whether a classification should act as a storage and retrieval system for information about the distribution of attributes over an organism or whether it should reflect the historical course of evolution of the organism concerned (Colless, 1970).

As with all polarized views, the real solution and major advances will be made down the middle. The phyleticists certainly can use the sophisticated techniques of the numerical taxonomists. At the same time, the pheneticists will eventually have to accept the biological species, as many already do, as the entity with which they must work. Their OTUs then can provide a better understanding of the species concept and the nature of the local population.

Summary

The classification of plants and animals may be an old facet of biology, but it is more necessary today than it has been in the past. Before any work in ecology can be done with accuracy, the organisms must be described, placed into recognizable groups, named, and their relationships worked out. The science of classifying organisms is taxonomy; the study of their relationships is systematics. Currently taxonomy and systematics is in a state of ferment, divided into two schools: the phyleticists, who think that classification should reflect the historical course of evolution, and the pheneticists, who think that classification should act as a storage and retrieval system of information on the characteristics of organisms. The dissent is leading to new approaches and new theories in taxonomy.

In the process of describing and classifying a plant or animal, it is usually treated as a morphological species, a discrete entity that exhibits little variation from the type of the species, the original specimen upon which the description is based. But organisms are not static; they are continuously changing through long periods of time. This fact has given rise to the concept of the biological species, a group of interbreeding individuals living together in a similar environment in a given region and under similar ecological relationships. This concept, limited as it is to bisexual organisms, is also filled with difficulties, such as the inability to distinguish, among some groups, just where one species begins and another ends. To get around this problem, some evolutionary biologists suggest that the species be retained to identify kinds of organisms, but as a concept the species should be regarded as evolutionary units.

It is through a long evolutionary process that species arise by an interaction of heritable variation, natural selection, and perhaps among most kinds, spatial isolation. Species formation under spatial isolation is known as geographical speciation. As one segment of the species is separated from another, it carries a somewhat different sample of the gene pool and faces different selective pressures. Eventually the population diverges so far from the original parent stock that interbreeding cannot take place even if the barriers are removed. At this point speciation is complete and the two populations could occupy the same geographical area (sympatry), yet remain distinct. This they accomplish by means of isolating mechanisms, which include any morphological character, behavioral trait, habitat selection, or genetic incompatibility that is species specific. If isolating mechanisms break down, hybridization results. Species may also arise in the absence of geographical isolation through disruptive selection in which the phenotypes of both extremes are favored at the expense of the average. Speciation in plants may also come about through alternation in the number of chromosomes—polyploidy.

Speciation often implies that selective pressures act only on individuals of the species. There is evidence that the local population, or deme, is also an evolutionary unit, that selective advantages exist between groups as well as between individuals. Other species may converge and become more similar either in response to selective pressures of the environment or to behavioral relations. The latter results in social mimicry, which in some situations facilitates foraging in mixed flocks, and in others increases aggressiveness between sympatric species in response to interspecific territoriality. Also apparently common is coevolution, in which the interacting species reciprocally influence the evolution of each other. Genetic diversity is the rule in natural populations, but man has so modified and purified the genetic constitution of domestic plants and has allowed so much genetic variability to disappear that the genetic base of crop plants is endangered.

Genetic variability and speciation all result from natural selection; essentially it is nonrandom reproduction. This can come

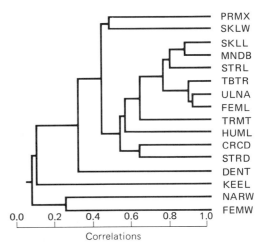

FIGURE 14-13

An example of a dendrogram that was developed from a matrix of correlations of male skeletal characters of North American house sparrows. Accuracy of the diagram in indicating inter-character correlations increases progressively to the right in the fine branching parts of the cluster. The letters indicate the skeletal character. For example, PRMX = premaxilla length; SKLL = skull length, etc. (From Johnston and Selander, 1971.)

*about in three ways: nonrandom mating,
nonrandom fecundity, and nonrandom survival. This nonrandomness lies in the variations contained in the gene pool of the
deme or local population of interbreeding
individuals, similar genotypically and pheno-
typically. Two sources of genetic variation
are mutation and a recombination of genes
provided by the parents in a bisexual population. Theoretically, the variations in
biparental populations, as reflected in gene
frequencies and genotypic ratios, remain in
equilibrium if the conditions of random
reproduction, equilibrium in mutation, and
a relatively large population size exist. In
nature such conditions do not occur, and
there is a departure from genetic equilibrium.
This departure is evolution, a continuing
process. The direction evolution takes depends upon the genetic characteristics of
those individuals in the population that
survive and leave behind viable progeny.
Thus natural selection, so often depicted
as a desperate "tooth and claw" and "law
of the jungle" sort of thing, is simply nonrandom reproduction. The successful are
not just those that survive, but rather those
that leave behind a sample of their genetic
constitution.*

IV

The organism and the ecosystem

One usually considers the organism within the framework of its habitat and its intra- and interspecific relations. But the organism is certainly coupled to and interacts with its physical environment. It has to contend with moisture problems; it has to cope with fluctuating environmental temperatures. Successful organisms have adapted to their physical environment by various means: morphologically, physiologically, and behaviorally.

The energy environment

Living organisms are intimately associated with an energy environment. The same influx of energy upon which photosynthesis depends is also the source of heat energy that characterizes the physical environment of life. What impact this heat influx or the thermal environment has on an organism depends on how it is adapted to cope with thermal budgets and thermal stresses. In any case the relationship involves the exchange of energy between the organism and the environment, the relationship between energy gained and energy lost. In effect energy absorbed from the environment plus energy produced must equal energy lost from the body and the energy stored. This statement can be clarified by considering the sources and processes of energy gains and energy losses.

The major source of absorbed energy is solar radiation (S). It may be direct, transmitted through the atmosphere; diffused radiation from the sky; and/or reflected radiation from objects in the environment. In addition there is infrared thermal radiation (IR) from the soil, rocks, vegetative surfaces, and the sky. There is also the heat of metabolism of the organism (M) (Figs. 15-1 and 15-2).

Just as the organism gains heat from the environment, so it loses heat to the environment. One source of loss is infrared radiation (IR_0). Organisms are continually emitting or giving off radiant energy, that which travels by electromagnetic waves. The amount of heat energy radiated from any object depends upon three values: (1) the emissivity (E) of the surface, (2) the temperature of the surface (T), and (3) the Stefan-Boltzmann constant (σ). The loss of heat is determined by

$$Q_r = E \sigma T^4$$

where Q_r = outward radiation flux in calories.

Because the infrared emissivity of a surface equals the absorption coefficient, an animal's surface, which emits as a black body or nearly so, also absorbs at the same rate. Thus an animal also absorbs radiant heat from the environment. The difference between the radiant energy lost from an animal and that absorbed from the environment represents the net energy exchange. Although it is simple to calculate net energy exchange for a one-dimensional flow between plane surfaces, as in a leaf, it is very complex for such an irregular object as an animal's body.

Another source of heat transfer is conduction (C), resulting from collision between oscillating molecules. The amount of heat transferred is given by

$$Q_c = kA \frac{(\Delta T)}{d}$$

where

Q_c = calories of heat conducted
k = thermal conductivity coefficient
A = area
ΔT = temperature difference between the two surfaces
d = distance between the two surfaces

Among animals the amount of heat lost by conduction varies depending upon whether the animal is standing or reclining.

Heat may also be transferred by convection, which takes place when molecules move from one place to another. Convection (G) may be natural, and result from density differences in the air or water surrounding an object. Or it may be forced, which involves external pressures from fluids passing by or over the object. The amount of heat transferred by convection is given by

$$Q_h = h_g A \Delta T$$

where

Q_h = calories transferred by convection
h_g = convection coefficient
A = area
ΔT = temperature difference between the surface of the convector and the fluid

The value of the convection coefficient depends upon the shape of the organism, the velocity of the fluid, and the physical properties of the fluid and the organism.

488

The organism and the physical environment

Forced convection can speed up another process, evaporation (LE). Evaporation depends upon a vapor pressure gradient that exists between the air and the object and boundary (surface) resistance. Every organism is coupled to air temperature by a boundary layer of air adhering to the surface. The resistance this layer offers to the absorbency and loss of heat depends upon the size, shape, texture, and orientation of the organism, and upon wind speed and the temperature gradient between surface of the organism and the air. In plants one has to add resistance by the leaf as influenced by the size and density of stomata, and the resistance of the cuticle to loss of water (see Gates, 1966).

When an animal possesses certain surface characteristics and assumes a certain posture, the various components discussed are not subject to further control by the animal, except to a limited extent through evaporation. For such a situation Newton's law of cooling applies:

$$Q = h_t A (T_s - T_e)$$

where

Q = calories of heat energy lost
h_t = proportionality factor dependent on the radiation and convection characteristics of the organism
A = surface area
T_s = effective surface temperature
T_e = effective environmental temperature.

The effective temperature of the surface is influenced by metabolic heat (M) produced by the organism. This includes basal heat production necessary to sustain life plus heat added by the physical act of digestion and by activity. In addition one has to consider gains and losses in energy in relation to changes in body temperature (s), or heat storage.

The relationship of all of these components can be expressed as

$$S + IR + M \pm s = IR_0 \pm C \pm LE \pm G$$

The net heat gain by solar radiation, infrared radiation, metabolism, and energy storage must equal the total heat lost by radiation, conduction, evaporation, and convection.

One aspect remains important. Heat produced continuously by plants and animals is lost passively to the environment. Such loss can take place only when the ambient or surrounding temperature is lower than the core body

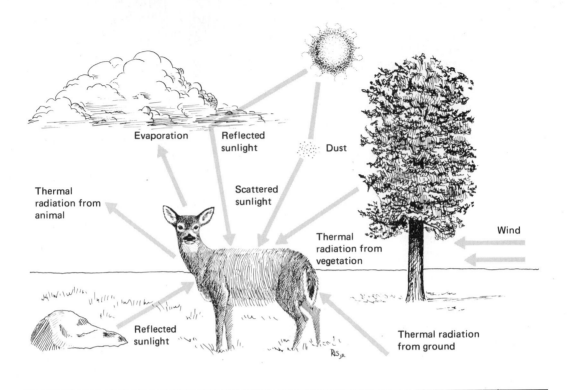

Evaporation   Reflected sunlight   Dust

Thermal radiation from animal

Scattered sunlight

Thermal radiation from vegetation

Wind

Reflected sunlight

Thermal radiation from ground

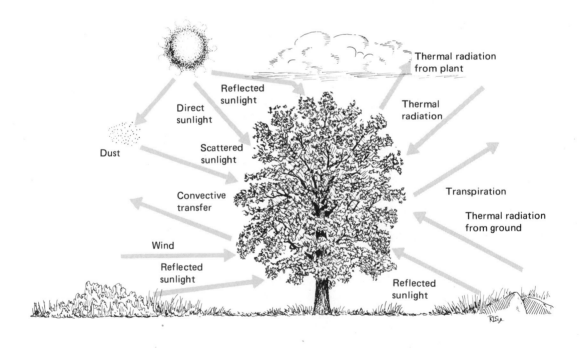

Reflected sunlight

Direct sunlight

Dust

Scattered sunlight

Convective transfer

Wind

Reflected sunlight

Thermal radiation from plant

Thermal radiation

Transpiration

Thermal radiation from ground

Reflected sunlight

temperature (Fig. 15-3). When the ambient temperature equals the core temperature, the route for passing heat off to the environment is lost. And when ambient temperature exceeds body temperature, the flow is reversed, and heat moves from the environment to the plant and animal.

Depending upon the values of the several components discussed, organisms are continually either emitting or absorbing radiation. Within a few minutes or even seconds an organism may be undergoing great temperature changes on the surface. Energy gained at one moment is lost at another. For example, Moen (1968b) measured the fluctuations on the surface of a ring-necked pheasant. Under a bright sun and free convection condition, the surface temperature of the bird was 45° C, 4° to 5° higher than the core temperature. Thus heat flow was from the outside surface to the interior of the animal. With a cloud cover and a slight breeze, the bird's surface temperature decreased nearly to air temperature (21° C), and heat flow was from the inside of the animal to the outside. These variations took place in a matter of seconds, while the air temperature remained the same. Extreme temperatures in the bird were buffered by the insulation of the feathers.

In animals evolutionary strategies (i.e., thermoregulatory adaptation) have evolved that exploit or decrease the effects by one or more of the components in the heat gain–heat loss equilibrium equation on the organism.

Plants and the energy environment

Although both plants and animals experience and exist within the same external energy environment, there are fundamental differences. Plants cannot move to escape adverse effects of heat or cold, and their metabolic heat is derived from absorbed solar radiation.

Except for the time when they sprout from seeds or vegetative cuttings, plants receive all of their energy from radiation in the environment and from convection. To balance the input, plants lose heat by radiation, convection, and evapotranspiration. They have some control over the latter by opening and closing the stomata and by changing the shape and position of the leaf. A notable example is the response of the evergreen leaves of rhododendron and laurel. In winter the leaves droop and curl.

FIGURE 15-1 [OPPOSITE TOP]
Energy exchange between an animal and its environment. (After Gates, 1962, Energy Exchange in the Biosphere, Harper & Row, New York.)

FIGURE 15-2 [OPPOSITE BOTTOM]
Energy exchange between a plant and its environment.

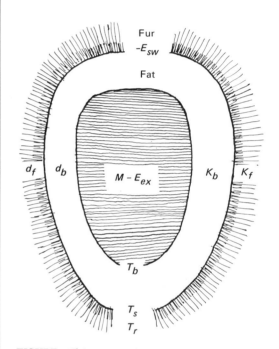

FIGURE 15-3
If one views an animal as a cylinder, the animal then consists of a body core, an insulating layer of fat bounded by skin, and an outer layer of fur or feathers that separates the skin from the radiating surface of the animal. These outer layers of fat and of feathers or fur enable the animal to balance its heat budget. (After Porter and Gates, 1969. © The Ecological Society of America.)

The colder the temperature, the tighter the curl. In fact it is possible to estimate winter temperature by the curl of the leaf.

The temperature of the leaf results from a combination of climatic variables such as wind speed, internal diffusion resistance, radiation absorbed, relative humidity, and air temperature. Wind speed as important. Winds as low as 1 mi/hour affect the temperature. Loss of water from the leaf increases if the air is relatively dry.

Heat affects the physiological processes of plants, particularly rate processes that depend upon temperature, such as photosynthesis and respiration. On a hot summer day a sunlit leaf can become too warm for photosynthesis. Following an early morning burst of activity, net photosynthesis ceases and respiration (Fig. 15-4) becomes dominant. Photosynthesis resumes later in the day after the leaf has cooled, but at this point light intensity has fallen to a suboptimal level. Because a plant's leaves face in different directions, photosynthesis of north-, south-, east- and west-facing leaves will differ throughout a summer day (see Gates, 1968). In arctic and alpine regions, plants have a low threshold for photosynthesis. Starting growth while snow is still on the ground, the plants possess basal leaves and prostrate forms that make use of the heat layer of air next to the soil surface. These organs start photosynthesis at air temperatures unfavorable for taller shoots (Scott and Billings, 1964).

In addition to possessing optimum temperatures for photosynthesis, plants also have an optimum temperature range for other aspects of their life cycle. To germinate, seeds of many plants in cold regions require chilling under moist conditions after a period of maturation. Fluctuation in temperatures is necessary for best germination of most seeds. Some plants require low temperatures during or shortly after germination. And the temperature necessary to stimulate flowering may be lower than the temperatures that favor flower development.

COLD INJURY

When temperatures drop below the minimum for growth, a plant becomes dormant even though respiration and photosynthesis may continue slowly. Low temperatures further affect the plant by precipitating the protein in leaves and tender twigs and by dehydrating the tissues. Intercellular ice draws water out of the cells, and rapid freezing causes ice to form within the protoplasts.

The ability to endure low temperature extremes varies among plants. They are not equally resistant at all stages of the life cycle. Flowers are more sensitive to low temperatures than fruits or leaves, and young leaves are more resistant than old ones. Trees may be more severely injured by frost than herbaceous plants.

Adaptations to endure low temperatures are primarily protoplasmic. Major exceptions are plants with waxy bloom or dense pubescence. Cold hardiness is acquired by a hardening process in which the protoplasm develops low structural viscosity. Free water content decreases and water-soluble sugars and proteins increase. Both of these changes lower the freezing point of tissues. Hardening of perennials of cool and cold climates takes place in the fall. In spring cold hardiness is lost with the renewal of activity. In summer tissues are killed by temperatures far higher than they endured in winter.

Animals and the energy environment

Animal response to the energy environment is more complex than that of plants. Not only do animals produce an often considerable amount of heat by their own metabolism, but they are able to move about to seek a favorable temperature regime, and in other ways, both behaviorally and physiologically maintain some control over their body temperature. The problems are the greatest for animals in terrestrial environments because air has a lower specific heat than water and absorbs less solar radiation. Because of incoming solar radiation, temperatures can be lethally high; and because of radiational loss to space, temperatures can be lethally low (see Chapter 2). Aquatic organisms, on the other hand, live in a more stable temperature environment that may range from −2° to 50° C, depending upon locality, but often they have a lower tolerance of temperature change.

SOME BASIC CONCEPTS

Body temperatures of animals are the result of heat gains and heat losses. Although all living things produce heat continuously, the magnitude and importance of this heat production depend on the particular evolutionary strategy of each species. Some animals exhibit high rates of thermal conductance and low rates of heat production (usually less than 2° C). Consequently these animals exploit sources of heat energy other than metabolism (solar radiation, reradiation, etc.). These animals are often called *poikilothermic* (changeable temperature) or more appropriately, *ectothermic* (heat from the outside). Within the range of temperatures that poikilotherms can tolerate, the rate of metabolism and thus oxygen consumption increases according to van't Hoff's rule. For every temperature rise of 10° C, the rate of oxygen consumption doubles. Oxygen consumption follows temperature, but its rate of change is exponential. Lacking any homeostatic devices, poikilotherms of terrestrial environments in particular must depend upon some behavioral control over body temperatures. Most aquatic poikilotherms generally encounter temperature fluctuations of lesser magnitude, and usually have more poorly developed behavioral and physiological thermoregulatory capabilities than terrestrial forms.

A small group of animals, largely birds and mammals, possess a sufficiently high rate of oxidative metabolism and a sufficiently low rate of thermal conductance that body temperature is a product of the animal's own oxidative metabolism (thermogenesis). Such animals are called *homeothermic* or *endothermic*. Their oxygen consumption decreases linearly with increasing temperature to some critical point where it is independent of environmental temperature. In other words, the lower the temperature, the more metabolic work the animal does to maintain a uniform temperature; but within a certain range of temperatures, the amount of energy needed to maintain body temperatures becomes minimal. This level of minimum work is the *zone of thermal neutrality*. The lower limit of this zone represents the critical air temperature, the lowest ambient temperature at which a bird or mammal can maintain its body temperature at basal metabolism at rest (BRM). At all ambient temperatures, homeotherms maintain their metabolic rate and body temperature by means of a closely regulated feedback system mediated by the hypothalamus (Fig. 15-5). The metabolic costs of such a system, however, are high. Homeotherms use some 80 to 90 percent of the oxidative energy to maintain thermal homeostasis.

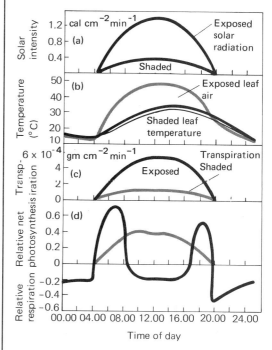

FIGURE 15-4

The effect of temperature on net photosynthesis is illustrated in these graphs: (a) solar radiation incident on an exposed and shaded leaf; (b) air temperature in relation to exposed and shaded leaf temperature; (c) transpiration rate of exposed and shaded leaf; (d) relative metabolic rates for an exposed and shaded leaf. If one follows the four graphs down by the hour of day, they show that the exposed leaf is inhibited photosynthetically when its leaf temperature is high. (From Gates, 1968b.)

493

THERMOREGULATION AND ACCLIMATION

How animals respond to the energy environment depends upon whether the animal is a poikilotherm or a homeotherm, whether it is living in a warm or cold environment, and whether it is in a terrestrial or aquatic situation.

Aquatic poikilotherms cannot maintain any significant difference between their body temperature and the environment. Their tissues are completely permeated by the circulatory system. Any heat produced in the muscles is transferred to the blood flowing through them and is carried away to the gills or skin, where it is lost to the water by convection and conduction. Fish are therefore ideal poikilotherms. Because of this close relationship between body temperature and environmental temperature, fish are readily victimized by any rapid changes in environmental temperatures. When temperature changes are not so rapid, the fish may escape to more favorable waters, for they can perceive minute differences in temperature— less than $1°$ C.

One can view the total range of temperature environment of a fish or any aquatic poikilotherm as a group of zones (Fig. 15-6). The central zone of *thermal tolerance* is the range of temperatures within which the fish is most at home. Within this range fish may have certain *preferred temperatures* that they will seek. The zone of thermal tolerance is bounded by an upper and lower zone of *thermal resistance*, temperature ranges within which the organism can survive for an indefinite period of time. The upper and lower bounds of thermal tolerance are marked by the upper and lower *incipient lethal temperature*. This is the temperature at which, when a fish is brought rapidly to it from a different temperature, will kill a stated fraction of the fish population (generally 50 percent) within an indefinitely prolonged exposure (Fry et al., 1946).

The incipient lethal temperature is not fixed. Its value is affected by the previous thermal history of the organism. Poikilothermic animals are, within limits, able to adjust or *acclimate* to higher and lower temperatures. If an organism lives at the higher end of the tolerable range, it acclimates itself sufficiently so that the lethal temperature will be somewhat higher; but its lower limits will be higher than if it were living within the cooler range of the tolerable environment. The reverse is true of animals living in cooler environments. Once acclimated to a given temperature, fish acclimate more readily to an increase than to a decrease in temperature. Acclimation to a higher temperature both increases the length of time an organism can survive an elevated temperature and raises the maximum temperature it can survive for a given period of time. However, a temperature ultimately will be reached that will be lethal to the organism. This is the *ultimate incipient lethal temperature* (for an excellent discussion of acclimation see Coutant, 1970).

Because acclimation takes place from one season to the next, a population is able to adjust to the prevailing temperatures each season. The brook trout exhibits a seasonal change in selected temperature from $12°$ C in summer and fall to $8°$ C from November to January (Sullivan and Fisher, 1953). On the other hand, experimental work indicates that when warm-water species of fish are acclimatized to a constant temperature level, the fish lose their ability to select their optimum temperature and show reduced precision in their reactions to high temperatures. Where such natural acclimatizations occur, the fish may become insensitive to high temperatures and high mortality may take place (see Norris, 1963).

Most aquatic cold-blooded animals remain active throughout periods of temperature extremes, but metabolism in general may be lowered greatly. Experiments have shown that large-mouth bass, acclimatized to $5°$ C, are slow to respond even to a slight current and are somewhat hampered in their maintenance or equilibrium. Food consumption, activity, and cruising speed all decreased with lowered temperatures (Fig. 15-7). Bass did not eat at all when the weekly mean temperature was $5°$ C; at $12°$ C, each bass fingerling consumed on the average 1.6 minnow/week, and at $20°$ C, an average of 4.1 minnows (Johnson and Charlton, 1960). Cruising speed of bass rose steadily from $5°$ C to a maximum at $25°$ C. Lowered activity at low temperatures, as well as lower metabolism, conserved oxygen, the concentration of which often drops dangerously low in ponds during winter. In fact the oxygen requirement of bass at $5°$ C is only about 10 percent of that at $29°$ C (Johnson and Charlton, 1960).

Increased metabolic activity increases oxygen consumption; and oxygen concentrations in heavily vegetated areas may be severely reduced

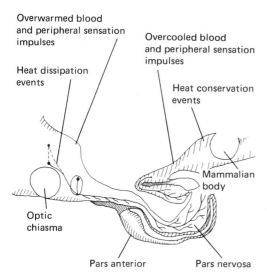

Anterior hypothalmus   Posterior hypothalmus

Overwarmed blood
and peripheral sensation
impulses

Overcooled blood
and peripheral sensation
impulses

Heat dissipation
events

Heat conservation
events

Mammalian
body

Optic
chiasma

Pars anterior   Pars nervosa

FIGURE 15-5
*Some regulatory events in the hypothalmus.
Shown are the probable physiological processes
associated with the control of temperature
regulation by the hypothalmus. (From G. E.
Folk, 1966,* Introduction to Environmental
Physiology, *Lea & Febiger, Philadelphia.)*

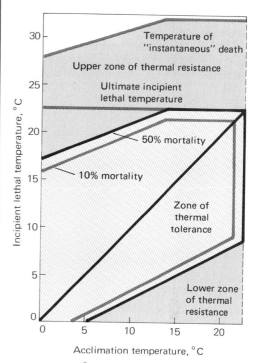

FIGURE 15-6
*Thermal tolerance of a hypothetical fish in
relation to thermal acclimation. (From C. C.
Coutant, 1970. © The Chemical Rubber Company,
1970. Used by permission of The Chemical Rubber
Company.)*

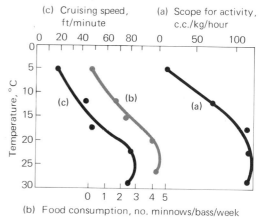

FIGURE 15-7
(a) *The scope of activity,* (b) *food consumption,
and* (c) *cruising speed of the large-mouthed
bass at several experimental temperatures. Scope
of activity is the difference between standard oxygen
uptake and active oxygen uptake. (After Johnson
and Charlton, 1960.)*

during the night by respiration of plants. When water becomes too warm (28° to 30° C), large-mouth bass either retreat to cooler water or remain in the shade of overhanging trees and refuse to eat. During this period of high body temperature, hemoglobin content increases in gills, the operculum pulse rate quickens, and the lactic-acid level, which increases the oxygen-combining power of blood, rises. By such physiological adaptability, the bass is able to cling to a precarious existence during periods of temperature stress (Denyes and Joseph, 1956).

Amphibians present a somewhat different situation. Permanently aquatic forms, such as many salamanders, maintain body temperatures in the same manner as fish. Their temperature control is primarily one of seeking preferred temperatures within their habitat. But for semiterrestrial frogs and salamanders the problem of temperature control is more complex. The avoidance of temperature extremes is probably the most important mechanism, but within this framework amphibians are able to exert considerable control over their body temperature, which does not, as is so often assumed, simply follow the air temperature (see also Brattstrom, 1963). By basking in the sun (heliothermism) frogs can raise their body temperature as much as 10° C above the ambient temperature. Because of associated evaporative water losses, such amphibians have to either be near water or sit partially submerged in it. Forms that live near water also use evaporative cooling through the skin to reduce body heat loads. By changing positions or locations, or by seeking a warmer or cooler substrate, amphibians are able to maintain body temperatures within a narrow range of variation (see Lillywhite, 1970) (Fig. 15-8).

Like fish, amphibians can acclimate to a temperature gradient. But among amphibians and reptiles the concept of the incipient lethal temperature is replaced by the *critical thermal maximum*. This is the temperature at which the animal's capacity to move becomes so reduced that it cannot escape from thermal conditions that will lead to its death (Cowles and Bogert, 1944). Like fish, if an amphibian has been acclimated to a relatively low temperature, its critical thermal maximum is lower than if it were acclimated to a higher temperature; and its speed of acclimation increases with increasing temperature.

Critical thermal maximum may actually represent two distinctly different critical temperatures. There may be a temperature at which an animal loses its capacity to escape lethal environmental temperatures, but if not exceeded, the animal survives. In addition there may be a body temperature that once achieved, although the animal recovers on cooling, renders the animal sterile. One need only ask which is more important, to survive extreme temperatures or to survive with reproductive ability intact.

The range of temperature requirements for amphibians varies through the life cycle. For the relatively cold-hardy wood frog, temperature limits that permit 50 percent survival of eggs through hatching is 6° to 24° C. Egg masses absorb and retain radiant energy so they have a higher temperature than the surrounding water. The preferred temperature range for tadpoles is 9° to 20° C. Temperatures above 24° C are lethal (Herreid and Kinney, 1967).

Relatively few reptiles are aquatic. Most are terrestrial and lack the buffering effects of water. Exposed to the widely fluctuating temperatures of the terrestrial environment, reptiles must possess more refined means of temperature regulation.

Although poikilothermic and ectothermic, reptiles exhibit little relationship between their core body temperature and the ambient temperature. By behavioral means they are able to maintain a relatively stable body temperature.

The *preferred body temperature* (PBT) is designated as the mean of those body temperature records taken during periods of minimal oscillations (Fig. 15-9). Another thermal characteristic often given by various authors for lizards is the *mean body temperature* (MBT). This temperature is represented by the average of all body temperatures recorded during the daylight hours. MBT represents a whole host of behavioral and physiological phenomena taking place throughout the day; thus care must be taken in interpreting MBT data. However, MBT often reflects elevational and latitudinal differences in the habitats occupied by reptilian species (Fig. 15-10). Because many reptiles bask in the sun to obtain sufficient heat energy, availability of light also influences MBT (Fig. 15-11). Reptiles utilize evaporative cooling by panting and by water loss through the skin to prevent body temperature from reaching the *critical thermal maximum* (CTM).

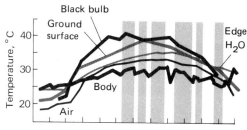

FIGURE 15-8

*The body temperature of a bullfrog measured
telemetrically. Dips in the black bulb temperature
indicate effects of cloud cover and/or convection.
Water temperature around the pond's edge varied
from one location to another by as much as
2 to 3°C. Thus while in shallow water a frog
may show a higher body temperature than that
recorded for the edge water. Note how
relatively uniform the bullfrog maintains its
temperature by moving in and out of water.
Shaded bars indicate periods out of
water. (From Lillywhite, 1970.)*

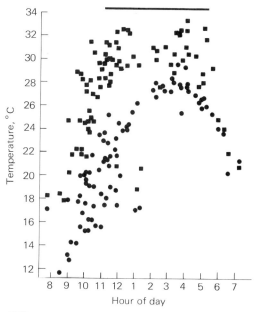

FIGURE 15-9

Body temperature records of Anolis carolinensis
*taken during spring at College Station, Texas.
The black line represents the period in which
preferred body temperature (PBT) was
achieved. (Courtesy of Donald R. Clark, Jr.,
and James C. Kroll.)*

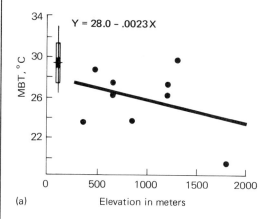

(a)

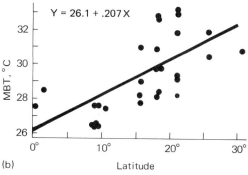

(b)

FIGURE 15-10

*The relationship of mean body temperature
(MBT) of anoles to elevation (a) and latitude
(b). (Data courtesy of Donald R. Clark, Jr.,
and James C. Kroll.)*

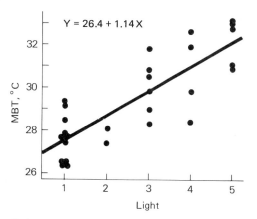

FIGURE 15-11

*Mean body temperatures (MBT) of various
species of tropical and temperate anoles in
relation to light intensity. The x axis represents
an artificial scale of light intensities, ranging
from dense tropical forest (1) to open
grassland (5). (Data courtesy Donald R.
Clark, Jr., and James C. Kroll.)*

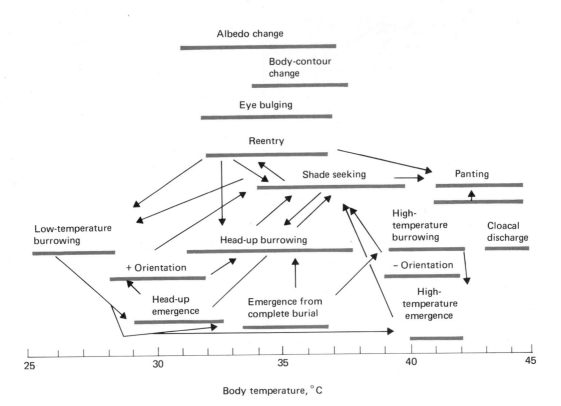

Thus reptiles possess some of the basic physiological mechanisms so characteristic of and so highly developed in the endotherms.

The reptile is able to regulate body temperature in a number of ways, the simplest of which is to shuttle back and forth between sun and shade. By basking or remaining in the open sun (heliothermism) the reptile is able to raise the core body temperature to the preferred level. At this point the animal moves back into the shade and remains there until the body temperature has dropped to some point, and then moves back out into the sun again. More elaborate behavior common to many lizards is proportional control. If the air temperature is lower than the preferred, the lizard can spread its ribs, flatten its body, and orient itself so that its body is at right angles to the sun. In this manner it gains the maximum amount of heat. But if the temperatures get too warm the lizard can appress its ribs and maintain the body parallel to the sun and decrease the surface area exposed to radiant energy. Other behavioral methods involve burrowing into the substrate, panting, and possibly changing color. Thus reptiles have at their command a variety of behavioral mechanisms useful for temperature regulation (Fig. 15-12).

In addition to behavior, reptiles possess some physiological mechanisms that permit them some degree of control over the maintenance of preferred body temperature. At least four families of lizards (Iguanidae, Gekkonidae, Varanidae, and Agamidae) can control the rate of change in body temperature by changing the rate of heartbeat and by varying the rate of metabolism. By accelerating the heartbeat they increase the metabolism and retard cooling. The heartbeat at a given temperature is higher during heating until the animal reaches the preferred temperature or above. Then it decreases. Though the lizard cannot stabilize its temperature in this manner, it can reduce daily oscillations.

Birds and mammals escape the constraints of the environment by becoming homeothermic and endothermic. Instead of depending upon the environment for heat, they produce it by metabolic oxidation. To do so they evolved a means of controlling heat exchange with the environment. Acclimation to changing temperatures by homeotherms depends upon their ability to regulate the gradient between body and air temperatures. They accomplish this mainly by changes in insulation, by the production of heat, by evaporative cooling, and by changing the body temperature.

Change in insulation is a major means by which homeotherms adapt to seasonal temperature changes. The degree of insulation is varied by changes in the thickness and type of fur, the degree of vascularization of the skin and limbs, and by posture and behavior. Insulation value of fur varies with thickness; in small mammals, there is a further correlation of fur thickness with body size (Scholander, Walters, Hock, and Irving, 1950). Fur of small animals is short and light; if it were heavy, the animals would have a difficult time moving about. Consequently in cold weather they must huddle together or remain in warm nests to conserve heat. Among mammals larger than a fox, there is no relationship between fur thickness and size. Arctic mammals have a thick underfur overlaid by a heavy fleece, a coat that provides excellent insulation. In summer the transfer of heat away from the body is increased when the heavy fur is shed. Arctic birds have longer feathers than birds of tthe temperate regions, and at the base of these feathers there is dense down as well (Irving, 1960). With numerous air spaces between them, feathers are excellent insulators; the bird can increase their effectiveness by erecting them. This increases the layer of warm air between the surface of the feathers and the surface of the skin. A heavy layer of fat, too, is an excellent insulator and is especially well developed in arctic marine mammals, such as the seal, walrus, and whale. Insulation can be controlled to some degree by posture and behavior. To conserve heat, the animals may curl up or wrap a furry tail about them. They may lose heat by stretching out and exposing thinly furred areas to the air.

Vascularization is another device among animals for conserving or getting rid of heat (Scholander, 1955). Many animals have extensive areas in the extremities where the arterial system joins the venous. The arteries carrying warm blood transfer the heat back to

FIGURE 15-12 [OPPOSITE]
Behavioral mechanisms involved in the regulation of body temperature by the horned lizard (Phrynosoma coronatum). (From Heath, 1965. Originally published by the University of California Press; reprinted by permission of The Regents of the University of California.)

499

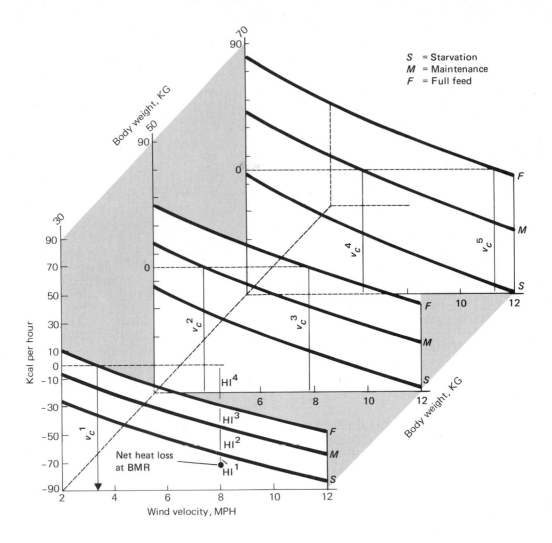

the returning venous blood, thus cooling extremities (Irving and Krogh, 1954). This is especially important among arctic animals. Seals, for example, have such a thick layer of fat and a pelage so dense that body heat can be dissipated only from the flippers, which are well supplied with sweat glands and are black in color. (Black bodies not only absorb all frequencies of radiation impinging upon them; they also emit all wavelengths back to the atmosphere.) This they accomplish by waving the flippers in the air and by swimming in cold water (Bartholomew and Wilke, 1956). Similar cooling is necessary among arctic animals to *prevent* excessive heat loss through the extremities, for legs and feet are exposed and the feet are in contact with the cold ground. If the extremities were too warm, the snow would melt and subsequently freeze; if they were too cold, the interior of the extremities would drop below the freezing point. Circulation in the

extremities is poor, which tends to cool the feet and check heat loss. Tissue temperature in arctic gulls declines rapidly along the tibia under the covering of feathers. The temperature change takes place there; the heat from the warm arterial blood is transferred to the cool venous blood returning from the foot, which may be close to 0° C (Irving, 1960). In addition, fat of low melting point is selectively deposited in those parts of the extremities subject to excessive cooling (Irving, 1960). The fat in the foot pads of arctic fox and caribou has a melting point of 0° C, about 30° C lower than the fat of the body core, yet the fat remains soft and flexible.

Within limits, warm-blooded animals can maintain their basal heat production by changing insulation. But with declining temperatures there is a point beyond which insulation is no longer effective and body heat must be maintained by increased metabolism. The temperature at which this takes place is called the *critical temperature*, and it varies greatly between tropical and arctic animals (Scholander, Hock, Walters, Johnson, and Irving, 1950). Tropical birds and mammals exposed to temperatures below 23.5° to 29.5° C increase their heat production. If the air temperature is lowered to 10° C, the tropical animal must triple its heat production; and if lowered to freezing, the animal is no longer able to produce heat as rapidly as it is being lost. Arctic small mammals, on the other hand, do not increase their heat production until the air temperature has fallen to −29° C. Large arctic mammals can sustain the coldest weather without heat beyond that produced by normal basal metabolism. Eskimo dogs and arctic foxes can sleep outdoors at temperatures of −40° C without stress. This is not due to any difference in metabolism itself but to effective insulation and cold acclimation. A number of mammalian species significantly increase their basal metabolism during cold acclimation without the intervention of shivering.

The production of metabolic heat by cold-acclimated birds and mammals depends upon an adequate high caloric diet (for review see Chaffee and Roberts, 1971). For example, the Alaskan redpoll, which commonly winters in areas where the environmental temperature may go as low as −60° C, increases its rate of food intake, selecting high-calorie food, and its digestive efficiency (W. S. Brooks, 1968). The role of diet in animals' ability to withstand cold is illustrated in studies of energy exchange of the white-tailed deer (Moen, 1968*a,b,c*). With heat production held constant at a full diet level, the lower limit of the thermoneutral range or critical temperature depends upon body size and surface area (Bergmann's rule). If the diet is at starvation level no body weight is sufficiently large to maintain a positive energy balance when the animal is standing in an open field under clear nocturnal skies with an air temperature of −30° C. On a maintenance diet a large deer of 70 kg reaches negative energy balance with a wind velocity of 6 mph; on a full diet it reaches a negative energy balance at 11 to 12 mph (Fig. 15-13).

When body temperature falls below the critical level the metabolism may be further increased by shivering. Among birds exposed to cold, shivering and muscular activity are primary sources of extra heat. In fact Arctic

FIGURE 15-13 [OPPOSITE]
The thermal balance (x axis) of deer exposed to wind velocities (y axis) from 2 to 12 mph. Body weights of 30, 50, and 70 kg are considered (z axis), and three dietary levels are used. Calculations are for a standing deer in an open snow field under clear skies at night, with an air temperature of —30 C. (From Moen, 1968c.)

and Temperate Zone birds weighing less than 300 to 400 mg must shiver when inactive at most cold temperatures normally encountered in their environment (G. C. West, 1965).

Mammals acclimated to cold temperatures can increase their heat production by non-shivering thermogenesis. This allows them to be active at lower temperatures than if they had to depend on shivering alone. Nonshivering thermogenesis and exercise are additive, so the two together can be important in maintaining body temperatures. Involved in nonshivering thermogenesis is brown fat, highly vascular brown adipose tissue found in the young of most species and in animals that hibernate (for detailed reviews see Smith and Horwitz, 1969; Chaffee and Roberts, 1971). Brown fat in-creases in mass when animals are chronically exposed to low temperatures. This fat is local-ized around the head, neck, thorax, and major blood vessels. Heat generated from the metabo-lism of this fat is transported to the heart and brain.

Physiologically it is more difficult for a homeotherm to adapt to high temperatures than to cold. Endogenously produced heat must be transferred to the atmosphere, and for the most part this involves an evaporative loss of water, largely through sweating and panting. Because birds possess a body temperature some 4° to 5° C higher than mammals they are more tolerant of intense heat. Birds do not sweat, and water loss through the skin is inhibited by the insulating covering of feathers. Body heat is lost largely by radiation, conduction, and convection. But when conditions demand it, birds can decrease their heat load by evapo-rative cooling through panting. Panting re-quires work that only adds more metabolic heat. Some groups of birds, particularly the goatsuckers, owls, pelicans, boobies, doves, and gallinaceous birds get arouund this by *gular fluttering* (see Fig. 15-14). This type of evapo-rative cooling uses less energy than the heavy breathing demanded by panting. Involved in gular fluttering are the gular area, the floor of the mouth, and parts of the esophagus. These highly vascular areas, in some species enlarged into a pouch, have little mass so they can be fluttered easily. By holding the bill open and fluttering the gular area, birds can evaporate water with little energy expenditure.

Another approach to heat problem by birds is *hyperthermia*. A rise in body temperature re-establishes the difference between the body and the environment, and thermal homeostasis is reestablished at a higher temperature. Hyper-thermia is not widespread among mammals but it is utilized by some of the African antelopes.

Among the artyodactyla, horns and antlers are richly vascular, and only incompletely shielded by hair. These function as thermal windows. For example, the horns of the goat possess a rich plexus of blood vessels that dilate in response to heat stress and constrict in the cold. At 30° C, goats can lose 3 percent of their basal heat by this avenue.

Just as animals' hair and fur act as an insula-tion against cold by keeping body heat in, so it acts to keep heat out. The heat is returned from the surface of the coat by reradiation and convection. A light breeze can carry away a large amount of heat. In addition animals that live in arid regions possess slick, glossy, light-colored pelage that reflects the wavelengths of sunlight.

Other animals of the arid country simply avoid the heat by adopting nocturnal habits and remaining underground or in the shade during the day. Some desert rodents that are active by day periodically seek burrows and passively lose heat through conduction by press-ing their bodies against the burrow walls. Some birds, such as the poorwill and certain hum-mingbirds, as well as bats, go into a daily torpor.

The large ears of such desert mammals as the kit fox and the jack rabbit may serve to reduce the need of water evaporation to regulate body heat (Fig. 15-15). The ears may function as efficient radiators to the cooler desert sky, which on clear days may have a radiation tem-perature of 25° C below that of the animal (Schmidt-Nielsen, 1964). By seeking shade where the ground temperatures are low and solar radiation is screened out, by sitting in depressions, where radiation from hot ground surface is obstructed, the jack rabbit through the two large ears (400 cm^2) could radiate 5 kcal/day. This is equal to one-third of the metabolic heat produced in a 3-kg rabbit. Such a radiation loss alone may be sufficient to take care of the necessary heat loss without much loss of water.

DORMANCY

One way to avoid both heat and cold is to go into a state of dormancy during a period of environmental stress. Many species of insects, as well as certain crustaceans, mites, and snails, enter *diapause*, a state of dormancy and arrested growth. Eggs, embryonic larvae, or pupal stages may be involved. In addition to tiding the animal over unfavorable periods, diapause also synchronizes the life cycle of a species with the weather and ensures that active stages will coincide with the climatic conditions and food supply that favor rapid development and high survival.

Some poikilothermic animals become dormant during periods of temperature extremes. Amphibians and turtles bury themselves in the mud of pond bottoms; snakes seek burrows and dens in rocky hill sides. There they remain in a state of suspended animation until the temperature warms again. Winter dormancy is called *hibernation*; summer dormancy is *estivation*.

Hibernation and estivation are not confined to poikilothermic animals. The phenomena also occur among a few homeothermic animals, particularly bats, ground squirrels, woodchucks, and jumping mice. Like dormancy in "cold-blooded" animals, hibernation in homeothermic animals is characterized by a reduction in the general metabolism to a degree never found in the deepest everyday sleep. But there is a difference. When reptiles and amphibians are exposed to cold, the animal cools because it has no way to stay warm. As the poikilotherm's system becomes cooler, the heart rate and metabolic rate decline. But when a mammal enters hibernation, the heart rate and metabolic rate decline, and *then* the body temperature drops.

The onset of dormancy in warm-blooded animals may take place with one decline in temperature, as in the hamster and pocket mouse, or it may come in a series of steps, as it does in the groundhog and ground squirrel. The temperature of the California ground squirrel drops a number of degrees, remains at this point for a while, then rises to the normal body temperature again (Strumwasser, 1960). The next day, when the animal again enters hiberation, its body temperature drops even lower. This continues daily, until the body temperature is just a few degrees above that of the environment. This apparently conditions the body and brain to lower temperatures. After a few such fitful starts at complete dormancy, the animal finally curls up to sleep in its retreat. In this state of suspended animation, only the vital life of the animal continues, maintained by shallow breathing, slow circulation, and a

503

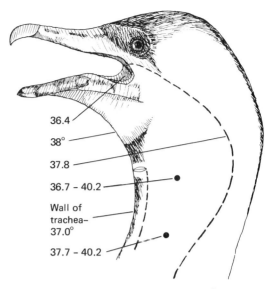

Body temperature–41.5°C

FIGURE 15-14
Representative evaporative surface temperatures from two 6.5-week-old double-crested cormorants (Phalacrocorax auritus). *Ambient temperature was held at body temperature* (41.5°C) *during measurements. The lower temperatures in the buccal and pharyngeal regions, tracheal walls, and a portion of the esophagus moved during gular fluttering indicate that these areas are sites of evaporation. (After Lasiewski and Snyder, 1969.)*

FIGURE 15-15
The jack rabbit, Lepus californicus, *lives in an extreme desert environment. On hot days the rabbit must rest in deep shade in a position where it can cool radiatively to the cold north sky.*

continual digestive absorption of stored fat. The reduced metabolism is characterized by great differences in body temperature, pulse rate, and breathing. The breathing rate, rapid in small mammals, is reduced to less than one breath a minute, and these breaths usually occur in a series of two or three gasps with long intervals between. The body temperature in the ground squirrel drops from 18.5° to 4.2° C. The heart rate is reduced to two to three beats a minute, although the blood pressure remains relatively high. The optimum environmental temperature for hibernators is 4.5° C (40° F); the body temperature passively follows the fluctuation in environmental temperatures between 3.3° to 12.8° C. If the environmental temperature reaches the freezing point of water, the metabolic rate of the hibernating animal increases. This increase may be able to keep the body temperature above freezing without the animal awakening. Mammalian hibernators, however, do not remain in a state of complete inertia all during the period of dormancy. They awake from time to time, eat a little, void, and return to their deep sleep.

Arousal from hibernation is explosive and certainly the most dramatic aspect of the hibernation cycle (Fig. 15-16). As the animal starts to come out of the sleep, its body temperature rises rapidly, perhaps from 4° to 17.5° C in an hour and a half (Mayer, 1960). In the case of the ground squirrel, when the anterior part of the body has reached about 36.5° C, the temperature of the hind parts rises rapidly, due to dilatation of the blood vessels. The animal's hair is erect, and its body shakes. At around 16° C, the animal tries to right itself, and at 17.5° C the shivering stops and the squirrel moves its tail. At 24° C the animal opens its eyes and suddenly sits up. At this time the temperature rises rapidly again. When the body temperature reaches 24° to 25.4° C, about 3 hours from the time of initial arousal, the ground squirrel is sitting up and flicking its tail. The animal is warm and active again.

The black bear (Fig. 15-17) and skunk are not true hibernators; in both, neither body temperature nor pulse is lowered significantly, and breathing is more frequent. During warm spells both will leave their dens and move about. Rarely are adult skunks inactive for over a month nor the young for more than 4 months. During these deep sleeps the animals may lose from 15 to 40 percent of body weight (Hock, 1960).

Estivation is common among some desert rodents. Not only does this prolong the period during which the animal can live on energy reserves through times of severe drought, but it also saves a considerable amount of water because the reduced ventilation of the lungs lowers the pulmonary evaporation of water. In addition, the lower body temperature of the animal reduces the amount of water which is needed to saturate the expired air (Schmidt-Nielsen, 1964).

Moisture

On a hot summer afternoon, a catbird sits perched in the shade of a shrub, its wings half-dropped, its bill open. At the same time the leaves of the shrub hang drooped and slightly curled on the stem. Both are responding to the stress of heat, which in turn is affecting the water economy of the organism. In a very practical sense it is difficult to separate the water budget from the heat budget, for in maintaining their temperature, animals utilize water to decrease their thermal load. But physiologically both plants and animals have evolved ways to maintain water and solute balance.

For plants the methods involve differential diffusion of ions across cell membranes into the roots, the ability to utilize by means of fine root hairs all but the hygroscopic water in the soil, and the ability to close the stomata and to curl leaves to reduce evapotranspiration.

Among animals the adaptations are more complex. A more or less universal mechanism is the excretory system, designed to rid the body of water and solute wastes or to conserve them. Maintenance of a water balance between an organism and its environment is not a serious problem in aquatic environments except where water levels fluctuate and where waters are salty. Aquatic organisms maintain their water balance by regulating osmotic pressure. Fresh-water organisms live in an environment where the surrounding water has a lower salt concentration than their bodies. The problem is to rid the body of excess water taken in and through the permeable membranes. Protozoans dispose of excess water through the contractile vacuole. Other organisms have either flame cells, nephridia, or kidneys. Fresh-water fish maintain their salt concentration by salt absorption through special cells in the gills. Marine and desert forms of plants and animals have other problems.

Beyond this animals have other strategies. Amphibians have skin that is permeable to water, and through it they take up water by osmosis. Water produced by the kidney flows to the bladder, where it is stored. If circumstances require it, aquatic amphibians cease urine production and conserve water and solutes for metabolic purposes by increasing reabsorption through the bladder.

Terrestrial animals have three means to gain water and solutes: through drinking, through food, and through production of metabolic water from food. They can lose water and solutes through urine, feces, evaporation over the skin, and respiration. If the animal is not to dehydrate, the input of water must equal the losses.

Birds and reptiles have three pathways through which they can physiologically control water losses: the gut, the kidney, and in some a salt gland. The kidneys of birds and reptiles are similar. Both possess a cloaca that modifies urinary volume and composition after it leaves the kidney. It converts the urine to a semisolid paste, consisting almost entirely of uric acid (but this is questioned in birds). Mammals have kidneys capable of lowering water losses by producing higher urinary osmotic and inorganic ion concentration. Although they lack salt glands and cloacas, many species of mammals do possess sweat glands. Among terrestrial animals a major source of water loss is through the skin and respiratory system. Birds and mammals have relatively high losses through this avenue and mammals through sweating and urine.

For the most part moisture becomes a problem in only two environments, the marine and saltwater environment and the desert. The former are physiological deserts, in which the concentration of salts outside the body of the organism can osmotically dehydrate it. In the latter an absolute lack of moisture exists. Interestingly, many organisms of the different environments use the same strategy to overcome the problem.

AQUATIC ENVIRONMENTS

In the marine and brackish environment the problem is one of inhibiting loss of water through the body wall by osmosis and preventing an accumulation of salts in the system.

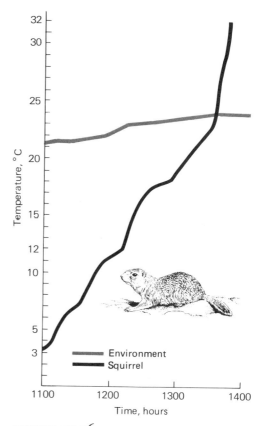

FIGURE 15-16

A graph showing a rapid rise in body temperature of the arctic ground squirrel upon arousal from hibernation. Upon the second rise, at 24°C, the animal opens its eyes and sits up. (After Mayer, 1960.)

FIGURE 15-17

A black bear that has just been aroused from its winter's sleep. The bear is not a true hibernator. (Photo courtesy Pennsylvania Game Commission.)

Algae and invertebrates get around the problem by possessing body fluids that are iso-osmotic with seawater. In a way this adds to the problem of marine vertebrates because they cannot obtain fresh water through their food. It is as salty as the water around them and the water they ingest with their food. Marine teleost fish absorb the water with the salt into the gut. The divalent ions are excreted by the kidneys and passed off as a partially crystalline paste of magnesium and calcium. The monovalent ions are excreted by means of active transport through special cells in the gills. Because active transport involves the movement of salts against a concentration gradient, energy is expended in the process. Sharks and other elasmobranch fish retain urea to maintain a slightly higher concentration of salt in the body than in the surrounding seawater. Pelagic sea birds are able to utilize seawater, for they possess special salt-secreting glands located on the surface of the cranium (Schmidt-Nielsen, 1960). Gulls, petrels, and other sea birds excrete from these glands fluids in excess of 5 percent salt. Petrels and other tube-nosed swimmers forcibly eject the fluid through their nostrils; in other species the fluid drips out of the internal or external nares. A gull given one-tenth of its weight in seawater excreted 90 percent of the salt through the salt gland within 3 hours.

Among the marine mammals the kidney is the main route for the elimination of salt. Porpoises have highly developed renal capacities in the kidney to eliminate salt loads rapidly (Malvin and Rayner, 1968). In marine mammals the urine is hyperosmotic to blood and seawater; the physiology is poorly understood.

Vertebrates in the Arctic and Antarctic have special problems. As seawater freezes it becomes colder and more salty (see Chapter 19). The only alternative for most organisms is to increase the solute concentration in the body fluids to lower the body temperature. Some species of fish in the Antarctic possess in their blood a glycoprotein antifreeze substance that enables the animals to exist in temperatures below the freezing point of blood.

Among the reptiles the diamondback terrapin, a common inhabitant of the salt marsh, lives in an environment in which the salinity is variable. It retains its osmotic pressure when the water is dilute, yet it possesses the ability to accumulate substantial amounts of urea in the blood when it finds itself in water more concentrated than 50 percent seawater. This it can achieve by switching on and off the functioning of the salt glands located in the head.

Although plants of the salt marsh are surrounded by an abundance of water, they grow in a physiologically dry substrate because the higher osmotic pressure outside their roots limits the water they can absorb. They compensate for an increase in sodium and chlorine uptake with a corresponding dilution of internal solutions with water stored in the tissues. In addition the plants exhibit a high internal osmotic pressure many times that of fresh-water and terrestrial plants, possess salt-secreting glands and heavy cutin on the leaves, and are succulent (Chapman, 1960).

ARID ENVIRONMENTS

In the other region of moisture stress, the desert, plants and animals are faced with the same problem of how to control water losses, this time by evaporation, and still get rid of solute wastes. They have several strategies available to them.

Plants can handle the moisture problem in one of several ways. As discussed in Chapter 17, they can adopt an annual life cycle by which the population can live through the dry periods as dormant seeds, ready to sprout quickly when the rains come. By completing their life cycle in a very short period before the moisture is completely gone, they can go into seed dormancy again. Or as the plants of the salt marsh do, they can adopt succulence as a method of conserving water. Notable are the cacti. Because their root systems are shallow, they are able to draw up considerable amounts of water during periods of rain. At that time the plants swell rapidly as they store water in the enlarged vacuoles of the cells. As they draw on this store of water during the dry season, they gradually shrivel. To further conserve water they open the stomata at night and take in and fix carbon dioxide for use during the day, when the stomata are closed.

The plants can absorb atmospheric moisture in the form of dew and fog through the cuticle and accumulate it in the shoots. This is relatively common among desert plants, particularly in the hygroscopic shoots of salt-secreting shrubs (but it is not confined to this type of plant).

Many desert shrubs are highly tolerant of

high salt concentrations in the soil. Among these, some exude salt or excrete high salt concentrations. Others grow and reproduce only during periods when the moisture content of the soil is high, and thereby avoid salt damage.

Plants of arid areas also meet moisture shortages by means of various leaf adaptations. In some, the leaf size is small, the cell walls are thickened, the vascular system is denser, and the palisade tissue is more highly developed than the spongy mesophyll. This increases the ratio between the internal exposed cellular surface area of the leaf and the external surface. If the plants still do not obtain enough water, they simply shed their leaves to reduce transpiration. Possessing small leaves, such plants can increase productive capacity by carrying out photosynthesis in the stem. Others combine the development of an extensive root system that is able either to penetrate deeply or to cover a large shallow surface area to secure the maximum amount of water when moisture is present.

Some desert plants are capable of selective ion transport. This allows the plant to absorb essential nutrients from the soil and help maintain internal osmotic pressures higher than those of the surroundings. Ion transport, like the elimination of salt in marine animals, involves active transport against a concentration and electrical potential gradient.

Animals in arid regions possess similar strategies. One is to adopt an annual life style and go either into estivation or some other stage of dormancy. For 8 or 9 months the spadefoot toad estivates in an underground cell lined with a gelatinous substance that reduces evaporative losses through the skin. It appears when rainfall saturates the earth, moves to the nearest puddle, mates, and lays eggs. Young tadpoles hatch in a day or two. They rapidly mature and metamorphose into functioning adults who are capable of digging their own retreats in which to estivate until the next rainy period. A flatworm, *Phagocytes vernalis*, which occupies ponds that dry up during the summer, encysts. As the water warms, this flatworm reacts by detaching small pieces from the posterior end of its body, until the entire animal is reduced to a number of fragments. Each fragment rounds up and secretes a layer of slime, which hardens into a cyst highly resistant to drying. These cysts remain in the debris of the pond bottom until the ponds fill again, when they

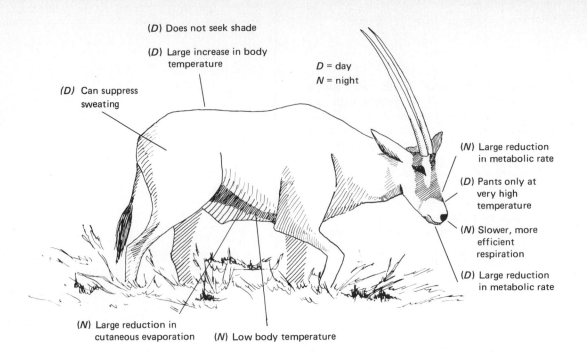

(D) Does not seek shade

(D) Large increase in body temperature

(D) Can suppress sweating

D = day
N = night

(N) Large reduction in metabolic rate

(D) Pants only at very high temperature

(N) Slower, more efficient respiration

(D) Large reduction in metabolic rate

(N) Large reduction in cutaneous evaporation

(N) Low body temperature

hatch. Other aquatic or semiaquatic organisms retreat deep into the soil until they reach groundwater level. Still others spend the dry season in estivation. Many insects undergo diapause, just as they do when confronted with unfavorable temperature.

The desert also contains many animals that are active the year round, all of which have evolved ways of circumventing the lack of water by physiological adaptations or by activity habits. The kangaroo rat, which seals its burrow by day and thereby keeps its chamber moist, lives all year without drinking water. It shares this trait with its ecological counterparts, the jerboas and gerbils of Africa and the Middle East and the marsupial kangaroo mice and pitchi-pitchi of Australia. These animals feed on dry seeds and dry plant material even when succulent green plants are available. The kangaroo rat obtains water from its own metabolic processes and from hygroscopic water in its food. To conserve water the animal remains in its burrow by day. It possesses no sweat glands; its urine is highly concentrated (24 percent urea and salt, twice the concentration of seawater) and its feces dry.

Large desert mammals, such as the camel, can use water effectively for evaporative cooling through the skin and respiratory system because their low surface area to body-size ratio and lower internal heat production result in slower accumulation of heat. The camel not only excretes highly concentrated urine but can withstand dehydration up to 25 percent of body

weight, and it loses water from body tissues rather than from the blood (Schmidt-Nielsen, 1959). The camel's body temperature is labile, dropping to 33.8° C overnight and rising to 40.6° C by day, at which point the animal begins to sweat. The camel accumulates its fat in the hump rather than over the body. This speeds heat flow away from the body, and its thick coat prevents the flow of heat inward to the body.

Perhaps a more extreme adaptation to aridity exists among some of the African antelope. Outstanding is the oryx (Fig. 15-18). Many African ungulates migrate to escape the heat and dryness, but the oryx remains. Its water requirements are low because it stores heat in its body during the day. This causes a substantial rise in body temperature (hyperthermia) (C. R. Taylor, 1969). The oryx further reduces daytime evaporative losses by suppressing sweating. It pants only at very high temperatures. It reduces its metabolic rate and conserves water by reducing the rate of internal production of calories, thus reducing the amount of evaporative cooling. These are daytime adaptations. By night the oryx reduces nonsweating cutaneous evaporation by 60 percent and reduces its metabolic rate by 60 percent. Its respiratory rates are proportional to its respiratory efficiency and inversely proportional to body temperature. A cool animal breathes more slowly than a warm one, thus using a greater proportion of inspired oxygen. With a lowered nighttime body temperature, the saturation level for water vapor in the exhaled air is lower. The animal normally does not drink water; it exists on metabolic water and by feeding on grasses and shrubs, many of them succulent. In fact the oryx can obtain all the water it needs by eating food containing an average of 30 percent water.

Some desert birds, like marine birds, utilize a salt gland to help maintain a water balance. Otherwise, both marine and desert birds and those of the salt marsh have the same basic adaptations for the conservation of water and the elimination of monovalent ions. The black-throated sparrow (*Amphispiza bilineata*, of the deserts of western North America) is not only restricted to arid regions and commonly occurs far from water in some of the most extreme desert habitat, but it also feeds largely on seeds. If the bird has some access to green vegetation it does not need to drink water (Smyth and Bartholomew, 1966). When it feeds largely on seeds, and water is available, it will drink up to 30 percent of its body weight per day. But if water is not available, the sparrow can survive indefinitely without drinking by reducing the water content of its excreta from about 81 percent to 57 percent. At the same time it is able to concentrate Cl⁻ in the urine. Among birds with similar adaptations are the budgerigar and zebra finch of Australia and the gray-backed finch lark of Africa (see Dawson and Bartholomew, 1968; Serventy, 1971).

Climate space

In order to survive, all animals (and for that matter plants) must live within a given set of climatic values, or a climate space (Porter and Gates, 1969). Climate space is a four-dimensional space in which the animal can endure. It is made up of radiation, air temperature, wind, and humidity, all acting simultaneously. If one knows the physiological characteristics and limits of an animal under a given set of conditions (body fat, insulation, body temperature, radiant surface, rate of water loss, and other properties), one can predict the climate space an animal can occupy (Fig. 15-19).

The climate space for the jack rabbit, *Lepus californicus*, of western North America is shown in Fig. 15-19. Except when moving from place to place, the jack rabbit cannot tolerate full sun when the air temperature is over 20° C. Although it can endure nighttime temperatures

FIGURE 15-18 [OPPOSITE]

The desert-dwelling oryx is well adapted to living in a hot and arid climate on limited water. The physiological adaptations are shown above. See text for details. (Adapted from M. S. Gordon, 1972, Animal Physiology, 2nd ed. With permission of Macmillan Publishing Co., Inc. Copyright © 1972 by Malcolm S. Gordon. Based on data from C. R. Taylor, 1969.)

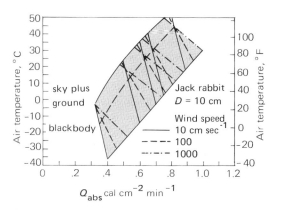

FIGURE 15-19

The climate space of the jack rabbit, showing relations between air temperature, radiation absorbed, and wind speed for constant body and radiant surface temperatures. A body diameter of 10 cm is assumed. Conditions vary with wind speed, given at 10, 100, and 1000 cm/sec. The left boundary of the climate represents the conditions the jack rabbit encounters in full sun over a range of air temperatures; the right boundary represents conditions of radiation encountered when sunlight is completely absorbed. The lines cutting across the climate space represent temperature conditions at the three different wind speeds and at five different temperature regimes. For example, in still air the jack rabbit can not be in full sun at air temperatures above 20°C except during transient conditions. During very hot days with temperatures above 40°C, the jack rabbit must be in deep shade. The rabbit may endure temperatures down to —15°C with considerable cover. (After Porter and Gates, 1969. © The Ecological Society of America.)

of 40° to 50° C, it cannot endure very hot days (temperature over 40° C) unless it remains in the deep shade and radiates body heat to the sky. With considerable cover the animal can endure temperatures down to −15° C. On a sunny day in winter it can withstand temperatures as low as −35° C.

Summary

Life exists in an energy or thermal environment. Plants and animals are continually receiving and losing heat to the environment. Heat gains come from solar radiation, diffuse and reflected radiation, and infrared thermal radiation. Heat losses result from infrared radiation from the organism, from conduction, convection, and evaporation.

Response to energy environment differs between plants and animals. Plants receive their energy from the environment. Heat affects the physiological processes of plants, especially photosynthesis and respiration. High temperatures inhibit photosynthesis and various phases of a plant's life cycle have their optimum range of largely protoplasmic temperature. Plants possess certain adaptations to low temperature, and other adaptations to heat, such as reduced transpiration and reduced size of leaves.

Animal responses are more complex. Although animals produce heat continuously, this heat has no effect on body temperature. Body temperatures of animals are maintained either from an external heat source or from metabolic regulation. Animals that depend upon environmental heat are poikilothermic and ectothermic. Animals that are heated metabolically are homeothermic and endothermic. The former gain most of their heat from the environment and regulate body temperature behaviorally by moving in and out of areas of different temperatures. The latter maintain body temperature by means of a closely regulated feedback system, mediated by the hypothalamus.

Both types respond to temperature changes by acclimation. Within limits, poikilotherms adjust or acclimate to higher or lower temperature either behaviorally (by avoiding temperature extremes), or seasonally (by adjusting tolerances to given temperature ranges, by basking in the sun to raise body temperature, and, in some instances, by evaporative cooling). Some reptiles change their body temperature by changing the rate of heart beat and metabolism. Acclimation in homeotherms is accomplished by changes in body insulation, by vascularization, evaporative heat loss, and nonshivering thermogenesis. Homeotherms find it more difficult to adapt to warm temperatures than to cold.

One method of evading periods of stress utilized by both poikilotherms and homeotherms is undergoing a period of dormancy. Animals going into dormancy in periods of heat and dryness are said to estivate. Those who become dormant in winter hibernate. Both involve a sharp reduction in physiological functions.

Plants and animals also face moisture problems, most acute in saline and arid environments. Similar strategies are employed in each. A number of animals secrete excess salts through salt glands, and thus conserve body moisture. Others concentrate their urine, use metabolic water, and reduce daytime evaporative losses by reducing the amount of sweating. Some plants and animals of arid regions possess annual life cycles, surviving dry periods as seeds, eggs, or cysts. Plants in both saline and arid environments may possess certain adaptations such as small leaves, waxy blooms, and succulence. In one way or another both plants and animals exhibit adaptations to handle variations in the thermal and moisture conditions of the environment.

Egocentric behavior: response to environment

A squirrel hurrying up a tree trunk and out across the limbs is a part of the forest in which it lives. It occupies a definite niche in the community; it is a member and a reproductive unit of a population of others of its own kind; and it contributes in its own small way to the gene pool of the population. The survival or nonsurvival of its young, indeed its own survival and ability to produce young over a period of several years, may influence in some small way the long-term evolution of the species. The ability of an animal to survive and leave offspring depends a good deal on how well the animal responds to its environment, to others of its own kind, to the whole world about it. How the squirrel reacts or attempts to adapt to sudden changes in its environment is called its *behavior*.

The behavior of an animal is directed toward itself, toward others of its own kind, and perhaps toward other species as well. The former is egocentric behavior; the latter is social behavior, already discussed (Chapter 13). *Egocentric behavior* includes activities that relate to the individual animal's welfare, such as maintenance, feeding, shelter seeking, and escape and defense behavior.

Maintenance behavior

Anyone who has ever watched a cat, even for a few hours, is well aware that the animal spends considerable time caring for its body. It not only licks, scratches, and bites its fur, but it also meticulously washes its face with pre-licked paws, especially after eating or rising from sleep. Similar maintenance activities are widespread among animals and include such behavior as bathing, dusting, grooming and preening, head scratching, sunning, and anting.

Body care is most conspicuous among birds and mammals, probably because they are the most easily observed, especially the birds. Grooming is an activity frequently indulged in, but descriptions of this behavior for our most common birds and mammals are very meager. Grooming in birds consists mainly of bill wiping, feather settling, scratching, and preening, in order of decreasing frequency (Iersel and Bol, 1958; Rowell, 1961). Common preening movements involve nibbling of the feathers and drawing the full length of the feather through the bill. Preening often is done with oil from the uropygial gland. Even here different groups of birds distribute oil over their plumage in different ways (Fickens, unpub.). Songbirds (oscines) take oil from the gland with the tip of the bill, then scratch, first touching the tip of the bill with the claw of the second toe and then rubbing the claw over the top of the head. Ducks (Anatids) bite the oil gland, press the oil into the feathers around the gland, and then rotate the head against these feathers.

Grooming activity frequently follows feeding, drinking, bathing, and other activities. Variations in the length and composition of grooming movements usually are associated with the intensity and type of the stimulus and available time. The longest periods of grooming contain the greatest variety of action. These usually are undisturbed periods following bathing, for when the bird is wet, the stimulus to groom is at its maximum. Preening movements in chaffinches often are absent at the beginning of grooming and reach their peak about a third of the way through the grooming activity (Rowell, 1961).

Grooming and preening are such common activities that it is not unusual for these behavior patterns to appear in other situations. One is the so-called displacement preening, which appears when two primary conflicting tendencies are opposing and equal (see Displacement, Chapter 13). Preening movements also have become incorporated in the courtship display of herons and ducks.

Grooming and comfort movements in mammals have received little attention. They are perhaps less involved, since the fur of mammals does not require the extensive care required by the feathers of birds. Rubbing, scratching, shaking, biting the fur, and licking movements are common to most mammals. Dogs, while lying down, spend some time licking the soles of their paws and the skin between the toes. Cows, horses, and dogs roll in the grass; buffaloes and pigs wallow either in wet or dry mud holes, a movement analogous to dust bathing in birds.

Comfort movements in wild moose have been described by Geist (1963). When they rise from their beds, moose often stretch in several different ways. When stretching the body the moose draws in its belly, arches its back downward, and swings its head up and

down (Fig. 16-1). These animals indulge in body shaking after exposure to water. The moose stands with its hind legs slightly spread, its head and neck held parallel to the ground to form a continuous line with the body, and raises the hair on the neck, withers, and rump. The shaking movement then begins at the head and spreads back. Rubbing new, growing antlers on the side of the hind leg is a common comfort movement in bulls. Moose scratch themselves with a raised hind leg on the nose, chin, cheeks, ears, and neck. The majority of these comfort movements, which vary with sex and season, take place mainly at the beginning or end of an activity period.

Sunbathing is common to both birds and mammals. During winter and spring, gray squirrels seek sunny areas where they can sun themselves. The squirrel will lie atop a limb and stretch outward from the trunk (Bakken, 1959). Among birds, two types of sunbathing are recognized (Hauser, 1957). One is voluntary, or normal, sunbathing, accompanied by preening, shaking, scratching, and the repeated resumption of the sunbathing posture. The second is compulsory sunning, exhibited when a bird is suddenly and apparently unexpectedly exposed to direct sunlight under extreme conditions of heat and humidity. This movement may be motivated by a sudden warming of the bird's immediate environment (Lanyon, 1958). Birds may assume a variety of positions in sunbathing (Hauser, 1957), but all are characterized by some form of feather fluffing, wing dropping, tail spreading, and often by panting.

An unusual form of body maintenance among birds is anting, the application of foreign substances to the plumage and possibly to the skin (see Simmons, 1957a; Whitaker, 1956). Anting may be active, as when the bird applies crushed ants to the underside of the wings and tail with the bill; or it may be passive, as when the bird postures among the thronging ants and allows them to crawl in the plumage. If ants are unavailable, then some substitute material, such as fruits, foliage, millipedes, beetles, grasshoppers, and even burned matches are used.

Numerous reasons have been advanced to explain anting, including the repulsion of ectoparasites by ant predation and acids, the attraction of birds to the odor of ants, and the conditioning of the plumage. Recently there has come to light a three-volume work on feather mites by a Russian parasitologist, V. B. Dubinin (Kelso and Nice, 1963), in which he describes some aspects and possible functions of anting. Birds typically are parasitized by feather mites (Analgesoidea), which lack mouthparts strong enough to chew feather substances. Apparently they feed instead on feather lipids, which exist in addition to those of the preen gland. Dubinin discovered that anting, which releases formic acid from the ants, may destroy the mites. He suggests that anting is performed to rid the bird of parasites and to treat or "nurse" the tail and flight feathers. Kelso and Nice (1963) suggest that "anting is an instinctive action present in many birds, perhaps aimed at the defense against feather mites. It appears to be 'triggered' by the acid or burning taste of ants and other substances and apparently may be performed *in vacuo*, i.e., in the absence of mite infestation."

Shelter seeking and habitat selection

All organisms, from the lowly paramecium to man, have one behavioral trait in common—*shelter seeking*. All seek optimum environmental conditions and avoid dangerous and injurious ones. Desert animals seek the shade of rock crevices or burrows in the sand during the heat of day. Deer and birds seek the cool of forest shade on a hot summer day. A bird escaping from a hawk darts into the thick underbrush. Shelter seeking can be divided into two broad groups, one in which the individual seeks a favorable situation for itself, and the other in which the individual seeks shelter from sudden danger.

Associated with shelter-seeking behavior is cover. Based on the behavior of the animal, Leopold (1933) recognized five different types (Table 16-1). Some time spent afield will reveal how well the behavior of animals is adjusted to their surroundings. The whitetailed deer, for example, has a predeliction for coniferous cover during winter (Hosley and Ziebarth, 1935; R. L. Smith, 1959).

Birds are rather selective of roosting cover, although this aspect of avian ecology has been somewhat neglected. The night-roosting habits of bobwhite quail have been rather intensively studied in southern Illinois (Klimstra and Ziccardi, 1963). The covies showed a strong preference for bare ground protected by low, sparse, open canopy of vegetation. The ma-

FIGURE 16-1
Comfort movements in moose: (left) stretching and (right) rubbing antler knobs, typical among yearling bulls. (After Geist, 1963.)

TABLE 16-1
Types of shelter or cover

Winter cover	Protection from snow and cold; a place to remain out of sight
Escape cover	Vegetation in which the animal is secure from predators and hunters
Loafing cover	Shade in summer; protection in winter
Nesting cover	Protection from both nest and parents
Roosting cover	Secure cover for sleeping

Source: Based on Leopold, 1933.

jority of roosting headquarters were located at medium to low elevations on well-drained soil, and faced to the south or southwest. Occasionally the covies shifted their roosting headquarters, and most of the shifts were made from lower to higher elevations. Shifts down to poorly drained sites and low elevations were made usually in response to wind and inclement weather.

Suggestive of roosting cover is loafing cover, a place where animals spend restful minutes during the day. In New York conifer plantations, white-tailed deer spent hot summer afternoons in thick spruce plantings where low-hanging branches offered protection as well as handy brushes against flies (R. L. Smith, 1959). Ducks congregate on open mud bars jutting out into the water. And sulfur butterflies gather about little pools of water along the summer roadside.

Basic to shelter-seeking behavior is *habitat selection*. The underlying question here is how the various species of animals select the environment they inhabit. There is some evidence, among birds at least, that imprinting is involved; yet other evidence suggests that habitat selection among birds and mice is innate and is independent of early experience. The whole question of habitat selection cannot be dissociated from competition and natural selection, the forces that "fit" the organism to the environmental niche it occupies (see Chapter 12).

Animals differ widely in their powers of locomotion and in their capacity to learn. Animals restricted in their ability to move any great distance or dispersed by wind and water survive only if the environment in which they arrive is favorable. Others with greater powers of locomotion move about until they locate a suitable area. This is particularly true of birds, flying insects, and mammals. Between these two extremes are many intermediates.

Habitat selection is partly a psychological process (Lack and Venerables, 1939; A. H. Miller, 1942). It is complex type of choice behavior, but is also partly instinctive. Species tend to select their ancestral habitat instinctively, recognizing it by conspicuous though not necessarily essential features. Salmon recognize their home stream from olfactory experiences gained as fry (Hasler, 1954). The Nashville warbler, a typical inhabitant of the open heath edges of northern bogs, selects open stands of aspen and balsam fir and forest openings of blackberry and sweet-fern in the southern part of its range. These habitats are visually suggestive of bog openings (R. L. Smith, 1956). MacArthur and MacArthur (1961) and later MacArthur and many others (see MacArthur, 1972) demonstrated a strong correlation between structural features of the vegetation and the species of birds present. Basically birds of forest and grassland choose their habitat on the basis of the density of leaves at different elevations above the ground, irrespective of plant species composition. In effect it is the overall aspect that is important—the type of terrain, whether rolling or flat, open or grown with woody vegetation, continuous or patchy. Woodcock, for example, will not utilize a singing ground unless it allows sufficient room for flight (Sheldon, 1967). Small openings surrounded by tall trees are unsuitable, but an opening of the same size surrounded by low shrubs is.

But there are other features of the habitat, the presence or lack of which will determine its suitability (Hilden, 1965). One is singing perches, the lack of which can prevent birds from colonizing an area. Their introduction can mean the colonization of an otherwise suitable area. For example, when telephone lines were strung across a treeless heath, they brought in tree pipits, birds that require an elevated singing perch.

An adequate nesting site is another requirement. Animals require sufficient shelter to protect parents and young against enemies and adverse weather. Selection of small island sites, such as muskrat houses, by geese provides protection against predators. The chukkar partridge selects its nesting site in response to microclimatic differences within the terrain and prefers places where its nest is protected against climatic extremes. Hole-nesting animals require suitable cavities or substrate in which they can construct such cavities. In areas where such sites are absent, populations of birds (Hartmann, 1956) and squirrels (Burger, 1969) can be increased dramatically by providing nesting boxes and den boxes.

There may be some relationship between food and habitat selection, although it is rarely a determinant. Some predatory birds nest only where food is abundant, which may account for local fluctuations in populations. Among such birds are owls, jaegers, and hawks. Perhaps the extension of the breeding range of the evening grosbeak has been encouraged by the widespread winter feeding of the birds south of their natural breeding range.

Once one or several animals have settled in an area, others may be stimulated to do likewise. Among colonial and semicolonial birds, the attractiveness of the colony is highly important. For herring gulls nesting for the first time, this is an important aspect in habitat selection (Dorst, 1958). Other birds are attracted to areas settled by different species. Thus the tufted duck, *Aythya fuligata*, has a strong social attraction to gull colonies. In fact this duck will not nest on small islets where gulls do not nest, and high-density populations have never been found where nesting gulls are few (Hildén, 1964). On the other hand, the presence of some animals has inhibited others from occupying otherwise suitable areas. Most pronounced in this respect is the presence of man.

Klopfer (1962, 1963) has been able to demonstrate by the use of a test chamber containing oak and pine foliage that adult chipping sparrows of the North Carolina Piedmont preferred pine foliage to oak. All environmental effects except foliage and light had been eliminated, including variations in perch space. In another experiment, chipping sparrows hand reared in isolation from either foliage type chose pine when given a choice. Conversely, isolates reared in the presence of oak foliage chose oak more often than pine.

Somewhat similar work has been done with the two subspecies of deer mice, *Peromyscus maniculatus bairdi*, restricted to open-field habitats, and *P. m. gracilis*, restricted to hardwood forests and brushy habitats. By presenting artificial habitats, one of forest, the other of grassland, to the two subspecies, V. T. Harris (1952) demonstrated that the two subspecies selected the appropriate artificial habitat. The physical conditions were essentially uniform throughout the experimental room, so he concluded that the mice were reacting to the form of artificial vegetation. And because laboratory-bred *Peromyscus*, who had no previous experience with either habitat, artificial or natural, exhibited a preference for the type of habitat they normally occupied in the wild state, he concluded that habitat selection was basically genetic.

This problem was probed more deeply by Wecker (1963), who investigated the role of early experience in determining the habitat preference of prairie deer mice under natural

515

conditions. Two strains of mice were used, live-trapped wild stock and laboratory-reared stock. The latter were descendents of 10 pairs of mice live-trapped by Harris and were 12 to 20 generations removed from any experience in the natural environment. In separate experiments they were released in a 1,600-ft² rectangular enclosure, constructed with its long axis lying halfway across a woods–field habitat boundary. This provided the introduced animals an opportunity to select one of the two distinct habitats. Movements of the mice were recorded by a sensitized treadle located in runways at various positions in the pen. Six groups of mice were tested in the enclosure: field-caught animals, laboratory animals, offspring of field-caught animals reared in the laboratory, offspring of laboratory animals reared in the field, offspring of field-caught animals reared in the woods, and offspring of laboratory animals reared in the woods. Without elaborating on details, the results of these experiments justified the conclusion that the choice of the grassland environment by the prairie deer mouse is normally predetermined by heredity. Early field experience can reinforce this innate preference, but it is not necessary for subsequent habitat selection. Early experience in other environments was not sufficient to reverse the normal selection of grassland by the experimental mice.

Confinement of prairie deer mice in the laboratory for 12 to 20 generations resulted in an apparent reduction of hereditary control over habitat selection and led to an increased variability in the behavior of these animals when tested in the enclosure. Although these mice lost their innate preference for the grassland habitat, they still retained an ability to learn from early field experience to respond to the stimuli associated with that environment. When these young laboratory-strain mice were reared in field, woods, and laboratory environment, only those with early experience in the field habitat chose the appropriate environment when released in the enclosure. Those mice with previous experience in woods and laboratory showed no preference for either the woods or the field habitat.

Habitat selection, however, is not extremely rigid. Most species exhibit some plasticity; otherwise these animals would not spread into "abnormal" habitats. This ability or trait of some members of a particular species to select habitats that deviate from that of their companions must exist on both a phenotypic and

genetic level. Hinde (1959) suggests that some animals may accept a normally inadequate habitat if some supernormal releaser, such as nest boxes, exists in the local environment. Plasticity in habitat selection is well illustrated by such birds as the chimney swift, which accepted chimneys over hollow trees as nest sites, and duck hawks, which find themselves as much at home in the man-made cliffs and canyons of our large cities as they do in natural cliff country. Pigeons become an adequate substitute for duck. And the magnolia warbler, partial to spruce and hemlock forests, is a typical inhabitant of the drier oak forests of West Virginia (M. B. Brooks, 1940; personal observations).

Feeding behavior

Feeding behavior of any animal is closely bound to its structural adaptation for acquiring food and to its niche in the community, both results of evolution. Moths and butterflies have extendible tongues that enable them to sip nectar. Mergansers have serrated margins on the bill that help them to hold onto fish. Herons have long legs, long necks, and long sharp bills, which permit them to spear fish in deep water. Cats have carnassial teeth, which allow them to shear flesh like scissors. Microtine rodents have teeth well adapted to a vegetable diet. There are nearly as many feeding adaptations as there are species of animals. This great variety has been discussed in one way or another throughout the pages of this book.

The diet of animals is also partly controlled by behavior patterns possessed by the animal. The possession of one or a series of behavior "programs" increases the possibility that the animal will take food that switches on this "program," rather than an item that does not. Among some animals these behavioral patterns are similar and suggest a common ancestry; among others, the "programs" from appetitive behavior to the consummatory act are as individualistic as the morphological structure of the species.

Feeding behavior is triggered by the internal state of the animal. Included are the general sensations of the body, such as hunger pangs and stomach contractions. The depletion of food material in the blood results in changes in blood glucose, which in turn may stimulate sensory nerves that affect centers in the brain. These functions are very closely related to the automatically self-stimulating physiological processes of the body. This, of course, is not the whole story, for food-seeking behavior can be triggered in the absence of any metabolic condition. A well-fed cat does not hesitate to chase a mouse immediately after consuming a meal. So other drives too may be involved in the release of the hunting behavior.

Once initiated, food-seeking behavior among many animals is relatively stereotyped and involves a series or chain of behavior patterns. Each pattern is dependent upon some stimulus different for each step or link and each is necessary for the behavior to proceed from one action to the other.

The solitary digger wasps (Sphecidae) offer a good example of stereotyped food-seeking behavior. Each species tends to be a specialist and hunts a particular prey, disregarding others similar in size and just as abundant, if not more so. Some hunting wasps prey only on a single genus or a single species. Very few find their prey among more than one order of arthropods. Among this interesting family of wasps, the affinity of the predator to the prey is as characteristic of each species as its anatomy. Thus there are such hunting wasps as the grasshopper and cricket killers, the cicada killers, the spider killers, and the like. This is a purely arbitrary classification, since grasshopper and cricket killers, for example, may be found among several subfamilies and tribes.

The unique fact about the predatory wasps is that they do not kill to feed themselves. The adults, male and female, feed on sugar in solution, supplied by nectar, ripe fruit, and the honeydew of aphids. Males are nonpredatory. The females hunt to provide food for the larvae, which are reared in galleries, chambers, or nests in the ground or in rotted wood or in nests built of mud above the ground.

To begin, the hunting behavior of the female predatory wasp is released by the completion of a burrow or nest that needs to be provisioned with food for the larvae before the eggs can be laid (Baerends, 1941). After the nest is completed, the digger wasp flies to the appropriate habitat to search for her prey. If the wasp sights a moving object, she moves in closer and hovers downwind. From this position, the wasp can scent the odor of the potential victim. If correct, the wasp moves in for the kill; but only if the victim has the feel of a bee, cicada, or whatever the specific prey must be, does the female immobilize the victim with a sting. Among the many different species of hunting wasps, this combination of visual, olfactory, and tactile cues guides them to their proper victims. The tendency to respond to the appropriate cue in each of these sensory areas is a part of their innate behavior. So accurate are the responses to these releasers or cues that the wasps rarely make a mistake.

The stinging behavior that follows a capture is also stereotyped. The motions differ from one major group to another. Each species is adapted to the anatomy and physiology of the prey it seeks. Caterpillar hunters sting their prey not only on the thorax but also along the underside of the abdomen, where a well-developed nervous system controls the prolegs, important in the caterpiller's locomotion. The bee killers sting the honeybee around the coxae or anchor segments of the front leg; from there the venom diffuses to the muscles controlling flight and movement of the legs.

Once the prey has been immobilized, the wasp has to carry it back to her nest. Here again clear-cut behavioral patterns have evolved, involving the size of prey, the methods of carrying it, and the distance from the nest that hunting takes place. The most primitive method is found among those wasps who install only a single victim in the brood nest (H. E. Evans, 1962). These wasps require a victim as large or larger than themselves to meet the food demands of the larvae, and they must of necessity drag the victim over the ground back to the nest. Even the method of dragging the prey across the ground varies. One group, regarded as the most primitive, seize the prey with the mandibles, often in some specific place as the base of the hind legs or the mouth parts, and drag it backwards to the nest. Others, somewhat more advanced and exhibiting structural or behavioral modifications that enable them to straddle their prey, move the victims forward over the ground. A few may make a short hopping flight.

Most of the digger wasps and a few vespid wasps use more than one paralyzed victim per cell (Evans, 1957, 1963). Thus the size of the prey is somewhat smaller. This allows the wasp to carry the prey in flight, holding onto the victim with the mandibles and often also with the legs. This ability enables such wasps as the grasshopper killer (*Sphex*) to search the grassy meadows some distance from their nests for suitable prey. All of these carry their prey with their mandibles, so they cannot dig while holding their load; they must lay the victim down at some obvious risk of loss while they scrape open the nest. Another group of digger wasps get around this by carrying their prey by the middle or hind legs alone, without the use of the mandibles. This is the method used by the well-known cicada killer, which carries its prey with the middle legs, although the great weight of the cicada often forces the wasp to drag the victim some of the distance across the ground. These wasps can clear away the nest entrance and perform other tasks without releasing the prey. The most advanced of the spechid hunting wasps, the Crabroninae and Philanthinae, carry their prey on their stings, which in some species is barbed. One genus, *Aphilanthops*, has its last abdominal segment modified for clamping the prey.

The stereotyped, highly specialized methods of the hunting wasps for seeking and handling prey represents a type of behavior that is adapted to specific prey. Feeding behavior, too, reduces competition among similar sympatric species less specific in food requirements. Such diversity in feeding behavior may be found among the herons, nine species of which may haunt common feeding grounds. The herons are somewhat separated in their feeding niches by the length of the legs and neck, which limits the depth of the water in which each can feed. The great blue heron can seek fish in deeper water than the green heron; but they differ too in the method they use to capture their meal.

All herons have one basic method of hunting: they stand and wait, usually hunched, with the head retracted (Meyerriecks, 1960). As the prey draws near, the heron flicks its tail up and down slightly, then strikes, darting its head forward and downward. The fish or frog, once seized, is held momentarily between the bill tips, then tossed head first into the mouth by a sharp backward motion of the head. Once the fish is swallowed, the heron dips its bill in the water and quickly shakes its head. The heron will perform this behavior even if it misses its prey, seemingly a necessary follow through of the strike. A method common to all herons, except the common egret, is to walk or wade slowly while holding the body erect and the head forward at a 45° angle. But there are variations. On sunny days when the surface glare is great, the little blue heron will tilt its extended head and neck sharply to one side and then the other as it peers into the water; and it will stop and quickly extend and withdraw its wings in barely perceptible flicks, which may startle the prey into activity. The Louisiana heron, improving on this, will extend one wing fully, whirl about, fold the wing, whirl again, extend the other, then rapidly pursue the startled prey that respond to the rapid shadow over the water. This same heron possesses an interesting variation in which it extends both wings, turns about, raises one wing higher than the other, then slowly tucks its head beneath the higher wing while peering into the water. As it turns again, the heron shifts the head beneath the other wing. Such action apparently shades the water, and by reducing surface glare, enables the heron to see down into the water. The reddish egret goes one step further. It extends both wings fully and while holding this pose stabs rapidly at the prey beneath (Fig. 16-2). This heron exhibits a variation that is most fully evolved in the black heron of Africa. The black heron brings both of its extended wings forward to form a canopy over the bird's body. The heron then tucks its head beneath this feathery umbrella. But the canopy formed by the reddish egret is not quite so complete. Prior to assuming the canopy pose, the reddish egret dashes about in the water, startling and dispersing the fish. Then the egret stops, spreads its wings, and waits for the frightened fish to take up cover beneath a false refuge. Another species, the snowy egret, chases fish out of cover by stirring the water with its feet either as it wades across the shallows or hovers above the surface (Fig. 16-2) (Meyerriecks, 1959).

These examples point out how fixed the food-seeking behavior of animals can be. They also illustrate how behavior, like structure, can evolve through natural selection, further adapting the animal to its niche in the community.

The preponderance of innate behavior patterns in food seeking does not eliminate learning. Like nest-building movements in birds, food seeking undoubtedly improves with experience. The young of precocial birds learn what to pick up and what to leave behind. Immature brown pelicans (Orians, 1969) and little blue herons (Recher and Recher, 1969) are less successful at hunting than adults under the same conditions and capture less food per unit time foraging. This may partially explain the delayed maturation and smaller clutch sizes of younger birds. Learning is probably most

FIGURE 16-2
Hunting behavior of reddish egret, in which the wings are fully extended. Foot-stirring behavior of the snowy egret. (After Meyerriecks, 1959.)

519

highly expressed in the food-seeking behavior of mammals, which often develops into something of an art, especially among the carnivores (see Fox, 1969).

Undoubtedly some individuals become more proficient than others. Even among mammals, however, one can still see basic patterns that some groups possess in common (Fig. 16-3). Among the canids, movement of the prey is a strong stimulus that elicits orientation, approach, attack, seizure, followed by a sequence of carrying prey to a safe or quiet place, unless capture is preceded by play (Fox, 1969). Learning is important and includes discovering the best time and place to hunt and the degree of social cooperation required. Young animals become experienced by aiding older animals in the chase, by observing, and by trial and error. Whatever the organism, those animals whose feeding behavior involves learning are much more flexible in their choice of food compared to those limited to stereotyped feeding, such as the hunting wasps.

Several diverse groups of animals, notably the bats and the dolphins, seek their food by echolocation. When hunting insects and flying about in the darkness of the night, bats are able to locate food and avoid objects by emitting through the mouth bursts of ultra-high-frequency pulses, which bounce off objects about them. The returning echo is picked up by their large ears. The time required for the return of the echo indicates the distance of the objects. When hunting insects, the bat engages in a fairly straight flight, during which it emits repetitive pulses at a relatively long interval of 50 to 100 msec (Griffin et al., 1960). If an insect is picked up by echolocation, the bat turns in that direction and increases the pulse rate to 10 to 50 msec. As the bat closes in, it increases the pulse rate to a very rapid 4 to 7 msec, a buzz. All this requires very little time. Bats have detected, located, and intercepted insects in flight within half a second.

Among animals whose young are highly dependent on parental care, food-seeking behavior is, in the young, care-soliciting behavior. This is reciprocal with the care-giving behavior of the parent. A newborn mammal instinctively seeks the nipple of the mother, yet unless the female is in proper psychological condition, the young animal may be rejected. Normally, however, when the young seeks to nurse the mother, the mother also seeks the young. Among altricial birds, the instinct of the young

to seek food and the stimuli that cause the adults to feed them involve a series of highly developed releasers. At the time the young birds hatch, the adults already exhibit a tendency to feed. In fact some birds, such as the starling, some warblers, and tanagers, may bring food to the nest before the eggs have hatched. At the other extreme, some late-nesting swallows may migrate south and leave the young in the nest to starve.

Young birds are fed only if they, to put it anthropomorphically, demand to be fed. The stimuli that elicit the feeding behavior of the adults to a certain extent depend upon the release of food-seeking behavior in the young. The typical response of the newly hatched young bird is to gape for food. Some newly hatched birds gape without any apparent external stimulus; others require some stimulus to elicit gaping. Altricial birds usually are born blind, so visual stimuli are not effective, but a slight jarring of the nest is. Later, after the eyes are open, the jarring of the nest may result in fear responses instead (Tinbergen and Kuenen, 1939). At this time visual and tactile releasers become important. Young thrushes, after their eyes have opened, gape at any object, provided it is not too small and is located above the horizontal of the nestlings' eyes (Tinbergen and Kuenen, 1939). Later, gaping is directed toward the head of the parent, suggesting that some learning is involved. Experiments show that food-begging responses of both the herring gull (Tinbergen and Perdeck, 1950) and the Franklin gull (Collias and Collias, 1957) are released by the sight of the parents' red-spotted beak and red bill, respectively. To be most effective the bill had to be held low, had to point downward, and had to have something (food) protruding from it. For the procedure involved in these experiments, which included the use of cardboard heads, variously colored bills, and in the case of the Franklin gull, variously colored flashing lights, the reader is referred to the original papers.

Just as the young require releasers to stimulate gaping or food begging, so do the adults require some sign stimuli to release the feeding response. Gaping is one releaser. Some species have evolved color patches, or "targets," at which the parents direct the food. Linings of the mouth, brightly colored in yellow, red, or orange, are one type of releaser, the bright yellow flanges on the margin of the beak another. Other species have bizarre color patterns in the mouth, such as the jet black spots and bars on the tongue in the desert horned lark. Wing fluttering, either horizontally or vertically, is another stimulus for the parent bird to feed, as is the food call of the young. Once fed, the young bird ceases to gape; and without gaping the young bird is not fed. Apparently when the young bird has swallowed food, the swallowing reflex becomes inoperative for a short while. Even if the bird should gape when the parent returns, it will refuse to swallow. The parent bird then removes the food and presents it to another. In such a manner food is rather equally distributed among all the young.

Escape and defense behavior

Of all the behavior patterns of animals, the most generalized and the least species-specific are those involving escape and defense. The responses—the escape movements, the distress calls and alarms—may release similar alarm or escape reactions in many different species. This behavior lacks specificity because no selective advantage is gained by having a behavior that emphasizes differences in alarm reactions between species. The importance is individual survival, and a whole effect is necessary.

ENEMY RECOGNITION

Recognition of some enemies is innate. A combination of visual characters elicit fear or alarm reactions from many birds. The classical experiments of Nice and Ter Pelwyk (1941) and of Hartley (1950) point out the importance of visual recognition. When small birds were presented with models of owls, they showed strong alarm or fright, but only if the models possessed one or a combination of four visual characteristics: (1) a large head, owllike in outline—short head, short neck, short tail; (2) an owllike pattern—barred, streaked, or spotted; (3) an owllike color—brown, gray, or tone contrasts of these colors; (4) a solid contour. The escape reaction of many birds to avian predators flying overhead is a response not only to movement but also to a definite outline involving a short neck. Lorenz and Tinbergen (1938) demonstrated that when a cardboard dummy with a characteristic outline of a short head was passed overhead, it released escape behavior in waterfowl and gallinaceous birds, regardless of the shape of wings and tail

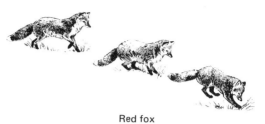

Red fox

FIGURE 16-3
Hunting behavior of a canid, as exemplified by the red fox. Note the pouncing movement on the prey. (Drawn from life.)

521

and regardless of color. When a generalized silhouette, which appears as a bird with a short neck and long tail when pulled in one direction, and as a bird with a long neck and short tail when pulled in the other, was presented, the birds reacted only when the dummy was moved in the direction of the short neck. As a result many birds react to swifts and doves as they do to hawks and owls. Movement, too, is involved. A swiftly moving object releases escape behavior even though an error in recognition is involved. Swiftly moving pigeons and swifts, for example, often release escape behavior in birds. In such situations, it is undoubtedly better for the animal to react swiftly and flee, even though in error, than to hesitate and perhaps die. More slowly moving predators permit distant recognition and allow more time for the release of appropriate behavior patterns, so that energy is not wasted on unnecessary flight or display of hostility.

Although enemy recognition is in many ways innate, the ability to distinguish between what to fear and what not to fear is learned. If an animal is faced with a possible enemy and nothing happens, its fear and alarm reactions decrease until the object no longer elicits a reaction. This is particularly true when an animal (especially a higher vertebrate) is exposed to something new and strange, yet harmless. Fear is the first reaction, but it gradually wanes until the animal accepts the object or other animal.

REACTION TO ENEMIES

How an animal reacts to an enemy depends upon the enemy, the behavior of the enemy, the circumstances under which the meeting takes place, and often the stage of the reproductive cycle.

When an animal faces the threat of an enemy or danger, it has five courses of action open to it. The animal may cry out a warning, if vocal; it may lead the enemy away from the nest or young; it may threaten, mob, or attack its adversary; it may hide; or it may flee.

Flight. If the enemy is formidable and resistance is useless or if the animal is unduly startled by swift approach, the animal flees. Often the animal needs no visual cues; scent, sound, or chemical stimuli are sufficient to produce an immediate response. At the sound of a man or dog moving through the woods, a deer will race to some safer spot. In response to the scent of the predatory starfish, sand dollars bury themselves in the sand. The rabbit runs a zigzag course ahead of pursuing hounds. In fact erratic display is common to many taxonomic groups such as mammals, birds, moths, grasshoppers, and cladoceran crustaceans. Flight taking the form of zigzagging, looping, spinning, and wild bouncing or leaping (Fig. 16-4) disorients the predator's attack. Flight behavior among some species and groups may involve flashing of colored surfaces (such as the white tail of the white-tailed deer, the white tail feathers of birds, the changing colors of fish). Such behavior may not only disorient the predator but may also stimulate escape tendencies in the predator. Game animals do not flee as soon as they see an enemy, but only after approached to within a certain flight distance. That distance will vary with different predators, and the age, sex, and social status of the intended prey. Stimuli for flight include sudden appearance, size, number, and course of approach of the enemy.

Flight reaction is contagious. If one animal flees, others may be startled into doing the same. This results in stampedes among the ungulates, and in waterfowl massed flight out of the area. If danger is from above, such as from a hawk, bird flocks tend to draw close together to form a dense flock and perform sharp, swift turns with coordination and precision. Such behavior makes picking off an individual much more difficult, an accomplishment achieved only by considerable maneuvering by the hawk.

Warning. Warning calls—a cry, a snort, a yelp—are not, in spite of our story books and our imaginations, given as a conscious alarm to other animals, although the effect may be the same. It is a cry of fear, and the intensity or nature of the call may vary with the intensity of fear. Among birds in particular, perhaps because they are as a group the most vocal of the vertebrates (other than man), the warning or alarm notes vary either in intensity or in kind according to the situation. They may vary with the type of predator or trespasser, with distance or nearness, or with the stage in the reproductive cycle. A hawk flying overhead may induce an animal to crouch or freeze and remain completely silent. A cat entering the territory of almost any bird elicits high-intensity alarm notes that spread rapidly to other birds in the area, yet man and dog may be ignored or be met with only mild alarm.

In her classic study on the song sparrow, Nice (1943) listed three stages of behavior in this species when alarmed. Although all three stages intergrade, each is characterized by a particular note. When alarmed, the bird utters a note *tchunk*, raises its crest, flips its wings, and raises and flips its tail. In the second, or fear, stage the bird utters *tik*, it compresses its body feathers, stretches its neck, and crouches on its perch. When frightened, the bird utters a rapid *tik-tik-tik* and flies and hides, and if unable to escape, pants with its bill open. The rate at which these notes are given indicates to some extent the intensity of the fear.

The tendency of birds to give a warning call increases as the reproductive cycle passes through the stages of nest building, incubation, and brooding the young. Just as the defense reaction of mammals is highest when the female is caring for the young, so the intensity of alarm is greatest among birds when the young are about to leave the nest or are under the care of the parents. The intensity of the response increases as the enemy comes nearer and nearer the young. During nest building and incubation, many birds are relatively unconcerned about most trespassing animals unless it is a cat or some other strong enemy. But later when young are in the nest, even human intrusion is met with alarm. The grasshopper sparrow, for example, utters only a sharp monosyllabic *tik* when humans or dogs enter its territory early in the nesting period (R. L. Smith, 1963). But if one closely approaches a nest containing young or if a cat enters the territory at any time, the bird raises its crest feathers, flicks its wings and tail, bobs up and down on its legs, and utters a sharp double note *chi-ip*. At high intensity the grasshopper sparrow gives this double note so rapidly that it runs into a trill. To dogs these birds react differently. They drop into the grass, crouch, and remain silent until the animal passes.

This emphasis on birds may overshadow the fact that mammals too have alarm notes. When a human approaches a fawn, the doe white-tailed deer, if nearby, reacts with a loud snort, the intent of which is unmistakable even to human ears. Squirrels are quite vocal and have a considerable repertoire of sound signals, including alarm. Six alarm signals have been re-

FIGURE 16-4

Stotting is a form of flight behavior and predator avoidance in Thompson's gazelle. In stotting, the gazelle holds its front and hind legs stiffly stretched downward and bounces, springing from the pastern joints. The high but not the highest level of flight excitation is caused by an approach of a predator. Stotting apparently conserves energy over other forms of flight behavior. At the same time a number of gazelles stotting at different intervals would tend to confuse the predator. Normal stotting gait (left). Paddling with hind legs in extremely high stotting (middle). Landing from high stotting (right). (After Walther, 1969.)

TABLE 16-2
Calls of the gray squirrel

Calls	Sex	Age	Description	Associated with
ASSOCIATED WITH MATING CHASES				
Buzzing	Male	Yearling and adult	Stridulating insect or partly stifled sneeze	Hunting for or chasing the female during mating chase
ASSOCIATED WITH WARNING OR ALARM				
Intense alarm chuck	Both	All	Rapid kuk, kuk, kuk	Imminent danger
General alarm chuck	Both	All	Drawn out ku-u-uk ca, one per two seconds	Danger
Warning chuck	Both	All	Slower, short kuk, kuk, kuk, kuk; or kuk, kuk, kuk, qua-a-a-a	Immediate past danger; usually follows general alarm call
Attention chuck	Female	Adult	Low chucking barely audible at 30 feet	Given to young as a warning (?) call
Juvenile scream	Both	Juvenile	Soprano scream; mouth at maximum gape	Handling in cone or removal from nest; probably signifies fear
Female scream	Female	Yearling and adult	Lower harsh scream, like combined scream and snarl	At males around her while cornered in mating chase (warning?)
ASSOCIATED WITH GROUPS				
Mew	Both	Adult and ?	Resembles meow of cat	?
Rapid clucking	Both	All ?	Kuk, kuk rolled together	Sometimes given before entering nest
Whistling	?	?	Resembles ground squirrel whistling	Sometimes given by occupants after nest is entered
Purr	Both	Young	Low purring like a cat	Play

Source: Bakken, 1959.

corded for the gray squirrel, each associated with a particular situation (Table 16-2).

Defensive secretions. Some species of fish (confined to the order Ostariophysi) release alarm substances from the skin into the water and induce a fright reaction in other members of the same or related species (Pfeiffer, 1962). This substance is produced in specialized epidermal cells, which do not open on the surface; they can release their contents only when the skin is broken. Fish in the vicinity receive the stimulus through the olfactory organs. Only minute quantities are needed to drive a school of minnows from a feeding area (0.002 mg in a 14-liter aquarium). The fright reaction is innate. Regardless of prior experience, fish exhibit this reaction only at a certain age—sometime after the alarm substances are produced in the skin. Fright reaction, which may be transferred visually to individuals not exposed to the substance, affords a twofold protection. It protects schools of young from cannibalistic attacks of adults, because once they break the skin of the young, the would-be cannibals are frightened away by the alarm substance, and it shields older fish from general predation. Fish possessing this alarm substance are mostly social, lack defensive structures such as spines, and generally are nonpredaceous.

Defensive secretions are commonly employed by arthropods to repel predators (Eisner and Meinwald, 1966; Eisner, 1970). Unlike pheromones, defensive secretions are produced in often copious quantities. Because they are strongly odorous they may be detected with ease. They are secreted by glands containing small openings but possessing large saclike reservoirs that are essentially infoldings of the body wall. There are three types of discharges: (1) one type, as in the millipedes, oozes onto the animal's body surface; (2) the secretions are aired by the evagination of the gland, as in beetles; (3) the animal discharges the contents as a spray for distances up to several feet. This is accomplished by such insects as grasshoppers, earwigs, and stink bugs. As a defensive mechanism, such secretions are very effective in repelling predators—birds, mammals, and insects alike. The secretions are discharged during the early phases of the attack to repel the assailant before bodily damage occurs. The secretions attack sensitive tissues about the face and mouth. Vertebrates quickly learn to discriminate against obnoxious prey (see Brower, 1969) a discrimination that has led to mimicry and warning coloration (see Chapter 12). Active components in the defensive secretions of arthropods occur as secondary substances of plants, which may act as deterrents against plant predators.

Detection by echolocation. Bats, as it has already been pointed out, detect obstacles and prey in complete darkness by echolocation. Several families of moths, a chief prey item of the bat, have evolved countermeasures enabling them to detect the bats. The Noctuidae, or owlet moths, in particular have a pair of ultrasonic ears located near the "waist" between the thorax and the abdomen (Roeder and Treat, 1961). Internal to the eardrum—a tympanum membrane directed obliquely backward and outward in a cavity of the body—is an air-filled cavity spanned by a thin strand of tissue. This contains a sound-detection apparatus of two acoustic sensory cells that eventually connect to the central nervous system by the way of the tympanic nerve. These ears respond to tones from 3 to 100 kc, but they are most sensitive near the middle range, the frequencies contained in the bat chirps. Acoustic stimulation of the moth ear by bat chirps releases erratic flight patterns in the moth, which appear to be highly effective in saving the moth. For every 100 moths that survived because they reacted to the bat chirps, only 60 nonreacting moths survived. This gives the reactors a very high survival rate over the nonreactors, which certainly must influence natural selection in favor of bat detection.

Distraction displays. When danger threatens the nest or young of birds, the adult may attempt to draw the attention of the intruder to itself. To do this some simulate injury. For this reason, such behavior has been termed injury feigning, although the pretense of being crippled is hardly a conscious one.

A predator near a nest releases in many birds, particularly the ground nesters, a distraction display in which the parent flutters along the ground, always keeping just one jump ahead of the enemy. If the predator follows, the bird continues the act until the predator is no longer near the nest. If the enemy remains, the stimulus is increased, and the bird returns to repeat the display until the response tires. When disturbed at the nest, the female grasshopper sparrow, for example, may dart off the nest, run for a short distance, arise in a short fluttering flight, then drop to the ground again where she spreads her tail and trails her wings, as if injured. At other times she may flutter off the nest as if crippled; or she may fly from the nest to a point 25 or 30 ft away, hide in the grass, and scold (R. L. Smith, 1963).

Distraction displays, like other behavior, vary from low to high intensity. Low-intensity distraction display is elicited in the golden plover when an enemy approaches. The bird holds the head high in alarm and may engage in displacement feeding (Drury, 1961). As the predator or man moves closer, the bird may run with head up, back ruffled, tail barely fanned and tilted, and cry an alarm. Occasionally it stops. stands up, and perhaps goes through displacement feeding again; or the bird may move diagonally to the side, crouched with both wings and tail partly spread. When near the intruder, the plover may stop, stand, and call. At high intensity the bird is hunched with its breast on the ground and it stamps its feet. The tail is often fanned, the head pulled in, the wings partly open and spread to resemble a threatening owl. If pursued, the plover runs off, crouching, with its wings open and down at the sides, with the tail partly fanned and held down, and beating and shuffling as if on four legs, creating an illusion of a four-footed animal. As the bird flees, it looks over its shoulder to watch the intruder.

The basic pattern of distraction displays among birds is similar, but there are some species-specific or at least generic-specific differences (Drury, 1961). The display of the black-bellied plover, closely related to the golden, differs in that at low intensity the bird frequently settles down with a fast squirming into a depression, as if settling on eggs, and it does not engage in a "rodent run" when pursued. Instead the black-bellied plover runs low, rolling from side to side, with tail spread and wings slightly lowered but folded. At high intensity the black-bellied plover is prostrate, with its breast on the ground in a hollow, its tail cocked and fanned, its wings almost fully spread and beaten spasmodically. In contrast, the distraction displays of the *Charadrius* plovers, including the ringed plover, little ringed plover, killdeer, and piping plover, involve wiggling along the ground, leaning on one side, and waving one or both wings in an uncoordinated manner. Distraction displays are not as highly developed among sandpipers, and the parent bird sits closer on the nest.

Distraction displays are quite interesting, for they do not appear as an expression of thwarted drives (Simmons, 1955; Drury, 1961). The postures are not solely an expression of conflict between attack and escape; aggression is a strong motivation. Distraction displays, like those of courtship, have become highly ritualized. Although there are variations in intensity, each stage grades uniformly into the other. As the drives of aggression, flight, brooding, fear, and concealment increase and decrease, the bird's behavior changes. Movement elicits fleeing, but it is a deliberate, slow, shuffling, flopping movement, which keeps the bird just ahead of the pursuer. All movements attract attention to the parent, who leads the intruder away. The movements have become ritualized and removed from their original context, including perhaps even the motivation. Compared with courtship, distraction displays are highly generalized, since here the "object" is to attract the attention of the predator. Strong interspecific differences would have no selective advantage and bear no relationship to the isolation of one species from another.

Threat. Threat rather than flight is used against some intruders and is probably more widely used among mammals than birds. Birds may use against their enemy the same or similar aggressive threats that they use against members of their own species (see Chapter 13). Distraction displays may grade into aggressive displays. The killdeer, for example, may run toward such intruders as a cow or a horse, fluff its feathers, trail its wings, then fly up and strike the animal in the face. Other birds, such as the owls, may remain on the nest, raise or spread the wings, and snap the bill. Some reptiles snap and hiss at intruders. Mammals become highly aggressive and lower the head, utter snorts or growls, snarl, bare the fangs, and so on. Other animals may change color, as do some fish, or increase in size, as do some frogs, toads, and snakes, in an attempt to intimidate the enemy.

Attack. When threat fails, attack may do the job. Active defense, especially of the young, is most highly developed among the mammals. Usually it is the function of the female, but in some species, such as the buffalo, musk-ox, and wolf, defensive attack is also highly developed among the males. Crocodiles vigorously protect eggs and young. Owls and hawks may directly attack intruders at the nest, and kingbirds are highly successful at routing cats from their territory. Small birds have a tendency to gang up on or mob an owl, hawk, snake, or even a cat. A warning call from a bird who has discovered some common enemy is sufficient to bring many birds to the scene, to fly about and scold the animal. Gulls, terns, and some shore birds will in the manner of small birds heckle the intruder, dive at it, and spray it with excrement.

Other means. Some animals assume a passive defense against predators and feign death. Such behavior is well known in the opossum and some species of snakes. Birds may crouch and hide at the approach of an enemy, small mammals may quickly take shelter. Squirrels are particularly adept at keeping a tree trunk between themselves and a hunter.

Summary

Survival is a chief preoccupation of an animal. To escape or defend itself from its enemies may be one of the more dramatic situations associated with the word survival. But the animal's ordinary day-by-day quest for food and shelter and its care for bodily comfort are just as important for the survival of an animal as escaping its enemies. Thus a good share of an animal's behavioral activities are directed toward its own welfare. Many of the egocentric behavior patterns are simple and relatively stereotyped. But with experience some in-

dividuals become more skilled than others at certain activities, such as hunting. Habitat selection too is fairly stereotyped, but animals recognize suitable areas for colonization more by general features than by plant composition. Although feeding, shelter-seeking, and main-tenance behavior may in some ways be species-specific, escape reactions—the alarms, the threat postures, defensive attitudes, and flight—are so generalized that the behavior pat-terns of one species are recognized by many others. Although escape and defense behavior of an individual may be largely directed to-ward its own survival, it also spreads the alarm to others. Involved is sort of a com-munication and relationship with other in-dividuals that places escape and defense be-havior one step beyond the egocentric and into the realm of social behavior. In fact, certain social behavior patterns of some species may be adaptations to reduce their vulnerability to predation.

A diversity of ecosystems

Man has always been interested in the plants and animals around him, but his knowledge was largely limited to life in his own immediate area. As adventurous explorers expanded man's horizons to new lands around the world, they brought back stories and specimens of new and often strange forms of life. As naturalists joined the explorers, they became more and more familiar with plant and animal life around the world, and began to note similarities and differences. But botanical explorers and zoological explorers looked at the world differently. Botanists in time noted that the world could be divided into great blocks of vegetation—deserts and grasslands, coniferous, temperate, and tropical forests. These divisions they called *formations* even though they had difficulty drawing sharp lines between them. In time plant geographers attempted to correlate the formations with climatic differences, and found that blocks of climate reflected the blocks of vegetation and their life form spectra (see Chapter 8).

At the same time zoogeographers studying the distributions of animals found them much more difficult to map. By the beginning of the twentieth century, naturalists had accumulated the basic facts of worldwide animal distribution. All they needed to do—and this was a great enough task—was to arrange the facts and draw some general conclusions. This was done for birds in 1878 by Philip Sclater, who mapped them into six regions roughly corresponding to continents. But the master work in zoogeography was done by Alfred Wallace, who is also known for reaching the same general theory of evolution as Darwin. The *realms* of Wallace, with some modification, still stand today.

Classification of regions

There are six biogeographical regions, each more or less embracing a major continental land mass and each separated by oceans, mountain ranges, or desert (Fig. 17-1). They are the Palearctic, Nearctic, Neotropical, Ethiopian, Oriental, and Australian. Because some zoogeographers consider the Neotropical and the Australian regions to be so different from the rest of the world, these two often are con-

sidered as regions or realms equal to the other four combined They are classified as Neogea (Neotropical). Notogea (Australia), and Metagea, the main part of the world. Each region possesses certain distinctions and uniformity in the taxonomic units it contains, and each to a greater or lesser degree shares some of the families of animals with other regions. Except for Australia, each at one time or another in the history of the earth has had some land connection with another. across which animals and plants could pass.

Two regions, the Palearctic and the Nearctic, are quite closely related; in fact the two are often considered as one, the Holarctic. The Nearctic contains the North American continent south to the Tropic of Cancer. The Palearctic region contains the whole of Europe, all of Asia north of the Himalayas, northern Arabia, and a narrow strip of coastal north Africa. The regions are similar in climate and vegetation. They are quite alike in their faunal composition, and together they share, particularly in the north, similar animals such as the wolf, the hare, the moose (called elk in Europe), the stag (called elk in North America), the caribou, the wolverine, and the bison.

Below the coniferous forest belt the two regions become more distinct. The Palearctic is not rich in vertebrate fauna, of which few are endemic. Palearctic reptiles are few and usually are related to those of the African and Oriental tropics. The Nearctic, in contrast, is the home of many reptiles and has more endemic families of vertebrates. The Nearctic fauna is a complex of New World tropical and Old World temperate families; the Palearctic is a complex of Old World tropical and New World temperate families.

South of the Nearctic lies the Neotropical, which includes all of South America, part of Mexico, and the West Indies. It is joined to the Nearctic by the Central American isthmus and is surrounded by the sea. Isolated until 15 million years ago, the fauna of the Neotropical is most distinctive and varied. In fact about half of the South American mammals such as the tapir and llama, are descendants of North American invaders, whereas the only South American mammals to survive in North America are the armadillo, opossum, and porcupine. Lacking in the Neotropical is a well-developed ungulate fauna of the plains, so characteristic of North America and Africa. The Neotropical however, is rich in endemic families of verte-

Terrestrial ecosystems

brates. Of 32 families of mammals, excluding bats, 16 are restricted to the Neotropical. In addition, 5 families of bats, including the famous vampire, are endemic.

The Old World counterpart of the Neotropical is the Ethiopian, which includes the continent of Africa south of the Atlas Mountains and Sahara Desert, and the southern corner of Arabia. It embraces tropical forests in central Africa and in the mountains of East Africa, savanna, grasslands, and desert. During the Miocene and the Pliocene, Africa, Arabia, and India shared a moist climate and a continuous land bridge, which allowed the animals to move freely between them. This accounts for some similarity in the fauna between the Ethiopian and the Oriental regions.

Of all the regions the Ethiopian contains the most varied vertebrate fauna; and in endemic families it is second only to the Neotropical. Lush forests cover much of the Oriental region, which includes India, Indochina, south China, Malaya, and the western islands of the Malay Archipelago. It is bounded on the north by the Himalayas and on the other sides by the Indian and Pacific Oceans. On the southeast corner, where the islands of the Malay Archipelago stretch out toward Australia, there is no definite boundary, although Wallace's line is often used to separate the Oriental from the Australian regions. This line runs between the Philippines and the Moluccas in the north, then bends southwest between Borneo and the Celebes, then south between the islands of Bali and Lombok. A second line, Weber's, has been drawn to the east of Wallace's line; it separates the islands with a majority of Oriental animals from those with a majority of Australian ones. Because the islands between these two lines are a transition between the Oriental and the Australian regions, some zoogeographers call the area Wallacea.

Of the tropical regions the Oriental possesses the fewest endemic species and lacks a variety of widespread families. It is rich in primate species, including two families confined to the region, the tree shrews and the tarsiers.

Perhaps the most interesting and the strangest region, and certainly the most impoverished in vertebrate species, is the Australian. This includes Australia, Tasmania, and New Guinea, and a few smaller islands of the Malay Archipelago. New Zealand and the Pacific Islands are excluded, for these are regarded as oceanic islands, separate from the major faunal regions. Partly tropical and partly south temperate, this region is noted for its lack of a land connection with other regions, the poverty of fresh-water fish, amphibians, and reptiles, the absence of placental animals, and the dominance of marsupials. Included are the Monotremes with the two egg-laying species, the duck-billed platypus and the spiny anteater. The marsupials have become diverse and have evolved ways of life similar to those of the placental animals of other regions.

By the turn of the century some biologists were attempting to combine the regional distribution of both plants and animals into one scheme. C. Hart Merriam, then chief and founder of the United States Bureau of Biological Survey (later to become the Fish and Wildlife Service), proposed the idea of *life zones*. Differences among these transcontinental belts running east to west, expressed by the animals and plants living there, supposedly are controlled by temperature. Merriam published his ideas first in 1894 in a paper entitled "Laws of Temperature Control of the Geographic Distribution of Terrestrial Plants and Animals," in the December *National Geographic* magazine. Later he summarized and expanded still further this idea in a U.S. Department of Agriculture bulletin, *Life Zones and Crop Zones of the United States*. Merriam divided the North American continent into three primary transcontinental regions, the Boreal, the Austral, and the Tropical. The Boreal region extends from the northern polar seas south to southern Canada, with extensions running down the three great mountain chains, the Appalachians, the Rockies, and the Cascade–Sierra Nevada Range. The Austral region embraces most of the United States and a large part of Mexico. The Tropical region clings to the extreme southern border of the United States and includes some of the lowlands of Mexico and most of Central America. Each of these regions Merriam further subdivided into life zones.

The Boreal region he subdivided into three zones. The Arctic-Alpine zone, characterized by arctic plants and arctic animals, lies north of the tree line and includes the arctic tundra as well as those parts of mountains further south that extend above the timber line. The Hudsonian zone, the land of spruce, fir, and caribou, embraces the northern coniferous forest and the boreal forests covering the high

mountain ranges to the south. The Canadian zone includes the southern part of the boreal forest and the coniferous forests that cloak the mountain ranges extending south.

The Austral region is split into five zones. First is the Transition zone, called the Alleghanian in the East, which extends across northern United States and runs south on the major mountain ranges. It is a zone in which the coniferous forest and the deciduous forest intermingle. Extending in a highly interrupted fashion across the country from the Atlantic to the Pacific is the upper Austral zone. It is further subdivided into the Carolinian area in the humid east and the upper Sonoran of the semiarid western North America. The lower Austral embraces the southern United States from the Carolinas and the Gulf States to California. In the humid southeast, it is known as the Austroriparian area and in the arid west as the Lower Sonoran.

Once widely accepted, the life zones are rarely used today, although they creep now and then into the literature on the vertebrates. In the first place a life zone is not a unit that can be recognized continent-wide by a characteristic and uniform faunal or vegetational component. There are wide differences in the Transition zones of the East, with its hardwoods, of the Rocky Mountains, with the yellow pine, and of northeastern California, with the redwoods. Thus the Transition covers too many types of vegetation and too many different animals to be useful. The life zones south of the Arctic and Canadian are not transcontinental; thus the Transition and Upper and Lower Austral of the East are totally different from those of the West. Then the temperatures at times of the year other than the season of growth and reproduction influence the distribution of plants and animals.

In spite of all this there is something evocative about the life-zone terminology. The Arctic-Alpine zone recalls the cold, windswept mountains above the timber line; and the name Sonoran, slowly spoken, sings of the sun-baked desert, cactus, mesquite, horned lizards, and roadrunners.

A third approach to the subdivision of the North American continent into geographical units of biological significance was the biotic-provinces concept defined and mapped by Dice in 1943. It differs from the others in that a province embraces a continuous geographic area that contains ecological associations distinguishable from those of adjacent provinces, especially at the species and subspecies level. Each biotic province is further subdivided into ecologically unique subunits, districts, or life belts, based largely on altitude, such as grassland belts and forest belts.

Basically the biotic-province concept is an attempt to classify the distribution of plants and animals, especially the latter, on the basis of ranges and centers of distribution of the various species and subspecies. But the regions themselves and their subdivisions are largely subjective. The boundaries more often than not coincide with physiographic barriers rather than with vegetation types; and the regions never occur as discontinuous geographic fragments. Although a number of species may be confined to some biotic province, others occur over several provinces, because their distribution is determined more by the presence of suitable habitat, which is rarely restricted to a single region. Because the boundaries of biotic provinces and the ranges of subspecies of animals with a wide geographic distribution do coincide, this system is used at times by mammalogists, ornithologists, and herpetologists in the study of a particular group.

As attempts at combining plant and animal distribution into one system, all of the above units of classification are unworkable, for plant and animal distributions do not coincide. Another approach, pioneered by Victor Shelford, was simply to accept plant formations as the biotic units and to associate animals with plants. This approach works fairly well because animal life does depend upon a plant base. These broad natural biotic units are called *biomes* (Fig. 17-2). Each biome consists of a distinctive combination of plants and animals in the fully developed or "climax" community; and each is characterized by a uniform life form of vegetation, such as grass or coniferous trees. It also includes the developmental stages, which may be dominated by other life forms. Because the species that dominate the seral stages are more widely distributed than those of the climax, they are of little value in defining the limits of the biome.

On a local and regional scale, communities are considered as gradients (Chapter 8) in which the combination of species varies as the individual species respond to environmental gradients. On a larger scale one can consider

FIGURE 17-1
Polar projection map of the world showing major biogeographical realms. (From M. F. Guyer and C. E. Lane, 1964, Animal Biology, 5th ed., Harper & Row, New York.)

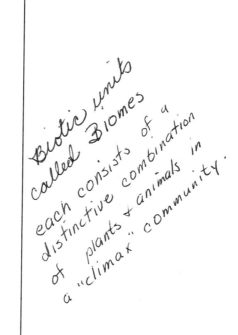

Biotic units
called Biomes
each consists of a
distinctive combination
of plants + animals in
a "climax" community.

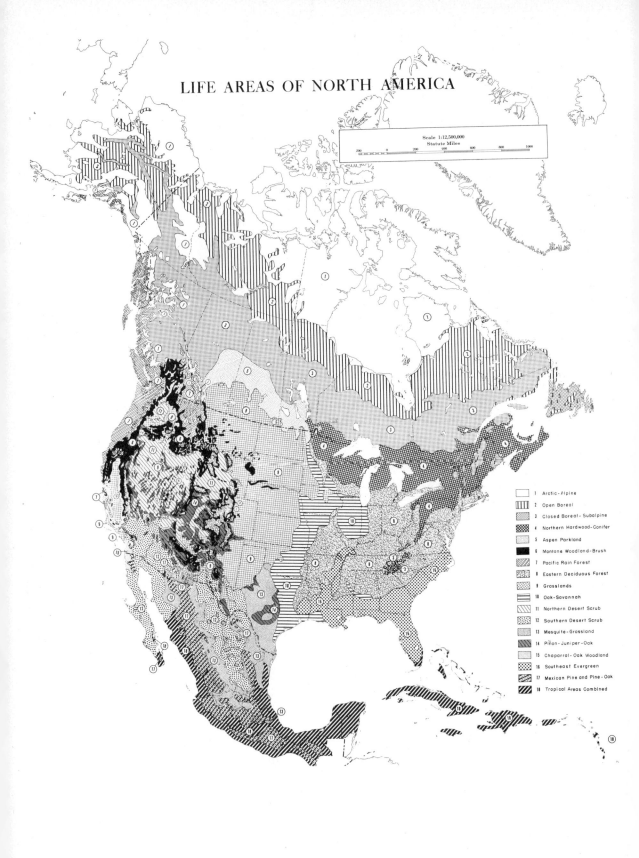

LIFE AREAS OF NORTH AMERICA

Scale 1:12,500,000
Statute Miles

1 Arctic-Alpine
2 Open Boreal
3 Closed Boreal-Subalpine
4 Northern Hardwood-Conifer
5 Aspen Parkland
6 Montane Woodland-Brush
7 Pacific Rain Forest
8 Eastern Deciduous Forest
9 Grasslands
10 Oak-Savannah
11 Northern Desert Scrub
12 Southern Desert Scrub
13 Mesquite-Grassland
14 Piñon-Juniper-Oak
15 Chaparral-Oak Woodland
16 Southeast Evergreen
17 Mexican Pine and Pine-Oak
18 Tropical Areas Combined

the terrestrial and even some aquatic ecosystems as gradients of communities and environments on a world scale. Such gradients of
ecosystems are *ecoclines*.

If one were to sample vegetation and associated animal life along a transect cutting across
a number of ecosystems, such as up a mountain
slope or north to south across a continent, one
would discover that communities change
gradually just as species change along community gradients. If one were to run this
transect across midcontinent North America
beginning in the moist, species-rich forests of
the Appalachians and ending in the desert, he
would follow a gradient of ecosystems on a
climatic moisture gradient—from the mesophytic forest of the Appalachians through oak-
hickory forests, oak woodlands with grassy
understory, tall-grass prairies (now cornland),
mixed prairies (wheatland), short-grass plains,
desert grasslands, and desert shrublands (Fig.
17-3a). Likewise, if one were to follow a
transect from southern Florida to the Arctic,
he would move along a climatic temperature
gradient that would take him from a subtropical forest through temperate deciduous
forest, temperate mixed forest, to boreal coniferous forest and tundra (Fig. 17-3b).

In addition to gradual changes in vegetation,
there are gradual changes in other ecosystem
characteristics. As one goes from highly mesic
and warm temperatures to xeric situations or
cold temperatures, productivity, species diversity, and the amount of organic matter decreases. There is a corresponding decline in
the complexity and organization of ecosystems,
in the size of the plants, and in the number of
strata to the vegetation. Growth forms change.
The tropical rain forest is dominated by
phanerophytes and epiphytes; the arctic tundra
by hemicryptophytes, geophytes, and thero-
phytes. Wherever similar environments exist
on the earth, the same growth forms exist even
though the species differences may be great.
Thus different continents tend to have communities of similar physiognomy.

The various biomes of the world fall into a
distinctive pattern when plotted on a gradient
of mean annual temperature and mean annual
precipitation (Fig. 17-4). The plots are obviously rough. Many types intergrade with one
another, and adaptations of various growth
forms may differ on several continents. Climate
alone is not responsible for biome types because
soil and exposure to fire can influence which

FIGURE 17-2 [OPPOSITE]
*Biomes of North America. (Map by John Aldrich,
courtesy U.S. Fish and Wildlife Service.)*

Climate
Soil   } responsible
exposure to fire for Biome Types

537

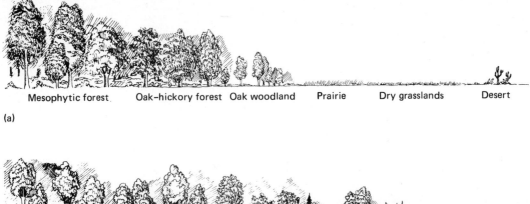

Mesophytic forest　Oak–hickory forest　Oak woodland　Prairie　Dry grasslands　Desert

(a)

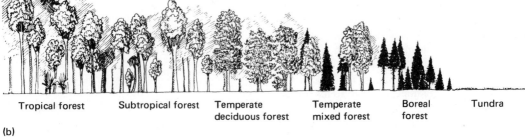

Tropical forest　Subtropical forest　Temperate deciduous forest　Temperate mixed forest　Boreal forest　Tundra

(b)

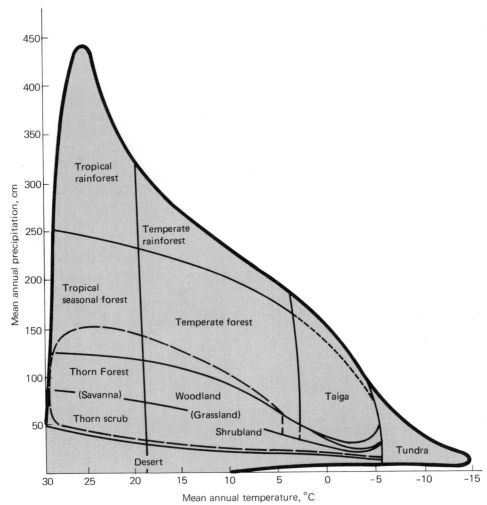

Tropical—Subtropical—Warm temperate—Cold temperate—Arctic-Alpine—

one of several biomes will occupy an area. Structure of biomes is further influenced by the nature of the climate, whether marine or continental. The same amount of rain, for example, can support either shrubland or grassland.

Life in the soil

Basic to all terrestrial ecosystems is soil (see Chapter 7). Populated by high numbers of species as well as individuals, the soil embraces another world with its whole chain of life, its predators and prey, its herbivores and carnivores, and its fluctuating populations. Because of their abundance, feeding habits, and ways of life, these small organisms have an important influence on the world a few inches above them. Because for all practical purposes it is a community separate from the one above, soil has been considered an ecosystem or biocenose, but it is not. Its energy source comes from dead bodies and feces from the community above (Fig. 17-5). It is but a stratum of the whole ecosystem of which it is a part (see Castri, 1970; Kuhnelt, 1970; Ghilarov, 1970).

THE SOIL AS AN ENVIRONMENT

The soil is a radically different environment for life than the one above the surface, yet the essential requirements do not differ. Like animals that live outside the soil, soil fauna require living space, oxygen, food, and water.

To the soil fauna, the soil in general possesses several outstanding characteristics as a medium for life. It is relatively stable, both chemically and structurally. Any variability in the soil climate is greatly reduced compared to above-surface conditions. The atmosphere remains saturated or nearly so, until soil moisture drops below a critical point. The soil affords a refuge from high and low extremes in temperature, wind, evaporation, light, and dryness. This permits soil fauna to make relatively easy adjustments to unfavorable conditions.

On the other hand, soil has low penetrability. Movement is greatly hampered. Except to such channeling species as earthworms, soil pore space is very important, for this determines the nature of the living space, the humidity, and the gaseous condition of the environment. The variability of these conditions creates a diversity of habitats, which is reflected by the diversity

FIGURE 17-3 [OPPOSITE TOP]
Gradients of vegetation in North America from east to west (a) and north to south (b). The east to west gradient runs from the mixed mesophytic forests of the Appalachians through the oak–hickory forests of the central states, the ecotone of burr oak and grasslands to the prairie, short-grass plains and the desert. The transect does not cross the Rocky Mountains. This gradient is largely a result of variations in precipitation. The north-south gradient reflects temperature. The transect cuts across the tundra, the boreal coniferous forest, the mixed northern hardwoods forest, the mixed mesophytic forests of the Appalachians, the subtropical forests of Florida, and the tropical forests of Mexico.

FIGURE 17-4 [OPPOSITE BOTTOM]
A pattern of world plant formation types in relation to climatic variables of temperature and moisture. In certain areas where the climate (maritime vs. continental) varies, soil can shift the balance between such types as woodland, shrubland, and grass. The dashed line encloses a wide range of environments in which either grassland or one of the types dominated by woody plants may form the prevailing vegetation in different areas. (After R. H. Whittaker, 1970, Communities and Ecosystems. *Reprinted with permission of Macmillan Publishing Co., Inc. © Copyright Robert H. Whittaker, 1970.*)

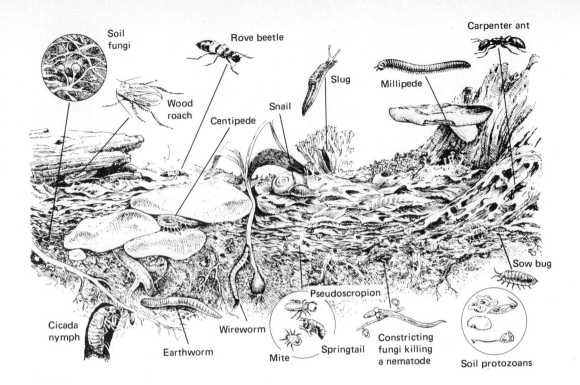

Soil fungi · Rove beetle · Slug · Carpenter ant · Wood roach · Snail · Millipede · Centipede · Pseudoscropion · Sow bug · Cicada nymph · Wireworm · Constricting fungi killing a nematode · Earthworm · Mite · Springtail · Soil protozoans

of species found in the soil (Birch and Clark, 1953). The number of different species, representing practically every invertebrate phylum, found in the soil is enormous. There are 250 species of Protozoa alone in English soils (Sandon, 1927). The number of species of soil animals exclusive of Protozoa found in a variety of habitats in Germany varied from 68 to 203 (Frenzel, 1936). In the soil of a beech woods in Austria live at least 110 species of beetles, 229 species of mites, 46 species of snails and slugs (Franz, 1950). E. C. Williams (1941) counted 294 species of soil animals, exclusive of Protozoa, in the Panama Rain Forest.

Only a part of the soil litter is available to most soil animals as living space. Spaces between the surface litter, cavities walled off by soil aggregates, pore spaces between individual soil particles, root channels and fissures—all are potential habitats. Most of the soil fauna are limited to pores and cavities larger than themselves. The distribution of these forms in different soils is often determined in part by the structure of the soil (Weis-Fogh, 1948), for there is a relationship between the average size of soil spaces and the fauna inhabiting them (Kuhnelt, 1950). Large species of mites inhabit loose soils with crumb structure, in contrast to smaller forms inhabiting compact soils. Larger soil species are confined to upper layers where the soil interstices are the largest (Haarløv, 1960).

Water in the spaces is essential because the majority of soil fauna is active only in this. Soil water is usually present as a thin film lining the surface of soil particles. This film contains, among other things, bacteria, unicellular algae, protozoa, rotifers, and nematodes. Most of these are restricted in their movements by the thickness and shape of the water film in which they live. Nematodes are less restricted, for they can distort the water film by muscular movements and thus bridge the intervening air spaces. If the water film dries up, these species encyst or enter a dormant state. Millipedes and centipedes, on the other hand, are highly susceptible to desiccation and avoid it by burrowing deeper into the soil (Kevan, 1962).

Excess water and the lack of aeration are detrimental to many soil animals. Excessive moisture, which typically occurs after heavy rains, often is disastrous to some soil inhabitants. Air spaces become flooded with deoxygenated water, producing a zone of oxygen shortage for soil inhabitants. If earthworms cannot evade this by digging deeper, they are forced to the surface, where they die from excessive ultraviolet radiation. The snowflea (a collembolan) comes to the surface in the spring to avoid excess soil water from melting snow (Kuhnelt, 1950). Many small species and immature stages of larger species of centipedes and millipedes may be completely immobilized by a film of water and unable to overcome the

540

surface tension imprisoning them. Adults of many species of these organisms possess a waterproof cuticle that enables them to survive some temporary flooding.

Soil acidity long has been regarded as having an important effect on soil fauna. But because pH is readily measured, it has been overplayed in an attempt to correlate soil characteristics with the fauna. Bornebusch (1930) regarded a pH of 4.5 as inimical to earthworms, yet some earthworms, such as *Lumbricus rubellus*, are quite tolerant of relatively acid conditions. Although every species of earthworm has its optimum pH, and although some species, such as the *Dendrobaenus*, are characteristic of acid conditions, most of them seem to be able to settle in most soils, provided they contain sufficient moisture (Petrov, 1946). In northern hardwood forests, earthworms are most abundant both in species and in numbers when the pH is between 5.5 and 4.1 (Stegeman, 1960). Mites and springtails (Collembola) can exist in very acid conditions (Murphy, 1953).

THE FAUNAL COMMUNITY

The interrelations of the organisms living in the soil are very complex; but within the upper layers of the soil, energy flows through a series of trophic levels similar to those of surface communities.

The primary source of energy in the soil community is the dead plant and animal matter and feces from the ground layer above. These are broken down by the microbial life—bacteria, fungi, protozoans. Upon this base rests a phytophagous consumer layer, which obtains its nourishment from assimilable substances of living plants, as do the parasitic nematodes and root-feeding insects; from fresh litter as do the earthworms; and from the exploitation of the soil microflora. Some members of this latter group, such as some protozoa and free-living nematodes, feed selectively on the microflora. Others, including most earthworms, pot worms, millipedes, and small soil arthropods, ingest large quantities of organic matter and utilize only a small fraction of it, chiefly the bacteria and fungi, as well as any protozoa and small invertebrates contained within the material. On the next trophic level are the predators— the turbellaria, which feed on nematodes and pot worms, the predatory nematodes and mites, insects, and spiders. In such a manner does the community in the soil operate on an energy

FIGURE 17-5 [OPPOSITE]
Life in the soil. This drawing shows only a tiny fraction of the organisms that inhabit the soil and litter. Note the fruiting bodies of the fungi, which in turn furnish food for animals.

541

source supplied by the unharvested organic material of the world above.

Prominent among the larger soil fauna are the Oligochaetes, which include two common families, the Lumbricidae (earthworms) and the Enchytraeidae (white or pot worms). The latter are small, whitish worms, which abound in the upper 3 in. of the soil if humidity is fairly constant. They are able to live under a greater variety of conditions than earthworms, but their numbers undergo violent fluctuations. Populations are at a maximum in winter and at a minimum in summer. They are not extensive burrowers and appear to divide the earth and humus more finely than earthworms. Little is known about their feeding biology beyond that they ingest organic debris, from which they may digest bacteria, protozoa, and other micro-organisms (C. O. Nielsen, 1961).

Earthworms may have considerable influence on the physical structure of the soil, an influence that depends in part on the species inhabiting the soil, the relative proportions of the species, and their numbers. Some species, such as *Lumbricus terrestris* and *Allolobophora longa,* work deep, even down to the C horizon, and develop well defined burrow systems 3 to 6 ft deep. Others, such as *Eisenia rosea* and *A. Chlorotica,* are shallow-working and are more or less confined to the top 6 in. These earthworms of shallow depths have no well-developed burrow systems and produce few casts. Only under unfavorable conditions, such as drought, do these earthworms burrow deep into the soil.

Earthworm activity in the soil consists mainly of burrowing, of ingestion and partial breakdown of organic matter, and of subsequent egestion in the form of surface or subsurface casts. Ingested soil is taken during burrow construction, mixed with intestinal secretions, and passed out either as aggregated castings on or near the surface or as a semiliquid in intersoil spaces along the burrow. Earthworms pull organic matter into their burrows and ingest some of this, which is then partially or completely digested in the gut. Casts of soil passed through the alimentary canal contain a larger proportion of soil particles less than 0.002 in. in diameter than uningested soil and a higher total nitrogen, organic carbon, exchangeable calcium and magnesium, available phosphorus, and pH.

Surface casting and burrowing slowly overturn the soil. Subsurface soil is brought to the top and organic matter is pulled down into and incorporated with the subsoil to form soil aggregates. These aggregates result in a more open structure in heavy soil and bind particles of light soil together.

The amount of soil worked over by earthworms is tremendous. A. C. Evans (1948) reports that the weight of worm casts produced per year on English fields varied from 1 to 5 tons. This variance is due primarily to the different ratios of only two of the several earthworms present, *Allolobophora noctura* and *A. longa.* He estimated that from 4 to 36 tons of soil passes through the alimentary tract of the total earthworm population living on an acre in a year.

Earthworms show a decided preference for certain leaves. The night crawler, *Lumbricus terrestris,* feeds on broadleaf litter rather than pine and soft-textured foliage. In one experimental study (J. W. Johnston, 1936) these earthworms accepted immediately large-toothed aspen, white ash, and basswood, took with less relish and did not entirely consume sugar maple and red maple, and did not eat red oak at all. In a European study (Lindquist, 1942) earthworms preferred the dead leaves of elm, ash, and birch, ate sparingly of oak and beech, and did not touch pine or spruce needles.

Millipedes probably are the next most important group of litter-feeders. They and their somewhat similar associates, the centipedes, are essentially animals of the woodland floor. Millipedes occupy essentially three woodland habitats, the floor and aerial parts of vegetation, the litter and upper soil layer, and the areas beneath bark and stones and in rotten logs and stumps. The three most common forms are the oval pill millipedes, the glomerids; the flat-backed polydesmids with flattened lateral expansions; and the large iuloids. The former two are not adapted to burrowing and they must find refuge against both floods and drought in surface retreats. Iuloids, however, burrow extensively in the soil. Millipedes ingest leaves, particularly those in which some fungal decomposition has taken place, for lacking the enzymes necessary for the breakdown of cellulose, they live on the fungi contained within the litter. Different species of millipedes ingest varying quantities of litter, depending upon the tree species (Van der Drift, 1951). *Iulus* consumes more red-oak litter, *Cylindroiulus* more pine.

The chief contribution of millipedes to soil development and to the soil ecosystem is the mechanical breakdown of litter, making it more vulnerable to microbial attack, especially by the saprophytic fungi.

Litter-feeders of importance are the snails and slugs, which among the soil invertebrates possess the widest range of enzymes to hydrolyze cellulose and other plant polysaccharides, possibly even lignin (C. O. Nielsen, 1962). In Australian rain forests, amphipods are a conspicuous part of the fauna and play a major part in the disintegration of the leaf litter (Birch and Clark, 1953).

Not to be ignored are the termites (Isoptera), white, wingless, social insects. Termites, together with some dipteran and beetle larvae, are the only larger soil inhabitants that are able to break down the cellulose of wood. This they accomplish with the aid of a symbiotic protozoan living in the termite's gut. The termite has a mouth structure adapted to ingest wood; the protozoan produces the enzymes that effectively digest cellulose into the simple sugars that the termite can use. Together, the two organisms function perfectly. Without the protozoan, the termite could not exist; without the termite the protozoan could not gain access to wood.

In spite of their role in the disintegration of wood, termites do not play a major role in the temperate soils, but in the tropics they dominate the soil fauna. In these regions they are responsible for the rapid removal of wood and other cellulose-containing materials, twigs, leaves, dry grass, structural timbers, etc., from the surface. In addition to removal of organic matter, the termites are important soil churners. They move considerable quantities of soil, perhaps as much as 5,000 tons/acre, in constructing their huge and complex mounds. In semidesert country the openings and galleries of subterranean termites allow the infrequent rains to penetrate deep into the subsoil rather than to run off the surface (Kevan, 1962).

Of all the soil animals, the most abundant and widely distributed are the mites (Acarina) and the springtails (Collembola). Both occur in nearly every situation where vegetation grows, from the tropical rain forest to the tundra. Flattened dorsoventrally, they are able to wiggle, squeeze, and even digest their way through tiny caverns in the soil. Here they browse on fungi or search for prey in the dark interstices and pores of the organic mass.

The most numerous of the two, both in species and in numbers, are the mites, tiny,

543

eight-legged arthropods from 0.1 to 2.0 mm in size. The most common mites in the soil and litter are the Orbatei. In the pine-woods litter of Tennessee, for instance, they make up 73 percent of all the litter mites (Crossley and Bohnsack, 1960). These mites live largely on fungal hyphae that attack dead vegetation as well as the sugars digested by this microflora, which these arthropods follow in evergreen needles.

The Collembola are the most generally distributed of all insects. Typically the springtails may be brightly colored, or completely white, and possess a remarkable springing organ at their posterior end, which enables them to leap comparatively great distances. From this they have earned their popular name "springtails." The springtails are small, from 0.3 to 1 mm. They consist of two groups, the round springtails, or Symphypleona, and the long springtails, or Arthropleona. Neither have specialized feeding habits. They consume decomposing plant materials, largely for the fungal hyphae they contain.

The small arthropods are the principal prey of spiders, beetles, especially the Staphylinidae, the pseudoscorpions, the mites, and the centipedes. The latter are one of the major invertebrate predators. The two most common groups are the nonburrowing, swift-running lithobiomorphs and the geophilomorphs that burrow, earthwormlike, into the soil. Predacious Mesostigmata mites prey on herbivorous mites, nematodes, enchytraeid worms, small insect larvae, and other small soil animals.

Most of the microorganisms of the soil, the protozoa and rotifers, myxobacteria and nematodes, feed on bacteria and algae. Nematodes are ubiquitous, found wherever their need for a film of water in which to move is met. Soil and fresh-water nematodes form one ecological group, with many species in common. But in the soil they exist at much higher densities than in fresh water, up to 20 million/m². They are most abundant in the upper 2 in. in the vicinity of roots, where they feed on plant juices, soil algae, and bacteria. A few are predaceous.

Also found in the organic layer of woodland and meadow soils, especially those with a cover of moss, are other representatives of fresh-water inhabitants, the rotifers. They feed largely on detritus and algae. Tartigrades or "water bears," too, are moss dwellers, for here they find the alternating wet and dry conditions they require for their existence.

These bacteria- and algae-feeders, in turn, are consumed by various predacious fungi. Among these, there are three groups: (1) the Zoopagales, an order of Phycomycetes that preys chiefly on protozoans, although a few species prey on nematodes; (2) the endozoic Hyphomycetes; and (3) the ensnaring Hyphomycetes. The latter two capture and digest nematodes, crustaceans, rotifers, and, to an extent, protozoans (Maio, 1958; Doddington, in Kevan, 1955). The Zoopagales possess sticky mycelia, which capture the prey like flypaper. The endozoic Hyphomycetes release spores, which stick to the integument of nematodes. Germ tubes penetrate the tube of the animal and develop into internal mycelium.

The most remarkable group of fungi, extremely common in the soil, are the nematode-trapping Hyphomycetes, which possess morphological adaptations that enable them to capture their prey. One of the commonest forms of traps is a network of highly adhesive loops, which catch and hold nematodes on contact. After a struggle, the animal dies and a narrow branch of hyphae penetrates and fills the body of the animal and absorbs its contents until only the skin is left. Others possess sticky, knoblike processes, to which the nematodes adhere. But the most unusual of all is the rabbit-snare trap, of which there are two types, nonconstricting and constricting. Both possess rings of filaments attached by short branches to the main filament. Each ring trap consists of three curved cells; and its inside diameter is just large enough to permit a nematode attempting to pass through to become wedged and unable to withdraw. In the constricting type ring, the friction of the nematode's body stimulates the ring cells to inflate to about three times their former volume and to grip the nematode in a stranglehold. The response is rapid; complete distention of the cells is accomplished within one-tenth of a second.

Other groups of animals, although feeding largely on the surface and contributing little to litter breakdown, are important as soil mixers. Ants are especially important as soil animals, for they are widely distributed, pioneer new sites, and bring up large quantities of soil from below ground. Mounds of the harvester ant dot large areas of the North American plains. The ants on one area moved an estimated 3,400 tons of soil/acre in constructing their mounds (J. Thorpe, 1949). Prairie dogs raise earth from lower levels and deposit it at the surface, where it is broken down by weathering and incorporated with organic material. They carry surface soil down to plug passageways, and on clay soils they increase the proportion of fine soil particles on the surface. On one area prairie dogs and badgers were credited with converting a silt-loam soil to a loam (J. Thorpe, 1949). The amount of soil moved by prairie dogs is large. In northern Colorado, the average volume of earth in dog mounds is 3 ft^2; with a burrow density of 25/acre, the soil in the mounds would weigh over 3 tons (C. Koford, 1958). In central Oklahoma, the total volume of 12 burrows with 25 entrances and 599 ft of tunnels was 95 ft^3, or roughly 4 tons of earth moved for each 25 entrance holes (M. Wilcomb, 1954). Moles, too, move considerable quantities of earth, although the amount has not been calculated. Their varied influences include improving the natural drainage and aeration of soil and increasing organic matter by burying surface vegetation and litter under their hills.

Abundance and distribution. If the diversity of species is one characteristic of soil fauna, then their enormous numbers is another.

When the Italian entomologist Antonio Berlese developed the funnel extraction method in 1905, quantitative studies of soil fauna became possible. Later others improved the method (see Kevan, 1955, 1962). Bornebusch in Denmark made early faunal counts in beech mor and mull, as did Ulrich (1933) in Germany. Because their samples were too large and their extraction methods poor, they underestimated the populations (Van der Drift, 1951). Bornebusch's maximum number obtained for a beech mull was 79 million/acre; Forrslund in 1947 came up with 4,410 million/acre. In fact Forrslund, by directly examining small samples from Swedish forest sites, obtained a population equivalent of 2,300,000 arthropods/m^2.

Because of the variability of sample depths and sizes and of differing efficiencies of extractions, accurate comparisons cannot be made; but they do give some indication of the numbers involved. Overgaard (1949), in a study of 31 localities in Denmark, found that nematodes varied from 708 million to 81 million/acre. Evans (1948) estimated the earthworm population in a 300-year-old pasture at Rothamsted at 167,000/acre, and Bornebusch obtained a population figure of 1,450,000 earthworms/acre in a beech mull forest. Mites and Collembola are the dominant soil arthropods and constitute 85 percent of the total number of animals in the soil (Salt et al., 1948). Mites accounted for 83 percent of all the soil fauna in a pine woods in Tennessee, where the number of soil animals was estimated at 102,000/m^2. Protozoans occur in tremendous numbers. Flagellates appear to be the most comon, ranging from 100,000 to 1 million/g of soil; amoebas from 50,000 to 500,000; and ciliates up to 1,000/g of soil (Waksman, 1952). A majority of these also inhabit fresh waters.

The numerical population of soil animals in mor is much larger than in mull, mainly because mites and springtails dominate the former. In Denmark the number of soil animals increased from oaks to beech and spruce (Bornebusch, 1930; Stevanovic, 1956), but on a biomass basis, mull supported the largest population. Bornebusch found that 50 to 80 percent of the fauna in mull humus of the broadleaf forest were earthworms. He found, however, a tremendous variation in biomass in 10 Danish forest localities. Spruce raw humus had but 0.90 g/m^2, whereas a rich mull had a biomass of 200 g/m^2, or 1,590 lb/acre. This is equivalent to the weight of livestock carried by a first-class Danish pasture. Evans and Guild (1948) obtained a considerably smaller biomass of 163 g/m^2 on a 300-year-old English pasture.

Soil fauna shows a marked zonation, with the densest populations concentrated in the upper surface layers. Grasslands do not show the same strong demarkation between surface and deep horizons, for the humus is more uniformly distributed. Eaton and Chandler (1942) found that the greatest number of arthropods always occurred in samples composed of litter in the process of fragmentation. Large numbers continued to be present in the upper humus layer, but a gradual reduction took place in the deeper parts. In natural heathland representing a mor humus, 96 percent of the population was concentrated in the upper 2.5 in. and the remaining 4 percent was scattered through the profile. In grasslands, 67 percent of the mites and springtails inhabited the 0- to 6-in. zone and 33 percent in the 6- to 12-in. zone (Murphy, 1953). Oribatei mite populations are greatest in the O layer, since most of them

feed on fungi in the litter. Larger species of these mites have their optimum in the O_1 layer because they feed on molds and algae growing in the youngest litter material. There appears, however, to be a correlation between the size of animals and the structure of the soil. Larger species are confined to the upper layers; the smaller, flatter species are most numerous in the deep soil layers. Vertical distribution of microarthropods is determined largely by water content, food, size of the soil spaces, and light in the litter layer (Kevan, 1962).

Soil fauna show marked and often violent fluctuations in numbers. These fluctuations are both seasonal and diurnal and are reflected in changes in temperature and moisture. The greatest fluctuations occur at the surface and decrease with depth. The daily vertical migration of the springtails appears to be conditioned chiefly by the degree of solar radiation and the degree of dew accumulation at night (Jacot, 1940). The downward migration of these and other microarthropods, however, is impeded, if not inhibited, by the mineral layer of the soil (Haarløv, 1960). Some soil invertebrates move deeper into the soil during the fall and winter and return to the surface the following spring (Dowdy, 1944). This migration often coincides with temperature overturns. During periods of drought, some species, such as earthworms, withdraw from the upper layers; or like the nematodes, they may enter an anabiotic state to become active again when conditions are favorable; others, such as the springtails, may die out. Still other species, as the Oribatei mites, seem to be adapted to extreme environmental conditions.

Relationship of soil populations to cover types. Distribution of the animals on the surface is influenced by vegetation types, and certain animals are characteristic of each. Much the same situation appears to exist beneath the surface of the soil. The species composition and age of the vegetation stand, the base status of the soil, and drainage, among others, affect soil faunal composition. Mixed stands are richer both in species and in numbers than pine and spruce stands (Pschorn-Walcher, 1952). Accumulated coniferous needles may provide suitable environmental conditions, whereas deciduous litter provides a sufficiently palatable food source to maintain the litter-feeding population of soil mites at a high level (Murphy, 1952). This mixture effect also seems to extend to mixed coniferous stands. White pine–hemlock stands supported a more varied soil fauna than pure white pine or red pine stands (Bellinger, 1954). Litter fauna under hemlock seems to be poorer than under white pine (Hope, 1943). The fact that white pine litter has a calcium content of 1 to 2 percent undoubtedly influences this increase, just as in the case of mixed stands of deciduous and coniferous forms.

Deserts

Deserts are defined by geographers as land where evaporation exceeds rainfall. No specific rainfall can be used as a criterion, but deserts may range from extremely arid ones to those with sufficient moisture to support a variety of life. Deserts occur in two distinct belts about the earth, one confined roughly about the Tropic of Cancer, the other about the Tropic of Capricorn.

Deserts are the result of several forces. One that leads to the formation of deserts and the broad climatic regions of the earth is the movement of air masses over the earth (see Chapter 2). High-pressure areas alter the course of rain. The high-pressure cell off the coast of California and Mexico deflects rainstorms moving south from Alaska to the east and prevents moisture from reaching the southwest. In winter high pressure areas move southward, allowing winter rains to reach southern California and parts of the North American desert. Winds blowing over cold waters become cold also; they carry very little moisture and produce little rain. Thus the west coast of California and Baja California, the Namib desert on coastal southwest Africa, and the coastal edge of the Atacama in Chile may be shrouded in mist yet remain extremely dry.

Mountain ranges also play a role in desert formation by causing a rain shadow on their lee side. The High Sierras and the Cascade Mountains intercept rain from the Pacific and help maintain the arid conditions of the North American desert. And the low eastern highlands of Australia effectively block the southeast trade winds from the interior of that continent. Other deserts, such as the Gobi and the interior of the Sahara, are so remote from the source of oceanic moisture that all the water has been wrung from the winds by the time they reach those regions.

547

CHARACTERISTICS

In spite of differences in physical appearances, all deserts have in common low rainfall, high evaporation (7 to 50 times as much as precipitation), and a wide daily range in temperature, from hot by day to cool by night. Low humidity allows up to 90 percent of solar insolation to penetrate the atmosphere and heat the ground. At night the desert yields the accumulated heat of the day back to the atmosphere. Rain, when it falls, is often heavy and, unable to soak into the dry earth, rushes off in torrents to basins below.

The topography of the desert, unobscured by vegetation, is stark and, paradoxically, partially shaped by water. Unprotected, the soil erodes easily during violent storms and is further cut away by the wind. Alluvial fans stretch away from eroded, angular peaks of more resistant rocks. They join to form deep expanses of debris, the *bajadas*. Eventually the slopes level off to low basins, or *playas*, which receive waters that rush down from the hills and water-cut canyons, or *arroyos*. These basins hold temporary lakes after the rains, but water soon evaporates and leaves behind a dry bed of glistening salt.

The aridity of the desert may seem inimical to life, yet in the deserts life does exist, surprisingly abundant, varied, and well adapted to withstand or circumvent the scarcity of water.

PLANT LIFE

The ways by which plants and animals of arid regions meet the water problem have been discussed. Essentially plants either are drought evaders or drought resisters. The evaders persist as seeds, ready to sprout and grow when moisture and temperature are favorable, to flower, produce seeds, and die. Drought resisters have evolved means of storing water, locating underground water, or reducing needs by shedding leaves and by reducing leaf size to cut down on transpiration. They may carry on photosynthesis in the stem. Desert plants may take in carbon dioxide by night and fix it, making it available during the day, when, because of stomatal closing, the intake is less. All this results in plants peculiar in habit and structure.

In sharp contrast to the forest, the struggle in the desert is not one of plant against plant for light and space, but one of all plants for moisture. Because little layering exists in the desert plant community, light is not a problem. Because moisture is limited, plants are mostly low; and because of competition for moisture, the plants are widely spaced.

Woody-stemmed and soft brittle-stemmed shrubs are characteristic desert plants. In the matrix of shrubs grows a wide assortment of other plants, the yuccas, cacti, small trees, and ephemerals. In the southwestern American desert large succulents rise above the shrub level and change the aspect of the desert all out of proportion to their numbers (Fig. 17-6). Like forest trees and prairie grasses, most desert species grow their best in certain topographical situations. The giant saguaro, the most massive of all cacti, grows on the bajadas of the Sonoran desert with other smaller, brilliant-flowered cacti. Other plants—ironwood, smoke tree, palo verde—grow best along the banks of intermittent streams, not so much because of their moisture requirements but because their hard-coated seeds must be scraped and bruised by the grinding action of sand and gravel during flash floods before they can germinate.

The ephemeral, or annual-flowering, plants supply seasonal brilliance to the desert. Surviving from one favorable period to another as seeds, they flourish on sandy soils, for here moisture easily penetrates the soil, and the ground warms rapidly, favoring germination of seeds. Although the annuals of the southwestern American desert are in so many ways similar to the annuals of more humid climates, they differ in one highly specialized way, the time of germination. This must correspond with periods of adequate rainfall if the population is to persist. There are two periods of flowering in the American desert, after the winter rains come in from the Pacific northwest and after the summer rains move up from the southwest out of the Gulf of Mexico. Some species flower only after the winter rains, others after the summer, while a few flower during both seasons.

Why these plants respond so unerringly to moisture has been the subject of considerable research (Went, 1955). Out of this have come the discoveries that none of the seeds would germinate unless the surface soil received an equivalent of 0.5 to 1 in. of rain, and that the seeds have on their coats a water-soluble growth inhibitor, which is washed away by a sufficient amount of percolating water. Thus the seeds are prevented from germinating after a trivial shower. The temperatures likewise are critical. Each species of desert annual has a definite and rather narrow limit of tolerance and will not grow in any quantity except when the moisture and temperatures are within these limits. For this reason the winter and summer annuals can be divided according to the temperatures at which they germinate and at which they survive. For winter annuals to germinate the first rains have to be well over 10 mm, and preferably over 20, and the night temperatures have to be above freezing.

ANIMAL LIFE

Under a harsh and shimmering sun, the deserts may appear void of animal life, yet animals do thrive there. Like plants, desert animals are either drought resisters or drought evaders. Drought evaders, like the ephemeral plants, make a sudden and dramatic appearance. For 8 or 9 months, perhaps even several years, eggs of insects and other invertebrates and insect pupae lie dormant. When the rains arrive and plants flourish, the deserts swarm with insects —crickets, grasshoppers, ants, bees and wasps, butterflies, moths, and beetles. Young bees emerge from underground cells at the very time the particular flowers on which they feed are in flower. Amphibians, such as the spadefoot toad, also make a brief appearance during the periods of winter and summer rains.

Birds nest during the rainy season when food is most abundant for the young. If extreme drought develops during the breeding season, some birds do not reproduce. Among some desert birds, the endocrine control of reproduction seems to depend upon rainfall rather than upon day length (Keast, 1959). A few birds, such as the swift, poorwill, and Allen and Anna's hummingbirds, become torpid when food is scarce (Bartholomew, Howell, and Cade, 1959). Small rodents, such as the kangaroo or pocket mice, estivate during periods of most severe drought.

But the desert also contains many animals that are active the year round, all of which have evolved ways of circumventing aridity and heat by physiological adaptations or by feeding and activity habits. By restricting their activity to the cooler parts of the day, early morning, late evening and night, these animals escape the heat of the desert. It is no coincidence that most desert animals are nocturnal. The most conspicuous daytime animals of the desert are

FIGURE 17-6
Saguaro dominates the aspect of this desert. Note the water-shaped topography. (Photo courtesy Arizona Fish and Game Department.)

549

lizards; snakes, much less abundant, are active by night. Rodents are chiefly burrowing and spend the hot day underground, where temperatures may be 70° F lower than on the sun-baked surface.

Most desert insects are abundant only during the rainy period, when their emergence is timed with the flowering of desert plants. But some are active throughout the year, the most notable of these being the harvester ant, which lives on seeds gathered from the desert floor and stored in underground granaries. During periods of drought these ants in the California desert gather mainly the seeds of two kinds of plants, wooly plantain and comb bur. When the winter rains come and the annual plants flower and seed, the ants gather the seeds of these other plants and ignore for the time being the seeds of plantain and the comb bur. But when the dry season returns, the ants again exploit their staple food source. This utilization of a constant food supply enables the insect to remain active through the year.

NORTH AMERICAN DESERTS

The desert is not the same everywhere. Differences in moisture, temperature, soil drainage, alkalinity, and salinity result either in the variation of dominants or in different groups of associated plants. Basically, the North American desert can be divided into two parts, the northern cool desert—the Great Basin—and the hot desert of the southwest—the Mohave, the Sonoran, and the Chihuahuan. The two, however, grade one into the other.

Sagebrush is the dominant plant of the cool desert, together with saltbush, shadscale, hop sage, winterfat, and greasewood. The plant communities of the Great Basin are simple, consist of essentially the same life forms, and form nearly the same stands of monotonous gray and gray-green for miles.

Hot deserts are dominated mostly by creosote bush accompanied by bur sage or burro bush. Together they often form a monotonous uniform growth broken only in areas of favorable moisture and soil by tall growths of acacia, saguaro, palo verde, and mesquite. Of the hot deserts the Mohave, transitional between the Great Basin and the Sonoran, is the poorest in species, has the simplest vegetational composition, and contains some species in common with the Great Basin. Mostly it is a rolling plain covered with creosote bush and sagebrush, its monotony broken by volcanic mountains and the curious Joshua trees, a spine-studded yucca found nowhere else.

South of the Mohave is the Sonoran desert, which contains a richly diversified vegetation of shrubs, trees, and succulent cacti (Shreve, 1951). Here the vegetation is the most dense of the American desert, the life forms—from small drought-deciduous trees to evergreen semishrubs and succulents—are most numerous, the number of species is the greatest, and the height most diverse. Although the physiognomy and structure of the vegetation throughout this desert is very similar, the dominants and subdominants differ. And although the life forms and structure are nearly the same, the floral composition is different. On higher slopes where more moisture is available grow the deep red-flowered ocotillo and the green-barked palo verde (Fig. 17-7). The drier plains are occupied by low-growing flowers, jatrophos, brittle bush, acacia, and organ-pipe cactus, while in the low washes smoke trees grow.

Within the Sonoran desert grows the richest variety of ephemerals, for here exists a variety of conditions not found elsewhere in the desert country. Many of these and other small plants grow in close association with or beneath the shade of old trees and shrubs, where conditions produced by an accumulation of wind-blown soil, organic debris of leaves, and broken plant stems about the bases of the large plants are more favorable for their germination and growth (C. H. Muller, 1953).

The Chihuahuan desert, which extends into Mexico, presents a different aspect. Intermediate between the Great Basin and the Sonoran, this desert is characterized by long-barreled cacti, yuccas, candelilla, and guayule. On the valley floors grow large patches of colo del zorro, a foxtail-like grass; and on the wind-ripple gypsum dunes a vigorous, aromatic, sumac squawbush, once used by Indians for baskets.

IMPACT OF MAN

Man, too, has had a hand in the creation of new deserts through poor land management, especially around the periphery of existing desert areas. Virgin lands, even in dry climates, are able to support some vegetation. The roots of trees, shrubs, and grasses tap the deeper water supply and bind the soil, but once these are eliminated, through fire, grazing, and cultivation, the destructive forces of erosion are

FIGURE 17-7 [OPPOSITE]
Palo verde–saguaro community. Rainfall here is about 14 to 16 in. In the distance the east-facing slopes are covered with interior chaparral. (Photo courtesy Arizona Fish and Game Commission.)

loosened. Fires are set to clear semiarid land for cultivation or to encourage the growth of grasses and eliminate the trees. Then, overcultivated and overgrazed, the land is exposed to wind and water erosion. Exhausted from cultivation, the lowlands are abandoned to grazing, first by cattle, then later by sheep and goats. Cultivation moves up to the high lands, and here the same cycle is repeated. Eventually the destruction is total. The vegetation and top soil are gone; bedrock is exposed; the land has reached the point of no return. In this way, the deserts are marching south through Africa into the African plains, and similarly they are encroaching on the semiarid grasslands of North America. The same land-use pattern is thought to have played a part in the downfall of the Mayan civilization and the decline of Greece and Rome. The once-rich lands of the southern Mediterranean that supported a great civilization are now barren, rock-filled, and impoverished. (For further discussion see Lowdermilk, 1953; Dale and Carter, 1955; Thomas, 1956; Stamp, 1961).

In recent years man has attempted, with great success, to bring some deserts back into production. But in doing so he is tapping reservoirs of water deep beneath the desert floor. Such "mining" of water is filled with danger, for once this water, like mineral wealth, is gone, it is irreplaceable. Thus the short-term agriculture and human occupancy of the desert may end with an even more sterile environment.

Grasslands

When the explorers looked out across the prairies for the first time, they witnessed a scene they had never before experienced. Nowhere in all western Europe had they seen anything similar. Lacking any other name to call them, the explorers named these grasslands "prairie," from the French, meaning "grassland." This was the North American prairie and plains, the climax grassland that occupied the midcontinent (Fig. 17-8). It was one of the several great grassland regions in the world, including the steppes of Russia, the pustza of Hungary, the South African veld, and the South American pampas. In fact at one time grasslands covered about 42 percent of the land surface of the world; but today much of it is under cultivation. All have in common a climate characterized by high rates of evaporation and periodically severe droughts, a rolling-to-flat terrain, and animal life that is dominated by grazing and burrowing species. They occur largely where rainfall is between 10 and 30 in./year, too light to support a heavy forest growth and too great to encourage a desert. Grasslands, however, are not exclusively a climatic formation—most of them require periodic fires for maintenance and renewal and for the elimination of incoming woody growth (see Chapters 7 and 9).

CHARACTERISTICS OF GRASSLANDS

The grasses that make up the haylands, pastures, and prairies are either sod formers or bunch grasses. As the names imply, the former develop a solid mat of grass over the ground, and the latter grow in bunches (Fig. 17-9), the space between which is occupied by other plants, usually herbs. Orchardgrass, broomsedge, crested wheatgrass, and little bluestem are typical bunch grasses, which form clumps by the erect growth of all shoots and spread at the base by tillers. Sod-forming grasses, which include such species as Kentucky bluegrass and western wheatgrass, reproduce and spread by underground stems. Some grasses may be either sod or bunch, depending upon the local environment. Big bluestem will develop a sod on a rich, moist soil and form bunches on a dry soil.

Associated with the grasses are a variety of legumes and forbs. Cultivated haylands and pastures usually are planted to a mixture of grasses and such legumes as alfalfa and red clover. With these may grow unwanted plants, such as mustard dandelion and daisy. Seral grasslands often consist of a mixture of native grasses, such as timothy and bluegrass, and an assortment of herbaceous plants, including cinquefoil, wild strawberry, daisy, dewberry, and goldenrod. On the prairie, legumes and forbs, particularly the composites, are important components of the climax grassland (J. E. Weaver, 1954). From spring to fall, the color and aspects of the grassland change from pasqueflower and buttercups to goldenrod.

STRATIFICATION

Grasslands possess essentially three strata, the roots, the ground layer, and the herbaceous layer (Fig. 17-10). The root layer is more pro-

FIGURE 17-8
Beyond the tall-grass prairie and mixed-grass prairies, now largely in crops, lie the short-grass plains dominated by buffalo grass and blue grama. (Photo by author.)

FIGURE 17-10
A cut through a hayfield showing the several strata in grasslands. Note the tall grass (timothy) and the denser understory of legumes (clover and alfalfa) and grass blades. (Photo by author.)

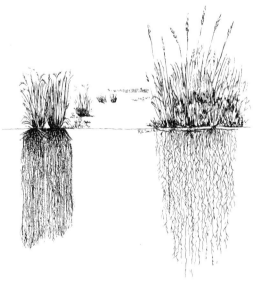

FIGURE 17-9
Growth form of a sod grass (right), crested wheatgrass, and a bunch grass (left), little bluestem. Also shown is root penetration and distribution in the soil (maximum depth: 7 ft.)

nounced in grasslands than in any other major community. Half or more of the plant is hidden beneath the soil; in winter this represents almost the total grass plant, a sharp contrast to the leafless trees of the forest. The bulk of the roots occupy rather uniformly the upper 0.5 ft or so of the soil profile and decrease in abundance with depth. The depth to which the roots of grasses extend is considerable. Little bluestem reaches 4 to 5.5 ft (1.3 to 1.7 m) and forms a dense mat in the soil to 2.5 ft (0.8 m) (J. E. Weaver, 1954). Roots of blue grama and buffalo grass penetrate vertically to 3 ft (1 m). In addition, many grasses possess underground stems, or rhizomes, that serve both to propagate the plants and to store food. On the end of the rhizome, which has both nodes and scalelike leaves, is a terminal bud, which develops into aerial stems or new rhizomes. Rhizomes of most species grow at shallow depths, not over 4 or 5 in. deep. The exotic quackgrass is notorious among farmers for its tough rhizomes. Forbs such as goldenrod, asters, and snakeroot possess large, woody rhizomes, and fibrous roots that add to the root mat in the soil. Some, such as snakeroot, have extensive taproots 16 ft (5 m) long, and rushlike lygodesmia, common in many prairies, extends down 21 ft (6+ m) in mellow soil. Among hayland plants, alfalfa possesses a taproot that grows to a considerable depth.

All the roots of grassland plants are not confined to the same general area of the soil but develop in three or more zones. Some plants are shallow-rooted and seldom extend much below 2 ft. Others go well below the shallow-rooted species but seldom more than 5 ft. Deep-rooted plants extend even further into the soil and absorb relatively little moisture from the surface soils. Thus plant roots absorb nutrients from different depths in the soil at different times, depending upon moisture (J. E. Weaver, 1954).

The ground layer is characterized by low light intensity during the growing season and by reduced windflow. Light intensity decreases as the grass grows taller and furnishes shade. Temperatures decrease as solar insolation is intercepted by a blanket of vegetation, and windflow is at a minimum. Even though the grass tops may move like waves of water, the air on the ground is calm. Conditions on grazed lands are different. Because the grass cover is closely cropped, the ground layer receives much higher solar radiation and is subject to higher temperatures and to greater wind velocity near the surface.

Grasslands, unmowed, unburned, and ungrazed, accumulate a layer of mulch on the ground surface. The oldest layer consists of decayed and fragmented remains of fresh mulch. Fresh mulch consists of residual herbage, leafy and largely undecayed. Three or 4 years must pass before natural grassland herbage will decompose completely (H. H. Hopkins, 1954). Not until mulch comes in contact with mineral soil does the decomposition process, influenced by compaction and depth, proceed with any rapidity. As the mat increases in depth, more water is retained, creating very favorable conditions for microbial activity (McCalla, 1943).

The amount of accumulated mulch often is enormous. On a relict of a climax prairie, organic matter and other humic materials amounted to 885.4 g/m², 581.1 g of it fresh mulch and 50.1 g fresh herbage (Dhysterhaus and Schmutz, 1947). Another prairie supported 461.2 g/m² of fresh mulch and 830.1 g/m² of humus (Dix, 1960).

Grazing reduces mulch, as do fire and mowing. Light grazing tends to increase the weight of humic mulch at the expense of fresh (Dix, 1960); moderate grazing results in increased compaction, which favors an increase in microbial activity and a subsequent reduction in both fresh and humic mulch. Heavy grazing of a stand of bluestem reduced mulch from 1014.7 g/m² to 100.8 g/m² (Zeller, 1961). An ungrazed North Dakota prairie averaged 441.2 g/m² of mulch compared to 241.5 g/m² for a grazed prairie. Burning reduces both, but the mulch structure returns 2 to 3 years after a fire on lightly grazed and ungrazed lands (Tester and Marshall, 1961; Hadley and Kieckhefer, 1963). Mowing greatly reduces fresh mulch and in a matter of time humic mulch also. An unmowed prairie accumulated 4.5 tons of humic mulch/acre; a similar prairie, mowed, had less than 1 ton (Dhysterhaus and Schmutz, 1947).

The influence of mulch on grasslands still is a point of controversy (see Tomanek, 1969). Mulch increases soil moisture through its effects on infiltration and evaporation; it decreases runoff, stabilizes soil temperature, and improves conditions for seed germination. In range management the question is how much natural mulch is needed for sustained yield of grass. Where mulch can accumulate, grassland maintains itself; but in areas of little or no accumulation, regression sets in, and the grassland deteriorates to weeds or mesquite. On the other hand, accumulation of mulch from one species can have a toxic effect on the germination and development of another. Heavy mulches can result in pure stands of some species or the invasion of forbs and woody vegetation. Some range ecologists maintain that a heavy mulch results in decreased forage production, smaller root biomass, lower caloric value of living shoots (Hadley and Kieckhefer, 1963), and affects the character and composition of grasslands by reducing understory plants (Weaver and Rowland, 1952).

The herb layer may vary from season to season and from year to year, depending upon the moisture supply. Essentially the layer consists of three or more strata, more or less variable in height, according to the grassland type (Coupland, 1950). Low-growing and ground-hugging plants, such as wild strawberry, cinquefoil, violets, dandelions, and mosses, make up the first stratum. All of these become hidden, as the season progresses, beneath the middle and upper layers. The middle layer consists of shorter grasses and such herbs as wild mustard, coneflower, and daisy fleabane. The upper layer consists of tall grasses and forbs, conspicuous mostly in the fall.

GRASSLANDS AND GRAZING

Because the prairie and plains once held the greatest concentration of herbivores in the world, it was only natural that, once settled, the same lands should support a grazing economy. Scattered widely across the plains were herds of buffalo and pronghorned antelope, animals that rarely stayed on one area long enough to overgraze the range. Their numbers were checked by wolves, coyotes, and Indians, who depended on the buffalo for their way of life. But in a short 80 years after the white man settled the plains, the buffalo was gone, and replaced first by the longhorns and later by modern beef cattle. Fences were built and the ranges were overstocked. Grazing too early in the spring to allow the plants to grow sufficient food reserves for winter soon upset the natural balance of the range. The grasslands degenerated.

Unable to withstand heavy grazing, such highly palatable plants as big and little bluestem, prairie and tall dropseed, and nitrogen-fixing legumes and forbs disappear, often within

2 to 3 years. Freed from competition, bluegrass and side-oats grama, daisy fleabane and ironweed increase. If the prairie continues to deteriorate, the area is invaded by such plants as weedy wheatgrass, brome grass, little barley, and annual dropseeds (Voigt and Weaver, 1951; Curtis, 1959). Degeneration is not confined above ground; it goes on underground as well. The dry weight of roots of palatable grasses such as bluestem decreases up to 75 percent as the grassland deteriorates from high-grade to low-grade pasture (J. E. Weaver, 1954). Grazing on desert grasslands increases the spread of mesquite because of the lessened competition from grass and because of seed dissemination by domestic stock (Humphrey, 1958). Protection from fire adds to this.

There are other consequences. Because no litter is added to the ground, mulch deteriorates and disappears. Water, once retained and fed into the soil, runs off the surface, taking the topsoil with it. Lacking moisture and nutrients, the original plants cannot maintain themselves, and the vegetative cover continues to decrease until only an erosion pavement remains (for a full discussion see Ellison, 1960). To maintain grasslands in good condition, at least one-third of the year's growth must be left to supply the annual addition to the mulch (Dhysterhaus and Schmutz, 1947).

GRASSLANDS AND DROUGHT

Dry weather for years at a time has always been a part of the plain's climate. Some grassland plants, such as blue grama, buffalo grass, and even bluegrass survive short periods of drought by becoming dormant; others, such as big bluestem, survive by deep rooting. A number of grasses have a short growing season during the cool part of the year. Some prairie grasses respond quickly to moisture at the end of a dry spell. Buffalo grass, although more sensitive to drought than blue grama, revives more quickly when the rains return. When drought is prolonged, however, the vegetational composition of the grassland changes. During the 7 years of drought during the 1930s, little bluestem was reduced up to 75 percent over much of its range, as were shallow-rooted forbs such as stiff sunflower and prairie coneflower (Coupland, 1958, 1959). Bare areas caused by their death were filled by the highly drought-resistant western wheatgrass, so successful a competitor for a scant supply of water that it caused the

death of more mesic species. In places, the depleted grasslands, particularly those covered by dust, were invaded by such weedy plants as many-flowered aster and daisy fleabane (Robertson, 1939). Big bluestem persisted and needlegrass, side-oats grama, and bluegrass spread eastward into the prairie. In the short-grass prairies blue grama and buffalo grass decreased from a basal area of 89 percent to 22 percent (Weaver and Albertson, 1956). Essentially the mesic tall-grass prairie became a mixed-grass prairie; the mixed changed to a short-grass plains and the latter became a dust bowl. When moisture returned to normal, the original communities were restored, except in the Dust Bowl, where so much of the original soil was lost through wind erosion in the droughts of the 1930s that the area has not yet fully recovered.

GRASSLAND ANIMAL LIFE

Seral or climax, grasslands support similar forms of life, although only in the prairies and plains do the distinctive grassland animals, such as buffalo, prairie chickens, and prairie dogs, live. Within the several strata of vegetation, the roots, ground layer, and herb cover, much of the animal life exists (Fichter, 1954). Invertebrates, particularly insects, which occur in an incredible number and variety of species, occupy all strata at some time during the year. Throughout the 12 months, there is a definite seasonal distribution. During winter, insect life is confined largely to the soil, litter, and grass crowns, where these organisms overwinter as pupae or eggs. Soil occupants in spring are chiefly earthworms and ants, the latter being the most prevalent. In some eastern meadows they make up to 26 percent numerically of the insect populations (Wolcott, 1937). The ground and litter layer harbors the scavenger carabid beetles and predaceous spiders, the majority of which are hunters rather than web-builders, since supports for strong webs are limited. Life in the herbaceous layer varies as the strata become more pronounced from summer to fall. Here invertebrate life is most diverse and abundant. Homoptera, Coleoptera, Orthroptera, Diptera, Hymenoptera, Hemiptera, all are represented. The cicadellid leafhoppers rise in swarms as one walks through the grass. Frothy secretions surrounding young spittlebugs cling to plant stems during the summer months. Grasshoppers and field crickets, insect singers of the fields, are most abundant in late summer and early fall. Insect life reaches two highs during the year, a major peak in summer and a less well-defined one in the fall.

Mammals are the most conspicuous vertebrates of the grasslands and the majority of these are herbivorous. On the grasslands evolved a large and rich ungulate fauna. The bison (Fig. 17-11) and antelope of North America were equaled in numbers only by the richer and more diverse ungulate fauna of the East African plains. Today herds of cattle have replaced the buffalo, and rodents and rabbits now have the distinction of being the most abundant native vertebrate herbivores, many of which are fossorial or burrowing.

Grassland animals all share some outstanding traits. Hopping or leaping is a common method of locomotion among grasshoppers, jack rabbits, and jumping mice. Long, strong hind legs enable these animals to rise above the level of the grass tops, where visibility is unimpeded. Speed, too, is well developed. Some of the world's fastest mammals, such as the prong-horned antelope, live in the grasslands. Because of the dense, thick grass and the lack of trees for singing perches, some grassland birds, especially the bobolink, meadowlarks, and horned larks have conspicuous flight songs, which advertise territory and attract a mate.

Across the prairies and plains a very close relationship exists between plants and animals. The two most typical and conspicuous plains animals are the prairie dog and the harvester ant, whose burrows and mounds dot the landscape. Both have several traits in common; they denude areas about the mound or burrow, they turn over and mix considerable quantities of soil, and they influence the nature of the grassland community. Occasionally prairie dogs will invade a clearing about an ant hill; and just as often the ants will establish hills on top of prairie-dog mounds (Koford, 1958).

The relationship of the prairie dogs to vegetation is very complex, yet typical of a range rodent's influence on plant life. Roundly condemned by many ranchers for the destruction of plant cover and reduction of grazing capacity, the prairie dog is regarded by others as instrumental in creating and maintaining the grassland. The virtues of the animal appear to change from one region of the plains to another. Because tall grass inhibits the activity of prairie dogs—they cannot see over it or move

557

through it with ease—the range of the rodent once ended at the edge of the tall and midgrass prairie. When the tall grasses were destroyed by overgrazing and replaced by short grasses and forbs, the prairie dogs moved in, a symptom of range deterioration (Osborn and Allen, 1949). Over much of the plains, however, the prairie dog appears to be instrumental in the development and maintenance of a short-grass community, especially when they have some assistance from grazing animals—the buffalo of the past and the cattle of today.

Prairie dogs, pocket gophers, and other burrowing rodents affect other life in the plains. Plain cottontails appear to be most abundant where the burrowing rodents are plentiful, since the rabbits use the burrows for cover and nesting sites. Camelback crickets, beetles, and mites live in the relatively constant temperature and humidity of the rodent burrows. Deer mice feed on the seeds of annual forbs that grow on the mounds and seek the supply stored by harvester ants. Coyotes, bobcats, badgers, hawks, owls, and snakes live on rodents and rabbits. Elimination of rodents checks the growth of forbs, the preferred food of pronghorned antelope; and the open ground, once inhabited by deer mice, supports a thick growth of grass and a heavy mulch that are attractive to meadow mice. Deep litter appears to be an important feature of bobolink and savanna sparrow habitat (Tester and Marshall, 1961). Grasshopper populations seemingly inhibited by deep litter are highest in areas of light-to-moderate litter depth. The interrelationship between animals and litter depth needs much more investigation.

The grassland grouse, like the prairie dog and buffalo, once were abundant on the plains. In the tall-grass country on the edge of the eastern forest lived the sharp-tailed grouse. Its habitat is a combination of grassland and brushland, maintained chiefly by burning. Once this land is protected from fire, the forest returns and the sharptails go (Hamerstrom et al., 1957). This is the high price being paid for overextended fire control and the reforestation of abandoned land. Prairie chickens, a subspecies of the eastern heath hen, occupy the grassland proper, although the ranges of the two grouse overlap (Fig. 17-12) and the prairie chicken utilizes some light shrub cover. At one time their low, rolling boom haunted the grasslands in spring, but overgrazing, plowing, and poor grassland management are destroying increasing amounts of prairie-chicken habitat. The species is in danger of following the heath hen into oblivion.

Scattered through the northern prairies and plains are small bodies of water (Fig. 17-13)—potholes, marshes, and sloughs, surrounded by cattails, bulrushes, and sedges. These potholes are highly attractive to ducks, coots, rails, gulls, and other water birds. In fact this area produces or produced the bulk of ducks in North America. Today a majority of these potholes have been and are being destroyed by drainage, reducing the waterfowl populations. This, like the plight of the prairie grouse, illustrates an ecological truism that without a place to live no species can exist.

Animal life in the seral and tame grasslands of eastern North America depends upon management by man for the maintenance of its habitat. But mowing, man's major management tool, also results in the destruction of habitat at a critical time of year. Nests of rabbits, mice, and birds are exposed at the height of the nesting season. Losses often are heavy from both mechanical injury and predation, although most species will remain on the area to complete or reattempt nesting activity. Increased popularity of grassland farming with emphasis on grass ensilage and hay further reduces the value of such lands for many species of bird and mammals. Grass for ensilage is cut early, at the very start of the breeding season, eliminating nesting cover and forcing most of the population elsewhere. This is one of the causes for the decline in numbers of such farm game species as bobwhite quail and cottontail rabbits.

Pasturelands, more often than not, are so badly overgrazed that they support little in the way of vertebrate life. The two most common inhabitants are the killdeer and horned lark. Rotation pastures, those broken down into small units that are grazed on a rotational basis, may support more grassland life, but this still needs to be determined.

Seral grassland, because of its infertility and plant cover of poverty grass and broomsedge, does not usually support as wide a variety of life as hay fields. Poverty grass–dewberry fields are inhabited by grasshopper sparrows, vesper sparrows, horned larks, and meadow mice, but deep-grass species, the meadowlark, bobolink, and Henslow sparrow, are few, if not entirely absent. Broomsedge fields contain grasshopper sparrows, meadowlarks, meadow mice, and cotton rats. Both seral types offer poor browse for herbaceous species such as cottontail rabbits

FIGURE 17-11 [OPPOSITE]
Buffalo, which once roamed the plains in countless numbers, epitomize the North American grasslands. (Photo courtesy South Dakota Fish and Game Department.)

FIGURE 17-12
An interspersion of brushy areas and grassland furnishes a habitat for both sharp-tailed grouse and prairie chicken. If the shrub growth increases, the prairie chicken will disappear. (Photo courtesy Michigan Conservation Department.)

and deer, although deer do feed on young broomsedge sprouts in early spring. Otherwise these dominant grasses are unpalatable to cattle and to native herbivores alike.

GRASSLAND TYPES

Eastern grasslands. Grasslands in the eastern forest region are either tame or seral. In highly developed agricultural areas, tame grasslands are the major representatives of this vegetation type and their development and maintenance enabled a number of grassland species—grasshopper sparrows, dickcissels, bobolinks, meadowlarks, prairie deer mice, and cottontail rabbits—to expand their range.

Tame grasslands usually are more rank and dense than seral ones and must be managed to be maintained. Management consists of fertilization, mowing, plowing, and reseeding when needed, or at regular intervals if hay and pasture are a part of the crop rotation. Haying removes cover and exposes the ground surface to high solar radiation in late spring and early summer. Seral grasslands usually contain a mixture of native and tame grasses, especially on more fertile soils. Because much of the land is abandoned because of its natural or man-made infertility, these grasslands are dominated either by poverty grass, cinquefoil, and dewberry, or by broomsedge.

Tall-grass prairie. The tall-grass prairie occupies, or rather occupied, a narrow belt running north and south next to the deciduous forest. In fact, it was well developed within a region that could support forests. Oak–hickory forests did extend into the grassland along streams and rivers, on well-drained soils, sandy areas, and hills. Prairie fires often set by Indians in the fall stimulated a vigorous growth of grass and eliminated the encroaching forest. When the settlers eliminated fire, oaks invaded and overtook the grassland (Curtis, 1959).

Big bluestem was the dominant grass of moist soils and occupied the valleys of rivers and streams and the lower slopes of the hills. The foliage stood two to three feet tall and the flower stalks 3 to 12 ft, so high that cattle were hidden in the grass. A sod former, big bluestem occupied perhaps only 17 percent of the soil surface, yet the foliage was so thick and spread so widely that few plants were able to grow in the understory. Associated with bluestem were a number of forbs, goldenrods, compass plants,

snakeroot, and bedstraw. Although grasses dominated the biomass, they were not numerically superior. Studies on remnant prairies in Wisconsin (Curtis, 1959) show that legumes comprised 7.4 percent of all species, grasses 10.2, composites 26.1. The high percentage of nitrogen-fixing legumes accounts in part for the annual production of 7,000 to 9,000 lbs of dry matter/acre (8,500 k/acre).

Drier uplands in the tall-grass country were dominated by the bunch-forming needlegrass, side-oats grama, and prairie dropseed. Like the lowland, the drier prairie contained many species other than grass. In Wisconsin, composites accounted for 27.5 percent of all species, butterfly weed and legumes 4.6 percent each, and grasses 13.7 (Curtis, 1959). The suggestion has been made that perhaps the xeric prairie might be more appropriately called "daisy-land."

Like the forest, the tall-grass prairie, as well as other types, is a continuously changing series of species ranging from those best adapted to wet, poorly aerated soils, such as slough grass, through those plants represented by big bluestem that flourish on mesic soils of high fertility, to those such as blue grama and a whole host of colorful forbs that dominate the xeric sites. Interestingly, no important genera of grasses or forbs have species that are at their optimum in all sections of the continuum nor have all their species with an optimum at any one particular point on the gradient (Curtis, 1959).

Mixed prairie. West of the tall grass is the mixed-grass prairie in which midgrasses occupy the lowland and short grasses the higher elevations. The mixed prairie, typical of the northern Great Plains, embraces largely the needlegrass–grama grass community, with needlegrass-wheatgrass dominating gently rolling soils of medium texture (Coupland, 1950). Because the mixed prairie is characterized by great annual extremes in precipitation, its aspect varies widely from year to year. In moist years midgrasses are prevalent, whereas in dry years short grasses and forbs are dominant. The grasses here are largely bunch and cool-season species, which begin their growth in early April, flower in June, and mature in late July and August.

Short-grass plains. South and west of the mixed prairie and grading into the desert is the short-grass plain, a country too dry for most midgrasses. The short-grass plains reflect a climate where the rainfall is light and infrequent (10 to 17 in. in the west, 20 in. in the east), the humidity low, the winds high, and the evaporation rapid. Shallow-rooted, the short grasses utilize moisture in the upper soil layers, beneath which is a permanent dry zone into which the roots do not penetrate. Sod-forming blue grama and buffalo grass dominate the short-grass plains, accompanied by such midgrasses as western wheatgrass, side-oats grama, and little bluestem. On wet bottomlands, switchgrass, Canada wild rye, and western wheatgrass replace grama and buffalo grass. Because of the dense sod, fewer forbs grow on the plains, but prominent among them is purple lupine.

Just as the tall-grass prairie was destroyed by the plow, so has much of the short-grass plains area been ruined by overgrazing and by plowing for wheat, which, because of low available moisture, the land could not support. Drought, lack of a tight sod cover, and winds turned much of the southern short-grass plains into the Dust Bowl, the recovery from which has taken years.

Desert grasslands. From southeastern Texas to southern Arizona and south into Mexico lies the desert grassland, similar in many respects to the short-grass plains except that triple-awn grass replaces buffalo grass (Humphrey, 1958). Composed largely of bunch grasses, the desert grasslands are widely interspersed with other vegetation types, such as oak savanna and mesquite. The climate is hot and dry. Rain falls only during two seasons, summer (July and August) and winter (December to February) in amounts that vary from 12 to 16 in. in the western parts to 20 in. in the east; but evaporation is rapid, up to 80 in./year. Vegetation puts on most of its annual growth in August. Annual grasses germinate and grow only during the summer rainy season, whereas annual forbs grow mostly in the cool winter and spring months.

Like the tall-grass prairies on the eastern rim of the grasslands, the desert grasslands on the west exist because of fires, which periodically swept across them and eliminated the mesquite, the cacti, and low trees. Without fire the desert grasslands, long before their discovery by white man, would have been a land of low trees with an understory of grasses and small shrubs.

Separated from the midcontinent grasslands

FIGURE 17-13
Midwest grasslands were once and in places still are pock-marked with shallow depressions called potholes. They are important as waterfowl breeding areas. (Photo courtesy U.S. Fish and Wildlife Service.)

by the Rocky Mountains is the California prairie, composed largely of needlegrass and bluegrass. A region of winter rains, much of the California prairie is either under cultivation or is overgrazed.

Tropical grasslands. Around the world is a belt of tropical monsoon grasslands that extend from western Africa to eastern China and Australia. Within this belt monsoon grasslands fall into ecoclimatic gradients of arid to semiarid grasslands; medium rainfall grasslands found mainly in India, Burma, and northern Australia; the high-rainfall monsoonal or equatorial grasslands of southeast Asia; and grasslands whose species are adapted to a hot monsoonal summer and cool-to-cold winters (Whyte, 1968). The tropical grasslands of South America do not fall into this group because geographical conditions do not promote true or false monsoonal conditions. Instead the grasslands of Latin America are largely steppe consisting almost entirely of bunch grass with no legumes, and very few herbs, bushes, or trees (McIlroy, 1972).

Within this broad belt the continental grasslands fall into their own ecoclines. In Africa for example, climatic climax grasslands are confined to desert areas with prolonged drought. Grass cover, dominated by *Aristida* is low and soil may be blown away by the wind. With a slight increase in wetness, desert scrub with scattered shrubs of the genera *Commiphora* and *Acacia* become the climax type. As rainfall increases, desert shrub gives way to desert-grass, Acacia savanna; tall-grass Acacia savanna; tall-grass low tree savanna, and finally humid forests. Thus low rainfall areas are characterized by quick-growing tufted types of low ground cover and high rainfall areas by tall coarse grasses.

Much tropical grassland exists because of fires that prevent the intrusion of woody vegetation. Grass in turn reacts to burning by putting out new shoots and drawing on reserves of moisture and food in the rhizomes and roots, which then are depleted before the arrival of rain. If the grass is overgrazed, the plants are weakened and deteriorate and may be replaced by annual or unpalatable species. On the other hand, the elimination of fire is just as disastrous.

There are a number of differences between tropical and temperate grasses (Steward, 1970). Tropical grasses are lower in crude protein and higher in crude fibers. Tropical grasses have maximum photosynthesis at 30° to 35° C, temperate grasses at 15° to 20° C. Tropical grasses have maximum photosynthetic rates of 50 to 70 mg^{-2}hour^{-1} compared with 20 to 30 mg for temperate grasses. Individual leaves of temperate grasses reach light saturation at relatively low levels, whereas leaves of tropical grasses become saturated only at much higher levels. Temperate grasses have high rates of photorespiration; tropical plants do not.

IMPACT OF MAN

Man had his beginnings in the savanna grasslands of Africa, and he has inhabited and utilized grasslands throughout his evolutionary history. Crop agriculture and pastoralism began there. It is only natural then that man has had an enormous impact on the grasslands of the world. He has modified grasslands intentionally or unintentionally by burning and grazing, too often converting productive grasslands into deserts. He has replaced native plant species with highly productive introduced forage plants accompanied by fertilization, plant control, pest control, and other intensive practices. He has broken up grasslands with the plow and planted the land to cereal grains and legumes. He has replaced wild ungulates with his domestic ones. He has transformed the most productive grasslands into breadbaskets of the world, dominated by a monoculture of cereal grains.

Shrubland

Covering large portions of the arid and semiarid world is climax shrubby vegetation. In addition climax shrubland exists in parts of temperate regions, because historical disturbances of landscapes have seriously affected their potential to support forest vegetation (Eyre, 1963). Among such shrub-dominated plagioclimaxes are the moors of Scotland and the macchia of South America. But outside of these regions, shrublands are seral, a stage in land's progress back to forest. There they exist as second-class citizens of the plant world (McGinnes, 1972), unfortunately, given little attention by botanists, who tend to emphasize dominant plants. As a result, little work has been done on the seral shrub communities.

CHARACTERISTICS OF SHRUBLANDS

Shrubs are difficult to characterize. They have, as McGinnes (1972) points out, a "problem in establishing their identity." They constitute neither a taxonomic nor an evolutionary category (Stebbins, 1972). One definition is that a shrub is a plant with woody persistent stems, no central trunk, and a height of up to 15 to 20 ft. But size does not set shrubs apart because under severe environmental conditions many trees will not exceed that size. Some trees, particularly coppice stands, are multiple-stemmed, and some shrubs have large single stems. Shrubs may have evolved either from trees or herbs (for detailed discussion on evolution of shrubs see Stebbins, 1972).

Shrub ecosystems, seral or climax, are characterized by woody structure, increased stratification over grasslands, dense branching on a fine scale, and low height. Dense growth of many shrub types, such as hawthorne and alder, develop nearly impenetrable thickets that offer protection for such animals as rabbits and quail, and nesting sites for birds. Deep shade beneath discourages any understory growth. Many seral shrubs have flowers attractive to insects (and to man) and seeds that are enclosed in palatable fruits and are easily dispersed by birds and mammals.

The success of shrubs depends largely on their abilities to compete for nutrients, energy, and space (West and Tueller, 1972). In certain environments shrubs have many advantages. They have less energetic and nutrient investment in aboveground parts than trees. Their structural modifications affect light interception, heat dissipation, and evaporative losses, depending upon species and environments involved. The more arid the site, the more common is drought deciduousness and the less common is evergreeness (Mooney and Dunn, 1970a). The multistemmed forms influence interception and stemflow of moisture, increasing or decreasing infiltration into the soil (Mooney and Dunn, 1970b). Because most shrubs can get their roots down quickly and form extensive root systems, they can utilize soil moisture deep in the profile. This feature gives them a competitive advantage over trees and grasses in regions where the soil moisture recharge comes during the nongrowing season. Because they do not have a high shoot-to-root ratio, shrubs draw less nutrient input into aboveground biomass and more into roots. Their perennial nature allows immobilization of limiting nutrients and slows the nutrient recycling, favoring further shrub invasion of grasslands. Subject to strong competition from herbs, some climax shrubs, such as chamise *Adenostoma fasciculatum* inhibit the growth of herbs by allelopathy (McPherson and Muller, 1969). Only when fire destroys mature shrubs and degrades the toxins do herbs appear in great numbers. As the shrubs recover, herbs decline. Seeds of herb species affected apparently have evolved the ability to lie dormant in the soil until released from suppression by fire.

SHRUBLAND ANIMAL LIFE

Shrub communities have their own distinctive animal life. Many occupants are common to shrubby edges of forest and shrubby borders of fields. In fact some species, such as the bobwhite quail, cottontail rabbit, and ruffed grouse, depend heavily on seral shrub communities and disappear if they are destroyed. Alder flycatchers, redwinged blackbirds, and swamp sparrows are typical of alder thickets; indigo buntings, field sparrows, and towhees of old-field communities.

Climax shrub communities have a complex of animal life that varies with the region. In North America the chaparral and sagebrush communities support deer and coyotes, a variety of rodents, as well as jackrabbits, kangaroo rats, mourning doves, and sagehens.

SHRUBLAND TYPES

Arid shrublands. In the semiarid country of western North America, in the regions bordering the Mediterranean, in Australia, and in parts of South America, western India, and central Asia are communities of xeric broadleaf evergreen shrubs and dwarf trees not over 8 ft tall. These sclerophyll communities are characteristic of regions where winters are mild and rainy and summers are long, hot, and dry (for detailed descriptions see McKell et al., 1972). Their vegetation varies with geographical location, altitude, and direction of slope.

These shrub communities go by different names in different parts of the world. In the Mediterranean region shrub vegetation results

from forest degradation and falls into three major types. The *garrigue* is low open shrubland on well-drained to dry calcareous soil, and results from the degradation of pine forests. The *maqui*, a higher type of thick shrubland, occurs where there is more rainfall, and replaces cork forests. The third type, the *matorral*, combines the other two types and appears to be equivalent to the North American chaparral. In South America the shrublands go by various names, depending upon the region. In central Brazil large areas of low evergreen woody plants are called *cerrados*; in northeastern Brazil shrublands, leafless in protracted dry periods, are the *catinga*. Along the Pacific ocean in Peru and Chile exists a highly diverse shrubland, again suggestive of chaparral (see Soriano, 1972). In southwestern Australia, the shrub country dominated by low-growing *Eucalyptus* is known as the *mallee*.

In North America the sclerophyllic shrub community is known as *chaparral*. There are three types: (1) the Mediterranean type in California, dominated by shrub oaks and chamise, where winters are wet and summers dry; (2) inland chaparral of Arizona and New Mexico, dominated by Gambel oak and other species but lacking chamise; and (3) the Great Basin sagebrush (Fig. 17-14). For the most part these communities lack an understory and ground litter, are highly inflammable, and range from light to heavy. Heavy seeders, many plant species of the chaparral require the heat and scarring action of fire to induce germination.

For centuries periodic fires have roared through the chaparral, clearing away the old growth, making way for the new, and recycling nutrients through the ecosystem. When man came to the chaparral, he changed the fire situation. Either he attempted to overcontrol fire in this type of vegetation, in which complete exclusion is impractical, thus seeding real disasters, or he allowed the chaparral to overburn. In the absence of fire, chaparral grows tall and dense and yearly adds more leaves and twigs to those already on the ground. During the dry season the shrubs, even though alive, will nearly explode when ignited. Once set on fire by lightning or man, an inferno follows. Sooner or later this is bound to happen in the chaparral.

Once fire swept, the land returns either to lush green sprouts coming up from buried root crowns or to grass, if a seed source is nearby. The grass and the vigorous young sprouts are excellent food for deer (Taber and Dasmann, 1958), sheep, and cattle. But as the sprout growth matures, chaparral becomes even more dense, the canopy closes, the litter accumulates, and the stage is set for another fire.

In lowlands near the grasslands, burned-over chaparral can revert to grass, which can be maintained by periodic controlled burns. At high elevations, where the pine forest begins, hot fires in the chaparral and pine result in the destruction of the forest and the encroachment of chaparral into higher country.

Seral shrublands. On drier uplands, shrubs rarely exert complete dominance over herbs and grass. Instead the plants are scattered or clumped in grassy fields, the open areas between filled with the seedlings of forest trees, which in the sapling stage of growth occupy the same ecological position as tall shrubs. Typical are the hazelnut, forming thickets in places, sumacs, chokecherry, and shrub dogwoods. But on wet ground the plant community often is dominated by tall shrubs and contains an understory intermediate between that of a meadow and a forest (Curtis, 1959). In northern regions the common tall shrub community found along streams and around lakes is the alder thicket, composed of alder, or alder and a mixture of other species such as willow and red-osier dogwood. Alder thickets are relatively stable and remain for some time before being replaced by the forest (Fig. 17-15). Out of the alder country, the shrub carr community (carr is an English name for wet-ground shrub communities) occupies the low places. Dogwoods are some of the most important species in the carr. Growing with them are a number of willows, which as a group usually dominate the community.

Shrub thickets (Fig. 17-16) are valued as food and cover for game; and many shrubs, such as hawthorn, blackberry, sweetbrier, and dogwoods, rank high as game food. However, the overall value of different types of shrub cover, its composition, quality, and the minimum amounts needed, have never been assessed. There is some evidence (Egler, 1953; Niering and Egler, 1955) that even where the forest is the normal end of succession, shrubs can form a stable community that will persist for many years. If incoming tree growth is removed either by selective spraying

FIGURE 17-14
Sagebrush in the Wasatch Mountains. Although considered an undesirable range plant, sagebrush supports an interesting assemblage of animal life. (Photo by author.)

FIGURE 17-15
Low wetland shrub communities or carrs are often dominated by alder, which forms relatively stable communities. (Photo by author.)

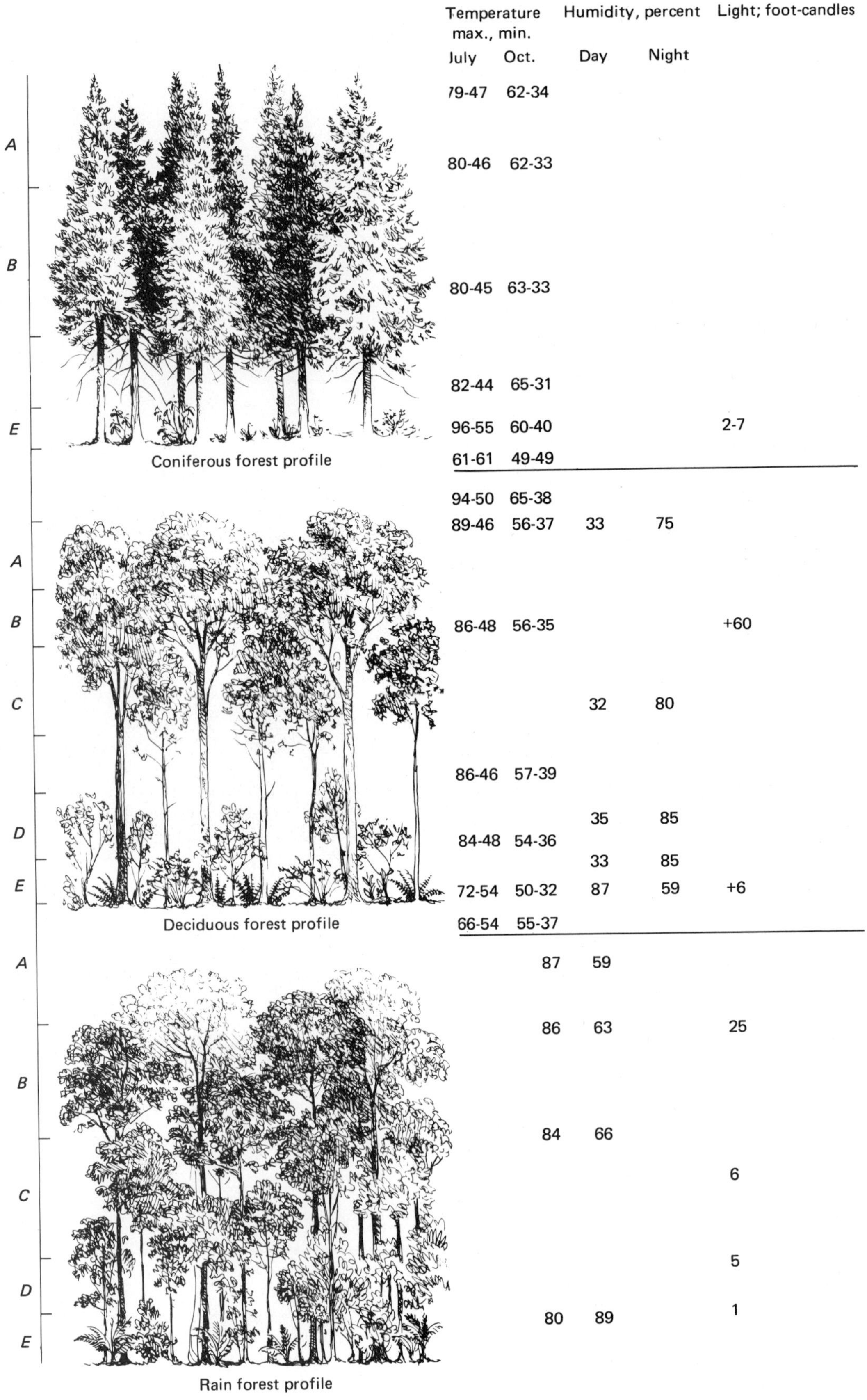

	Temperature max., min.		Humidity, percent		Light; foot-candles
	July	Oct.	Day	Night	
	79-47	62-34			
A					
	80-46	62-33			
B					
	80-45	63-33			
	82-44	65-31			
E	96-55	60-40			2-7
	61-61	49-49			

Coniferous forest profile

	94-50	65-38			
	89-46	56-37	33	75	
A					
B	86-48	56-35			+60
C			32	80	
	86-46	57-39			
			35	85	
D	84-48	54-36			
			33	85	
E	72-54	50-32	87	59	+6
	66-54	55-37			

Deciduous forest profile

A			87	59	
			86	63	25
B					
			84	66	
C					6
					5
D					1
E			80	89	

Rain forest profile

or by cutting, the shrubs eventually form a closed community resistant to further invasion by trees. This could have wide application to power line rights-of-way.

IMPACT OF MAN

In many regions of the world, shrubland exists because of disturbance or degradation of other vegetation types by man. Once established, they are in their own way valuable as grazing lands, as wildlife habitat, as a source of food, and for pharmaceutical and industrial raw materials (for extensive review see McKell et al., 1972). At the same time shrubland is subject to destruction by man. Too often unappreciated and regarded as worthless wastelands, shrublands are converted into housing developments and destroyed by land reclamation projects. Wetland carrs are drowned by flood control and hydroelectric projects. Hedgerows, once a familiar part of the landscapes of the United States and western Europe, are being torn out rapidly to make larger agricultural fields required by mechanization (see Moore et al., 1967). In Africa the Mediterranean shrublands, themselves a result of human disturbance, are in turn being destroyed by overgrazing, cultivation, and uprooting the plants for fire wood. The latter alone is destroying approximately 3 million acres of shrubland a year and allowing the desert to encroach upon them (C. K. Pearce, 1970).

The forest

Of all the vegetational types of the world, probably none are more widespread (at least at one time) or more diverse than the forest. In spite of a wide range of forest types, all possess certain characteristics in common. Dominated by large aboveground biomass, they possess a multilayered structure that provides a habitat for a wide assemblage of species.

STRATIFICATION OF THE FOREST

Because of the variety of life forms of forest vegetation, forests are often highly stratified (Fig. 17-17). The mixed tropical rain forests consist of five layers above the soil. The uppermost, or A, stratum, consists of trees 35

FIGURE 17-17 [OPPOSITE]
Stratification of a coniferous, deciduous, and tropical rain forest. (Sources: Coniferous forest: Atkins, 1957; Fowells, 1948; Vezina, 1961. Deciduous forest: Christy, 1952; McCormick, 1959. Rain forest: Davis and Richards, 1933.)

FIGURE 17-16
In eastern North America shrublands are usually successional communities, but if dense, shrub communities may persist for a very long time. This slope is claimed by St. Johnswort (Hypericum virginicum) and wild indigo (Baptisia tinctoria). (Photo by author.)

to 42 m (116 to 138 ft) high, whose deep crowns rise above the rest of the forest to form a discontinuous canopy. Below this is the B stratum, also discontinuous, of lower trees 20 m (66 ft) high. Strata A and B are not clearly separated from one another and together form an almost complete canopy. The lowest tree stratum, C, is continuous and is often the deepest layer and, unlike A and B, very well defined. The D layer, usually poorly developed in the dense shade, consists of shrubs, young trees, tall herbs, and ferns. The E, or ground layer, is made up of tree seedlings and low herbaceous plants.

Highly developed, uneven-aged, deciduous forests usually consist of four strata. The A, or upper canopy, consists of dominant or co-dominant trees and corresponds to the A and B layers of the tropical rain forest. Beneath this and corresponding to the C layer of the rain forest is the lower tree canopy. Below is the D, or shrub layer. The ground (E) or field layer consists of herbs, ferns, and mosses. Its composition varies with the seasons, from spring with the trilliums and hepaticas to the fall with its asters and goldenrods.

Even-aged stands, the results of fire, clear-cut logging, and other disturbances, often have poorly developed strata beneath the canopy because of the dense shade. The low tree and shrub strata are thin and the ground layer often is poorly developed, except in small, open areas (Fig. 17-18). Stratification in coniferous forests is somewhat similar to that in even-aged deciduous stands, for here too the lower strata are poorly developed and the ground layer consists largely of ferns, mosses, and very few herbs. However, old-stand pitch pine may have three strata, an upper canopy, a shrub layer of blueberries, and a thin herbaceous layer. Jack pine may have four strata: an upper canopy of jack pine, a lower tree canopy of incoming hardwoods, a low shrub and tall herb layer, and a low herb layer of grasses and sedges.

ENVIRONMENTAL STRATIFICATION

A forest is like a blanket over the land, the top of which receives the full impact of the wind, rain, and sun. The various strata, determined in part by light and moisture, modify the environment from the forest canopy to the forest floor. Stratification of light and tempera-

ture are as pronounced here as they are in some aquatic communities.

Light. Bathed in full sunlight, the uppermost layer of the canopy is the brightest part of the forest. From here on down through the forest strata, the light intensity is progressively reduced to only a fraction of full sunlight (Fig. 17-17). In an oak forest, only about 6 percent of the total midday sunlight reaches the forest floor, and the brightness of the light is about 0.4 percent of that of the upper canopy. Pines generally form a dense upper canopy that excludes so much sunlight that the lower strata cannot develop. Pitch pine, Virginia pine, and jack pine have rather open crowns, which allow more light to reach the forest floor than oak, hickory, beech, or maple. Conditions are different with spruce and fir. The upper crown, a zone of widely spaced narrow spires, is open and well lighted, while the lower crown is most dense and intercepts most of the light (Fig. 17-19).

There are seasonal differences of light intensity within the forest. The greatest extremes exist in the deciduous forest. The forest floor receives its maximum illumination during March and April, before the leaves appear; and a second, lower peak occurs in the fall. The darkest period is in midsummer. Light in the coniferous forest is reduced approximately the same throughout the year, since the trees retain their foliage. Here illumination is the greatest in mid-summer, when the sun's rays are most direct and the lowest in winter when the intensity of incident sunlight is the lowest.

Humidity and moisture. The forest interior is a humid place, a fact soon discovered by those who spend midsummer working in the woods. The humidity is high because of plant transpiration and poor air circulation. Variation of humidity inside the forest is influenced in part by the degree to which the lower strata are developed. Within these layers, leaves are adding moisture to the immediate surrounding air, increasing the humidity just above each. Thus layers of increasing and decreasing humidity may exist from the forest floor to the canopy. During the day, when the air warms and its water-holding capacity increases, relative humidity is the lowest. At night, when temperatures and moisture-holding capacities are low, relative humidity in the forest rises. The lowest humidity in the whole forest is a few feet above the canopy, where air circulation is the best.

Highest humidity is found near the forest floor, the result of evaporation of moisture from the ground and the settling of cold air from the strata above.

Temperature. The highest temperatures, in the deciduous forest at least, are in the upper canopy because this stratum intercepts solar radiation. Temperatures tend to decrease through the lower strata. The most rapid decline occurs from the leaf litter down through the soil.

The temperature profile of coniferous forests tends to be somewhat the reverse of those of deciduous forests, particularly in those containing sprucelike trees. Here the coolest layer is in the upper canopy, perhaps because of greater air circulation, and the temperature increases down through the several strata to the forest floor.

The temperature profile changes through the 24-hour period. At night, temperatures are more or less uniform from the canopy to the floor. This is due to the fact that radiation takes place most rapidly in the canopy; as the air cools it sinks and becomes slightly heated by the warmer air beneath the canopy. During the day, the air heats up, and by midafternoon temperature stratification becomes most pronounced. On rainy days the temperatures are more or less equalized because water absorbs heat from warmer surfaces and transfers it to the cooler surfaces.

Temperature stratification varies seasonally (Christy, 1952). In fall, when the leaves drop and the canopy thins, temperatures fluctuate more widely at the various levels. Maximum temperatures decrease from the canopy downward but rise again at the litter surface. The soil, no longer shaded by an overhead canopy, absorbs and radiates more heat than in summer. Below the insulating pavement of litter, temperatures decrease again through the soil. Thus there may be two temperature maximums in the profile, one in the canopy, the other on the surface of the litter. Winter temperatures decrease from the canopy down to the small tree layer, where in some forests they rise and then drop at the litter surface. From here the temperature increases rapidly down through the soil. During spring, conditions are highly variable. Maximum temperatures are found on the leaf-litter surface, which at this season of the year intercepts solar radiation, and temperatures decrease upward toward the canopy.

FIGURE 17-18
A managed 65-year-old even-aged stand of Appalachian hardwoods on an excellent site. Two intermediate cuttings (thinnings) have been made to date. (Photo courtesy U.S. Forest Service.)

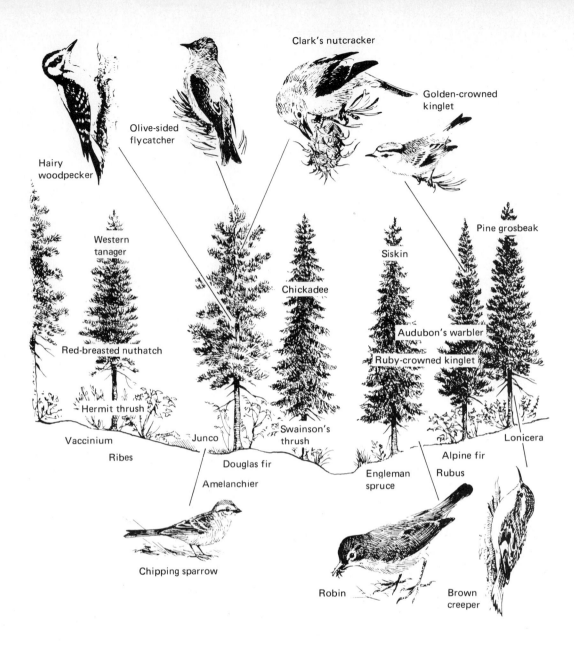

Clark's nutcracker

Golden-crowned kinglet

Olive-sided flycatcher

Hairy woodpecker

Western tanager

Pine grosbeak

Siskin

Chickadee

Red-breasted nuthatch

Audubon's warbler

Ruby-crowned kinglet

Hermit thrush

Vaccinium

Junco

Swainson's thrush

Lonicera

Ribes

Douglas fir

Alpine fir

Engleman spruce

Rubus

Amelanchier

Chipping sparrow

Robin

Brown creeper

Wind. Anyone who has entered a woods on a windy day is well aware of how effectively the forest reduces wind velocity. The influence of forest cover on wind velocity varies with the height and density of the stand and the size and density of the crown. Overall wind velocities inside the forest may be reduced by 90 percent, and velocity near the ground usually ranges from 1 to 2 percent of wind velocity outside. Velocities in the open and cutover stands and in wintertime deciduous forests are greater than in dense stands and in coniferous forests, the latter of which are most effective in reducing the flow of wind.

The forest edge deflects the wind upward and over the trees, where the roughness of the canopy surface reduces the velocity. The velocity of wind in the forest is not a constant percentage of the speed of the wind above the canopy. During midday, for example, when the temperature decreases with height, this percentage decreases as the wind velocity above the crown increases.

FOREST ANIMAL LIFE

Deciduous and coniferous forests. Although the deciduous and coniferous forests involve, with some exceptions, different species of animal life, the ecological niches and the adaptations of animals are similar. In general the diversity of animal life is associated with the stratification of and growth forms of plants. Some animals are associated with or spend the major part of their life in a single stratum; others may range over two or more. Arthropods, in particular, confine their activities to one stratum (Dowdy, 1951). Sixty-seven percent of all the arthropod species in Missouri oak–hickory forests inhabited one stratum only, and 78 percent of a red cedar forest were so restricted. None of the arthropod species were common to all strata and only 2 percent ranged through as many as four strata.

The greatest concentration and diversity of life are found on and just below the ground layer. Many animals, the soil invertebrates in particular, remain in the subterranean strata. Others, the mice, shrews, ground squirrels, and foxes burrow into the soil for shelter or food but spend considerable time aboveground. The larger mammals live on the ground layer and feed on herbs, shrubs, and low trees. Birds move rather freely between several strata, but even here there is some restriction. Ruffed grouse, spruce grouse, hooded warblers, and ovenbirds occupy essentially the ground layer but may move up into the tree to feed, to roost, or to advertise territory. Some invertebrates, such as the millipedes and spiders, move up into the upper strata at night when humidity is favorable, a vertical migration that is somewhat similar to that of the plankton organisms in lakes and seas (Dowdy, 1944).

Other species occupy the upper strata—the shrub, low tree, and canopy (Fig. 17-20). Red-eyed vireos inhabit the lower tree stratum of the eastern deciduous forest, the wood peewee the lower canopy. Blackburnian warblers and scarlet tanagers dwell in the upper canopy, where they are very difficult to observe. Flying squirrels and tree mice are mammalian inhabitants of the canopy, and the woodpeckers, nuthatches, and creepers live in the open space of tree trunks between the shrubs and the canopy. Most of the intensive work on stratification of bird life in the temperate deciduous forest has been done in European woodlands, where one biologist (Turcek, 1951) found that 15 percent of the bird species in an oak–hornbeam forest nested on the ground, 25 percent in the herb and shrub strata, and 29 percent in the canopy. Thirty-three percent of the total population occupied the forest canopy, where more niches apparently were available; but its biomass was less than the ground and shrub populations. The ground layer, however, was

FIGURE 17-20 [OPPOSITE]
Foraging niches of some birds in a spruce–fir forest in Wyoming. (After Salt, 1957.)

FIGURE 17-19
Spire-shaped, the spruce has a fairly open upper canopy; the crown is densest in the lower one-half to one-third. (Photo by author.)

the feeding area for 52 percent of the species; herb and shrubs, 9 percent; tree trunk, 10 percent; tree foliage, 23 percent; and the open spaces, 6 percent. In deciduous forests, the diversity of bird species seems to be related to the height and density of foliage and resulting stratification, rather than to plant species composition (MacArthur and MacArthur, 1961).

As habitats, deciduous and coniferous forests differ chiefly in foliage arrangement and retention. The stiff, needlelike leaves of conifers, arranged spirally around the branches, restrict the movements of some birds. And the poor decomposition of accumulated litter in coniferous forests is not favorable to small animal life, especially the soil invertebrates. Coniferous forests provide better winter cover for wildlife. The presence of such shelter often is decisive in the survival of white-tailed deer in winter deer yards.

Differences also exist within deciduous and coniferous forests, even within a given region. In Wisconsin, xeric forests and south-facing slopes are inhabited by such birds as the scarlet tanager, blue jay, and red-eye towhee. Mesic forests are occupied by wood thrushes, redstarts, and least flycatchers (Bond, 1957). The dry, shaded forest floor is the haunt of the ovenbird, whereas well-shaded ground is preferred by the Canada warbler (Kendeigh, 1945). Moisture regimes also strongly influence the distribution of litter fauna. In extensive reforestation areas in New York State, where pine and spruce have been planted over thousands of acres, a definite species preference for certain conifer growth forms exist among birds (R. L. Smith, 1956). The pine forests attracted the lowest number of species—8—and were preferred by the black-throated green warbler. Spruce plantings held 22 species, of which the magnolia and Nashville warblers were the most common. The greater diversity of species in spruce undoubtedly reflected the greater abundance of niches offered by the spirelike crowns and the branches retained close to the ground. The low branches also provided excellent winter cover for grouse, snowshoe hares, and cottontail rabbits. But because of the difficulty they had in moving through the dense growth of spruce in winter, white-tailed deer sought the pine.

Some animals commonly associated with the deciduous forest—the white-tailed deer, the black bear, and the mountain lion, for example—actually range over both the deciduous and coniferous forests. Their north–south distribution appears to be limited more by climate, especially temperature, than by vegetation. The red squirrel, commonly associated with coniferous and mixed hardwood–coniferous forests, is quite common in deciduous woodlands, including oak, in the northern part of its range. But in the southern Appalachians this squirrel is restricted to the northern hardwood–coniferous forests of the high elevations and does not inhabit the oak forests at lower elevations, where the climate is warmer. On the other hand, warblers characteristic of northern coniferous forests, the magnolia and the black-throated green, are common in purely deciduous northern hardwood and oak forests of the Appalachian Mountains in West Virginia; and a northern mammal associate, the red-backed mouse, lives on the forest floor (R. L. Smith, 1966). Possibly these animals were associated with the hemlock, and after the hemlock was cut, they adjusted to the purely hardwood situation.

Tropical rain forest. Contrasting sharply with life in the temperate deciduous and coniferous forests is life in the tropical rain forests. Animal species are more diverse because there is a greater variety of niches. Stratification of life is most pronounced here, where six distinct feeding strata are recognized (Harrison, 1962). (1) An insectivorous and carnivorous feeding group, consisting largely of bats and birds, works the upper air above the canopy. (2) Within the A—C canopy, a wide variety of birds, fruit bats, and mammals feed on leaves, fruit, and nectar. A few are insectivorous and mixed feeders. (3) Below the canopy, in the middle zone of tree trunks, is a world of flying animals, birds, and insectivorous bats. (4) Here too are the scaunsorial mammals, which range up and down the trunk, entering both the canopy and the ground zones to feed on the fruit of epiphytes growing on the tree trunks, on insects, and other animals. (5) Large ground animals make up the fifth feeding group. This includes large mammals and, rarely, birds, living on the ground and lacking climbing ability, that are able to reach up into the canopy or cover a large area of the forest. They include the plant-feeders that either browse on the leaves, eat fallen fruit or root tubers, and their attendant carnivores. (6) The final feeding stratum includes the small ground animals, birds, and small mammals capable of some climbing, which search the ground litter and lower parts of tree trunks for food. This includes insectivorous, herbivorous, carnivorous, and mixed feeders.

The tropical rain forest has been compared to a lake. The canopy, area of primary production, represents the phytoplankton, exploited directly by insects, comparable to the zooplankton, and the large animals, comparable to the nekton. Food carried down into the deeper layer, the middle zone, and the ground, as fallen leaves, fruit, insect bodies, compares to the bodies of plankton organisms. This food source is utilized by the middle-zone birds and bats, corresponding to the nekton organisms of the deeper layer; and the middle-zone mammals, which might be considered as periphytic organisms. The small ground mammals and birds are equivalent to the benthic organisms.

Animal life in the tropical rain forest is largely hidden, either by the dense foliage of the upper strata or by the cover of night. Birds are largely arboreal, and although brightly colored, remain hidden in the dense foliage. Ground birds are small and dark-colored, difficult to see. Mammals appear scarcer than they really are, for they are largely nocturnal or arboreal. Ground-dwelling mammals are small and secretive. Tree frogs and insects are most conspicuous at evening, when their tremendous choruses are at full volume. Insects are most diverse about forest openings and along streams and forest margins, where light is more intense, where temperatures fluctuate and air circulates freely. Highly colored butterflies, beetles, and bees are common. Among the unseen invertebrates, hidden in loose bark and in axils of leaves, are snails, worms, millipedes, centipedes, scorpions, spiders, and land planarians. Termites are abundant in the rain forest and play a vital role in the decomposition of woody plant material. Together with ants, they are the dominant insect life. Ants are found everywhere in the rain forest, from the upper canopy to the forest floor, although in common with other rain-forest life, the majority tend to be arboreal.

FOREST ECOSYSTEMS

Tropical rain forest. The rain forest or some variation of it—the monsoon forest, the evergreen savanna forest, the montane rain forest—forms a worldwide belt about the equator. The largest continuous rain forest is found in the Amazon Basin of South

America. West and central Africa and the Indo-Malayan region are other major areas of tropical rain forest. Although the rain forest is remote and incompletely studied (see Richards, 1952; Cain and Castro, 1959; Odum, 1971), its contrasts and similarities to other American forests deserve some comment.

The rain forest, so-called because of its constant high humidity, grows where seasonal changes are minimal. The mean annual temperature is about 26° C (79° F), the mean minimum rarely goes below 25° C (77° F), and heavy rainfall occurs through much of the year. Under such perpetual midsummer conditions, plant activity continues uninterrupted, resulting in very luxurious growth. Tree species number in the thousands, none of which are usually dominant and the majority of which are represented by a very few individuals. Communities with single dominants usually are limited to certain soils and to areas of particular combinations of soils and topography. The tree trunks are straight, smooth, and slender, often buttressed, and reach 25 to 30 m (82 to 98 ft) before expanding into large, leathery, simple leaves.

Climbing plants, the lianas, long, thick, and woody, hang from trees like cables, and epiphytes grow on the trunks and limbs. The undergrowth of the dark interior is sparse and consists of shrubs, herbs, and ferns. Litter decays so rapidly that the clay soil, more often than not, is bare. The tangled vegetation, popularly known as the jungle, is a second-growth forest that develops where the primary forest has been despoiled.

The monsoon forests and other deciduous and semideciduous forests grow where rainfall diminishes and where a pronounced dry season occurs. They differ from the rain forest in that they lose their leaves during the dry season. Such forests, most common in southeastern Asia and India, are commonly known as tropical seasonal forests. Similar stands occur along the Pacific side of Mexico and Central America.

Tropical savanna. Where forests merge with grasslands, a distinctive type of vegetation, the savanna, may exist. Natural savannas, the halfway world between grassland and forest (Fig. 17-21) occur on an environmental gradient between a pure grassland and a pure forest climate. Savannas are dominated by grasses and sedges, and contain open stands of widely spaced short trees, usually thorny.

True savanna exists only where climate is favorable. Elsewhere it has been created, either by fire, by grazing, or by a combination of both. Best known are the tropical savannas of Africa, dominated by grasses and sedge, which put out new growth in the rainy season and die down in the dry. The short, flat-crowned trees are uniformly scattered or form a gallery forest along the river banks. Just as in the Temperate Zone, the tropical savanna in places is increasing as the forest cover is destroyed; in other areas trees and shrubs invade overgrazed grassland. The African savanna is inhabited by the richest and most diverse ungulate fauna in the world, including the gazelle, the impala, the eland, the buffalo, the giraffe, the zebra, and the wildebeest. In contrast, the seral and geologically more recent South American savannas lack these conspicuous mammalian herds.

Temperate evergreen forest. In the subtropical regions of North America and the Caribbean Islands one finds the temperate evergreen forest. Restricted to a region of warm maritime climate, the temperate or subtropical evergreen forest is best developed on the North American continent along the Gulf Coast and in the hammocks of the Florida Everglades and the Keys. Depending upon location, these forests are characterized by live oaks, magnolias, gumbo limbo, royal and cabbage palms, and bromeliads.

Temperate deciduous forest. The temperate deciduous forest once covered large areas of Europe, China, parts of South American and Middle American highlands, and eastern North America. In eastern North America, it consists of a number of forest types, which intergrade into one another. The northern segment of the deciduous forest complex is the hemlock–white pine–northern hardwoods forest, which occupies southern Canada, and extends southward through northern United States and along the high Appalachians into North Carolina and Tennessee. Beech, sugar maple, basswood, yellow birch, black cherry, red oak, and white pine are the chief components. White pine was once the outstanding tree of the forest. But because most of it was cut before the turn of the century, it now grows only as a successional tree on

FIGURE 17-21 [OPPOSITE]
The tropical savanna is made up of widely scattered trees in a tropical grassland. The tropical savanna is the heart of East Africa big game country. (Photo courtesy Bernheim, Woodfin Camp & Associates.)

abandoned land and as scattered trees through the forest. On relatively flat, glaciated country with its deep, rich soil grow two somewhat similar forests, the beech–sugar maple forest, restricted largely to southern Indiana and central Minnesota and east to western New York; and the sugar maple–basswood forest, found from Wisconsin and Minnesota south to northern Missouri. South of this is the extensive central hardwood forest. The central hardwood can be divided into three major types. (1) The first is the cove, or mixed mesophytic, forest, which consists of an extremely large number of species, dominated by yellow poplar. This forest, which reaches its best development on the northern slopes and deep coves of the southern Appalachians, is one of the most magnificent in the world. Much of its original grandeur has been destroyed by high-grading and fire, but even in second- and third-growth stands, its richness is apparent. (2) On more xeric sites, the southern slopes and drier mountains, grows the oak–chestnut forest. The chestnut, killed by blight, has been replaced by additional oaks. (3) The western edge of the central hardwoods in the Ozarks and the forests along the prairie river systems are dominated by oak and hickory.

The southern pine forests of the coastal plains of the South Atlantic and Gulf States are considered part of the temperate, deciduous forest because they represent a seral rather than a final stage. Unless maintained by fire and cutting, these forests are succeeded by such hardwoods as oak, hickory, and magnolia. Magnolia and live oak dominate the climax forest of the southern Gulf States and much of Florida.

Northern coniferous forest. North of the eastern deciduous forest is the continent-wide belt of coniferous forest, which extends from New England, northern New York, and southern Canada north to the tundra, westward to the Pacific, and southward through the Rockies and Sierras into Mexico. The northern coniferous forest starts out as pine and hemlock mixed with hardwoods, a gradient or ecotone of the northern hardwood forest. Eastern hemlock, jack pine, red pine, white pine, and white cedar are characteristic. Originally the pine forests were most highly developed about the Great Lakes, but they were destroyed by exploitative logging in the 1880s and 1900s. The coniferous forest extends southward from New England through the high Appalachians. Here at high elevations the spruce and fir, the major forest type of the north woods, end. In the southern Appalachians, red spruce and Frasier fir dominate; but north, in the Adirondacks and White Mountains on into Canada and across the continent to Alaska, white spruce, black spruce, and balsam fir form the matrix of the forest. Occupying, for the most part, glaciated land, the northern coniferous forest is a region of cold lakes, bogs, rivers, and alder thickets. This same boreal forest encircles the northern part of the globe.

Temperate rain forest. South of Alaska, the coniferous forest differs from the northern boreal forest, both floristically and ecologically. The reasons for the change are climatic and topographic. Moisture-laden winds move inland from the Pacific, meet the barrier of the Coast Range, and rise abruptly. Suddenly cooled by this upward thrust into the atmosphere, the moisture in the air is released as rain and snow in amounts up to 635 cm (250 in.). During the summer, when the winds shift to the northwest, the air is cooled over chilly northern seas. Though the rainfall is low, the cool air brings in heavy fog, which collects on the forest foliage and drips to the ground to add perhaps 127 cm (50 in.) more of moisture. This land of superabundant moisture, high humidity, and warm temperatures supports the "temperate rain forest," a community of lavish vegetation dominated by western hemlock, western red cedar, Sitka spruce, and Douglas fir. Further south, where the precipitation still is high, grows the redwood forest, occupying a strip of land about 724 km (450 mi) long.

Montane coniferous forest. The air masses that dropped their moisture on the western slopes of the Coast range descend down the eastern slopes, heat up, and absorb moisture, creating the conditions that produce the Great Basin deserts already discussed. The same air rises up the western slopes of the Rockies, cools, and drops moisture again, although far less than it did on the Coast Range. Here in the Rockies develop several coniferous forest associations, influenced to a great extent by elevation. At high elevation, where the winters are long and snow is heavy, grows the subalpine forest, characterized by Engelmann spruce, alpine fir, and white-barked and bristlecone pines (Fig. 17-22). Lower elevations support Douglas fir and ponderosa pine. The aridity-tolerant ponderosa pine more often than not has an understory of grass and shrubs.

Forests similar to those of the Rocky Mountains grow in the Sierras and Cascades. Alpine forests there consist largely of mountain hemlock, red fir, and lodgepole pine. At lower elevations grow the huge sugar pine, incense cedar, and the largest tree of all, the giant sequoia, which grows only in scattered groves on the west slopes of the California Sierras.

A deciduous seral stage, common to much of the western coniferous forest and the northern coniferous forest as well, is the aspen parkland, supporting trembling aspen, the most widespread tree of North America. The aspen is an important segment of the coniferous forest, for it is utilized by deer, grouse, bear, snowshoe hare, and beaver.

Temperate woodland. In western parts of North America where the climate is too dry for montane coniferous forests, one finds the temperate woodlands. These ecosystems are characterized by open growth small trees beneath which grows a well-developed stand of grass or shrubs. There are a number of types of temperate woodlands, which may consist of needle-leaved trees, deciduous broadleaved trees, or sclerophylls, or any combination of these. An outstanding example is the pinyon–juniper woodland in which two dominant genera, *Pinus* and *Juniperus* are always associated. This ecosystem is found from the front range of the Rocky Mountains to the eastern slopes of the Sierra Nevada foothills. One of the best examples stands on the Kaibab Plateau of northern Arizona. In southern Arizona, New Mexico, and northern Mexico occur oak juniper and oak woodlands, and in the Rocky Mountains, particularly in Utah (Fig. 17-23), one finds oak-sagebrush woodlands. In the Great Valley of California grows still another type—evergreen oak woodlands with a grassy undergrowth.

IMPACT OF MAN

Although man originally developed in a grassland ecosystem, he achieved his greatest economic and cultural development in the temperate forest regions. Destruction of the temperate forest began early in Europe and was largely complete by the Middle Ages (see Tubbs, 1968; Berglund, 1969). Over most of Europe the forests that do remain, except for

FIGURE 17-22

The higher slopes of the Rocky Mountains support mixed stands of alpine fir and Engelmann spruce. Lower slopes support ponderosa pine and lodgepole pine. (Photo by author.)

the boreal ones, have been highly modified by man. In North America great expanses of natural forest still exist in spite of mismanagement, logging, and land clearing. Logging wiped out the magnificent virgin stands of eastern temperate hardwoods, destroyed the vast stands of pines in New England and the Lake states. Fires and subsequent conversion to now-abandoned agriculture degraded most of the region. Most of the eastern forest land was cleared for agriculture; but with a decline in eastern agriculture, the forest is returning to abandoned land. But as forest returns in one place, it is being destroyed in another for housing, highways, or recreational development. Large areas of forested land in the Northeast are gradually becoming urban forest, owned by suburbanites and maintained for aesthetic reasons. There is some question of how natural they can remain. Large reserves of strippable coal, coupled with energy demands, threaten the very existence of the rich mesic forests of the southern Appalachians (Fig. 17-24), with their great diversity of tree species and endemic salamanders. As the demand for timber increases, intensive management of forests with its emphasis on monoculture and hybrid species, and wood fibers rather than board feet, will change the character of the American forest and its animal life.

The tropical forest is the most gravely threatened forest ecosystem. Although man has occupied the tropical forest for a long time, only in recent times has he had any real impact on it. Now there is great pressure to settle tropical forests. Roads are cutting through the tropical jungle, and large areas are being cleared for questionable agriculture (see Chapter 4). Once destroyed, tropical forest species, unlike temperate species, cannot recolonize large areas (see Gomez-Pompa et al., 1972). Thus millions of years of evolution of plant and animal species in the most complex biotic community in the world are threatened with extinction.

The tropical savanna of Africa is also being degraded, and the unique ungulate fauna and its predators are disappearing under pressure from domestic cattle. Increasing populations of cattle and overgrazing are converting parts of it to desert conditions; and increasing human populations, the decline in nomadic life, and the concentration of both people and cattle in small areas are further aggravating a deteriorating situation.

The tundra

North of the coniferous forest belt lies a frozen plain, clothed in sedges, heaths, and willows, which encircles the top of the world. This is the arctic tundra. At lower latitudes similar landscapes, the alpine tundra, occur in the mountains of the world. But in the Antarctic a well-developed tundra is lacking. Arctic or alpine, the tundra is characterized by low temperatures, a short growing season, and low precipitation (cold air can carry very little water vapor).

The tundra—the word comes from the Finnish *tunturi*, meaning a treeless plain—is a land dotted with lakes and transected by streams. Where the ground is low and moist, there are extensive bogs. On high, drier areas and places exposed to the wind, vegetation is scant and scattered, and the ground is bare and rock-covered. These are the fell-fields, an anglicization of the Danish *fjoeld-mark*, or rock deserts. Lichen covered, the fell-fields are most characteristic of the highly exposed alpine tundra.

CHARACTERISTICS

Frost molds the tundra landscape. Alternate freezing and thawing, and the presence of a permanent frozen layer in the ground, the permafrost, create conditions unique to the tundra. The sublayer of soil is subject to annual thawing in spring and summer and freezing in fall and winter. The depth of thaw may vary from a few inches in some places to 1 to 2 ft in others. Below this, the ground is always frozen solid and is impenetrable to both water and roots. Because the water cannot drain away, flat lands of the Arctic are wet and covered with shallow lakes and bogs. This reservoir of water lying on top of the permafrost enables plants to exist in the driest parts of the Arctic.

The symmetrically patterned land forms so typical of the tundra result from frost. As the surface freezes, the fine soil materials and clays, which hold more moisture, expand while freezing and then contract when they thaw. This action tends to push the larger

FIGURE 17-23
*Oak–sagebrush communities comprise much of
the chaparral of the western mountains.
(Photo by author.)*

FIGURE 17-24
*Strip-mining for coal in Appalachia is destroying
the heart of mixed mesophytic forest region
with its rich diversity of plant life. (Photo by
author.)*

579

material upward and outward from the mass to form the patterned surface.

Typical nonsorted patterns associated with seasonally high water tables are frost hummocks, frost boils, and earth stripes (Fig. 17-25). Frost hummocks are small earthen mounds up to 5 ft in diameter and 4 ft high, which may or may not contain peat. Frost boils are formed when the surface freezes across the top, trapping the still unfrozen muck beneath. As this chills and expands, the mud is forced up through the crust. Raised earth stripes, found on moderate slopes, appear as lines or small ridges flowing downhill. They apparently are produced by a downward creep or flow of wet soil across the surface of the permafrost.

Sorted patterns are characteristic of better-drained sites. The best known of these are the stone polygons, the size of which is related to frost intensity and the size of the material (Johnson and Billings, 1962). The larger stones are forced out to a peripheral position, and the smaller and finer material, either small stones or soil, occupies the center. The polygon shape may result from an accumulation of rocks in desiccation cracks formed during drier periods. These cracks ap-

pear as the surface of the soil dries out, in much the same way as cracks appear in bare, dry, compacted clay surfaces in temperate regions. On the slopes, creep, frost-thrusting, and downward flow of soil change polygons into sorted stripes running downhill. Mass movement of supersaturated soil over the permafrost forms solifluction terraces, or "flowing soil." This gradual downward creep of soils and rocks eventually rounds off ridges and other irregularities in topography. This molding of the landscape by frost action is called *cryoplanation* and is far more important than erosion in wearing down the arctic landscape.

Permafrost in the alpine tundra exists only at very high elevations and in the far north, but the frost-induced processes—small solifluction terraces and stone polygons—are still present. The lack of a permafrost results in drier soils; only in alpine wet meadows and bogs do soil-moisture conditions compare with the Arctic. Precipitation, especially snowfall and humidity, are higher in the alpine region than in the arctic tundra, but the steep topography results in a rapid runoff of water.

The tundra, arctic or alpine, has a number of features that, although shared independently with other ecosystems, collectively make the tundra unique. The vegetation of the tundra is structurally simple, dominated by low-growing grasses, sedges, heaths, willows, and birches. Although the vegetation appears homogeneous, small variations in microtopography result in steep gradients of moisture that in turn influence vegetative cover. Species diversity is low, and most of the biomass and functional activity are contained in relatively few groups (Fig. 17-26). The ecosystem experiences a short growing and reproductive season. Most of the vegetation is perennial, and vegetative reproduction is common. Animals are characterized by extended life cycles with long periods of dormancy or by strong migratory habits. Annual productivity is low, partly because of the short growing season, partly because of low temperatures, and partly because of a poor nutrient supply (especially of nitrogen, due to inhibition of microbial activity by the cold) (Warren-Wilson, 1957). Annual increment to biomass is low, but because of a low rate of decomposition there is a large accumulation of energy and nutrients in dead organic matter. Animal populations are characterized by large oscillations, especially among rodents.

ARCTIC AND ALPINE VEGETATION

Tundra vegetation begins at the tree line (Fig. 17-27). In appearance, it suggests a short-grass plain, but it is actually a land of lichens and mosses, sedges and rushes, grasses and low-growing shrubs, chiefly heaths. Although the vegetation may appear homogeneous over wide areas of the tundra, the pattern is often complex. Large differences exist across the tundra, which reflect changes in the macroclimate and microclimate, land form, history, time, and substrate. The relatively permanent instability of the substrate plays a primary role in influencing tundra vegetation, physically dislodged by the movements of the surface soil. In fact the timber line may be regarded as a zone of transition from relatively stable to instable soils (Raup, 1951) rather than a purely climatic phenomenon. Thus climate would have only an indirect effect through its influence on soil movements.

Although the environments of the arctic and alpine regions are somewhat similar, the

FIGURE 17-26 [OPPOSITE]
The wide expanse of the arctic tundra. This photograph shows an area near the Sadlerochit River on the Arctic National Wildlife Range, 5 mi from the Arctic Ocean. Note the frost polygons in the foreground. The caribou, a major arctic herbivore, are a part of the Porcupine herd. (Photo courtesy U.S. Fish and Wildlife Service.)

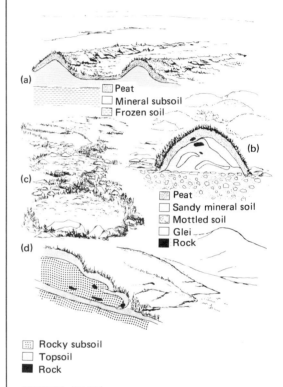

(a) Peat
Mineral subsoil
Frozen soil

(b)

(c) Peat
Sandy mineral soil
Mottled soil
Glei
Rock

(d)

Rocky subsoil
Topsoil
Rock

FIGURE 17-25
Patterned ground forms typical of the tundra region: (a) unsorted earth stripes; (b) frost hummock; (c) sorted stone nets and polygons; (d) solifluction terrace. (Diagrams adapted from Johnson and Billings, 1963. © The Ecological Society of America.)

vegetation of the two differs considerably (Bliss, 1956). However, they do have some features in common. Because the environment is especially rigorous, the number of species tends to be few, and their growth form is low. The plants are rarely able to take a sufficient hold on the surface to have any sort of control over the immediate soil environment. The complications of snowbanks, microrelief, aspect, snowmelt, frost heaving, and so on produce an almost endless change in plant associations from spot to spot (Polunin, 1955). Because of frequent and drastic disturbances, plants are often short-lived, and successional cycles are fragmentary.

The growth rate of both arctic and alpine species is reduced and appears to be controlled to some degree by soil and air temperatures. In Greenland, for example, plant growth begins when the mean daily soil-surface temperatures pass 0° C (32° F) (Krog, 1955). The influence of temperature near the surface of the ground is illustrated by a study of the alder, *Alnus crispus* (Bliss, 1956). This shrub leafed out 2 to 4 days earlier at the base of the clump than at the top, 1 m above; and leaves opened on the south side of the clump 6 days earlier than on the north side. Another plant, *Arctagrostis latifolia*, increased and decreased its growth with corresponding temperature changes. The greatest stem elongation took place during high-temperature periods. The cessation of growth in tundra plants, however, is completely independent of temperature and is probably due to physiological changes during the latter part of the growing season.

There are adaptive differences between the plants of the arctic and alpine regions, associated in part with the photoperiod. The alpine sorrel, found both in the arctic and alpine tundras, produces an increasing number of flowers in the southern portions of its range and a decreased production of rhizomes (Mooney and Billings, 1960). Northern populations of the plant have a higher photosynthetic rate at lower temperatures and attain a maximum rate at a lower temperature. Alpine plants reach the saturation point for light at higher intensities than arctic plants, which are adapted to lower light intensities. Arctic plants require longer periods of daylight than alpine plants. The further north the geographic origin of the plant, the more slowly the plant grows under short photoperiod. Arctic plants propagate themselves almost entirely by vegetative means; alpine plants propagate themselves by seedlings. The adventitious roots of arctic plants are short and parallel to the rhizomes and short-lived. Those of the alpine species are long, penetrate to considerable depths, and are long-lived (see Billings and Mooney, 1968).

Arctic vegetation. In the Arctic only those species able to withstand constant disturbance of the soil, buffeting by the wind, and abrasion from wind-carried particles of soil and ice can survive. On well-drained sites, heath shrubs, dwarf willows and birches, dryland sedges and rushes, herbs, mosses, and lichens cover the land. On the driest and most exposed sites—the flat-topped domes, the rolling hills, and low-lying terraces, all usually covered by coarse, rocky material and subject to extreme action by the frost—vegetation is sparse and often confined to small depressions. Plant cover consists of scattered heaths and mats of mountain avens, as well as crustose and foliose lichens growing on the rocks. Willows, birch, and heath occupy well-drained soils of finer material, and between them grow grasses, sedges, and herbs.

But over much of the Arctic the typical vegetation is a cotton grass–sedge–dwarf heath complex (H. C. Hanson, 1953). Hummocks may support growths of lichens, willow, blueberry, and Laborador tea. Depressions are covered with sedge-marsh vegetation, and over the rest grows tussocks of cotton grass. The spaces between the tussocks may be filled with sphagnum, on top of which dwarf shrubs grow; in other places sphagnum may overgrow the sedges and cotton grass. On mounds and hummocks in fresh-water marshes, on well-drained knolls and slopes, in areas of late-melting snowbanks, along streams and on sandy and gravelly beaches, grassland types develop.

Topographic location and snow cover delimit a number of arctic plant communities. Steep, south-facing slopes and river bottoms support the most luxurious and tallest shrubs, grasses, and legumes, whereas cotton grass dominates the gentle north and south slopes, reflecting higher air and soil temperatures and greater snow depth. Pockets of heavy snow cover create two types of plant habitats, the snow patch and the snowbed. Snow-patch communities occur where wind-driven snow collects in shallow depressions and protects the plants beneath, particularly the aerial stems of shrubs. As a result, snow patches support rather tall growths of willow, birch, and heaths, in contrast to the shorter vegetation surrounding them. Snowbeds, typical of both arctic and alpine situations, are found where large masses of snow accumulate because of certain topographic peculiarities. Not only does the deep snow protect the plants beneath, but the meltwater from the slowly retreating snowbank provides a continuous supply of water throughout the growing season. Snowbed plants, usually found only here, have an extremely short growing season but are able to break into leaf and flower quickly because of the advanced stage of growth beneath the snow. In unfavorable seasons, the snowbank may melt so late that the plants do not have sufficient time to ripen the fruit or perhaps to open the flower. Thus vegetative propagation is important in these species. In fact the gradual retreat of the snow may result in zonation of vegetation across the snowbed, with lichens along the exposed edge, to sedges and hairgrass, to shrubs such as willow. Where the snow cover is very deep and extremely late in melting, only mosses grow (Hanson, 1953; Polunin, 1934–35).

Because of the cold climate, short summers, low soil temperatures, sparse vegetation, and slow decay of organic matter, arctic soils are poor and deficient in nitrates, phosphates, and other salts. Where these nutrients are added to the soil, as around bird cliffs, animal burrows, goose nesting grounds, and present and past areas of human habitation, the vegetation is lush and green, in contrast to the less colorful surrounding expanse. These local areas are the "arctic oases," for they contain not only species common to the region but also some species confined to these habitats (see Bank, 1953; Wiggins and Thomas, 1962).

One of the outstanding features of the arctic ecosystem is the relationship of permafrost, heat transfer, and vegetation, which interact to create conditions unique to the Arctic. Two characteristics of permafrost have a pronounced effect: (1) it is sensitive to thermal changes—any natural or manmade disturbances, however slight, can greatly affect the thermal equilibrium and (2) relatively impervious to moisture, the permafrost forces all water to move above it. Thus surface water becomes quite conspicuous even though pre-

FIGURE 17-27

The tree line in the Rocky Mountains is sharply defined dwarf spruce and fir growing in narrow pockets that hold snow in winter. Although the low growth form is partly genetic, parts of the plants exposed above snow are broken and killed by wind and cold. (Photo by author.)

cipitation is low (see Brown and Johnson, 1964; Brown, 1970).

Vegetation protects the permafrost and influences the heat exchanged between the soil and the air. Vegetation by transpiration dries the soil and reduces its temperature. It further influences heat exchange by reducing evaporation from the soil surface, by holding precipitation on the surface of the plant, by holding wind-blown snow, and by condensing moisture from the air (Tyrtikov, 1959). Vegetation decreases solar radiation, resulting in less warming of the soil and thus in retardation of the warming and thawing of soil in summer and in increasing its average temperature in winter. If vegetation is removed, the depth of thaw is 1.5 to 3 times that of areas still retaining vegetation. The accumulation of organic matter and dead plant material retards the warming of soil in summer more than a living vegetative cover. Thus vegetation impedes thawing of permafrost and acts to conserve it (see Pruitt, 1970).

In turn permafrost chills the soil, retards the general growth of both aboveground and underground parts of plants, and slows the activity of soil microorganisms. It hinders aeration and impoverishes nutrients in the soil (Tyrtikov, 1959). The effect is more pronounced the closer the permafrost comes to the surface, when it contributes to the formation of shallow root systems and reduces the stability of trees against wind. In spite of the name *permafrost*, it is not permanent. The upper boundary moves upward or downward, resulting in widely varying site conditions for plants. In fact the effect of permafrost on vegetation is so pronounced that vegetation can be used to map areas of permafrost (Pewe, 1957). Black spruce, for example, grows best on areas of permafrost, whereas white spruce grows on areas free of it. Cottonwood and white spruce grow where permafrost is 4 ft or more below the ground; where it is less than 4 ft below, stunted black spruce grows; and where permafrost is within 18 in. of the surface, the area is wet and treeless, with a growth of grasses and low shrubs.

Alpine vegetation. In general the alpine tundra is a more severe environment for plants than the arctic tundra; and the adaptation of plants to the physical environment is probably more important than the interrelations of one species with another. The alpine tundra is a land of strong winds, snow, and of cold and widely fluctuating temperatures. During the summer the temperature on the surface of the soil ranges from 40° to 0° C (104° to 32° F) (Bliss, 1956). The atmosphere is thinner in the alpine tundra, and because of this, light intensity, especially ultraviolet, is high on clear days.

The alpine tundra of the Rocky Mountains is a land of rock-strewn slopes, bogs, and alpine meadows, and shrubby thickets (Fig. 17-28). In spite of the similarity of conditions, only about 20 percent of the plant species of the Arctic and of the Rocky Mountain alpine tundra are the same, and these are different ecotypes. Heaths are lacking in the tundras of the Rockies, as well as the heavy growth of lichens and mosses between other plants. Lichens are more or less confined to the rocks, and the ground is bare between plants.

Cushion- and mat-forming plants, rare in the Arctic, are important in the alpine tundra. Low and hugging the ground, they are able to withstand the buffeting of the wind. The cushionlike blanket traps heat and the interior of the cushion may be 20° warmer than the surrounding air, a microclimate that is utilized by insects. Thick cuticles, which increase plants' resistance to desiccation, and the abundance of epidermal hairs and scales are characteristic of alpine plants. The significance of this is still debated. These hairs appear to absorb and reflect the bright light of the alpine environment. At the same time, the hairs may act as a heat trap and perhaps prevent cold injury when air temperatures drop to freezing (Krog, 1955), and enable the plants to develop and bloom while the air is still cold.

Alpine vegetation and its associated soils vary on a rather complex gradient or continuum controlled by topographic site and snow cover, both of which interact with wind (Fig. 17-29). The vegetational pattern has been worked out for the Beartooth Plateau of Wyoming (Johnson and Billings, 1962) and serves as an example of the high alpine areas of the Rocky Mountains. The high, windswept areas, rocky and free of snow, support only lichens, which may completely cover the sheltered side; but on the windward side they are short, no higher than the depth of snow, and they may be completely lacking on the most exposed sites. Below the lichen growth are the xeric cushion-plant communities, which extend further downslope on the windward side than they do on the lee. This land of rock, lichens, and cushion plants is the alpine rock desert. In somewhat more protected sites grows the geum turf, a sodlike covering of geum and associated plants, such as sedges, lupines, polygonums, and mountain avens. Alpine meadows develop on well-drained soils of sheltered uplands and lower mesic slopes and basins. Hairgrass, *Deschampsia*, often growing in pure stands, is the dominant species. These meadows are subject to considerable disturbance both from frost activity and from pocket gophers. Alpine bogs, communities quite similar to those of the arctic tundra, support a growth of sedge and cottongrass. Willow thickets, dense and uniform in height, grow along drainage channels and in alpine valley bottoms.

The alpine tundra of the high Appalachians is not nearly so cold and windswept as that of the Rockies. Tundra areas are small and lack the diversity of species found in the western mountains. Indeed little floristic similarity exists between the regions. There is a much closer affinity between the flora of the eastern alpine tundra and that of the arctic and the alpine communities of Scandinavia and central Europe.

Nine plant communities are recognized in the tundras of the Presidential Range in New Hampshire (Bliss, 1963). These occur on two gradients, one of increasing snow depth, the other of increasing moisture (Fig. 17-30). On exposed windswept sites where winter snow cover is thin or nonexistent, *Diapensia*, a dwarf, tussock-forming shrub, grows. Over those widespread areas where snow cover is variable a dwarf-shrub-heath–rush community occupies the sites. Dwarf-shrub heaths—bearberry, bilberry, Lapland rosebay—dominate where the deep snow cover melts early. Snowbank communities are most prevalent on the east- and southeast-facing slopes, in the lee of the prevailing winds. The second gradient, one of increasing summer atmospheric and soil moisture and fog, is largely restricted to north- and west-facing slopes on the higher peaks. Sedge meadows at the highest elevations give way downslope to a sedge–dwarf-shrub–heath community. At lower elevations this is replaced by a sedge–rush–dwarf-shrub–heath type. Two other communities, the streamside and the bog, are common at low elevations.

(a)

(b)

FIGURE 17-28

Alpine tundra in the Rocky Mountains: (a) *the tundra just beginning to come to life in the spring,* (b) *stone polygons. (Photos by author.)*

585

Krumholtz. At the tree line, where the forest gives way to the tundra, lies an area of stunted, wind-shaped trees, the krumholtz or "crooked wood." The krumholtz in the North American alpine region is best developed in the Appalachians. In the west it is much less marked, for there the timber line ends almost abruptly with little lessening of height; the trees for the most part are flagged—branches remain only on the lee side. On the high ridges of the Appalachians, particularly in the White and Adirondack Mountains, the trees begin to show signs of stunting far below the timber line. As the trees climb upward, stunting increases until spruces and birches, deformed and semiprostrate, form carpets 2 to 3 ft high, impossible to walk through but often dense enough to walk upon. Where strong winds come in from a constant direction, the trees are sheared until the tops resemble close-cropped heads, although the trees on the lee side of the clumps grow taller than those on the windward side. Though the wind and cold generally are regarded as the cause of the dwarf and misshapen condition of the trees, Clausen (1965) has demonstrated that the ability of some species of trees to show a "krumholtz" effect is genetically determined. Eventually conditions become too severe even for the prostrate forms, and the trees drop out completely except for those that have taken root behind the protection of some high rocks, and tundra vegetation then takes over completely. On some slopes the trees might be able to grow on better sites at higher elevations, but they appear to be eliminated by competition from sedges (Griggs, 1946).

ANIMALS OF THE TUNDRA

Arctic tundra. The tundra world holds some fascinating animals, even though the diversity of species is small. Like the plants, the animals of the arctic and the alpine tundras are of two distinct groups. Except for the caribou and the pipit, there are no important species in common. The animals of the arctic tundra are mostly circumpolar, and those species of the North American tundra that are not, have close relatives in the Eurasian tundra. The North American barren ground caribou, for example, has the European reindeer, and the North American wolverine is matched by the glutton. Musk-ox, arctic hare, arctic fox, and arctic ground squirrels· are or have been

common to both. In addition, some 75 percent of the birds of the North American tundra are common to the European tundra (Udvardy, 1958).

The arctic tundra, with its wide expanse of ponds and boggy ground, is the haunt of myriads of waterfowl, sandpipers, and plovers, which arrive when the ice is out, nest, and return south before winter sets in.

Invertebrate life is scarce, as are amphibians and reptiles. Some snails are found in the arctic tundra about the Hudson Bay. Insects, reduced to a few genera, are nevertheless abundant, especially in mid-July. The insect horde is composed of black flies, deer flies, and mosquitos (Shelford and Twomey, 1941).

White is the dominant color of the animals of the arctic tundra. In the high Arctic, white coloration is universal and maintained the year round, as in the polar bear, arctic fox, and greater snow goose. Others are dimorphic. The gray falcon is white in the high Arctic, gray in the low Arctic, and dark in the sub-Arctic. Some animals alternate a white winter color with a dark summer color, and again there is a relation between the winter climate and the degree of whiteness. The rock ptarmigan of the low boreal mountains, for example, never quite acquires a full white winter plumage, for the back is marked with dark; the ptarmigans of the high and low Arctic are pure white in winter. The snow bunting, too, exhibits the same relationship. Those birds wintering in Siberia have creamy white edges to the feathers; those that winter in central Europe and North America have the feathers edged in buff. The reason for the white color is not fully known. It does conceal both predator and prey in winter and was believed capable of reducing the amount of heat radiating from the body, but recent experiments indicate that this apparently is not true (Hammel, 1956).

Animal activity in the arctic tundra is geared to short summers and long winters. The only hibernator is the ground squirrel, although the female polar bear does den up in the snow where she gives birth to her cub. The ground squirrel is active only from May to September. It mates almost as soon as it emerges from the burrow in spring, and the young are born in mid-June, after a 25-day gestation period. The young are self-sufficient by mid-July, attain adult weight and are ready to hibernate by late September and early October (W. Mayer, 1960). A similar speed-up in the life cycle is found among some of the arctic birds. Because of the long days, the northern robin feeds the young for 21 hours/day, and the young may leave the nest when a little over 8 days old, in contrast to 13 days or more in the temperate region (Karplus, 1949). The larger mammals and the few overwintering birds, such as the ptarmigan and the redpoll, and the marine mammals have heavy layers of fat beneath the skin, long dense fur, or long dense feathers. Ptarmigan and snowy owls have feathers on their tarsi; the caribou and the musk-ox have winter coats of extremely heavy, wooly underfur overlaid with a thick fleece that is almost airtight. Small rodents and weasels tunnel beneath the snow, itself a good insulator. Those species unable to withstand the severe cold migrate to warmer or more protected areas. Few birds remain behind over winter, and even the caribou leaves the high arctic tundra for the southern portions, even south of the tree line. Invertebrates usually pass the winter in the larval and pupal stages. A few, however, such as rotifers and diving beetles, may be frozen in the ice, while others pass the season as adults.

Alpine tundra. The alpine tundra, which extends upward like islands in the mountain ranges of the world, is small in area and contains few characteristic species. The alpine regions of western North America are inhabited by pikas, marmots, mountain goat—not a goat at all but related to the South American chamois—mountain sheep, and elk. The sheep and the elk spend their summers in the high alpine meadows and winter on the lower slopes. The marmot, a mountain woodchuck, hibernates over winter, while the pika cuts grass and piles it in tiny haycocks to dry for winter. Some rodents, such as the vole and the pocket gopher, remain under the ground and snow during winter. The pocket gopher is important, for its activities influence the pattern of alpine vegetation. The tunneling gophers kill the sedge and cushion plants by eating the roots and by throwing the soil to the surface, smothering plant life. Other plants then take over on the wind-blown, gravelly soil. These pioneering plants are rejected by the gopher. The rodent moves on, the cushion plants move back in, organic matter accumulates again. Slowly the sedges recover, and

FIGURE 17-29
Relationship of major vegetation types to slope cover and slope position in the western alpine tundra on Beartooth Plateau, Wyoming. (After Johnson and Billings, 1962. © The Ecological Society of America.)

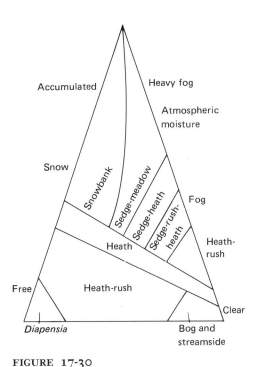

FIGURE 17-30
Relationship of alpine communities to snow and atmospheric moisture in the Presidential Range of New Hampshire. (After Bliss, 1963. © The Ecological Society of America.)

when they do, the gophers return once more.

The alpine regions contain a fair representation of insect life. Flies and mosquitos are scarce, but springtails, beetles, grasshoppers, and butterflies are common. Because of the ever-present winds, butterflies fly close to the ground; other insects have short wings or no wings at all. Insect development is slow; some butterflies may take 2 years to mature, and grasshoppers 3.

A major problem for mammals of the alpine tundra is low oxygen pressure. Birds obviously are able to live at high altitudes where oxygen pressure is low, but mammals that wander into high places are not, unless they live there permanently and have become adapted to rarefied air. Normal air pressure for most mammalian life is 15 lb/in.² At high alpine elevations this pressure may drop to 10 lb at 10,000 ft and even lower higher up; at 18,000 ft, air pressure is reduced by half. In such a rarefied atmosphere the mammal, including man, has a difficult time securing sufficient oxygen. Mammals can temporarily adapt to high altitudes by increasing the heart beat and the rate of respiration. Mammals permanently adapted to high altitudes have blood richer in red cells and hemoglobin.

IMPACT OF MAN

For centuries the Arctic, a remote and forbidding environment, has escaped the ravages of western man. Now that the resources of the Arctic, especially oil, are in demand, the region faces environmental degradation. The impact can be grave, for the tundra is a fragile ecosystem sensitive to disturbance.

Because the permafrost has a controlling influence over the arctic ecosystem, any disturbance to soil cover can affect the permafrost, and that in turn affects the tundra ecosystem. The activities of man tend to reduce the effectiveness of vegetation as an insulation for permafrost, leading to its degradation and thawing. Over wide areas of the Arctic, permafrost is in such a delicate balance that even minor disturbances can have disastrous consequences. Removal of mosses and lichens can increase the depth of thaw 20 to 50 percent (Tyrtikov, 1959). Footpaths used for only one season may persist unchanged for decades (Mackay, 1966). Frequent vehicular traffic slices up the tundra

(Fig. 17-31). In regions of Alaska, where relatively light track-laying vehicles designed for the Arctic have traveled, the tracks have thawed and eroded into gullies 10 ft deep. Trails made by Navy equipment in the 1940s on the Arctic Slope have eroded into gullies 20 ft wide and 10 to 15 ft deep and miles long.

There is concern over the impact of oil pipelines across the tundra. Pipelines buried 6 ft deep in the permafrost and heated to 80° C (176° F) would thaw a cylindrical region 20 to 30 ft in diameter in a few years. In 20 years permafrost would be thawed to a depth of 40 to 50 ft in southern Alaska, and 30 to 40 ft in northern Alaska. If the temperature were held to 30° C, the thaw would be reduced only by 30 to 40 percent (Lachenbruch, 1970). Insulating the pipe only increases the temperature of the oil rather than reducing the thaw.

Arctic wildlife is particularly vulnerable to human activities (Holloway, 1970; Pruitt, 1970; Uspenski, 1970). Land development, especially hydroelectric schemes and mineral exploitation, threaten nesting and wintering grounds. Pipelines, railroads, and highways can obstruct migratory movements of caribou (Klein, 1971). Oil pollution on land and sea can create special problems. Increased disturbance by human activity and increased hunting (legal or illegal) can reduce game population. Fires may become more frequent. Already a problem in the Arctic, fire has devastated large areas of lichen ranges, upon which the caribou depend. The destruction of lichen ranges has been associated with the alarming reduction of caribou in Canada. The situation is aggravated by the notoriously slow recovery of lichens from fire.

Because of the circumpolar nature of the arctic ecosystem, the worldwide threat of exploitation, and the international distribution of its wildlife, the maintenance of the arctic ecosystem is an international problem. Uspenski (1970) writes:

The specificity of the arctic ecosystems, their simplicity and instability, their extreme vulnerability to economic activities, the peculiarities of the geographical distribution of its mobile components naturally emphasize the necessity of a much wider profile of the international-legal protection of the entire natural complex of the Arctic, and especially those species the conservation of which exceeds the limit of the ability of individual countries.

Summary

The alpine tundra of the high mountain ranges in lower altitudes and the arctic tundra that extends beyond the tree line of the far north are at once similar and dissimilar. Both have low temperature, low precipitation, and a short growing season. Both possess a frost-molded landscape and plant species whose growth form is low and whose growth rate is slow. The arctic tundra has a perma-frost layer; rarely does the alpine tundra. Arctic plants require longer periods of day-light than alpine plants and reproduce vege-tatively; alpine plants progagate themselves by seedlings. Over much of the Arctic, the dominant vegetation is cotton grass, sedge, and dwarf heaths. In the alpine tundra, cushion- and mat-forming plants, able to withstand buffeting by the wind, dominate the exposed sites; cotton grass and other tundra plants are confined to protected sites. At the tree line lies the krumholtz, a land of stunted, wind-shaped trees, a growth form that may be genetically determined. Animal life in the arctic tundra, except for the caribou and the pipit, is distinct from that of the alpine and is circumpolar in distribution. White is the dominant color of tundra birds and mammals, especially in winter. For mammals of the alpine tundra a major prob-lem is low oxygen pressure, a situation that animals of the arctic tundra generally do not face. The tundra region, arctic and alpine, is a rigorous environment for plants and animals; but in spite of it, the tundra, con-trary to its implied barrenness, is a land rich in life—if not in variety, then in seasonal abundance.

Another harsh environment is the desert. As in the Arctic, plants and animals have evolved ways to circumvent aridity and high temperatures. Desert animals avoid the heat by becoming nocturnal in habit, by seeking shady places, or by spending the day in underground burrows. They obtain drinking water from succulent plants, from the blood and body fluids of their prey, or from the metabolic oxidation of carbohydrates and

FIGURE 17-31
A seismographic exploration trail has left its marks on the fragile tundra ecosystem. This trail is near the Canning River, and about 20 mi south of the Arctic Ocean. Removal of the vegetation mat with bulldozers to create the trail is no longer permitted by the Bureau of Land Management or the State Divisions of Lands. Note the thaw erosion on the trail. (Photo courtesy U.S. Fish and Wildlife Service.)

fats. Further water conservation involves reabsorption of water from urine and feces. Plants evade aridity by living through the dry period as seeds, which sprout when sufficient rainfall arrives, by storing water in the plant body, or by shedding leaves and possessing small leaves that reduce transpiration. In North America there are two major types of deserts, the cool desert of the Great Basin, dominated by sagebrush, and the hot desert of the Southwest, dominated by creosote bush and cacti. Deserts occupy about one-seventh of the land surface of the earth and are largely confined to a worldwide belt between the Tropic of Cancer and the Tropic of Capricorn. Deserts are largely the result of the climatic patterns of the earth, as well as the locations of mountain ranges and the remoteness of land areas from sources of oceanic moisture. For centuries man has lived on the periphery of the desert, and few peoples have made it their home. Today modern man is looking to the desert as potential agricultural land and a place to live. Settlement of the desert is being achieved by tapping deep but unreplenishable water reserves beneath the desert floor. Successful though such attempts may be now, the danger exists that man's occupancy of the desert may so "mine" the area of water and nutrients that even more arid conditions will result. Before deserts are exploited, man should first understand their ecology, about which we know so little.

Semiarid regions support climax shrub ecosystems, characterized by densely branched woody structure and low height. The success of shrubs depends largely on their ability to compete for nutrients, energy, and space. In semiarid situations the shrubs have numerous competitive advantages, including structural modifications that affect light interception, heat losses, and evaporative losses. Shrublands, which go by different names in different regions of the world, are most characteristic of places where winters are mild and rainy and summers are long, hot, and dry. Seral shrublands occupy land going back to forest. Although a successional stage, many seral shrublands may remain stable for long periods of time. Shrublands, an important wildlife habitat, are often the result of disturbances to forest and grassland by man and are themselves being destroyed by man.

Coniferous, deciduous, and tropical rain forests are the three dominant kinds of forest cover of the earth. Confined to the northern hemisphere, the coniferous forest is typical of regions where the summers are short and the winters long and cold. The deciduous forest, which is richly developed in North America, western Europe, and eastern Asia, grows in a region of moderate precipitation and mild temperatures during the growing season. The tropical rain forest, which grows in equatorial regions, is the vegetative cover where the humidity is high, the rainfall is heavy, especially during at least one season of the year, seasonal changes are minimum, and the annual mean temperature is about 80° F. Of the three, the tropical rain forest is the most highly stratified and contains the greatest diversity of ecological niches. A well-developed deciduous forest is the next most highly stratified. Accompanying this vegetative stratification is a stratification of light, temperature, and moisture. The canopy receives the full impact of the climate and intercepts light and rainfall; the forest floor is shaded through the year in most coniferous and tropical rain forests, and in late spring and summer in the deciduous forest. The coniferous and deciduous forests hold different species of animal life, but animal adaptations are similar. The greatest concentration and diversity of life is on and just below the ground layer. Other animals live in various strata from low shrubs to the canopy. The tropical rain forest has pronounced feeding strata from above the canopy to the forest floor, and many of its animals are strictly aboreal. Whatever the forest, the different trees that compose it create different environments, which ultimately dictate the kind of plants and animals that can live within it.

Grasslands, once covering extensive areas of the globe, including the midcontinent of North America, have shrunk to a fraction of their original size because of man's requirements for crop and grazing lands. With them also declined or disappeared many of the large grazing herbivores, replaced in part by cattle and sheep. Clearing of the forest, the planting of hayfields, and the development of seral grasslands on disturbed sites extended the range of some grassland animals into once-forested country. Seral or climax grasslands consist of sod formers, bunch grasses, or both, the latter of which often provide excellent nesting sites for birds and small mammals. When grasslands are undisturbed by grazing, mowing, or burning, they accumulate a layer of mulch, which retains moisture, influences the character and composition of plant life, and provides shelter for some animals. Insects occur in great numbers and in a wide variety of species, and mammals are the conspicuous vertebrates of the grasslands. The prairie dog of the plains, now much reduced in its range, is typical of the burrowing animals common to grasslands. Because of its feeding and tunneling activities, the prairie dog prevents the encroachment of shrubs and annuals and allows the perennial grasses and forbs to increase. Among the birds, the prairie chicken epitomizes the grassland, but its rolling boom is almost silent, as overgrazing and plowing have destroyed its habitat. Even in the seral and managed grasslands of the forested regions, habitat for animals is declining as agricultural practices change and as more and more abandoned land reverts back to forest.

All ecosystems support and are supported by a heterotrophic soil community. Organisms present in the soil, like all other animal populations, reflect their environment. Their abundance and faunal composition depend upon the nature of the soil, its nutrient status and the vegetation present, the kind of litter it produces, and the ability of the plants to return calcium and other nutrients to the soil. The soil animals, in turn, play their role in influencing the future development of the upper soil layers. The direct decomposition of the plant litter is accomplished by the microflora, bacteria, and fungi. The soil invertebrate fauna make the organic matter more readily available to the microflora by the mechanical breakdown of the litter, by exposing new areas for fungal invasion, by spreading fungal spores through their feces, and by increasing surface area exposed to attack by bacteria and fungi. But the soil fauna also consume great quantities of fungi and depress bacterial and fungal populations. The predaceous species in turn influence the population levels of the litter-feeding and decomposer organisms. Such is the chain of life in the world beneath the surface of the ground.

Through the years, a number of attempts have been made to classify life into mean-

ingful distributional units. Most of these attempts are faunistic in their approach, but one considers both plants and animals together. The first division of the world into distributional units was the biogeographical or faunal regions and realms. Of the latter, three are recognized and are further subdivided into six regions. Each region is separated by a barrier of oceans, mountain ranges, or deserts, which prevent the free dispersal of animals; and each possesses its own distinctive forms of life. Each region is further subdivided by secondary barriers, such as vegetation types and topography. These subdivisions have been recognized as life zones, biotic provinces, and biomes. The life-zone concept, restricted to North America, divides the continent into broad transcontinental belts, the plant and animal differences between which are governed chiefly by temperature. The biotic-provinces approach divides the North American continent into continuous geographic units that contain ecological associations different from those of adjacent units, especially at the species and subspecies levels. The biome system groups the plants and animals of the world into integral units, characterized by distinctive life forms in the climax, the stage of development at which the community is in approximate equilibrium with its environment. Boundaries of biomes, or "major life zones," as they are known in Europe, coicide with the boundaries of the major plant formations of the world. By including both plants and animals as a total unit that evolved together, the biome permits the recognition of the close relationship that exists among all living things. The biomes today have been greatly disturbed by man. The preservation of samples of the original climax communities of each biome as indicators of the natural potential and limitations of a given region would provide a reference point against which man's influence on the natural environment can be assessed.

lobal aquatic ecosystems fall into broad classes definable by salinity—the fresh-water ecosystem and the saltwater ecosystem. The latter may include inland brackish water, as well as marine and estuarine habitats. Fresh-water ecosystems, the study of which is known as limnology, are conveniently divided into two groups, *lentic* or standing water habitats, and *lotic* or running water habitats. Both can be considered on an environmental gradient. The lotic follows a gradient from springs to mountain brooks to streams and rivers. The lentic involves a gradient from lakes to ponds to bogs, swamps, and marshes.

Lentic ecosystems

Lakes are inland depressions containing standing water. They may vary in size from small ponds of less than an acre to large seas covering thousands of square miles. They may range in depth from a few feet to over 5,000 ft. Ponds, however, are considered as small bodies of standing water so shallow that rooted plants can grow over most of the bottom. Most ponds and lakes have outlet streams; and both are more or less temporary features on the landscape because filling, no matter how slow, is inevitable.

Lakes and ponds arise in many ways. Some North American lakes were formed by glacial erosion and deposition and a combination of the two. Glacial abrasion of slopes in high mountain valleys carved basins, which filled with water from rain and melting snow to produce tarns. Retreating valley glaciers left behind crescent-shaped ridges of rock debris, which dammed up water behind them. Numerous shallow kettle lakes and potholes were formed on the glacial drift sheets that cover much of northeastern North America and northwestern Europe. Lakes are also formed by the deposition of silt, driftwood, and other debris in the beds of slow-flowing streams. Loops of streams that meander over flat valleys and flood plains often become cut off, forming crescent-shaped oxbow lakes. Shifts in the earth's crust, either by the uplifting of mountains or by the breaking and displacement of rock strata, causing part of the

Fresh-water ecosystems

valley to sink, develop depressions that fill with water. Craters of extinct volcanos, too, may fill with water; and landslides can block off streams and valleys to form new lakes and ponds. In a given area, however, all natural lakes and ponds have the same geological origin and the same general characteristics. But because of varying depths at the time of origin, they may represent several stages of succession.

Many lakes and ponds are formed by nongeological activity. Beavers dam up streams to make shallow but often extensive ponds. Man intentionally creates artificial lakes by damming rivers and streams for power, irrigation, and water storage, or by constructing small ponds and marshes for water, fishing, and wildlife. Quarries and strip mines fill with water to form other ponds.

CHARACTERISTICS OF LAKES

The relatively still waters of lakes and ponds offer environmental conditions that contrast sharply with running water. Light penetrates only to a certain depth, depending upon turbidity. Temperatures vary seasonally and with depth. Because only a relatively small proportion of the water is in direct contact with the air and because decomposition is taking place on the bottom, the oxygen content of lake water is relatively low compared to that of running water. In some lakes, oxygen may become less with depth, but there are many exceptions. These gradations of oxygen, light, and temperature profoundly influence life in the lake, its distribution, and adaptations.

Temperature stratification. Each year the waters of many lakes and ponds undergo seasonal changes in temperature (Fig. 18-1). As the ice melts in early spring, the surface water, heated by the sun, warms up. When it reaches 4° C and becomes more dense, a slight temporary stratification develops, which sets up convection currents. Aided by the strong winds of spring, these currents mix the water throughout the basin until the water in the lake is uniformly 4° C. Now even the slightest winds can cause a complete circulation of the water between the surface and the bottom. This is the spring overturn, when the nutrients on the bottom, the oxygen on the top, and the plankton within are mixed throughout the lake.

With the coming of summer, the sun's

593

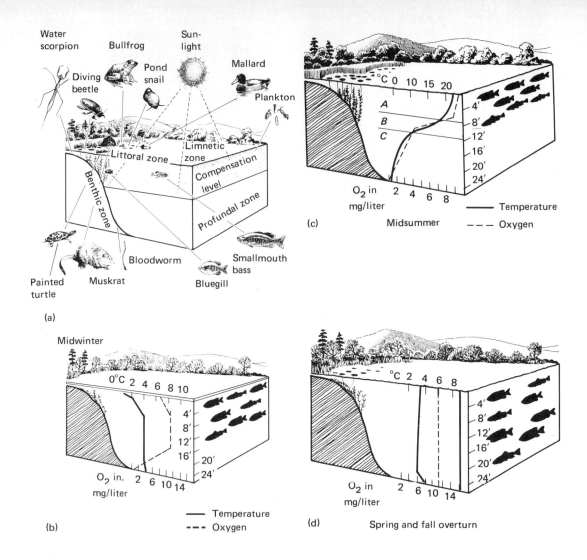

Water
scorpion

Bullfrog

Sun-
light

Mallard

Diving
beetle

Pond
snail

Plankton

Limnetic
zone

Littoral zone

Compensation
level

Benthic zone

Profundal zone

Painted
turtle

Muskrat

Bloodworm

Smallmouth
bass

Bluegill

(a)

°C 0 10 15 20

A
B
C

4'
8'
12'
16'
20'
24'

O₂ in
mg/liter

2 4 6 8

(c)   Midsummer

——— Temperature
‐ ‐ ‐ Oxygen

Midwinter

0°C 2 4 6 8 10

4'
8'
12'
16' 20'
24'

O₂ in.
mg/liter

2 6 10 14

——— Temperature
‐ ‐ ‐ Oxygen

(b)

°C 2 4 6 8

4'
8'
12'
16'
20'
24'

O₂ in
mg/liter

2 6 10 14

(d)   Spring and fall overturn

intensity increases, and the temperature of the surface water rises. The higher the temperature of the surface water, the greater is the difference in the density between the surface and deeper layers. Because the thermal density gradient opposes the energy of the wind, it becomes more difficult for the waters to mix. As a result, a mixing barrier or wall is established. The warm, freely circulating surface water, with a small but variable temperature gradient, is the *epilimnion* (Fig. 18-1c). Below this is the wall, the *metalimnion*, a zone characterized by a very steep and rapid decline in temperature. Within the metalimnion is the *thermocline*, the plane at which the temperature drops most rapidly—1° C for each meter of depth. Below these two layers is the *hypolimnion*, a deep, cold layer, in which the temperature drop is gentle.

With the coming of autumn, the air temperature falls. The surface water loses heat to the atmosphere through evaporation, convection, and conduction. The temperature of the surface water drops and the thermocline sinks; the epilimnion increases until it includes the entire lake. The temperature once again is uniform from top to bottom, the lake waters circulate, and oxygen and nutrients are recharged throughout the lake. This is the fall overturn, which, through stirring actions caused by the slightest wind, may last until ice forms.

As the surface water cools below 4° C, it becomes lighter, remains on the surface, and, if the climate is cold enough, freezes; otherwise, it remains very close to 0° C. Now a slight inverse temperature stratification may develop, in which the water becomes warmer up to 4° C with depth. The water immediately beneath the ice may be warmed by solar radiation through the ice. Because this increases its density, this water subsequently flows to the bottom, where it mixes with water warmed by heat conducted from bottom mud. The result of this is a higher temperature at the bottom, although the overall stability of the water is undisturbed. As the ice melts in spring, the surface water again becomes warm, currents pass unhindered, and the spring overturn takes place.

This is a general picture of seasonal changes in temperature stratification and must not be considered as a uniform condition in all deep bodies of water. In shallow lakes and ponds temporary stratification of short duration may occur; in others stratification may exist but no thermocline develops. In some very deep lakes the thermocline may simply descend during periods of overturn and not disappear at all. In such lakes the bottom water never becomes mixed with the top layers. But some form of thermal stratification occurs in all very deep lakes, including those of the tropics.

Based on temperature, lakes can be broadly classified into three basic groups (Hutchinson, 1957). Lakes that have two overturns during the year are called *dimictic*. Dimictic lakes are inversely stratified in winter and directly stratified in summer. These lakes are characteristic of temperate regions and of high mountains in subtropical regions. In warm oceanic climates and in the mountains of subtropical latitudes are lakes whose waters never cool sufficiently to allow complete circulation, except at the very coldest time of the year. Such lakes, which undergo but one overturn in a year, are called *warm monomictic*. The water is never below 4° C at any level; and in fact the temperature of the water in the hypolimnion is never lower than the mean temperature of the air at the time of the last circulation. These lakes too are directly stratified in summer. Lakes in polar and arctic regions may never rise above 4° C, and circulation takes place only in summer. These are *cold monomictic* lakes. Freely circulating in summer, these lakes are inversely stratified in winter. There are a few other types. Lakes in the high mountains of equatorial regions, where there is little seasonal change in temperature, have waters that are continually circulating at a little above 4° C. These lakes, called *polymictic*, always lose just enough heat to prevent stable stratification. On the other hand there are some lakes in the humid tropics whose waters, always well above 4° C, circulate only at irregular intervals.

FIGURE 18-1 [OPPOSITE]

These drawings summarize life in the lake as it is influenced by the seasons and seasonal stratification of oxygen and temperature. (a) A generalized picture of the lake in midsummer showing the major zones—the littoral, the limnetic, the profundal, and the benthic. The compensation level is the point at which the light is too low for photosynthesis. Surrounding the lake is a variety of organisms typical of the lake community. (b) The distribution of oxygen and temperature in a lake during midwinter and its effect on the distribution of fish life. The narrow fish are trout, representing the cold-water species; the bass silhouette, warm-water species. (c) In midsummer there is a pronounced stratification: A the epilimnion; B, the metalimnion; C, the hypolimnion. (d) During the spring and fall overturns, the oxygen and temperature curves are almost straight lines. The graphs are general, and represent no particular lake.

595

Oxygen stratification. Oxygen stratification during summer nearly parallels that of temperature (Fig. 18-1c), although here again there are exceptions. In general the amount of oxygen is greatest near the surface, where there is an interchange between the water and atmosphere and some stirring by the wind. The quantity decreases with depth, a decrease caused in part by the respiration of decomposer organisms feeding on the organic matter dropping down from the layers above. In some lakes oxygen may vary little from top to bottom; every layer will be saturated relative to temperature and pressure. Water in a few lakes may be so clear that light penetrates below the depth of the thermocline and permits the development of phytoplankton. Here, because of photosynthesis, the oxygen content may be greater in deep water than on the surface.

During the spring and fall overturns, when water recirculates through the lake, oxygen is replenished in the deep water and nutrients are returned to the top. In winter the reduction of oxygen in unfrozen water is slight because bacterial decomposition is reduced and water at low temperatures holds a maximum amount of oxygen. Under ice, however, oxygen depletion may be serious and result in a heavy winter kill of fish.

Currents and seiches. Oxygen and thermal stratification, depth and position of the thermocline, circulation of nutrients, and distribution of organisms all are influenced by currents. The most conspicuous water movements observed are traveling surface waves, the result of wind pressures on the surface of lakes and ponds. Except for the effects they have on the shore line and shore organisms, surface waves are not too important biologically. More important are standing waves, or *seiches* (sāches), a term that comes from the French and means dry, exposed shore line. These seiches are produced by an oscillation of a structure of water about a point or node.

There are two kinds of seiches, surface and internal. Both are produced by the wind blowing across the water's surface, by heavy rain showers, or perhaps even by changes in atmospheric pressure (see Bryson and Ragotzkie, 1960; Vallentyne, 1957).

When the wind blows across a lake or pond, it piles up water on the leeward end and creates a depression on the windward end. When the wind subsides, the current flows back; but because the momentum of the returning currents is not broken on the shore, a depression is created on the former leeward side, and the water flows back again. Thus an oscillation or rocking motion is established about a stationary node (Fig. 18-2). This continues until it is finally halted by the lake basin proper or by such meteorological forces as an opposite wind or rain. Although these surface seiches occur on all lakes, they can be observed visually on larger lakes.

Internal seiches, not observable on the surface, occur during the summer in thermally stratified lakes. They are much more pronounced and exert a greater influence on life in the lake than surface seiches. Internal seiches are caused not only by the action of wind on the surface waters but also by density differences between warm and cold water. When the wind piles the water of the epilimnion up on the leeward side, the weight and circulation of the lighter water over the denser cold water tilts the thermocline and raises the hypolimnion on the windward side. When the wind stops, the raised hypolimnion, pushed down on and toward the leeward side, causes the epilimnion water to flow back toward the windward side. Thus an oscillation is established between the lighter layer of the epilimnion and the denser water of the hypolimnion (Fig. 18-2). In time, the oscillations are slowed by friction on the lake basin or by the wind from an opposite direction.

These oscillations move about a point in the center of the lake where resistance is highest. Here vertical displacement movement is the least; the greatest displacement is on the windward and leeward ends. The position of internal seiches can best be determined by charting the variations of the depth of a particular temperature over a period of hours. This movement makes difficult the determination of the true position of the thermocline because its position at any one point is determined by the amplitude of the oscillation at that point.

Ecologically, internal seiches are important because they distribute heat and nutrients vertically in the lake and transport them into shallow water. The seiche sets up two rhythmic but opposite current systems above and below the thermocline, whose speeds are greatest when the thermocline is level (Fig. 18-2). A turbulence develops in the hypolimnion; without this turbulence the hypolimnion would become stagnant. In addition, plankton are moved up or down with the water mass. Even fish and other organisms are influenced by internal seiches.

The mixing of water and the transfer of nutrients also are carried on by *eddy currents*. These are small, turbulent currents, whose energy is dissipated at right angles to the major currents. Because these largely are horizontal and the turbulence is vertical, an interchange of adjacent water masses takes place. Here oxygen may be transferred downward and heat upward; nutrients and even plankton are intermixed. The degree of intermixing depends upon the intensity of the turbulence, which changes gradually from one depth to another.

LIFE IN LAKES AND PONDS

So far, lakes and ponds have been considered only as if they were natural laboratories in which chemical processes and demonstrations of physical laws could be observed. This was necessary in order to understand the biology of lakes and ponds, because the abundance, distribution, and diversity of lake and pond life are influenced by light, temperature, oxygen, and nutrients.

The energy source of the lake and pond ecosystem is sunlight. The depth to which light can penetrate is limited by the turbidity of the water and the absorption of light rays. On this basis, lakes and ponds can be divided into two basic layers—the *trophogenic zone*, roughly corresponding to the epilimnion, in which photosynthesis dominates; and the second, the lower *tropholytic zone*, where decomposition is most active. This zone is about the same as the hypolimnion. The boundary between the two zones is the *compensation depth*—the depth at which photosynthesis balances respiration and beyond which light penetration is so low that it is no longer effective. Generally the compensation depth occurs where light intensity is about 100 footcandles, or approximately 1 percent of full noon sunlight incident to the surface (see Edmondson, 1956).

The region of photosynthetic activity can be divided into two subzones. First is the *littoral*, or shallow-water zone, where light penetrates to the bottom. This area is occupied by rooted plants such as water lilies, rushes, and sedges. Beyond this is the *lim-*

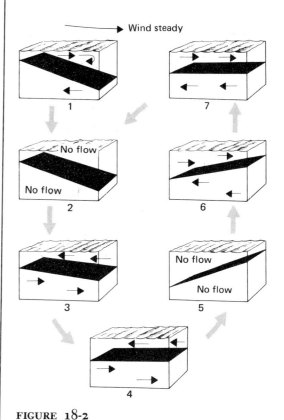

FIGURE 18-2

A diagrammatic representation of a thermocline slope and subsequent seiches after the wind dies down.

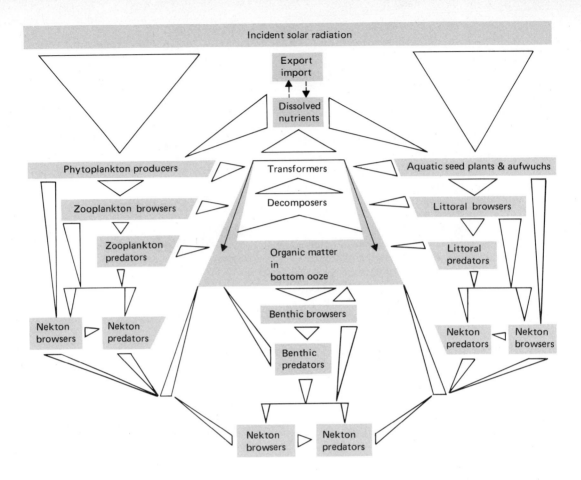

Incident solar radiation

Export import

Dissolved nutrients

Phytoplankton producers

Transformers

Aquatic seed plants & aufwuchs

Zooplankton browsers

Decomposers

Littoral browsers

Zooplankton predators

Organic matter in bottom ooze

Littoral predators

Nekton browsers

Nekton predators

Benthic browsers

Nekton predators

Nekton browsers

Benthic predators

Nekton browsers

Nekton predators

netic, or open-water zone, which extends to the depth of effective light penetration. It is inhabited by plant and animal plankton and the *nekton,* free-swimming organisms such as fish, which are capable of moving about voluntarily. Common to both the tropholytic and trophogenic zones is the *benthic,* or bottom region. Although these areas are named and often described separately, all are closely dependent upon one another in nutrient cycles and energy flow (Fig. 18-3). The organisms that inhabit the benthic zone are known collectively as the *benthos.*

Limnetic zone. The open water is a world of minute suspended organisms, the plankton. Dominant are the phytoplankton, among them the diatoms, desmids, and the filamentous green algae. Because these tiny plants alone carry on photosynthesis in open water, they are the base upon which the rest of limnetic life depends. Suspended with the phytoplankton are the animal, or zooplankton organisms, which graze upon the minute plants. These animals form an important link in the energy flow in the limnetic zone. Most characteristic are the rotifers, copepods, and cladocerans.

Rotifers, the so-called wheelbearers, have at the front end of the body a circlet of moving cilia, which look like two rapidly rotating wheels. These cilia aid the organisms to pull algae and protozoa into the pharynx, where the food is crushed and ground. The crustaceans, cladocerans, and copepods filter phytoplankton, bacteria, and detritus out of the water by means of comblike setae on the thoracic appendages.

Movements and distribution of phytoplankton and most of the zooplankton are influenced largely by the physical forces already described. Unable to determine their own position in the water, these organisms must either float or sink. The ability to float depends upon the specific gravity of the organisms in relation to the specific gravity of the medium. If the specific gravity of an organism is the same as water, it will float; and if its specific gravity is greater, it will sink. The rate at which small bodies sink varies directly with their weight in excess of an equal volume of surrounding water, and their form, and indirectly with viscosity. Provided viscosity and specific gravity remain unchanged, a sphere of 0.01 mm will sink 100 times more slowly than one of 0.1 mm. Also involved is the ratio of total surface area to the volume of the body. The greater the ratio, the greater the friction and thus the resistance to sinking (see Brooks and Hutchinson, 1950).

Many plankton organisms have adaptations for staying afloat or remaining suspended. Fresh-water diatoms and desmids have thin, siliceous shells, which reduce their weight. Some plankton organisms may live in a gelatinous envelope, which has about the same specific gravity as water. Others may have gas vacuoles or oil droplets in the body. Many plankton organisms have surfaces set with spines, ridges, horns, or setae or possess an elongated body, all of which increase the total surface area in relation to weight.

But eddy diffusion currents more than any floating device keep plankton from sinking. The currents maintain a mixing action that not only holds the organisms in suspension but also prevents a persistent stratification of plankton in the epilimnion.

Vertical distribution or stratification of plankton organisms is influenced by the physicochemical properties of water, especially temperature, oxygen, light, and current. Light, of course, sets the lower limit at which phytoplankton can exist. Because animal plankton feeds on these minute plants, most of these too are concentrated in the trophogenic zone. Phytoplankton, by its own growth, limits light penetration and thus reduces the depth at which it can live. As the zone becomes more shallow, the phytoplankton can absorb more light, and organic production is increased. But within these limits the depths at which the various species live is influenced by the optimum conditions for their development. Some phytoplankton live just beneath the water's surface; others are more abundant a few feet beneath, while those requiring colder temperatures live deeper still. Cold-water plankton, in fact, is restricted to those lakes in which phytoplankton growth is scarce in the upper region and in which the oxygen content of the deep water is not depleted by the decomposition of organic matter. Many of these cold-water species never move up through the metalimnion.

Because many of them are capable of independent movement, animal plankton exhibits stratification that often changes seasonally. In winter some plankton forms are

FIGURE 18-3 [OPPOSITE]
A diagrammatic relationship between the littoral, limnetic, and benthic organisms and the trophic structure of a lake. The heavy arrows of various sizes indicate the relative quantities of energy flow. (This diagram was originally designed by Arnold Benson, West Virginia University.)

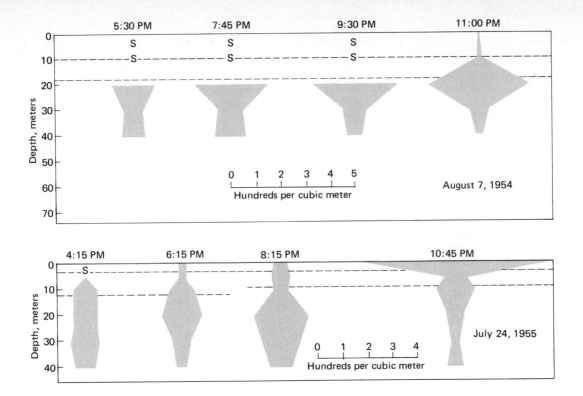

5:30 PM   7:45 PM   9:30 PM   11:00 PM

0 1 2 3 4 5
Hundreds per cubic meter

August 7, 1954

4:15 PM   6:15 PM   8:15 PM   10:45 PM

0 1 2 3 4
Hundreds per cubic meter

July 24, 1955

spread evenly to considerable depths; in summer they concentrate in the layers most favorable to them and to their stage of development. At this season animal plankton undertakes a vertical migration during some part of the 24-hour period. Depending upon the species and their stage of development, zooplankton spends the night or day in the deep water or on the bottom and moves up to the surface during the alternate period to feed on phytoplankton (Fig. 18-4).

During the spring and fall overturns, plankton is carried down, but at the same time nutrients released by decomposition in the tropholytic zone are carried upward into the impoverished upper layers. In spring, when the surface water warms and stratificacation again develops, phytoplankton has access to both nutrients and light. The spring bloom develops, followed, especially in shallow water, by a rapid depletion of nutrients and a reduction in plankton populations.

Free, to a large extent, from the action of weak currents and capable of moving about at will are nekton organisms, the fish, and some invertebrates. In the limnetic zone, fish make up the bulk of the nekton. Their distribution is influenced mostly by food supply, oxygen, and temperature (Fig. 18-1). During the summer, largemouth bass, pike, and muskellunge inhabit the warmer epilimnion

waters, where food is abundant. In winter they retreat to deeper water. Lake trout, on the other hand, move to greater depths as summer advances. During the spring and fall overturn, when oxygen and temperature are fairly uniform throughout, both warm-water and cold-water forms occupy all levels.

Profundal zone. Below the depth of effective light, beneath the thermocline, lie the profundal waters. Diversity and abundance of life here are influenced by oxygen and temperature; and the organisms depend upon the rain of organic material from the layers above as their energy source. In highly productive waters, decomposer organisms so deplete the profundal waters of oxygen that little aerobic life can exist there. The profundal zone of a deep lake is relatively much larger, so the productivity of the epilimnion is low in comparison to the volume of water, and decomposition does not deplete the oxygen. Here the profundal zone supports some life, particularly fish, some plankton, and such organisms as certain cladocerans, which live in the bottom ooze. Other zooplankton may occupy this zone during some part of the day but migrate upward to the surface to feed. Only during the spring and autumn overturns, when organisms from the upper layers enter this zone, is life abundant in the profundal waters.

Easily decomposed substances floating down

through the profundal zone are partly mineralized while sinking (Kleerekoper, 1953). The remaining organic debris, the dead bodies of plants and animals of the open water and decomposing plant matter from shallow-water areas, settle on the bottom. This, together with quantities of material washed in by inflowing water, makes up the bottom sediments, the habitat of benthic organisms.

Benthic zone. The bottom ooze is a region of great biological activity, so great that oxygen curves for lakes and ponds show a sharp drop in the profundal water just above the bottom. Because organic muck lacks oxygen completely, the dominant organisms there are anaerobic bacteria. Under anaerobic conditions, decomposition cannot proceed to inorganic end-products. When the amounts of organic matter reaching the bottom are greater than can be utilized by the bottom fauna, odoriferous muck rich in hydrogen sulfide and methane results. Thus lakes and ponds with highly productive limnetic and littoral zones have an impoverished fauna on the profundal bottom. Life in the bottom ooze is most abundant in lakes with a deep hypolimnion in which oxygen is still available.

Some of the bottom species, such as the flatworm (Rhabdocoela), live on the surface of the ooze, but others burrow into the bottom mud to feed on decaying organic matter. Prominent among these are the rhizopods (Rhizopoda), amoebalike protozoans encased in shells, the clam *Psidium*, small crustaceans such as isopods and cladocerans, phantom midges, which rise to the surface at night and return to the bottom at dawn, and tardigrades or water bears.

Many of these organisms live relatively deep in the ooze. They meet their limited oxygen demands by constructing tubes up through the ooze to the water. One of these, the bloodworm, or midge larva (*Tendipes*), sets up a current of water by movement within the tube. It utilizes the small quantity of oxygen through a hemoglobinlike pigment in its blood. The very abundant annelid worm, *Tubifex*, lives with its anterior end buried in the mud and its posterior end, which it undulates briskly, extended up in the water.

Most organisms of the profundal bottom are not unique to this zone. They represent for the most part a few species of the larger littoral bottom fauna that can tolerate se-

FIGURE 18-4 [OPPOSITE]
The vertical distribution of the copepod Limnocalanus macurus *on two midsummer days. On August 7, 1954, the sunset was at 8:00 P.M. (EST); on July 24, 1955, it was 8:15 P.M. (EST). The maximum number reach the surface 1.5 to 4 hours before sunset. Note that this organism inhabits the deeper water. The broken lines represent the metalimnion. (After Wells, 1961.)*

601

vere stagnation. Even these, if subject to continuous stagnation, will be eliminated wholly or in part. Bottom fauna is restocked from egg stages carried over during periods of stress or by downward migration from the littoral bottom during seasonal turnover (Welch, 1952). But there are a few species found only on the profundal bottom. Among these are the amphipod *Pontoporeia affinis*, and the opossum shrimp *Mysis oculata*, var. *relicta*, which inhabit only deep, cold-water lakes, where the oxygen may be less than 7 percent of saturation.

As the water becomes more shallow, the benthos changes. The bottom materials—stones, rubble, gravel, marl, clay—are modified by the action of water, by plant growth, by drift materials, and by recent organic deposits. Increased oxygen, light, and food result in a richness of species and an abundance not found on the profundal bottom. Here on the bottom of the littoral zone live, in addition to the tube worms, the midges, and the water bears, numerous other plants and detritus-feeders.

Closely associated with the benthic community are the *periphyton* or *aufwuchs*, those organisms that are attached to or move upon a submerged substrate but do not penetrate it. Small aufwuchs communities are found on the leaves of submerged aquatics, on sticks, rocks, and other debris. The organisms found there depend upon the movement of the water, temperature, kind of substrate, and depth.

Periphyton found on living plants are fast growing, lightly attached, and consist primarily of algae and diatoms. Because the substrate is so short-lived, these rarely exist for more than one summer. Aufwuchs on stones, wood, and debris form a more crustlike growth of blue-green algae, diatoms, water mosses, and sponges. Burrowing into and living up in this crust is a host of associated animals—rotifers, hydras, copepods, insect larvae, and a wide variety of protozoans, such as *Stentor* and *Vorticella*. Unlike those on plants, the periphyton on more substantial substrate are more persistent. Periphyton found in moving water, such as currents and waves on the lake shore, adhere tightly. In fact the various means of attachment of periphyton to the substrate are as varied as the methods phytoplankton use to remain afloat. Some aufwuchs organisms have gelatinous sheaths, tubes, or cups, and are attached to the substrate by gelatinous stalks or basal disks.

Littoral zone and ponds. Aquatic life is the richest and most abundant in the shallow water, where sunlight can reach the bottom. This is the littoral zone found about the edges of lakes and usually throughout the pond. The plants and animals found here vary with water depth, and a distinct zonation of life exists from deeper water to shore.

A blanket of duckweed, supporting a world of its own, may cover the surface of the littoral water. Desmids and diatoms, protozoans and minute crustaceans, hydras and snails live on the undersurface of the blanket; mosquitos and collembolans live on the top.

Submerged plants such as the alga *Chara* and the pondweeds or potamogetons, one of the largest genera of rooted aquatic plants, are found at water depths beyond that tolerated by submergent vegetation. They serve as supports for small algae and as cover for swarms of minute aquatic animals.

Submerged plants are highly modified in structure. Living, as they do, completely immersed in water, they lack cuticle, which reduces transpiration in leaves exposed to the air. Lacking cuticle, submerged plants can absorb nutrients and gases directly from the water through thin and finely dissected, or ribbonlike, leaves. On many, the roots are small, poorly branched, and lack root hairs. The buoyance and protection of water eliminates the need for supporting tissues in the stems. Air chambers and passages are common in leaves and in the cortex of stems, where oxygen produced by photosynthesis is stored for use in respiration. For the most part, vegetative reproduction is highly developed.

As sedimentation and accumulation of organic matter increases toward the shore and the depth of water decreases, floating rooted aquatics, such as pond lilies and smartweeds, appear. Many of these floating plants have poorly developed root systems but highly developed aerating systems. Supporting tissue is greatly reduced and the spongy tissue in the stems and leaves is filled with large air spaces. The upper surfaces of floating leaves are heavily waxed to prevent clogging of the stomata by water. Underwater leaves of these same floating plants are like those of submerged plants, thin and small, while the floating leaves usually are large and different in form. The leaves and stems are leathery and tough and able to withstand the action of waves.

Floating plants offer food and support for numerous herbivorous animals that feed both on phytoplankton and the floating plants. The undersides and stems of plants support a highly interesting assemblage of organisms, among them protozoans, hydras, snails, and sponges.

In the shallow water beyond the zone of floating plants grow the emergents, plants whose roots and lower stems are immersed in water and whose upper stems and leaves stand above the water. Large areas of such plants make up the marsh, to be discussed later. Among these emergents are plants with narrow, tubular or linear leaves, such as the bulrushes, reeds, and cattails. With these are associated such broadleaf emergents as pickerelweed and arrowhead. The distribution and variety of plants vary with the water depth and fluctuation of the water level.

Life is abundant about the sheltering beds of emergent plants. Damselflies and dragonflies lay their eggs on submerged stems just below the water line. The water scorpion, nearly 2 in. long and resembling a walking stick, has a long air tube on the abdomen. Through this it draws air from the atmosphere while it hangs its head and raptorial forelegs 2 to 3 in. below the surface and waits to seize any prey that ventures near. Conspicuous diving insects are the backswimmers, the diving beetles, and the water boatmen. The first two are predacious. The abundant backswimmers have their backs keeled like boat bottoms and navigate upside down on the surface; beneath the surface they swim right side up. The diving beetles, usually black or brownish black in color, include some of the largest aquatic insects and some of the smallest, almost microscopic in size. Both the backswimmers and the diving beetles hang head downward in the water and dive swiftly after prey. Both feed on tadpoles, fish fry, and insects. The backswimmers, however, feed on the juices, which they secure by piercing the prey with their piercing-sucking mouth parts. The water boatmen feed on algae, protozoa, and microscopic metazoa, which they gather by sweeping the fine material on the bottom into their mouths. They feed on algae filaments also, which they pierce with protrusible stylets to suck out the contents.

All three can remain underwater for some time because they carry with them their own "scuba tanks" of air in the form of bubbles. The backswimmers carry a film of air on the abdomen; the diving beetles and water scavenger beetles carry a large bubble of air in the subelytral air space (beneath the wings), into which their spiracles open. The water boatmen, on the other hand, are wrapped in air and carry an additional supply trapped beneath the wings; because of this they must cling to plant stems and debris to remain submerged.

The air bubbles that these insects carry act as physical gills. The air in the bubble is composed of approximately 21 percent oxygen and 78 percent nitrogen, about the same as that of the atmosphere. When the insects are submerged, they withdraw oxygen from the bubble and give off carbon dioxide. But carbon dioxide dissolves readily in water, so it does not accumulate in the bubble. At the same time the nitrogen dissolves very slowly into the water. As the oxygen supply decreases, the tension or partial pressure of the nitrogen increases and maintains the surface of the bubble. As the oxygen tension inside the bubble decreases, it reaches a level at which it is less inside the bubble than outside. Unless the oxygen content of the water is sharply limited, oxygen will diffuse into the bubble almost as fast as it is used. Thus as long as sufficient nitrogen remains to provide an adequate surface for oxygen diffusion, the animal can remain submerged. When the oxygen is reduced to about 1 percent of the total gas inside the bubble, the insect rises to the surface for a recharge of air.

Fish, such as pickerel and sunfish, find shelter, food, and protection among the emergent and floating plants; and the ubiquitous catfish, often tolerant of extreme conditions of turbidity, feeds close to the bottom. Fish of lakes and ponds lack the strong lateral muscles characteristic of fish living in swift water; and some, such as the sunfish, have compressed bodies, which permit them to move with ease through masses of aquatic plants.

PRODUCTIVITY CLASSES OF LAKES

At this point one fact emerges—that production is greatest in lakes and ponds that are relatively shallow and that are rich in organic matter and nutrients. Because of the heavy accumulation of organic matter, oxygen depletion in the hypolimnion occurs during the summer. Such lakes are termed *eutrophic*. These contrast sharply with deep lakes which often have steep sides. Littoral plants are scarce, and plankton growth is low in proportion to the total volume of water. The quantity of organic matter produced in relation to water volume is much less and the deep waters are continuously oxygenated (see the discussion earlier in this chapter). Because the amount of nutrients recycled is low per unit volume of water, deep lakes often are poor in phosphorus, nitrogen, and calcium. Such lakes are termed *oligotrophic*.

A third type of lake to come under this classification is *dystrophic*. Characteristic of bog situations, dystrophic lakes are low in calcium carbonate, high in humus content, and very poor in nutrients.

TEMPORARY PONDS

Many are the shallow depressions and the remnants of old ponds that fill with water shortly after the first spring thaw. For a while they sparkle in the spring sunshine, but by early summer they are gone. Few give these ponds a second thought; some consider them a nuisance or a breeding place for mosquitos. Actually they merit close observation, for few places are more interesting.

Because of extreme environmental conditions, the *temporary pond* is a difficult place for living organisms to inhabit. For a time the area is submerged; then follows a period of progressively drier conditions, until finally the bottom is dry and covered with woody or herbaceous growth. In winter the depression may hold some ice-covered water, collected from melted snow, but most of the area is frozen and snow-covered. When the early spring thaw sets in, the depression fills with water and the cycle begins again. During the few months that open water is present, the inhabitants of temporary ponds must complete their reproductive activity and meet in one way or another the problem of species survival during the dry period (see Chapter 7).

Animal life in the temporary pond is at low ebb during late winter and early spring. If the depression holds some water during the winter, the flatworm, *Phagocytes vernalis*, may be present, in company with the copepod *Cyclops*. Shortly after the snow melts, the fairy shrimp appears, the most typical inhabitant of the temporary pond. Fairy shrimp hatch during February and March from eggs laid the spring before. Alternate drying and freezing of eggs on the exposed pond bottom stimulates the processes of development and hatching (see C. R. Weaver, 1943). Because such conditions do not usually occur in permanent ponds, fairy shrimp are confined to temporary ponds or ponds that undergo considerable drawdown of water during the summer.

The population of fairy shrimp reaches its height in April and disappears nearly as rapidly as it appeared. The sudden decline seems to be associated with an increase in predatory insects (Kenk, 1949). Even early in spring, while ice still clings to the edges of the pond and the water is very cold, individuals of predaceous beetles appear. They lay their eggs in April, and young larvae hatch to feed on fairy shrimp. Slow swimmers, the fairy shrimp soon fall prey to the beetles. As the fairy shrimp population declines, another temporary pond crustacean may appear, the clam shrimp, which has a bivalved carapace enclosing a compressed body.

As the water warms, activity increases. New pond residents appear, either by hatching from dormant eggs or cysts or by migrating to the pond. Frogs and salamanders arrive in March and April, court, mate, deposit their gelatinous eggs, and depart. Fingernail clams are common from February on.

By April the waters of temporary ponds are alive with small crustaceans, including the aquatic sow bug, water fleas, ostracods, and copepods. In addition to these minute crustaceans, vernal pools support abundant plankton organisms, perhaps encouraged by decaying vegetation on the pond bottom. Browsing on this plankton growth, algae, and organic debris are the scud and the crayfish. As the water level sinks, the crayfish resort to burrows that reach groundwater level, and stays until the spring thaw. In the pocket of water filling the burrow a great concentration of other small crustaceans find refuge.

The pond waters begin to drop in May, and by June the water is nearly gone. The crayfish has retreated to its burrow, insects have hatched from eggs, matured, and flown away. Eggs of shrimp, cysts of flatworms, and dormant mollusks lie buried in the bottom

mud. But for many animals, the drying of the pond means death. Nymphs of insects not yet matured and some unhatched eggs are destroyed. Larvae of frogs and salamanders that fail to reach adult form before the pond dries perish.

Pond life has ended and in its place is mesophytic growth of willow and dogwood, cattail and sedge. Nine months later the pond will fill again and the cycle start anew.

THE MARSH AND SWAMP

Along the shallow margins of lakes and ponds and in low, poorly drained lands where water stands for several months of the year, wetland vegetation appears. Mostly this represents the last filling-in stages of lake and pond succession. Generally such areas are saturated or covered by water during the growing season. At times, particularly in early fall, the substrate may be exposed, a condition necessary for the germination of many wetland plants. The substrate is soft muck, rich in decaying organic matter mixed with mineral soil. As time passes, drainage gradually im-

proves, for the vegetation continuously builds up the bottom with debris. The vegetation is discontinuous and distributed singly in small stands or clumps separated from one another (Fig. 18-5).

These wetlands are marshes and swamps. Biologically, they are among the richest and most interesting communities. Yet they are also the least appreciated and the first to be destroyed by filling and drainage.

Marshes are wetlands in which the dominant vegetation consists of reeds, sedges, grasses, and cattails; essentially they are wet prairies. *Swamps*, on the other hand, are wooded wetlands and in some cases represent a successional step from marsh to mesic forest. In other regions the swamp may develop without a preceding marsh stage.

Marsh. The floating plants of the premarsh stages were both protected and buoyed up by the water. But the emergent marsh plants, rising clear of the water, lack the protection and buoyancy of the water. Instead the emergent plants are wandlike and flexible, able to bend before the stress of wind and water. Such plants require firm bases. Into the soft ooze, cattails, sedges, and associated plants send tough, fibrous rhizomes and roots. These develop a firm mat, which resists both waves and the pulling of the plant tops by wind and water. Thus marsh vegetation is restricted to plants that can tolerate submerged or water-logged organic soil and which form firm mats or tussocks in the ooze.

Marshes vary in depth from deep water to shallow (Table 18-1). The maximum tolerable depth for emergent vegetation is about 3 ft. At this depth such plants as phragmites and wild rice grow. Shallow fresh-water marshes, those whose soil is waterlogged or submerged under as much as 6 in. of water, are dominated by cattails, sedges, and rushes. Wet meadows develop on soils raised enough to prevent standing water from remaining throughout the growing season but still waterlogged within a few inches of the surface.

Although grasses are the dominant marsh vegetation, other plants are there. In shallow, open water between the cattails and the reeds, algae are abundant, especially the green, coarse, threadlike *Spirogyra* and the net-forming *Hydrodicton*. Floating on the water's surface may be the silt-covered, jellylike alga, *Nostoc*. Other plants of the shallows include arrowheads and pickerel weed.

Plant life of the marshes supports a rich and abundant animal life. Snails, one of the most common marsh animals, feed on ooze and dead animal matter and in turn are consumed by birds and fish. Birds add the most color to marshlands. Here is found an avian richness hard to equal in other temperate communities. Characteristic of marshes are ducks; in fact the two are inseparable. The most typical mammalian member of the marsh community is the muskrat. This rodent uses cattails, sedges, and other plants to build its moundlike lodge, and it subsists on their roots and leaves.

Swamp. Swamps, like marshes, can be divided into two broad groups, deep-water and shallow-water; but here much of the similarity ends.

Deep-water swamps occur extensively on the flood plains of the larger southern river systems, especially in the Mississippi River drainage system and on the uplands of the coastal plain. Flood-plain swamps usually consist of southern cypress and tupelo gum, whereas upland swamps often are dominated by black gum and pond cypress; but all four species may

FIGURE 18-5 [OPPOSITE]
Marshlands, too often drained and filled, are important to many forms of wildlife. Loss of wetlands is responsible for the decline in waterfowl populations. (Photo courtesy U.S. Fish and Wildlife Service.)

TABLE 18-1
A classification of wetlands

<table>
<tr><td colspan="2" align="center">*Fresh areas*</td></tr>
</table>

INLAND FRESH AREAS

1. Seasonally flooded basins or flats	Soil covered with water or waterlogged during variable periods but well drained during much of the growing season. In upland depressions and bottomlands. Bottomland hardwoods to herbaceous growth.
2. Fresh meadows	Without standing water during growing season; waterlogged to within a few inches of surface. Grasses, sedges, rushes, broadleaf plants.
3. Shallow fresh marshes	Soil waterlogged during growing season; often covered with 6 in. or more of water. Grasses, bulrushes, spike rushes, cattails, arrowhead, smartweed, pickerelweed. A major waterfowl-production area.
4. Deep fresh marshes	Soil covered with 6 in. to 3 ft of water. Cattails, reeds, bulrushes, spike rushes, wild rice. Principal duck-breeding area.
5. Open fresh water	Water less than 10 ft deep. Bordered by emergent vegetation: pondweed, naiads, wild celery, water lily. Brooding, feeding, nesting area for ducks.
6. Shrub swamps	Soil waterlogged; often covered with 6 in. or more of water. Alder, willow, buttonbush, dogwoods. Ducks nesting and feeding to limited extent.
7. Wooded swamps	Soil waterlogged; often covered with 1 ft of water. Along sluggish streams, flat uplands, shallow lake basins. North: tamarack, arborvitae, spruce, red maple, silver maple; South: water oak, overcup oak, tupelo, swamp black gum, cypress.
8. Bogs	Soil waterlogged; spongy covering of mosses. Heath shrubs, sphagnum, sedges.

COASTAL FRESH AREAS

9. Shallow fresh marsh	Soil waterlogged during growing season; at high tides as much as 6 in. of water. On landward side, deep marshes along tidal rivers, sounds, deltas. Grasses and sedges. Important waterfowl areas.
10. Deep fresh marshes	At high tide covered with 6 in. to 3 ft of water. Along tidal rivers and bays. Cattails, wild rice, giant cutgrass.
11. Open fresh water	Shallow portions of open water along fresh tidal rivers and sounds. Vegetation scarce or absent. Important waterfowl areas.

<table>
<tr><td colspan="2" align="center">*Saline areas*</td></tr>
</table>

INLAND SALINE AREAS

12. Saline flats	Flooded after periods of heavy precipitation; waterlogged within few inches of surface during the growing season. Vegetation: seablite, salt grass, saltbush. Fall waterfowl-feeding areas.
13. Saline marshes	Soil waterlogged during growing season; often covered with 2 to 3 ft of water; shallow lake basins. Vegetation: alkali hard-stemmed bulrush, wigeon grass, sago pondweed. Valuable waterfowl areas.
14. Open saline water	Permanent areas of shallow saline water. Depth variable. Sago pondweed, muskgrasses. Important waterfowl-feeding areas.

COASTAL SALINE AREAS

15. Salt flats	Soil waterlogged during growing season. Sites occasionally to fairly regularly covered by high tide. Landward sides or islands within salt meadows and marshes. Salt grass, seablite, saltwort.
16. Salt meadows	Soil waterlogged during growing season. Rarely covered with tide water; landward side of salt marshes. Cord grass, salt grass, black rush. Waterfowl-feeding areas.
17. Irregularly flooded salt marshes	Covered by wind tides at irregular intervals during the growing season. Along shores of nearly enclosed bays, sounds, etc. Needlerush. Waterfowl-cover area.
18. Regularly flooded salt marshes	Covered at average high tide with 6 in. or more of water; along open ocean and along sounds. Salt-marsh cord grass along Atlantic. Pacific: Alkali bulrush, glassworts. Feeding area for ducks and geese.
19. Sounds and bays	Portions of salt-water sounds and bays shallow enough to be diked and filled. All water landward from average low tide line. Wintering areas for waterfowl.
20. Mangrove swamps	Soil covered at average high tide with 6 in. to 3 ft of water. Along coast of southern Florida. Red and black mangroves.

Source: Adapted from S. P. Shaw and C. G. Fredine, 1956, *Wetlands of the United States*, U.S. Fish and Wildlife Serv. Circ. 39.

grow in mixed stands. In more shallow water, sweet bay, slash pine, and pond pine are common associates. These deep-water swamp forests lack herbaceous plants but support abundant epiphytes, plants that grow perched on other plants. Outstanding among these epiphytes are bromalids, Spanish moss, and the more specialized tank epiphytes, whose leaves are shaped like gutters and collect water in the axils. These little tanks are occupied by a variety of small invertebrates, including mosquitos. Orchids are abundant in swamps of the deep south. The best known is the cigar or cowhorn orchid, whose flowers—yellowish splotched with brown—are borne on long stalks.

Shallow-water swamps range from shrubby willows, alders, and buttonbush to oaks and maples. Where water is nearly permanent through the year, willow thrives, together with buttonbush, a shrub capable of growing in deep water. Alder likewise is tolerant of permanent shallow-water areas and is common throughout North America. Shallow-water swamp forests include such associates as elm, silver maple, red maple, white pine, and northern white cedar. Common southern swamp associates are overcup oak and water hickory. In the deep south, the understory of these swamps usually is a tangle of shrubs, vines, ferns, and palmetto.

Swamp vegetation, like that of the marsh, must meet the problem of anchorage in soft muck saturated with water. The root systems of trees are often massive, but they develop superficially and do not penetrate deeply. They spread widely and develop buttresses. These are exaggerated by deep-water swamp trees such as bald cypress and tupelo gum. The height and development of these swollen bases depend on the water depth at the start of the growing season. In addition, all deep-swamp dominant trees develop so-called pneumatophores from the superficial lateral root system that extends outward like cables. Most familiar are the knees of the swamp black gum, tupelo, and cypress; those of the latter may rise 8 to 10 ft above the low-water mark (Fig. 18-6). Once it was thought that these knees acted in respiratory gas exchange, but their real function is unknown. They may serve to support the massive trees in the soft substrate.

An outstanding characteristic of shallow-water swamps is the rapid and uneven elevation of the land. The floor of the swamp is

FIGURE 18-6
Wooded wetlands are swamps. Some of the best developed are the cypress swamps of the southern United States. (Photo by author.)

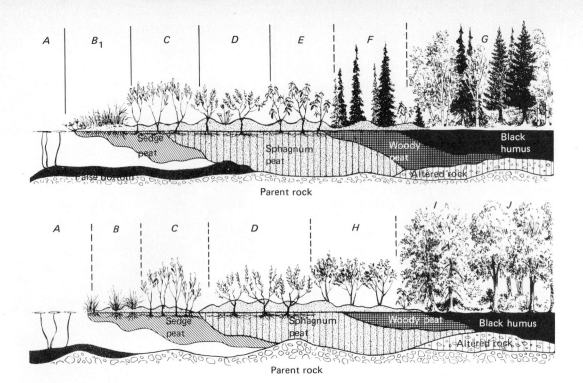

Parent rock

Parent rock

a series of depressions and rises created by fallen logs and upturned roots. On the rises grow ferns, vines, shrubs, and more mesic trees. Logs and tree bases are covered with mosses and liverworts, lichens and fungi. In spring and other periods of high water, the depressions are small pools; in dry weather they are little hollows, grown to mosses, liverworts, and marsh annuals. Thus across the swamp there is rapid and marked differentiation of microclimates, with varying moisture conditions and temperatures.

Variations in environmental conditions within the swamp provide a diversity of microhabitats for animals. The inhabitants range from aquatic and semiaquatic insects, turtles, snakes, alligators, and crocodiles, ducks and herons to animals more closely associated with purely terrestrial situations, such as the deer, bear, and squirrel. Like the marsh, the swamp contains a richness of life too often little appreciated by man.

BOGS

Of all aquatic communities, none are more intriguing than bogs (Fig. 18-7). Although bogs are rather widely distributed over the humid regions of the earth, they are best developed and most abundant in the cold northern forested regions in North America, Europe, and Asia. In physical and chemical characteristics

and community composition, they differ sharply from lakes, ponds, and marshes.

Regardless of species composition or location, all bogs have several features in common. They usually develop where drainage is blocked, all have cushionlike vegetation, and all have an accumulation of peat. Most at some time have a marginal semifloating mat of vegetation, usually sphagnum moss and heaths.

Lakes, ponds, and abandoned stream meanders typically accumulate organic sediments on the bottom. If drainage is inadequate, the area is saturated with water the year around and little organic matter is carried away or mixed with mineral substrate. Through time, drainage is further congested by plant growth and the circulation of water in these lakes and ponds is reduced, creating anaerobic conditions. Because biological activity is slowed down by a combination of anaerobic conditions and cold temperatures, the organic debris is only partially decomposed to form peat. These peats release humic acids, which tinge the water brown. Bog-lake waters are low in electrolytes, especially calcium carbonate, and range from strongly acid to alkaline, a phenomenon imperfectly understood but apparently influenced by plant cover and peat. Water adjacent to bog vegetation may be acid or alkaline, but water with sphagnum moss, the dominant plant of bogs, is always acid.

Sphagnum and peat can absorb bases from dissolved salts and thus free acids. Bog waters also are low in available nitrogen, potassium, and phosphorus because these nutrient elements are tied up in the peat. These waters have a high carbon dioxide content and possess traces of hydrogen sulfide, the result of activity by sulfur bacteria. Lakes and ponds exhibiting these characteristics are termed *dystrophic*.

The decidedly acid condition of sphagnum bogs accounts for their uniqueness. The harsh conditions limit the kind of plants able to survive there. Thus a bog community, consisting of plants incapable of growing in alkaline situations, develops.

Bog development begins much like succession in other lakes (Fig. 18-8). Open water is colonized by submerged and then floating plants, especially the yellow pond lily. The accumulation of organic matter raises the lake or pond floor. Many of these sediments, especially the finer materials, may oxidize, decompose, or remain suspended as colloids during their movement from the water's surface to the bottom. Often they will remain in solution until precipitated by bacterial, chemical, or photosynthetic activity to form a fine, soft deposit called a "false bottom."

Meanwhile reeds and sedges, cotton grass, buckbean, and marsh cinquefoil grow in the shallows. Sphagnum creeps into the area and fills the open spaces between the plants. When a consolidated mass of peat develops or when jutting rocks, logs, or other solid objects allow a foothold, a mat extends outward over the water (Fig. 18-8). Leading edges of plants, particularly buckbean, send out a buoyant network of rhizomes, which sprout new shoots above the water level. On this loose tangle, *Sphagnum* and *Hypinium* mosses grow, filling in the open spaces and consolidating the mass. As the mat thickens, other plants, such as the cranberry, sweet gale, and bog rosemary appear (Fig. 18-9). Sphagnum mosses grow in rounded cushions about the bases of shrubs and annually add new growth on the accumulating remains of past moss generations. As the mat thickens and raises, other shrubs that are intolerant of very wet conditions invade it. Typical are leatherleaf and Labrador tea. Leatherleaf, found in practically all boreal bogs, is exclusively a bog species and is confined to that habitat. Usually it is the domi-

FIGURE 18-8 [OPPOSITE]

Transects through two typical bogs showing the zones of vegetation, sphagnum mounds, and the nature of peat deposits and floating mat. A, Pond lily in open water; B₁, buckbean and sedge zones; B₂, sedge zones; C, sweet-gale zone; D, leatherleaf; E, Laborador tea; F, black spruce; G, birch, balsam, black spruce forest; H, mountain holly; I and J, silver and red maple. Tamarack is often associated with black spruce but is soon replaced by the spruce. (Redrawn from Dansereau and Segadas-Vianna, 1952.)

FIGURE 18-7

A bog. This is the "muskeg furthest south," Cranberry Glades, Pocahontas County, West Virginia. Tamarack is absent, as is Laborador tea. But this is the southernmost point in the range of bog rosemary. Note the hummock effect of moss and lichens. (Photo by author.)

nant shrub, but in some bogs Labrador tea occupies this position.

On the sides of sphagnum hummocks, in depressions between, and on the meadowlike floating mat of sedges and buckbean grow two outstanding bog inhabitants, the pitcher plant and sundew, both predacious plants. The flat leaves of the small sundew bear tentacles tipped with sticky globules, which ensnare insects. The pitcher plants have a rosette of bronzy, tubular leaves, which contain water. Insects that inadvertently wander into the leaves find themselves unable to crawl over the down-pointed hairs and fall into the water to drown. Here they apparently are digested by the plants, although the exact nature of this process and its function in the plant are not clearly understood. In spite of this digestive action and the low pH—approximately 4.5 —of the fluid in the leaves, the mosquito, *Wyeomyua smithii*, passes through the egg, larval, and pupal stages within the cavity of the plant.

Along the edge of the bog where the peat mat is thick and reaches the bottom, the forest invades the shrub and sphagnum growth. Usually the first trees are scattered clumps and individuals of larch and black spruce, beneath which develop such plants as partridge berry and bunchberry. Eventually the forest thickens and advances as the mat closes toward the center of the lake. Because of its thin crown, tamarack allows a dense shrub growth to remain, but finally even these are replaced by black spruce, arborvitae, and balsam fir. In regions where deciduous forest is the climax, spruce is replaced by maples, birch, and balsam fir.

Peat accumulates and consolidates beneath the mat of sedges and heaths. The weight causes the mat to sink gradually to the bottom, where further decomposition is arrested. Occasionally the peat mat becomes infiltrated with alkaline water, which allows fermentation and decomposition of the peat. Pockets form beneath the surface; the peat mat sags and caves in. Patches of open water appear again. In the bog forest the evaporative power of the vegetation may become greater than the water-retaining capacity of the peaty material. Then succession tends toward a drier, more mesic condition. But lumbering, windthrow, or any action that destroys tree growth can reverse this process and convert the area back to early bog conditions. Thus succession in a bog may regress a number of times, delaying the time when stable vegetation will claim the area.

Life in the bog lake is restricted in the number of species, but the organisms present may be abundant. Phytoplankton is more abundant and richer in species than the zooplankton. Dominating the bog waters are varied and beautiful forms of desmids and a few blue-green algae. Among the zooplankton the testaceous rhizopods and the rotifers are the richest in species, and some forms are restricted to bog waters. Protozoans, too, are common, particularly *Sarcodina*.

More characteristic of bogs, perhaps, than the presence of a few species is the absence of many. Mollusks, except for certain Sphaeridae, are missing because of the lack of calcium. Among the aquatic insects are representatives of Odonata, Hemiptera, Trichoptera, Diptera, and Coleoptera. Fish are few in number or absent, but amphibians, especially the Ranidae, are universally present. Warblers are the most abundant birds of the bog forests, and the bog lemming is a common small mammal.

The moors. On drier ridges surrounding the dystrophic lake and bogs, and on flat, poorly drained areas, raised bogs or high moors develop (Dansereau and Segadas-Vianna, 1952). Sphagnum moss creeps into depressions up on the drier ridges. The spongy sphagnum with its tremendous water-holding capacity retains more and more precipitation in the ground. As a result mesophytic trees are replaced by trees and shrubs that live in a wet, acidic environment. The sphagnum accumulates, often to a height of several feet, and rises above the adjoining ground in extensive cushionlike mounds. As the mounds increase in size, they join, until the whole area becomes a raised bog surrounded by a swampy moat. Like the flat bog, the area eventually is occupied by concentric zones of vegetation interrupted by bog puddles and pools.

IMPACT OF MAN

Lakes, being settling basins, receive a heavy input of sewage, industrial wastes, and silt, which change the chemical conditions of water and fill up the basin. The impact of pollution on lakes and the eutrophication of water have already been discussed. Adding to eutrophication and destroying the floating and emergent vegetation about lakes is the proliferation of permanent and summer homes and marinas on the shores. Such developments and accompanying activities ruin both fisheries and wetland habitat. Oil pollution from motor boats is a major environmental problem on lakes. Wasting 10 to 40 percent of their fuel, motor boats discharge an oily mixture with gas exhausts beneath the surface of the water, where it escapes immediate detection. One gal of oil/million of water imparts an odor to lake waters; 8 gal/million taint the flesh of fish. A mixture of 50 gal of water to 1 of oil requires 30 million gal of lake water for decomposition of the hydrocarbons. Such oily discharges can lower oxygen levels and have an adverse effect on the growth and longevity of fish.

Marshes and swamps are major targets for drainage projects, flood control, sanitary land fills, and real estate development. A prime cause of wetland losses is drainage for agricultural development. For example, 90 percent of all the Delta woodland swamps drained in Arkansas has been done to increase soybean production. Some 150,000 acres of such habitat have been lost during the 1960s in that state alone. Throughout the United States 45 million acres of primitive swamp, marshes, and seasonally flooded land have been lost. The remaining 75 million are disappearing under agricultural drainage, and flood control and hydroelectric projects. These growing losses have affected populations of waterfowl and other wetland wildlife. Continuing losses can bring wetland wildlife to a point of extinction.

FIGURE 18-9
A close-up view of sphagnum moss and bog rosemary in a New York State bog. (Photo by author.)

Lotic ecosystems

Current or continuously moving water is the outstanding feature of streams and rivers. Current cuts the channel, molds the character of the stream, and influences the life and ways of organisms inhabiting flowing waters.

Streams may begin as outlets of ponds or lakes, or they may arise from springs and seepage areas. Added to this in varying quantities is surface runoff, especially after heavy or prolonged rains and rapid snow melt. Because precipitation, the source of all runoff and subsurface water, varies seasonally, the rate and

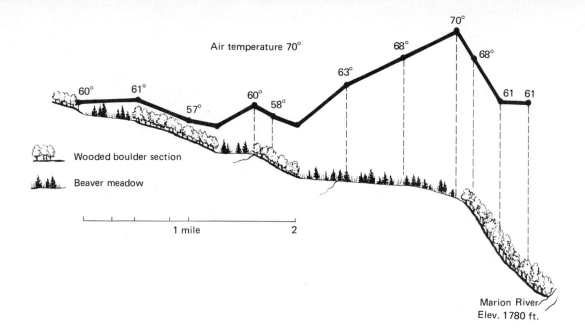

Air temperature 70°

70°

68°

68°

63°

60° 61°

57°

60° 58°

61 61

Wooded boulder section

Beaver meadow

1 mile   2

Marion River
Elev. 1780 ft.

volume of stream flow also fluctuates widely from flood conditions to a nearly dry stream bed.

As water drains away from its source, it flows in a direction dictated by the lay of the land and the underlying rock formations. Its course may be determined by the original slope and its regularities; or the water, seeking the least resistant route to lower land, may follow the joints and fissures in bedrock near the surface and shallow depressions in the ground. Whatever its direction, water is concentrated into rills that erode small furrows, which soon grow into gullies. Water, moving downstream, especially where the gradient is steep, carries with it a load of debris that cuts the channel wider and deeper and that sooner or later is deposited within or along the stream. At the same time, erosion continues at the head of the gully, cutting backward into the slope and increasing its drainage area. Just below its source, the stream may be small, straight, and often swift, with waterfalls and rapids. Further downstream, where the gradient is less and the velocity decreases, meanders become common. These are formed when the current, deflected by some obstacle on the floor of the channel, by projecting rocks and debris, or by the entrance of swifter currents, strikes the opposite bank. As the water moves downstream, it is thrown back to the other side again. These abrasive forces create a curve in the stream, on the inside of which the velocity is slowed and the water drops its load. Such cutting and desposition often cause valley

streams to change course and to cut off the meanders to form oxbow lakes. When the water reaches level land, its velocity is greatly reduced, and the load it carries is deposited as silt, sand, or mud.

At flood time the material carried by the stream is dropped on the level lands, over which the water spreads to form flood-plain deposits. These flood plains on which man has settled so extensively are a part of the stream or river channel used at the time of high water, a fact that few people recognize. The current at flood time is swiftest in the normal channel of the stream and slowest on the flood plain. Along the margin of the channel and the flood plain, where the rapid water meets the slow, the current is checked and all but the fine sediments are dropped on the edges of the channel. Thus the deposits on the flood plain are higher on the immediate border and slope off gradually toward the valley side.

When a stream or river flows into a lake or sea, the velocity of the water is suddenly checked and the load of sediment is deposited in a fan-shaped area about the mouth of the river to form a delta. Here the course of the stream is broken into a number of channels, which are blocked or opened with subsequent deposits. As a result the delta is characterized by small lakes and swampy or marshy islands. Material not deposited at the mouth is carried further out to sea, where it settles on the bottom. Eventually the sediments build up above the water to form a new land surface.

CHARACTERISTICS OF THE STREAM

The character of a stream is molded by the velocity of the current. This velocity varies from stream to stream and within the stream itself; and it depends upon the size, shape, and steepness of the stream channel, the roughness of the bottom, the depth, and the rainfall.

The velocity of flow influences the degree of silt deposition and the nature of the bottom. The current in the riffles is too fast to allow siltation, but coarser silt particles drop out in the smooth or quiet sections of the stream. High water increases the velocity; it moves bottom stones, scours the stream bed, and cuts new banks and channels. In very steep stream beds, the current may remove all but very large rocks and leave a boulder-strewn stream.

Flowing water also transports nutrients to and carries waste products away from many aquatic organisms, and may even sweep them away. Balancing this depletion of bottom fauna, the current continuously reintroduces bottom fauna from areas upstream. Similarly, as nutrients are washed downstream, more are carried in from above. Because of this, the productivity of primary producers in streams is 6 to 30 times that of those in standing water (Nelson and Scott, 1962). The transport and removal action of flowing water benefits such continuous processes as photosynthesis.

The temperature of the stream is not constant. In general, small shallow streams tend to follow, but lag behind, air temperatures, warming and cooling with the seasons but never falling below freezing in winter. Streams with large areas exposed to direct sunlight are warmer than those shaded by trees, shrubs, and high steep banks (Fig. 18-10). This is ecologically important because temperature affects the composition of the stream community.

The constant swirling and churning of stream water over riffles and falls result in greater contact with the atmosphere. Thus the oxygen content of the water is high at all levels and often is near the saturation point for the existing temperature. Only in deep holes or in polluted waters does dissolved oxygen show any significant decline.

Free carbon dioxide in rapid water is in equilibrium with that of the atmosphere, and the amount of bound carbon dioxide is influ-

FIGURE 18-10 [OPPOSITE]
Profile of Bear Brook (Adirondack Mountains) and a graph of its water temperatures, showing the warming effect of open beaver meadows and the cooling effects of wooded boulder stream. (From the Biological Survey of Raquette Watershed, New York State Conservation Department, 1934.)

enced by the nature of the surrounding terrain and the decomposition taking place in
pools of still water. Most of the carbon dioxide
in flowing water occurs as carbonate and bicarbonate salts. Streams fed by groundwater
from limestone springs receive the greatest
amount of carbonates in solution. Because of
a coating of algae and ooze on the bottom,
little calcium carbonate is added by the action of carbonic acid on a limestone stream
bed (Neel, 1951).

The degree of acidity and alkalinity, or pH,
of the water reflects the carbon dioxide content as well as the presence of organic acids
and pollution. The higher the pH of stream
water, the richer the natural waters generally
are in carbonates, bicarbonates, and associated
salts. Such streams support more abundant
aquatic life and larger trout populations than
streams with acid waters, which generally are
low in nutrients.

Most streams receive a part of their basic
energy supply from land, as well as from connected ponds and backwaters. Many of the
first-level consumer organisms in the stream
are detritus-feeders, much of whose food
comes from the terrestrial vegetation along the
banks (Marshall, 1967; Cummins et al.,
1966). The primary consumers in a Piedmont
stream in Georgia obtained 66 percent of their
energy from leaf material and other organic
debris in the water (Nelson and Scott, 1962).
Marginal stream vegetation also supplies a
major portion of food for secondary consumers, especially in late spring and summer.
Brook trout in northern West Virginia streams
feed largely on aquatic organisms from March
through May, when these invertebrates are
most abundant. In June the fish showed a
pronounced shift to invertebrates. From July
to September emergent and terrestrial organisms comprised 60 to 80 percent of the total
food the fish consumed (Redd and Benson,
1962). The supply of organic nutrients has
a pronounced influence on stream productivity. A low concentration of sucrose (4 mg/liter)
continuously introduced to the riffles of a
stream was sufficient to increase productivity
markedly yet not change the quality of the
water. The input of sucrose represented an
additional energy input of 130,000 kcal/m²
to the riffle, twice that supplied by the sun to
the forest-shaded riffles (Warren, 1971). The
additional energy input resulted in a lush
growth of the bacteria *Sphaerotilus*, a doub-

ling of the herbivorous insects, a quadrupled
increase in carnivorous insects, and an increase
in the production of cutthroat trout from 0.2
kcal/m² to 4.3 kcal/m².

FAST STREAMS

Fast or swiftly flowing streams are, roughly,
all those whose velocity of flow is 50 cm/
second or higher (A. Nielsen, 1950). At this
velocity the current will remove all particles
less than 5 mm in diameter and will leave behind a stony bottom. The fast stream is often
a series of two essentially different but interrelated habitats, the turbulent riffle and the
quiet pool (Figs. 18-11 and 18-12). The waters
of the pool are influenced by processes occurring in the rapids above, and the waters of
the rapids are influenced by events in the pool.

The riffles are the sites of primary production in the stream (see Nelson and Scott,
1962). Here the aufwuchs assume dominance
and occupy a position of the same importance
as the phytoplankton of lakes and ponds. The
aufwuchs consist chiefly of diatoms, blue-
green and green algae, and water moss. Extensive stands of algae grow over rocks and
rubble on the stream bed and form a slippery
covering familiar and often dangerous to
fishermen and others who wade the stream.
Growth during favorable periods may be so
rapid that the stream bottom is covered in
10 days or less (Blum, 1960). Many small
algal species are epiphytes and grow on the
tops of or in among other algae.

The outstanding feature of much of this
algal growth is its ephemeral nature. Scouring
action of water and the debris it carries tears
away larger growth, epiphytes and all, and
sends the algae downstream. As a result there
is a constant contribution from upstream to
the downstream sequence.

Above and below the riffles are the pools.
Here the environment differs in chemistry, intensity of current, and depth. Just as the
riffles are the sites of organic production, so
the pools are sites of decomposition. They are
the catch basins of organic materials, for here
the velocity of the current is reduced enough
to allow a part of the load to settle out. Pools
are the major sites for free carbon dioxide
production during the summer and fall, which
is necessary for the maintenance of a constant supply of bicarbonate in solution (Neel,
1951). Without pools, photosynthesis in the

riffles would deplete the bicarbonates and result in smaller and smaller quantities of available carbon dioxide downstream.

Overall production in a stream is influenced in part by the nature of the bottom. Pools with sandy bottoms are the least productive because they offer little substrate for either aufwuchs or animals. Bedrock, although a solid substrate, is so exposed to currents that only the most tenacious organisms can maintain themselves (Fig. 18-13). Gravel and rubble bottoms support the most abundant life because they have the greatest surface area for the aufwuchs, provide many crannies and protected places for insect larvae, and are the most stable (see Figs. 18-13 and 18-14). Food production decreases as the particles become larger or smaller than rubble. Insect larvae, on the other hand, differ in abundance on the several substrates. Mayfly nymphs are most abundant on rubble, caddisfly larvae on bedrock, and Diptera larvae on bedrock and gravel (Pennak and Van Gerpen, 1947).

The width of the stream also influences overall production. Bottom production in streams 20 ft wide decreases by one-half from the sides to the center, and in streams 100 ft wide it decreases one-third (Pate, 1933). Streams 6 ft or less in width are four times as rich in bottom organisms as those 19 to 24 ft wide. This is one reason why headwater streams make such excellent trout nurseries.

Life in a fast stream. Although the pools and the riffles are more or less distinct, no clear-cut differences exist in the animals that inhabit them. Some animals of the riffles are carried by the current to the pools; others move back and forth between the two at will, as do the trout. For these, the riffles furnish food and the pools shelter. A good trout stream should be about 50 percent in pools and 50 percent in riffles. But the majority of stream inhabitants live in riffles, on the underside of rubble and gravel, where they are sheltered from the current. Characteristic of the riffle insects are the nymphs of mayflies, caddisflies, true flies, stoneflies, and alderflies or dobsons. In the pools the dominant insects are the burrowing mayfly nymphs, dragonflies and damselflies, and water striders.

A major reason for the richer aquatic life in the riffles is the current. Stream animals depend upon flowing water to aid their respiration and to bring them food. If riffle fauna

617

FIGURE 18-11
Two different but related habitats in a stream, the riffles (foreground) and the pool. (Photo by author.)

FIGURE 18-12
A segment of the stream where the transect in Fig. 18-15 was taken. (Photo courtesy Arnold Benson.)

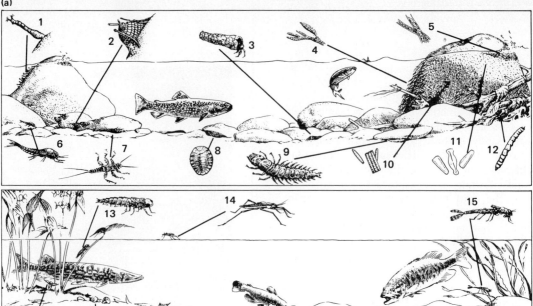

are transplanted to still water, many suffocate in a few hours. Although quiet water may contain more than a sufficient amount of oxygen to meet their needs, the organisms become surrounded by a closely adhering film of liquid that forms a sort of cloak and is impoverished of substances, including oxygen. In fast water, such a cloak cannot form, and the absorbing and respiratory surfaces are in continuous contact with oxygenated water (Ruttner, 1953).

A streamlined form, which offers lessened resistance to water current, is typical of many animals of fast streams, such as the black-nosed dace and the brook trout (Fig. 18-14). Among the insect larvae, the "howdy" mayfly nymph is unique for its streamlined body, its strongly plumed cerci, and its ability to dart among stones in swift water. Also able to move from stone to stone are the mayflies of the genus *Baetis*, which are fishlike in form.

Other insect larvae possess extremely flattened bodies and broad, flat limbs, with which they cling to the undersurface of stones where the current is very weak. Typical among these are the mayflies of the genera *Stenonema* and *Iron*. Their eyes are dorsally located and their gills are platelike, a characteristic of mayflies that do not have to agitate the water around them to secure circulation. Similarly flattened are the stonefly nymphs, with two cerci and the tufts of filamentous gills on the bases of the

legs below the thorax. Another example of body flattening even more extreme is the larvae of the water penny beetle. Viewed from the top the water penny appears as an encrustation on the stone. But underneath the widened, protective, sucker-forming covering of the back is the body of the beetle larva. The flattening of the body enables the animals to inhabit the thin crevices and the undersides of stones away from the strong current (A. Nielsen, 1950).

Other forms attach themselves temporarily in one way or another to the substrate. The blackfly larvae, *Simulium*, occur in such numbers on the down-current sides of stones that they have gained the name "black moss." They attach themselves to rocks by means of a circlet of rows of outwardly directed hooks on a swollen posterior. Attached to the center of the circlet are muscles that contract to pull the hooks inward (Nielsen, 1950). When the muscles are relaxed, the hooks move outward, releasing the hold. This allows the larvae to move about slightly on a silken web, secreted by the salivary glands and placed on the rock by the larva. Head end downstream, the larvae hang swaying in the water. They strain out the food brought by the current by means of feeding fans on the head.

The larvae of certain species of caddisflies construct cases of sand or small pebbles (Fig. 18-15), which protect them from the wash of the current. Some species have portable houses, the weight of which increases with the velocity of the current. The thickened walls act as ballast to hold the case on the bottom. Other species have cases firmly cemented to the sides and bottoms of stones. The net-spinning caddisflies firmly cement to stones funnel-shaped nets, the open ends of which face upstream. The larval inhabitants feed on the minute plants and animals swept into the meshes of the nets. Free-living caddisfly larvae, *Rhyacophila*, roam over the stones. Equipped with hooks on the anal prolegs and strong anal claws, these caddis flies grasp irregularities on the stone. They creep along, lengthening and shortening their bodies and clinging alternately with their anal prolegs and strong anal claws. These caddisflies, however, form stone cases for pupation (H. H. Ross, 1944).

Algae and water moss also are attached permanently to the substrate (Fig. 18-14). Water moss and heavily branched filamentous algae are held to rocks by strong holdfasts and are aligned with the current. In addition, the algae are covered with a slippery, gelatinous coating. Other algae grow in spheric, wartlike or cushionlike colonies with smooth, gelatinous surfaces. Some species are reduced to simplified platelike forms, which grow in closely appressed sheets that follow the contours of the rocks. In limestone water some algae, such as *Phormidium* and *Androinella*, secrete calcium carbonate to cover the entire growth in a crust. These may be overgrown at certain seasons by such algae as *Cladophera*. Algae growth in streams often exhibits zonation on rocks; the growth is influenced by both depth and current (Fig. 18-14) (Blum, 1960).

Sticky undersurfaces aid snails and planarians to remain stationary in the current. Snails can cling tightly to rubble in flowing water, and often such bottoms in hard-water streams support large populations.

The few ciliates living in the stream differ little from those of still water and exhibit no adaptation for current. This is possible because of the rapid decrease in the current near the bottom. In the submicroscopic distances from the bottom, current is almost zero. Here in this slowly moving water the ciliates live.

In spite of these adaptations, many bottom organisms tend to drift downstream to form a sort of traveling benthos. This is a normal process in streams, even in the absence of

FIGURE 18-14 [OPPOSITE]
Life in a fast stream compared to that in a slow stream. (1) *Blackfly larva;* (2) *net-spinning caddisfly;* (3) *stone case of caddisfly;* (4) *water moss* (Fontinalis); (5) *alga* Ulothrix; (6) *mayfly nymph* (Isonychia); (7) *stonefly nymph* (Perla); (8) *water penny;* (9) *hellgrammite;* (10) *diatoms* (Diatoma); (11) *diatoms* (Gomphonema); (12) *cranefly larva;* (13) *dragonfly nymph;* (14) *water strider;* (15) *damselfly nymph;* (16) *water boatman;* (17) *fingernail clam* (Sphaerium); (18) *burrowing mayfly nymph* (Hexagenia); (19) *bloodworm;* (20) *crayfish. The fish in the fast stream* (a) *are* (left) *brook trout,* (right) *redbelly dace. The fish in the slow stream* (b) *are* (left to right) *northern pike, bullhead, and small-mouth bass.*

FIGURE 18-13
A fast mountain stream in a deep woods. The bottom is largely bedrock. (Photo by author.)

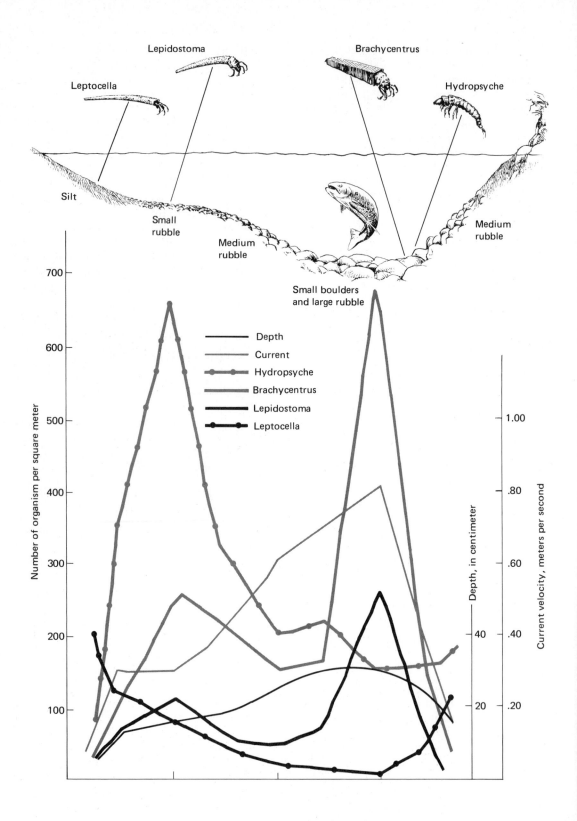

Leptocella

Lepidostoma

Brachycentrus

Hydropsyche

Silt

Small
rubble

Medium
rubble

Medium
rubble

Small boulders
and large rubble

Depth
Current
Hydropsyche
Brachycentrus
Lepidostoma
Leptocella

Number of organism per square meter

700

600

500

400

300

200

100

Depth, in centimeter

40

20

Current velocity, meters per second

1.00

.80

.60

.40

.20

high water and abnormal currents (for detailed review see Hynes, 1970; T. F. Waters, 1972). This drift consists mainly of organisms that are free-ranging such as the mayfly larva, *Baetis*, and the scud, *Gammarus*. (K. Muller, 1966; Waters, 1965). Drift is so characteristic of streams that a mean rate of drift can serve as an index of the production rate of a stream.

Drift of stream insects is not a haphazard incident. It is relatively constant and exhibits diurnal variations through the year (Muller, 1966). It is greatest at night, especially soon after sunset, and is lowest during the day. Illumination of an artificial stream at night reduces the amount of drift to a daytime volume (Elliot, 1965); continuous light, whether artificially applied (Elliot, 1967) or maintained through a 24-hour period by bright moonlight (N. H. Anderson, 1966), inhibits drift. Drift is also influenced by temperature, different species reacting differently during different seasons (Hynes, 1970; Waters, 1972).

Drift, however great, unless catastrophic, rarely depletes the upstream population of drift organisms. There is some question about this. Waters (1961, 1965) suggests that drift represents excessive production from the upstream habitat. However, there is some evidence that the depleted areas are repopulated either by upstream flight of adult insects or upstream movements of larvae through bottom rubble (Hynes, 1970). Drift is a means of colonizing depleted downstream sites (K. Muller, 1957; Waters, 1964). But a large portion of organisms is consumed by fish, and an even greater number undoubtedly perish.

SLOW STREAMS AND RIVERS

As the current slows, a noticeable change takes place in streams (Fig. 18-14). Silt and decaying organic matter accumulate on the bottom, and fine detritus from upstream is the main surce of energy (Fig. 18-16). Faunal organisms are able to move about to obtain their food, and a plankton population of a sort develops. The composition and configuration of the stream community approaches that of standing water.

With increasing temperatures, decreasing current, and accumulating bottom silt, organisms of the fast water are replaced gradually by organisms adapted to these conditions. The brook trout and the sculpin give way to

FIGURE 18-15 [OPPOSITE]
Distribution of four species of caddisfly in relation to water depth and current across a 12-m transect of a high mountain stream in Colorado. Note that two species, Hydropsyche and Brachycentrus, are in deep, fast water, while the remaining two are in shallow, less swift water where silt is deposited on the bottom. (Data and graphs from Arnold Benson, West Virginia University.)

FIGURE 18-16
A slow stream reflects the sky of a summer afternoon. (Photo by author.)

621

small-mouth bass and rock bass, the dace to shiners and darters. With current at a minimum, many resident fish lack the strong lateral muscles typical of the trout and have compressed bodies that permit them to move with ease through masses of aquatic plants. Mollusks, particularly *Sphaerium* and *Pisidium*, and pulmonate snails, crustaceans, and burrowing mayflies replace the rubble-dwelling insect larvae (Fig. 18-14). Only in occasional stretches of fast water in the center of the stream are remnants of headwater-stream organisms found.

As the volume of water increases, as the current becomes even slower and the silt deposits heavier, detritus-feeders increase. Tube-dwelling annelids and midges are common (Fig. 18-14), as are the bottom-feeding catfish, suckers, and the introduced carp. Backswimmers, water boatmen, and diving beetles inhabit the sluggish stretches and backwaters of rivers; and where water conditions are suitable, muskies, pike, and turtles are common. Rooted aquatics appear. Emergent vegetation grows along the river banks, and duckweeds float on the surface. Indeed the whole aspect approaches that of lakes and ponds, even to zonation along the river margin.

The higher water temperature, weak current, and abundant decaying matter promotes the growth of protozoan and other plankton populations. Scarce in fast water, plankton increase in numbers and species in slow water. Rivers have no typical plankton of their own. Those found there originated mainly from backwaters and lakes. In general, plankton populations in rivers are not nearly as dense as those in lakes. Time is too short for much multiplication of plankton because relatively little time is needed for a given quantity of water to flow from its source to the sea. Also occasional river rapids, often some distance in length, kill many plankton organisms by violent impact against suspended particles and the bottom. Aquatic vegetation filters out this minute life as the current sweeps it along. Plankton populations at the beginning of the Shenango River, which arises out of Pymatuning Reservoir in northwestern Pennsylvania, are the same as those found in the reservoir itself. But 11 mi downstream, the plankton populations decreased by 73 percent (Hartman and Himes, 1961). Interestingly, the proportion of silica-cased diatoms increased downstream.

INTERMITTENT STREAMS

All streams at some time experience very low flow and may nearly dry up. Some streams, except for occasional pools, dry up seasonally every year. These are intermittent streams. Although they cease to exist as streams during the summer and fall, they are still inhabited by aquatic fauna that does not have an active aquatic life stage during the dry period.

In a general way the fauna of intermittent streams can be divided into two groups (Clifford, 1966). One is the summer–fall association of the dry stream. Species are few and consist of those organisms that survive in the remaining pool in spite of their rising temperature and small organisms that can burrow into the moist interstitial spaces below the stream bed where some water still remains. The other group is the winter–spring association, composed of organisms characteristic of flowing water. They appear each year and make up a fairly stable permanent population. By possessing certain features in their life cycle most members are preadapted to escape the unfavorable dry period. For example two crustaceans, *Lircens fontinalis* and *Crangonyx forbesi*, breed in March and April. The old generation of the former dies when the stream dries up; the females of the latter die before the water goes. The population survives as young small individuals existing in the moist substrate. A species of caddisfly, *Ironoquia punctatissima* estivates in the pupal or larval stage in litter along the edge of the stream. Other invertebrates survive the dry period in the egg stage. Species whose life cycle does not include inactive stages during the dry period either are forced from the area or die as the streams dry.

SPRINGS

Many streams are born as springs back in the woods or in pasture fields. A spring is a concentrated flow of groundwater issuing from openings in the ground. Springs may range from tiny seep holes, through which the water oozes to form wet spots in the ground, to large fissures in rocks or openings in the ground that are cleaned out and enlarged by percolating water. If the rate of flow is great enough, pools of water nearly devoid of suspended matter form around the point of discharge. Runoff water erodes a channel away, which is the beginning of a stream, while precipitated mineral matter is deposited in the vicinity of the pool.

From an ecological viewpoint, the spring pool is important as a natural constant aquatic environment. Compared to lakes and rivers, its temperature is relatively constant, as is its chemical composition and water velocity. The organisms do not modify the pool environment, for almost as rapidly as the water is altered by photosynthesis and aquatic organisms, it is replaced by fresh water from the ground.

Springs, from a production standpoint, can be classified as autotrophic or heterotrophic. Large springs, those of first magnitude, such as the famous Silver Springs of Florida, are usually autotrophic and support a standing crop of such producer organisms as algae and submerged aquatic plants. Plankton is usually absent. This base supports a consumer pyramid of aquatic insects, snails, fish, and turtles (see H. T. Odum, 1957a, 1957b). Pasture springs are heterotrophic. For the community as a whole, a large portion of the assimilated energy is supplied by imports of organic debris. Primary production is small, whereas herbivore production is large because detritus-feeders comprise the bulk of the spring fauna (Teal, 1957).

LONGITUDINAL ZONATION

In a general way lotic ecosystems exhibit a longitudinal biotic zonation from the upper reaches or source to the slow-moving river far downstream. There is no precise boundary between one zone and another. The upper reaches of swift cold water are characterized by certain combinations of species such as brook trout, sculpin, and stoneflies. Downstream, where the gradient is reduced and the temperatures are warmer, the small-mouthed bass replaces the brook trout, and in the slow-running deep rivers, catfish, largemouth bass, and pickerel replace the small-mouthed bass (see Fig. 18-1). Longitudinal zonation is something of a continuum, but one characterized by a number of discontinuities. Communities of fast water may appear and disappear downstream as conditions change in the watercourse.

Although there have been a number of attempts to classify the longitudinal zones of a

623

stream, all largely unsuccessful, specific ecological entities are identifiable. Illes and
Botosaneanu (1963) have suggested that these
entities can be defined at points where the
largest faunal changes take place. They divide
the lotic ecosystem into the spring region
(*eucrenon*); the spring brook (*hypocrenon*);
the fast stream to small river (*rhithron*),
where the mean temperature rises to 20° C;
and the slow large rivers (*potamon*), where
the temperature rises to over 20° C, and the
bottom is mainly sand and mud.

IMPACT OF MAN

Few streams exist that in some way have not
been affected by man through pollution from
silt, sewage, and industrial wastes. The magnitude of ecological changes produced depends
upon the kind of pollutant and the quantity
in both time and space.

Industrial pollution, most serious in larger
streams and rivers, is very complex. Water
withdrawn from streams and used for cooling
in certain manufacturing processes and then
returned again raises the stream temperature
and lowers dissolved oxygen. Disposed water,
used for flushing and chemical treatment, imparts bad tastes and odors and even toxic substances to stream water. Discharges from
chemical plants and sulfurous wastes from
pulp and paper mills are highly poisonous to
aquatic life. Sterile stream bottoms and brown
water are mute testimony of this. Many chemical wastes, perhaps harmless alone, react
with other chemicals to produce highly toxic
conditions. Two ppm of copper or eight of
zinc alone will not harm fish, but as little as
one-tenth part of the two combined will
eliminate fish in a stream. Many spectacular
and tragic kills of fish and other aquatic organisms result from a sudden influx of such
chemical pollutants. More recently, detergents
have become an added problem. Acid water
from both deep and surface coal mines has
destroyed stream life in coal regions and so
reduced bacterial activity that biological puri-
fication of sewage and other organic wastes in
water is impossible.

With the coming of the atomic age, radioactive wastes from uranium mills and atomic-
energy plants present a serious problem.
Streams cannot purify themselves naturally of
radioactive wastes, as they can of organic
wastes. Radioactive materials, however, are
increasingly diluted as they are carried downstream; some are deposited on the bottom;
others are taken up by aquatic organisms, both
plant and animal.

Raw sewage dumped into streams sharply
changes biological conditions, and even sewage disposal plant effluent, rich in nutrients,
can upset ecological stability. Streams can
purify themselves by natural processes, the
breakdown of organic matter by bacterial activity. The time required depends upon the
degree of pollution and the character of the
stream. A fast-flowing stream constantly saturated with oxygen will purify itself much
faster than a slow stream. As sewage enters a
stream, it is dispersed, and the solids settle
to the bottom, where they are attacked by
aerobic bacteria. This depletes the oxygen, but
this loss is offset by the absorption of more
oxygen from the air into the stream. The carbon dioxide and hydrogen sulfide content of
the water in the discharge area is high. Normal stream life, particularly vertebrates and
mollusks, is absent, and the dominant organisms consist of a number of protozoans, mosquito larvae, and tubifex worms. Below this
zone of active decomposition, flowing waters
dilute the pollutants. Although conditions are
improved, the stream still is far from normal.
Green algae are present but reduced in numbers; bacteria still are abundant and oxygen is
low. Downstream the pollutants are diluted
further, dissolved oxygen is higher, and organisms tolerant of such conditions, such as
carp, catfish, chiromonid larvae, and protozoans, inhabit the area. Eventually the water
becomes clean and fresh again and normal
populations of fish and invertebrates reappear.

In far too many streams, however, conditions become worse downstream. No sooner
has the stream somewhat recovered from its
polluted condition than another town dumps
in its sewage. As a result, the stream carries
a greater load than it can recover from, aerobic
conditions no longer exist, and putrifaction
occurs. Aerobic bacteria are replaced by anaerobic, and normal stream life is completely
destroyed. Putrifying bacteria alone remain,
and the stream becomes a foul-smelling open
sewer.

Although water pollution affects aquatic organisms, the organisms themselves also produce changes in the stream. In badly polluted
water, populations of tubifex worms, *Asellus*,
and *Chironomus* larvae are so high that they

too reduce the oxygen supply (Westlake, 1959). *Chironomus* larvae alone can lower the oxygen content by over 1.0 ppm/mi of river. Dense plant populations, stimulated by nutrient content of the water, cause additional, wide diurnal and seasonal changes in the oxygen content of rivers.

Siltation, caused by erosion of farmlands, roadsides, construction, and other forms of soil disturbance, is the most insidious form of pollution, for it is widespread, often goes unnoticed, and the damage it does is often permanent. Silt can change a cold, clear trout stream to a warm, murky one, inhabited by fish tolerant of turbid waters and muddy bottom.

Silt destroys stream habitats (Fig. 18-17), changes the environment, and kills aquatic organisms. Clay soils suspended in water block out light and prevent the growth of aquatic plants. Silt settles on the stream bottom, covering sites for insect larvae and smothering mussels and other bottom organisms. Silt also blankets sewage and other organic material and retains it in place. Here it is decomposed, depleting the oxygen supply (Ellis, 1936). Aquatic insects characteristic of rubble disappear, and burrowing forms appear (Eustis and Hiller, 1954). Caddisflies and mayflies are replaced by bloodworms. Evidence indicates that high turbidity kills fish by clogging the opercular cavities and gill filaments with silt. Because water cannot reach the gill filaments, blood aeration is impeded, and sooner or later the fish die from carbon dioxide retention, anoxemia, or both. Similarly the mantle and gills of mollusks are either clogged or injured by soil particles (Cordone and Kelly, 1961). Silty water flowing through the gravel nests or redds of trout and salmon causes heavy mortality of eggs. Silt clogs the spaces between the gravel, reducing the water flow through the redd, and settles on the eggs. With insufficient water washing them, the eggs die from lack of oxygen. Thousands of miles of trout and salmon water have been destroyed by siltation, which more than any other cause limits the natural reproduction of these fish.

One of the most recent threats to natural streams is that of channelization. The dredging and straightening of streams under the guise of flood control and agricultural development is converting hundreds of miles of meandering productive fish-filled streams into

FIGURE 18-17
Silt carried to the stream by erosion and destruction of streambanks is deposited in the stream where the current slows. (Photo by author.)

sterile, unattractive drainage ditches. The dredging and straightening of stream channels destroys the bottom habitat, produces great uniformity along the watercourse, and increases laminar flow. Newly cut channels support little bottom fauna, and fish move out because of a lack of food, shelter, and breeding sites.

Dams, whatever their purpose, convert large portions of streams into lakes. They eliminate migratory fish and increase the plankton content of the water. If the discharge is from the upper stratified layers of the reservoir, the effect of the dam downstream is similar to that of a natural lake. But if the discharge is from the cold hypolimnion, the summer temperature of the river or stream is kept cold and the downstream stretches possess cold-water conditions and hold a cold-water biota. If the reservoir is used for irrigation or power, the fluctuating water level, both above and below the dam, is such that many organisms cannot survive.

However attractive and pleasant free-flowing rivers and meandering streams may be, their days appear to be numbered.

Summary

Fresh-water ecosystems are of two general types, the lotic or flowing water environment, and the lentic or standing water environment.

From its source to its mouth, the lotic ecosystem exhibits changes in its character. Up in the hills, the headwater streams are small, shallow, usually swift, and cold. Production is high, but the nutrients are carried downstream by the current. To exist in the current, the organisms of fast-flowing waters may be streamlined in shape or flattened to conceal themselves in the crevices of and underneath rocks; or they may attach themselves in one fashion or another to the substrate. Downstream, the volume of flow, augmented by tributaries, increases, the channel becomes wider and deeper, and the waters are not nearly so swift. In lowlands, the flow is slow, and the bottom is soft with silt and mud. Aquatic plants and animals characteristic of ponds and lakes replace life of the swift headwaters. But there is no clear-cut boundary between the swift and slow water communities. Like the vegetation continuum, one gradually blends into the other, reflecting

a longitudinal gradient in temperature, velocity of current, and often pH. Certain conditions, such as changes in current velocity, may reappear along the gradient, and with this a return of some fast-water populations. But life in the streams of most settled parts of the country has been affected by pollution from industrial wastes, sewage, and siltation. Organisms of natural waters are replaced by those tolerant of turbid waters, low in oxygen. Most of our streams and rivers face destruction as natural systems from the building of dams and the dredging and straightening of stream channels.

Lentic ecosystems are characterized by lakes and ponds, standing bodies of water that fill depressions in the earth. Lakes and ponds are ephemeral features of the landscape, geologically speaking. In time they fill in, grow smaller, and are finally replaced by a terrestrial community. Before its death, the lake was a nearly self-contained ecosystem. As with most lakes, it probably exhibited gradients in light, temperature, and dissolved gases. In summer its waters were stratified. It had a surface layer of warm, circulating water, the epilimnion, and a middle zone, the metalimnion, in which the temperature rapidly dropped. Below this was the hypolimnion, a bottom layer of denser water, approximately 4° C, often low in oxygen and high in carbon dioxide. When the surface water cooled in autumn, the difference in density between the layers decreased, and the waters circulated throughout the lake. A similar mixing of waters took place in spring when the lake warmed up. These seasonal overturns were important in recirculating the nutrients and in mixing the bottom waters with the top. Areas where light penetrated to the bottom of the lake, a zone known as the littoral, were occupied by rooted plants. Beyond this was open water, or the limnetic zone, inhabited by plant and animal plankton and fish. Below the depth of effective light penetration was the profundal region, where the diversity of life varied with the temperature and oxygen supply. The bottom, or benthic zone, was a place of intense biological activity, for here decomposition of organic matter was taking place. Anaerobic bacteria were dominant on the bottom beneath the profundal water, while the benthic zone of the littoral was rich in decomposer organisms and detritus-feeders. Activities of

*plants and animals and sedimentation built
up the lake bottom. Little by little, the zones
of vegetation so characteristic of the littoral,
the floating plants and emergents, pushed
outward toward the middle.*

*The late fill-in stages of a lake create bogs,
marshes, and swamps, but swamps also develop on the flood plains of the larger river
systems of southern North America. Each is
distinctively different. Bogs, confined to northern regions, are characterized by blocked
drainage conditions, an accumulation of peat,
cushionlike vegetation, a marginal semifloating mat of plant growth, and acidic conditions, created largely by sphagnum moss. Only
those plants tolerant of acidic conditions
occupy the bog. Life is restricted in the
number of plant and animal species, but the
organisms present are often abundant.
Marshes, on the other hand, are wetlands in
which the grass lifeform is dominant; swamps
are wooded. Both may range from deep to
shallow water, and both embrace a richness
and diversity of life that is hard to equal in
other temperate communities. Yet marshes
and swamps too frequently are considered
more as places to be drained or filled than as
areas to be managed and preserved.*

Fresh-water rivers eventually empty into the oceans, and terrestrial ecosystems end abruptly at the edge of the sea. For some distance there is a region of transition. The river enters the saline waters of oceans, creating a gradient of salinity. That gradient provides a habitat for organisms uniquely adapted to exist in the half-world between saltwater and fresh. The inhabitants of the region between land and sea are adapted to live in often severe environments dominated by tides. Beyond all this lies the open ocean—the shallow seas overlying the continental shelf and the deep ocean.

Some characteristics of the marine environment

The marine environment is marked by a number of extreme differences from the fresh-water world. It is large, occupying 70 percent of the earth's surface. The volume of surface area lighted by the sun is small in comparison to the total volume of water involved. This and the dilute solution of nutrients limits production. It is deep, in places nearly 4 mi. All of the seas are interconnected by currents, dominated by waves, influenced by tides, and characterized by saline waters.

SALINITY

The salinity of the open sea is fairly constant, averaging about 35 ppt (o/oo). This probably has not changed greatly since the earth was formed. Although quantities of salts are carried to the sea by rivers, they are removed at about the same rate as supplied by means of complex chemical reactions with sediments and particulate matter.

Salinity is due to two elements, sodium and chlorine, which make up some 86 percent of sea salt. These along with other such major elements as sulfur, magnesium, potassium, and calcium comprise 99 percent of sea salts. Seawater, however, differs from a simple sodium chloride solution in that the equivalent amounts of cations and anions are not balanced against each other. The cations exceed the anions by 2.38 milliequivalents. As a result, seawater is weakly alkaline (pH 8.0

to 8.3) and strongly buffered, a condition that is biologically important.

The amount of dissolved salt in seawater is usually expressed as chlorinity or salinity. Because oceans are usually well mixed, sea salt has a constant composition; that is, the relative proportions of major elements change little. Thus the determination of the most abundant element, chlorine (Table 19-1), can be used as an index of the amount of salt present in a given volume of seawater. Chlorine expressed in o/oo is the amount of chlorine in grams in a kilogram of seawater. Chlorinity can be converted to salinity, the total amount of solid matter in grams per kilogram of seawater. The relationship of salinity to chlorinity is expressed by

$$S (o/oo) = 1.80655 \times chlorinity$$

The salinity of parts of the ocean is variable because physical processes change the amount of water in the seas. Salinity is affected by evaporation, precipitation, movement of water masses, mixing of water masses of different salinities, formation of insoluble precipitates that sink to the ocean floor, and diffusion of one water mass to another. Salinities are most variable near the interface of sea and air. Elements so affected by these physical processes are conservative elements not involved in biological processes. The most variable elements in the sea, such as phosphorus and nitrogen, are the nonconservative ones because their concentration is related to biological activity. Because they are taken up by organisms (see Chapter 5), these elements are usually depleted near the surface and are enriched at lower depths. In parts of the ocean some of these nutrients are returned to the surface by upwelling.

Salinity imposes certain restrictions on life that inhabits the oceans (see Kinne, 1971). Fish and marine invertebrates that inhabit marine, estuarine, and tidal environments have to maintain osmotic pressure under conditions of changing salinities (see Chapter 15). Most marine species are adapted to living in high salinities, and the number declines as salinity is reduced.

TEMPERATURE AND PRESSURE

What has already been written about temperature and fresh-water relations also applies to the sea (Fig. 19-1). The range of

Marine ecosystems

temperature is far less than that on land, although as one would expect, the range of variation in temperature over the oceans is considerable. Arctic waters at 27° F are much colder than tropical waters at 81° F, and currents are warmer or colder than the waters through which they flow. Seasonal and daily temperature changes are larger in coastal waters than on the open sea. The surface of coastal waters is the coolest at dawn and the warmest at dusk. In general, seawater is never more than 2° to 3° below the freezing point of fresh water or higher than 81° F. At any given place the temperature of deep water is almost constant and cold, below the freezing point of fresh water. Seawater has no definite freezing point, although there is a temperature for seawater of any given salinity at which ice crystals form. Thus pure water freezes out, leaving even more saline water behind. Eventually it becomes a frozen block of mixed ice and salt crystals. With rising temperatures the process is reversed.

Unlike fresh water, seawater (with a salinity of 24.7 o/oo or higher) becomes heavier as it cools and does not reach its greatest density at 4° C. Thus the limitation of 4° C as the temperature of bottom water does not apply to the sea. In spite of this, bottom temperatures of the sea rarely go below the freezing point of fresh water, generally averaging around 2° C, even in the tropics if the water is deep enough. The temperature of the ocean floor over 1 mi deep is 3° C.

Another aspect of the marine environment is pressure. Pressure in the ocean varies from 1 atm at the surface to 1,000 atm at the greatest depth. Pressure changes are many times greater in the sea than in terrestrial environments and have a pronounced effect on the distribution of life. Certain organisms are restricted to surface waters where the pressure is not so great, whereas other organisms are adapted to life at great depths. Some marine organisms, such as the sperm whale and certain seals, can dive to great depths and return to the surface without difficulty.

ZONATION AND STRATIFICATION

Just as lakes exhibit stratification and zonation, so do the seas. The ocean itself is divided into two main divisions, the pelagic, or whole body of water, and the benthic, or

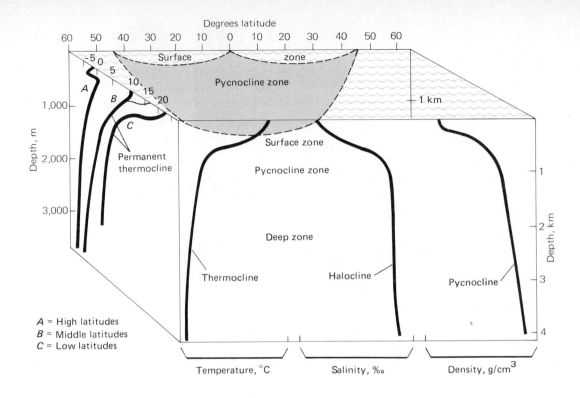

A = High latitudes
B = Middle latitudes
C = Low latitudes

Temperature, °C   Salinity, ‰   Density, g/cm³

bottom region (Fig 19-2). The pelagic is further divided into two provinces: the neritic, water that overlies the continental shelf, and the oceanic provinces. Because conditions change with depth, the pelagic is divided into three vertical layers or zones: from the surface to about 200 m is the photic zone, in which there are sharp gradients of illumination, temperature, and salinity. From 200 to 1,000 m, where very little light penetrates and where the temperature gradient is more even and gradual and without much seasonal variation, is the mesopelagic zone. It contains an oxygen-minimum layer and often the maximum concentrations of nitrate and phosphate. Below the mesopelagic is the bathypelagic zones, where darkness is virtually complete except for bioluminescence, and where temperature is low and pressure very great.

The upper layers of ocean water exhibit a stratification of temperature and salinity. Depths below 300 m are usually thermally stable. In high and low latitudes temperatures remain fairly constant throughout the year (Fig. 19-1). In middle latitudes temperatures vary with the season, associated with climatic changes. In summer the surface waters become warmer and lighter, forming a temporary seasonal thermocline. In subtropical regions the surface waters are constantly heated, developing a marked permanent ther-

mocline. Between 500 and 1,500 m a permanent but relatively slight thermocline exists.

Associated with a temperature gradient is a salinity gradient or *halocline*, especially at the higher latitudes (Fig. 19-1). There the abundant precipitation reduces surface salinity and causes a marked change in salinity with depth. Thus in the middle latitudes in particular, the two produce a marked gradient in density. Water masses form density layers with increasing depth. Because density of seawater does increase in depth and does not reach its greatest density at 4° C as with fresh water, there is no seasonal overturn. This results in a normally stable stratification of density known as the *pycnocline* (Fig. 19-1). Because there are marked changes in temperature with depth in the open ocean and thus with density, the pycnocline often coincides with the thermocline. Below the pycnocline is the deep zone, which comprises 80 percent of the ocean. Because of the contact of the deep ocean water with the cold surface waters of the high latitudes (Fig. 19-1), the deep ocean is cold and contains appreciable amounts of oxygen.

WAVES AND CURRENTS

Waves on the surface of the sea are stirred by the wind. The frictional drag of the wind on the surface of smooth water ripples the water. As the wind continues to blow, it applies more pressure to the steep side of the ripple, and wave size begins to grow. As the wind becomes stronger, short, choppy waves of all sizes appear, and as they absorb more energy, they continue to grow in size. When the waves reach a point at which the energy supplied by the wind is equal to the energy lost by the breaking waves, they become the familiar whitecaps. Up to a certain point, the stronger the wind, the higher the waves.

Waves are generated on the open sea. The stronger the wind, the longer the fetch (the distance the waves can run without obstruction under the drive of the wind blowing in a constant direction) and the higher the waves. As the waves travel out of the fetch, or if the winds die down, sharp-crested waves change into smooth long-crested waves or swells that can travel great distances because they lose little energy as they travel. Swells are characterized by troughs and ridges, the height of which is measured from the bottom of the trough to the crest. The length of the wave is measured from the crest to the following wave and its period by the time required for successive crests to pass a fixed point. None of these are static, for all depend upon the wind and the depth of the wave.

The waves that break up on a beach are not composed of water driven in from distant seas. Each particle of water remains largely in the same place and follows an elliptical orbit with the passage of the wave form. As the wave moves forward with a velocity that corresponds to its length, the energy of a group of waves moves with a velocity only half that of individual waves. The wave at the front loses energy to the waves behind and disappears, its place taken by another. Thus the swells that break on a beach are distant descendents of waves generated far out at sea.

As the waves approach land they advance into increasingly shallow water. The height of the wave rises higher and higher until the wave front grows too steep and topples over. As the waves break onshore they dissipate their energy against the coast, pounding rocky shores or tearing away at sandy beaches at one point and building up new beaches elsewhere.

Surface waves are the most obvious ones. But in the ocean there are also internal waves. Similar to surface waves, internal waves appear at the interface of layers of waters of

FIGURE 19-1 [OPPOSITE]

A composite graph of the temperature and salinity profiles of the sea. The side panel illustrates in a general way the temperature profiles of the ocean. The arctic or high latitudes, A. Note that the low-salinity colder surface waters overlie the warmer, more saline water. At the middle latitudes the ocean exhibits a seasonally varying surface temperature and a permanent thermocline, B. Waters of the lower latitudes possess stable surface temperatures and a permanent thermocline, C. (After R. V. Tait, 1968, Elements of Marine Ecology, Plenum, New York.) The front panels show the relationship between temperature and salinity and the pycnocline. The back panel shows a schematic representation of the horizontal layering of ocean waters. Note the latitudinal differences. (After M. Gross, 1972, Oceanography, Prentice-Hall, Englewood Cliffs, N.J.)

TABLE 19-1

Composition of seawater of 35 o/oo salinity, major elements (original)

Elements	g/kg	Millimole/kg	Milli-equiva-lent/kg
CATIONS			
Sodium	10.752	467.56	467.56
Potassium	0.395	10.10	10.10
Magnesium	1.295	53.25	106.50
Calcium	0.416	10.38	20.76
Strontium	0.008	0.09	0.18
			605.10
ANIONS			
Chlorine	19.345	545.59	545.59
Bromine	0.066	0.83	0.83
Fluorine	0.0013	0.07	0.07
Sulphate	2.701	28.12	56.23
Bicarbonate	0.145	2.38	—
Boric acid	0.027	0.44	—
			602.72

Source: K. Kalle, 1971, in O. Kinne, 1971, *Marine Ecology*, Vol. 2, Part 1. Wiley-Interscience, New York.

Note: Surplus of cations over strong anions (alkalinity): 2.38.

different densities. In addition there are also stationary waves or seiches, already described.

Just as there are internal waves, so there are internal currents in the sea. Surface currents are produced by wind, heat budgets, salinity, and the rotation of the earth (see Chapter 2). Water moving in surface currents must be replaced by a corresponding inflow from elsewhere. Because the surface waters are cooled and salinity changes, high-density water formed on the surface, largely at high latitudes, sinks and flows toward the low latitudes. These currents, subject to the Coriolis effect, are deflected or obstructed by submarine ridges and are modified by the presence of other water masses. The result is three main systems of subsurface water movements—the bottom, the deep, and the intermediate ocean currents—each of which runs counter to the other.

Also influencing subsurface currents is the Eckman spiral. As the wind sets the surface layer of water in motion, that layer in turn sets in motion a layer of water beneath, and that layer in turn another. Each layer moves more slowly than the layer above and is deflected to the right of it because of the Coriolis effect. At the base of the spiral the movement of water is counter to the flow on the surface, although the average flow of the spiral is at right angles to the wind.

In coastal regions the Eckman transport of surface water can bring deep waters up to the surface, a process called upwelling. Wind blowing parallel to a coast causes surface waters to be blown offshore. This is replaced by water moving upward from the deep. Although cold and containing less dissolved oxygen, upwelling water is rich in nutrients that support a rich growth of phytoplankton. For this reason regions of upwelling are highly productive.

TIDES

One of the fundamental laws of physics is Newton's law of universal gravitation. The law states that every particle of matter in the universe attracts every other particle with a force that varies directly as the product of their masses and inversely as the square of the distance between them. This is the reason why apples fall to the earth; it is also, in part, why the planets revolve about the sun. The earth attracts the apple, the apple in turn attracts the earth; but because the mass of the earth is so much greater, the apple falls to the ground. Likewise, the sun and its planets exert an attraction on each other; but the sun, being the largest, exerts the most powerful force. There is also an attraction between the earth and the moon. The moon is much closer to the earth than the sun, and it exerts a force twice as great as that of the sun. The gravitational pull of the sun and the moon each cause two bulges in the waters of the oceans. The two caused by the moon occur at the same time on opposite sides of the earth on an imaginary line extending from the moon through the center of the earth. The tidal bulge on the moon side is due to gravitational attraction; the bulge on the opposite side occurs because the gravitational force there is less than at the center of the earth. As the earth rotates eastward on its axis, the tides advance westward. Thus any given place on the earth will in the course of one daily rotation pass through two of the lunar tidal bulges, or high tides, and two of the lows or low tides, at right angles to the high. Because the moon revolves in a 29.5-day orbit around the earth, the average period between successive high tides is approximately 12 hours, 25 minutes.

The sun also causes two tides on opposite sides of the earth; and these tides have a relation to the sun such as the lunar tides have to the moon. Because they are less than half as high, solar tides are partially masked by lunar tides except for twice during the month, when the moon is full and when it is new. At these times, the earth, the moon, and the sun are nearly in line, and the gravitational pulls of the sun and the moon are additive. This causes the high tides of those periods to be exceptionally large, with maximum rise and fall. These are the fortnightly spring tides, a name derived from the Saxon *sprungen*, which refers to the brimming fullness and active movement of the water. When the moon is at either quarter, the pull of the moon is at right angles to the pull of the sun, and the two forces interfere with each other. At this time the differences between high and low tide are exceptionally low. These are the neap tides, from an old Scandinavian word meaning "barely enough."

Tides are not entirely regular nor are they the same all over the earth. They vary from day to day in the same place, following the waxing and the waning of the moon. They may act differently in several localities within the same general area. In the Atlantic, semi-daily tides are the rule. In the Gulf of Mexico, the alternate highs and lows more or less efface each other, and flood and ebb follow one another at about twenty-four hour intervals to produce one daily tide. Mixed tides are common in the Pacific and Indian oceans. These are combinations of the other two, with different combinations at different places. Local inconsistencies of tides are due to many variables. The elliptical orbit of the earth about the sun and of the moon about the earth influence the gravitational pull, as does the declination of the moon—the angle of the moon in relation to the axis of the earth. Latitude, barometric pressure, offshore and onshore winds, depth of water, contour of shore, and natural periods of oscillation, or internal waves (see Currents and Seiches, Chapter 18), modify tidal movements.

Estuaries

Waters of all streams and rivers eventually drain into the sea; the place where this fresh water joins the salt is called an *estuary*. Estuaries are semienclosed parts of the coastal ocean where the seawater is diluted and partially mixed with water coming from land. Estuaries differ in size, shape, and volume of water flow, all influenced by the geology of the region in which they occur. As the river reaches the encroaching sea, the stream-carried sediments are dropped in the quiet water. These accumulate to form deltas in the upper reaches of the mouth and shorten the estuary. When silt and mud accumulations become high enough to be exposed at low tide, tidal flats (Fig. 19-3) develop, which divide and braid the original channel of the estuary. At the same time, ocean currents and tides erode the coast line and deposit material on the seaward side of the estuary, also shortening the mouth. If more material is deposited than is carried away, then barrier beaches, islands, and brackish lagoons appear.

Current and salinity, both very complex and variable, shape life in the estuary, where the environment is neither fresh water nor salt. Estuarine currents result from the interaction of a one-direction stream flow, which varies with the season and rainfall, with oscil-

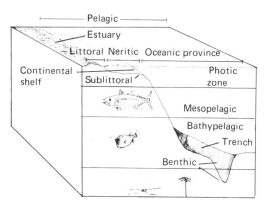

FIGURE 19-2
The various regions of the ocean visualized.

FIGURE 19-3
A muddy tidal flat covered with blue mussels.
(Photo by author.)

lating ocean tides, and with the wind (Ketchum, 1951; Burt and Queen, 1957). Because of the complex nature of the currents, generalizations about estuaries are difficult to make (see Lauff, 1967).

Salinity varies vertically and horizontally, often within one tidal cycle. Vertical salinity may be the same from top to bottom or it may be completely stratified, with a layer of fresh water on top and a layer of dense saline water on the bottom. Salinity is homogenous when currents, particularly eddy currents, are strong enough to mix the water from top to bottom. The salinity in some estuaries may be homogeneous at low tide, but at flood tide a surface wedge of seawater moves upstream more rapidly than the bottom water. Salinity is then unstable and density is inverted. The seawater on the surface tends to sink, as lighter fresh water rises, and mixing takes place from the surface to the bottom. This phenomenon is known as tidal overmixing. Strong winds, too, tend to mix salt water with the fresh (Barlow, 1956) in some estuaries; but when the winds are still, the river water flows seaward on a shallow surface over an upstream movement of seawater that only gradually mixes with the salt. Horizontally, the least saline waters are at the river entrance, and the most saline at the mouth of the estuary (Fig. 19-4). The configuration of the horizontal zonation is determined mainly by the deflection caused by the incoming and outgoing currents. In all estuaries of the northern hemisphere, outward-flowing fresh water and inward-flowing seawater are deflected to the right because of the earth's rotation. As a result, salinity is higher on the left side (Fig. 19-4).

Because the quantity of fresh water pouring into the estuary varies through the year, salinity also varies. Salinity is highest during the summer and during periods of drought, when less fresh water flows into the estuary. It is lowest during the winter and spring, when rivers and streams are discharging their peak loads. At times this change in salinity can happen rather rapidly. For example, early in 1957 a heavy rainfall broke the most severe drought in the history of Texas. The resultant heavy river discharge reduced the salinities in Mesquite Bay on the central Texas coast by over 30 ppt in a 2-month period. At the height of the drought, salinities ranged from 35.5 to 50 ppt, but after the drought was broken, the salinities ranged from 2.3 to 2.9 ppt (Hoese, 1960). Such rapid changes have a profound impact on the life of the estuary.

The salinity of seawater is about 35 o/oo; that of fresh water ranges from 0.065 to 0.30 o/oo. Because the concentration of metallic ions carried by rivers varies from drainage to drainage, the salinity and chemistry of estuaries differ. The proportion of dissolved salts in the estuarine waters remains about the same as that of seawater, but the concentration varies in a gradient from fresh water to sea (Fig. 19-4).

Exceptions to these conditions exist in regions where evaporation from the estuary may exceed the inflow of fresh water from river discharge and rainfall (a negative estuary). This causes the salinity to increase in the upper end of the estuary, and horizontal stratification is reversed.

Temperatures in estuaries fluctuate considerably diurnally and seasonally. Waters are heated by solar radiation and inflowing and tidal currents. High tide on the mud flats may heat or cool the water, depending on the season. The upper layer of estuarine water may be cooler in winter and warmer in summer than the bottom, a condition that, as in a lake, will result in a spring and autumn overturn.

Mixing waters of different salinities and temperatures acts as a nutrient trap (Fig. 19-5). Inflowing river waters more often than not impoverish rather than fertilize the estuary, except for phosphorus. Instead, nutrients and oxygen are carried into the estuary by the tides. If vertical mixing takes place, these nutrients are not soon swept back out to sea but circulate up and down between organisms, water, and bottom sediments.

LIFE IN THE ESTUARY

Organisms inhabiting the estuary are faced with two problems—maintenance of position and adjustment to changing salinity. The bulk of estuarine organisms are benthic and are securely attached to the bottom, buried in the mud, or occupy crevices and crannies about sessile organisms. Motile inhabitants chiefly are crustaceans and fish, largely young of species that spawn offshore in high-salinity water. Planktonic organisms are wholly at the mercy of the currents. Because the seaward movement of stream flow and ebb tide transport plankton out to sea, the rate of circulation or flushing time determines the nature of the plankton population. If the circulation is too vigorous, then the plankton population may be relatively small. Phytoplankton in summer are most dense near the surface and in low-salinity areas. In winter phytoplankton are more uniformly distributed. For any planktonic growth to become endemic in an estuary, reproduction and recruitment must balance the losses from physical processes that disperse the population (Barlow, 1955).

Salinity dictates the distribution of life in the estuary. Essentially the organisms of the estuary are marine, able to withstand full seawater. Except for andromonous fishes, no freshwater organisms live there. Some estuarine inhabitants cannot withstand lowered salinities, and these species decline along a salinity gradient. Sessile and slightly motile organisms have an optimum salinity range within which they grow best. When salinities vary on either side of this range, either up or down, populations decline. Two animals, the clam worm and the scud, illustrate this situation. Two species of clam worm, *Nereis occidentalis* and *Neanthes succinea*, inhabit the estuaries of the southern coastal plains of North America. *Nereis* is more numerous at high salinities, and *Neanthes* is most abundant at low salinities. In European estuaries the scud *Gammarus* is an important member of the bottom fauna. Two species, *Gammarus locusta* and *G. marina*, are typical marine species and cannot penetrate very far into the estuary. Instead they are replaced by a typical estuarine species, *G. zaddachi*. This species, however, is broken down into three subspecies, separated by salinity tolerances. *G. zaddachi* lives at the seaward end, *G. z. salinesi* occupies the middle, and *G. z. zaddachi*, which can penetrate up into fresh water for a short time at least, lives on the landward end (Spooner, 1947; Segerstrale, 1947).

Size influences the range of motile species within the estuarine waters. This is particularly pronounced among estuarine fish. Some, such as the striped bass, spawn near the interface of fresh and low-salinity water (Fig. 19-6). The larvae and young fish move downstream to more saline waters as they mature. Thus for the striped bass the estuary serves both as a nursery and feeding ground for the young. Andromonous species, such as the shad, spawn in fresh water, but the young

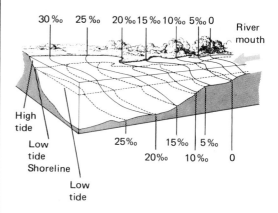

FIGURE 19-4
A generalized diagram of an estuary showing the vertical and horizontal stratification of salinity from the river mouth to the estuary at both high and low tide. At high tide the incoming sea water increases the salinity toward the river mouth; at low tide the salinity is reduced. Note also how salinity increases with depth, since the lighter fresh water flows over the denser salt.

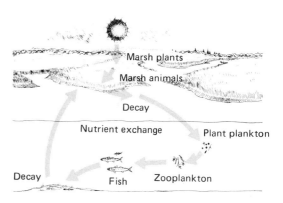

FIGURE 19-5
The relationship of the salt marsh to the estuary in nutrient cycling. (Adapted from C. N. Schuster, Jr., 1966.)

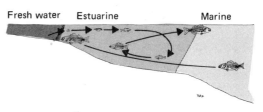

FIGURE 19-6
The relationship of a semi-anadromous fish, the striped bass, to the estuary. See text for details. (From Cronin and Mansueti, 1971.)

635

fish spend the first summer in the estuary, then move out to the open sea. Other species, such as the croaker, spawn at the mouth of the estuary, but the larvae are transported upstream to feed in the plankton-rich low-salinity areas. Others, such as the bluefish, move into the estuary to feed. In general marine species drop out toward fresh water and are not replaced by fresh-water forms. In fact the mean number of species progressively decreases from the mouth of the estuary to upstream stations (H. W. Wells, 1961).

Because salinities in estuaries vary considerably, organisms must be able to cope with changes in osmotic pressures. Some excrete excess water as fast as it enters the body, by means of kidneys, flame cells, and contractile vacuoles. Others increase internal hydrostatic pressure, especially by strong contractions of the body wall. Salt retention (discussed in Chapter 15) is still another method. A fourth solution common among estuarine animals such as snails, clams, shrimp, and crabs is an impermeable covering. Clams close up during periods of low salinity and function on stored glycogen available for anaerobic utilization; once this energy source is gone, the clam must open and suffer the consequences if the salinity is not back in the tolerable range.

Salinity changes often affect larval forms more severely than adults. Larval veligers of the oyster drill *Thais* succumb to low salinity more easily than the adults. A sudden influx of fresh water, especially after hurricanes or heavy rainfall, sharply lowers the salinity and causes a high mortality of oysters and their associates. When the drought-breaking heavy rainfall sharply reduced the salinities of Mesquite Bay in Texas, the high-salt-tolerant marine sessile and infaunal mollusks were completely wiped out. The high-salinity community of the oyster *Ostrea equestris* and the mussel *Brachidontes exustus* was replaced by the oyster *Crassostrea virginica* and the mussel *Brachidontes recurvus*. The rapid lowering of the salinity did not kill fish or other motile forms, which apparently moved out of the area (Hoese, 1960).

The oyster bed and the oyster reef are the outstanding communities of the estuary. The oyster is the dominant organism about which life revolves. Oysters may be attached to every hard object in the intertidal zone or they may form reefs, areas where clusters of living oysters grow cemented to the almost-buried shells of past generations. Oyster reefs usually lie at right angles to tidal currents, which bring planktonic food, carry away wastes, and sweep the oysters clean of sediment and debris.

Closely associated with oysters are encrusting organisms such as sponges, barnacles, and bryozoans, which attach themselves to oyster shells and are dependent on the oyster or algae for food. The oyster crab strains food from the oyster's gills (Christensen and McDermott, 1958), and a pramidellid snail lives an ectoparasitic life by feeding on body fluids and tissue debris from the oyster's mouth (Hopkins, 1958). Beneath and between the oysters live polychaete worms, decapods, pelecypods, and a host of other organisms. In fact, 303 different species have been collected from the oyster bed (H. W. Wells, 1961).

TIDAL MARSHES

On the alluvial plains about the estuary and in the shelter of the spits and offshore bars and islands exists a unique community, the tidal marsh (Fig. 19-7). Although to the eye tidal marshes appear as waving acres of grass, they are instead a complex of distinctive and clearly demarked plant associations. The reasons for this complex are again the tides and salinity. The tides perhaps play the most significant role in plant segregation, for twice a day the salt-marsh plants on the outermost tidal flats are submerged in salty water and then exposed to the full insolation of the sun. Their roots extend into poorly drained, poorly aerated soil, in which the soil solution contains varying concentrations of salt. Only plant species with a wide range of salt tolerance can survive such conditions (Fig. 19-8). Thus from the edge of the sea to the highlands, zones of vegetation, each recognizable by its own distinctive color, develop.

Tidal salt marshes begin in most cases as mud or sand flats, first colonized by algae and, if the water is deep enough, by eelgrass. As organic debris and sediments accumulate, eelgrass is replaced by the first salt-marsh colonists—the sea poa, *Puccinellia*, on the European coast, and salt-water cord grass, *Spartina alterniflora* var. *glabra*, on the coast of eastern North America. Stiff, leafy, up to 10 ft tall, and submerged in salt water at every high tide, salt-water cord grass forms a marginal strip between the open mud flat to the front and the higher grassland behind (Fig. 19-9). The pattern of vegetation behind the tall *Spartina* reflects microelevations, depth of tidal coverage, and salinity (Fig. 19-10). *Spartina al-*

FIGURE 19-7 [OPPOSITE]
A view of a tidal marsh on the Virginia coast.

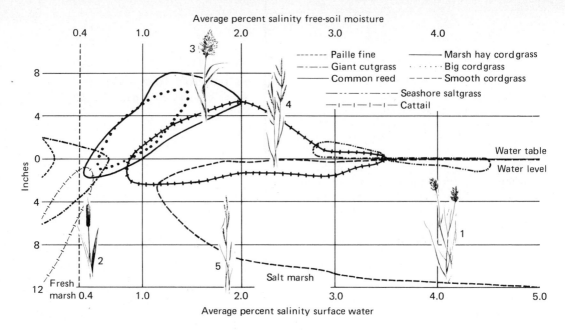

Average percent salinity free-soil moisture

0.4　　1.0　　　2.0　　　3.0　　　4.0

----- Paille fine　　——— Marsh hay cordgrass
—·—·— Giant cutgrass　　· · · · · Big cordgrass
——— Common reed　　— — — Smooth cordgrass
——·—·—·—— Seashore saltgrass
—ı—ı—ı—ı— Cattail

Water table
Water level

Inches

Fresh marsh 0.4　　1.0　　　2.0　　　3.0　　　4.0　　　5.0

Salt marsh

Average percent salinity surface water

terniflora may follow the winding channels of tidal creeks. On higher ground, where tidal waters are shallow, grows the short, green, fine-cut marsh-hay cord grass, *Spartina patens*. It grows so heavy and forms such a tight mat that few other plants can grow in this zone. At higher microelevations, where ordinary high tides rarely reach and where the salinity of the soil is relatively high, *Spartina patens* is replaced by or shares the site with the salt-tolerant *Distichlis spicata*, spike grass.

Where the salinity is even higher and the area is covered with even less tidal water, glasswort, *Salircornia*, with fleshy succulent pointed stems dominates the marsh (Fig. 19-11). On the rising ground level beyond may be a zone of rushes, the black grass, *Juncus*, so called because of its very dark green color, which becomes almost black in the fall. Rarely are rushes covered by ordinary high tides, but often they are submerged by the neap tides of spring and fall. Beyond the black grass a narrow belt of tussock-forming grasses may mark a transition between the marsh below and the upland vegetation.

Two conspicuous physiographic features of the salt marshes are the meandering creeks and the salt pans. The creeks are the drainage channels that carry the tidal waters back out to sea. The formation of these creeks is a complex process. In some cases, the channels are formed by water deflected by minor irregularities on the surface. In estuarine marshes, the river itself forms the main channel. Once formed, the channels are deepened by scouring and heightened by a steady accumulation of organic matter and silt. At the

same time, the heads of the creeks erode backward and small branch creeks develop. Where lateral erosion and undercutting take place, the banks may cave in, blocking or overgrowing the smaller channels. The distribution and pattern of the creek system plays an important role in the drainage of the surface water and the drainage and movement of the water in the subsoil.

Across the tidal marshes are many circular to elliptical patches of vegetation, different from the dominant plants about them and concentrically arranged about small depressions in the earth. These depressions are salt pans. At high tide they are flooded; at low tide the depressions remain filled with salt water. If shallow enough, the water may evaporate completely and leave an accumulating concentration of salt on the mud.

Pans come about in several ways. The great majority of them are formed as the marsh develops. Early plant colonization is irregular and bare spots on the flat become surrounded by vegetation. As the level of the vegetated marsh rises, the bare spots lose their water outlet. If such a pan eventually becomes attached to a creek, normal drainage is restored and the pan eventually becomes vegetated. Other pans, especially in sandy marshes, are derived from creeks. Marsh vegetation may grow across the creek bottom and dam a portion of it; or lateral erosion may block the channel. With drainage no longer effective, water remains behind after the flood tide, inhibiting the growth of plants. Often a series of such pans may form on the upper reaches of a single creek. Still another type is the

638

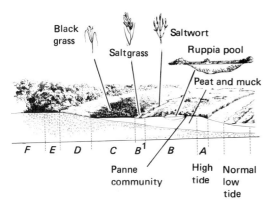

Black grass

Saltgrass

Saltwort

Ruppia pool

Peat and muck

F E D C B¹ B A

Panne community

High tide

Normal low tide

FIGURE 19-9

A generalized view of a tidal marsh on the northeastern Atlantic Coast of the United States. Note the braiding of the tidal streams and the zonation of marsh vegetation, reflecting the gradients of water depth and salinity. (A) Zone of salt-water cord grass; (B) marsh-hay cord grass; (B'), saltwort and salt grass; (C) black grass, Juncus; (D) grassland—switch grass, fresh-water cord grass; (E) shrub zone; (F) forest. For details, see text. (Drawing based in part on data from Miller and Egler, 1950; Chapman, 1960; and author's photographs and sketches.)

FIGURE 19-8 [OPPOSITE]

Approximate ranges of species dominance among fresh-water and salt marsh plants in relation to water and salinity. (Adapted from Water, USDA Yearbook of Agriculture, 1955, p. 450.)

FIGURE 19-11

A closer view of the floor of the salt marsh. Note the absence of debris, the holes of the fiddler crab, and the new growth of glasswort. (Photo by author.)

FIGURE 19-10

A view of a salt marsh on the Virginia coast showing the pattern of salt-marsh vegetation. The mosaic is influenced by both water depth and salinity. The three dominant grasses are the coarse salt-marsh cord grass, Spartina alterniflora, the fine-leafed Spartina patens, salt-marsh hay grass, and Distichlis spicata, spike grass. (Photo by author.)

rotten-spot pan, caused by the death of small patches of vegetation from one cause or another. Perhaps it may be due to small depressions, which result in the eventual waterlogging of the soil, or it may be due to patches of trash that remain behind after snowmelt in the spring. Retardation of vegetative growth under these patches induces further accumulation in the same spot the following winter. Eventually the vegetation is completely killed and a bare area results (see Miller and Egler, 1950; Chapman, 1960).

Pans support a distinctive vegetation, which varies with the depth of the water and salt concentration. Pools with a firm bottom and sufficient depth to retain tidal water support dense growths of wigeon grass, *Ruppia maritima*, with long, threadlike leaves and small, black, triangular seeds relished by waterfowl (Fig. 19-9). The pools are usually surrounded by those forbs such as sea lavender that add so much color to the sea coast. Shallow depressions in which water evaporates are covered with a heavy algal crust and crystallized salt. At best, they support only saltwort, pale green in summer and scarlet in fall. Some such pans are occasionally invaded by forbs and a yellow, stunted growth of salt-water cord grass, considered by some botanists to be a variety of the species occupying the first major zone of estuarine vegetation (see Chapman, 1960).

Ecologically, the estuary, its sounds, creeks, marshes, and mud and sand flats are all one production unit. The marshes produce an excess of organic material. In fact less than 5 percent of the net production of a *Spartina* marsh is consumed by insects and other herbivores; and two-thirds of that consumed by the major herbivore, the salt-marsh grasshopper, passes through the gut undigested and available to other organisms (Smalley, 1960). Excess organic matter in the marsh is transported into estuarine waters, where it is available to a whole host of decomposers and other detritus feeders, such as the horse mussel, important in the phosphorus cycle. While filtering water to obtain its food, the mussel excretes large quantities of organic particles as pseudofeces, which sink to the bottom (Kuenzler, 1961).

As soon as organic debris is buried more than a few millimeters, either through the movements of animals or by further deposition on top, an anaerobic environment is created. Here bacteria and nematodes live on organic matter, utilizing it by parallel oxidizing and reducing reactions (Teal and Kanwisher, 1961). This results in the accumulation of such end products as methane, hydrogen sulfide, and ferrous compounds. Increasing degrees of reduction suppress biological activity. In fact if the bacteria of the mud are supplied with oxygen, their rate of energy degradation increases 25 times. Thus the tidal marsh is a horizontally stratified system in which free oxygen is abundant in the surface and absent in the mud. Between these two extremes is a zone of diffusion and mixing of oxygen.

The exposed banks of tidal creeks that braid through the salt marshes support a dense population of mud algae, the diatoms and dinoflagellates, photosynthetically active all year. Photosynthesis in summer is highest during high tides; in winter it is highest during low tides, when the sun warms the sediments (Pomeroy, 1959). Some of the algae are washed out at ebb tide and become a temporary part of the estuarine plankton, available to such filter feeders as the oysters. Thus the salt marsh functions both as a source of food and of fertilizer for the estuary.

MANGROVE SWAMPS

Mangrove swamps replace tidal marshes in tropical regions, and in North America they reach their best development along the southwest coast of Florida and southward (Fig. 19-12). As with the salt marsh, the vegetation is influenced by salinity and the tides. The pioneering red mangrove colonizes the submerged soft-bottomed shoals protected from the full beat of the surf. It has a peculiar system of branching prop roots that extend downward like stilts from the trunks and lower branches. The seeds germinate on the tree to form club-shaped hypocotyls 30 cm long. They hang by two cotyledons until they drop into the water or the mud below. If the mud is deep enough, the seedling may take root. If not, the seedling rises to the surface and for a while floats horizontally, often drifting with the wind and tidal currents far from its point of origin. Eventually the root end becomes waterlogged and the seedling floats in a vertical position until the tip of the root touches bottom in shallow water. If the bottom is soft, the roots form rapidly, the young plant is anchored, and a new mangrove blossoms above water. As the plant grows, new prop roots sprout from the limbs and grow down to the water. The tangle of roots block tidal currents, which drop their load of organic debris among the mangroves. This, together with falling leaves and the droppings of birds, gradually builds up the soil to high-tide level (J. H. Davis, 1940).

At the upper limits of high tide, the black or honey mangrove replaces the red. This mangrove does best in a sandy or less organic soil and tolerates a shorter period of flooding. It has pencil-like pneumatophores rising through the substrate from shallow, horizontal roots. The flowers produce abundant nectar used by bees, and the leaves exude excess salts on the leaf surface, where it is washed away by the rain. Behind the black mangrove and on firm ground at the edge of the tide line may be a zone of white mangrove. The white mangrove also possesses pneumatophores, but they are fewer in number and smaller. Above the normal high tides grows a relative of the white mangrove, the button mangrove, with loose bark and twisted and often prostrate trunk.

Associated with the mangroves are a number of vines and shrubs. The ivy, *Cissis*, climbs through the mangrove crowns and sends long, cordlike, aerial roots down to the ground. A spreading, succulent-leafed shrub, the saltwort, grows on the wet, salty mud.

Birds are the most colorful and conspicuous inhabitants of the mangroves. In the water about the feet of mangrove roots and in the shallow mangrove pools live an assortment of small fish. Fiddler crabs crawl through the tangle of mud and roots, and small coon oysters grow attached to the roots.

IMPACT OF MAN

Because they are semienclosed, provide natural harbors, and act as doorways to inland commerce, estuaries have long been sites for the development and growth of cities and industrial activities. Because they receive the drainage from vast inland areas, estuaries accept all pollution—from chemicals, pesticides, and sewage to silt from mining, agriculture, and lumbering. The commercial and recreational value of estuaries stimulates bulkheading, dredging, and filling to create waterfront, industrial, recreational, and residential sites. Thermal power plants pour heated effluents into the water, raising the temperature. All of

FIGURE 19-12
Mangrove swamps replace tidal marshes in the tropical regions. (Photo by author.)

these activities change the nature of the estuary and the circulation of estuarine water. In doing so they destroy both life in the estuary and the estuary as an environment for life. Oyster beds disappear, and fish are no longer able to utilize the estuary as a spawning ground and nursery.

Regarded by man as wasteland, the associated tidal marshes have been ditched, drained, and filled for real estate and industrial development, or used as garbage dumps. Others have been ditched for mosquito control. Any ditching destroys the normal drainage pattern and may produce the conditions it was designed to destroy (Miller and Egler, 1950). Loss of tidal marshes is the greatest in urban and suburban areas along the northeast coast. From 1954 to 1969 New York lost 30 percent of its wetlands, Connecticut 22 percent, and Maryland, 10 percent. Continuing destruction of the estuary and the tidal marsh can only hasten the disappearance of wetland wildlife, a number of commercially important fish, and other offshore and estuarine life.

The seashore

Where the edge of the land meets the edge of the sea there exists the fascinating and complex world of the seashore. Rocky, sandy, muddy, protected, exposed to the pounding of incoming swells, all shores have one feature in common: they are alternatingly exposed and submerged by the tides. Roughly, the region of the seashore is bounded on one side by the height of the extreme high tides and on the other by the height of extreme low tides. Within these confines, conditions change from hour to hour with the ebb and flow of the tides. At flood tide the seashore is a water world; at ebb tide it belongs to the terrestrial environment, with its extremes in temperature, moisture, and solar radiation. In spite of all this, the seashore inhabitants are essentially marine, adapted to withstand some degree of exposure to the air for varying periods of time.

THE ROCKY SHORE

As the sea recedes at ebb tide, rocks glistening and dripping with water begin to appear. Life hidden by tidal water emerges into the open air, layer by layer. The uppermost layers of life are exposed to the air, wide temperature fluctua-

tions, intense solar radiation, and desiccation for a considerable period of time, while the lowest fringes on the intertidal shore may be exposed only briefly before the flood tide submerges them again. These varying conditions result in one of the most striking features of the rocky shore, the zonation of life (Fig. 19-13). Although this zonation may be strikingly different from place to place as a result of local variations in aspect, nature of the substrate, wave action, light intensity, shore profile, exposure to prevailing winds, climatic differences, and the like, it possesses everywhere the same general features. All rocky shores have three basic zones, characterized by the dominant organisms occupying them (Fig. 19-14).

Where the land ends and seashore begins is difficult to fix. The approach to a rocky shore from the landward side is marked by a gradual transition from lichens and other land plants, the maritime zone, to marine life, dependent in part at least on the tidal waters. The first major change from land shows up on the supralittoral fringe (Fig. 19-15), where the salt water comes only every fortnight on the spring tides. It is marked by the black zone, a patchlike or beltlike encrustation of Verrucaria-type lichens and Myxophyceae algae such as *Calothrix* and *Entrophsalis*. Capable of existing under conditions so difficult that few other plants could exist, these blue-green algae, enclosed in slimy, gelatinous sheaths, and their associated lichens represent an essentially nonmarine community, on which graze basically marine animals, the periwinkles (Doty, in Hedgpeth, 1957). Common to this black zone is the rough periwinkle that grazes on the wet algae covering the rocks. On European shores lives a similarly adapted species, the rock periwinkle, the most highly resistant to desiccation of all the shore animals.

Below the black zone lies the littoral, a region covered and uncovered daily by the tides. It is universally characterized by the barnacles, although often they are hidden under a dense growth of fucoid seaweeds or kelp (a brown alga, Phaeophyceae) and in the more northern reaches of the North American and European coasts by the red algae (Rhodophyceae). The littoral tends to be divided into subzones. In the upper reaches the barnacles are most abundant. The oyster, blue mussel, and limpet appear in the middle and lower portions of the littoral, as does the common periwinkle, the second of the three periwinkles common to the rocky shore.

Of all the animals of the littoral, the common periwinkle, which may range from the high water of neap tides to the low water of spring tides, is the most adaptable. It may live on bare rock, among barnacles, among seaweeds, and on soft mud; it is at home equally on shores exposed to open sea and in sheltered bays.

Barnacles (Figs. 19-15 and 19-16) are the most distinctive organisms of the littoral. At low tide their chiseled limestone whiteness forms a distinctive belt across the rocky shore. Two kinds are involved, *Chthamalus*, which occupies the upper limits of the littoral, and the larger *Balanus*, which lives lower downshore (see Chapter 12). Barnacles are sedentary crustaceans, glued to stones and dependent upon tides to bring them their sustenance. They are cone-shaped, with six fitted plates for their sides and with a door of four plates covering the top. At low tide, the door plates are closed, which protects the barnacles from drying. As the first tidal waters flow over them, the doors open; and the barnacles thrust out six pairs of branched, slender, wandlike appendages to sweep in the diatoms and other microscopic life in the water.

Found among the barnacles is the limpet, a rock-clinging mollusk that is able to move freely about. The limpets, in common with the barnacles, are cone-shaped, but they grow much larger. And like their associates they, too, lead a planktonic existence before they settle on the shore. When the tide is in, they move about from their home "base" to browse on the algae covering the rock. This they efficiently remove by their scraping organ, the radula; in fact they may ingest particles of rock and leave striations on the stone. Their grazing area, which extends about 3 ft from "home," is kept nearly barren of algae. If the limpets are removed, the surface of the stone is soon colonized by *Entromorpha*, *Porphyra*, and *Fucus* algae. When it has finished grazing, the limpet by some means returns to its "home," where it clings during ebb tide.

The dog whelk, a highly colored predatory snail about the size of the common periwinkle, moves about on the intertidal shore to prey on mussels, barnacles, and to a limited extent, the browsing snails. Diet greatly influences the color of whelks. Those feeding primarily on

FIGURE 19-13
The rocky shore of the Bay of Fundy shows the broad zones of life exposed at low tide. Note the heavy growth of bladder wrack or rockweed on the lower portion, the white zone of barnacles above. (Photo by author.)

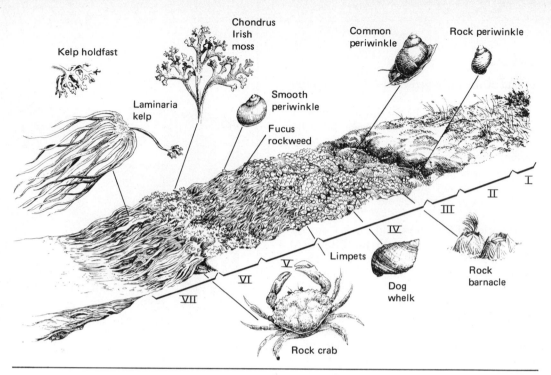

Kelp holdfast

Chondrus
Irish
moss

Laminaria
kelp

Smooth
periwinkle

Fucus
rockweed

Common
periwinkle

Rock periwinkle

I

II

III

IV

Limpets

Dog
whelk

Rock
barnacle

V

VI

VII

Rock crab

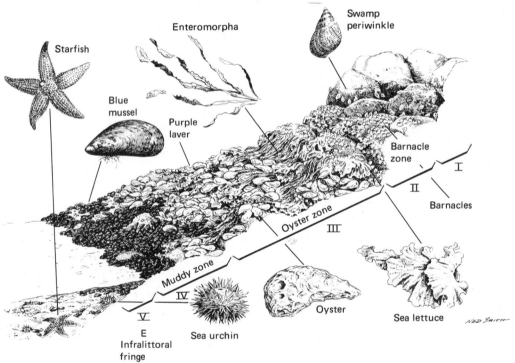

Starfish

Enteromorpha

Swamp
periwinkle

Blue
mussel

Purple
laver

Barnacle
zone

I

II

Barnacles

Oyster zone

III

Muddy zone

IV

Sea urchin

Oyster

Sea lettuce

NED SMITH

V

E
Infralittoral
fringe

mussels have brown, black, and mauve-pink
shells; those feeding on barnacles have white
shells.

Occupying the lower half of the littoral zone
of colder climates and in places overlying the
barnacles are an ancient group of plants, the
brown algae, more commonly known as rockweeds, or wrack (*Fucus*). Rockweeds attain

their finest growth on protected shores, where
they may grow 7 ft long; on wave-whipped
shores they are considerably shorter. The rockweeds that live furthest up on the tidelands are
channeled rockweeds, or wrack, a species found
on the European but not the American shore.
It is replaced here by another, the spiral rockweed, a low-growing, orange-brown alga, whose

short, heavy fronds end in turgid, roundish swellings. The spiral rockweeds on the northeastern coast form only a very narrow band, if present at all, for it is replaced by the more abundant bladder rockweed and, in sheltered waters only, the knotted rockweed. The bladder rockweed can withstand the heavy surf best because it is shorter, has a strengthening midrib, and possesses great tensile strength. Both the bladder and knotted rockweeds have gas-filled bladders, which tend to buoy them up at high tide. Rockweeds have no roots—they draw their nutrients from the seawater surrounding them—but they cling to the rocks by means of disk-shaped holdfasts.

At low tide the rockweeds lie draped over the rocks. Limpets and smooth periwinkles browse on the fronds, while underneath lie starfish and young green crabs. In the sediments live scarlet ribbon worms, minute bristle worms, and the predaceous nereids, or clam worms. Minute copepods and mites are abundant. On the fronds of rockweeds are attached encrusting animals, such as the flowerlike, pink-colored hydroid *Clava*. Also attached to the fronds of rockweed are the coiled tubes of the small polychaete worm, *Spirorbis*.

The lower reaches of the littoral zone may be occupied by blue mussels (Fig. 19-16) instead of rockweeds. This is particularly true on shores where the hard surfaces have been covered in part by sand and mud. No other shore animal grows in such abundance; the blue-black shells packed closely together may blanket the area (Fig. 19-17). Although mussels, in common with other bivalves, possess a flattened foot, it is not suited for gripping. So this mollusk attaches itself to the substrate by a byssus, the threads of which are formed initially as a thick fluid by a gland in the foot.

Near the lower reaches of the littoral zone, mussels may grow in association with a red alga, *Gigartina*, a low-growing, carpetlike plant. The algae and the mussels together often form a tight mat over the rocks. Here, too, grows another seaweed, Irish moss, which grows in carpets some 6 in. deep. Its color is variable, ranging from purple to yellow and green; its fronds are branched and tough (Fig. 19-15), and when covered by water the plant has an iridescent sheen. Here, well protected in the dense growth from the waves, live infant starfish, sea urchins, brittle stars, and the bryozoan sea mats or sea lace, *Membranipora*. These bryozoans are almost microscopic, tentacled,

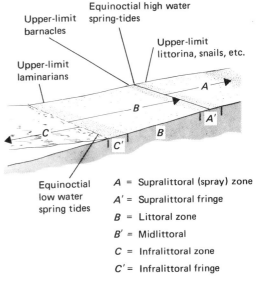

FIGURE 19-14
A diagram of the basic zones on a rocky shore. Use this as a guide when studying the subsequent drawings of zonation on rocky shores. (Adapted from Stephenson, 1949.)

colonial animals, which form thin films in which each chamber is visible, giving the colony the appearance of lace.

The lowest part of the littoral zone, uncovered only at the spring tides and not even then if the wave action is strong, is the sublittoral fringe. This zone, exposed for relatively short periods of time, consists of forests of the large brown alga, *Laminaria*, one of the kelps (Fig. 19-15), with a rich undergrowth of smaller plants and animals among the holdfasts. Limpets, horse mussels, and other bivalves are abundant here, together with encrusting organisms, copepods, isopods, amphipods, sea squirts, crabs, segmented bristle worms, and starfish. Starfish wait out the low tide beneath the seaweed, beneath rocks, and in pools. At high tide they range out over the mussel beds to feed on the bivalves. Moving onto the mussel, the starfish humps its body and grips the valves of the shells with its tube feet. By continuously pulling on the valves and gradually inserting its thin stomach into the mussel, the shell finally opens. The starfish then digests the bivalve externally with its protruding stomach. A related echinoderm, the globular, spiny sea urchin also is common in the sublittoral fringe.

Beyond the sublittoral fringe is the sublittoral zone, the open sea. This zone is principally neretic and benthic and contains a wide variety of fauna, depending upon the substrate, the presence of protruding rocks, gradients in turbulence, oxygen tensions, light, and temperature. This offshore zone is extremely interesting to explore by skin diving. Appropriately, this area is becoming known to marine biologists as the scuba zone.

VARIATIONS IN ZONATION

The zonation just described is essentially that of the North Atlantic Coast north of Cape Cod. This coast is duplicated in part by the rocky coasts of the British Isles, for 85 percent of the plants and 65 percent of the animals are shared in common (Stephenson and Stephenson, 1949). Although adhering to the same general pattern, the details of zonation vary on other rocky shores wherever they exist. This variation in eastern North America is due to temperature influenced by latitude and to that great river of the sea, the Gulf Stream.

Southern Atlantic coast. From the Florida Straits to Cape Hatteras, the Gulf Stream fol-

lows the edge of the continental shelf. On the east coast of Florida the shelf is narrow; where the waters flow close to shore, especially in extreme southern Florida and the Keys, tropical fauna flourish. North of Cape Kennedy, the Gulf Stream swings out from the coast, but at Cape Hatteras in the Carolinas the shelf narrows and the Gulf Stream comes closer to shore. This area marks an indefinite temperature barrier whose effects are more pronounced in winter than in summer. During the summer, the fauna of these warm temperate waters range north to Cape Cod. A huge sandspit jutting out into sea, Cape Cod deflects the Gulf Stream eastward away from North America and cups the cold Laborador currents in its northern shores (a fact well recognized by anyone who has plunged into the waters along both shores of the Cape in summer). Because distribution of life in the seas is strongly influenced by temperature, Cape Cod marks the northernmost distribution of many warm, temperate-water species and the southern-most distribution of arctic species. Within recent years even this has been changing, for with a general warming trend occurring in the arctic waters, southern species such as the green crab and the mantis shrimp are extending their range northward. Cape Cod also marks a radical change in the substrate of the shores from sand to rock. Thus from Cape Cod northward the intertidal regions are characterized by a heavy growth of seaweeds, vast numbers of periwinkles, and millions of barnacles and mussels.

South of the Cape, rocky shores are scarce, but where a rock substrate does exist, as in jetties and breakwaters, rocky-shore fauna appear, although the composition of the community differs somewhat from that of the northern coast. In the warm, temperate waters from Cape Kennedy to Cape Hatteras, some species appear and disappear on a north–south gradient (Stephenson and Stephenson, 1952). The shore populations include tropical forms distributed as far north as North Carolina; the cold-water species extend somewhat south of Hatteras. Mixed with these two are the more cosmopolitan species that occupy longer or shorter stretches of the Atlantic Coast from Cape Cod to Cape Kennedy. Within this wide range of coastline, seasonal variations exist, with a strengthening of the southern element in summer, the northern element in winter. But in spite of the physical and biological variations, the intertidal zone exhibits the same common features. The supralittoral fringe possesses the typical black zone of encrusting algae and lichens, including such species as *Enterophysalis deusta* and *Calothrix pulvenata*, but the periwinkles are extremely scarce, a sharp contrast to the northern rocky coast (Fig. 20-17). Faunal inhabitants include the rapid isopod, or sea roach, ghost crabs, and fiddler crabs. The littoral is characterized by a barnacle zone, occupied in the upper limits by *Chthamalus* and on the seaward side by the larger *Balanus*. Below, but overlapping the barnacles, is a belt of oysters, mussels, and a red seaweed, the ribbonlike purple laver (Fig. 19-17). Here too grow the green seaweeds, *Ulva* and *Enteromorpha*, and the brown bladder rockweed. Beyond this is a zone of mud claimed by the mussels, mixed algae, the orange sponge, limpets, predaceous whelks, the anemones, and sea urchins.

The rocky, clinkerlike coraline shores of the Florida Keys is a tropical world, yet even here the major intertidal zones are discernible (Stephenson, 1950, 1971). The supralittoral fringe is especially well developed and is divided into subzones, a white, a gray, and a black. The white zone is the meeting place of land and sea and is occupied by such maritime species as the purple-clawed hermit crab, rapid crabs, and the sea roach. The gray zone occupies rocks wetted only part of the time, often only during the high waters of spring tides. This is the first zone of macroscopic marine algae, characterized by the mosslike *Bostrychia*. The periwinkle *Littorina ziczac*, the beaded periwinkle *Tectarius*, and other small, boldly colored snails graze on the algal film. The black zone claims areas completely submerged by the high waters of spring tides and is coated with an encrusting growth of the blue-green algae *Entyophysalis* and *Brachytrichia*, utilized by the periwinkles. Below the black zone is the littoral, characteristically yellowish in color. Only in the upper littoral do the barnacles live, and they vary in numbers from complete absence to crowded colonies. They include the small barnacle *Chthamalus stellatus* and the large barnacle *Tetraclita squamosa*, which occupies the lower barnacle zone. Associated with the barnacles are the neritid snails, the ribbed mussel, chitons, and the everpresent predatory whelks. Large algae, so characteristic of the littoral of the northern shores, are absent; the seaweeds here are short, mossy, turflike, or encrusting. The lower part of the

FIGURE 19-17
A close-up view of a blue mussel bed. (Photo by author.)

647

littoral is inhabited by two dominant colonial organisms, sheets of small, bubblelike green alga, *Valonia*, and masses of sedentary tube-dwelling gastropods, *Spiroglyphus*. Below the yellow zone lies the lower reef, the most distinctive feature of which is the extensive growth of the yellow, carpetlike alga, *Laurencia*. Beyond this are the reef flats, a world of corals rich in many marine plants and animals. This, however, is not a part of the intertidal region.

Pacific coast. The intertidal regions of the Atlantic Coast of North America fall into tropical, warm temperate, and cold temperate regions. On the intertidal Pacific Coast, these regions are obscured by complex arrangements of currents and water masses. As a result, the intertidal flora and fauna possess many common features from Alaska to the outer coast of California. Typical zonation, suggestive of the cold temperate Atlantic Coast, exists. There is a black, supralittoral fringe occupied by the periwinkle, *Littorina scutulata*. The littoral is inhabited by barnacles, the upper zone by *Chthamalus*, the lower by *Balanus*. The lower littoral supports both mussels and kelp. The growth of kelp often is prolific, but the conspicuous east coast species, the bladder, serrated, and spiral rockweeds, are missing. The Pacific Coast rockweeds lack bladders and serrations, and all are hermaphroditic. They vary widely in size and general appearance, in the shape of the receptacles, and in habitat, yet taxonomically they still appear to represent one species, *Fucus distichus*, also found along the Atlantic Coast (Stephenson and Stephenson, 1961). The sublittoral possesses two subzones, a strip of bare rocks and below this a zone of mossy, beardlike algae and a colorful array of large starfish.

Perhaps the most striking feature of the Pacific Coast intertidal region is the contrast between north-facing and south-facing shores (Stephenson and Stephenson, 1961). The supralittoral fringe on the north-facing shore is narrower than on the south and is located much higher on the shore. There is also a correspondingly higher shift in the barnacle zone, which is much deeper on the north-facing shores. The barnacle zone on the south-facing shores involves two genera, an upper zone of *Chthamalus* and a lower *Balanus* zone. *Chthamalus* is missing entirely on north-facing shores. The main zone of kelp is up to 4 ft higher on the north-facing coast, and its sublittoral has a somewhat different faunal composition than

the south-facing shores. Particularly characteristic of the north is the tube worm, *Serpula vermicularis*, a weak fringe of kelp, and a rich group of echinoderms and anemones, all of which seemingly replace the band of short, carpetlike algae of the south-facing shore.

TIDE POOLS

When the tide flows out it leaves behind (in rock crevices, in rocky basins, and in depressions) pools of water that dot the rocky shores (Fig. 19-18). These are tide pools, "microcosms of the sea," as Yonge (1949) describes them. They represent distinct habitats, which differ considerably from the exposed rock and the open sea, and even differ among themselves. At low tide all the pools are subject to wide and sudden fluctuations in temperature and salinity, but these changes are most marked in shallow pools. Under the summer sun the temperature may rise above the maximum many organisms can tolerate. As the water evaporates, especially in the smaller and more shallow pools, salinity increases and salt crystals may appear around the edges. When rain or land drainage brings fresh water to the pool, salinity may decrease. In deep pools such fresh water tends to form a layer on the top, developing a strong salinity stratification in which the bottom layer and its inhabitants are little affected. Obviously pools near low tide are influenced least by the rise and fall of the tides; those that lie near and above the high-tide line are exposed the longest and undergo the widest fluctuations. Some may be recharged with seawater only by the splash from breaking waves or occasional high spring tides. Regardless of their position on the shore, most pools suddenly return to sea conditions on the rising tide and experience drastic and instantaneous changes in temperature, salinity, and pH. Life in the tidal pools must be able to withstand wide and rapid fluctuations in their environment.

Tidal pools are fascinating places, one as interesting as the other, for often they differ in the organisms they contain. Small pools high up on the shore and subject to rapid evaporation may hold little more than the reflection of the sky or perhaps the little copepod *Tigriopus*, which can survive the high temperatures and salinities. Pools of the upper shore may have thin, pink encrustations of a calcareous alga, *Lithothamnion*, or support a bright-

green growth of a tubelike alga, *Entromorpha*.
Or they may be lined with a crustose growth
of a brown alga, *Ralfsia*, which holds a wide
variety of microscopic life. Periwinkles may
live in the pools, as well as barnacles and
mussels on whose shell may grow threadlike
colonies of hydroids. The common seashore
insect, *Anurida*, runs on the surface film. Like
some of its fresh-water counterparts, this insect
holds a film of air about its body when it enters
the water. This air bubble enables the insect
to remain submerged during high tide, ready to
roam over the rocks at low tide in search of
dead animal matter, on which it feeds. Pools
on the lower shore may contain more diverse
and more colorful forms—green crumb-of-bread
sponges, which obtain their green color from
symbiotic algae scattered through the tissues,
yellow sponges, mussels, starfish, and anemones.
One small tidal pool investigated contained
5,185 individuals representing 45 macroscopic
species (Bobvjerg and Glynn, 1960).

All this life in the pools adds still another
dimension to the environment—oxygen balance. If algal growth is considerable, oxygen
content of the water varies through the day.
Oxygen will be high during the daylight hours
but will be low at night, a situation that rarely
occurs at sea. The rise of carbon dioxide at
night means a lowering of pH. Only organisms
well adapted to these extreme short-term fluctuations in the environment can colonize the
pools.

LOOSE ROCKS AND BOULDERS

The rocky shore represents one extreme of the
intertidal habitat; the sandy beach, soon to be
discussed, represents the other. Lying between
these two are those shores covered with loose
stones and small boulders mixed in a matrix of
mud and sand. Like the undersides of stream-
bottom stones, the rocks and boulders of the
seashore also support an abundance of life
(Fig. 19-19). Stones embedded in sand and
gravel on the lower zones are subject to abrasion great enough to prevent the growth of
algae and barnacles. But unless buried in the
mud, the undersides of these stones support
populations of those organisms requiring protection from sunlight, desiccation, and violent
water movements. These animals live in almost
perpetually calm seawater, which remains behind long after the tide has gone out. Because
of low illumination, algae are scarce. The fauna

649

(a)

(b)

FIGURE 19-18

*Tidal pools: (a) A rock pool nestled in a canyon
on a rocky shore; (b) A tiny rock pool on the
ledge of a large rock. Note the rockweed, the
barnacles, and the transparency of the water.
(Photos by author.)*

utilize the food suspended in the water about them and organic debris dropped about the stones, or they leave the area by night to feed. The nature of the faunal community depends upon the stability of the rocks. If the rocks are easily overturned by the waves, the populations beneath will die from exposure. If rock movement is greatest in the winter, then the summer population will consist largely of summer annuals. Because small stones of pebble size are so unstable, beaches covered with them are usually barren of life.

The undersurfaces of rocks may support a few periwinkles (Fig. 19-19) and a host of small crustaceans such as the shore skipper *Orchestia* and the amphipod *Gammarus*. Other amphipods live in tubular nests made from the fragments of weeds. The small, coiled *Spirorbis*, so common on seaweeds, also attaches itself to the undersides of stones, as do the serpulid worms with limy tubes and the broad, flattened scale worms. Rock-boring worms, such as species of *Potamilla* and *Polydora*, may also live here. Crabs seek shelter beneath the rocks, together with starfish, sea urchins, and worms of many kinds. Where organic debris has accumulated, the omnivorous bristle worms, *Nereis*, and related genera abound. These worms live in irregular burrows beneath the stones and feed on plant and animal debris. Associated with them and feeding upon them are the ribbon worms, or nemertines. Where organic matter has accumulated to such a point that it creates a rich and odoriferous muck, the substrate beneath the stones may be covered with thin, twisting masses of redthreads. Examining the undersurfaces of stones on the intertidal beach can be rewarding; only be sure to return the stones to their original positions if the small communities they support are to be preserved.

SANDY SHORES

The sandy shore at low tide appears barren of life, a sharp contrast to the life-studded rocky shores. But the beach is not as dead as it seems, for beneath the wet and glistening sand, life exists, waiting for the next high tide.

The sandy shore in some ways is a harsh environment; indeed the very matrix of this seaside environment is a product of the harsh and relentless weathering of rock, both inland and along the shore. Through eons the ultimate products of rock weathering are carried away by rivers and waves to be deposited as sand along the edge of the sea. The size of the sand particles deposited influences the nature of the sandy beach, water retention during low tide, and the ability of animals to burrow through it. Beaches with relatively steep slopes usually are made up of larger sand grains and are subject to more wave action. Beaches exposed to high waves are generally flattened, for much of the material is transported away from the beach to deeper water, and left behind is fine sand (Fig. 19-20). Sand grains of all sizes, especially the finer particles in which capillary action is greatest, are more or less cushioned by a film of water about them, reducing further wearing away. The retention of water by the sand at low tide is one of the outstanding environmental features of the sandy shore.

Existence of life on the surface of the sand is almost impossible. It provides no surface for attachments of seaweeds and their associated fauna; and the crabs, worms, and snails so characteristic of rock crevices find no protection here. Life then is forced to exist beneath the sand (Fig. 19-21).

Life on the sandy beach does not experience the same violent fluctuations in temperature as that of the rocky shores. Although the surface temperature of the sand at midday may be 10° C or more higher than the returning seawater, the temperature a few inches below remains almost constant throughout the year. Nor is there a violent fluctuation in salinity, even when fresh water runs over the surface of the sand. Below 10 in., salinity is little affected. Organic matter accumulates within the sand, especially in sheltered areas. This detritus offers food for some inhabitants of the sandy beach, but where it accumulates in large amounts, it prevents the free circulation of water. A region of stagnation and oxygen deficiency results, characterized by anaerobic bacteria and by the formation of ferrous sulfides. The iron sulfides cause a zone of blackening, the depth of which varies with the exposure of the beach.

Most animals of the sandy beach either occupy permanent or semipermanent tubes within the sand, or they can burrow rapidly into the sand when they need to do so. Except for a few beach hoppers and a few plant-feeding insects, the true sand dwellers feed on particulate and suspended organic matter carried to the beach by the tides or supplied by the decomposition of other animal and plant debris on the beach. For this reason, sandy shores containing the greater amounts of organic matter support more life. The multicellular invertebrates of the beach obtain oxygen either by gaseous exchange with the water through the outer covering or by breathing through gills and elaborate respiratory siphons.

Zonation of animal life exists on the beach but is hidden and must be discovered by digging. Based on the animal organisms, the sandy shores can be divided roughly into supralittoral, littoral, and sublittoral zones, but a universal pattern similar to that for the rocky shores is lacking (Dahl, 1953).

Pale, sand-colored ghost crabs (Fig. 19-21) and beach hoppers, scavengers of the drift line, occupy the upper beach, the supralittoral. Like the periwinkles of the upper rocky shores, these animals are more terrestrial than marine. The beach flea, or sand hopper, too, is a marine form that has nearly forsaken the sea. It seldom, if ever, enters the water, and if submerged for very long it may drown; yet it requires the dampness and salinity of the shore.

The intertidal beach, the littoral, is the zone where true marine life really appears. Although it lacks the variety found in the intertidal rocky shores, the populations of individual species often are enormous. The energy base for sandy-beach fauna is organic matter, much of it made available by bacterial decomposition, which goes on at the greatest rate at low tide. The bacteria are concentrated about the organic matter in the sand, where they escape the diluting effects of water. The products of decomposition are dissolved and washed into the sea by each high tide, which in turn brings in more organic matter for decomposition. Thus the sand beach is an important site of biogeochemical cycling, supplying the offshore waters with phosphates, nitrogen, and other compounds.

Some animals of the sandy beach obtain their food by actively burrowing through the sand and ingesting the substrate to obtain the organic matter it contains. The commonest of all these sandworms is the lugworm (Fig. 19-21), responsible for the conspicuous coiled and cone-shaped casts on the beach.

Two animals that sort through the sands for food are the sand-burrowing amphipod *Haustorius* and the ghost shrimp (Fig. 19-21), but the methods they use are different. *Haustorius* swims upside down in the tidal waters like fairy shrimp but rights itself to burrow in the sand. Here it feeds on minute organic

FIGURE 19-19

The underside of a small boulder exposed at low tide. Note the barnacles, the periwinkles, the dog whelk. Scud were abundant in the water beneath the stone. (Photo by author.)

FIGURE 19-20

A long stretch of sandy beach pounded by the surf. Although the beach appears barren, life is abundant just beneath the sand. (Photo by author.)

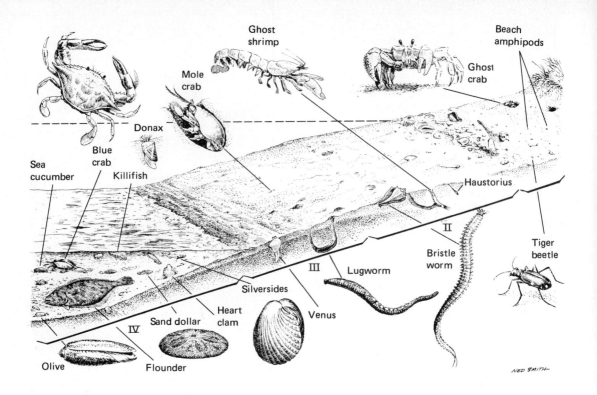

Ghost shrimp

Mole crab

Beach amphipods

Ghost crab

Donax

Blue crab

Sea cucumber

Killifish

Haustorius

II

Bristle worm

Tiger beetle

III

Lugworm

Silversides

Heart clam

Venus

IV

Sand dollar

Olive

Flounder

NED SMITH

particles clinging to the sand grains and floating between them. The ghost shrimp found on the low-tide end of the littoral zone is a burrower. It excavates a vertical burrow several feet deep in the sand, where it stays at low tide. At high tide, the shrimp moves up to the mouth of the burrow and sifts through the sand grains for organic detritus, bacteria, and diatoms.

Other sandy beach animals obtain their food by sorting and filtering particles of organic matter from the tidal waters. Two of these are "surf-fishers," which advance and retreat up and down the beach with the flow and ebb of the tide. One is the mole crab, which lies partially buried within the sand and allows the surf to roll over it. Then, as the surf subsides and the waves recede, it peers through the sand and extends its antennae into the current. It then draws the antennae through the feeding appendages to remove the captured food. As the tide advances, whole populations of mole crabs will leave the sand, ride the surf up the beach, and then quickly burrow into the sand as the waves fall back. The other is a bivalve, *Donax*, the coquina clam, found on the Atlantic, Gulf, and Pacific coasts. At flood tide, the coquina emerges from the sand and is carried shoreward. As the waves lose force, the coquina, digging into the substrate with its stout foot, buries into the sand. Once settled, it extends its divided siphon to the surface to

pick up detritus and to discharge wastes as pseudofeces. As the tide recedes, the clam moves seaward with the outgoing current and spends the low tide in the sublittoral. This movement with the tides is characteristic of coquina of the Atlantic and Gulf coasts. The Pacific coast species appears to be more sedentary (Hedgpeth, 1957).

The sublittoral zone and the sublittoral fringe are inhabited by still other bivalves and clams. One of the most characteristic is the heart clam, or common cockle, which may occur in tremendous numbers. It is globular in shape, with conspicuous radial ridges on the shell and a large powerful foot. Like the coquina, this clam strains out microscopic phytoplankton. It moves freely across the surface but can burrow rapidly into the sand with the aid of its powerful foot.

Associated with these essentially herbivorous animals are the predators, ever present whether the tide is in or out. Near and below the low-tide line live predatory gastropods, which, like the dog whelks of the rocky shore, prey on the bivalves that they hunt beneath the sand. In the same area lurk the predatory portunid crabs, such as the blue crab (Fig. 19-21) and the green, which feed on mole crabs, clams, and other organisms. They, like the snails, move back and forth with the tides. The incoming tides bring with them other predators, such as the killifish and silversides. As the tides recede, gulls and shore birds scurry across the sand to snatch up or probe the sand for food.

Within the beach sand live vast numbers of microscopic copepods, ostracods, nematodes, and gastrotrichs, all making up the interstitial life (Pennak and Zinn, 1943; Pennak, 1951). The interstitial fauna are in general elongated forms with the setae, spines, or tubercles greatly reduced. The great majority do not have pelagic larval stages. These animals feed largely on microscopic algae, bacteria, and detritus. Interstitial life, best developed on more sheltered beaches made up of larger sand particles, shows seasonal variations, reaching its maximum development in the summer months.

An array of other animals can be found just above the low-tide line and in the sublittoral. Among them are the starfish and the related sand dollar (Fig. 19-21). The latter animals, disk-shaped and wafer thin, have supporting pillars inside the shell and a five-pointed-star design faintly marked on the disk. In the sand lives the heart urchin, which inhabits a chamber six or more inches deep. It remains in contact with the surface by a mucus-lined channel. A detritus feeder, it collects organic debris from the sand about the mouth of the channel with modified tube feet, which the urchin extends up through the tunnel.

Although much of the interest on the sandy shores may be directed to beach animals themselves, the drift line should not be ignored. Here one may find strays from the open sea washed up on shore. Common in this tidal debris may be the thin, papery shell of the paper nautilus, actually its elaborate egg case, the horned egg cases of skates, beadlike egg chains of the whelks, and a variety of bivalve shells. Jellyfish may be washed ashore in numbers, and on southern shores they may include the dangerous Portuguese man-of-war. Even pieces of driftwood washed ashore should be examined, for these may harbor such organisms as the gooseneck barnacle and the burrowing shipworms, of which there are many species. More often than not, the shipworms themselves will be gone, but the long, cylindrical tunnels penetrating all parts of the wood remain.

The coral reef

In warm, sunlit subtropical and tropical waters, one finds structures of biological rather than

FIGURE 19-21 [OPPOSITE]
Life on a sandy ocean beach along the Atlantic coast. Although strong zonation is absent, organisms still change on a gradient from land to sea. (I) Supratidal zone: ghost crabs and sand fleas; (II) flat beach zone: ghost shrimp, bristle worms, clams; (III) intratidal zone: clams, lugworms, mole crabs; (IV) subtidal zone. The dotted line indicates high tide.

TABLE 19-2
Division of the ocean into provinces according to their level of primary organic production

Province	Percentage of ocean	Area (km^2)	Mean productivity (g of carbon/ m^2/year)	Total productivity (10^9 ton of carbon/year)
Open ocean	90	326×10^6	50	16.3
Coastal zone[a]	9.9	36×10^6	100	3.6
Upwelling areas	0.1	3.6×10^5	300	0.1
Total				20.0

Source: J. Ryther, 1969. © 1969 American Association for the Advancement of Science.
[a] Includes offshore areas of high productivity.

geological origin, coral reefs. Coral reefs are built by carbonate-secreting organisms of which coral (Coelenterata) may be the most conspicuous but not always the most important organisms. Also contributing heavily are the corraline red algea (*Porolithon*) as well as foraminifera and mollusks. Reefs are formed by intergrown skeletons that withstand both the action of waves and the attack of coral-eating animals. Built only underwater, usually to a maximum depth of tens of meters below low tide, coral reefs need a stable foundation to permit them to grow. Such foundations are provided by shallow continental shelves and submerged volcanos.

There are three kinds of coral reefs with many gradations among them: (1) *fringing reefs*, which grow along the rocky shores of islands and continents; (2) *barrier reefs*, which parallel shore lines along continents; and (3) *atolls*, horseshoe-shaped reefs surrounding a lagoon. Such lagoons are about 40 m deep and are usually connected to the open sea by breaks in the reef. Reefs build up to sea level. To become islands or atolls, the reefs have to be exposed by a lowering of the sea level or be built up by the accumulation of sediments and the piling up of reef material by the action of wind and waves.

Ecologically, coral reefs are interesting for the life they support and the ecosystems they represent. Coral reefs are complex ecosystems involving close relationships between coral and algae. In the tissue of coral live zooxanthellae, endozoic dinoflagellate algae; and on the calcareous skeletons live still other kinds, both filamentous and calcareous. At night the coral polyps feed, extending their tentacles to capture zooplankton from the water, thus securing phosphorus and other elements needed by the coral and its symbiotic algae. During the day the algae absorb sunlight and carry on photosynthesis and directly transfer organic material to coral tissue. Thus nutrients tend to be recycled between the coral and the algae (Johannes, 1967; Pomeroy and Kuenzler, 1969). In addition the algae, by altering the carbon dioxide concentration in animal tissue, enable the coral to extract the calcium carbonate needed to build the skeletons (Goreau, 1963). Living on the coral are filamentous algae (Chlorophyta) that add to the primary production. Thus the symbiotic relationship between algae and coral, as well as the nutrient recharging by tidal waters, makes the coral reef one of the most productive ecosystems in the world.

Reflecting this high productivity and a wide variety of habitats provided by the coral structures is the great diversity of life about the coral reef. Thousands of kinds of exotic invertebrates, some of which feed on the coral animals and algae, hundreds of highly colored herbivorous fish, and a large number of predatory fish that lie in ambush for prey in the caverns that honeycomb the reef swarm about the coral. In addition there is a wide array of symbionts, such as the cleaning fish and crustaceans that pick parasites and detritus from larger fish and invertebrates.

The open sea

Beyond the estuaries and the rocky and sandy shores, the mangrove swamps and coral reefs, lies the open sea (Fig. 19-22). As in lakes, the dominant form of plant life or primary producers is phytoplankton, small microscopic plants drifting with the currents. There is a reason for the smallness of the plants. Surrounded by a chemical medium that contains in varying quantities the nutrients necessary for life, they absorb nutrients directly from the water (see Chapter 5). The smaller the organism, the greater is the surface area exposed for the absorption of nutrients and the utilization of the energy of the sun. The density of seawater is such that there is no need for well-developed supporting structures. In coastal waters and areas of upwelling, phytoplankton 100 μ or more in diameter may be common, but in general plant life is much smaller and widely dispersed.

Because of their requirements for light, phytoplankton are restricted to the upper surface waters, which, determined by depth of light penetration, may vary from tens to hundreds of feet. Because of seasonal, annual, and geographic variations in light, temperature, and nutrients, and grazing by zooplankton, the distribution and species of phytoplankton vary from ocean to ocean and from place to place.

Each ocean and region within an ocean appears to have its own dominant forms. Littoral and neritic waters and regions of upwelling are richer in plankton than mid oceans. In regions of downwelling, the dinoflagellates, a large diverse group characterized by two whiplike flagellae, concentrate near the surface in areas of low turbulence. They attain their greatest abundance in warmer waters. In summer they may so concentrate in the surface waters that they color it red or brown. Often toxic to other marine life, such concentrations of dinoflagellates are responsible for red tides. In regions of upwelling, the dominant forms of phytoplankton are diatoms. Enclosed in a silica case, diatoms are particularly abundant in arctic waters. Smaller than diatoms are the coccolithophoridae, so small they pass through plankton nets (and so are classified as *nanoplankton*). Their minute bodies are protected by calcareous plates or spicules embedded in a gelatinous sheath. Universally distributed in all waters except the polar seas, the Coccolithophoridae possess the ability to swim. They are characterized by droplets of oil, which aid in buoyancy and serve as a storage of food. In the equatorial currents and in shallow seas, the concentration of phytoplankton is variable. Where both lateral and vertical circulation of water is rapid, the composition reflects in part the ability of the species to grow, reproduce, and survive under local conditions.

Because of the impoverished nutrient status

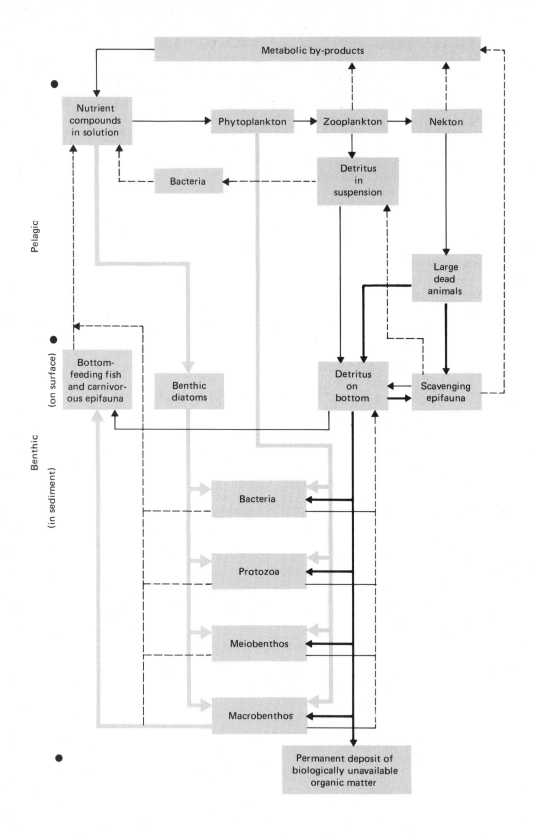

of ocean water, productivity is low. Least productive is the open sea (Table 19-2). The most productive zones are zones of upwelling where phosphorus-enriched cold deep waters come to the surface and shallow coastal waters. In the presence of an adequate nutrient supply, phytoplankton become relatively abundant, utilizing the bulk of the nutrients on the surface. As the phytoplankton die or are consumed by zooplankton, the nutrient supply is depleted. As both phytoplankton and zooplankton settle toward the bottom, nutrients become trapped below the level of light. But in relatively shallow coastal waters and over shallow banks or shelves, light penetrates to greater depths, phytoplankton are distributed through the water column, and the nutrients are more rapidly circulated.

ZOOPLANKTON CONSUMERS

Grazing on the phytoplankton are the herbivorous zooplankton, largely copepods, planktonic arthropods that are the most numerous animals of the sea, and the shrimplike euphausids, commonly known as krill. Other planktonic forms are the larval stages of such organisms as gastropods, oysters, and cephalopods. Feeding on the herbivorous zooplankton are the carnivorous zooplankton, which include such organisms as the larval forms of combjellies (Ctenophora) and arrowworms (Chaetognatha).

Because their food in the ocean is so small and widely dispersed, zooplankton have adapted ways more efficient and less energy-demanding to harvest phytoplankton than filtering water through pores. They have evolved webs, bristles, rakes, combs, cilia, sticky structures, and even bioluminescence.

Like phytoplankton, the composition of zooplankton varies from place to place, season to season, year to year. In general zooplankton fall into two main groups, the larger forms characteristic of shallow coastal waters and the generally smaller forms characteristic of the deeper open ocean. Zooplankton of the continental shelf contain a large proportion of larvae of fish and benthic organisms. They have a greater diversity of species, reflecting a greater diversity of environmental and chemical conditions. The open ocean being more homogenous and nutrient-poor, supports less diverse zooplankton. Zooplankton species of polar waters, having spent the winter in a dormant state in the deep water, rise to the surface dur-

FIGURE 19-22 [OPPOSITE]
A simplified marine food chain. The upper portion represents the pelagic zone, below that the epifauna and bottom-feeding nekton. The bottom of the diagram includes the infauna and marine humus. (After Raymont, 1963. From M. Gross, 1972, Oceanography, Prentice-Hall, Englewood Cliffs, N.J.)

ing short periods of diatom blooms to reproduce. In temperate regions distribution and abundance depend upon temperature conditions. In tropical regions, where temperature is nearly uniform, zooplankton are not so restricted, and reproduction occurs throughout the year.

Also like phytoplankton, zooplankton live mainly at the mercy of the currents. But possessing sufficient swimming power, many forms of zooplankton exercise some control. Most species of zooplankton migrate vertically each day to arrive at a preferred level of light intensity. As darkness falls, zooplankton rapidly rise to the surface to feed on phytoplankton. At dawn the forms move back down to preferred depths.

There are advantages to this migration. By feeding in the darkness of night and hiding in the darkened waters by day, zooplankton avoid heavy predation, and by remaining in cooler water by day, they conserve energy during the resting period. Surface currents move zooplankton away from their daytime location during feeding, but they can return home by countercurrents present in the deeper layers. Response to changing light conditions is useful in another way. As clouds of phytoplankton pass over the water above, zooplankton respond to the shadow of food by moving upward. This takes them out of the deep current drift and nearer the surface. At night they can move directly up to the food-rich surface water. Zooplankton that lack a vertical migration, and even some of those that do, drift out of breeding areas with surface currents. Survival of a breeding population is assured by a complex cycle of seasonal migrations.

NEKTON

Feeding on zooplankton and passing energy along to even higher trophic levels are the small fishes, squids, and ultimately the large carnivores of the sea. Some of the predatory fish, such as herring and tuna, are more or less restricted to the photic zone. Others are found in the deeper mesopelagic and bathypelagic zones. Living in a world that lacks any sort of refuge as a defense against predation or as a site for ambush, inhabitants of the pelagic zone have evolved various means of defense and of securing prey. Among them are the stinging cells of the jellyfish, the remarkable streamlined shapes that allow speed both for escape

and for pursuit, unusual coloration, advanced sonar, a highly developed sense of smell, and a social organization involving schools or packs. Some animals, such as the baleen whale, have specialized structures that permit them to strain krill and other plankton from the water. Others, such as the sperm whale and certain seals, have the ability to dive to great depths to secure food. Phytoplankton light up darkened seas, and fish take advantage of that bioluminescence to detect their prey.

The dimly lighted regions of the mesopelagic and the dark regions of the bathypelagic zone depend upon a rain of detritus as an energy source. Such food is limited. The rate of descent of organic matter, except for larger items, is so slow that it is consumed, decayed, or dissolved before it reaches the deepest water or the bottom. Other sources include saprophytic plankton, which exist in the darker regions, particulate organic matter (see Chapter 5), and import of such material as wastes from the coastal zone, garbage from ships, and large dead animals. Because food is so limited, species are few and the populations low.

Residents of the deep also have special adaptations for securing food. Some, like the zooplankton, swim to the upper surface to feed by night; others remain in the dimly lighted or dark waters. Darkly pigmented and weakbodied, many of the deep-sea fish depend upon luminescent lures, mimicry of prey, extensible jaws, and expandable abdomens (which enable them to consume large items of food) as means of obtaining sustenance. Although most fish are small (usually 6 in. or less in length), the region is inhabited by rarely seen large species such as the giant squid. In the bathypelagic region bioluminescence reaches its greatest development, where two-thirds of the species produce light. Bioluminescence is not restricted to fishes. Squid and euphausiids possess searchlightlike structures complete with lens and iris; and squid and shrimp discharge luminous clouds to escape predators. Fish have rows of luminous organs along their sides and lighted lures that enable them to bait prey and recognize other individuals of the same species.

BENTHIC REGION

Benthal refers to the floor of the sea and the benthos to plants and animals that live there (Fig. 19-22). There is a gradual transition of life from the benthos that exists on the rocky and sandy shores and that which exists in the ocean's depths. From the tide line to the abyss, organisms that colonize the bottom are influenced by the nature of the substrate. If the bottom is rocky or hard, the populations consist largely of organisms that live on the surface of the substrate, the *epifauna* and the *epiflora*. Where the bottom is largely covered with sediment, most of the inhabitants, chiefly animals, live within the deposits and are known collectively as the *infauna*. The kind of organism that burrow into the substrate is influenced by the particle size because the mode of burrowing is often specialized and adapted to a certain type of substrate.

The substrate varies with the depth of the ocean and the relationship of the benthic region to land areas and continental shelves. Near the coast, bottom sediments are derived from the weathering and erosion of land areas along with organic matter from marine life. Sediments of deep water are characterized by fine textured material, which varies with depth and the type of organisms in the overlying waters. Although these sediments are termed organic, they contain little decomposable carbon, consisting largely of skeletal fragments of planktonic organisms. In general, with regional variations, organic deposits down to 4,000 m are rich in calcareous matter. Below 4,000 m hydrostatic pressure causes some form of calcium carbonate to dissolve. At 6,000 m and lower, sediments contain even less organic matter and consist largely of red clays, rich in aluminum oxides and silica.

Within the sediments are layers that relate to oxidation-reduction reactions. The surface or oxidized layer, yellowish in color, is relatively rich in oxygen, ferric oxides, nitrates, and nitrites. It supports the bulk of the benthic animals, such as polychaete worms and bivalves in shallow water, and flatworms, copepods, and others in deeper water, and throughout a rich growth of aerobic bacteria. Below this is a grayish transition zone to the black reduced layer, characterized by a lack of oxygen, iron in the ferrous state, nitrogen in the form of ammonia, and hydrogen sulfide. It is inhabited by anaerobic bacteria, chiefly reducers of sulfates and methane.

In the deep benthic regions variations in temperature, salinity, and other conditions are negligible. In this world of darkness there is no photosynthesis, so the bottom community is strictly heterotrophic, depending entirely on what organic matter finally reaches the bottom as a source of energy. Estimates suggest that the quantity of such material amounts to only 0.5 g/m²/year (H. B. Moore, 1958). Bodies of dead whales, seals, and fish may contribute another 2 or 3 g.

Among the bottom organisms there are four feeding strategies: (1) they may filter suspended material from the water, as do stalked coelenterates; (2) they may collect food particles that settle on the surface of the sediment, as do the sea cucumbers; (4) they may be selective or unselective deposit-feeders, such as the polychaetes; or (4) they may be predatory as are the brittlestars and the spiderlike pycnogonids.

Important in the benthic food chain are bacteria (see Fig. 19-22) of the sediments. Common where large quantities of organic matter are present, bacteria may reach several tens of grams per square meter in the topmost layer of silt. Bacteria synthesize protein from dissolved nutrients and in turn become a source of protein, fats, and oils for deposit-feeders.

Summary

The marine environment is characterized by salinity, waves, tides, and vastness. Salinity is due largely to sodium and clorine, which make up 86 percent of sea salt. Although sea salt has a constant composition, salinity varies throughout the ocean. It is affected by evaporation, precipitation, movement of water masses, and the mixing of water masses of different salinities. Because of its salinity, seawater does not reach its greatest density at 4° C, but becomes heavier as it cools.

Like lakes, the marine environment exhibits stratification and zonation. Because there is no seasonal overturn, a normally stable stratification occurs, known as the pycnocline. It often coincides with the thermocline. The marine environment also exhibits surface waves produced by winds, internal waves at the interface of layers of water, surface currents, and internal currents that run counter to the surface.

Estuaries, where the fresh water meets the sea, and their associated tidal marshes and swamps are a unit in which the nature and distribution of life are determined by salinity. In the estuary itself, salinity declines from the mouth back up the river. This decrease in salinity is accomplished by a decline in estu-

arine fauna because the fauna consists chiefly of marine species. The estuary serves as a nursery for marine organisms, for here the young can develop protected from predators and competing species unable to withstand the lower salinity. Tidal marshes add to the productivity of the estuary. Composed of salt-tolerant grasses and flooded by daily tides, most of the primary production goes unharvested by herbivores. The excess organic matter is carried out to the estuary by the tides, where it is utilized by decomposers and detritus-feeders. If vertical mixing between the fresh water and the salt occurs in the estuary, then the nutrients are circulated up and down between the organisms and bottom sediments. The importance of the tidal marsh as a nutrient source to the estuary and the role of the estuary as a nursery for such marine species as the flounder and oyster should be reason enough for their protection. Yet the pollution of the estuaries and the filling in of tidal marshes, unless the trend is stopped, may have a deleterious effect on the productivity of the shallow seas.

The drift line marks the furthest advance of tides on the sandy shore; on the rocky shore, the tide line is marked by the zone of black algal growth on the stone. The most striking feature of the rocky shore, its zonation of life, is the result of alternate exposure and submergence by the tides. The black zone marks the supralittoral, the upper part of which is flooded only every 2 weeks by the spring tides. Submerged daily by the tides is the littoral, characterized by barnacles, periwinkles, mussels, and fucoid seaweeds. Uncovered only at the spring tides is the sublittoral, dominated by large, brown, laminarian seaweeds, Irish moss, and starfish. Left behind by outgoing tides are the tidal pools, "microcosms of the sea." These are distinct habitats, subject over a 24-hour period to wide fluctuations in temperature and salinity, and inhabited by a varying number of marine organisms, depending upon the amount of submergence and exposure. Sandy shores, in contrast to the rocky ones, appear barren of life at low tide; but beneath the sand, conditions for life are more amenable than on the rocky shore. Inhabitants of the sandy beach occupy permanent or semipermanent tubes in the sand and feed on organic matter carried to the shore by the tides. Zonation of life is hidden beneath the sand, where the variety of life is less than on the rocky shore. But whether hidden beneath the sand or exposed on the rocks, all forms of life at the edge of the sea exhibit a remarkable adaptation to the ebb and flow of the tides.

Beyond the estuary and the rocky shore lies the open sea. There the dominant plant life is phytoplankton, and the chief consumers are zooplankton. Depending upon these for an energy base are the nekton organisms dominated by the fishes. Because of the impoverished nutrient status of ocean water, productivity is low. Most productive are the coastal waters and zones of upwelling.

The open sea can be divided into three main regions. The bathypelagic is the deepest, void of sunlight and inhabited by darkly pigmented, weak-bodied animals possessing luminescence. Above it lies the dimly lighted mesopelagic region, inhabited by its own characteristic species, such as certain sharks and the squid. Both the mesopelagic and bathypelagic regions depend upon a rain of detritus from the upper region, the epipelagic, for their energy source.

Because of currents and tides, the open sea can be considered one vast interconnected ecosystem that has its variants in the sandy beaches and rocky shores below the high-tide mark.

LA MACHINE

Située sur la Riviere de Seine près St. Germain en Laye et encore
la charmante et Magnifique Maison Royale de Marly dont elle po.
Cette Machine eleve 200. pouces d'eau, Soixante et deux Toises de
fournit au fameux Versailles qui n'en est éloigné que d'une bonne heu.
par N. de Fer.

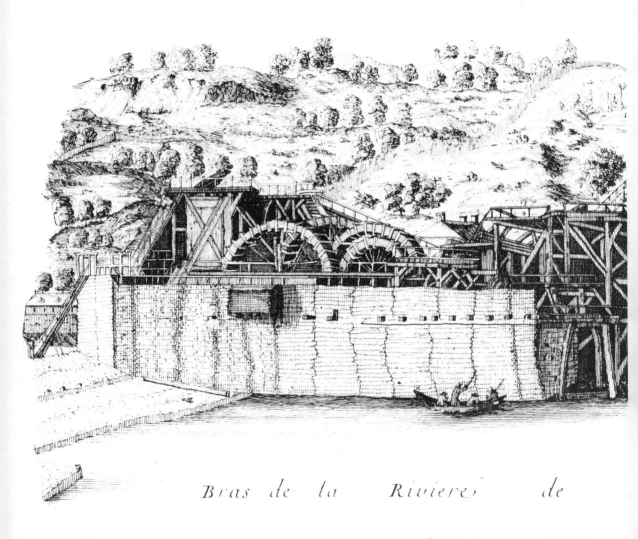

Bras de la Riviere de

Management
of ecosystems

plus pres de
te le nom
haut qu'elle
re de Chemin.

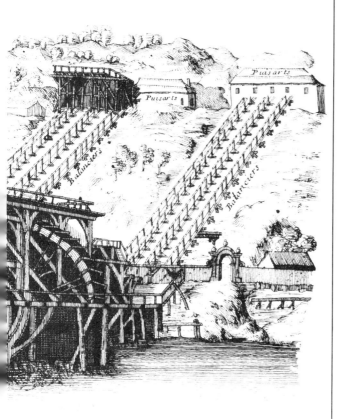

Seine

By now two facts should be evident. One is that man is the dominant organism on earth, imposing his influence on all other forms of life. Intentionally or unintentionally, he has exterminated numbers of plants and animals, locally or worldwide. He has destroyed some ecosystems and so modified others to meet his own needs that they are highly artificial or at best seminatural. Indeed one must question whether any natural ecosystems now exist. The oceans have received such heavy quantities of man's wastes that they are polluted far beyond any expectation. Air pollution affects ecosystems and interferes with their natural functions. To obtain minerals man has destroyed mountains and released toxic elements and silt in waterways. He has drained swamps and marshes, cleared forest lands for agriculture, and replaced agricultural lands with urban and suburban expansions. Even the so-called wilderness areas, no matter how remote, are affected. By excluding such natural processes as fire, man is managing or in some way interfering with natural processes, turning future vegetational development in directions it would not otherwise take. In one way or another man has changed the world's ecosystems to fit his own immediate needs.

The other fact is that man is a part of the world's ecosystems. Any study or management of those ecosystems must take man into account. In spite of his culturally constructed economic and social systems, man still depends upon and is very much involved in the natural cycles and energy flows of ecosystems.

These two basic facts have been largely ignored by many ecologists and resource managers. Ecologists, past and present, have been reluctant to fit man into the conceptual framework of the ecosystem. Many ecologists are so involved in working out ecological theory that they fail to relate ecological principles to resource management. When some do become involved in applied ecology, their solutions to growing ecological problems often seem unreasonable and unrealistic because they are unfamiliar with resource management.

On the other hand, those involved in resource management—the agriculturalists, the foresters, and the wildlife and range managers—often are so poorly trained in ecology that they ignore the fact that ecosystems do exist and that man is an integral part of them. Only recently has the concept of ecosystems begun to creep into resource management.

Compounding these problems is the view of nature held by the public, more specifically, urban society. Typically city-dwellers have no concept of what is involved in the production of agricultural crops or the lumber used for building homes and furniture. To them nature lies untouched and unspoiled. Timber is not to be cut, wildlife is not to be hunted. Ecosystems are for viewing, and man is an intruder in them.

These factors create a dilemma in resource management. Ecologists may understand ecosystems, but they are naive about resource management; resource managers are naive about ecosystems. Their efforts are geared only toward the development and exploitation of resources. They work for a society that is geared to an exploitative economy, taking from today with no thought for tomorrow. Finally, there is urban man, who is so detached from the land that he expects to use resources, yet to leave the world untouched. Essentially man has approached the planet earth as an open ecosystem. He has assumed there is a constant supply of energy and raw materials and that all the products and wastes he creates are ultimately lost to the system. But all the while man has been treating the earth as an open system, he has in fact been operating within a closed system that he has been constantly short-circuiting.

While working within the open-system approach, man has dealt with the land and its products as discreet, noninteracting resources. For centuries man's basic relationship to his environment was founded on what uses he could make of the various components of the natural world for food, clothing, and shelter. As long as his numbers were few, man was able to get away with this approach. But with increasing populations and advanced technology, he is making changes in his environment at a much faster rate. These radical changes are upsetting longstanding ecological balances. All of a sudden an earth that has been regarded as infinite is now being seen as finite. If man is to exist in a finite world, he must work with nature's rules, under which he evolved. He has to operate within the complex of dynamic interrelated systems of which he is an integral part and whose existence he is just beginning to comprehend. In short, man needs to develop

Ecosystem approach to resource management

an ecosystem approach to the management of his world.

In order to develop an ecosystem approach man must take an ecosystem point of view. However, this runs counter to the way agriculturists, foresters, bankers, speculators, and even governments generally tend to view land and resources. The ecosystem point of view means a total systems approach and includes many things omitted in a less comprehensive view. It negates a single-purpose approach to the environment and the treatment of natural resources and land as discrctc commodities (for more on this idea see R. L. Smith, 1972). An ecosystems approach demands an application of systems thinking to the relationship between artificial and natural environments and man. But it does not, as many argue, demand a return to nature. An ecosystem approach to artificial ecosystems means that man should benefit to the fullest extent from natural processes and substitute wherever possible the economy of nature for an exploitive economy.

Ecosystem approach to agriculture

Of all of man's exploitative uses of land resources, urbanization excepted, none has changed ecosystems so radically on a large scale as agriculture. As practiced today, particularly in the United States, modern technological agriculture is the antithesis of the ecosystem approach.

Crop agriculture involves the complete replacement of a natural ecosystem (with its balanced nutrient cycles, diversity of species, and complexity of organization) with a highly simplified artificial ecosystem composed of plants and animals wholly dependent upon man's care and protection, both involving an input of energy (Fig. 20-1). Historically, as agriculture developed, diversity decreased and artificiality increased. Early swidden agriculture involved the growth of several crops in the same fields that usually were surrounded by forest growth (see Chapter 4). Traditional agriculture as practiced in the United States until World War II involved a variety of monocultural crops grown in small separate fields often separated by hedges and pasturelands that supported cattle, horses, and sheep. There was a certain stability to an area that consisted of diverse vegetational types. After World War II hedgerows, small fields, and family

farms gave way to large farms with fields of considerable acreage devoted to single crops. The diversity of traditional farming was replaced by monocultural systems that were highly vulnerable to pests and diseases because of the low genetic diversity of the crops.

In addition to being characterized by low diversity monoculture, technological agriculture is carried on largely outside of natural nutrient cycling. In traditional agriculture a considerable portion of the crops were fed into livestock for milk, meat, and their energy supply. Crop and animal wastes were also returned to the soil (see Fig. 6-2). Today cattle are held in feedlots, fed with grain and hay raised elsewhere, and their wastes are stockpiled rather than returned to the land (see Chapter 6). Nutrients are removed when crops grown in one place are transported to another. Calcium and phosphorus from the wheat fields of the Midwest go to India and Russia, and those from the pineapple fields of Hawaii wind up in the offshore waters of New York harbor. Every item of food purchased in a supermarket represents a quantity of mineral matter pumped out of some nutrient cycle and never recycled in that ecosystem again. The full effects of this impoverishment are prevented only by costly input of chemical fertilizers.

Because agricultural crops need protection from insects and disease, pesticides are applied liberally to the landscape. These enter global biogeochemical cycles, often with disastrous results (see Chapter 6). Herbicides are sprayed both before and after crops begin to grow to reduce competition from weeds and to channel the maximum amount of energy into crops.

Technology adds further stress. No one can

foresee the long-term effects of heavy machinery moving across fields. Will it affect soil structure and the future productivity of the land? The need for a constant supply of water in both semiarid and humid regions results in the interference in a number of ways with the water cycle and subsequently with the movement and translocation of mineral matter in the soil (see Chapter 6). To produce crops that can be harvested by machinery rather than by men, plant geneticists are shaping crops to machines rather than to a more palatable and nutritious food for man.

Increased populations and the corresponding increased demand for food have in part been responsible for the technological revolution in agriculture (see Chapter 4). Because of these demands and the total separation of the consumer from production, a return to the old ecosystem-oriented subsistence agriculture is impossible.

Agricultural production and consumption can be brought at least partly back to an ecological orientation in a number of ways. Crop production can become more diversified, at least to the extent of growing a variety of crops in a given area. Experimental work can be advanced on ways of producing more than one crop in a field. There is some effort along this line involving the no-tillage method of planting corn in sod fields, thereby reducing erosion, reducing the need for cultivation, and providing forage for livestock. More important, the waste products of animals could be recycled back to fields, and the waste water, nutrients, and solid wastes from urban sewage disposal plants could provide both irrigation water and fertilizer for croplands surrounding urban areas.

Such a scheme has already been worked out experimentally (Parizek et al., 1967). Effluents are piped from sewage treatment plants and sprayed on croplands, hayfields, and forests at a rate commensurate with the soil's ability to absorb water. Not only does this increase crop production, but the plants and microorganisms act as living filters, removing phosphates and nitrates from the water and returning them to groundwater in a purified state. (The ecology of agriculture is discussed in some detail in Chapter 4.)

Range management

Range management is the area that has made the greatest advances in the application of ecological concepts. There are reasons why this is so. Rangelands are essentially wildlands, uncultivated areas that support shrubby or herbaceous vegetation. Because a high degree of control over range ecosystems is neither possible nor economical, the range manager must fit his management practices within the natural ecosystem in order to maintain productivity. There was also an early interest in ranges as ecosystems even before the concept was defined. As far back as 1898 Jared Smith (1899) described the ecological reasons why rangelands deteriorated. In 1920 Frederick Clements, the pioneering plant ecologist, outlined a complete system of range improvement that is still appropriate today. Early definitions of range management had a strong ecological ring to them. For example, Jardine and Anderson (1919) stated that "grazing on the National Forests is regulated with the object of using the grazing resources to the fullest extent possible consistent with the protection, development and use of other resources." Thus range management has an early history of ecological orientation.

That appreciation for ecology carried through the development of range management. It is summarized in the definition by Heady (1967), who views range management as

. . . a land management discipline which depends upon basic sciences; limits its activities to uncultivated lands in subhumid, semiarid, and arid regions; centers its activities on grazing animals and forage; and is concerned with the production of animal products, water, timber,

FIGURE 20-1 [OPPOSITE]
Agriculture has become corporate business with large land holdings and heavy mechanization. As small farms disappear the close bond between agriculture and the land is broken and sound ecological principles are often difficult to apply. (Photo courtesy U.S. Department of Agriculture.)

wildlife, and recreation which are useful to mankind.

Another definition is supplied by Lewis (1969), who states:

Range ecosystems are natural pastures or derived pastures managed extensively on the basis of ecological principles. The optimum combination of goods and services is determined by the capabilities of the ecosystem, the level of technology, economic demands, and social pressures. The objectives may include any of the values which the ecosystem is capable of producing.

Most grassland ecosystems were in natural equilibrium when civilization moved in. At that point in history man replaced some natural components with ones of his own that would provide him with removable products. Usually he reduced the native herbivores, shortened food chains, and added domesticated animals in large numbers. Thus, the carrying capacity of the range was overtaxed, and the ecosystem began to deteriorate.

Early range managers recognized the regressive effects of overgrazing. They observed that certain plants declined as grazing intensity increased. These "decreasers" as they called them, were replaced by other species, the "increasers," which were able to persist under close grazing or drought conditions. The behavior of the increasers or decreasers depends upon the nature of the site, the kind of stock, and the season of use. Under conditions of overgrazing, unpalatable plants or invaders begin to take over the range (Figs. 20-2, 20-3). Usually the increasers are capable of rapid migration, are adapted to drier habitats than the dominant plants of the original ecosystem, and escape grazing because of spines or low palatability. In the final stages of deterioration, wind and water erosion may be severe. Thus the range manager can use the plants as indicators of the health of the ecosystem.

The range manager can manipulate a number of ecological variables. But to do this he needs to know what kind of plants are being grazed; the degree, time, and uniformity of grazing; the life cycles and environmental relationships of plants involved; as well as the character, nutritional requirements, and behavior of grazing animals, domesticated or wild. With this knowledge he can manipulate, within limits, certain variables that will result in a

FIGURE 20-2 [OPPOSITE TOP]
Heavy grazing by sheep has left an insufficient residue of leafy growth for the range to recover. (Photo courtesy Soil Conservation Service.)

FIGURE 20-3 [OPPOSITE BOTTOM]
A well-managed range on a high mountain meadow in Sawtooth Valley, Idaho. (Photo courtesy Soil Conservation Service.)

progressive return of the range to a productive state or will maintain the range in a state of high productivity.

One tool the range manager can use is the control of the number and kinds of grazing animals allowed on the range. Because cattle, sheep, and wild grazing animals usually have different food preferences, the range manager can allow common-use grazing. That will make maximum use of the forage and reduce competition among different plants that could result if several species of plants are preferred over others by one kind of animal. If certain plants have declined, the manager may not want certain kinds of animals to graze on the range until the plants recover. Often competition arises between cattle, deer, and elk, and choices have to be made. The optimum proportions of each are usually determined by economic and social pressures. In areas where hunting leases on private rangelands bring in a substantial income, deer may get preference over cattle. On public lands there is constant pressure to make more range available to wild animals.

Another option is to control the distribution of grazing animals by the use of fences, by the distribution of watering places, by enticing animals to new or little-used parts of the range with supplemental feeding or range fertilization, and by herding. The range manager can achieve similar results by selecting the grazing season and keeping the animals off certain areas until the vegetation is sufficiently mature to withstand grazing and trampling. Or he can use a system of continuous, deferred, or rotational grazing, in which the cattle are moved and restricted to one grazing area, allowing one to recover while the other is being used. Such a system reduces damage resulting from selective grazing by allowing highly palatable plants to recover.

The range manager may choose to manipulate vegetation. He can accomplish this by selecting the kinds of grazing animals and grazing pressure that will result in the supression of some plants and the encouragement of others. Or he can selectively use herbicides for plant control. When range regression has proceeded too far, the range manager may resort to reseeding or some type of mechanical treatment of the soil. If undesirable woody vegetation invades the range, he may resort to prescribed burning to change the composition of the vegetation (see Chapter 7).

Unfortunately, too many studies have approached the range from the standpoint of its vegetation or animals rather than as an ecosystem. This lack of deep understanding of grassland systems is an obstacle to truly effective range management.

Forest management

Although range management developed with an ecosystem point of view, forestry developed with an economic and utilitarian point of view. Only in very recent years has the forestry profession begun to recognize ecosystem concepts and to attempt to apply them to forest management. In the past the area of forest ecology (once called silvics) was more concerned with the autecology of forest trees and had little to do with ecosystem concepts.

Because of its emphasis on economic production of timber, forestry has made some serious ecological mistakes. A good example involves the national controversy over clearcutting on the national forests both in the East and West, as well as on large private holdings. This issue provides insight into the consequences of ignoring and misunderstanding ecological principles.

Economic returns aside, the objective of timber harvesting is to regenerate the stand. Four basic strategies are available. These range from single-tree selection cut at one extreme and clear-cutting at the other. Between the two there are shelterwood and seed-tree cuttings. The selection cut provides for all-aged management; the other three provide for even-aged management.

Briefly, the *selection method* (Fig. 20-4) involves the removal of mature trees either singly or in small groups at intervals, permitting the continuous establishment of regeneration. The objective of this method is the maintenance of an uneven-aged stand with trees of different ages and sizes.

The *shelterwood method* involves the removal of the mature stand in a series of cuts. It provides for the development of a new stand at one time under the cover of a partial forest canopy. The final cut removes the shelterwood and allows the new stand to develop in the open as an even-aged forest.

The *seed-tree* method involves the harvesting of nearly all of the timber on the selected area in one cut. But a few of the better trees of usually economically desired species are left uniformly distributed on each acre to reseed the site. Once they have served their purpose, the seed trees are cut.

In the *clear-cutting* method (Fig. 20-5) all of the trees on the area are cut for the purpose of creating a new, even-aged stand. There are many variations to the clear cut. The area cut may involve a whole stand, a group, a patch, or a strip.

The final harvest results in the destruction of one stand of trees and starts another. As the new forest grows and matures, trees are cut periodically to reduce overcrowding, to stimulate the development of an understory, and to improve the species composition and quality of the remaining stands. Collectively known as intermediate cuts, the methods employed are *thinning* and *improvement* cuttings.

Until the 1950s American foresters embraced the idea that the selection method was the ideal and only way of managing a forest. It came about partly because of the depressed timber markets of the time, the need to extend badly depleted growing stock, and a reaction to the timber barons' policy of clear-cutting. The public was sold on the idea that the selection method was the only way to manage the nation's forests.

The selection method is based on the premise that most timber stands, some southern pine and western coniferous forests excepted, are uneven-aged. That accounted for the different sizes of trees and the vertical structure of the forest. As the dominant or overstory trees mature, they are to be cut, allowing younger trees to develop into merchantable timber. Thus a forest cover was always maintained and the stand had a certain esthetic appeal.

But in time foresters discovered that the stands weren't uneven-aged after all. The size differences in the trees reflected different growth rates among species, different degrees of tolerance, competitive influences and site conditions. Smaller understory trees were really stagnated trees of the same age. When the mature trees were removed, the remaining trees failed to respond to added space and sunlight. In effect selection cutting became another form of high grading.

Then there were ecological problems. When a forest is subject to selection cutting, it remains much the same structurally as it was prior to any timber harvesting. Where the trees are removed, young growth responds,

FIGURE 20-4
A new clear cut in the dormant season. The large and small materials have been completely utilized and 1- to 5-in. stems and snags have been felled. A new forest will become established by natural succession. (Photo courtesy U.S. Forest Service.)

FIGURE 20-5
A 7-year-old stand of hardwood reproduction following clear-cutting on a fair hardwood growing site. (Photo courtesy U.S. Forest Service.)

671

much of it consisting of intolerant species. But unless the openings are large, the intolerant trees eventually succumb to shade and competition, and tolerant trees claim the opening. Thus the ecological effect of selection cutting is to increase the proportion of tolerant trees in the stand, such as sugar maple, and reduce the proportion of yellow poplar, black cherry and red oak, all commercially important.

So the dogma of the selection method died in the 1960s only to be replaced by the dogma of clear-cutting, a practice roundly condemned earlier as an example of forestry at its worst. With an objective of growing high-quality timber of high-value species, foresters concluded that even-aged management was desirable on many stands of mature timber, including eastern hardwoods. As they systematically cleared large areas of fine old timber, foresters did not fully understand or expect two things —the ecological impacts of heavy clear-cutting and the esthetic impact on the summer tourists.

The increased amount of clear-cutting in the late 1960s and the 1970s coincided with the public's concern over environmental protection. Clear-cut forests look devastated after cutting. The loss of cool forest in familiar places encouraged many to protest any timber cutting at all. The majority of opposition came from those who believed, as they were once taught, that clear-cutting is inherently a bad practice associated with exploitation. Outcries against forest-cutting practices led to legislative investigations at state and federal levels, moratoriums on cutting on public lands, and the like.

Clear-cutting is a rather drastic practice. When a forest or a part of a forest is clear-cut, the original community is destroyed. When the forest cover is removed, the site is exposed to the full impact of sunlight. In an uncut hardwood forest in West Virginia, the net energy available was 425 kcal/cm² on a clear day in July. Of this about 8 kcal were used in photosynthesis, 317 were used in evapotranspiration, and about 100 in heating (Hornbeck, 1970; Reifsnyder and Lull, 1965). On a clear-cut area about 291 kcal went into heating and 121 into evapotranspiration. The energy that went into heating raised the temperature of the soil by as much as 60° F higher than soils shaded by forest trees (Marquis, 1966). Such high soil temperatures may reduce the germination and survival of some tree species and favor

the germination of others. It increases the rate of humus decomposition and has a marked effect on the water budget. Because clear-cutting markedly reduces the vegetation for a few years, transpiration is reduced. Thus, more water flows out through the soil to streams. In northeastern hardwood forests, clear-cutting will divert about half of the water normally used in evapotranspiration into stream flow. Because this increased flow comes is the summer, it can have a beneficial effect on stream ecosystems (see Table 20-1).

Increased runoff resulting from clear-cutting may carry with it soil particles or dissolved nutrients (see Chapter 5). The greatest losses are of nitrogen, as emphasized by the Hubbard Brook experiments. However, the losses there were exaggerated, for all vegetative regrowth was treated with a herbicide for 3 years. These results have been used to argue against clear-cutting, in spite of the fact that the treatment was extreme. Generally vegetative growth recovers within 3 to 7 years, and nitrogen returns to its precutting level (Table 20-2).

The sharp change in environmental conditions and the flood of light stimulates the plants that grow best in full sunlight, the so-called intolerant species. These species do best at 50 to 100 percent of full sunlight, and can grow at 30 percent of full sunlight. To get 30 percent of full sunlight on the forest floor, about 60 percent of the basal area must be removed, and to obtain 50 percent, 80 percent of the basal area must be cut. Therefore the stand density must be reduced considerably in order to regenerate these intolerant species. A forester who wants to favor oaks and yellow poplar is forced to use an even-aged management system.

But there is also an economic reason why clear-cutting became popular in forest management. Even-aged management allows a more efficient organization of the forest and regulation of the cut. It reduces harvesting costs by concentrating all the cutting activity to a limited area and by reducing the amount of roads needed at any one time.

As a result, foresters used clear-cutting with the same dedication in the 1960s as they did selection cutting in the 1940s and 1950s. Ignoring ecological considerations and concentrating on timber production, both private and federal forestry agencies overdid clear-cutting. They disregarded site differences, failing to recognize that while clear-cutting may be desirable on one site, it might not be on another. New growth failed to regenerate on sites where trees should never have been cut. Cutting of extremely large blocks of 100 acres or more on mountainsides and the poor placement of roads resulted in erosion and degradation of the forest site in the eastern hardwoods regions. Clear-cutting was done on areas where a partial cut would have been more desirable. No consideration was given to its effects on climax forest, wildlife, or esthetics. In the West, clear-cutting of ponderosa pine and Douglas fir was often followed by such drastic site preparations as terracing, and the cutting of contour benches on which a tree-planting machine can be operated. Not only is this esthetically unappealing, but it also encourages erosion.

The influences of these two extremes in cutting practices on forest development and the kinds of trees that make up the forest stand follow the general pattern described. But within this general pattern are wide variations from site to site. The response of forest vegetation on moist cool slopes is different than the response on drier south-facing slopes. The type of vegetation that follows clear-cutting is influenced by the amount of slash left behind, the presence or absence of residual trees, the amount of soil disturbance, and other factors. In both management modes the incoming vegetation is influenced by the kind of reproduction already present in the understory and the seed source.

In a given forest area the forester has available to him not one but many different cutting practices that he can adapt to the site, the kind of forest, the species composition, and his management objectives. Pressures from an esthetically oriented public are forcing foresters to abandon some of their strong economic points of view for an ecosystem point of view. Timber harvests have to be planned around the timber, the structure and function of the forest ecosystem, the wildlife it holds, and esthetics. Some stands should not be cut at all. Other stands, so badly mismanaged in the past, might better be replaced by a complete removal of trees, allowing nature to start all over again. In any case the forest ecosystem should be viewed as a dynamic system. By studying that system, the nature of the forest stand, its location, the site conditions, and other demands, largely social, upon the forest land, the truly professional forest manager can adapt or modify the cut as the situation requires.

TABLE 20-1
Water budget for hardwood forest in eastern region

	Water budget	
Item	Uncut forest (in.)	Clear-cut forest (in.)
Precipitation	44	44
Evapotranspiration	20	10
Runoff (stream flow)	24	34

Source: Lull, 1971.

TABLE 20-2
Nitrogen budget

	Nitrogen budget	
Item	Uncut forest	Clear-cut forest
Total N capital of site (lb/acre)	3,500	3,500
Input (lb/acre)	7	7
Losses (lb/acre)	3	32
Net loss or gain (lb/acre)	+4	−23
Total loss over 3–7 years (lb/acre)	—	50–100
Proportionate loss	—	2–3
Time required for recovery (years)	—	10–25

Source: Adapted from Pierce et al., 1970; Marbut, 1935; and unpubl. file rep. of Apr. 9, 1971, Northeast. For. Exp. Stn., Parsons, W. Va.

673

Wildlife management

Wildlife management, an outgrowth of forestry, developed largely under the influence of Aldo Leopold. Historically, wildlife management is strongly based on ecological principles that for various reasons are difficult to apply.

Wildlife in America is owned by the public and controlled by state and federal governments. Thus its management unfortunately is subject both to public pressure and political whims rather than to ecological principles. Everyone from the president of the local garden club to sportsmen and legislators consider themselves wildlife experts. Even some wildlife biologists are so narrowly trained that they fail to see wildlife in the context of the ecosystem and from the viewpoint of the garden club. For some time many wildlife biologists have placed emphasis on single-species management and have failed to consider all the values of wildlife.

Wildlife is a product of the land. The number and kinds of wildlife are dictated by the land-use decisions made by forestry, agriculture, reclamation bureaus, land speculators, and realtors. Rarely is wildlife given major consideration in land-use planning (see Nat. Acad. Sci. 1970).

A diversity of wildlife requires a diversity of habitats. But as agriculture becomes more industrialized, as small fields are incorporated into larger ones, as hedgerows disappear, as forests are cleared for crop production, and as potholes and marshes are drained, this diversity declines. In addition, the loss of land to road construction and to suburban and urban developments further reduces habitat available to wildlife. With these developments both the abundance and variety of wildlife are declining, and more and more species become endangered (for example, see Cal. Fish and Game Comm., 1972).

Because many people associate the decline in wildlife with sport hunting, an antihunting sentiment is rapidly developing that could have a serious effect on the future of wildlife. These people do not consider the fact that because of sport hunting many forms of wildlife still exist today. Revenue derived from hunting and fishing licenses and from taxes on hunting and fishing equipment has supported wildlife research, habitat restoration and maintenance, and the reintroduction of wildlife on depleted ranges. This work in turn has bene-

fitted many nongame species. No animal has ever been exterminated or seriously reduced by regulated sport hunting. Commercial hunting for profit, however, has driven some wildlife to the edge of extinction.

Based on habitat, wildlife can be divided into four groups: wilderness, late successional, midsuccessional, and early successional. Wilderness species, among them the caribou, grizzly bear, wolf, and prairie chicken occupy climax vegetational types—large areas of mature forest, tundra, undisturbed deserts, and grasslands. Highly sensitive to human disturbance, these species show the sharpest decline. As the few remaining wilderness areas are invaded by recreation, urban growth, roads, or as their structure and composition are changed by the deliberate exclusion of fire, these animals slowly head down the road to extinction. They are largely incompatible with man. The late successional species, such as the gray squirrel and wild turkey, require mature forests that produce an abundance of acorns and nuts upon which these animals depend. Midsuccessional species occupy areas created by grazing, burning, logging, and land abandonment by agriculture. This habitat is a temporary feature of the landscape and unless maintained by intensive management practices, such as logging and fire, it will advance into late successional stages. If this occurs, its characteristic species of wildlife such as the deer, moose, snowshoe hare, and ruffed grouse will decline. The abundance of deer and grouse in the country today is a result of poor logging practices and unproductive agriculture of past years. Early successional species such as the bobwhite quail and cottontail rabbit depend upon weed stages of succession and an interspersion of grass and shrubby growth. Once abundant in the heyday of traditional agriculture, these species find little in the way of habitat in this age of industrialized farming. Thus the key to wildlife abundance is the maintenance and manipulation of habitat.

Manipulation of vegetation is one aspect of wildlife management. The other is regulation of wildlife populations. For many species, such as ruffed grouse and cottontail rabbit, natural regulatory mechanisms (see Chapter 11) take care of surplus individuals. Hunting acts as a substitute for natural mortality. In other species, such as waterfowl, hunting is added to natural mortality, and regulations must be carefully controlled. In such game animals as deer regulatory mechanisms are only weakly operative (see Chapter 11). As the numbers of animals exceed the carrying capacity of the range, the habitat deteriorates, and animals die of winter starvation. Such animal populations should be carefully controlled by regulated hunting. It is this group of animals that causes so much concern among antihunting groups.

To appreciate the complex problem of managing wildlife, one might examine the case of the white-tailed deer, an animal whose management has aroused so much emotion down through the years that the result has usually been no management at all. Prior to settlement by white man, the North American continent supported relatively low populations of white-tailed deer. The animals were confined mostly to the forest edge, to burned-over areas, and to the vicinity of Indian settlements. As the timber was cut, the country opened up, and the vegetation was set back to an earlier successional stage. The deer responded, and their numbers increased dramatically. The abundance of game encouraged its commercial exploitation. All manner of methods were used to take deer and other game that found a ready sale in such cities as New York, Philadelphia, and Chicago. Game was so depleted that it became evident it was headed for extinction. Game laws evolved to protect these animals (see Leopold, 1933). To build up the deer herds states instituted buck laws, which restricted the hunting of deer to antlered males. With excellent habitat and such protection the deer herds recovered. Deer actually became more abundant than they had ever been in presettlement times. In time the new forests matured and deer cover and food declined. The range could no longer support a high deer population. The habitat deteriorated and deer died of starvation during the winter (Fig. 20-6). Yet the general acceptance of the buck law and the reluctance of society to permit the hunting of does made deer management almost impossible. In many parts of the range conditions grew steadily worse and the number of deer sharply declined (see, for example, Dahlberg and Guettinger, 1956; Silver, 1957).

To maintain a high number of deer, one must hold a substantial portion of forest lands in early successional stages. The only way to accomplish this is to harvest timber annually in relatively small clear-cut blocks. Timber

FIGURE 20-6

A doe deer that died of starvation, the result of a population too high for available food. (Photo courtesy of Michigan Conservation Department.)

cutting, however, is governed by market conditions for wood and wood products. If the demand for wood is low, timber remains uncut, and the amount of available food and cover declines. If the demand is heavy, timber cutting is increased. The more forest lands that are cut, the more deer habitat is created. Thus the very type of cutting practices so deplored by the public is the very kind needed to support a high deer population.

In many parts of the deer range the numbers of deer are so high and so close to the starvation level that even if cuttings are extensive, the deer consume new growth as fast as it develops. This, of course, has a serious impact on the regeneration of a new forest. In places where deer numbers are lower, the blocks of timber cut may be so large that deer are unable to utilize the great amount of food available. By the time the deer herd has responded, the forest is back in the pole stage and deer food becomes scarce again. Thus the development and maintenance of deer habitat is controlled by economic forces and forestry, and is outside the influence of the wildlife manager.

The wildlife biologist could control the deer population size if allowed to do so by politicians. This he can do by regulating the length of the hunting season, the number of hunters permitted in an area, and the age and sex of the deer to be harvested. By such regulation the wildlife biologist can keep the herd in balance with the range. If the range will support a high number of deer and the area is understocked, he can allow the herd to build up by enforcing the buck law. When the herd reaches the carrying capacity he can hold it there by periodically removing a portion of the does. Because deer are polygamous, a large number of bucks can be removed without noticeably affecting the productivity of the herd (see Table 20-3).

An ecosystem approach to deer herd management requires public acceptance of deer herd management, acceptance by the hunter for the need to hold the size of the deer herd in balance with available food, and absence of political interference. Once these social requirements are met, then the wildlife manager must take into account the constraints imposed by economic and recreational demands placed on forest lands, as well as the energy requirements of deer. As has already been pointed out (Chapter 3), young animals can convert energy to flesh more efficiently than older animals.

Mature deer (those 2 years of age and older) utilize most of their energy intake for body maintenance. How well the conversion of plant energy to deer biomass takes place depends, of course, on the quality of food available.

By modifying the sex and age structure of the deer herd, the wildlife biologist can modify energy requirements of the herd and thus productivity. One deer management strategy is to set season regulations and hunting pressure so that the animals are removed when relatively young. The other is to set the hunting season and pressure so that the animals are taken after they have reached maximum development. By emphasizing late-age harvesting, the wildlife biologist can produce larger deer with well-developed antlers. But he exchanges this for high food consumption, high sustained biomass, and low production of new biomass. With an early-age harvest the deer herd consumes less food and produces more biomass per unit of forage consumed and a greater number of animals. But this, of course, does not mean more total biomass supported per unit of range.

The differences between the two strategies can be illustrated by two different deer herds in the United States, one in the Llano Basin of Texas, the other on the George Reserve in Michigan. The former herd is characterized by conservation hunting, the latter by intensive harvesting. The Llano Basin herd consists of a dense population of relatively small deer (Teer et al., 1965). The sex ratio is one buck to two does; the production rate is 0.6 fawn per doe, a low annual increment. Hunters remove less than 50 percent of the annual increment and only about 14 percent of the total fall population. On the George Reserve (Chase and Jenkins, 1962) the density of deer is half that of Llano Basin; the sex ratio is 1:1 because both bucks and does are hunted. In spite of a smaller number of animals, the total fawn crop is only 25 percent less than that of the Texas herd because each doe produces twice as many fawns. Hunters remove 40 percent of the total fall population, 100 percent of the annual increment, holding the herd within the limits of the carrying capacity of the habitat. Although the biomass is nearly the same, the interpopulation energy flow is four times greater in the George Reserve herd.

If the management objective is to produce trophy animals, then the late-age harvest is appropriate. If the objective is to provide the maximum amount of hunting, then the early-age harvest is appropriate. In the United States hunting regulations favor the early-age harvest. The strategy used is also influenced by the nature of the habitat. There is, for example, little use in employing a late-aged harvest in a region where the soil fertility is so low that the vegetation can not support deer of excellent quality.

Another aspect to be taken into account in managing a deer herd is the increasing public demands made on public forest lands for recreation. The public may not be willing to exchange increased timber cutting for more deer. Too large a deer herd may cause excessive damage to farm crops and orchards, inhibit forest reproduction, and perhaps reduce the esthetic qualities of woodlands.

One point about wildlife management still needs to be emphasized, a point rarely understood by the public. Different species of wildlife, game and nongame, have different habitat requirements. If the objective is an abundance of wild turkey and gray squirrels, then the mature forest must go uncut. If the objective is deer, grouse, or moose, then the forest must be cut periodically to maintain early successional stages. In doing so, one must sacrifice old-age timber stands. It is impossible on any given piece of land to manage for high populations of all forms of wildlife. But by interspersing blocks of open land with forest, and blocks of clear-cut timber with blocks of uncut and selectively cut timber, one can create a diversity of habitats that will support a diversity of wildlife with perhaps minimal populations of each.

Systems analysis

Ecosystems are extremely complex, and so are the problems of resource management, which involve not only ecosystems but economics as well. The approaches to these problems and their solutions can no longer be handled on an empirical level. Ecosystems, natural or artificial, consist of many interacting components. The state of one influences the state of others. This calls for a more sophisticated approach to ecological problems. One such approach is that of systems analysis.

In using systems analysis, ecologists and resource managers have borrowed some analytical tools from business and industry used to solve

TABLE 20-3

Populations and composition of the deer harvest in the Llano Basin, Texas, and the George Reserve, Michigan

Criteria	George Reserve	Llano Basin
	number/100 acres	
Peak population	7.8	14.4
Nonfawn population	4.7	10.1
Bucks (all ages)	2.3[a]	3.2
Does (all ages)	2.4[a]	6.9
Fawn population	3.1	4.3
Fawns/all nonfawn does	1.3	0.6
Total removal	3.1	2.0
Composition of killed animals		
Mature bucks	1.3	1.1
Immature bucks	0.3	—
Mature does	1.1	0.9
Immature does	0.3	—
Unknown	0.1	—
Killed animals	percent	
Increment	100	47
Total population	40	14

Source: H. Short, 1972.
[a] Assumes 1:1 sex ratio for nonfawn population.

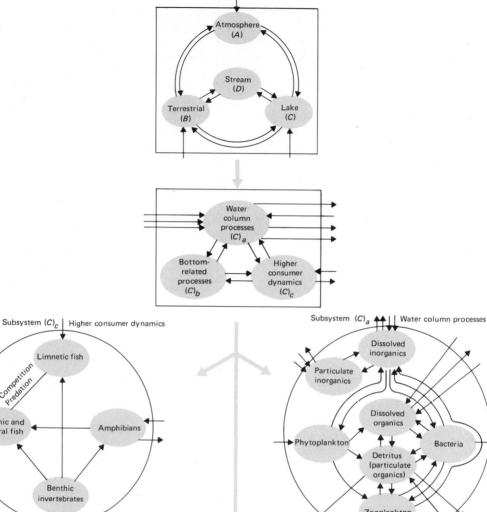

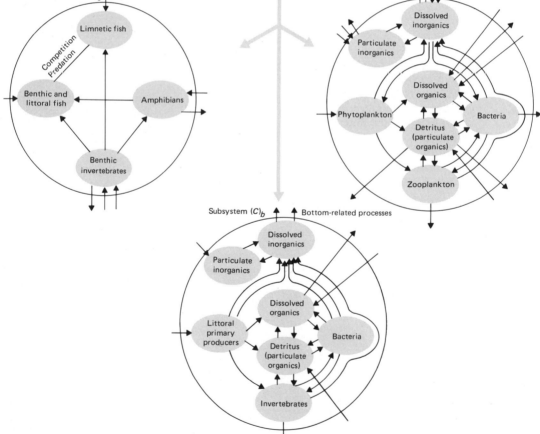

complex managerial problems. Systems analysis involves the translation of selected physical and biological characteristics of ecosystems and populations into sets of mathematical relationships or systems, called models. These models obviously are limited in their application; they are abstract, simplified approximations of real-world conditions. However abstract, systems analysis does provide some predictive insight into the functioning of ecosystems and some predictive basis for decision making in resource management. Whether the systems approach is used within a purely ecological or a resource management framework, the approaches to the problems are similar.

The first step is to define the components of the system to be studied. Each component in some way must be related to other components, either by sharing some common feature or involving some significant transfers between them. Such components of ecosystems might be various trophic levels as described by biomass, calories, or nutrients, and the abiotic elements. Or they might be such structural elements as subsoil, soil, litter, and trees; or roots, mulch, live vegetation, and standing dead vegetation. In resource management the components may be populations of predators and prey, year classes, exploitation pressures, market prices of resources, fixed expenses for harvesting, and the like. Once the components are identified they can be varied by lumping or splitting into larger or smaller categories. Some boundary to the system must be set; and the flow or interactions between components or compartments must be defined. Also to be determined are the inputs or driving forces that affect the system but are not affected by it. Once the system, its boundary, its compartments, and their interactions are placed in some hierarchical arrangement, the result is a picture model of the system (Figs. 20-7 and 20-8). Such models have been used throughout the text without referring to them as systems models.

The picture model can be translated into a tabular arrangement or matrix of functions showing the existence of transfers between compartments (Fig. 20-9). Again the transfer functions can be expanded or contracted as needed. This matrix of transfer functions sets the stage for the development of mathematical models.

Once the picture model and the matrix have been constructed, the next step is to determine

679

FIGURE 20-8 [OPPOSITE]

A complex model of an aquatic ecosystem in which a major compartment is further subdivided into a number of smaller subcompartments for study. This model shows how complex ecological studies can become. (Adapted from Taub et al., 1972.)

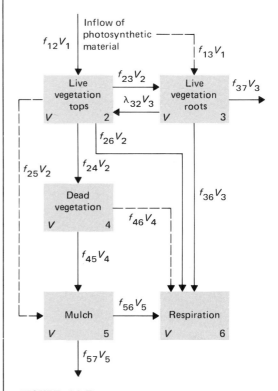

FIGURE 20-7

A picture model of a system showing compartments and transfers between them that can be studied.

the state conditions for the compartments. What measurable characteristics are involved and how shall they be measured? Ecosystem compartments might be measured in terms of biomass in grams or kilocalories per square meter, or in quantities of nutrients. Resource compartments might be measured in terms of stocking rates, year class, amount of catch or harvest, costs, and so on. The state conditions are translated into sets of system variables that are ordered into lists already defined by the matrix of transfer functions, called the system state vector. What you now have is a compartmental system, V, made up of n compartments:

$$V = \begin{matrix} v_1 \\ v_2 \\ v_3 \\ \cdot \\ \cdot \\ \cdot \\ v_n \end{matrix}$$

in which v_1, v_2, etc. represent at any time the value of system variable 1, 2, etc. This might be the amount of nutrient per unit area in that compartment, the amount of organic matter, kilocalories of energy, and so on.

The matrix represents the transfer function between the compartments:

$$F = \begin{matrix} f_{11} \ f_{12} \cdots f_{1n} \\ f_{21} \ f_{22} \cdots f_{2n} \\ \cdot \\ \cdot \\ \cdot \\ f_{n1} \ f_{n2} \cdots f_{nn} \end{matrix}$$

Thus f_{12} might represent the transfer function controlling the movement of a nutrient or energy from compartment 1 to compartment 2. The f_{ij} represents the x's +'s in the picture matrix (Fig. 20-9).

Because the values in each compartment are variables changing with time, the change is represented by

$$\dot{V} = \begin{matrix} \dfrac{dv_1}{dt} \\ \dfrac{dv_2}{dt} \\ \cdot \\ \cdot \\ \cdot \\ \dfrac{dv_n}{dt} \end{matrix}$$

The general form of the equation then becomes

$$\dot{V} = Fv$$

and it is solved by matrix algebra, differential equations, and obviously by the use of a computer.

After constructing the model and perhaps testing it by using values drawn from the literature, the investigator must measure the variables in the field. He may measure the rate of input and output of energy across a transfer boundary from one level to another, or perhaps the quantities of nutrients transferred. Once these processes have been measured, the values obtained are inserted into the matrix. Because the flow or transfer between the compartments is a rate, the system must be observed at different times, when the compartments have different values. To obtain values for the rates the investigator may modify the biomass by introducing or removing organisms, or he may harvest some of the material or use any technique necessary to obtain the values needed. After all the values have been obtained the full solution of the model requires the use of correlation and multiple regression.

One of the best ways to appreciate systems analysis is to examine an example. The one chosen is a study of seasonal primary productivity in two old-field ecosystems, one dominated by the widely planted Kentucky-31 tall fesque, *Festuca elatior*, the other by broomsedge, *Andropogon virginicus*. The procedures are as clear cut as the analysis, yet the study does point out how complex the analysis of even the simplest function of an ecosystem can become.

The boundaries of the system are described by the picture model Fig. 20-7 as well as by the flow between compartments. The transfer matrix is described in Fig. 20-9. The flow between compartments is expressed as percentage of organic matter, representing a series of losses from one compartment and gains by the receiving compartments.

Kelly et al. (1969) measured a full-year cycle of growth. For each species they measured the aboveground and below-ground live vegetation and standing dead vegetation, as well as the total amount of mulch and roots. They made several independent estimates of transfer rates by such methods as use of litter bags (see Appendix B) and removal of litter from plots.

They also measured soil moisture, precipitation, solar energy influx, and temperature.

To analyze the data they constructed a seven-compartment system (Fig. 20-9) with eight nonzero transfer coefficients that were later reduced to seven (Table 20-4). Two of these coefficients were derived from separate experiments, two were abstracted from the literature, and four were determined empirically. These coefficients assumed a constant rate of transfer between compartments during all the seasons. They were used along with some values measured in the field to come up with a predictive model against which field data could be compared. These comparisons are shown in Figs. 20-10 and 20-11. The dashed lines represent the results of the model. The points represent the mean values for each compartment and their standard errors. In general the actual data agree with the model, except in the early growing season. Thus the constant coefficients had their limitations, and the assumption of a constant transfer rate during all seasons was not realistic.

The transfer functions from one compartment to another were not a constant fraction as assumed by the model. The compartments and the transfers between them varied through the season. For example there was a large seasonal increase in the root compartment of both *Festuca* and *Andropogon* communities and a rapid decrease in the standing dead compartment of the *Andropogon* (see Figs. 20-10 and 20-11). To make the model more realistic, the investigators encorporated it to seasonally varying coefficients, one set for the cool season, another for the warm (for details see Kelly et al., 1969), as given in Tables 20-5 and 20-6. The use of seasonally varying coefficients resulted in a closer agreement between the predictive model and the field data as shown in Figs. 20-12 and 20-13.

The major similarity in the two models was the relatively constant value of the ground litter. The model predicted an increased decay rate at the times of increased input, and this was observed in the field. Differences in the live top compartment reflected the different periodicity from the source compartment, as well as transfers to the root and to the standing dead compartments. In the *Festuca* community the amount of standing dead vegetation did not fluctuate greatly during the growing seasons because there was a gradual dying of the leaves and a continual transfer to the dead compart-

TABLE 20-4
The transfer coefficients (%/day) for the constant coefficient form of the Festuca *and* Andropogon *models*

Coefficient	Festuca	Andropogon
Source to live top (λ_{12})	1.0	1.0
Live top to live root (λ_{23})	0.005	0.004
Live top to standing dead (λ_{24})	0.005	0.002
Live top to respiration (λ_{26})	0.0014	0.0014
Live root to live top (λ_{32})[a]	combined in transfer coefficient λ_{23}	
Live root to respiration (λ_{36})	0.0007	0.0007
Standing dead to mulch (λ_{45})	0.0018	0.001
Mulch to respiration (λ_{56})	0.005	0.002

[a] Specifying an upward flow proportional to root mass would alter the systems behavior and require a compensating increase in λ_{23}. Both would require seasonally varying regulators and were not attempted at this stage.

Component	(1)	(2)	(3)	(4)	(5)	(6)	(7)
Inflow (1)	—	X	O	O	O	O	O
Live vegetation tops (2)	X	+	X	X	X	X	X
Live vegetation roots (3)	X	X	+	O	O	X	X
Dead vegetation (4)	O	X	O	O	X	X	O
Mulch (5)	O	X	O	X	O	X	X
Respiration (6)	O	X	X	X	X	O	O
Organic matter soil (7)	O	O	O	O	X	O	O

FIGURE 20-9
A transfer matrix table in which the picture model, Fig. 20-7, has been translated into a tabular arrangement or matrix of functions.

TABLE 20-5
Seasonally varying transfer coefficients (%/day) for the Andropogon *model* ($1 \le t \le 365$)

Coefficient	Equations
Source to live tops (λ_{12})	1.0
Live top to live root (λ_{23})	$[0.002 + 0.002 \sin(2t - 0.7)1.4]\,4.2$
Live top to standing dead (λ_{24})	$0.00027\, e^{0.012t}$
Live top to respiration (λ_{26})	0.0014
Live root to live top (λ_{32})	combined in transfer coefficient λ_{23}
Live root to respiration (λ_{36})	$[0.0005 \pm 0.01 \sin(t + 2)]1.1 \quad 1 \le t \le 280$
	$[0.0005 + 0.01 \sin(t + 2)](365 - t)0.01 \quad 281 \le t \le 365$
Standing dead to mulch (λ_{45})	$0.0018[1 + \sin(2t - 1.56)]$
Mulch to respiration (λ_{56})	$(\lambda_{45}V_4 - 180 + V_5)/V_5$

TABLE 20-6
Seasonally varying coefficients (%/day) for the Festuca *model* ($1 < t \le 365$)

Coefficient	Equations
Source to live top (λ_{12})	1.0
Source to live root (λ_{13})	$0.169\, t - 0.0014\, t^2 \quad t < 120; 0.0\, t \ge 120.$
Live top to standing dead (λ_{24})	0.005
Live top to respiration (λ_{26})	0.0014
Live root to live top (λ_{32})	combined in transfer coefficient λ_{23}
Live root to respiration (λ_{36})	0.0007
Standing dead to mulch (λ_{45})	0.0018
Mulch to respiration (λ_{56})	$(\lambda_{45}V_4 - 117 + V_5)/V_5$

ment. In the *Andropogon* community death came at the end of the growing season. Because a large part of the live mass of *Andropogon* is in the flower stalk, there was a rapid increase in the dead compartment soon after flowering. In the *Festuca* community there was a large increase in the root community in the spring before the live vegetation, and a constant value for the rest of the growing season. In the *Andropogon* community increases in the root compartment took place later in the year along with live vegetation.

The model and the field data also permitted the investigators to estimate net primary production. They did this by using the estimates of input to the seasonal coefficient models (gross production) and subtracting from them preliminary estimates of respiration that occurred in the live vegetation (roots and shoots), information they derived from the transfer coefficients for the live top and root respiration.

The gross primary production of the *Festuca* was an estimated 1220 g/m²/year, and net primary production ranged from 921 g/m² to 1116 g/m²/year, depending upon whether the turnover in the root compartment was due entirely to the death of the roots or only to respiration. The calculated value for positive biomass increments, determined from field data, was 992 g/m²/year, a value that fell within the limits set by the model. The gross primary production of the *Andropogon* community was 1145 g/m²/year, and the range of net production was from 853 g/m² to 1060 g/m²/year. Both were considered to be underestimates.

This example points up some of the problems involved in constructing a model. If one attempts to make the model too realistic, it becomes too complex and almost impossible to handle. However, a model that can be handled with some ease may not represent any real-world situations.

Many ecologists believe that systems analysis has its widest application in natural resource management, where predictive answers are required. The variables, such as population size and growth rates of both organisms and populations, are clear cut. In such instances the model is designed to give specific information, such as the optimum yield or rate of harvest.

Often such ecosystem models have to be wedded to economic models, which add further complexity, as illustrated in Fig. 20-14. In this instance the ecosystem model is a simple sustained-yield decision model in forestry, in which the timber growth is equal to timber

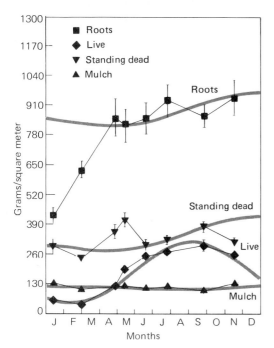

FIGURE 20-10

Comparison of the final constant coefficient model (dashed lines) with the 1967 field data (connected points) for the Festuca *community (g/m² ± S.E.). (Courtesy Oak Ridge National Laboratory.)*

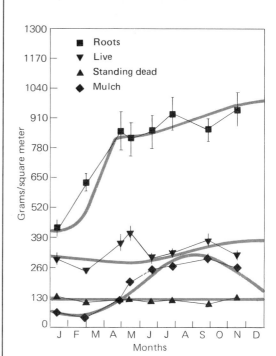

FIGURE 20-12

Comparison of the final seasonally varying coefficient model (dashed lines) with the 1967 field data (connected points) for the Festuca *community (g/m² ± S.E.). (Courtesy Oak Ridge National Laboratory.)*

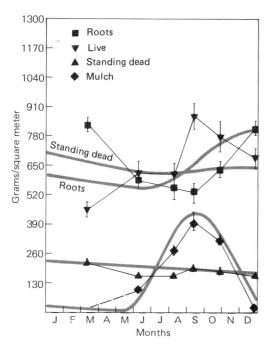

FIGURE 20-11

Comparison of the final constant coefficient model (dashed lines) with the 1967 field data (connected points) for the Andropogon *community (g/m² ± S.E.). (Courtesy Oak Ridge National Laboratory.)*

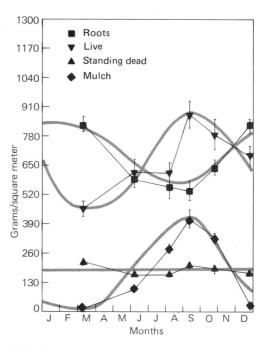

FIGURE 20-13

Comparison of the final seasonally varying coefficient model (dashed lines) with the 1967 field data (connected points) for the Andropogon *community (g/m² ± S.E.). (Courtesy Oak Ridge National Laboratory.)*

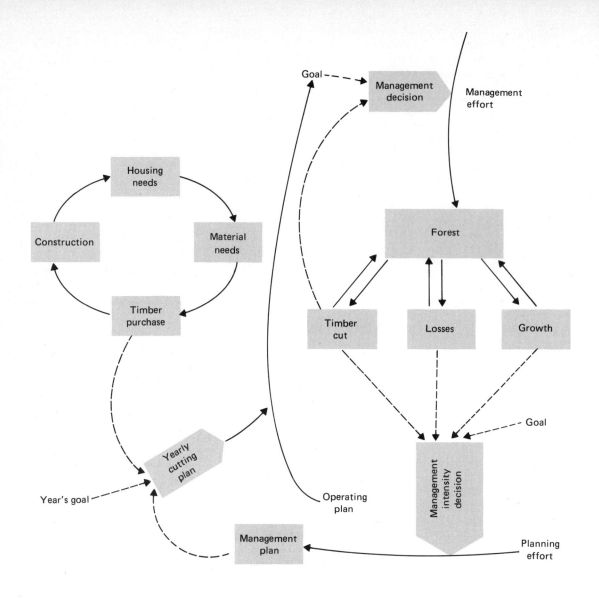

cut. But the model is too simplistic because it fails to take into account that timber cut and growth rates are interdependent. So a more complex model is needed (Fig. 20-15). This model builds in a plan for intensity of management. There are now three state variables, the forest, timber cut, and losses, and two decision-making (rate) variables.

But no matter how well this model is constructed it is still unrealistic because cutting and management obviously are going to be influenced by timber prices and timber demand, which in turn are going to be influenced by demands for housing and construction. Now an even more complex model is required that involves the relationship of economics to ecosystems. Thus the resource manager, in this case the forest manager, must face the fact that economic demands are not geared to biological

forest growth. Although it is not added in, the fact that the forest is not just a commodity to be harvested but also relates to other human needs (such as esthetics, recreation, and watershed protection) further complicates the model.

Similar problems and approaches arise in range management. By means of computer simulation the range manager can forecast the effect on both vegetation and herbivores of any proposed grazing intensity, of allowable differences in grazing intensity, of differential utilization of different grasses by different kinds of grazing animals, and of uneven distribution of moisture.

Also susceptible to computer simulation are the management and exploitation of fisheries resources (Royce et al., 1963), forest insect control (K. E. F. Watt, 1964), and even huntable wildlife species (Miller et al., 1972).

Ecology and the future

At this point it should be obvious that if man is to live in harmony with his environment, he must approach the earth as a finite, closed ecosystem. The only real input comes from the energy of the sun. This energy powers the photosynthesis needed to sustain life, circulates air masses, mixes oxygen and carbon dioxide, recycles and circulates water, and drives the cycling of nutrients and materials, all of which are available only on earth. To ensure his existence man must live within the broad global cycles that characterize ecosystems. To do this he must incorporate an ecosystems viewpoint into his technological and cultural life.

To adopt such an approach is difficult, for it implies a whole new way of organizing man's relations with the natural world and a restructuring of his society. It means that the cost of pollution and environmental restoration would be added to the cost of production. It would demand less dependence on raw materials and more on recycling salvageable materials. It would mean planning for a lower growth rate and stabilizing the Gross National Product, ideas that are hardly acceptable by governments and industries that gauge progress by the annual growth rate. The transition from a growing state to a stabilized state could be highly frustrating, for it would mean fewer jobs temporarily and a decreased demand for raw materials produced by developing countries. An ecosystem approach would mean slowing down population growth and eventually holding the world population within the carrying capacity of the earth. That carrying capacity would have to be determined by the quality of life we are willing to accept (see Wagar, 1970; R. L. Smith, 1972).

An ecosystem approach means redefining our priorities. To date our priorities have been out of line ecologically. We have been expending money and resources on huge defense establishments and on space exploration, areas noncritical to environmental quality. We give low-level priorities to such important expenditures as pollution abatement, recycling of wastes, removal of sulfur emissions from power plant and industrial stacks, environmental restoration, improvement of the urban environment, and the elimination of malnutrition. Such priorities obviously do not have wide

685

FIGURE 20-15 [OPPOSITE]
A more complex model of sustained-yield production in which managerial decisions and in turn forest ecosystems are influenced by outside economic and market forces. The impact of these influences is too little understood by the lay public and the ecologists alike. (After Gould, 1972.)

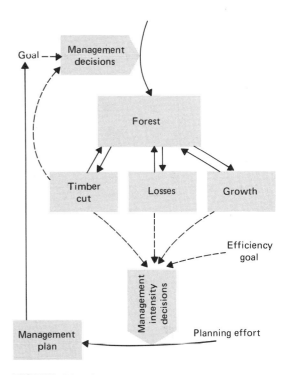

FIGURE 20-14
A simple sustained-yield production model for a forest. The basic model consists of the central compartments, but the model is modified because decisions in the intensity of management influence the forest. In turn, growth, loss, and timber cut influence management decisions. This is a good example of the interplay between ecology and economics in a forest ecosystem. (After Gould, 1972.)

appeal nor do they add significantly to the Gross National Product.

An ecosystem approach means a new approach to agricultural land use. Industrial agriculture would have to be modified to work within natural ecological systems. Animal and human wastes would be recycled, and the nutrients they contain would stimulate terrestrial plant production instead of being lost to the sea. This would also reduce further eutrophication of rivers, lakes, and estuaries. Agriculture would have to depend less on pesticides and more on the biological control of pests and diseases. Crop production would become more diversified locally, reducing the damaging effects of plant diseases, stabilizing crop production, and maintaining soil fertility.

An ecosystem approach would require a new outlook in resource management. The emphasis would have to shift, as it is beginning to do now, from a single-objective type of management for economic exploitation to multiple-use objectives. New objectives would consider not only the major product of the land such as meat or timber, but such other products as wildlife, watershed protection, and recreation. Management practices would have to conform to sound ecological principles, especially those relating to successional stages and nutrient cycling.

It would mean slowing down our pace of life. Within the past 25 years the changes have been too rapid for man to comprehend and for natural ecosystems to absorb. Change is inherent in natural ecosystems and in life itself, but those changes are relatively slow, providing time for adjustment. But in the last century the speed of travel has increased by a factor of 10^2, our speed of communication by a factor of 10^7, our speed of data handling by 10^6, our use of energy resources by 10^3, and our population over the past few thousand years by 10^3 (Platt, 1969). This pace has to level off.

If man is to live within the framework of ecological principles, he must have greater understanding and appreciation of them. We need to increase our studies of ecosystems. A beginning has been made under the International Biological Program. Teams of ecologists are studying the grassland, temperate forest, coniferous forest, tropical forest, desert, and tundra biomes. To apply basic principles to the utilization of these various biomes as well as to artificial ecosystems we will need well-trained ecologists. They will have to want to get involved in applied areas of agriculture, forestry, range management, wildlife management, and mineral exploitation. Generalists as well as specialists will be needed. Lewis (1969), for example, suggests that to manage such resources as forests and range there should be a team of specialists headed by a resource ecologist. That person would not necessarily have a deep understanding of all aspects of the field but he would be able to wed ecological principles to sociological and economic needs.

An ecosystem approach would require a more ecological orientation in education at all levels and in all fields. Sociology, medicine, engineering, and law should be taught within the context of man's place in the ecosystem. Greater emphasis will have to be placed on ecology at the applied level. Although it is of interest to study ecological theories or the interrelationships of a single species of environment, the environmental situation is serious enough to demand that most of our educational energies be directed in the applied areas. This means that ecologists in general are going to have to become much more familiar with basic problems in agriculture and resource management than they are now.

Such changes are difficult to make. But somewhere and somehow high priority must be given to such changes, not just on a local or national level but on an international level (see Ward and Dubos, 1972). Ecological approaches involve global strategies. This means international recognition of problems and international cooperation in their solution. Among such problems are global air pollution, the pollution of the ocean, and the uncontrolled exploitation of the resources of the sea. We can go stumbling along as we have been, but as Lynton Caldwell (1970) notes, "Today there may be higher political priorities, but, ecologically, tomorrow may be too late."

Appendixes

An ecologist friend of mine once described ecology as quantified natural history. In many respects this is an acceptable definition. It does emphasize that modern studies of plant and animal life and environmental studies must be quantified, in contrast to the purely descriptive studies of an earlier day. Even the most elementary studies demand a quantified approach. Because much of the work involves sampling of one sort or another, statistics are involved. But in the hands of those who know little about it, statistics can be dangerous. They are often abused rather than used, and misapplied to the problems at hand. Too often data are collected and forced to fit some statistical procedure; instead the research should be planned with a particular statistical approach in mind. To aid the investigator the following list of selected references is appended. Quantified approaches also can employ computers and computer programming. Because programming is not only specialized but also constantly changing, I have not included any specific references on its use. Instead the investigator should consult local experts. For an introduction to the usefulness of the computer see Watt (1966, 1968).

ANDREWARTHA, H. G.: 1970, *Introduction to the Study of Animal Behavior*, University of Chicago Press, Chicago. Parts of this book contain helpful and easily understood discussions on the use and application of statistics to the study of animal populations: sampling, tests for nonrandomness, analysis of variance, etc.

BAILEY, N. T.: 1959, *Statistical Methods in Biology*, Wiley, New York. A good readable elementary book on statistics, but limited in scope. Treats variation, statistical significance tests, testing methods for homogeneity, correlation, regression, factorial experiments, random sampling, handling a calculating machine, and a summary of statistical formulas.

BATSCHELET, E.: 1965, *Statistical Methods for the Analysis of Problems in Animal Orientation and Certain Biological Problems*, American Institute Biological Sciences, Washington, D.C. Prepared for the use of biologists working in the area of animal migration and homing, biological clocks and periodic activity. Contains solved examples.

COCHRAN, W. G., AND G. M. COX: 1957, *Experimental Designs*, Wiley, New York. Application and interpretation of experimental design. Requires a background of statistics. Should be consulted if the problem requires some experimental design.

COX, D. R.: 1958, *Planning of Experiments*, Wiley, New York. Describes basic ideas underlying statistical aspects of experimental design. Emphasis is on planning. Probably the best general introduction to the design of experiments now available to nonstatisticians.

GOLDSTEIN, A.: 1964, *Biostatistics: An Introductory Text*, Macmillan, New York. An introductory text written from an applied point of view. Discusses and illustrates quantitative data, enumeration data, and linear regression. Some simple nonparametric techniques are treated briefly.

GRIEG-SMITH, P.: 1964, *Quantitative Plant Ecology*, Butterworth, Washington, D.C. A statistical approach to plant ecology. This book is a must for all ecologists concerned with vegetation studies.

HOGG, R. V., AND A. T. CRAIG: 1970, *An Introduction to Mathematical Statistics*. A clear mathematical exposition of statistics. Treats binominal and Poisson distributions, density functions, etc.

KERSHAW, K. A.: 1964, *Quantitative and Dynamic Ecology*, Elsevier, New York. Oriented toward plant ecology, this book considers simpler statistical procedures and their application to vegetation studies. Strong on correlation of positive and negative association between species, Poisson series, and the detection of nonrandomness and natural groupings.

LEWIS, T., AND L. R. TAYLOR: 1967, *An Introduction to Experimental Ecology*, Academic, New York. Useful but not as detailed or comprehensive as Southwood's *Ecological Methods*.

PATTEN, B. C. (ED.): 1971, 1972, *Systems Analysis and Simulation in Ecology* (2 vols., 3d in preparation), Academic, New York. Application of systems science and computer technology to ecology. Invaluable reference on modeling.

PEARCE, S. C.: 1963, *Biological Statistics: An Introduction*, McGraw-Hill, New York. Presents statistical methods to biologists who have a minimum of mathematical background, but need to use the techniques. Discusses statistical analysis, experimental design, and interpretation of data.

PIELOU, E. C.: 1969, *An Introduction to Mathematical Ecology*, Wiley, New York. Excellent introduction to statistical and mathematical approaches to ecology. A graduate level book.

SCHULTZ, V.: 1961, An annotated bibliography on the uses of statistics in ecology—A Search of 31 periodicals, Publ. TID 3908, *U.S. Atomic Energy Commission*, Office of Technical Information, Environmental Science Branch, Division of Biology and Medicine, Washington, D.C. An extremely useful compendium. Summarizes the uses of statistics in ecology.

An annotated bibliography of statistical methods

SIEGEL, S.: 1956, *Nonparametric Statistics for the Behavioral Sciences*, McGraw-Hill, New York. A collection of nonparametric tests in common use. Procedures are explained and illustrated by mathematical examples.

SIMPSON, G. G.; A. ROE; AND R. C. LEWONTIN: 1960, *Quantitative Zoology*, rev. ed., Harcourt Brace Jovanovich, New York. As the title implies, this book discusses zoology quantified. Statistical concepts and procedures are applied to taxonomy, distribution, populations, etc. Invaluable for the ecologist involved with animals and animal populations.

SNEDECOR, G. W.: 1957, *Statistical Methods*, Iowa State College Press, Ames. Excellent reference for those with some familiarity with statistics.

SOKAL, R. R., AND F. J. ROHLF: 1969, *Biometry, The Principles and Practices of Statistics in Biological Research*, Freeman, San Francisco. Perhaps the best introduction to biometrics and applied statistics.

SOKAL, R. R., AND F. J. ROHLF: 1973, *Introduction to Biostatistics*, Freeman, San Francisco. Excellent introduction. Requires minimal background in math and covers nonparametric tests.

SOUTHWOOD, T. R.: 1966, *Ecological Methods: With Particular Reference to the Study of Insect Populations*, Barnes & Noble, New York. An excellent handbook on statistics in ecology. Especially good on population statistics.

WATT, K. E. F. (ED.): 1966, *Systems Analysis in Ecology*, Academic, New York. Surveys problems and techniques of systems analysis in ecology. See in particular Chap. 1.

WATT, K. E. F.: 1968, *Ecology and Resource Management, A Quantitative Approach*, McGraw-Hill, New York. A systems-oriented text, useful mostly for graduate students, but much material useful to other students.

Abstracts of current biometrical literature useful to ecologists appear in *Wildlife Reviews* (see Bibliography).

O

ne of the major problems in ecology is the determination of population distribution, size, and changes in abundance. The problem involves sampling to estimate some characteristic of the population. By taking into account the variability within the population, one can make some general inferences about the population as a whole. To be valid the samples must be completely random; that is all combinations of sampling units must have an equal probability of being selected. To characterize the population as a whole, certain *parameters* are used. The proportion of males in a population is a parameter; so is the mean value per plot.

The objective of sampling is to estimate some parameter or a function of some parameter. The value of the parameter as estimated from a sample is the *estimate*, which is hoped to be *accurate*—close to the true value. But often the estimate is *biased*. This is a systematic distortion due to some flaw in the measurement or to the method of collecting the sample. It should be avoided, but often it cannot be. In any case the important thing is to recognize the source of the bias and take it into account. A biased account can never be accurate, although it may be *precise*. Precision refers to a clustering of sample values about their own mean (Fig. B-1).

SAMPLING PLANT POPULATIONS

Methods

Methods of analyzing the vegetation occupying a given site are numerous and the literature discussing them, the underlying philosophies, and the statistical treatments is extensive. Basic references are given at the end of this appendix, and they should be consulted. Because a major decision in ecological studies involves the methods to be used, a number of ways of handling the vegetation are given here with some comments on their advantages and disadvantages. Again these are simply personal selections, based on experience, and they are not complete by any means.

QUADRATS OR SAMPLE PLOTS

Strictly speaking, the quadrat applies to a square sample unit or plot. It may be a single sample unit or it may be divided into subplots. Quadrats may vary in size, shape, number, and arrangement, depending upon the nature of the vegetation and the objectives of the study.

The size of the quadrat must be adapted to the characteristics of the community. The richer the flora, the larger or more numerous the quadrats must be. To sample forest trees the fifth-acre plot is a popular size, but it may be too large if trees are numerous or if many species are involved. Smaller plots can be used to study shrubs and understory trees. For grass and herbaceous plants, 1 m^2 is the usual size. The quadrat is usually square, but rectangular or circular ones may work better. Circular plots are the easiest to lay out because one needs only a center stake and string of desired length.

The number of sample units to be employed always presents some problems. The number will vary with the characteristics of the community, objectives of the investigation, degree of precision, and so on. The final number more often than not is arbitrary, but by using statistical methods the reliability of the sample and the number of samples needed for any desired degree of accuracy can be determined, once a normal distribution around a mean has been established.

Species-area curves. A second approach to this problem is through the species-area curve (Fig. B-2). This is obtained by plotting the number of species found in plots of different sizes (vertical axis) against the sample size area (horizontal axis). The curve rises sharply at first because the number of new species found is large. As the sample plot size or number is increased, the quantity of new species added declines to a point of diminishing returns where there is little to be gained by continuing the sampling. This curve can be plotted on an arithmetic or a logarithmic base (see Vestal, 1949). The method can be employed to determine the largest size of a single plot (minimal area) needed to survey the community adequately. In this case the sampling should be done by using a geometric system of nested plots (Fig. B-3). Or the curve can be used to determine the minimum

Sampling plant and animal populations

number of small multiple plots needed for a satisfactory sample. In addition the species-area curve can be used to compare one community with another (Fig. B-4).

Kinds of quadrats. Quadrats are often labeled according to the uses of or data derived from them.

1. List quadrat. The organisms found are listed by name. A series of list quadrats gives a floristic analysis of the community and allows an assignment of a frequency index, but nothing else.
2. Count quadrat. The numbers as well as the names of each species encountered are recorded. This method is widely used in forest survey work; but here additional information, such as height, volume, basal area, etc., also is taken. Quadrats used in deer-browse studies also fall into this category.
3. Cover quadrat. The actual or relative coverage is recorded, usually as a percentage of the area of the ground surface covered or shaded by vegetation.
4. Chart quadrat. A quadrat that is mapped to scale to show the location of individual plants. This is a tedious job, but where long-range studies of vegetation changes are being made, this method provides the "big picture."

Location of plots. For statistically reliable estimates, the plots must be randomized. This is rather easily done. On an area photo or map of the study area, lay out a series of grid lines. Number the lines on a horizontal and vertical axis. Write the numbers of the vertical and horizontal grid lines on small squares of paper, place in two separate bowls, mix, and draw out pairs of numbers. Then locate the position of each plot by plotting the point given by the paired numbers.

THE BELT TRANSECT

A variation of the quadrat method is the belt transect. A *transect* is a cross section of an area used as a sample for recording, mapping, or studying vegetation. Because of its continuity through an area, the transect can be used to relate changes in vegetation along the line or strip with changes in the environment. As a sample unit, the measurements

693

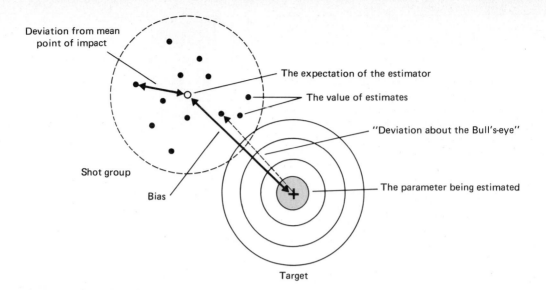

Deviation from mean
point of impact

The expectation of the estimator

The value of estimates

"Deviation about the Bull's-eye"

The parameter being estimated

Shot group

Bias

Target

within a transect are pooled and each transect is treated as a single observation.

The belt transect can be set up as follows.

1. Determine the total area of the site to be sampled; then divide by 5 or 10 to obtain the total sample area.
2. Lay out a series of belt transects of a predetermined width and length, sufficient to embrace the area to be sampled. Then divide the belts into equal-sized segments. These are sometimes called quadrats or plots, but they differ from true quadrats in that each represents an observational unit rather than a sampling unit.
3. The vegetation in each unit is measured for some attribute, depending upon the problem at hand: abundance, sociability, frequency, stem counts, etc.

A variation of the segmented-belt transect consists of taking observations only on alternate segments. The precision seems to be affected very little (Oosting, 1956). For example, 10 quadrats alternately spaced on a 20-ft belt are nearly twice as efficient statistically as 10 quadrats on a 10-ft belt.

Advantages and disadvantages. The quadrat method has its advantages and disadvantages. It is a popular method, easily employed. If the individual organisms are randomly distributed, then the accuracy of the sample and the estimate of the density depend upon the size of the sample. But individuals seldom are randomly dispersed, so the accuracy of quadrat sampling may be low, unless a great number of plots are involved. The quadrat method is tedious and time-consuming.

The belt transect is well adapted to estimate abundance, frequency, and distribution. But for estimating the frequency index, it has the disadvantage that frequency by classes is related to the size of the plot. If one wishes to compare one area with another, the segment size used in sampling must be the same for both areas.

Analysis. There are several ways to record data from quadrats and belt transects. One is simply to record the presence of a species. This perhaps is the most objective method but it limits analysis to frequency and relative frequency. A second method is to record the number of individuals of the various species found in the quadrat. This gives a density figure. Because of the variations in growth forms among the various species, numbers mean little. Counts are most useful in certain situations such as counting the number of stems of shrubby plants available for deer browse or counting the amount of forest reproduction. The samples are broken down into such classes 1 ft high, 1 to 12 ft, or 1 to 3 in., 4 to 9 in., and so on.

A third method is that of Braun-Blanquet (1951). This involves a total estimate based on abundance and cover. If the number of individuals in a plant community is estimated but not counted, the data are referred to as *abundance*. Abundance implies a number of individuals, but number does not necessarily reflect dominance or *cover*. Cover is the result of both numbers and massiveness. Although abundance and coverage are separate and distinct, they can be combined in a community description as the total estimate. For many field studies this method works very well. But it is subjective and the data are difficult to handle statistically. However, this method does provide a useful general picture of the plant community. The scales are given in Table B-1. Along with total estimates should go an estimate of sociability of each species (Table B-2)—whether the plant grows singly, in clumps, mats, and so on.

By determining both the total estimate and the sociability, each species has paired values. These are usually written as plant species A, 4.3, in which the first figure is the total estimate, the second the sociability. Once a number of stands have been surveyed, the community characteristics can be combined in a sort of an association table. The plant species usually are listed on the basis of fidelity or presence, the characteristic species of the community often heading the list. Examples are given in Table B-3. It is far from being complete and is given here only as an example of how such tables are constructed.

Data so collected describe individual stands. By the use of another attribute, presence, one can compare stands of a community type or of related types. Presence refers to the degree of regularity with which a species recurs in different examples of a community type. It is commonly expressed as a percentage that can be assigned to one of a limited number of presence classes, as given in Table B-4. Presence is determined by dividing the total number of stands in which the species is found by the total number of stands investigated. Species that have a high percentage of presence or which fall within presence class 5 often are regarded as more or less characteristic of that community.

LINE INTERCEPT

The line transect is one-dimensional. Most useful for sampling shrub stands and woody understory of the forest, the line transect or line intercept method consists of taking observations on a line or lines laid out randomly or systematically over the study area.

1. Stretch a metric steel tape, steel chain, or a tape between two stakes 50 to 100 m or ft apart.
2. Subdivide the line into predetermined intervals such as 1 m, 5 m, etc.
3. Move along the line, and for each interval record the plant species found and the

FIGURE B-1 [OPPOSITE]

A graphic model of the various statistical concepts involved in estimating populations. The solid bulls-eye represents the parameter being estimated, in this instance population size. The dotted target represents the sampling estimated. The distance of the samples or shots from the true target or bulls-eye is the "deviation." Value of estimates gives one the mean value or expectation of the estimate. The distance of the shots from the mean point of impact are the deviations. The distance between the mean point of impact or expectation of the estimator and the true value, the bulls-eye (which one may never know) is the bias. *A tight shot or a clumping of estimates around a mean value indicates a high precision but the estimates are not necessarily accurate. A shot group close to the parameter being estimated increases accuracy. The variance is the mean squared deviation about the mean point of impact. (From Giles, 1969,* Wildlife Management Techniques.)

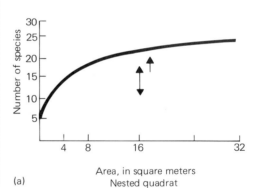

(a) Area, in square meters
Nested quadrat

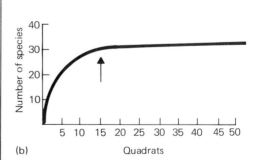

(b) Quadrats

FIGURE B-2

Species area curves: (a) *for miminal area of quadrat;* (b) *for miminal number of quadrats. Arrows indicate minimal areas.*

TABLE B-3
Stand composition, Cumberland Plateau, West Virginia

Herbaceous species	Plot number										Frequency %
	1	2	3	4	5	6	7	8	9	10	
Polystichum aerostichoides	2.2	+0.1	1.2	·1.2	2.2	2.2	2.2	2.2	2.1	1.1	100
Cimicifuga racemosa	3.2	2.2			2.2	3.2	2.2	1.2	2.2	2.2	80
Geranium maculatum		+0.1	+0.1	+0.1	1.2	2.2	2.2	+0.2	1.1	+0.1	90
Disporum lanquinosium	3.2	3.3		3.3	1.1	2.2	1.1	+0.1	+0.2	2.2	90
Galium circaezans	+0.2	+0.2			1.2	+0.1	2.2	+0.1			60
Thalictrum dioicum		+0.2			1.2	1.1	2.1			+0.2	50
Sanicula canadensis	+0.1	2.2	1.1		+0.1						40

Note: Selected species only, to illustrate table construction.

distance it covers along that portion of the line intercept. Consider only those plants touched by the line or lying under or over it. Treat each stratum of vegetation separately, if necessary,

   a. For grasses, rosettes, and dicot herbs, measure the distance along the line at ground level.

   b. For shrubs and tall dicot herbs, measure the shadow or distance covered by a downward projection of the foliage above.

4. Usually 20 to 30 such lines are sufficient.

The data can be summarized as follows.

1. Number of intervals in which each species occurs along the line.
2. Frequency of occurrence for each species in relation to total intervals sampled.
3. Total linear distance covered by each species along the transect.
4. Total length of line covered by vegetation and total "open" length.
5. Total number of individuals, if they can be so recorded. Because of the nature of branching and size variations it is difficult to count individual plants on line transect.

Advantages and disadvantages. The advantages and disadvantages can be summarized as follows. This method is rapid, objective, and relatively accurate. The area may be determined directly from recorded observations. The lines can be randomly placed and replicated to obtain the desired precision. The method is well adapted for measuring changes in vegetation if the ends of the lines are well marked. Generally it is more accurate in mixed plant communities than quadrat sampling and is especially well suited for measuring

low vegetation. On the debit side, the method is not well adapted for estimating frequency or abundance because the probability of an individual being sampled is proportional to its size. Nor is it suited where vegetation types are intermingled and the boundaries indistinct.

Analysis. From the data the following measurements may be calculated.

$$\text{relative density} = \frac{\text{total individuals species A}}{\text{total individuals all species}} \times 100$$

(Calculation of relative density may not be possible if individual plants can not be identified.)

$$\text{dominance (cover)} = \frac{\text{total intercept length, species A}}{\text{total transect length}} \times 100$$

$$\text{relative dominance} = \frac{\text{total intercept length, species A}}{\text{total intercept length, all species}} \times 100$$

$$\text{frequency} = \frac{\text{intervals in which species occurs}}{\text{total number of transect intervals}} \times 100$$

$$\text{relative frequency} = \frac{\text{frequency value, species A}}{\text{total frequency value, all species}} \times 100$$

POINT-QUARTER METHOD

There have been several variations of the variable plot or "plotless" method developed for ecological work. These methods arose from the variable radius method of forest sampling developed in Germany by Bitterlich. He used it to determine timber volume without establishing plot boundaries. The method was introduced into the United States by Grosenbaugh (1952, 1958).

One of the most useful of the plotless methods is the point-quarter method (see Cottam and Curtis, 1956; Greig-Smith, 1964). It is most useful in sampling communities in which individual plants are widely spaced or in which the dominant plants are large shrubs or trees.

The procedure is as follows.

1. Locate a series of random points within the stand to be sampled, or pick random points along a line transect passing through the stand.
2. At each station make a point in the ground.
3. Divide the working area into four quarters or quadrants by visualizing a grid line, predetermined by compass bearing, and a line crossing it at right angles, both passing through the point (Fig. B-5).
4. Select the closest tree (or plant) to the point in each of the four quarters, measure and record the distance of each from the point, obtain the diameter breast height, and record the species. Four trees are measured at each point. The tally sheet should indicate their species, diameter (diameter tapes can be obtained at forestry supply houses) and distance from the point.
5. Tally at least 50 such points.

The computations are as follows.

1. Add all distances in the samples and divide this by the total number of distances to obtain a mean distance of point to plant.

$$\text{mean distance} = \frac{\text{total distance}}{\text{number of distances}}$$

2. Square the mean distance to obtain the mean area covered on the ground per plant.
3. Divide the mean area per plant into the unit area on which density is expressed. If the area is in feet then divide the mean distance squared into 43,560 ft^2 to obtain the total number of trees per acre.
4. Determine basal area for each tree (see Tables B-5 and B-6) from diameter

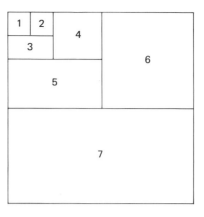

FIGURE B-3
An example of nested quadrats.

TABLE B-1
Total estimate scale (abundance plus coverage)

+ Individuals of a species very sparsely present in the stand; coverage very small
1 Individuals plentiful, but coverage small
2 Individuals very numerous if small; if large, covering at least 5% of area
3 Individuals few or many, collectively covering 6–25% of the area
4 Individuals few or many, collectively covering 26–50% of the area
5 Plants cover 51–75% of the area
6 Plant species cover 76–100% of the area

TABLE B-2
Sociability classes of Braun-Blanquet

Class 1   Shoots growing singly
Class 2   Scattered groups or tufts of plants
Class 3   Small, scattered patches or cushions
Class 4   Large patches or broken mats
Class 5   Very large mats of stands or nearly pure populations that almost blanket the area

TABLE B-5
Areas of circles in square feet for diameters in inches

Diameter	0.0	0.1	0.2	0.3	0.4	0.5	0.6	0.7	0.8	0.9
0	0.000	0.000	0.000	0.000	0.001	0.001	0.002	0.003	0.003	0.004
1	0.005	0.007	0.008	0.009	0.011	0.012	0.014	0.016	0.018	0.020
2	0.022	0.024	0.026	0.029	0.031	0.034	0.037	0.040	0.043	0.046
3	0.049	0.052	0.056	0.059	0.063	0.067	0.071	0.075	0.079	0.083
4	0.087	0.092	0.096	0.101	0.106	0.110	0.115	0.120	0.126	0.131
5	0.136	0.142	0.147	0.153	0.159	0.165	0.171	0.177	0.183	0.190
6	0.196	0.203	0.210	0.216	0.223	0.230	0.238	0.245	0.252	0.260
7	0.267	0.275	0.283	0.291	0.299	0.307	0.315	0.323	0.332	0.340
8	0.349	0.358	0.367	0.376	0.385	0.394	0.403	0.413	0.422	0.432
9	0.442	0.452	0.462	0.472	0.482	0.492	0.503	0.513	0.524	0.535
10	0.545	0.556	0.567	0.579	0.590	0.601	0.613	0.624	0.636	0.648
11	0.660	0.672	0.684	0.696	0.709	0.721	0.734	0.747	0.759	0.772
12	0.785	0.799	0.812	0.825	0.839	0.852	0.866	0.880	0.894	0.908
13	0.922	0.936	0.950	0.965	0.979	0.994	1.009	1.024	1.039	1.054
14	1.069	1.084	1.100	1.115	1.131	1.147	1.163	1.179	1.195	1.211
15	1.227	1.244	1.260	1.277	1.294	1.310	1.327	1.344	1.362	1.379
16	1.396	1.414	1.431	1.449	1.467	1.485	1.503	1.521	1.539	1.558
17	1.576	1.595	1.614	1.632	1.651	1.670	1.689	1.709	1.728	1.748
18	1.767	1.787	1.807	1.827	1.847	1.867	1.887	1.907	1.928	1.948
19	1.969	1.990	2.011	2.032	2.053	2.074	2.095	2.117	2.138	2.160
20	2.182	2.204	2.226	2.248	2.270	2.292	2.315	2.337	2.360	2.382
21	2.405	2.428	2.451	2.474	2.498	2.521	2.545	2.568	2.592	2.616
22	2.640	2.664	2.688	2.712	2.737	2.761	2.786	2.810	2.835	2.860
23	2.885	2.910	2.936	2.961	2.986	3.012	3.038	3.064	3.089	3.115
24	3.142	3.168	3.194	3.221	3.247	3.274	3.301	3.328	3.355	3.382
25	3.409	3.436	3.464	3.491	3.519	3.547	3.574	3.602	3.631	3.659
26	3.687	3.715	3.744	3.773	3.801	3.830	3.859	3.888	3.917	3.947
27	3.976	4.006	4.035	4.065	4.095	4.125	4.155	4.185	4.215	4.246
28	4.276	4.307	4.337	4.368	4.399	4.430	4.461	4.493	4.524	4.555
29	4.587	4.619	4.650	4.682	4.714	4.746	4.779	4.811	4.844	4.876
30	4.909	4.942	4.974	5.007	5.041	5.074	5.107	5.140	5.174	5.208
31	5.241	5.275	5.309	5.343	5.378	5.412	5.446	5.481	5.515	5.550
32	5.585	5.620	5.655	5.690	5.726	5.761	5.796	5.832	5.868	5.904
33	5.940	5.976	6.012	6.048	6.084	6.121	6.158	6.194	6.231	6.268
34	6.305	6.342	6.379	6.417	6.454	6.492	6.529	6.567	6.605	6.643
35	6.681	6.720	6.758	6.796	6.835	6.874	6.912	6.951	6.990	7.029
36	7.069	7.108	7.147	7.187	7.227	7.266	7.306	7.346	7.386	7.426
37	7.467	7.507	7.548	7.588	7.629	7.670	7.711	7.752	7.793	7.834
38	7.876	7.917	7.959	8.001	8.042	8.084	8.126	8.169	8.211	8.253
39	8.296	8.338	8.381	8.424	8.467	8.510	8.553	8.596	8.640	8.683
40	8.727	8.770	8.814	8.858	8.902	8.946	8.990	9.035	9.079	9.124
41	9.168	9.213	9.258	9.303	9.348	9.393	9.449	9.484	9.530	9.575
42	9.621	9.667	9.713	9.759	9.805	9.852	9.898	9.945	9.991	10.038
43	10.085	10.132	10.179	10.226	10.273	10.321	10.368	10.416	10.463	10.511
44	10.559	10.607	10.655	10.704	10.752	10.801	10.849	10.898	10.947	10.996
45	11.045	11.094	11.143	11.192	11.242	11.291	11.341	11.391	11.441	11.491
46	11.541	11.591	11.642	11.692	11.743	11.793	11.844	11.895	11.946	11.997
47	12.048	12.100	12.151	12.203	12.254	12.306	12.358	12.410	12.462	12.514
48	12.566	12.619	12.671	12.724	12.777	12.830	12.882	12.936	12.989	13.042
49	13.095	13.149	13.203	13.256	13.310	13.364	13.418	13.472	13.527	13.581
50	13.635	13.690	13.745	13.800	13.854	13.909	13.965	14.020	14.075	14.131

measurements. (Basal area is the area of a plane passed through the stem of a tree at right angles to its longitudinal axis and at breast height. Because the cross section approximates a circle, its area can be computed from a standard formula.)

5.
$$\text{relative density} = \frac{\begin{array}{c}\text{individual of}\\\text{species } A\end{array}}{\begin{array}{c}\text{total individuals}\\\text{of all species}\end{array}} \times 100$$

6.
$$\text{density} = \frac{\begin{array}{c}\text{relative density}\\\text{of species } A\end{array}}{100} \times \begin{array}{c}\text{total density}\\\text{of all species}\end{array}$$

7.
$$\begin{array}{c}\text{relative}\\\text{dominance}\end{array} = \frac{\begin{array}{c}\text{total basal area}\\\text{of species } A\end{array}}{\begin{array}{c}\text{total basal area}\\\text{of all species}\end{array}} \times 100$$

8.
$$\text{frequency} = \frac{\begin{array}{c}\text{no. of points at which}\\\text{species } A \text{ occurs}\end{array}}{\text{total no. of points sampled}} \times 100$$

9.
$$\begin{array}{c}\text{relative}\\\text{frequency}\end{array} = \frac{\begin{array}{c}\text{frequency value}\\\text{for species } A\end{array}}{\begin{array}{c}\text{total frequency value}\\\text{for all species}\end{array}} \times 100$$

10. Absolute values for the number of trees per unit area of any species and the basal area per unit area of any species are determined by multiplying the relative figures for density by the total trees per acre to determine density and by the total basal area per acre to determine absolute dominance.

Advantages and disadvantages. The point-quarter method is simple and rapid and works quite well. The underlying assumption is that individuals of all species together are randomly dispersed. Although this assumption may not be true, it does not seem to produce significant error, except where a deviation from overall randomness is quite obvious. In spite of this, relative density and relative dominance are valid even if dispersion is not random. Error would appear in the calculation of absolute density and absolute dominance.

699

TABLE B-4
Presence classes

Presence class	Stands of one community type studied in which species occur %
1	1–20
2	21–40
3	41–60
4	61–80
5	81–100

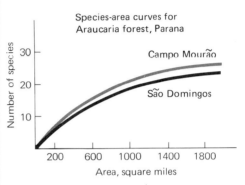

FIGURE B-4
Species-area curve for a stand of Araucaria *forest on different sites near Campo Mourão, Parana, showing the use of species-area curves to compare one community with another. (From Cain and Castro, 1959, Manual of Vegetation Analysis, Harper & Row, New York.)*

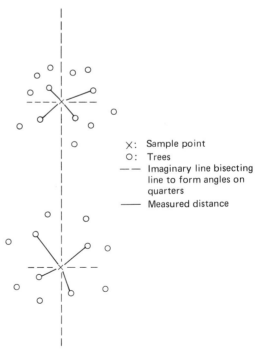

Point quarter method

FIGURE B-5
The point-quarter method of sampling forest stands. See text for details.

Appendix B
SAMPLING POPULATIONS

TABLE B-6

Areas of circles in square feet for circumferences and diameters in inches and in square meters for circumferences and diameters in centimeters

Diameter (in./cm)	Circumference (in./cm)	Area (ft²)	Area (m²)	Diameter (in./cm)	Circumference (in./cm)	Area (ft²)	Area (m²)
1	3.14	0.005	—	51	160.22	14.186	0.240
2	6.28	0.022	—	52	163.36	14.748	0.212
3	9.42	0.049	0.001	53	166.50	15.321	0.221
4	12.57	0.087	0.001	54	169.65	15.904	0.229
5	15.71	0.136	0.002	55	172.79	16.499	0.238
6	18.85	0.196	0.003	56	175.93	17.104	0.246
7	21.99	0.267	0.004	57	179.07	17.721	0.255
8	25.13	0.349	0.005	58	182.21	18.348	0.264
9	28.27	0.442	0.006	59	185.35	18.986	0.273
10	31.42	0.545	0.008	60	188.50	19.635	0.283
11	34.56	0.660	0.010	61	191.64	20.295	0.292
12	37.70	0.785	0.011	62	194.78	20.966	0.302
13	40.84	0.922	0.013	63	197.92	21.648	0.312
14	43.98	1.069	0.015	64	201.06	22.340	0.322
15	47.12	1.227	0.018	65	204.20	23.044	0.332
16	50.26	1.396	0.020	66	207.34	23.758	0.342
17	53.41	1.576	0.023	67	210.49	24.484	0.352
18	56.55	1.767	0.025	68	213.63	25.220	0.363
19	59.69	1.969	0.028	69	216.77	25.967	0.374
20	62.83	2.182	0.031	70	219.91	26.725	0.385
21	65.97	2.405	0.035	71	223.05	27.494	0.396
22	69.12	2.640	0.038	72	226.19	28.274	0.407
23	72.26	2.885	0.042	73	229.34	29.065	0.418
24	75.40	3.142	0.045	74	232.48	29.867	0.430
25	78.54	3.409	0.049	75	235.62	30.680	0.442
26	81.68	3.687	0.053	76	238.76	31.503	0.454
27	84.82	3.976	0.057	77	241.90	32.338	0.466
28	87.96	4.276	0.062	78	245.04	33.183	0.478
29	91.11	4.587	0.066	79	248.18	34.039	0.490
30	94.25	4.909	0.071	80	251.33	34.907	0.503
31	97.39	5.241	0.075	81	254.47	35.785	0.515
32	100.53	5.585	0.080	82	257.61	36.674	0.528
33	103.67	5.940	0.086	83	260.75	37.574	0.541
34	106.81	6.305	0.091	84	263.89	38.484	0.554
35	109.96	6.681	0.096	85	267.04	39.406	0.567
36	113.10	7.069	0.102	86	270.18	40.339	0.581
37	116.24	7.467	0.108	87	273.32	41.282	0.594
38	119.38	7.876	0.113	88	276.46	42.237	0.608
39	122.52	8.296	0.119	89	279.60	43.202	0.622
40	125.66	8.727	0.126	90	282.74	44.179	0.636
41	128.81	9.168	0.132	91	285.88	45.166	0.650
42	131.95	9.621	0.138	92	289.03	46.164	0.665
43	135.09	10.085	0.145	93	292.17	47.173	0.679
44	138.23	10.559	0.152	94	295.31	48.193	0.694
45	141.37	11.045	0.159	95	298.45	49.224	0.709
46	144.51	11.541	0.166	96	301.59	50.266	0.724
47	147.65	12.048	0.173	97	304.73	51.318	0.739
48	150.80	12.566	0.181	98	307.88	52.382	0.754
49	153.94	13.095	0.189	99	311.02	53.456	0.770
50	157.08	13.635	0.196	100	314.16	54.542	0.785

IMPORTANCE VALUES

In regions where the plant communities are highly heterogeneous, the classification of communities on the basis of dominants or codominants becomes impractical. Therefore, Curtis and McIntosh (1951) came up with the index "importance value" to develop a logical arrangement of the stands. This index is based on the fact that most species do not normally reach a high level of importance in the community, but those that do serve as an index, or guiding species. Once importance values have been obtained for species within a stand, the stands can then be grouped by their leading dominants according to importance values, and the groups are then placed in a logical order based on the relationships of several predominant species. In Table B-7, for example, are four species that were the leading dominants in 80 of 95 forest stands in southern Wisconsin. Note that the dominants are arranged in order of decreasing importance value, from stands dominated by black oak to those dominated by sugar maple. Such an arrangement also shows increasing values for sugar maple. Trees intermediate in dominance can be handled in the same way.

CONTINUUM INDEX

The importance value can be expanded further into a continuum index, a composite figure that can be used to compare a large number of stands. Curtis and others in Wisconsin found in their study of importance values that each species reaches its best development in stands whose position bears a definite relationship to that of other species. In other words the stands varied continuously along a gradient; thus the word *continuum*. The continuum index is extremely useful because it can be employed to investigate environmental relationships of component communities, to designate the position of a stand on a gradient, and to provide a background for studies of other organisms.

The continuum index for a stand is obtained by first assigning a climax adaptation number to the various tree species involved. Curtis and McIntosh (1951) established a 10-part scale for Wisconsin, in which the highly tolerant sugar maple was given a value of 10, because it was best adapted to maintain itself in a stand. The intolerant bur oak

Appendix B
SAMPLING POPULATIONS

TABLE B-7
The average importance-value index of trees in stands
with four species as the leading dominants

Species	Leading dominant in stand				Ecological sequence number
	Quercus velutina	*Quercus alba*	*Quercus rubra*	*Acer saccharum*	
Black oak (*Quercus velutina*)	165.1	39.6	13.6	0	2
Shagbark hickory (*Carya ovata*)	0.3	8.8	5.2	5.9	3.5
White oak (*Quercus alba*)	69.9	126.8	52.7	13.7	4
Black walnut (*Juglans nigra*)	1.5	1.2	2.2	1.9	5
Red oak (*Quercus rubra*)	3.6	39.2	152.3	37.2	6
American basswood (*Tilia americana*)	0.3	5.9	19.0	33.0	8
Sugar maple (*Acer saccharum*)	0	0.8	11.7	127.0	10

Source: Adapted from Curtis and McIntosh, 1951.

and trembling aspen were given a value of 1 because they are early successional species. Cain and Castro (1959) suggest that the term *ecological sequence number* be used in place of the *climax adaptation number* to get rid of that questionable term *climax*. Next the importance value of each species is multiplied by its ecological sequence number, and the values for all the species in the stand are added. The sum is the stand continuum index. This is used to place that stand on a continuum scale that runs from 300 to 3,000. After the stand indices have been calculated, the position of the individual species can be plotted in relation to the position of the stand on the index (Fig. B-6).

Periphyton and phytoplankton

In aquatic communities algae are the dominant vegetation. There are two kinds of growth involved: the plankton suspended on the water and the periphyton growing attached to some substrate.

PHYTOPLANKTON

The phytoplankton can be obtained by drawing water samples from several depths (see Appendix D). Cell counts of algae present in each sample, either normal or concentrated, can be made with a Sedgewich-Rafter counting chamber and a Whipple ocular. If necessary, the samples can be concentrated by centrifugation in a Foerst plankton centrifuge. The centrifuged samples are then diluted to a suitable volume (100 to 200 ml) in a volumetric flask.

When counting the cells, a separate tally should be kept for each species. This will permit an analysis of community structure at each station. Record the number of cells for single-celled forms, and the number of colonies for colonial forms. The latter should then be multiplied by an appropriate factor for each species to convert the colonies into cells. These factors will have to be predetermined by averaging the cell counts from a large number of typical colonies from the area in question.

Another method of handling the phytoplankton is by filtration (see McNabb, 1960; Clark and Sigler, 1963). The organisms in the sample are first fixed by the addition of 4 parts 40 percent aqueous solution of formaldehyde to each 100 parts of the sample. The analysis is as follows.

1. Thoroughly agitate the sample; withdraw a fraction with a pipette large enough to hold a sample that will provide an optimum quantity of suspended matter on the filter.
2. This sample is placed in the tube of a filter apparatus designed to accommodate a 1-in.-diameter membrane filter. The water is drawn through the filter with a vacuum pump.
3. Remove the filter and place it on a glass slide. Put two or three drops of immersion oil over the residue, and store the slide in the dark to dry (about 24 hours). The oil replaces the water in the pores of the filter and makes it transparent.
4. Place a cover slip over the transparent filter.
5. Determine the most abundant species by scanning and then choose a quadrat size

702

that will contain individuals of this species approximately 80 percent of the time.
6. Move the mechanical stage so that approximately 30 random quadrats are
   viewed. The presence or absence of individual species is noted. There is no need
   to count.
7. When 30 quadrats have been surveyed, the
   percentage frequency can be calculated by:

$$\text{frequency} \atop (\%) = \frac{\text{total no. of occurrences of a species}}{\text{total no. of quadrats examined}} \times 100$$

PERIPHYTON

The periphyton has not received quite the
same attention from ecologists as the phytoplankton, particularly in a quantitative way.
Methods for studying the periphyton are
given in detail by Sladeckova, 1962.

Epiphyton, the periphyton growing on
living plants and animals, can be observed
in place on the organism if the substrate is
thin or transparent enough to allow the transmission of light. If the leaves are thin and
transparent, the task is relatively easy, but
the growth on one side must be scraped away.
If the leaf is opaque, the chlorophyll can be
extracted by dipping the leaf in chloral hy-
drate. Small leaves can be examined over the
whole area. Large leaves can be sampled in
strips marked by grids on a slide or by an
ocular micrometer. With large aquatic plants
a square will have to be cut from the leaf or
stem. If the leaf is too thick to handle under
the microscope, scrape off the periphyton
quantatively and mount in a counting cell for
examination. The results can be related back
to the total surface area.

Algae growing on such aquatic animals as
turtles, mollusks, etc., and on stones must be
removed for study. The quantitative scraping
and transfer is difficult, but there are several
techniques available.

One method employs a simple hollow square
instrument with a sharpened edge, which is
pressed closely or driven into the substrate.
This separates out a small area of given size
around which the periphyton is washed away.
The instrument is then raised and the periphyton remaining in the sample square is scraped
into a collecting bottle.

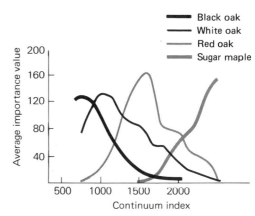

FIGURE B-6
*Distribution of four forest tree species on a
continuum index. (From Curtis and McIntosh,
1951.)*

If the stones can be picked up from the bottom, then the periphyton can be removed with an apparatus consisting of a polyethylene bottle with the bottom cut out and a brush with nylon bristles. A section of the stone is delimited by the neck of the bottle held tightly on the surface. The periphyton is scraped loose by the brush and then washed into a collecting bottle with a fine-jet pipette.

The periphyton can be counted in a Sedgewich-Rafter cell recording a predetermined number, usually 100 to 1,000, as they appear in the field of view. The results can be expressed as a percentage; or the algae can be checked for frequency in the field of view, using the Braun-Blanquet scale of total estimate.

Some of the difficulties can be avoided by growing the periphyton on an artificial substrate, usually glass or transparent plastic slides. These are attached in the water in a variety of ways. In lentic situations they can be placed on sand or stones in the water. Or they can be placed in saw-cuts on boards, in holes in bricks, clipped to a rope, attached to a wooden frame, or tucked into rubber corks, if the aquatic habitat is lotic.

In fresh water, the glass slides are placed in a vertical or horizontal position. In shallow water the plates are usually laid horizontally, directly on the bottom, especially if the influence of light on the composition of periphyton is one of the objectives of a study. Or they can be hung vertically in the water; in this case both sides will be covered with periphyton. Plates placed horizontally collect true periphyton, detritus, etc., on the upper surface, while the bottom will be colonized by heterotrophic organisms. In running water the vertical position with plates parallel to the current is best since the surfaces will not become too badly filled with mud and debris.

For algae and protozoa, the plates should be exposed for one to two weeks, and for hydras, sponges, and the like, about one month. At the end of this time the slides are collected and placed in wide-mouthed jars filled with water from the locality. In the lab, the periphyton can be observed directly on the slides under the microscope. For counting, the slides should be marked in a grid. To do this one can use a Whipple ocular micrometer, or clean the bottom of the slide and place it on an auxiliary slide marked off in a grid. If the

growth is very dense, it can be scraped off and examined in a Sedgewich-Rafter cell.

Biomass and production of periphyton can be determined by methods described below: dry weight, loss through ignition to determine organic carbon, pigment extraction, light and dark bottles containing pieces of plant with and without periphyton attached.

SAMPLING STREAM-BOTTOM ORGANISMS

Samples of stream-bottom organisms can be taken with a modified Surber bottom-fauna sampler (Fig. B-7). This consists of a brass frame with stainless-steel side pieces and a current baffle. To this is attached on a removable brass frame a fine net of 74 meshes/linear in. and a coarse net with 19 meshes/linear in. The latter is fitted in the sampler in front of the fine net to produce a cone or cone effect. Flanges on the insert prevent its being forced into the fine net in the rear.

This modified sampler picks up many small organisms that might otherwise be lost. In fact collection of virtually all macroorganisms is assured. In addition one obtains two subsamples with respect to size, and the small organisms are associated with fine detritus only.

This sampler encloses a specified area of stream (500 cm²), which is the sample unit. Organisms, detritus, and trash are scrubbed free from the substrate, and the current washes it into the net. Transfer the contents to a container and take back to the lab for examination and sorting.

Artificial stream habitat. Just as it is often desirable to observe pond and lake inhabitants in an indoor aquarium, so it is with stream organisms. But the problem of maintaining a stream population under artificial conditions is more difficult. However, a setup that works fairly well consists of a length of gutter, preferably wood, set on an angle with water from an outlet running in at the top and draining into a sink at the bottom. Water depth can be controlled by inserting a series of wooden partitions, each notched in the top, through which water pours like little waterfalls from one compartment to the next. The bed of the "stream" should be lined with small stones and gravel.

Conditions in a slow stream or pool can be simulated by arranging a series of shallow pans, one above the other. Each is filled with an overflow pipe covered with a fine wire filter. If the overflow pipes are arranged about 9 in. above the next pan, the oxygen concentration will approach saturation (see Warren and Davis, 1971).

Dendrochronology

Dendochronology is the science that treats the accurate dating of past events through the aging of trees. As such it is a valuable tool for the ecologist. It has been used in a number of studies—to age trees for management information, to establish dates of past forest fires, insect outbreaks, glaze, periods of suppression and release in the life history of forest trees. It has been involved in hydrological and archaeological studies and even in legal cases involving boundary disputes in which specimens are taken from fence posts and witness trees. An outstanding example of growth-ring analysis in an ecological investigation is Spencer's study (1958) of porcupine fluctuations in the Mesa Verde National Park.

Dendrochronology is based on the variation of growth rings. Growth rings, despite popular belief, are not regular, nor are they all necessarily laid down annually. Because of the failure of the cambium to form a sheath of xylem the entire length of the bole, rings may be omitted, especially near the base. This may be caused by the lack of food manufacture in the crown, by drought, fire, extreme cold, insects, and so on. At the other extreme are multiple rings produced by multiple waves of cambial activity during the growing season. These are caused by temporary interruptions in the normal growth, such as a late spring frost, or by regrowth taking place after normal seasonal growth has ended. At any rate the growth rings reflect the interaction of woody plants and their environment.

The fundamental principle of dendrochronology is crossdating, the correlation of distinctive patterns of growth between trees for a given sequence of years. Because no two plants have exactly the same growing conditions and life history (although the broad features are common to all trees involved) the similarities are relative rather than quantitative. The relative widths of corresponding rings are the same in relation to adjoining rings. By lining

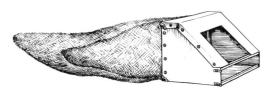

(a)

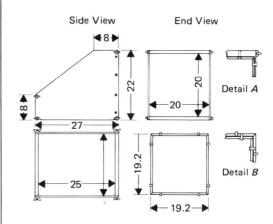

(b)

FIGURE B-7
(a) *A modified Surber bottom sampler. (Redrawn from Withers and Benson, 1962.)* (b) *Construction details for the Surber bottom sampler.*

up these similarities, the investigator can establish the relative identity of any rings in sequence and aberrant rings in the individual specimen. A great number of specimens must be cross-dated before each ring with a sequence can be dated.

COLLECTION OF MATERIAL

A recently logged-over area can provide an abundance of material, but new sections must be cut from the stump. Sections cut at a 30° angle, clean and sharp, may be ready for examination. Or the stump section can be smoothed with a carpenter's plane or by machine sanding. If the study involves shrubs, sample stems should be cut close to the ground and the entire cross section used.

Increment boring. The usual method of obtaining samples from forest trees is with the increment borer, available from forestry supply houses. This is an instrument designed to bore a core from a tree. It consists of a T handle, a hollow bit, and an extractor. Increment borers are fairly easy to use, but without care they can be damaged or broken. Here are a few hints.

1. For growth and age studies, remove the core as near to the base of the tree as the instrument handle will allow.
2. Coat the screw with heavy-grade oil.
3. To start the borer, combine a strong pushing and twisting motion until the borer is engaged in the wood.
4. Line the borer on the radius, keep the borer straight, and attempt to reach the center of the trunk.
5. When the core is drilled, insert the extractor and press firmly to cut the core from the trunk.
6. Remove the borer with reverse rotation.
7. Paint the wound with tree paint; a small cork can be inserted.
8. Store the cores in soda straws or polyethylene tubing.

Personally I find the large-diameter soda straws very satisfactory. Be sure to label each sample fully, including the directional side of the tree from which it was removed.

To obtain a freshly cut edge for examination, the core is held firmly in a core holder. The groove in a plastic foot ruler is fine if the ruler is clamped to a table and the end is stopped. With a razor blade make a transverse cut the length of the top of the core. It can then be brushed with water or kerosene to make the rings stand out better. When the core is ready, clip it to the stage of a microscope for examination. One-hundred-power magnification usually is sufficient.

With the use of a graduated mechanical stage, a stage micrometer, or a dial micrometer, measure the distance of each ring. The total distance included in the layers observed can be measured and then compared with the accumulated individual measurements. Any error should be distributed over the individual measurements. For serious research a dendrochronometer, a special device with a microscope and precise measuring devices, should be used.

CROSS-DATING

Although the more involved methods and problems of cross-dating cannot be described in detail here, the basic procedure is as follows.

1. On graph paper write down a series of numbers horizontally from left to right to represent growth layers. You can begin with one or with the years, the first number being the season preceding. This gives a series of numbers starting with the present and leading backward through the tree's life. A number of such blanks should be made up.
2. Set up a scale on the graph in thousands-of-an-inch so that the largest bars represent the *narrowest* widths.
3. Make a small bar graph for each year of the tree's life.
4. Make such a coded summary of all wood samples available.
5. Compare these visually, two at a time, sliding them along each other. Keep looking for corresponding groups of years with the same pattern of ring sequences. By such a technique, multiple rings can be checked, or extremely narrow growth rings previously missed can be picked up.

A simpler method but lacking the quantitative precision is to place an underline beneath the year that has a ring slightly less than the rings adjacent to it. If the decrease is more pronounced, draw two lines, and if very narrow, three lines. For very wide rings, draw two lines above the year.

STATISTICAL ANALYSIS

The data can be reduced to average values, and then compared to weather data covering the principal growing season for the species involved. Comparisons can be made between rainfall and the current year's growth, rainfall and the previous year's growth, monthly evapotranspiration deficits and growth, with frost-free periods and so on.

Data will have to be analyzed by simple or multiple regression, depending upon the variables, using partial correlations and standard errors for tests of significance. For sample analysis and interpretations, see Fretts, 1962.

Palynology

Palynology is the study of past plant communities by the analysis of pollen profiles. These studies are especially enlightening if they are coupled with carbon-14 dating.

PEAT COLLECTION

Peat cores are bored at one to several stations in the bog. They are taken with a peat borer, available commercially. (A Hiller-type borer is manufactured by the Deans Manufacturing Company, Deans, New Jersey.) At each station two separate borings should be drilled, several feet apart. Then by taking successive samples from alternate borings (example: first foot sample in core number one; second foot sample from core number two) contamination of one sample with another can be prevented. Collect two 6-in. samples, one from the lower part of the cylinder, the other from the upper, at each boring; then place the samples in glass vials. If the vials are completely filled and tightly sealed, no preservative should be needed (Walker and Hartman, 1960).

TREATMENT OF SAMPLES

Back in the laboratory, the samples can be treated as follows.

1. Thoroughly mix each 6-in. sample and remove a pea-size lump for deflocculation.
2. Boil the peat for a few minutes in a dilute solution of NaOH, gently breaking it apart with a wooden cocktail stirrer.

Use only one stirrer for each sample to avoid contamination.

3. While boiling add several drops of gentian-violet stain.
4. Stir vigorously and strain through fine wire mesh. Then stir again and draw up a 0.5-ml sample into a pipette.
5. Mix with a very small amount of warm glycerine jelly added to this sample.
6. Mount several drops on a slide and add cover slip.

EXAMINATION

The samples, now transferred to slides, should be examined under a microscope equipped with a mechanical stage.

1. Tally 100 or 200 pollen grains as they are encountered by systematically moving the slide.
2. Each kind of pollen grain must be identified, if only by code, and tallied separately. Identification should be made as far as possible from a reference pollen collection made up beforehand.
3. Record the results from each slide directly as a percentage for each kind of pollen.

PLOTTING THE POLLEN PROFILE

The pollen profile can be constructed by plotting a graph for each species or kind of pollen. The vertical scale is set up for depth in feet or meters; or the horizontal scale is percentage, based on counts of 100 or 200 pollen grains for each spectrum level.

SAMPLING ANIMAL POPULATIONS

The study of animals involves considerably more problems than the study of plants. Animals are harder to see and most are not stationary—they are here one minute and gone the next. When it comes to sampling, the animals have something to say about getting caught, and they are more liable to mortality than plants. The following methods of estimating animal numbers, determining age structure, mortality, home range, and so on will enable the field biologist to make some measurements, however rough, of animal populations in the ecosystem.

Trapping and collecting

The sampling of an animal population involves collecting animals, either alive for marking and release, or dead. Because so much information on collecting and trapping is available in other publications (see references at the end of this appendix), I will deal with this topic very briefly.

FLYING INSECTS

For diurnal insects, use aerial nets and heavy-duty sweep nets designed to withstand the hard wear encountered when put through grass and woody vegetation. For nocturnal insects, use traps containing ultraviolet light or a mercury-vapor light, or use an old sheet fitted on a slant against some support with a strong light above it. Insects can then be picked off the sheet. If the insects are to be killed, place them in a killing jar containing a layer, either on the bottom or in a deep lid, of plaster of paris and potassium cyanide. Insert thin layers of tissue or light cloth in the jar to prevent damage to moths and butterflies.

AQUATIC ORGANISMS

For aquatic organisms use dip nets for organisms in the water, bottom nets for scraping along the bottom of ponds, wire-basket scraper nets, and plankton towing nets. For aquatic collecting from the shore, aquatic throw nets are useful. To collect bottom organisms in deep water, use a bottom dredge lowered from a boat. Fish, tadpoles, and large crustaceans can be collected with seines. A set of assorted widths will be necessary.

SOIL ORGANISMS

The most difficult component of soil fauna to study are the soil arthropods. They are the most numerous, the most difficult to identify, and possibly the most difficult to sample accurately. Nematodes, white or pot worms, and protozoans are the worst, but these are not considered here, for they require specialized techniques for their extraction that will not be discussed. Those interested are referred to the book *Progress in Soil Zoology* (Murphy, 1962).

Soil arthropods can be extracted from the

soil by means of a Tullgren funnel, an improved version of the Berlese funnel, the construction of which is simple (Fig. B-8). Essentially it consists of a heat source and a smooth funnel, preferably glass, and a shelf of hardware cloth on which to place the sample. The heat from the lamp and then dessication drives the arthropods downward, until they fall into the collecting bottle beneath the funnel.

The procedure is rather simple. Place the sample of litter or soil on the hardware cloth shelf, so fitted in the funnel that air space is present between the wire and the wall of the funnel. To begin extraction, open the lid of the funnel 90° and turn on the 100-watt bulb. After about 16 hours, depending on sample size and moisture content, change to a 15-watt bulb and shut the lid. There will be two periods of arthropod exodus, the first wave due to heat, the second due to desiccation. The collecting bottle beneath the funnel may contain alcohol, formalin, or water. Water may be preferable, since it increases the humidity gradient toward which the animals move. The animals are then sorted and identified under the microscope.

These funnels are adequate for introductory soil biology. For more efficient extraction, necessary for serious studies in soil zoology, a better extractor is required. A new extractor for woodland litter has been described by Kempson, Lloyd, and Ghelardi (1963) in the journal *Pedobiologia*. The funnels are replaced by wide-mouthed bowls filled with an aqueous solution of picric acid. The acid not only preserves the specimens but also produces by evaporation a high humidity in the air just under the sample. The humid air is cooled by conduction from a cold-water bath in which the bowls are immersed.

Another method of extraction is floatation. Procedures, although relatively simple, are too lengthy to include here. Refer to Jackson and Raw (1966) and Andrews (1972).

Larger soil animals, such as spiders and beetles, can be taken in traps made from funnels and cans set in the soil to ground level. Boards placed on the ground may attract millipedes, centipedes, and slugs. Meat bait in small wire traps will attract scavenger insects.

Sampling earthworm populations presents some difficulties, for no really successful method has been devised to extract the ani-

(a)

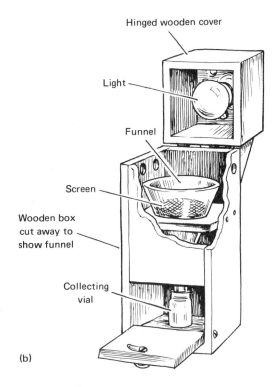

(b)

FIGURE B-8

Although the Berlese and other types of funnels can be purchased, they can be constructed easily in the workshop: (a) *relatively simple, sufficient for introductory work;* (b) *more elaborate and more efficient.*

mals from lower layers of the soil. One of the better methods is a combination of formalin and a shovel. Apply a dilute solution of formalin (25 ml of 40 percent formalin to 1 gal of water) to a quadrat 2 ft². Within a few minutes worms will come to the surface. After earthworm movement to the surface stops, pour on a second application. When worms cease to come to the surface the second time, dig out the quadrat as deep as necessary. The soil must be hand-sorted for maximum recovery. Earthworm cocoons can be extracted by the floatation method.

SMALL ANIMALS IN VEGETATION

Sweep nets with stout frames to withstand sweeps close to the ground and in woody growth are useful for collecting many types of insects and even some arboreal amphibians and reptiles. Drag nets consisting of light tubular frames, to which are attached canvas bags, are useful on flat ground. Overhead vegetation can be sampled by beating the limbs with sticks to dislodge the animals, which should fall into canvas collecting trays beneath.

BIRDS AND MAMMALS

Birds can be trapped for banding in specially constructed traps, cannon nets for larger game birds, and mist nets. Both federal and state permits are required for such work. Once a permit is granted, the operator of the banding station will be furnished plans for suitable traps. For mammals, live traps of wood or wire and snap traps are used. Both are available commercially, but live traps are easily constructed (see references). Traps can be baited with natural foods, dripping water, etc. For small mammals, a mixture of peanut butter and oatmeal works well. Also useful are grain, apple, meat, and appropriate scents.

Marking animals

Marking individuals in an animal population is necessary if you wish to distinguish certain members of a population at some future date, to recognize individuals from their neighbors, to study movements, or to estimate populations by the mark–recapture method.

Arthropods and snails are best marked with a quick-drying cellulose paint. It is easily applied with any pointed object. Marking butterflies is a two-man operation. One has to hold the wings together with a pair of forcepts; the other does the marking on the side of the wing exposed at rest. For aquatic mollusks and insects better results are obtained through the use of ship-fouling paint because the acetate paints do not hold up well in water.

Fish are usually marked by tagging in several ways. Strap tags of monel metal may be attached to the jaw, the preopercle, or the operculum. Streamer or pennant tags attached to various parts of the body, usually at the base of the dorsal fin, are used in some studies. Another method is to insert a plastic tag into the body cavity. To place the tag, a narrow incision is made in the side of the abdominal wall. Once the incision heals, which it does quickly, the fish carries the tag for life. Tags can be recovered only when the fish is cleaned. Fish can also be marked by clipping the fins, but this does not permit the individual recognition of a very large number of fish.

Frogs, toads, salamanders, and most lizards can be marked by some system of toe clipping, which involves the removal of the distal part of one or more toes. One method worked out by Martof (1953) is as follows: Number the toes on the left hind foot 1 to 5, the toes on the right foot 10 to 50. Up front, the left forefoot toes are numbered 100 to 400, and the toes on the right forefoot 800, 1,600, 2,400, 3,200. Thus one can mark up to 6,399 individuals by clipping no more than 2 toes.

Snakes and lizards can be marked by removing scales or patches of scales in certain combinations.

Birds are usually marked by serially numbered aluminum bands and by cellulose and aluminum colored bands. The latter are necessary for individual recognition in the field. In some specialized studies, the plumage is dyed a conspicuous or contrasting color.

Small mammals may be marked by toe clipping in combinations similar to those given for amphibians, or by notching the ear.

A number of other methods have been devised for marking mammals. Fur clipping and tattooing may be employed. Bear, deer, elk, moose, rabbits, and hares can be marked with strap tags or plastic discs attached to the ear. Aluminum bands similar to those used on birds can be attached to the forearm of bats. Dyes can be used to mark both large and small mammals. Small amphibians and reptiles can be marked by branding. For details see Clark (1968).

RADIOACTIVE TRACERS

The use of radioactive tracers is a particularly useful method for studying animals that are secretive in habits, live in dense cover, spend part or all of their lives underground, or that have radically different phases in their life cycle, such as the moths and butterflies.

If animals are fed small traces of gamma-emitting radioactive material, then the radioactive materials are metabolically incorporated into the tissue. The tracer becomes a part of the animal and is passed along to egg or offspring. Radioactive larvae remain so as they transform to adults, and the material is passed on to the egg. The same is true for birds. This technique is useful for studying dispersal, for the identification of specific broods or litters, for obtaining data on population dynamics and natural selection.

Another method involves the application of a radioactive tracer in or on an animal in such a way that the animal is not seriously injured and behaves in a normal way. This is usually done by fastening a radioactive wire to the animal or inserting it under the skin of the abdomen with a hypodermic needle. The movements of the tagged individual are then followed with a Geiger counter.

Although these techniques have their merits, they also have their disadvantages. The greatest is the potential radioactive hazard to the investigator himself, to other humans, and to the ecosystem. Another disadvantage is the impossibility of separating one animal from another. Because most work with radioactive tracers requires an AEC license, details are not given here. Specific techniques can be reviewed in Tester, 1963; Godfrey, 1954; Pendleton, 1956; and Graham and Ambrose, 1967.

Aging techniques

Information on the age structure of wild populations is not easily obtained. During the past several decades, a number of aging techniques have been developed, mostly for game and

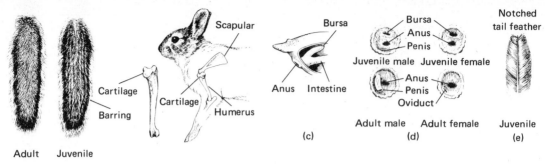

(a) Gray squirrel tail pelage   (b) Rabbit

fish. Fish aging began first when Hoffbauer (1898) pubished his studies on the scale markings of known-age carp. Since then the technique has been developed further and refined. It is based on the fact that a fish scale starts as a tiny plate and grows as the fish grows. A number of microscopic ridges, the circuli (Fig. B-9), are laid down about the center of the scale each year. When the fish is growing well in summer, the ridges are spaced wide apart. During winter, when growth slows down, the ridges are close together. This annual check on growth enables the biologist to determine the age of a fish by counting the number of areas of closed rings, the annuli (Fig. B-9). Salmon and some species of trout spend 1 or 2 years in streams before migrating out to sea or into lakes. Because stream growth is slower than lake or sea growth, the scales show when the fish migrate. When salmonid fish spawn, reabsorption of scales occurs, eroding the margins of the scales and interrupting the pattern of circular ridges. This erosion leaves a mark that can be detected in later years (Fig. B-9). Because the growth of a scale continues throughout the life of a fish, it also provides information on the growth rate. This is obtained by measuring the total radius of the scale, the radius to each year's growth ring, and the total body length of the fish. Then by simple proportion, the yearly growth rate can be determined.

Other techniques in aging fish include the length–frequency distribution, vertebral development, and rings or growth layer in the otolith or ear stone.

Because of the large number of year classes (animals born in a population during a particular year) that can be identified, one can determine dominant year classes, learn the age when fish reach sexual maturity, estimate production mortality, and the effects of fish harvest.

Aging techniques for mammals and birds have developed more slowly and are not as refined as those for fish, but a number of methods are in common use. These involve handling all or a part of the animal. Among birds, plumage development is frequently used. Until molted, the tail feathers of juvenile waterfowl are notched at the tip, in contrast with the normally contoured feather of the winter plumage (Fig. B-10). The shape of the primary wing feather separates adults from young among many gallinaceous game birds. When the whole bird is available, the presence or depth of the bursa of Fabricus, a blind pouch lying dorsal to the cecum and opening into the cloaca (Fig. B-10), indicates juvenile birds.

Among mammals the examination of reproductive organs is useful because the majority do not breed until the second year. This method can be used only during the breeding season. The presence of the epiphyseal cartilage in rabbits and squirrels identifies juveniles up to 6 or 7 months (Fig. B-10). Black bars on the pelage of the underside of the tail of juvenile gray squirrels separate the young from the adults (Fig. B-10). Primeness of pelt on the inside of skins is a good means of aging muskrat during the trapping season. Dark pigmentation on the flesh side of the pelt indicates areas of growing hair. This pigmentation in adults appears in irregular, scattered dark areas, whereas in immature animals it is more or less symmetrical and linear. Skull measurements are useful in beavers and muskrats. Annual growth rings on the roots of canine teeth indicate age for the first few years of life in the fur seal and other pinnipeds and in canids. Growth rings also show up in the horns of mountain sheep. The wear and replacement of teeth in deer and elk permit the determination of different age classes in these mammals (see Giles, 1969, *Wildlife Management Techniques*).

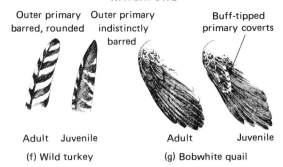

WATERFOWL

Outer primary
barred, rounded

Outer primary
indistinctly
barred

Buff-tipped
primary coverts

Adult   Juvenile

(f) Wild turkey

Adult   Juvenile

(g) Bobwhite quail

LENS-WEIGHT TECHNIQUE

Because the lens of the eye of most mammals (and possibly birds) grows continuously throughout life, and because there is only slight variation between individuals in lens size and growth, the measurement of the lens is a feasible method for aging a number of mammals. It has been done successfully for the cottontail rabbit, the raccoon, the black bear, and the fur seal.

The technique involves the weighing of the dry lens and comparing its weight against a chart of lens weights of known-age individuals (Fig. B-11). This may require the development of such a table by the investigator, who will have to rear young animals in captivity and sacrifice them week by week for their eyes. A table has been prepared for rabbits by Lord (1959, 1963).

The technique is as follows.

1. Remove eyes as soon as possible after the animal is killed and place in a solution of 10 percent formalin. The formalin will harden the lens so that it can be removed from the vitreous humor.
2. Fix for 1 week, but the longer the better.
3. After fixing, remove the lens from the eye and roll it a few minutes on a paper towel to remove excess moisture.
4. Place the lens in an oven to dry at 80° C.
5. Lenses are considered dry when repeated weighing after intervals of drying results in no additional loss of weight. This will usually require 24 to 36 hours.
6. Weigh immediately after removal from oven, since the dried lens are hygroscopic and take on water. Electric scales that read weights rapidly are preferred over other types of balances.

The lens-growth curve permits a rather close approximation of the age of the mammal. For cottontail rabbits, the method permits the

FIGURE B-10 [OPPOSITE]
Age determination in some game birds and mammals. (a) Regular barring on the underside of the tail distinguishes the juvenile gray squirrel from the adult. (b) Presence of the epiphyseal cartilage on the humerus of the juvenile cottontail rabbit separates that age class from the adult. This method is useful when the biologist wishes to collect data from a wide area by requesting that successful rabbit hunters turn in these bones. (c) The bursa of Fabricus (enlarged). Its presence or greater depth indicates a juvenile bird. The depths vary with the species. This method is useful in both waterfowl and some gallinaceous game birds. (d) A method of sexing and aging water fowl by examining the cloaca. Note the presence of the bursal opening on juvenile waterfowl, its absence on the adult. (e) The notched tail feather of juvenile water fowl. (f) The number X (ten) primary in juvenile gallinaceous birds is sharply pointed; in adults it is rounded. The juvenile wild turkey in addition has its outer primary indistinctly barred. (g) The juvenile bobwhite quail, in addition to having a pointed number X primary, also possess buff-tipped primary coverts. (For a complete discussion on aging and sexing techniques, see R. Giles, Wildlife Investigational Techniques.)

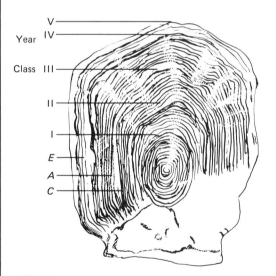

FIGURE B-9
The age of fish can be determined by the growth rings on the scale. C is the circuli; A the annuli; E, erosion of the scale from spawning.

determination of the month of birth of young rabbits and the year of birth of rabbits over one year of age.

Sex determination

The sex of mammals in most instances can be determined by examining external genitalia, and the sex of birds by plumage differences. For example, the male ruffed grouse can be distinguished from the female by the length of the central tail feather (Fig. B-12); and females of prairie chickens can be distinguished from the males by strong mottling or barring of all the tail feathers (Fig. B-12). Male wild turkeys have body feathers black-tipped, whereas the feathers of the female are brown-tipped (Fig. B-12). For detailed information see Giles, 1969, *Wildlife Management Techniques*.

Determination of home range and territory

There are a number of methods available for obtaining an approximation of the size of home range. Some methods are offered here.

CALCULATING HOME RANGE

On a map, outline and measure the area that includes all the observations made on the movements of individuals. If the observations are obtained by trapping, then one must assume that the animal could have gone half-way toward an adjacent trap, especially if the traps were set in a regular grid.

CENTER OF ACTIVITY

Arrange the recaptures on a grid and determine the values on an XY axis. An average of these locations will give the center of activity (see Hayne, 1949b). This method has the advantages that the information is relatively easy to summarize and the calculations are not complicated. However, a map of the area must be made, and many recaptures of the same individual are required before the extent of the home range can be obtained. As the number of recaptures increases, the area of

the known range increases. At least 15 recaptures or more are necessary. This method is unsuited for mammals that follow paths or tunnel underground.

FREQUENCY OF CAPTURE

Record the distances between captures in live traps set randomly or in grids. Record the number of captures as a frequency distribution according to the distance between them. Distances between captures are then tallied and proportions calculated for each distance by sex or age categories.

The distances can be measured in two ways. They can be taken from the place where the animal was first marked or observed, or from each successive location.

This method has the advantage that the traps can be set out haphazardly, avoiding the labor of setting them out in grids. The recaptures of all individuals can be used. And information can be obtained during a short period of time because the data from animals captured only two or three times can be used. The disadvantages of this method are that short movements are favored and that no definite boundaries of home ranges can be given. Home range is described as a frequency of distances observed.

DYED BAIT

The fact that small mammals come readily to bait can be exploited in a manner that does not involve trapping. Bait boxes are made from quart-size paper milk containers and are set out in a 50-ft grid over the study area. These boxes serve as containers for the bait and as receptacles for the feces. If food at certain stations is stained with a dye, then the visitation by animals that fed on the dyed bait to other bait boxes can be traced through colored droppings. The distance moved by a particular animal from a station with dyed bait can be determined when colored scats are recovered at various stations in the grid. The results with this method compare favorably with those obtained by other methods, and it does not involve the time and work of trapping. Complete details, as well as a list of suitable dyes, can be found in a paper by New, 1958.

TERRITORY

Because territorial boundaries are rather rigidly maintained by birds during the breeding season, territorial boundaries can be mapped by observing the movements of the birds during the day, by plotting singing perches, by observing locations of territorial disputes, and on occasion by chasing the bird. When the bird arrives at the boundary of its territory, it generally will double back.

RADIO TRACKING

The development of transistors and other miniature electronic devices has made possible the construction of small transmitters that can be attached to animals usually by means of a specially designed collar or harness. Mercury cells are used as the source of power. The transmitter is a transistor-crystal-controlled oscillator with the tank coil for the oscillator acting as the magnetic-dipole transmitting antenna. The antenna is constructed of copper or aluminum and has a figure-8 directional pattern. The receiver is a portable battery-powered unit, whose basic components are the receiver, a radio range filter, and two transistorized radio-frequency converters. Positions of stationary animals can be obtained by a single portable direction-finding receiver. Running animals are best located by triangulation, using at least two direction-finding receivers. Details of construction are not given here. In fact, radiotracking equipment is available commercially.

Although highly useful in obtaining data on animals that could not be acquired otherwise, radiotracking has its disadvantages for most investigators. The most serious disadvantage is the need for electronic expertise. Unless the investigator is an electronics whiz or has the assistance of an electronics technician, he should not attempt radiotracking. Other disadvantages are cost and the need for an FCC license. Useful references are Giles, 1969, *Wildlife Management Techniques*, Chapter 9; Mackay, 1970; and *Bioscience*, Feb. 1965, Vol. 15, No. 2.

Estimating the numbers of animals

Basic to the study of animal populations is the estimation of their numbers, no small task

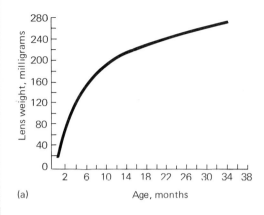

(a)

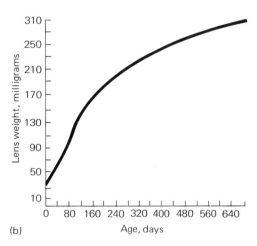

(b)

FIGURE B-11

Growth rate curves for the lens of (a) cottontail rabbit (from Lord, 1961) and (b) black-tailed jack rabbits. (After Tiemeier and Plenert, 1964.)

in wild populations. During the past several decades, much work has gone into the development of techniques and statistical methods to arrive at some estimates of animal populations. Basically the methods of estimating the numbers of animals can be put into three categories: a *true census,* a count of all individuals on a given area; *sampling estimates,* derived from counts on sample plots; and *indices,* in which the trends of populations from year to year or from area to area are obtained through roadside counts, animal signs, and the like. In this category everything is relative, and to be meaningful, the data must be compared with those from other areas or other years.

TRUE CENSUS

A true census implies a direct count of all individuals in a given area. This is rather difficult to do for most wild populations, but there are situations where a total count can be made.

Many territorial species are easily seen and heard and can be located in their specific area. Such a census is regularly used for birds. The spot-map method is probably the best approach. A sample plot of at least 25 acres (10 hectares) is marked out in a grid with numbered stakes or tree tags placed at intervals of 50 m. Five or more daily counts are made throughout the breeding season. Each time a bird is observed, it is marked on a map of the plot. At the end of the census period all the spots at which a species is observed are placed on one map. The spots should fall into groups, with each group indicating the presence of a breeding pair. The groups for each species can then be counted in order to arrive at the total population for the given area. Results are usually expressed as animals per acre or per hectare.

Direct counts can be made in areas of concentration. Deer in open country, herds of elk and caribou, waterfowl on wintering grounds, rookeries, roosts, breeding colonies of birds and mammals permit direct counting usually either from the air or from aerial photographs. Covies of bobwhite quail can be located and counted with the aid of a well-trained bird dog.

ESTIMATES FROM SAMPLING

Estimations of population from sampling involve two basic assumptions: (1) that mortality and recruitment during the period the data are being taken are negligible or can be accounted for and (2) that all members have an equal probability of being counted—that the members of the population are not trap-shy or trap-addicted, that they are distributed randomly through the population if marked and released, and that they do not group by age, sex, or some other characteristic.

Sampling also involves one major general consideration. The method employed in taking the sample must be adapted to the particular species, the time, the place, the purpose.

Sample plots. Relatively immobile forms, such as barnacles, mollusks, and cicada emergence holes, can be estimated by the quadrat method, similar to that employed for plants. The data can be analyzed for presence, frequency, etc., or the results can be converted to a density per acre, etc. The size and shape of the quadrat will depend upon the density of the population, the diversity of the habitat, the nature of the organism involved. Make a few preliminary surveys before settling on a quadrat size.

Foliage arthropods may be sampled by a number of strokes with a standard sweep net over a 10-m² area. The number of strokes needed to secure the sample must be predetermined. It will vary with the type of vegetation.

Estimates of zooplankton, obtained by pulling a plankton net through a given distance of water at several depths, can be made by filtering a known volume of sample through a funnel using a filter pump. The filter paper should be marked off in equal squares. With the aid of a hand lens or a binocular microscope, count the organisms in each square. The numbers then can be related back to the total volume of water sampled.

If the organisms are too small to be counted in this manner, a Rafter plankton-counting cell can be used. This consists of a microscope slide base plate ruled into ten 1-cm squares. The slides are made from strips of microscope glass slides cemented to the base with Canada balsam. This should hold 1 cc of liquid. After a small volume of water is introduced, cover with a long cover glass, place under the microscope, and count the organisms square by square. Record the number of each form per square until at least 100 observations have been made. The occurrence of individual species can be recorded as percentage frequency. (Note: Plankton-counting cells and eyepiece micrometers can be purchased commercially, but they are expensive.)

Mark–recapture method. This method is based on trapping, banding, or marking, and then later recapturing sample individuals. It involves the release of a known number of marked animals turned loose in the original area. Then after an appropriate interval of time (approximately 1 week for rabbits), a sample of the population is taken. An estimate of the total population is then computed from the ratio of marked to unmarked individuals.

$$N : T :: n : t$$

or

$$N = \frac{T}{t/n} \quad \text{or} \quad \frac{nT}{t}$$

where

T = number marked in the precensus period
t = number of marked animals trapped in the census period
n = total animals trapped in the census period
N = the population estimate

Suppose that in a precensus period a biologist tags 39 rabbits. Then during the census period he traps 15 tagged rabbits and 19 unmarked ones, a total of 34. The following ratio is set up:

$$N : 39 :: 34 : 15$$

or

$$N = \frac{39}{15/34} = 88 \text{ rabbits}$$

The confidence limits at the 95 percent level may be calculated from

$$\text{S.E.} = N \sqrt{\frac{(N-T)(N-n)}{Tn(N-1)}}$$

To determine the limits within which the population lies, add and subtract two standard errors from the estimate. A large standard error and rather wide confidence limits are the result of a small number of recaptures.

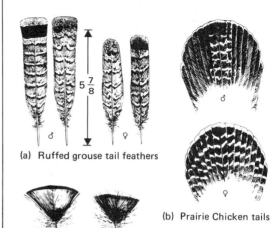

(a) **Ruffed grouse tail feathers**

(b) **Prairie Chicken tails**

(c) **Wild turkey breast feathers**

(d) **Ring-necked pheasant**

FIGURE B-12
Sex determination in game birds. (a) *Length of the central tail feather separates male and female ruffed grouse. The male tail feather is 5⅞ in. long or more; the female tail is shorter.* (b) *The female prairie chicken has heavily barred tail feathers; barring is absent on the outer tail feather of the male.* (c) *The breast feathers of the male wild turkey are black-tipped; those of the female are brown tipped.* (d) *Sexual dimorphism distinguishes males from females in many species of birds.*

Appendix B
SAMPLING POPULATIONS

TABLE B-8
The Schnabel method of estimating populations

P Period (date)	A Number trapped	Number marked	B Marked animals in area	(A)×(B)	(A)×(B) Sum	Recaptures	C Sum of recaptures	(A)×(B) (C) Estimated population
1	4	4	—	00	0	—	—	—
2	4	4	4	16	16	0	0	—
3	2	2	8	16	32	0	—	—
4	6	6	10	60	92	0	—	—
5	10	7	16	160	252	3	3	—
6	4	4	23	92	344	0	3	—
7	8	6	27	216	560	2	5	—
8	4	2	33	132	692	2	7	—
9	5	4	35	175	867	1	8	—
10	7	6	39	273	1140	1	9	—
11	7	6	45	315	1455	1	10	145
12	9	7	51	459	1914	2	12	159
13	6	3	58	348	2262	3	15	150
14	10	6	61	610	2872	4	19	151
15	8	5	67	536	3408	3	22	154
16	6	1	72	432	3840	5	27	142
17	4	2	73	292	4132	2	29	142
18	12	7	75	900	5032	5	34	148
19	8	4	82	656	5688	4	38	149

For this example the

$$S.E. = 88 \sqrt{\frac{(88-39)(88-34)}{(34)(39)(87)}}$$
$$= (0.1513)(88) = 13.31$$

Upper limits:

$88 + 26 = 114$

Lower limits:

$88 - 26 = 62$

The chances are 95 out of a hundred that the population of rabbits lies between 62 and 114. This wide spread is typical in wildlife studies.

A variation of this procedure is to accumulate the captures and recaptures. There are several ways in which this can be done, but only the Schnabel method will be illustrated here (for others, see Giles, 1969, *Wildlife Management Techniques*). All animals captured are tagged or marked and released daily. A record is kept of the total animals caught each day, the number of recaptures, and the number of animals newly tagged. An example is given in Table B-8. The method of calculating the population is the same as that already given. However, in the Schnabel method, T becomes progressively larger. Population esti-

mates can be calculated daily later in the period, or the season can be divided into periods of, say, a week, and the population computed for each period. True confidence limits cannot be determined, and the calculation of the standard error becomes rather involved. For details see Ricker, 1958. An excellent approach using the regression method to estimate population size from mark–recapture data is described by Marten (1970).

Another variation of the mark–recapture method is the multiple recapture, in which all animals caught on any particular day are marked and released including the recaptures. Thus an animal caught on day 1 and again on day 2 will bear the marks of both days. By such a method one not only can keep account of total marks recaptured each day of trapping, but can also relate recaptures to the initial day of marking. Known as the Jolly method (Jolly, 1965), this technique is useful in following population trends. The basic equation is

$$P_i = \frac{\hat{M}_i n_i}{r_i}$$

where

P_i = estimate of population on day i
$\hat{M}_i$ = estimate of total number of marked animals in population on day i

718

r_i = total number of marked animals on day i

n_i = total number captured on day i

The formula is essentially that of the Lincoln index.

For bookkeeping, the trellis diagram is most convenient (Fig. B-13). In the trellis diagram the marginal column on the left running downward from left to right contains the totals captured for each day. The marginal column on the right contains the totals released as marked animals each day. In the body of the table are the figures for the recaptures as they relate to the day they were originally marked. For example, on trapping day 5, 220 animals were caught, 214 released. Of the 220 captured, 30 had been marked on day 4, 13 on day 3, 8 on day 2, and 2 on day 1. The data can provide two different population estimates. A column starting at any date and running downward to the left gives the necessary information for determining the population on the day of recapture. Columns starting at any date and running downward to the right give the necessary information to determine population size on the day of release. To use this method, trapping does not have to be done on consecutive days, but it is necessary that it be done at equally spaced times.

Although raw data can be used to estimate population (for details see Southwood, 1966, pp. 83–87), it is more meaningful if the recapture values are corrected to the number of marked recaptures per 100 marked on day i and per 100 in the recapture sample. This must be done for each recapture value in the body of the table. The corrected values are then substituted for the raw values. The formula is

$$y = \frac{X_i}{1} \times \frac{100}{M} \times \frac{100}{n_i}$$

where

X_i = number recaptured on day n_i, sample i

M = total number of marked animals initially released (prior to time of sample)

n_i = total number of marked and unmarked animals in sample i

y = corrected recaptures in sample i

For example, on day 5, there were four sets of recaptures: 30, 13, 8, and 2, out of 220 captures. To correct these values:

recaptures from day 4: $y = \dfrac{30}{1} \times \dfrac{100}{202} \times \dfrac{100}{220}$

$$= 6.7$$

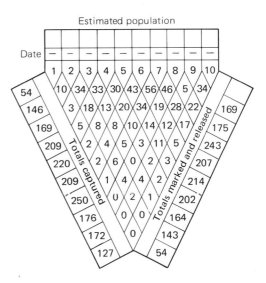

FIGURE B-13
Trellis diagram.

recaptures day 3: $y = \dfrac{13}{1} \times \dfrac{100}{164} \times \dfrac{100}{202}$
$= 3.6$, and so on

The next step is to calculate a weighted ratio, r, to show the rate of decrease of recapture values. Determine r for each column running to the right, which gives you $r+$ values, size of the population on the day of release. Do the same for each column running to the left, which gives you $r-$ values for the size of the population on the day of recapture. The formulas are

1. For three or four values for y:

$$r = \frac{y_2 + y_3 + \cdots y_n}{y_1 + y_2 + \cdots y_{n-1}}$$

2. For more than four values for y:

$$r = \frac{y_3 + y_4 + \cdots y_n}{y_1 + y_2 + \cdots y_{n-2}}$$

With this weighted ration for each method you can calculate the theoretical number of recaptures that would have been obtained on the day of release:

$$y_0 = \frac{y_1 + y_2 + \cdots y_{n-1}}{r}$$
$$-(y_1 + y_2 + \cdots + y_{n-2})$$

With the theoretical values for recaptures at the time of release, you can estimate population size for each day by the following:

$$N \text{ (on day } i) = \frac{100}{1} \times \frac{100}{y_0}$$

For convenience and for comparison of population estimates by the two methods, positive and negative, you may tabulate your calculations in a table headed as follows.

Day	$r+$	y_0	N	$r-$	y_0	N

This method works quite well where relatively large numbers of animals are involved, such as insect populations.

Removal method. The removal method of estimating populations has been widely used in small-mammal studies, although the assumptions on which it is based are open to question. This method assumes (1) that the population is essentially stationary; (2) that the probability of capture during the trapping period is the same for each animal exposed to capture;

(3) that the probability of capture remains constant from trapping to trapping (not trapshy, etc.). The trapping program consists of setting the same number of traps for several nights. The number caught the first night is expected to exceed that caught the second, and the second night's catch should exceed the third. In other words, the population becomes depleted.

The field procedure is as follows.

1. Set the traps (snap) out in a grid system, three traps to a station; or preferably set the traps in two parallel lines 50 ft apart, 20 stations to a line, 3 traps at each station. The stations are spaced 25 or 50 ft apart, depending upon the nature of the vegetation.
2. Trap for three nights. Prebait for best success.

There are several ways of handling the data to obtain the estimate. The most popular is to plot the size of the daily catch against the number of animals previously caught (Fig. B-14). A straight line is drawn through the plotted points to cut the horizontal axis. The point at which the horizontal axis is cut represents the population estimate.

A simpler method of estimating the population based on two captures is suggested by Zippin (1956). The population N equals the number caught the first day (y_1) squared, divided by the difference between the number caught the first day (y_1) and the number caught the second (y_2). Using the same data as on the graph,

$$N = \frac{(y_1)^2}{y_1 - y_2} = \frac{36}{2} = 18$$

The two methods give approximately the same answer.

A second variation of the removal method is the proposed IBP Standard Minimum method. This method involves a 16 by 16 grid of points or stations spaced 15 m apart, each station containing two snap traps. The stations are prebaited for 3 days, a procedure that attracts small mammals to the area and usually ensures the largest catch on the first day. The prebaiting period is followed by a 5-day trapping period. Again the daily catch is plotted against accumulated catch, and the population is estimated at the point where the axis crosses the horizontal line. But a better fit is obtained by calculating a simple regression line from the 5-day catch data after the method suggested by Hayne (1949*b*).

As with all trapping procedures, there is disagreement over the Standard Minimum method. Some argue for a smaller grid of, say, 8 by 8. Traps on the outer stations usually capture more animals than those in the center of the grid because the animals in the border zone react to the sudden removal of the animals in the center. These immigrants are picked up by the border traps. Thus outside invaders can significantly contribute to the catch on the margin zones and influence population estimates. Some attempt should be made to assess the effect of movement into the grid by estimating the size of the border area. Although nearly all animals are removed in the 5-day period, some animals still avoid capture.

The removal method is useful where one desires a relative measure or index figure for small-mammal populations in order to compare one habitat with another. The data will be more valuable if details on vegetation and litter are recorded for each station. Often some association can be obtained between vegetation and trapping success at the various stations.

INDICES

Indices are estimates of animal populations derived from counts of animal signs, calls, roadside counts, and so on. The results do not give estimates of absolute populations, but they do indicate trends of populations from year to year and from habitat to habitat. Often this type of information is all that is needed.

Call counts. Call counts are used chiefly to obtain population trends of certain game birds such as the mourning dove, bobwhite quail, woodcock, and pheasant. A predetermined route is established along country roads; it should be no longer than can be covered in 1 hour's time. Stations are located at quarter- or half-mile intervals, depending upon the terrain and the species involved. The route is run around sunrise for gallinaceous birds and for doves, and around sunset for woodcock. The exact time to start must be determined for each area by the investigator. At each station the vehicle is stopped, the observer listens for a minute or two (standardize), records the calls heard, and goes on to the next station. Routes should be run several times and an

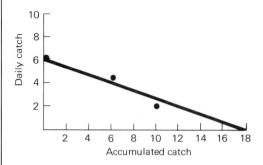

FIGURE B-14
Daily catch on a 3-day removal trap line plotted against accumulated catch to estimate the exposed population.

average taken. The number of calls divided by the number of stops gives a call-index figure.

Roadside counts. This is similar to a call count, with the exception that the number of animals observed along the route is recorded and the results divided by the number of miles. Other variations include counting of animal tracks, browse, signs, active dens and lodges, and so on.

Pellet counts. Counting pellet or fecal groups is widely used to estimate big-game populations. This method involves the counting of pellet groups in sample plots or transects located in the study area. It may be used for estimating the relative intensity of use of the range by one or more kinds of animals, to determine trends in animal populations, or in rarer cases to estimate the total population. The latter is possible only when an entire herd is known to occupy a given area for a definite period. Intensity of use is usually expressed as the number of pellets or pellet groups per unit area.

The accuracy of estimating populations by this method depends upon some knowledge of the rate of defecation by the animals involved. Herein lies the weakness of the technique, because pellet groups vary with the diet, season, age, sex, rate of decomposition, and the type of vegetation (that is, the plants can cover the pellet groups). Usually rates vary with the region; so some preliminary observations will have to be made to arrive at some useful figure. For deer, a pellet-group figure of 15 (per deer) is satisfactory on good range, and 13 for poor range. Rabbits vary too widely in their pellet groups for the technique to have much value with that group.

The field procedure is as follows.

1. On randomly located transect lines, establish a number of rectangular plots of 1/50th acre, 12 × 72.6 ft divided in half longitudinally for ease in counting.
2. Count pellet groups at the most favorable time, when plant growth, leaf fall, and so on are least likely to interfere.
3. Mark the plots permanently and clear or paint the pellet groups at the beginning of the study where age determination of pellets is difficult.
4. Then let

$$t = \frac{1}{na'}\, y$$

where

y = sum of pellet groups counted over the plots
a' = area of one plot
n = number of plots
t = pellet groups per unit area

5. Determine the value for t. To translate t to total deer days of use:
   a. Assume a defecation rate of 13 pellet groups/deer/day for poor range, 15/day for adequate range.
   b. Determine the period, the number of days, over which the pellet groups were deposited (for example, since the last count, time plots cleared, etc.)
   c. Divide t by the defecation rate to obtain days of utilization by deer per acre.
   d. Divide the above, the number of days of utilization, by the number of days in the period. This gives you the number of deer per acre (assuming a constant population.)
6. Multiply result by 640 to obtain the number of deer per square mile.

Dropping boards. A modification of the pellet-group method for small mammals is the dropping-board method (see Emlen et al., 1957).

1. Set out 4-in. squares of weatherproof plywood (in natural color) in lines or grids. Use at least 100 boards. This will cover 1.2 acres if spaced 25 ft apart, 4.7 acres with a 50-ft spacing, and 18.8 acres with a 100-ft spacing. Be sure the squares are level and placed firmly on the ground.
2. Number each station.
3. The boards may or may not be baited depending upon local conditions.
4. Make a series of at least three visits. The time of day the visits are made and their frequency will depend upon local conditions, such as coprophagous insect activity. Daily visits may be necessary.
5. At each station record the presence of droppings by species, and brush the board clean for the next visit. The droppings of small mammals are distinctive and with some experience can be identified. See Murie, 1954, *Field Guide to Animal Tracks.*

6. Results can be expressed as incidence of droppings for each species.

Figures obtained from the record of usage are indices of the population useful in comparative studies of interspecific, interseasonal, and interregional abundance. The dropping-board technique can be used in studies of population trends and fluctuations, local distribution, species association, activity, rhythms, effects of weather and environmental conditions on activity, and movements if the animal is tagged with a radioactive tracer or with dyed bait.

Measuring mortality

The measurement of mortality in natural populations is imperative in the analysis of animal populations. Coupled with a knowledge of age structure, mortality data can be used to construct life tables, life equations, survivorship curves, growth curves, and the like. Knowledge of the mortality rate is basic in the management of game and fish populations.

Several methods of determining mortality in wild populations will be discussed here. Further details on these methods and the description of others, including the construction of life tables and the conversion of monthly rates of probability of dying to annual rates and to death rates and estimations of life expectancy can be found in Giles, 1969, *Wildlife Management Techniques*; Davis, 1960; Davis and Golley, 1963; and Ricker, 1958.

KNOWN DEATHS

If a number of marked animals are found dead a known time after marking, the percentage dying can be plotted against time and a curve drawn through the points. The probability of dying can be read directly from the chart, or it can be calculated by dividing the number of individuals that die during a period of time by the initial population. At best, however, this is a very rough measurement.

If the animals can be readily aged, as is true with deer, rabbits, mountain sheep, and others, and if sufficient lower jaws, eye lenses, or horns can be obtained, then mortality and the probability of dying can be expressed on an age basis.

If the size of a population can be determined

723

at two different times, if immigration and emigration are held to a minimum, and if there is no recruitment from births, then the difference in population size between time A and time B is the result of deaths. (Winter deer yards offer an excellent opportunity to determine mortality in this way.) The probability of dying during the period under observation can be calculated by dividing the number dying over the period (population time A − population time B) by the initial population. Davis and Golley (1963) give formulas for changing such data to an annual basis. However, because the probability of dying may vary with the season, as it does with deer, an annual probability of dying based on a seasonal figure may be erroneous.

References

Vegetation analysis

BRAUN-BLANQUET, J.

1951 *Pflanzensoziologie: Grundzuge der Vegetationskunde*, 2d ed., Springer, Vienna.

1932 *Plant Sociology*, McGraw-Hill, New York. (An English translation of the 1st edition of above.)

CAIN, S. A., AND G. M. DEO. CASTRO

1959 *Manual of Vegetation Analysis*, Harper & Row, New York.

GREIG-SMITH, P.

1964 *Quantitative Plant Ecology*, 2d ed., Academic, New York.

KERSHAW, K. A.

1964 *Quantative and Dynamic Ecology*, Edward Arnold, London.

OOSTING, H. J.

1956 *The Study of Plant Communities*, Freeman, San Francisco, Calif.

Methodology

BECKER, D. A., AND J. J. CROCKETT

1973 Evaluation of sampling techniques on tall-grass prairie, *J. Range Management*, **26**:61–65.

BRAY, R. AND J. T. CURTIS

1957 An ordination of upland forest communities of southern Wisconsin, *Ecol. Monographs*, **27**:325–349.

COTTAM, G., AND J. T. CURTIS

1956 The use of distance measures in phytosociological sampling, *Ecology*, **37**:451–460.

CURTIS, J. T., AND R. P. MCINTOSH

1951 The upland forest continuum in the prairie–forest border region of Wisconsin, *Ecology*, **32**:476–496.

GROSENBAUGH, L. R.

1952 Plotless timber estimates—new, fast, easy, *J. Forestry*, **50**:32–37.

1958 Point-sampling and line-sampling: probability theory, geometric implications, synthesis, *U.S. Forest Serv., Southern Forest Expt. Sta. Occ. Paper* 160.

HYDER, D. N., AND F. A. SNEVA

1960 Bitterlich's plotless method for sampling basal ground cover of bunch grasses, *J. Range Management*, **13**:6–9.

LONG, G. A.; P. S. POISSONET; J. A. POISSONET; PH. M. DAGET; AND M. P. GODRON

1972 Improved needle point frame for exact line transects, *J. Range Management*, **25**:228.

POISSONET, P. S.; J. A. POISSONET; M. P. GODRON; AND G. A. LONG

1973 A comparison of sampling methods in dense herbaceous pasture, *J. Range Management*, **26**:65–67.

U.S. FOREST SERVICE

1958 Techniques and methods of measuring understory vegetation, a symposium, *Southern Forest Expt. Station*.

1962 Range research methods; a symposium, *U.S. Dept. Agr. Misc. Publ.* 940.

VESTAL, A. G.

1949 Minimum areas for different vegetations, *Illinois Biol. Monographs*, **20**:1–129.

Phytoplankton

CLARK, W. J., AND W. F. SIGLER

1963 Method of concentrating phytoplankton samples using membrane filters, *Limnol. Oceanog.*, **8**:127–129.

HARTMAN, R. T.

1958 Studies of plankton centrifuge efficiency, *Ecology*, **39**:374–376.

LUND, J. W. G., AND J. F. TALLING

1957 Botanical limnological methods with special reference to algae, *Botan. Rev.*, **23**:489–583.

MCNABB, C. D.

1960 Enumeration of fresh water phytoplankton concentrated on the membrane filter, *Limnol. Oceanog.*, **5**:57–61.

SCHWOERBEL, J.

1970 *Methods of Hydrobiology (Freshwater Biology)*, Pergamon, Elmsford, N.Y.

SLADECKOVA, ALENA

1962 Limnological investigation methods for the periphyton (aufwuchs) community, *Botan. Rev.*, **28**:286–350.

WELCH, P. S.

1948 *Limnological Methods*, McGraw-Hill–Blakiston, New York.

WOOD, E. J. F.

1962 A method for phytoplankton study, *Limnol. Oceanog.*, **7**:32–35.

Dendrochronology

FRETTS, H. C.

1960 Multiple regression analysis of radial growth in individual trees, *Forest Sci.*, **6**:334–349.

1962 An approach to dendrochronology—screening by means of multiple regression techniques, *J. Geophys. Res*, **67**:1413–1420.

ROUGHTON, R. D.

1962 A review of literature on dendrochronology and age determination of woody plants, *Tech. Bull.* 15, *Colorado Dept. Fish Game,* Denver.

SPENCER, D. A.

1958 Porcupine population fluctuations in past centuries revealed by dendochronology, PhD thesis, University of Colorado, Boulder, Colo.

TAYLOR, R. F.

1936 An inexpensive increment core holder, *J. Forestry*, **34**:814–815.

Palynology

BROWN, C. A.

1960 *Palynological Techniques*, published by author, 1180 Stanford Ave., Baton Rouge 8, La.

ERDTMANN, G.

1954 *An Introduction to Pollen Analysis*, Ronald, New York.

FAEGRI, K., AND J. IVERSON

1964 *Textbook of Pollen Analysis*, 2d ed., Hafner, New York.

FELIX, C. F.

1961 An introduction to palynology, in H. N. Andrews, *Studies in Paleobotany*, Wiley, New York.

WALKER, P. C., AND R. T. HARTMAN

1960 The forest sequence of the Hartstown bog area in western Pennsylvania, *Ecology,* **41**:461–474.

WODEHOUSE, R. P.

1935 *Pollen Grains*, McGraw-Hill, New York.

Animal populations, general

ANDREWARTHA, H. G.

1970 *An Introduction to the Study of Animal Populations*, 2d ed., University of Chicago Press, Chicago.

ANDREWS, W. A.

1972 *A Guide to the Study of Freshwater Ecology*, Prentice-Hall, Englewood Cliffs, N.J.

1972 *A Guide to the Study of Soil Ecology*, Prentice-Hall, Englewood Cliffs, N.J.

GILES, R. H., JR.

1969 *Wildlife Management Techniques*, The Wildlife Society, Washington, D.C.

HOLME, N. A., AND A. D. MCINTYRE

1971 *Methods for the Study of Marine Benthos*, IPB Handbook No. 16, Blackwell, Oxford.

JACKSON, R. M., AND F. RAW

1966 *Life in the Soil*, St. Martin, New York.

MADSEN, R. M.

1967 *Age Determination of Wildlife, A Bibliography*, Biblio. No. 2, U.S. Department of the Interior, Dept. Library, Washington, D.C.

MACKAY, R. F.

1970 *Bio-medical Telemetry*, Wiley, New York.

725

Appendix B
SAMPLING POPULATIONS

PARKINSON, D.; T. R. G. GRAY; AND S. T. WILLIAMS
1971 *Methods for Studying the Ecology of Soil Microorganisms*, IBP Handbook 19, Blackwell, Oxford.

PHILLIPSON, J. (ED.)
1970 *Methods of Study in Soil Ecology*, UNESCO, Paris.
1971 *Methods of Study in Quantitative Soil Ecology*, IBP Handbook No. 18, Blackwell, Oxford.

RICKER, W. E.
1958 Handbook of computations for biological statistics of fish populations, Fishery Res. Board Can. Bull., **119**:1–300
1971 *Methods for Assessment of Fish Production in Fresh Waters*, IBP Handbook No. 3, Blackwell, Oxford.

SCHWOERBEL, J.
1970 *Methods of Hydrobiology (Freshwater Biology)*, Pergamon, Elmsford, N.Y.

SOUTHWOOD, T. R. E.
1966 *Ecological Methods*, Methuen, London.

TEPPER, E. E.
1967 *Statistical Methods in Using Mark-Recapture Data for Population Estimation (A Compilation)*, Bibliography No. 4, U.S. Department of Interior, Department Library, Washington, D.C.

U.S. DEPARTMENT OF AGRICULTURE
1962 "Range research methods," *U.S.D.A. Misc. Publ. 940*, Supt. of Documents, Washington, D.C.

WELCH, P. S.
1948 *Limnological Methods*, McGraw-Hill–Blakiston, New York.

Animal populations, other

ADAMS, L.
1951 Confidence limits from the Petersen or Lincoln index in animal population studies, *J. Wildlife Management*, **15**:13–19.

CLARK, D. R., JR.
1968 Branding as a marking technique for amphibians and reptiles, *Copia*, **1968**:148–151.

COCHRAN, W. W., AND R. D. LORD, JR.
1963 A radio-tracking system for wild animals, *J. Wildlife Management*, **27**:9–24.

DAVIS, D. E.
1960 A chart for estimation of life expectancy, *J. Wildlife Management*, **24**:344–348.

DAVIS, D. E., AND F. B. GOLLEY
1963 *Principles in Mammalogy*, Reinhold, New York.

ELLIS, R. J.
1964 Tracking raccoons by radio, *J. Wildlife Management*, **28**:363–368.

EMLEN, J. T.; R. HINE; W. A. FULLER; AND P. ALFONSO
1957 Dropping boards for population studies of small mammals, *J. Wildlife Management*, **21**:300–414.

GODFREY, G. K.
1954 Tracing field voles (*Microtus agrestis*) with a Geiger-Muller counter, *Ecology*, **35**:5–10.

GRAHAM, W. J., AND H. W. AMBROSE, III
1967 A technique for continuously locating small mammals in field enclosures. *J. Mammal.*, **48**:639–642.

GRODZINSKI, W.; Z. PUCEK; AND L. RYSZKOWSKI
1966 Estimation of rodent numbers by means of prebaiting and intensive removal. *Acta Theriol.*, **11**:297–314

HAYNE, D. W.
1949a An examination of the strip census method for estimating animal populations, *J. Wildlife Management*, **13**:145–157.
1949b Two methods for estimating populations of mammals from trapping records, *J. Mammal.*, **30**:399–411.

JOLLY, G. M.
1965 Explicit estimates from capture–recapture data with both death and immigration—stochastic model, *Biometrika*, **52**:225–247.

KAYE, S. V.
1960 Gold-198 wires used to study movements of small mammals, *Science*, **13**:824.

LORD, R. D., JR.
1959 The lens as an indicator of age in cottontail rabbits. *J. Wildlife Management*, **23**:358–360.
1963 The cottontail rabbit in Illinois, *Illinois Dept. Conserv. Tech. Bull. 3*.

MACLULICH, D. A.
1951 New techniques of animal census, with examples, *J. Mammal.*, **32**:318–328.

MARTEN, G. C.
1970 A regression method for mark–recapture estimate of population size with unequal catchability, *Ecology*, **51**:257–312.

MARTOF, B. S.
1953 Territoriality in the green frog *Rana clamitans*, *Ecology*, **34**:165–174.

NEVILLE, A. C.
1963 Daily growth layers for determining the age of grasshopper populations, *Oikos*, **14**:1–8.

NEW, J. G.
1958 Dyes for studying the movements of small mammals, *J. Mammal.*, **39**:416–429.
1959 Additional uses of dyes for studying the movements of small mammals, *J. Wildlife Management*, **23**:348–351.

PENDLETON, R. C.
1956 Uses of marking animals in ecological studies: Labelling animals with radioisotopes, *Ecology*, **37**:686–689.

ROBINETTE, W. L.; R. B. FERGUSON; AND J. S. GASHWEILER
1958 Problems involved in the use of deer pellet group counts, *Trans. North Am. Wildlife Conf.*, **23**:411–425.

SCHULTZ, V.
1972 *Ecological Techniques Utilizing Radionuclides*

and Ionizing Radiation, A Selected Bibliography,
U.S. Atomic Energy Comm., RLO-2213
(Suppl.1)

STICKEL, LUCILLE F.

1946 Experimental analysis of methods for
measuring small mammal populations, *J. Wildlife
Management,* 10:140–158.

TABER, R. D.

1956 Marking of mammals: standard methods and
new developments, *Ecology,* 37:681–685.

TESTER, J. R.

1963 Techniques for studying movements of
vertebrates in the field, in *Radioecology,*
Reinhold, New York, pp. 445–450.

TIEMEIER, O. W., AND M. L. PLENERT

1964 A comparison of three methods for determining the age of black-tailed jackrabbits, *J. Mammal.*
45:409–416.

VAN ETTEN, R. C., AND C. L. BENNET, JR.

1965 Some sources of error in using pellet group
counts for censusing deer, *J. Wildlife Management,* 29:723–729.

VERTS, B. J.

1963 Equipment and techniques for radio-tracking
skunks, *J. Wildlife Management,* 27:325–339.

ZIPPIN, C.

1958 The removal method of population estimation,
J. Wildlife Management, 22:325–339.

Collecting animals

ANDERSON, R. M.

1948 Methods of collecting and preserving
vertebrate animals, *Nat. Museum Can. Bull.,* 69.

KNUDSEN, J.

1966 *Biological Techniques,* Harper & Row, New
York.

NEEDHAM, J. G. (ED.)

1937 *Culture Methods for Invertebrate Animals,*
reprint, Dover, New York.

OMAN, P. W., AND A. D. CUSHMAN

1948 Collection and preservation of insects, *U.S.
Dept. Agr. Misc. Publ.* 60.

PETERSEN, A. M.

1953 A *Manual of Entomological Techniques,*
published by the author, Ohio State University,
Columbus, Ohio.

WAGSTAFFE, R. J., AND J. H. FIDLER

1955 *The Preservation of Natural History Specimens,* vol. 1, *The Invertebrates,* Philosophical
Library, New York.

stimating the rates of energy fixation and storage in an ecosystem is difficult, and the task of measuring all aspects of energy flow is enormous. Several major techniques for determining primary production are in current use. All estimate energy fixation indirectly by relating the amounts of materials, oxygen, or carbon dioxide released or used.

Methods of estimating production

LIGHT AND DARK BOTTLES

The light-and-dark-bottle method, commonly used in aquatic environments, is based on the assumption that the amount of oxygen produced is proportional to gross production, since one molecule of oxygen is produced for each atom of carbon fixed. Two bottles containing a given concentration of phytoplankton are suspended at the level from which the samples were obtained. One bottle is black to exclude light; the other is clear. In the light bottle, a quantity of oxygen proportional to the total organic matter fixed (gross production) is produced by photosynthesis. At the same time the phytoplankton is using some of the oxygen for respiration. Thus the amount of oxygen left is proportional to the amount of fixed organic matter remaining after respiration or net production. In the dark bottle, oxygen is being utilized but is not being produced. Thus the quantity of oxygen utilized, obtained by subtracting the amount of oxygen left at the end of the run (usually 24 hours) from the quantity at the start, gives a measure of respiration. The amount of oxygen in the light bottle added to the amount used in the dark provides an estimate of total photosynthesis or gross production.

Pratt and Berkson (1959) point out two sources of error in this method. At a temperature range of 11° to 21° C, bacteria are responsible for 40 to 60 percent of the total respiration customarily attributed to the phytoplankton. Failure to adjust gives a low estimate of production by plants. Then in a 2-day experiment large changes in the plankton population occurred in the light bottle. Thus a difference existed in the concentration of phytoplankton inside and outside the bottle. The increase in

the plankton inside the bottle was caused by an accelerated regeneration of nutrients by bacteria attached to the bottle walls.

There are other shortcomings. Estimations of production are confined to that portion of the plankton community contained within the sample bottles. The method fails to take into account the metabolism of the bottom community. And the procedure is based on the assumption that respiration in the dark is the same as that in the light, which is not necessarily true.

A modification of this method involves the whole aquatic ecosystem, which becomes the light and dark bottles, the daytime representing the light bottle, the nighttime the dark. The oxygen content of the water is taken every 2 to 3 hours during a 24-hour period. The rise and fall of the oxygen during the day and night can be plotted as a diurnal curve. To obtain a correct estimate for the oxygen production of plants, the oxygen exchanged between air and water and between the water and bottom must be estimated and deducted. Details for this method, adaptable to the study of flowing waters, can be found in Odum, 1956.

CARBON DIOXIDE

Although the light-and-dark-bottle method is the most useful in aquatic ecosystems, the measurement of the uptake of carbon dioxide and its release in respiration is better adapted to the study of terrestrial ecosystems. In this method a sample of the community, which may be a twig and its leaves, a segment of the tree stem, the ground cover and soil surface, or even a portion of the total community, is enclosed in a plastic tent. Air is drawn through the enclosure, and the carbon dioxide concentration of the incoming and outgoing air is measured with an infrared gas analyzer. The assumption is that any carbon dioxide removed from the incoming air has been incorporated into organic matter. A similar sample may be enclosed within a dark bag. The amount of carbon dioxide produced in the dark bag is a measure of respiration. In the light bag, the quantity of carbon dioxide would be equivalent to photosynthesis minus respiration. The two results added together indicate gross production.

Because the enclosure of a segment of a community necessarily alters the environment of the sample, new methods that produce relatively little disturbance to the normal environment have been sought. One technique is the so-called aerodynamic method (Lemon, 1960; Monteith, 1960). It is based on the assumption that the reduction of carbon dioxide in a canopy of vegetation is equivalent to net photosynthesis. The method involves the periodic measurement of the vertical gradient of carbon dioxide concentration by means of sensors arranged in a series from the ground to a point above the canopy of the vegetation. The difference in the flux or flow between each layer represents the amount of carbon dioxide unitlized by the foliage of each layer. The rate of diffusion or flux within the canopy is proportional to the gradient of carbon dioxide and to the wind speed and turbulence within the canopy (see Inoue, 1968; Lemon, 1968). The technique in many ways is still in the developmental phase, particularly in the design of instruments sensitive to short-term responses. One shortcoming of this and other techniques is the inability to distinguish between the soil and plant contributions to the carbon dioxide flux and to determine the carbon dioxide uptake by the roots and its transport to the chloroplast site.

CHLOROPHYLL

An estimate of the production of some ecosystems can be obtained from chlorophyll and light data. This technique evolved from the discoveries by plant physiologists that a close relationship exists between chlorophyll and photosynthesis at any given light density. This relationship remains constant for different species of plants and thus communities, even though the chlorophyll content of organisms varies widely, as a result of nutritional status and duration and intensity of the light to which the plant is exposed.

This method, best adapted to aquatic ecosystems, involves the determination of chlorophyll *a* content of the plant per gram or per square meter, which under reasonably favorable conditions is the same. Because the quantity of chlorophyll in aquatic (and terrestrial) communities tends to increase or decrease with the amount of photosynthesis (which varies at different light intensities), the chlorophyll per square meter indicates the food manufacturing potential at the time. This photosynthesis : chlorophyll ratio remains rather constant even in cells whose chlorophyll content

Measuring productivity and community structure

varies widely because of nutrition or the duration of the intensity of light to which they were previously exposed.

Chlorophyll *a* can be extracted by filtering natural water through a membrane filter and then extracting the pigments with acetone. The light absorption of the acetone extract is measured at selected wave lengths in the spectrophotometer, the chlorophyll *a* content is computed from this information and expressed in grams per square meter. Nomographs for converting the plankton pigment into chlorophyll biomass can be found in a paper "Plankton pigment nomographs" by A. C. Duxbury and C. S. Yentsch, *Journal of Marine Research,* **15**:92–101 (1956).

To estimate production one must know, in addition to the chlorophyll content of the water, the total daily solar radiation reaching the water's surface and the extinction coefficient of light. The following expression is used to determine daily photosynthesis.

$$P = \frac{R}{k} \times C \times p(\text{sat})$$

where

$P =$ the photosynthesis of the phytoplankton population in grams of carbon per square meter per day

$R =$ relative photosynthesis determined from the curve in Fig. C-1 for the appropriate value of surface radiation

$k =$ extinction coefficient, per meter, as measured

$p(\text{sat}) =$ photosynthesis of a sample of the population at 2,000 foot-candles, as measured in grams of carbon per cubic meter per hour; a rough value for this is 3.7.

$C =$ grams of chlorophyll per cubic meter in a sample of a homogeneously distributed population

The chlorophyll method is well suited for survey work of aquatic ecosystems. It eliminates the need to enclose a sample of the community in artificial containers, or to make time studies of photosynthesis. The method is less useful in terrestrial ecosystems. Productivity is limited more by the lack of nutrients than by the quantity of chlorophyll. The concentration of chlorophyll is greater in plants of shady habitats than in plants growing in sunlight, yet the productivity of such plants is low, although the efficiency of the utilization of light is high. In terrestrial ecosystems, the amount of chlorophyll per unit of leaf and stem is useful in determining the size of photosynthetic systems and in comparing one photosynthetic system with another (Newbould, 1967).

HARVEST METHOD

Widely used in terrestrial ecosystems is the harvest method. It is most useful for estimating the production of cultivated land and range, where production starts from zero at seeding or planting time, becomes maximum at harvest, and is subject to minimal utilization by consumers. A modification of this method is used to estimate forest productivity (see Newbould, 1967).

Briefly the techniques involve the clipping or removal of vegetation at periodic intervals, drying to a constant weight, and expressing that weight as biomass in grams per square meter per year. The caloric value of the material can also be determined. Then the biomass is converted to calories and the harvested material expressed as kilocalories per square meter per year. To be accurate, plant material must be sampled throughout the growing season and the contribution of each individual species determined. Different species of plants reach their peak production at different times during the growing season. To estimate community production from a sample taken only at the end of the growing season seriously underestimates production.

In some determinations only the aboveground portions are harvested, and the root matter goes unsampled. Other studies attempt to estimate the root biomass. Sampling the root biomass is difficult at best. Although the roots of some annual and crop plants may be removed from the soil, the problem becomes more difficult with grass and herbaceous species and even more difficult with forest trees. The task is laborious; except for annual plants, the investigator faces the almost impossible task of separating new roots from older ones. He has the added problem of estimating the turnover of short-lived small roots and root materials and the variability of the sample.

Because plants of different age, size, and species make up the forest and shrub ecosystems, a modified harvest technique known as dimension analysis is used. This involves the measurement of height, the diameter at breast height (DBH), and the diameter growth rate of the trees in a sample plot. A set of sample trees is cut, weighed, and measured, usually at the end of the growing season. The height to the top of the tree, DBH, depth and diameter of the crown, and other parameters are taken. Total weight, both fresh and dry, of the leaves and branches are determined, as is the weight of the trunk and limbs. Roots are excavated and weighed. By various calculations the net annual production of wood, bark, leaves, twigs, roots, flowers, and fruits is obtained. From this information the biomass and production of the trees in the sample unit are estimated and then summed for the whole forest. The biomass of the ground vegetation is also determined, as well as litter fall, estimated from material collected in litter traps. By the use of such techniques, the biomass and nutrient content of various trees and forest stands can be obtained. Because energy utilized by plants and plant material consumed by animals and microorganisms is not accounted for, the harvest method estimates net community rather than net primary production.

RADIOACTIVE TRACERS

The most recent method of determining production involves the measurement of the rate of uptake of radioactive carbon (^{14}C) by plants. This is the most sensitive technique now available to measure net photosynthesis under field conditions.

Basically the method involves the addition of a quantity of radioactive carbon as a carbonate ($^{14}CO_3$) to a sample of water containing its natural phytoplankton population. After a short period of time, to allow photosynthesis to take place, the plankton material is strained from the water, washed, and dried. Then radioactivity counts are taken, and from them calculations are made to estimate the amount of carbon dioxide fixed in photosynthesis. This estimate is based on the assumption that the ratio of activity of $^{14}CO_3$ added to the activity of phytoplankton is proportionate to the ratio of the total carbon available to that assimilated.

In common with other techniques the ^{14}C method has certain inherent weaknesses. The method does not adequately measure changes in the oxidative states of the carbon fixed.

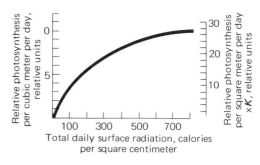

FIGURE C-1

Radiation between total daily surface radiation and daily relative photosynthesis (R) beneath a unit of sea surface. (Adapted from Ryther and Yentsch, 1957.)

731

All of the carbon fixed is not retained by the producers. Some tends to seep out of algal cells as water-soluble organic compounds used by bacteria. The various primary producers have different abilities to utilize available light. The amount of carbon fixed is influenced by the species composition of the plant community.

Radioactive phosphorus, ^{32}P, also has been used, but because phosphorus tends to be absorbed by sediments and organisms more rapidly than it is assimilated by organisms, estimates obtained from phosphorus are inaccurate. This method is much more useful for determining the direction of energy flow than for estimating the fixation of energy. This has been done rather successfully for an old-field ecosystem by Odum and Kuenzler (1963). These investigators labeled three dominant species of plants, *Heterotheca*, *Rumex*, and *Sorgum*, by spraying a solution of ^{32}P on the crowns of the plants; each of the three was isolated in separate quadrats. The solution was soon absorbed and incorporated into the plant biomass. If an animal ate the plant, it in turn became radioactive. These investigators found that two animals—the cricket, *Oecanthus*, and the ant, *Dorymyrmex*—were actively feeding on the plants. Radiation showed up in these two in 1 to 2 weeks, whereas in other grazing herbivores, including the grasshopper, it appeared later, from 2 to 5 weeks. Among the predators, such as the spider, the radioactive tracer did not reach its maximum level until 4 weeks after the initial labeling. Late, too, in showing any concentration of ^{32}P were the detritus- or litter-feeders, the snails. Thus radioactive tracers not only aid in the separation of animals of the community into their appropriate trophic levels but also make possible the determination of habitat niches. In the aforementioned experiment, for example, Odum and Kuenzler found that the most common grazing herbivores in the old-field community fed freely on several dominant species of plants without any marked preference, whereas some of the rarer species were quite selective.

Estimating consumer production

Estimating consumer production also has its problems and difficulties. Methods involve the determination of food consumption, energy assimilation, heat production, maintenance requirements, and growth.

The first step involves some estimation of food consumption. This can be determined in the laboratory or estimated in the field. Laboratory determinations involve feeding the animal a known quantity of its natural foods, allowing it to eat over a period of time, usually 24 hours, then removing the food and weighing the remains. The amount of food consumed equals the amount fed minus the amount removed. The caloric value of the food consumed can be determined by burning a sample in the calorimeter or by obtaining the caloric value for the foods involved from a table—if one exists. If the activity periods of the animal and the weight of the food its stomach will hold are known, then consumption can be rather accurately determined by multiplying the activity periods by the mean weight of observed stomach contents from a sample of animals from the population. Activity periods are used because most animal activity usually is concerned with feeding.

Once consumed, the food must be assimilated. Assimilation can be determined by subtracting the energy voided in feces from energy consumed. The assimilation of natural foods by animals is still largely unknown.

The energy assimilated is used for maintenance and growth. Energy used for maintenance is lost. The cost of maintenance can be determined by confining the animal to a calorimeter and measuring the heat production directly, or the energy used in maintenance can be determined indirectly by placing the animal in a respirometer and measuring the oxygen consumed or the carbon dioxide produced. These results are then converted to calories of heat. But to do this one must know the respiratory quotient, the ratio of the volume of carbon dioxide produced to the oxygen consumed. The respiratory quotient varies with the type of food utilized in the body. To estimate accurately the heat production of a population from laboratory determinations, one must also know the daily activity periods, the weight distribution of the population, and the environmental temperature.

Production or storage of energy is estimated by weighing individuals fed on a natural diet in the laboratory or by weighing animals each successive time they are caught in the field. An indirect and usually more useful method is based on the age distribution of a population, the growth curve for the species, and the caloric value of the animal tissue. Growth curves must be obtained for each population under investigation and for each season under study. Once a growth curve is available and the age distribution of the population is known, the weight of the tissue produced in a given period can be estimated for each age category. The weight gain is then converted to caloric equivalents.

Radioisotopes are also used to determine secondary production, particularly of insects. The rate at which a radioisotope is ingested can be converted to food consumption as long as the concentration of the isotope in the food is known. The method is based on the fact that insects feeding on plant material tagged with a radioactive tracer ^{137}Cs accumulate the isotope in their tissues until there is a steady-state equilibrium. The intake through the consumption of plants is balanced by loss through biological elimination. The balance is given as

$$I = \frac{kQ}{a}$$

where

$I =$ the feeding rate in units of radioactivity
$a =$ the proportion of ingested isotope assimilated
$k =$ the elimination coefficient (0.693/biological half-life)
$Q =$ the equilibrium whole-body burden of the isotope

By using such a method, Reichle and Crossley (1969) determined that forest geometrid caterpillars in a tulip poplar (*Liriodendron*) forest consumed 3.39 mg of foliage/day or about 56 percent of their dry body weight; the tree cricket, *Oecanthus*, consumed 7.35 mg dry weight of foliage/individual/day or about 81 percent of its body weight. Maintenance energy flow amounted to 20.34 kcal/m²/year for the geometrids and 12.39 kcal/m²/year for the tree crickets.

Community structure

POPULATION DISPERSION

One of the problems associated with community structure is the spatial pattern of organisms in the community as it relates to the

733

interaction of organisms with various aspects of the environment. Data collected from quadrats, point quadrats, etc., may be used to determine intrapopulation dispersion, as long as one remembers that the analysis can be influenced by the size of the sampling unit.

As pointed out in Chapter 10, population dispersion may be uniform, random, or clumped. In situations where the density of individuals is low relative to the available surface area or volume, the Poisson method is useful to determine types of dispersion. The Poisson distribution furnishes values expected on the basis of a random dispersion pattern and approximates an extremely asymmetrical distribution. Because the mean of the Poisson is equal to its variance, the Poisson distribution is completely specified by the mean. Thus the theoretical Poisson distribution corresponding to the observed distribution can be constructed from the sample mean alone.

To calculate the Poisson one must know the number of sample units, the number of organisms in each sample unit, and the probability or chance that the organism is located in the sample unit or area. The Poisson series is expressed as

$$P_x = e^{-x}\left(1, x, \frac{x^2}{2!}, \cdots, \frac{x^i}{i!}\right)$$

The steps for setting up the Poisson distribution are as follows:

1. Determine the sample means obtained by the equation:

$$x = \frac{OX}{N}$$

where

O = the observed frequencies
X = the frequency class
N = total frequency

This is an estimate of np, which is then substituted into the general expression for Poisson probability, $e^{-\bar{x}}$, where e is the base of natural logarithms.

2. Determine from the table of exponential functions the value of $e^{-\bar{x}}$.
3. Calculate the Poisson probabilities (see example below).
4. Multiply the probability distribution by the total frequency N to convert it to absolute frequency, so that the probabilities are comparable with the observed distribution.

As an example we can use data from Gabbutt (1961) on the distribution of fleas on mice. The problem here is to determine whether or not the fleas are randomly distributed through a population of mice. The information for *Microtis* is lumped into five classes as follows:

Number of fleas per mouse	0	1	2	3	4+
Mice	44	8	9	3	4

Calculating for the mean:

$$\bar{x} = \frac{fX}{N}$$

$$= \frac{(0 \times 44) + (1 \times 8) + (2 \times 9) + (3 \times 3) + (4 \times 4)}{68}$$

$$= 0.75$$

To determine the Poisson probability:

$X = 0 : e^{-\bar{x}} = 0.472$

where $e^{-\bar{x}} = e^{-.75} = 0.472$

$\quad \bar{x} = 0.75$

$X = 1 : \bar{x}e^{-\bar{x}} = 0.472 \times 0.75 = 0.354$

$X = 2 : \dfrac{\bar{x}}{2} e^{-\bar{x}} = 0.472 \times \dfrac{(0.75)^2}{2} = 0.133$

$X = 3 : \dfrac{\bar{x}^3}{6} e^{-\bar{x}} = 0.472 \times \dfrac{(0.75)^3}{6}$

$$= 0.0345 \quad (0.04)$$

$X = 4 : \dfrac{\bar{x}^4}{24} e^{-\bar{x}} = 0.472 \times \dfrac{(75)^4}{24}$

$$= 0.00623 \quad (0.01)$$

$X = 5 : 1 - 1.00 = 0$

To obtain the theoretical frequency or distribution multiply N, the total number of observations by the Poisson probability for each class (Table C-1). For example, the theoretical distribution for class 0 is $68 \times 0.47 = 31.96$ or 32.

Once the theoretical frequencies have been obtained, the next step is to determine how well the Poisson distribution fits the data. To do this we have to set up a chi-square table in which the observed distribution, O, can be compared with the expected value, E. This is done in Table C-2.

There are five classes after lumping the data and two constants, $\bar{x}$ and N. Thus the degrees of freedom are $5 - 2 = 3$. The high value of chi-square (see statistics books for tabled values of chi-square) with a probability lying well below 0.005 indicates that there is no agreement between the observed and expected distribution. Thus the population of fleas is not randomly dispersed among the mice.

A further test for randomness is the ratio of the variance to the mean. In this example, the ratio is 1.925, which further indicates nonrandomness. In general, if the ratio of the variance to the mean varies around the value of 1, individuals are randomly distributed. If the value is less than 1, the individuals tend toward uniform distribution; and if the values arc much greater than 1, the individuals tend toward a clumped distribution.

SPECIES DIVERSITY

Species diversity implies both the number of species and the number of individuals in a community. But one also has to consider how the individuals are apportioned among the species. For example, a community consisting of five species and 100 individuals with the individuals equally divided among all five species would be more equitable than a community in which 80 individuals were of one species and the remaining 20 were allotted to the other four species.

Two approaches to species diversity are widely used today: Simpson's index (Simpson, 1949) and the Shannon-Wiener formula (Shannon and Wiener 1963). Both are sensitive to changes in the number of species and to changes in the distribution of individuals among the species. However, the index value of both is influenced by sample size. Thus if the index is to be used to compare the diversity among communities, the sample sizes must be equal. Or if complete census data are used, the areas sampled must be of equal size.

Simpson's index of diversity is based on the number of samples of random pairs of individuals that must be drawn from a community to provide at least a 50 percent chance of obtaining a pair with both individuals of the same species. The index is calculated by the formula

$$D = \frac{N(N-1)}{\Sigma\, n(n-1)}$$

where

D = the diversity index
N = total number of individuals of all species
n = number of individuals of a species

TABLE C-1

Poisson probability and theoretical frequency

Number of fleas per mouse X	Observed frequency f	fx²	Poisson probability	Theoretical frequency
0	44	0	.47	32.0
1	8	8	.35	23.8
2	9	36	.13	8.9
3	3	27	.04	2.7
4+	4	64	.01	0.6
Total	68	135		68

MEASURING STRUCTURE

TABLE C-2
Comparison of observed distribution with expected distribution

Number of fleas per mouse	Observed distribution	Expected distribution	O-E	$(O-E)^2$	$\dfrac{(O-E)^2}{E}$
0	44	32.0	12.0	144	4.5
1	8	23.8	−15.8	249.64	10.48
2	9	8.9	0.1	.01	.001
3	3	2.7	0.3	.09	.033
4	4	0.6	3.4	11.56	19.266
Total	68	68			$\chi^2 = 34.28$

$$x^2 = 34.28$$

$$\text{variance: } s^2 = \frac{f(X - \bar{x})^2}{N - 1} = \frac{fX^2 - N(\bar{x}^2)}{N - 1} = \frac{135 - 68(0.56225)}{67} = 1.444$$

$$\frac{s^2}{x} = \frac{1.444}{0.75} = 1.925 \qquad P > .005$$

A community containing only one species would have a value of 1.0. From this values would increase to an infinite one in which every individual belongs to a different species.

The Shannon-Wiener formula comes from information theory. In ecological use the function describes the degree of uncertainty of predicting the species of a given individual picked at random from the community. As the number of species increases and as the individuals are more equally distributed among the species present, the more the uncertainty increases. This function has been criticized as being biologically meaningless (Hurlbutt, 1971). The Shannon-Wiener formula in a general form is

$$D = -\Sigma \, p_i \log_2 p_i$$

where p_i = decimal fraction of total individuals belonging to the ith species.

In words the formula states that the probability that any one individual belongs to species i is p_i. This in turn is equal to the ratio n_i/n, where n_i is the number of individuals in the ith species and n is the total number of individuals of all species. Diversity is greatest if each individual belongs to a different species, the least if all individuals belong to one species. A working formula is

$$H = -\sum_{i=1}^{s} \left(\frac{n_i}{N}\right) \log_2 \left(\frac{n_i}{N}\right)$$

where s = the total number of species collected; $\log_2 = 3.322 \log_{10}$

For calculation of the index use

$$D = 3.322 \left[\log_{10} n - (1/n \, \Sigma \, n_i \log_{10} n_i)\right]$$

To compute:

1. Obtain appropriate $\log_{10} n$.
2. For each species calculate $\log_{10} n_i$.
3. Calculate $n_i \log_{10} n_i$, which means multiply the number of individuals in each species by $\log_{10} n_i$.
4. Sum the $n_i \log_{10} n_i$ and divide the value by n, the total number of individuals of all the species.
5. Subtract this value from $\log_{10} n$ and multiply by 3.322.

ASSOCIATION BETWEEN SPECIES

Some species in a community may occur together more frequently than by chance. This may result from a symbiotic relationship, food-chain coactions, or from similarities in adaptation and response to environmental conditions. Some measurement of this association provides an objective method for recognizing natural groupings of species. Negative associations may indicate interactions detrimental to one or both species, such as interspecific competition or adaptations to different sets of environmental conditions.

The data are obtained by sampling quadrats or point-centered quadrats. Presence or absence data for pairs of species are arranged in a 2 × 2 contingency table:

Species A			
	+	−	
+	a	b	$a + b$
−	c	d	$c + d$
	$a + c$	$b + d$	$a + b + c + d = n$

Species B (left axis label)

Species A (top label)

where

a = samples containing both A and B
b = samples containing only species B
c = samples containing only species A
d = samples containing neither species

From these data a coefficient of association, C, can be calculated. It will vary from a $+1.0$ for maximum possible association to -1.0 for negative association. A value of 0 suggests that frequency of association is that expected by chance.

If $bc > ad$ and $d \gtreqless a$

then

$$C = \frac{ad - bc}{(a + b)(a + c)}$$

If $bc > ad$ and $a > d$

then

$$C = \frac{ad - bc}{(b + d)(c + d)}$$

If $ad \gtreqless bc$

then

$$C = \frac{ad - bc}{(a + b)(b + d)}$$

To determine whether the coefficient of association is significant, one can by a chi-square test determine the significant level of the deviations between the observed values of the contingency table and the expected based on chance association. This requires a slight modification of the contingency table above, as indicated in Table C-3. The expected number of samples containing both species can be found by

$$\frac{(a + b)(a + c)}{n}$$

The remaining expected values can be calculated in a similar manner or obtained by subtraction from the observed marginal totals.

The chi-square value, with one degree of

TABLE C-3
Contingency table for determining the degree of association

	Species A			
	+		−	
	Observed	Expected	Observed	Expected
+				
−				

Species B (left axis label)

freedom, can be found from the following formula:

$$\text{chi square} = \sum \frac{(\text{observed} - \text{expected})^2}{\text{expected}}$$

References

Productivity

In the past few years a number of excellent books on methods of estimating productivity have appeared. Outstanding are the International Biological Program (IBP) handbooks, which belong in all ecological libraries.

ECKARDT, F. E. (ED.)
1968 *Functioning of Terrestrial Ecosystems at the Primary Production Level*, UNESCO, Paris.

ELLENBERG, H. (ED.)
1971 *Integrated Experimental Ecology*, Springer-Verlag, New York.

INOUE, E.
1968 The CO_2 concentration profile within crop canopies and its significance for the productivity of plant communities, in Eckardt (ed.), *Functioning of Terrestrial Ecosystems*, pp. 359–366.

LEMON, E. R.
1960 Photosynthesis under field conditions, II. An aerodynamic method for determining the turbulent carbon dioxide exchange between the atmosphere and a corn field, *Agron. J.*, 52:697–703.
1968 The measurement of height distribution of plant community activity using the energy and momentum balances approaches, in Eckardt (ed.), *Functioning of Terrestrial Ecosystems*, pp. 381–389.

MILNER, C., AND R. E. HUGHES
1968 *Methods for the Measurement of the Primary Production of Grassland*, IBP Handbook No. 6, Blackwell, Oxford.

MONTEITH, J. L., AND G. SZEICZ
1960 The carbon-dioxide flux over a field of sugar beets, *Quart. J. R. Met. Soc.*, 86:205–214.

NEWBOULD, P. J.
1967 *Methods in Estimating the Primary Production of Forests*, IBP Handbook No. 2, Blackwell, Oxford.

ODUM, H. T.
1956 Primary production in flowing waters, *Limnol. Oceanog.*, 1:102–117.

PRATT, D. M., AND BERKSON
1959 Two sources of error in the oxygen light and dark bottle method, *Limnol. Oceanog.*, 4:328–334.

RYTHER, J. H., AND C. S. YENTSCH
1957 The estimation of phytoplankton production in the ocean from chlorophyll and light data, *Limnol. Oceanog.*, 2:281–286.

U.S. FOREST SERVICE
1962 Range research methods: a symposium. *U.S.D.A. Misc. Publ 940.* Vollenweider, R. A.
1969 *A Manual on Methods for Measuring Primary Production in Aquatic Environments*, IBP Handbook No. 12, Blackwell, Oxford.

Secondary production

EDMONDSON, W. T., AND G. G. WINBERG
1971 A *Manual on Methods for the Assessment of Secondary Production in Fresh Waters*, IBP Handbook No. 17, Blackwell, Oxford.

ODUM, E. P., AND E. J. KUENZLER
1963 Experimental isolation of food chains in old-field ecosystem with the use of phosphorus-32, in Schultz and Klements (eds.), *Radioecology*, pp. 113–120.

PETRUSEWICZ, K. (ED.)
1967 *Secondary Productivity of Terrestrial Ecosystems*, 2 vols., Pantsworve Wydawnictwo Naukowe, Warsaw.

PETRUSEWICZ, K., AND A. MACFADYEN
1970 *Productivity of Terrestrial Animals—Principles and Methods*, IBP Handbook No. 13, Blackwell, Oxford.

PHILLIPSON, J. (ED.)
1970 *Methods of Study in Soil Ecology*, UNESCO, Paris.
1971 *Methods of Study in Quantative Soil Ecology*, IBP Handbook No. 18, Blackwell, Oxford.

REICHLE, D. E., AND D. A. CROSSLEY, JR.
1969 Trophic level concentrations of cesium-137, sodium, and potassium in forest arthropods, in D. J. Nelson and F. E. Evans (eds.), *Symposium on Radioecology*, CONF-670503, pp. 678–686.

SCHWOERBEL, J.
1970 *Methods of Hydrobiology: Freshwater Biology*, Pergamon, New York.

SOROKIN, Y., AND H. KADOTA (EDS.)
1972 *Handbook on Microbial Production*, IBP Handbook No. 23, Blackwell, Oxford.

UNESCO
1969 *Soil Biology*, UNESCO, Paris.

WINBERG, G. G. (ED.)
1971 *Methods for the Estimation of Production of Aquatic Animals*, Academic, New York.

Community structure

ANDREWARTHA, H. G.
1970 *Introduction to the Study of Animal Populations*, University of Chicago Press, Chicago.

GABBUTT, P. D.
1961 The distribution of some small mammals and their associated fleas from Central Laborador, *Ecology* 42:518–525.

GREIG-SMITH, P.
1964 *Quantitative Plant Ecology*, Butterworth, London.

HURLBUTT, S. H.
1971 The nonconcept of species diversity: a

critique and alternative parameters, *Ecology*
52:577–586.

LLOYD, M., AND R. J. GHELARDI
1964 A table for calculating the "equitability"
component of species diversity, *J. Animal
Ecology*, **33**:217–225.

MACARTHUR, R. H., AND J. W. MACARTHUR
1961 On bird species diversity, *Ecology* **42**:594–598.

PIELOU, E. C.
1969 *An Introduction to Mathematical Ecology*,
Wiley, New York.

SHANNON, C. E., AND W. WIENER
1963 *The Mathematical Theory of Communication*,
University of Illinois Press, Urbana.

SIMPSON, E. H.
1949 Measurement of diversity, *Nature*, **163**:688.

WILLIAMS, C. B.
1964 *Patterns of Balance in Nature*, Academic, New
York.

739

he methods suggested here are for the most part the simplest and least expensive for the field student to use. Some of them have a little of the homemade quality about them. They may not give the precise results of techniques using more refined instrumentation, but if performed carefully, these methods will give results accurate enough for most studies. Fortunately there are now on the market some prepackaged kits for analyzing water and air pollution. For air pollution testing equipment and material write:

National Environmental Instruments
P.O. Box 590 Pilgrim Station
Warwick, R.I. 02888

For excellent water analysis kits write:

Hach Chemical Co.
Box 907
Ames, Iowa 50010

Lamotte Chemical Co.
P.O. Box 329
Chestertown, Md. 21620

Lamotte also supplies a good soil testing kit for field use.

For graphic analysis of the aquatic environment see Kaill and Frey (1973) for plotting a unique environmental profile.

Collecting water samples

Collecting samples of water from various depths of lakes, ponds, and estuaries can be a difficult task. Water samplers designed to collect waters from various depths can be purchased from scientific apparatus companies or they can be made by assembling a Meyer sampler. This latter apparatus consists of a heavy bottle fitted with a weight heavy enough to sink it when full of air and attached to a wire harness. A stopper with an eyebolt attached is loosely inserted. A cord is attached both to the eyebolt and to the wire harness. When the bottle is lowered to the desired depth (indicated by a marked line), the cord is jerked to remove the cork. Allow sufficient time for the bottle to fill, and then pull it to the surface. Because the bottle cannot be recorked, there will be some admixture with

water from other layers, but this will be minimal.

Water samples must be obtained in 200-ml or 250-cc glass bottles with tight-fitting glass stoppers and taken so that no bubbling or splashing occurs. Be sure that the sample bottle is completely filled, so that there is no air space between the neck and the stopper. Special water samples are available that make it possible to secure water without introducing atmospheric oxygen. If a sampler is used, then when the water is transferred to the bottle, it should be allowed to overflow the container two or three times to eliminate atmospheric oxygen. The water is then ready for analysis in the field by the rapid Winkler method.

Oxygen determination

REAGENTS NEEDED

The following reagents are needed:

1. Alkaline sodium solution. Dissolve 500 g of NaOH and 135 g of NaI to 1 liter of water.
2. Manganous sulfate. Dissolve 480 g $MnSO_4 \cdot 4H_2O$ in distilled water; dilute to 1 liter. Where water is organically enriched, add to the solution 10 g of sodium azide, NaN_3 dissolved in 40 ml of water. This effectively removes interference from nitrites in the water sample.
3. Sodium thiosulfate. Prepare a solution (compute first) of $0.025N$ $Na_2S_2O_3 \cdot 5H_2O$. When dissolved, add a good drop of chloroform to act as a stabilizer, or add 3 g borax/liter, since this solution does not store. Next standardize a 250-ml sample by adding 25 ml of $0.025N$ potassium dichromate, $K_2Cr_2O_7$. Then add 1 ml of the alkaline NaI solution and 5 ml of concentrated hydrochloric acid (HCl). Titrate this solution with the thiosulfate until the solution turns a faint yellow. Add 10 drops of starch solution (see next paragraph) and continue titration until the blue color disappears. The solution is now ready for use; it should be stored in a brown bottle.
4. Starch reagent. Into 350 ml of water stir 2 g of powdered potato starch. To this add 30 ml of 20 percent NaOH, mix, and let stand 1 hour. Neutralize with

HCl, using litmus as an indicator. Then add 1 ml of glacial acetic acid.

OXYGEN DETERMINATION

To determine oxygen follow this procedure:

1. Add 1 ml of the manganous sulfate solution so that it sinks to the bottom. (Slowly and carefully add all reagents deep into the sample with a narrow pipette. When all reagents have been added, the bottle will overflow.) Then with a clean pipette add 1 ml of alkaline sodium iodine solution. Close the bottle, quickly upend, and shake for 20 seconds. A glass bead added to the bottle speeds up mixing. A yellow precipitate will form and settle to the bottom.

2. Add 2 ml of concentrated sulfuric acid (H_2SO_4) and invert several times to mix the acid with the solution. This will dissolve the precipitate (manganese hydroxide, in which is fixed the oxygen in the solution) and leave a yellowish-brown color, the deepness of which is proportional to the amount of dissolved oxygen in the sample.

3. After inverting the bottle several times, place the sample into a 400-ml beaker and titrate with the 0.025N sodium thiosulfate solution until a pale straw yellow is reached. Before doing this, *be sure* to read the level of the sodium thiosulfate in the burette and record.

4. Place a white background against the beaker. Then add 10 drops of the starch indicator. Continue titration until the blue color disappears. Read the new level of the sodium thiosulfate in the burette. Record and then calculate the amount used in titration. The number of milliliters used represents the quantity of dissolved oxygen in parts per million.

Free carbon dioxide concentration

In general the amount of carbon dioxide varies inversely with the dissolved oxygen content. Free carbon dioxide occurs only in acid waters.

1. Obtain a water sample and from it fill a Nessler tube to the 100-cc mark. Do not

Environmental measurements

agitate or splash, because CO_2 will easily diffuse into the air.

2. Add 10 drops of phenolphthalein solution. This can be made by dissolving 5 g of phenolphthalein in 1 liter of 50 percent alcohol, then neutralizing with $0.02N$ NaOH.
3. Titrate this with NaOH, but be sure to record the level of the solution in the burette before starting. When a pink color appears for a few seconds under agitation, quit. Read the new level and record the amount of NaOH used. Multiply amount used by 10. The answer equals the amount of CO_2 in parts per million in the sample.

Temporary hardness

Temporary hardness of water is caused by the carbonates of calcium and magnesium. Because any carbon dioxide present in alkaline and neutral waters will be in the form of bicarbonate, the determination of temporary hardness and the determination of combined carbon dioxide are one and the same.

REAGENTS NEEDED

1. Hydrochloric acid, $0.01N$.
2. Indicator solution consisting of a mixture of 0.02 percent methyl red and 0.1 percent brome cresol green mixed in 95 percent alcohol.

DETERMINATION

1. To the water sample (100 ml) add a couple of drops of indicator.
2. Titrate the HCl into solution until the first appearance of pink.
3. Record the amount in cubic centimeters of HCl needed per 100 ml of water.
4. Multiply this by 5 and the answer will be the amount of CO_2 expressed as parts per million of $CaCO_3$.

Total hardness

Total hardness or permanent hardness generally is caused by the chlorides and sulfates of calcium and magnesium.

There are several ways of determining total hardness, an old-fashioned way, with the soap method, and a colorimetric method. The latter is the most accurate and once the stock solution is mixed, the easiest. The colorimetric method is based on the ability of sodium versenate (sodium diethylenediamine tetracetate, $Na_2C_{10}H_{14}O_8N_2 \cdot 2H_2O$) to form un-ionized complexes with calcium and magnesium. If eriochrome black T, a dark blue dye, is added to this solution containing Ca^{2+} and Mg^{2+} ions, a complex, pink in color, is formed. By adding sodium versenate solution to this complex, the solution can be turned back to blue again by removing Ca^{2+} and Mg^{2+} from the dye complex to the versenate complex again. The end point—when the pink changes back to blue—can be used as a measure of total hardness.

REAGENTS NEEDED

1. Standard calcium chloride solution. Dissolve 55 g of $CaCl_2 \cdot 6H_2O$ in distilled water; then add enough to make 1 liter of solution.
2. Sodium versenate solution. This can be purchased as a standard solution or it can be mixed as follows. Dissolve 2.5 g of sodium versenate in 2 liters of distilled water. To this add 13.5 cc of N NaOH solution, prepared by mixing 40 g in 1 liter of water, and make up to 2500 cc. This solution must be adjusted against the standard chloride solution so that 1 cc equals 0.1 mg of Ca^{2+}. Use eriochrome black T as the indicator.
3. Indicator. Eriochrome black T. To 30 cc of distilled water add 1 g of the eriochrome black T and 1 cc of N Na_2CO_3. Mix and make up to 100 cc with isopropyl alcohol.
4. Buffer solution. Dissolve 40 g of borax in 800 cc of water and 10 g of NaOH and 5 g of sodium sulfide ($Na_2S_2 \cdot 9H_2O$) in 100 cc of distilled water. Mix the two solutions together and dilute to 1 liter. This is used to control pH at the 8 to 10 level and to eliminate the effects of copper, iron, and manganese ions.

DETERMINATION

1. Take a 100-cc sample of water and slightly acidify with $0.1N$ HCl; boil for a few minutes.

2. Add 0.5 cc of the buffer solution.
3. Add 5 drops of eriochrome black *T* indicator solution.
4. Titrate with the standard sodium versenate, the end point being reached when the blue color appears. Each cubic centimeter of sodium versenate needed to titrate the 100-cc sample equals 1 ppm of Ca^{2+} and Mg^{2+}.

Chlorides: salinity

Salinity of coastal waters varies little, but in the estuary it varies considerably, both vertically and horizontally. Because of its biological importance, the determination of chlorides is necessary in any estuarine study. The most accurate method is titration.

DETERMINATION

1. Add a few drops of potassium chromate solution to a 100-cc sample of water.
2. Titrate with 0.01N silver nitrate solution, stirring constantly. When a faint red color of silver chromate appears, the end point has been reached. Determine the number of cubic centimeters of 0.01N silver nitrate used and multiply this by 0.000355. The answer equals chlorides in parts per million.

Acidity

Colorimetric determination of pH is the quickest and simplest method in the field and fairly accurate. For aquatic work there are the narrow-range pH papers. For soils work there are a number of soil-acidity test kits on the market that use color indicators. But be sure to obtain a kit with a selection of narrow-range indicators.

For more sophisticated measurement and research, the electronic pH meter should be used. Most of these are expensive, although moderately priced pocket meters are available.

Current

Estimates of stream flow are necessary in any study of the flowing-water community. Fortunately a short, accurate method for esti-

743

mating the volume of flow is available without resorting to current meters (Robins and Crawford, 1954). It works very well for most stream studies.

Choose a cross section of the stream where the current and depth are most uniform, and measure the width. Then divide this width into three equal segments, marked or separated by pushing sticks into the bottom, coloring the surface of stones, and so on. Next record the depth at the midpoint of each segment and determine the velocity of the surface current. The velocity can be found by dropping a fisherman's float (without projecting arms) attached to 5 ft of limp, monofilament nylon fishing line (0.005–0.01-in. diameter). Record with a stopwatch the time required for the float to travel the 5 ft. This should be done several times and the average figure recorded and converted into feet per second.

Next determine the volume of flow R for each segment of the cross section by the following formula:

$$R = WDaV$$

where

$a =$ a bottom factor constant (0.8 for rocks and coarse gravel; 0.9 for mud, sand, hardpan, bedrock)
$W =$ width of the segment
$D =$ depth at the midpoint of the segment
$V =$ surface current velocity taken at the midpoint of the segment

Total flow is determined by adding the flows for the three segments.

ESTUARY

The study of currents in the estuary is more complicated because of the flow of fresh water to the sea and the incoming flow of seawater at high tide. A knowledge of the circulation set up by the incoming and outgoing flow is of major importance in the study of both the physical and biological aspects of the estuary.

A rapid technique for obtaining current profiles in estuarine waters has been developed by Prichard and Burt (1951). It involves the use of a current indicator consisting of a submerged biplane-shaped drag and a device for reading the angle made by the suspending wire and the vertical. The device is made from two half-inch (five ply) fir plywood panels 4 ft long and 3 ft high, assembled so that each panel bisects the other. This gives four planes 2 by 3 ft. To this is added a 30-lb weight, and the entire biplane is attached to hydrographic wire and suspended into the water from a block extending as far out as possible from the anchored ship. The current will swing this drag out in the direction of the current, whose speed of flow is then computed by the angle of the wire from the vertical. The velocity of flow is proportional to the square root of the tangent of the angle made by the supporting line with the vertical, so the current velocity can be solved by the following formula:

$$v = k\sqrt{\tan \theta}$$

The value of k, a proportionality factor to take care of the opposing forces of frictional drag of water flowing past the drag and the restoring force of gravity, for this device is 1.04. For those interested, the formula for k can be found in the original paper. This method works well for the determination of current velocity down to 50 ft, even where the velocity is as low as 1 cm/second (less than 0.02 knot).

Light

AQUATIC

A commonly employed method of estimating the depth of light penetration into a body of water is the Secchi disk, which is easily constructed. Take a firm metal disk about 8 in. in diameter and paint it with several coats of white enamel. Then divide the disk into quadrants and paint the alternate sectors black or red. Attach the disk by the center to a rod, cord, or chain calibrated for depth. Be sure the disk is fitted to ride level when it is lowered into the water. To read, lower the disk over the side of the boat and note the depth at which the disk disappears. Sink the disk several more feet, then raise it, noting the depth at which the disk reappears. The average of the two observations gives a single light-penetration reading. This method gives an *estimate* of the compensation level.

For precision readings of light in aquatic environments photometers adapted for underwater work are necessary.

The usual method of obtaining light-intensity data in terrestrial communities is the use of a photometer, a wide variety of which are available today. Preferable are ones that are calibrated in foot-candles.

Although the photometer works satisfactorily in open situations, its use in the forest community leaves a good deal to be desired. Because the photometer measures light intensity at a given spot at a given minute, it fails to yield information on the changing intensity of light on the forest floor. The only alternative with this instrument is to take a large number of readings over a period of time, which is time-consuming, inaccurate, and even impossible when more than a few locations are involved.

Superior to the exposure meter is the chemical light meter described briefly in a paper by W. G. Dore (1958). It offers several advantages: (1) it measures the cumulative amount of light reaching a particular location during a period of time; (2) it is comparatively inexpensive; (3) a great number of locations can be sampled at the same time; and (4) the measurement can be made in almost any location where a small vial of the light-sensitive chemical can be placed.

The principle of the method is based on the fact that anthracene $(C_{14}H_{10})$ in benzene will polymerize into insoluble dianthracene $(C_{14}H_{10})_2$ upon exposure to light. This property can be used to measure the amount of light that enters an environment over a period of time. In use, vials of the anthracene-benzene solution are exposed to the light of a particular environment for a period of time, then are analyzed to determine the amount of unconverted anthracene remaining in solution. The analysis is made with a spectrophotometer. A standardization curve that relates the percent transmittance from the spectrophotometer to concentration of anthracene, and a calibration curve that relates the concentration of anthracene to light exposure are required to convert the chemical reaction into standard units of light.

Details of the procedure are too lengthy to describe here, but they are given in step-by-step pictorial detail by Marquis and Yelenosky, 1962, in a U.S. Forest Service publication, *A Chemical Light Meter for Forest Research*.

Temperature

The temperature of air and water at any one given time can be recorded with a standard mercury thermometer or with a dial thermometer. The latter is preferable for ecological work. The cost is moderate, they are not easily broken, and they can be used to take temperatures of streams, of crevices in bark, etc., underneath litter and in the soil.

A semirecording instrument is the maximum-minimum thermometer. It consists of two thermometers, each with a metallic float that in the one thermometer lodges at the highest temperature, in the other at the lowest. For new readings the floats are reset at the top of the mercury columns, preferably with a small magnet. Temperatures usually are recorded on a 24-hour basis. For ecological purposes any time for resetting can be chosen, preferably at some time away from the extremes. In terrestrial communities, maximum-minimum thermometers should be protected in slotted wooden shelters and should be placed at several levels, including the ground level. In animal studies the ground level is best.

The maximum-minimum thermometer can be used to record the water temperatures at different depths in lakes and ponds. Attach one end of the thermometer to a measured line and to the other end attach a weight. Take readings vertically from the surface to the bottom at 0.5- or 1.0-m intervals.

In the absence of a maximum-minimum thermometer, the temperatures of deep water can be taken as soon as the water samples are raised to the surface.

Electronic temperature-recording devices with battery power and with a sensing probe or lead especially designed to read instantaneously the temperature of water at various depths are now available at reasonable cost. Even some mail-order houses have them available as fish-finding devices.

Highly accurate temperature measurements of microlocations—soil, leaf litter, tree wood, bark, leaf, etc.—can be taken with resistance thermometers or thermistors and microprobes that are buried in the material or object. The thermistors have leads projecting above the surface. To take a reading, the wires are attached to the terminals of a potentiometer, on which temperatures are read directly. Compact sensory units are available with which one can take simultaneous readings of both the soil-moisture content and the soil temperature.

Available for continuous recording of temperatures are recording thermographs. A number of types are available, but all work with a simple clock within a cylinder drum upon which temperatures are recorded in graph form by pen and ink.

Humidity

The usual method of determining atmospheric moisture involves the use of the sling psychrometer. This can be purchased rather reasonably from scientific supply houses. When readings are to be taken, the fabric over the wet bulb must be soaked in distilled water. Then the instrument is rapidly swung so that the evaporation cools the wet bulb. When the readings of the two bulbs, wet and dry, are constant, the temperatures on both are recorded. Relative humidity is obtained from the dry-bulb reading and the difference between the dry-bulb reading and the wet-bulb reading. To convert these readings to relative humidity, a conversion table prepared for this purpose must be consulted. Vapor-pressure deficits can also be obtained from the original readings with the aid of suitable tables.

ATMOMETER

For certain studies the natural rates of evaporation may mean more than the relative humidity.

A useful instrument for measuring total evaporative power is the atmometer, or more specifically the Livingston (or James) atmometer, a standardized and calibrated wet surface constant in size, shape, and texture. It consists of a hollow unglazed sphere connected to a reservoir or graduated cylinder.

For long time measurements, over a period of hours or days, the atmometer is connected to a large reservoir of water (about 500 ml in a flask or bottle) by ordinary glass tubing sealed at the end but with a small hole in the end about 1 in. from the top. This in turn is covered with a tightly fitting rubber tubing, which acts as a valve to allow the water to pass from the reservoir into the porous chamber but not back again. The tube is then inserted into the reservoir through one hole in a two-hole stopper. The bottom of the tube should reach nearly to the bottom of the reservoir. In the other hole, place a hooked vent pipe with the external opening pointing downward. Comparisons are made by weighing the whole apparatus and then determining the loss of weight over a period of time. Multiply the raw data by the conversion factor on the glazed neck of each atmometer. Use distilled water at all times.

A simple short-term atmometer can be assembled by filling an inverted atmometer with distilled water and inserting a one-hole rubber stopper in the neck. Then carefully insert a 1-ml graduated pipette through the stopper and into the atmometer far enough to cause the water to rise in the graduated pipette. Expose the atmometer for a constant period of time in the habitat under study and note the amount of water evaporated.

Humidity also can be recorded continuously with the use of a recording hygrograph, which plots the humidity on a graph attached to a revolving drum. In fact the thermograph and hygrograph are often combined into one (rather expensive) instrument, the hygrothermograph.

SOIL MOISTURE

Soil moisture is best determined through the use of moisture blocks, thermistors, and a potentiometer. This has been discussed briefly under temperature. The whole setup can be purchased from agricultural and forestry supply houses.

An alternate but less accurate method is to collect soil samples in the field, place them in plastic bags, and bring them back to the laboratory. If quite wet, the samples should be allowed to drain. Soil moisture content is determined by comparing the weight of the samples at the time of collection with their weight in oven-dry condition. Weigh over 10 g of a sample, place on an evaporating dish, and put in an oven at 105° C until the weight is constant. Cool in the desiccator and weigh. The loss in weight from fresh to dry represents the moisture content or actual amount of water lost. The loss in weight expressed as a percentage is the percent moisture content. It is found as follows:

fresh weight − oven dry weight =
$$\text{moisture content, M.C.}$$

$$\frac{M.C. \times 100}{O.D.} = P.M.C.$$

747

Soil atmosphere

The soil atmosphere, its fluctuations in composition between oxygen and carbon dioxide, and its influence on animal distribution in the soil have received little attention. It is a fertile field for investigation, although its study requires some equipment and is rather time-consuming. The collection of the soil gases is relatively simple. An ingenious procedure for obtaining samples of soil atmosphere from the litter has been devised by Haley and Brierley (1953). This procedure might be modified to study atmosphere of deeper soil layers. The apparatus used (Fig. D-1) consists of polyvinyl plastic tubing 30 cm long and 3 cm in diameter. Each end is fitted with a rubber stopper. One is fitted with a small piece of glass capillary tubing, which in turn is fitted with a stopper made by fitting a piece of glass rod inside a short rubber tube. The other stopper is fitted with wider glass tubing, to which is attached a short piece of rubber tubing with a screw clamp attached. A length of wire is soldered to the clamping screw.

These tubes, which will hold two 25-cc gas samples, are then buried in the surface soil or the litter with the location wire projecting upward. Leave the nozzles of the tubes closed in location for several days until the disturbed soil conditions have returned to normal. Then open the tubes by turning the wire handles on the screw clamps. Gaseous diffusion between the tubes and the soil atmosphere will take place until equilibrium is established, usually in about 24 hours. For best results, however, wait a week, return, close the nozzle, and remove the tubes.

The gases within the tube are then tested in the laboratory for oxygen and carbon dioxide. Analysis procedure can be obtained in analytical chemistry books. A required piece of equipment is the Bancroft-Haldane or Haldane-Gutherie gas-analysis apparatus, obtainable from scientific supply houses. A more accurate field method involving more equipment is described by Wallis and Wilde, 1957.

Soil organic matter

Accurate estimate of soil organic matter is difficult without special techniques, but a reasonable estimate can be obtained by loss of weight on ignition. Take about 5 g of oven-dried soil, weigh carefully, and then heat red hot in a crucible for about a half hour. Cool in a desiccator, reweigh, and heat again. Continue this until weight becomes constant. The loss in weight, expressed as a percentage, will represent the amount of oxidizable organic matter present. Correction for the accompanying decomposition of carbonates can be made by adding some ammonia carbonate to the cooled sample, then heating in the oven to $105°$ C to drive off the excess solution. Cool and reweigh. The gain of weight indicates the amount of CO_2 lost by the carbonates during the previous heating. This gain in weight can then be subtracted from the initial weight loss before percentage of organic matter is determined.

Soil texture and composition

Examine samples of sand, silt, clay, and gravel with a hand lens. Moisten each of these samples and feel them between your fingers. Note the difference between the four, both as to their feel and the way they behave when moist.

Collect samples of soils from various areas —forests, fields, and gardens. Rub a sample of each between your fingers. If gritty, sand particles are present. If some or all of the soil feels like flour or talcum powder, then it is silty. Clay soils have enough clay particles to give them a harsh feeling when dry. Dampen the samples by sprinkling them with water until they have the consistency of workable putty. From this sample make a ball about half an inch in diameter. Hold the ball between thumb and forefinger. Gradually press your thumb forward, forming the soil into a ribbon. If the ribbon forms easily and remains long and flexible, the soil is probably a clay or silty clay; if it breaks easily under its own weight, it probably is a clay loam or a silty clay loam. If a ribbon is not formed, the soil is probably a silt loam, sandy loam, or sand.

Soil textures are best determined in the laboratory by passing a known weight of oven-dry soil from several layers through various-size sieves (U.S. Department of Agriculture standard sieves with decreasing aperture sizes), separating the various components. By determining the proportion of each, the soil texture can be classified. A similar though less accurate method can be used to demonstrate soil texture. Collect samples of various soils, enough to fill a pint jar about half full of each sample. Add water to fill the jar to the shoulder. Put the cap on and shake the jar. Then let the soil settle out; allow plenty of time, for small particles will be slow in settling. The heavier particles will settle out first, the silt next, and the clay last. These components will build up several layers. Hold up a card or heavy piece of paper against the side of the jar and draw a diagram showing the different layers. Label each layer. The relative amounts of each will help you determine the texture of the soil. The ecologist is more concerned with combined effects than with detailed particle analysis, so this rough estimate may be sufficient.

Sampling the soil

The most efficient device for sampling the soil is a soil augur, which enables you to extract soil from known depths and horizons. After extraction the soil is laid out to dry before determining its composition.

Soil profile

Because the distribution of soil inhabitants and the vegetation is influenced by the nature of the soil, some attention must be given to the soil profile (Fig. 7-31). To become familiar with a soil profile, make a vertical cut through a well-drained soil. A good place to do this is along a roadside cut, but be careful to select an area where no debris or fill has been added to the surface soil. Observe the depth of the various horizons, their texture, color, and pH. Measure the pH of the three major horizons with a soil-pH test kit. Make a sketch of the soil profile, indicating the horizons, their depth, color, and pH. If possible, make a similar cut in a wooded section and note the difference between disturbed and undisturbed soil. Then sample the profile of the soil in an area where water stands most of the year. Note the difference between the vegetation growing on the well-drained soil and that growing on the wet areas. Select an area where the slope of the land is variable and possesses a number of drainage conditions. Dig into the subsoil of each drainage class and compare the color of

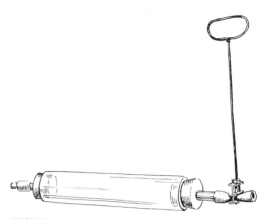

FIGURE D-1
Tube for sampling the soil atmosphere.

749

the subsoil on the top of the ridge with that lower on the slope and at the bottom of the slope. Notice the differences in the type or condition of the vegetation growing on these various drainage situations. Dig a pit to expose A and B horizons under a deciduous and a coniferous woodland, or under an old, established grassland and a woods. Compare the distribution of organic matter. Learn from your local Soil Conservation Service Office the major soil groups found in your area and their location.

Mapping the study area

In some studies an accurate map is desirable and often necessary. It may be needed to show the location of sample plots, animal concentrations and movements, vegetative cover, and the like. The details of the map, what it should show, all depend upon the nature of the study.

The simplest method of mapping is with the use of aerial photographs. If the area to be mapped is located within the center of the photograph and the terrain is not too rough, then the major vegetational units and features of the landscape can be traced directly from a photograph enlarged to a scale of 660 ft to 1 in. (Photographs of most areas are available from the Agricultural Stabilization and Conservation Service, U.S. Department of Agriculture, Washington, D.C.). Scale can be corrected by pacing a given distance along a road, field boundary, or other straight-line feature and comparing it with the distance scale of the photograph. Make any necessary adjustments. For rough or hilly country, or for areas requiring several photographs, the map should be prepared by radial triangulation. For details see the aerial photo references at the end of this appendix.

The area can also be mapped by tape and compass. This will give accurate enough maps for most work. More precise surveying instruments require knowledge of land-surveying techniques.

THE COMPASS

The box-type compass is the most satisfactory for mapping work. It requires the use of a tripod, so that the compass can be leveled at each survey point.

The face of the compass is graduated into quadrants. The graduations at the North and South Poles are 0 and increase through 90° from each pole to east and west. The greater the number of degrees, the more the bearing approaches east or west. The bearing is always read from north or south to east or west. For example, S50E means a point 50° away from south toward the east. If the needle points directly east or west, it is read so. Or the compass may be graduated into 360°. North is 0 and the degrees increase clockwise. East is 90, south 180, west 270, and north 360. Both 0 and 360 are the same—north.

The compass needle on a central pivot always points to magnetic north. The angle this line makes with a line pointing to true north is known as the angle of declination. This declination varies in different parts of the country. The degrees of declination are usually indicated on geologic survey topographic maps. If the survey lines are to be changed to true north, they must be corrected by degrees of declination. This is best done by adjusting the dial of the compass to true north before mapping.

The east and west directions on the box-type compass are printed on opposite sides. In other words, east is where one normally would expect to find west and vice versa. There is a reason for this. The direction is indicated by means of a sight or line on the compass lid. The bearing is always read on the graduated dial at the point where the north end of the needle comes to rest. If the sighting is to the magnetic north, the needle will rest at north; if the compass is shifted to the west, the needle will still point north, but the dial is shifted so that west is at the tip of the needle.

DISTANCE BY PACING

Distance can be measured in two ways: (1) by the use of a steel tape, marked in meters, feet, or chains or (2) by pacing. Pacing is faster but not as accurate, and it is more difficult to correct for slope. However, it will suffice for most mapping, if the terrain is fairly level. First one must know the distance of one's average step or pace. Distance by pacing is measured by counting the steps and multiplying the number of steps by the distance covered in each step. Usually only one foot, the left, is counted. Thus the pace is the average distance taken in two steps. Pacing should be practiced by stepping off measured distances in different terrains and dividing the distance by the number of paces or steps taken.

THE SURVEY TRAVERSE

A traverse is connected by a series of survey lines. Each line is determined by its bearing and distance. A traverse begins at some established point, follows all exterior limits of the area, and closes at the starting point. Traverses that follow roads and streams will not close and their direction will be determined by the route the road or stream takes.

The compass is set up at the starting point, A, and a sight is taken on the next station, B. The bearing is read and entered in the notebook. The distance between A and B is then measured or paced and the distance recorded. When the line AB is completed, the compass is set up at B, a bearing taken on C, recorded, and the distance to C is then measured. The work progresses in this manner until completed. Accurate and complete notes must be taken as the survey progresses. From these notes the map will be drawn. Notes must be entered so they can be interpreted accurately.

After the exterior lines have been surveyed, interior details can be determined by meandering traverses, as the course of a road or stream. The starting point is located at the exact point on the exterior boundary where the road or stream enters the area. The course will be determined by plotting the bearings and distances. The point where the traverse leaves the area should be determined on the opposite boundary traverse.

PLOTTING THE TRAVERSE

A map is plotted from the traverse notes. A protractor, a ruler or engineer's scale, and a straight edge are needed. Before transferring the notes to the map paper, the scale must be determined. The scale will depend upon the extent to which the area can be reduced to fit the size of the map paper. For small areas, the scale of 1 in. on the paper to equal 660 ft on the ground is satisfactory. For larger areas the scale can be reduced; for smaller areas it can be increased.

The protractor is used to transfer the bearings of the individual lines of the map. The

751

protractor is graduated in degrees, used in plotting the angles.

Drawing the map. Select a point on the map to represent the starting point of the survey. Through this point draw a straight line for the meridian, or north–south line. The protractor is centered over the initial point and carefully aligned to the meridian. The bearing of the initial line is marked by a pencil dot according to the number of degrees it bears east or west from north or south. A line is drawn between the two points, A, the starting station and the point located by the protractor. The appropriate distance is then scaled off on this line. The protractor is then centered at *B* and the same operation repeated to locate station *C*. The plotting is continued until all stations and lines are located on the map.

Adjusting the errors. More often than not, the error in plotting the bearings and distance will be great enough so that the map will not close. The error of closure is determined by measuring the distance connecting the last point plotted with the starting point. If the length of line or error of closure is 0.5 percent of the total length of the perimeter, the error is not excessive and will not seriously affect area determinations. If the error is larger than this, the field work should be checked.

Small errors must be corrected if the map is to close. This can be done as follows.

1. The line representing the closing error is drawn and measured.
2. Through each angle, lines parallel to the closing error are drawn.
3. The length of each line between stations is divided by the total perimeter and the results multiplied by the length of the closing error.
4. On the parallel lines the respective amounts of closing error are marked off, increasing cumulatively at each successive corner.
5. The newly located points are connected and the adjusted boundary to the map is completed.

Interior details. Interior details are now added from the notes to the outline. Care must be taken to have the traverse of vegetation boundaries, roads, etc., end on or leave the area at a point on the map that corresponds to the measured distance in the field notes.

Area determination. If a planimeter is not available, the area can be determined as follows.

1. Divide the map into triangles.
2. Compute in square inches, square feet, or square meters the areas of each triangle by the formula—area = base × one-half altitude. If the calculations are made in feet, divide by 43,560 to obtain the area in acres; if in chains, divide by 10; if in meters, divide by 10,000 to obtain the area in hectares.

Mapping lakes and ponds

Mapping is essential to the study of any natural habitat, and lakes and ponds are no exception. One does not need to spend a long time on this, but something fairly accurate is necessary. The map needs to show the shape of the pond or lake, and the surrounding features, including the major zones of vegetation. In addition it should include the depth contours and the bottom vegetation (Fig. D-2). Depth can be roughly determined by taking soundings with a weighted, marked line lowered from a boat. This can be done more accurately with a hydrograph or sonar, if available. Mapping will require a number of transects across the pond. The contours can be plotted and drawn on the map, or depths can be indicated by inserting the figures at appropriate places on the map. The bottom vegetation (weed beds) should be mapped and emergents as well as submerged vegetation indicated. The result should be a base map on which can be placed transparent overlays of transects, temperature and oxygen profiles, fish cover, spawning beds, waterfowl and other bird nests, etc.

Stream mapping

As in the study of other habitats, a map is almost a necessity. A section of stream, if not too wide, can be laid out in a grid and the surface plan of the bottom mapped on paper. This should show such major features as: current flow, large stones, current deflection around the stones, depth contours (far easier to determine here than in lakes and ponds), distribution and nature of vegetation, and bottom substrate. By using transparent overlays, one can plot in the distribution of aquatic organisms in relation to current, depth, bottom substrate, etc. Line transects run across the stream will be helpful in determining the distribution of aquatic organisms.

Remote sensing

Most techniques used in the study of the environment involve the handling and measurement of some component of the ecosystem. Of growing importance is remote sensing, techniques of learning about ecosystems and the environment without direct contact. These methods involve the use of electromagnetic and sound waves, radiant energy, and ionizing radiation.

Remote sensing is not new. One method, the use of aerial photography, has been employed for years. Black and white stereoscopic photos have been used for some time to determine the volume of timber stands, to count wild animals, to record vegetational and man-made changes in the landscape, and to prepare topographic maps. More recently color photography has been widely employed, for its visual contrasts make the identification of tree species, insect damage, plant disease, and details of aquatic environment much easier.

One of the newer techniques is the use of infrared radiation. Infrared photography is very useful in the study of vegetation because the molecules of pigments in plants do not absorb infrared wavelengths. Instead the infrared is either transmitted through the leaf or is reflected by the cell walls. Cells of one species of plant have a different reflectivity than those of another, and normal cells have a different reflectivity than abnormal ones. Therefore infrared photography can be used to distinguish species and to detect unhealthy plants. In fact diseases and environmental stresses in plants can be detected by infrared photography before they can be detected visually.

Infrared photography is also used to detect differences in environmental temperatures. Such pictures or thermographs, in which differences in temperature appear as contrasting bands of color, permit the detection of thermal pollution.

Because more information from aerial photography can be obtained if different types

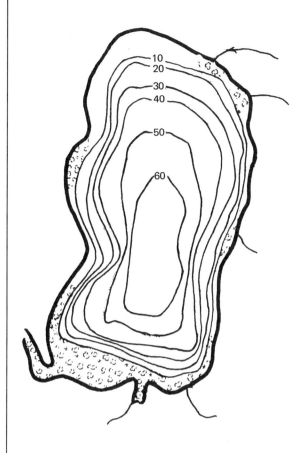

FIGURE D-2
A map of a lake showing depth profiles and a bed of aquatic vegetation.

753

of photography are employed, a new technique called multispectral scanning has been developed. With this method several or many spectral bands from ultraviolet to infrared are recorded. The visual data are gathered by telemetry or are recorded on film or magnetic tape. The image data are fed into a multi-channeled sensor, which sorts out the various displays or images. Because all the channels were recorded simultaneously, each image is in perfect register. This enables the interpreter to pick out and study an object of interest with different types of images taken at the same instant.

Because of its technical nature, remote sensing beyond the use of black and white and color aerial photographs is beyond the range of most field biologists and ecologists.

References

General

AMERICAN PUBLIC HEALTH ASSOCIATION
1971 *Standard Methods for the Examination of Water and Wastewater*, 3rd ed., American Public Health Assoc., Washington, D.C.

AMERICAN SOCIETY FOR PHOTOGRAMMETRY
1966 *Manual of Photogrammetry*, American Society of Photogrammetry, Falls Church, Va.

ANDREWS, W. A. (ED.)
1972 *A Guide to the Study of Fresh Water Ecology*, Prentice-Hall, Englewood Cliffs, N.J.

1972 *A Guide to the Study of Soil Ecology*, Prentice-Hall, Englewood Cliffs, N.J.

1972 *A Guide to the Study of Terrestrial Pollution*, Prentice-Hall, Englewood Cliffs, N.J.

1972 *A Guide to the Study of Environmental Pollution*, Prentice-Hall, Englewood Cliffs, N.J.

BARNES, H.
1963 *Oceanography and Marine Biology: A Book of Techniques*, Macmillan, New York.

GOLTERMAN, H. L. (ED.)
1969 *Methods for Chemical Analysis of Fresh Water*, IBP Handbook No. 8, Blackwell, Oxford.

JACKSON, M. L.
1958 *Soil Chemical Analysis*, Prentice-Hall, Englewood Cliffs, N.J.

JOHNSON, P. L.
1969 *Remote Sensing in Ecology*, University of Georgia Press, Athens.

KAILL, W. M., AND J. K. FREY
1973 *Environments in Profile*, Canfield, San Francisco.

KUCHLER, A. W.
1967 *Vegetation Mapping*, Ronald, New York.

PLATT, R. B., AND J. F. GRIFFITHS
1964 *Environmental Measurement and Interpretation*, Reinhold, New York.

SCHWOERKEL, J.
1970 *Methods of Hydrobiology*, Pergamon, Elmsford, N.Y.

SOIL SURVEY STAFF
1951 "Soil Survey Manual," *U.S. Dept. Agriculture Handbook No. 18*.

STROBBE, M. A.
1972 *Environmental Science Laboratory Manual*, Mosby, St. Louis.

WADSWORTH, R. M. (ED.)
1968 *The Measurement of Environmental Factors in Terrestrial Ecology*, Blackwell, Oxford.

WELCH, P. S.
1948 *Limnological Methods*, McGraw-Hill–Blakiston, New York.

WHITE, W., JR. (ED.)
1972 *North American Reference Encyclopedia of Ecology and Pollution*, North American, Philadelphia.

Other

AVERY, G.
1957 Foresters' guide to aerial photo interpretation, *Southern Forest Exp. Sta. Occ. Paper*, 156.

1962 *Interpretation of Aerial Photographs*, Burgess, Minneapolis.

DORE, W. G.
1958 A simple chemical light meter, *Ecology*, 39:151–152.

GERKING, S. D.
1959 A method of sampling the littoral macrofauna and its application, *Ecology*, 38:219–226.

HALEY, J. L., AND J. K. BRIERLEY
1953 A method of estimation of oxygen and carbon dioxide concentrations in the litter layer of beech woods, *J. Ecology*, 41:385–387.

JOHNSTON, R.
1964 Recent advances in the estimation of salinity, in H. Barnes (ed.), *Oceanography and Marine Biology*, ann. rev. no. 2, Allen and Unwin, London, pp. 97–120.

MARQUIS, D. A., AND G. YELENOSKY
1962 A chemical light meter for forest research, *Northeast Forest Exp. Sta. Paper* 165.

PRITCHARD, D. W., AND W. V. BURT
1951 An inexpensive and rapid technique for obtaining current profiles in estuarine waters, *J. Marine Res.*, 10:180–189.

ROBINS, C. R., AND R. W. CRAWFORD
1954 A short, accurate method for estimating the volume of stream flow, *J. Wildlife Management*, 18:366–369.

SPURR, S. H.
1960 *Photogrammetry and Photo-interpretation*, 2d ed., Ronald, New York.

WALLIS, G. W., AND S. A. WILDE
1957 Rapid method for the determination of carbon dioxide evolved from forest soils, *Ecology*, 38:359–361.

Suggested readings

In this list of suggested readings, I have aimed entirely at books. Back in 1966, when the first edition appeared, books in ecology were not overly abundant. Since 1970 there has been an avalanche of ecologically oriented books. In fact one might call it ecological pollution. For the most part these books consist either of a wide range of readings of all sorts or short paperback texts on ecology. Of the former most will vanish through a process of natural selection. The latter have their greatest use as a review or summary of selected ecological principles. With few exceptions I have not included these shorter texts in the bibliography because they are not specifically reference publications; the readings have been highly selected.

There has also appeared an overwhelming number of books in specialized areas. Here the choice of selections is wholly personal. Any book listed could be replaced by another, but space limits the number. Although I have not indicated so, many of the books are available in paperback. The books are listed by parts and then broken down into smaller subject areas.

General

Reviews

In recent years a number of review volumes have come on the market. Some of the most valuable are the *Annual Review of Entomology, Annual Review of Plant Physiology*, and *Annual Review of Ecology and Systematics*, published by Annual Reviews, Palo Alto, Calif., as well as *Advances in Ecological Research*, published by Academic Press, New York.

Readings

CONNELL, J. H.; D. B. MERTZ; AND W. W. MURDOCK (EDS.): 1970, *Readings in Ecology and Ecological Genetics*, Harper & Row, New York. Thirty selections in areas of community and ecosystems, population dynamics, and genetics. Differs in that editors provide some illuminating commentary. Additional bibliographies.

DAWSON, P. S., AND C. E. KING (EDS.): 1971, *Readings in Population Biology*, Prentice-Hall, Englewood Cliffs, N.J. Covers a wide spectrum of populations from genetics to population dynamics. Forty-one papers.

DETWYLER, T. R. (ED.): 1971, *Man's Impact on Environment*, McGraw-Hill, New York. An

excellent assortment of readings on man's inhumanity to the environment, although not strongly oriented ecologically.

EHRLICH, P. R.; J. P. HOLDREN; AND R. W. HOLM (EDS.): 1971, *Man and the Ecosphere*, Freeman, San Francisco. Informative articles from *Scientific American*.

FORD, R. F., AND W. E. HAZEN (EDS.): 1972, *Readings in Aquatic Ecology*, Saunders, Philadelphia. From physiology and behavior of aquatic pollution, Twenty-nine papers.

HAZEN, W. W. (ED.): 1970, *Readings in Population and Community Ecology*, Saunders, Philadelphia. Contains a full set of papers of debate on balance of nature between Ehrlich and Birch, Slobodkin, etc.

NYBAKKEN, J. W. (ED.): 1971, *Readings in Marine Ecology*, Harper & Row. An excellent collection of papers on marine ecology. Thirty-two papers; some overlap with Ford.

SCHULTZ, V., AND F. W. WHICKER (EDS.): 1972, *Ecological Aspects of the Nuclear Age: Selected Readings in Radiation Ecology*, TID-25978, National Technical Information Service, Springfield, Va. The best general introduction to radiation ecology.

SMITH, R. L. (ED.): 1972, *Ecology of Man: An Ecosystem Approach*, Harper & Row, New York. Papers and lengthy commentaries view man and the environment from an ecosystem point of view.

Texts

ALEXANDER, M.: 1971, *Microbial Ecology*, Wiley, New York. An excellent new book on microbial ecology by a well-known soil microbiologist.

ALLEE, W. C.; A. C. EMERSON; O. PARK; T. PARK; AND K. P. SCHMIDT: 1949, *Principles of Animal Ecology*, Saunders, Philadelphia. A dated but still valuable text and reference.

BOUGHEY, A. S.: 1971, *Man and the Environment*, Macmillan, New York. An interesting combination of human evolution, ecology, and environmental pollution.

BROCK, T.: 1966, *Principles of Microbial Ecology*, Prentice-Hall, Englewood Cliffs, N.J. The first book on the subject.

COLLIER, B. D.; G. W. COX; A. W. JOHNSON; AND P. C. MILLER: 1973, *Dynamic Ecology*, Prentice-Hall, Englewood Cliffs, N.J. A graduate-level text in ecology, theoretically oriented.

EHRLICH, P. R., AND A. H. EHRLICH: 1972, *Population, Resources, Environment*, 2d ed., Freeman, San Francisco. An important book that considers all three items in the title from an ecological viewpoint. Well written.

EMLEM, J. M.: 1972, *Ecology, An Evolutionary Approach*, Addison Wesley, Reading, Mass. A graduate-level text, mathematically oriented, that looks at ecology from an evolutionary viewpoint. Probably the best advanced text.

KREBS, C. J.: 1972, *Ecology: Experimental Analysis of Distribution and Abundance*, Harper & Row, New York. Experimental in approach, this intermediate text is largely a book on population ecology.

MACARTHUR, R. H.: 1972, *Geographical Ecology*, Harper & Row, New York. Largely theoretical, this is one of the most important ecology books of the 1970s.

NAUMOV, N. P.: 1972 (Russia, 1963), *The Ecology of Animals*, University of Illinois Press, Urbana. Translation of the leading Russian text in ecology. Interesting contrast. Russian ecology is much more applied than American. Should be read by all ecologists for a different viewpoint.

ODUM, E. P.: 1971, *Fundamentals of Ecology*, 3d ed., Saunders, Philadelphia. The original ecology text updated. Chapters on remote sensing, modeling.

SCIENTIFIC AMERICAN: 1970, *The Biosphere*, Freeman, San Francisco. An excellent series of articles on the ecosystem.

WARREN, C. E.: 1971, *Biology and Water Pollution Control*, Saunders, Philadelphia. Although too much of the book is devoted to general ecology, it is a good source on the biology of water pollution.

WATTS, M. T.: 1967, *Reading the Landscape*, Macmillan, New York. A delightful book that should be read by all ecologists.

WHITTAKER, R. H.: 1970, *Communities and Ecosystems*, Macmillan, New York. A somewhat advanced small paperback that is a rich storehouse of concepts in community ecology.

Part I

BATES, M.: 1960, *The Forest and the Sea*, Random House, New York (several editions available). Probably the finest general approach to ecology ever written.

BATES, M.: 1961, *The Nature of Natural History*, Scribner, New York. A well-written, thought-provoking, humanistic approach to ecology.

FARB, P.: 1964, *Ecology*, Life Nature Library, Time-Life Books, New York. An enjoyable and concise introduction to ecology, heavily illustrated with outstanding color photographs.

LEOPOLD, A.: 1949, *A Sand County Almanac*, Oxford University Press, New York (several editions available). A classic in ecological writing, rediscovered in the 1970s.

MURDOCH, W. W. (ED.): 1971, *Environment: Resources, Pollution, Society*, Sinauer, Stamford, Conn. An excellent source book on the environment.

WAGNER, R. H.: 1971, *Environment and Man*, Norton, New York. A clearly written discussion of man's environmental problems.

Part II

Energy and Biogeochemical cycles

BARDACH, J.: 1968, *Harvest of the Sea*, Harper & Row, New York. Overview of the sea as a source of food.

BARTHOLOMEW, W. V., AND F. F. CLARK (EDS.): 1963, *Soil Nitrogen*, American Society of Agronomists, Madison, Wis. A review of nitrogen in the soil and role in nitrogen cycle.

BRADY, N. C. (ED.): 1967, *Agriculture and the Quality of Our Environment*, American Association for the Advancement of Science, Washington, D.C. Excellent review of pollution from agriculture and the effects of pollution on agriculture.

BROWN, L. R., AND G. W. FINSTERBUSCH: 1972, *Man and His Environment: Food*, Harper & Row, New York. A superb account of the environmental consequences of man's quest for food.

CRAWFORD, M. A. (ED.): 1968, "Comparative Nutrition of Wild Animals," *Symposium Zoological Society, London*, No. 21, Academic, New York. Important reference on nutrition of wild animals; contains information often hard to find.

CRISP, D. J. (ED.): 1965, *Grazing in Terrestrial and Marine Environments*, Blackwell, Oxford. Energy flow at the herbivore level.

DAVIS, D. E., AND F. B. GOLLEY: 1963, *Principles in Mammalogy*, Reinhold, New York. Considers energy flow at the population level. (See Chap. 9, Metabolism of Populations.)

DUVIGNEAUD, P. (ED.): 1971, *Productivity of Forest Ecosystems*, UNESCO, Paris. Invaluable source of information on primary production in the forest. Worldwide in scope.

ECKARDT, F. E. (ED.): 1968, *Functioning of Terrestrial Ecosystems at the Primary Production Level*, UNESCO, Paris. A comprehensive look at primary production of terrestrial ecosystems. Includes research and methods.

GOLDMAN, C. R. (ED.): 1966, *Primary Productivity in Aquatic Environments*, University of California Press, Berkeley. Contains some technical but basic papers.

HICKLING, C. F.: 1968, *The Farming of Fish*, Pergamon, Elmsford, N.Y. The commercial rearing of fish for food.

KLEIBER, M.: 1961, *The Fire of Life: An Introduction to Animal Energetics*, Wiley, New York. Energy metabolism in animals considered; dated.

LEEDS, A., AND A. P. VAYDA (EDS.): 1965, *Man, Culture, and Animals*, American Association for the Advancement of Science, Washington, D.C. Ecological relationship between man and his domestic animals.

MARGALEF, R.: 1968, *Perspectives in Ecological Theory*, University of Chicago Press, Chicago. Thought-provoking ideas for advanced students.

Suggested readings

MATSUMUIA, F.; G. M. BOUSH; AND T. MISATO (EDS.):
1972, *Environmental Toxocology of Pesticides*,
Academic, New York. A symposium volume that
reviews recent findings.

MILLER, G. T., JR.: 1971, *Energy, Kinetics, and
Life*, Wadsworth, Belmont, Calif. An outstanding
elementary approach to energy and ecology.

MOROWITZ, H. J.: 1968, *Energy Flow in Biology*,
Academic, New York. Physicist's view of energy
in the biological world; advanced.

NATIONAL ACADEMY OF SCIENCE: 1969, *Eutrophication: Causes, Consequences, and Corrections*,
National Academy of Sciences, Washington, D.C.
An excellent overview of eutrophication. Worldwide in scope.

ODUM, H. T. (ED.): 1970, *A Tropical Rain Forest*,
TID-24270, National Technical Information
Service, Springfield, Va. A massive compendium
on productivity of tropical rain forest.

ODUM, H. T.: 1971, *Energy, Power, and Environment*, Wiley, New York. Energy viewed from the
aspect of man and his culture and activities. A
curious and controversial book.

PETRUSEWICZ, K. (ED.): 1967, *Secondary Productivity of Terrestrial Ecosystems*, 2 vols., Inst. Ecol.
Polish Acad. Sci. Int. Biol. Program, Warsaw. A
wide range of papers on secondary production.

PHILLIPSON, J.: 1966, *Ecological Energetics*, St.
Martin, New York. A short, lucid introduction to
the subject.

PIMENTEL, D. D.: 1971, *Ecological Effects of Pesticides on Non-Target Species*. Executive Office of
the President, Office of Science and Technology,
Washington, D.C. Excellent summary and
bibliography of subject.

PURDOM, P. W. (ED.): 1971, *Environmental Health*,
Academic, New York. Effects of pollution on
health.

RAYMONT, J. E. C.: 1963, *Plankton and Productivity
in the Oceans*, Macmillan, New York. Primary
production well covered in a readable fashion in
Chaps. 6 through 10.

REICHLE, D. E. (ED.): 1970, *Analysis of Temperate
Forest Ecosystems*, Springer-Verlag, New York.
Good synthesis of productivity in temperate
forests.

RUSSEL-HUNTER, W. D.: 1970, *Aquatic Productivity*,
Macmillan, New York. Poorly organized, but
presents current knowledge of aquatic production
from phytoplankton to harvest of sea by man.

SAN PIETRO, A.; F. A. GREER; AND T. J. ARMY (EDS.):
1967, *Harvesting the Sun*, Academic, New York.
A good overview of photosynthesis from its
mechanisms to importance to man.

SINGER, S. F. (ED.): 1970, *Global Effects of
Environmental Pollution*, AAAS Symposium,
Springer-Verlag, New York. Contains a number
of significant papers on pollution, especially
atmospheric.

STEELE, J. H. (ED.): 1970, *Marine Food Chains*,

University of California Press, Berkeley. Secondary
productivity in marine environment.

STERN, A. C. (ED.): 1968, *Air Pollution, A Comprehensive Treatise*, 3 vols., 2d ed., Academic, New
York. Covers all aspects of air pollution in detail.

STROBBEM, M. A. (ED.): 1971, *Understanding
Environmental Pollution*, Mosby, St. Louis. A
good selection of readings devoted wholly to
pollution.

STUDY OF CRITICAL ENVIRONMENTAL PROBLEMS:
1970, *Man's Impact on the Global Environment*,
MIT Press, Cambridge, Mass. An excellent
reference source.

STUDY OF MAN'S IMPACT ON CLIMATE: 1971,
Inadvertent Climate Modification, MIT Press,
Cambridge, Mass. An excellent introduction to
climate and man's impact on it.

THE INSTITUTE OF ECOLOGY: 1971, *Man in the
Living Environment*, University of Wisconsin
Press, Madison. An excellent assessment of man's
impact on the environment. Considers man's
effects on biogeochemical cycles, man and agriculture, and resource management.

Environmental influences

BECK, S. D.: 1963, *Animal Photoperiodism*, Holt,
Rinehart, & Winston, New York. A short introduction to the subject.

BUNNING, E.: 1964, *The Physiological Clock*, 2d ed.,
Academic, New York. A short book by an
authority in the field of periodicity.

BROWN, F. A., ET AL.: 1970, *Biological Clocks, Two
Views*, Academic, New York. Discusses the
controversy over exogenous and endogenous
control over circadian rhythms.

GATES, D. M.: 1962, *Energy Exchange in the
Biosphere*, Harper & Row, New York. Read in
particular Chap. 1.

GEIGER, R.: 1957, *The Climate Near the Ground*,
2d ed., Harvard University Press, Cambridge,
Mass. Emphasizes the microclimate, which makes
this the climatology book for ecologists.

GOLTERMAN, H. L., AND R. S. CLYMO: 1967,
Chemical Environment in Aquatic Habitats, an
IBP symposium, N. V. Noord-Hollandsche
Vitgevers, Maatsschappy, Amsterdam.

HENDERSON, L. J.: 1913, *The Fitness of the
Environment* (reprint), Beacon, Boston. An old
classic that ecologists still should read.

HUNT, C. A., AND R. M. GARRELS: 1972, *Water, The
Web of Life*, Norton, New York. A short
elementary introduction to water.

JOHNSON, C. G., AND L. P. SMITH (EDS.): 1965,
*Biological Significance of Climatic Changes in
Britain*, Academic, New York. Some ecological
information.

KRENKEL, P. A., AND F. L. PARKER (EDS.): 1969,
Biological Aspects of Thermal Pollution, Vander-
bilt University Press. Symposium. Good source,
but literature changing rapidly.

LOWERY, W. P.: 1969, *Weather and Life*, Academic, New York. An introductory text on weather and its effects on plants and animals, including man.

NELSON, D. J. (ED.): 1973, *Radionuclides in Ecosystems*, Proc. 3d National Symposium on Radioecology, 2 vol., CONF-710501-Pl, National Technical Information Service, Springfield, Va. A wealth of information on radioecology up to 1971.

NELSON, D. J., AND F. C. EVANS (EDS.): 1969, *Symposium on Radioecology*, Proc. 2nd Symposium. CONF 670503, National Technical Information Service, Springfield, Va. A major reference source.

SCHROEDER, M. J., AND C. C. BUCK: 1970, "Fire Weather," U.S.D.A. *Agricultural Handbook 360*, GPO, Washington, D.C. Although written as a forest fire control handbook, this book is a superb reference on climatology.

SCHULTZ, V., AND A. W. KLEMENT (EDS.): 1963, *Radioecology*, Reinhold, New York. Proceedings of First Symposium on Radioecology. An important sourcebook.

SOLLBERGER, A.: 1965, *Biological Rhythm Research*, Elsevier, Amsterdam. An exhaustive survey of the literature to 1964. A superb bibliography.

U.S. FOREST SERVICE: 1971, *Fire in the Northern Environment—A Symposium*, Northwest Forest and Range Exp. Station, Portland, Oreg. The role of fire in the northern coniferous forest and tundra.

Community and succession

AGER, D. V.: 1963, *Principles of Paleoecology*, McGraw-Hill, New York. A survey of the subject of communities in the past.

ASHBY, M.: 1969, *Introduction to Plant Ecology*, St. Martin, New York. An elementary text that is both interesting and stimulating.

BROOKHAVEN NATIONAL LABORATORY: 1969, "Diversity and Stability in Ecological Systems," *Brookhaven Symp. in Biology No. 22*, National Technical Information Service, Springfield, Va. Summary of current status of theory.

DAUBENMIRE, D.: 1968, *Plant Communities: A Textbook of Synecology*, Harper & Row, New York. A book strong on succession.

DICE, L. C.: 1952, *Natural Communities*, University of Michigan Press, Ann Arbor. A good book that has never been fully appreciated.

ELTON, C. E.: 1927, *Animal Ecology*, Sidgwick & Jackson, London (Chapman & Hall; University of Washington Press).

ELTON, C. E.: 1958, *The Ecology of Invasion by Plants and Animals*, Wiley, New York. This and the previous book, classics in the best sense of the word, are excellent introductions to community ecology. Short, clearly written.

GRAHAM, E. H.: 1944, *Natural Principles of Land Use*, Oxford University Press, New York (Green-wood). An excellent book stressing the biological concepts applied to land management. Should be required reading for all. Contains photos of the Harvard College Forest Models, dioramas illustrating forest succession in mid-New England.

KOMAREK, E. V. (ED.): 1962–1973, *Annual Tall Timbers Fire Ecology Conference*, Tallahassee, Fla. The finest source of information on forest fire ecology available.

OOSTING, H. S.: 1956, *The Study of Plant Communities*, Freeman, San Francisco. A general introduction to plant ecology.

THOMAS, W. L., JR. (ED.): 1956, *Man's Role in Changing the Face of the Earth*, University of Chicago Press, Chicago. A symposium volume that presents a thorough discussion of man's impact on the biosphere.

U.S. DEPARTMENT OF AGRICULTURE: 1958, "Land," *Yearbook of Agriculture*, GPO, Washington, D.C. A survey of the American land, past and present, and the problems of the future.

WEST, R. G.: 1968, *Pleistocene Geology and Biology*, Wiley, New York. A general introduction of interest as background to paleoecology.

Part III

Populations

ANDREWARTHA, H. G.: 1970, *Introduction to the Study of Animal Populations*, University of Chicago Press, Chicago. Basically a summary of the following volume.

ANDREWARTHA, H. G., AND L. C. BIRCH: 1954, *The Distribution and Abundance of Animals*, University of Chicago Press, Chicago. A stimulating and controversial text.

CHENG, T. C.: 1964, *The Biology of Animal Parasites*, Saunders, Philadelphia. A sound introduction to animal parasitology.

CHENG, T. C.: 1971, *Aspects of the Biology of Symbiosis*, University Park Press, Baltimore. Papers from AAAS Symposium. Wide range of topics.

CLARK, L. R., ET AL. (EDS.): 1967, *The Ecology of Insect Populations in Theory and Practice*, Methuen, London. Introduction to population ecology from entomological point of view. Pedantic but important reference.

DORST, J.: 1963, *The Migration of Birds*, Houghton Mifflin, Boston. A survey of bird migration the world over.

EASTWOOD, E.: 1967, *Radar Ornithology*, Methuen, London. An account of bird movements as observed by radar.

ELTON, C. E.: 1958, *Voles, Mice and Lemmings*, Methuen, London. A classic book on animal populations.

ERRINGTON, P. L.: 1963, *Muskrat Populations*, Iowa State University Press, Ames. The results of a

761

20-year study of muskrats. A model of field investigations.

ERRINGTON, P. L.: 1967, *Of Predation and Life*, Iowa State University Press, Ames. A philosophical view toward predation.

FENNER, F., AND F. M. RATCHIFFE: 1965, *Myxomatosis*, Cambridge University Press, New York. The story of myxomatosis, rabbit populations, and the adaptations of hosts and parasites.

HARRISON, G. A., AND A. J. BOYCE (ED.): 1971, *The Structure of Human Populations*, Clarendon Press, London. Demography, ecology, psychology, anthropology, and sociology brought together in an analysis of human populations.

HENRY, I. S. M.: 1966, *Symbiosis*, vols. 1 and 2, Academic, New York. A sweeping survey of symbiosis. Important reference.

JOHNSON, C. G.: 1969, *Migration and Dispersal of Insects by Flight*, Methuen, London. Exhaustive review of subject. Basic reference.

JONES, F. R. H.: 1968, *Fish Migration*, St. Martin, New York. Exhaustive account of fish migration over the world.

KEYFITZ, N., AND W. FLIEGER: 1971, *Populations, Facts and Methods of Demography*, Freeman, San Francisco. Fundamentals of demography and a compendium of the world's human populations.

LACK, D.: 1954, *The Natural Regulation of Animal Numbers*, Oxford University Press, New York. Emphasizes the role of food.

LACK, D.: 1966, *Population Studies of Birds*, Oxford University Press, New York. Case history studies.

LE CREN, E. D., AND M. W. HOLDGATE (EDS.): 1962, *The Exploitation of Natural Animal Populations*, Wiley, New York. A symposium volume, which explores the scientific basis of the response of natural animal populations to exploitation.

MACFADYEN, A.: 1963, *Animal Ecology*, 2d ed., Pitman, London. A more technical introduction, with emphasis on soil organisms.

ORR, R. T.: 1970, *Animals in Migration*, Macmillan, New York. An excellent popular account of migration among animals.

SLOBODKIN, L. B.: 1962, *Growth and Regulation of Animal Populations*, Holt, Rinehart & Winston, New York. A valuable publication on population dynamics, but laboratory rather than field populations dominate the text.

WATSON, A.: 1970, *Animal Populations in Relation to Their Food Resources*, Blackwell, Oxford. Investigations into animal–food interactions. A number of interesting papers.

Genetics and speciation

BEADLE, G., AND M. BEADLE: 1966, *The Language of Life*, Doubleday, New York. An engaging and lucid account of genetics.

BLAIR, W. F. (ED.): 1961, *Vertebrate Speciation*, University of Texas Press, Austin. A symposium

volume that contains a tremendous amount of compressed information.

COLD SPRING HARBOR SYMPOSIA ON QUANTITATIVE BIOLOGY NO. 20: 1955, *Population Genetics: The Nature and Causes of Genetic Variability in Populations*, The Biological Laboratory, Cold Spring Harbor, N.Y. A volume that well summarizes the field up to the mid-1950s.

CREED, R. (ED.): 1971, *Ecological Genetics and Evolution*, Blackwell, Oxford. Essays in honor of E. B. Ford that stress polymorphism and melanism.

DOBZHANSKY, T.: 1951, *Genetics and the Origin of Species*, 3d ed. Columbia University Press, New York. Hard reading in places but one of the most important books on evolution.

DOBZHANSKY, T., ET AL.: 1968, *Evolutionary Biology*, Appleton-Century-Crofts, New York. Review volumes on the subject.

DRAKE, F. T. (ED.): 1968, *Evolution and Environment*, Yale University Press, New Haven, Conn. Symposium papers.

EATON, T. H., JR.: 1970, *Evolution*, Norton, New York. A concise, lucid account with a strong ecological approach.

FORD, E. B.: 1970, *Ecological Genetics*, 3d ed., Methuen, London. The ecological aspects are well developed in this book.

GRANT, V.: 1971, *Plant Speciation*, Columbia University Press, New York. Speciation phenomena in higher plants.

LACK, D.: 1947, *Darwin's Finches: An Essay on the General Biological Theory of Evolution*, Cambridge University Press, London. A classic of biology. Probably the best introduction to the actual origin of new species.

LACK, D.: 1971, *Ecological Isolation in Birds*, Blackwell, London. A valuable book on interspecific competition and isolating mechanisms.

LOWE-MCCONNELL, R. H. (ED.): 1967, Speciation in tropical environments, *Biol. J. Linn. Soc. Lond.* 1:1–2. Academic, New York. Excellent papers on speciation and species diversity in the tropics.

MAYR, E.: 1970, *Population, Species, and Evolution*, Harvard University Press, Cambridge, Mass. An abridgment of *Animal Species and Evolution*, and much more usable. Excellent reference.

METTLER, L., AND T. G. GREGG: 1969, *Population, Genetics and Evolution,*" Prentice-Hall, Englewood Cliffs, N.J. An excellent elementary introduction.

MOODY, P. A.: 1970, *Introduction to Evolution*, 3d ed., Harper & Row, New York. A time-tested introduction.

SAVORY, T.: 1963, *Naming the Living World*, Wiley, New York. An introduction to the principles of biological nomenclature.

STEBBINS, G. L.: 1971, *Processes of Organic Evolution*, Prentice-Hall, Englewood Cliffs, N.J. An introductory text, strong on speciation.

Behavior (including Part IV)

ARMSTRONG, E. A.: 1964, *Bird Display and Behavior*, rev. ed., Dover, New York. Overall survey of bird behavior. Excellent bibliography.

CROOK, J. H. (ED.): 1970, *Social Behavior in Birds and Mammals*, Academic, New York. Good on feeding dispersal of birds, breeding adaptations, and parental care.

DARLING, F. F.: 1937, *A Herd of Red Deer*, Oxford University Press, New York. A classic in the study of animal behavior.

EIBL-EIBESFELDT, I.: 1970, *Ethology, The Biology of Behavior*, Holt, Rinehart & Winston, New York. An excellent introductory text on behavior.

ETKIN, W. (ED.): 1964, *Social Behavior and Organization Among Vertebrates*, University of Chicago Press, Chicago. A rather good summary by a variety of contributors.

EVANS, H. E.: 1963, *Wasp Farm*, Natural History Press, New York. An engaging account of the comparative behavior of wasps.

GEIST, V.: 1971, *Mountain Sheep, A Study in Behavior and Evolution*, University of Chicago Press, Chicago. A good field study of behavior, well illustrated.

HAFEZ, E. S. (ED.): 1962, *The Behavior of Domestic Animals*, Williams & Wilkins, Baltimore. A thorough reading of this book would be a valuable beginning for the study of big game animals.

HINDE, R. A.: 1970, *Animal Behavior*, 2d ed., McGraw-Hill, New York. A synthesis of ethology and comparative psychology. Concentrates on the causes of behavior.

JEWELL, P. A., AND C. LORZOS (EDS.): 1966, "Play, Exploration, and Territory in Animals," *Symposium Zool. Soc. London*, Academic, New York. Contains a group of informative papers.

JOHNSGARD, P. A.: 1965, *Handbook of Waterfowl Behavior*, Cornell University Press, Ithaca, N.Y. Excellent descriptions of waterfowl displays.

JOHNSGARD, P. A.: 1971, *Animal Behavior*, 2d ed., Brown, Dubuque, Iowa. Best introduction to animal behavior available.

KENDEIGH, S. C.: 1952, Parental care and its evolution in birds. *Illinois University Biological Monographs 22*, University of Illinois Press, Urbana, Ill. Summarizes parental care by birds with special reference to the house wren.

KLOPFER, P. H.: 1973, *Behavioral Aspects of Ecology*, 2d ed. Prentice-Hall, Englewood Cliffs, N.J. A behavioral approach to ecology. Highly speculative in places, but a refreshing point of view.

KLOPFER, P. H., AND J. P. HAILMAN: 1967, *An Introduction to Animal Behavior; Ethology's First Half Century*, Prentice-Hall, Englewood Cliffs, N.J. An introductory text.

LORENZ, K.: 1952, *King Solomon's Ring*, Crowell, New York. A highly popular book in which this great ethologist describes his experiences in studying animal behavior.

MAIER, R. A., AND B. M. MAIER: 1970, "Comparative Animal Behavior," Brooks-Cole, Belmont, Calif. A comparative look at animal behavior throughout the animal kingdom.

MARLER, P., AND W. J. HAMILTON, III: 1967, *Mechanisms in Animal Behavior*, Wiley, New York. Basic behavior, but little on behavioral ecology.

RHEINGOLD, H.: 1963, *Maternal Behavior in Mammals*, Wiley, New York. The first general review of maternal behavior in mammals.

TRAVOLGA, W. N.: 1969, *Principles of Animal Behavior*, Harper & Row, New York. Superb bibliography of animal behavior.

THORPE, W. H.: 1961, *Bird Song: The Biology of Vocal Communication and Expression in Birds*, Cambridge University Press, London. An important work on vocal communications in birds.

THORPE, W. H.: 1963, *Learning and Instinct in Animals*, rev. ed., Methuen, London. A basic work in the field of ethology.

TINBERGEN, N.: 1951, *The Study of Instinct*, Oxford University Press, New York. A major work in the comparative study of behavior.

TINBERGEN, N.: 1960, *The Herring Gull's World*, Basic Books, New York. Tinbergen's finest book, an excellent example of how basic research can be presented as good literature.

TINBERGEN, N.: 1965, *Animal Behavior*, Life Nature Library, Time-Life Books, New York. A highly readable and richly illustrated introduction to ethology.

WILSON, E. O.: 1971, *Insect Societies*, Harvard University Press, Cambridge, Mass. An outstanding book on social insects and their behavior.

Part IV

Physiological ecology

BENTLEY, P. J.: 1971, *Endocrines and Osmoregulation*, Springer-Verlag, New York. A comparative account of regulation of water and salt in vertebrates.

DAUBENMIRE, R. F.: 1959, *Plants and Environment*, 2d ed., Wiley, New York. A basic reference.

FOLK, G. E., JR.: 1966, *Introduction to Environmental Physiology*, Lea & Febiger, Philadelphia. Well illustrated, this book is an excellent introduction to environmental physiology of mammals.

GORDON, M. S.: 1971, *Animal Functions, Principles and Adaptations*, Macmillan, New York. Perhaps the finest book on animal physiology. Valuable because it is an ecological approach.

IRVING, L.: 1972, *Arctic Life of Birds and Mammals, Including Man*, Springer-Verlag, New York. A review of adaptive physiology of arctic homeotherms by an authority.

NEWELL, R. C.: 1970, *Biology of Intertidal Animals,* Elsevier, New York. A thorough review of physiology of intertidal animals.

NICOL, J. A. C.: 1967, *Biology of Marine Animals,* Wiley, New York. Somewhat similar to Newell, but considers animals of open water.

TRESHOW, M.: 1970, *Environment and Plant Response,* McGraw-Hill, New York. Discusses plants under environmental stress, especially pollution.

VERNBERG, F. J., AND A. B. VERNBERG: 1970, *The Animal and the Environment,* Holt, Rinehart & Winston, New York. Animal ecology from a physiological point of view.

WITTOW, G. C.: 1970, 1971, *Comparative Physiology of Thermoregulation:* vol. 1, *Invertebrates and Non-mammalian Vertebrates,* vol. 2, *Mammals.* A thorough review of the literature. A basic reference.

Part V

A popular yet authoritative treatment of the material covered in this part is contained in the *Our Living World of Nature* series published by McGraw-Hill. Volumes include forest, seashore, African plains, and the far north. Fourteen volumes available, all authoritatively written and exceptionally well illustrated.

Biogeography

The best treatment of biogeography for the student and lay reader alike is contained in the six-volume series on land and wildlife, a part of the Life Nature Library, Time-Life Books, New York. These volumes include South America, Australia, Eurasia, Africa, North America and Tropical Asia.

In addition the books below should serve as standard references:

ALLEN, D. L.: 1967, *The Life of Prairie and Plains,* Our Living World of Nature Series, McGraw-Hill, New York. An excellent introduction to life and ecology of North American grasslands. Well illustrated.

CAIN, S. A.: 1944, *The Foundations of Plant Geography,* Harper & Row, New York.

DARLINGTON, P. J., JR.: 1957, *Zoogeography: The Geographical Distribution of Animals,* Wiley, New York.

UDVARDY, M. D. F.: 1969, *Dynamic Zoogeography. With Special Reference to Land Animals,* Van Nostrand Reinhold, New York.

Soil

BUCKMAN, H. O., AND N. C. BRADY: 1969, *The Nature and Properties of Soils,* 7th ed., Macmillan, New York. A classic college textbook that covers most aspects of soil science.

BUOL, S. W., ET AL.: 1973, *Soil Genesis and Classification,* Iowa State University Press, Ames. A new text on this subject.

EYRE, S. R.: 1968, *Vegetation and Soils, A World Picture,* 2d ed., Aldine, Chicago. A broad picture of soils of the world and their development.

FARB, P.: 1959, *The Living Soil,* Harper & Row, New York (in several editions). A popular account of life in the soil; scientifically sound and beautifully written.

KEVAN, D. K. MCE. (ED.): 1955, *Soil Zoology,* Butterworth, Washington, D.C. Symposium proceedings. Excellent; notable are the keys to soil invertebrates.

KEVAN, D. K. MCE.: 1962, *Soil Animals,* Philosophical Library, New York. Simplified introduction to life in the soil. Good on the ecology of soils.

LUTZ, H. J., AND R. F. CHANDLER: 1946, *Forest Soils,* Wiley, New York. Out of print, but a basic reference on the subject.

UNESCO: 1969, *Soil Biology,* UNESCO, Paris. An outstanding collection of papers on all aspects of soil biology.

U.S. DEPARTMENT OF AGRICULTURE: 1957, "Soils," *Yearbook of Agriculture,* GPO, Washington, D.C. Principles of soils, soil fertility, and soil management.

WALLWORK, J. A.: 1970, *Ecology of Soil Animals,* McGraw-Hill, New York. An up-to-date reference with a good review of world literature.

Grassland

The finest reference on grasslands is the relatively unavailable "The Grassland Ecosystem, A Preliminary Synthesis," *Range Sci. Dept. Sci. Ser. No. 2,* 1969, Colorado State Univ. Fort Collins, Colo., edited by R. L. Dix and R. G. Beidleman.

CARPENTER, J. R.: 1940, The grassland biome, *Ecol. Monographs,* 10:617–684. Included among the books because it is an excellent review of North American grasslands.

HUMPHREY, R. R.: 1962, *Range Ecology,* Ronald, New York. An ecology text with grassland orientation.

SPEDDING, C. R. W.: 1971, *Grassland Ecology,* Clarendon Press, Oxford. Synthesis of ecology and agriculture. A fresh modern approach to grassland management.

SPRAGUE, H. E. (ED.): 1959, *Grasslands,* American Association for the Advancement of Science, Washington, D.C. A symposium volume with a number of papers on grassland ecology.

WEAVER, J. E.: 1954, *North American Prairie,* Johnsen, Lincoln, Nebr. A detailed description of the original prairie.

WEAVER, J. E.: 1968, *Prairie Plants and Their Environment,* University of Nebraska Press, Lincoln. A summary of many years' work on prairie grasslands.

WEAVER, J. E., AND F. W. ALBERTSON: 1956, *Grasslands of the Great Plains,* Johnsen, Lincoln, Nebr. A sound reference on the grasslands of the midcontinent.

Shrubland

FRIEDLANDER, C. P.: 1961, *Heathland Ecology*, Harvard University Press, Cambridge, Mass. A good introduction to the ecology of shrub communities, although oriented toward British heathland.

MCKELL, C. M., ET AL. (EDS.): 1972, "Wildland Shrubs, Their Biology and Utilization," *U.S.D.A. Forest Service General Tech. Rept. INT-1*. Intermountain Forest and Range Experiment Station. A definitive and much-needed reference on shrubs, a subject on which little literature exists. Good bibliography.

Desert

BUXTON, P. A.: 1923, *Animal Life in Deserts*, Edward Arnold, London. An old book on animals in the desert that still has its interest.

CLOUDSLEY-THOMPSON, J. L.: 1954, *Biology of the Deserts*, Hafner, New York. A general book on desert life.

HODGE, C., AND P. C. DUISBERG: 1963, *Aridity and Man*, American Association for the Advancement of Science, Washington, D.C. Man and the desert, with emphasis on the arid zones of the United States.

HOWES, P. G.: 1954, *The Giant Cactus Forest and Its World*, Duell, Sloan & Pearce, New York. A pleasant introduction to the southwestern desert.

JAEGER, E. C.: 1957, *The North American Deserts*, Stanford University Press, Stanford, Calif. A general survey; excellent reference.

KIRMIZ, J. P.: 1962, *Adaptation to Desert Environment; A Study On the Jerboa, Rat and Man*, Butterworth, Washington, D.C. A detailed treatment of the adaptations of mammalian life in the desert.

KRUTCH, J. W.: 1952, *The Desert Year*, Sloane, New York. A classic on the desert that should be read for both the description and the philosophy.

LEOPOLD, A. STARKER: 1961, *The Desert*, Life Nature Library, Time-Life Books, New York. An excellent general account of the desert. Well illustrated.

SHREVE, F., AND I. L. WIGGINS: 1963, *Vegetation and Flora of the Sonoran Desert*, Stanford University Press, Stanford, Calif. This two-volume work is a monumental study of the desert vegetation of the Southwest.

WHITE, G. F. (ED.): 1956, *Future of Arid Lands*, American Association for the Advancement of Science, Washington, D.C. A symposium volume that discusses the future of deserts in human affairs.

Forest

AUBERT, DE LA RUE; E. F. BOURLIERE; AND J. HARROY: 1957, *The Tropics*, Knopf, New York. A good introduction to the tropical world.

BRAUN, E. L.: 1950, *Deciduous Forests of Eastern North America*, McGraw-Hill–Blakiston, New York. The ecology of the forests of eastern North America, their development, composition, and distribution.

CURTIS, J. T.: 1959, *The Forest of Wisconsin*, University of Wisconsin Press, Madison. A sound regional ecology of forests that should be a model for future studies.

FARB, P.: 1961, *The Forest*, Life Nature Library, Time-Life Books, New York. An excellent introduction to forest ecology; beautifully written and illustrated.

MCCORMICK, J.: 1959, *The Living Forest*, Harper & Row, New York. A short introduction to the forest—ecology, disease, insects, harvesting, types.

NEAL, E.: 1958, *Woodland Ecology*, Harvard University Press, Cambridge, Mass. A gem; written with the British woodland in mind, but overall presentation applicable to all temperate forests.

PLATT, R.: 1965, *The Great American Forest*, Prentice-Hall, Englewood Cliffs, N.J. A sweeping look at the forest in all its aspects. A popular introduction.

RICHARDS, P. W.: 1952, *The Tropical Rain Forest*, Cambridge University Press, London. The book on the tropical rain forest.

U.S. DEPARTMENT OF AGRICULTURE: 1949, "Trees," *Yearbook of Agriculture*, Supt. of Documents, Washington, D.C. An overall view of the forest—the trees, forest regions, forest management.

Tundra

BRITTON, M. E.: 1957, Vegetation of the arctic tundra, *18th Biol. Colloq. Oregon State Chapter Phi Kappa Phi*, Oregon State College, Corvallis, pp. 26–61. A detailed summary.

BROOKS, M. B.: 1967, *The Life of the Mountains*, Our Living World of Nature Series, McGraw-Hill, New York. Beautifully written survey of mountain life in North America.

FREUCHEN, P., AND F. SALMOSEN: 1958, *The Arctic Year*, Putnam, New York. Excellent reading, scientifically sound.

LEY, WILLY: 1962, *The Poles*, Life Nature Library, Time-Life Books, New York. A description of life and exploration in the polar regions. Well illustrated.

MILNE, L., AND M. MILNE: 1962, *The Mountains*, Life Nature Library, Time-Life Books, New York. An account of alpine life.

PEARSALL, W. H.: 1960, *Mountains and Moorlands*, Collins, London. European mountains and moorlands well described: emphasis on Britain.

POLUNIN, N.: 1948, "Botany of the Canadian Eastern Arctic; III, Vegetation and Ecology," *Bulletin 104, National Museum of Canada*, Ottawa. A basic reference by an authority of the tundra.

STONEHOUSE, B.: 1971, *Animals of the Arctic; the Ecology of the Far North*, Holt, Rinehart &

Winston, New York. A colorful and informative book on arctic animal life.

WASHBURN, A. L.: 1956, Classification of patterned ground and review of suggested origins, *Geol. Soc. Am. Bull.* 67, pp. 823–865. A detailed discussion of patterned ground in the Arctic.

ZWINGER, A. H., AND B. E. WILLARD: 1972, *Land Above the Trees*, Harper & Row, New York. An excellent guide to American alpine tundra.

Fresh-water ecosystems

AMOS, W. H.: 1970, *The Infinite River*, Random House, New York. Story of the river from the headwaters to the sea.

BARDACH, J. E.: 1964, *Downstream: A Natural History of the River*, Harper & Row, New York. Follows development of the river from headwaters, with practical considerations.

BENNETT, G. W.: 1962, *Management of Artificial Lakes and Ponds*, Van Nostrand Reinhold, New York, Limnology and fishery biology from a practical approach. Deals only with artificial impoundments, especially small ponds, popular in ecology laboratory work.

COKER, R. E.: 1954, *Streams, Lakes, and Ponds*, University of North Carolina Press, Chapel Hill. An account for the general reader.

FREY, D. G. (ED.): 1963, *Limnology in North America*, University of Wisconsin Press, Madison. A thorough account.

HUTCHINSON, G. E.: 1957, 1967, A *Treatise on Limnology*, vol. 1, *Geography, Physics and Chemistry*; vol 2, *Introduction to Lake Biology and Limnoplankton*, Wiley, New York. An invaluable reference on lake ecosystems.

HYNES, H. B. N.: 1960, *The Biology of Polluted Waters*, Liverpool University Press, Liverpool. The effects of pollution on fresh-water ecosystems.

HYNES, H. B. N.: 1970, *The Ecology of Running Water*, University of Toronto Press, Toronto. A major work on running-water ecosystems. Indispensable reference.

MACAN, T. T.: 1963, *Fresh Water Ecology*, Longmans, Essex. A revised edition of a major work on aquatic life.

MACAN, T. T.: 1970, *Biological Studies of English Lakes*, Elsevier, New York. Good summary of lake biology.

MACAN, T. T., AND E. B. WORTHINGTON: 1952, *Life in Lakes and Rivers*, Collins, London. Excellent limnology; the finest general book on the subject in spite of the date.

NEEDHAM, P. R.: 1969, *Trout Streams*, Winchester, New York. A reprint of a classic published in 1938. With annotations to update the text.

POPHAM, E. J.: 1961, *Life in Fresh Water*, Harvard University Press, Cambridge, Mass. An excellent introduction to fresh-water ecology. Contains an account of a long-term survey of an actual pond that can serve as a model for practical field work.

REID, G. K.: 1961, *Ecology of Inland Waters and Estuaries*, Reinhold, New York. Good summary of fundamentals.

RUTTNER, F.: 1963, *Fundamentals of Limnology*, rev. ed., University of Toronto Press, Toronto, a basic reference, but somewhat dated.

Estuaries

COWELL, E. B. (ED.): 1971, *The Ecological Effects of Oil Pollution on Littoral Communities*, Institute of Petroleum, London. Papers dealing with all aspects of oil spills on life along the coast.

GREEN, J.: 1968, *Biology of Estuarine Animals*, University of Washington Press, Seattle. From vegetation to vertebrates, an overview of the estuary.

LAUFF, G. (ED.): 1967, *Estuaries*, American Association for the Advancement of Science, Washington, D.C. A comprehensive look at estuaries. An important basic reference.

RANWELL, D. S.: 1972, *Ecology of Salt Marshes and Sand Dunes*, Chapman and Hall, London. A good introduction to salt marshes and dunes. More technical than Teal; plant oriented, little on animal life.

REMANE, A., AND C. SCHLIEPER: 1971, *Biology of Brackish Water*, Wiley, New York. An important reference that covers not only the estuary but all brackish water habitats.

TEAL, J., AND M. TEAL: 1969, *Life and Death of a Salt Marsh*, Little, Brown, Boston. Finest reference on the salt marsh.

Seashore

BERRILL, N. J.: 1951, *The Living Tide*, Dodd, Mead, New York. A popular book on life in the tidal zone.

CARSON, R.: 1955, *The Edge of the Sea*, Houghton Mifflin, Boston. A beautifully written book that deals with marine life along the eastern coast of North America.

ELTRINGHAM, S. K.: 1971, *Life in Mud and Sand*, Crane, Russak & Co., Inc., New York. A fresh, concise introduction to physical environment and life of muddy and sandy seashores.

MACGINITIE, G. E., AND N. MACGINITIE: 1968, *Natural History of Marine Animals*, 2d ed., McGraw-Hill, New York. Excellent introduction to animal life between the tide marks.

RICKETTS, E. F.; J. CALVIN; AND J. HEDGPETH (EDS.): 1968, *Between Pacific Tides*, 4th ed., Stanford University Press, Stanford, Calif. Contains a wealth of information on Pacific coast intertidal life. A must.

RUDLOE, J.: 1971, *The Erotic Ocean: Handbook for Beachcombers*, Harcourt Brace Jovanovich, New York. An excellent handbook for seaside ecologists also.

SILVERBERG, R.: 1972, *The World within the*

Tidepool, Weybright & Talley, New York. A well-illustrated popular account of the marine microcosm.

STEPHENSON, T. A., AND A. STEPHENSON: 1972, *Life Between Tidemarks on Rocky Shores*, Freeman, San Francisco. Excellent descriptions of rocky shores over the world.

Open sea

BARNES, H. (ED.): 1963, *Oceanography and Marine Biology*, Allen and Unwin, London. An annual review publication.

CARSON, R.: 1961, *The Sea Around Us*, Oxford University Press, New York. A classic; best popular account of the sea.

COKER, R. E.: 1947, *This Great and Wide Sea*, University of North Carolina Press, Chapel Hill. A brief but comprehensive book on the ocean.

EKMAN, S.: 1953, *Zoogeography of the Sea*, Sidgwick and Jackson, London. An indispensable reference on the "shelf fauna" around the world.

ENGEL, L.: 1961, *The Sea*, Life Nature Library, Time-Life Books, New York. A brief introduction to the sea with an excellent collection of illustrations.

FREDRICH, H.: 1969, *Marine Biology*, University of Washington Press, Seattle. Technical introduction to marine world.

HARDY, A.: 1971, *The Open Sea: Its Natural History*, Houghton Mifflin, Boston. The original two-volume work now in one volume. One of the great writings on the sea.

HEDGPETH, J. W. (ED.): 1957, *Treatise on Marine Ecology and Paleoecology*, vol. 1, *Ecology*, Geol. Soc. Amer., Mem. 67, New York. A book immense in scope—a bible for the marine ecologist. Covers the open sea, the sea shore, the estuary, etc., but unfortunately out-of-print.

HILL, M. N. (ED.): 1962–1964, *The Seas: Ideas and Observations on Progress in the Study of the Seas*, 3 vols., Wiley, New York. The three volumes cover physical oceanography, seawater, comparative and descriptive oceanography, etc. A major reference on the sea.

HOOD, D. W. (ED.): 1971, *Impingement of Man on the Oceans*, Wiley, New York. Ocean pollution and other impacts of man on the sea.

KINNE, O.: 1971, *Marine Ecology*, vol. 1, *Environmental Factors*, 3 parts. A proposed five-volume treatise on marine ecology. Technical but invaluable. Work far from completed.

OLSEN, T. A., AND F. J. BURGESS (EDS.): 1967, *Pollution and Marine Ecology*, Wiley, New York. Studies on the effects of pollution on marine life and ecosystems.

RUSSEL, F. S. (ED.): 1963, *Advances in Marine Biology*, Academic, New York. Annual review volumes.

SCIENTIFIC AMERICAN: 1971, *Oceanography* Freeman, San Francisco. A collection of articles

arranged to provide an excellent introduction to the ocean and its use by man.

SEARS, M. (ED.): 1961, *Oceanography*, American Association for the Advancement of Science, Washington, D.C. A symposium volume.

TAIT, R. V.: 1968, *Elements of Marine Ecology*, Plenum, New York. An elementary and readable introduction to marine ecology. Recommended.

WEINS, H. J.: 1962, *Atoll Environment and Ecology*, Yale University Press, New Haven, Conn. An excellent and complete book on ecology of coral reefs.

Part VI

The number of books relating directly to an eco-system approach to resource management are very few, indeed. In this group of references I have included a few books that contain ideas that can be related to such an approach. The list of such books is far from complete.

DARLING, F. F., AND J. P. MILTON (EDS.): 1966, *Future Environments of North America*, Natural History Press, Garden City, N.Y. A look into the future of the North American continent by a number of experts.

DASMANN, R. F.: 1968, *A Different Kind of Country*, Macmillan, New York. Arguments for diversity and a slowdown of progress.

FLAWN, P. T.: 1970, *Environmental Geology: Conservation, Land-use Planning and Resource Management*, Harper & Row, New York. Application of geological principles to environmental management. A new approach for geology. Should be a part of an ecologist's library.

UDALL, S. A.: 1963, *The Quiet Crisis*, Holt, Rinehart & Winston, New York. A history of the conservation movement in the United States and outlook for the future.

VALLENTINE, J. F.: 1972, *Range Development and Improvements*, Brigham Young University Press, Provo, Utah. Application of ecological principles to management of the range.

VAN DYNE, G. (ED.): 1970, *The Ecosystem Concept in Natural Resource Management*, Academic, New York. A pioneering volume; a few articles fall rather far from the mark but it is a beginning.

WARD, B., AND R. DUBOS: 1972, *Only One Earth*, Norton, New York. Examines our environmental problems from a global, social, economic, and political perspective. A book for all ecologists.

WATT, K. E. F.: 1968, *Ecology and Resource Management*, McGraw-Hill, New York. Not a book for the beginner. An attempt to apply ecological concepts and computer technology to resource management with some success.

WATT, K. E. F.: 1973, *Principles of Environmental Science*, McGraw-Hill, New York. Application of ecological principles to land and resource planning, pollution problems.

General references

The following is a list of recommended guides to various groups of organisms. The list is not complete, but the volumes represent one man's opinion about what are the best and most acceptable guides to plants and animals. The regional works included are those applicable to a wider region than the state borne in the title. Except in rare instances, technical monographs are not included.

Lower plants

BODENBERG, E. T.
1954 *Mosses: A New Approach to the Identification of Common Species*, Burgess, Minneapolis.

COBB, B.
1956 *A Field Guide to the Ferns*, Houghton Mifflin, Boston.

CONRAD, H. J.
1956 *How to Know the Mosses and Liverworts*, Brown, Dubuque, Iowa.

DAWSON, E. Y.
1956 *How to Know the Seaweeds*, Brown, Dubuque, Iowa.

FINK, B.
1935 *The Lichen Flora of the United States*, University of Michigan Press, Ann Arbor.

FRYE, T., AND L. CLARK
1937–1947 *Hepaticae of North America*, 2 vols., University of Washington Press, Seattle.

GRAHAM, V. O.
1970 *Mushrooms of the Great Lakes Region*, reprint, Dover, New York.

GROOT, A. J.
1947 *Mosses with a Hand Lens*, 4th ed., published by the author, 1 Vine Street, New Brighton, Staten Island, N.Y.

HALE, M. E.
1969 *How to Know the Lichens*, Brown, Dubuque, Iowa.

KAUFFMAN, C. H.
1971 *The Gilled Mushrooms (Agaricaceae) of Michigan and the Great Lakes Region*, 2 vols., Reprint of 1918 ed., Dover, New York.

KREIGER, L. C. C.
1947 *The Mushroom Handbook*, 1967 reprint, Dover, New York.

PARSONS, F. T.
1961 *How to Know the Ferns: A Guide to Names, Haunts, and Habits* (reprint), Dover, New York.

PRESCOTT, G. W.
1964 *How to Know the Fresh Water Algae*, Brown, Dubuque, Iowa.

SHAVER, J. M.
1954 *Ferns of the Eastern Central States, with*

special Reference to Tennessee, 1970 reprint, Dover, New York.

SMITH, A. H.
1951 *Puffballs and Their Allies in Michigan*, University of Michigan Press, Ann Arbor.

SMITH, A. H.
1963 *The Mushroom Hunter's Field Guide*, rev. ed., University of Michigan Press, Ann Arbor.

SMITH, G.
1950 *The Fresh Water Algae of the United States*, McGraw-Hill, New York.

SMITH, G.
1951 *Manual of Phycology: An Introduction to the Algae and Their Biology*, Chronica Botanica, Waltham, Mass.

TAYLOR, W. R.
1937 *Marine Algae of the Northeastern Coast of North America*, University of Michigan Science Series 13, University of Michigan Press, Ann Arbor.

TAYLOR, W. R.
1960 *Marine Algae of the Eastern Tropical and Subtropical Coasts of the Americas*, University of Michigan Press, Ann Arbor.

THOMAS, W. C.
1936 *Field Book of Common Mushrooms*, Putnam, New York.

WATSON, E. V.
1955 *British Mosses and Liverworts*, Cambridge University Press, London. (Most species treated also occur in northeastern North America.)

WHERRY, E. T.
1942 *Guide to Eastern Ferns*, 2d ed., Science Press, Lancaster, Pa.

Grasses and wildflowers

BRITTON, N. L., AND J. N. ROSE
1937 *The Cactaceae*, 2 vols., reprint, Dover, New York.

CRAIGHEAD, J.; F. C. CRAIGHEAD; AND R. J. DAVIS
1963 *Field Guide to Rocky Mountain Wildflowers*, Houghton Mifflin, Boston.

CUTHBERT, M.
1943, 1948 *How to Know the Spring Wildflowers; How to Know the Fall Wildflowers*, Brown, Dubuque, Iowa.

DANA, MRS. WM. STARR
1962 *How to Know the Wildflowers*, rev. modernized ed. of 1900 volume, Dover, New York.

DAWSON, E. Y.
1963 *How to Know the Cacti*, Brown, Dubuque, Iowa.

DAYTON, W. A.
1960 "Notes on Western Range Forbs," U.S.D.A. *Handbook*, 161, Supt. of Documents, Washington, D.C.

DEGENER, O.

1946–1957 *New Illustrated Flora of the Hawaiian Islands*, published by the author, Makuleia Beach, Waialua, Oahu, Hawaii.

FASSETT, N. C.

1940 *A Manual of Aquatic Plants*, McGraw-Hill, New York.

FERNALD, M. L.

1950 *Gray's Manual of Botany*, 8th ed., American Book, New York.

GLEASON, H. A.

1952 *The New Britton and Brown Illustrated Flora of the Northeastern United States and Adjacent Canada*, 3 vols., New York Botanical Garden, New York.

HITCHCOCK, A. S.

1950 rev. by Agnes Chase, *Manual of Grasses of the United States*, Supt. of Documents, Washington, D.C., reprint, Dover, New York.

HITCHCOCK, C. L.; A. CRONQUIST; M. OWNBEY; J. W. THOMPSON

1964 *Vascular Plants of the Pacific Northwest*, 5 vols., University of Washington Press, Seattle.

HOTCHKISS, N.

1972 *Common Marsh, Underwater and Floating-leaved Plants*, Dover, New York.

HULTEN, E.

1960 *Flora of the Aleutian Islands*, 2d ed., Hafner, New York.

JAQUES, H. E.

1949 *Plant Families, How to Know Them*, Brown, Dubuque, Iowa.

JAQUES, H. E.

1972 *How to Know the Weeds*, Brown, Dubuque, Iowa.

KUMMER, A. P.

1951 *Weed Seedlings*, University of Chicago Press, Chicago.

MARTIN, A. C.; H. S. ZIN; AND A. L. NELSON

1951 *American Wildlife and Food Plants*, McGraw-Hill, New York (new paperback 1961 edition, Dover, New York).

MATTHEWS, F. S.

1955 ed. by Norman Taylor, *Field Book of American Wildflowers*, rev. ed., Putnam, New York.

MUENSCHER, W. C.

1944 *Aquatic Plants of the United States*, Comstock, Ithaca, N.Y.

MUENSCHER, W. C.

1955 *Weeds*, Macmillan, New York.

PETERSON, R. T., AND M. MCKENNY

1968 *A Fieldguide to the Wildflowers of Northeastern and Northcentral North America*, Houghton Mifflin, Boston.

POHL, R. W.

1954 *How to Know the Grasses*, Brown, Dubuque, Iowa.

PORSILD, A. E.

1957 Illustrated flora of the Canadian Arctic

Aids to identification

Archipelago, *Bulletin 146, National Museum of Canada*, Ottawa.

PRESCOTT, G. W.

1969 *How to Know the Aquatic Plants*, Brown, Dubuque, Iowa.
Iowa.

RICKETT, H. W.

1966–1971 *Wild Flowers of the United States*, 5 vols., 1, *Northeastern States*; 2, *The South-eastern States*; 3, *Texas*; 4, *The Southwestern States*; 5, *The Northwestern States*, McGraw-Hill, New York.

RYDBERG, P. A.

1969 (1922 reprint) *Flora of the Rocky Mountains and Adjacent Plains*, Hafner, New York.

WHERRY, E. T.

1948 *The Wildflower Guide*, Doubleday, New York.

WIGGANS, I. L., AND J. H. THOMAS

1962 *Flora of the Alaskan Arctic Slope*, University of Toronto Press, Toronto.

Trees and shrubs

BAERS, H.

1955 *How to Know the Western Trees*, Brown, Dubuque, Iowa.

BENSON, L. D., AND R. A. DARROW

1945 *A Manual of Southwestern Desert Trees and Shrubs*, University of Arizona Press, Tucson.

BROCKMAN, C. F.

1968 *Trees of North America*, Golden, New York.

GRAVES, A. H.

1956 *Illustrated Guide to Trees and Shrubs*, Harper & Row, New York.

GRIMM, W. C.

1960 *Recognizing Trees*, Stackpole, Harrisburg, Pa.

GRIMM, W. C.

1966 *How to Recognize Shrubs*, (*Recognizing Native Shrubs*) Stackpole, Harrisburg, Pa.; Castle, New York.

HARLOW, W. M.

1957 *Trees of Eastern and Central United States and Canada; Fruit Key and Twig Key to Trees and Shrubs*, reprint ed., Dover, New York.

HARRAR, E. S., AND J. G. HARRAR

1946 *Guide to Southern Trees*, reprint, Dover, New York.

HAYES, DORIS W.

1960 "Key to important woody plants of eastern Oregon and Washington," *U.S.D.A. Handbook* 148, Supt. of Documents, Washington, D.C.

HOSIE, R. C.

1973 *Native Trees of Canada*, 7th ed., Canadian Forestry Service, Department of Environment Ottawa.

JAQUES, H. E.

1946 *How to Know the Trees*, rev. ed., Brown, Dubuque, Iowa.

KEELER, H.

1969 *Our Northern Shrubs and How To Identify Them*, reprint, Dover, New York.

MCMINN, H. E., AND E. MAINO

1946 *An Illustrated Manual of Pacific Coast Trees*, University of California Press, Berkeley.

PEATTIE, D. C.

1953 *Natural History of Western Trees*, Houghton Mifflin, Boston.

1966 *Natural History of Trees of Eastern and Central North America*, Houghton Mifflin, Boston.

PETRIDIES, G.

1972 *A Field Guide to Trees and Shrubs*, Houghton Mifflin, Boston.

SARGENT, C. S.

1922 *Manual of Trees of North America*, reprint, 2 vols., Dover, New York.

TRELEASE, W.

1967 *Winter Botany: An Identification Guide to Native Trees and Shrubs*, reprint, Dover, New York.

VINES, R. A.

1963 *Trees, Shrubs and Woody Vines of the Southwest*, University of Texas Press, Austin.

Invertebrates

General

DAVIS, C. C.

1955 *The Marine and Fresh Water Plankton*, Michigan State University Press, East Lansing.

EDMONDSON, W. T. (ED.)

1959 *Fresh-water Biology*, 2d ed., Wiley, New York.

JOHNSON, M. E., AND H. J. SNOOK

1967 *Seashore Animals of the Pacific Coast*, reprint, Dover, New York.

KLOTS, E. B.

1966 *New Field Book of Freshwater Life*, Putnam, New York.

LIGHT, S. F.; R. I. SMITH; F. A. PITELKA; D. P. ABBOTT; AND F. M. NEESNER

1957 *Intertidal Invertebrates of the Central California Coast*, University of California Press, Berkeley.

MINER, R. W.

1950 *Field Book of Seashore Life*, Putnam, New York.

NEEDHAM, J. G., AND P. R. NEEDHAM

1962 *A Guide to the Study of Fresh Water Biology*, Holden-Day, San Francisco.

PENNAK, R.

1953 *Fresh Water Invertebrates of the United States*, Ronald, New York.

PRATT, H. S.

1953 *Manual of the Common Invertebrate Animals*, McGraw-Hill–Blakiston, New York.

Protozoa

JAHN, T. L.

1949 *How to Know the Protozoa*, Brown, Dubuque, Iowa.

Colenterata

SMITH, F. G. W.

1948 *Atlantic Reef Corals*, University of Miami Press, Coral Gables.

Earthworms

EATON, T. H., JR.

1942 Earthworms of the northeastern United States, *J. Wash. Acad. Sci.*, **32**:242–249.

OLSON, H. W.

1928 The earthworms of Ohio, *Ohio Biol. Survey*, 4:45–90.

Mollusca

ABBOT, R. T.

1954 *American Seashells*, Van Nostrand, Reinhold, New York.

BAKER, F. C.

1939 Field book of Illinois Land snails, *Manual 2, Illinois Natural History Survey*, Urbana, Ill.

BURCH, J. B.

How to Know the Eastern Land Snails, Brown, Dubuque, Iowa.

KEEN, A. M.

1963 *Marine Molluscan Genera of Western North America*, Stanford University Press, Stanford, Calif.

KEEP, J.

1935 rev. by J. L. Bailey, Jr., *West Coast Shells*, Stanford University Press, Stanford, Calif.

MORRIS, P.

1951 *Field Guide to the Shells of our Atlantic and Gulf Coast*, Houghton Mifflin, Boston.

MORRIS, P.

1952 *Field Guide to the Shells of the Pacific Coast and Hawaii*, Houghton Mifflin, Boston.

PILSBRY, H. A.

1939–1948 Land mollusca of North America, *Academy of Natural Sciences Monograph No. 3*, Philadelphia, Pa.

Insects

BORROR, D. J., AND R. E. WHITE

1970 *A Field Guide to the Insects of North America*, Houghton Mifflin, Boston.

CHU, H. F.

1949 *How to Know the Immature Insects*, Brown, Dubuque, Iowa.

DILLON, E., AND L. S. DILLON

1961 *A Manual of Common Beetles of Eastern North America*, Harper & Row, New York. (Dover reprint.)

EHRLICH, P. R.

1970 *How to Know the Butterflies*, Brown, Dubuque, Iowa.

ESSIG, E. O.

1958 *Insects and Mites of Western North America*, Macmillan, New York.

HELFER, J. R.

1963 *How to Know the Grasshoppers, Cockroaches, and Their Allies*, Brown, Dubuque, Iowa.

HOLLAND, W. J.

1949 *The Butterfly Book*, Doubleday, Garden City, N.Y.

1968 *The Moth Book*, reprint, Dover, New York.

JAQUES, H. E.

1947 *How to Know the Insects*, Brown, Dubuque, Iowa.

JAQUES, H. E.

1951 *How to Know the Beetles*, Brown, Dubuque, Iowa.

KLOTS, A. K.

1951 *Field Guide to the Butterflies*, Houghton Mifflin, Boston.

LUTZ, F. E.

1935 *Field Book of Insects*, 3d rev. ed., Putnam, New York.

NEEDHAM, J. G., AND M. J. WESTFALL, JR.

1955 *A Manual of the Dragonflies of North America*, University of California Press, Berkeley.

SWAIN, R. B.

1948 *The Insect Guide*, Doubleday, Garden City, N.Y.

USINGER, R. I. (ED.)

1956 *Aquatic Insects of California, with Keys to North American Genera and California Species*, University of California Press, Berkeley.

Spiders

COMSTOCK, J. H.

1940 ed. by W. J. Gertsch, *The Spider Book*, 2d rev. ed., Doubleday, Garden City, N.Y.

EMERTON, J. H.

1961 *The Common Spiders of the United States*, reprint, Dover, New York.

KATSON, B. J., AND E. KATSON

1972 *How to Know the Spiders*, Brown, Dubuque, Iowa.

Vertebrates

General

BLAIR, W. F.; A. P. BLAIR; P. BRODKROB; F. R. CAGLE; AND G. A. MOORE

1957 *Vertebrates of the United States*, McGraw-Hill, New York.

Fishes

BIGELOW, H. B., AND W. C. SCHRODER

1953 Fishes of the gulf coast of Maine, *U.S. Fish and Wildlife Service Fishery Bull. No. 74*.

BREDER, C. M.

1948 *Field Book of Marine Fishes*, Putnam, New York.

EDDY, S.
1957 *How to Know the Freshwater Fishes*, Brown, Dubuque, Iowa.

HARLAN, J. R., AND E. B. SPEAKER
1956 *Iowa Fish and Fishing*, 3d ed., Iowa Conservation Commission, Des Moines.

HUBBS, C., AND K. LAGLER
1958 Fishes of the Great Lakes region, *Cranbrook Institute of Science Bulletin 26*, Bloomfield, Ill.

PERLMUTTER, A.
1961 *Guide to Marine Fishes*, New York University Press, New York.

SCOTT, W. B.
1967 *Freshwater Fishes of Eastern Canada*, University of Toronto Press, Toronto.

TRAUTMAN, M. B.
1957 *The Fishes of Ohio*, Ohio State University Press, Columbus.

Amphibians and reptiles

BISHOP, S.
1943 *Handbook of Salamanders*, Comstock, Ithaca, N.Y.

CARR, A.
1952 *Handbook of Turtles*, Comstock, Ithaca, N.Y.

CONANT, R.
1958 *A Field Guide to the Reptiles and Amphibians of the United States and Canada East of the 100 Meridian*, Houghton Mifflin, Boston.

DITMARS, R.
1949 *Fieldbook of North American Snakes*, Doubleday, Garden City, N.Y.

ERNST, C. H., AND R. W. BARBOUR
1972 *Turtles of the United States*, University of Kentucky Press, Lexington.

PICKWELL, G.
1972 *Amphibians and Reptiles of the Pacific States*, reprint, Dover, New York.

POPE, C.
1939 *Turtles of the United States and Canada*, Knopf, New York.

SCHMIDT, K., AND D. D. DAVIS
1941 *Field Book of Snakes of the United States and Canada*, Putnam, New York.

SMITH, H. M.
1946 *Handbook of Lizards of the United States and Canada*, Comstock, Ithaca, N.Y.

STEBBINS, R. C.
1954 *Amphibians and Reptiles of Western North America*, McGraw-Hill, New York.

STEBBINS, R. C.
1966 *A Field Guide to Western Reptiles and Amphibians*, Houghton Mifflin, Boston.

WRIGHT, A. H., AND A. A. WRIGHT
1949 *Handbook of Frogs and Toads*, 3d ed., Comstock, Ithaca, N.Y.

WRIGHT, A. H., AND A. A. WRIGHT
1957 *Handbook of Snakes of the United States and Canada*, vols. 1, 2, Comstock, Ithaca, N.Y.

Birds

BOND, J.
1971 *A Field Guide to the Birds of the West Indies*, Houghton Mifflin, Boston.

DE SCHAUENSEE, R. M.
1970 *A Guide to the Birds of South America*, Livingston, Wynnewood, Pa.

HEADSTROM, R.
1970 *A Complete Field Guide to Nests in the United States*, Washburn, N.Y.

JAQUES, H. E.
1947 *How to Know the Land Birds*, Brown, Dubuque, Iowa.

KORTRIGHT, F. H.
1942 *Ducks, Geese and Swans of North America*, Stackpole, Harrisburg, Pa.

PALMER, R.
1962 *Handbook of North American Birds*, vol. I, Yale University Press, New Haven, Conn.

PETERSON, R. T.
1947 *A Field Guide to the Birds*, Houghton Mifflin, Boston.

PETERSON, R. T.
1961 *A Field Guide to Western Birds*, Houghton Mifflin, Boston.

POUGH, R. H.
1946–1951 *Audubon Bird Guides*, vol. 1, *Eastern Land Birds*; vol. 2, *Water Birds*. Doubleday, Garden City, N.Y.

POUGH, R. H.
1957 *Audubon Western Bird Guide*, Doubleday, Garden City, N.Y.

ROBBINS, C. S.; B. BRUNN; AND H. S. ZIM
1966 *Birds of North America*, Golden, New York.

SAUNDERS, A. A.
1959 *A Guide to Bird Song*, Doubleday, New York.

Note: For identification of bird song see the Peterson's Field Guide Records, Dover Records, and records produced by The Federation of Ontario Naturalists.

Mammals

BOOTH, E. S.
1961 *How to Know the Mammals*, Brown, Dubuque, Iowa.

BURT, W. H.
1957 *Mammals of the Great Lakes Region*, University of Michigan Press, Ann Arbor.

BURT, W. H., AND R. P. GROSSENHEIDER
1963 *A Field Guide to the Mammals*, 2d ed., Houghton Mifflin, Boston.

HALL, E. R., AND K. R. KELSON
1959 *Mammals of North America*, Ronald, New York.

HAMILTON, W. J., JR.
1943 *The Mammals of Eastern United States*, Comstock, Ithaca, N.Y.

INGLES, L. G.
1965 *Mammals of the Pacific States,* Stanford
University Press, Stanford, Calif.

JACKSON, H. H. T.
1961 *The Mammals of Wisconsin,* University of
Wisconsin Press, Madison.

LEOPOLD, A. S.
1959 *Wildlife in Mexico,* University of California
Press, Berkeley.

MURIE, O. J.
1954 *A Field Guide to Animal Tracks,* Houghton
Mifflin, Boston.

PALMER, R. S.
1954 *The Mammal Guide,* Doubleday, Garden
City, N.Y.

SCHWARTZ, C. W., AND E. R. SCHWARTZ
1960 *The Wild Mammals of Missouri,* University
of Missouri Press, Columbia.

he following list is by no means complete. The number of new journals has increased considerably during the past few years. The scope is international. A number of journals in foreign languages also publish many papers in English or have English abstracts.

Acta Botanica Neerlandica: 1952 on, Royal Botanical Society of the Netherlands. Wide diversity of papers, mostly in English. Important.

American Birds: 1946 on, Formerly *Audubon Field Notes.* Devoted to the reporting of distribution and abundance of birds. Important information on continental trends in bird populations.

American Journal of Botany: 1914 on, Botanical Society of America. Technical, of interest chiefly to the professional.

American Midland Naturalist: 1909 on, University of Notre Dame, Notre Dame, Ind. Specializes in papers in field natural history.

American Naturalist: 1867 on, American Society of Naturalists. Concerned largely with morphology, evolution, physiology, but of late ecology creeping in.

American Scientist: Society of Sigma Xi. Contains in every volume several papers of importance to ecologists.

American Zoologist: Quarterly Publication of American Society of Zoologists. Most papers published originate in symposia of society and its annual refresher course.

Animal Behavior: 1952 on, Baillière, London. Contains a wide range of papers on animal behavior.

Annales Zoologici Fennici: 1963 on, Societas Biologica Fennica Vanamo, Helsinki. An important journal; many papers on ecology.

Annals of Botany: 1887 on, Annals of Botany Co., Clarendon, Oxford. Heavy stress on physiological ecology.

Annals of the Entomological Society of America: 1908 on, Entomological Society of America. The major entomological journal in the United States. Strong on taxonomy and morphology.

The Auk: 1884 on, American Ornithologists Union. Leading American ornithological journal.

Australian Journal of Zoology: 1958 on, Commonwealth Scientific and Industrial Research Organization. General zoological papers, many of ecological interest.

Behavior Genetics: 1970 on, Plenum, New York. Papers on inheritance of behavior in animals and men.

Behaviour: 1947 on, E. Brill. Leiden, Netherlands.

The journal of animal behavior; strongly ethological in viewpoint.

Biological Bulletin: 1927 on, Marine Biological Laboratory, Woods Hole, Mass. General papers with emphasis on marine biology.

Biological Conservation: 1970 on, Applied Science Publishers, Barking, Essex, England. Important new journal on scientific protection of plants and animals and management of natural resources.

Bioscience: 1951 on, American Institute of Biological Sciences. A general biological journal with many papers devoted to ecology and environment.

Bird-Banding: 1930 on, Bird Banding Organizations. A small quarterly. Noted for its excellent coverage of current literature.

Botanical Gazette: 1862 on, University of Chicago Press, Chicago. Covers all departments of botanical science including plant ecology.

Botanical Review: 1935 on, New York Botanical Garden. Excellent review papers, often on plant ecology.

British Birds: H. F. and G. Witherby, London. An important ornithological journal issued monthly. Contains some first-rate papers on the ecology of birds.

Brittonia: 1949 on, American Society of Plant Taxonomists. Papers deal largely with plant taxonomy.

The Bryologist: 1899 on, American Bryological and Lichenological Society. Only journal devoted to mosses and lichens. Many papers of ecological interest.

Bulletin of Environmental Contamination and Toxicology: 1966 on, Springer-Verlag, New York. Short, current papers dealing largely with pesticides and heavy metal pollution.

California Fish and Game: 1914 on, State of California, Department of Fish and Game. Papers largely devoted to fish and game management. Contains many important papers on western wildlife.

Cambridge Philosophical Society: Biological Reviews, Cambridge University Press. General review papers, many of importance to ecology and behavior.

Canadian Entomologist: 1868 on, Entomological Society of Ontario. A major entomological journal devoted to insects of Canada.

Canadian Field Naturalist: 1887 on, Ottawa Field-Naturalist's Club. Papers on all phases of natural history; of strong interest to the field biologist.

Canadian Journal of Botany: 1921 on, National Research Council of Canada, Ottawa. General botanical papers, a few of ecological interest.

Canadian Journal of Zoology. 1923 on, National Research Council of Canada, Ottawa. General zoological papers but many of ecological interest.

Chesapeake Science: State of Maryland, Department

of Research and Education, Chesapeake Biological Laboratory, Solomons. A regional publication, but papers of wide interest.

Condor: 1899 on, Cooper Ornithological Society. An ornithological journal with excellent scientific papers.

Copeia: 1913 on, American Society of Ichthyologists and Herpetologists. Papers on fishes, amphibians, and reptiles.

East African Wildlife Journal: East African Wildlife Society. Covers general ecology and wildlife and park management in East Africa.

Ecological Monographs: 1930 on, Ecological Society of America. Contains papers too long for *Ecology.*

Ecology: 1920 on, Ecological Society of America. Indispensable to ecologists.

Ecology Law Quarterly: 1971 on, School of Law, University of California, Berkeley. Papers on law and the environment. Informative new journal.

Economic Botany: 1947 on, Society of Economic Botany. Many interesting papers on man and plant relationships. More ecology than economics.

Environment: Formerly *Scientist and Citizen.* Scientists Institute for Public Information, St. Louis. Included among journals because it does contain significant articles well documented.

Environment and Behavior: 1969 on, Sage Publications, New York. Papers on relationship of man's environment and his behavior. Interesting.

Environment and Planning: 1969 on, Pion Ltd., London. Papers on environment and ecological approach to urban and suburban planning.

Environment, Science and Technology: 1967 on, American Chemical Society. Environmental articles written from the technological point of view.

Environmental Affairs: 1971 on, Boston College Environmental Law Center, Brighton, Mass. New journal devoted to environmental law. Similar to *Ecology Law Quarterly.*

Environmental Letters: 1970 on, Dekker, New York. Short papers quickly published on current environmental pollution problems.

Environmental Pollution: 1970 on, Applied Science Publ. Ltd., Barking, Essex, England. Important journal with papers on pollution written from ecological point of view.

Evolution: 1947 on, Society for Study of Evolution. Papers on evolution and natural selection.

Fishery Bulletin: U.S. Department of Commerce. A wide range of well-edited papers on ecology and life history of commercial marine species, marine environment, and population dynamics.

Forest Science: 1955 on, Society of American Foresters. Technical, but occasionally contains papers of interest to ecologists.

Forestry: Society of Foresters of Great Britain, Oxford University Press. Frequently contains good papers on forest ecology.

Journals of interest to ecologists

The Geographical Journal: Royal Geographical Society. Many articles on man–environment relations. Geographical journals a good source of material on human ecology.

Geographical Review: American Geographical Society of New York. An outstanding journal with articles of strong ecological interest.

Geography: The journal of the Geographical Association, Sheffield, England. Another source of papers relating to ecology and man.

Herpetologica: 1945 on, Herpetologists League. A Journal with papers on all aspects of herpetology. Has improved with age.

Human Ecology: 1972 on, Plenum, New York. A new journal devoted to the ecology of man.

Hydrobiologica: Dr. W. Junk, Publisher, The Hague, Netherlands. An important journal on fresh-water biology.

Hydrobiological Journal: American Fisheries Society and Scripta Publishing Co., New York. Translation of Russian journal. Valuable source on fresh-water biology.

Ibis: British Ornithological Society. The leading British ornithological journal.

Journal of Animal Ecology: 1933 on, British Ecological Society. Blackwell, Oxford. A journal devoted exclusively to animal ecology. Contains many valuable papers.

Journal of Applied Ecology: 1966 on, British Ecological Society. Journal devoted to the applied aspects of ecology. Many useful papers on wildlife and man-dominated ecosystems.

Journal of Ecology: 1912 on, British Ecological Society. Papers on general ecology but with major emphasis on plants. An important reference journal.

Journal of Environmental Quality: 1972 on, American Society of Agronomy, Crop Science Society of America, and Soil Science Society of America, 677 South Segal Road, Madison, Wisconsin. Papers on environmental quality in natural and agricultural ecosystems. A significant new journal.

Journal of Environmental Systems: 1971 on, Baywood Publ. Co. Papers on analysis, design, and management of environmental problems. Applied systems analysis.

Journal of Experimental Marine Biology and Ecology: 1963 on, North Holland Publishing Co. All aspects of marine ecology.

Journal of Fish Biology: 1969 on, Fisheries Society of the British Isles, Academic, New York. Papers devoted to fishery biology and fishery management.

Journal of the Fisheries Rsearch Board of Canada: 1943 on, Fisheries Research Board of Canada. A very valuable source of papers on fish, fish management, and ecology.

Journal of Forestry: Society of American Foresters. Papers on general forestry. Some of interest to ecologists.

Journal of Herpetology: 1967 on, Society of the Study of Amphibians and Reptiles. General papers on herpetology.

Journal of Mammalogy: 1919 on, American Society of Mammalogists. Only journal in English devoted to mammals.

Journal of the Marine Biological Association of the United Kingdom: 1920 on, Marine Biological Association of the United Kingdom. Contains papers on marine biology.

Journal of Marine Research: Sears Foundation for Marine Research. Bingham Oceanographic Laboratory, Yale University Press, New Haven, Conn. A major source of papers on oceanography, heavy on the physical side.

Journal of Parasitology: 1914 on, American Society of Parasitology. All aspects of animal parasitism.

Journal of Range Management: 1947 on, American Society of Range Management. Advances in the science and art of grazing land management, understanding of practical and scientific range and pasture problems.

Journal of Water Pollution Control Federation: Water Pollution Control Federation. An important source of material on water pollution from causes and effects to solutions.

Journal of Wildlife Management: 1937 on, The Wildlife Society. Devoted to wildlife research and management. Contains excellent material for the field biologist.

Limnology and Oceanography: 1956 on, American Society of Limnology and Oceanography. The major journal in the field. All papers of strong interest to ecologists.

Marine Biology: 1967 on, Springer-Verlag, New York. An international journal devoted to life in the oceans and coastal waters.

Natural Resources Journal: 1961 on, The University of New Mexico School of Law. The original environmental law journal, in print long before a general interest in law and the environment developed.

Nature: Macmillan, London. A weekly journal of science since 1869. Contains a number of important papers of interest to ecologists and field biologists.

New York Fish and Game Journal: 1953 on, State of New York, Conservation Department. An important regional publication with a number of major papers in fish and wildlife field.

Oecologica: 1963 on, Springer-Verlag, New York. In cooperation with International Association for Ecology. An international journal containing general ecological papers.

Oikos, Acta Oecologica Scandinavica: Munksgaard, Copenhagen. A major ecological journal. Papers are in English.

Pesticide Monitoring Journal: 1967 on, **Federal** Workers Group on Pesticide Management, EPA, Chamblee, Georgia. Important current source on pesticide information.

Physiological Zoology: 1927 on, University of Chicago Press, Chicago. Contains many papers on physiological ecology.

Proceedings of the Zoological Society of London: 1822 on, Zoological Society of London. Contains a number of major papers on ecology and behavior of animals.

Quarterly Review of Biology: American Institute of Biological Sciences. A review journal occasionally publishing papers of interest to field biologists. Excellent book reviews.

Radiation Botany: 1961 on, Pergamon, Elmsford, N.Y. An international journal with papers on effects of ionizing radiation on plants.

Science: 1833 on, American Association for the Advancement of Science. Covers whole field of science, but some papers of interest to ecologists. Good book reviews.

Systematic Zoology: 1952, Society of Systematic Zoologists. Although papers are largely taxonomic, many are of strong ecological interest.

Transactions of American Fisheries Society: American Fisheries Society. Began as a quarterly in 1959. Contains a number of papers on fish and fresh-water ecology.

Viltrevy. Swedish Wildlife: 1964 on, Swedish Sportsmen's Association, Stockholm. Long papers on wildlife, covering both ecology and behavior.

Wilson Bulletin: 1888 on, Wilson Ornithological Society. A midwestern ornithological journal that is especially oriented to the field.

n recent years a number of specialized bibliographies have appeared. However useful, they are too narrow to list here. The following list, also incomplete, is confined to current general bibliographies. These are excellent gateways to the literature.

Bibliography of Agriculture: 1942 on, U.S. Department of Agriculture. Monthly; covers wide variety of biological subjects. Especially useful for coverage of publications of state agricultural experiment stations.

Biological Abstracts: 1926 on, Philadelphia. Most comprehensive biological abstracting journal in North America. A primary reference, it includes brief abstracts available in groups on microfilm.

Forestry Abstracts: 1940 on, Commonwealth Agricultural Bureau, Oxford, England. Abstracts of forestry subjects compiled from world literature. Covers all phases of forestry including ecology, soils, general and systematic zoology, animal ecology, general botany, etc. A major bibliography for the ecologist.

KLEMENT, A. W., JR., AND V. SCHULTZ: 1962 on, *Terrestrial and Freshwater Radioecology. A Selected Bibliography.* Seven supplements through 1971, Report TID-3910 (and supplements). U.S. Atomic Energy Commission, Division of Technical Information. An invaluable source of materials on radio-ecology.

Pesticides Documentation Bulletin: 1963–1969, National Agricultural Library. U.S. Department of Agriculture. An excellent comprehensive bibliography on all aspects of pesticides. Discontinued in 1969 because of budget cuts.

ROBERTS-PICHETTE: 1972, *Annotated Bibliography of Permafrost, Vegetation, and Wildlife and Landform Relationships.* Forest Management Institute, Ottawa Canada. Inf. Rept. FRM-X-43. Included because it is a rich source of material on the Arctic and impact of oil and development.

SCHULTZ, V.: 1972, *Ecological Techniques Utilizing Radionuclides and Ionizing Radiation. A Selected Bibliography.* RLO-2213, Suppl. 1. U.S. Atomic Energy Commission, Tech. Inform. Center. Sourcebook for techniques.

SCHULTZ, V., AND F. W. WHICKER: 1971, A *Selected Bibliography of Terrestrial Freshwater and Marine Radiation Ecology.* TID-25650. U.S. Atomic Energy Commission. Available from National Technical Information Service, Springfield, Va. A highly useful and valuable bibliography for ecologists. Broken down into subject matter.

Sport Fishery Abstracts: 1956 on, U.S. Fish and Wildlife Service. Covers sport fishery research and management, limnology, ecology, and natural history of fishes. Indispensable to the fishery biologist, ecologist, and limnologist. Contains short abstracts of each paper.

Water Resource Abstracts: 1968 on, U.S. Department of Commerce. Comprehensive. Covers water-related aspects of life, physical and social sciences, as well as related engineering and legal aspects of water.

Wildlife Abstracts: 1935–1951; 1952–1955; 1956–1960. An annotated bibliography of the publications abstracted in *Wildlife Review.* A major reference source.

Wildlife Review: 1935 on, U.S. Fish and Wildlife Service. Coverage much wider than title indicates. Invaluable to field biologists; a necessity for those in wildlife management.

Zoological Record: 1864 on, Zoological Society of London. The major reference for zoologists. Worldwide coverage; essential.

General
bibliographies

Bibliography

eference information for sources cited throughout the text is presented here. Any point of particular interest that has been stressed in the text can be further researched.

ABBOT, B. C., AND D. BALLANTINE
1957 The toxin from *Gymnodinium veneficum* Ballantine, *J. Marine Biol. Assoc. U.K.*, **36**:169–189.

ADAMS, C. L.
1909 An ecological survey of Isle Royale, Lake Superior, *Rept. Bd. Geol. Surv. for 1908*.

ADAMS, L.
1959 An analysis of a population of snowshoe hares in northwestern Montana, *Ecol. Monographs*, **29**:141–170.

ADKISSON, P. L.
1964 Action of the photoperiod in controlling insect diapause, *Am. Naturalist*, **98**:357–374.
1966 Internal clocks and insect diapause, *Science*, **154**:234–241.

AHLGREN, I. F., AND C. E. AHLGREN
1960 Ecological effects of forest fires, *Botan. Rev.*, **26**:483–533.

ALDRICH, S. R.
1972 Some effects of crop production technology on environmental quality, *Bioscience*, **22**:90–95.

ALEXANDER, MARTIN
1965 Nitrification, in W. V. Bartholomew and F. E. Clark (eds.), *Soil Nitrogen, Am. Soc. Agron. Monogr. No. 10*, pp. 307–343.

ALEXANDER, M. M.
1958 The place of aging in wildlife management, *Am. Sci.*, **46**:123–137.

ALEXANDER, R. D.
1960 Sound communication in *Orthoptera* and *Cicadidae*, in Lanyon and Tavolga (eds.), *Animal Sounds and Communication*, pp. 38–96.
1961 Aggressiveness, territoriality and sexual behaviour in field crickets (Orthoptera: Gryllidae), *Behaviour*, **17**:130–223.
1962 The role of behavioral study in cricket classification, *System. Zool.*, **11**:53–72.

ALEXANDER, R. D., AND T. E. MOORE
1962 The evolutionary relationships of 17-year and 13-year cicadas, and three new species (Homoptera, Cicadidae, Magicicada), *Museum Zool. Misc. Publ. Univ. Mich.*, **121**:1–59.

ALLEE, W. C.
1926 Measurements of environmental factors in the tropical rain forest of Panama, *Ecology*, **7**:273–302.

ALLEE, W. C.; A. E. EMERSON; O. PARK; T. PARK; AND K. P. SCHMIDT
1949 *Principles of Animal Ecology*, Saunders, Philadelphia.

ALLEE, W. C.; B. GREENBERG; G. M. ROSENTHAL; AND P. FRANK
1948 Some effects of social organization on growth in the green sunfish *Lepomis cyanellas*, *J. Exp. Zool.*, **108**:1–19.

ALLEN, D. L.
1943 Michigan fox squirrel management, *Mich. Dept. Conserv. Game Div. Publ. No. 100*, Lansing, Mich., p. 404.

ALLEN, K. R.
1960 Effect of land development on stream bottom faunas, *Proc. New Zealand Ecol. Soc.*, **7**:20–31.

ALLISON, F. E.
1966 The fate of nitrogen applied to soils, *Adv. Agron.*, **18**:219–258.

ALTMANN, M.
1952 Social behaviour of elk, *Cervus canadensis nelsoni*, in the Jackson Hole area of Wyoming, *Behaviour*, **4**:116–143.
1956a Patterns of social behavior in big game of the United States and Europe, *Trans. North Am. Wildlife Conf.*, **21**:538–545.
1956b Patterns of herd behavior in free-ranging elk of Wyoming, *Cervus canadensis nelsoni*, *Zoologica*, **41**:65–71.
1960 The role of juvenile elk and moose in the social dynamics of their species, *Zoologica*, **45**:35–40.

ALTSCHUL, A. M.
1967 Food proteins: new sources from seeds, *Science*, **158**:221–226.

AMADON, D.
1947 Ecology and evolution of some Hawaiian birds, *Evolution*, **1**:63–68.
1950 The Hawaiian honey creepers (Aves, Drepanididae), *Bull. Am. Mus. Nat. Hist.*, **100**:397–451.
1959 Behavior and classification: some reflections, *Naturforschenden Gesellschaft*, Zurich, **104**:73–78.

AMERICAN CHEMICAL SOCIETY
1969 *Cleaning Our Environment: The Chemical Basis for Action*, American Chemical Society, Washington.

AMMANN, G. A.
1957 *The Prairie Grouse of Michigan*, Michigan Department of Conservation, Lansing.

ANDERSON, C. C.
1958 The elk of Jackson Hole, *Wyoming Fish and Game Comm. Bull. No. 10*, Cheyenne.

ANDERSON, D. W.; J. J. HICKEY; R. W. RISEBROUGH; D. F. HUGHES; AND R. E. CHRISTENSEN
1969 Significance of chlorinated hydrocarbon residues to breeding pelicans and cormorants, *Can. Field Naturalist*, **83**:91–112.

ANDERSON, G. C.
1973 The built-in population control mechanism. *W. Va. Agr. and Forestry*, **5**(1):10–13.

ANDERSON, K. L.

1965 Fire ecology: some Kansas prairie forbs, *Proc. 4th Tall Timbers Fire Ecology Conf.*, pp. 153–159.

ANDERSON, N. H.

1966 Depressant effect of moonlight on activity of aquatic insects, *Nature*, 209:319–320.

ANDERSON, W. L., AND P. L. STEWART

1969 Relationships between inorganic ions and the distribution of pheasants in Illinois, *J. Wildlife Manage.*, 33:254–270.

ANDREW, R. J.

1956 Normal and irrelevant toilet behaviour in *Emberiza, Supp. Brit. J. Anim. Behav.*, 4:85–91.

1961 The displays given by passerines in courtship and reproductive fighting: a review, *Ibis*, 103a:315–348.

ANDREWARTHA, H. G.

1961 *Introduction to the Study of Animal Populations*, University of Chicago Press, Chicago.

ANDREWARTHA, H. G., AND L. C. BIRCH

1954 *The Distribution and Abundance of Animals*, University of Chicago Press, Chicago.

ANTOINE, L. H., JR.

1964 Drainage and best use of urban land, *Public Works*, 95:88–90.

ARDREY, R.

1966 *The Territorial Imperative*, Atheneum, New York.

ARMELAGOS, G. J., AND J. R. DEWEY

1970 Evolutionary response to human infectious disease, *Bioscience*, 20:271–275.

ASCHOFF, J.

1958 Tierische Periodik unter dem Einfluss von Zeitgebern, *Z. F. Tierpsychol.*, 15:1–30.

1962 Comparative physiology: diurnal rhythms, *Ann. Rev. Physiol.*, 25:581–601.

1965 Circadian rhythms in man, *Science*, 148:1427–1432.

1966 Circadian activity pattern with two peaks, *Ecology*, 47:657–662.

1969 Desynchronization and resynchronization of human rhythms, *Aerospace Med.*, 40:844–849.

ASCHOFF, J., AND J. MEYER-LOHMANN

1954 Angeborene 24-Stunden-Periodik beim Kucken, *Pfluegers Arch. Ges. Physiol.*, 260:170–176.

ASCHOFF, J., AND R. WEVER

1962 Spontanperiodik des Menschen bei Ausschluss aller Zeitgeber, *Naturwissenschaften*, 49:337ff.

ATKINS, E. S.

1957 Light measurement in a study of white pine reproduction, *Can. Dept. Northern Affairs Nat. Resources, Forestry Branch, For. Res. Div. Tech. Note No. 60.*

ATLAVINYTE, O., AND J. D. DACIULYTE

1969 The effect of earthworms on the accumulation of vitamin B_{12} in the soil, *Pedobiologia*, 9:165–170.

AUMANN, G. D.

1965 Microtine abundance and soil sodium levels, *J. Mammal.*, 594–604.

AUMANN, G. D., AND J. T. EMLEN

1965 Relation of population density to sodium availability and sodium selection by microtine rodents, *Nature*, 208:198–199.

AYALA, F. J.

1968a Environmental factors limiting the productivity and size of experimental populations of *Drosophila serrata* and *D. birchii, Ecology*, 49:562–565.

1968b Genotype, environment, and population numbers, *Science*, 162:1453–1459.

BACKIEL, T., AND E. D. LECREN

1967 Some density relationships for fish population parameters, in S. D. Gerking (ed.), *The Biological Basis of Freshwater Fish Production*, Wiley, New York, pp. 261–293.

BAER, J. G.

1951 *Ecology of Animal Parasites*, University of Illinois Press, Urbana, pp. 1–224.

BAERENDS, G. P.

1941 Fortpflanzungsverbalten und Orientierung der Grabwespe *Ammophila campestris, Jur. Tijd voor Entom.*, 84:71–275.

1959 The ethological analysis of incubation behaviour, *Ibis*, 101:357–368.

BAERENDS, G. P., AND J. M. BAERENDS-VAN ROON

1950 An introduction to the study of the ethology of cichlid fishes, *Behaviour Suppl.*, 1:7–242.

BAILEY, R. E.

1952 The incubation patch of passerine birds, *Condor*, 54:121–136.

BAKKEN, A.

1959 Behavior of gray squirrels, *Symp. Gray Squirrel Cont. 162, Maryland Dept. Res. Ed.*, pp. 393–407.

BALL, R. C., AND F. F. HOOPER

1963 Translocation of phosphorus in a trout stream ecosystem, in Schultz and Klement (eds.), 1963, *Radioecology*, pp. 217–228.

BALLINGER, R. E.; K. R. MARION; AND O. J. SEXTON

1970 Thermal ecology of the lizard *Anolis limifrons* with comparative notes on three additional Panamanian anoles, *Ecology*, 51:246–254.

BALSER, DONALD S.; H. H. DILL; AND H. K. NELSON

1968 Effect of predator reduction on waterfowl nesting success, *J. Wildlife Management*, 32:669–682.

BANFIELD, A. W. F.

1961 Migrating caribou, *Nat. Hist.*, 70(5):57–62.

BANK, T. P., II

1953 Ecology of prehistoric Aleutian village sites, *Ecology*, 34:246–264.

BARBER, H. S.

1951 North American fireflies of the genus *Photuris, Smithsonian Inst. Misc. Collections*, 117:1–58.

BARDACH, J. E.
1968 Aquaculture, *Science*, **161**:1098–1106.

BARKALOW, F. S., JR.; R. B. HAMILTON; AND R. F.
SOOTS, JR.
1970 The vital statistics of an unexploited gray
squirrel population, *J. Wildlife Manage.*,
34:489–500.

BARLOW, J. C., AND D. M. POWER
1970 An analysis of character variation in red-eyed
and Philadelphia vireos (Aves: Vireonidae) in
Canada, *Can. J. Zool.*, **48**:673–678.

BARLOW, J. P.
1955 Physical and biological processes determining
the distribution of zooplankton in a tidal estuary,
Biol. Bull., **109**:211–225.
1956 Effect of wind on salinity distribution in an
estuary, *J. Marine Res.*, **15**:193–203.

BARRY, R. G.
1969 The world hydrological cycle, in R. J. Chorley
(ed.), *Water, Earth, and Man*, Methuen,
London, pp. 11–29.

BARTHOLOMEW, G. A.
1959 Mother–young relations and the maturation of
pup behaviour in the Alaskan fur seal, *Anim.
Behav.*, **7**:163–171.

BARTHOLOMEW, G. A., AND T. J. CADE
1957 Temperature regulation, hibernation and
aestivation in the little pocket mouse *Perognathus
longimembris*, *J. Mammal.*, **38**:60–72.

BARTHOLOMEW, G. A., AND N. E. COLLIAS
1962 The role of vocalization in the social behaviour
of the northern elephant seal, *Anim. Behav.*,
10:7–14.

BARTHOLOMEW, G. A.; T. R. HOWELL; AND T. J. CADE
1959 Torpidity in the white-throated swift, Anna
hummingbird, and poor-will, *Condor*, **59**:145–
155.

BARTHOLOMEW, G. A., AND F. WILKE
1956 Body temperature in the northern fur seal
Callorhinus ursinus, *J. Mammal.*, **37**:327–337.

BASSAM, J. A.
1965 Photosynthesis: the part of carbon, in
J. Bonner and J. E. Varnea (eds.), *Plant
Biochemistry*, Academic, New York, pp. 875–902.

BASTOCK, M.; D. MORRIS; AND M. MOYNIHAN
1953 Some comments on conflict and thwarting in
animals, *Behaviour*, **6**:66–84.

BATCHELDER, R. B.
1967 Spatial and temporal patterns of fire in the
tropical world, *Proc. 6th Tall Timbers Fire
Ecology Conf.*, pp. 171–208.

BATES, H. W.
1863 *The Naturalist on the River Amazon*, Murray,
London.

BATESON, G.
1963 The role of somatic change in evolution,
Evolution, **17**:529–539.

BATZLI, G. O., AND F. A. PITELKA
1970 Influence of meadow mouse populations on
California grassland, *Ecology*, **51**:1027–1039.

BAUMGARTNER, A.
1968 Ecological significance of the vertical energy
distribution in plant stands, in F. E. Eckardt
(ed.), *Functioning of Terrestrial Ecosystems at
the Primary Production Level*, Proceedings of the
Copenhagen Symposium, Natural Resources
Research V, Unesco, pp. 367–374.

BEAL, H. W.
1934 The penetration of rainfall through hardwood
and softwood forest canopy, *Ecology*, **15**:412–
415.

BEALL, G.
1941a The monarch butterfly *Danaus archippus*
Fab.: I, General observations in southern Ontario,
Can. Field Naturalist, **55**:123–129.
1941b The monarch butterfly *Danaus archippus*
Fab.: II, The movement in southern Ontario,
Can. Field Naturalist, **55**:133–137.
1946 Seasonal variation in sex proportion and
wing length in the migrant butterfly *Danaus
plexippus* L. (Lep. Danaidae), *Trans. Royal
Entomol. Soc.*, London, **97**:337–353.

BEALS, E. W.
1968 Spatial pattern of shrubs on a desert plain in
Ethiopia, *Ecology*, **49**:744–746.

BEATLEY, J. C.
1969 Dependence of desert rodents on winter
annuals and precipitation, *Ecology*, **50**:721–724.

BECKWITH, S. L.
1954 Ecological succession on abandoned farm lands
and its relationship to wildlife management,
Ecol. Monographs, **24**:349–376.

BEEBE, F. L.
1960 The marine peregrines of the northwest Pacific
coast, *Condor*, **62**:145–189.

BEER, J. R., AND W. TRAUX
1950 Sex and age ratios in Wisconsin muskrats,
J. Wildlife Manage., **14**:323–331.

BEETON, A. M.
1965 Eutrophication of the St. Lawrence Great
Lakes, *Limnol. Oceanogr.*, **10**:240–254.
1969 Changes in the environment and biota of
the Great Lakes, in *Eutrophication, Causes,
Consequences, Correctives*, National Academy of
Sciences, Washington, D.C., pp. 150–187.

BEHLE, W. H.
1956 A systematic review of the mountain
chickadee, *Condor*, **58**:51–70.

BELLINGER, P. F.
1954 Studies of soil fauna with special reference to
the Collembola, *Conn. Agr. Exp. Sta. Bull.
No. 583*.

BELLROSE, F. C.
1967 Orientation in waterfowl migration, in R. M.
Storm (ed.), *Animal Orientation and Navigation*,
Oregon State University Press, Corvallis, pp.
73–98.

BELLROSE, F. C.; T. G. SCOTT; A. S. HAWKINS; AND
J. B. LOW

1961 Sex ratios and age ratios in North American ducks, *Illinois Nat. Hist. Surv. Bull.*, **27**:391–474.

BENNETT, R.

1951 Some aspects of Missouri quail and quail hunting, 1938–1948, *Missouri Conserv. Comm. Tech. Bull. No. 2, Jefferson City.*

BENNINGHOFF, W. S.

1952 Interaction of vegetation and soil frost phenomena, *Arctic*, **5**:34–44.

BENSON, L.

1962 *Plant Taxonomy, Methods and Principles,* Ronald, New York.

BENTLEY, P. J.

1966 Adaptations of amphibia to arid environments, *Science*, **152**:619–623.

BERG, W.; A. JONNELS; B. SJOSTRAND; AND
T. WESTERMARK

1966 Mercury content in feathers of Swedish birds from the past 100 years, *Oikos*, **17**:71–83.

BERGERUD, A. T.

1971 The population dynamics of Newfoundland caribou, *Wildlife Monograph No. 25,* Wildlife Society, Washington, D.C.

BERGERUD, A. T., AND F. MANUEL

1969 Aerial census of moose in central Newfoundland, *J. Wildlife Manage.*, **33**:910–916.

BERGH VAN DEN, J. P.

1969 Distribution of pasture plants in relation to chemical properties of the soil, in J. H. Rorison (ed.), *Ecological Aspects of the Mineral Nutrition of Plants*, pp. 11–23.

BERGH VAN DEN, J. P., AND C. T. DEWIT

1960 Concurrentie tussen Timothee en Reukgras, *Meded. Inst. biol. scherk. Onderz. Lanlb. Gewass.*, **121**:155–165.

BERNIER, B.

1961 *Forest Humus, a Consequence and Cause of Local Ecological Conditions,* mimeo, Northeast Forests Soils Conference.

BERRY, J. L. B.; J. W. SIMMONS; AND R. J. TENNANT

1963 Urban population densities: structure and change, *Geogr. Rev.*, **53**:389–405.

BIEL, E. R.

1961 Microclimate, bioclimatology, and notes on comparative dynamic climatology, *Am. Scientist*, **49**:326–357.

BIGGAR, J. W., AND R. B. COREY

1969 Agricultural drainage and eutrophication, in *Eutrophication, Causes, Consequences, Correctives*, National Academy of Sciences, Washington, D.C., pp. 404–445.

BILLINGS, W. D., AND L. C. BLISS

1959 An alpine snowbank and its effects on vegetation, plant development, and productivity, *Ecology*, **40**:388–397.

BILLINGS, W. D., AND H. A. MOONEY

1968 The ecology of arctic and alpine plants, *Biol. Rev.*, **43**:481–529.

BIRCH, L. C.

1948 The intrinsic rate of natural increase of an insect population, *J. Anim. Ecol.*, **17**:15–26.

BIRCH, L. C., AND D. P. CLARK

1953 Forest soil as an ecological community with special reference to the fauna, *Quart. Rev. Biol.*, **28**(1):13–36.

BITMAN, J.

1970 Hormonal and enzymatic activity of DDT, *Ag. Sci. Rev.*, **7**(4):6–12.

BLACK, R. F.

1954 Permafrost: a review, *Geol. Soc. Am. Bull.*, **65**:839–856.

BLAIR, A. P.

1942 Isolating mechanisms in a complex of four species of toad, *Biol. Symp.*, **6**:235–249.

BLAIR, W. F.

1943 Populations of the deer mouse and associated small mammals in southern New Mexico, *Contrib. Lab. Vert. Biol.*, University of Michigan, **21**:1–40.

1955 Mating call and stage of speciation in the *Microhyla olivacea— M. carolinensis* complex, *Evolution*, **9**:469–480.

1958 Mating call in the speciation of Anuran amphibians, *Am. Naturalist*, **92**:27–51.

1960 *The Rusty Lizard: A Population Study,* University of Texas Press, Austin.

1961 Calling and spawning seasons in a mixed population of Anurans, *Ecology*, **42**:99–110.

BLEST, A. D.

1957 The function of eyespot patterns in the Lepidoptera, *Behaviour*, **11**:209–256.

BLISS, L. C.

1956 A comparison of plant development in microenvironments of arctic and alpine tundras, *Ecol. Monographs*, **26**:303–337.

1963 Alpine plant communities of the Presidential Range, New Hampshire, *Ecology*, **44**:678–697.

BLUM, J. L.

1960 Algae populations in flowing waters, in *The Ecology of Algae, Spec. Publ. No. 2, Pymatuning Lab. of Field Biology*, pp. 11–21.

BLYDENSTEIN, J.

1968 Burning and tropical American savannas, *Proc. 8th Tall Timbers Fire Ecology Conf.*, pp. 1–14.

BOBVJERG, R. V., AND P. W. GLYNN

1960 A class exercise on a marine microcosm, *Ecology*, **41**:229–232.

BODE, I. H. R.

1958 Beiträge zur Kenntnis allelopathischen Eischeinungen bei einigen Juglandacean, *Planta*, **51**:440–480.

BOGERT, C. M.

1960 The influence of sound on the behavior of amphibians and reptiles, in W. E. Lanyon and W. N. Tavolga (eds.), *Animal Sound and Communications*, pp. 137–320.

BOND, R. R.

1957 Ecological distribution of breeding birds in

the upland forests of Southern Wisconsin, *Ecol. Monographs*, 27:351–384.

BONNER, J.

1962 The upper limit of crop yield, *Science*, 137:11–15.

BORG, K.; H. WANNTROP; K. ERNE; AND E. HANKO

1969 Alkyl mercury poisoning in terrestrial Swedish wildlife, *Viltrevy*, 6(4):301–379.

BORMANN, F. H.; G. E. LIKENS; D. W. FISHER; AND R. S. PIERCE

1968 Nutrient loss accelerated by clear-cutting of a forest ecosystem, in H. E. Young (ed.), *Symposium on Primary Productivity and Mineral Cycling in Natural Ecosystems*, University of Maine Press, Orono.

BORNEBUSCH, C. H.

1930 *The Fauna of Forest Soils*, Nielsen and Lydiche, Copenhagen.

BORROR, D. J.

1961 Intraspecific variation in passerine bird songs, *Wilson Bull.*, 73:57–78.

BOTT, T. L., AND T. D. BROCK

1969 Bacterial growth rates above 90°C in Yellowstone hot springs, *Science*, 164:1411–1412.

BOX, T. W.; J. POWELL; AND D. L. DRAWE

1967 Influence of fire on south Texas chaparral, *Ecology*, 48:955–961.

BRAEMER, W.

1959 Versuche zu der im Richlungsfinder der Fische enthaltenen Zeitschatzung, *Verh. d. deutschen Zool. Ges. Zool. Anz. 23, Supplementband 276*.

1960 A critical review of the sun-azimuth hypothesis, *Cold Spring Harbor Symp. Quant. Biol.*, 25:413–427.

BRAGG, A. N.

1950 Observations on *microhyla* (Salientia: Microhylidae), *Wasmann J. Biol.*, 8:113–118.

BRAMBLE, W. C., AND R. H. ASHLEY

1955 Natural revegetation of spoil banks in central Pennsylvania, *Ecology*, 36:417–423.

BRANT, D. H.

1962 Measures of the movements and population densities of small rodents, *Univ. Calif. Pub. Zool.*, 62:105–184.

BRATTSTROM, B. H.

1963 A preliminary review of the thermal requirements of amphibians, *Ecology*, 44:238–255.

1965 Body temperature of reptiles, *Am. Midland Nat.*, 73:376–422.

BRAUN, E. LUCY

1950 *Deciduous Forests of Eastern North America*, McGraw-Hill, Blakiston, New York.

BRAWN, V. M.

1961 Reproductive behaviour of the cod (*Gadus callarias* L.), *Behaviour*, 18:177–198.

BRAY, J. R.

1961 Measurement of leaf utilization as an index of minimum level of primary consumption, *Oikos*, 12:70–74.

BRAYTON, R. O., AND G. M. WOODWELL

1966 Effects of ionizing radiation and fire on *Gaylussica baccata* and *Vaccinium vacillans*, *Am. J. Bot.*, 53:812–820.

BREDER, C. M., JR.

1951 Studies on the structure of the fish school, *Bull. Am. Mus. Nat. Hist.*, 98:1–28.

BRETT, J. R., AND D. MACKINNON

1954 Some aspects of olfactory perception in migrating adult coho and spring salmon, *J. Fisheries Res. Board Can.*, 11:310–318.

BROCK, T. R.

1966 *Principles of Microbial Ecology*, Prentice-Hall, Englewood Cliffs, N.J.

BROCK, V. E., AND R. H. RIFFENBURGH

1960 Fish schooling, a possible factor in reducing predation, *J. Du Conseil*, 25:307–317.

BROECKER, W. S.

1970 Man's oxygen reserves, *Science*, 168:1537–1538.

BRONSON, F. H.

1969 Pheromonal influences on mammalian reproduction, in M. Diamond (ed.), *Perspectives in Reproduction and Sexual Behavior*, Indiana University Press, Bloomington.

BROOKS, J. L., AND S. I. DODSON

1965 Predation, body size, and composition of plankton, *Science*, 150:28–35.

BROOKS, J. L., AND G. E. HUTCHINSON

1950 On the rate of passive sinking of Daphnia, *Proc. Nat. Acad. Sci. U.S.*, 36:272–277.

BROOKS, M. G.

1940 The breeding warblers of the central Allegheny Mountain region, *Wilson Bull.*, 52:249–266.

1951 Effect of black walnut trees and their products on other vegetation, *W. Va. Univ. Agr. Exp. Stat. Bull.*, 347:1–31.

BROOKS, W. S.

1968 Comparative adaptations of the Alaskan redpolls to the arctic environment, *Wilson Bull.*, 80:253–280.

BROWER, J. V. Z.

1958 Experimental studies of mimicry in some North American butterflies: I, The monarch, *Danaus plexippus*, and Viceroy, *Limenitis archippus*; II, *Battus philenor* and *Papilio troilus*, *P. polyxenes* and *P. glaucus*; III, *Danaus gilippus berenice* and *Limenitis archippus floridensis*, *Evolution*, 12:32–47, 123–136, 273–285.

BROWER, J. V. Z., AND L. P. BROWER

1962 Experimental studies of mimicry: 6, The reaction of toads (*Bufo terrestris*) to honeybees (*Apis mellifera*) and their dronefly mimics (*Eristalis vinetorum*), *Am. Naturalist*, 97:297–307.

BROWER, L. P.

1961 Studies on the migration of the monarch butterfly: I, Breeding populations of *Danaus plexippus* and *D. gilippus berenice* in south central Florida, *Ecology*, 42:76–83.

1969 Ecological chemistry, *Sci. American*, **220**(2): 22–29.

BROWER, L. P.; J. V. Z. BROWER; AND F. P. CRANSTON

1965 Courtship behavior of the Queen butterfly, *Danaus gilippus berenice* Cramer, *Zoologica*, **50**:1–39.

BROWER, L. P.; J. V. Z. BROWER; AND P. W. WESTCOTT

1960 Experimental studies of mimicry: 5, The reactions of toads (*Bufo terrestris*) to bumblebees (*Bombus americanorum*) and their robberfly mimics (*Mallophora bomboides*) with a discussion of aggressive mimicry, *Am. Naturalist*, **94**:343–355.

BROWN, E. R.

1961 The black-tailed deer of western Washington, *Washington State Game Dept. Biol. Bull. No. 13*, Olympia.

BROWN, F. A., JR.

1959 Living clocks, *Science*, **130**:1535–1544.

BROWN, F. A., JR.; J. W. HASTINGS; AND J. D. PALMER

1970 *The Biological Clock: Two Views*, Academic, New York.

BROWN, J. H., AND E. H. TRYON

1960 Establishment of seeded black locust on spoil banks, *W. Va. Univ. Agr. Exp. Sta. Bull. No. 440*.

BROWN, J. L.

1963 Aggressiveness, dominance and social organization in the Stellar jay, *Condor*, **65**:460–484.

1969a The buffer effect and productivity in tit populations, *Am. Naturalist*, **103**:347–354.

1969b Territorial behavior and population regulation in birds: a review and reevaluation, *Wilson Bull.*, **81**:293–329.

BROWN, L. R.

1967 The world outlook for conventional agriculture, *Science*, **158**:604–611.

1968 The agricultural revolution in Asia, *Foreign Affairs*, **46**:688–698.

BROWN, R. J. E.

1970 *Permafrost in Canada*, University of Toronto Press, Toronto.

BROWN, R. J. E., AND G. H. JOHNSON

1964 Permafrost and related engineering problems, *Endeavour*, **23**:66–72.

BROWN, W. L., AND E. O. WILSON

1956 Character displacement, *System. Zool.*, **5**:49–64.

BRUCE, V. G.

1960 Environmental entrainment of circadian rhythms, *Cold Spring Harbor Symp. Quant. Biol.*, **25**:29–47.

BRUSH, J. E.

1968 Spatial patterns of populations in Indian cities, *Geogr. Rev.*, **58**:362–391.

BRYSON, R. A., AND R. A. RAGOTZKIE

1960 On internal waves in lakes, *Limnol. Oceanogr.*, **5**:397–408.

BUCKNER, C. H., AND W. J. TURNOCK

1965 Avian predation on the larch sawfly, *Pristiphora erichsonii* (Htg) (Hymenoptera: Tenthredinidae), *Ecology*, **46**:223–236.

BUDOWSKI, G.

1966 Fire in tropical American lowland areas, *Proc. 5th Tall Timbers Fire Ecology Conf.*, pp. 5–22.

BUDYKO, M. I.

1963 *The Heat Budget of the Earth*, Hydrological Publishing House, Leningrad.

BUECHNER, H. K.

1961 Territorial behavior in the Uganda kob, *Science*, **133**:698.

BUELL, M. F., AND R. E. WILBUR

1948 Life form spectra of the hardwood forests of the Itasca Park region, Minnesota, *Ecology*, **29**:352–359.

BÜNNING, E.

1935a Zur Kenntnis der erblichen Tagesperiodizität bei den Primärblättern von *Phaseolus multifloris*, *Jb. Wiss. Botany*, **81**:411ff.

1935b Zur Kenntnis der endogenen Tagesrhythmik bei Insekten und bei Pflanzen, *Ber. Deut. Botan. Ges.*, **53**:594–623.

1959 Physiological mechanism and biological importance of the endogenous diurnal periodicity in plants and animals, in R. B. Withrow (ed.), *1959, Photoperiodism and Related Phenomena*, pp. 507–530.

1964 *The Physiological Clock*, 2d ed., Academic, New York.

BÜNNING, E., AND G. JOERRENS

1960 Tagesperiodische antagonistische Schwankungen de Blauviolettund Gelbrot-Empfindlichkeit als Grundlage der photoperiodischen Diapause-Induktion bei *Pieris brassicae*, *Z. Naturforsch.*, **15b**:205–223.

BÜNNING, E., AND D. MÜLLER

1961 Wie messen Organismen lunare Zyklen?, *Z. Naturforsch.*, **16b**:391–395.

BÜNSOW, R.

1953 Endogene Tagesrhythmik und Photoperiodismus bei *Kalanchoe blossfeldiana*, *Planta*, **42**:220–252.

BURDICK, G. E.; E. J. HARRIS; H. J. DEAN; T. M. WALKER; J. SKEA; AND D. COLBY

1964 The accumulation of DDT in lake trout and the effect of reproduction, *Trans. Am. Fisheries Soc.*, **93**:127–136.

BURGER, G. V.

1969 Response of gray squirrels to nest boxes at Remington Farms, Maryland, *J. Wildlife Manage.*, **33**:796–801.

BURGES, A.

1963 The microbiology of a podzol profile, in J. Doeksen and J. van der Drift (eds.), *Soil Organisms*, North Holland Publishing Co., Amsterdam, pp. 151–157.

BURGHARDT, G. M., AND E. H. HESS
1966 Food imprinting in the snapping turtle, *Chelydra serpentinea, Science*, **151**:108–109.

BURT, W. H.
1940 Territorial behavior and populations of some small mammals in southern Michigan, *Mus. Zool. Misc. Publ. Univ. Mich.*, **45**:1–58.

BURT, W. V., AND J. QUEEN
1957 Tidal overmixing in estuaries, *Science*, **126**:973–974.

BUSTARD, H. R.
1969 The population ecology of the gekkonid lizard (*Gehyra variegata* Dumeril and Bibrou) in exploited forests in northern New South Wales, *J. Anim. Ecol.*, **38**:35–51.
1970 The role of behavior in the natural regulation of numbers in the gekkonid lizard *Gehyra variegata, Ecology*, **51**:724–728.

BUTLER, P. A.
1964 Commercial fisheries investigations in *Pesticide–Wildlife Studies, U.S. Fish and Wildlife Serv. Circ.* 226, pp. 11–25.

BYERLY, T. C.
1967 Efficiency of feed conversion, *Science*, **157**:890–895.

CAIN, A. J., AND P. M. SHEPPARD
1950 Selection in the polymorphic land snail *Cepaea nemoralis, Heredity*, **4**:275–294.
1954 Natural selection in *Cepaea, Genetics*, **39**:89–116.

CAIN, S. A., AND G. M. CASTRO
1959 *Manual of Vegetation Analysis*, Harper & Row, New York.

CALDWELL, L. K.
1970 The ecosystem as a criterion for public land policy, *Nat. Resources J.*, **10**:203–220.

CALHOUN, J. B.
1952 The social aspects of population dynamics, *J. Mammal.*, **33**:139–159.

CALIFORNIA FISH AND GAME COMMISSION
1972 *At the Crossroads: A Report on California's Endangered and Rare Fish and Wildlife*, California Fish and Game Comm., Sacramento.

CAMPBELL, R. W.
1969 Studies on gypsy moth population dynamics, in *Forest Insect Population Dynamics, USDA Res. Paper NE-125*, pp. 29–34.

CANTLON, J. E.
1953 Vegetation and microclimates on north and south slopes of Cushetunk Mountain, New Jersey, *Ecol. Monographs*, **23**:241–270.

CARL, E. A.
1971 Population control in arctic ground squirrels, *Ecology*, **52**:395–413.

CARLISLE, A.; A. H. F. BROWN; E. J. WHITE
1966 The organic matter and nutrient elements in the precipitation beneath a sessile oak canopy, *J. Ecol.*, **54**:87–98.

CARPENTER, C. R.
1934 A field study of the behavior and social relations of howling monkeys (*Aloatta palliata*), *Comp. Psychol. Monographs No. 2.*

CARR, A.
1962 Orientation problems in the high seas travel and terrestrial movements of marine turtles, *Am. Scientist*, **50**:359–374.

CASTRI, F. DI
1970 Les grands problems qui se posent aux ecologistes pour l'étude des écosystemes du sol, in J. Phillipson (ed.), *Methods of Study in Soil Biology*, UNESCO, Paris, pp. 15–31.

CAUGHLEY, G.
1966 Mortality patterns in mammals, *Ecology*, **47**:906–918.
1970 Eruption of ungulate populations with emphasis on Himalayan thar in New Zealand, *Ecology*, **51**:53–72.

CAUGHLEY, G., AND L. C. BIRCH
1971 Rate of increase, *J. Wildlife Manage.*, **35**:658–663.

CHAFFEE, R. R. J., AND J. C. ROBERTS
1971 Temperature acclimation in birds and mammals, *Ann. Rev. of Physiol.*, **33**:155–202.

CHAMBERS, G. D.; K. C. SADLER; AND R. P. BREITENBACK
1966 Effects of dietary calcium levels on egg production and bone structure of pheasants, *J. Wildlife Manage.*, **30**:65–73.

CHAMBERS, R. E., AND W. M. SHARP
1958 Movement and dispersal within a population of ruffed grouse, *J. Wildlife Manage.*, **22**:231–239.

CHANGNON, S. A.
1968 La Porte weather anomaly: Fact or fiction?, *Bull. Amer. Metero. Soc.*, **49**:4–11.

CHAPMAN, V. J.
1960 *Salt Marshes and Salt Deserts of the World*, Wiley, New York.

CHAPPEL, H. G.
1963 The effect of ionizing radiation on *Smilax* with special reference to the protection afforded by their production of underground vegetative structures, in Schultz and Klement (eds.), *Radioecology*, pp. 289–294.

CHASE, W., AND D. H. JENKINS
1962 Productivity of the George Reserve deer herd, in *Proc. 1st Nat. White-tailed Deer Disease Symposium*, pp. 78–88.

CHEATUM, E. L., AND C. W. SEVERINGHAUS
1950 Variations in fertility of white-tailed deer related to range conditions, *Trans. North Am. Wildlife Conf.*, **15**:170–189.

CHITTY, D.
1952 Mortality among voles (*Microtus agrestus*) at Lake Vyrnwy, Montgomeryshire, in 1936–9, *Phil. Trans. Roy. Soc. London*, Ser. B, **236**:505–552.
1960 Population processes in the vole and their reference to general theory, *Can. J. Zool.*, **38**:99–113.

CHITTY, D., AND E. PHIPPS
1966 Seasonal changes in survival in mixed popula-

tions of two species of vole, *J. Anim. Ecol.*, 35:313–331.

CHOW, T. J.

1970 Lead accumulation in roadside soil and grass, *Nature*, 225:295–296.

CHOW, T. J., AND J. L. EARL

1970 Lead aerosols in the atmosphere: increasing concentrations, *Science*, 169:577–580.

CHRISTENSEN, A. M., AND J. J. MCDERMOTT

1958 Life history of the oyster crab *Pinnotheres ostreum*, *Biol. Bull.*, 114:146–179.

CHRISTIAN, J. J.

1963 Endocrine adaptative mechanisms and the physiologic regulation of population growth, in W. V. Mayer and R. G. Van Gelder (eds.), *Physiological Mammalogy*, Vol. 1, *Mammalian Populations*, Academic, New York, pp. 189–353.

CHRISTIAN, J. J., AND D. E. DAVIS

1964 Endocrines, behavior and populations, *Science*, 146:1550–1560.

CHRISTIAN, J. J.; V. FLYGER; AND D. E. DAVIS

1960 Factors in the mass mortality of a herd of Sika deer, *Cervus nippon, Chesapeake Science*, 1:79–95.

CHRISTY, H. R.

1952 Vertical temperature gradients in a beech forest in central Ohio, *Ohio J. Sci.*, 52:199–209.

CHURCHILL, E. D., AND H. C. HANSON

1958 The concept of climax in arctic and alpine vegetation, *Botan. Rev.*, 24:127–191.

CHURCHMAN, C. W.

1968 *The Systems Approach*, Dell, New York.

CLARK C.

1967 *Population Growth and Land Use*, Macmillan, London and St. Martins Press, New York.

CLARK, F. E.

1969a The microflora of grassland soils and some microbial influences on ecosystem functions, in R. L. Dix and R. G. Beidleman (eds.), *The Grassland Ecosystem, Range Sci. Dept. Sci. Ser. No. 2*, Colorado State University.

1969b Ecological associations among soil micro-organisms, in *Soil Biol.*, UNESCO, Paris, pp. 125–161.

CLARK, F. W.

1972 Influence of jackrabbit density on coyote population change, *J. Wildlife Manage.*, 36:343–356.

CLARK, L. B.

1938 Observations on the Atlantic palolo, *Yearbook Carnegie Inst. Wash.*

CLARKE, J. F.

1969 Nocturnal urban boundary layer over Cincinnati, Ohio, *Monthly Weather Rev.*, 97:582–589.

CLAUSEN, J.

1965 Population studies of alpine and subalpine races of conifers and willows in the California High Sierra Nevada, *Evolution*, 19:56–68.

CLAUSEN, J.; D. D. KECK; AND W. M. HIESEY

1948 Experimental studies on the nature of species:

III, Environmental responses of climatic races of Achillea, *Carnegie Inst. Wash. Publ. No. 581.*

CLAWSON, M.; H. H. LANDSBERG; AND L. T. ALEXANDER

1969 Desalted seawater for agriculture: Is it economic?, *Science*, 164:1141–1148.

CLAWSON, S. G.

1958–59 Fire ant eradication and quail, *Alabama Conserv.*, 30(4):14–15, 25.

CLAY, W. M.

1953 Protective coloration in the American sparrow hawk, *Wilson Bull.*, 65:129–134.

CLEMENTS, F. C.

1916 *Plant Succession, Carnegie Inst. Wash. Publ.* 242, Washington, D.C.

1920 *Plant Indicators: The Relation of Plant Communities to Process and Practice, Carnegie Inst. Wash. Publ.* 290, pp. 1–388.

CLEMENTS, F. E., AND V. E. SHELFORD

1939 *Bio-ecology*, Wiley, New York.

CLIFFORD, H. F.

1966 The ecology of invertebrates in an intermittent stream, *Inv. Indiana Lakes and Streams*, 7(2):57–98.

CLOUDSLEY-THOMPSON, J. L.

1956 Studies in diurnal rhythms: VII, Humidity responses and nocturnal activity in woodlice (Isopoda), *J. Exp. Biol.*, 33:576–582.

1960 Adaptive functions of circadian rhythms, *Cold Spring Harbor Symp. Quant. Biol.*, 25:345–355.

CLOUGH, G. C.

1965a Lemmings and population problems, *Am. Scientist*, 53:199–212.

1965b Viability of wild voles, *Ecology*, 46:119–134.

COBB, W. F., JR., AND R. W. STARK

1970 Decline and mortality of smog-injured ponderosa pine, *J. Forestry*, 68:147–149.

COCHRANE, C. R.

1968 Fire ecology in southeastern Australian schlerophyll forests, *Proc. 8th Tall Timbers Fire Ecology Conf.*, pp. 15–40.

CODY, M. L.

1968 On the methods of resource division in grassland bird communities, *Am. Naturalist*, 102:107–147.

1969 Convergent characteristics in sympatric species: a possible relation to interspecific competition and aggression, *Condor*, 71:222–239.

CODY, M. L., AND J. H. BROWN

1970 Character divergence in Mexican birds, *Evolution*, 24:304–310.

COKER, R. E.

1947 *This Great and Wide Sea*, University of North Carolina Press, Chapel Hill.

Cold Spring Harbor Symposia on Quantitative Biology

1957 Population studies: animal ecology and demography, Vol. 22, Biological Lab., Cold Spring Harbor, New York.

1960 Biological clocks, Vol. 25, Biological Lab., Cold Spring Harbor, New York.

COLE, D. W.; S. P. GESSEL; AND S. F. DICE

1967 Distribution and cycling of nitrogen, phosphorus, potassium and calcium in a second-growth Douglas-fir ecosystem, in *Symposium on Primary Productivity and Mineral Cycling in Natural Ecosystems*, pp. 197–232.

COLE, L. C.

1946 A study of the crypotozoa of an Illinois woodland, *Ecol. Monographs*, 16:49–86.

1951 Population cycles and random oscillations, *J. Wildlife Manage.*, 15:233–252.

1954a The population consequences of life history phenomena, *Quart. Rev. Biol.*, 29:103–137.

1954b Some features of random cycles, *J. Wildlife Manage.*, 18:107–109.

1957 Sketches of general and comparative demography, *Cold Spring Harbor Symp. Quant. Biol.*, 22:1–15.

COLLESS, D. H.

1970 The relationship of evolutionary theory to phenetic taxonomy, *Evolution*, 24:721–722.

COLLIAS, ELSIE R., AND N. E. COLLIAS

1957 The response of chicks of the Franklin's gull to parental bill color, *Auk*, 74:371–375.

COLLIAS, N. E.

1956 Analysis of socialization in sheep and goats, *Ecology*, 37:228–239.

COLLIAS, N. E., AND R. D. TABER

1951 A field study of some grouping and dominance relations in ring-necked pheasants, *Condor*, 53:265–275.

COLLIER, C. C.

1962 Influence of strip mining on the hydrological environment of parts of Beaver Creek Basin, Ky., 1955–1959, *U.S. Geol. Surv. Prof. Paper*, 427-B.

COMMONER, B.

1970 Threats to the integrity of the nitrogen cycle: nitrogen compounds in soil, water, atmosphere, and precipitation, in S. F. Singer (ed.), *Global Effects of Environmental Pollution*, Springer-Verlag, New York.

CONNELL, J. H.

1961 The influence of interspecific competition and other factors on the distribution of the barnacle *Chthamalus stellatus*, *Ecology*, 42:710–723.

CONNELL, J. H., AND E. ORIAS

1964 The ecological regulation of species diversity, *Am. Naturalist*, 98:399–414.

COOK, C. W.; L. A. STODDARD; AND L. E. HARRIS

1954 The nutritive value of winter range plants in the Great Basin, *Utah Agr. Exp. Sta. Bull. No. 372.*

COOKE, F., AND F. G. COOCH

1968 The genetics of polymorphism in the goose *Anser caerulescens*, *Evolution*, 22:289–300.

COOKE, F., AND J. R. RYDER

1971 The genetics of polymorphism in the Ross Goose (*Anser rossii*), *Evolution*, 25:483–496.

COOKE, G. D.

1967 The pattern of autotrophic succession in laboratory microcosms, *Bioscience*, 17:717–721.

COOPER, C. F.

1960 Changes in vegetation, structure and growth of southwestern pine forests since white settlement, *Ecol. Monographs*, 30:129–164.

COOPER, E. L.

1959 Trout stocking as an aid to fish management, *Penn. State Univ. Agr. Exp. Sta. Bull. No. 663.*

COPE, O. B.

1971 Interaction between pesticides and wildlife, *Ann. Rev. Entomology*, 16:325–364.

CORBETT, E. S., AND R. P. CROUSE

1968 Rainfall interception by annual grass and chaparral, *USDA For. Serv. Res. Paper PSW-48.*

CORDONE, A. L., AND W. KELLY

1961 The influence of inorganic sediment on the aquatic life of streams, *Calif. Fish Game*, 47:189–228.

CORNFORTH, I. S.

1970a Leaf-fall in a tropical rain forest, *J. Appl. Ecol.*, 7:603–608.

1970b Reafforestation and nutrient reserves in the humid tropics, *J. Appl. Ecol.*, 7:609–615.

CORY, L.; P. FJELD; AND W. SERAT

1970 Distribution patterns of DDT residues in the Sierra Nevada mountains, *Pesticides Monitoring J.*, 4:8–10.

COSTONIS, A. C., AND W. A. SINCLAIR

1969 Relationships of atmospheric ozone to needle blight of eastern white pine, *Phytopathology*, 59:1566–1574.

COTT, H. B.

1940 *Adaptive Coloration in Animals*, Methuen, London.

COTTAM, G., AND J. T. CURTIS

1956 The use of distance measures in phytosociological sampling, *Ecology*, 37:451–460.

COUPLAND, R. T.

1950 Ecology of mixed prairie in Canada, *Ecol. Monographs*, 20:217–315.

1958 The effects of fluctuations in weather upon the grassland of the Great Plains, *Botan. Rev.*, 24:273–317.

1959 Effect of changes in weather conditions upon grasslands in the northern Great Plains, in Sprague (ed.), 1959, *Grassland*, 291–306.

COUTANT, C.

1970 Biological aspects of thermal pollution: I, Entrainment and discharge canal effects, *CRC Critical Reviews in Environ. Control*, Nov. 1970, 341–381.

COWAN, I., AND V. GEIST

1961 Aggressive behavior in deer of the genus *Odocoileus*, *J. Mammal.*, 42:522–526.

COWAN, R. L.

1962 Physiology of nutrition as related to deer, *Proc. 1st Natl. White-tailed Deer Disease Symp.*, pp. 1–8.

COWLES, H. C.

1899 The ecological relations of the vegetation on the sand dunes of Lake Michigan, *Botan. Gaz.*, **27**:95–117, 167–202, 281–308, 361–391.

COWLES, R. B., AND C. M. BOGERT

1944 A preliminary study of the thermal requirements of desert reptiles, *Bull. Amer. Museum Nat. Hist.*, **83**:265–296.

CRAGG, J. B. (ED.)

1962 *Advances in Ecological Research*, Vol. 1, Academic, New York.

CRAIG, W.

1918 Appetites and aversions as constituents of insects, *Biol. Bull.*, Woods Hole, Massachusetts, **34**:91–107.

CRAIGHEAD, F., AND J. CRAIGHEAD

1956 *Hawks, Owls and Wildlife*, Stackpole, Harrisburg, Pa.

CROMBIE, A. C.

1947 Interspecific competition, *J. Anim. Ecol.*, **16**:44–73.

CRONIN, L. E., AND A. J. MANSUETI

1971 The biology of the estuary, in P. A. Douglas and R. H. Stroud (eds.), *A Symposium on the Biological Significance of Estuaries*, Sport Fishing Institute, Washington, D.C., pp. 14–39.

CROSBY, G. T.

1972 Spread of the cattle egret in the Western hemisphere, *Bird-banding*, **43**:205–212.

CROSSLEY, D. A., JR., AND K. K. BOHNSACK

1960 Long-term ecological study in the Oak Ridge area: III, Oribatid mite fauna in pine litter, *Ecology*, **41**:628–638.

CROSSMAN, E. J.

1959a Distribution and movements of a predator, the rainbow trout, and its prey, the redside shiner, in Paul Lake, British Columbia, *J. Fisheries Res. Board Can.*, **16**:247–267.

1959b A predator-prey interaction in freshwater fish, *J. Fisheries Res. Board Can.*, **16**:269–281.

CUMMINGS, B. G., AND F. WAGNER

1968 Rhythmic processes in plants, *Annual Rev. Plant Physiol.*, **19**:381–416.

CUMMINS, K. W.; W. P. COFFMAN; AND P. A. ROFF

1966 Trophic relationships in a small woodland stream, *Verheindlung der Internationalen Vereinigung fur Theoretische und Angewandte Limnologie*, **16**:627–637.

CURRY, L. L.

1965 A survey of environmental requirements for the midge (*Diptera tendipedidae*), in *Hearings Before the Subcommittee on Air and Water Pollution of the Committee on Public Works, United States Senate*, pp. 127–141.

CURRY-LINDAHL, K.

1962 The irruption of Norway lemming in Sweden during 1960, *J. Mammal.*, **43**:171–184.

CURTIS, J. T.

1959 *The Vegetation of Wisconsin*, University of Wisconsin Press, Madison, p. 657.

CURTIS, J. T., AND R. P. MCINTOSH

1951 An upland forest continuum in the prairie-forest border region of Wisconsin, *Ecology*, **32**:476–496.

DAHL, E.

1953 Some aspects of the ecology and zonation of the fauna of sandy beaches, *Oikos*, **4**:1–27.

DAHLBERG, B. L., AND R. C. GUETTINGER

1956 The white-tailed deer in Wisconsin, *Tech. Wildlife Bull. No. 14*, Wisconsin Conserv. Dept., Madison.

DALE, T., AND V. G. CARTER

1955 *Topsoil and Civilization*, University of Oklahoma Press, Norman, p. 270.

DAMBACK, C. A., AND E. GOOD

1943 Life history and habits of the cicada killer in Ohio, *Ohio J. Sci.*, **43**:32–41.

DANIEL, C. P.

1963 Study of succession in fields irradiated with fast neutron and gamma radiation, in Schultz and Klement (eds.), *Radioecology*, pp. 277–282.

DANSEREAU, P.

1945 Essae de correlation sociologique entre les plantes superieures et les poissons de la Beine du Lac Saint-Louis, *Rev. Can. Biol.*, **4**:369–417.

1959a Vascular aquatic plant communities of southern Quebec: a preliminary analysis, *Trans. 10th Northeast Wildlife Conf.*, pp. 27–54.

1959b *Biogeography: An Ecological Perspective*, Ronald, New York.

DANSEREAU, P., AND F. SEGADAS-VIANNA

1952 Ecological study of the peat bogs of eastern North America, *Can. J. Bot.*, **30**:490–520.

DARLING, F. F.

1937 *A Herd of Red Deer*, Oxford, London.

1960 *Wildlife in an African Territory*, Oxford University Press, London.

DARLINGTON, P. J., JR.

1957 *Zoogeography: The Geographical Distribution of Animals*, Wiley, New York.

DARNELL, R. M.

1961 Trophic spectrum of an estuarine community based on studies of Lake Pontchartrain, Louisiana, *Ecology*, **42**:553–568.

DARWIN, C.

1881 *The Formation of Vegetable Mould Through the Action of Worms, with Observations on Their Habits*, Murray, London.

DASMANN, R. F.

1964 *Wildlife Biology*, Wiley, New York.

DAUBENMIRE, R. F.

1959 *Plants and Environment: A Textbook of Plant Autecology*, Wiley, New York.

1966 Vegetation: identification of typal community, *Science*, **151**:291–298.

1968a Soil moisture in relation to vegetation distribution in the mountains of northern Idaho, *Ecology*, **49**:431–438.

1968b Ecology of fire in grasslands, *Adv. in Ecological Res.*, **5**:208–266.

1968c *Plant Communities: A Textbook of Plant Synecology*, Harper & Row, New York.

DAVIDSON, J., AND H. G. ANDREWARTHA

1948a Annual trends in a natural population of *Thrips imaginis* (Thysanoptera), *J. Anim. Ecol.*, 17:193–199.

1948b The influence of rainfall, evaporation and atmospheric temperature on fluctuations in size of a natural population of *Thrips imaginis* (Thysanoptera), *J. Anim. Ecol.*, 17:200–202.

DAVIS, D. E.

1960 A chart for estimation of life expectancy, *J. Wildlife Manage.*, 24:344–348.

DAVIS, J. H.

1940 The ecology and geologic role of mangroves in Florida, *Papers Tortugas Lab.*, 32:302–412.

DAVIS, T. A. W., AND P. W. RICHARDS

1933 The vegetation of Moraballi Creek, British Guiana: An ecological study of a limited area of tropical rain forest, *Ecology*, 19:503–514.

DAWSON, P. S.

1968 Xenocide, suicide and cannabalism in flour beetles, *Am. Naturalist*, 102:97–105.

DAWSON, W. R., AND G. A. BARTHOLOMEW

1968 Temperature regulation and water economy of desert birds, in G. W. Brown (ed.), *Desert Biology*, Academic, New York.

DAY, R. K., AND D. DENUYL

1932 The natural regeneration of farm woods following the exclusion of livestock, *Purdue Univ. Agr. Sta., Res. Bull. No. 368*.

DECOURSEY, PATRICIA J.

1960a Phase control of activity in a rodent, *Cold Spring Harbor Symp. Quant. Biol.*, 25:49–54.

1960b Daily light sensitivity rhythm in a rodent, *Science*, 131:33–35.

1961 Effect of light on the circadian activity rhythm of the flying squirrel, *Glaucomys volans*, *Z. Vergleich. Physiol.*, 44:331–354.

DEEVEY, E. S.

1947 Life tables for natural populations of animals, *Quart. Rev. Biol.*, 22:283–314.

DELACOUR, J., AND C. VAURIE

1950 Les mesanges charbonnières (révision de l'espece *Parus major*), *L'Oiseau*, 20:91–121.

DELONG, K. T.

1966 Population ecology of feral house mice: interference by *Microtus*, *Ecology*, 47:481–484.

DENYES, H. A., AND J. M. JOSEPH

1956 Relationships between temperature and blood oxygen in the large-mouth bass, *J. Wildlife Manage.*, 20:56–64.

DETHIER, V. G.

1970 Chemical interactions between plants and insects, in E. Sondheimer and J. B. Simone (eds.), *Chemical Ecology*, Academic, New York, pp. 83–102.

DEVOS, A.

1969 Ecological conditions affecting the production of wild herbivore mammals on grasslands, *Adv. in Ecol. Res.*, 6:137–183.

DEWIT, C. T.

1968 Photosynthesis: its relationship to overpopulation, in A. San Pietro et al. (eds.), *Harvesting the Sun*, Academic, New York, pp. 315–320.

DHYSTERHAUS, E. J., AND E. M. SCHMUTZ

1947 Natural mulches or "litter of grasslands"; with kinds and amounts on a southern prairie, *Ecology*, 28:163–179.

DICE, L. R.

1943 *The Biotic Provinces of North America*, University of Michigan Press, Ann Arbor.

1947 Effectiveness of selection by owls on deer-mice (*Peromyscus maniculatus*) which contrast in color with their background, *Contrib. Lab. Vert. Biol.*, University of Michigan, 34:1–20.

DIETZ, D. R.; R. H. UDALL; AND L. E. YEAGER

1962 Chemical composition and digestibility by mule deer of selected forage species, Cache le Poudre Range, Colorado, *Colorado Dept. Game and Fish Tech. Bull. 14*.

DILGER, W. C.

1956a Adaptative modifications and ecological isolating mechanisms in the thrush genera *Catharus* and *Hylocichla*, *Wilson Bull.*, 68:171–199.

1956b Hostile behavior and reproductive isolating mechanisms in the avian genera *Catharus* and *Hylocichla*, *Auk*, 73:313–353.

1960 Agonistic and social behavior of captive redpolls, *Wilson Bull.*, 72:115–132.

DILGER, W. C., AND P. JOHNSGARD

1959 Comments on species recognition with special reference to the wood duck and mandarin duck, *Wilson Bull.*, 71:46–53.

DIAMOND, J. B., AND J. A. SHERBURNE

1969 Persistence of DDT in wild populations of small mammals, *Nature*, 221:481–487.

D'ITRI, F.

1971 Mercury accumulation in the aquatic environment, *Proc. 162d American Chemical Society Meeting*, Washington, D.C.

DIX, R. L.

1960 The effects of burning on the mulch structure and species composition of grassland in western North Dakota, *Ecology*, 41:49–56.

DIX, R. L., AND F. E. SMEINS

1967 The prairie, meadow, and marsh vegetation of Nelson County, North Dakota, *Can. J. Bot.*, 45:21–58.

DIXON, A. F. G.

1970 Quality and availability of food for a sycamore aphid population, in A. Watson (ed.), *Animal Populations in Relation to Their Food Resources*, Blackwell, Oxford.

DIXON, K. L.

1949 Behavior of the plain titmouse, *Condor*, 51:110–136.

DOBZHANSKY, T.

1947 Effectiveness of intraspecific and interspecific matings in *Drosophila pseudoobscura* and *D. persimilis*, *Am. Naturalist*, 81:66–73.

1950 Evolution in the tropics, *Am. Sci.*, 38:209–221.

1951 *Genetics and the Origin of Species*, 3d ed., Columbia University Press, New York.

DOCHINGER, L. S.

1968 The impact of air pollution on eastern white pine: the chlorotic dwarf disease, *J. Air Pollution Control Assoc.*, 18:814–816.

DODD, A. P.

1940 The biological campaign against prickly pear, *Comm. Prickly Pear Board*, Brisbane, Queensland, Australia.

1959 The biological control of prickly pear in Australia, *Biogeography and Ecology in Australia, Monog. Biologicae*, 8:565–577.

DODDS, D. G.

1960 Food competition and range relationships of moose and snowshoe hare in Newfoundland, *J. Wildlife Manage.*, 24:52–60.

DORN, H. F.

1962 World population growth: an international dilemma, *Science*, 125:280–283.

DOTY, M. S.

1957 Rocky intertidal surfaces, in J. Hedgpeth, *Treatise on Marine Ecology and Paleoecology: I, Ecology*, pp. 535–585.

DOWDY, W. W.

1944 The influence of temperature on vertical migration of invertebrates inhabiting different soil types, *Ecology*, 25:449–460.

1951 Further ecological studies on stratification of the arthropods, *Ecology*, 32:37–52.

DREW, J. V.; J. C. TEDROW; R. E. SHANKS; AND J. J. KORANDA

1958 Rate and depth of thaw in arctic soils, *Trans. Am. Geophys. Union*, 39:697–701.

DRIFT, J. VAN DER

1951 Analysis of the animal community in a beech forest floor, *Tijdschrift voor Entomologie*, 94:1–168.

1971 Production and decomposition of organic matter in an oakwood in the Netherlands, in *Productivity of Forest Ecosystems*, UNESCO, Paris, pp. 631–634.

DRURY, W. H., JR.

1961 The breeding biology of shorebirds on Bylot Island, Northwest Territories, Canada, *Auk*, 78:176–219.

DRURY, W. H., JR.; I. C. T. NISBET; AND R. E. RICHARDSON

1961 The migration of "angels," *Nat. Hist.*, 70(8):11–16.

DUCE, R. A., ET AL.

1972 Enrichment of heavy metals and organic compounds in the surface microlayer of Narragansett Bay, Rhode Island, *Science*, 176:161–163.

DUDDINGTON, C. L.

1955 Interrelations between soil microflora and soil nematodes in Kevan, *Soil Zool.*, Butterworth, Washington, D.C., pp. 284–301.

DUGDALE, R. C., AND J. J. GOERING

1967 Uptake of new and regenerated forms of nitrogen in primary productivity, *Limnol. and Oceanogr.*, 12:196–206.

DUNAWAY, P. B., ET AL.

1969 Radiation effects in the Soricidae, Cricetidae, and Muridae, in Nelson and Evans (eds.), *Symposium in Radioecology*, pp. 173–184.

DUVIGNEAUD, P., AND S. DENAEYER-DESMET

1967 Biomass, productivity and mineral cycling in deciduous forests in Belgium, in *Symposium on Primary Productivity and Mineral Cycling in Natural Ecosystems*, University of Maine Press, Orono, pp. 167–186.

1970 Biological cycling of minerals in temperate deciduous forests, in D. Reichle (ed.), *Analysis of Temperate Forest Ecosystems*, Springer-Verlag, New York, pp. 199–225.

DYBAS, H. S., AND M. LLOYD

1962 Isolation by habitat in two-synchronized species of periodical cicadas (Homoptera, Cicadidae, Magicada), *Ecology*, 43:444–459.

EASTWOOD, E.

1971 *Radar Ornithology*, Methuen, London.

EATON, T. H., JR., AND R. F. CHANDLER, JR.

1942 The fauna of the forest humus layers in New York, *Cornell Univ. Agr. Exp. Sta. Mem. No. 247.*

EDEBURN, R.

1947 A study of the breeding distribution of birds in a typical upland area, *Proc. W. Va. Acad. Sci. V 18, W. Va. Univ. Bull. Ser. 47*, No. 9-I pp. 34–47.

EDMONDSON, W. T.

1956 The relation of photosynthesis by phyloplankton to light in lakes, *Ecology*, 37:161–174.

1969 Eutrophication in North America, in *Eutrophication: Causes, Consequences, Correctives*, National Academy of Sciences, Washington, D.C., pp. 124–149.

1970 Phosphorus, nitrogen, and algae in Lake Washington after diversion of sewage, *Science*, 169:690–691.

EDWARDS, C. A.

1969 Effects of gamma radiation on populations of soil invertebrates, in D. J. Nelson and F. C. Evans (eds.), *Symposium on Radioecology: A.E.C. Report CONF-670503*, National Clearinghouse for Technical Information, Springfield, Va., pp. 68–77.

EDWARDS, C. A., AND G. W. HEATH

1963 The role of soil animals in breakdown of leaf material, in J. Doeksen and J. van der Drift (eds.), *Soil Organisms*, North Holland Publishing Co., Amsterdam, pp. 76–84.

EGLER, F. E.

1953 Vegetation management for rights-of-way and roadsides, *Smithsonian Inst. Ann. Rep. 1953*, pp. 299–322.

EHRET, C. F., AND T. TRUCCO

1967 Molecular models for the circadian clock: I, The chronon concept, *J. Theoret. Biol.*, 15:240–262.

EHRET, C. F., AND J. J. WILLE

1970 The photobiology of circadian rhythms in protozoa and other eukaryotic microorganisms, in P. Holldal (ed.), *Photobiology of Microorganisms*, Wiley, New York.

EHRLICH, P. R., AND L. C. BIRCH

1967 The "balance of nature" and "population control," *Am. Naturalist*, 101:97–107.

EHRLICH, P. R.; D. E. BREEDLOVE; P. F. BRUSSARD; AND M. A. SHARP

1972 Weather and the "regulation" of subalpine populations, *Ecology*, 53:243–247.

EHRLICH, P. R., AND P. H. RAVEN

1964 Butterflies and plants: a study in coevolution, *Evolution*, 18:586–608.

EISENBERG, R. M.

1970 The role of food in the regulation of the pond snail, *Lymnaea elodes*, *Ecology*, 51:680–684.

EISNER, E.

1960 The relationship of hormones to the reproductive behavior of birds, referring especially to parental behavior: a review, *Anim. Behav.*, 8:155–179.

EISNER, T.

1970 Chemical defense against predation in arthropods, in E. Sondheimer and J. Simeone (eds.), *Chemical Ecology*, Academic, New York, pp. 157–217.

EISNER, T., AND J. MEINWALD

1966 Defensive secretions of arthropods, *Science*, 153:1341–1350.

ELLIOT, J. M.

1965 Daily fluctuations of drift invertebrates in a Dartmoor stream, *Nature*, 285:1127–1129.

1967 Invertebrate drift in a Dartmoor stream, *Arch. Hydrobiol.*, 63:202–237.

ELLIS, M. M.

1936 Erosion silt as a factor in aquatic environments, *Ecology*, 17:29–42.

ELLISON, L.

1954 Subalpine vegetation of the Wasatch Plateau, Utah, *Ecol. Monographs*, 24:89–184.

1960 Influence of grazing on plant succession on rangelands, *Botan. Rev.*, 26:1–78.

ELSTER, H. J.

1965 Absolute and relative assimilation rates in relation to phytoplankton populations, in C. R. Goldman (ed.), *Primary Productivity in Aquatic Environments*, Mem. 1st Ital. Idrobiol. 18 Suppl., University of California Press, Berkeley, pp. 79–103.

ELTON, C. S.

1927 *Animal Ecology*, Sidgwick & Jackson, London.

1958 *The Ecology of Invasions by Animals and Plants*, Methuen, London.

EMANUELSSON, A.; E. ERIKSSON; AND H. EGNER

1954 Composition of atmospheric precipitation in Sweden, *Tellus*, 3:261–267.

EMLEN, J. M.

1973 *Ecology: An Evolutionary Approach*, Addison-Wesley, Reading, Mass.

EMLEN, J. T., JR.

1940 Sex and age ratios in the survival of California quail, *J. Wildlife Manage.*, 4:92–99.

1952 Social behavior in nesting cliff swallows, *Condor*, 54:177–199.

EMLEN, S. T.

1967 Migratory orientation in the indigo bunting *Passerina cyanea*, *Auk*, 84:309–342.

1972 Migration: orientation and navigation, in D. S. Farner and J. R. King (eds.), *Avian Biology*, Vol. 3, Academic, New York.

ENGELMANN, M. D.

1961 The role of soil arthropods in the energetics of an old field community, *Ecol. Monographs*, 31:221–238.

ENGLE, L. G.

1960 Yellow poplar seedfall pattern, *Central States For. Expt. Stat. Note 143*.

ENRIGHT, J. T.

1963a The tidal rhythm of activity of a sand beach amphipod, *Z. Vergleich Physiol.*, 46:276–313.

1963b Endogenous tidal and lunar rhythms, *Proc. XVI Inter. Congress Zool.*, pp. 355–359.

1965 Entrainment of a tidal rhythm, *Science*, 147:864–867.

1966 Influences of seasonal factors on the activity onset of the house finch, *Ecology*, 47:662–666.

EPLING, C.

1947 Natural hybridization of *Salvia apicna* and *S. mellifera*, *Evolution*, 1:69–78.

ERIKISSON, E.

1952 Composition of atmospheric precipitation: I, Nitrogen compounds, *Tellus*, 4:214–232.

1963 The yearly circulation of sulfur in nature, *J. Geophys. Res.*, 68:4001–4008.

ERRINGTON, P. L.

1937a Habitat requirements of stream-dwelling muskrats, *Trans. North Am. Wildlife Conf.*, 2:411–416.

1937b Drowning as a cause of mortality in muskrats, *J. Mammal.*, 18:497–500.

1939 Reactions of muskrat populations to drought, *Ecology*, 20:168–186.

1940 Natural restocking of muskrat-vacant habitats, *J. Wildlife Manage.*, 4:173–185.

1943 An analysis of mink predation upon muskrats in the north-central United States, *Iowa Agr. Exp. Sta. Res. Bull.*, 320:797–924.

1946 Predation and vertebrate populations, *Quart. Rev. Biol.*, 21:144–177, 221–245.

1951 Concerning fluctuations in populations of the prolific and widely distributed muskrat, *Am. Naturalist*, 85:273–292.

1957 Of population cycles and unknowns, *Cold Spring Harbor Symp. Quant. Biol.*, 22:287–300.

1963 *Muskrat Populations*, Iowa State University Press, Ames.

ESCHMEYER, P. H., AND W. R. CROWE

1955 The movement and recovery of tagged walleyes in Michigan, 1929–1953, *Misc. Publ. Inst. Fishery Research*, University of Michigan, 3:1–32.

ESSER, A. H.

1965 Territoriality of patients on research wards, in J. Wortis (ed.), *Recent Advances in Biological Psychiatry*, vol. 8, Plenum Press, New York.

EUSTIS, A. B., AND R. H. HILLER

1954 Stream sediment removal by controlled reservoir releases, *Progressive Fish Culturalist*, 16:30–35.

EVANS, A. C.

1948 Studies on the relationships between earthworms and soil fertility: II, Some effects of earthworms on soil structure, *Ann. Appl. Biol.*, 35:1–13.

EVANS, A. C., AND W. J. MCLAREN GUILD

1948 Studies on the relationships between earthworms and soil fertility: V, Field populations, *Ann. Appl. Biol.*, 35:485–493.

EVANS, F. C., AND S. A. CAIN

1952 Preliminary studies on the vegetation of an old-field community in southeastern Michigan, *Contrib. Lab. Vert. Biol.*, University of Michigan, 51:1–17.

EVANS, H. E.

1957 *Studies on the Comparative Ethology of Digger Wasps of the Genus Bembex*, Comstock, Ithaca, New York.

1962 A review of the nesting behavior of the digger wasps of the genus *Aphilanthops*, with special attention to the mechanics of prey carriage, *Behaviour*, 19:239–260.

1963 Predatory wasps, *Sci. Am.*, 208(4):144–154.

EWER, R. F.

1968 *Ethology of Mammals*, Plenum, New York.

EYRE, S. R.

1963 *Vegetation and Soils: A World Picture*, Aldine, Chicago.

FARNER, D. S.

1955 The annual stimulus for migration: experimental and physiologic aspects, in A. Wolfson (ed.), *Recent Studies on Avian Biology*, pp. 198–237.

1959 Photoperiodic control of animal gonodal cycles in birds, in Withrow (ed.), *Photoperiodism and Related Phenomena*, pp. 717–758.

1964a The photoperiodic control of reproductive cycles in birds, *Am. Scientist*, 52:137–156.

1964b Time measurement in vertebrate photoperiodism, *Am. Naturalist*, 98:375–386.

FELTON, P. M., AND H. W. LULL

1963 Suburban hydrology can improve watershed conditions, *Public Works*, 94:93–94.

FENNER, F.

1953 Host-parasite relationships in myxomatosis of the Australian wild rabbit, *Cold Spring Harbor Symp. Quant. Biol.*, 18:291–294.

FICHTER, E.

1954 An ecological study of invertibrates of grassland and deciduous shrub savanna in eastern Nebraska, *Am. Midland Naturalist*, 51:321–439.

FICKEN, M. S.

1963 Courtship of the American redstart, *Auk*, 80:307–317.

FICKEN, M. S., AND R. W. FICKEN

1962 The comparative ethology of the wood warblers: a review, *Living Bird*, 1:103–122.

FICKEN, R. W.

1963 Courtship and behavior of the common grackle *Quiscalus quiscula*, *Auk*, 80:52–72.

FISCHER, A. G.

1960 Latitudinal variation in organic diversity, *Evolution*, 14:64–81.

FISCHER, K.

1960 Experimentelle Beinflussung der inneren Uhr bei der Sonnenkompassorientierung und der Laufaktivitat von *Lacerta viridis Laur*, *Naturwissenschaften*, 47:287–288.

FISCHER, K. C.

1958 An approach to organ and cellular physiology of adaptation to adaptation to temperature in fish and small mammals, in Prosser (ed.), *Physiological Adaptations*, American Physiological Society, Washington, D.C.

FISCHER, R. A.

1929 *The Genetical Theory of Natural Selection*, 2d rev. ed., Dover, New York.

FITCH, H. S., AND J. R. BENTLEY

1949 Use of California annual plant forage by range rodents, *Ecology*, 30:306–321.

FLOOK, D. R.

1970 Causes and implications of an observed sex differential in the survival of wapitie, *Can. Wildlife Serv. Rept. Ser. No. 11*

FLORIN, J.

1965 The advance of frontier settlement in Pennsylvania 1638–1850, M.A. dissertation, Pennsylvania State University.

FOGG, G. E.

1968 *Photosynthesis*, American Elsevier, New York.

FOLK, G. E., JR.

1966 *Introduction to Environmental Physiology*, Lea & Febiger, Philadelphia.

FONS, W. L.

1940 Influence of forest cover on wind velocity, *J. Forestry*, 38:481–486.

FORSSLUND, K. H.

1947 Nagot om insamlingsmetodiken vid markfaunaundersokingar, *Medd Skogsforsoksanst*, Stockholm, 37:1–22.

FORTESQUE, J. A. C., AND G. C. MARTIN
1970 Micronutrients: forest ecology and systems analysis, in D. E. Richle (ed.), *Analysis of Temperate Forest Ecosystems*, Springer-Verlag, New York, pp. 173–198.

FOWELLS, H. A.
1948 The temperature profile in a forest, *J. Forestry*, 46:897–899.

FOX, M. W.
1969a Ontogeny of prey-killing behavior in Canidae, *Behaviour*, 35:259–272.

1969b The anatomy of aggression and its ritualization in Canidae: A developmental and comparative study, *Behaviour*, 35:242–258.

1970 A comparative study of the development of facial expressions in canids wolf, coyote, and foxes, *Behaviour*, 36:4–73.

FRANCIS, W. J.
1970 The influence of weather on population fluctuations in California quail, *J. Wildlife Manage.*, 34:249–266.

FRANK, F.
1957 The causality of microtine cycles in Germany, *J. Wildlife Manage.*, 21:113–121.

FRANK, P. W.
1952 A laboratory study of intraspecies and interspecies competition in *Daphnia puliceria* and *Simocephalus vetulus*, *Physiol. Zool.*, 25:178–204.

FRANZ, H.
1950 *Bodenzoologie als Grundlage der Bodenpflege*, Akademie-Verlag, Berlin, pp. 316ff.

FREDERIKSEN, H.
1969 Feedbacks in economic and demographic transition, *Science*, 166:837–847.

FRENCH, C. E.; L. C. MCEWEN; N. C. MAGRUDER; R. H. INGRAM; AND R. W. SWIFT
1955 Nutritional requirements of white-tailed deer for growth and antler development, *Penn. State Univ. Agr. Exp. Sta. Bull. No. 600.*

FRENCH, N. R.; B. G. MAZA; AND H. W. KAAZ
1969 Mortality rates in irradiated rodent populations, in D. J. Nelson and F. E. Evans (eds.), *Symposium on Radioecology*, 670–503, National Technical Information Service, Springfield, Va., pp. 46–60.

FRENZEL, G.
1936 *Untersuchungen uber die Tierwelt des Weisenbodens*, Gustav Fischer, Jena, E. Germany.

FRETWELL, S. D., AND H. L. LUCAS
1969 On territorial behavior and other factors influencing habitat distribution in birds, *Acta Biotheoretica*, 19:16–36.

FRINK, C. R.
1969 Water pollution potential estimated from farm nutrient budgets, *Agron. J.*, 61:550–553.

1970 The nitrogen cycle of a dairy farm, in *Relationship of Agriculture to Soil and Water Pollution*, Cornell University, Ithaca, pp. 127–133.

1971 Plant nutrients and water quality, *Agr. Sci. Rev.*, 9(2):11–25.

FRISCH, K. VON
1954 *The Dancing Bees*, Methuen, London.

FRY, F. E. J.
1947 Effects of the environment on animal activity, *Univ. Toronto Stud. Biol.*, 55:1–62.

GALLE, O. R.; W. R. GROVE; AND J. M. MCPHERSON
1972 Population density and pathology: What are the relations for man?, *Science*, 176:23–30.

GASHWILER, J. S.
1970a Plant and mammal changes on a clearcut in West Central Oregon, *Ecology*, 51:1018–1026.

1970b Further study of conifer seed survival in a western Oregon clearcut, *Ecology*, 51:849–854.

GATES, D.
1962 *Energy Exchange in the Biosphere*, Harper & Row, New York.

1965 Radiant energy: its receipt and disposal, *Meteorological Monographs*, 6:1–26.

1966 Spectral distribution of solar radiation at the earth's surface, *Science*, 151:523–528.

1968a Energy exchange and ecology, *Bioscience*, 18:90–95.

1968b Toward understanding ecosystems, *Adv. in Ecol. Research.*, 5:1–35.

GAUFIN, A. R.
1957 The use and value of aquatic insects as indicators of organic enrichment, *Biol. Problems in Water Pollution*, Robert A. Taft Sanitary Engineering Center, Cincinnati, Ohio, pp. 136–143.

1965 Environmental requirements for Plecoptera, in *Biological Problems in Water Pollution*, 3d Seminar, *Public Health Serv. Publ. No. 999-WP-25*, pp. 105–110.

GAUSE, G. F.
1934 *The Struggle for Existence*, Williams & Wilkins, Baltimore.

GEIS, A. D.; R. I. SMITH; AND J. P. ROGERS
1971 Black duck distribution, harvest characteristics, and survival, *U.S. Fish and Wildlife Service Spec. Sci. Rept., Wildlife No. 139.*

GEIST, V.
1963 On the behaviour of the North American moose (*Alces alces andersoni* Peterson, 1950) in British Columbia, *Behaviour*, 20:377–416.

1971 *Mountain Sheep: A Study in Behavior and Evolution*, University of Chicago Press, Chicago.

GERKING, S. D.
1959 The restricted movement of fish populations, *Biol. Rev.*, 34:221–242.

GHILAROV, M. S.
1970 Soil biocoenosis, in J. Phillipson (ed.), *Methods of Study in Soil Ecology*, UNESCO, Paris, pp. 67–77.

GIBSON, J. B., AND J. M. THODAY
1962 Effects of disruptive selection: VI, A second chromosome polymorphism, *Heredity*, 17:1–26.

GILBERTSON, C. B., ET AL.

1970 The effect of animal density and surface slope on the characteristics of runoff, solid waste, and nitrate movement on unpaved feedyards, *Nebraska Agri. Exp. Sta. Bull.*, 508.

GILMOUR, J. S. L.

1951 The development of taxonomic theory since 1851, *Nature*, 168:400–402.

GISBORNE, H. T.

1941 How the wind blows in the forest of northern Idaho, *Northern Rocky Mt. Forest Range Expt. Sta.*

GLEASON, H. A.

1926 The individualistic concept of the plant association, *Bull. Torrey Botan. Club*, 53:7–26.

GLOOSCHENKO, W.

1968 Thermal pollution, in *Hearings before the Subcommittee on Air and Water Pollution of the Committee on Public Works*, United States Senate, p. 752.

GOLDSMITH, J. R.

1969 Epidemiological bases for possible air quality criteria for lead, *Air Poll. Contr. Assoc. J.*, 19:714–719.

GOLDSMITH, J. R., AND A. C. HEXTER

1967 Respiratory exposure to lead: epidemiological and experimental dose–response relationship, *Science*, 158:132–134.

GOLDSMITH, J. R., AND S. A. LANDAW

1968 Carbon monoxide and human health, *Science*, 162:1352–1359.

GOLLEY, F. B.

1960 Energy dynamics of a food chain of an old-field community, *Ecol. Monographs*, 30:187–206.

GOMEZ-POMPA, A.; C. VASQUEZ-YANES; AND S. GUEVARA

1972 The tropical rainforest: a nonrenewable resource, *Science*, 177:762–765.

GOOD, E. E., AND C. A. DAMBACH

1943 Effect of land use practices on breeding bird populations in Ohio, *J. Wildlife Manage.*, 7:291–297.

GOODHART, C. B.

1962 Variation in a colony of the snail *Cepaea nemoralis* L., *J. Anim. Ecol.*, 31:207–237.

GOREAU, T. F.

1963 Calcium carbonate deposition by coralline algae and corals in relation to their roles as reef builders, *Ann. N.Y. Acad. Sci.*, 109:127–167.

GOTTLIEB, G.

1963 "Imprinting" in nature, *Science*, 139:497–498.

GOULD, E. M., JR.

1972 Toward a new forest decision policy, in R. D. Nyland (ed.), *A Perspective on Clearcutting in a Changing World*, Misc. Rept. No. 4, Applied Forestry Research Institute, Syracuse, N.Y., pp. 103–107.

GOULD, S. J., AND R. F. JOHNSTON

1972 Geographic variation, *Ann. Rev. Ecol. and Systematics*, 3:457–498.

GRAHAM, S. A.

1958 Results of deer exclosure experiments in the Ottawa National Forest, *Trans. North Am. Wildlife Conf.*, 23:478–490.

GRANT, V.

1963 *The Origins of Adaptations*, Columbia University Press, New York.

GREENBANK, D. O.

1956 The role of climate and dispersal in the initiation of outbreaks of the spruce budworm in New Brunswick: I, The role of climate, *Can. J. Zool.*, 34:453–476.

GREENBERG, B.

1947 Some relations between territory, social hierarchy and leadership in the green sunfish (*Lepomis cyanellus*), *Physiol. Zool.*, 20:267–299.

1963 Parental behaviour and imprinting in cichlid fishes, *Behaviour*, 21:127–144.

GREIG-SMITH, P.

1964 *Quantitative Plant Ecology*, 2d ed., Butterworth, Washington, D.C.

GRETSCHY, G.

1952 Die Suksession der Bodentiere auf Fitchlenschlagen, *Veroffent lichungen der Bundes anstalt fur Alpine Landwirtsclaft in Admont.*, 6:25–85.

GRIFFIN, D. R.; F. A. WEBSTER; AND C. R. MICHAEL

1960 The echolocation of flying insects by bats, *Anim. Behav.*, 8:141–154.

GRIGGS, R. F.

1946 The timberlines of northern America and their interpretation, *Ecology*, 27:275–289.

GRINNELL, J.

1924 Geography and evolution, *Ecology*, 5:225–229.

GUHL, A. M., AND W. C. ALLEE

1944 Some measurable effects of social organization in flocks of hens, *Physiol. Zool.*, 17:320–347.

GULLION, G. W.

1953 Territorial behavior of the American coot, *Condor*, 55:169–186.

1970 Factors influencing ruffed grouse populations, *Trans. 35th N. Am. Wildlife and Natural Resource Conf.*

GUNTER, G.

1961 Some relations of estuarine organisms to salinity, *Limnol. Oceanogr.*, 6:182–190.

GYSEL, L. W.

1960 An ecological study of the winter deer range of elk and mule deer in the Rocky Mountain National Park, *J. Forestry*, 58:696–703.

HAARLOV, N.

1960 Microarthropods from Danish soils, *Oikos*, Supplement No. 3, pp. 1–176.

HAARTMAN, L. VON

1956 Territory in the pied flycatcher, *Muscicapa hypoleuca*, *Ibis*, 98:460–475.

1957 Adaptation in hole nesting birds, *Evolution*, 11:339–347.

HACSKAYLO, E.

1971 Metabolite exchanges in ectomycorrhizae, in

E. Hacskaylo (ed.), *Mycorrhizae, USDA For. Serv. Misc. Publ. No. 1189.*

HADLEY, E. B., AND B. J. KIECKHEFER
1963 Productivity of two prairie grasses in relation to fire frequency, *Ecology*, 44:389–395.

HAGAN, D. C.
1960 Interrelationships of logging, birds, and timber regeneration in Douglas fir region of northwestern California, *Ecology*, 41:116–125.

HAHN, H. C., AND W. P. TAYLOR
1950 Deer movements in the Edwards Plateau, *Texas Game and Fish*, 8:4–9, 31.

HAIRSTON, N. G.
1949 The local distribution and ecology of the plethodontid salamanders of the southern Appalachians, *Ecol. Monographs*, 19:47–73.
1969 On the relative abundance of species, *Ecology*, 50:1091–1094.

HAIRSTON, N. G., ET AL.
1968 The relationship between species diversity and stability: an experimental approach with protozoa and bacteria, *Ecology*, 49:1091–1101.

HAIRSTON, N. G., AND C. H. POPE
1948 Geographic variation and speciation in Appalachian salamanders (*Plethodon jordani* group), *Evolution*, 2:266–278.

HAIRSTON, N. G.; F. E. SMITH; AND L. B. SLOBODKIN
1960 Community structure, population control and competition, *Am. Naturalist*, 94:421–425.

HALE, N.
1971 *Biology of Lichens*, Edward Arnold, London.

HAMERSTROM, F. N., JR.
1939 A study of Wisconsin prairie chicken and sharp-tailed grouse, *Wilson Bull.*, 51:105–120.

HAMERSTROM, F. N., JR.; O. E. MATTSON; AND F. HAMERSTROM
1957 A guide to prairie chicken management, *Tech. Wildlife Bull. No. 15*, Wisconsin Conserv. Dept., Madison.

HAMILTON, W. J., III
1959 Aggressive behavior in migrant pectoral sandpipers, *Condor*, 61:161–179.
1962 Bobolink migratory pathways and their experimental analysis under night skies, *Auk*, 79:208–233.

HAMMEL, H. T.
1956 Infrared emissivities of some arctic fauna, *J. Mammal.*, 37:375–378.

HAMMOND, A. L.
1972 Chemical pollution: polychlorinated biphenyls, *Science*, 175:155–156.

HAMMOND, J.
1953 Periodicity in animals: the role of darkness, *Science*, 177:389–390.

HAMNER, W. M.
1963 Diurnal rhythm and photoperiodism in testicular recrudescence of the house finch, *Science*, 142:1294–1295.
1968 The photorefractory period of the house finch, *Ecology*, 49:211–227.

HANDLEY, C. O., JR.
1969 Fire and mammals, *Proc. 9th Tall Timbers Fire Ecology Conf.*, pp. 151–159.

HANSON, H. C.
1953 Vegetation types in northwestern Alaska and comparisons with communities in other arctic regions, *Ecology*, 34:111–140.
1958 Principles concerned in the formation and classification of communities, *Botan. Rev.*, 24:65–125.

HANSON, H. C., AND E. D. CHURCHILL
1961 *The Plant Community*, Reinhold, New York.

HANSON, W. C.
1971 Seasonal patterns in native residents of three contrasting Alaskan villages, *Health Physics*, 20:585–591.

HARDIN, G.
1960 The competitive exclusion principle, *Science*, 131:1292–1297.

HARLAN, J. R.
1972 Genetics of disaster, *J. Environ. Quality*, 1:212–215.

HARLEY, J. L.
1959 *The Biology of Mycorrhiza*, Plant Science Monographs, Leonard Hill, London.

HARPER, J. A.
1964 Calcium in grit consumed by pheasants in east-central Illinois, *J. Wildlife Manage.*, 28:265–270.

HARPER, J. A., AND R. F. LABISKY
1964 The influence of calcium in the distribution of pheasants in Illinois, *J. Wildlife Manage.*, 28:722–731.

HARPER, J. L.
1967 A Darwinian approach to plant ecology, *J. Ecology*, 55:247–270.
1969 The role of predation in vegetational diversity, *Brookhaven Symposia in Biology*, No. 22, pp. 48–62.

HARRINGTON, R. W., JR.
1959 Photoperiodism in fishes in relation to the annual sexual cycle, in Witherow (ed.), *Photoperiodism and Related Phenomena*, pp. 651–667.

HARRIS, D. R.
1967 New light on plant domestication and the origins of agriculture: a review, *Geog. Review*, 57:90–107.
1969 Agricultural systems, ecosystems, and the origins of agriculture, in P. Ucko and G. Dimbleby (eds.), *The Domestication and Exploitation of Plants and Animals*, Aldine, Chicago, pp. 5–15.

HARRIS, T. M.
1958 Forest fire in the Mesozic, *J. Ecology*, 46:447–453.

HARRIS, V. T.
1952 An experimental study of habitat selection by prairie and forest races of the deermouse, *Peromyscus maniculatus*, *Contrib. Lab. Vert. Zool.*, University of Michigan, 56:1–53.

HARRISON, J. L.

1962 Distribution of feeding habits among animals in a tropical rain forest, *J. Anim. Ecol.*, **31**:53–63.

HART, J. S.

1951 Photoperiodicity in the female ferret, *J. Expt. Biol.*, **28**:1–12.

HARTESVELDT, R. J., AND H. T. HARVEY

1967 The fire ecology of sequoia regeneration, *Proc. California Tall Timbers Fire Ecology Conf.*, pp. 65–77.

HARTLEY, P. H. T.

1950 An experimental analysis of interspecific recognition, *Symp. Soc. Expt. Biol.*, **4**:313–336.

HARTMAN, R. T., AND C. L. HIMES

1961 Phytoplankton from Pymatuning Reservoir in downstream areas of the Shenango River, *Ecology*, **42**:180–183.

HARTMAN, W. C.

1957 Finger Lakes rainbows: Part III, A chronicle of their progress from egg to adult, N.Y. *State Conserv.*, **11**(6):20–22, 34.

HARTMAN, W. L.

1972 Lake Erie: effects of exploitation, introductions, and eutrophication on the salmonid community, *J. Fish Res. Bd. Can.*, **29**:899–912.

HARVEY, H. W.

1950 On the production of living matter in the sea off Plymouth, *J. Marine Biol. Assoc. U.K.*, n.s., **29**:97–137.

HASLER, A. D.

1954 Odour perception and orientation in fishes, *J. Fisheries Res. Board Can.*, **11**:107–129.

1960 Guideposts of migrating fishes, *Science*, **132**:785–792.

1969 Cultural eutrophication is reversible, *Bioscience*, **19**:425–431.

HASLER, A. D., AND H. O. SCHWASSMAN

1960 Sun orientation of fish at different latitudes, *Cold Spring Harbor Symp. Quant. Biol.*, **25**:429–441.

HASSELL, M. P.

1966 Evaluation of parasite or predator response, *J. Anim. Ecol.*, **35**:65–75.

HASTINGS, J. W.

1959 Unicellular clocks, *Ann Rev. Microbiol.*, **13**:297–312.

1970 Cellular-biochemical clock hypothesis, in F. Brown et al., *The Biological Clock*, Academic, New York, pp. 61ff.

HAUENSCHILD, C.

1960 Lunar periodicity, *Cold Spring Harbor Symp. Quant. Biol.*, **25**:491–497.

HAUSER, D. C.

1957 Some observations on sun-bathing in birds, *Wilson Bull.*, **69**:78–90.

HAYNE, D. W., AND R. C. BALL

1956 Benthic productivity as influenced by fish predation, *Limnol. Oceanogr.*, **1**:162–175.

HAYNE, D. W., AND D. Q. THOMPSON

1965 Methods for estimating microtine abundance, *Trans. 30th North Am. Wildlife Conf.*, pp. 393–400.

HAYS, H., AND R. W. RISEBROUGH

1972 Pollutant concentrations in abnormal terns from Long Island Sound, *Auk*, **89**:19–35.

HEADY, H. F.

1967 *Practices in Range Forage Production*, University of Queensland Press, Brisbane, Australia.

HEATH, J. E.

1965 Temperature regulation and diurnal activity in horned lizards, *University of California Publication on Zoology*, **64**:97–136.

HEATWOLE, H., AND K. LIM

1961 Relation of substrate moisture to absorption and loss of water by the salamander, *Plethodon cinereus*, *Ecology*, **42**:814–819.

HEDGPETH, J. W. (ED.)

1957a Treatise in marine ecology and paleocology: I, Ecology, *Memoir 67*, Geological Soc. Am.

1957b Sandy beaches, in Hedgpeth (ed.), *Treatise in Marine Ecology and Paleoecology: I*, pp. 587–608.

HEILMAN, P. E.

1966 Change in distribution and availability of nitrogen with forest succession on north slopes in interior Alaska, *Ecology*, **47**:825–831.

HEINROTH, O.

1911 Beitrange zur Biologie, namentlich Ethologie und Psychologie der Anatiden, *Verhl. 5 Int. Orn. Kongr.*, pp. 589–702.

HEINSELMAN, M. L.

1963 Forest sites, bog processes, and peatland types in the glacial Lake Agassiz region, Minnesota, *Ecol. Monographs*, **33**:327–374.

1971 The natural role of fire in northern coniferous forests, in *Fire in the Northern Environment: A Symposium*, Pacific Northwest Forest and Range Expt. Station, Portland, Oregon, pp. 61–72.

HELBAEK, H.

1959 Domestication of food plants in the Old World, *Science*, **130**:365–375.

HELLER, H. C.

1971 Altitudinal zonation of chipmunks (*Eutamias*): interspecific aggression, *Ecology*, **52**:312–319.

HELLER, H. C., AND D. GATES

1971 Altitudinal zonation of chipmunks (*Eutamias*): energy budgets, *Ecology*, **52**:424–433.

HELLMERS, H., AND J. BONNER

1959 Photosynthetic limits of forest tree yields, *Proc. Soc. Am. Foresters*, pp. 32–35.

HENDERSON, NANCY E.

1963 Influence of light and temperature on the reproductive cycle of the eastern brook trout *Salvelinus fontinalis*, *J. Fisheries Res. Board Can.*, **20**:859–897.

Bibliography

HENNING, W.

1966 *Phylogenetic Systematics*, University of Illinois Press, Urbana.

HENRIKSSON, E.

1971 Nitrogen fixation by lichens, *Oikos*, **22**:119–121.

HENRY, S. M. (ED.)

1966 *Symbiosis*, Academic Press, New York.

HENSLEY, M. M., AND J. B. COPE

1951 Further data on removal and repopulation of the breeding birds in a spruce-fir forest community, *Auk*, 68:483–493.

HERREID, C. F., AND S. KINNEY

1967 Temperature and development of wood frog *Rana sylvalica* in Alaska, *Ecology*, 48:579–590.

HESS, E. H.

1959 Imprinting, *Science*, 130:133–141.

1964 Imprinting in birds, *Science*, 146:1129–1139.

1971 The imprinting process, in *Topics in the Study of Life*, BIOS: *The Bio Source Book*, Harper & Row, New York, pp. 314–320.

HICKEY, J. J.

1952 *Survival Studies of Banded Birds*, U.S. Fish and Wildlife Serv. Spec. Sci. Rep. 15.

HILDÉN, O.

1964 Ecology of duck populations in the island group of Valassarret, Gulf of Bothnia, *Ann. Zool. Fennici*, 1:153–279.

1965 Habitat selection in birds: A review, *Ann. Zool. Fennica*, 2:53–75.

HINDE, R. A.

1954 Factors governing the changes in strength of a partially inborn response as shown by the mobbing behaviour of the chaffinch (*Fringella coelebs*): I, The nature of the response and the examination of its course, *Proc. Roy. Soc.*, Ser. B., London, 142:306–331.

1955–1956 A comparative study of the courtship of certain finches (*Fringillidae*), *Ibis*, 97:706–745; 98:1–23.

1959 Behavior and speciation in birds and lower vertebrates, *Biol. Rev.*, 34:85–128.

1970 *Animal Behaviour: A Synthesis of Ethology and Comparative Psychology*, McGraw-Hill, New York.

HINDE, R. A.; W. H. THORPE; AND M. A. VINCE

1956 The following response of young coots and moorhens, *Behaviour*, 9:214–242.

HIRTH, H. F.

1959 Small mammals in old field succession, *Ecology*, 40:417–425.

HJORTH, I.

1970 Reproductive behaviour in Tetraonidae, *Viltrevy*, 7:184–596.

HOCHBAUM, H. A.

1955 *Travels and Traditions of Waterfowl*, University of Minnesota Press, Minneapolis.

HOCK, R. J.

1960 Seasonal variations in physiological functions of arctic ground squirrels and black bears, in Lyman and Dawe (eds.), *Mammalian Hibernation*, pp. 155–169.

HOESE, H. D.

1960 Biotic changes in a bay associated with the end of a drought, *Limnol. Oceanogr.*, 5:326–336.

HOFFMAN, K.

1959 Die aktivitalsperiodik von im 18-und-36 studen-tag eibruteten eidechsen, *Z. Vergleich. Physiol.*, 42:422–432.

1960 Versuche zur analyse der tagesperiodik: 1, Der einfluss der licktintensitat, *Z. Vergleich. Physiol.*, 43:544–566.

1963 Zur Beziehung zwischen Phasenlage und Spontanfrequenz bei der endogenen Tagesperiodik, *Z. Naturforsch.*, 18b:154–157.

1965 Clock-mechanisms in celestial orientation of animals, in J. Aschoff (ed.), *Circadian Clocks*, North Holland Publishing Co., Amsterdam.

HOLDEN, C.

1971 Fish flour: Protein supplement has yet to fulfill expectations, *Science*, 173:410–412.

HOLLING, C. C.

1959 The components of predation as revealed by a study of small mammal predation of the European pine sawfly, *Can. Entomologist*, 91:293–320.

1961 Principles of insect predation, *Ann. Rev. Entomol.*, 6:163–182.

1966 The functional response of invertebrate predators to prey density, *Mem. Entomol. Soc. Canada*, No. 48, pp. 1–86.

HOLLOWAY, C. W.

1970 Threatened vertebrates in northern circumpolar regions, in W. A. Fuller and P. G. Kevan (eds.), *Productivity and Conservation in Northern Circumpolar Lands*, IUCN Publ., n.s. 16:175–192.

HOLMGREN, R. C.

1956 Competition between annuals and young bitterbrush (*Purshia tridentata*) in Idaho, *Ecology*, 37:370–377.

HOOVER, E. E., AND H. E. HUBBARD

1937 Modifications of the sexual cycle in trout by control of light, *Copeia*, 4:206–210.

HOPE, J. G.

1943 An investigation of the litter fauna of two types of pine forest, *Bull. Wagner Free Inst. Sci.*, Philadelphia, 18:1–7.

HOPKINS, H. H.

1954 Effects of mulch upon certain factors of the grassland environment, *J. Range Manage.*, 7:255–258.

HOPKINS, S. H.

1958 The planktonic larvae of *Polydora websteri* Hartman (Annelida, Polychaeta) and their settling on oysters, *Bur. Marine Sci. of Gulf and Caribbean*, 8:268–277.

HORN, M. H.; J. M. TEAL; AND R. M. BACKUS

1970 Petroleum lumps on the surfaces of the sea, *Science*, 168:245–246.

HORNBECK, J. W.

1970 The radiant energy budget of clearcut and forested sites in West Virginia, *For. Sci.*, **16**:139–145.

HORNOCKER, M. G.

1969 Winter territoriality in mountain lions, *J. Wildlife Manage.*, **33**:457–464.

1970 An analysis of mountain lion predation upon mule deer and elk in the Idaho primitive area, *Wildlife Monograph No. 21*, Wildlife Society, Washington, D.C.

HOSLEY, N. W., AND R. K. ZIEBARTH

1935 Some winter relations of the white-tailed deer to the forests in north-central Massachusetts, *Ecology*, **16**:535–553.

HOWARD, E.

1920 *Territory in Bird Life*, Dutton, New York.

HOWARD, W. E.

1960 Innate and environmental dispersal of individual vertebrates, *Am. Midland Naturalist*, **63**:152–161.

HUBER, B.

1952 Der Emfluss der Vegetation auf die Schwankungen des CO_2-Gehaltes der Atmosphäre, *Arch. Met. Geophys. Bioklim.*

HUDSON, H. J., AND J. WEBSTER

1958 Succession of fungi on decaying stems of *Agropyron repens*, *Brit. Mycol. Soc. Trans.*, **41**:165–177.

HUFFAKER, C. B.

1958 Experimental studies on predation: dispersion factors and predator-prey oscillations, *Hilgardia*, **27**:343–383.

HULL, D. L.

1970 Contemporary systematic philosophies, *Ann. Rev. Ecol. and System.*, **1**:19–54.

HUMPHREY, R. R.

1958 The desert grassland, a history of vegetational changes and an analysis of causes, *Botan. Rev.*, **24**:193–252.

HUMPHRIES, D. A., AND P. M. DRIVER

1967 Erratic display as a device against predators, *Science*, **156**:1767–1768.

HUNT, E. G., AND A. I. BISCHOFF

1960 Inimical effects on wildlife of periodic DDT applications to Clear Lake, *Calif. Fish Game*, **46**:91–106.

HUNT, L. B.

1960 Songbird breeding populations in DDT-sprayed Dutch elm disease communities, *J. Wildlife Manage.*, **24**:139–146.

HUNTER, G. W., III, AND WANDA S. HUNTER

1934 Further studies on fish and bird parasites, *Suppl. 24th Ann. Rep. N.Y. State Dept. Conserv.*, No. 9, *Rept. Biol. Surv. Mohawk-Hudson Watershed*, pp. 267–283.

HURD, L. E.; M. V. MELLINGER; L. L. WOLF; AND S. T. MCNAUGHTON

1971 Stability and diversity at three trophic levels in terrestrial successional ecosystems, *Science*, **173**:134–136.

HURLBERT, S. H.

1971 The nonconcept of species diversity: a critique and alternative parameters, *Ecology*, **52**:577–586.

HUSCH, B.; C. I. MILLER; AND T. W. BEERS

1972 *Forest Mensuration*, Ronald, New York.

HUTCHINSON, G. E.

1957 *A Treatise on Limnology*, Vol. 1, *Geography, Physics, Chemistry*, Wiley, New York.

1969 Eutrophication, past and present, in *Eutrophication: Causes, Consequences, Correctives*, National Academy Sciences, Washington, D.C., pp. 12–26.

HUXLEY, J. S.

1934 A natural experiment on the territorial instinct, *Brit. Birds*, **27**:270–277.

HYDER, D. N.

1969 The impact of domestic animals on the structure and function of grassland ecosystems, in R. L. Dix and R. G. Beidleman, *The Grassland Ecosystem: A Preliminary Synthesis*, Colorado State University Range Sciences Dept., Sci. Ser. No. 2, pp. 243–260.

HYNES, H.

1970 *Biology of Running Water*, University of Toronto Press, Toronto, Ontario.

IDE, F. P.

1935 The effect of temperature on the distribution of the mayfly fauna of a stream, *Univ. Toronto Biol. Ser.*, **31**:1–76.

IKUSIMA, I.

1965 Ecological studies on the productivity of aquatic plant communities; I, Measurement of photosynthetic activity, *Bot. Mag. Tokyo*, **78**:202–211.

ILLIES, J., AND L. BOTOSANEANU

1963 Problemes et methodes de la classification et de la zonation écologique des eaux courantes, considerées surtout du point de vue faunistique, *Mitt. int. Verein theor. Angew. Limnol.*, **12**:1–57.

IRVING, I., AND J. KROGH

1954 Body temperatures of arctic and subarctic birds and mammals, *J. Appl. Physiol.*, **6**:667–680.

IRVING, L.

1960 Birds of Anaktuviik Pass, Kobuk and Old Crow: a study in Arctic adaptation, *U.S. Natl. Mus. Bull.*, **217**.

JACKSON, A. S.

1965 Wildfires in the Great Plains grasslands, *Proc. 4th Tall Timbers Fire Ecology Conf.*, pp. 241–259.

JACOBSON, J. S., AND A. C. HILL

1970 *Recognition of Air Pollution Injury to Vegetation: A Pictorial Atlas*, Air Pollution Control Association, Pittsburgh, Pa.

JACOT, A. P.

1939 Reduction of spruce and fir litter by minute animals, *J. Forestry*, **37**:858–860.

1940 The fauna of the soil, *Quart. Rev. Biol.*, 15:38–58.

JAFFE, L. S.

1970 The global balance of carbon monoxide, in S. F. Singer (ed.), *Global Effects of Environmental Pollution*, Springer-Verlag, New York, pp. 34–49.

JAHN, E.

1950 Bodentieruntersuchungen in den Flugsandgebieten des Marchfeldes, *Z. Angew. Entomol.*, 32:208–274.

JAMES, F. C.

1970 Geographic size variation in birds and its relationship to climate, *Ecology*, 51:365–390.

JANZEN, D. H.

1969 Seed-eaters versus seed size, number, toxicity, and dispersal, *Evolution*, 23:1–27.

1970 Herbivore and the number of tree species in tropical forests, *Am. Naturalist*, 104:501–528.

1971 Seed predation by animals, *Ann. Rev. Ecol. and System.*, 2:465–492.

JARDINE, J. T., AND M. ANDERSON

1919 Range management on the National Forests, *USDA Agr. Bull.*, 790:1–98.

JAWORSKI, N. A., AND L. J. HELLING

1970 Relative contributions of nutrients to the Potomac River Basin from various sources, in *Relationship of Agriculture to Soil and Water Pollution*, Cornell University Press, Ithaca, N.Y.

JENKINS, D.; A. WATSON; AND G. R. MILLER

1963 Population studies on red grouse *Lagopus lagopus scoticus* (Lath) in northeast Scotland, *J. Anim. Ecol.*, 32:317–376.

JENKINS, D. W.

1944 Territory as a result of despotism and social organization in geese, *Auk*, 61:30–47.

JENNINGS, C. D., AND C. OSTERBERG

1969 Sediment radioactivity in the Columbia River Estuary, in D. Nelson and F. Evans (eds.), *Symposium on Radioecology*, pp. 300–318.

JENNY, H.

1933 Soil fertility losses under Missouri conditions, *Missouri Agri. Exp. Sta. Bull.*, 324.

1958 Role of the plant factor in podogenic functions, *Ecology*, 39:5–16.

JENSEN, A. C.

1970 Thermal pollution in the marine environment, *The Conservationist*, 25(2):8–13.

JENSEN, D. D.

1971 Learning and behavior, in *Topics in the Study of Life*, Harper & Row, New York, pp. 307–314.

JEPSEN, G. L.; E. MAYR; AND G. G. SIMPSON (EDS.)

1949 *Genetics, Paleontology, and Evolution*, Princeton University Press, Princeton, N.J.

JOENSUU, O. I.

1971 Fossil fuels as a source of mercury pollution, *Science*, 172:1027–1028.

JOHANNES, R. E.

1964 Uptake and release of dissolved organic phosphorus by representatives of a coastal marine ecosystem, *Limnol. Oceanogr.*, 9:224–234.

1965 Influence of marine protozoa on nutrient regeneration, *Limnol. Oceanogr.*, 10:434–442.

1967 Ecology of organic aggregates in the vicinity of coral reef, *Limnol. Oceanogr.*, 12:189–195.

1968 Nutrient regeneration in lakes and oceans, in M. R. Droop and E. J. Ferguson (eds.), *Advances in the Microbiology of the Sea*, Vol. 1, Academic, New York, pp. 203–213.

JOHANNES, R. E., AND K. L. WEBB

1970 Release of dissolved organic compounds by marine and fresh water invertebrates, *Proc. Conf. on Organic Matter in Natural Waters*, Inst. Marine Sci., Univ. Alaska Occas. Publ. No. 1.

JOHNSGARD, P. A.

1960 Hybridization in the Anatidae and its taxonomic implications, *Condor*, 62:25–33.

1961 The sexual behavior and systematic position of the hooded merganser, *Wilson Bull.*, 73:227–236.

JOHNSON, C. G.

1969 *Migration and Dispersal of Insects by Flight*, Methuen, London.

JOHNSON, D. W.

1968 Pesticides and fishes: A review of selected literature, *Trans. Am. Fish. Soc.*, 97:398–424.

JOHNSON, F. S.

1970 The oxygen and carbon dioxide balance in the earth's atmosphere, in F. S. Singer (ed.), *Global Effects of Environmental Pollution*, Springer-Verlag, New York, pp. 4–11.

JOHNSON, L.

1972 Keller Lake: characteristics of a culturally unstressed salmonid community, *J. Fish Res. Bd. Can.*, 29:731–740.

JOHNSON, M. G., AND W. H. CHARLTON

1960 Some effects of temperature on the metabolism and activity of the large mouth bass *Micropterus salmoides Lacepede*, *Progressive Fish Culturist*, 22:155–163.

JOHNSON, N. M.; R. C. REYNOLDS; AND G. E. LIKENS

1972 Atmospheric sulphur: its effect on the chemical weathering of New England, *Science*, 177:514–515.

JOHNSON, P. L., AND W. O. BILLINGS

1962 The alpine vegetation of the Beartooth Plateau in relation to cryopedogenic processes and patterns, *Ecol. Monographs*, 32:105–135.

JOHNSON, W. K., AND G. J. SCHROEPFER

1964 Nitrogen removal by nitrification and denitrification, *J. Water Pollut. Control Fed.*, 36:1011–1036.

JOHNSTON, D. W., AND E. P. ODUM

1956 Breeding bird populations in relation to plant succession on the Piedmont of Georgia, *Ecology*, 37:50–62.

JOHNSTON, J. W.

1936 The macrofauna of soils as affected by certain

coniferous and hardwood types in the Harvard Forest, Ph.D. dissertation, Harvard University Library, Cambridge, Mass.

JOHNSTON, R. F.

1956a Predation by short-eared owls in a Salicornia salt marsh, *Wilson Bull.*, 68:91–102.

1956b Population structure in salt marsh song sparrows: Part II, Density, age structure and maintenance, *Condor*, 58:254–272.

JOHNSTON, R. F., AND R. K. SELANDER

1971 Evolution in the house sparrow: II, Adaptative differentiation in North American populations, *Evolution*, 25:1–28.

JOHNSTON, V. R.

1947 Breeding birds of the forest edge in Illinois, *Condor*, 49:45–53.

KABAT, C.; N. E. COLLIAS; AND R. C. GUETTINGER

1953 Some winter habits of white-tailed deer and the development of census methods in the flag yard of northern Wisconsin, *Tech. Wildlife Bull. No. 7*, *Wisconsin Conserv. Dept.*, Madison.

KABAT, C., AND D. R. THOMPSON

1963 Wisconsin Quail, 1834–1962: Population dynamics and habitat management, *Tech. Bull. No. 30*, *Wisconsin Conserv. Dept.*, Madison.

KALININ, G. R., AND V. D. BYKOV

1969 The world's water resources, present and future, *Impact of Science on Society*, 19:135–150.

KALLE, K.

1971 Salinity: general introduction, in O. Kinne (ed.), *Marine Ecol.*; *Vol. 1*, *Environmental Factors*, Part 2.

KALLEBERG, H.

1958 Observations in a stream tank of territoriality and competition in juvenile salmon and trout (*Salmo salar* L. and *S. trutta* L.), *Report No. 39*, *Inst. Freshwater Research*, Sweden, pp. 55–98.

KANWISHER, J.

1963 The effect of wind on CO_2 exchange across the sea surface, *J. Geophysical. Res.*, 73:4543.

KARPLUS, M.

1949 Bird activity in continuous day-light of arctic summer, *Bull. Ecol. Soc. Am.*, 30:60.

KEAST, A.

1959 Australian birds: their zoogeography and adaptation to an arid continent, *Biogeography and Ecology in Australia*, 8:89–114.

KEITH, J. O.; L. A. WOODS; AND E. G. HUNT

1970 Reproductive failure in brown pelican, *Trans. 35th North Am. Wildlife and Nat. Resource Conf.*, pp. 56–63.

KEITH, L. B.

1963 *Wildlife's Ten-year Cycle*, University of Wisconsin Press, Madison.

KELLOGG, C. E.

1936 Development and significance of the great soil groups of the United States, *USDA Misc. Publ.* 229.

KELLOGG, W. W., ET AL.

1972 The sulfur cycle, *Science*, 175:587–599.

KELLY, J. M., ET AL.

1969 Models of seasonal primary productivity in eastern Tennessee *Festuca* and *Andropogon* ecosystems, *ORNL-4310*, Oak Ridge National Lab., Oak Ridge, Tenn.

KELSO, L., AND M. M. NICE

1963 A Russian contribution to anting and feather mites, *Wilson Bull.*, 75:23–26.

KEMP, G. A., AND L. B. KEITH

1970 Dynamics and regulation of red squirrels (*Tamiasciurus hudsonicus*) populations, *Ecology*, 51:765–779.

KEMP, W. B.

1971 The flow of energy in a hunting society, *Sci. Amer.*, 224(3):104–115.

KENDEIGH, S. C.

1932 A study of Merrian's temperature laws, *Wilson Bull.*, 44:129–143.

1945 Community selection by birds on the Helderberg Plateau of New York, *Auk*, 62:418–436.

KENDEIGH, S. C.; G. C. WEST; AND G. W. COX

1960 Annual stimulus for spring migration in birds, *Anim. Behav.*, 8:180–185.

KENK, R.

1949 The animal life of temporary and permanent ponds in southern Michigan, *Misc. Publ. Mus. Zool.*, University of Michigan, No. 71, pp. 1–66.

KERSTER, H. W.

1968 Population age structure in the prairie forb, *Liatris aspera*, *Bioscience*, 18:430–432.

KESSEL, B.

1953 Distribution and migration of the European starling in North America, *Condor*, 55:49–67.

KETCHUM, B. H.

1951 The exchanges of fresh and salt waters in tidal estuaries, *J. Marine Res.*, 10:18–38.

1954 Relation between circulation and planktonic populations in estuaries, *Ecology*, 35:191–200.

1967 The phosphorus cycle and productivity of marine phytoplankton, in *Symposium on Primary Productivity and Mineral Cycling in Natural Ecosystems*, University of Maine Press, Orono, pp. 32–51.

KETTLEWELL, H. B. D.

1961 The phenomenon of industrial melanism in Lepidoptera, *Ann. Rev. Entomol.*, 6:245–262.

1965 Insect survival and selection for pattern, *Science*, 148:1290–1296.

KEVAN, D. K. MCE.

1955 *Soil Zoology*, Butterworth, Washington, D.C.

1962 *Soil Animals*, Philosophical Library, New York.

KEYFITZ, N., AND W. FLIEGER

1971 *Population Facts and Methods of Demography*, Freeman, San Francisco.

KIESTER, A. R.

1971 Species density of North American amphibians and reptiles, *Systematic Zool.*, 20:127–137.

805

KILHAM, L.

1959 Early reproductive behavior of flickers, *Wilson Bull.*, 71:299–408.

KING, J. A.

1955 Social behavior, social organization and population dynamics in a black-tailed prairie dog town in the Black Hills of South Dakota, *Contrib. Lab. Vert. Biol.*, University of Michigan, No. 67, pp. 1–123.

KINNE, O. (ED.)

1970 *Marine Ecology: A Comprehensive Integrated Treatise on Life in Oceans and Coastal Waters: Vol. 1, Environmental Factors*, Wiley, New York.

KIRKPATRICK, R. L., AND D. P. KIBBE

1971 Nutritive restriction and reproductive characteristics of captive cottontail rabbits, *J. Wildlife Manage.*, 35:332–337.

KLEEREKOPER, H.

1953 The mineralization of plankton, *J. Fisheries Res. Board Can.*, 10:283–291.

KLEIN, D. R.

1968 The introduction, increase, and crash of reindeer on St. Matthews Island, *J. Wildlife Manage.*, 32:350–367.

1970a Food selection by North American deer and their response to over-utilization of preferred plant species, in A. Watson (ed.), *Animal Populations in Relation to Their Food Resources*, Blackwell, Oxford, pp. 25–46.

1970b The impact of oil development in Alaska, in *Productivity and Conservation in Northern Circumpolar Lands*, IUCN Publ., n.s., 16:209–243.

1970c Tundra ranges north at the boreal forest, *J. Range Manage.*, 23:8–14.

1971 Reaction of reindeer to obstructions and disturbances, *Science*, 173:393–398.

KLIMSTRA, W. D., AND V. C. ZICCARDI

1963 Night-roosting habitat of bobwhites, *J. Wildlife Manage.*, 27:202–214.

KLOMP, H.

1964 Intraspecific competition and the regulation of insect numbers, *Ann. Rev. Entomol.*, 9:17–40.

KLOPFER, P. H.

1959 An analysis of learning in young Anatidae, *Ecology*, 40:90–102.

1962 *Behavioral Aspects of Ecology*, Prentice-Hall, Englewood Cliffs, N.J.

1963 Behavioral aspects of habitat selection: the role of early experience, *Wilson Bull.*, 75:15–22.

1964 Parameters of imprinting, *Am. Naturalist*, 98:175–182.

1965 Imprinting: a reassessment, *Science*, 147:302–303.

KLOPFER, P. H., AND J. P. HAILMAN

1964 Perceptual preferences and imprinting in chicks, *Science*, 145:1333–1334.

KLOPFER, P. H., AND R. H. MACARTHUR

1961 On the causes of tropical species diversity and niche overlap, *Am. Naturalist*, 95:223–226.

KLUIJVER, H. N., AND L. TINBERGEN

1953 Territory and the regulation of density in titmice, *Arch. Neerl. Zool.*, 10:265–289.

KNOX, E. G.

1952 Jefferson County (N.Y.) soils and soil map, N.Y. *State College Agr.*, Cornell University, Ithaca, N.Y.

KOFORD, C. B.

1957 The vicuna and the puna, *Ecol. Monographs*, 27:153–219.

1958 Prairie dogs, white faces, and blue grama, *Wildlife Monograph*, No. 3, pp. 1–78.

KOK, B.

1965 Photosynthesis: the pathway of energy, in J. Bonner and J. E. Varner (eds.), *Plant Biochemistry*, Academic, New York, pp. 904–960.

KOLENOSKY, G. B.

1972 Wolf predation on wintering deer in east central Ontario, *J. Wildlife Manage.*, 36:357–369.

KOLMAN, W. A.

1960 The mechanism of natural selection for the sex ratio, *Am. Naturalist*, 94:373–377.

KOMAREK, E. V., SR.

1964 The natural history of lightning, *Proc. 3d Tall Timbers Fire Ecology Conf.*, pp. 139–183.

1965 Fire ecology: Grasslands and man, *Proc. 4th Tall Timbers Fire Ecology Conf.*, pp. 169–220.

1966 The meteorological basis for fire ecology, *Proc. 5th Tall Timbers Fire Ecology Conf.*, pp. 85–125.

1967a The nature of lightning fires, *Proc. California Tall Timbers Fire Ecology Conf.*, pp. 5–41.

1967b Fire and the ecology of man, *Proc. 6th Tall Timbers Fire Ecology Conf.*, pp. 143–170.

1968 Lightning and lightning forces as ecological forces, *Proc. 8th Tall Timbers Fire Ecology Conf.*, pp. 169–197.

KOPEC, R. J.

1970 Further observations of the urban heat island of a small city, *Bull. Amer. Meteorological Soc.*, 51:602–606.

KOPLIN, J. R.

1972 Measuring predator impact of woodpeckers on spruce beetles, *J. Wildlife Manage.*, 36:308–320.

KORSTIAN, C. F.

1931 Southern white cedar, *USDA Tech. Bull. No. 251.*

KOWAL, N. E.

1966 Shifting agriculture, fire, and pine forest in the Cordillera Central, Luzon, Philippines, *Ecol. Monographs*, 36:389–419.

KOZLOVSKY, D. C.

1968 A critical evaluation of the trophic level concept: I, Ecological efficiencies, *Ecology*, 49:48–60.

KREBS, C.

1963 Lemming cycle at Baker Lake, Canada, during 1959–62, *Science*, 146:1559–1560.

1964 The lemming cycle at Baker Lake, Northwest

Territories, during 1959–1962, *Tech. Paper No. 15, Arctic Institute of North America.*

KREBS, C. J., AND K. T. DELONG

1965 A *Microtus* population with supplemental food, *J. Mammal.,* **46**:566–572.

KREBS, C. J.; M. S. GAINES; B. L. KELLER; J. H. MYERS; AND R. H. TAMARIN

1973 Population cycles in small rodents, *Science,* **179**:35–41.

KREBS, C. J.; B. L. KELLER; AND J. H. MYERS

1971 *Microtus* population densities and soil nutrients in southern Indiana grasslands, *Ecology,* **52**:660–663.

KREBS, J. R.

1971 Territory and breeding density in the great tit *Parus major, Ecology,* **52**:2–22.

KROG, J.

1955 Notes on temperature measurements indicative of special organization in arctic and subarctic plants for utilization of radiated heat from the sun, *Physiol. Plantarum,* 8:836–839.

KUCERA, C. L.; R. C. DAHLMAN; AND M. R. KOELLING

1967 Total net productivity and turnover on an energy basis for tallgrass prairie, *Ecology,* **48**:536–541.

KUENZLER, E. J.

1958 Niche relations of three species of Lycosid spiders, *Ecology,* **39**:494–500.

1961 Phosphorus budget of a mussel population, *Limnol. Oceanogr.,* 6:400–415.

KUHNELT, W.

1950 *Bodenbiologie mit besonderer Berucksichtigung der Tierwelt,* Herold, Vienna, pp. 368ff.

1970 Structural aspects of soil-surface-dwelling biocoenoses, in J. Phillipson (ed.), *Methods of Study in Soil Biology,* UNESCO, Paris, pp. 45–56.

KURCHEVA, G. F.

1964 Wirbellose Tiere abs Faktor der Zersetzung von waldstreu, *Pedobiologia,* **4**:8–30.

LACHENBRUCH, A. H.

1970 Some estimates of the thermal effects of a heated pipeline in permafrost, *Geol. Surv. Circ.,* **632**.

LACK, D. L.

1945 The Galápagos finches, a study in variation, *California Acad. Sci. Occas. Paper No. 21,* San Francisco.

1947 The significance of clutch size, *Ibis,* 89:30–52; 90:25–45.

1953 *The Life of the Robin,* Penguin, London.

1954 *The Natural Regulation of Animal Numbers,* Clarendon Press, Oxford.

1964 A long-term study of the great tit (*Parus major*), *J. Anim. Ecol.,* Suppl., **33**:159–173.

1966 *Population Studies of Birds,* Clarendon Press, Oxford.

1971 *Ecological Isolation in Birds,* Harvard University Press, Cambridge, Mass.

LACK, D., AND L. S. V. VENERABLES

1939 The habitat distribution of British woodland birds, *J. Anim. Ecol.,* 8:39–71.

LANDSBERG, H. E.

1970 Man-made climatic changes, *Science,* **170**: 1265–1274.

LANGFORD, A. N., AND M. F. BUELL

1969 Integration, identity, and stability in the plant association, *Adv. in Ecol. Res.,* 6:83–135.

LANYON, W. E.

1958 The motivation of sun-bathing in birds, *Wilson Bull.,* 70:280.

LANYON, W. E., AND W. N. TAVOLGA (EDS.)

1960 *Animal Sounds and Communication,* AIBS *Symposium Series, Publ. No. 7.*

LARSON, F.

1940 The role of bison in maintaining the short grass plains, *Ecology,* **21**:113–121.

LASIEWSKI, R. C., AND G. K. SNYDER

1969 Responses to high temperatures in nestling double-crested and pelagic cormorants, *Auk,* 86:529–540.

LAVE, L. B., AND E. P. SESKIN

1970 Air pollution and human health, *Science,* 169:723–733.

LAWRENCE, D. B.

1958 Glaciers and vegetation in southeastern Alaska, *Am. Scientist,* **46**:89–122.

LAY, D. W.

1958 Fire ant eradication and wildlife, *Proc. 12th Annual Conf. Southeast Assoc. Game Fish Comm.,* pp. 22–24.

LEAHY, M. G., JR., AND G. B. CRAIG, JR.

1967 Barriers to hybridization between *Aedes aegypti* and *Aedes albopictus* (Diptera *Culicidae*), *Evolution,* **21**:41–58.

LEE, R. B.

1966 !Kung Bushman subsistence: an input-output analysis, in D. Damas (ed.), *Ecological Essays: Proceedings of the Conference of Cultural Ecology, National Mus. Can. Bull. No. 230.*

LEES, A. D.

1955 *The Physiology of Diapause in Arthropods,* Cambridge University Press, London.

1960 Some aspects of animal photoperiodism, *Cold Spring Harbor Symp. Quant. Biol.,* **25**:261–268.

LEHRMAN, D. S.

1959 Hormonal responses to external stimuli in birds, *Ibis,* **101**:478–496.

LEITH, HELMUT

1963 The role of vegetation in the carbon dioxide content of the atmosphere, *J. Geophys. Res.,* 68:3887–3898.

LEOPOLD, A.

1933 *Game Management,* Scribner, New York.

1949 *A Sand County Almanac,* Oxford University Press, New York.

LEOPOLD, A., AND A. EYNON

1961 Avian daybreak and evening song in relation to time and light intensity, *Condor,* **63**:269–293.

LEVINE, R. P.

1969 The mechanism of photosynthesis, *Sci. Amer.*, **221**:58–70.

LEVINS, R.

1968 *Evolution in Changing Environments*, Princeton University Press, Princeton, N.J.

LEWIS, J. K.

1969 Range management viewed in the ecosystem framework, in G. Van Dyne (ed.), *The Ecosystem Concept in Natural Resources Management*, Academic, New York, pp. 97–187.

LEWONTIN, R. C.

1965 Selection in and of populations, in J. A. Moore (ed.), *Ideas in Modern Biology*, XVI *Inter. Congr. Zoology.*

1969 The meaning of stability, in *Brookhaven Symp. Biol.*, **22**:13–24.

LIBBY, W. F.

1961 Radiocarbon dating, *Science*, **133**:621–629.

LICHT, L. E.

1967 Growth inhibition in crowded tadpoles: intraspecific and interspecific effects, *Ecology*, **48**:736–745.

LIDICKER, W. Z., JR.

1966 Ecological observations on a feral house mouse population declining to extinction, *Ecol. Monographs*, **36**:27–50.

LIEBIG, J.

1940 *Chemistry in Its Application to Agriculture and Physiology*, Taylor and Walton, London.

LIETH, H.

1960 Patterns of change within grassland communities, in J. L. Harper (ed.), *The Biology of Weeds*, Oxford, pp. 27–39.

LIGNON, J. D.

1968 Sexual differences in foraging behavior in two species of *Dendrocopos* woodpeckers, *Auk*, **85**:203–215.

LIKENS, G. E., ET AL.

1967 The calcium, magnesium, potassium, and sodium budgets for a small forested ecosystem, *Ecology*, **48**:772–785.

LIKENS, G. E.; F. H. BORMANN; AND N. M. JOHNSON

1969 Nitrification: importance to nutrient losses from a cutover forest ecosystem, *Science*, **163**:1205–1206.

LILLYWHITE, H. B.

1970 Behavioral temperature regulation in the bullfrog, *Rana catesbeiana*, *Copeia*, **1970**:158–168.

LINDQUIST, B.

1942 Experimentelle Untersuchungen uber die Bedeutung einiger Landmollusken fur die Zersetgung der Waldstreu, *Kgl. Fysiograf. Sallskap Lund Forh.*, **11**:144–156.

LINDUSKA, J. P.

1947 Winter den studies of the cottontail in southern Michigan, *Ecology*, **28**:448–454.

LITTLE, E. L., JR.

1971 *Atlas of United States Trees: Vol. 1, Conifers and Important Hardwoods*, *USDA Misc. Publ. No. 1146.*

LLOYD, M.

1968 Self-regulation of adult numbers by cannabalism in two laboratory strains of flour beetles (*Trilobium castaneum*), *Ecology*, **49**:245–259.

LLOYD, M., AND H. S. DYBAS

1966 The periodical cicada problem: 1, Population ecology, *Evolution*, **20**:133–149.

LLOYD, M., AND R. J. GHELARDI

1964 A table for calculating the "equitability" component of species diversity, *J. Anim. Ecol.*, **33**:217–225.

LOFTUS, K.

1958 Studies on river-spawning populations of lake trout in eastern Lake Superior, *Trans. Am. Fishery Soc.*, **87**:259–277.

LONGHURST, W. M.; H. K. OH; M. B. JONES; AND R. E. KEPNER

1968 A basis for the palatability of deer forage plants, *Trans. N. Am. Wildlife Nat. Res. Conf.*, **33**:181–189.

LOOMIS, R. S., ET AL.

1967 Community architecture and the productivity of terrestrial plant communities, in A. San Pietro et al. (eds.), *Harvesting the Sun*, Academic, New York, pp. 291–308.

LORD, R. D., JR.

1961a A population study of the gray fox, *Am. Midland Naturalist*, **66**:87 109.

1961b Magnitudes of reproduction in cottontail rabbits, *J. Wildlife Manage.*, **25**:28–33.

1961c Mortality rates of cottontail rabbits, *J. Wildlife Manage.*, **25**:33–40.

LORENZ, K.

1931 Beitrage zur Ethologie der sozialer corviden, *J. fur Ornith.*, **46**:67–127.

1935 Der Kumpan in der Umwelt des Vogels, *J. fur Ornith.*, **83**:137–213; 289–413.

1937 The companion in the bird's world, *Auk*, **54**:245–273.

1966 *On Aggression*, Harcourt Brace Jovanovich, New York.

LORENZ, K., AND N. TINBERGEN

1938 Taxis und Instinkthandlung in der Eirollbewegung der Grangans I, *Z. Tierpsychol.*, **2**:1–29.

LOTKA, A. J.

1925 *Elements of Physical Biology*, Williams & Wilkins, Baltimore.

LOUCKS, O. L.

1970 Evolution of diversity, efficiency, and community stability, *Am. Zoologist*, **10**:17–25.

LOWDERMILK, W. C.

1953 Conquest of the land through 7000 years, *USDA Conserv. Serv. Agri. Information Bull.*, **99**.

LOWE, V. P. W.

1969 Population dynamics of red deer (*Cervus elaphus* L) on Rhum, *J. Animal Ecol.*, **38**:425–457.

LOWE-MCCONNELL, R. H.

1969 Speciation in tropical freshwater fishes, *Biol. J. Linn. Soc.*, 1:51–75.

LULL, H. W.

1967 Factors influencing water production from forested watersheds, *Municipal Watershed Mgmt. Symp. Proc. 1965*, Univ. Mass. Coop. Ext. Serv. Publ., 446:2–7.

1971 Effects of trees and forest on water relations, in *Forests and Trees in an Urbanizing Environment*, Univ. Mass. Coop. Ext. Serv., *Planning and Resource Devl. Ser.*, 17:65–69.

LULL, H. W., AND W. E. SOPPER

1969 Hydrologic effects from urbanization of forested watersheds in the northeast, *USDA For. Serv. Res. Paper*, NE-146.

LUNK, W. A.

1952 Notes on variation in the Carolina chickadee, *Wilson Bull.*, 64:7–21.

LUTZ, H. J.

1956 Ecological effects of forest fires in the interior of Alaska, *USDA Tech. Bull. No. 1133*.

LUTZ, H. J., AND R. F. CHANDLER

1946 *Forest Soils*, Wiley, New York.

LYMAN, C. P., AND A. R. DAWE

1960 Mammalian hibernation, *Museum Comp. Zool.*, Harvard, Cambridge, Mass.

MCARDLE, R. E.; W. H. MEYER; AND D. BRUCE

1949 The yield of Douglas-fir in the Pacific Northwest, *USDA Tech. Bull. No. 201* (rev.).

MACARTHUR, R. H.

1958 Population ecology of some warblers of northeastern coniferous forests, *Ecology*, 39:599–619.

1960 On the relative abundance of species, *Am. Naturalist*, 94:25–36.

1961 Population effects of natural selection, *Am. Naturalist*, 95:195–199.

1972 *Geographical Ecology*, Harper & Row, New York.

MACARTHUR, R. H., AND J. W. MACARTHUR

1961 On bird species diversity, *Ecology*, 42:594–598.

MACARTHUR, R. H., AND E. O. WILSON

1963 An equilibrium theory of insular zoogeography, *Evolution*, 17:373–387.

1967 *The Theory of Island Biogeography*, Princeton University Press, Princeton, N.J.

MCATEE, W. L.

1939 Wildlife of the Atlantic Coast salt marshes, *USDA Circ.*, 520:1–28.

MCBEE, R. H.

1971 Significance of intestinal microflora in herbivory, *Ann. Rev. Ecology and Systematics*, 2:165–176.

MCBRIDE, G.; I. P. PARE; AND F. FOENANDER

1969 The social organization and behavior of the feral domestic fowl, *Anim. Behav. Monographs*, 2:127–181.

MCCALLS, T. M.

1943 Microbiological studies of the effects of straw used as a mulch, *Trans. Kansas Acad. Sci.*, 43:52–56.

MCCANN, G. D.

1942 When lightning strikes, *Sci. American*, 49:23–25.

MCCORMICK, F.

1963 Changes in a herbaceous plant community during a three-year period following exposure to ionizing radiation gradients, in Schultz and Klements (eds.), *Radioecology*, pp. 271–276.

1969 Effects of ionizing radiation on a pine forest, in D. Nelson and F. Evans (eds.), *Symposium on Radioecology*, USAEC Rept., CONF-670503, pp. 78–87.

MCCORMICK, J.

1959 *The Living Forest*, Harper & Row, New York.

MACFAYDEN, A.

1963 *Animal Ecology: Aims and Methods*, 2d ed., Pitman, London.

MCGINNES, W. G.

1972 North America, in C. M. McKella et al. (eds.), *Wildland Shrubs: Their Biology and Utilization*, USDA For. Serv. Gen. Tech. Rept., INT-1, pp. 55–66.

MCILROY, R. J.

1972 *An Introduction to Tropical Grassland Husbandry*, 2d ed., Oxford University Press, London.

MCINTOSH, R. P.

1958 Plant communities, *Science*, 128:115–120.

1967 An index of diversity and the relation of concepts to diversity, *Ecology*, 48:392–403.

MACKAY, J. R.

1966 Tundra and tiaga, in F. Darling and J. P. Milton (eds.), *Future Environments of North America*, Natural History Press, New York, pp. 156–171.

MCKELL, C. M., ET AL. (EDS.)

1972 *Wildland Shrubs: Their Biology and Utilization*, USDA For. Serv. Gen. Tech. Rept. INT-1.

MACLULICH, D. A.

1947 Fluctuations in the numbers of varying hare (*Lepus americanus*), Univ. Toronto Biol. Ser. No. 43.

MCMILLAN, C.

1959 The role of ecotypic variation in the distribution of the Central Grassland of North America, *Ecol. Monographs*, 29:285–308.

MCNAB, B. K.

1963 Bioenergetics and the determination of home range size, *Am. Naturalist*, 97:133–140.

1971 On the ecological significance of Bergmann's rule, *Ecology*, 52:845–854.

MCNAUGHTON, S. J.

1968 Structure and function in California grassland, *Ecology*, 49:962–972.

MCPHERSON, J. K., AND C. H. MULLER

1969 Allelopathic effects of *Adenostoma fascicu-*

latum "chamise" in the California chaparral,
Ecol. Monographs, **39**:177–179.

MADGWICK, H. A. I., AND J. D. OVINGTON

1959 The chemical composition of precipitation
in adjacent forest and open plots, *Forestry,*
32:14–22.

MAGUIRE, B., JR.

1971 Phytotelmata: biota and community structures
determination in plant-held waters, *Ann. Rev.
Ecol. and System.,* **2**:439–464.

MAIO, J. J.

1958 Predatory fungi, *Sci. American,* **199**:67–72.

MAISUROW, D. K.

1941 The role of fire in the perpetuation of virgin
forests of northern Wisconsin, *J. Forestry,*
39:201–207.

MALTHUS, T. R.

1798 *An Essay on Principles of Population,* Johnson,
London (numerous reprints).

MALVIN, R. L., AND M. RAYNER

1968 Renal function and blood chemistry in
Cetacea, *Am. J. Physiol.,* **214**:187–191.

MANGELSDORF, P. C.

1958 Ancestor of corn, *Science,* **128**:1313–1320.

MARE, M. F.

1942 A study of the marine benthic community with
special reference to microorganisms, *J. Marine
Biol. Assoc.,* **25**:517–554.

MARGALEF, R.

1963 On certain unifying principles in ecology, *Am.
Naturalist,* **47**:357–374.

1968 *Perspectives in Ecological Theory,* University
of Chicago Press, Chicago.

1969 Diversity and stability: a practical proposal
and a model of interdependence, *Brookhaven
Symp. Biol.,* **22**:25–37.

MARKS, P. L., AND F. H. BORMANN

1972 Revegetation following forest cutting: mechanisms for return to steady-state nutrient cycling,
Science, **176**:914–915.

MARLER, P. R.

1956 Behaviour of the chaffinch, *Fringilla coelebs,
Behaviour Suppl.,* **5**:1–184.

MARLER, P. R., AND W. J. HAMILTON, III

1966 *Mechanisms of Animal Behavior,* Wiley, New
York.

MARQUIS, D. A.

1972 Effect of forest clearcutting on ecological
balances, in R. D. Nyland (ed.), *A Perspective
on Clearcutting in a Changing World, Applied
Forestry Research Institute, Misc. Rept. No. 4,*
Syracuse, New York, pp. 47–66.

MARSDEN, H. M., AND N. R. HOLLER

1964 Social behavior in confined populations of the
cottontail and swamp rabbit, *Wildlife Monographs No. 13.*

MARSH, G. P.

1864 *Man and Nature: or Physical Geography as
Modified by Human Action,* reprinted 1965 by
Harvard University Press, Cambridge, Mass.

MARSH, J. A.

1970 Primary productivity of reef-building calcareous
red algae, *Ecology,* **51**:255–263.

MARSHALL, J. T., JR.

1948 Ecological races of song sparrows in the San
Francisco Bay region: Part II, Geographic
variation, *Condor,* **50**:233–256.

MARSHALL, S. M., AND A. P. ORR

1955 On the biology of *Calanus finmarchicus:* 8,
Food uptake, assimilation, and excretion in
stage V *Calanus, J. Marine Biol. Assoc. U.K.,*
34:495–529.

MARSHALL, W. H.

1951 Pine marten as a forest product, *J. Forestry,*
49:899–905.

MARTIN, M. M.

1970 The biochemical basis of the fungus-attine ant
symbiosis, *Science,* **169**:16–20.

MARTIN, N. D.

1960 An analysis of bird populations in relation to
forest succession in Algonquin Provincial Park,
Ontario, *Ecology,* **41**:126–140.

MARX, D. H.

1971 Ectomycorrhizae as biological deterrents to
pathogenic root infections, in E. Hacskaylo (ed.),
Mycorrhizae, USDA Misc. Publ. 1189, pp. 81–96.

MATHER, K.

1955 Polymorphism as an outcome of disruptive
selection, *Evolution,* **9**:52–61.

MATTHEWS, G. V. T.

1968 Bird navigation, *Cambridge Monogr. Exp.
Biology No. 3,* 2d ed., Cambridge University
Press, London.

MAY, J. M.

1960 The ecology of human disease, *Ann. N.Y.
Acad. Sci.,* **84**:789–794.

MAYANDON, J., AND P. SIMONART

1959 Étude de la décomposition de la matière
organique dans le sol au moyen de carbon radio-
actif: V, Décomposition de cellulose et de lignine,
Plant Soil, **11**:176–181.

MAYER, H. M.

1969 *The Spatial Expression of Urban Growth,*
Commission on College Geography Resource
Paper No. 7, Association of American Geog-
raphers, Washington, D.C.

MAYER, W. V.

1960 Histological changes during the hibernation
cycle in the arctic ground squirrel, in Lyman and
Dawe (eds.), 1960, *Mammalian Hibernation,*
pp. 131–148.

MAYFIELD, H. F.

1960 *The Kirtland's Warbler, Cranbrook Inst. Sci.
Bull. No. 40.*

MAYR, E.

1942 *Systematics and the Origin of Species,*
Columbia University Press, New York.

1963 *Animal Species and Evolution,* Harvard
University Press, Cambridge, Mass.

1969 *Principles of Systematic Zoology*, McGraw-Hill, New York.

1970 *Population, Species, and Evolution*, Harvard University Press, Cambridge, Mass.

MEANLEY, B.

1957 Notes on the courtship behavior of the king rail, *Auk*, 74:433–440.

MECH, L. D.

1970 *The Wolf*, Natural History Press, Garden City, New York.

MENDALL, H. L.

1958 The Ring-necked duck in the Northeast, *Univ. Maine Studies, Second Series, No. 73.*

MENDEL, G.

1966 *Mendel's Principles of Heredity*, translated by J. A. Peters, 1959, Harvard University Press, Cambridge.

MENTZER, L. W.

1951 Studies on plant succession in true prairie, *Ecol. Monographs*, 21:255–267.

MENZIE, C. M.

1969 Metabolism of pesticides, *USDI Fish and Wildlife Serv. Sp. Sci. Rept. Wildlife No. 127.*

MERGEN, FRANCOIS

1963 Ecotypic variation in *Pinus strobus* L, *Ecology*, 44:716–727.

MERRIAM, C. H.

1898 Life zones and crop zones of the United States, *Bull., U.S. Bureau Biol. Survey*, 10:1–79.

MERRIAM, H. G.

1960 Problems in woodchuck population ecology and a plan for telemetric study, Ph.D. dissertation, Cornell University, Ithaca, N.Y.

MESLOW, E. C., AND L. B. KEITH

1968 Demographic parameters of a snowshoe hare population, *J. Wildlife Manage*, 32:812–835.

MEYERRIECKS, A. J.

1959 Foot stirring behavior in herons, *Wilson Bull.*, 71:153–158.

1960 *Comparative Breeding Behavior of Four Species of North American Herons, Publ. Nuttall Ornithological Club, No. 2.*

MICHENER, C. D.

1970 Diverse approaches to systematics, *Evolutionary Biol.*, 4:138.

MIELKE, L. N., ET AL.

1970 Groundwater quality and fluctuation in a shallow unconfined aquifer under a level feedlot, in *Relationship of Agriculture to Soil and Water Pollution*, Cornell University Press, Ithaca, N.Y., pp. 31–40.

MIHURSKY, J. A., AND V. S. KENNEDY

1967 Water temperature criteria to protect aquatic life, in *Symp. on Water Quality Criteria, Amer. Fish. Soc. Spec. Publ. No. 4*, pp. 20–32.

MILLER, A. H.

1942 Habitat selection among higher vertebrates and its relation to intraspecific variation, *Am. Naturalist*, 76:25–35.

MILLER, C. A.

1963 The spruce budworm: 1, The bionomics of the spruce budworm, in Morris (ed.), 1963, *The dynamics of epidemic spruce budworm population*, pp. 12ff.

MILLER, G. R.

1968 Evidence for selective feeding on fertilized plots by red grouse, hares and rabbits, *J. Wildlife Manage.*, 32:849–853.

MILLER, P. R.; J. R. PARMENTER; O. C. TAYLOR; AND E. A. CARDIFF

1963 Ozone injury to the foliage of *Pinus ponderosa*, *Phytopathology*, 53:1072–1076.

MILLER, R., AND J. RUSCH

1960 Zur Frange der Kohlensaureversorgung des Waldes, *Forstwiss Cbl.*, 79(1–2):42–64.

MILLER, R. S.

1967 Pattern and process in competition, *Adv. in Ecol. Res.*, 4:1–74.

MILLER, R. S.; G. S. HOCHBAUM; AND D. B. BOTKIN

1972 A simulation model for the management of sandhill cranes, *Yale Univ. School of Forestry and Environmental Studies Bull. No. 80.*

MILLER, W. R., AND F. E. EGLER

1950 Vegetation of the Wequetequock-Pawcatuck tidal marshes, Connecticut, *Ecol. Monographs*, 20:141–172.

MILLICENT, E., AND J. M. THODAY

1960 Gene flow and divergence under disruptive selection, *Science*, 131:1311–1312.

1961 Effects of disruptive selection: IV, Gene flow and divergence, *Heredity*, 16:199–217.

MILTON, W. E. J.

1940 The effect of manuring, grazing and cutting on the yield; botanical and chemical composition of natural hill pastures: I, Yield and botanical composition, *J. Ecol.*, 28:326–356.

MINSHALL, G. W.

1967 Role of allochthonous detritus in the trophic structure of a woodland spring brook community, *Ecology*, 48:139–149.

MISHUSTIN, E. N., AND V. K. SHILNIKOVA

1969 The biological fixation of atmospheric nitrogen by free-living bacteria, in *Soil Biology*, UNESCO, Paris, pp. 65–124.

MITCHELL, R. D.

1969 A model accounting for sympatry in water mites, *Am. Naturalist*, 103:311–346.

MOBIUS, K.

1883 The oyster and oyster culture, *Rept., U.S. Fish Comm. 1880*, pp. 683–824.

MOEN, A. M.

1968a Energy exchange of white-tailed deer, western Minnesota, *Ecology*, 49:676–681.

1968b The critical thermal environment, *Bioscience*, 18:1041–1043.

1968c Surface temperatures and radiant heat from white-tailed deer, *J. Wildlife Manage.*, 32:388–344.

MOHLER, L. L.; J. H. WAMPOLE; AND E. FICHTER
1951 Mule deer in Nebraska National Forest, *J. Wildlife Manage.*, **15**:129–157.

MOIR, W. H.
1969a Energy fixation and the role of primary producers in the energy flux of grassland ecosystems, in R. L. Dix and R. G. Beidleman (eds.), *The Grassland Ecosystem, Range Sci. Dept. Colorado State Univ. Sci. Ser. No. 2*.
1969b Steppe communities in the foothills of the Colorado Front Range and their relative productivities, *Am. Midland Naturalist*, **81**:331–340.

MOLCHANOV, A. A.
1960 *The Hydrological Cycle of Forest*, Israel Program for Scientific Publication, Jerusalem.

MONK, C. A.
1967 Tree species diversity in the eastern deciduous forest with particular reference to northcentral Florida, *Am. Naturalist*, **101**:173–187.
1970 An ecological significance of energetics, *Ecology*, **47**:504–505.

MONRO, J.
1967 The exploitation and conservation of resources by populations of insects, *J. Anim. Ecol.*, **36**:531–347.

MONSI, M.
1968 Mathematical models of plant communities, in F. E. Eckardt (ed.), *Functioning of Terrestrial Ecosystems at the Primary Production Level*, Proceedings of the Copenhagen Symposium, Natural Resources Research V, Unesco, pp. 131–149.

MONTEITH, J. L.
1962 Measurement and interpretation of carbon dioxide flues in the field, *Netherlands J. Agr. Sci.*, **10**(sp. issue):334–346.

MOOK, L. J.
1963 Birds and spruce budworm, in R. Morris, *The dynamics of epidemic spruce budworm populations*, pp. 244–248.

MOONEY, H. A., AND W. O. BILLINGS
1960 The annual carbohydrate cycle of alpine plants as related to growth, *Am. J. Bot.*, **47**:594–598.

MOONEY, H. A., AND E. L. DUNN
1970a Convergent evolution of Mediterranean climate evergreen sclerophyll shrubs, *Evolution*, **24**:292–303.
1970b Photosynthetic systems of Mediterranean climate shrubs and trees of California and Chile, *Am. Midland Naturalist*, **104**:447–453.

MOORE, C. W. E.
1959 The competitive effect of *Danthonia* spp. on the establishment of *Bothriochloa ambigua*, *Ecology*, **40**:141–143.

MOORE, H. B.
1934 The relation of shell growth to environment in *Patella vulgata*, *Proc. Malacological Soc.*, London, **21**:217–222.

1958 *Marine Ecology*, Wiley, New York.

MOORE, J. A.
1949a Geographic variation of adaptive characters in *Rana pipiens* Schreber, *Evolution*, **3**:1–24.
1949b Patterns of evolution in the genus *Rana*, in Jepsen, Mayr, and Simpson, *Genetics Paleontology and Evolution*, pp. 315–338.

MOORE, N. W.; M. D. HOOPER; AND B. N. K. DAVIS
1967 Hedges: I, Introduction and reconnaissance studies, *J. Appl. Ecol.*, **4**:201–220.

MOROWITZ, H. J.
1968 *Energy Flow in Biology*, Academic, New York.

MORRIS, D.
1954 The reproductive behaviour of the river bullhead (*Cottus gobio* L.) with special reference to the fauning activity, *Behaviour*, **7**:1–32.

MORRIS, R. F. (ED.)
1963 The dynamics of epidemic spruce budworm populations, *Mem. Entomol. Soc. Can. No. 31*.

MORRIS, R. F.
1963 Predictive population equations based on kep factors, *Entomol. Soc. Can. Mem. 32*, pp. 16–21.

MORRIS, R. F.; W. F. CHESHIRE; C. A. MILLER; AND D. G. MOTT
1958 The numerical response of avian and mammalian predators during a gradation of the spruce budworm, *Ecology*, **39**:487–494.

MORRISON, P. R., AND F. A. RYSER
1952 Weight and body temperature in mammals, *Science*, **116**:231–232.

MORTIMER, G. H.
1952 Water movements in lakes during summer stratification: evidence from the distribution of temperature in Windermere, *Phil. Trans. Roy. Soc.*, Ser. B., London, pp. 236–355.

MOSS, R.
1972 Food selection by red grouse (*Lagopus lagopus scoticus* Lath.) in relation to chemical composition, *J. Anim. Ecol.* **41**:411–428.

MOSSER, J. L.; N. S. FISHER; T. TENG; AND C. F. WURSTER
1972 Polychlorinated biphenyls toxicity to certain phytoplankton, *Science*, **175**:191–192.

MOUNT, A. B.
1969 Eucalypt ecology as related to fire, *Proc. 9th Tall Timbers Fire Ecology Conf.*, pp. 75–108.

MOYNIHAN, M.
1955a Some aspects of reproductive behaviour in the blackheaded gull (*Larus ridibundus ridibundus* L.) and related species, *Behaviour Suppl.*, **4**:1–201.
1955b Types of hostile display, *Auk*, **72**:247–259.
1955c Remarks on the original sources of displays, *Auk*, **72**:240–246.
1962 Display patterns of tropical American "Nine-primaried" songbirds: I, *Chlorospingus*, *Auk*, **79**:310–344.
1968 Social mimicry: character convergence versus character displacement, *Evolution*, **22**:315–331.

MOYNIHAN, M., AND M. F. HALL

1954 Hostile, sexual and other social behaviour patterns of the spice finch (*Lonchura punctulata*) in captivity, *Behaviour*, 7:3–76.

MUIR, R. C.

1965 The effect of sprays on fauna of apple trees, *J. Appl. Ecol.*, 2:31–41, 43–57.

MULLER, C. H.

1953 The association of desert annuals with shrubs, *Am. J. Bot.*, 40:53–59.

MULLER, C. H.; R. B. HANAWALT; AND J. K. MCPHERSON

1968 Allelopathic control of herb growth in the fire cycle of California chaparral, *Bull. Torrey Bot. Club*, 95:225–231.

MULLER, K.

1954a Investigations on the organic drift in north Swedish streams, *Dept. Inst. Freshwater Research*, 35:133–148.

1954b Die drift in fliessenden Gewässern, *Arch. Hydrobiol.*, 49:539–545.

1966 Die Tagesperiodik von Fleisswasserorganismem, *Z. Morpho. Ohol. Tiere*, 56:93–142.

MULLER, P. E.

1889 Becherches sur les formes naturelles de l'humus, *Ann. Sci. Agron.*, Paris, 6:85–423.

MULLER, W. H.; P. LORBER; AND B. HALEY

1968 Volatile growth inhibitors produced by *Salvia leucophylla*: effect on seedling growth and respiration, *Bull. Torrey Bot. Club*, 95:415–422.

MÜLLER-SCHWARZE, D.

1971 Pheromones in black-tailed deer, *Anim. Behav.*, 19:141–152.

MURATORI, A., JR.

1968 How outboards contribute to water pollution, *The Conservationist*, 22(6):6–8, 31.

MURDOCH, W. W.

1969 Switching in general predators: experiments on predator specificity and stability of prey populations, *Ecol. Monographs*, 39:335–354.

MURIE, A.

1944 *The Wolves of Mount McKinley*, USDI Natl. Park Serv., *Fauna Ser.*, 5:1–238.

MURIE, O. J.

1954 *A Field Guide to Animal Tracks*, Houghton-Mifflin, Boston, Mass.

MURPHY, D. A., AND J. A. COATES

1966 Effects of dietary protein on deer, *Trans. N. Am. Wildlife and Nat. Res. Conf.*, 31:129–139.

MURPHY, G. I.

1966 Population biology of the Pacific sardine (*Sardinops caerulea*), *Proc. Calif. Acad. Sci. 4th Series*, 34:1–84.

1967 Vital statistics of the Pacific sardine (*Sardinops caerulea*) and the population consequences, *Ecology*, 48:731–736.

MURPHY, P. W.

1952 Soil faunal investigations, in Report on forest research for the year ending March, 1951, *Forestry Comm.*, London, pp. 130–134.

1953 The biology of the forest soils with special reference to the mesofauna or meiofauna, *J. Soil Sci.* 4:155–193.

MUSCATINE, L., AND H. M. LENHOFF

1963 Symbiosis: on the role of algae symbiotic with hydra, *Science*, 142:956–958.

MUSSER, J. J.

1963 Description of the physical environment and of strip mining operations in parts of Beaver Creek Basin, Kentucky, *U.S. Geol. Surv. Prof. Paper 427A*.

MUUL, I.

1965 Daylength and food caches, *Nat. Hist.*, 74(3):22–27.

1969 Photoperiod and reproduction flying squirrels, *Glaucomys volans*, *J. Mammal.*, 50:542–549.

MYERS, J. H., AND C. J. KREBS

1971 Genetic, behavioral, and reproductive attributes of dispersing field voles *Microtus pennsylvanicus* and *Microtus ochrogaster*, *Ecol. Monographs*, 41:53–78.

MYERS, K.; C. S. HALE; R. MYKYTOWYCZ; AND R. L. HUGHES

1971 The effects of varying density and space on sociality and health in animals, in A. H. Esser, *Behavior and Environment: The Use of Space by Animals and Men*, Plenum, New York, pp. 148–187.

MYERS, K., AND W. E. POOLE

1967 A study of the biology of the wild rabbit *Oryctolagus cuniculus* L. in confined populations: IV, The effects of rabbit grazing on sown pastures, *J. Ecology*, 55:435–451.

NACE, R. L.

1969 Human uses of ground water, in R. J. Chorley (ed.), *Water, Earth, and Man*, Methuen, London, pp. 285–294.

NAGEL, W. V.

1943 How big is a coon? *Missouri Conservation*, 4(7):6–7.

NAPIER, J. R.

1966 Stratification and primate ecology, *J. Anim. Ecol.*, 35:411–412.

NATIONAL ACADEMY OF SCIENCES

1970 *Land Use and Wildlife Resources*, National Academy of Sciences, Washington, D.C.

NAYLOR, E.

1958 Tidal and diurnal rhythms of locomotory activity in *Carcinus maenus* L., *J. Exptl. Biol.*, 35:602–610.

1965 Effects of heated effluents upon marine and estuarine organisms, in F. M. Russel (ed.), *Adv. in Marine Biol.*, Vol. 3, Academic, New York, pp. 68–103.

NEAVE, F.

1944 Racial characteristics and migratory habits in *Salmo gairdneri*, *J. Fisheries Res. Board Can.*, 6:245–251.

NEEL, J. K.

1951 Interrelations of certain physical and chemical features in headwater limestone streams, *Ecology*, 32:368–391.

NELLIS, C. H.

1972 Lynx-prey interactions in central Alberta, *J. Wildlife Manage.*, 36:320–329.

NELSON, D. J.

1962 Clams as indicators of strontium-90, *Science*, 138:38–39.

NELSON, D. J., ET AL.

1971 Hazards of mercury, *Environmental Research*, 4:3–69.

NELSON, D. J., AND D. C. SCOTT

1962 Role of detritus in the productivity of a rock-outcrop community in a Piedmont stream, *Limnol. Oceanogr.*, 3:396–413.

NELSON, P. R., AND W. T. EDMONDSON

1955 Limnological effects of fertilizing Bare Lake, Alaska, *U.S. Fish Wildlife Serv. Fishery Bull.*, 56:415–436.

NERO, R. W.

1956a A behavior study of the red-winged blackbird: I, Mating and nesting activities, *Wilson Bull.*, 68:5–37.

1956b A behavior study of the red-winged blackbird: II, Territoriality, *Wilson Bull.*, 68:129–150.

NESTLER, R. B.

1949 Nutrition of bobwhite quail, *J. Wildlife Manage.*, 13:342–358.

NEUWIRTH, R.

1957 Some recent investigations into the chemistry of air and of precipitation and their significance for forestry, *Allg. Forst-v. Jagdztg*, 128:147–150.

NEWBOULD, P. J.

1967 *Methods for Estimating the Primary Production of Forests*, IBP Handbook No. 2, Blackwell, Oxford.

1968 Methods for estimating root production, in F. E. Eckardt (ed.), *Functioning in a Terrestrial Ecosystem at the Primary Production Level*, UNESCO, Paris, pp. 187–190.

NEWELL, R. C.

1965 The role of detritus in the nutrition of two marine deposit feeders, the prosobranch *Hydrobia ulvae* and the bivalve *Macoma balthica*, *Proc. Zool. Soc. London*, 144:25–45.

NEWLING, B. E.

1966 Urban growth and spatial structure: mathematical models and empirical evidence, *Geogr. Rev.* 56:213–225.

NEWMAN, M. A.

1956 Social behavior and interspecific competition in two trout species, *Physiol. Zool.*, 29:64–81.

NEWMAN, M. T.

1962 Ecology and nutritional stress in man, *Am. Anthropologist*, 64:22–34.

NICE, M. M.

1941 The role of territory in bird life, *Am. Midland Naturalist*, 26:441–487.

1943 Studies in the life history of the song sparrow: II, *Trans. Linn. Soc.*, New York, 6:1–329.

1950 Development of a redwing: *Agelaius phoeniceus*, *Wilson Bull.*, 62:87–93.

1962 Development of behavior in precocial birds, *Trans. Linn. Soc.*, New York, Vol. 8.

NICE, M. M., AND W. E. SCHANTZ

1959 Head-scratching in Passerines, *Ibis*, 101:250–251.

NICE, M. M., AND J. J. TER FELWYK

1941 Enemy recognition by the song sparrow, *Auk*, 58:195–214.

NICHOLSON, A. J.

1954 An outline of the dynamics of animal populations, *Aus. J. Zool.*, 2:9–65.

1958 The self-adjustment of populations to change, *Cold Spring Harbor Symp. Quant. Biol.*, 22:153–173.

NICHOLSON, A. J., AND V. A. BAILEY

1935 The balance of animal populations: Part 1, *Proc. Zool. Soc. London*, pp. 551–598.

NICOL, J. A. C.

1967 The luminescence of fishes, in N. B. Marshall (ed.), *Aspects of Marine Zoology*, *Symp. Zool. Soc. London No. 19*, Academic, New York, pp. 27–55.

NIELSEN, A.

1950 The torrential invertebrate fauna, *Oikos*, 2:176–196.

NIELSEN, C. O.

1961 Respiratory metabolism of some populations of enchytraeid worms and free-living nematodes, *Oikos*, 12:17–35.

1962 Carbohydrates in soil and litter invertebrates, *Oikos*, 13:200–215.

NIERING, W. A., AND F. E. EGLER

1955 A shrub community of *Viburnum lentago* stable for twenty-five years, *Ecology*, 36:356–360.

NOBLE, G. K.

1936 Courtship and sexual selection of the flicker (*Colaptes auratus luteus*), *Auk*, 53:269–282.

NOBLE, G. K., AND H. T. BRADLEY

1933 The mating behavior of lizards: its bearing on the theory of sexual selection, *Ann. N.Y. Acad. Sci.*, 35:25–100.

NOLAN, V., JR.

1958 Anticipatory food-bringing in the prairie warbler, *Auk*, 75:263–278.

NOMMIK, H.

1965 Ammonium fixation and other reactions involving nonenzymatic immobilization, in W. Bartholomew and F. Clark (eds.), *Soil Nitrogen*, Amer. Society Agronomy, Madison, Wis., pp. 198–258.

NORRIS, K. S.

1963 The functions of temperature in the ecology of the percoid fish *Girella nigricans* Ayres, *Ecol. Monographs*, 33:23–62.

NYDAL, R.

1968 Further investigation on the transfer of

radiocarbon in nature, *J. Geophysical Res.*, 2:1617–1635.

ODEN, S., AND T. AHL

1970 Forsurningen av skandinaviska vatten [The acidification of Scandinavian lakes and rivers], Ymer, Arsbok, pp. 103–122.

ODUM, E. P.

1941 Annual cycle of the black-capped chickadee, *Auk*, 53:314–333, 518–535; 59:499–531.

1959 *Fundamentals of Ecology*, 2d ed., Saunders, Philadelphia.

1960 Organic production and turnover in old field succession, *Ecology*, 41:34–49.

1969 The strategy of ecosystem development, *Science*, 164:262–270.

1971 *Fundamentals of Ecology*, 3d ed., Saunders, Philadelphia.

ODUM, E. P., AND E. J. KUENZLER

1963 Experimental isolation of food chains in an old-field ecosystem with the use of phosphorus-32, in V. Schultz and A. Klement (eds.), 1963, *Radioecology*, pp. 113–120.

ODUM, H. T.

1956 Primary production in flowing water, *Limnol. Oceanogr.*, 1(2):102–117.

1957a Trophic structure and productivity of Silver Springs, Florida, *Ecol. Monographs*, 27:55–112.

1957b Primary production measurements in eleven Florida springs and a marine turtle-grass community, *Limnol. Oceanogr.*, 2:85–97.

1971 *Environment, Power, and Society*, Wiley-Interscience, New York.

ODUM, H. T., AND E. P. ODUM

1955 Trophic structure and productivity of a windward coral reef community on Eniwetok Atoll, *Ecol. Monographs*, 25:291–320.

ODUM, H. T., AND R. C. PINKERTON

1955 Time's speed regulator, the optimum efficiency for maximum output in the physical and biological systems, *Am. Scientist*, 43:331–343.

OGAWA, H.; K. YODA; AND T. KIRA

1961 A preliminary survey of the vegetation of Thailand, *Nature and Life in Southeast Asia*, 1:21–157.

OKE, T. R., AND C. EAST

1971 The urban boundary layer in Montreal, *Boundary Layer Meteorology*, 1:411.

OLSON, J. S.

1958 Rates of succession and soil changes on southern Lake Michigan sand dunes, *Botan. Gaz.*, 119:125–170.

1963 Energy storage and balance of decomposers in ecological systems, *Ecology*, 44:322–332.

1970 Carbon cycles and temperate woodlands, in D. Reichle (ed.), *Analysis of Temperate Forest Ecosystems*, Springer-Verlag, New York, pp. 226–241.

OOSTING, H. J., AND L. E. ANDERSON

1937 The vegetation of a bare-faced cliff in western North Carolina, *Ecology*, 18:280–292.

1939 Plant succession on granite rock in eastern North Carolina, *Botan. Gaz.*, 100:750–768.

OPHEL, I. L.

1959 Investigation of the effects of radioactive material on aquatic life in Canada, *Trans. 2nd Seminar Biol. Problems; Water Pollution, Robert A. Taft Sanitary Engineering Center Tech. Report W-60*, pp. 3–21.

1963 The fate of radiostrontium in a freshwater community, in V. Schultz and A. Klement (eds.), *Radioecology*, pp. 213–216.

ORIANS, G. H.

1969a Age and hunting success in the brown pelican, *Anim. Behav.*, 17:316–319.

1969b The number of bird species in some tropical forests, *Ecology*, 50:783–801.

ORIANS, G. H., AND G. COLLIER

1963 Competition and blackbird social systems, *Evolution*, 17:449–459.

ORIANS, G. H., AND M. F. WILLSON

1964 Interspecific territories of birds, *Ecology*, 45:736–745.

OSBORN, B., AND P. F. ALLEN

1949 Vegetation of an abandoned prairie-dog town in tall grass prairie, *Ecology*, 30:322–332.

OVERGAARD, C.

1949 Studies on the soil microfauna: II, The soil-inhabiting nematodes, *Nat. Jutland*, 2:131ff.

OVERLAND, L.

1960 Endogenous rhythms in opening and odor of flowers of *Cestrum nocturnum*, *Am. J. Bot.*, 47:378–382.

OVINGTON, J. D.

1957 Dry matter production by *Pinus sylvestris* L, *Ann. Bot. London*, n.s. 21:287–314.

1959 Mineral content of plantations of *Pinus sylvestris* L, *Ann. Bot. London*, n.s. 23:75–88.

1961 Some aspects of energy flow in plantation of *Pinus sylvestris* L, *Ann. Bot. London*, n.s. 25:12–20.

1962a Quantitative ecology and the woodland ecosystem concept, in J. Cragg (ed.), *Advances in Ecological Research I*, pp. 103–192.

1962b The application of ecology to multi-purpose use of woodlands, *Proc. Lockwood Conf. on the Suburban Forest and Ecology Bull. 652*, Conn. Agr. Expt. Sta., pp. 76–89.

OVINGTON, J. D.; D. HEITKAMP; AND D. B. LAWRENCE

1963 Plant biomass and productivity of prairie, savanna, oakwood and maize field ecosystems in Central Minnesota, *Ecology*, 44:52–63.

OVINGTON, J. D., AND H. A. I. MADGWICK

1959 The growth and composition of natural stands of birch: 1, Dry matter production, *Plant Soil*, 10:271–283.

PAINE, R. T.

1966 Food web complexity and species diversity, *Am. Naturalist*, 100:65–75.

1969 The Pisaster-Tegula interaction: Prey patches,

predator food preference and intertidal community structure, *Ecology*, 50:950–961.

PALMER, H. E.; W. C. HANSON; B. I. GRIFFIN; AND W. C. ROESCH

1963 Cesium-137 in Alaskan Eskimos, *Science*, 142(3588):64–65.

PALMER, R. W.

1941 A behavior study of the common tern (*Sterna hirundo hirundo* L.), *Proc. Boston Soc. Nat. Hist.*, 43:1–119.

PALMGREN, P.

1949 Some remarks on the short-term fluctuations in the numbers of northern birds and mammals, *Oikos*, 1:114–121.

PALOHEIMO, J. E., AND L. M. DICKIE

1970 Production and food supply, in J. H. Steele (ed.), *Marine Food Chains*, University of California Press, Berkeley.

PAPI, F.

1955 Astronomische Orientierung bei der Wolfspinne *Arctosa perita*, *Z. Vergleich. Physiol.*, 37:230–233.

PAPI, F.; L. SERRETTI; AND S. PARRINI

1957 Nuove richerche sull'orientamento e il senso del tempo di *Arctosa perita*, *Z. Vergleich. Physiol.*, 39:531–561.

PARENTI, R.

1968 Inhibitional effects of *Digitaria sanguinalis* and possible role in old field succession, Ph.D. dissertation, University of Oklahoma, Norman.

PARIS, O. H.

1969 The function of soil fauna in grassland ecosystems, in R. L. Dix and R. G. Beidleman (eds.), *The Grassland Ecosystem: A Preliminary Synthesis*, Colorado State Univ. Range Sci. Dept., Sci. Ser. No. 2, pp. 331–360.

PARIZEH, R. R., ET AL.

1967 Waste water renovation and conservation, *Penn. State Univ. Studies No. 23*, University Park, Pa.

PARK, T.

1948 Experimental studies of interspecies competition: I, Competition between populations of the flour beetles, *Trilobium confusum* Duval and *Trilobium castaneum* Herbst, *Ecol. Monographs*, 18:265–308.

1954 Experimental studies of interspecies competition: II, Temperature, humidity, and competition in two species of *Trilobium*, *Physiol. Zool.*, 27:177–238.

1955 Experimental competition in beetles with some general implications, in J. B. Cragg and N. W. Pirie (eds.), *The Numbers of Man and Animals*, Oliver & Boyd, London.

PARKER, J.

1969 Further studies of drought resistance in woody plants, *Botan. Rev.*, 35:317–371.

PATE, V. S. L.

1933 Studies on fish food in selected areas: a biological survey of Raquette Watershed, N.Y.,

State Conserv. Dept. Biol. Survey No. 8, 136–157.

PATRICK, R.

1949 A proposed biological measure of stream conditions based on a survey of the Conestoga Basin, Lancaster County, Pennsylvania, *Proc. Acad. Nat. Sci. Phila.*, 101:277–341.

PAYNE, R.

1968 Among wild whales, N.Y. *Zoological Society Newsletter*, November, 1968.

PEAKALL, D. B.

1970 Pesticides and the reproduction of birds, *Sci. American*, 222:72–78.

PEAKALL, D. B., AND J. L. LINCER

1970 Polychlorinated biphenyls: another long-life widespread chemical in the environment, *Bioscience*, 20:958–964.

PEARCE, C. K.

1970 Range deterioration in the Middle East, XI *Int. Grassland Congr. Proc.*, Queensland, Australia, pp. 26–30.

PEARCE, R. B.

1967 Photosynthesis in plant communities as influenced by leaf angle, *Crop Sci.*, 7:321–326.

PEARCE, R. B., ET AL.

1965 Relationships between leaf area index, light interception and net photosynthesis in orchardgrass, *Crop Sci.*, 6:15–18.

PEARCY, W.

1962 Ecology of an estuarine population of winter flounder *Pseudopleuronectes americanus* (Walbaum), Parts I–IV, *Bull. Bingham Oceanographic Collection*, Vol. 18, Art. 1, pp. 1–78.

PEARL, R., AND L. J. REED

1920 On the rate of growth of the population of the United States since 1790 and its mathematical representation, *Proc. Nat. Acad. Sci. U.S.*, 6:275–288.

PEARSE, A. S.; H. J. HUMM; AND G. W. WHARTON

1942 Ecology of sand beaches at Beaufort, North Carolina, *Ecol. Monographs*, 12:136–190.

PEARSON, D. L.

1971 Vertical stratification of birds in a tropical dry forest, *Condor*, 73:46–55.

PEARSON, O. A.

1946 Scent glands of the short-tailed shrew, *Anat. Record*, 94:615–629.

PEARSON, O. P.

1966 The prey of carnivores during one cycle of mouse abundance, *J. Anim. Ecol.*, 35:217–233.

1971 Additional measurements of the impact of carnivores on California voles, *J. Mammal.*, 52:41–49.

PEARSON, P. G.

1959 Small mammals and old field succession on the Piedmont of New Jersey, *Ecology*, 40:249–255.

PELTON, M. R., AND E. E. PROVOST

1969 Effects of radiation on survival of wild cotton rats (*Sigmodon hispidus*) in enclosed areas of natural habitat, in D. J. Nelson and F. C. Evans (eds.), *Symposium on Radioecology*, pp. 39–45.

PENNAK, R. W.

1942 Ecology of some copepods inhabiting intertidal beaches near Woods Hole, Massachusetts, *Ecology*, **23**:446–456.

1951 Comparative ecology of the interstitial fauna of fresh-water and marine beaches, *Ann. Biologique*, **27**:217–248.

PENNAK, R. W., AND E. D. VAN GERPEN

1947 Bottom fauna production and physical nature of a substrate in a northern Colorado trout stream, *Ecology*, **28**:42–48.

PENNAK, R. W., AND D. J. ZINN

1943 Mystacocarida, a new order of Crustacea from intertidal beaches in Massachusetts and Connecticut, *Smith's Misc. Coll.*, Vol. 103, No. 9.

PEQUEGNAT, W. E.

1961 Life in the scuba zone: II, *Natural History*, **70**(5):46–54.

PETERLE, T. J.

1969 DDT in Antarctic snow, *Nature*, **224**:620.

PETROV, V. S.

1946 Aktevnaia reaktsiia pochvy pH kah faktor rasprpstraneniia dozhdevykh chorvei (Lumbricidae, Oligochaetae), *Zool. Zh.*, **25**:107–110.

PEWE, T. L.

1957 Permafrost and its effect on life in the north, in H. P. Hansen (ed.), *Arctic Biology, 18th Biol. Colloq.*, Oregon State College, pp. 12–25.

PFEIFFER, W.

1962 The fright reaction of fish, *Biol. Rev.*, **37**:495–511.

PHILLIPS, J.

1965 Fire—as master and servant: its influence in the bioclimatic regions of trans-Saharan Africa, *Proc. 4th Tall Timbers Fire Ecology Conf.*, pp. 7–109.

PHILLIPS, W. S.

1963 Depth of roots in soil, *Ecology*, **44**:242.

PIANKA, E. R.

1966 Latitudinal gradients in species diversity: a review of concepts, *Am. Naturalist*, **100**:33–46.

1967 On lizard species diversity: North American flatlands desert, *Ecology*, **48**:333–351.

PIELOU, E. C.

1966a Shannon's formula as a measure of specific diversity: its use and misuse, *Am. Naturalist*, **100**:463–465.

1966b The measurement of diversity in different types of biological collections, *J. Theoret. Biol.*, **13**:131–144.

1969 *An Introduction to Mathematical Ecology*, Wiley, New York.

PIKE, G.

1962 Migration and feeding of the gray whale, *J. Fisheries Res. Board Can.*, **19**:815–838.

PIMENTEL, D.

1961 Animal population regulation by the genetic feedback mechanism, *Am. Naturalist*, **95**:65–79.

1968 Population regulation and genetic feedback, *Science*, **159**:1432–1437.

1971a *Ecological Effects of Pesticides on Non-target Species*, Executive Office of the President, Office of Science and Technology, Washington, D.C.

1971b Evolutionary and environmental impact of pesticides, *Bioscience*, **21**:109.

PIMENTEL, D.; J. E. DEWEY; AND H. H. SCHWARDT

1951 An increase in the duration of the life cycle of DDT-resistant strains of the house fly, *J. Econ. Entomol.*, **44**:477–481.

PIMENTEL, D.; E. H. FEINBERG; P. W. WOOD; AND J. T. HAYES

1965 Selection, spacial distribution, and the coexistance of competing fly species, *Am. Naturalist*, **99**:97–109.

PIMENTEL, D.; W. P. NAGEL; AND J. L. MADDEN

1963 Space-time structure of the environment and the survival of the parasite-host system, *Am. Naturalist*, **97**:141–167.

PIMLOTT, D. H.

1967 Wolf predation and ungulate populations, *Am. Zoologist*, **7**:267–278.

PITELKA, F. A.

1957a Some aspects of population structure in the short term cycle of the brown lemming in northern Alaska, *Cold Spring Harbor Symp. Quant. Biol.*, **22**:237–251.

1957b Some characteristics of microtine cycles in the Arctic, *18th Biology Coll. Proc.*, Oregon State College, Corvallis, pp. 73–88.

1964a The nutrient recovery hypothesis for Arctic microtine cycles: I, Introduction, in *Grazing in Terrestrial and Marine Environments*, Blackwell, Oxford, pp. 55–56.

1964b Predation in the lemming cycle at Barrow, Alaska, *Proc. 16th Inter. Cong. Zool.*, **1**:265.

PITTENDRIGH, C. S.

1954 On temperature independence in the clock system controlling emergence time in Drosophila, *Proc. Natl. Acad. Sci. U.S.*, **40**:1018–1029.

PLATT, J.

1969 What we must do, *Science*, **166**:1115–1121.

PLATT, R. B.

1963 Ecological effects of ionizing radiation on organisms, communities and ecosystems, in V. Schultz and A. Klement (eds.), *Radioecology*, pp. 243–255.

POLT, J. M., AND E. H. HESS

1964 Following and imprinting effects of light and social experience, *Science*, **143**:1185–1187.

POLUNIN, N.

1934–1935 The vegetation of Akpatok Island: I–II, *J. Ecol.*, **22**:337–395; **23**:161–209.

1948 Botany of the Canadian eastern Arctic: III, Vegetation and ecology, *Nat. Mus. Can. Bull.* 104.

1955 Aspects of arctic botany, *Am. Scientist*, **43**:307–322.

1960 *Introduction to Plant Geography*, McGraw-Hill, New York.

POMEROY, L. R.

1959 Algae productivity in salt marshes of Georgia, *Limnol. Oceanogr.*, 4:386–397.

POMEROY, L. R., R. E. JOHANNES ET AL.

1969 The phosphorus and zinc cycles and productivity of a salt marsh, in D. J. Nelson and F. E. Evans (eds.), *Symposium on Radioecology*, Proc. 2nd Symposium, National Technical Information Service, Springfield, Va. pp. 412–419.

POMEROY, L. R., AND E. J. KUENZLER

1969 Phosphorus turnover by coral reef animals, in D. Nelson and F. Evans (ed.), *Symposium on Radioecology*, pp. 478–482.

POMEROY, L. R.; H. M. MATHEWS; AND H. SHIK MIN

1963 Excretion of phosphate and soluble organic phosphorus compounds by zooplankton, *Limnol. Oceanogr.*, 4:50–55.

POORE, M. E. D.

1962 The method of successive approximation in descriptive ecology, *Adv. in Ecol. Res.*, 2:35–68.

PORTER, W. P., AND D. M. GATES

1969 Thermodynamic equilibria of animals with environment, *Ecol. Monographs*, 39:245–270.

POTZGER, J. E.

1956 Pollen profiles as indicators in the history of lake filling and bog formation, *Ecology*, 37:476–483.

POULSON, T. L.

1969 Salt and water balance in seaside and sharp-tailed sparrows, *Auk*, 86:473–489.

POWER, J. F.

1970 Leaching of nitrate-nitrogen under dryland agriculture in the northern Great Plains, in *Relationship of Agriculture to Soil and Water Pollution*, Cornell University Press, Ithaca, N.Y., pp. 111–122.

PREISTER, L. E.

1965 The accumulation in metabolism of DDT, parathion, and endrin by aquatic food chain organisms, Ph.D. dissertation, Clemson University.

PRESCOTT, G. W.

1960 Biological disturbances resulting from algal populations in standing water, in *The Ecology of Algae, Pymatuning Lab. Field Biol. Spec. Publ. No. 2.*

PRESTON, F. W.

1948 The commonness and rarity of species, *Ecology*, 29:254–283.

1962 The canonical distribution of commonness and rarity: Part I, Part II, *Ecology*, 43:185–215, 410–432.

PRITCHARD, D. W.

1952 Salinity distribution and circulation in the Chesapeake Bay estuarine system, *J. Marine Res.*, 11:106–123.

PROSSER, C. L. (ED.)

1958 *Physiological Adaptations*, American Physiological Society, Washington, D.C.

PRUITT, W. O., JR.

1960 Behavior of the barren-ground caribou, *Biol. Paper No. 3*, University of Alaska.

1970a Some aspects of interrelationships of permafrost and tundra biotic communities, in *Productivity and Conservation in Northern Circumpolar Lands*, IUCN Publ., n.s. 10:33–41.

1970b Tundra animals: What is their future? *Roy. Soc. Can. Trans. Ser. 4*, 8:373–385.

PSCHORN-WALCHER, H.

1952 Vergleich der Bodenfauna in Mischwaldern und Fichtenmonokulturen der Nordostalpen, *Mitt. forstl. Versanst. Mariabrunn.*, 48:44–111.

PUTWAIN, P. D.; D. MACHIN; AND J. L. HARPER

1968 Studies in the dynamics of plant populations: II, Components and regulation of a natural population of *Dumex acetosella* L., *J. Ecol.*, 56:421–431.

RABB, G. B.; J. H. WOOLPY; AND B. E. GINSBURG

1967 Social relationships in a group of captive wolves, *Am. Zoologist*, 7:305–311.

RAINEY, R. C., AND Z. WALOFF

1948 Desert locust migrations and synoptic meteorology in the Gulf of Aden area, *J. Anim. Ecol.*, 17:101–112.

1951 Flying locusts and convection currents, *Bull. Anti-locust Research Centre*, London, 9:51–70.

RAND, A. L.

1961 Some size gradients in North American birds, *Wilson Bull.*, 73:46–56.

RANDALL, JOHN E.

1965 Grazing effect on sea grasses by herbivorous reef fishes in the West Indies, *Ecology*, 46:255–260.

RANSAY, A. O.

1956 Seasonal patterns in the epigamic displays of some surface-feeding ducks, *Wilson Bull.*, 68:275–281.

RAPOPORT, E. H.

1969 Gloger's rule and pigmentation of Collembola, *Evolution*, 23:622–626.

RASMUSSEN, D. I.

1941 Biotic communities of Kaibab Plateau, Arizona, *Ecol. Monographs*, 3:229–275.

RAUP, H. M.

1951 Vegetation and cyroplanation, *Ohio J. Sci.*, 51:105–116.

RAUSCH, R. A.

1967 Some aspects of the population ecology of wolves, *Am. Zoologist*, 7:253–256.

RAVELING, D. G.

1970 Dominance relationships and agonistic behavior of Canada geese in winter, *Behaviour*, 37:291–319.

RAYMONT, J. E. G.

1963 *Plankton and productivity in ocean*, Pergamon, Elmsford, N.Y.

READ, R. A., AND L. C. WALKER
1950 Influence of eastern red cedar on soil in Connecticut pine plantations, *J. Forestry*, 48:337–339.

RECHER, H. F., AND J. A. RECHER
1969 Comparative foraging efficiency of adult and immature little blue herons (*Florida caerulae*), *Anim. Behav.*, 17:320–322.

REDD, B. L., AND A. BENSON
1962 Utilization of bottom fauna by brook trout in a northern West Virginia stream, *Proc. West Va. Acad. Sci.*, 34:21–26.

REGIER, F. E., AND E. O. WILSON
1968 The alarm defense system of the ant, *Acanthomyops claviger*, *J. Insect Physiol.*, 14:955–970.

REGIER, H. A., AND E. B. COWELL
1972 Application of ecosystem theory, succession, diversity, stability, stress and conservation, *Biol. Cons.*, 4:83–93.

REGIER, H. A., AND K. H. LOFTUS
1972 Effects of fisheries exploitation on salmonid communities in oligotrophic lakes, *J. Fish.Res. Bd. Canada*, 29:959–968.

REICHLE, D. E., ET AL.
1973 Carbon flow and storage in a forest ecosystem, in *Brookhaven Biology Colloquium, Carbon in the Biosphere*.

REICHLE, D. E., AND D. A. CROSSLEY, JR.
1967 Investigation of heterotrophic productivity in forest insect communities, in K. Petrusewicz (ed.), *Secondary Productivity of Terrestrial Ecosystems*, Polish Acad. Sciences, Warsaw, pp. 563–587.

REICHLE, D. E.; P. B. DUNAWAY; AND D. J. NELSON
1970 Turnover and concentration of radionuclides in food chains, *Nuclear Safety*, 11:43–56.

REIFSNYDER, W. E., AND H. W. LULL
1965 Radiant energy in relation to forests, *USDA Tech. Bull. No. 1344*.

REINERS, W. A.; I. A. WORLEY; AND D. B. LAURENCE
1971 Plant diversity in a chronosequence at Glacier Bay, Alaska, *Ecology*, 52:55–69.

REITEMEIER, R. F.
1957 Soil potassium and fertility, *Yearbook of Agriculture*, USDA, Washington, D.C., pp. 101–106.

RENFRO, W. C., AND C. OSTERBERG
1969 Radiozinc decline in starry flounders after temporary shutdown of Hanford reactors, in D. J. Nelson and F. E. Evans (ed.), *Second National Symposium on Radioecology*, pp. 372–379.

RENNER, M.
1960 The contribution of the honey bee to the study of the time sense and astronomical orientation, *Cold Spring Harbor Symp. Quant. Biol.*, 25:361–367.

RETZER, J. L.
1956 Alpine soils of the Rocky Mountains, *J. Soil Sci.*, 7:22–32.

RHOADES, W. A., AND R. B. PLATT
1971 Beta radiation damage to vegetation from close-in fallout from two nuclear detonations, *Bioscience*, 21:1121–1125.

RICE, E. L.
1964 Inhibition of nitrogen fixing and nitrifying bacteria by seed plants, *Ecology*, 45:824–837.
1965 Inhibition of nitrogen-fixing and nitrifying bacteria by seed plants: II, Characterization and identification of inhibitors, *Physiol. Plant*, 18:255–268.

RICE, T. R., AND J. W. ANGELOVI
1969 Radioactivity in the sea: effects in fisheries, in F. Firth (ed.), *The Encyclopedia of Marine Resources*, Van Nostrand Reinhold, New York, pp. 574–578.

RICE, E. L.; W. T. PENFOUND; AND L. M. ROHRBAUGH
1960 Seed dispersal and mineral nutrition in succession in abandoned fields in central Oklahoma, *Ecology*, 41:224–228.

RICHARDS, C. M.
1958 The inhibition of growth in crowded *Rana pipiens* tadpoles, *Physiol. Zool.*, 31:138–151.
1962 The control of tadpole growth by algal-like cells, *Physiol. Zool.*, 35:285–296.

RICHARDS, P. W.
1952 *The Tropical Rain Forest*, Cambridge University Press, London.

RICKER, W. E.
1940 On the origin of Kokanee, a freshwater type of sockeye salmon, *Trans. Roy. Soc. Can. Sect. V*, 34:121–135.
1954 Stock and recruitment, *J. Fisheries Res. Board Can.*, 11:559–623.
1958a Maximum sustained yields from fluctuating environments and mixed stocks, *J. Fisheries Res. Board Can.*, 15:991–1006.
1958b Handbook of computations for biological statistics of fish populations, Bull 119, *J. Fisheries Res. Board Can.*, pp. 1–300.

RODIN, L. E., AND N. I. BAZILEVIC
1964 The biological productivity of the main vegetation types in the Northern Hemisphere of the Old World, *Forestry Abstracts*, 27:369–372.
1967 *Production and Mineral Cycling in Terrestrial Vegetation* (transl. from Russian by Scripta Technica), G. E. Fogg (ed.), Oliver & Boyd, London.

ROE, A., AND G. G. SIMPSON
1958 *Behavior and Evolution*, Yale University Press, New Haven, Conn.

ROEDER, K. D., AND A. E. TREAT
1961 The detection and evasion of bats by moths, *Am. Scientist*, 49:135–148.

RORISON, I. H. (ED.)
1969 *Ecological Aspects of Mineral Nutrition of Plants*, Blackwell Scientific Publications, Oxford.

ROSENWEIG, M. L., AND R. H. MACARTHUR
1963 Graphical representation and stability

conditions of predator–prey interactions, *Am. Naturalist*, **97**:209–223.

ROSS, B. A.; J. R. BRAY; AND W. H. MARSHALL
1970 Effects of a long-term deer exclusion on a *Pinus resinosa* forest in north-central Minnesota, *Ecology*, **51**:1088–1093.

ROSS, H. H.
1944 The caddisflies or Trichoptera of Illinois, *Bull. Illinois Nat. Hist. Surv.*, **23**:1–326.
1962 *A Synthesis of Evolutionary Theory*, Prentice-Hall, Englewood Cliffs, N.J.

ROWAN, W. R.
1925 Relation of flight to bird migration and developmental changes, *Nature*, **115**:494–495.
1929 Experiments in bird migration: I, Manipulation of the reproductive cycle, seasonal histological changes in the gonads, *Proc. Boston Soc. Nat. Hist.*, **39**:151–208.

ROWE, P. B., AND T. M. HENDRIX
1951 Interception of rain and snow by second growth ponderosa pine, *Trans. Am. Geophys. Union*, **32**:903–908.

ROWELL, C. H. F.
1961 Displacement grooming in the chaffinch, *Anim. Behav.*, **9**:38–63.

ROWLAND, F. S.
1973 Mercury levels in swordfish and tuna, *Biol. Conserv.*, **5**:52–53.

ROYCE, W. F., ET AL.
1963 Salmon gear limitation in northern Washington waters, *Univ. Wash. Publ. in Fish.*, n.s. **2**(1):1–123.

RUDD, R. L.
1964 *Pesticides and the Living Landscape*, University of Wisconsin Press, Madison.

RUDD, R. L., AND R. E. GENELLY
1956 Pesticides: their use and toxicity in relation to wildlife, *Calif. Fish and Game Bull. No. 7*.

RUINEN, J.
1962 The phyllosphere: 1, An ecologically neglected region, *Plant Soil*, **15**:81–109.

RUSCH, D. H.; E. C. MESLOW; P. D. DOERR; AND L. B. KEITH
1972 Response of great horned owl populations to changing prey densities, *J. Wildlife Manage.*, **36**:282–296.

RUTTNER, F.
1953 *Fundamentals of Limnology*, University of Toronto Press, Toronto.

RYTHER, J. H.
1956 Photosynthesis in the ocean as a function of light intensity, *Limnol. Oceanogr.*, **1**:61–70.
1960 Organic production by plankton algae and its environmental control, in *Ecology of Algae*, *Pymatuning Lab. Field Biol. Spec. Publ. No. 2*, University of Pittsburgh Press, Pittsburgh, pp. 72–83.
1969 Photosynthesis and fish production in the sea, *Science*, **166**:72–75.

RYTHER, J. H., AND W. M. DUNSTAN
1971 Nitrogen, phosphorus and eutrophication in the coastal marine environment, *Science*, **171**:1008–1013.

RYTHER, J. H.; W. M. DUNSTAN; K. R. TENORE; AND J. E. HUGUENIN
1972 Controlled eutrophication: increasing food production from the sea by recycling human wastes, *Bioscience*, **22**:144–152.

RYTHER, J. H., AND C. S. YENTSCH
1957 The estimation of phytoplankton production in the ocean from chlorophyll and light data, *Limnol. Oceanogr.*, **2**:281–286.

SABINE, W. S.
1959 The winter society of the Oregon junco: intolerance, dominance, and the pecking order, *Condor*, **61**:110–135.

SALT, G.; F. S. J. HOLLICK; F. RAW; AND M. V. BRIAN
1948 The arthropod population of pasture soil, *J. Anim. Ecol.*, **17**:139–150.

SALT, G. W.
1952 The relation of metabolism to climate and distribution in three finches of the genus *Carpodacus*, *Ecol. Monographs*, **22**:121–152.
1957 An analysis of avifaunas in the Teton Mountains and Jackson Hole, Wyoming, *Condor*, **59**:373–393.

SAMUEL, D. E.
1967 A review of the effects of plant estrogenic substances on animal reproduction, *Ohio J. Sci.*, **67**:308–312.

SANDERS, H. L.
1968 Marine benthic diversity: a comparative study, *Am. Naturalist*, **102**:243–282.

SANDON, H.
1927 *The Composition and Distribution of the Protozoan Fauna of the Soil*, Oliver & Boyd, London.

SAUER, E. G. F., AND ELENORE SAUER
1955 Zur Frage der Nachtlichen Zugorientierung von Grasmuken, *Rev. Suisse Zool.*, **62**:250–259.
1960 Star navigation of nocturnal migrating birds, *Cold Spring Harbor Symp. Quant. Biol.*, **25**:463–473.

SAUNDERS. J. K., JR.
1963 Movements and activities of the lynx in Newfoundland, *J. Wildlife Manage.*, **27**:390–400.

SCHALLER, G. B.
1972 *Serengeti: A Kingdom of Predators*, Knopf, New York.

SCHARLOO, W.
1971 Reproductive isolation by disruptive selection: Did it occur? *Am. Naturalist*, **105**:86.

SCHELDERUP-EBBE, T.
1922 Beitrage zur Socialpsychologie des Haushuhns, *Zeitschr. Phychol.*, **88**:225–252.

SCHENKEL, R.
1948 Ausdruckstudien an Wolfen, *Behaviour*, **1**:81–130.

SCHMIDT, K. P.
1954 Faunal realms, regions and provinces, *Quart. Rev. Biol.*, **29**:322–331.

SCHMIDT-NIELSEN, K.
1956 Animals and arid conditions: physiological aspects of productivity and management, in White (ed.), *Future of Arid Lands*, pp. 368–389.
1959 Physiology of the camel, *Sci. American*, **201**:140–151.
1960 The salt secreting gland of marine birds, *Circulation*, **21**:955–967.
1964 *Desert Animals: Physiological Problems of Heat and Water*, Oxford University Press, London.

SCHMIDT-NIELSEN, K., AND B. SCHMIDT-NIELSEN
1952 Water metabolism of desert mammals, *Physiol. Rev.*, **32**:135–166.

SCHNELL, J. H.
1963 The effect of neutron-gamma radiation on free-living small mammals at the Lockheed Reactor Site, in V. Schultz and A. Klement (eds.), 1963, *Radioecology*, pp. 339–344.
1968 The limiting effects of natural predation on experimental cotton rat populations, *J. Wildlife Manage.*, **32**:698–711.

SCHOENER, T. W.
1968 Sizes of feeding territories among birds, *Ecology*, **49**:123–141.

SCHOLANDER, P. F.
1955 Evolution of climatic adaptation in homeotherms, *Evolution*, **9**:15–26.

SCHOLANDER, P. F.; R. HOCK; V. WALTERS; F. JOHNSON; AND L. IRVING
1950 Heat regulation in some arctic and tropical birds and mammals, *Biol. Bull.*, **99**:237–258.

SCHOLANDER, P. F.; V. WALTERS; R. HOCK; L. IRVING; AND F. JOHNSON
1950 Body insulation of some arctic and tropical mammals and birds, *Biol. Bull.*, **99**:225–236.

SCHORR, A.
1963 *Slums and Social Insecurity*, Government Printing Office, Washington, D.C.

SCHUCK, H.
1943 Survival, population density, gorwth and movement of the wild brown trout in Crystal Creek, *Trans. Am. Fisheries Soc.*, **73**:209–230.

SCHULTZ, A. M.
1964 The nutrient-recovery hypothesis for arctic microtine cycles: II, Ecosystem variables in relation to arctic microtine cycles, in *Grazing in Terrestrial and Marine Environments*, Blackwell, London.

SCHULTZ, V., AND A. W. KLEMENT (EDS.)
1963 *Radioecology*, Van Nostrand Rheinhold, New York.

SCHUSTERMAN, R. J., AND R. G. DAWSON
1968 Barking, dominance and territoriality in male sea lions, *Science*, **160**:434–436.

SCHWARTZ, C. W.
1944 *The Prairie Chicken in Missouri*, Missouri Conserv. Comm., Jefferson City.

SCLATER, P. L.
1858 On the general geographical distribution of the members of the class Aves, *J. Proc. Limnol. Soc. (Zool.)*, **2**:130–145.

SCOTT, D., AND W. O. BILLINGS
1964 Effects of environmental factors on standing crop and productivity of an alpine tundra, *Ecol. Monographs*, **34**:243–270.

SCOTT, J. P.
1958 *Animal Behavior*, University Chicago Press, Chicago.
1962 Critical periods in behavioral development, *Science*, **138**:949–958.

SCOTT, J. W.
1942 Mating behavior of the sage grouse, *Auk*, **59**:472–498.
1950 A study of the phylogenetic or comparative behavior of three species of grouse, *Ann. N.Y. Acad. Sci.*, **51**:1062–1073.

SCOTT, T. C.
1943 Some food coactions of the northern plains red fox, *Ecol. Monographs*, **13**:427–479.
1955 An evaluation of the red fox, *Ill. Nat. Hist. Surv., Biol. Notes No. 35*, pp. 1–16.

SEBA, D., AND E. COCHRANE
1969 Surface slicks as a concentrator of pesticides, *Pesticide Monitoring J.*, **3**:190–193.

SEGERSTRALE, S. G.
1947 New observations on the distribution and morphology of the amphipod *Gammarus zaddachi* Sexton, with notes on related species, *J. Marine Biol. Assoc. U.K.*, **27**:219–244.

SELANDER, R. K., AND D. R. GILLER
1961 Analysis of sympatry of great-tailed and boat-tailed grackles, *Condor*, **63**:29–56.

SELANDER, R. K., AND R. F. JOHNSTON
1967 Evolution in the House Sparrow: I, Intrapopulation Variation in North America, *Condor*, **69**:217–258.

SELLECK, G. W.
1960 The climax concept, *Botan. Rev.* **26**:534–545.

SERVENTY, D. L.
1971 Biology of desert birds, in D. S. Farner and J. R. King (eds.), *Avian Biology*, Academic, New York, pp. 287–339.

SEVENSTER, P.
1961 *A Causal Analysis of Displacement Activity in Fanning in Gasterosteus aculeatus* L., E. Brill, Leiden.

SEVERINGHAUS, C. W.
1972 Weather and the deer population, *The Conservationist*, **27**(2):28–31.

SEVERINGHAUS, C. W., AND ROSALIND GOTTLEIB
1959 Big deer vs. little deer, N.Y. *State Conservationist*, **14**(2):30–31.

SHAW, S. P., AND C. G. FREDINE
1956 Wetlands of the United States, *U.S. Fish and Wildlife Circ. 39.*

SHELDON, W. G.
1967 *The Book of the American Woodcock,* University of Massachusetts Press, Amherst.

SHELFORD, V. E.
1913 *Animal Communities in Temperate America,* University of Chicago Press, Chicago.
1926 *Naturalists Guide to the Americas,* Williams and Wilkins Co., Baltimore.
1932 Life zones, modern ecology, and the failure of the temperature summing, *Wilson Bull.,* 44:144–157.

SHELFORD, V. E., AND A. C. TWOMEY
1941 Tundra animal communities in the vicinity of Churchill, Manitoba, *Ecology,* 22:47–69.

SHEPPARD, P. M.
1951 Fluctuations in the selective value of certain phenotypes in the polymorphic land snail *Cepaea nemoralis* L., *Heredity,* 5:125–134.
1959 *Natural Selection and Heredity,* Hutchinson, London.

SHETTER, D. S.
1937 Migration, growth rate, and population density of brook trout in the north branch of the Au Sable, Michigan, *Trans. Am. Fisheries Soc.,* 66:203–210.

SHORT, H. L.
1972 Ecological framework for deer management, *J. Forestry,* 72:200–203.

SHREVE, F.
1951 Vegetation of the Sonoran Desert, *Carnegie Inst. Wash. Publ. 591,* Washington, D.C.

SHUSTER, C. N., JR.
1966 The nature of a tidal marsh, *The Conservationist,* 21(1):22–29, 36.

SIBLEY, C. G.
1957 The evolutionary and taxonomic significance of sexual dimorphism and hybridization in birds, *Condor,* 59:166–191.
1960 The electrophoretic patterns of avian egg-white proteins as taxonomic characters, *Ibis,* 102:215–284.

SIEBURTH, J. MCN., AND A. JENSEN
1970 Production and transformation of extracellular organic matter from marine littoral algae: a resume, in D. E. Hood (ed.), *Organic Matter in Natural Waters,* Univ. Alaska, pp. 203–223.

SIGAFOOS, R. S.
1951 Soil instability in tundra vegetation, *Ohio J. Sci.,* 51:281–298.
1952 Frost action as a primary physical factor in tundra plant communities, *Ecology,* 33:480–487.

SILVER, H.
1957 A History of New Hampshire Game and Furbearers, *Survey Rept. No. 6,* New Hampshire Fish and Game Dept., Concord.

SIMMONS, K. E. L.
1955 The nature of predator-reactions of waders towards humans, with special reference to the role of aggressive-, escape-, and brooding-drives, *Behaviour,* 8:130–173.
1957 A review of the anting behaviour of passerine birds, *Brit. Birds,* 50:401–424.

SIMMS, E.
1965 The effects of cold weather of 1962/63 on the blackbird population of Dallis Hill, *Brit. Birds,* 58:33–43.

SIMPSON, G. G.
1964 Species density of North American recent mammals, *Syst. Zool.,* 13:57–73.

SINGH, J. S., AND R. MISRA
1968 Diversity, dominance, stability and net production in the grasslands at Varanasi, India, *Can. J. Botany,* 47:425–427.

SINGH, R. N.
1961 *Role of Blue-Green Algae in the Nitrogen Economy of Indian Agriculture,* Indian Council of Agricultural Research, New Delhi.

SKINNER, W. A.; R. D. MATHEWS; AND R. M. PARKHURST
1962 Alarm reaction of the topsmelt *Atherinops affinia* (Ayers), *Science,* 138:681–682.

SLOBODKIN, L. B.
1962 *Growth and Regulation of Animal Populations,* Holt, Rinehart & Winston, New York.

SLOBODKIN, L. B., AND H. L. SANDERS
1969 On the contribution of environmental predictability to species diversity, *Brookhaven Symp. Biol.,* 22:82–95.

SLOBODKIN, L. B.; F. SMITH; AND N. HAIRSTON
1967 Regulation in terrestrial ecosystems and the implied balance of nature, *Am. Naturalist,* 101:109–124.

SLUSS, R. R.
1967 Population dynamics of the walnut aphid *Chromaphes juglandicola* (Kalt.) in northern California, *Ecology,* 48:41–58.

SMALLEY, A. E.
1960 Energy flow of a salt marsh grasshopper population, *Ecology,* 41:672–677

SMITH, B. E., AND G. COTTAM
1967 Spatial relationships of mesic forest herbs in southern Wisconsin, *Ecology,* 48:546–557.

SMITH, C. C.
1968 The adaptative nature of social organization in the genus of tree squirrels *Tamiasciurus, Ecol. Monographs,* 38:31–63.
1970 The coevolution of pine squirrels (*Tamiascurius*) and conifers, *Ecol. Monographs,* 40:349–371.

SMITH, D. G.
1972 The role of the epaulets on the red-winged blackbird (*Agelaius phoeniceus*) social system, *Behaviour,* 41:251–268.

SMITH, F. E.

1952 Experimental methods in population
dynamics: a critique, *Ecology*, **33**:441–450.

1961 Density dependence in the Australian thrips,
Ecology, **42**:403–407.

SMITH, J. G.

1899 Grazing problems in the Southwest and how
to meet them, *USDA Div. Agrost. Bull.*, **16**:1–47.

SMITH, M. C.

1968 Red squirrel responses to spruce cone failure in
interior Alaska, *J. Wildlife Manage.*, **32**:305–317.

1971 Food as a limiting factor in the population
ecology of *Peromyscus polionotus* (Wagner),
Ann. Zool. Fennici, **8**:109–112.

SMITH, R. E., AND B. A. HORWITZ

1969 Brown fat and thermogenesis, *Physiological
Rev.*, **49**:330–425.

SMITH, R. H., AND S. GALLIZIOLI

1965 Predicting hunter success by means of a spring
call count of Gambel quail, *J. Wildlife Manage.*,
29:806–813.

SMITH, R. L.

1956 An evaluation of conifer plantations as
wildlife habitat, Ph.D. dissertation, Cornell
University, Ithaca, N.Y.

1959a Conifer plantations as wildlife habitat, *N.Y.
Fish Game J.*, **5**:101–132.

1959b The songs of the grasshopper sparrow, *Wilson
Bull.*, **71**:141–152.

1962 Acorn consumption by white-footed mice
(*Peromyscus leucopus*), *Bull. 482T*, *West Va.
Univ. Agr. Expt. Sta.*

1963 Some ecological notes on the grasshopper
sparrow, *Wilson Bull.*, **75**:159–165.

1966 Animals and the vegetation of West Virginia,
in E. L. Core, *Vegetation of West Virginia*,
McClain, Parsons, W. Va., pp. 17–24.

1972 *The Ecology of Man: An Ecosystem Approach*,
Harper & Row, New York.

SMITH, S. H.

1972 Factors of ecologic succession in oligotrophic
fish communities of the Lawrentian Great Lakes,
J. Fish. Res. Bd. Canada, **29**:717–730.

SMYTH, M., AND G. A. BARTHOLOMEW

1966 The water economy of the black-throated
sparrow and the rock wren, *Condor*, **68**:447–458.

SNYDER, R. L.

1962 Reproductive performance of a population of
woodchucks after a change in sex ratio, *Ecology*,
43:506–515.

SOKAL, R. R., AND R. C. RINKEL

1963 Geographic variation of alate *Pemphigus
populi-transversus* in eastern North America,
Univ. Kansas Sci. Bull., **44**:467–507.

SOKAL, R. R., AND P. H. SNEATH

1963 *The Principles of Numerical Taxonomy*,
Freeman, San Francisco.

1966 Efficiency in taxonomy, *Taxonomy*, **15**:1–21.

SOLBRIG, O.

1970 *Principles and Methods of Plant
Biosystematics*, Macmillan, New York.

SOLLBERGER, A.

1965 *Biological Rhythm Research*, Elsevier,
Amsterdam.

SOLOMON, M. E.

1957 Dynamics of insect populations, *Ann. Rev.
Entomol.*, **2**:121–142.

1964 Analysis of processes involved in the natural
control of insects, in Cragg (ed.), *Advances in
Ecological Research*, Vol. 2, pp. 1–58.

SOMMER, R.

1968 Spatial parameters in naturalistic social
research, in A. H. Esser (ed.), *Behavior and
Environment: The Use of Space by Animals and
Men*, Plenum, New York, pp. 281–289.

SORENSEN, T.

1954 Adaptation of small plants to deficient
nutrition and a short growing season, *Botan.
Tidsskr.*, **51**:339–361.

SORIANO, A.

1972 South America, in *Wildland Shrubs, Their
Biology and Utilization*, USDA For. Serv. Gen.
Tech. Rept. INT-1, pp. 51–54.

SOUTHERN, H. N., AND V. P. W. LOWE

1968 The pattern of distribution of prey and
predation in tawny owl territories, *J. Anim.
Ecol.*, **37**:75–97.

SOUTHERN, W. E.

1972 Magnets disrupt the orientation of juvenile
ring-billed gulls, *Bioscience*, **22**:476–479.

SOWLS, L. K.

1960 Results of a banding study of Gambels quail in
southern Arizona, *J. Wildlife Manage.*, **24**:185–190.

SPAETH, J. N., AND C. H. DIEBOLD

1938 Some interrelations between soil characteristics,
water tables, soil temperature, and snow cover
in the forest and adjacent open areas in
south-central New York, *Cornell Univ. Agr.
Expt. Sta. Mem.*, **213**.

SPAGNOLI, J. J.

1971 What's happening to our salt marshes? *The
Conservationist*, **25**(5):22–27.

SPARROW, A. H.; L. A. SCHAIRER; AND R. C. SPARROW

1963 Relationship between nuclear volumes,
chromosome numbers and radiosensitivities,
Science, **141**:163–166.

SPARROW, A. H., AND G. M. WOODWELL

1963 Prediction of the sensitivity of plants to
chronic gamma radiation, in Schultz and Klement
(eds.), *Radioecology*, pp. 257–270.

SPECHT, R. L.; P. RAYSON; AND M. E. JACKSON

1958 Dark Island Heath (Ninety-mile Plain, South
Australia): VI, Pyric succession: Changes in
composition, coverage, dry weight, and mineral
nutrient status, *Aust. J. Botany*, **6**:59–88.

823

SPINAGE, C. A.

1972 African ungulate life tables, *Ecology*, 53:645–652.

SPOONER, G. M.

1947 The distribution of *Gammarus* species in estuaries: Part I, *J. Marine Biol. Assoc. U.K.*, 27:1–52.

SPRAGUE, H. B. (ED.)

1959 *Grasslands*, American Association for the Advancement of Science, Washington, D.C.

SPURR, S. H.

1952 Origin of the concept of forest succession, *Ecology*, 33:426–427.

1957 Local climate in the Harvard Forest, *Ecology*, 38:37–56.

STAMP, L. D.

1961 A history of land use in arid regions, *Arid Zone Research No. 17*.

STARK, N.

1972 Nutrient cycling pathways and litter fungi, *Bioscience*, 22:355—360.

STEBBINS, G. L., JR.

1950 *Variation and Evolution in Plants*, Columbia, New York.

1972 Evolution and diversity of arid-land shrubs, in *Wildland Shrubs, Their Biology and Utilization*, *USDA For. Serv. Gen. Tech. Rept.*, INT-1, pp. 111–116.

STEGEMAN, L. C.

1960 A preliminary survey of earthworms of the Tully Forest in central New York, *Ecology*, 41:779–782.

STEHLI, F. G.; R. G. DOUGLAS; AND N. D. NEWELL

1969 Generation and maintenance of gradients in taxonomic diversity, *Science*, 164:947–949.

STENGER, J., AND J. B. FALLS

1959 The utilized territory of the ovenbird, *Wilson Bull.*, 71:125–140.

STEPHENSON, T. A., AND A. STEPHENSON

1949 The universal features of zonation between tide-marks on rocky coasts, *J. Ecology*, 37:289–305.

1950 Life between tide-marks in North America: I, The Florida Keys, *J. Ecology*, 38:354–402.

1952 Life between tide-marks in North America: II, North Florida and the Carolinas, *J. Ecology*, 40:1–49.

1954 Life between the tide-marks in North America: IIIA, Nova Scotia and Prince Edward Island: description of the region; IIIB, Nova Scotia and Prince Edward Island: The geographical features of the region, *J. Ecology*, 42:14–45, 46–70.

1961 Life between tide-marks in North America: IVA, IVB: Vancouver Island, I, II, *J. Ecology*, 49:1–29, 229–243.

1971 *Life Between the Tide-marks on Rocky Shores*, W. H. Freeman, San Francisco.

STERN, W. L., AND M. F. BUELL

1951 Life-form spectra of New Jersey pine barren forest and Minnesota jack pine forest, *Torrey Botan. Club Bull.*, 78:61–65.

STEVANOVIC, D.

1956 Populations of collembola in forest associations on Mt. Kopaonik, *Zbornik Radov Inst. Ekol. Biogeogr.*, Beograd, 7:16ff.

STEWARD, G. A.

1970 High potential productivity of the tropics for cereal crops, grass forage crops, and beef, *J. Aust. Institite Agr. Sci.*, 36:85.

STEWART, B. A.; L. K. PORTER; AND F. G. VIETS

1966a Effect of sulfur content of straws on rates of decomposition and plant growth, *Soil Sci. Soc. Amer. Proc.*, 30:355–358.

1966b Sulfur requirements for decomposition of cellulose and glucose in soil, *Soil Sci. Soc. Amer. Proc.*, 30:453–456.

STEWART, B. A.; F. G. VIETS, JR.; AND G. L. HUTCHISON

1968 Agricultural effects on nitrate pollution of ground water, *J. Soil Water Conserv.*, 23:13–15.

STEWART, P. A.

1952 Dispersal, breeding behavior and longevity of banded barn owls in North America, *Auk*, 69:227–245.

STEWART, R. E., AND J. W. ALDRICH

1952 Ecological studies of breeding bird populations in northern Maine, *Ecology*, 33:226–238.

STEWART, R. E.; A. D. GEIS; AND C. D. EVANS

1958 Distribution of populations and hunting kill of the canvasback, *J. Wildlife Manage.*, 22:333–370.

STEWART, W. D. P.

1967 Nitrogen-fixing plants, *Science* 158:1426–1432.

STOECKLER, J. H.

1962 Shelterbelt influence on Great Plains field environment and crops, *USDA Prod. Res. Rept. No. 62.*

STOKES, A. W.

1955 Population studies of the ring-necked pheasants on Pelee Island, Ontario, *Ont. Dept. Lands Forests, Tech. Bull. Wildlife Ser. No. 4*, Ottawa.

1961 Voice and social behavior of the chukar partridge, *Condor*, 63:111–127.

STONE, W. B.; E. PARKS; AND B. L. WEBER

1972 The little foxes, *The Conservationist*, 26(4):8–9, 47.

STOTT, D. H.

1962 Cultural and natural checks on population growth, in Montagu (ed.), *Culture and the Evolution of Man*, Oxford University Press, New York, pp. 355–376.

STOUT, B. B., AND R. J. MCMAHON

1961 Throughfall variation under tree crowns, *J. Geophys. Res.*, 66:1839–1843.

STOUT, J. F.

1963 The significance of sound production during the reproductive behaviour of *Notropis analostanus* (Family Cyprinidae), *Anim. Behav.*, 11:83–92.

STOUT, P. R., AND R. G. BURAN

1967 The extent and significance of fertilizer buildup in soils as revealed by vertical distribution of nitrogenous matter between soils and underlying water reservoirs, in N. Brady (ed.), *Agriculture and the Quality of the Environment*, AAAS Publ. No. 85, pp. 283–310.

STRICKLAND, J. D. H.

1965a Production of organic matter in the primary stages of the marine food chain, in Riley and Skirrow (eds.), *Chem. Oceanogr.*, vol. 1, Academic, New York, p. 477–610.

1965b Phytoplankton and marine primary production, *Ann. Rev. Microbiol.*, 19:127–162.

STRINGER, G. E., AND W. S. HOAR

1955 Aggressive behavior of underyearling Kamloops trout, *Can. J. Zool.*, 33:148–160.

STRUMWASSER, F.

1960 Some physiological principles governing hibernation in *Citellus beecheyi*, in Lyman and Dawe (eds.), *Mammalian Hibernation*, pp. 285–318.

STUDY OF CRITICAL ENVIRONMENTAL PROBLEMS

1970 *Man's Impact on the Global Environment*, MIT Press, Cambridge, Mass.

STUDY OF MAN'S IMPACT ON CLIMATE

1971 *Inadvertant Climate Modification*, Massachusetts Institute of Technology Press, Cambridge, Mass.

SUKACHEV, V., AND N. DYLIS

1968 *Fundamentals of Forest Biogeocoenology*, Oliver & Boyd, London. Sullivan, C. M., and K. C. Fisher

1953 Seasonal fluctuations in the selected temperature of speckled trout *Salvelinus fontinalis* (Mitchell), *J. Fisheries Res. Board Can.*, 10:187–195.

SWANK, W. G.

1959 The mule deer in Arizona chaparral, *Wildlife Bull. No. 3, Arizona Game Fish Dept.*, Phoenix, Ariz.

SWEENEY, B. M.

1969 *Rhythmic Phenomena in Plants*, Academic, New York.

SWEENEY, B. M., AND J. W. HASTINGS

1957 Characteristics of the diurnal rhythm of luminescence in *Gonyaulax polyedra*, *J. Cellular Comp. Physiol.*, 49:115.

SWIFT, R. W.

1948 Deer select most nutritious forages, *J. Wildlife Manage.*, 12:109–110.

TABER, R. D.

1949 Observations on the breeding behavior of the ring-necked pheasant, *Condor*, 51:153–175.

1953 Studies of black-tailed deer reproduction on three chaparral cover types, *Calif. Fish Game*, 39:177–186.

TABER, R. D., AND R. F. DASMANN

1957 The dynamics of three natural populations of the deer *Odocoileus hemionus columbianus*, *Ecology*, 38:233–246.

1958 The black-tailed deer of the chaparral, *Calif. Dept. Fish Game, Game Bull. No. 8*, Sacramento, Calif.

TALBOT, L. M., AND MARTHA H. TALBOT

1963a The wildebeest in western Masailand, East Africa, *Wildlife Monographs No. 12*, Wildlife Society, Washington, D.C.

1963b The high biomass of wild ungulates on East African Savanna, *Trans. N. Amer. Wildlife Conf.*, 28:465–476.

TAMARIN, R. H., AND C. J. KREBS

1969 *Microtus* population biology: II, Genetic changes at the transferrin locus in fluctuating populations of two vole species, *Evolution*, 23:183–211.

TAMM, C. O.

1951 Removal of plant nutrients from tree crowns by rain, *Physiol. Plant.*, 4:184–188.

TANNER, J. T.

1966 Effects of population density on growth rates of animal populations, *Ecology*, 47:733–745.

TANSLEY, A. G.

1935 The use and abuse of vegetational concepts and terms, *Ecology*, 16:284–307.

TARRANT, R. F.

1971 Persistence of some chemicals in Pacific Northwest forests, in *Pesticides, Pest Control, and Safety on Forest Lands*, Continuing Education Books, Corvallis, Oregon, pp. 133–141.

TARRANT, R. F.; D. G. MOORE; W. B. BOLLEN; AND B. F. LOPER

1972 Pesticides in soil, *Pesticides Monitoring J.*, 6:65–72.

TATUM, L. A.

1971 The southern corn leaf blight epidemic, *Science*, 171:1113–1116.

TAUB, F. B.; R. L. BURGNER; E. B. WELCH; AND D. E. SPYRIDAKES

1972 A comparative study of four lakes, in J. F. Franklin et al. (eds.), *Proceedings on Research in Coniferous Forest Ecosystems*, N. W. For. and Range Exp. Stat., Portland, Oregon.

TAYLOR, C. R.

1969 The eland and the oryx, *Sci. American*, 220(1):88–95.

1970a Strategies of temperature regulation: effect of evaporation on East African ungulates, *Am. J. Physiol.*, 219:1131–1135.

1970b Dehydration and heat: effects on temperature regulation of East African ungulates, *Am. J. Physiol.*, 219:1136–1139.

TAYLOR, W. P. (ED.)

1956 *The Deer of North America*, Stackpole, Harrisburg, Pa.

825

TEAL, J. M.
1957 Community metabolism in a temperate cold spring, *Ecol. Monographs*, 27:283–302.
1962 Energy flow in the salt marsh ecosystem of Georgia, *Ecology*, 43:614–624.

TEAL, J. M., AND J. KANWISHER
1961 Gas exchange in a Georgia salt marsh, *Limnol. Oceanogr.*, 6:388–399.

TEDROW, J. F. C.; J. V. DREW; D. E. HILL; AND L. A. DOUGLAS
1958 Major genetic soils of the arctic slope of Alaska, *J. Soil Sci.*, 9:33–45.

TEDROW, J. F. C., AND H. HARRIES
1960 Tundra soil in relation to vegetation, permafrost and glaciation, *Oikos*, 11:237–249.

TEER, J. G.; J. W. THOMAS; AND E. A. WALKER
1965 Ecology and management of white-tailed deer in the Llano Basin of Texas, *Wildlife Monographs No. 15*, Wildlife Society, Washington, D.C.

TESTER, J. R., AND W. H. MARSHALL
1961 A study of certain plant and animal interrelations on a native prairie in northwestern Minnesota, *Minnesota Museum Natural Hist.*, *Occasional Paper No. 8*

TEVIS, L., SR.
1956 Responses of small mammal populations to logging of Douglas fir, *J. Mammal.*, 37:189–196.

THIELCKE, GERHARD
1966 Ritualized distinctiveness of song in closely related sympatric species, *Philosophical Transactions of the Royal Society of London*, B, Vol. 251, pp. 493–497.

THODAY, J. M.
1958 Effects of disruptive selection: the experimental production of a polymorphic population, *Nature*, 181:1124–1125.
1959 Effects of disruptive selection: I, Genetic flexibility, *Heredity*, 13:187–203.
1960 Effects of disruptive selection: III, Coupling and repulsion, *Heredity*, 14:35–49.

THODAY, J. M., AND T. G. BOAM
1959 Effects of disruptive selection: II, Polymorphism and divergence without isolation, *Heredity*, 13:205–218.

THODAY, J. M., AND J. B. GIBSON
1962 Isolation by disruptive selection, *Nature* 193:1164–1166.
1970 The probability of isolation by disruptive selection, *Am. Naturalist*, 104:219–230.
1971 Reply to Scharloo, *Am. Naturalist*, 105:86–88.

THOMAS, W. L., JR. (ED.)
1956 *Man's Role in changing the Face of the Earth*, University of Chicago Press, Chicago.

THOMPSON, D. Q., AND R. H. SMITH
1970 The forest primeval in the northeast: a great myth?, *Proc. 10th Tall Timbers Fire Ecology Conf.*, pp. 255–265.

THOMPSON, H. V.
1953 The grazing behavior of the wild rabbit, *Brit. J. Anim. Behav.*, 1:16–20.
1954 The rabbit disease, myxomatosis, *Ann. Appl. Biol.*, 41:358–366.

THOMPSON, W. L.
1960 Agonistic behavior in the house finch: Part I, Annual cycle and display patterns, *Condor*, 62:245–271.

THOREAU, H. D.
1860 Succession of forest trees: address read to the Middlesex Agricultural Society, Sept., 1860, in *Excursion*, 1891, Houghton Mifflin, Boston, Mass.

THORPE, J.
1949 Effects of certain animals that live in soils, *Sci. Monthly*, 68:180–191.

THORPE, W. H.
1945 The evolutionary significance of habitat selection. *J. Animal Ecol.*, 14:67–70.
1951 The learning abilities of birds, *Ibis*, 93:1–52, 252–296.
1958 The learning of song patterns by birds, with special reference to the song of the chaffinch *Fringilla coelebs*, *Ibis*, 100:535–570.
1963 *Learning and Instinct in Animals*, 2d ed., Methuen, London.

THORSON, G.
1950 Reproductive and larval ecology of marine bottom invertebrates, *Biol. Rev.*, 25:1–45.
1957 Bottom communities in Hedgpeth (ed.), *Treatise on marine ecology and peleoecology*, 1:461–534.

THURSTON, J. M.
1969 The effect of liming and fertilizers on the botanical composition of permanent grassland, and on the yield of hay, in I. H. Rorison (ed.), *Ecological Aspects of the Mineral Nutrition of Plants*, Blackwell, Oxford, England, pp. 3–10.

TILLY, L. J.
1968 The structure and dynamics of Cone Spring, *Ecol. Monographs*, 38:169–197.

TINBERGEN, L.
1960 The natural control of insects in pinewoods: I, Factors influencing the intensity of predation by songbirds, *Arch. Neerl. Zool.*, 13:265–343.

TINBERGEN, L., AND H. KLOMP
1960 The natural control of insects in pinewoods: II, Conditions for Nicholson oscillations in parasite-host systems, *Arch. Neerl. Zool.*, 13:344–379.

TINBERGEN, N.
1951 *The Study of Instinct*, Oxford University Press, New York.
1952 Derived activities, their causation, biological significance, origin and emancipation during evoluton, *Quart. Rev. Biol.*, 27:1–32.
1953 *The Herring Gull's World*, Collins, London.
1960 Comparative studies of the behaviour of gulls (Laridae): a progress report, *Behaviour*, 15:1-70.

TINBERGEN, N., AND D. J. KUENEN

1939 Uber die austosenden und die richtungge-
benden Reizsituationen der Sperrbewegung von
jengen Drosseln (*Turdus m. merula* L. und *T. e.
ericetorum* Tuxton), *Z. Tierpsychol.*, 5:182–226.

TINBERGEN, N., AND M. MOYNIHAN

1952 Head flagging in the black-headed gull: its
function and origin, *Brit. Birds*, 45:19–22.

TINBERGEN, N., AND A. C. PERDECK

1950 On the stimulus situation releasing the
begging response in the newly hatched herring
gull chick (*Larus argentatus*), *Behaviour*,
3:1–39.

TOMANEK, G. W.

1969 Dynamics of mulch layer in grassland
ecosystems, in R. Dix and R. Beidleman (eds.),
*The Grassland Ecosystem, Colorado State Univ.
Range Sci. Dept. Sci. Ser. No. 2*, pp. 225–240.

TOMPA, F. S.

1962 Territorial behavior: the main controlling
factor of a local song sparrow population, *Auk*,
79:687–697.

1964 Factors determining the numbers of song
sparrows *Melospiza melodia* (Wilson) on
Mandarte Island, B. C., Canada, *Acta Zool.
Fennica*, 109:73ff.

1971 Catastrophic mortality and its population
consequences, *Auk*, 88:753–759.

TORDOFF, H. B.

1954 Social organization and behavior in a flock of
captive, non-breeding red crossbills, *Condor*,
56:346–358.

TORNABENE, T. G., AND H. W. EDWARDS

1972 Microbial uptake of lead, *Science*,
176:1334–1335.

TRANSEAU, E. N.

1926 The accumulation of energy by plants, *Ohio
J. Sci.*, 26:1–10.

TREMBLEY, F.

1965 Effects of cooling water from steam electric
power plants on stream biota, in *Biological
Problems in Water Pollution, Public Health Serv.
Publ. No. 999-WP-25*, pp. 334–345.

TRESHOW, M.

1970 *Environment and Plant Response*, McGraw-
Hill, New York.

TRIMBLE, G. R., JR., AND S. WEITZMAN

1954 Effect of a hardwood forest canopy on rainfall
intensities, *Trans. Amer. Geophys. Union*,
35:226–234.

TRUE, R. P.; H. L. BARNETT; ET AL.

1960 Oak Wilt in West Virginia, *West Va. Univ.
Agr. Expt. Sta. Bull. 448T.*

TSIVOGLOU, E. C., AND W. W. TOWNE

1957 Sources and control of radioactive water
pollutants, *Sewage Ind. Wastes*, 29:143–156.

TURCEK, F. J.

1951 On the stratification of the avian populations
of the Querceto-Carpinetum forest communities

in southern Slovakia (English summary), *Sylvia*,
13:71–86.

TYRTIKOV, A. P.

1959 Perennially frozen ground, in *Principles of
Geocryology; Part I, General Geocryology* (transl.
from Russian by R. E. Brown), *Nat. Res. Coun.
Canada Tech. Trans.*, 1163(1964): 399–421.

UCKO, P. J., AND G. W. DIMBLEBY (EDS.)

1969 *The Domestication and Exploitation of Plants
and Animals*, Aldine, Chicago.

UDVARDY, M. D. F.

1958 Ecological and distributional analysis of North
American birds, *Condor*, 60:50–66.

UEBELMESSER, E. R.

1954 Uber den endonomen Rhythmus der Sporan-
gientrager-Bildung von Pilobolus, *Arch.
Mikrobiol.*, 20:1–33.

UGENT, D.

1970 The potato, *Science*, 170:1161–1166.

UHLIG, H.

1955 The gray squirrel: its life history, ecology, and
population characteristics in West Virginia,
*Conserv. Comm. W. Va., Final Rept. PR
Project 31-R.*

ULRICH, A. T.

1933 Die Macrofauna der Waldstreu, *Mitt.
Forstwirstch Fortwiss*, 4:283–323.

URBAN, D.

1970 Raccoon populations, movement patterns, and
predation on a managed waterfowl marsh, *J.
Wildlife Manage.*, 34:372–382.

USPENSKI, S. M.

1970 Problems and forms of fauna conservation in
the Soviet arctic and subarctic, in *Productivity
and Conservation in Northern Circumpolar
Lands*, IUCN Publ., n.s. 16:199–207.

VALLENTYNE, J. R.

1957 The principles of modern limnology, *Am.
Scientist*, 45:218–244.

VANCE, B. D., AND W. DRUMMOND

1969 Biological concentration of pesticides by algae,
J. Am. Water Works Assoc., 61:360–362.

VAN HOOKE, R. I.

1971 Energy and nutrient dynamics of spider and
orthropteran populations in a grassland ecosystem,
Ecol. Monographs, 41:1–26.

VAN IERSEL, J. J. A.

1953 An analysis of the parental behaviour of the
three-spined stickleback, *Behaviour, Suppl. No. 3.*

VAN IERSEL, J. J. A., AND A. C. A. BOL

1958 Preening of two tern species: a study on
displacement activities, *Behaviour*, 13:1–88.

VAN LAWICK-GOODALL, J.

1968 The behaviour of free-living chimpanzees in
the Gombe Stream Reserve, *Anim. Behav.
Monographs*, 1(3):161–311.

VAN VALEN, L.

1971 Group selection and the evolution of dispersal,
Evolution, 25:591–598.

VAYDA, A. P.
1961 Expansion and warfare among swidden agriculturists, *Am. Anthropologist*, **63**:346–358.

VEALE, P. T., AND H. L. WASCHER
1956 Henderson County soils, *Illinois Univ. Agr. Expt. Stat. Soil Report No. 77.*

VERDUIN, J.
1956 Energy fixation and utilization by natural communities in western Lake Erie, *Ecology*, **37**:40–50.

VESEY-FITZGERALD, D.
1972 Fire and animal impact in Tanzania National Parks, *Proc. 11th Tall Timber Fire Ecology Conf.*, pp. 297–318.

VEZINA, P. E.
1961 Variation in total solar radiation in three Norway spruce plantations, *Forest Sci.*, **7**:257–264.

VIESSMAN, W., JR.
1966 The hydrology of small impervious areas, *Water Resources Res.*, **2**:405–412.

VIETS, F. G., JR.
1971a The mounting problem of cattle feedlot pollution, *Agr. Science Rev.*, **9**:1–8.
1971b Water quality in relation to farm use of fertilizers, *Bioscience*, **21**:460–467.

VIMMERSTED, J. P., AND P. H. STRUTHERS
1968 Influence of time and precipitation on chemical composition of spoils and drainage, *Second Symposium on Coal Mine Drainage Research*, Mellon, Pittsburgh, Pa.

VIRO, P. J.
1953 Loss of nutrients and the natural nutrient balance of the soil in Finland, *Comm. Inst. Forest. Fenn.*, **42**:1–50.

VOGL, R. J.
1967 Fire adaptations of some southern California plants, *Proc. California Tall Timbers Fire Ecology Conf.*, pp. 79–109.
1969 One hundred and thirty years of plant succession in a southeastern Wisconsin lowland, *Ecology*, **50**:248–255.

VOIGT, G. K.
1960 Distribution of rain under forest stands, *Forest Sci.*, **6**:2–10.
1971 Mycorrhizae and nutrient mobilization, in E. Hacskaylo (ed.), *Mycorrhizae, USDA Forest Serv. Mscl. Pub. No. 1189*, pp. 122–131.

VOLTERRA, V.
1926 Variazione e fluttazioni de numero d'individiu in specie animali conviventi, *Mem. Accad. Lincei*, **2**:31–113, translated in R. N. Chapman, 1931, *Animal Ecology*, McGraw-Hill, New York.

VOSE, R. N., AND D. G. DUNLAP
1968 Wind as a factor in the local distribution of small mammals, *Ecology*, **49**:381–386.

WADDINGTON, C. H.
1957 *The Strategy of the Genes*, Allen, London.

WADLEIGH, C. H.
1968 Wastes in relation to agriculture and forestry, *USDA Misc. Pub. No. 1065.*

WAGAR, J. A.
1970 Growth versus the quality of life, *Science*, **168**:1179–1184.

WAGNER, F. H.; C. D. BESADNY; AND C. KABAT
1965 Population ecology and management of Wisconsin pheasants, *Wisconsin Cons. Dept. Tech. Bull. No. 34.*

WAGNER, F. H., AND L. C. STODDART
1972 Influence of coyote predation on black-tailed jack rabbit populations in Utah, *J. Wildlife Manage.*, **36**:329–342.

WAGNER, F. H., AND A. W. STOKES
1968 Indices to overwinter survival and productivity with implications for population regulation in pheasants, *J. Wildlife Manage.*, **32**:32–36.

WALKER, P. C., AND R. T. HARTMAN
1960 Forest sequence of the Hartstown bog area in western Pennsylvania, *Ecology*, **41**:461–474.

WALKER, T. J., AND A. D. HASLER
1949 Detection and discrimination of odors of aquatic plants by the bluntnosed minnow (*Hyborhynchus notatus*), *Physiol. Zool.*, **22**:45–63.

WALLACE, A. R.
1876 *The Geographical Distribution of Animals*, 2 vols., Macmillan, London.

WALLACE, G. J.
1959 Insecticides and birds, *Audubon Mag.*, **61**:10–12.

WALLRAFF, H. G.
1960 Does celestial navigation exist in animals? *Cold Spring Harbor Symp. Quant. Biol.*, **25**:451–460.

WALTER, E.
1934 Grundlagen der allegmeinen fisherielichen Produktionslehre, *Handb. Binnenfisch. Mitteleur.*, **14**:480–662.

WALTHER, F. R.
1969 Flight behaviour and avoidance of predators in Thomson's gazelle (*Gazella thomsoni* Guenther, 1884), *Behaviour*, **34**:184–221.

WARD, B., AND R. DUBOS
1972 *Only One Earth, the Care and Maintenance of a Small Planet*, Norton, New York.

WARD, R. R.
1971 *The Living Clocks*, Knopf, New York.

WARD, W. W., AND T. W. BOWERSOX
1970 Upland oak response to fertilization with nitrogen, phosphorus, and calcium, *Forest Sci.*, **16**:113–120.

WARNER, R. E.
1968 The role of introduced diseases in the extinction of the endemic Hawaiian avifauna, *Condor*, **70**:101–120.

WARREN, C. E.
1971 *Biology and Water Pollution Control*, Saunders, Philadelphia, Pa.

WARREN, C. E., AND G. E. DAVIS
1971 Laboratory stream research: objectives, possibilities, constraints, *Ann. Rev. Ecol. and System.*, 2:111–144.

WARREN-WILSON, J.
1957 Arctic plant growth, *Adv. Sci.*, 53:383–387.

WARRINGER, J. E., AND M. L. BREHMER
1966 The effects of thermal pollution on marine organisms, *Air and Water Pollution Int. J.*, 10:277–289.

WASSINK, E. C.
1959 Efficiency of light energy conversion in plant growth, *Plant Physiol.*, 34:356–361.

1968 Light energy conversion in photosynthesis and growth of plants, in F. Echardt (ed.), *Functioning in Terrestrial Ecosystems at the Primary Production Level*, UNESCO, Paris, pp. 53–66.

WATERHOUSE, F. L.
1955 Microclimatological profiles in grass cover in relation to biological problems, *Quart. J. Roy. Meterol. Soc.*, 81:63–71.

WATERS, T. F.
1961 Standing crop and drift of stream bottom organisms, *Ecology*, 42:532–537.

1964 Recolonization of denuded stream bottom areas by drift, *Trans. Am. Fish. Soc.*, 91:243–250.

1965 Interpretation of invertebrate drift of day-active stream invertebrates, *Ecology*, 46:327–334.

1968 Diurnal periodicity in the drift of day-active stream invertebrates, *Ecology*, 49:152–153.

1972 The drift of stream insects, *Ann. Rev. Entomology*, 17:253–272.

WATERS, W. E.
1969 The life table approach to analysis of insect impact, *J. Forestry*, 67:300–304.

WATSON, A., AND D. JENKINS
1968 Experiments on population control by territorial behavior in red grouse, *J. Anim. Ecol.*, 37:595–614.

WATSON, A., AND R. MOSS
1970 Dominance, spacing behavior, and aggression in relation to population limitation in vertebrates, in A. Watson (ed.), *Animal Populations in Relation to Their Food Resources*, Blackwell, Oxford, pp. 167–218.

WATSON, G. E.
1962a Three sibling species of *Alectoris* partridge, *Ibis*, 104:353–367.

1962b Sympatry in Palearctic *Alectoris* partridges, *Evolution*, 16:11–19.

WATT, K. E. F.
1964 The use of mathematics and computers to determine optimal strategy and tactics for a given insect pest control problem, *Can. Entomol.*, 96:202–220.

1968 *Ecology and Resource Management: A Quantitative Approach*, McGraw-Hill, New York.

WEAVER, C. R.
1943 Observations of the life cycle of the fairy shrimp, *Eubranchipus vernalis*, *Ecology*, 24:500–502.

WEAVER, H.
1967 Fire and its relation to ponderosa pine, *Proc. California Tall Timbers Fire Ecology Conf.*, pp. 127–149.

WEAVER, J. E.
1954 *North American Prairie*, Johnson, Lincoln, Nebr.

WEAVER, J. E., AND F. W. ALBERTSON
1956 *Grasslands of the Great Plains: Their Nature and Use*, Johnson, Lincoln, Nebr.

WEAVER, J. E., AND N. W. ROWLAND
1952 Effects of excessive natural mulch on development, yield and structure of native grassland, *Botan. Gaz.*, 114:1–19.

WEBBER, H. H.
1968 Mariculture, *Bioscience*, 18:940–945.

WEBSTER, D. W.
1954 Smallmouth bass *Micropterus dolomieui* in Cayuga Lake: I, Life history and environment, *Cornell Univ. Agr. Expt. Sta. Mem.*, 327:3–39.

WEBSTER, J.
1956–1957 Succession of fungi on decaying cocksfoot culms: I, II, *J. Ecol.*, 44:517–544; 45:1–30.

WECKER, S. C.
1963 The role of early experience in habitat selection by the prairie deer mouse, *Peromyscus maniculatus bairdi*, *Ecol. Monographs*, 33:307–325.

WEIR, J. S.
1972 Spatial distribution of elephants in an African National Park in relation to environmental sodium, *Oikos*, 23:1–13.

WEIR, W. C., AND D. T. TORELL
1959 Selective grazing by sheep as shown by a comparison of the chemical composition of range and pasture forage obtained by hand clipping and that collected by esophageal-fistulated sheep, *J. Anim. Sci.*, 18:641–649.

WEIS-FOGH, T.
1948 Ecological investigations of mites and collembola in the soil, *Nat. Jutland*, 1:135–270.

WELCH, P. S.
1952 *Limnology*, McGraw-Hill, New York.

WELLER, M. W.
1959 Parasitic egg laying the redhead (*Aythya americana*) and other North American Anatidae, *Ecol. Monographs*, 29:333–365.

WELLINGTON, W. G.
1960 Qualitative changes in natural populations during changes in abundance, *Can. J. Zool.*, 38:238–314.

WELLS, H. W.
1961 The fauna of oyster beds, with special reference to the salinity factor, *Ecol. Monographs*, 31:239–266.

WELLS, H. W., AND I. E. GRAY
1960 Some oceanic subtidal oyster populations,

Nautilus, 73:139–146.

WELLS, L.

1960 Seasonal abundance and vertical movements of planktonic crustaceans in Lake Michigan, *USDI Fishery Bull.*, 60(172):343–369.

WENT, F. W.

1955 The ecology of desert plants, *Sci. American*, 192:68–75.

WENT, F. W., AND L. O. SHEPS

1969 Environmental factors in regulation of growth and development: ecological factors, in F. Steward (ed.), *Plant Physiology: A Treatise*, Academic, New York, pp. 299–406.

WENT, F. W., AND N. STARK

1968 Mycorrhiza, *Bioscience*, 18(11):1935–1039.

WEST, D. A.

1962 Hybridization in grosbeaks (Pheucticus) of the Great Plains, *Auk*, 79:399–424.

WEST, G. C.

1965 Shivering and heat production in wild birds, *Physiol. Zool.*, 38:111–120.

WEST, N. E., AND P. T. TUELLER

1972 Special approaches to studies of competition and succession in shrub communities, in C. M. McKell et al. (eds.), *Wildland Shrubs: Their Biology and Utilization*, USDA For. Serv. Gen. Tech. Rept., INT-1, pp. 172–181.

WESTLAKE, D. F.

1959 The effects of organisms on pollution, *Proc. Linn. Soc., London*, 170:171–172.

WHITAKER, L. M.

1957 A resume of anting, with particular reference to a captive oriole, *Wilson Bull.*, 69:195–262.

WHITE, E. J., AND F. TURNER

1970 A method of estimating income of nutrients in a catch of airborne particles by a woodland canopy, *J. Appl. Ecol.*, 7:441–461.

WHITE, G. F. (ED.)

1956 The future of arid lands, *Amer. Assoc. Adv. Sci. Publ. No. 43*, Washington, D.C.

WHITTAKER, R. H.

1951 A criticism of the plant association and climatic climax concept, *Northwest Sci.*, 25:17–31.

1952 A study of summer foliage insect communities in the Great Smoky Mountains, *Ecol. Monographs*, 22:1–44.

1953 A consideration of the climax theory: the climax as a population and pattern, *Ecol. Monographs*, 23:41–78.

1956 Vegetation of the Great Smoky Mountains, *Ecol. Monographs*, 26:1–80.

1961 Estimation of net primary production of forest and shrub communities, *Ecology*, 42:177–183.

1962 Classification of natural communities, *Botan. Rev.*, 28:1–239.

1963 Net production of heath balds and forest heaths in the Great Smoky Mountains, *Ecology*, 44:176–182.

1965 Dominance and diversity in land plant com-munities, *Science*, 147:250–260.

1966 Forest dimensions and production in the Great Smoky Mountains, *Ecology*, 47:103–121.

1967 Gradient analysis of vegetation, *Biol. Rev.*, 42:207–264.

1970a *Communities and Ecosystems*, Macmillan, New York.

1970b The biochemical ecology of higher plants, in E. Sondheimer and J. B. Simeone (eds.), *Chemical Ecology*, Academic, New York, pp. 43–70.

WHITTAKER, R. H., AND P. R. FEENEY

1971 Allelochemics: chemical interactions between species, *Science*, 171:757–770.

WHITTAKER, R. H., AND G. M. WOODWELL

1969 Structure, production, and diversity of the oak-pine forest at Brookhaven, New York, *J. Ecol.*, 57:155–174.

WHYTE, R. O.

1968 *Grasslands of the Monsoon*, Praeger, New York.

WIEGERT, R. G., AND D. F. OWEN

1971 Trophic structure, available resources, and population density in terrestrial vs. aquatic ecosystems, *J. Theoret. Biol.*, 30:69–81.

WIGGINS, I. L., AND J. H. THOMAS

1962 A flora of the Alaskan arctic slope, *Publ. Arctic Inst. North America No. 4*, University of Toronto Press, Toronto.

WILCOMB, M. J.

1954 A study of prairie dog burrow systems and the ecology of their arthropod inhabitants in central Oklahoma, Ph.D. dissertation, University of Oklahoma, Norman.

WILKES, H. G.

1972 Maize and its wild relatives, *Science*, 177:1071–1077.

WILLARD, W. K.

1963 Relative sensitivity of nestlings of wild passerine birds to gamma radiation, in Schultz and Klement (eds.), *Radioecology*, pp. 345–349.

WILLIAMS, C. B.

1964 *Patterns in the Balance of Nature*, Academic, New York.

WILLIAMS, C. E.

1965 Soil fertility and cottontail body weights: a reexamination, *J. Wildlife Manage.*, 28:329–337.

WILLIAMS, C. E., AND A. L. CASHEY

1965 Soil fertility and cottontail fecundity in south-eastern Missouri, *Am. Midland Naturalist*, 74:211–224.

WILLIAMS, E. C.

1941 An ecological study of the floor fauna of the Panama rain forest, *Bull. Chicago Acad. Sci.*, 6:63–124.

WILLIAMSON, P.

1971 Feeding ecology of the red-eyed vireo (*Vireo olivaceus*) and associated foliage-gleaning birds, *Ecol. Monographs*, 41:129–152.

WILLIS, A. J.
1963 Braunton burrows: the effects on vegetation of the addition of mineral nutrients to the dune soils, *J. Ecol.*, **51**:353–374.

WILLSON, M. F., AND G. H. ORIANS
1963 Comparative ecology of red-winged and yellow-headed blackbirds during the breeding season, *Proc. XVI Inter. Zool. Congress*, **3**:242–346.

WILSON, E. O.
1971 Competitive and aggressive behavior, in J. Eisenberg and W. Dillon (ed.), *Man and Beast: Comparative Social Behavior*, Smithsonian Institute Press, Washington, D.C.

WILSON, R. E., AND E. H. RICE
1968 Allelopathy as expressed by *Helianthus annus* and its role in old field succession, *Bull. Torrey Botan. Club*, **95**:432–448.

WILTSCHKO, W., AND H. HOCK
1972 Orientation behavior of night migrating birds (European robins) during late afternoon and early morning hours, *Wilson Bull.*, **84**:149–163.

WILTSCHKO, W.; H. HOCK; AND F. W. MERKEL
1971 Outdoor experiments with migrating robins (*Erithacus rubecula*) in artificial magnetic fields, *Z. Tierpsychol.*, **29**:409–415.

WING, L. D., AND I. D. BUSS
1970 Elephants and forests, *Wildlife Monographs No. 19*, Wildlife Society, Washington, D.C.

WINN, H. E.
1958a Comparative reproductive behavior and ecology of fourteen species of darters (Pisces-Percidae), *Ecol. Monographs*, **28**:155–191.
1958b Observations on the reproductive habits of darters (Pisces-Percidae), *Am. Midland Naturalist*, **59**:190–212.

WINN, H. E., AND J. STOUT
1960 Sound production by the satinfin shiner, *Notropis analostanus*, and related fishes, *Science*, **132**:222–223.

WINSTON, F. W.
1956 The acorn microsere with special reference to arthropods, *Ecology*, **37**:120–132.

WITHERS, J. D., AND A. BENSON
1962 Evaluation of a modified Surber bottom sampler, *Proc. West Va. Acad. Sci.*, **37**:16–20.

WITHERSPOON, J. P.; S. I. AVERBACH; AND J. S. OLSON
1962 Cycling of Cesium-134 in white oak trees on sites of contrasting soil type and moisture, *Oak Ridge Natl. Lab.*, **3328**:1–143.

WITHROW, R. B. (ED.)
1959 Photoperiodism and related phenomena in plants and animals, *Publ. No. 55*, *Amer. Assoc. Adv. Science*, Washington, D.C.

WITKAMP, M.
1963 Microbial populations of leaf litter in relation to environmental conditions and decomposition, *Ecology*, **44**:370–377.

WITKAMP, M., AND D. A. CROSSLEY
1966 The role of arthropods and microflora on the breakdown of white oak litter, *Pedabiologia*, **6**:293–303.

WITKAMP, M., AND J. S. OLSON
1963 Breakdown of confined and nonconfined oak litter, *Oikos*, **14**:138–147.

WOLCOTT, G. N.
1937 An animal census of two pastures and a meadow in northern New York, *Ecol. Monographs*, **7**:1–90.

WOLF, D. D., AND D. SMITH
1964 Yield and persistence of several legume-grass mixtures as affected by cutting frequency and nitrogen fertilization, *Agron. J.*, **56**:130–133.

WOLFE, J. N.; R. T. WAREHAM; AND H. T. SCOFIELD
1949 Microclimates and macroclimates of Neotoma, a small valley in central Ohio, *Ohio Biol. Survey Bull. No. 41*.

WOLFSON, A.
1955 *Recent Studies on Avian Biology*, University of Illinois Press, Urbana.
1959 The role of light and darkness in the regulation of spring migration and reproductive cycles in birds, in Withrow (ed.), *Photoperiodism and Related Phenomena*, pp. 679–716.
1960 Regulation of annual periodicity in the migration and reproduction of birds, *Cold Spring Harbor Symp. Quant. Biol.*, **25**:507–514.

WOOD, O. M.
1937 The interception of precipitation in an oak-pine forest, *Ecology*, **18**:251–254.

WOOD-GUSH, D. G. M.
1955 The behaviour of the domestic chicken: a review, *Brit. J. Anim. Behav.*, **3**:81–110.

WOODWELL, G. M.
1962 Effects of ionizing radiation on terrestrial ecosystems, *Science*, **138**:572–577.
1963 The ecological effects of radiation, *Sci. American* **208**(6):40–49.
1967 Radiation and the pattern of nature, *Science*, **156**:461–470.

WOODWELL, G. M.; P. P. CRAIG; AND H. A. JOHNSON
1971 DDT in the biosphere: where does it go? *Science*, **174**:1101–1107.

WOODWELL, G. M., AND W. R. DYKEMAN
1966 Respiration of a forest measured by carbon dioxide accumulations during temperature inversions, *Science*, **154**:1031–1034.

WOODWELL, G. M., AND T. G. MARPLES
1968 The influence of chronic gamma radiation on the production and decay of litter and humus in an oak-pine forest, *Ecology*, **49**:456–465.

WOODWELL, G. M., AND A. W. SPARROW
1965 Effects of ionizing radiation on ecological systems, in G. Woodwell (ed.), *Ecological Effects of Nuclear War*, Brookhaven National Lab. USAEC Rept. BNL-917 (C-43).

WOODWELL, G. M., AND R. H. WHITTAKER
1968 Primary productivity in terrestrial ecosystems, *Amer. Zool.*, **8**:19–30.

WOODWELL, G. M.; C. F. WURSTER, JR.; AND
P. A. ISAACSON
1967 DDT residues in an east coast estuary: a
case of biological concentration of a persistant
pesticide, *Science*, 156:821–823.

WOOLHOUSE, W. W. (ED.)
1969 *Dormancy and Survival, Symp. Soc. Expt.
Biol. No. 23*, Academic, New York.

WOOLPY, J. H.
1968 The social organization of wolves, *Nat. Hist.*
77(5):46–55.

WRIGHT, J. H.
1970 Electric power generation and the environment, *Westinghouse Engineer*, 30(3):66–80.

WRIGHT, R. T.
1970 Glycollic acid uptake by plankton bacteria, in
D. Wood (ed.), *Organic Matter in Natural
Waters, Occ. Publ. No. 1*, Institute Marine
Science, University of Alaska.

WRIGHT, S.
1931a Evolution in Mendelian populations, *Genetics*,
16:97–159.
1931b Statistical theory of evolution, *Amer.
Statistical J. March Suppl.*, pp. 201–208.
1935 Evolution in population in approximate
equilibrium, *J. Genetics*, 30:243–256.
1943 Evolution in Mendelian populations, *Genetics*,
16:97–159.
1955 Classification of the factors of evolution,
Cold Spring Harbor Symp. Quant. Biol.,
20:16–24.
1964 Stochastic processes in evolution, in J. Gurland
(ed.), *Stochastic Models in Medicine and Biology*,
University of Wisconsin Press, Madison, pp.
199–241.

WURSTER, C. F., JR.
1968 DDT reduces photosynthesis by marine
plankton, *Science*, 159:1474–1475.
1969 Chlorinated hydrocarbon insecticides and the
world ecosystem, *Biol. Conser.*, 1:123–129.

WYNNE-EDWARDS, V. C.
1962 *Animal Dispersion in Relation to Social
Behavior*, Hafner, New York.
1963 Intergroup selection in the evolution of social
systems, *Nature*, 200:623–628.
1965 Self-regulating system in populations of
animals, *Science*, 147:1543–1548.

YEAGLEY, H. L.
1947 A preliminary study of a physical basis of
bird navigation, *J. Appl. Physics*, 18:1035–1063.

YEATTER, R. E., AND D. H. THOMPSON
1952 Tularemia: weather and rabbit populations,
Illinois Nat. Hist. Surv. Bull. No. 25, pp. 351–
382.

YONGE, C. M.
1949 *The Sea Shore*, Collins, London.

ZAK, B.
1964 Role of mycorrhizae in root disease, *Ann. Rev.
Phytopathol.*, 2:377–392.

ZELLER, D.
1961 Certain mulch and soil characteristics of major
range sites in western North Dakota as related to
range conditions, M.A. dissertation, North Dakota
State University, Fargo.

ZIMMERMAN, J. L.
1971 The territory and its density dependent effect
in *Spiza americana*, *Auk*, 88:591–612.

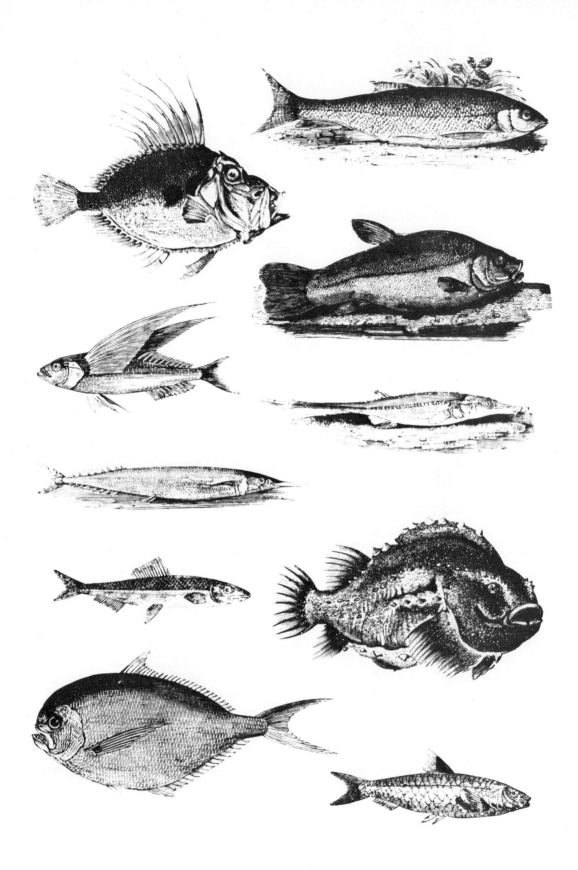

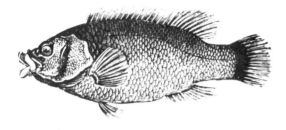

Index

74 75 76   9 8 7 6 5 4 3

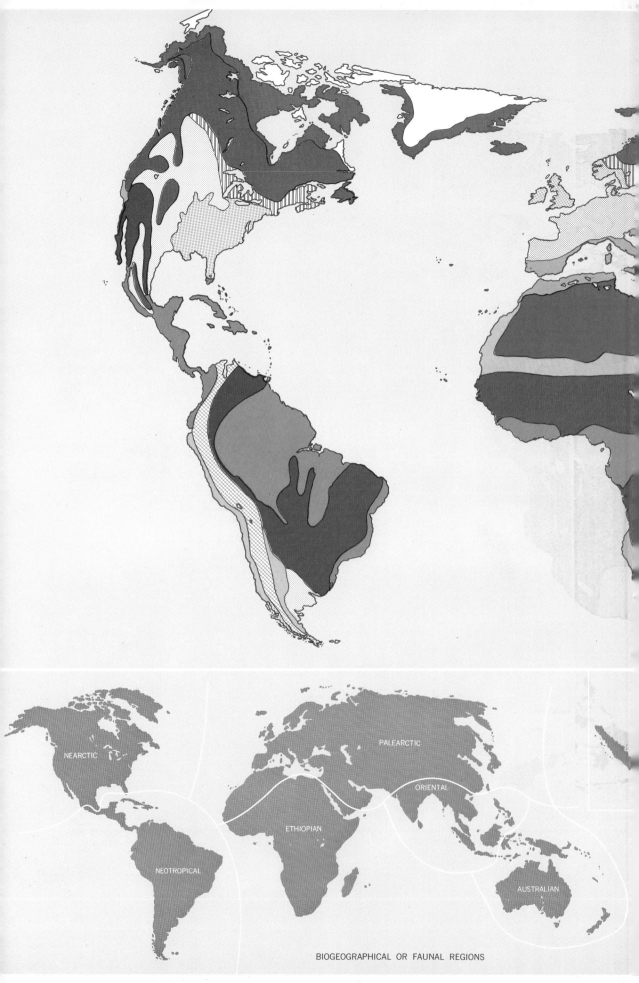

BIOGEOGRAPHICAL OR FAUNAL REGIONS

NEARCTIC

PALEARCTIC

ORIENTAL

ETHIOPIAN

NEOTROPICAL

AUSTRALIAN